America's National Park Roads and Parkways

THE ROAD AND AMERICAN CULTURE

Drake Hokanson, *Series Editor*
George F. Thompson, *Series Founder and Director*
Karl Raitz, *Consulting Editor*

Published in cooperation with the Center for American Places,
Santa Fe, New Mexico, and Staunton, Virginia

America's National Park

Drawings from the Historic American Engineering Record

Roads and Parkways

Edited by Timothy Davis, Todd A. Croteau, and Christopher H. Marston

With an introductory essay by Timothy Davis *and a foreword by* Eric DeLony

The Johns Hopkins University Press *Baltimore & London*

Printed in China on acid-free paper

2 4 6 8 9 7 5 3

The Johns Hopkins University Press
2715 North Charles Street
Baltimore, Maryland 21218-4363
www.press.jhu.edu

Book design: Teresa Bonner

Library of Congress Cataloging-in-Publication Data

America's national park roads and parkways : drawings from the Historic American Engineering Record / Timothy Davis, Todd A. Croteau, and Christopher H. Marston, editors ; introductory essay by Timothy Davis ; and foreword by Eric DeLony.
p. cm.—(The road and American culture)
Includes bibliographical references.
ISBN 0-8018-7878-0 (alk. paper)
1. National parks and reserves—United States—Pictorial works. 2. Roads—United States—Pictorial works. 3. Parkways—United States—Pictorial works. 4. Bridges—United States—Pictorial works. 5. Historic sites—United States—Pictorial works. 6. Roads—United States—Design and construction. 7. Parkways—United States—Design and construction. 8. Bridges—United States—Design and construction. 9. United States—Pictorial works. 10. United States—History, Local. I. Davis, Timothy, 1958– II. Croteau, Todd A. III. Marston, Christopher. IV. Historic American Engineering Record. V. Series.
E160.A578 2004
388.1'0973—dc22 2003023211

A catalog record for this book is available from the British Library.

To all of the talented researchers, writers, delineators,
photographers, and editors who produced the documentation.
Their individual efforts have contributed
to a greater understanding of the American landscape.

Contents

Foreword, by Eric DeLony — *ix*
Acknowledgments — *xi*

Drawing on the Road, by Timothy Davis — *1*
- A Brief History of National Park Roads and Parkways — *3*
- The HAER NPS Park Roads and Bridges Program — *13*

National Park Roads — *21*
- Acadia National Park — *25*
- Crater Lake National Park — *49*
- Glacier National Park (Going-to-the-Sun Road) — *56*
- Great Smoky Mountains National Park — *74*
- Marsh-Billings-Rockefeller National Historical Park — *87*
- Mount Rainier National Park — *95*
- Rocky Mountain National Park — *110*
- Scotts Bluff National Monument — *115*
- Sequoia National Park (Generals Highway) — *123*
- Shenandoah National Park (Skyline Drive) — *131*
- Yellowstone National Park — *144*
- Yosemite National Park — *168*
- Zion National Park — *201*

Parkways — *211*
- Baltimore-Washington Parkway — *214*
- Blue Ridge Parkway — *219*
- Colonial Parkway — *247*
- George Washington Memorial Parkway — *259*
- Natchez Trace Parkway — *272*
- Rock Creek and Potomac Parkway — *283*

National Military Parks — *299*
- Chickamauga and Chattanooga National Military Park — *302*
- Gettysburg National Military Park — *310*
- Shiloh National Military Park — *322*
- Vicksburg National Military Park — *326*

Park Road Precedents — *337*
- Bronx River Parkway — *340*
- Columbia River Highway — *363*

Appendix: The Historic American Engineering Record National Park Service Park Roads and Bridges Projects — *373*
Selected References — *377*

Foreword

In 1986, Randy Biallas, chief historical architect for the National Park Service, asked me if I might be interested in a collection of inspection reports for the bridges in the National Park System. This offer intrigued me, so over the next few months Randy and I scheduled time to review a selection of the hundreds of inspection reports completed by the Federal Highway Administration to assist the Park Service in managing its tremendous network of roads and bridges. Many of these reports detailed simple concrete culverts and austere steel-girder spans, but others depicted beautiful stone-veneered concrete arches and picturesque rustic wooden structures. The artistic, technological, and historic qualities of these bridges were so impressive that we began making plans to augment the basic inspection reports with more extensive documentation compiled to the standards of the Historic American Engineering Record (HAER), a division of the National Park Service charged with documenting America's engineering and industrial heritage.

Fortunately, we found a sympathetic sponsor in John Gingles, the servicewide federal lands highway coordinator for park facilities management, who was responsible for overseeing bridge rehabilitation and replacement. Since the Park Service was contemplating a comprehensive program of road and bridge improvements, a systematic survey of this nature would dovetail nicely with the agency's broader practical and regulatory needs. Gingles saw the merit of this collaboration and agreed to fund a pilot project so that HAER could begin to develop a programmatic approach to the daunting challenge of documenting the complex and wide-ranging resources of America's national park roads and parkways. In the summer of 1988, we organized a small team to look at a selection of the bridges on park roads in the National Capitol Region in Washington, D.C.

From this modest beginning emerged one of the most comprehensive and exciting projects ever conducted by the Historic American Engineering Record. For the next twelve years, HAER fielded teams of young architects, landscape architects, historians, and engineers, who documented roads, bridges, and associated landscapes throughout the national parks. Though the project started out exclusively focused on bridges, it soon became apparent that documenting bridges in isolation from their surroundings was shortsighted. Park Service landscape architects and Bureau of Public Roads engineers had integrated the roads and bridges into their natural settings so carefully and subtly that it would have been shameful not to include the roadside landscape and associated features as part of the documentation package. By the third year of the program, HAER and its sister agency, the Historic American Buildings Survey (HABS), had begun to conceptualize the documentation process in broader terms as a means of recording both engineered structures and the broader landscapes in which they were situated. The HAER Park Roads and Bridges Program went on to pioneer new methods of recording and interpreting park roads as complex cultural landscapes encompassing a wide range of natural, technological, perceptual, and historical attributes. The following portfolio provides ample evidence of the value of this expansive and innovative approach.

I have always viewed the HAER documentation of park roads and bridges as a means of making park managers and maintenance staff more aware of the unique attributes and cultural significance of the historic bridges and culverts, elegant and sinuous rights of way, and other charismatic features associated with traditional park roads. The detailed historical reports and attractive, information-packed graphics can play an important role in helping to convince park road managers and designers of the need to retain the historic fabric and feeling of roads and related features when they undergo rehabilitation or replacement. The partnership between the engineers and landscape architects that produced these wonderful park roads and parkways continues with our friends and colleagues at the Federal Highway Administration. This portfolio and its sequel volume, providing additional details on the history of park roads and bridges, testify to the strength of this partnership and to the possibilities that are within reach when engineers, historians, park managers, preservationists, and landscape architects combine their talents to ensure that park roads—and indeed all highways and byways—are integrated with their surroundings so that they "lie lightly on the landscape" and showcase America's bountiful natural and historic resources.

Eric DeLony
Chief, Historic American Engineering Record (retired)
National Park Service

Acknowledgments

Like America's national park roads and parkways themselves, this book reflects the diverse skills, strenuous labor, and steadfast commitment of many individuals and federal programs over an extended period of time. The Historic American Engineering Record (HAER) National Park Service (NPS) Park Roads and Bridges Program was the brainchild of HAER Chief Eric DeLony and NPS Chief Historical Architect Randall J. Biallas, who set the project in motion in 1988. Former Historic American Buildings Survey (HABS) Historian Sara Amy Leach initiated the parkway component of the documentation with the endorsement of HABS Chief Paul Dolinksy. Former HABS/HAER Chief Robert Kapsch supported the project in its early years. His successor, E. Blaine Cliver, saw it through to completion. This ambitious documentation effort would not have been possible without the generous backing of the Federal Highway Administration's Federal Lands Highway Office, which underwrote the program's expenses throughout the duration of the project. Former Federal Lands Highway Program Administrator Tom Edick was an early and enthusiastic proponent, along with his staff members Donald Patrick and Al Burden. Current Federal Lands Highway Program Administrator Art Hamilton continued this support. The National Park Service's Facilities Management Division was the direct sponsor of the project, disbursing federal highway monies and providing administrative support and technical guidance. Division Chief Dale Wilking and retired Deputy Chief John Gingles expressed unwavering commitment to HAER's multiyear effort. NPS Federal Lands Highway Program managers Lou DeLorme and Mark Hartsoe served as liaisons with HAER and provided ongoing advice and encouragement. The superintendents and staffs of individual parks lent invaluable technical and administrative support, sharing their knowledge of park resources and history. National Register of Historic Places Historian Linda McClelland generously shared her research on Skyline Drive and other aspects of the history of park roads. The National Park Foundation was a key partner in the publication of this volume. Marjorie Thomas served as the foundation's chief liaison with the Park Service throughout this process.

The drawings themselves were produced by more than a decade's worth of HAER summer teams working long hours under tremendous pressure in frequently adverse conditions. The number of architects, landscape architects, and other illustrators participating in the program is too great to permit individual acknowledgment, but the artist responsible for each drawing is identified in the title block to the right or bottom of the image. Among the illustrators who returned to HAER for numerous years and deserve special recognition are Pete Brooks, Robert Harvey, and Ed Lupyak. The drawing component was markedly enriched through the participation of interns from the International Council on Monuments and Sites (ICOMOS) U.S. summer exchange program, administered throughout most of the program's duration by Ellen Delage. A particularly valuable source of landscape documentation expertise was provided by representatives of the Polish National Center for Research and Documentation of Historical Monuments, participating through the ICOMOS program. The project historians are also too numerous to list, but their contributions are acknowledged on the cover sheets of individual projects. Richard Quin, Dawn Duensing, and Timothy Davis lent their historical talents to multiple projects, producing an extraordinary amount of insightful research and analysis. The photographic component of the documentation, comprising more than four thousand large-format black-and-white negatives and more than two hundred large-format color transparencies, was created by HAER photographer Jet Lowe, HABS photographer Jack Boucher, and contractors Brian

Grogan, David Haas, and William Faust. HAER Park Roads and Bridges Program historians Justine Christianson and Kelly Young performed extraordinary service in organizing and finalizing a tremendous amount of documentation and preparing it for transmittal to the Library of Congress; both contributed significantly to the preparation of this volume. George F. Thompson, president of the Center for American Places, enthusiastically embraced this book project and shepherded it from concept to reality.

America's National Park Roads and Parkways

Tundra Curves, Trail Ridge Road, Rocky Mountain National Park, 1993 *Photographer: Brian Grogan, HAER*

Drawing on the Road

Timothy Davis

America's national park roads and parkways are outstanding design achievements that exemplify the harmonious integration of highway engineering and landscape architecture. The challenge of building roadways through remote and rugged terrain inspired some of the most spectacular feats in the history of American engineering, yet even in the most demanding locations, designers went to great lengths to make sure that park roads would "lie lightly on the land," impinging as little as possible on their natural and cultural surroundings. By designing roadways to showcase park scenery and employing graceful curves, naturalistic landscaping, and attractive rustic features, the National Park Service (NPS) created a world-renowned road system that provides access to America's most treasured scenery while standing as a remarkable social, artistic, and technological achievement in its own right.

In recognition of the importance of America's national park roads and parkways, the Historic American Engineering Record (HAER) undertook an ambitious documentation project that produced the drawings compiled in this volume. Since its founding in 1969, HAER has recorded a wide range of sites and structures that exemplify America's technological and industrial heritage. Like its companion program, the Historic American Buildings Survey (HABS), HAER is a key component of the Cultural Resources Division of the National Park Service, which promotes the preservation and stewardship of cultural resources both within the national parks and throughout the country as a whole. HABS/HAER teams produce detailed records of historic sites ranging from individual structures to industrial complexes and from residential districts to expansive cultural landscapes. This documentation, consisting of large-format photographs, historical reports, and measured and interpretive drawings, is

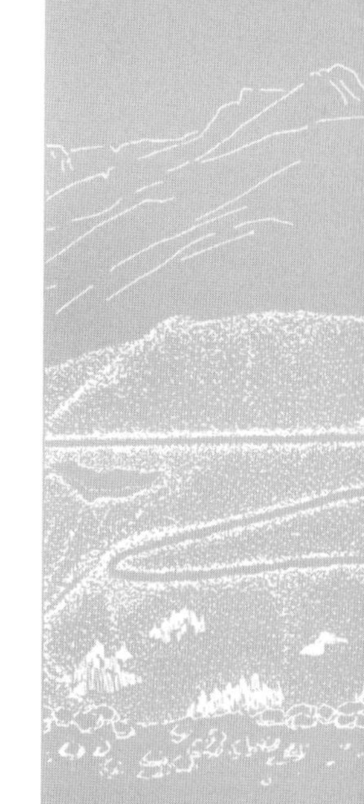

Grand Loop Road, Yellowstone National Park, 1916 *Photographer: J. E. Haynes; Haynes Foundation Collection, Montana Historical Society*

housed in the Library of Congress, where it is available to the public in both print and electronic forms. With the completion of this project, America's national park roads and parkways take their place alongside such icons of America's architectural and engineering heritage as the Golden Gate Bridge, the White House, and the Washington Monument.

The drawings in this book articulate the unique characteristics of America's national park roads and demonstrate the complex social, technological, and aesthetic influences that produced them. With innovative combinations of graphics and text, the drawings detail the characteristics of individual sites and structures; explicate landscape design strategies, construction processes, and environmental conditions; and show how these forces combine to shape the motorist's view from the road. While not every park could be included, a wide range appears in the following pages: from western giants like Yellowstone, Glacier, and Yosemite to eastern gems such as Acadia, Great Smokies, and the diminutive Marsh-Billings-Rockefeller National Historical Park. Separate sections cover national parkways and national military parks. A final section presents two independent projects that significantly influenced NPS policies: New York's Bronx River Parkway and Oregon's Columbia River Highway. A complete collection of the drawings produced by the HAER NPS Park Roads and Bridges Program would require a substantially larger volume. This compendium provides a representative overview of park road resources along with a broad sampling of related documentation techniques.

In addition to providing a rich source of information about park road development, this volume is intended to serve as a source book for historic site documentation. The challenge of documenting park road landscapes led the Historic American Engineering Record to experiment with a wide range of interpretive strategies and representational techniques. Park roads are continually evolving landscapes encompassing a complex array of physical features and cultural attributes—from engineering, architecture, and landscape architecture to ecological conditions, historical developments, and human perceptions. Capturing these qualities through the traditional documentary medium of two-dimensional ink-on-Mylar drawings was a daunting task. The drawings in this volume represent a range of responses to this challenge, combining conventional documentation with a broad array of innovative interpretations. Engineered structures such as bridges and culverts are generally represented through traditional methods such as sections, elevations, and axonometric projections. Additional technical details are revealed through construction process drawings and multilayered "peel-away" views that illustrate the various components of engineered structures. Natural systems are portrayed through maps, profiles, and botanical sketches. A wide variety of creative diagrams illustrate evolving design strategies and landscape development techniques. The motorist's perspective is conveyed through skillful renditions of views and vistas. These techniques were developed to communicate the complexities of park roads, but they can easily be adapted to document and interpret a broad array of historic sites and cultural landscapes. Whether applied to traditional site documentation or to the rapidly growing field of cultural landscape analysis, these drawings should be seen not as blueprints for identical efforts but as sources of inspiration for future attempts to communicate the essential qualities of the places that give form and meaning to the American experience.

A Brief History of National Park Roads and Parkways

The first national park roads were relatively primitive affairs constructed by private entrepreneurs and the U.S. Army Corps of Engineers. Some of the more progressive road builders sought to incorporate aesthetic concerns in their designs, but meager budgets and the challenges of construction in mountainous terrain necessitated a pragmatic focus on providing basic circulation systems to afford access to remote western wonderlands. Crude and dangerous pack trails slowly gave way to narrow, winding dirt roads designed to accommodate the colorful stagecoaches that provided the primary means of transportation. Construction crews labored with picks, shovels, and horse-drawn equipment to carve rudimentary roadbeds out of the steep mountainsides. Black powder was occasionally used to blast away obstacles, but road builders were generally forced to follow the dictates of the terrain, zigzagging up treacherous slopes with unsightly switchbacks and hazardous hairpin turns. Unimproved dirt or gravel surfaces produced choking clouds of dust under the heavy pounding of stagecoaches and four-horse teams.

This historical essay draws on the author's research on park roads and parkways and the many historical reports produced by the HAER NPS Park Roads and Bridges Program. The individual reports, drawings, and photographs can be found in the HABS/HAER collection of the Library of Congress's Prints and Photographs Division or accessed on line through the Library of Congress's American Memory website.

Despite modest resources, primitive equipment, and difficult conditions, early park road builders accomplished many impressive feats. In Yellowstone, America's first national park (established 1872), the Army Corps of Engineers completed entrance roads from surrounding railroad depots and constructed a Grand Loop Road providing access to the park's major features. The engineers built or improved approximately three hundred miles of roadway, contending with rugged canyons, steep mountain passes, troublesome swamps, and treacherous hot spring formations. Most of this work was completed between 1883 and 1905. In Yosemite, which was administered by the state of California from 1864 until 1890, private toll road companies provided the first vehicular access. Rival concerns from the outlying communities of Big Oak Flat, Coulterville, and Mariposa constructed steep stage roads that zigzagged through the mountains to the valley floor, where visitors could tour the primary sights on a relatively smooth and level carriage drive. In Mount Rainier National Park, the Army Corps of Engineers built the scenic Nisqually Road between 1903 and 1910, upgrading an existing toll road and extending it farther up the mountain to the subalpine meadows of Paradise Valley. The corps also constructed some of the first roads at Crater Lake National Park and assumed control of road building in Yosemite Valley when it was declared a national park in 1890.

Most early park bridges were utilitarian spans of wood and steel, but several notable structures were constructed during the first decades of national park development. Yellowstone's Golden Gate Viaduct carried stagecoaches across a sheer cliff on the way from Mammoth Hot Springs to the geyser basins. The first version, constructed in 1884 by Army Corps of Engineers Lieutenant Dan Kingman, was a rickety wooden trestle that caused considerable consternation among faint-hearted tourists. In 1900, Kingman's successor Captain Hiram Chittenden built a more substantial concrete viaduct in the same location, carefully moving a twenty-four-ton stone monolith that had stood at the lower end of the original structure. (The stone pillar was relocated again in 1933, when Chittenden's viaduct was widened for automobile traffic.) Chittenden was also responsible for the impressive gateway arch erected at the park's northern entrance and for the graceful concrete bridge located just upstream from the Grand Canyon of the Yellowstone, both constructed in 1903. The bridge's elegant bow-like profile was achieved by using the Melan arch system, an early form of steel-reinforced concrete construction. Another unusual structure, the Corkscrew, or

Old Big Oak Flat Road, Yosemite National Park, 1903

Photographer: J. T. Boysen; National Park Service Historic Photograph Collection (NPSHPC)

Bad road conditions, Yellowstone National Park, 1922
Courtesy Yellowstone National Park

Spiral, Bridge on the East Entrance Road, was completed in 1904, easing passage on the steep grade of Sylvan Pass. Yosemite had no such elaborate structures in its early years, though the covered bridge at Wawona provided a touch of picturesque charm. The deep ravines cutting into the flanks of Mount Rainier forced engineers to construct several impressive timber trestles, which were replaced by more substantial steel-and-concrete structures in the 1920s and 1930s to accommodate the demands of automobile traffic.

The arrival of the automobile at the beginning of the twentieth century presented park managers with new challenges and opportunities. A few pioneering motorists made their way into Yosemite as early as 1900, but automobiles were rapidly banned from national parks as menaces to public safety on the narrow, winding stagecoach roads. Aggressive lobbying by automobile clubs and tourism interests, combined with the widespread desire to make the national parks more accessible to middle-class Americans, ensured that this prohibition would be short-lived. Automobiles were legalized in Mount Rainier in 1907, Crater Lake in 1911, Glacier in 1912, and Yosemite and Sequoia in 1913. Yellowstone, the last to admit cars, opened its gates to motorists in 1915. The stagecoach companies quickly adapted to the situation, chauffeuring visitors around the parks in gaily colored touring cars.

The motorization of park transportation was generally heralded as a positive development. Not only did it eliminate the tedious and uncomfortable experience of clattering along in slow and dusty stagecoaches, but by significantly reducing the time and expense of park vacations, automobiles made the national park experience available to a broader segment of the public. While long-distance motor tourism was largely a middle-class pursuit for many years, it was significantly cheaper than the traditional railroad, stagecoach, and hotel vacation—especially if motorists stayed in modest cabins or engaged in the rapidly growing pastime of auto camping.

The National Park Service adopted an enthusiastically pro-automobile philosophy when the agency was formed in 1916 to improve the administration of park affairs. NPS leaders believed that embracing the automobile was both good for the country—because it allowed more visitors to enjoy the parks—and good for the Park Service—because it created broader support for park protection and expansion. The new agency proclaimed that one of its most pressing tasks was to upgrade the National Park System to accommodate motorists. Narrow, one-way roads had to be widened, sharp turns eliminated, and grades made more gentle and consistent. Road surfaces also had to be improved, as dirt and gravel wagon roads that were barely sufficient for stagecoach travel rapidly deteriorated under the added stress of automobile traffic. Guardrails also became more important, since vehicles would be moving at higher rates of speed and more people would be traveling on their own rather than relying on professional drivers. A variety of new facilities would have to be developed to accommodate the growing crowds of motorists. Gas stations, garages, parking lots, scenic turnouts, expanded lodging and dining options, and enhanced entrance stations and sign systems would be needed. New roads would

Auto Stages, Mount Rainier National Park, early 1920s *NPSHPC*

also have to be constructed, both in new areas the Park Service was bringing into the system and in older parks where it was difficult or impossible to upgrade historic routes to modern standards.

Although the National Park Service was determined to accommodate the influx of motorists, the agency's leaders insisted that the resulting development should not compromise the parks' natural and historic resources. Roads and related features were to be designed to harmonize with their surroundings, and great care was taken to ensure that the parks were not gridironed by excessive road construction. A limited number of high-quality motor roads would provide access to major attractions, but the vast majority of most parks would remain off-limits to automobiles. The preferred strategy was to provide one or two roads linking prominent features and affording views of a cross section of park scenery while leaving large sections of backcountry accessible only on foot or by horseback.

The daunting task of upgrading the national park road system required strong leadership, a highly skilled team of design and engineering professionals, and a substantial commitment of federal funds over an extended period. NPS Director Stephen Mather provided able and enthusiastic leadership, as did his successor Horace Albright and subsequent Park Service directors. Technical expertise and aesthetic direction were provided by the NPS's growing staff of landscape designers, working closely with highway engineers from the U.S. Bureau of Public Roads (BPR). Congress was initially reluctant to commit significant funding, but Mather finally secured a $7.5 million road-building appropriation in 1924, followed by an additional $2.5 million in 1928. New Deal programs such as the Public Works Administration and the Civilian Conservation Corps channeled millions in much-needed funding and thousands of workers into the effort during the 1930s. Because of the tremendous amount of high-quality work accomplished during the 1920s and 1930s, this period has been widely regarded as the Golden Age of national park road building. The character and extent of the national park road system enjoyed by visitors today were largely established during this era.

The interbureau agreement between the NPS and the BPR played a fundamental role in guaranteeing the aesthetic quality and technical excellence of these park roads. Under this arrangement, which began in the early 1920s and was formalized in 1926, the NPS outlined the general goals, location, and aesthetic considerations of each project. BPR engineers then conducted detailed surveys and developed specific plans. After securing NPS approval, the BPR supervised the actual construction process. NPS personnel also monitored the work for conformance with guidelines developed to ensure that park road construction exhibited high standards of design and execution. This arrangement, which still governs the relationship between the NPS and the BPR's successor agency, the Federal Highway Administration, was hailed by both parties as an exemplary model of interprofessional and interagency collaboration.

NPS Chief Landscape Architect Tom Vint (second from left) **and design staff, 1934** *NPSHPC*

The BPR's engineering expertise enabled the NPS to undertake a far more ambitious construction program than it could have accomplished on its own, including projects of extreme technical difficulty, such as Glacier National Park's Going-to-the-Sun Highway and Zion's Mount Carmel Tunnel. The NPS, in turn, encouraged the BPR to pay greater attention to the aesthetic aspects of road construction. BPR engineers embraced the challenge and played a major role in translating the NPS's "lying lightly on the land" philosophy from idea to reality. While the NPS established the general parameters of the projects and prepared most of the architectural treatments for bridges, guardwalls, and related structures, BPR engineers ensured that proposals were technically feasible and suggested many attractive and innovative design solutions of their own. The spectacular but visually unobtrusive cliffside location of Glacier's Going-to-the-Sun Road, for instance, was developed by BPR engineers after NPS designers had proposed a technically simpler but much more conspicuous route zigzagging up the heart of a pristine mountain valley. It was this insight, in fact, that convinced Mather to make the BPR a full-time partner in national park road development.

The NPS-BPR collaboration produced a distinctive "park road" aesthetic that is explicated in detail in the following portfolio of drawings. NPS landscape architects and BPR engineers drew heavily on nineteenth-century carriage road design techniques but updated these practices to accommodate the demands of automobile traffic and the geographic diversity of the national parks. The alignments of park roads were subtly configured to showcase park landscapes in the most attractive and engaging manner. Turnouts were provided at particularly scenic locations so that tourists could pull off without endangering themselves or other motorists. Where roadside vegetation obstructed potentially appealing views, trees were cut to create carefully calculated vistas. Variety of scenery was sought to maintain the motorist's interest. Rather than follow a ridgeline for miles, a park road or parkway drive might drop down to a hillside location for a while or wind back and forth across the crest to provide views in both directions. In forested areas, the distance between the edge of the road and the treeline could be manipulated to achieve a variety of different effects, from a narrow, wooded aisle to a broad, green corridor or an alternating series of smaller and larger "rooms" carved out of the surrounding woodlands. Selective cutting to create an undulating treeline produced a more natural and informal effect. Trees, shrubs, and charismatic rock formations were generally allowed to remain much closer to the pavement than along conventional roadways, bringing motorists into intimate contact with their surroundings.

Although hazardous hairpin curves were eliminated wherever possible and most curves were straightened somewhat to accommodate the shift to automobile traffic, park roads remained much more circuitous than conventional highways. Prolonged straightaways were avoided as much as possible. Sinuous curves were thought to be more attractive. Curvilinear alignments also allowed road builders to follow the contours of the land more closely, significantly reducing the need for expensive, visually unappealing, and environmentally disruptive excavations. Road widths were narrower than contemporary highways, as well, rarely exceeding twenty-two feet, with no shoulders to speak of and curbs only on the most highly developed

Going-to-the-Sun Road, Glacier National Park, 1932 *Photographer: George Grant; NPSHPC*

parkways and heavily used areas. The NPS realized that it could not afford to equip the vast amount of road mileage under its control with modern concrete or bituminous "asphalt" paving during this initial improvement campaign, so the agency tried to stabilize as much of its road network as possible with various oil-treated gravel or macadam surfaces. More substantial road-mixed asphalt pavements were applied in some situations. Whenever possible, local rock was used for the crushed-stone component of park road pavements, saving on material and hauling expenses and helping the roads blend in with their surroundings. The results of this policy were most striking in the southwestern parks, where the red-tinted pavements harmonized with local rock formations.

Where excavations were unavoidable, park road builders did their best to minimize disruptions and rehabilitate areas disturbed during the construction process. Instead of leaving steep, raw banks alongside newly graded roadways, as was the common practice in contemporary highway construction, the NPS insisted that roadside embankments be sloped back gradually and rounded to resemble natural contours. Sodding and planting programs helped stabilize disturbed roadsides and naturalize their appearance. Revegetation policies promoted the use of native species arranged in naturalistic groupings. On particularly steep and prominent mountainsides, tunnel excavation afforded a means of minimizing the visual impact of road construction while improving safety and often reducing costs as well. Excavating a tunnel required specialized equipment and expertise, but it eliminated unsightly hillside cuts and expensive retaining walls while reducing the risks of rock fall and snowslides. Designers could also employ tunnels to direct the motorist's gaze, either through windows providing lateral views, as in Zion's Mount Carmel Tunnel, or by using exit portals to frame striking scenes, as was done to spectacular effect with Yosemite's Wawona Tunnel.

The desire to protect motorists on steep hillsides led the Park Service to develop an array of attractive guardwall designs, which were eventually codified and sent to parks as standardized plans. Constructed of locally quarried, hand-laid stone masonry in most cases, the walls had a pleasingly rugged appearance and mirrored the hues and texture of neighboring outcrops. Considerable variation was possible within the NPS rustic design framework. Different sizes and shapes of rocks were laid in various patterns ranging from seemingly random to relatively regular, although every effort was made to avoid a formal rectilinear appearance. Both flat-topped and crenelated walls were popular. Further variation could be achieved by changing the height, width, shape, and spacing of the crenelations. Rugged log guardrails were used in many locations, and a few hybrid designs of stone and timber could be found in Yellowstone and elsewhere. Traditional split-rail fences flanked many southern park roads and parkways, but these were intended more for scenic and sentimental purposes than for safety. While the subtle variations between different styles may not be apparent to casual observers, the sensation of driving along winding mountain roadways flanked by rugged stone or timber barriers has long been one of the classic components of the national park experience.

NPS bridge construction followed a similar design ethos during this period. While the BPR ensured that new bridges embraced the latest technology in reinforced concrete and steel construction, the architectural designs and surface treatments reflected the NPS's naturalistic design philosophy. Most bridges were faced with rustic stone veneers. Longer spans were occasionally left undecorated, especially if their elevations were unlikely to be viewed by the general public. In addition to favoring locally quarried stone and random masonry patterns, NPS guidelines stipulated that great care should be taken to ensure that drill marks and other signs of quarrying be turned inward so that only the weathered surfaces of rocks were exposed to view. NPS bridges from this period tended to have simple lines with subtly accented arches. In Acadia National Park, however, designers produced a more eclectic array of architectural treatments, with rustic stone enlivened by a variety of picturesque embellishments. Where steel girders were used to support bridge spans, they were often covered with heavy timbers to maintain the rustic design aesthetic and harmonize better with their surroundings. On Colonial Parkway, several modern reinforced-concrete bridges were faced with red brick laid in traditional bond patterns to harmonize with the architecture of nearby Williamsburg. Portals for the tunnel carrying the parkway under Colonial Williamsburg were given similar treatments. In mountainous areas, tunnel portals were generally left in their natural state or, where stabilization with concrete was necessary, faced with random rubble masonry. Culvert headwalls were also given rustic stone treatments where they were likely to be viewed by the public.

Entrance stations similarly reflected the determination to harmonize structures with their natural and cultural surroundings. Again, considerable variation could be found within the reigning rustic aesthetic as designers sought to employ local materials and evoke regional building traditions. Oversized log structures were popular in the western parks. Adobe, or concrete made to mimic adobe, was favored in the Southwest. Vernacular building motifs were adapted for many park structures in the Appalachian parks. Entrance stations were often the most prominent buildings encountered by motorists, but similar policies shaped the design of a wide range of visitor facilities, from comfort stations and concessions to water fountains and wayside exhibits. The iconic brown wooden signs familiar to generations of park visitors became standard features of the national park landscape during this period.

The intent behind these design strategies was to showcase park scenery, minimize the negative effects of development, and ensure that improvements harmonized with their surroundings. The visual appeal, intimate scale, and rustic character of roads constructed during this Golden Age of national park road building profoundly shaped the ways that visitors viewed their national parks. The sensation of motoring along sinuous curves, crossing attractive rustic bridges, winding beside rugged guardwalls, and being led through a series of sensational views rapidly became an integral component of the national park experience. For most visitors, in fact, the park road experience *was* the national park experience. By encouraging motorists to slow down and enjoy their surroundings instead of racing blindly along to their next destination, moreover, park road designers were implicitly engaged in the process of traffic calming, an approach to transportation management that is becoming increasingly popular in contemporary planning circles. The narrow, winding roads, enfolding trees, looming chasms, and encroaching vegetation brought motorists into intimate contact with their surroundings while simultaneously preventing them from driving too fast. A leaflet handed out to early park motorists encapsulated this sentiment, advising that "park roads are for leisurely driving only. If you are in a hurry, you might do well to take another route, and come back when you have more time" (cited in U.S. Department of the Interior, NPS, *Park Road Standards,* 1968).

A large number of important projects were completed during this period, producing some of the most spectacular scenic drives in the world and firmly establishing the distinctive character of America's national park roads and parkways. Glacier's Transmountain Highway, later renamed the Going-to-the-Sun Road, was one of the most significant accomplishments of this era. Not only was it an impressive example of high-altitude road construction and the progenitor of the historic NPS-BPR design collaboration, but also, by providing an attractive motor route across the park, it completed the last link in the western Park-to-Park Highway when it was officially opened in

Christine Falls Bridge, Mount Rainier National Park, 1992
Photographer: Jet Lowe, HAER

1933. The Zion–Mount Carmel Highway Tunnel, completed in 1930, surmounted another imposing obstacle by enabling motorists to climb from the valley floor to the surrounding rim country. Concealing the road in a tunnel prevented the much-needed roadway from scarring the park's striking sandstone cliffs. This technically complex project remains the longest tunnel in the National Park System. A similarly ambitious tunnel was constructed in Yosemite during the relocation of the Wawona Road, which was rerouted significantly to make it safer for automobile traffic. The lower tunnel opening framed a spectacular view of Yosemite Valley. Three shorter tunnels were used to reconfigure the Big Oak Flat for automobile traffic without undue scarring of the mountainsides. Both of these projects were completed during the early 1930s, along with improvements to the El Portal Road so that it could provide year-round access to the valley floor.

Ricksecker Point Overlook, Mount Rainier National Park

Photographer: Jack E. Boucher, 1960; NPSHPC

Rocky Mountain National Park's Trail Ridge Road was another impressive new project. It was designed to replace the nearby Fall River Road, which, though scarcely a decade old, was already considered unsuited for automobile traffic. Trail Ridge Road followed more moderate grades and wound its way over the Continental Divide in gentle and graceful curves. The highest sections were constructed over areas of permafrost that posed novel challenges to the NPS-BPR design team. Generals Highway, developed in stages between 1921 and 1935, led motorists through the towering trees and impressive scenery of Sequoia and General Grant National Parks. Yellowstone did not receive any major new roads or dramatic relocations, but grades and curves were eased, bridges were upgraded, and road surfaces were stabilized with oil treatments and asphaltic pavements. At Mount Rainier, the NPS widened, paved, and straightened the Nisqually Road and undertook the construction of several new routes designed to provide access to previously remote sections of the park. Numerous other parks received vital improvements to their road systems during the 1920s and 1930s.

In addition to upgrading roadways in the traditional western parks, the Park Service made a major effort to acquire and develop additional national parks, particularly in the eastern United States. Before the creation of Lafayette (later Acadia) National Park in 1919, there were no national parks east of the Mississippi River. By the outbreak of World War II, a number of eastern parks had been established to preserve priceless natural and historical resources and provide recreational and educational opportunities within easy reach of the bulk of the nation's population. Recognizing that the vast majority of these visitors would be traveling by automobile, the NPS and the BPR began outfitting the new parks with modern motor roads. The engineers and landscape architects combined the lessons learned in western park construction with new approaches to recreational roadway design developed by eastern park system officials to serve the needs of major metropolises. The result was an impressive collection of scenic roadways located within a day's drive of millions of Americans. This tremendous expansion of eastern national park opportunities enabled motorists to enjoy picturesque seaside drives, solemn Civil War battlefields, striking mountain scenery, and a vast array of natural and historic sites without taking extended vacations or having to endure the rigors of cross-country journeys over marginal interstate highways.

Acadia, the first East Coast addition, had a significant head

The newly completed Transmountain Highway, or Going-to-the-Sun Road, Glacier National Park, circa 1933
Vintage postcard: Timothy Davis

Cadillac Mountain Road, Acadia National Park, 1995
Photographer: Jet Lowe, HAER

start in the road-building department. John D. Rockefeller Jr., who maintained a summer residence nearby, had already outfitted the island with an impressive network of carriage drives using the same design aesthetic that informed the national park road experience. To maintain the sanctity of this network for horse-drawn traffic, Rockefeller helped the Park Service develop a complementary system for automobiles, contributing considerable funds and hiring the noted landscape architect Frederick Law Olmsted Jr. to ensure their scenic quality. The NPS-BPR collaboration also shaped Acadia's road system, resulting in such exemplary projects as the popular Cadillac Mountain Road, completed in 1932.

Shenandoah National Park's Skyline Drive exemplified the Park Service's determination to bring the national park experience within easy reach of eastern motorists. Constructed between 1931 and 1936, Skyline Drive was intended to provide an attractive day trip for motorists from Washington, D.C., and surrounding regions. The winding ridgetop route provided the same sorts of scenic views and rustic features that western tourists had grown to associate with the national park experience. Skyline Drive also epitomized the manner in which a park's road could become synonymous with the park itself. While Shenandoah National Park encompassed nearly 200,000 acres of mountainous terrain, most visitors' perceptions were limited to the narrow corridor surrounding the seventy-mile-long scenic drive. Great Smoky Mountains National Park, a much larger reservation developed at roughly the same time in the heart of the southern Appalachians, had its own signature scenic drive in the Newfound Gap Road, along with additional tour routes providing access to the park's varied features. Numerous stone-faced bridges were required to span the park's many streams, but the most striking road-related structure remains the Loopover at Newfound Gap, a

Blue Ridge Parkway, 1950

Photographer: T. W. Kines; U.S. Bureau of Public Roads Collection, National Archives

Civilian Conservation Corps crew constructing guardrails, Yosemite National Park

Yosemite National Park

stone-faced corkscrew bridge that replaced two dangerous and unsightly switchbacks.

Parkways played an important role in the NPS's effort to reach a broader audience during the 1930s. Originally conceived as sedate carriage drives linking nineteenth-century urban parks, parkways were adapted to the automobile age by the staff of New York's Westchester County Park Commission, several of whom went on to work for the National Park Service. The NPS realized that parkways could be used both to provide attractive links between individual park units and to serve as destinations themselves, with recreational features catering to day-trippers and overnight visitors. After developing Colonial Parkway as a relatively modest scenic drive between the historic sites of Jamestown, Yorktown, and Williamsburg, Virginia, the NPS expanded the parkway concept to an unprecedented scale. The Blue Ridge Parkway, following the crest of the Appalachian Mountains between Skyline Drive and Great Smoky Mountains National Park, was 469 miles long, while the Natchez Trace Parkway, extending most of the way from Nashville, Tennessee, to Natchez, Mississippi, would eventually contain more than four hundred miles of rolling two-lane blacktop. Along with the traditional emphasis on natural scenery, these parkways were designed to showcase cultural landscapes. Split-rail fences, restored log cabins, and service facilities designed to mimic vernacular building traditions celebrated regional culture and exposed urban motorists to America's agrarian heritage. The location, acquisition, design, and development of these two extraordinary linear parks was such a daunting task that the Blue Ridge Parkway was not completed until 1987 and the Natchez Trace remains under construction today.

As part of a major reorganization of government bureaus in 1933, the Park Service inherited several suburban parkways in the Washington, D.C., area, along with other parks and historic properties in the nation's capital. Rock Creek and Potomac Parkway, linking Rock Creek Park in northwest Washington with the National Mall, was completed in 1936 with the help of New Deal funds. Rock Creek Park, with its collection of historic drives and bridges dating to the early twentieth century, also became a Park Service property. Mount Vernon Memorial Highway, stretching along the Potomac from Washington to Mount Vernon, was constructed by the BPR between 1928 and 1932. It was immediately incorporated into the proposed George Washington Memorial Parkway, which was designed to preserve both sides of the Potomac River in the national capital region. Minor progress was made on

Construction crew, Going-to-the-Sun Road, circa 1932
Glacier National Park

this broader project before World War II intervened, and it was not completed until the 1960s. The Baltimore-Washington Parkway, while planned during this era, was not developed until the mid-1950s, by which point it functioned largely as an unusually attractive commuter thoroughfare.

The same government reorganization that placed Washington's parks under Park Service control gave the NPS authority over most of the nation's military parks. These unique reservations, intended to preserve and commemorate significant battlefields such as Gettysburg, Shiloh, and Vicksburg, were previously under the jurisdiction of the War Department. Their road systems had been developed by a variety of special commissions, veterans groups, and military custodians and were designed to showcase historic battlefield landscapes and associated commemorative elements. Bridges, road surfaces, and related features were generally constructed to very high standards because of the reverence accorded these symbolically important sites. The Park Service attempted to preserve the character of these reservations as much as possible while selectively upgrading aspects of their road systems to accommodate automobile traffic and increased visitation.

The impressive extent and quality of the roads completed during this Golden Age of park road development owed much to the skill and determination of the construction forces that labored on them, often in adverse and trying conditions. Most of the primary road work—grading, paving, drainage, and bridge construction—was performed by private contractors whose employees were experienced with specialized construction techniques and the operation of heavy equipment. The invention of dynamite and the advent of gasoline- and diesel-powered machinery enabled these crews to overcome obstacles that had stymied earlier road builders. Paradoxically, more powerful technology enabled the NPS to produce roadways that often seemed to lie more lightly on the land. Large amounts of earth and rock could be moved with relative ease, making it possible to create longer, more graceful curves and gentler grades. The Civilian Conservation Corps (CCC) and its predecessor, the Emergency Conservation Work (ECW) program played an indispensable role in park road development between 1933 and 1942. The labor-intensive but relatively low-skilled work of regrading slopes and planting trees and ground cover provided employment for thousands of CCC enrollees, without whom the NPS would never have been able to undertake its ambitious roadside rehabilitation programs. Many miles of guardwall and guardrail were constructed by CCC workers, along with campgrounds, picnic shelters, and other visitor facilities. America's entry into World War II ended the CCC program and park road development in general, as men and resources were shifted to more vital national needs.

By the late 1940s, when Americans had begun to put the war behind them and resume recreational travel, the NPS realized that it was facing major problems. Not only were new facilities needed to accommodate the ever-increasing numbers of park visitors, but years of low budgets had created a maintenance backlog that was turning into a growing embarrassment. The situation became steadily worse during the 1950s, as widespread automobile ownership, rising incomes, expanded leisure time, and rapidly improving interstate highways brought unprecedented crowds into already overburdened national parks. Prominent commentators and politicians began to decry the condition of America's national parks and call for action to improve them. The NPS's response to this outcry was to initiate a comprehensive ten-year improvement program, which Congress funded with approximately $1 billion of appropriations over the ensuing decade. Designated Mission 66, this program was intended to bring the National Park Service in line with contemporary needs in time for the agency's fiftieth anniversary in 1966.

Building on the prewar Park Service's determination to promote and accommodate motor tourism, Mission 66 resulted in a comprehensive upgrading of roads and related facilities, along with a significant increase in the number of parks and a resulting expansion in total park road mileage. Few new roads were built in existing parks, but many prewar roads were widened, straightened, and paved to accommodate larger, faster, and more numerous automobiles. Parking lots, campgrounds, and visitor facilities were expanded, and many older bridges were replaced with new structures better suited to modern traffic demands. Yellowstone's beloved Chittenden Bridge was torn down in 1961 and replaced with a more substantial but decidedly less graceful structure. While some efforts were made to evoke traditional park road aesthetics, many new bridges and visitor facilities were unabashedly modern structures exemplifying the unadorned form-follows-function ethos of Modernist design. The cost of labor and materials also factored into the appearance of Mission 66 structures, since the hand-hewn rustic quality of prewar park roads had become prohibitively expensive. Advances in highway engineering made it easier to calculate and construct complex serpentine curves, however, improving the safety and visual harmony of parkways and park roads designed for higher speeds and heavier traffic volumes.

Although most visitors appeared to enjoy these improvements, Mission 66 had many detractors. In addition to objecting to the Modernist aesthetics, critics contended that the Park Service was building too many roads and "improving" others to the point where they lost their scenic beauty, intimacy, and primitive appeal. By making it easier for ever-increasing crowds of visitors to motor through the parks at higher speeds, moreover, the Park Service seemed to be encouraging a trend that many traditional park supporters found deeply disturbing. By the mid-1950s, a growing chorus of environmentalists and wilderness advocates began protesting Park Service road improvement efforts. One of the most controversial Mission 66 projects was the modernization of Yosemite's Tioga Road, an old mining trace over the Sierras that had been upgraded several times during the prewar era but remained such a slow and tortuous trip that many motorists were reluctant to explore the beauties of the High Sierra. While wilderness advocates embraced the road's primitive nature as a means of keeping the area from becoming over-

Overcrowded conditions leading to Mission 66: Yellowstone National Park, circa 1954
Yellowstone National Park

Mission 66 Road Project, Teton National Park, circa 1960
NPSHPC

crowded, NPS Director Conrad Wirth and other prominent officials insisted that the road should be widened, paved, and straightened to encourage greater use. Environmentalists were more successful in other locations, stopping or curtailing the extension of George Washington Memorial Parkway and the creation of a proposed parkway along the historic Chesapeake and Ohio Canal from Washington, D.C., to Cumberland, Maryland.

By the mid 1960s—and even earlier in some regions—the Park Service was beginning to reexamine its road-building policies. The earlier strategy of bisecting each park with a major scenic artery was dropped in the development of most new parks in the Pacific Northwest and Alaska, where road construction was limited to minimal approach roads. The long-contemplated plan of encircling Mount Rainier with an all-weather motor road was also abandoned. In 1967 the Park Service assembled a blue-ribbon panel, including prominent landscape photographer and Tioga Road critic Ansel Adams, to evaluate the status of roads in the national parks. The panel concluded that additional road development should be kept to a minimum and issued a series of guidelines intended to ensure that future improvement and rehabilitation programs would reflect the spirit of prewar park road design. The panel also suggested that the NPS should begin looking at ways to facilitate different methods of transportation, such as shuttle buses and alternative fuel vehicles, especially in the most crowded parks. The 1969 National Environmental Policy Act added regulatory restrictions to these proposals by mandating a rigorous procedure of environmental review for major construction projects.

NPS park road managers and designers rose to the challenge in many locations, devising alternative transportation systems for congested areas like Yosemite Valley and striving to ensure that new projects were both attractive and environmentally sensitive. The BPR's modern-day successor, the Federal Highway Administration (FHWA), continued to provide state-of-the-art technical support and design guidance. Blue Ridge Parkway's Linn Cove Viaduct epitomizes the blending of modern design and technology with traditional park road values and greater ecological awareness. Constructed between 1979 and 1983 as the final link in the 469-mile parkway, the 1,243-foot-long concrete structure carries the parkway drive high above the scenic and ecologically sensitive slopes of North Carolina's Grandfather Mountain. To avoid scarring the hillside, builders lowered individually precast concrete segments into place by crane from the previously assembled portion of the structure. No conventional access roads were permitted, and the only ground-level construction was the drilling of foundations for seven piers supporting the superstructure. The complex engineering and construction combined with the graceful curves of the elevated roadway to earn the project accolades on technological, ecological, and aesthetic grounds. The Double Arch Bridge on the Natchez Trace Parkway provides another example of innovative engineering in the service of park road development. The Double Arch Bridge was completed in 1994 as a means of spanning a broad valley, stream, and highway. Its designers used a somewhat similar process to produce a light, soaring structure that barely seems to touch the ground as it stretches more than 1,600 feet above the rural Tennessee landscape.

Even as the NPS and FHWA were employing increasingly sophisticated technologies to accommodate the ever-expanding crowds of motorists who poured into the parks in the closing decades of the twentieth century, a growing awareness of the historical importance of the nation's classic park roads and parkways began to spread throughout the park road community. Cultural resource managers began to emphasize the importance of preserving and interpreting historic roads and related resources, particularly those associated with the Golden Age of park road building between the two world wars. There was greater recognition that these roadways bore witness to an important era in Park Service history and reflected broader developments in highway engineering and landscape architecture. For many park professionals and visitors alike, they also epitomized the role of the national parks in twentieth-century American life. While nineteenth-century parks had catered to a small number of well-heeled tourists, who rattled along rutted dirt roads in dusty horse-drawn stagecoaches, the twentieth-century national park experience was rooted in the more expansive vision of middle-class motorists winding through the wilderness along smoothly paved and sensitively designed scenic drives. By balancing the desire for democratic access with a firm commitment to scenic preservation and conscientious stewardship, America's national park roads and parkways represent a social, aesthetic, and technological achievement without par in the history of park development and recreational planning.

The HAER NPS Park Roads and Bridges Program

This book is a direct result of the growing appreciation for the historic stature and cultural significance of America's park roads and parkways. Recognizing the need to document and interpret these irreplaceable national landmarks, the Historic American Engineering Record worked with current NPS facilities managers

Linn Cove Viaduct, Blue Ridge Parkway, 1997
Photographer: David Haas, HAER

A family enjoying the Blue Ridge Parkway, circa 1950
Photographer: Abbie Rowe; NPSHPC

and the FHWA's Federal Lands Highway Program to develop a multiyear program to record national park roads and parkways through large-format photographs, historical narratives, and measured and interpretive drawings. After a 1988 pilot project focusing on park bridges in the nation's capital, HAER directed its attention to classic western parks such as Yellowstone, Yosemite, Glacier, and Mount Rainier. The program's scope soon expanded to encompass eastern parks and parkways along with national military parks and additional bridges in the Washington area. The Historic American Building Survey assisted in the effort, documenting George Washington Memorial Parkway and Rock Creek and Potomac Parkway, the two oldest parkways in the national capital region. By the time the project drew to a close in 2002, the combined programs had documented dozens of park roads and parkways along with hundreds of individual bridges, producing more than four thousand photographs, 476 drawings, and over ten thousand pages of written history. Although not every park could be surveyed in detail, the oldest and most prominent park road systems were documented along with many significant but less well-known roads and bridges. Several non–Park Service scenic roads were also documented using similar techniques and funding from state and local sources. These include New York's Bronx River and Taconic Parkways, Connecticut's Merritt Parkway, California's Arroyo Seco Parkway, and Oregon's Historic Columbia River Highway.

Most of the HAER NPS Park Roads and Bridges documentation was completed through the HABS/HAER summer internship program. This popular program provides students and young professionals with hands-on training while enabling HABS/HAER to produce significantly more documentation than could be accomplished by the division's full-time staff. The HABS/HAER summer program has long served as a primary training ground for preservation professionals, with many participants returning year after year before moving on to full-time positions with state, local, and federal preservation agencies. After a highly competitive selection process, interdisciplinary teams comprising students and recent graduates in architecture, landscape architecture, architectural history, and related disciplines typically devoted twelve weeks to researching and documenting each individual parkway or park road system. Architects and historians from the Washington-based HABS/HAER office familiarized the team with the physical resources of the site, helped identify the defining characteristics of each park road or parkway, explained the program's broader goals, and provided instruction in documentation techniques. Team members usually spent the next few weeks surveying the park road system, conducting preliminary measurements, sketching, and studying archival materials such as original development plans, construction reports, and historic photographs. They identified potential topics for drawings, presenting thumbnail sketches of prospective graphic layouts, which were reviewed by team members and project leaders and then developed in greater detail through further research, discussion, and on-site examination.

At several points during the course of each summer, HAER staff

HAER documentation team surveying a historic bridge, Mount Rainier, 1992

Photographer: Todd Croteau, HAER

visited the teams to provide further direction and help refine the drawings. Key park employees, such as cultural resource managers, engineers, and landscape architects, were invited to participate in these reviews. After one last round of consultation with HAER program leaders, the final drawings were executed in archival ink on Mylar sheets measuring either 24 by 36 inches or 33 by 44 inches in accordance with the Secretary of the Interior's Guidelines for Recording Historic Structures. The basic HABS/HAER field manual, *Recording Historic Structures* (Washington, D.C.: American Institute of Architects Press, 1989) provided additional guidance in historic site measurement and documentation, as did the growing corpus of park road drawings completed by HAER teams over previous summers. While computer-assisted drawing (CAD) technologies were employed in a few instances, nearly all the NPS Park Roads and Bridges documentation was hand drawn. Despite the increasing sophistication of CAD technologies, this traditional approach seemed best suited to capturing the nuances of park road landscapes, especially for the more interpretive drawings that came to play an increasingly prominent role in the documentation effort. In a few experiments, historic photographs were scanned onto Mylar to depict construction processes and similar issues, but hand drawing generally produced better results in this realm as well. Not only did the scanned photographs frequently produce disturbing contrasts of tone and content, but the creative capabilities of hand drawing afforded greater clarity and concision by enabling delineators to combine components of multiple images and edit out distracting or superfluous visual material. The hand-drawn, hand-lettered appearance also helped establish a sense of stylistic continuity among the wide-ranging drawing sets while evoking the handcrafted rustic quality of park roads themselves. The unique characteristics of the many different parks generated a tremendous variety of conceptual topics, landscape scenes, and structural detail, but the overall appearance of the HAER NPS Park Roads and Bridges documentation displays a remarkable sense of character and consistency, especially for a project that relied on dozens of individual artists and took more than a decade to complete.

Despite the overall similarity in tone and format, the drawing component of the documentation evolved significantly over the course of the program. Early projects focused on engineered structures such as bridges, tunnels, culverts, and viaducts, but the program's scope gradually expanded to address broader issues of highway engineering, landscape design, environmental perception, ecology, and cultural history. This transformation reflected both the inherent complexity of park road environments and the growing influence of interdisciplinary cultural landscape studies. The detailed bridge drawings that dominated early documentation efforts at Yellowstone, Glacier, and Yosemite provided a wealth of technical information, but it rapidly became apparent that new conceptual strategies and more creative methods of graphic representation were needed to convey the complex spatial, temporal, and experiential aspects of park roads and parkways. HAER's project managers worked with field teams to move beyond traditional site plans, sections, and elevations and develop interpretive drawings of construction processes, landscape design techniques, topographical experiences, and evolving historical processes. By promoting an interdisciplinary perspective on historic road research, the cultural landscape approach helped expand the NPS Park Roads and Bridges Program from a narrow technical investigation into a more ambitious social, environmental, and conceptual exploration.

The benefits of this expanded approach became obvious in the next group of projects conducted at parks such as Great Smokies,

HAER delineator Daniela Trettel preparing a drawing for the Mount Rainier documentation project, 1992

Photographer: Todd Croteau, HAER

Pohono Bridge, Yosemite National Park. Documented by HAER in 1991
Photographer: Brian Grogan, HAER

Mount Rainier, and Acadia, where conventional bridge documentation was augmented by drawings depicting such diverse themes as the evolution of road networks over time, the geological and vegetative characteristics of park road corridors, the ways in which roads and related features conformed to the lay of the land, and the motorist's changing view of the roadside landscape. Process drawings and cutaway views provided insights into the ways in which park roads were constructed and landscaped to harmonize with their natural surroundings. Additional sheets chronicled the topographical details of selected sites along with characteristic roadside features such as guardwalls, drainage systems, signs, and entrance stations.

This broad-ranging approach became increasingly important as HAER expanded its efforts to encompass parkways and military parks. Complex conceptual agendas, historical and interpretive elements from various eras, and elaborately orchestrated views and vistas imbue these park road landscapes with tremendous richness and complexity. Military parks were intended both to preserve and commemorate hallowed scenes of conflict. They contain historic and designed landscapes, commemorative and interpretive features, and circulation networks comprising battle-era historic roads and historically significant tour routes; all of these elements are often overlapping and intertwining. HAER conveyed these concepts with diagrams of design strategies, comparisons of key landscapes during various time periods, and depictions of commemorative and interpretive elements such as signs, monuments, wayside exhibits, and road alignments tracing troop movements or other historically significant routes. Parkway designers were similarly concerned with shaping both physical landscapes and cultural values, employing classic landscape design techniques to combine naturalistic scenery with celebrations of America's colonial and agrarian heritage. Although it was possible to document Washington's 2½-mile-long Rock Creek and Potomac Parkway in minute detail using digitized maps produced by aerial reconnaissance, the tremendous length of most other parkways and park road systems made such a comprehensive approach impossible, or at least inordinately expensive. Instead, team members identified distinctive landscape zones and represented them through a mixture of site plans, impressionistic vignettes, renderings of significant details, and illustrations of dominant botanical and topographical characteristics. Additional drawings articulated key management policies such as scenic easements and agricultural leases, which play essential roles in establishing and maintaining the character of many parkway and park road landscapes.

The final phase of the HAER Park Roads and Bridges Program involved diversifying the collection with examples from additional parks and parkways throughout the country and returning to several of the parks surveyed early in the process to augment the initial bridge drawings with additional documentation incorporating the broader cultural landscape approach that had evolved over the course of the project. This further broadening of the research scope produced excellent historical reports on relatively undeveloped parks such as Lassen, Redwood, Hawaii Volcanoes, and Haleakala, along with innovative drawing projects at Rocky Mountain National Park—where the Trail Ridge and Fall River Roads were compared in terms of grade, width, curvature, and landscape setting—and at Scotts Bluff National Monument—where attempts to convey the symbolic resonance of the site and the role that the Depression-era road-building project played in the social history of local communities complemented more traditional technical drawings of construction processes and engineered features.

The return visits to Yellowstone, Yosemite, and Glacier were extremely productive. The Glacier team worked closely with park staff to develop an array of drawings specifically designed to augment the park's efforts to interpret the history and significance of the Going-to-the-Sun Road, which had recently been designated a National Historic Landmark—the first national park road to receive this honor. Drawings traced the history of transportation in the park; articulated crucial location decisions, design considerations, and environmental factors that shaped the route; chronicled the construction process; and outlined contemporary management techniques such as native plant revegetation and snow removal. Additional drawings detailed variations in guardwalls and visitor facilities, including a demonstration of the evolution of auto camp designs—a vital but often undervalued aspect of the national park landscape. At Yosemite, the follow-up team documented the diversity of guardwalls, entrance stations, and landscape types and also developed drawings showing the historical evolution and key characteristics of individual park roads. A set of historically focused bridge drawings traced the evolution in design and technology from the nineteenth century to the present. The final Yellowstone team was also more historically oriented, documenting the evolution of individual bridges and viaducts over time and tracing changes in transportation practices, road widths and composition, entrance stations, and interpretive features. The team historian supplemented the original construction narrative with a more interpretive overview of design philosophies and brief reports on several smaller side roads that had been overlooked during the initial project.

Although the historical component of the documentation program was generally more broad-based from the start, the individual reports also tended to become more complex and multifaceted as the project progressed, expanding from basic technical and administrative matters to situate park road development practices in larger social, cultural, and conceptual contexts. The historians usually worked somewhat independently from the drawing teams, producing narrative reports that varied in length from a page or two for minor bridges to fifty to five hundred pages for comprehensive overviews of entire park road systems or parkways. For convenience, many of the more complex park road histories were broken down into an overview of broader development patterns, narrative accounts of the evolution of individual roads, and short reports on significant bridges and other engineered structures. The historical development of national parkways was related through comprehensive narratives augmented by separate reports on individual bridges. Historians frequently played an active role in the drawing component of the projects, as well, both by helping shape the content of the visual documentation and by providing texts for individual drawings. The historians and delineators worked together to develop preliminary itineraries for the large-format photographers. HABS/HAER staff photographers or specialized contractors used these suggestions to guide their efforts, which were designed to capture a representative sample of landscape views, engineered structures, and construction details rather than to provide blanket coverage of park road resources—an impossible or, at least, exorbitantly expensive task, given the vast mileage of America's national park road and parkway system.

At the end of each project, the teams made final presentations to park staff and other interested parties. These events proved to be extremely valuable, both by providing detailed information to assist park road management efforts and, perhaps more importantly, by fostering greater appreciation for the value of park roads as significant national resources in need of preservation and interpretation. Many employees who had worked in parks for years reported that these sessions changed their perspectives on park roads, making them more aware of historically significant features and fostering greater respect for the contributions of road development to national park history. Park visitors and local media representatives also attended many of these presentations. Such public participation, combined with sympathetic newspaper articles, helped further the program's goal of promoting greater appreciation for the history of park roads and the roles they have played in shaping the national park experience.

The ultimate value of the HAER NPS Park Roads and Bridges Program is as wide-ranging, multifaceted, and historically significant as America's park roads and parkways themselves. The basic HABS/HAER mission has long been to create an archive of American design while serving as a training ground for preservation professionals. The HAER NPS Park Roads and Bridges Program produced impressive results in both of these realms. Several thousand miles of park roads, hundreds of bridges, and a wide variety of associated features were documented over the course of the program; field team members have gone on to careers in academia, private practice, and local, state, and federal preservation agencies. Another longstanding function of HABS/HAER documentation has been to serve as a means of "preserving on paper" aspects of the built environment that are threatened by natural disasters, human actions, and other historical forces. This baseline documentary role has already proven to be an important benefit of the HAER NPS Park Roads and Bridges Program. Several bridges documented by HAER have been demolished, and others are currently slated for replacement or removal. Guardrails and guardwalls have been altered and extended, pavements have been widened and straightened, landscape settings have changed because of both natural and cultural factors, and significant road relocations have been enacted or proposed to address a variety of social, technological, and environmental concerns. Additional alterations to park roads and related resources are inevitable as structures and landscapes age, visitation continues to increase, and management priorities evolve to accommodate changing social practices, technological influences, and cultural concerns.

While the documentary function of the HAER Park Roads and Bridges Program is invaluable, the drawings, research, and photographs can also play significant proactive roles in the realm of design, management, and public education. This material can be used by engineers, landscape architects, maintenance personnel, and cultural resource specialists to inform decisions about preservation, rehabilitation, and other management issues. By calling attention to the historic importance of park roads and providing detailed baseline information, HAER documentation projects have already influenced management decisions in several parks. The HAER NPS Roads and Bridges Program has also contributed significantly to the growing commitment on the part of the National Park Service and the Federal Highway Administration to embrace policies geared toward maintaining the character of classic park roads and employing traditional park road development techniques in new projects.

This type of documentation can be particularly valuable in the area of interpretation, where the attractive graphics and detailed historical narratives can be transformed into engaging and informative publications, exhibits, and wayside panels. HAER has developed interpretive brochures for many parks, which present the history of park road development in concise and accessible terms. These brief summaries combine photographs, text, and drawing excerpts in a convenient format designed to be distributed at park entrances and visitor centers. The most ambitious public outreach effort to date was a major exhibition hosted by the National Building Museum in Washington, D.C., from June 1997 through March 1998. This exhibition, titled Lying Lightly on the Land: Building America's National Park Roads and Parkways, told the park road story through historic images and artifacts, HAER drawings and photographs, rare con-

Lying Lightly on the Land: Building America's National Park Roads and Parkways, exhibition at the National Building Museum, Washington, D.C., 1997 *HAER*

Samples of the brochure series developed by HAER to disseminate research to the public
HAER

struction films, contemporary video footage, and interactive features including computerized virtual park road tours and create-your-own-park-road design simulations. Lying Lightly on the Land was highly praised by visitors and received several national awards for articulating the complexities of park road development in an engaging and accessible manner and calling attention to the ways in which national park roads and parkways embodied broader trends in American social history. The exhibition will serve as the basis for a companion volume to the current portfolio of drawings. By combining additional photographic documentation with compelling historical imagery and a more extensive historical narrative, this treatment will make the park road story accessible to an even broader audience.

The HAER NPS Park Roads and Bridges Program has made major contributions to the study of America landscape architecture, engineering, recreation management, and social history. By documenting historical development patterns and articulating key technological factors, design features, and landscape characteristics, HAER has chronicled the development of America's park roads and parkways while providing a solid basis for their preservation and interpretation. Stewardship of historic road resources has become an increasingly prominent concern at all levels of national park management. The HAER Park Roads and Parkways Program has played a prominent role in advancing the cause of historic road preservation outside the parks as well. Just as the original park road design ethic spread from the national parks to the country at large, the historic road preservation ethos is beginning to take root and flourish at local, state, and national levels. From a broader methodological perspective, the program has markedly advanced the field of historic site documentation. By combining traditional methods of architectural and engineering description with innovative strategies for communicating an increasingly complex array of social, environmental, and conceptual factors, HAER has helped transform the study of park roads and other designed landscapes into a more broad-based and engaging interdisciplinary endeavor. Above all, the HAER NPS Park Roads and Bridges Program has created a permanent and compelling record of an important aspect of America's cultural heritage that continues to be enjoyed and appreciated by the millions of visitors to America's national park roads and parkways.

Recently rehabilitated road, Yellowstone National Park, 1996

Photographer: Timothy Davis, HAER

National Park Roads

From the rugged coast of Maine to the Appalachian highlands, Rocky Mountains, southwestern deserts, and Hawaiian islands, national park roads afford access to much of America's most spectacular scenery and many of its hallowed historic sites. Park roads are more than just means of arriving at exotic destinations, however. With their attractively landscaped surroundings, charming rustic features, and imaginatively engineered alignments, they are enormously appealing environments in their own right. Millions of visitors a year enjoy the sensation of motoring along their gentle curves, crossing over picturesque bridges, and partaking of an endlessly fascinating panorama of scenic beauty and cultural diversity.

Since one of the key goals of park road design is to disguise the artifice of their construction, most motorists have little appreciation for the tremendous amount of hard work and creative thinking that goes into their development. Even trained professionals are often unaware of the subtle design strategies and sophisticated engineering practices involved in the creation of these seemingly simple yet extraordinarily complex cultural landscapes. The following drawings show how park roads were built, trace their evolution over time, and delineate the design techniques that allow park roads to provide safe and stimulating access to a wide range of natural and cultural features without unduly affecting the landscapes through which they pass.

While not every national park could be included in this survey, the following examples provide a representative sample of the richness and diversity of America's park road landscapes while articulating the complex social, technological, aesthetic, and environmental forces that have made them cherished components of America's cultural heritage.

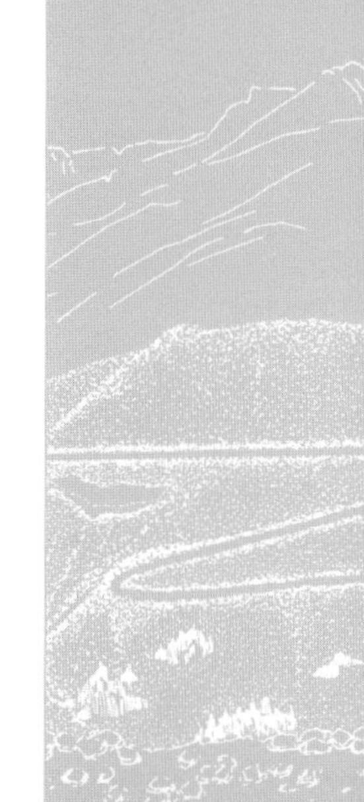

East Side Highway, Mount Rainier National Park, 1992

Photographer: Jet Lowe, HAER

National Park Road Systems

Acadia National Park Roads and Bridges

Cover Sheet PAGE 25
Geology of the Landscape 26
Road Design Similarities 27
System Separation 28
Grade Separations 29
Motor Road Development 30
Cadillac Mountain Road and Park Loop Road 31
Ocean Drive 32
Sieur de Monts Spring 33
The Bridges of Acadia National Park 34
Around the Mountain 39
Bubble Pond Carriage Road 40
Otter Creek Cove Bridge and Causeway 41
Duck Brook Motor Road Bridge, 1953 42
Bubble Pond Bridge, 1928 43
Stanley Brook Bridge, 1933 44
Brown Mountain Gatelodge, 1931 45
Jordan Pond Gatelodge, 1931 47

Crater Lake National Park Road System, 1919/1933

Cover Sheet 49
Evolution of Road 50
Road Design Principles 51
Comprehensive Design 52
Orientation Features 53
Historic Bridges and Conduits 54
Goodbye Creek Bridge, 1954 55

Going-to-the-Sun Road, Glacier National Park, I

Cover Sheet 56
Systems Map: Structures, 1926–1936 57
Bridges 58
Triple Arches, circa 1927 59
Baring Creek Bridge, 1931 60
Axonometric of Construction 61

Going-to-the-Sun Road, Glacier National Park, II

Cover Sheet 62
History of Transportation 63
Choosing a Route 64
Road Construction 65
Vista Design 66
Across the Divide: Experiencing the Landscape 67
Why a Tunnel? 68
Guard Rails and Retaining Walls 69
Park Entrances 70
The Loop 71
Auto Camp Design 72
Snow Removal 73

Great Smoky Mountains National Park Roads and Bridges

Cover Sheet 74
Reference Map 75
Evolution of Major Roads 76
Design Principles 77
Motorist Experience 78
Stonework 79
Roadside Barriers 80
Bridge Comparison 81
The Loop Over 83
Elkmont Vehicle Bridge 84
Cades Cove 85
Roaring Fork Motor Nature Trail 86

Marsh-Billings-Rockefeller National Historical Park Carriage Roads

Cover Sheet 87
Carriage Road Site Map 88
Bird's Eye View from the Northeast 89
The Recovering Landscape 90
Forest Experience 91
Road Anatomy 92
The Pogue Loop 93
A Working Landscape 94

Mount Rainier National Park Roads and Bridges

Cover Sheet 95
Historical Base Map, 1899–1992 96
Nisqually River Suspension Bridge, 1924/1952 97
Nisqually Construction Details 98
Nisqually Structural Subsystems 99
Christine Falls Bridge, 1928 100
Christine Falls Bridge 1992 Site Plan 101
Saint Andrews Creek Bridge, 1931 102
Laughingwater Creek Bridge, 1935 103
Chinook Pass Entrance, 1936 104
Construction Details 105
Deer Creek Bridge, 1939 106
Stevens Creek Bridge, 1941 107
Muddy Fork River Bridge at Box Canyon, 1952 108
Box Canyon Site Plan 109

Rocky Mountain National Park Roads

Cover Sheet 110
Fall River Road, 1920 111
Trail Ridge Road, 1932 112
Roadside Edging 113
Snow Removal 114

(continued on next page)

Scotts Bluff National Monument Summit Road
Cover Sheet 115
Historic Migration Corridor 116
Hiding the Road 117
Building a Road Up a Rock 118
Putting Men to Work 119
Tunnel Construction 120
Summit Road Design 121
Summit Development 122

Sequoia National Park, Generals Highway
Cover Sheet 123
Generals Highway 124
Tunnel Rock 125
Hospital Rock 126
Four Guardsmen 127
Stonework 128
Approach Bridges 129
Historic Bridges 130

Shenandoah National Park, Skyline Drive
Cover Sheet 131
Entrance Stations 132
Road Construction 133
Parapet Construction 134
Landscape Techniques 135
Road Details 136
Waysides and Picnic Areas 137
Overlooks, North District 138
Overlooks, Central District 139
Marys Rock Tunnel 140
Skyland 141
Big Meadows 142
Changing Nature of Skyline Drive 143

Yellowstone Roads and Bridges, I
Cover Sheet 144
Obsidian Creek Bridge, 1910 145
Obsidian Creek Bridge, Plan/Elevation 146
Fishing Bridge, 1936 147
Fishing Bridge, Plan/Elevation 148
Fishing Bridge, Isometric 149
Crawfish Creek Bridge, 1936 150
Crawfish Creek Bridge, Plan/Elevation 151
Crawfish Creek Bridge, Axonometric 152
Gardner River Bridge, 1939 153
Gardner River Bridge, Plan/Elevation 154
Gardner River Bridge, Details 155

Yellowstone Roads and Bridges, II
Cover Sheet 156
Experiencing Wonderland 157
Evolution of Yellowstone National Park Roadways 158
Roadway Construction: Methods and Evolution 159
Entrance Checking Stations 160
Experiencing the Landscape 161
Scenic Drives and Bypass Roads of the Grand Loop 162
Waysides and Pullouts 163
Golden Gate Viaduct 164
Corkscrew Bridge 165
Canyon Bridge over Jay Creek, 1895/1915 166
Chittenden Memorial Bridge, 1903/1962 167

Yosemite National Park Roads and Bridges, I
Cover Sheet 168
Historical Base Map 169
Wawona Covered Bridge, 1868 170
Wawona Covered Bridge, Isometric 171
Bridalveil Fall Bridges, 1913 172
Sentinel Bridge, 1919 173
Sentinel Bridge Construction Details 174
Yosemite Creek Bridge, 1921 175
Stoneman Bridge, 1932 176
Stoneman Bridge Construction Details 177
El Capitan Bridge, 1933 178
El Capitan Bridge Construction Details 179
Wawona Tunnel, 1933 180
Wawona Tunnel Construction Details 181
Cascade Creek Bridge, 1939 182

Yosemite National Park Roads and Bridges, II
Cover Sheet 183
Map of Historic and Current Roads 184
Race to the Valley 185
Coulterville Road, 1874 186
Glacier Point Road, 1882 187
Big Oak Flat Road, 1874/1940 188
Big Oak Flat Road Tunnels 189
Wawona Road, 1875/1933 190
El Portal Road 191
Mariposa Grove Roads 192
Tioga Road, 1883/1961 193
Valley Roads 194
Entrance Stations 195
Arch Rock Entrance 196
Guard Walls 197
Roadside Landscape 198
Bridge Evolution, Nineteenth Century–1920s 199
Bridge Evolution, 1920s–Present 200

Zion National Park Roads and Bridges
Cover Sheet 201
Zion National Park Roads and Bridges 202
Zion–Mount Carmel Highway Tunnel 203
Stages of Tunnel Construction 204
Site Schematic 205
Pine Creek Bridge 206
Pine Creek Bridge Construction Details 207
North Fork Virgin River Bridge, 1930 208
Construction Details 209

ACADIA NATIONAL PARK

ROADS AND BRIDGES

Located primarily on Mount Desert Island, Acadia National Park preserves all or portions of seven offshore islands and a peninsula jutting into the Gulf of Maine. Its approximately 35,000 acres include forty-one miles of coastline, the only fjord on the east coast, and the highest point on the American Atlantic coast.

The precursor to Acadia National Park, the Sieur de Monts National Monument, was established on July 3, 1916. When the monument was upgraded to park status three years later, it became the first national park east of the Mississippi and the only park in the country created solely from donations of private land. Initially called Lafayette in honor of its colonial heritage it was renamed Acadia in 1929.

Although present-day visitors traditionally drive to the top of Cadillac Mountain before dawn to witness the first sunlight striking the American continent, the roads of Mount Desert Island were not always so accommodating. Long before the park was established, Abnaki Indians hunted throughout the island on a sophisticated network of footpaths. When white settlers arrived in 1761, they too used these paths, widening many to make room for horses and buckboard wagons. The Native American trail to the top of Cadillac Mountain, for instance, was widened into a toll road during the 1850s. Early white settlers to Mount Desert Island also quickly began constructing their own roads in order to enhance economic and social interaction. As early as 1777 residents had connected the island communities of Bar, Bass, and Southwest Harbors with crude roads though they were often impassable during inclement weather.

With the rise of the automobile during the final decades of the 19th century, the beauty of Mount Desert Island became increasingly accessible to tourists. Sensing economic opportunity, some residents promoted road construction as a means of drawing visitors to the island. In 1888, for example, locals built a road to the shore of Eagle Lake in an effort to promote the island as a vacation destination. Other islanders were less enthused. Summer resident John D. Rockefeller, Jr., whose family fortune resulted from the Standard Oil Company, ironically saw the increase in the number of cars as a threat to his idyllic island getaway. In reaction to this symbol of urban industrial society, he began constructing carriage roads, first on his Seal Harbor estate and then by 1941 he had built more than forty miles of such roads throughout the park.

This project is part of the Historic American Engineering Record (HAER), a long-range program to document historically significant engineering and industrial works in the United States. The HAER program is administered by the National Park Service (NPS), US Department of the Interior. Conducted during the summers of 1994-95, the Acadia National Park Roads and Bridges Recording Project was jointly sponsored by HAER; Robert J. Kapsch, Chief and Eric DeLony, Principal Architect, the NPS Park Roads and Parkways Program; Mark Hartsoe, Program Manager, and Acadia National Park; Paul Haertel, Superintendent. Todd Croteau, HAER staff architect, and Richard Quin, HAER historian, gave overall supervision and direction to the project.

The 1994 recording team consisted of David Haney, supervising architect (Yale), landscape architectural technicians Joseph Korzeniewski (VPI), Kate Curtis (University of Leeds, England [International Council Of Monuments and Sites -US-ICOMOS]); architectural technician Shannon Barras (VPI); historian Richard Quinn; and Jet Lowe, HAER photographer.

The 1995 recording team consisted of harlen d. Groe, supervising landscape architect (ISU); landscape architectural technicians Edward Lupyak (PSU), Joseph Korzeniewski (VPI), architectural technician Sarah Desbiens (University of Montreal, Quebec, Canada [International Council of Monuments and Sites -US-ICOMOS]); historian Neil Maher (NYU); and Jet Lowe, HAER photographer.

In building these carriage roads, Rockefeller went to great lengths to ensure their aesthetic character. He personally oversaw much of their construction and hired landscape gardener Beatrix Farrand to recommend roadside plantings to hide unattractive scarring. Along with the beauty of the roads themselves, Rockefeller was equally concerned with the scenery they exhibited. He consciously designed his carriage routes to present the park visitor with a sequence of scenic experiences including views across meadows, alongside streams, and out from mountain tops. This desire to construct roads that complimented their natural surrounding is perhaps best illustrated by Rockefeller's sixteen stone carriage bridges. Painstakingly constructed from native materials that allow the bridges to blend in with the natural landscape, most also provide scenic vistas.

During the early 20th century as motorists continued to visit Mount Desert Island, they began urging park officials to open up Acadia's carriage roads to automobiles. Attempting to alleviate this public pressure, Rockefeller decided to build scenic motor roads for the park separate from the carriage system. He began in 1922 by constructing the "Mountain Road" motor road from the northern end of Eagle Lake to the Jordan Pond Tea House, and continued with similar projects during the following three decades. Here again Rockefeller maintained high aesthetic standards, this time by hiring landscape architect Frederick Law Olmsted, Jr. to consult on many of the routes. Although these roads were widely praised for their scenic quality, Rockefeller's overall plan to construct a twenty-three mile motor loop around the eastern portion of the park aroused considerable local disapproval from those fearing such a project would threaten the wilderness quality of several key regions. While Rockefeller was ultimately successful in seeing his plan carried out, when the Bureau of Public Roads completed the final segment in 1958 this community opposition had significantly altered the shape of his Park Loop Road.

In many ways the carriage and motor roads of Acadia National Park appear quite distinct. Signs, grade separations, and two picturesque gatehouses remind today's visitors that these two systems were intended to be kept physically separate. Yet the tension between Acadia's carriage and motor roads also lies at the heart of their shared history. Both were built in response to the rise of the automobile and each was influenced by the aesthetic vision of John D. Rockefeller, Jr. It is this common heritage that visitors atop Cadillac Mountain must keep in mind when they watch the sun rise over the horizon from the front seats of their automobiles.

DELINEATED BY: harlen d. Groe, Ed Lupyak, Sarah Desbiens, Joe Korzeniewski, and Neil Maher, 1995

NATIONAL PARK SERVICE
ACADIA ROADS & BRIDGES RECORDING PROGRAM
NATIONAL PARK SERVICE
UNITED STATES DEPARTMENT OF THE INTERIOR

ACADIA NATIONAL PARK ROADS & BRIDGES
BAR HARBOR VICINITY · HANCOCK COUNTY · MAINE

SHEET 1 OF 19

HISTORIC AMERICAN ENGINEERING RECORD ME-12

IF REPRODUCED, PLEASE CREDIT: HISTORIC AMERICAN ENGINEERING RECORD, NATIONAL PARK SERVICE, NAME OF DELINEATOR, DATE OF THE DRAWING

ACADIA NATIONAL PARK

Geology of the Landscape

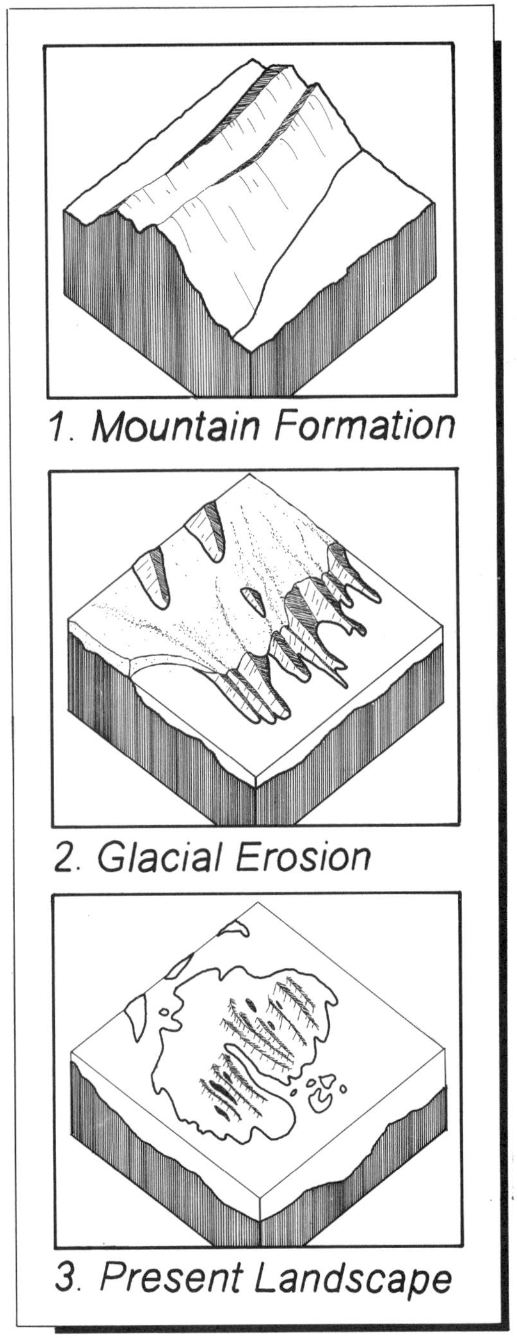

Acadia's mountains had their beginning long before visitors began watching the sunrise on Cadillac Summit. The first stage in their formation occurred approximately 400 million years ago, when a huge mass of molten rock moved upward beneath what is today Mount Desert Island. Eventually this magma broke through the older surface rock and cooled to form a coarse-grained granite ridge running east to west. As geologic time wore on, rivers, rain, and wind carried away much of the older surface rock while leaving behind the more resistant granites. As a result, V-shaped valleys were carved into this east-west lateral ridge.

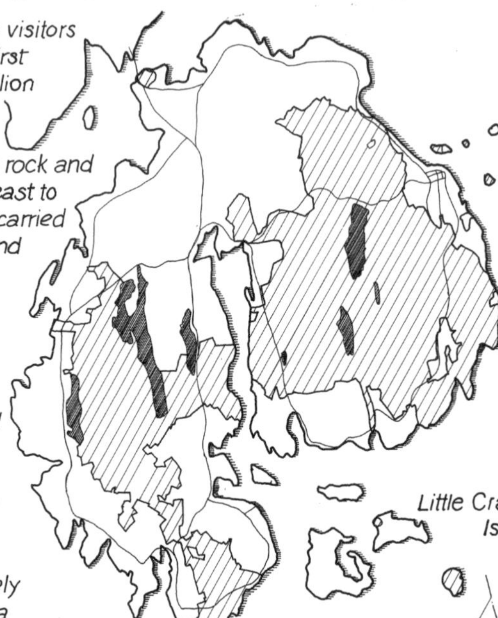

When glaciers descended out of Canada across New England two to three million years ago, this lateral ridge on Mount Desert Island acted as a barrier. As the ice sheet built up on the north flank of the ridge, it eventually spilled through the low places in the divide to form long finger-like glaciers which descended the southern face. The streaming of glacial ice across the divide cut deep valleys running north to south completely across the lateral ridge, which was now dissected into a series of detached peaks. This erosion continued until a rise in global temperatures halted and reversed these glaciers approximately 18,000 years ago. Left behind were the mountains, valleys, and glacial lakes of the present day park. As the ice receded further, rising sea levels separated Mount Desert Island from the mainland.

Although volcanic fire and glacial ice dominated the early geologic history of Mount Desert Island, the ocean has more recently shaped Acadia's landscape. Every headland, bay, and inlet has been influenced by the power of the sea. The constant pounding of waves against the island's cliffs, for instance, dislodges rocks to create scenic wonders such as Thunder Hole. These same waves also grind these rocks into fine particles and deposit them along the shore at locations like Sand Beach. The geologic forces that continually eroded and built up the mountains of Mount Desert Island more than 400 million years ago, therefore, are still at work today in Acadia National Park.

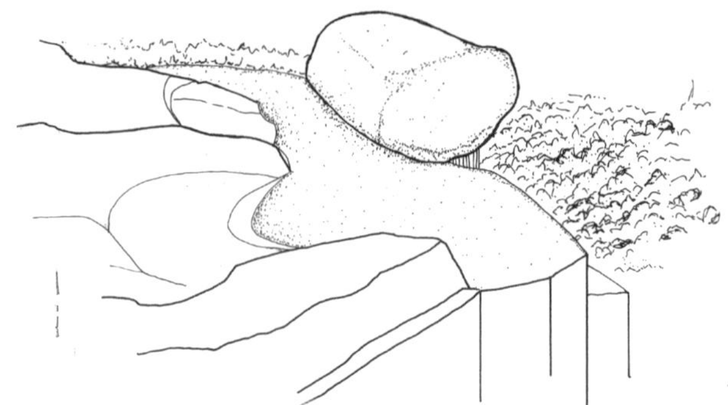

Glacial erratic, like this boulder near the summit of the South Bubble, were transported to Mount Desert Island by ice sheets three million years ago

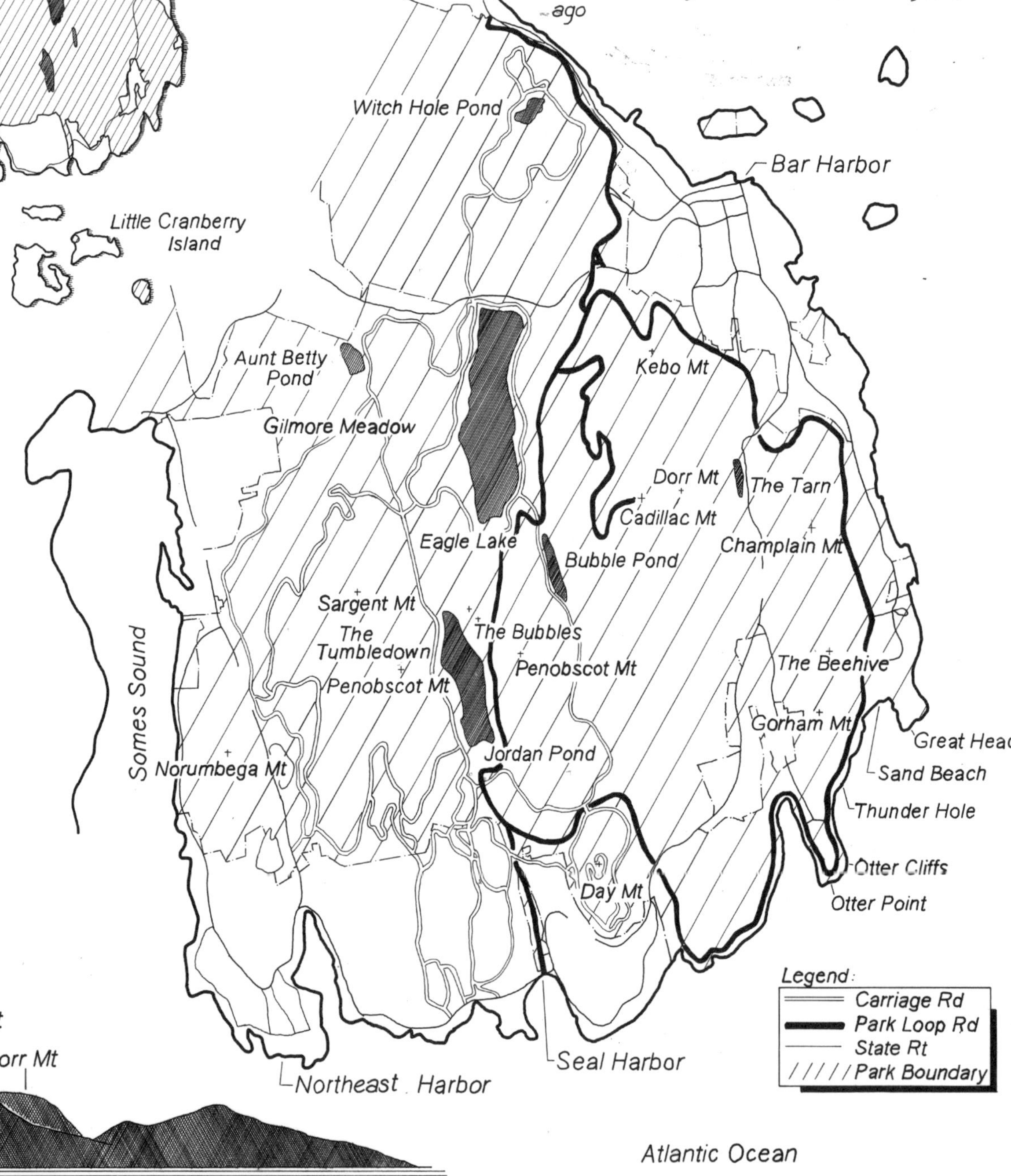

The Mountains of Acadia as Seen from Little Cranberry

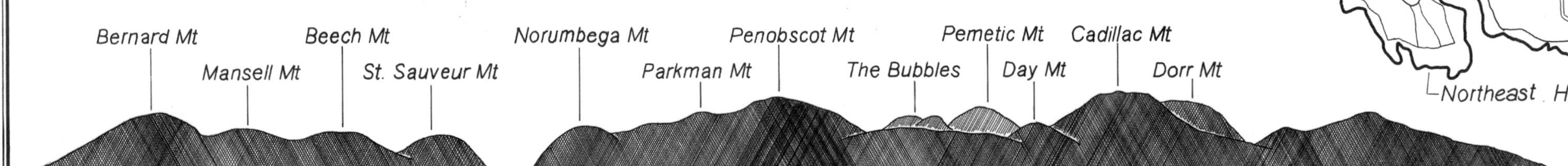

DELINEATED BY: Joe Korzeniewski, 1995

NATIONAL PARK SERVICE
ACADIA ROADS & BRIDGES RECORDING PROGRAM

ACADIA NATIONAL PARK ROADS & BRIDGES
BAR HARBOR VICINITY
HANCOCK COUNTY
MAINE

SHEET 2 OF 19

HISTORIC AMERICAN ENGINEERING RECORD
ME-12

IF REPRODUCED, PLEASE CREDIT: HISTORIC AMERICAN ENGINEERING RECORD, NATIONAL PARK SERVICE, NAME OF DELINEATOR, DATE OF THE DRAWING

ROAD DESIGN SIMILARITIES

1 Aunt Betty Pond Road

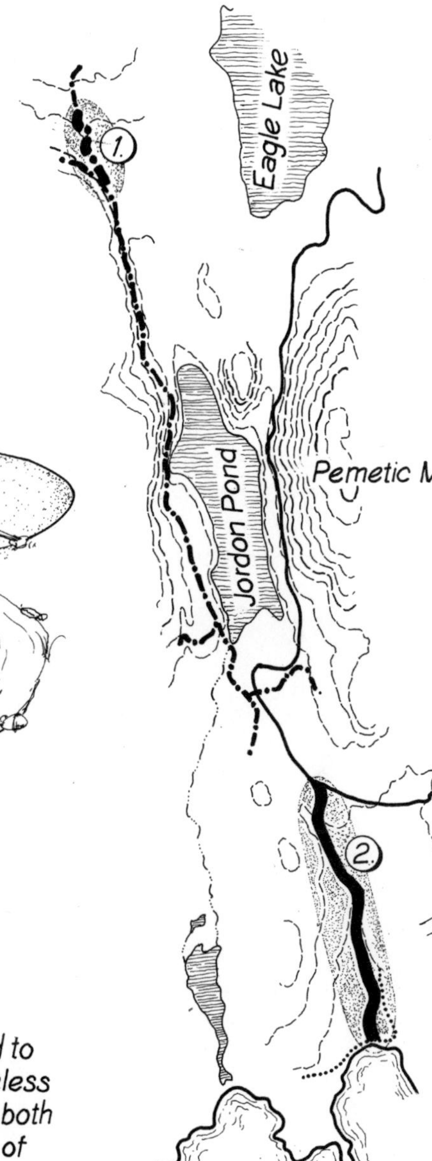

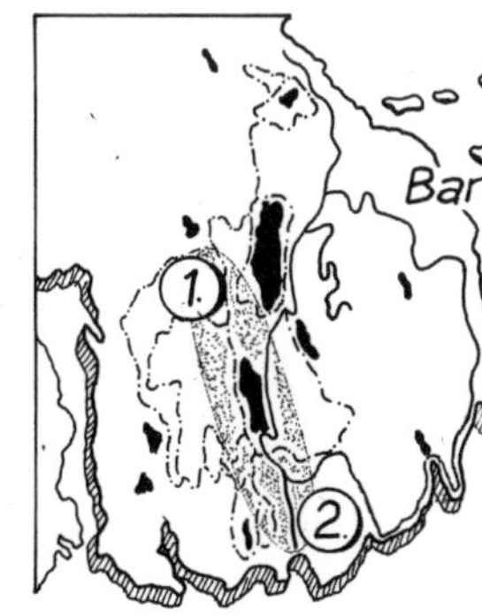

Rockefeller first used coping stones, known locally as "Rockefeller's teeth," on his carriage roads. In an attempt to enhance the consistency between the two systems, similar stone guardrails were also placed along the park's motor roads.

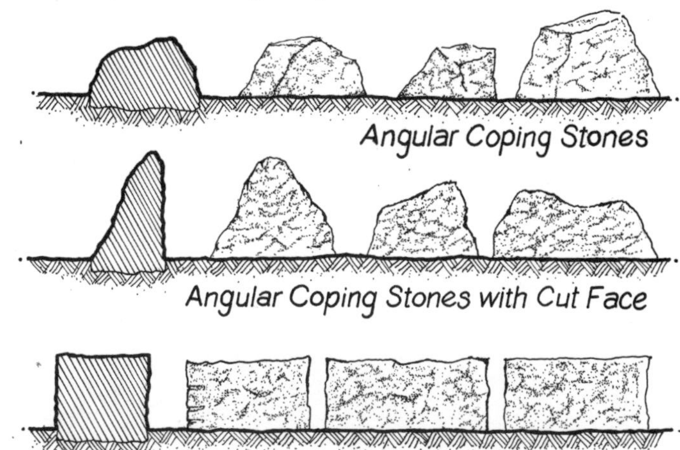

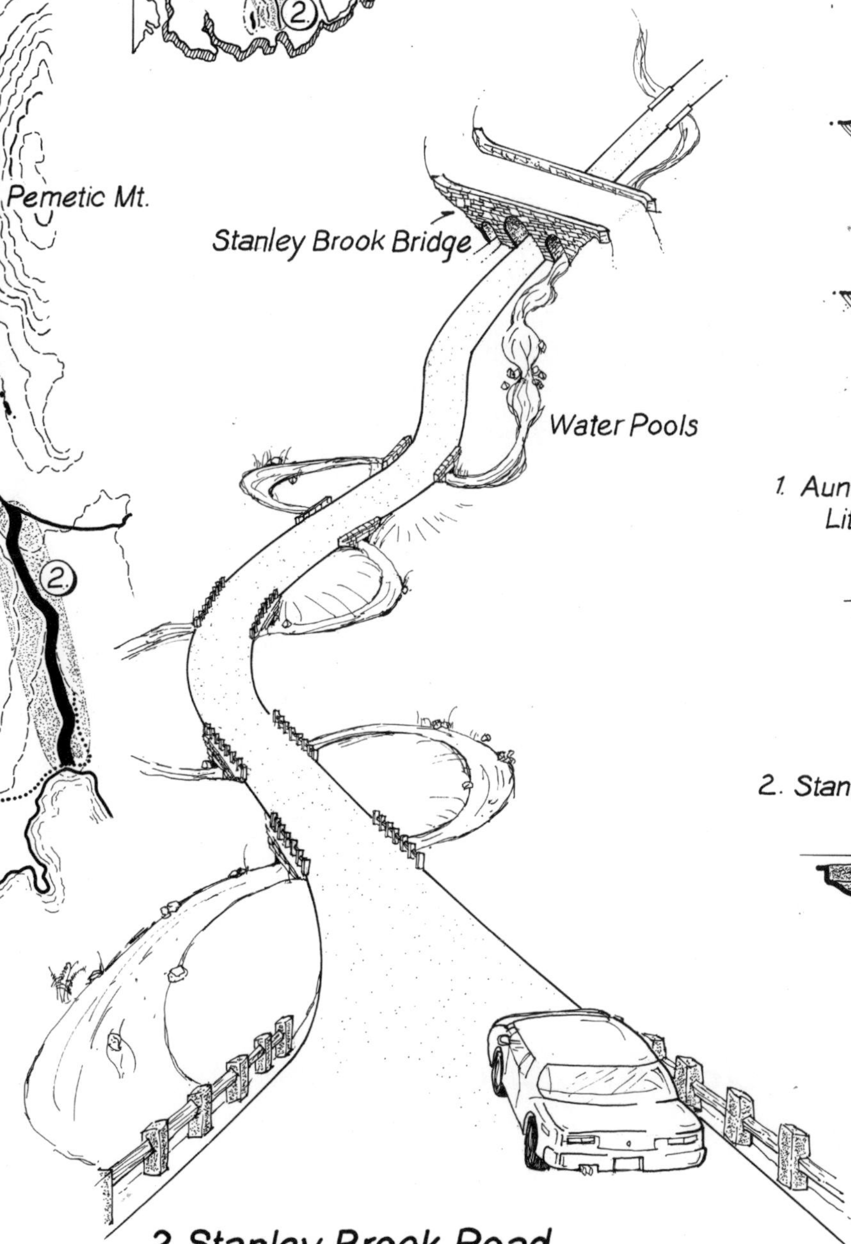

2. Stanley Brook Road

1. Aunt Betty Pond Road Little Bridges

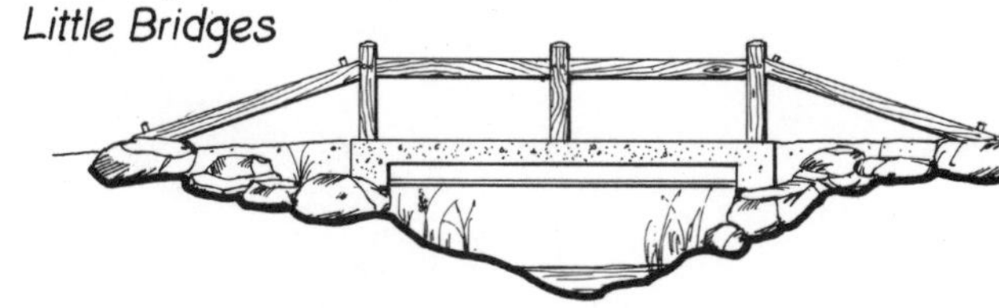

2. Stanley Brook Road Bridges

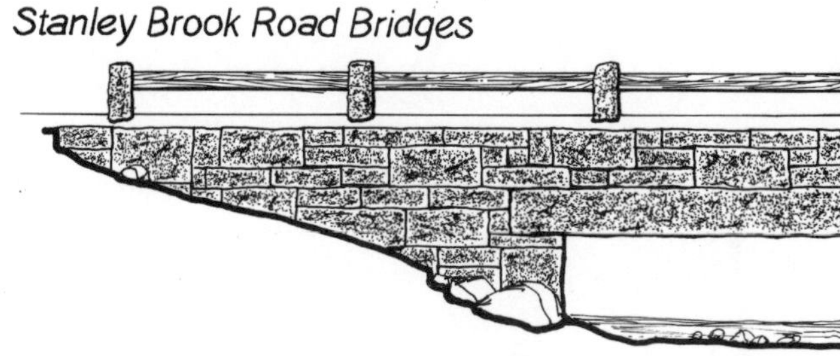

Feet 0 1 2 3 4 5 10

Meters 0 1 2 3

Although the carriage and motor roads of Acadia were intended to remain separate, distinct transportation systems, they nevertheless share a variety of design similarities. For instance, planners of both systems attempted to lay out their roads to take full advantage of scenic areas within the park, especially along watercourses. Acadia's carriage roads, like this one to Aunt Betty Pond, were therefore often manipulated with tight turning radii in order to cross the stream beds numerous times. The designers of the park's motor roads, who had to incorporate straighter alignments in order to account for faster speeds, accomplished a similar design by manipulating the stream bed itself. The course of the Stanley Brook was altered to make it pass under the archway of the Stanley Brook Bridge. The flow of the brook was also changed when a number of scenic pools were created along the edge of the motor road.

DELINEATED BY: Ed Lupyak, 1995

NATIONAL PARK SERVICE
ACADIA ROADS & BRIDGES RECORDING PROGRAM
NATIONAL PARK SERVICE
UNITED STATES DEPARTMENT OF THE INTERIOR

ACADIA NATIONAL PARK ROADS & BRIDGES

BAR HARBOR VICINITY HANCOCK COUNTY MAINE

SHEET 3 OF 19

HISTORIC AMERICAN ENGINEERING RECORD ME-12

SYSTEM SEPARATION

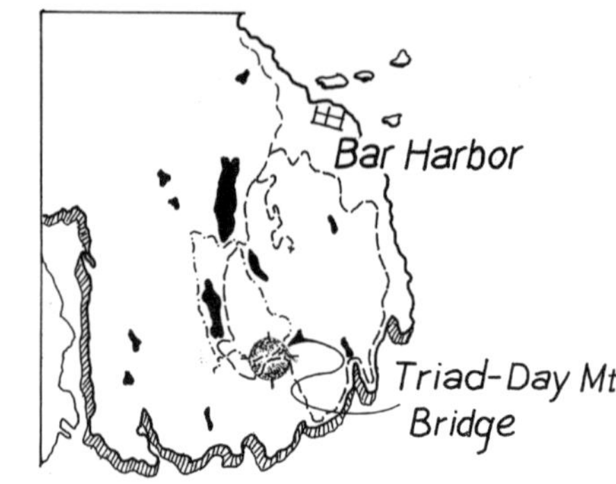

The transportation network of Acadia National Park is comprised of three distinct systems. Hiking trails, carriage roads, and motor roads enable visitors to enjoy the park by a variety of means. To ensure the safe and efficient flow of visitors, as well as to protect the natural beauty of the park, designers of this network have attempted to maintain the separateness of each of these three systems wherever possible. On a physical level this is achieved by keeping systems from coming into contact with one another. Where this is unavoidable, such as at this intersection near the Wildwood Riding Stables, bridges, underpasses, and gates are used to allow hikers, carriage drivers, bicyclists, and motorists to bypass one another with minimal contact.

Acadia's signs also serve to direct the flow of visitors within these three transportation systems by providing directional and usage information. Due partly to the varying speeds at which visitors pass them, each system's signage is also distinctive. The more detailed, rustic quality of the trail and carriage road signs, as well as the more visible character of the signs along the park's motor roads, serve as psychological reminders of the separation of these three systems.

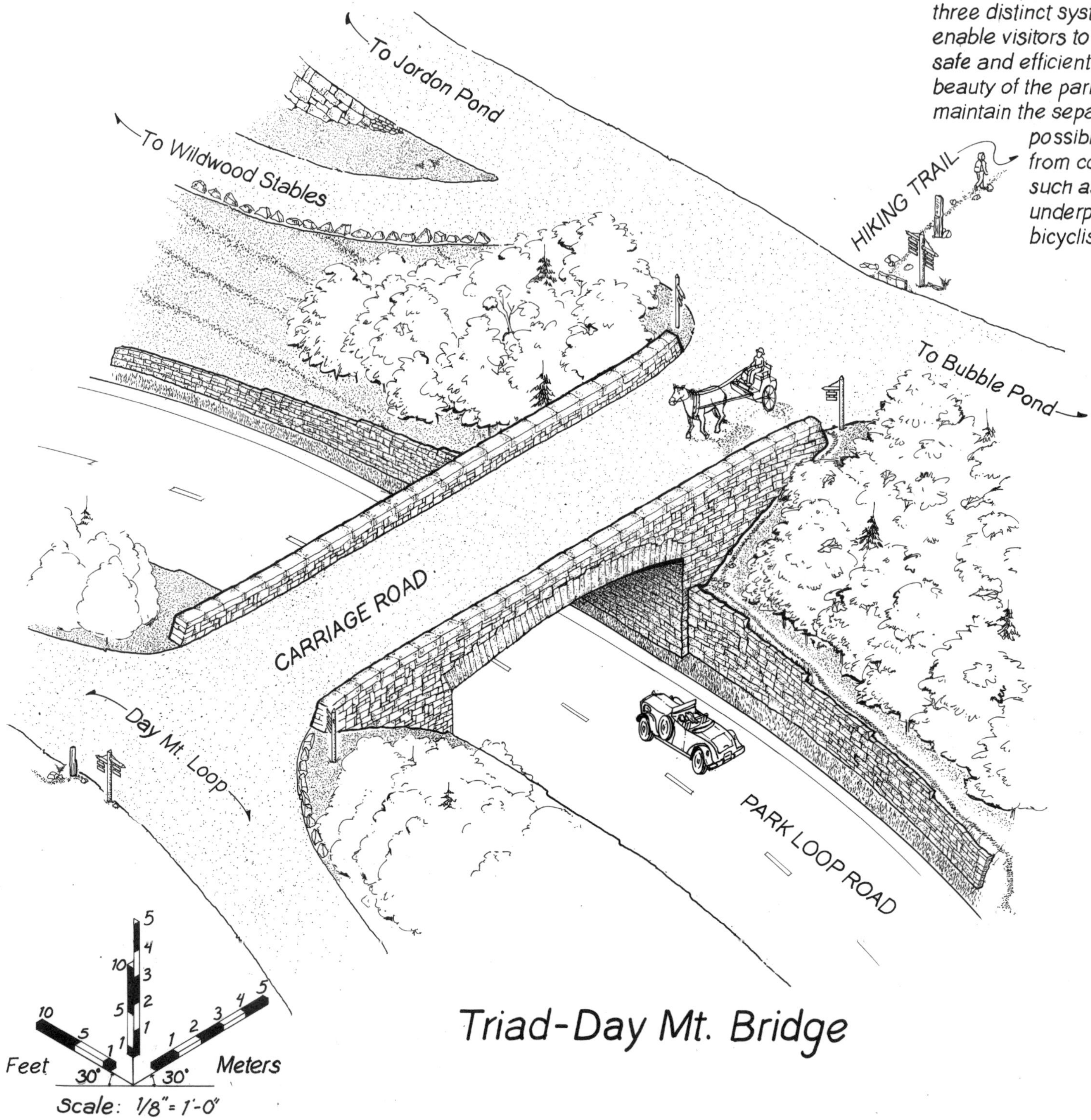

Triad-Day Mt. Bridge

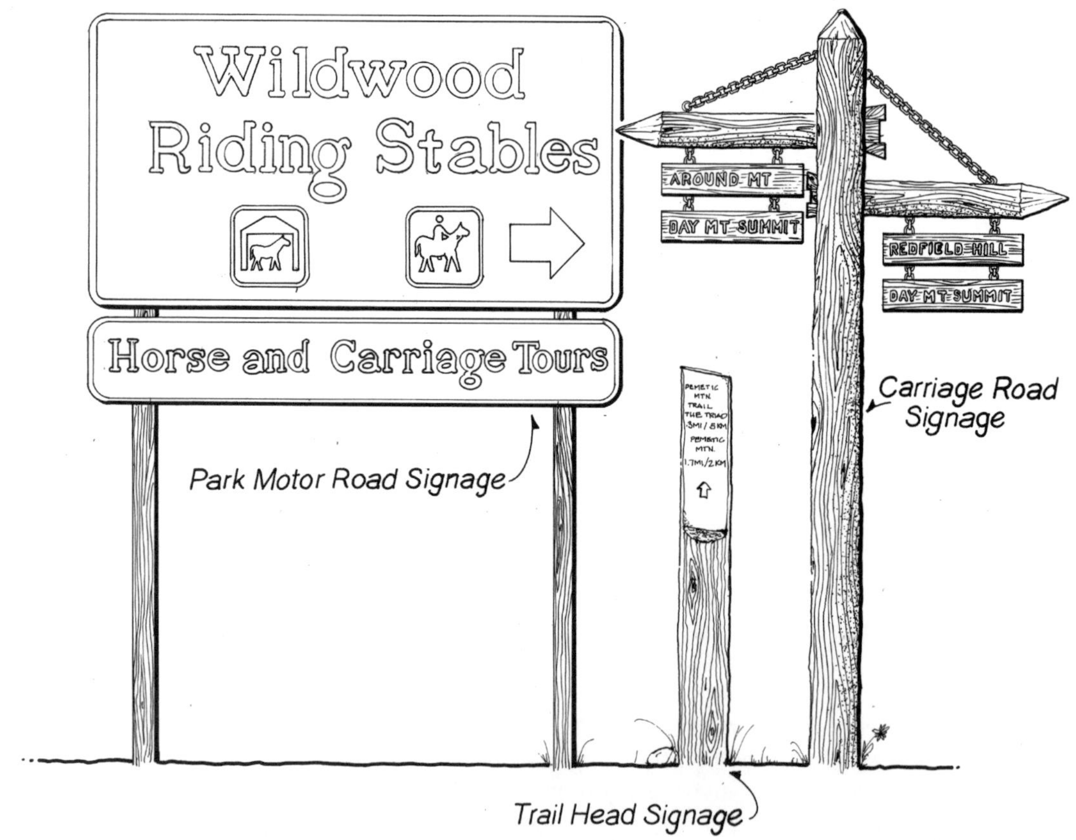

DELINEATED BY: Ed Lupyak, 1995

NATIONAL PARK SERVICE
ACADIA ROADS & BRIDGES RECORDING PROGRAM
NATIONAL PARK SERVICE
UNITED STATES DEPARTMENT OF THE INTERIOR

ACADIA NATIONAL PARK ROADS & BRIDGES
BAR HARBOR VICINITY HANCOCK COUNTY MAINE

SHEET 4 of 19

HISTORIC AMERICAN ENGINEERING RECORD ME-12

IF REPRODUCED, PLEASE CREDIT: HISTORIC AMERICAN ENGINEERING RECORD, NATIONAL PARK SERVICE, NAME OF DELINEATOR, DATE OF THE DRAWING

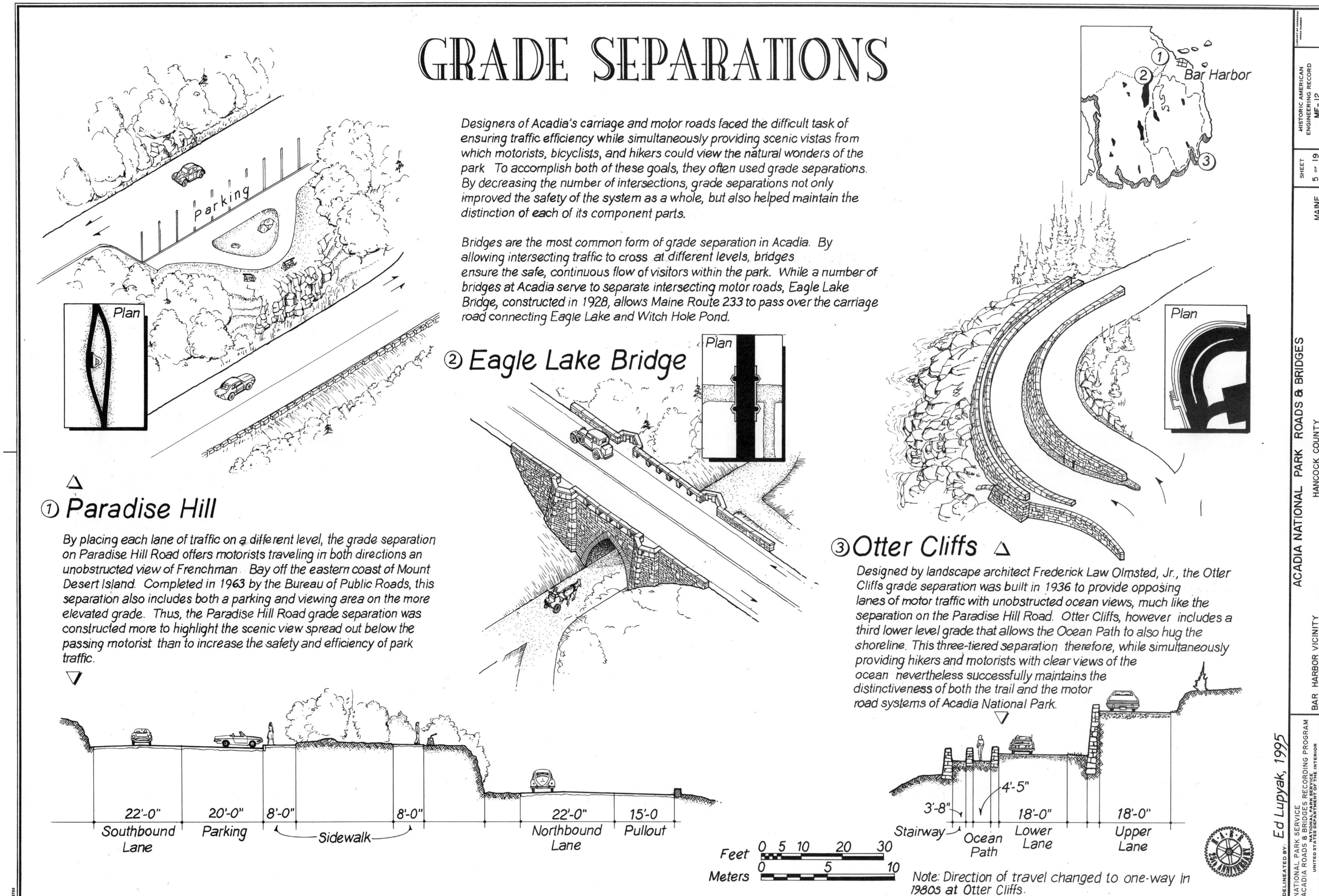
GRADE SEPARATIONS
Designers of Acadia's carriage and motor roads faced the difficult task of ensuring traffic efficiency while simultaneously providing scenic vistas from which motorists, bicyclists, and hikers could view the natural wonders of the park. To accomplish both of these goals, they often used grade separations. By decreasing the number of intersections, grade separations not only improved the safety of the system as a whole, but also helped maintain the distinction of each of its component parts.
Bridges are the most common form of grade separation in Acadia. By allowing intersecting traffic to cross at different levels, bridges ensure the safe, continuous flow of visitors within the park. While a number of bridges at Acadia serve to separate intersecting motor roads, Eagle Lake Bridge, constructed in 1928, allows Maine Route 233 to pass over the carriage road connecting Eagle Lake and Witch Hole Pond.
Parking
Plan
① Paradise Hill
By placing each lane of traffic on a different level, the grade separation on Paradise Hill Road offers motorists traveling in both directions an unobstructed view of Frenchman Bay off the eastern coast of Mount Desert Island. Completed in 1963 by the Bureau of Public Roads, this separation also includes both a parking and viewing area on the more elevated grade. Thus, the Paradise Hill Road grade separation was constructed more to highlight the scenic view spread out below the passing motorist than to increase the safety and efficiency of park traffic.
② Eagle Lake Bridge
Plan
Bar Harbor
1
2
3
Plan
③ Otter Cliffs
Designed by landscape architect Frederick Law Olmsted, Jr., the Otter Cliffs grade separation was built in 1936 to provide opposing lanes of motor traffic with unobstructed ocean views, much like the separation on the Paradise Hill Road. Otter Cliffs, however includes a third lower level grade that allows the Ocean Path to also hug the shoreline. This three-tiered separation therefore, while simultaneously providing hikers and motorists with clear views of the ocean nevertheless successfully maintains the distinctiveness of both the trail and the motor road systems of Acadia National Park.
22'-0" Southbound Lane
20'-0" Parking
8'-0"
Sidewalk
8'-0"
22'-0" Northbound Lane
15'-0 Pullout
Feet 0 5 10 20 30
Meters 0 5 10
3'-8"
4'-5"
18'-0"
18'-0"
Stairway
Ocean Path
Lower Lane
Upper Lane
Note: Direction of travel changed to one-way in 1980s at Otter Cliffs.
DELINEATED BY: Ed Lupyak, 1995
NATIONAL PARK SERVICE
ACADIA ROADS & BRIDGES RECORDING PROGRAM
BAR HARBOR VICINITY
ACADIA NATIONAL PARK ROADS & BRIDGES
HANCOCK COUNTY
MAINE
SHEET 5 OF 19
HISTORIC AMERICAN ENGINEERING RECORD ME-12
IF REPRODUCED, PLEASE CREDIT: HISTORIC AMERICAN ENGINEERING RECORD, NATIONAL PARK SERVICE, NAME OF DELINEATOR, DATE OF THE DRAWING

MOTOR ROAD DEVELOPMENT

A) Jordan Pond/Eagle Lake (1923-27): Park Superintendent George Dorr originally planned this road in 1922 to open up the interior of the park to visitors and to protect against wildfire. When federal appropriations for the road fell through, John D. Rockefeller, Jr. agreed to fund the project. The Jordan Pond/Eagle Lake road connects Rt. 233 near Eagle Lake with Seal Harbor.

B) Sieur de Monts (planned 1929, built 1940): Although not represented here, this section of the Park Loop motor road was the most controversial to construct (see Sieur de Monts Spring sheet). Originally designed to run alongside The Tarn, the final layout went around Champlain Mt., connecting Ocean Drive and the Kebo Mt. section with the old park headquarters near downtown Bar Harbor (on Ledgelawn Ave.).

C) Ocean Drive (1929-58): Because much of its proposed route ran across privately owned land, Ocean Drive took longer to complete than any other section of the Park Loop motor road. As Acadia obtained these properties, more segments of Ocean Drive were added. The Ocean Path, a trail running along the road's southern portion, was reconstructed during the Depression by workers from President Roosevelt's Civilian Conservation Corps (see Ocean Drive sheet).

D) Stanley Brook (1934-36): Rockefeller constructed the Stanley Brook motor road on his own land and then donated it to Acadia. In order to maintain the scenic quality of both the brook and the narrow valley, as well as to alleviate local opposition to the road, Rockefeller hired landscape architect Frederick Law Olmsted, Jr. to help design it. The final route crosses Stanley Brook six times on small, unadorned bridges. (Extensively landscaped by Beatrix Farrand).

E) Otter Point (1938-39): The 1932 decision, relocating the Otter Cliff Naval Radio Station to nearby Schoodic Peninsula made way for the extension of the Park Loop motor road through Otter Point. Rockefeller donated the land for the Otter Cove Causeway and confirmed the design with F. L. Olmsted, Jr. A memorial to Allesandro Fabbri, commander of the radio station, marks the station's former location.

F) Paradise Hill (1940-41): One of the last sections of the park motor road to be completed, the Paradise Hill Road was built so that motorists entering the park from Rt. 3 could avoid Bar Harbor traffic. Although it was constructed by the Bureau of Public Roads, the land for the road was donated by Rockefeller.

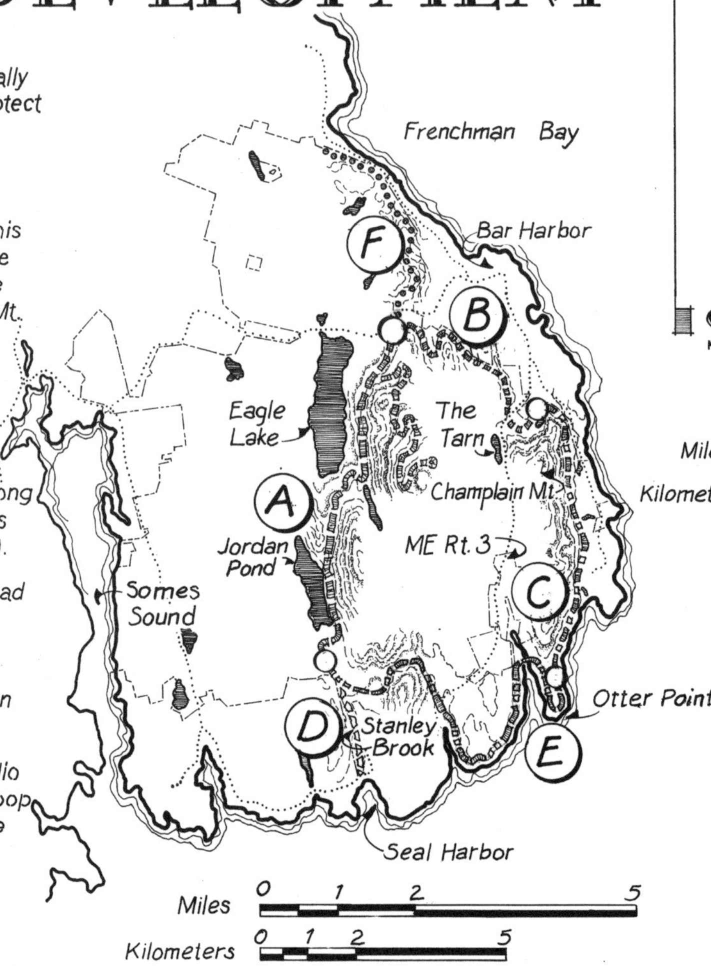

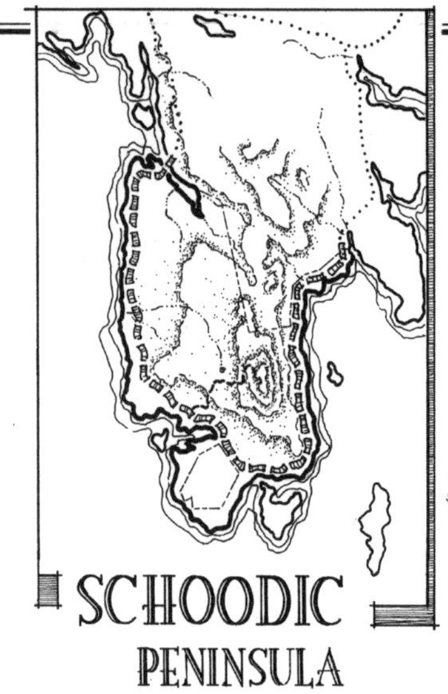

(F) Granite Median (rumble strips)

lookout area — Paradise Hill

scale 1/2" = 1'—0

Otter Creek Cove Bridge and Causeway

(E)

Fabbri Memorial — planting

(A) Jordan Pond / Eagle Lake Road "Mountain Road"

(C) Ocean Drive with Pulloff Parking

(D) Stanley Brook Road

concept drawing — F. L. Olmsted Jr. never implemented

DELINEATED BY: harlen d. groe, 1995

NATIONAL PARK SERVICE
ACADIA ROADS & BRIDGES RECORDING PROGRAM
NATIONAL PARK SERVICE
UNITED STATES DEPARTMENT OF THE INTERIOR

BAR HARBOR VICINITY — ACADIA NATIONAL PARK ROADS & BRIDGES — HANCOCK COUNTY — MAINE

SHEET 6 OF 19

HISTORIC AMERICAN ENGINEERING RECORD ME-12

IF REPRODUCED, PLEASE CREDIT: HISTORIC AMERICAN ENGINEERING RECORD, NATIONAL PARK SERVICE, NAME OF DELINEATOR, DATE OF THE DRAWING

CADILLAC MT. ROAD

and Park Loop Road

Historic Road Sign c. 1932.

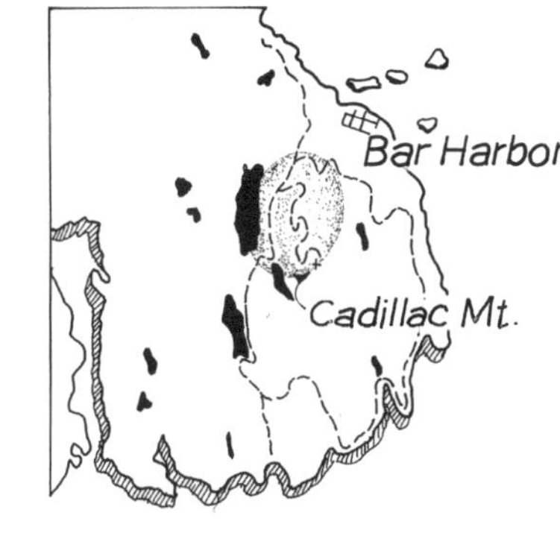

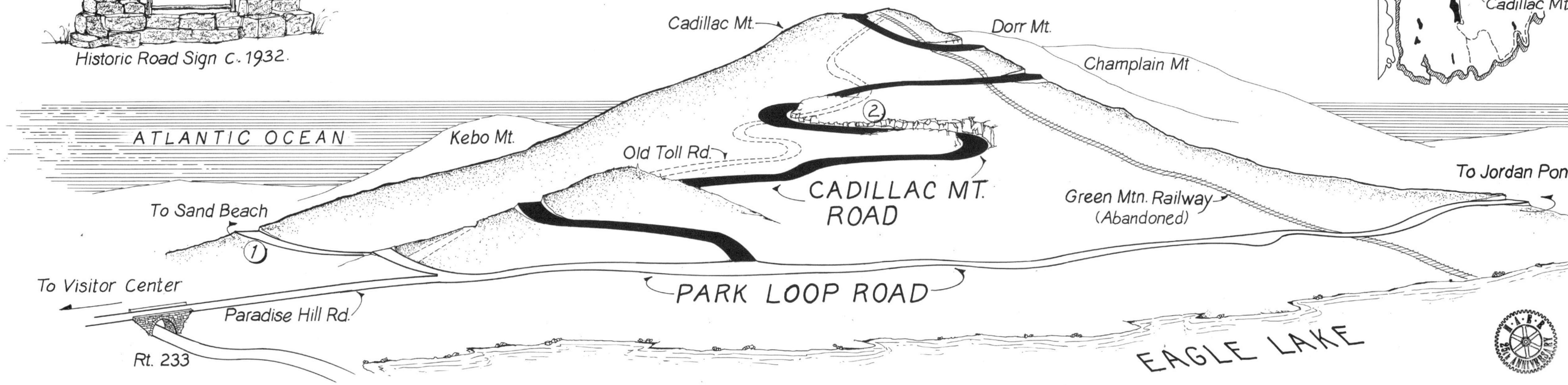

The present Cadillac Mountain Road was not the first route to the top of Acadia's highest mountain. Long before it was completed by the Bureau of Public Roads in 1932, Abnaki Indians blazed a footpath to Cadillac's Summit. By widening much of this Native American trail, early white settlers were able to lay out the first buckboard toll road up the mountain during the 1850s. Because of increasing tourism on Mount Desert Island, a cog railway was also constructed up Cadillac from the shore of Eagle Lake in 1883. The present Cadillac Mountain Road, which connects with the Park Loop Road near Eagle Lake, allows visitors to drive continuously from the highest point on America's Atlantic seaboard down to the ocean's edge.

Designed by the Bureau of Public Roads between 1929 and 1932, Cadillac Mountain Road is an extraordinary example of construction in mountainous terrain. The nearly four-mile road maintains a positive grade for its entire length and includes spiral transition curves, which were an adaptation from railroad construction, to give a smoother ride between the road's straight and curved sections. Guardrails were fashioned by spacing three to four foot angular coping stones along the edge of the road approximately four feet apart.

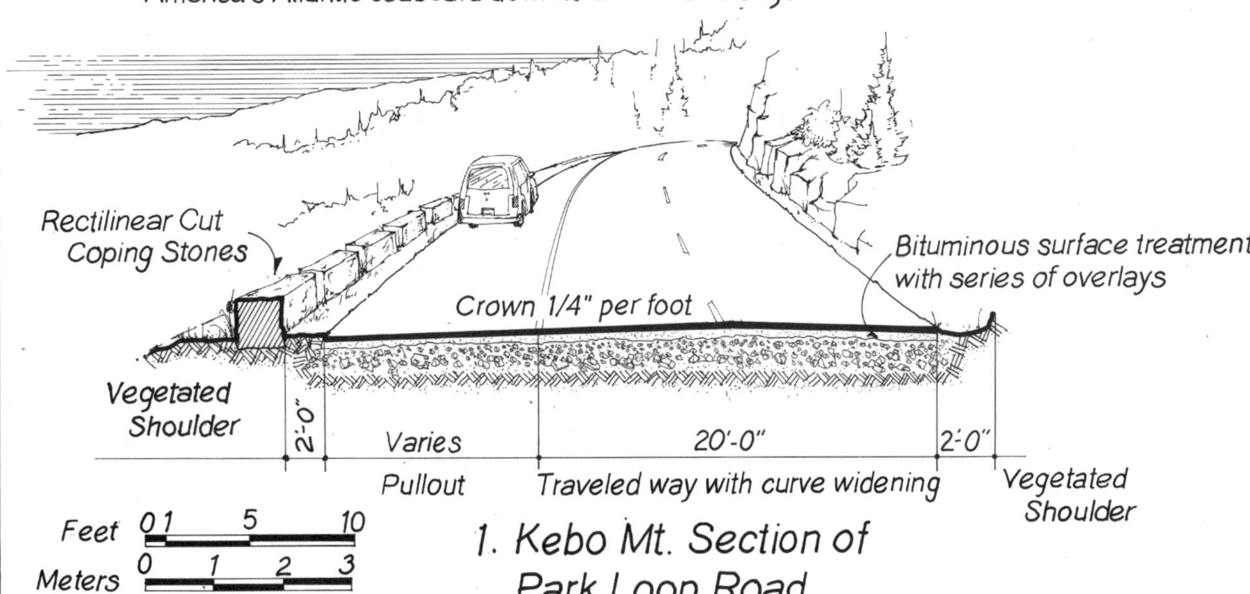

1. Kebo Mt. Section of Park Loop Road

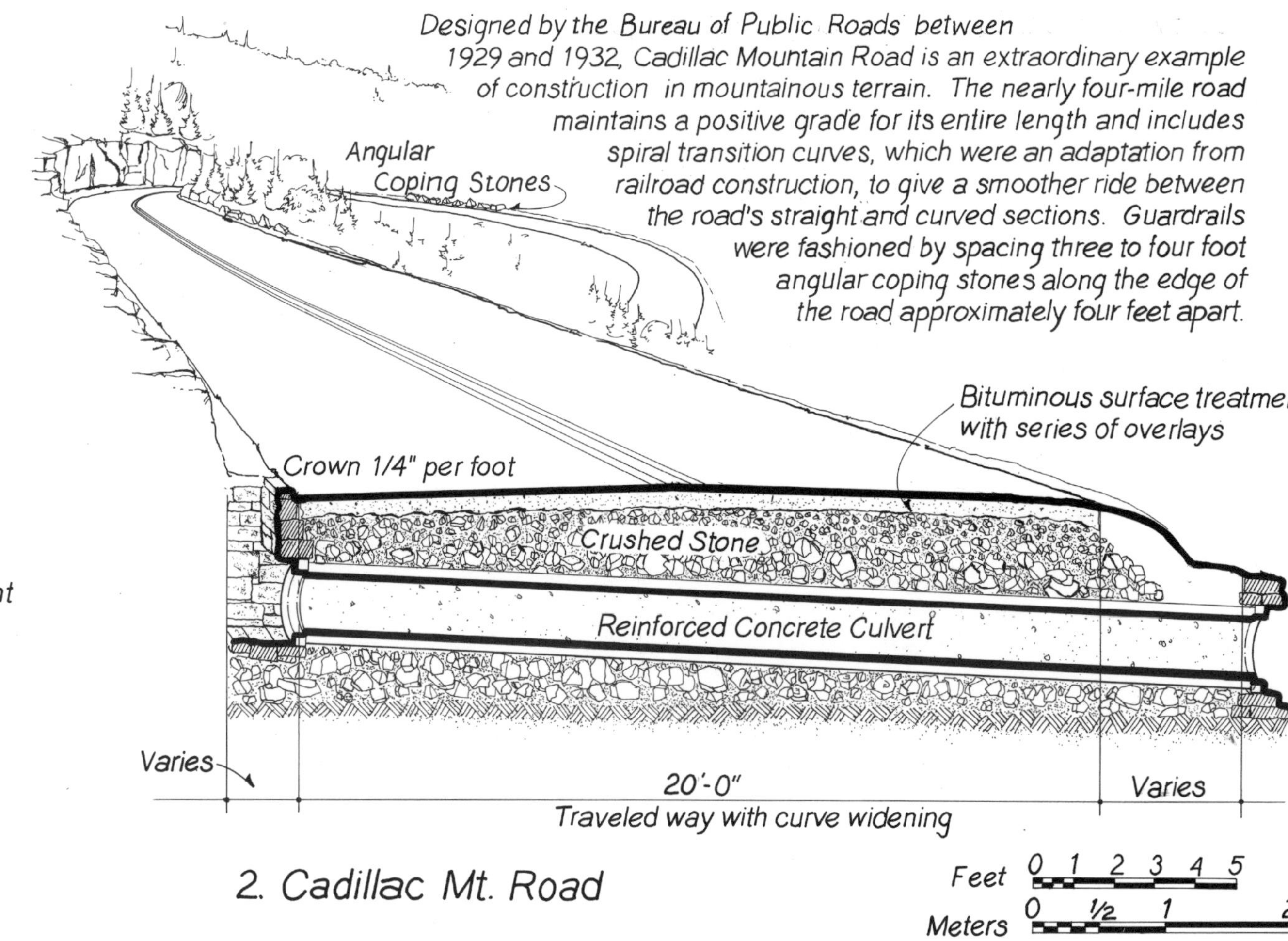

2. Cadillac Mt. Road

DELINEATED BY Ed Lupyak, 1995

NATIONAL PARK SERVICE
ACADIA ROADS & BRIDGES RECORDING PROGRAM

BAR HARBOR VICINITY | HANCOCK COUNTY | MAINE

ACADIA NATIONAL PARK ROADS & BRIDGES

SHEET 7 OF 19

HISTORIC AMERICAN ENGINEERING RECORD ME-12

OCEAN DRIVE

Hugging the eastern shoreline of Mount Desert Island from the Sieur de Monts Spring park entrance to the grade separation at Otter Cliffs, Ocean Drive is one of the most breathtaking roads on the Atlantic seaboard. It was designed specifically to present motorists with a sequence of scenic experiences, including views of Sand Beach, Thunder Hole, and the cliffs of Otter Point. For almost its entire four mile length motorists can glimpse the ocean from their automobiles, or are given access to viewing areas at which they can park and look out unobstructed over the sea.

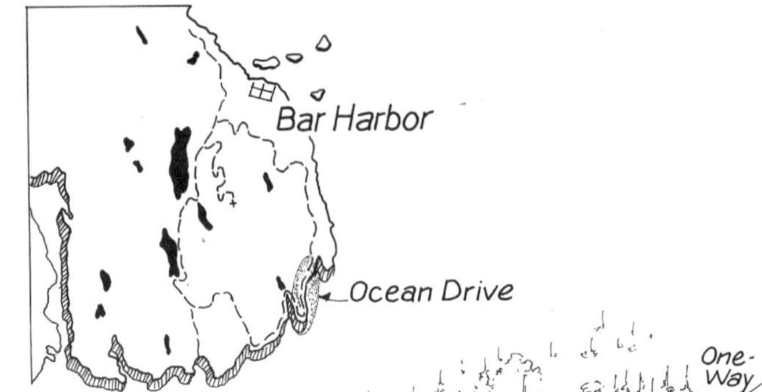

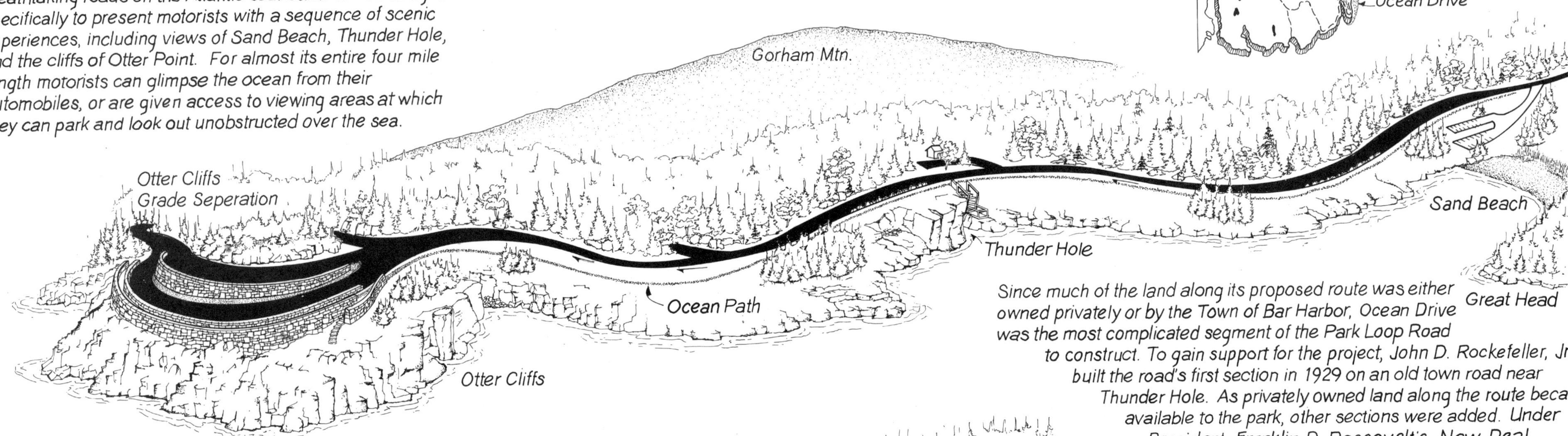

Since much of the land along its proposed route was either owned privately or by the Town of Bar Harbor, Ocean Drive was the most complicated segment of the Park Loop Road to construct. To gain support for the project, John D. Rockefeller, Jr. built the road's first section in 1929 on an old town road near Thunder Hole. As privately owned land along the route became available to the park, other sections were added. Under President Franklin D. Roosevelt's New Deal, Civilian Conservation Corps workers reconstructed the Ocean Path, a scenic trail running alongside the southern portion of Ocean Drive. The Bureau of Public Roads constructed the final section of Ocean Drive in 1958 completing the entire Park Loop Road in the process.

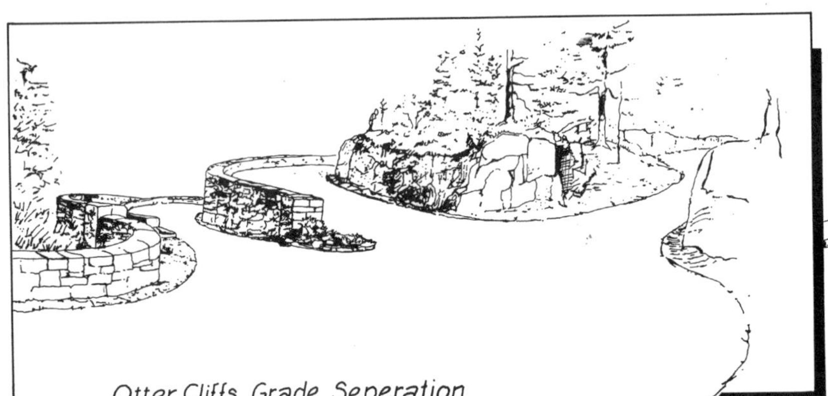

Otter Cliffs Grade Seperation

Landscape architect Frederick Law Olmsted, Jr. first began work with motor roads in Acadia in 1929, when John D. Rockefeller, Jr. hired him to settle a dispute concerning a planned realignment in the Sieur de Monts Spring area of the park. His continual involvement throughout the following decade ensured much of the aesthetic character of Acadia's motor road system. In this 1935 sketch Olmsted laid out his design for the Otter Cliffs grade separation.

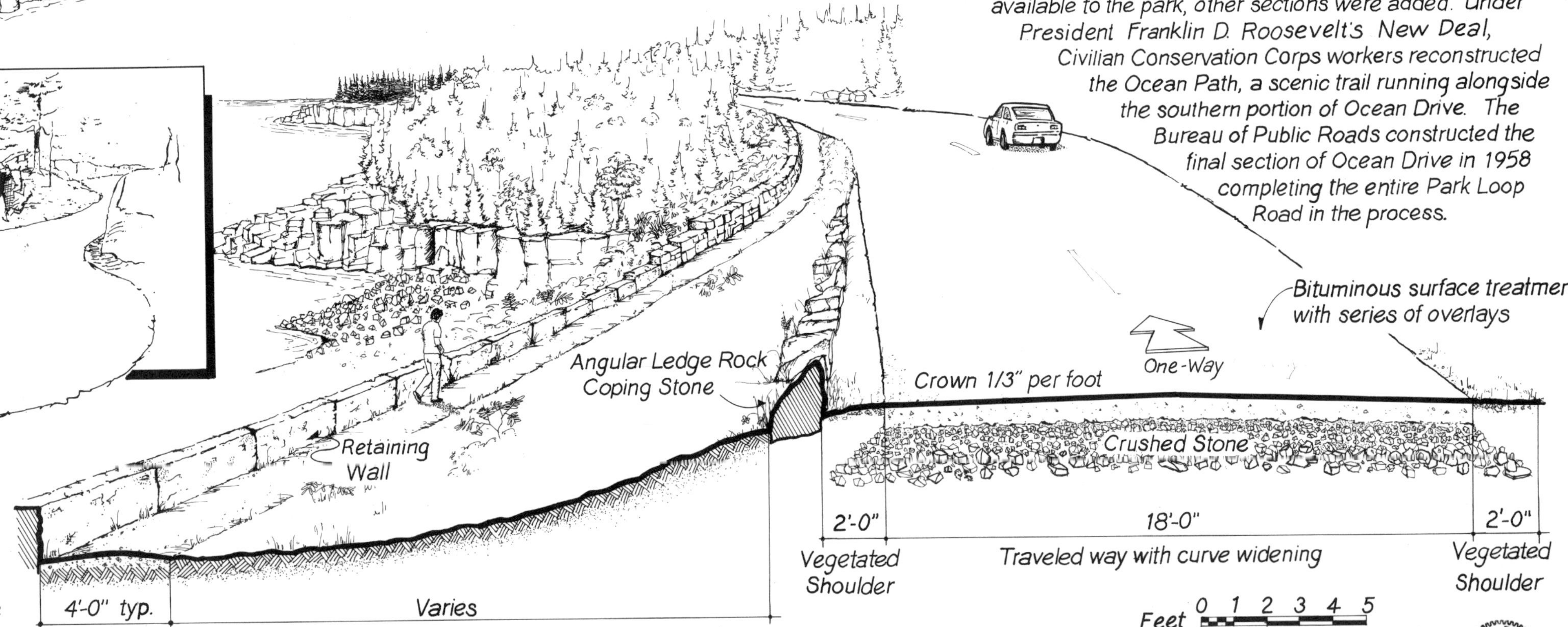

Section thru Ocean Drive and Path

Note: Ocean Drive in its present form is a redesign/reconstruction of a road by the same name built ca. 1890.

Delineated by: Ed Lupyak, 1995

National Park Service
Acadia Roads & Bridges Recording Program
National Park Service
United States Department of the Interior

Historic American Engineering Record ME-12

Sheet 8 of 19

Acadia National Park Roads & Bridges
Bar Harbor Vicinity
Hancock County
Maine

If reproduced, please credit: Historic American Engineering Record, National Park Service, name of delineator, date of the drawing

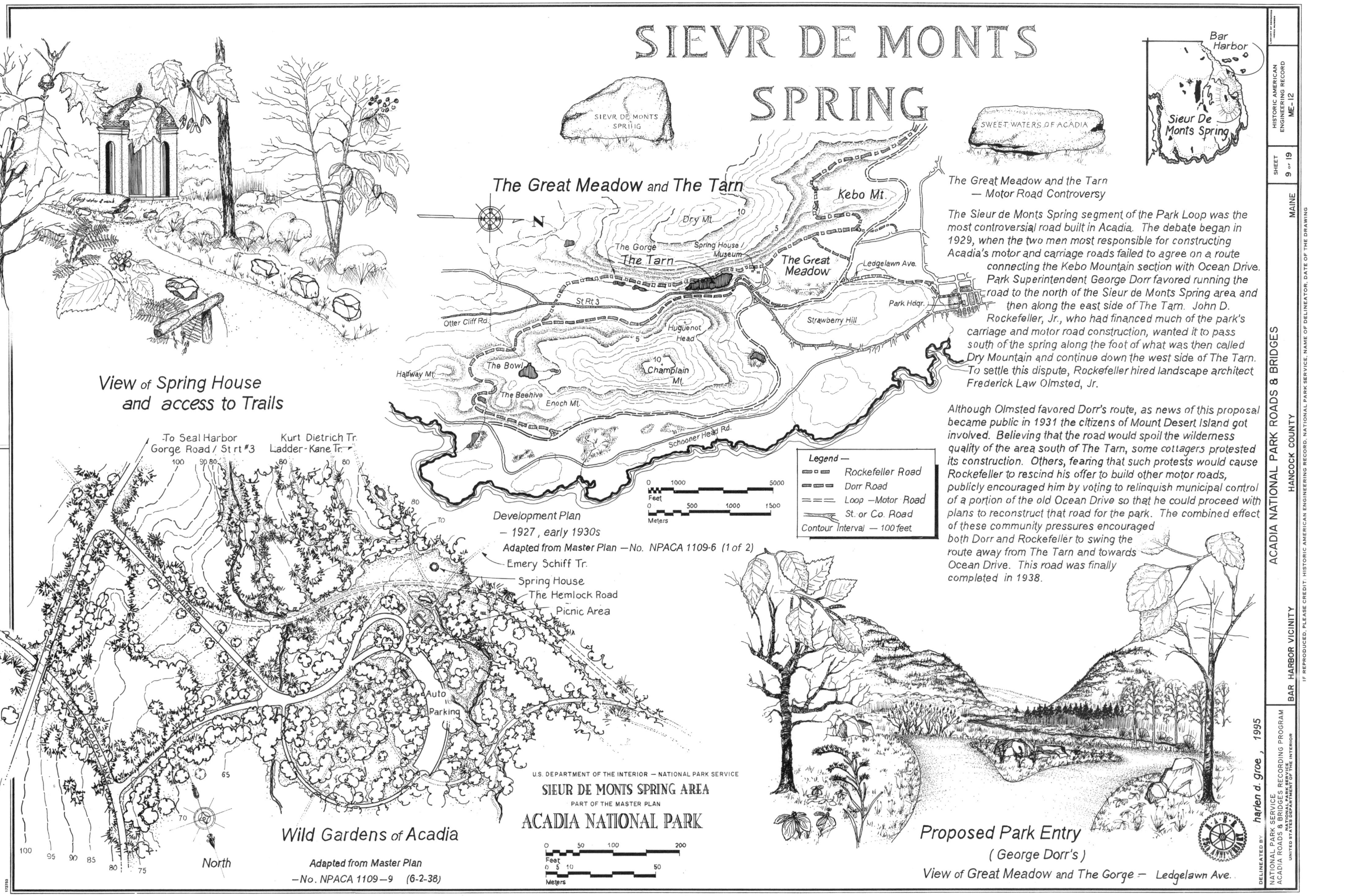

SIEVR DE MONTS SPRING
SIEVR DE MONTS SPRING
SWEET WATERS OF ACADIA
Bar Harbor
Sieur De Monts Spring
HISTORIC AMERICAN ENGINEERING RECORD
ME-12
SHEET 9 of 19
View of Spring House and access to Trails
The Great Meadow and The Tarn
N
Dry Mt.
Kebo Mt.
The Gorge
The Tarn
Spring House / Museum
The Great Meadow
Ledgelawn Ave.
St Rt 3
Park Hdqr.
Otter Cliff Rd.
Huguenot Head
Strawberry Hill
The Bowl
Champlain Mt.
Halfway Mt.
The Beehive
Enoch Mt.
Schooner Head Rd.
Legend —
Rockefeller Road
Dorr Road
Loop — Motor Road
St. or Co. Road
Contour Interval — 100 feet
Feet
Meters
Development Plan — 1927, early 1930s
Adapted from Master Plan — No. NPACA 1109-6 (1 of 2)
The Great Meadow and the Tarn — Motor Road Controversy
The Sieur de Monts Spring segment of the Park Loop was the most controversial road built in Acadia. The debate began in 1929, when the two men most responsible for constructing Acadia's motor and carriage roads failed to agree on a route connecting the Kebo Mountain section with Ocean Drive. Park Superintendent George Dorr favored running the road to the north of the Sieur de Monts Spring area and then along the east side of The Tarn. John D. Rockefeller, Jr., who had financed much of the park's carriage and motor road construction, wanted it to pass south of the spring along the foot of what was then called Dry Mountain and continue down the west side of The Tarn. To settle this dispute, Rockefeller hired landscape architect Frederick Law Olmsted, Jr.
Although Olmsted favored Dorr's route, as news of this proposal became public in 1931 the citizens of Mount Desert Island got involved. Believing that the road would spoil the wilderness quality of the area south of The Tarn, some cottagers protested its construction. Others, fearing that such protests would cause Rockefeller to rescind his offer to build other motor roads, publicly encouraged him by voting to relinquish municipal control of a portion of the old Ocean Drive so that he could proceed with plans to reconstruct that road for the park. The combined effect of these community pressures encouraged both Dorr and Rockefeller to swing the route away from The Tarn and towards Ocean Drive. This road was finally completed in 1938.
To Seal Harbor
Gorge Road / St rt #3
Kurt Dietrich Tr.
Ladder-Kane Tr.
Emery Schiff Tr.
Spring House
The Hemlock Road
Picnic Area
Auto Parking
North
Wild Gardens of Acadia
Adapted from Master Plan — No. NPACA 1109—9 (6-2-38)
U.S. DEPARTMENT OF THE INTERIOR — NATIONAL PARK SERVICE
SIEUR DE MONTS SPRING AREA
PART OF THE MASTER PLAN
ACADIA NATIONAL PARK
Proposed Park Entry (George Dorr's)
View of Great Meadow and The Gorge — Ledgelawn Ave.
DELINEATED BY harlen d. groe, 1995
NATIONAL PARK SERVICE
ACADIA ROADS & BRIDGES RECORDING PROGRAM
BAR HARBOR VICINITY
ACADIA NATIONAL PARK ROADS & BRIDGES
HANCOCK COUNTY
MAINE

THE BRIDGES OF ACADIA NATIONAL PARK

	Bridge Name	Date	HAER #
1	Cobblestone Bridge	1917	ME-31
2	Jordan Stream Little Bridges	1919	ME-48a-c
3	Little Harbor Brook Bridge	1919	ME-32
4	Jordan Pond Dam Bridge	1920	ME-33
5	Hemlock Bridge	1924	ME-34
6	Waterfall Bridge	1925	ME-35
7	Hadlock Brook Bridge	1926	ME-37
8	Chasm Brook Bridge	1926	ME-28
9	Bubble Pond Bridge	1928	ME-39
10	Eagle Lake Bridge	1928	ME-55
11	Deer Brook Bridge	1929	ME-36
12	Duck Brook Bridge	1929	ME-40
13	Eagle Lake Little Bridges	1931	ME-49a-b
14	Amphitheatre Bridge	1931	ME-41
15	West Branch Jordan Stream Bridge	1931	ME-42
16	Aunt Betty Pond Little Bridges	1931	ME-50a-g
17	Cliffside Bridge	1932	ME-43
18	Jordan Pond Road Bridge	1932	ME-44
19	Stanley Brook Bridge	1933	ME-45
20	Stanley Brook Bridges (Stanley Brook Road)	1933	ME-24a-f
21	Otter Creek Cove Bridge And Causeway	1938	ME-19
22	Kebo Brook Bridge	1938	ME-20
23	Route 3 Bridge (Blackwoods)	1939	ME-15
24	Fish House Bridge	1939	ME-16
25	Little Hunters Beach Brook Bridge	1939	ME-21
26	Hunters Beach Brook Bridge	1939	ME-22
27	Sieur de Monts Spring Bridge	1949	ME-14
28	Triad-Day Mountain Bridge	1941	ME-46
29	Wildwood Farm Bridge	1941	ME-47
30	Duck Brook Bridge (Paradise Hill Road)	1950	ME-30
31	Route 233 Bridge (Paradise Hill Road)	1952	ME-17
32	New Eagle Lake Road Bridge	1952	ME-18

Acadia National Park contains the most diverse and detailed collection of historic bridges within the National Park Service. Since 1917, when John D. Rockefeller forces constructed the first of his carriage road bridges, a legacy of spans have been created that heighten the aesthetic experience of the park's carriage and motor roads. Constructed mainly of reinforced concrete with stone facing, these spans range in size from small slabs to monumental arches, spanning various brooks, chasms and valleys. Acadia's bridges illustrate two distinct design philosophies; Rockefeller and federally funded creations differ in their attention to details, site location and overall character. Rockefeller influenced structures maintain a higher level of detailing, which includes gothic arches, antiquated masonry surfaces, protruding scuppers and reliefs. Rockefeller spans exclaimed the craft in the creation. Federally designed and engineered structures of the NPS and Bureau of Public Roads continued the traditions of building with stone; however, the level of detail is reduced. Function and efficiency was of primary concern, with appearance receiving less attention. The later motor road bridges are consistent with the national tradition of NPS spans, while Rockefeller's survive as sculptural inspirations.

(11) Deer Brook Bridge

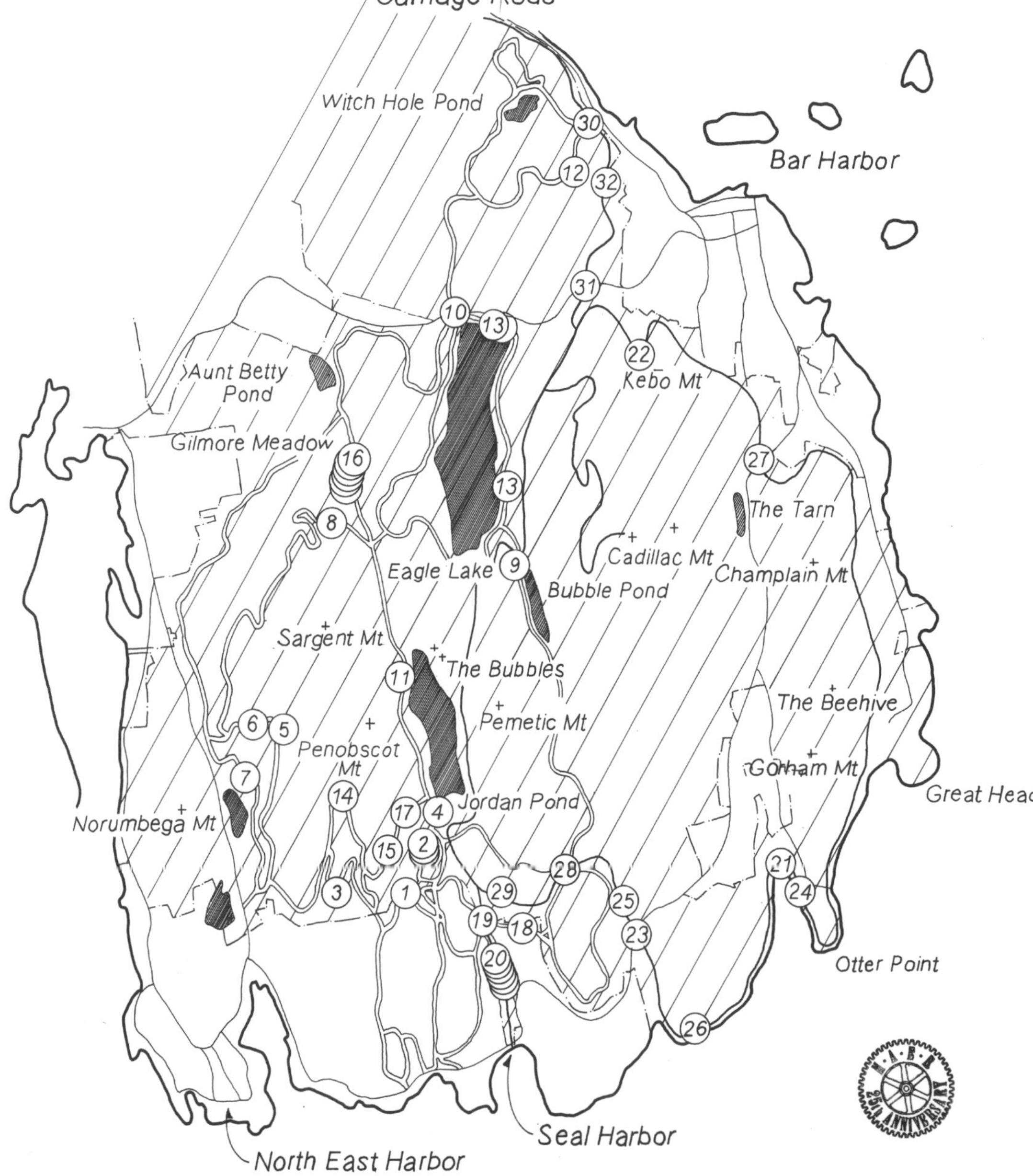

HISTORIC AMERICAN ENGINEERING RECORD ME-12

SHEET 10 of 19

ACADIA NATIONAL PARK ROADS & BRIDGES

MAINE

HANCOCK COUNTY

BAR HARBOR VICINITY

NATIONAL PARK SERVICE
ACADIA ROADS & BRIDGES RECORDING PROGRAM
NATIONAL PARK SERVICE
UNITED STATES DEPARTMENT OF THE INTERIOR

DELINEATED BY: Joe Korzeniewski, 1995

IF REPRODUCED, PLEASE CREDIT: HISTORIC AMERICAN ENGINEERING RECORD, NATIONAL PARK SERVICE, NAME OF DELINEATOR, DATE OF THE DRAWING

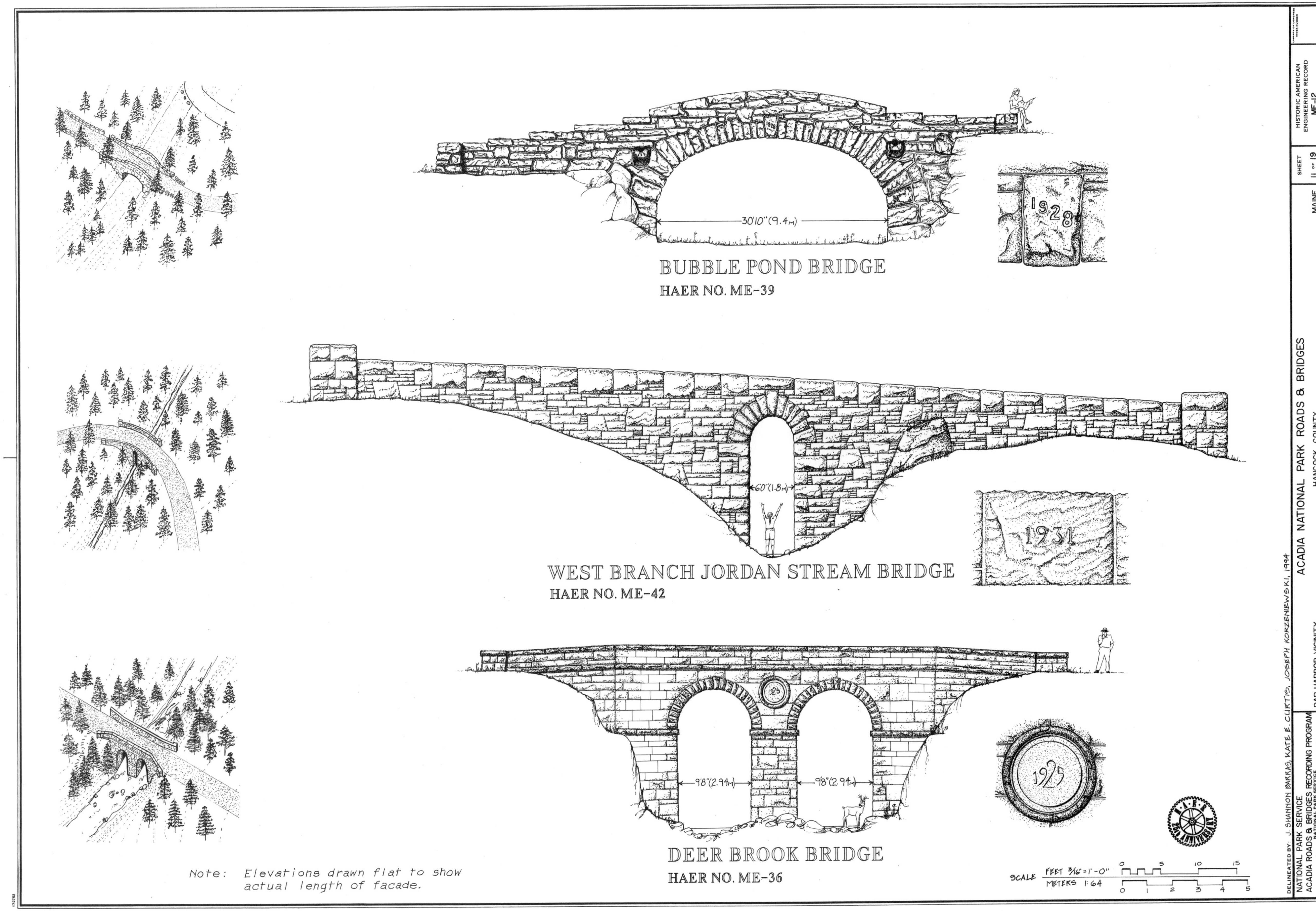

30'10" (9.4m)
1928
BUBBLE POND BRIDGE
HAER NO. ME-39
6'0" (1.8m)
1931
WEST BRANCH JORDAN STREAM BRIDGE
HAER NO. ME-42
9'8" (2.94m)
9'8" (2.94m)
1925
DEER BROOK BRIDGE
HAER NO. ME-36
Note: Elevations drawn flat to show actual length of facade.
SCALE FEET 3/16" = 1'-0"
METERS 1:64
0 5 10 15
0 1 2 3 4 5
DELINEATED BY: J. SHANNON BARKAS, KATE E. CURTIS, JOSEPH KORZENIEWSKI, 1994
NATIONAL PARK SERVICE
ACADIA ROADS & BRIDGES RECORDING PROGRAM
NATIONAL PARK SERVICE
UNITED STATES DEPARTMENT OF THE INTERIOR
ACADIA NATIONAL PARK ROADS & BRIDGES
BAR HARBOR VICINITY
HANCOCK COUNTY
MAINE
SHEET
11 OF 19
HISTORIC AMERICAN ENGINEERING RECORD
ME-12
IF REPRODUCED, PLEASE CREDIT: HISTORIC AMERICAN ENGINEERING RECORD, NATIONAL PARK SERVICE, NAME OF DELINEATOR, DATE OF THE DRAWING

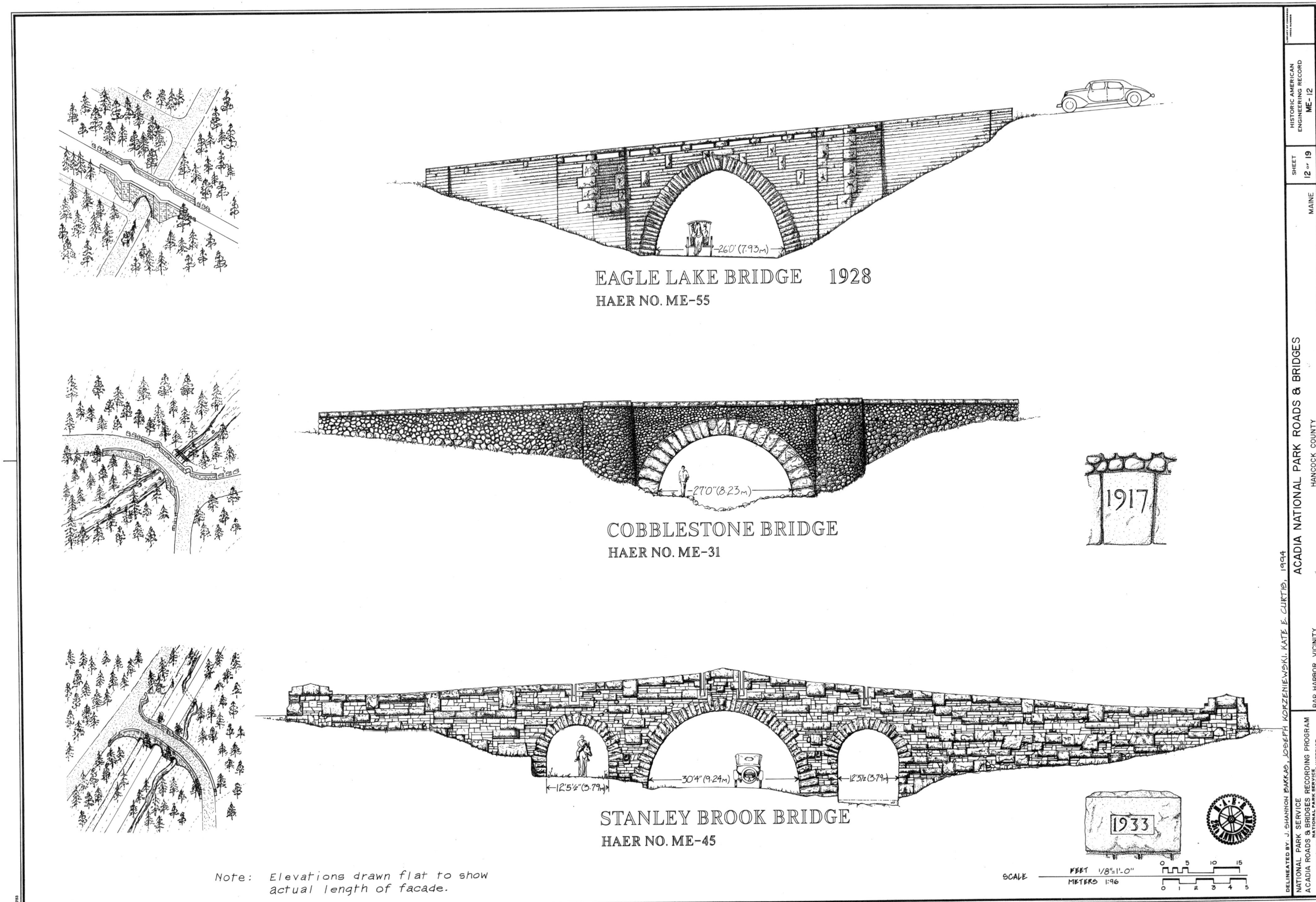

260' (7.93m)
EAGLE LAKE BRIDGE 1928
HAER NO. ME-55
27'0" (8.23m)
1917
COBBLESTONE BRIDGE
HAER NO. ME-31
12'5½" (3.79m)
30'4" (9.24m)
12'5½" (3.79m)
1933
STANLEY BROOK BRIDGE
HAER NO. ME-45
Note: Elevations drawn flat to show actual length of facade.
SCALE
FEET 1/8"=1'-0"
METERS 1:96
0 5 10 15
0 1 2 3 4 5
DELINEATED BY: J. SHANNON BAKKAS, JOSEPH KORZENIEWSKI, KATE E. CURTIS, 1994
NATIONAL PARK SERVICE
ACADIA ROADS & BRIDGES RECORDING PROGRAM
NATIONAL PARK SERVICE
UNITED STATES DEPARTMENT OF THE INTERIOR
ACADIA NATIONAL PARK ROADS & BRIDGES
BAR HARBOR VICINITY
HANCOCK COUNTY
MAINE
SHEET
12 OF 19
HISTORIC AMERICAN ENGINEERING RECORD
ME-12
IF REPRODUCED, PLEASE CREDIT: HISTORIC AMERICAN ENGINEERING RECORD, NATIONAL PARK SERVICE, NAME OF DELINEATOR, DATE OF THE DRAWING

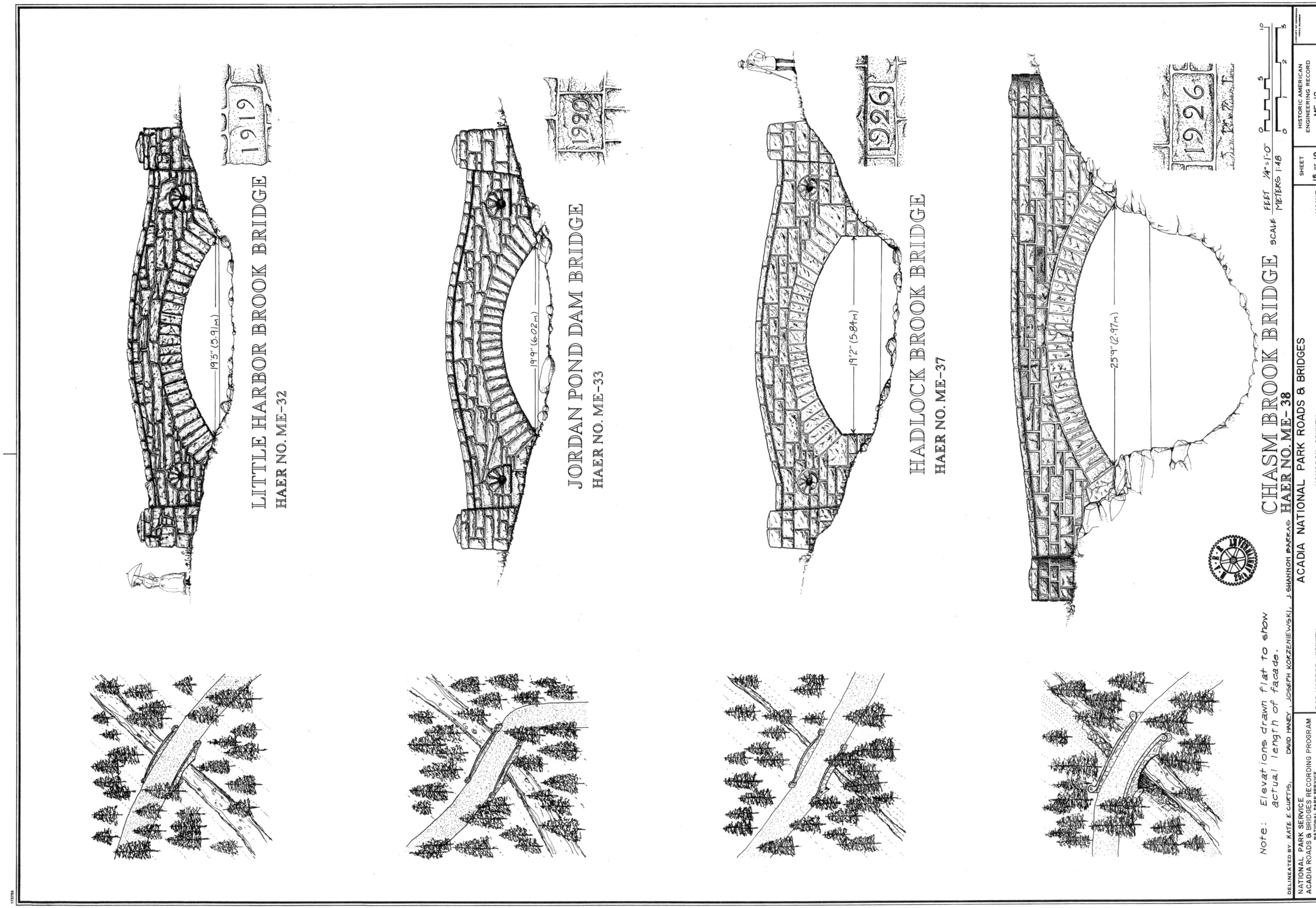
LITTLE HARBOR BROOK BRIDGE
HAER NO. ME-32
19'5" (5.91m)
1919
JORDAN POND DAM BRIDGE
HAER NO. ME-33
19'9" (6.02m)
1920
HADLOCK BROOK BRIDGE
HAER NO. ME-37
19'2" (5.84m)
1926
CHASM BROOK BRIDGE
HAER NO. ME-38
25'9" (2.97m)
1926
SCALE FEET 1/4"=1'-0" METERS 1:48
Note: Elevations drawn flat to show actual length of facade.
DELINEATED BY: KATE K. CURTIS, DAVID HANEY, JOSEPH KORZENIEWSKI, J. SHANNON BARRAS
NATIONAL PARK SERVICE
ACADIA ROADS & BRIDGES RECORDING PROGRAM
UNITED STATES DEPARTMENT OF THE INTERIOR
BAR HARBOR VICINITY
ACADIA NATIONAL PARK ROADS & BRIDGES
HANCOCK COUNTY
MAINE
SHEET 15 OF 19
HISTORIC AMERICAN ENGINEERING RECORD
ME-12
IF REPRODUCED, PLEASE CREDIT: HISTORIC AMERICAN ENGINEERING RECORD, NATIONAL PARK SERVICE, NAME OF DELINEATOR, DATE OF THE DRAWING

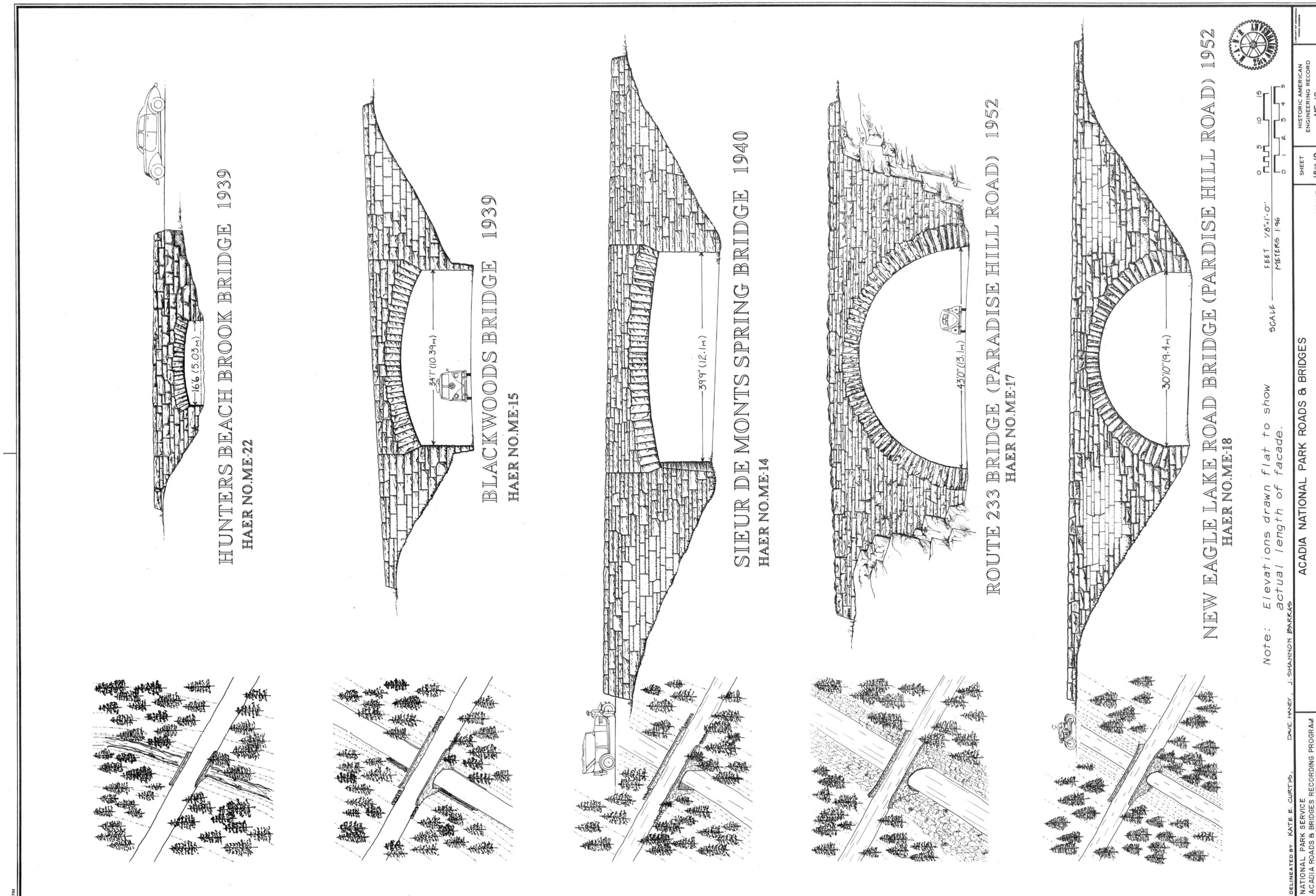
HUNTERS BEACH BROOK BRIDGE 1939
HAER NO.ME-22
16'6 (5.03m)
BLACKWOODS BRIDGE 1939
HAER NO.ME-15
34'1" (10.39m)
SIEUR DE MONTS SPRING BRIDGE 1940
HAER NO.ME-14
39'9" (12.1m)
ROUTE 233 BRIDGE (PARADISE HILL ROAD) 1952
HAER NO.ME-17
43'0" (13.1m)
NEW EAGLE LAKE ROAD BRIDGE (PARDISE HILL ROAD) 1952
HAER NO.ME-18
30'10" (9.4m)
Note: Elevations drawn flat to show actual length of facade.
SCALE
FEET 1/8"=1'-0"
METERS 1:96
DELINEATED BY: KATE E. CURTIS, DAVE HANEY, J. SHANNON BARRAS
NATIONAL PARK SERVICE
ACADIA ROADS & BRIDGES RECORDING PROGRAM
NATIONAL PARK SERVICE
UNITED STATES DEPARTMENT OF THE INTERIOR
ACADIA NATIONAL PARK ROADS & BRIDGES
BAR HARBOR VICINITY
HANCOCK COUNTY
MAINE
SHEET 18 OF 19
HISTORIC AMERICAN ENGINEERING RECORD
ME-12
IF REPRODUCED, PLEASE CREDIT: HISTORIC AMERICAN ENGINEERING RECORD, NATIONAL PARK SERVICE, NAME OF DELINEATOR, DATE OF THE DRAWING

AROUND-THE-MOUNTAIN

Jordan & Sargent Mountain Carriage Road

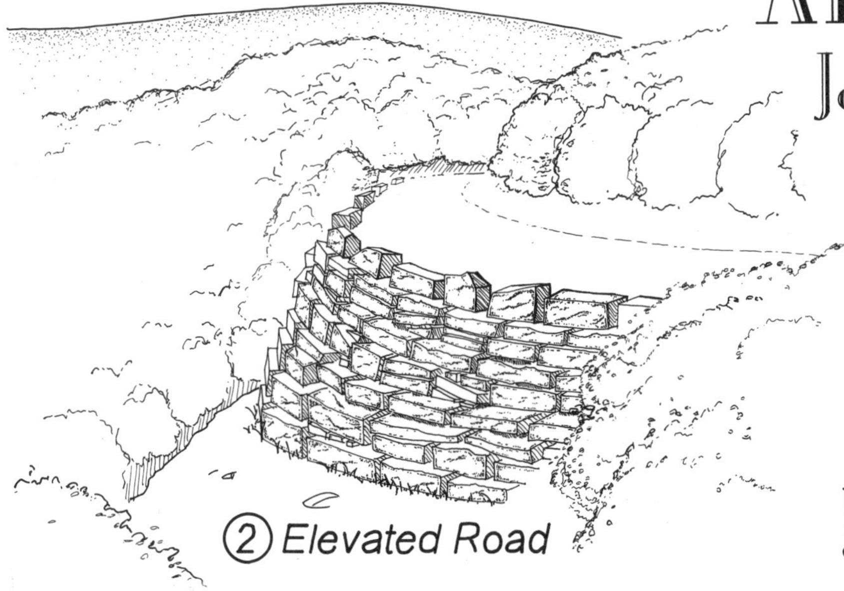

Engineers used retaining walls not only to maintain the gradual grade of Acadia's carriage roads, but also to provide visitors with scenic overlooks. This wall on the northern portion of the Around-the-Mountain loop allows unobstructed views of Eagle Lake, Cadillac Mountain and the Atlantic Ocean.

To keep them from washing out, all of Acadia's carriage roads were constructed with culverts, which safely transport runoff water underneath the roadbed. This culvert on the Around-the-Mountain loop, one of the most elaborate within the entire carriage system, was intentionally built in this manner in order to make the intersection between the trail and carriage road more scenic.

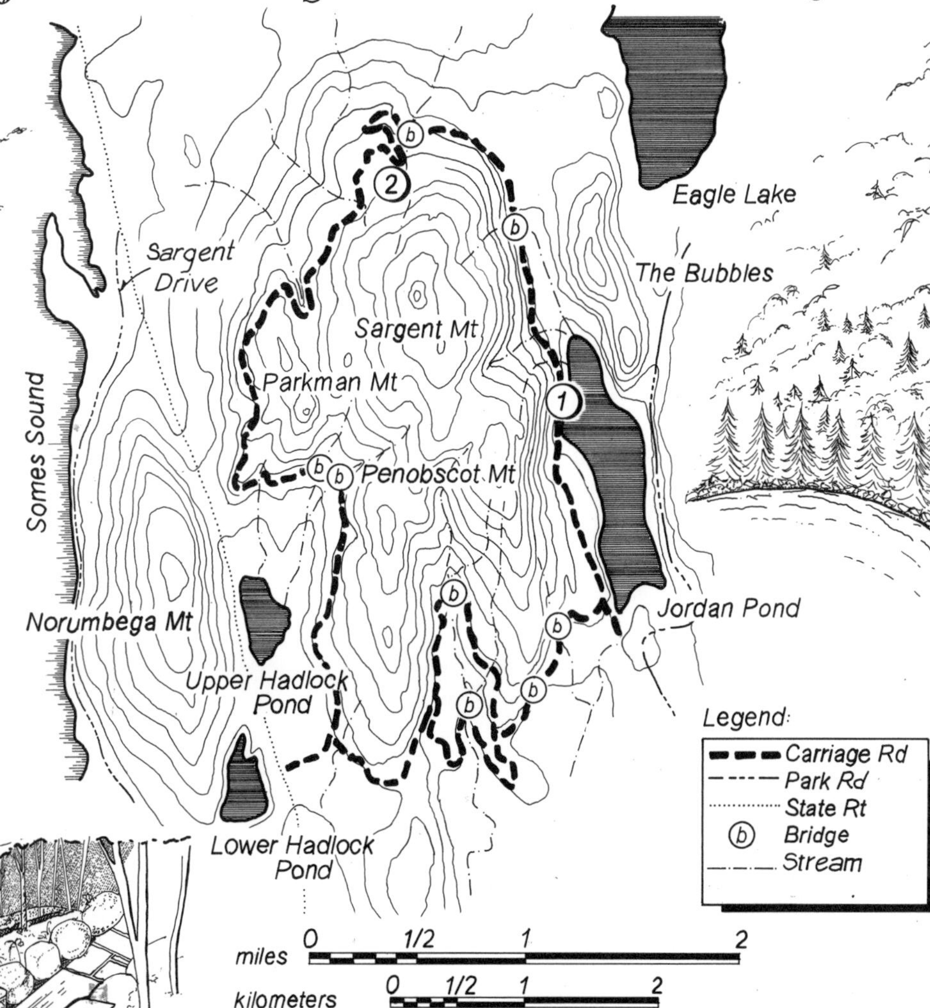

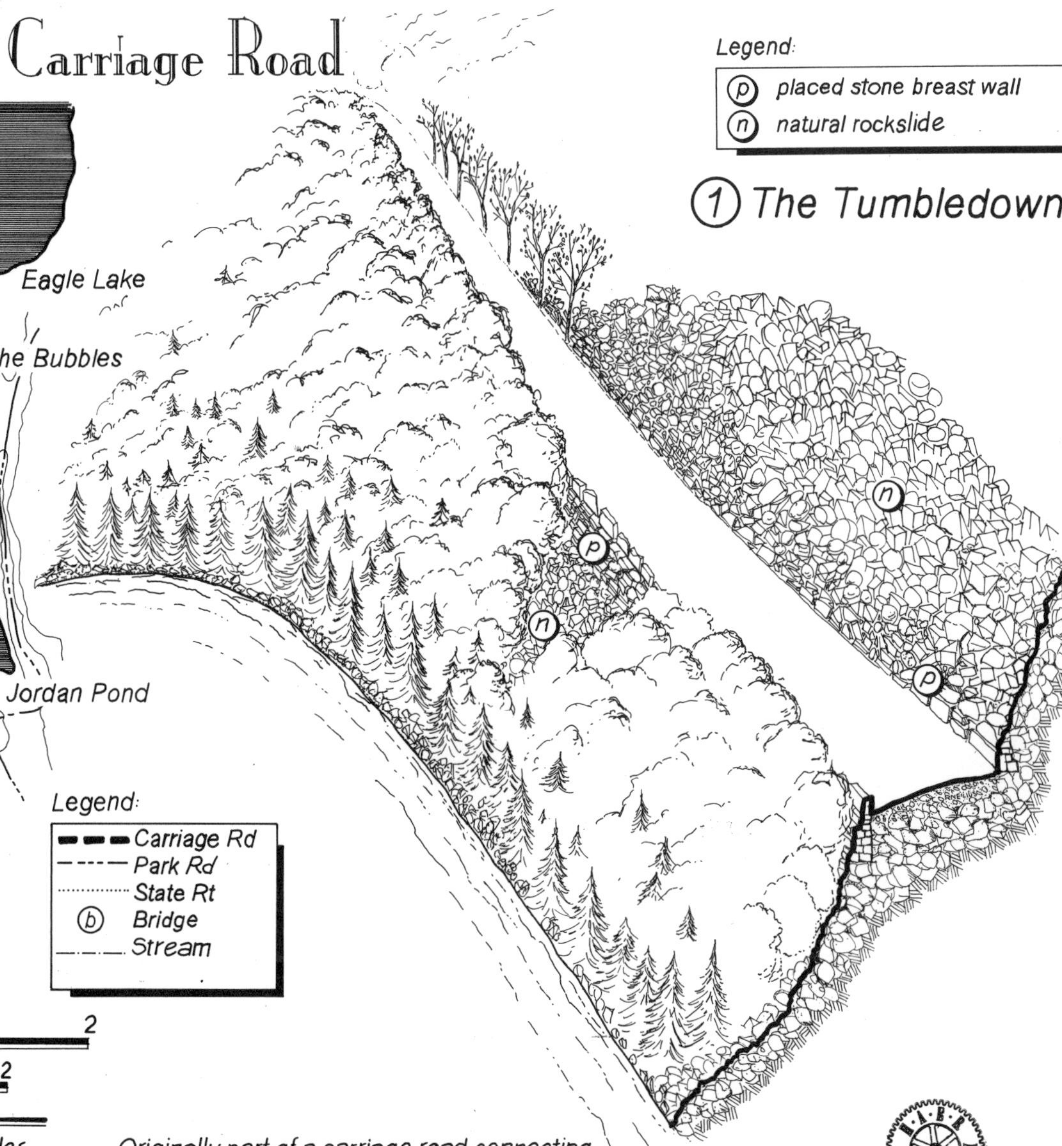

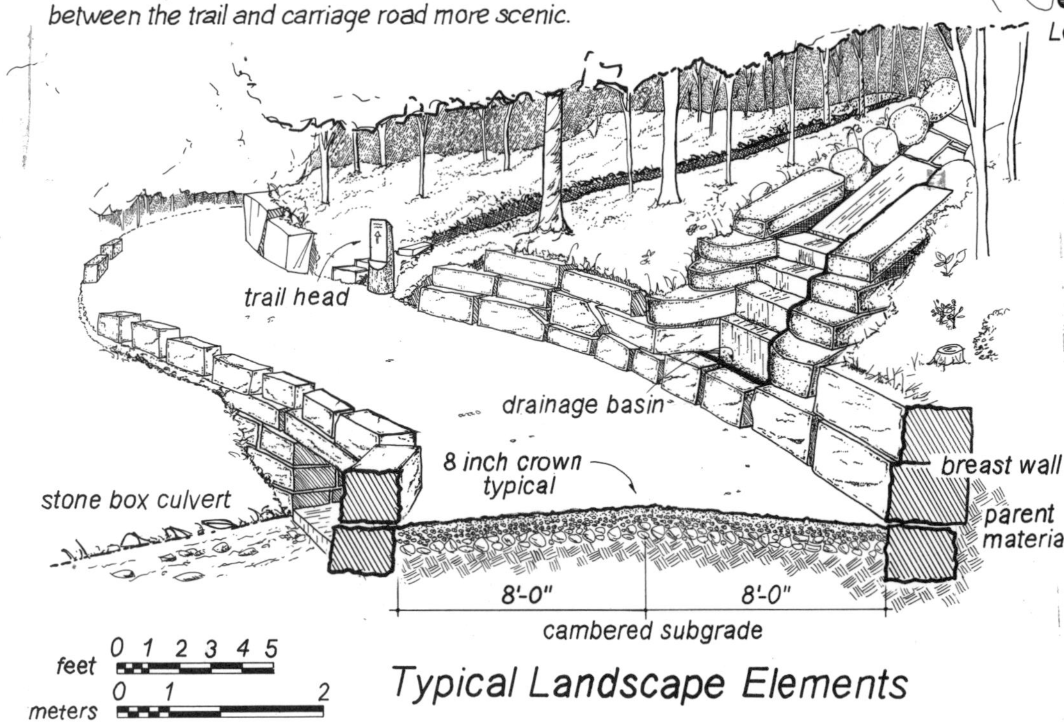

Built in stages by John D. Rockefeller, Jr. between 1921 and 1932 the Around the Mountain loop attains the highest elevation of any carriage road in Acadia. Like many of Rockefeller's motor roads, this carriage road presents the visitor with a sequence of scenic experiences including views of Gilmore Meadow to the north, Cadillac Mountain in the east, and the Cranberry Islands and Somes Sound in the south and west respectively. The nine large stone bridges along the Around-the-Mountain carriage road are just one illustration of its highly engineered design. Other examples include the widespread use of retaining and breast walls, elaborate culverts, and planned vistas.

Originally part of a carriage road connecting the northern and southern regions of the park, this section of the Around the Mountain loop, known locally as "The Tumbledown," was built by Rockefeller ca1920. When completed it was touted as an engineering feat, primarily because designers maintained the existing rockslide when cutting the road bed into the side of Penobscot Mountain. Although large rectilinear coping stones act as guardrails along the shoulder, the stonework becomes less formalized as it recedes from the road, blending into the more natural rockslide above and below the cut. Care was even taken to place as much weathered rock near the roadside as possible. The result, many argued at the time, was the near invisibility of the road when viewed from across Jordan Pond.

Landscape gardener Beatrix Farrand worked on many of Acadia's carriage roads, including this section of the Around the Mountain loop. The birches shown above lining the road just south of "The Tumbledown," recommended by Farrand, are yet another example of its highly planned design.

DELINEATED BY: Joe Korzeniewski, 1995

NATIONAL PARK SERVICE
ACADIA ROADS & BRIDGES RECORDING PROGRAM

NATIONAL PARK SERVICE
UNITED STATES DEPARTMENT OF THE INTERIOR

ROCKEFELLER CARRIAGE ROADS
BAR HARBOR VICINITY
HANCOCK COUNTY
MAINE

SHEET 1 OF 4

HISTORIC AMERICAN ENGINEERING RECORD
ME-13

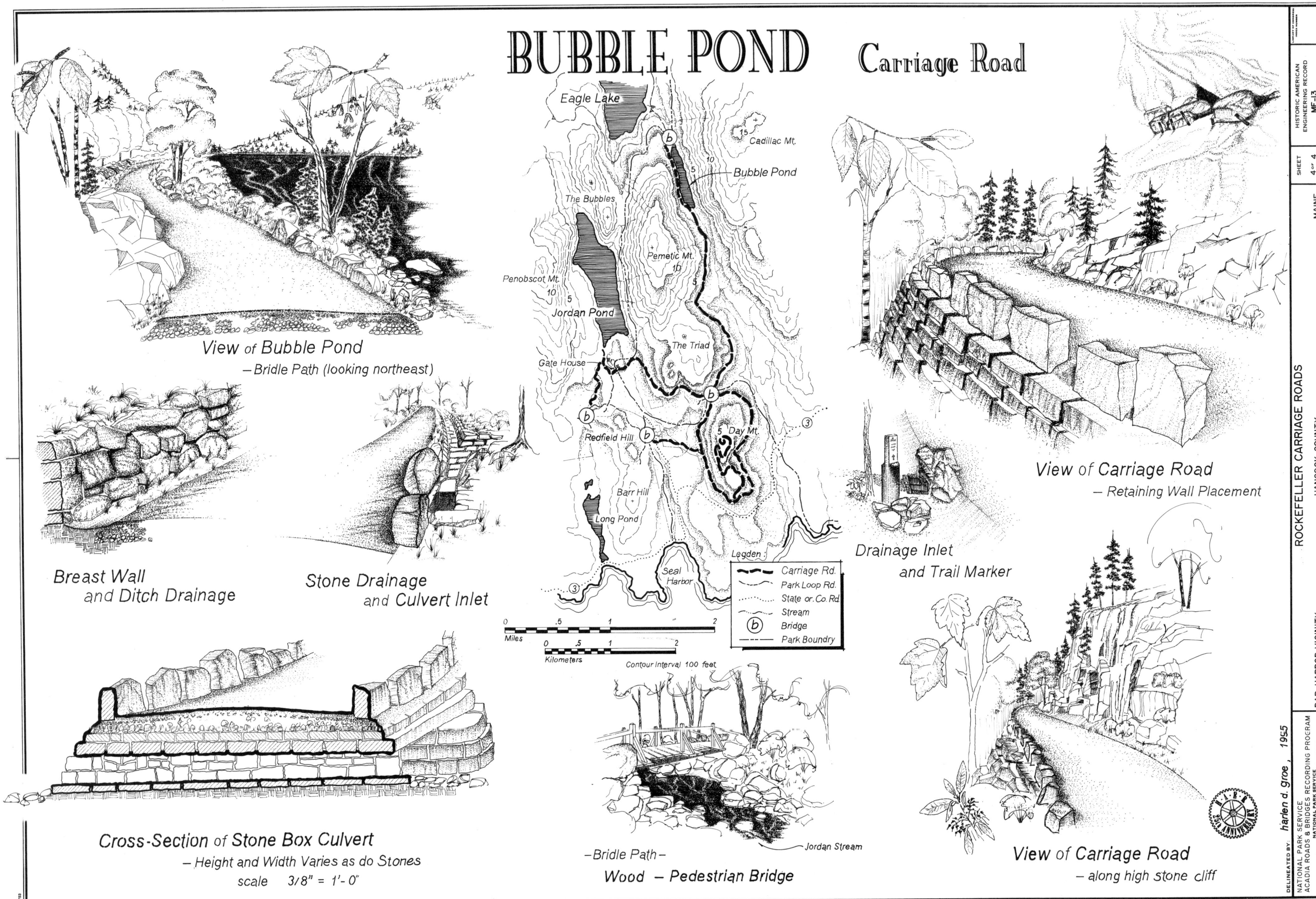
BUBBLE POND Carriage Road
View of Bubble Pond
– Bridle Path (looking northeast)
Breast Wall
and Ditch Drainage
Stone Drainage
and Culvert Inlet
Cross-Section of Stone Box Culvert
– Height and Width Varies as do Stones
scale 3/8" = 1'- 0"
Eagle Lake
Cadillac Mt.
Bubble Pond
The Bubbles
Pemetic Mt.
Penobscot Mt.
Jordan Pond
The Triad
Gate House
Day Mt.
Redfield Hill
Barr Hill
Long Pond
Seal Harbor
Carriage Rd.
Park Loop Rd.
State or Co. Rd.
Stream
Bridge
Park Boundry
Miles
Kilometers
Contour Interval 100 feet
–Bridle Path–
Wood – Pedestrian Bridge
Jordan Stream
View of Carriage Road
– Retaining Wall Placement
Drainage Inlet
and Trail Marker
View of Carriage Road
– along high stone cliff
DELINEATED BY: harlen d. groe, 1995
NATIONAL PARK SERVICE
ACADIA ROADS & BRIDGES RECORDING PROGRAM
NATIONAL PARK SERVICE
UNITED STATES DEPARTMENT OF THE INTERIOR
ROCKEFELLER CARRIAGE ROADS
BAR HARBOR VICINITY
HANCOCK COUNTY
MAINE
SHEET 4 of 4
HISTORIC AMERICAN ENGINEERING RECORD
ME-13
IF REPRODUCED, PLEASE CREDIT: HISTORIC AMERICAN ENGINEERING RECORD, NATIONAL PARK SERVICE, NAME OF DELINEATOR, DATE OF THE DRAWING

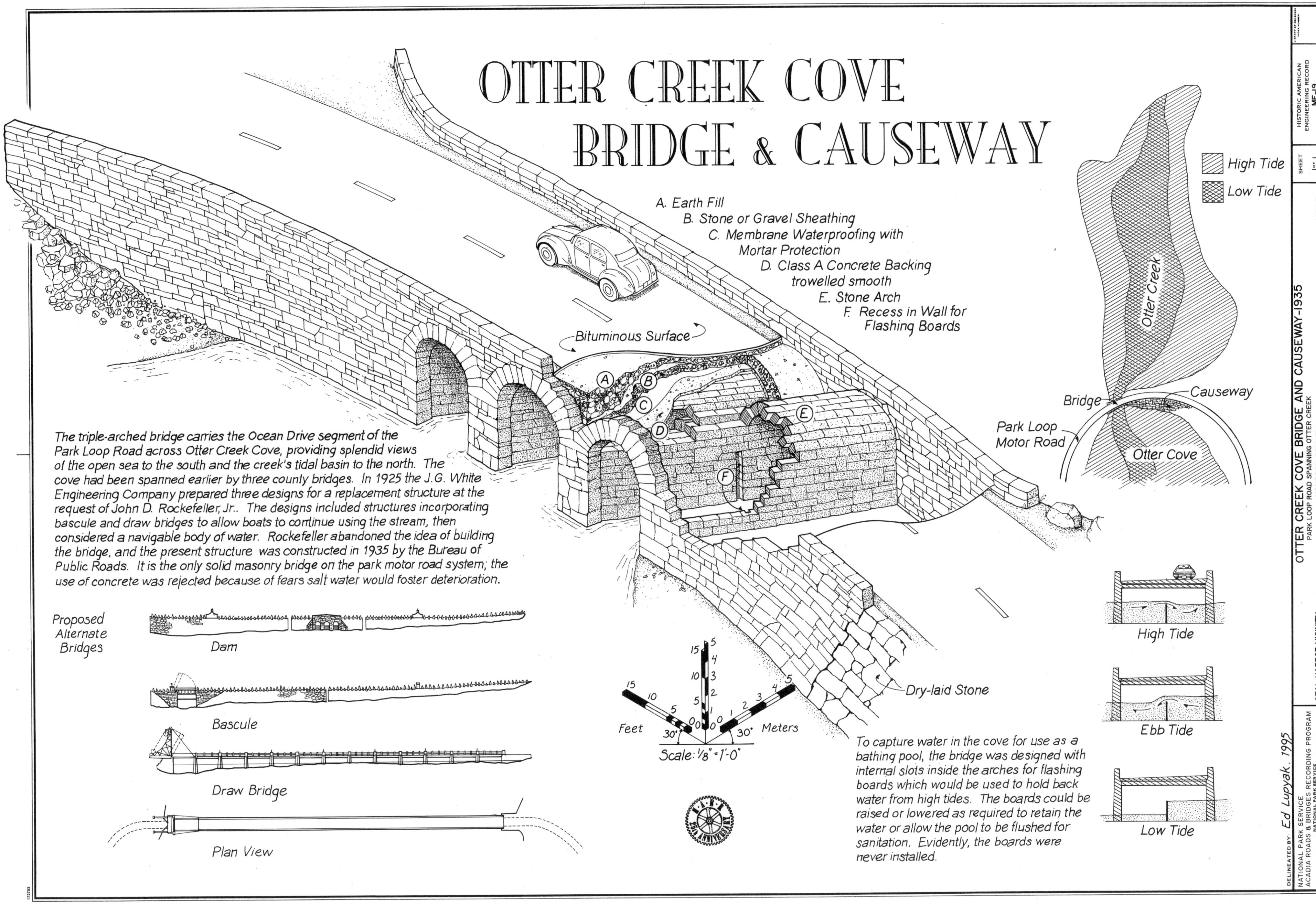
OTTER CREEK COVE
BRIDGE & CAUSEWAY
A. Earth Fill
B. Stone or Gravel Sheathing
C. Membrane Waterproofing with Mortar Protection
D. Class A Concrete Backing trowelled smooth
E. Stone Arch
F. Recess in Wall for Flashing Boards
Bituminous Surface
A
B
C
D
E
F
High Tide
Low Tide
Otter Creek
Bridge
Causeway
Park Loop Motor Road
Otter Cove
The triple-arched bridge carries the Ocean Drive segment of the Park Loop Road across Otter Creek Cove, providing splendid views of the open sea to the south and the creek's tidal basin to the north. The cove had been spanned earlier by three county bridges. In 1925 the J.G. White Engineering Company prepared three designs for a replacement structure at the request of John D. Rockefeller, Jr.. The designs included structures incorporating bascule and draw bridges to allow boats to continue using the stream, then considered a navigable body of water. Rockefeller abandoned the idea of building the bridge, and the present structure was constructed in 1935 by the Bureau of Public Roads. It is the only solid masonry bridge on the park motor road system; the use of concrete was rejected because of fears salt water would foster deterioration.
Proposed Alternate Bridges
Dam
Bascule
Draw Bridge
Plan View
Feet
Meters
30°
30°
Scale: 1/8" = 1'-0"
Dry-laid Stone
High Tide
Ebb Tide
Low Tide
To capture water in the cove for use as a bathing pool, the bridge was designed with internal slots inside the arches for flashing boards which would be used to hold back water from high tides. The boards could be raised or lowered as required to retain the water or allow the pool to be flushed for sanitation. Evidently, the boards were never installed.
DELINEATED BY Ed Lupyak, 1995
NATIONAL PARK SERVICE
ACADIA ROADS & BRIDGES RECORDING PROGRAM
SEAL HARBOR VICINITY
OTTER CREEK COVE BRIDGE AND CAUSEWAY-1935
PARK LOOP ROAD SPANNING OTTER CREEK
HANCOCK COUNTY
MAINE
SHEET 1 OF 1
HISTORIC AMERICAN ENGINEERING RECORD
ME-19
IF REPRODUCED, PLEASE CREDIT: HISTORIC AMERICAN ENGINEERING RECORD, NATIONAL PARK SERVICE, NAME OF DELINEATOR, DATE OF THE DRAWING

DUCK BROOK MOTOR ROAD BRIDGE 1953

The last major bridge constructed in Acadia National Park, Duck Brook Bridge is the largest road-related structure in the park and the largest stone-faced concrete arch bridge east of the Mississippi. The structure was planned in the early 1940s as part of the Paradise Hill Road, which would provide a scenic connecting link to the Park Loop Road from Hulls Cove west of Bar Harbor. However, construction was delayed because of World War II, and the bridge work did not begin until 1950. The contractor was the "M&M Construction Company," comprised of W. Robinson Martin of Manchester, Vermont and Harold MacQuinn of Hulls Cove. The project was under construction supervision by the Bureau of Public Roads, U.S. Department of Commerce.

The bridge spans Duck Brook's deep ravine on three massive semicircular arches. An internal concrete frame stiffens the arches and supports the concrete deck. To support the arches during construction, the Timber Structure Company of New York shipped 42 knocked-down prefabricated wooden trusses to the project. The trusses were erected in pairs by a crane without the use of traditional centering. Due to the size of the arches, the trusses had to support weights of up to 800 tons. A construction railway, called the "M&M Railroad," was employed to convey hopper cars carrying concrete to the site. The pink granite face stone was obtained from the famed Halls Quarry near Somesville. Some 4,000 tons of concrete, 2,000 tons of reinforcing steel, and 1,100 cubic yards of stone was used to build the bridge. The bridge was completed in 1953. At a cost of $336,000, it was the park's most expensive road-related structure.

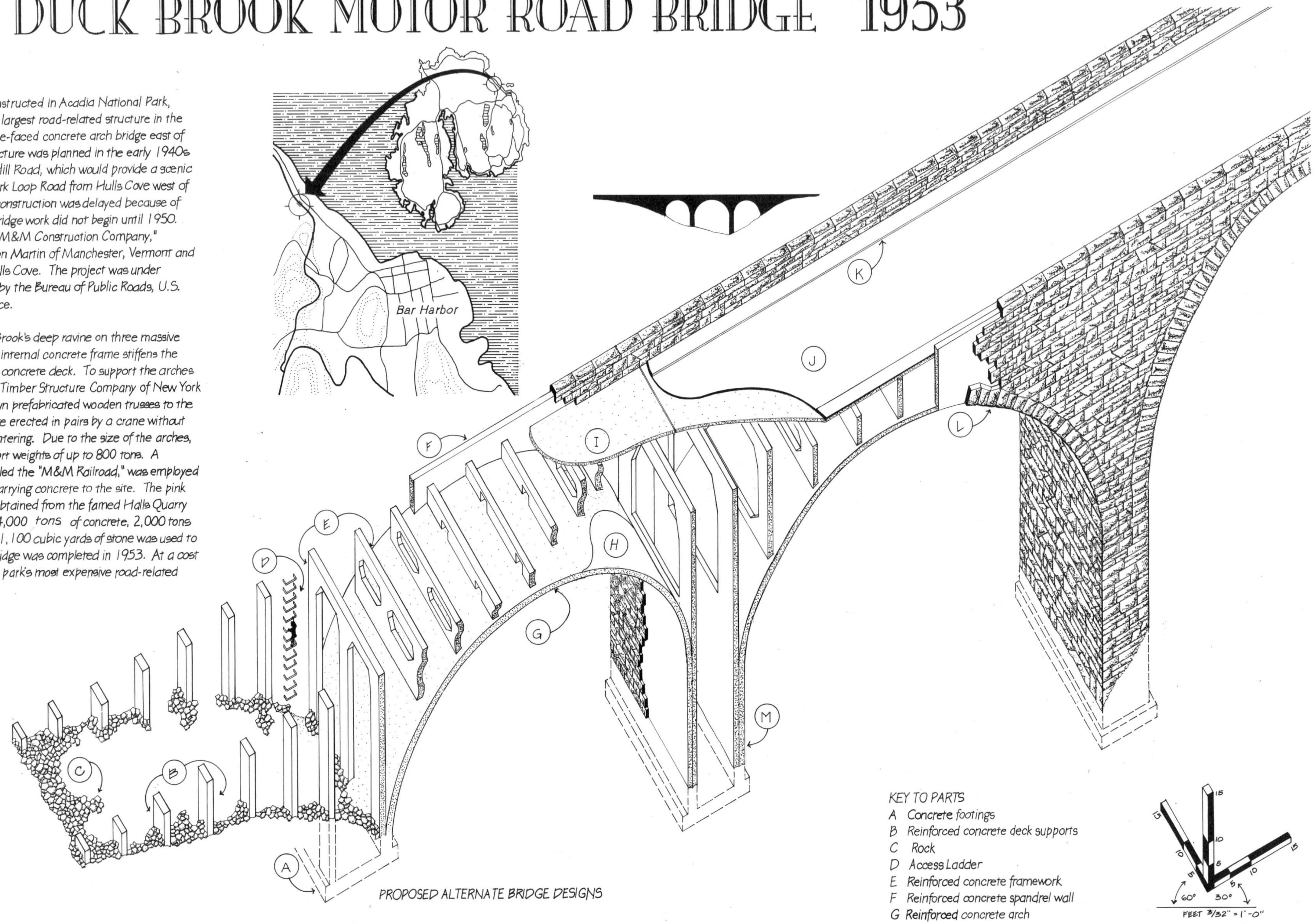

PROPOSED ALTERNATE BRIDGE DESIGNS

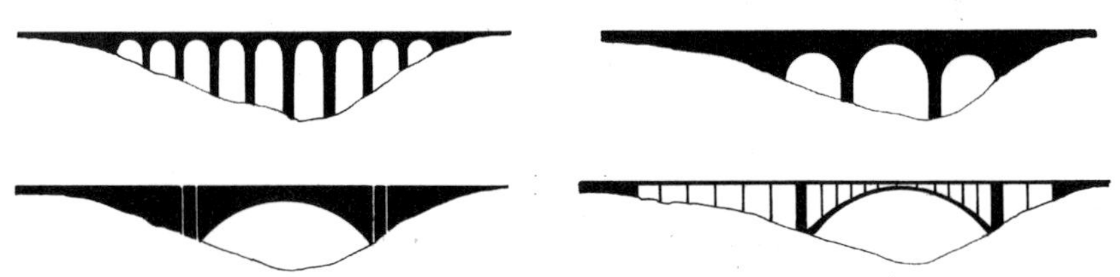

NOTE: This drawing is a conjectural interpretation based on construction drawings and field measurements.

KEY TO PARTS
- A Concrete footings
- B Reinforced concrete deck supports
- C Rock
- D Access Ladder
- E Reinforced concrete framework
- F Reinforced concrete spandrel wall
- G Reinforced concrete arch
- H Bituminous waterproofing
- I Reinforced concrete deck with expansion joints
- J Asphalt roadway surface
- K Pedestrian sidewalk
- L Granite facing stone
- M Reinforced concrete pier

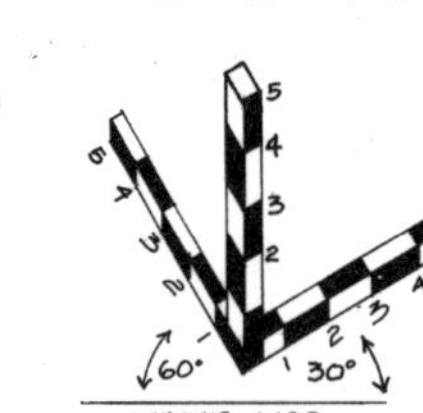

DELINEATED BY J. SHANNON BARRAS, 1994

NATIONAL PARK SERVICE
ACADIA ROADS & BRIDGES RECORDING PROGRAM
NATIONAL PARK SERVICE
UNITED STATES DEPARTMENT OF THE INTERIOR

BAR HARBOR VICINITY

DUCK BROOK BRIDGE - 1953
PARADISE HILL ROAD SPANNING DUCK BROOK
HANCOCK COUNTY

MAINE

SHEET 1 of 1

HISTORIC AMERICAN ENGINEERING RECORD
ME-30

BUBBLE POND BRIDGE 1928

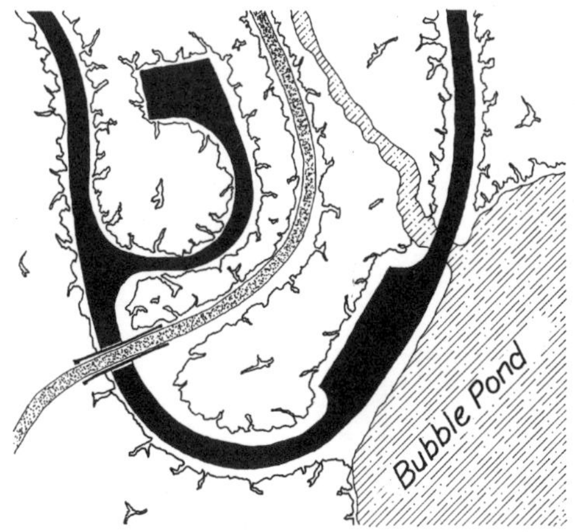

1928 Appearance

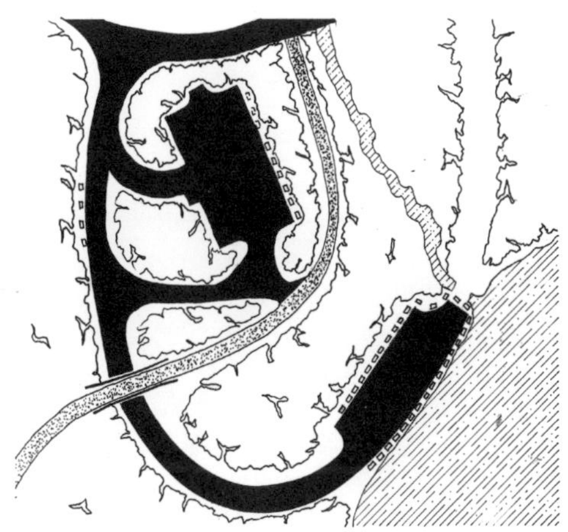

1960 Appearance

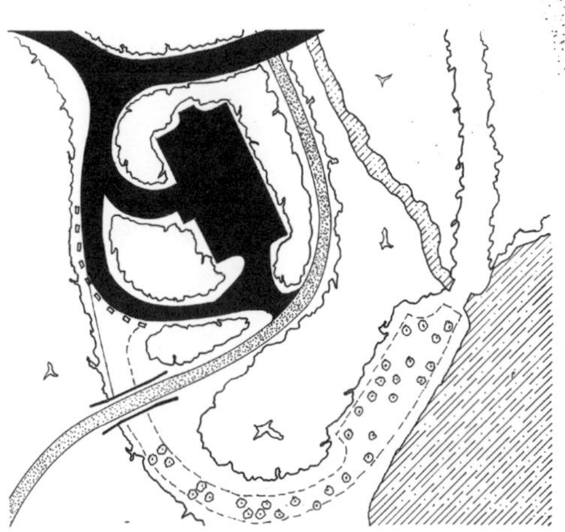

1994 Appearance

SITE DEVELOPMENT

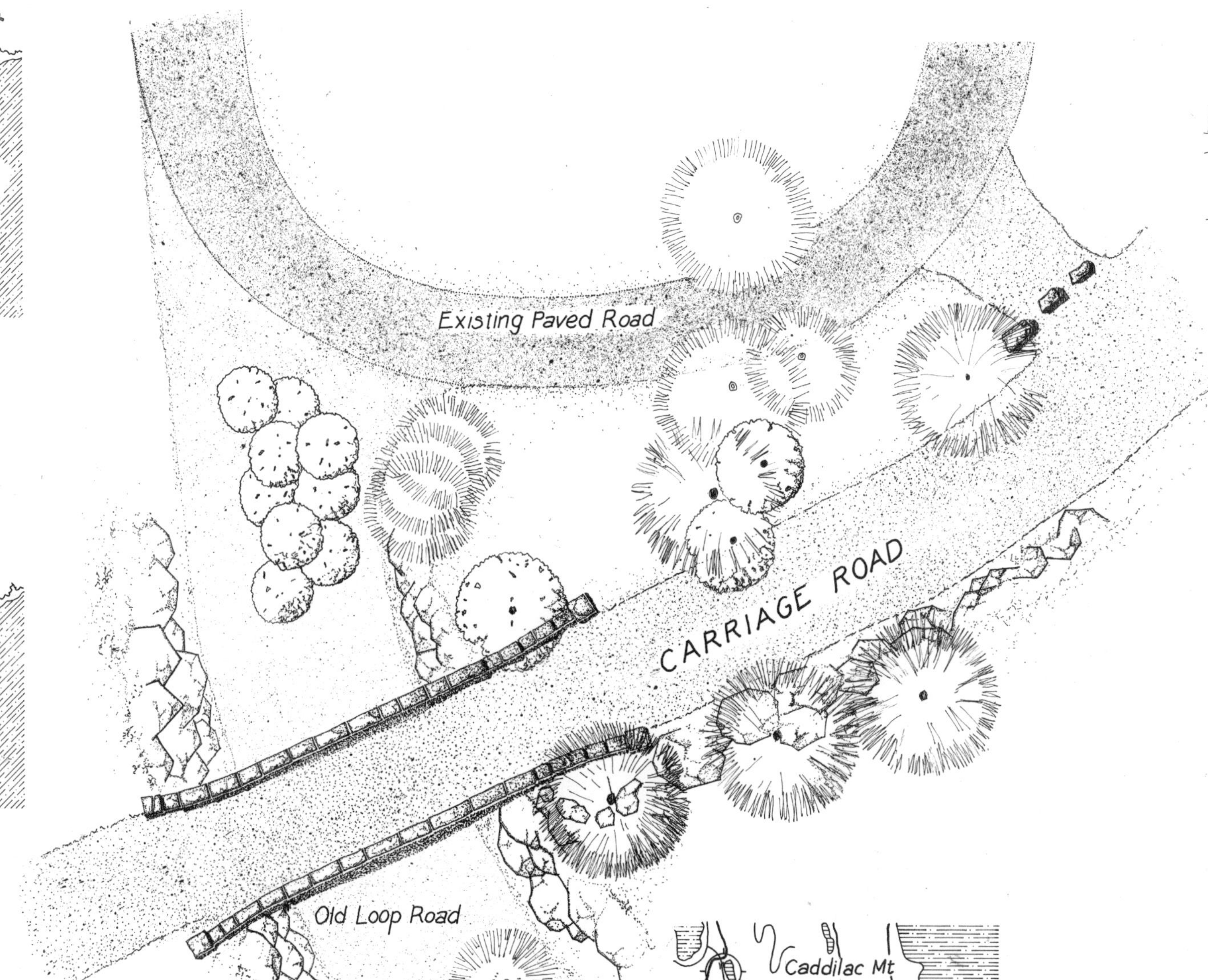

NOTE: This drawing is a conjectural interpretation based on construction drawings and field measurements.

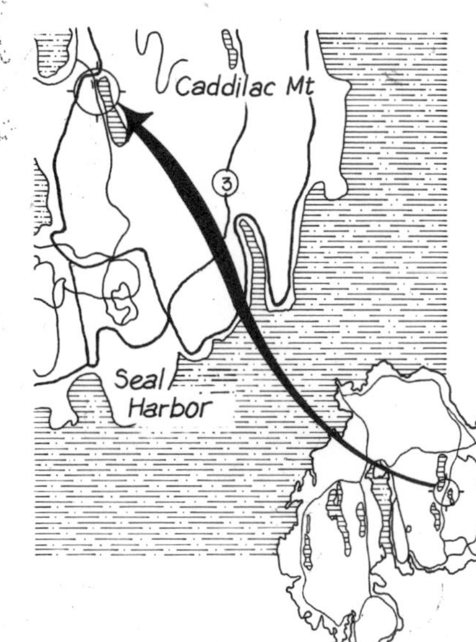

BOSWORTH'S ORIGINAL PROPOSAL

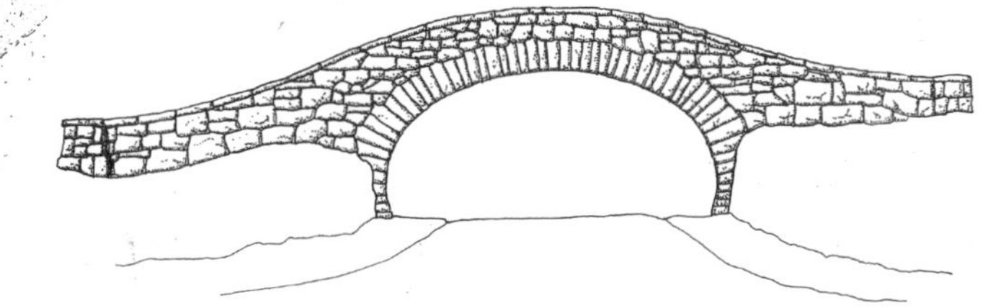

BOSWORTH'S REVISED PROPOSAL

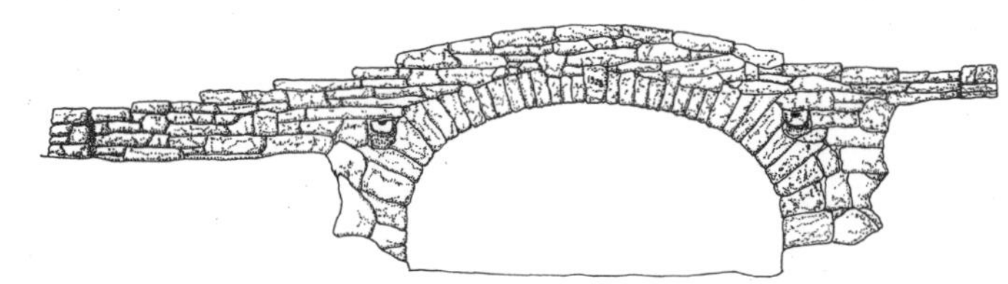

NATIONAL PARK SERVICE DESIGN

As John D. Rockefeller, Jr. planned his new Bubble Pond-Triad carriage road, he determined it should cross the new "Mountain Road" motor road (which he funded) on a bridge in order to avoid an unsafe grade crossing. He turned to New York architect William Welles Bosworth, who responded with a formal multi-centered arch bridge with open balustrade and finely-cut voussoirs. Although Mr. Rockefeller accepted the design, the NPS Assistant Director Arno Cammerer objected to it as being too formal for a national park. Bosworth prepared a second design, eliminating the open balustrade, but retaining the carefully cut stone and the formal voussoirs. This design was rejected as being "too citified". The Park Service then designed a more "rustic" structure of roughly-quarried stone, making it possibly the only carriage road bridge of solid masonry construction. Although disappointed, Mr. Rockefeller agreed to fund the construction of this design. His engineer, Paul D. Simpson, prepared the working drawings, and the bridge was constructed in 1928 by Philadelphia mason Pringle Borthwick. The bridge site was landscaped under the direction of landscape gardener Beatrix Farrand, who had worked with Mr. Rockefeller in other areas with the carriage roads.

Mr. Rockefeller's site plan was changed in the 1960s, when a segment of the "Mountain Road" (now a segment of the Park Loop Road) was relocated away from the bridge to eliminate a sharp curve. In the 1980s the parking area was relocated and the road under the bridge was abandoned, leaving the bridge as it stands today.

DELINEATED BY: J. Shannon Barras, 1994

NATIONAL PARK SERVICE
ACADIA ROADS & BRIGDGES RECORDING PROGRAM
NATIONAL PARK SERVICE
UNITED STATES DEPARTMENT OF THE INTERIOR

BAR HARBOR VICINITY

BUBBLE POND BRIDGE-1928
BUBBLE POND CARRIAGE ROAD SPANNING ABANDONED MOTOR ROAD
HANCOCK COUNTY

MAINE

SHEET 1 OF 1

HISTORIC AMERICAN ENGINEERING RECORD
ME-39

IF REPRODUCED, PLEASE CREDIT: HISTORIC AMERICAN ENGINEERING RECORD, NATIONAL PARK SERVICE, NAME OF DELINEATOR, DATE OF THE DRAWING

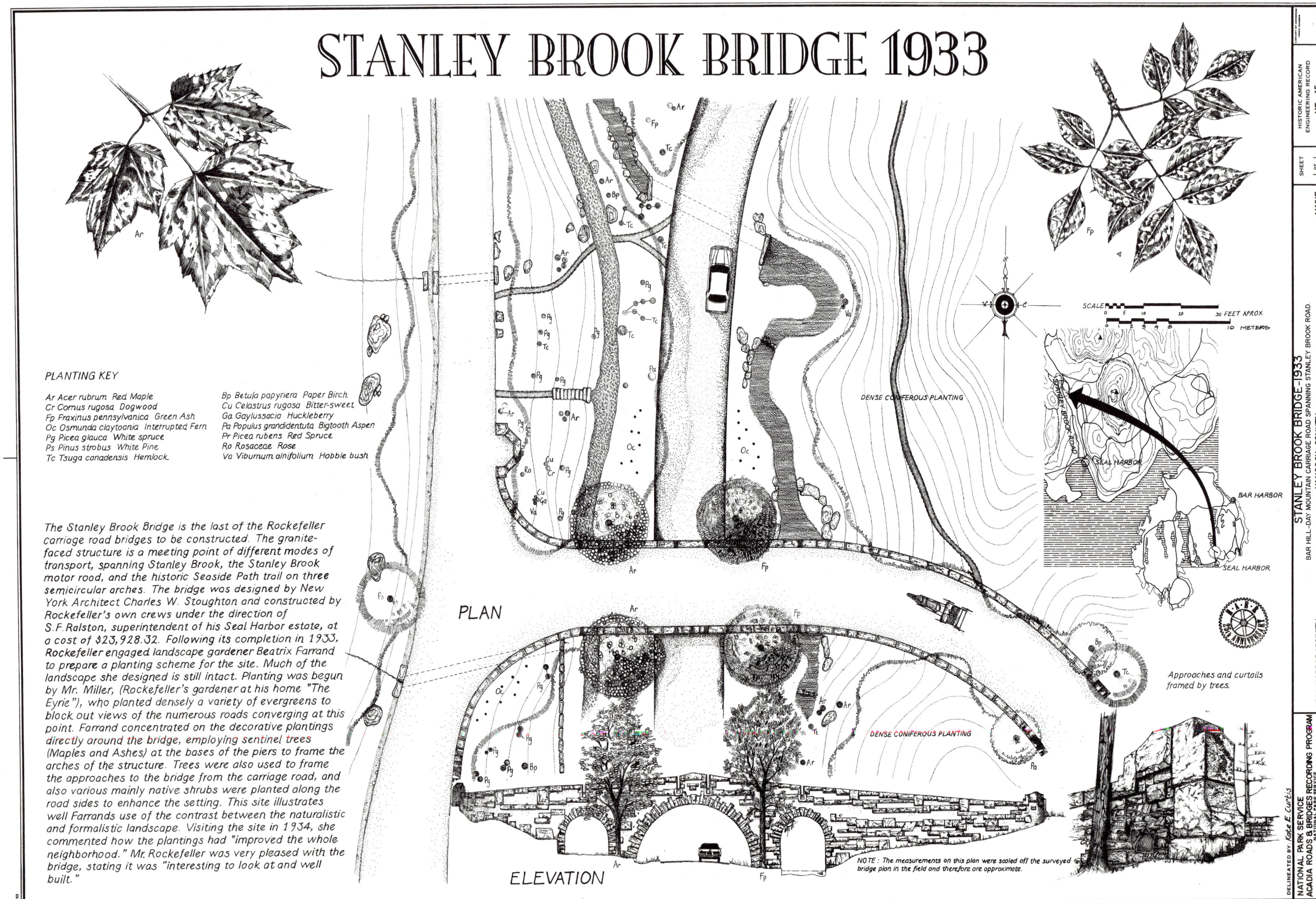
STANLEY BROOK BRIDGE 1933
PLANTING KEY
Ar Acer rubrum Red Maple
Cr Cornus rugosa Dogwood
Fp Fraxinus pennsylvanica Green Ash
Oc Osmunda claytoania Interrupted Fern
Pg Picea glauca White spruce
Ps Pinus strobus White Pine
Tc Tsuga canadensis Hemlock
Bp Betula papyriera Paper Birch
Cu Celastrus rugosa Bitter-sweet
Ga Gaylussacia Huckleberry
Pa Populus grandidentuta Bigtooth Aspen
Pr Picea rubens Red Spruce
Ro Rosaceae Rose
Va Viburnum alnifolium Hobble bush
The Stanley Brook Bridge is the last of the Rockefeller carriage road bridges to be constructed. The granite-faced structure is a meeting point of different modes of transport, spanning Stanley Brook, the Stanley Brook motor road, and the historic Seaside Path trail on three semicircular arches. The bridge was designed by New York Architect Charles W. Stoughton and constructed by Rockefeller's own crews under the direction of S.F. Ralston, superintendent of his Seal Harbor estate, at a cost of $23,928.32. Following its completion in 1933, Rockefeller engaged landscape gardener Beatrix Farrand to prepare a planting scheme for the site. Much of the landscape she designed is still intact. Planting was begun by Mr. Miller, (Rockefeller's gardener at his home "The Eyrie"), who planted densely a variety of evergreens to block out views of the numerous roads converging at this point. Farrand concentrated on the decorative plantings directly around the bridge, employing sentinel trees (Maples and Ashes) at the bases of the piers to frame the arches of the structure. Trees were also used to frame the approaches to the bridge from the carriage road, and also various mainly native shrubs were planted along the road sides to enhance the setting. This site illustrates well Farrands use of the contrast between the naturalistic and formalistic landscape. Visiting the site in 1934, she commented how the plantings had "improved the whole neighborhood." Mr. Rockefeller was very pleased with the bridge, stating it was "interesting to look at and well built."
PLAN
ELEVATION
DENSE CONIFEROUS PLANTING
DENSE CONIFEROUS PLANTING
SCALE 0 5 10 20 30 FEET APROX
10 METERS
SEAL HARBOR
BAR HARBOR
SEAL HARBOR
STANLEY BROOK ROAD
Approaches and curtails framed by trees.
NOTE: The measurements on this plan were scaled off the surveyed bridge plan in the field and therefore are approximate.
DELINEATED BY: Kate E. Curtis
NATIONAL PARK SERVICE
ACADIA ROADS & BRIDGES RECORDING PROGRAM
NATIONAL PARK SERVICE
UNITED STATES DEPARTMENT OF THE INTERIOR
SEAL HARBOR VICINITY
STANLEY BROOK BRIDGE-1933
BAR HILL-DAY MOUNTAIN CARRIAGE ROAD SPANNING STANLEY BROOK ROAD.
HANCOCK COUNTY
MAINE
SHEET 1 OF 1
HISTORIC AMERICAN ENGINEERING RECORD
ME-45
IF REPRODUCED, PLEASE CREDIT: HISTORIC AMERICAN ENGINEERING RECORD, NATIONAL PARK SERVICE, NAME OF DELINEATOR, DATE OF THE DRAWING

BROWN MOUNTAIN GATELODGE 1931

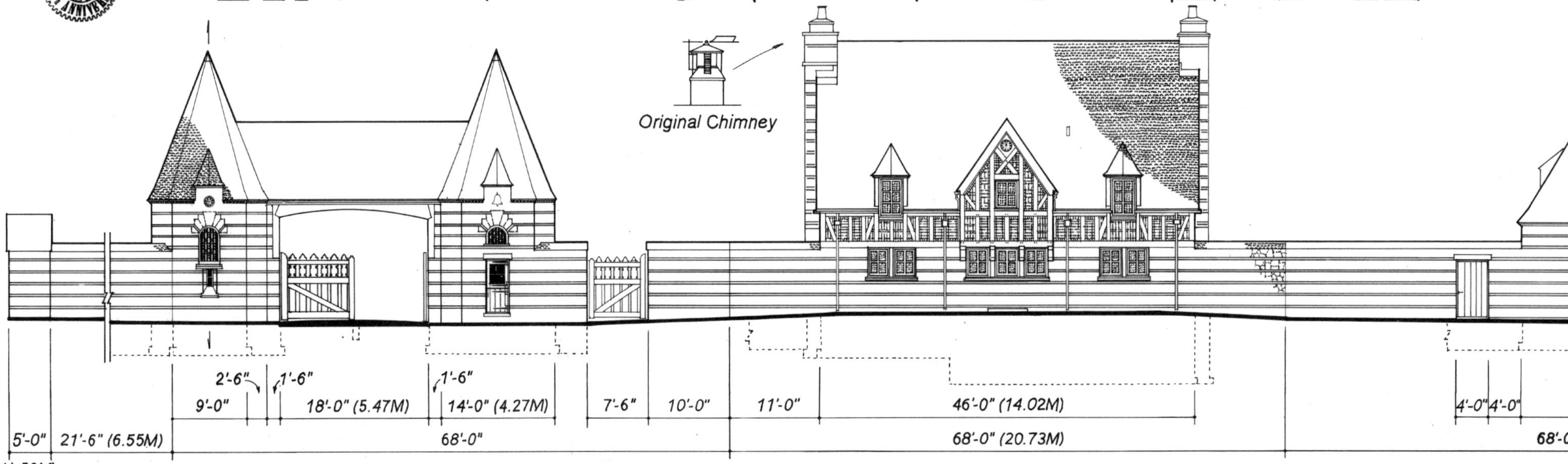

West Elevation

scale: 1/8" = 1'-0"

Built by John D. Rockefeller, Jr. to deter motorists from entering his carriage road system, Acadia's gate lodges are two of the most unique structures on Mount Desert Island. They were designed by prominent New York architect Grosvenor Atterbury, who in attempting to recall the island's colonial heritage chose a highly picturesque style reminiscent of the provincial architecture of the Le Puis district in France. After their completion in 1932, Rockefeller donated both gatehouses to the park.

Located just north of Northeast Harbor near Lower Hadlock Pond, the Brown Mountain Gate Lodge rests on a semi-octagonal foundation that reflects the shape of its monumental gate-towers. The high masonry wall bordering the front of the structure, as well as the positioning of the entrance and garage behind it, result in a nearly unbroken facade intended to draw the visitor's attention towards the gate.

The original design for the Brown Mountain Gate Lodge located the entranceway in front facing Maine Route 3 and included a bell post that was eventually incorporated into the plans for the Jordan Pond Gate Lodge.

Entrance and Keeper's Lodge (First proposal December 31, 1930)

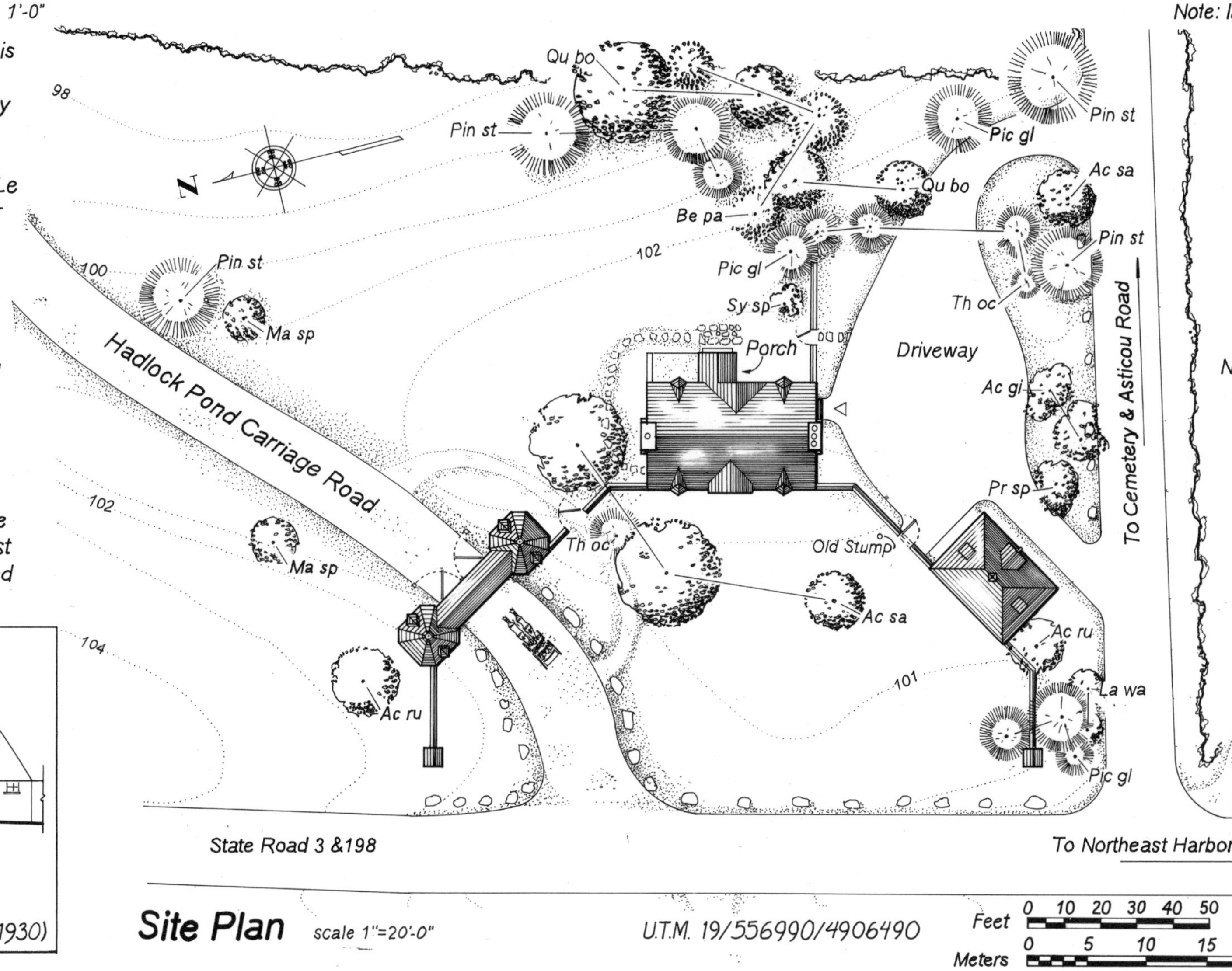

Site Plan

scale 1"=20'-0"

U.T.M. 19/556990/4906490

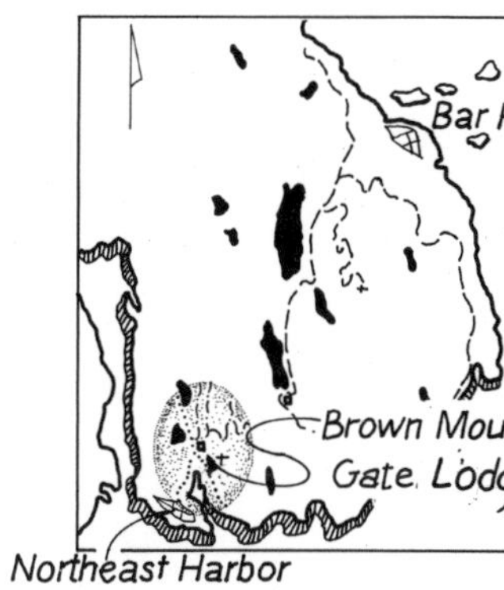

Rockefeller asked landscape gardener Beatrix Farrand for advice concerning the plantings around the Brown Mountain Gate Lodge. Although she at first advocated screening the driveway and entrance area on the south side of the complex from view with a grassy mound, she later decided, along with Atterbury, to use plantings instead.

Key	Botanic Name	Common Name
Ac gi	Acer ginnala	Amur Maple
Ac ru	" rubrum	Red Maple
Ac sa	" saccharum	Sugar Maple
Be pa	Betula papyrifera	Paper Birch
La wa	Laburnum waterari	Walterer Laburnum
Ma sp	Malus species	Crabapple sp.
Pic gl	Picea glauca	White Spruce
Pin st	Pinus strobus	Eastern White Pine
Pr sp	Prunus species	Cherry sp.
Qu bo	Quercus borealis	Red Oak
Sy sp	Syringa species	Lilac sp.
Th oc	Thuja occidantalis	Eastern Arborvitae

DELINEATED BY: Sarah F. Desbiens, 1995

NATIONAL PARK SERVICE
ACADIA ROADS & BRIDGES RECORDING PROGRAM
NATIONAL PARK SERVICE
UNITED STATES DEPARTMENT OF THE INTERIOR

BROWN MOUNTAIN GATE LODGE-1931
MAINE ROUTE 3, ONE HALF MILE NORTH OF ASTICOU
HANCOCK COUNTY
NORTHEAST HARBOR VICINITY
MAINE

SHEET 1 OF 2

HISTORIC AMERICAN ENGINEERING RECORD
ME-53

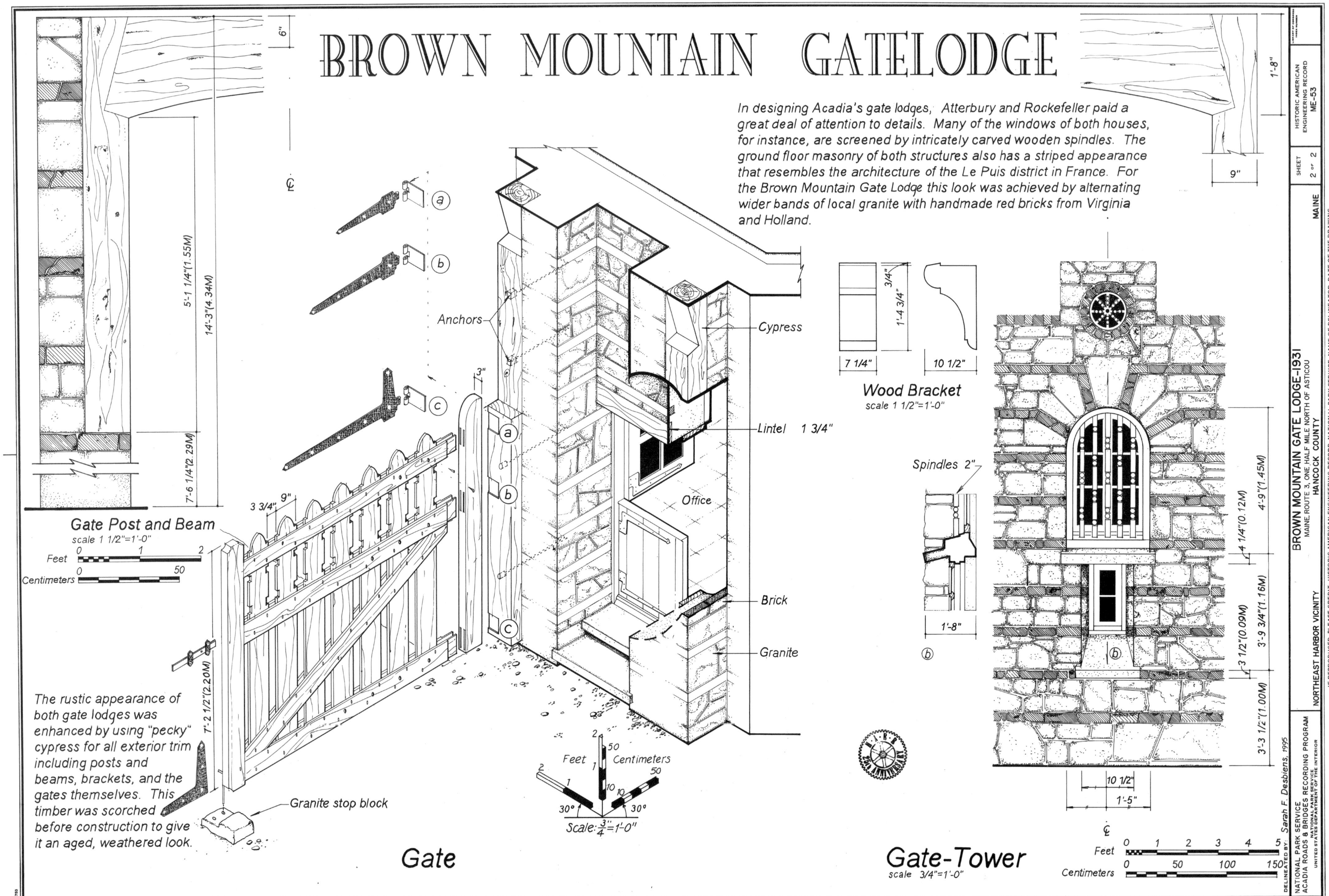
BROWN MOUNTAIN GATELODGE
In designing Acadia's gate lodges, Atterbury and Rockefeller paid a great deal of attention to details. Many of the windows of both houses, for instance, are screened by intricately carved wooden spindles. The ground floor masonry of both structures also has a striped appearance that resembles the architecture of the Le Puis district in France. For the Brown Mountain Gate Lodge this look was achieved by alternating wider bands of local granite with handmade red bricks from Virginia and Holland.
The rustic appearance of both gate lodges was enhanced by using "pecky" cypress for all exterior trim including posts and beams, brackets, and the gates themselves. This timber was scorched before construction to give it an aged, weathered look.
Gate Post and Beam
scale 1 1/2"=1'-0"
Feet
Centimeters
5'-1 1/4"(1.55M)
14'-3"(4.34M)
7'-6 1/4"(2.29M)
7'-2 1/2"(2.20M)
Anchors
Cypress
Lintel 1 3/4"
Office
Brick
Granite
Granite stop block
Wood Bracket
scale 1 1/2"=1'-0"
7 1/4"
10 1/2"
1'-4 3/4"
Spindles 2"
1'-8"
9"
Scale: 3/4"=1'-0"
Gate
Gate-Tower
scale 3/4"=1'-0"
4'-9"(1.45M)
4 1/4"(0.12M)
3'-9 3/4"(1.16M)
3 1/2"(0.09M)
3'-3 1/2"(1.00M)
10 1/2"
1'-5"
HISTORIC AMERICAN ENGINEERING RECORD ME-53
SHEET 2 of 2
BROWN MOUNTAIN GATE LODGE-1931
MAINE ROUTE 3, ONE HALF MILE NORTH OF ASTICOU
NORTHEAST HARBOR VICINITY
HANCOCK COUNTY
MAINE
NATIONAL PARK SERVICE
ACADIA ROADS & BRIDGES RECORDING PROGRAM
DELINEATED BY: Sarah F. Desbiens, 1995

JORDAN POND GATELODGE 1931

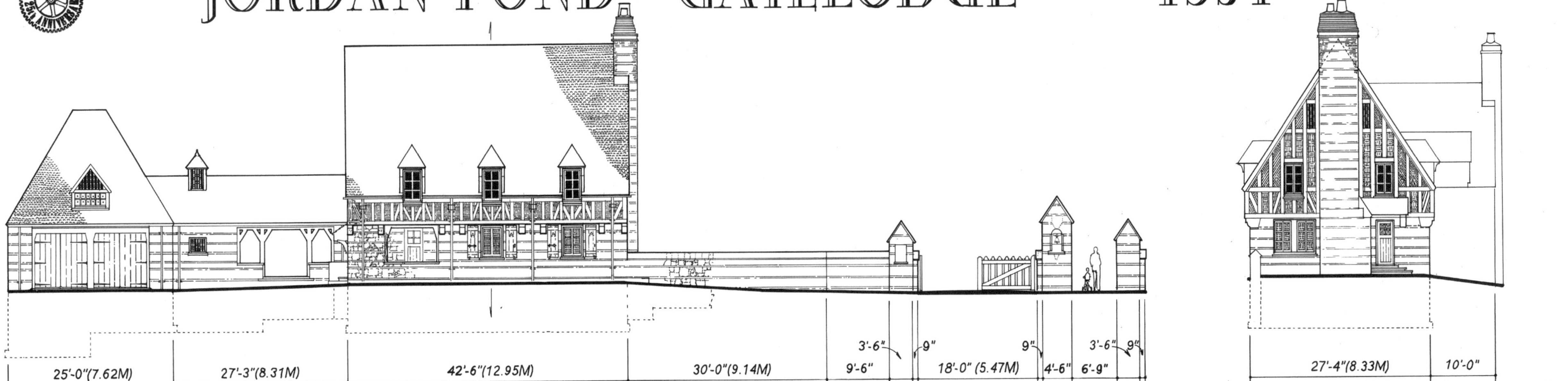

West Elevation Note: laid flat scale 1/8"= 1'-0"

South Elevation

Like its counterpart near what used to be called Brown Mountain, the Jordan Pond Gate Lodge was built by John D. Rockefeller, Jr. in 1931. It too was designed by Grosvenor Atterbury in the same French style, and includes a main house, garage, and carriage road gate all connected by a high masonry wall. Yet, whereas the layout of the Brown Mountain Gate Lodge was octagonal, the one near Jordan Pond is linear. The facade of the Jordan Pond Gate Lodge complex, with its breezeway and porch, also emphasizes the house and garage rather than the gate itself, which is mirrored by another located across the Park Loop Road. After its completion, the Jordan Pond Gate Lodge was occupied first by the family of Rockefeller's chief engineer, Paul D. Simpson, and later the structure was donated to Acadia by Rockefeller.

The Jordan Pond Gate Lodge was built in the vicinity of the Jordan Pond Tea House. This popular island eatery was converted from a farmhouse in 1896, and destroyed by fire in 1979 only to be rebuilt. The outbuildings of the original farm were razed when the Park Loop Road was realigned in 1964.

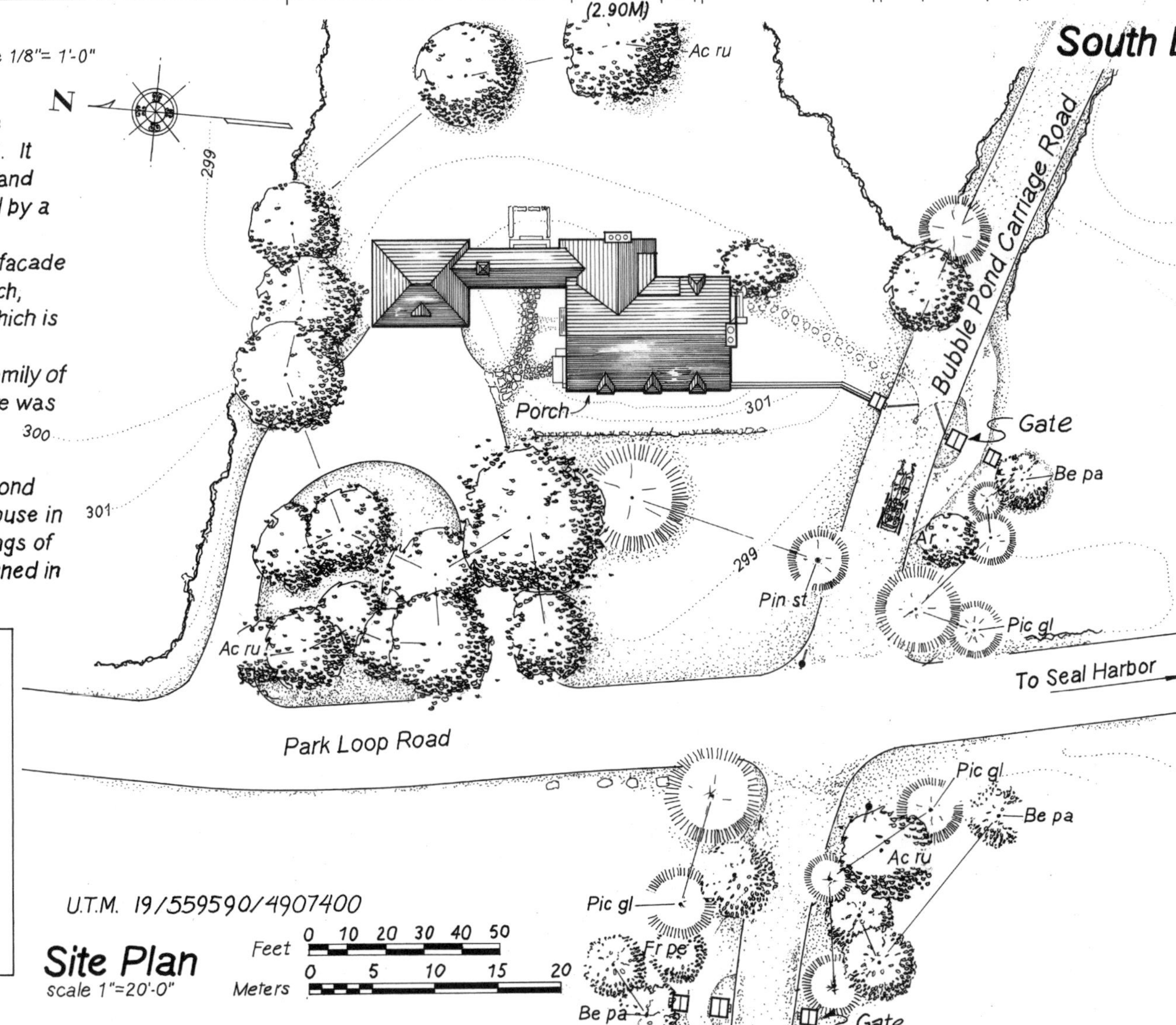

Site Plan
scale 1"=20'-0"

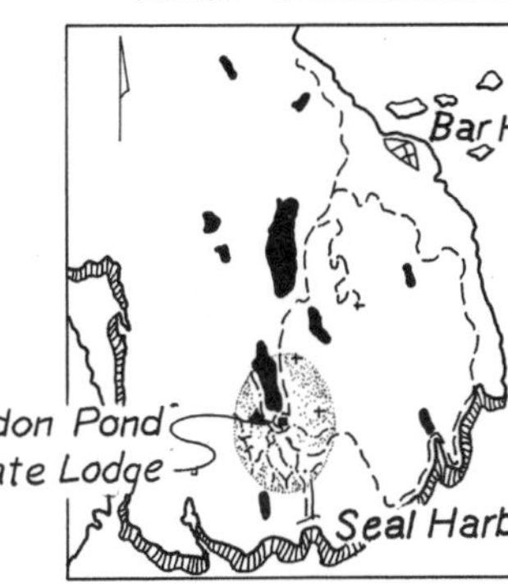

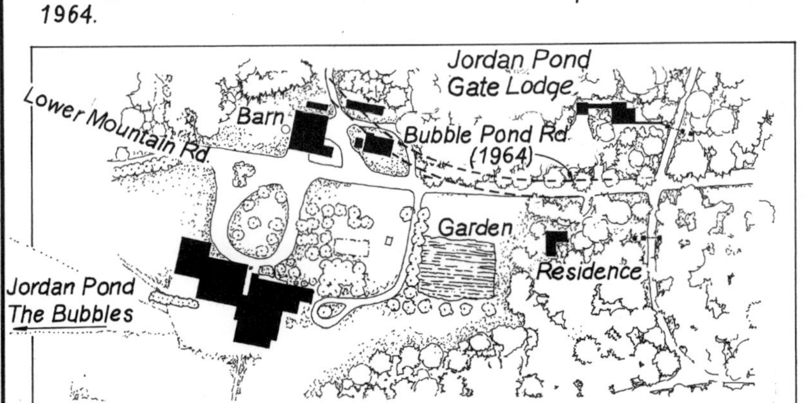

Jordan Pond House, Part of the Master Plan (1937)

Landscape gardener Beatrix Farrand was also involved with the plantings around the Jordan Pond Gate Lodge Like her other gate lodge work, Farrand attempted to blend the Jordan Pond structure in with its natural surroundings. To accomplish this she recommended that vines be placed on the exterior of the facade, shrubs and ferns be located near sharp corners of the buildings, and a terraced lawn be constructed in front of the house in order to salvage trees indigenous to the area.

Key	Botanic Name	Common Name
Ac ru	Acer rubrum	Red Maple
Be pa	Betula papyrifera	Paper Birch
Fr pe	Fraxinus pennsylvanica lanceolata	Green Ash
Pic gl	Picea glauca	White Spruce
Pin st	Pinus strobus	Eastern White Pine

HISTORIC AMERICAN ENGINEERING RECORD ME-54
SHEET 1 of 2
MAINE
JORDAN POND GATE LODGE-1931
PARK LOOP ROAD, ONE MILE NORTH OF SEAL HARBOR
HANCOCK COUNTY
SEAL HARBOR VICINITY
DELINEATED BY: Sarah F. Desbiens 1995
NATIONAL PARK SERVICE
ACADIA ROADS & BRIDGES RECORDING PROGRAM
NATIONAL PARK SERVICE UNITED STATES DEPARTMENT OF THE INTERIOR
IF REPRODUCED, PLEASE CREDIT: HISTORIC AMERICAN ENGINEERING RECORD, NATIONAL PARK SERVICE, NAME OF DELINEATOR, DATE OF THE DRAWING

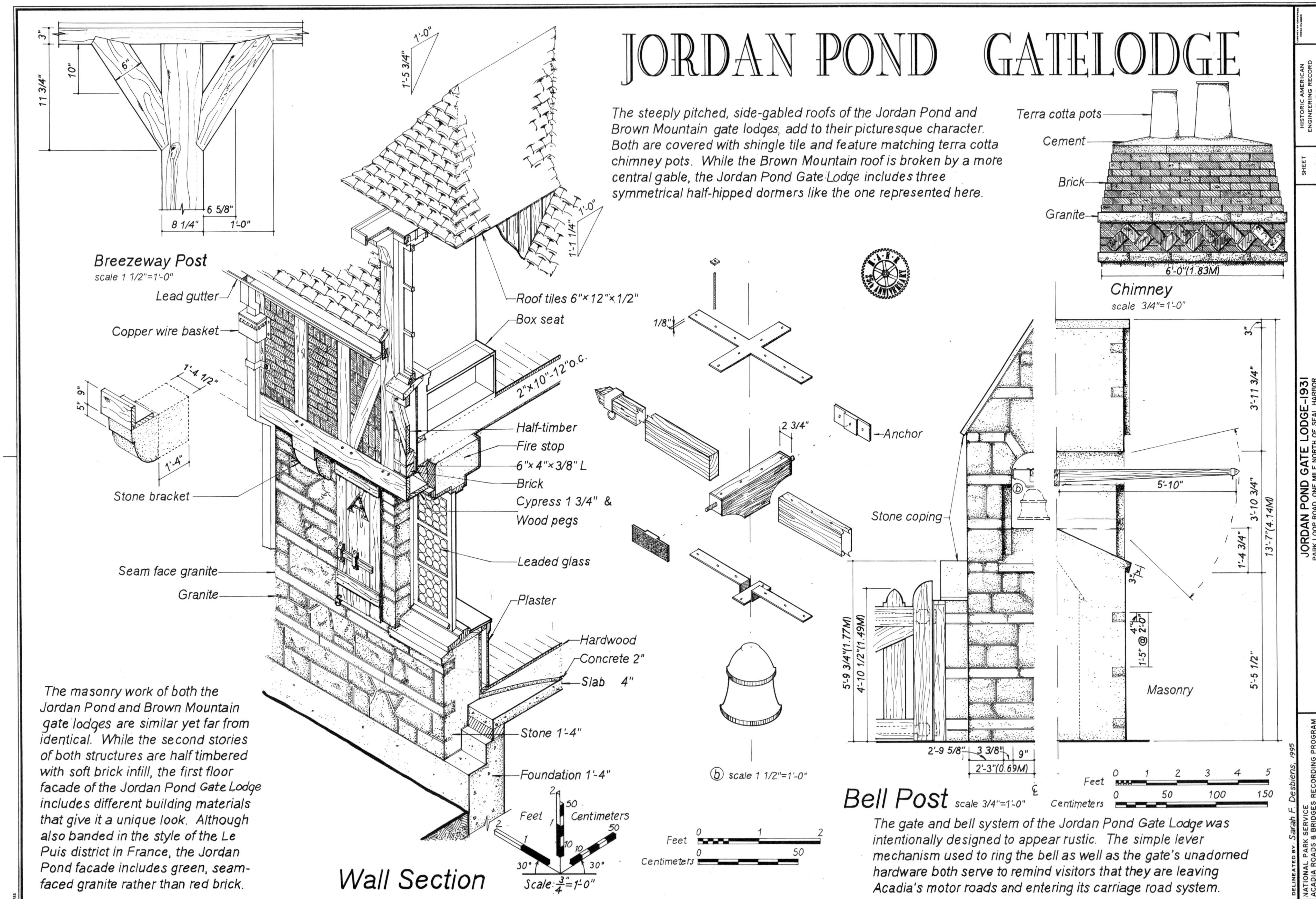
JORDAN POND GATELODGE
The steeply pitched, side-gabled roofs of the Jordan Pond and Brown Mountain gate lodges, add to their picturesque character. Both are covered with shingle tile and feature matching terra cotta chimney pots. While the Brown Mountain roof is broken by a more central gable, the Jordan Pond Gate Lodge includes three symmetrical half-hipped dormers like the one represented here.
Breezeway Post
scale 1 1/2"=1'-0"
3"
10"
11 3/4"
6"
6 5/8"
8 1/4"
1'-0"
1'-0"
1'-5 3/4"
1'-1 1/4"
Lead gutter
Copper wire basket
Roof tiles 6"×12"×1/2"
Box seat
2"×10"-12"o.c.
Half-timber
Fire stop
6"×4"×3/8" L
Brick
Cypress 1 3/4" & Wood pegs
Leaded glass
Plaster
Hardwood
Concrete 2"
Slab 4"
Stone 1'-4"
Foundation 1'-4"
1'-4 1/2"
9"
5"
1'-4"
Stone bracket
Seam face granite
Granite
The masonry work of both the Jordan Pond and Brown Mountain gate lodges are similar yet far from identical. While the second stories of both structures are half timbered with soft brick infill, the first floor facade of the Jordan Pond Gate Lodge includes different building materials that give it a unique look. Although also banded in the style of the Le Puis district in France, the Jordan Pond facade includes green, seam-faced granite rather than red brick.
Feet
Centimeters
Scale: 3/4"=1'-0"
Wall Section
1/8"
2 3/4"
Anchor
ⓑ scale 1 1/2"=1'-0"
Feet 0 1 2
Centimeters 0 50
Terra cotta pots
Cement
Brick
Granite
6'-0"(1.83M)
Chimney
scale 3/4"=1'-0"
Stone coping
5'-9 3/4"(1.77M)
4'-10 1/2"(1.49M)
2'-9 5/8"
3 3/8"
9"
2'-3"(0.69M)
5'-10"
3"
3'-11 3/4"
3'-10 3/4"
1'-4 3/4"
13'-7"(4.14M)
4" 1'-5" @ 2'-0"
Masonry
5'-5 1/2"
Feet 0 1 2 3 4 5
Centimeters 0 50 100 150
Bell Post scale 3/4"=1'-0"
The gate and bell system of the Jordan Pond Gate Lodge was intentionally designed to appear rustic. The simple lever mechanism used to ring the bell as well as the gate's unadorned hardware both serve to remind visitors that they are leaving Acadia's motor roads and entering its carriage road system.
HISTORIC AMERICAN ENGINEERING RECORD ME-54
SHEET 2 OF 2
MAINE
JORDAN POND GATE LODGE-1931
PARK LOOP ROAD, ONE MILE NORTH OF SEAL HARBOR
HANCOCK COUNTY
SEAL HARBOR VICINITY
NATIONAL PARK SERVICE
ACADIA ROADS & BRIDGES RECORDING PROGRAM
DELINEATED BY: Sarah F. Desbiens, 1995
IF REPRODUCED, PLEASE CREDIT: HISTORIC AMERICAN ENGINEERING RECORD, NATIONAL PARK SERVICE, NAME OF DELINEATOR, DATE OF THE DRAWING

CRATER LAKE NATIONAL PARK

ROAD SYSTEM

1919/1933

The Park-to-Park Highway route, as published in the 1920 Park Service Annual Report.

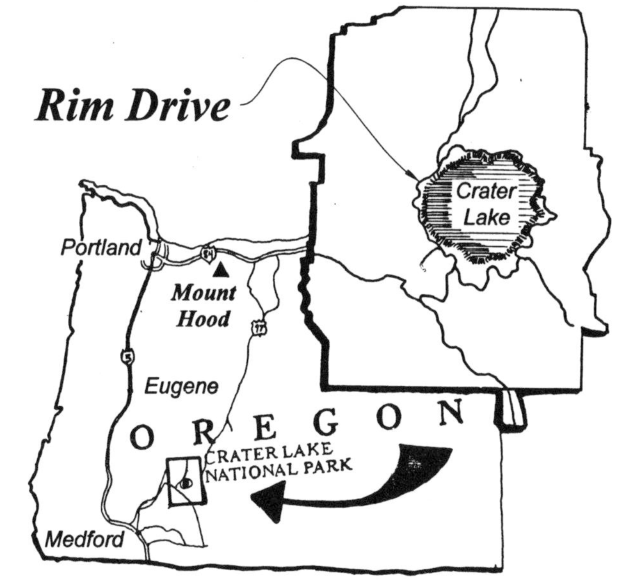

The "Crater" of Crater Lake is actually a deep caldera that was formed thousands of years ago by the violent volcanic eruption and collapse of Mount Mazama. Today Crater Lake is primarily experienced by motorists along roads integrated into the natural terrain of the surrounding wilderness. Rim Drive is one of the most scenic drives in the National Park System, and affords visitors a varied series of views amidst a stunning natural spectacle.

The lake and its environs have always been a sacred landscape for Native Americans, especially for vision quests to the spirit world. Nineteenth-century travelers and local settlers were fascinated by the "exquisite beauty" and "awful grandeur" of the lake. In 1885 one of those admirers, William G. Steel, began his own quest to protect the natural beauty of the area. On May 22, 1902, Congress recognized Crater Lake as the nation's sixth national park. Initial development efforts were devoted to making it accessible to visitors who were hindered by the park's isolated location.

In 1913, after surveying a system of park roads, the centerpiece of which was a scenic road encircling the caldera, the U.S. Army Corps of Engineers (COE) began construction of the first Rim Drive. This project continued for the next six years. Though some visitors considered it "crude" and "only partially satisfactory", this Rim Drive soon became popular with many motorists. Approaching the edge of the caldera at numerous points, it afforded magnificent views to the lake at various elevations and angles as well as vistas out to the Cascade Range. In 1916 the National Park Service assumed responsibility for the park, though the COE continued its roadwork until 1919.

In 1928 the redevelopment of the original Rim Drive began in conjunction with a program for improvement and reconstruction of the roads, trails and bridges throughout the National Park System. Rim Drive was the central focus of the park's master plan. Road construction was started by the Bureau of Public Roads (BPR) contract labor in 1931 and augmented by the Civilian Conservation Corps (CCC) in 1933. This program eventually brought Rim Drive up to contemporary standards. Special attention was paid to the protection of the natural landscape. Landscape architects Merel Sager and Francis Lange planned a route by which visitors could perceive and enjoy the natural beauty and power of the site. They literally constructed the motorist's experience of Crater Lake by presenting a sequence of different views.

The new Rim Drive followed a slightly different alignment, but still respected the topography of the caldera. Observation towers, viewpoints, parking areas, and pull-outs were skillfully located and designed to be subordinate to the natural features of the park which had inspired their creation. Taking advantage of new construction techniques, the designers attempted to make Rim Drive relate harmoniously in scale, color, and texture to the natural context. The master plan of the 1930s is still evident in the general pattern of the roads, but over the years some additions and changes have occurred.

This documentation interprets the Crater Lake road system's evolution and design including trails, historical structures and small scale details. It reveals artificial relationships that have been incorporated into the natural context of the Crater Lake National Park.

This project is part of the Historic American Engineering Record (HAER), a long-range program dedicated to documenting the historically significant engineering, industrial and transportation heritage of the United States. The HAER Program is administered by the HABS/HAER Division of the National Park Service (NPS), U.S. Department of the Interior. The project was sponsored by the NPS Park Roads and Parkways Program through Federal Lands Highway funds.

The field work, measured drawings, historical report and photographs were prepared under the direction of NPS Roads & Bridges Program Manager Todd Croteau and historian Tim Davis. The recording team consisted of field supervisor and historian Christian Carr (Bard Graduate Center), architectural technicians Sarah Lehman (University of Oregon), Walton Stowell (SCAD Savannah, Georgia), and Simona Stoyanova (ICOMOS/Bulgaria). Accompanying large-format photography was prepared by Jet Lowe, HAER Staff photographer.

The team worked in active collaboration, and with the generous assistance of the Crater Lake National Park Cultural Resource personnel, especially Historian Steve Mark, Master Carpenter David Harry, and Chief of Interpretation Kent Taylor.

DELINEATED BY: *Walton D. Stowell II, 1999*

NPS ROADS & BRIDGES RECORDING PROGRAM

CRATER LAKE

CRATER LAKE NATIONAL PARK ROADS – 1919/1933
CRATER LAKE NATIONAL PARK
KLAMATH COUNTY

OREGON

SHEET 1 of 10

HISTORIC AMERICAN ENGINEERING RECORD
OR-107

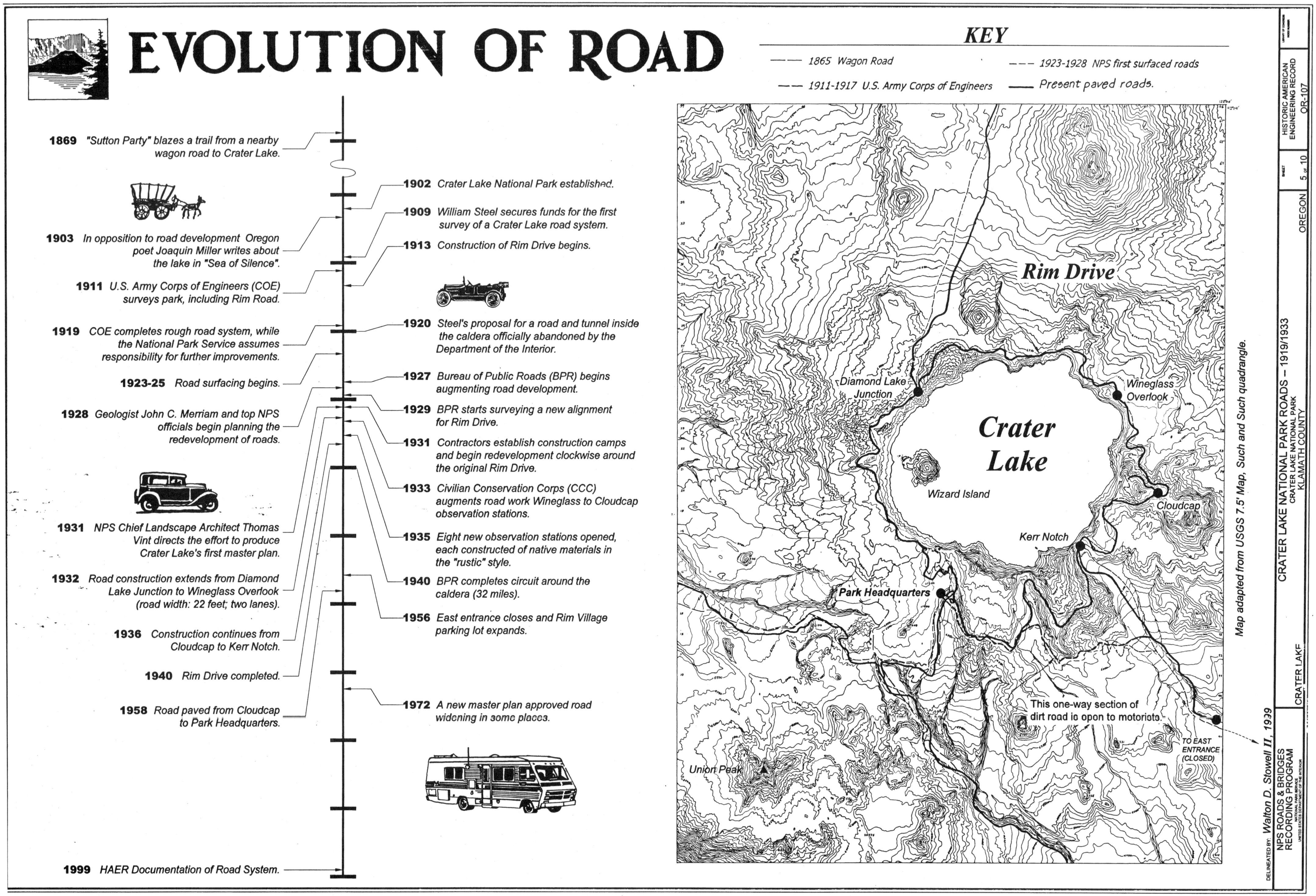
EVOLUTION OF ROAD
1869 "Sutton Party" blazes a trail from a nearby wagon road to Crater Lake.
1902 Crater Lake National Park established.
1909 William Steel secures funds for the first survey of a Crater Lake road system.
1903 In opposition to road development Oregon poet Joaquin Miller writes about the lake in "Sea of Silence".
1913 Construction of Rim Drive begins.
1911 U.S. Army Corps of Engineers (COE) surveys park, including Rim Road.
1919 COE completes rough road system, while the National Park Service assumes responsibility for further improvements.
1920 Steel's proposal for a road and tunnel inside the caldera officially abandoned by the Department of the Interior.
1923-25 Road surfacing begins.
1927 Bureau of Public Roads (BPR) begins augmenting road development.
1928 Geologist John C. Merriam and top NPS officials begin planning the redevelopment of roads.
1929 BPR starts surveying a new alignment for Rim Drive.
1931 Contractors establish construction camps and begin redevelopment clockwise around the original Rim Drive.
1933 Civilian Conservation Corps (CCC) augments road work Wineglass to Cloudcap observation stations.
1931 NPS Chief Landscape Architect Thomas Vint directs the effort to produce Crater Lake's first master plan.
1935 Eight new observation stations opened, each constructed of native materials in the "rustic" style.
1932 Road construction extends from Diamond Lake Junction to Wineglass Overlook (road width: 22 feet; two lanes).
1940 BPR completes circuit around the caldera (32 miles).
1956 East entrance closes and Rim Village parking lot expands.
1936 Construction continues from Cloudcap to Kerr Notch.
1940 Rim Drive completed.
1958 Road paved from Cloudcap to Park Headquarters.
1972 A new master plan approved road widening in some places.
1999 HAER Documentation of Road System.
KEY
1865 Wagon Road
1923-1928 NPS first surfaced roads
1911-1917 U.S. Army Corps of Engineers
Present paved roads.
Rim Drive
Diamond Lake Junction
Wineglass Overlook
Crater Lake
Wizard Island
Cloudcap
Kerr Notch
Park Headquarters
This one-way section of dirt road is open to motorists.
TO EAST ENTRANCE (CLOSED)
Union Peak
Map adapted from USGS 7.5' Map, Such and Such quadrangle.
DELINEATED BY: Walton D. Stowell II, 1939
NPS ROADS & BRIDGES RECORDING PROGRAM
CRATER LAKE
CRATER LAKE NATIONAL PARK ROADS – 1919/1933
CRATER LAKE NATIONAL PARK
KLAMATH COUNTY
OREGON
5 of 10
HISTORIC AMERICAN ENGINEERING RECORD
OR-107

ROAD DESIGN PRINCIPLES

First Rim Drive (1910 - 1919)

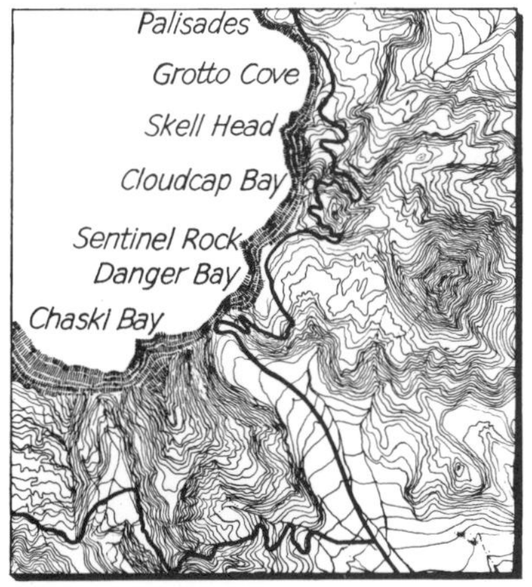

Old and new road alignments
Drawing based on historical photo

Road crew at Devil's Backbone
Drawing based on historical photo

The East Rim Drive near Kerr Notch
Drawing based on historical photo

Because of the exceedingly rough nature of the lava rock terrain around Crater Lake and the short working seasons it was a difficult task to build the Rim Drive of standard width, with a minimum number of curves and a gentle grade. The First Rim Drive (35.4 miles), completed in 1919, was a typical mountain road, rocky in some places, dusty in others and crudely surfaced most of the way.

In conjunction with the program for improvement and reconstruction of roads throughout the National Park system, the redevelopment of Rim Drive commenced in 1928. This program aimed to make the road system appropriate to modern standards. The Second Rim Drive followed the old route wherever possible in order to provide similar views of the lake and reduce construction costs.

In an attempt to preserve the primitive picture of the natural setting, the new road was located so that it could not be seen from any point of the lake and, with a few exceptions, from any other point on the road. Special attention was paid to the protection of the landscape and the integration of the road within the natural context. Viewpoints, observation stations and parking lots became part of the road design. The construction of the Second Rim Drive involved blasting and hauling rock, grubbing, burning and clearing vegetation, grading and surfacing the road bed, building stone walls and planting native vegetation in the disturbed areas. The two-lane, two-way road was designed with an average width of 22 feet. The completed road had a paved bituminous surface from the Park Headquarters to Cloudcap, and an oiled surface from Cloudcap back to the Park Headquarters. The completed length of the Second Rim Drive was 32.6 miles.

Second Rim Drive (1928 - 1940)

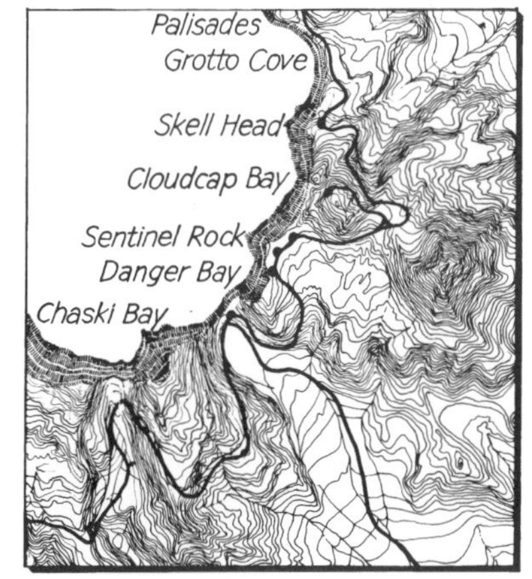

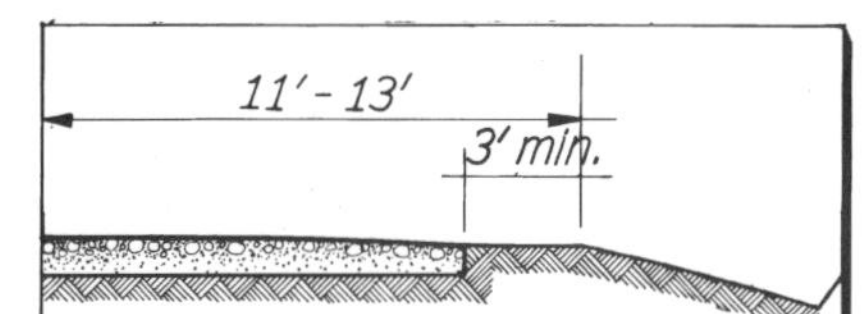

Roadbed section:
1. Bituminous surfacing
2. Crushed rock base
3. Compacted sub-base

Steamroller
Drawing based on historical photo

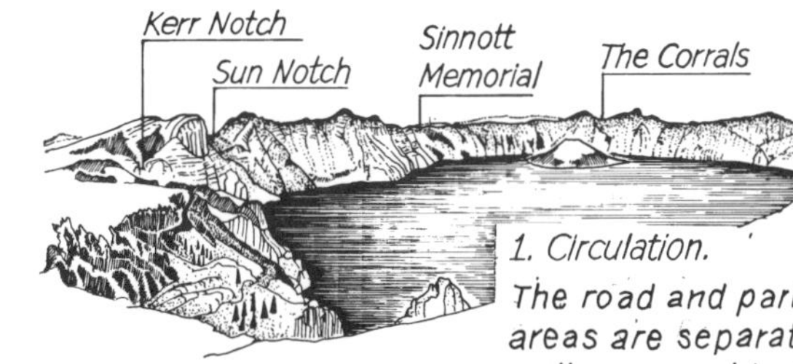

1. Circulation.
The road and parking areas are separated from walkways and trails.

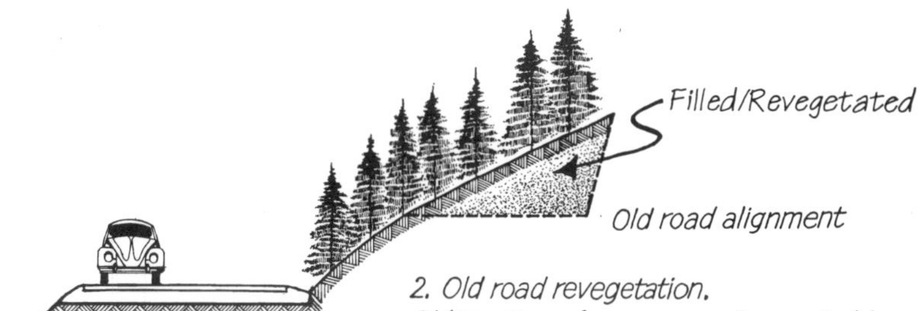

2. Old road revegetation.
Obliteration of any scar or trace of old roads visible from major points of interest, through planting or application of sod and filling, was one of the landscape design principles involved in the construction of the Second Rim Drive.

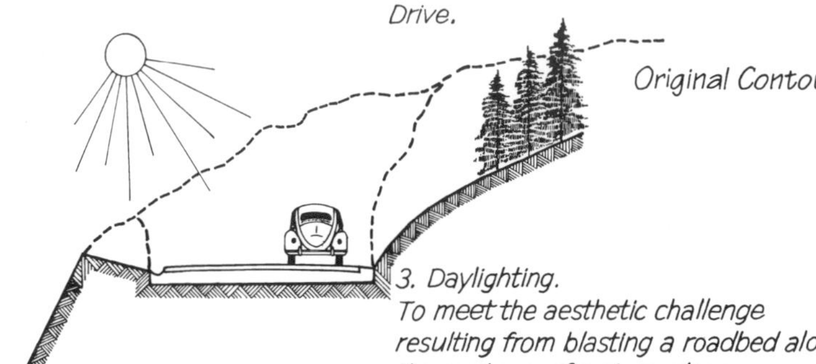

3. Daylighting.
To meet the aesthetic challenge resulting from blasting a roadbed along the contours of a steep slope, "daylighting" was applied. The slope was cut back, leaving a broad shoulder at the side of the road, which contributed to a feeling of greater spaciousness.

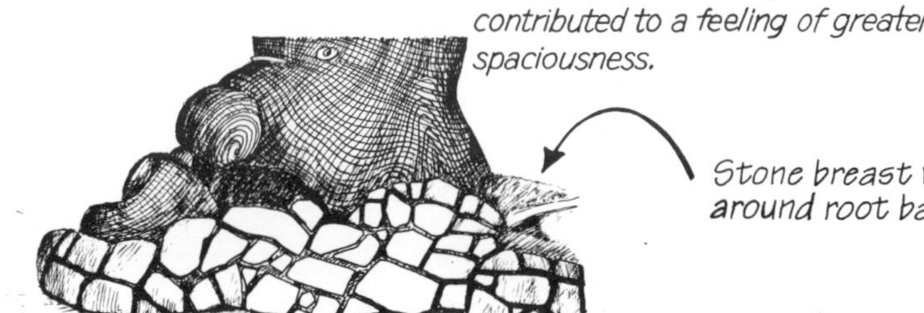

4. Design with nature.
Inspired by the natural setting, the Rim Drive elements were aesthetically designed and harmoniously integrated within its surroundings.

DELINEATED BY: Simona Stoyanova, 1999
NATIONAL PARK SERVICE ROADS & BRIDGES RECORDING PROGRAM
NATIONAL PARK SERVICE
UNITED STATES DEPARTMENT OF THE INTERIOR
CRATER LAKE
Crater Lake National Park Roads - 1919/1933
CRATER LAKE NATIONAL PARK
KLAMATH COUNTY
OREGON
SHEET 6 of 10
HISTORIC AMERICAN ENGINEERING RECORD
OR-107

IF REPRODUCED, PLEASE CREDIT: HISTORIC AMERICAN ENGINEERING RECORD, NATIONAL PARK SERVICE, NAME OF DELINEATOR, DATE OF THE DRAWING

COMPREHENSIVE DESIGN

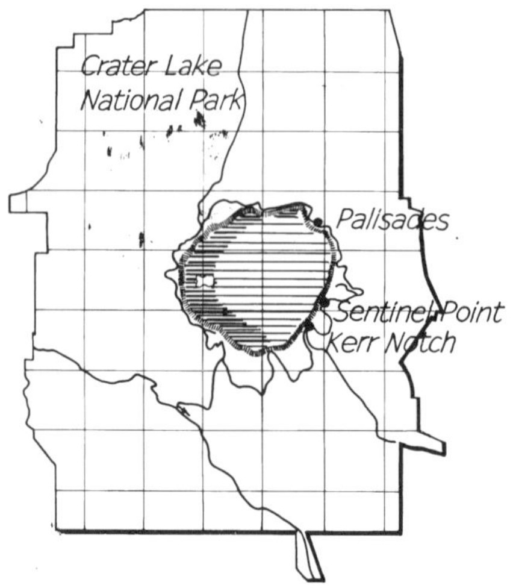

Present view of Sentinel Point

The unity in design of the Rim Drive and all its supporting elements is a major characteristic of the Rim system. Construction drawings prepared by the NPS Branch of Plans and Design in 1936 depict the design scale, form and materials of the observation stations. Designed as scenic overlooks, where visitors could leave their cars and enjoy the view, some stations were large enough to hold 50 cars, parked at an angle or perpendicular to the rim. The semi-elliptical pullouts were adjacent to the road and substantial in size, averaging 400-500 feet in length and 150-160 feet in width. Typically, the bituminous surface of the parking area was separated from that of the roadbed by an island plant bed. Within the paved area an attempt was made to separate pedestrian and vehicular circulation. A raised bituminous pathway, approximately 4 feet wide, was located as a promenade along the edge of the rim. The pathway was edged on the rim-side by a stone guardrail and on the parking side by a stone curb. At a promontory point, the path widened into a larger gathering area 10 feet wide – the so called "observation platform". The numerous small scale features add a subtle level of detail to the constructed landscape. Most of them – drinking fountains, directional and interpretive signs, guardrails – were built of local stone or wood, designed with rustic forms to integrate with the surrounding natural setting. This harmonious relationship among the various elements within the Rim Drive system is a significant feature of this site.

Steps from Sentinel Point
Drawing based on historical photo

Interpretive sign from the Palisades

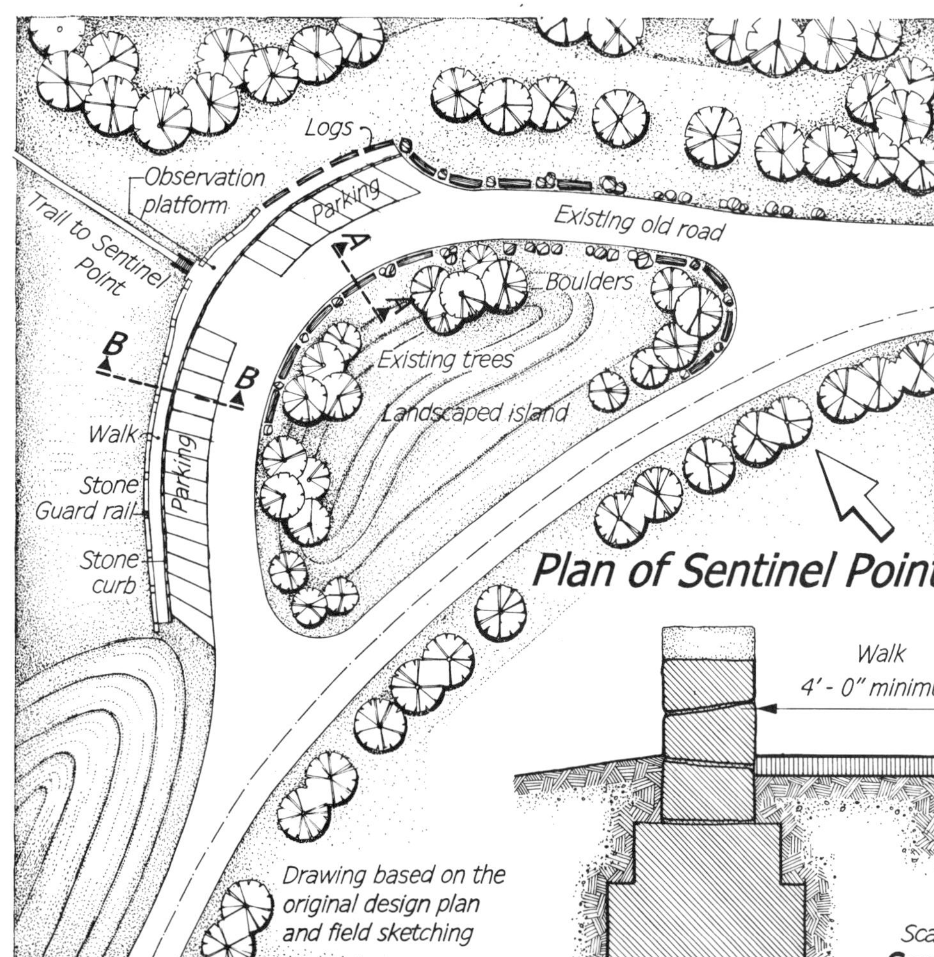

Plan of Sentinel Point

Drawing based on the original design plan and field sketching

Drinking fountain from Kerr Notch

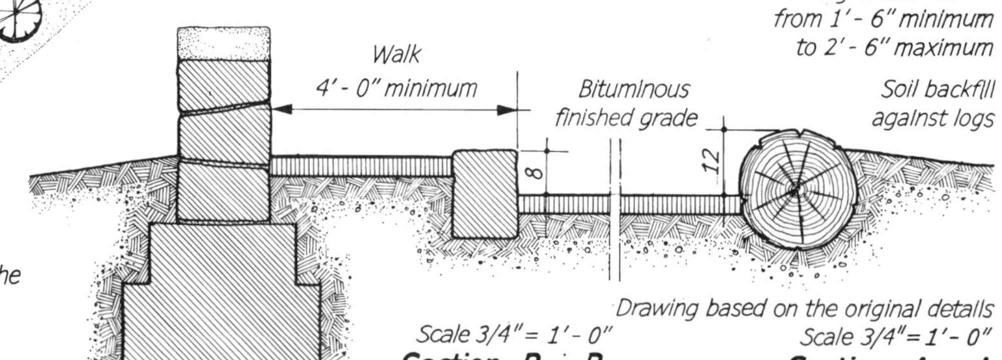

Drawing based on the original details

Scale 3/4" = 1' - 0"
Section B – B

Scale 3/4" = 1' - 0"
Section A – A

DELINEATED BY: Simona Stoyanova, 1999

NATIONAL PARK SERVICE ROADS & BRIDGES RECORDING PROGRAM
NATIONAL PARK SERVICE
UNITED STATES DEPARTMENT OF THE INTERIOR

Crater Lake National Park Roads - 1919/1933
CRATER LAKE NATIONAL PARK
KLAMATH COUNTY

CRATER LAKE

OREGON

SHEET 7 of 10

HISTORIC AMERICAN ENGINEERING RECORD
OR-107

ORIENTATION FEATURES

North Entrance Sign

North Entrance Checking Kiosk

East Entrance Station

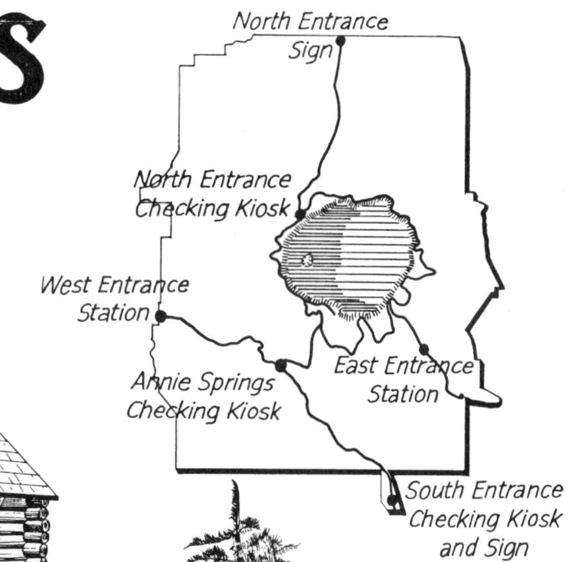

South Entrance Sign

During the 1920s and 1930s, visitors entered Crater Lake National Park through one of five entrance stations: Annie Springs, East, South, North or West. in 1925, log gateways were built alongside two park entrances . Situated near the entrance signs, the south and west stations were located along Highway 62. In 1951, the South and West checking areas were combined with Annie Springs into one southern checking area. The North entrance station was originally positioned at the junction of the Diamond Lake Road and Rim Road. In 1956 this location was abandoned and relocated to the northern boundary. The East entrance served visitors who entered from Highway 232. Access from this direction was later closed and the station was demolished soon after.

Four entrance signs located along the park's access routes are the initial orientation features for visitors. The south and west entrance signs are located at the park's boundaries along Highway 62. Both of these signs were originally timber construction and demolished in the 1970s only to be reconstructed through careful analysis of historic drawings and photographs in 1998. The design of the north and east entrance signs called for the based to be constructed of masonry. The north sign located near the junction of Highway 138 and the Diamond Lake Road was demolished in 1970 but reconstructed in 2000. the original base of the east sign along the Pinnacles Road still remains intact although the board was replaced in 1994.

West Entrance Station

Drawings and Reference Maps adapted from Historical Sources.

Annie Springs Checking Kiosk

South Entrance Checking Kiosk

HISTORIC AMERICAN ENGINEERING RECORD OR-107

SHEET 9 of 10

OREGON

CRATER LAKE NATIONAL PARK ROADS – 1919/1933
CRATER LAKE NATIONAL PARK
KLAMATH COUNTY

CRATER LAKE

NPS ROADS & BRIDGES RECORDING PROGRAM

DELINEATED BY: Sarah Lehman, 1999

HISTORIC BRIDGES AND CONDUITS

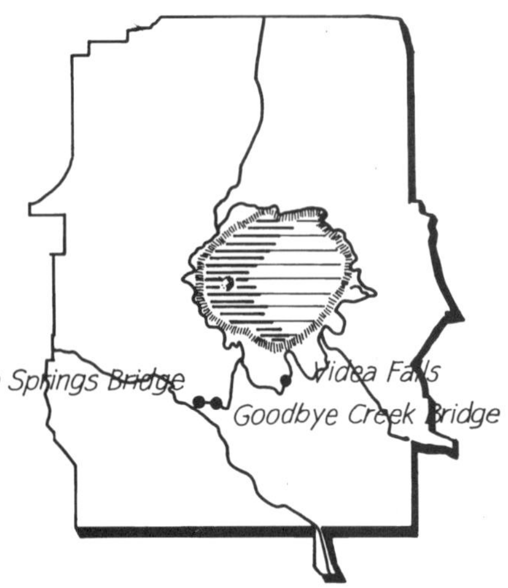

Annie Springs Bridge
- 1927 -

Goodbye Creek Bridge
- 1929 -

Over the years, Annie Creek and Goodbye Creek have been spanned by several different bridges. The Army Corps of Engineers built the first crossing in the 1910s. The National Park Service spanned Annie Creek in 1927 and Goodbye Creek in 1929 with three bridges of lodgepole pine construction, typical of this era of park design. The Annie Creek bridge spanned seventy-eight feet and the Goodbye Creek Bridge spanned two hundred and forty feet. The railing was made of balusters and rounded posts, while the floor was comprised of a two by six laminated deck. Both of these bridges collapsed in April 1942. Traffic was rerouted above the two springs on a hazardous narrow route for fourteen years before present day laminated beam bridges were constructed in the 1950s.

Videa Falls Fill

During the 1938 and 1939 construction seasons the road alignment near Vidae Falls was modified to allow a higher alignment. Rather than designing a bridge to cross the waterfall, a large fill was used, covering up approximately two-fifths of the fall's horizontal distance. The fill has been replanted and the carefully sculpted views once visible from the old road bed are no longer present.

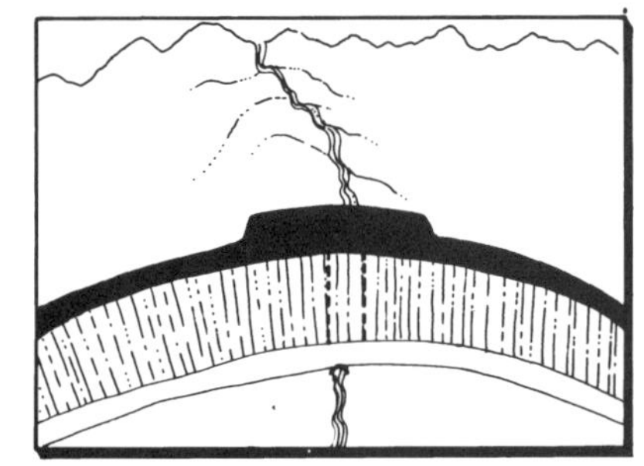

Mosaic Waterways

Only a few mosaic waterways exist along the Rim Road. Both of these examples are located near Vidae Falls on the East Rim Drive and closely follow the principle of design with nature. Stonework blends with natural exposed rock.

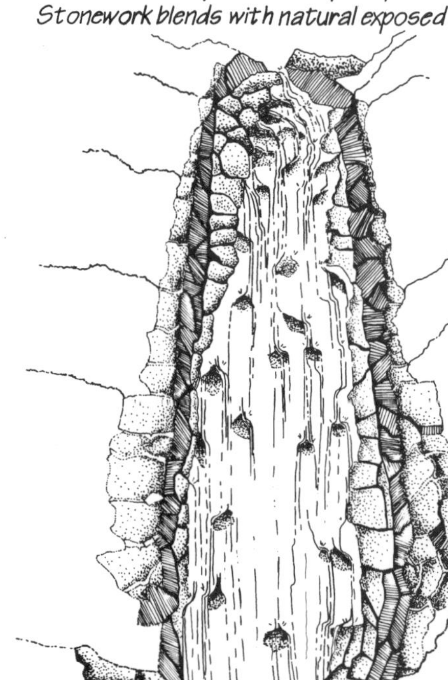

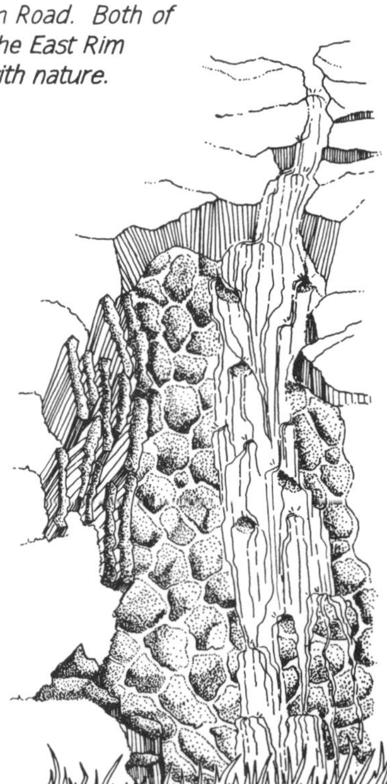

Typical Culverts

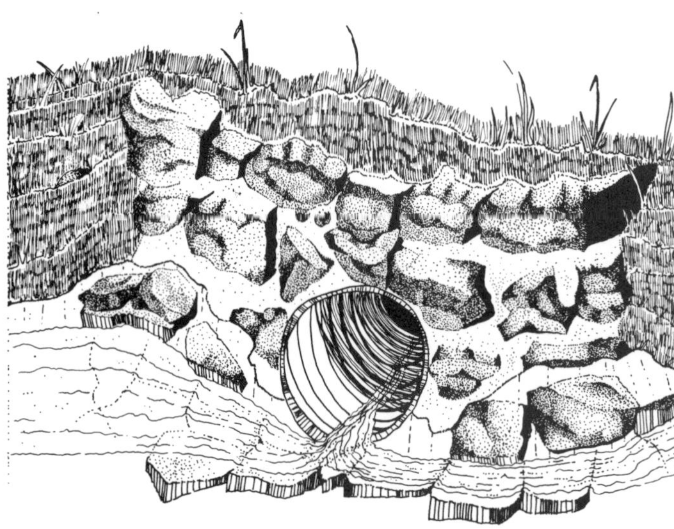

Flat stone culverts are the primary type of drainage feature evident along the Rim Drive. They were designed of materials that are inconspicuous viewed in conjunction with the existing landscape. Variations in culvert type were realized by modifying the orientation of the metal pipe as depicted in the two illustrations.

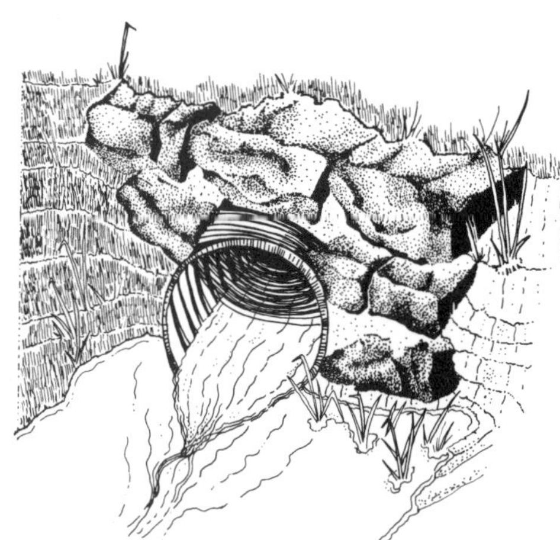

DELINEATED BY Sarah Lehman 1999

NATIONAL PARK SERVICE ROADS & BRIDGES RECORDING PROGRAM
NATIONAL PARK SERVICE
UNITED STATES DEPARTMENT OF THE INTERIOR

CRATER LAKE

Crater Lake National Park Roads - 1919/1933
CRATER LAKE NATIONAL PARK
KLAMATH COUNTY

OREGON

SHEET 10 of 10

HISTORIC AMERICAN ENGINEERING RECORD
OR-107

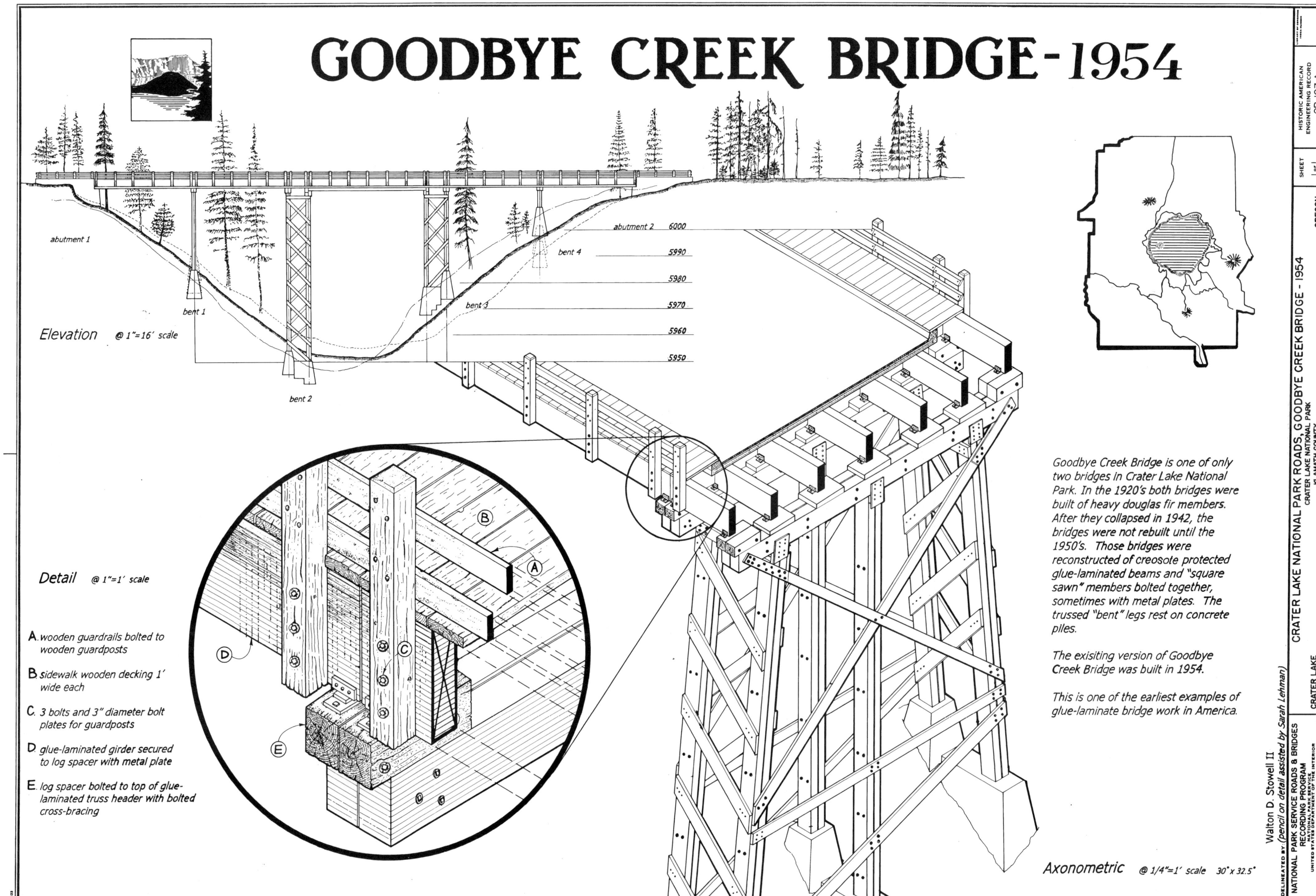

GOODBYE CREEK BRIDGE - 1954
Elevation @ 1"=16' scale
abutment 1
bent 1
bent 2
bent 3
bent 4
abutment 2
6000
5990
5980
5970
5960
5950
Detail @ 1"=1' scale
A. wooden guardrails bolted to wooden guardposts
B. sidewalk wooden decking 1' wide each
C. 3 bolts and 3" diameter bolt plates for guardposts
D. glue-laminated girder secured to log spacer with metal plate
E. log spacer bolted to top of glue-laminated truss header with bolted cross-bracing
Goodbye Creek Bridge is one of only two bridges in Crater Lake National Park. In the 1920's both bridges were built of heavy douglas fir members. After they collapsed in 1942, the bridges were not rebuilt until the 1950's. Those bridges were reconstructed of creosote protected glue-laminated beams and "square sawn" members bolted together, sometimes with metal plates. The trussed "bent" legs rest on concrete piles.
The exisiting version of Goodbye Creek Bridge was built in 1954.
This is one of the earliest examples of glue-laminate bridge work in America.
Axonometric @ 1/4"=1' scale 30" x 32.5"
Walton D. Stowell II
DELINEATED BY: (pencil on detail assisted by Sarah Lehman)
NATIONAL PARK SERVICE ROADS & BRIDGES RECORDING PROGRAM
NATIONAL PARK SERVICE
UNITED STATES DEPARTMENT OF THE INTERIOR
CRATER LAKE
CRATER LAKE NATIONAL PARK ROADS, GOODBYE CREEK BRIDGE - 1954
CRATER LAKE NATIONAL PARK
KLAMATH COUNTY
OREGON
SHEET 1 OF 1
HISTORIC AMERICAN ENGINEERING RECORD
OR-107-A
IF REPRODUCED, PLEASE CREDIT: HISTORIC AMERICAN ENGINEERING RECORD, NATIONAL PARK SERVICE, NAME OF DELINEATOR, DATE OF THE DRAWING

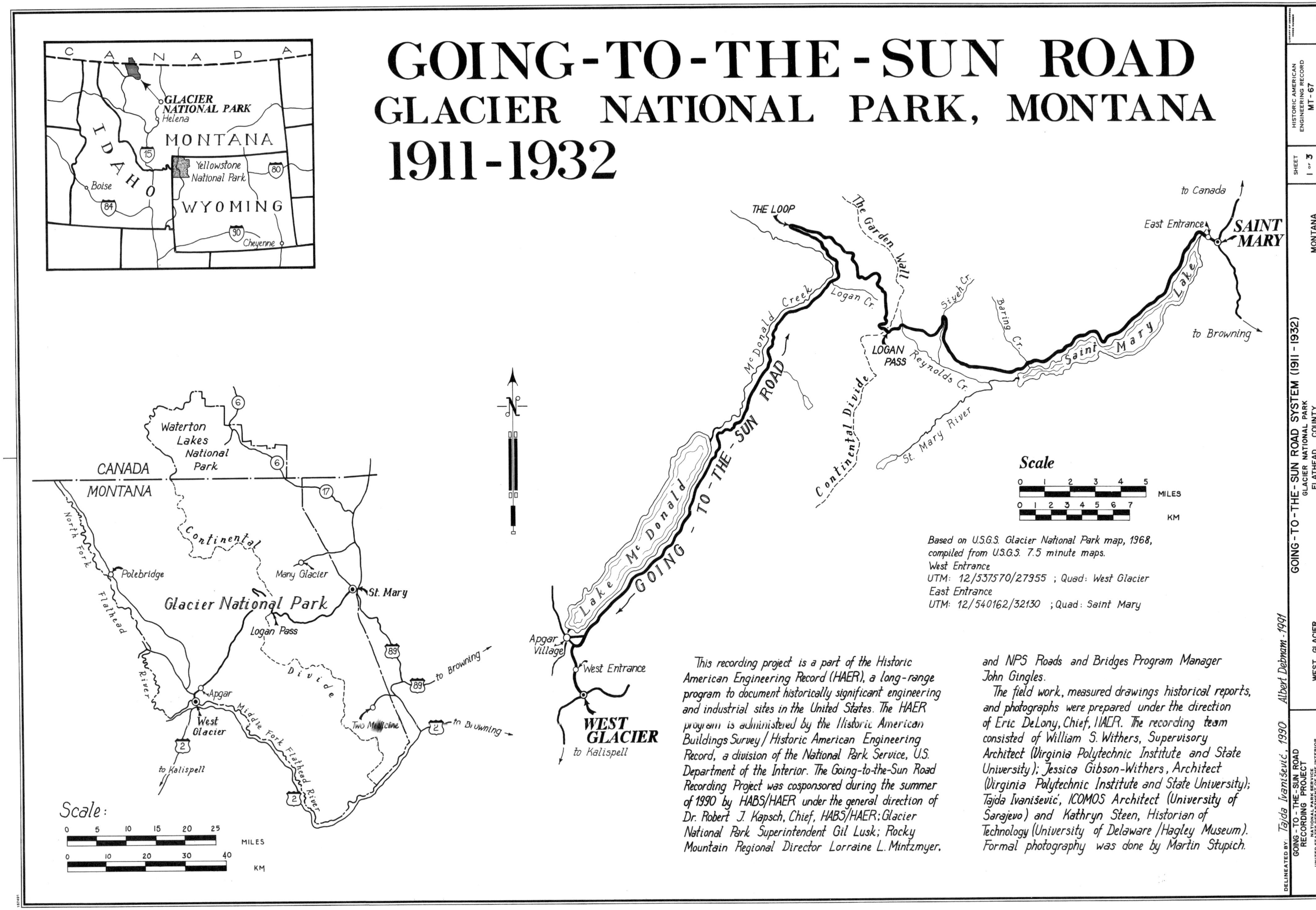
GOING-TO-THE-SUN ROAD
GLACIER NATIONAL PARK, MONTANA
1911-1932
CANADA
GLACIER NATIONAL PARK
Helena
MONTANA
IDAHO
15
Yellowstone National Park
80
Boise
84
WYOMING
90
Cheyenne
THE LOOP
The Garden Wall
to Canada
East Entrance
SAINT MARY
Logan Cr.
Siyeh Cr.
Baring Cr.
Saint Mary Lake
to Browning
McDonald Creek
LOGAN PASS
Reynolds Cr.
GOING-TO-THE-SUN ROAD
Continental Divide
St. Mary River
Lake McDonald
Scale
0 1 2 3 4 5 MILES
0 1 2 3 4 5 6 7 KM
Based on U.S.G.S. Glacier National Park map, 1968, compiled from U.S.G.S. 7.5 minute maps.
West Entrance
UTM: 12/537570/27955 ; Quad: West Glacier
East Entrance
UTM: 12/540162/32130 ; Quad: Saint Mary
Apgar Village
West Entrance
WEST GLACIER
to Kalispell
Waterton Lakes National Park
6
CANADA
MONTANA
17
North Fork
Continental
Polebridge
Many Glacier
Glacier National Park
St. Mary
Flathead
Logan Pass
Divide
89
to Browning
River
Apgar
West Glacier
Middle Fork Flathead River
Two Medicine
2
to Browning
to Kalispell
Scale:
0 5 10 15 20 25 MILES
0 10 20 30 40 KM
This recording project is a part of the Historic American Engineering Record (HAER), a long-range program to document historically significant engineering and industrial sites in the United States. The HAER program is administered by the Historic American Buildings Survey / Historic American Engineering Record, a division of the National Park Service, U.S. Department of the Interior. The Going-to-the-Sun Road Recording Project was cosponsored during the summer of 1990 by HABS/HAER under the general direction of Dr. Robert J. Kapsch, Chief, HABS/HAER; Glacier National Park Superintendent Gil Lusk; Rocky Mountain Regional Director Lorraine L. Mintzmyer, and NPS Roads and Bridges Program Manager John Gingles.
The field work, measured drawings historical reports, and photographs were prepared under the direction of Eric DeLony, Chief, HAER. The recording team consisted of William S. Withers, Supervisory Architect (Virginia Polytechnic Institute and State University); Jessica Gibson-Withers, Architect (Virginia Polytechnic Institute and State University); Tajda Ivanišević, ICOMOS Architect (University of Sarajevo) and Kathryn Steen, Historian of Technology (University of Delaware / Hagley Museum). Formal photography was done by Martin Stupich.
HISTORIC AMERICAN ENGINEERING RECORD MT-67
SHEET 1 of 3
GOING-TO-THE-SUN ROAD SYSTEM (1911-1932)
GLACIER NATIONAL PARK
FLATHEAD COUNTY
MONTANA
WEST GLACIER
DELINEATED BY: Tajda Ivanišević, 1990 Albert Debnam-1991
GOING-TO-THE-SUN ROAD RECORDING PROJECT
NATIONAL PARK SERVICE
UNITED STATES DEPARTMENT OF THE INTERIOR
IF REPRODUCED, PLEASE CREDIT: HISTORIC AMERICAN ENGINEERING RECORD, NATIONAL PARK SERVICE, NAME OF DELINEATOR, DATE OF THE DRAWING

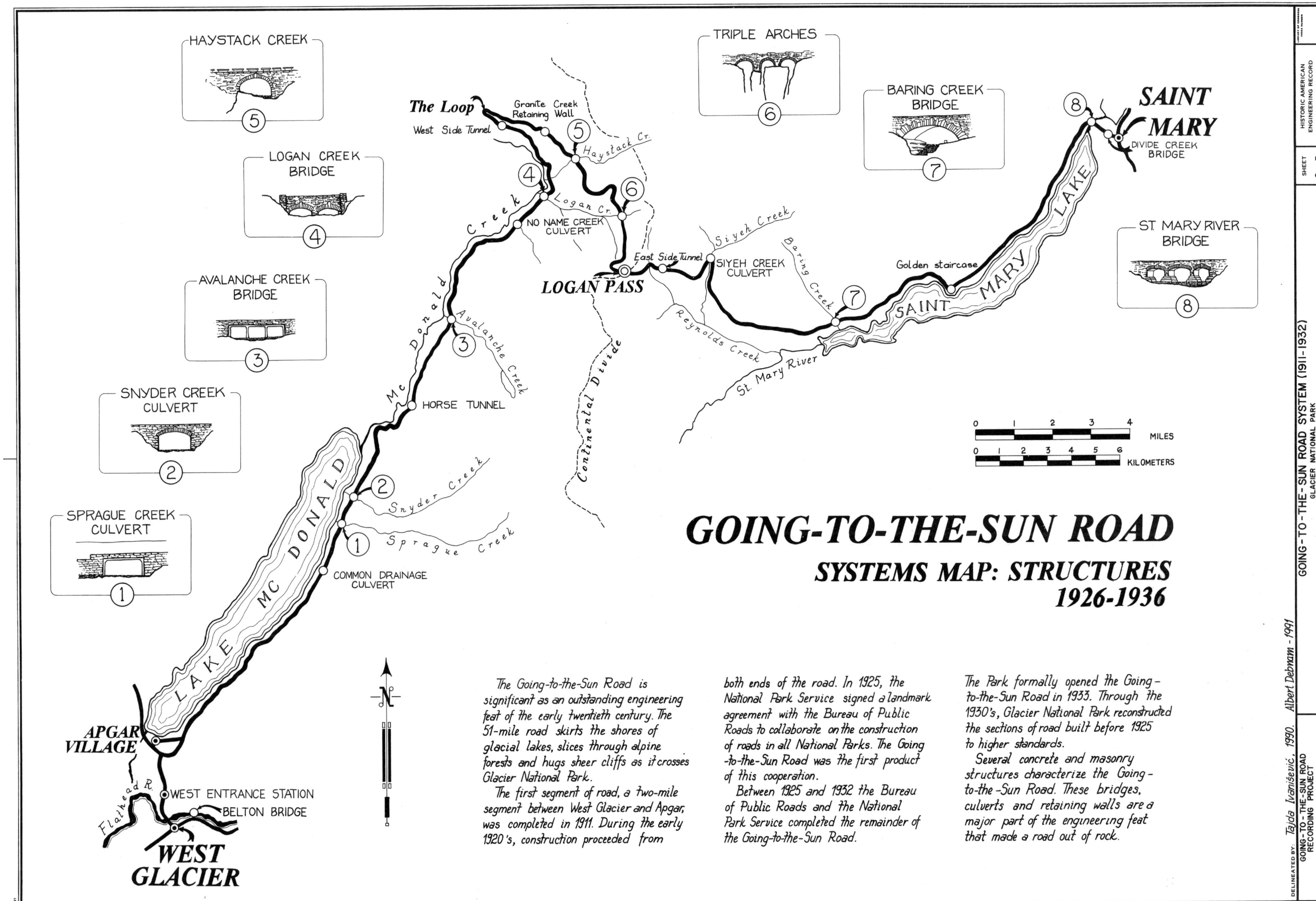
GOING-TO-THE-SUN ROAD
SYSTEMS MAP: STRUCTURES
1926-1936
HAYSTACK CREEK
5
LOGAN CREEK BRIDGE
4
AVALANCHE CREEK BRIDGE
3
SNYDER CREEK CULVERT
2
SPRAGUE CREEK CULVERT
1
TRIPLE ARCHES
6
BARING CREEK BRIDGE
7
ST. MARY RIVER BRIDGE
8
The Loop
Granite Creek Retaining Wall
West Side Tunnel
Haystack Cr.
Logan Cr.
NO NAME CREEK CULVERT
Mc Donald Creek
Avalanche Creek
HORSE TUNNEL
LOGAN PASS
East Side Tunnel
SIYEH CREEK CULVERT
Siyeh Creek
Baring Creek
Reynolds Creek
St. Mary River
Golden staircase
SAINT MARY LAKE
SAINT MARY
DIVIDE CREEK BRIDGE
Continental Divide
0 1 2 3 4 MILES
0 1 2 3 4 5 6 KILOMETERS
Snyder Creek
Sprague Creek
COMMON DRAINAGE CULVERT
LAKE MC DONALD
APGAR VILLAGE
Flathead R.
WEST ENTRANCE STATION
BELTON BRIDGE
WEST GLACIER
N
The Going-to-the-Sun Road is significant as an outstanding engineering feat of the early twentieth century. The 51-mile road skirts the shores of glacial lakes, slices through alpine forests and hugs sheer cliffs as it crosses Glacier National Park.
The first segment of road, a two-mile segment between West Glacier and Apgar, was completed in 1911. During the early 1920's, construction proceeded from both ends of the road. In 1925, the National Park Service signed a landmark agreement with the Bureau of Public Roads to collaborate on the construction of roads in all National Parks. The Going-to-the-Sun Road was the first product of this cooperation.
Between 1925 and 1932 the Bureau of Public Roads and the National Park Service completed the remainder of the Going-to-the-Sun Road.
The Park formally opened the Going-to-the-Sun Road in 1933. Through the 1930's, Glacier National Park reconstructed the sections of road built before 1925 to higher standards.
Several concrete and masonry structures characterize the Going-to-the-Sun Road. These bridges, culverts and retaining walls are a major part of the engineering feat that made a road out of rock.
DELINEATED BY: Tajda Ivanisevic, 1990. Albert Debnam - 1991
GOING-TO-THE-SUN ROAD RECORDING PROJECT
NATIONAL PARK SERVICE
UNITED STATES DEPARTMENT OF THE INTERIOR
GOING-TO-THE-SUN ROAD SYSTEM (1911-1932)
GLACIER NATIONAL PARK
FLATHEAD COUNTY
WEST GLACIER
MONTANA
SHEET 2 of 3
HISTORIC AMERICAN ENGINEERING RECORD
MT-67
IF REPRODUCED, PLEASE CREDIT: HISTORIC AMERICAN ENGINEERING RECORD, NATIONAL PARK SERVICE, NAME OF DELINEATOR, DATE OF THE DRAWING

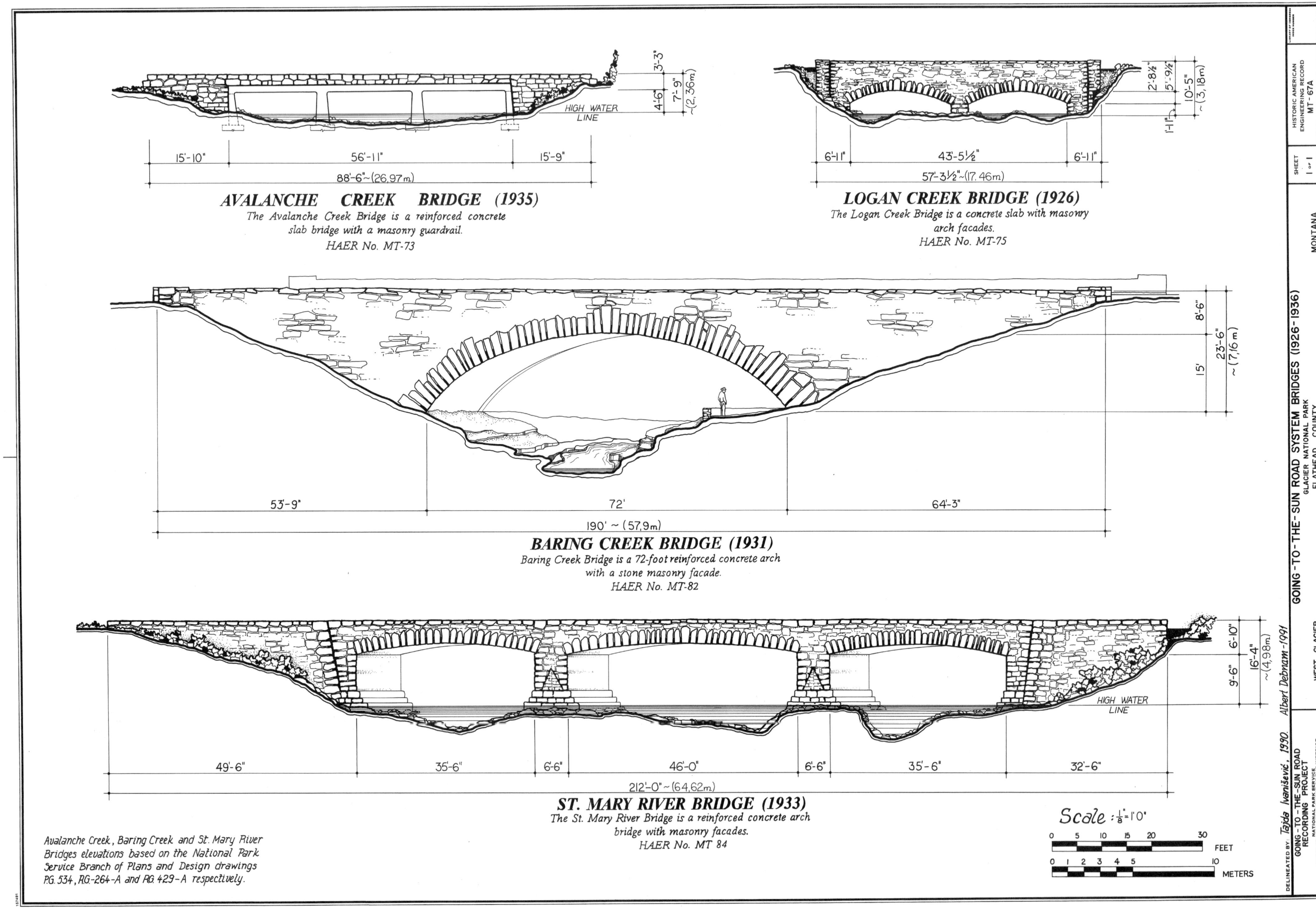

3'-3"
4'-6"
7'-9"
~(2,36m)
HIGH WATER LINE
15'-10"
56'-11"
15'-9"
88'-6"~(26,97m)
AVALANCHE CREEK BRIDGE (1935)
The Avalanche Creek Bridge is a reinforced concrete slab bridge with a masonry guardrail.
HAER No. MT-73
2'-8½"
5'-9½"
10'-5"
~(3,18m)
1-1"
6'-11"
43'-5½"
6'-11"
57'-3½"~(17,46m)
LOGAN CREEK BRIDGE (1926)
The Logan Creek Bridge is a concrete slab with masonry arch facades.
HAER No. MT-75
8'-6"
15'
23'-6"
~(7,16m)
53'-9"
72'
64'-3"
190' ~ (57,9m)
BARING CREEK BRIDGE (1931)
Baring Creek Bridge is a 72-foot reinforced concrete arch with a stone masonry facade.
HAER No. MT-82
6'-10"
9'-6"
16'-4"
~(4,98m)
HIGH WATER LINE
49'-6"
35'-6"
6'-6"
46'-0"
6'-6"
35'-6"
32'-6"
212'-0"~(64,62m)
ST. MARY RIVER BRIDGE (1933)
The St. Mary River Bridge is a reinforced concrete arch bridge with masonry facades.
HAER No. MT 84
Avalanche Creek, Baring Creek and St. Mary River Bridges elevations based on the National Park Service Branch of Plans and Design drawings P.G. 534, P.G.-264-A and P.G. 429-A respectively.
Scale: 1/8"=1'0"
0 5 10 15 20 30 FEET
0 1 2 3 4 5 10 METERS
DELINEATED BY: Tajda Ivanišević, 1990. Albert Debnam-1991
GOING-TO-THE-SUN ROAD RECORDING PROJECT
NATIONAL PARK SERVICE
UNITED STATES DEPARTMENT OF THE INTERIOR
GOING-TO-THE-SUN ROAD SYSTEM BRIDGES (1926-1936)
GLACIER NATIONAL PARK
FLATHEAD COUNTY
WEST GLACIER
MONTANA
IF REPRODUCED, PLEASE CREDIT: HISTORIC AMERICAN ENGINEERING RECORD, NATIONAL PARK SERVICE, NAME OF DELINEATOR, DATE OF THE DRAWING
SHEET 1 of 1
HISTORIC AMERICAN ENGINEERING RECORD
MT-67A

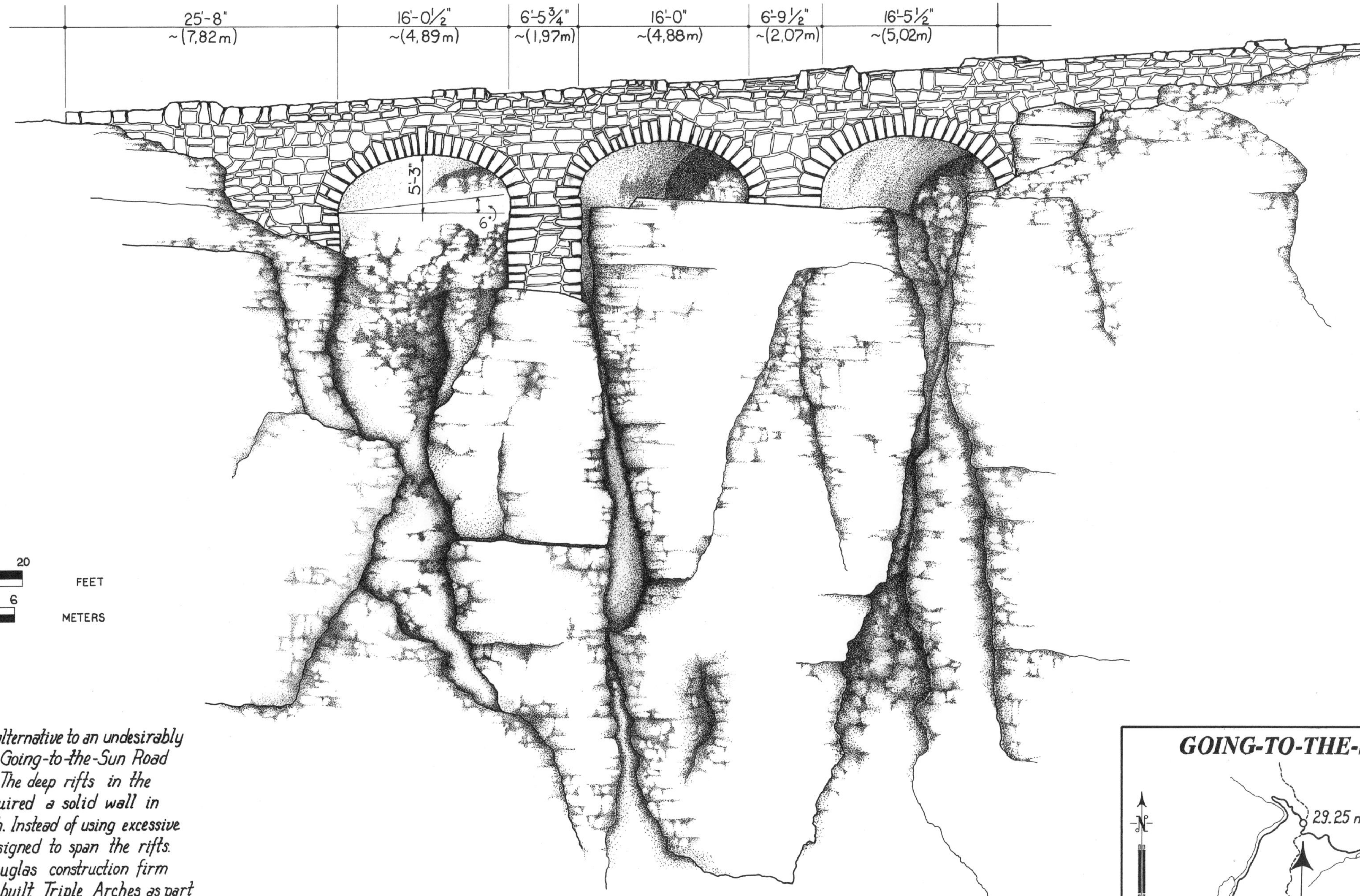

Scale: 3/16" = 1'0"

0 5 10 20 FEET

0 1 2 3 4 5 6 METERS

Triple Arches was an alternative to an undesirably large retaining wall on the Going-to-the-Sun Road in Glacier National Park. The deep rifts in the mountain would have required a solid wall in excess of 20 feet in depth. Instead of using excessive fill, three arches were designed to span the rifts.

The Williams and Douglas construction firm of Tacoma, Washington, built Triple Arches as part of their 12-mile contract on the Going-to-the-Sun Road. With the approval of the National Park Service and Bureau of Public Roads, the contractors revised the original N.P.S. plans on site.

Mistakenly, the contractors constructed the western-most arch with the spring line parallel to the six percent grade of the road. They corrected this error, however, in the remaining two arches. These two arches have horizontal spring lines.

TRIPLE ARCHES - ca. 1927

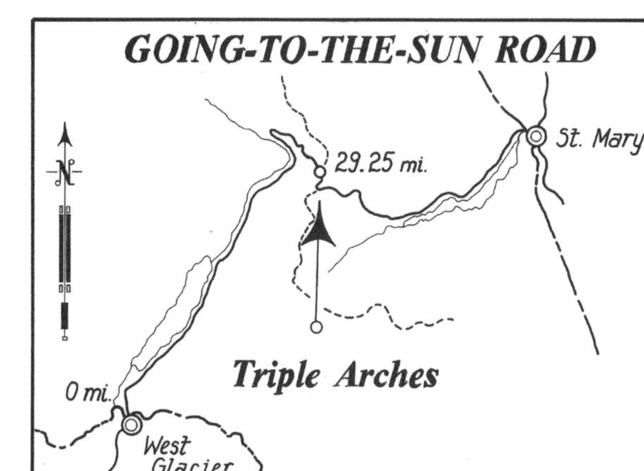

Scale: 0 3 6 12 16 MILES

UTM: 12/539930/30010
Quad: Logan Pass

DELINEATED BY: William Withers, 1990 Albert Debnam, 1991

GOING-TO-THE-SUN ROAD RECORDING PROJECT
NATIONAL PARK SERVICE
UNITED STATES DEPARTMENT OF THE INTERIOR

GOING-TO-THE-SUN ROAD: TRIPLE ARCHES (ca. 1927)
29 MILES EAST OF WEST GLACIER IN GLACIER NATIONAL PARK
WEST GLACIER FLATHEAD COUNTY MONTANA

SHEET 1 of 1

HISTORIC AMERICAN ENGINEERING RECORD
MT-79

LOCATION MAP

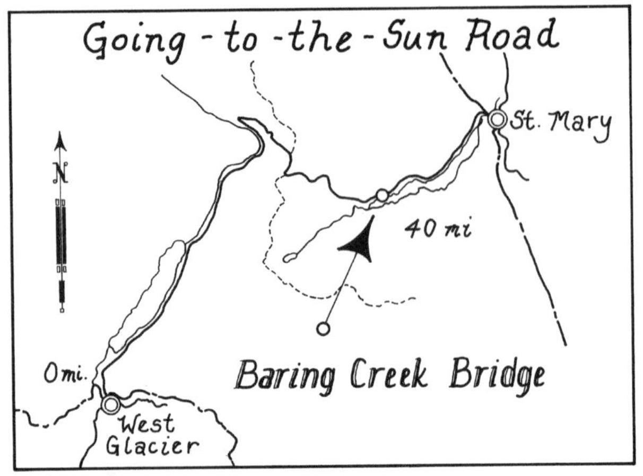

Scale: 1" = 6 MILES

0 3 6 12 18 MILES

0 5 10 15 25 KILOMETRES

UTM: 12/539470/30899
Quad: Rising Sun

Baring Creek Bridge is a 72-foot reinforced concrete arch with a stone masonry facade. A.R. Guthrie and Co. of Portland, Oregon, built the bridge in 1931 as a part of a 4.5-mile contract on one of the final links of the Going-to-the-Sun Road.

The Landscape Architecture Division of the National Park Service influenced the design and specifications of all structures on the Going-to-the-Sun Road in a attempt to make the structures unobtrusive to the natural setting. According to the landscape architect's recommendations, Baring Creek Bridge was designed unusually high and long to minimize the destruction of the natural landscape and to allow the existing hiking trail from Sunrift Gorge to Baring Creek Falls to pass underneath the bridge.

AREA MAP

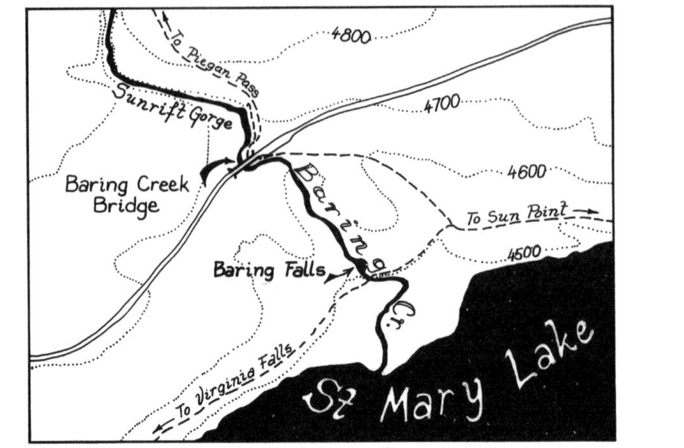

Scale: 1" = 200'

0 200 400 600 FEET

0 50 100 150 METERS

BARING CREEK BRIDGE - 1931

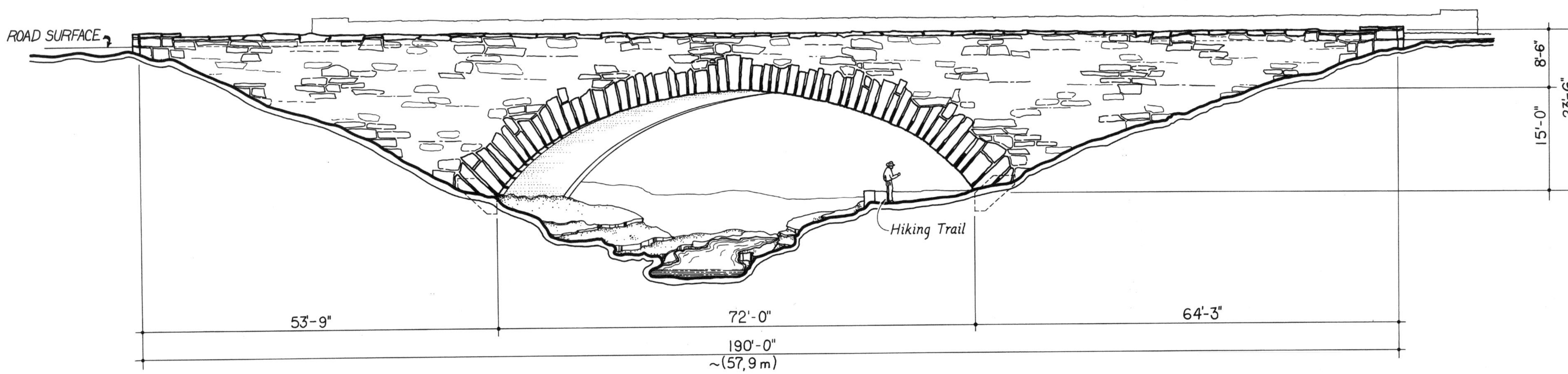

DOWNSTREAM ELEVATION

Scale: 1/8" = 1'-0"

0 5 10 15 20 30 FEET

0 1 2 3 4 5 10 METERS

Baring Creek Bridge elevation based on the National Park Service Branch of Plans and Design drawings RG-264-A and B.

DELINEATED BY: *Jessica Gibson-Withers, 1990. Albert Debnam-1991*

GOING-TO-THE-SUN ROAD RECORDING PROJECT
NATIONAL PARK SERVICE
UNITED STATES DEPARTMENT OF THE INTERIOR

GOING-TO-THE-SUN ROAD: BARING CREEK BRIDGE (1931)
40 MILES EAST OF WEST GLACIER IN GLACIER NATIONAL PARK
WEST GLACIER FLATHEAD COUNTY MONTANA

SHEET 1 of 3

HISTORIC AMERICAN ENGINEERING RECORD
MT-82

LIBRARY OF CONGRESS INDEX NUMBER

IF REPRODUCED, PLEASE CREDIT: HISTORIC AMERICAN ENGINEERING RECORD, NATIONAL PARK SERVICE, NAME OF DELINEATOR, DATE OF THE DRAWING

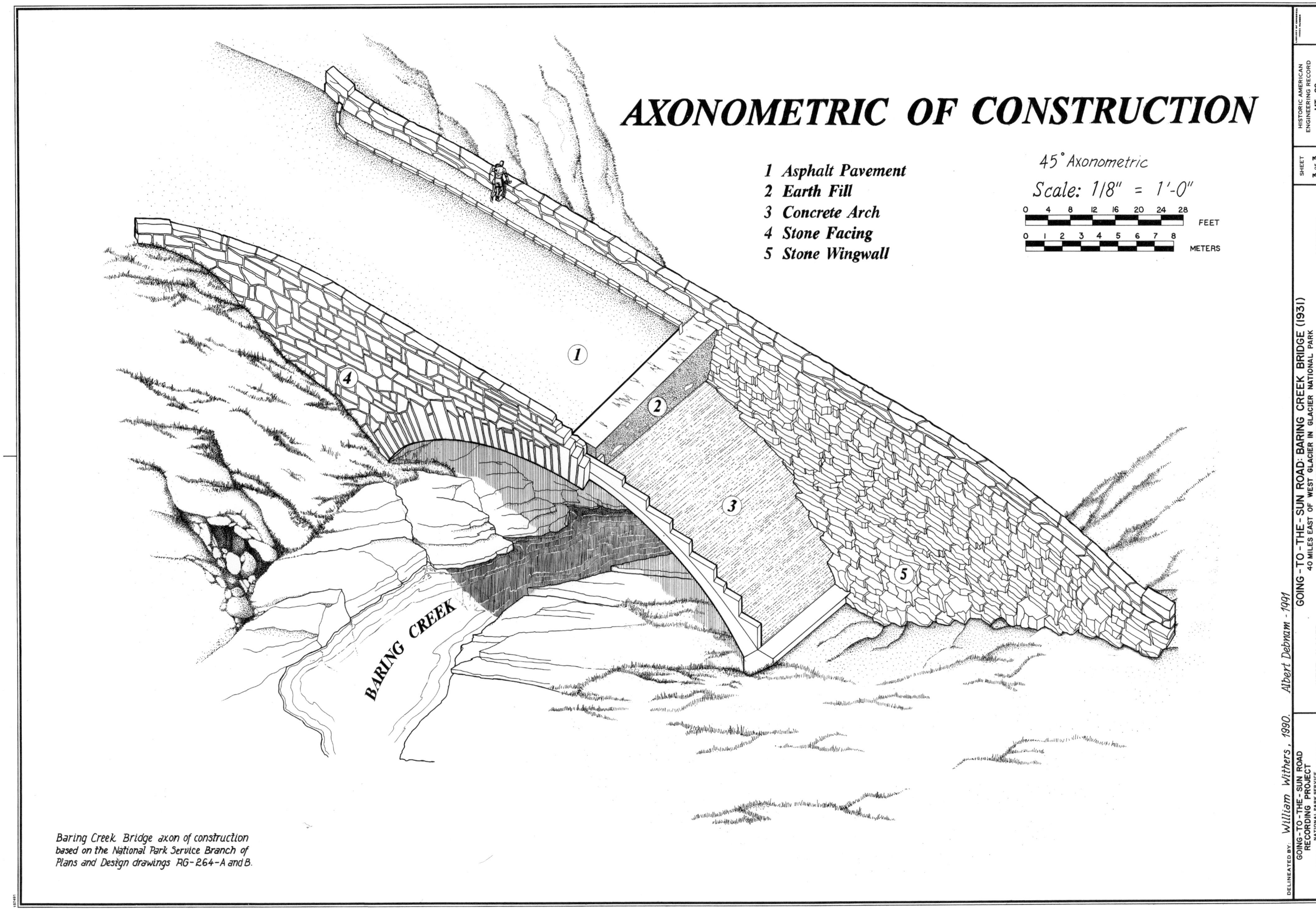

AXONOMETRIC OF CONSTRUCTION
1 Asphalt Pavement
2 Earth Fill
3 Concrete Arch
4 Stone Facing
5 Stone Wingwall
45° Axonometric
Scale: 1/8" = 1'-0"
0 4 8 12 16 20 24 28
FEET
0 1 2 3 4 5 6 7 8
METERS
1
2
3
4
5
BARING CREEK
Baring Creek Bridge axon of construction based on the National Park Service Branch of Plans and Design drawings RG-264-A and B.
DELINEATED BY: William Withers, 1990. Albert Debnam - 1991
GOING-TO-THE-SUN ROAD RECORDING PROJECT
NATIONAL PARK SERVICE
UNITED STATES DEPARTMENT OF THE INTERIOR
GOING-TO-THE-SUN ROAD: BARING CREEK BRIDGE (1931)
40 MILES EAST OF WEST GLACIER IN GLACIER NATIONAL PARK
WEST GLACIER
FLATHEAD COUNTY
MONTANA
IF REPRODUCED, PLEASE CREDIT: HISTORIC AMERICAN ENGINEERING RECORD, NATIONAL PARK SERVICE, NAME OF DELINEATOR, DATE OF THE DRAWING
SHEET 3 of 3
HISTORIC AMERICAN ENGINEERING RECORD
MT-82

GOING · TO · THE · SUN · ROAD

GLACIER NATIONAL PARK

Going-to-the-Sun Road is a landmark in the history of National Park road construction. The spectacular 51-mile long route through the heart of Glacier National Park is regarded as one of the most scenic and technologically impressive mountain roads in the world.

When Congress authorized Glacier National Park in 1910 there were no roads connecting the east and west sides of the park. The rapid rise in automobile tourism over the next decade made road construction an imperative, as motorists were forced to take wide detours or ship their vehicles around the park via the Great Northern Railway. To remedy this situation and provide access to the spectacular scenery in the heart of Glacier National Park, National Park Service (NPS) officials began considering several alternative routes. One option, proposed by NPS Chief Engineer George Goodwin, called for the roadway to skirt the east side of Lake McDonald, parallel McDonald Creek for approximately eight miles, and then climb sharply toward Logan Pass, ascending the steep, secluded valley with a series of closely spaced switchbacks. NPS landscape architect Thomas Vint objected that the multiple switchbacks would deface one of the prettiest sections of the park. Vint proposed replacing Goodwin's convoluted alignment with one long graceful curve that would swing nearly two miles further to the west and then reverse direction, traversing a series of rugged cliffs known as the Garden Wall as it ascended gradually toward Logan Pass. Vint acknowledged that constructing such a road would be difficult, but insisted it was worth the expense. With its relatively gentle grade and minimal curvature, the Garden Wall road would be easier to drive and maintain. It would also preserve and display the attractive Logan Creek Valley.

NPS Director Stephen Mather favored Vint's proposal and engaged U.S. Bureau of Public Road's engineer Frank Kittredge to determine its feasibility. Kittredge endorsed Vint's alternative on both practical and aesthetic grounds. Impressed by Kittredge's detailed and insightful report, Mather established an official partnership between the NPS and the Bureau of Public Roads (now the Federal Highway Administration) that still governs park road-building today. NPS landscape architects guide the basic location and design of park road projects while federal highway engineers provide technical expertise and oversee construction contracts.

The construction of Going-to-the-Sun Road was an epic process. Severe technical challenges, harsh working conditions, and chronic funding shortages stretched the construction period over several decades. The lower portions were completed by the mid-1920s, but Going-to-the-Sun Road was not officially opened until July 15, 1933. Work on the guard walls continued for several more years and the road was not completely paved until 1952. Maintaining the road's scenic character and historic integrity has remained a constant challenge.

Going-to-the-Sun Road's dramatic beauty and status as a seminal contribution to the history of park road development have earned it many accolades. The American Society of Civil Engineers named it a National Civil Engineering Landmark in 1985. In 1997 Going-to-the-Sun Road was declared a National Historic Landmark - the first park road in America to receive this prestigious designation.

The Glacier National Park Roads Recording Project was undertaken during the summer of 2000 and is part of the Historic American Engineering Record (HAER), a long-range program to document historically significant engineering and industrial works in the United States. HAER (Eric DeLony, Chief) is administered by the Historic American Buildings Survey/Historic American Engineering Record (E. Blaine Cliver, Chief), a division of the National Park Service, U.S. Department of the Interior. The project was funded by the U.S. Department of Transportation's Federal Lands Highway Program (Art Hamilton, Administrator) through the NPS Park Roads and Parkways Program (Lou DeLorme, Manager) and cosponsored by Glacier National Park (Randy Jones, Superintendent) and the NPS Cooperative Program at Montana State University (Barry Sulam, Manager).

GOING-TO-THE-SUN-ROAD

The field work, measured drawings, and historical reports were completed under the direction of Todd A. Croteau, Project Leader; and Tim Davis, Program Historian. The recording team consisted of Field Supervisor Brandy Dubs (Montana State University); Architects Arin Streeter (University of Tennessee), Eszter Vogel (US/ICOMOS, Hungary), Lucas Dupuis and Nathan Junkert (Montana State University), Christopher Boldt (University of Washington); and Landscape Architect Magdalena M. Lisowska (US/ICOMOS, Poland).

DELINEATED BY: Brandy Dubs & Magdalena M. Lisowska, 2000

NPS PARK ROADS RECORDING PROGRAM
NATIONAL PARK SERVICE
UNITED STATES DEPARTMENT OF THE INTERIOR

ADDENDUM TO GOING-TO-THE-SUN ROAD - 1933
GLACIER NATIONAL PARK
FLATHEAD COUNTY

WEST GLACIER VICINITY | MONTANA

SHEET 1 OF 14

HISTORIC AMERICAN ENGINEERING RECORD MT-67

HISTORY OF TRANSPORTATION

1891- The Great Northern Railway completes its track between Midvale (East Glacier) and Belton (West Glacier). encouraging settlement and tourism in West Glacier.
1893-The Great Northern Railway completes its mainline to the Pacific Coast, thus placing the Glacier region on it's transcontinental line and making it more accessible to potential tourists and homesteaders.

1895-Early transportation into the area that would become the park consisted of a rough buck board line from Belton to the foot of Lake McDonald on the West side. 1911-a rough road was constructed from East Glacier Station to Saint Mary. By 1912 an early stage coach line ran from East Glacier Park to Many Glacier via Saint Mary.

1895-The first passenger boat service begins on Lake McDonald with a 40' steam boat. Due to lack of roads, boat service continued to be an integral part of the transportation system on Lake McDonald and St Mary Lake until the opening of the Going-to-the-Sun Road.

1915-A number of small saddle-horse outfits combine to form The Park Saddle Horse Company which at this time offered the only means of transportation, aside from by foot, into the vast backcountry of the park. The company would continue to operate until 1942, at that time accommodating 10,000 visitors a season.

1891-1933

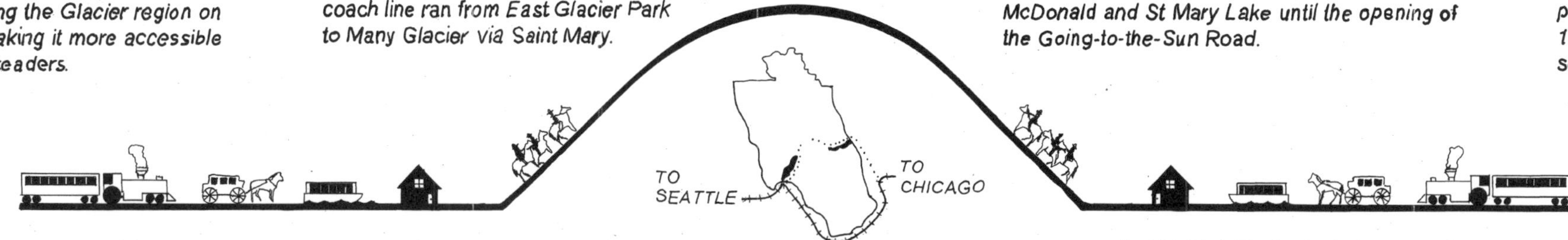

Early visitors used a number of different forms of transportation to tour the Glacier area. Most visitors would arrive by train. Wagons, stagecoaches, and eventually motorbuses would then carry them from the train stations to hotels located near the edges of the park. In the absence of good roads motor launches provided access to accomodations on the shores of Lake McDonald and Saint Mary Lake. The vast backcountry of the park was accessible only by foot or the saddle horse tours that originated from the various hotels.

In 1914 the Glacier Park Transport Company purchased its first motor fleet consisting of six 7 passenger touring cars, ten 11 passenger buses and one 1 1/2 ton truck from the White Motor Company. These vehicles quickly replaced stage service to Many Glacier Chalet and served as the major form of land transportation in and around the park until they were replaced in 1936.

1933-2000

When Going-to-the-Sun Road opened in 1933 it dramatically changed the Glacier National Park visitor experience transforming both what people saw and how they saw it. Not only did the road provide access to the backcountry and unite the East and West sides of the park over Logan Pass, it made automotive travel the primary means of experiencing the landscape Visitors could ride over the pass in the concessionaire's open topped buses, or they could enjoy the scenery from their own automobiles. This newfound convenience caused a rapid rise in park visitation, which rose from 53,202 in 1932 to 76615 during the road's opening season. Today the Going-to-the-Sun Road carries two million visitors each season Guided tours still operate today, though most people opt to drive their own vehicles.

In the years 1936-1939 new buses were purchased from the White Motor Company to replace the aging 1914 fleet. According to the 1949 drivers manual for Glacier Park Transportation Company "these custom built buses fit exactly the requirements of sightseeing in Glacier Park. The company sees no important factor of obsolescence because if additional buses are required the present type will be ordered." Over 60 years of service on the Going-to-the-Sun Road has made these red buses an icon of the Glacier National Park experience.

Today the Going-to-the-Sun Road carries about two million visitors each season. Glacier Park Inc. still offers guided scenic bus tours, however most visitors opt to drive their own vehicles.

HISTORIC AMERICAN ENGINEERING RECORD MT-67

SHEET 2 OF 14

MONTANA

ADDENDUM TO GOING-TO-THE-SUN ROAD - 1933
GLACIER NATIONAL PARK
FLATHEAD COUNTY
WEST GLACIER VICINITY

DELINEATED BY: Lucas Dupuis 2000

NPS PARK ROADS RECORDING PROGRAM
NATIONAL PARK SERVICE
UNITED STATES DEPARTMENT OF THE INTERIOR

CHOOSING A ROUTE

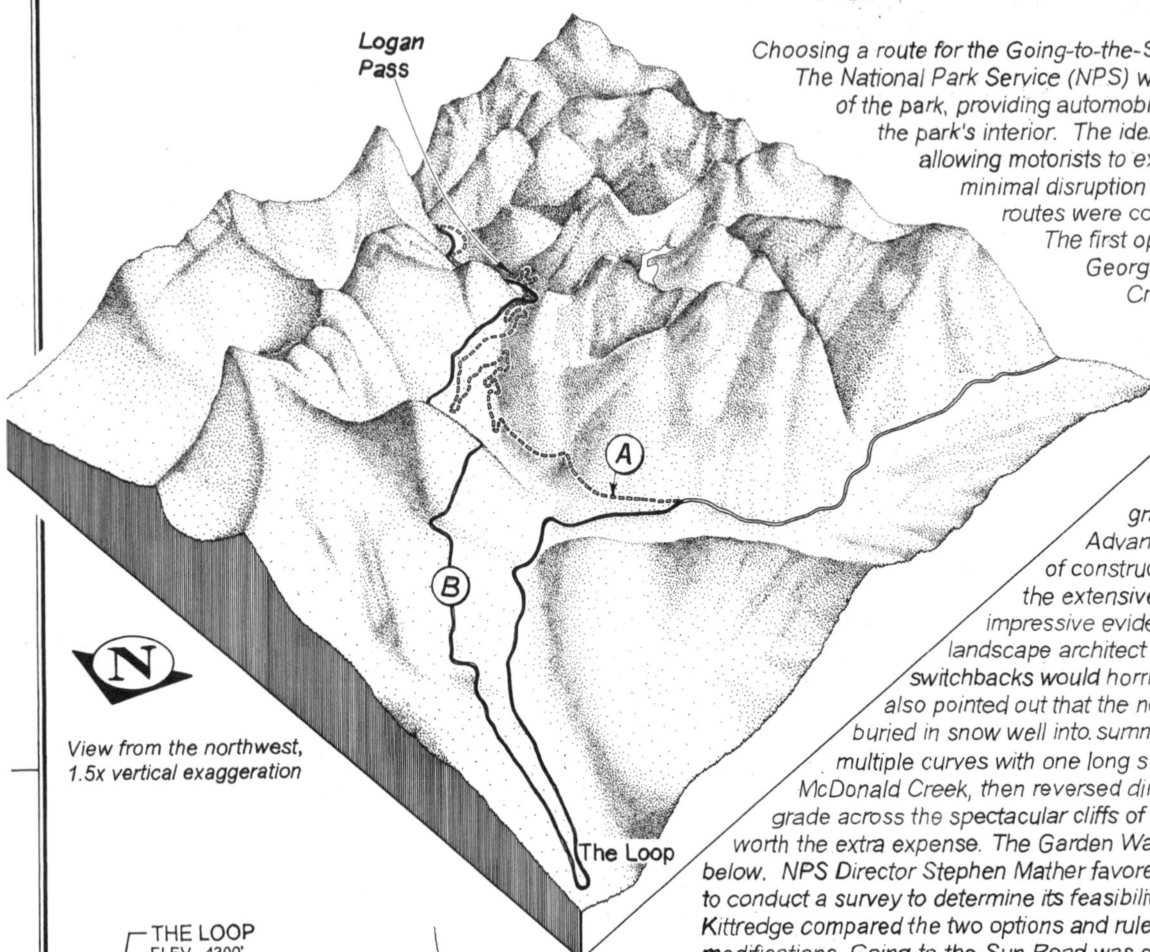

View from the northwest, 1.5x vertical exaggeration

Choosing a route for the Going-to-the-Sun Road was a momentous decision. The National Park Service (NPS) wanted to connect the east and west sides of the park, providing automobile access to the spectacular scenery of the park's interior. The ideal roadway would "lie lightly on the land," allowing motorists to experience the grandeur of the park with minimal disruption to the natural environment. Two alternate routes were considered.

The first option, proposed by NPS Chief Engineer George Goodwin in 1918, followed McDonald Creek until it reached the Logan Creek valley, where the road angled right and climbed toward Logan Pass, ascending to the summit in a series of fifteen switchbacks. Another set of switchbacks would carry the road down the east side of Logan Pass to the head of St. Mary Lake.

This route required an 8% maximum grade and sharp 50-foot radius curves. Advantages of this proposal were relative ease of construction and economy. Goodwin also believed the extensive switchbacks would be viewed as impressive evidence of man's mastery over nature. NPS landscape architect Thomas Vint protested that Goodwin's switchbacks would horribly scar the scenic Logan Creek valley. He also pointed out that the north-facing valley location would remain buried in snow well into summer. Vint proposed replacing Goodwin's multiple curves with one long switchback that ascended gradually from McDonald Creek, then reversed direction sharply and ran in a constant easy grade across the spectacular cliffs of the Garden Wall to Logan Pass. Construction would be more difficult, but Vint insisted it was worth the extra expense. The Garden Wall route would blend into the surrounding cliffs while offering spectacular views into the valley below. NPS Director Stephen Mather favored Vint's proposal. In 1924 he enlisted U.S. Bureau of Public Roads engineer Frank Kittredge to conduct a survey to determine its feasibility. This route, with a lower 6% grade and wider curves, would be easier to drive and maintain. Kittredge compared the two options and ruled in favor of the Garden Wall route on practical and aesthetic grounds. With only slight modifications, Going-to-the-Sun Road was constructed along the lines proposed by Vint and Kittredge.

"Its alignment or route should include both the practical means of making this grandeur accessible and the dramatic values in displaying it."

Thomas Vint, 1925

A 1918 Goodwin Survey

B 1924 Kittredge Survey

(Route Chosen)

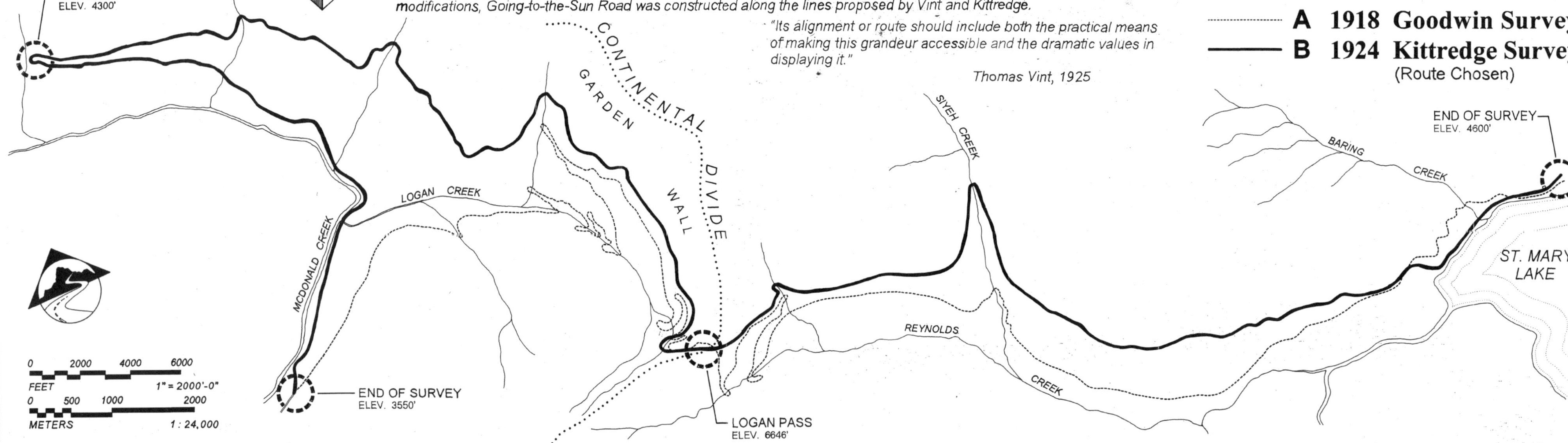

DELINEATED BY: Arin Streeter, 2000

NPS PARK ROADS RECORDING PROGRAM
NATIONAL PARK SERVICE
UNITED STATES DEPARTMENT OF THE INTERIOR

ADDENDUM TO GOING-TO-THE-SUN ROAD - 1933
GLACIER NATIONAL PARK
WEST GLACIER VICINITY FLATHEAD COUNTY MONTANA

SHEET 3 OF 14

HISTORIC AMERICAN ENGINEERING RECORD
MT-67

LIBRARY OF CONGRESS INDEX NUMBER

ROAD CONSTRUCTION

The construction of Going-to-the-Sun Road was a formidable task. From the initial surveys to the final paving and landscaping, work crews had to contend with steep and hazardous terrain, severe mountain weather, and short construction seasons. Painstaking design criteria and strict regulations designed to minimize the road's negative impact on the surrounding terrain added to the complexity of the process. During the early years of the project, chronic funding shortages also hampered progress. By the time the road was completed, however, it was widely hailed as an engineering triumph, an esthetic masterpiece, and a construction tour-de-force.
Summer 1911: William Logan supervises construction of road from Belton to Lake McDonald. Summer 1921: NPS Labor clears section of road along east side of Lake McDonald. 1921-Aug. 15,1922: 10-mile section of road to the Glacier Hotel is completed. (Lake McDonald Lodge) June 1922–July 20, 1924: Section of road from Lake McDonald Lodge to Avalanche Creek is finished. 1924: An 8-mile section of road from the end of the steel bridge contract to two miles east of Going to the Sun Point is finished. 1925: The section of road between Avalanche Creek and Logan Creek is finished. 1925 – October 1928: The final 12 miles of road on the west side, from Logan Creek to Logan Pass, is finished. 1931- 1932: The east side of the Going-to-the-Sun Road is finished. July 15, 1933: Formal ceremony at Logan Pass to dedicate the completion of the Going-to-the-Sun Road.

1) Surveyors laying out the route had to scale precarious cliffs and watch out for loose rocks and other dangers. The initial location survey was begun in September 1924. By the time the crews finished in early November they were wading through waist deep snows. 2) Many sections of the road had to be blasted out of solid rock. This was especially true on the Garden Wall, where benches for the roadway had to be carved into the cliffs. Small blasts were used to avoid damage to the landscape. 3) Steam shovels and other heavy equipment were used to clear away material left by blasting. The blasted material was sorted to be used as fill material or in the construction of retaining walls. To prevent damage to the surrounding terrain, "side-casting", or pushing unwanted material below the road, wasn't allowed. 4) Extensive retaining walls were used to stabilize precarious areas and bench the road out from the cliffs. Many of the retaining walls were topped with attractively crenelated stone guard walls, which was added when necessary to ensure safe and comfortable travel. Dozens of experienced masons were employed to construct the walls but they could barely keep up with the demand. 5) The surface of the roadway was made up of a layer of crushed rock on top of the rocky roadbed. Because of the limestone present in Glacier, binder for the surfacing was not required. A blade pulled by a caterpillar was used for ditching and grading where it was required. 6) The road was improved and resurfaced numerous times between 1933 and 1950, and was finally completely paved in 1952.

SURVEYING ①

BLASTING ②

CLEARING/SORTING ③

RETAINING WALLS ④

GRADING/SURFACING ⑤

FINISHED ROADWAY ⑥

HISTORIC AMERICAN ENGINEERING RECORD MT-67
SHEET 4 OF 14
MONTANA
ADDENDUM TO GOING-TO-THE-SUN ROAD - 1933
GLACIER NATIONAL PARK
FLATHEAD COUNTY
WEST GLACIER VICINITY
NPS PARK ROADS RECORDING PROGRAM
NATIONAL PARK SERVICE
UNITED STATES DEPARTMENT OF THE INTERIOR
DELINEATED BY: Nathan Junkert 2000

VISTA DESIGN

Motorists on Going-to-the-Sun Road experience a carefully choreographed series of striking views and stimulating vistas. Road designers employed a variety of sophisticated techniques to display natural scenery along the route and produce a memorable motoring experience. The road's location in relationship to features such as lakes, mountains, cliffs, and streams, together with changes in the bordering vegetation and subtle variations in alignment and grade produce a constantly changing succession of visual effects and environmental perceptions

PANORAMAS

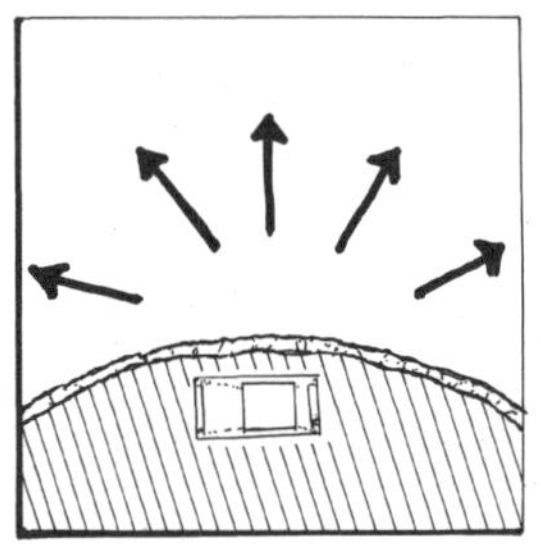

Many places along the Going-to-the-Sun Road provide expansive views of the distant landscape. Small pullouts were constructed at many of these points to provide visitors opportunities to stop and enjoy the sweeping vistas.

AXIAL VIEWS

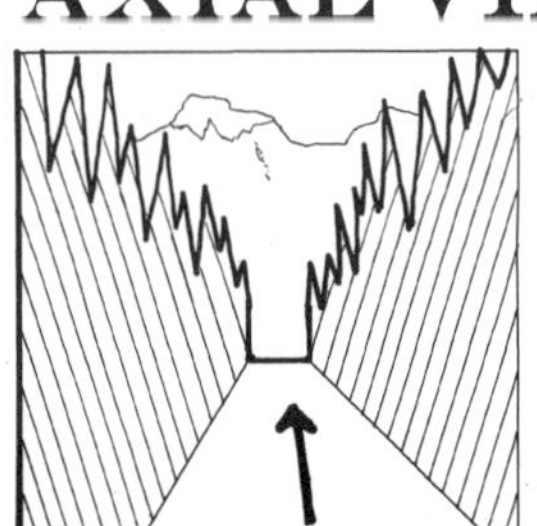

This directional view is created by employing long, relatively straight road alignments to focus the traveler's attention and provide sufficient time to enjoy significant views. Road designers employed this technique to highlight striking features such as the Garden Wall and Jackson Glacier.

VISTA MANAGEMENT

Lake McDonald

Interior View

Filtered View

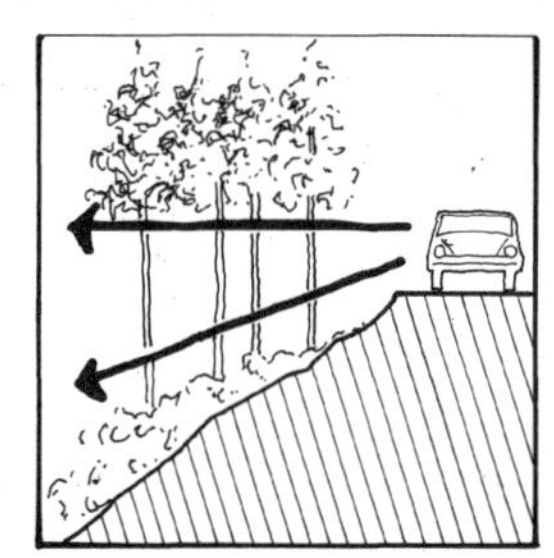

French Cut

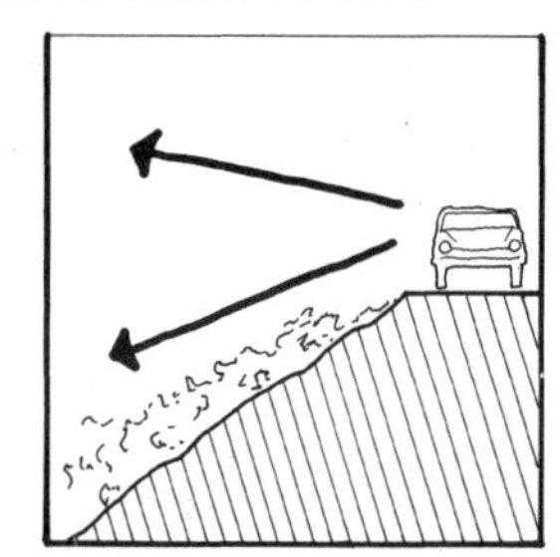

Canopy Cut

Managing roadside vegetation to enhance views has long been a standard landscape design technique. Selective cutting can be used to improve existing views or create outlooks onto scenic highlights obscured by dense vegetation. Many vistas that appear to be natural are actually carefully designed by park road managers. Glacier National Park has developed a series of vista management guidelines to reduce undesirable vegetation and enhance the visitor's experience. The four standard approaches include: the removal of lower limbs and brush along the immediate roadside to provide views into the forest; selective cutting of lower limbs and brush to create filtered views of background landscapes; a narrow window or "French Cut" that gives the visitor a selective view through the trees, and the "Canopy Cut," which removes all obstructions and completely opens the view. Managed vistas are often located at curves and turnouts where motorists have better opportunities to enjoy the designed views.

ROCK LEDGES

Alignment

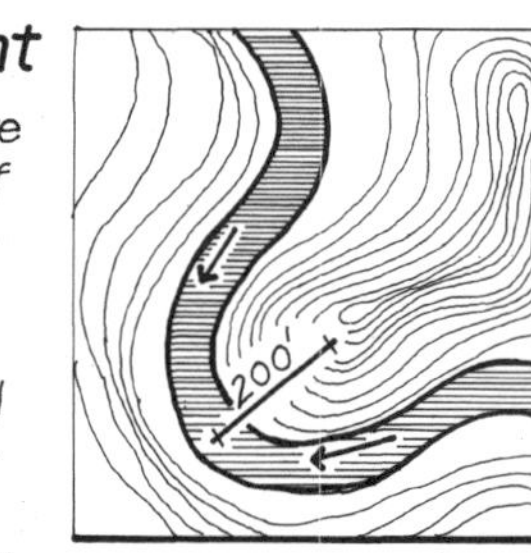

Variations in the curvature or horizontal alignment of a road produce distinct scenic effects. The rugged terrain of Going-to-the-Sun Road resulted in numerous short tight curves. These curves reflect the desire to follow the mountain's natural contours with a minimum of unsightly excavation. The sharp curves create a striking motoring experience as the road seems to disappear around corners and visitors are left staring over steep cliffs into thin air.

Grade

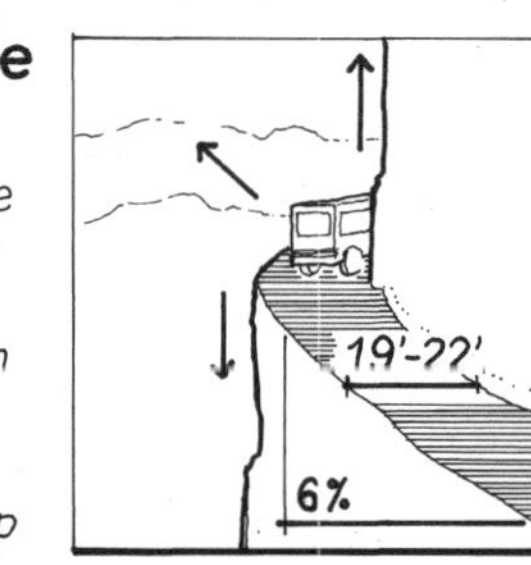

Changes in the road's vertical alignment and the level of the surrounding terrain also affect the motoring experience. On the Garden Wall and other cliff sections, the sharp drop-offs and steep 6% road grade combine with the narrow width of the pavement to heighten the feeling of exposure. The low rock barriers allow motorists to peer down into dizzying depths as they look out over the valleys far below.

HISTORIC AMERICAN ENGINEERING RECORD MT-67

SHEET 5 of 14

MONTANA

ADDENDUM TO GOING-TO-THE-SUN ROAD - 1933
GLACIER NATIONAL PARK
FLATHEAD COUNTY

WEST GLACIER VICINITY

NPS PARK ROADS RECORDING PROGRAM
NATIONAL PARK SERVICE
UNITED STATES DEPARTMENT OF THE INTERIOR

DELINEATED BY: Brandy Dubs, 2000

Driving the Going-to-the-Sun Road is the primary way in which visitors experience Glacier National Park. While enjoying the road's ever-changing scenery, attentive motorists can gain an appreciation of a variety of distinct environmental zones and ecological associations. Going-to-the-Sun Road passes through six regions that are ecologically and visually distinct and provide different but related experiences. From west to east these include, a densely forested area, stream valley, rock ledges, a sub-alpine zone, a dry forest, and a meadow of grasslands. From the bleak summits of the highest slopes where only hardy alpine plants can thrive, to the luxuriant Pacific-type forests of west side valleys and the expanse of grassy, flower-dotted plains on the eastern approach, the profusion of flora and fauna provide a spectacular lesson in natural history.

Rock Ledge

Along the Garden Wall and other higher sections of the road, the visitor has spectacular views across steep glacial valleys, precipitous cliffs, and craggy mountains. Snow fields, dramatic waterfalls, and green tapestries created by varied vegetation add to the alpine effect. Beyond the outer edge of the road, steep cliffs drop precipitously into the valley far below, adding excitement to the driving experience.

Dry Forest

The east side of Logan Pass transitions to a dry forested area, primarily Douglas Fir, Spruce, and Pine. This creates a varied pattern as it blends from somewhat open woodlands, with a grassy understory, to quite dense forests that are virtually opaque. The changing ages and sizes of the trees gives the road varied scenic character. The traveler must proceed slowly to experience the views because the openings are short and only quick glimpses of lateral scenes are possible.

Stream Valley

The temperate rainforest ecology, rushing mountain streams and dramatic upward vistas along this segment of the Going-to-the-Sun Road provide opportunities to enjoy classically picturesque scenery while observing the dynamics of mountain streams. The creek passes through stretches of cascades and pools to roaring McDonald Falls.

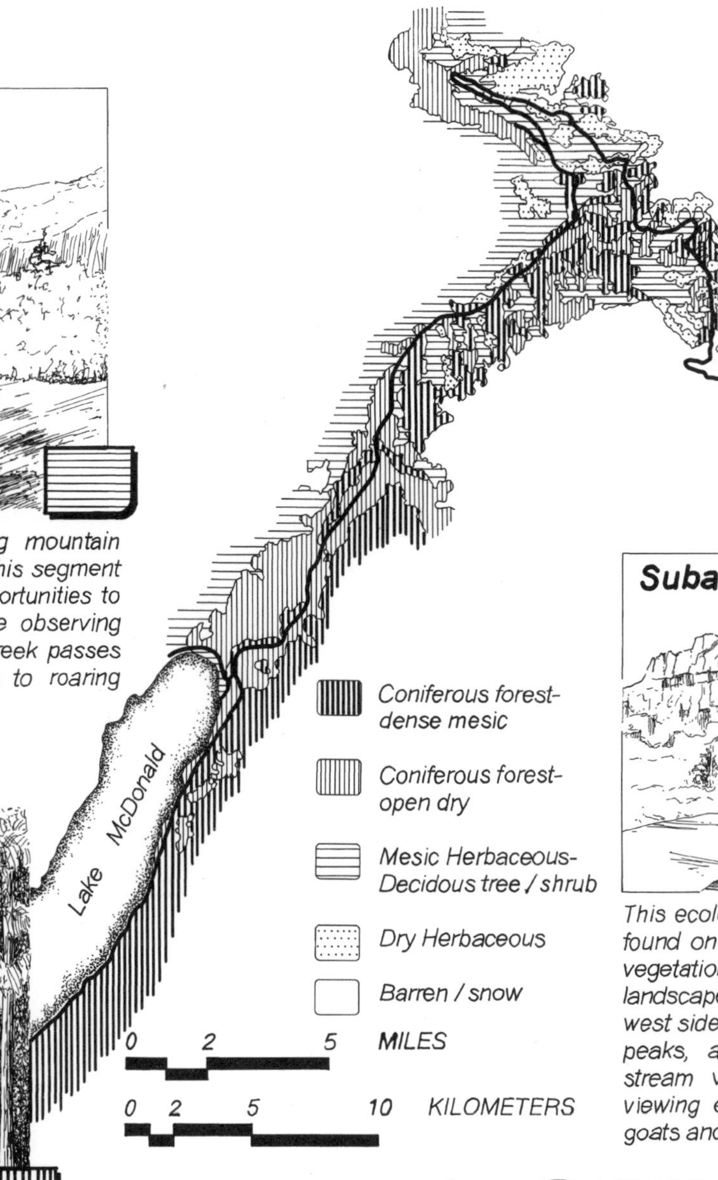

Aspen Grasslands

This broad valley with St. Mary Lake at the bottom and the rugged mountains in the background produces a unique environment. Dry prairie grasslands dominate the valley floor while groves of aspen climb the surrounding slopes. This abrupt transition between prairies and mountains provides a habitat ideal for wildlife. In the early mornings and late evenings it is not uncommon to see elk and bear in the meadows.

Subalpine

This ecological zone straddles the Continental Divide and is found on both the east and west sides of the park. Low vegetation with islands of spruce and fir dominate the landscape. The trees are larger and in denser stands on the west side of Logan Pass. The expansive views of mountain peaks, alpine tundra, wildflowers, waterfalls, and lower stream valleys are spectacular. The opportunities for viewing elusive high mountain animals such as mountain goats and big horn sheep are outstanding.

Dense Forest / Lakeshore

This segment of the road is thickly vegetated, passing through cool, evergreen forests that include Douglas Fir, White Pine and large Cedar and Hemlock trees, which take centuries to develop and grow. Roadside vegetation is lush, giving a feeling of enclosure within the forest that is interrupted by intermittent views of the lake, mountains and forest floor.

ACROSS THE DIVIDE

EXPERIENCING THE LANDSCAPE

DELINEATED BY: *Brandy Dubs & Magdalena M. Lisowska, 2000*

NPS PARK ROADS RECORDING PROGRAM
NATIONAL PARK SERVICE
UNITED STATES DEPARTMENT OF THE INTERIOR

ADDENDUM TO GOING-TO-THE-SUN ROAD - 1933
GLACIER NATIONAL PARK
FLATHEAD COUNTY
WEST GLACIER VICINITY
MONTANA

SHEET 6 OF 14

HISTORIC AMERICAN ENGINEERING RECORD
MT-67

WHY A TUNNEL?

EAST TUNNEL - AS BUILT

Channel Cut Scheme

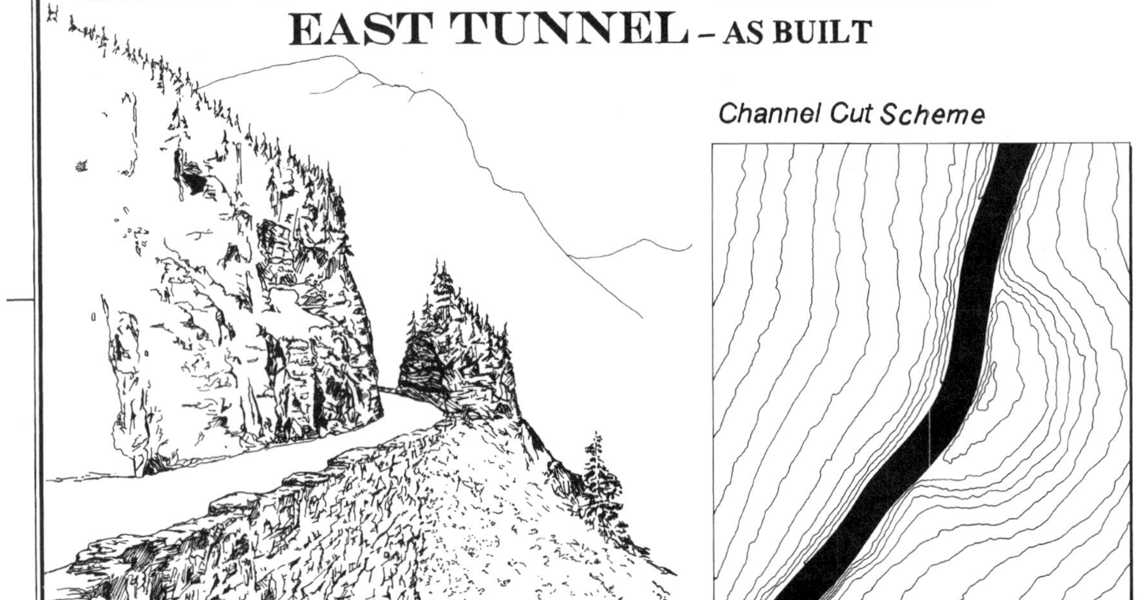

Cleared Bench Scheme

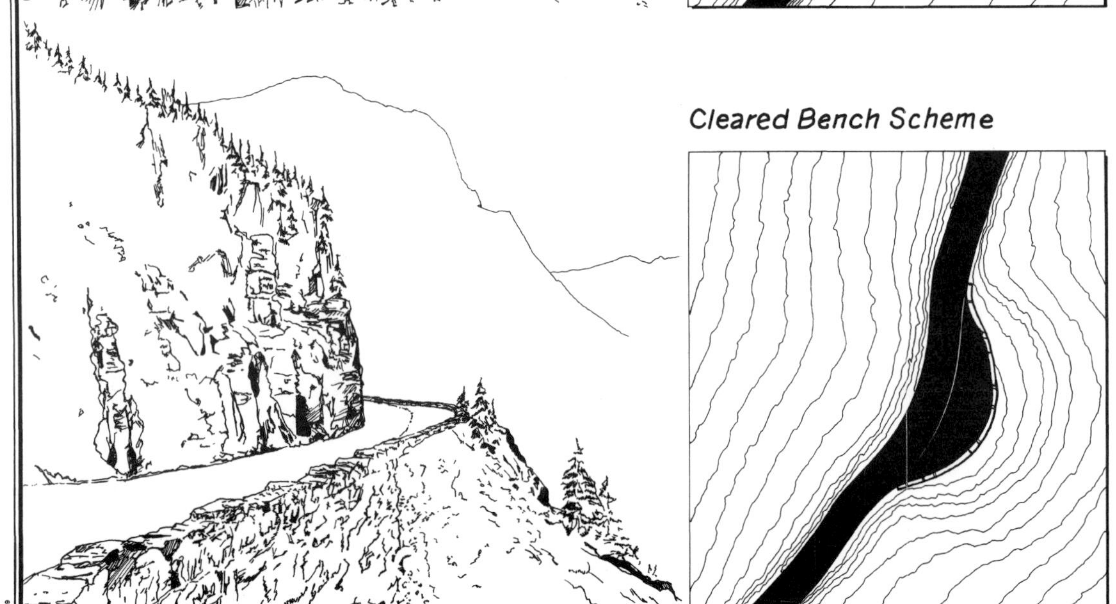

"The finished road should minimize the effects of the work of the hand of man that the effects of the work of nature will predominate in the picture."

Thomas Vint, 1925

When constructing roads across steep slopes, designers have several options. The simplest approach is to clear off a level bench for the roadway, but this requires extensive excavation and leaves a highly visible construction scar. The steep rock faces left behind can be very unstable, endangering motorists and creating maintenance problems. Another approach is to make a narrow rock cut, saving the outer portion of the slope to retain scenic interest and reduce excavation costs. Tunnels, while sometimes more complex to construct, help minimize the road's negative impact on fragile ecosystems and spectacular mountain scenery. By protecting the road from falling rock, they also reduce driving hazards and maintenance concerns.

The East Tunnel, illustrated to the left, is bored straight through a solid rock outcropping, its portals framing mountain views along the road. The West Tunnel is cut through steeper, more fractured cliffs, making water a problem but allowing dramatic, pedestrian windows oriented toward the sideways views. This opportunity makes the West tunnel one of the major stopping places along the road.

WEST TUNNEL GALLERY

(View from automobile)

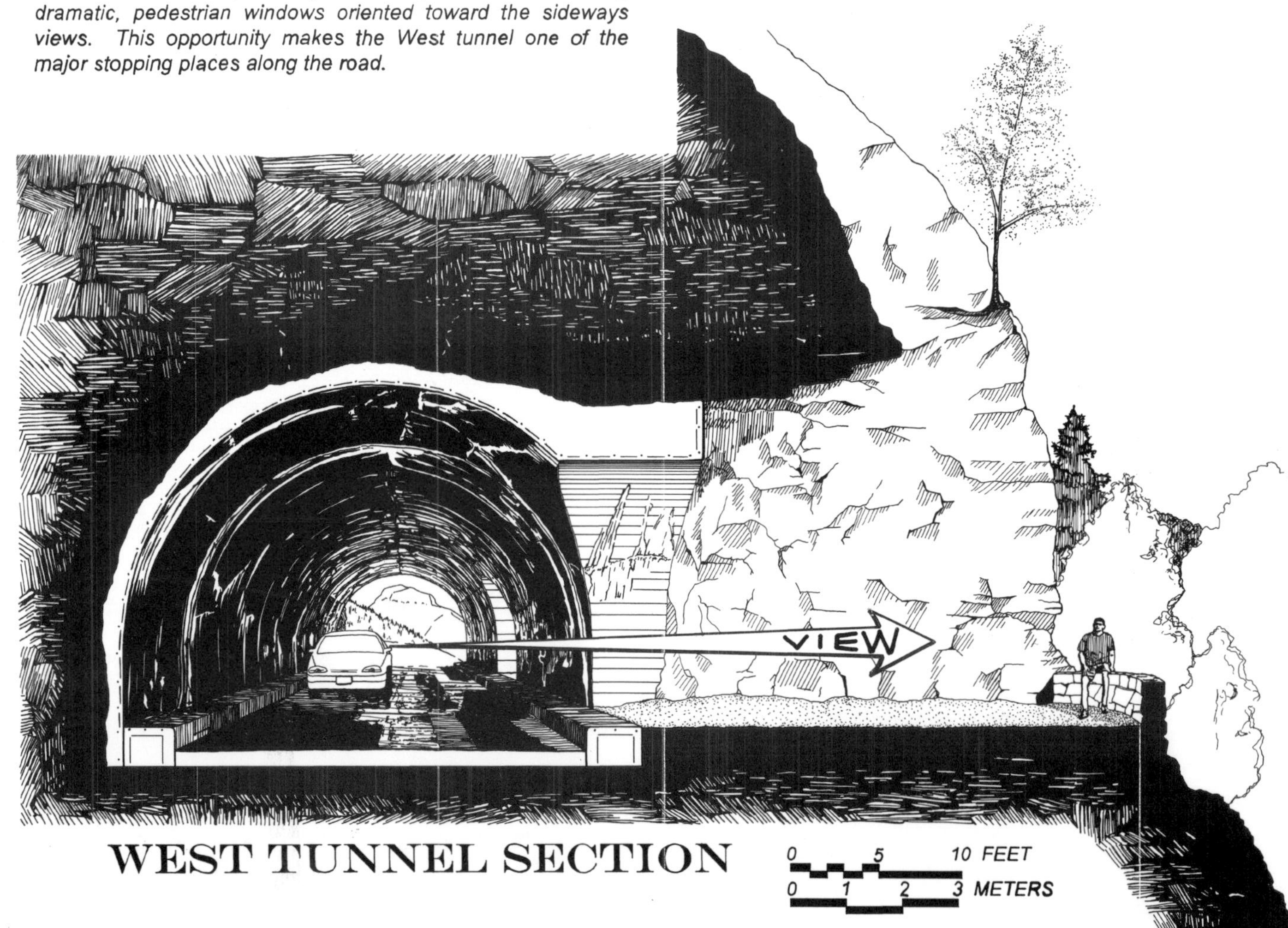

WEST TUNNEL SECTION

0 5 10 FEET
0 1 2 3 METERS

HISTORIC AMERICAN ENGINEERING RECORD MT-67

SHEET 8 of 14

MONTANA

ADDENDUM TO GOING-TO-THE-SUN ROAD - 1933
GLACIER NATIONAL PARK
FLATHEAD COUNTY

WEST GLACIER VICINITY

DELINEATED BY: Christopher A. Boldt, 2000

NPS PARK ROADS RECORDING PROGRAM
NATIONAL PARK SERVICE
UNITED STATES DEPARTMENT OF THE INTERIOR

GUARD RAILS & RETAINING WALLS

PERMANENT

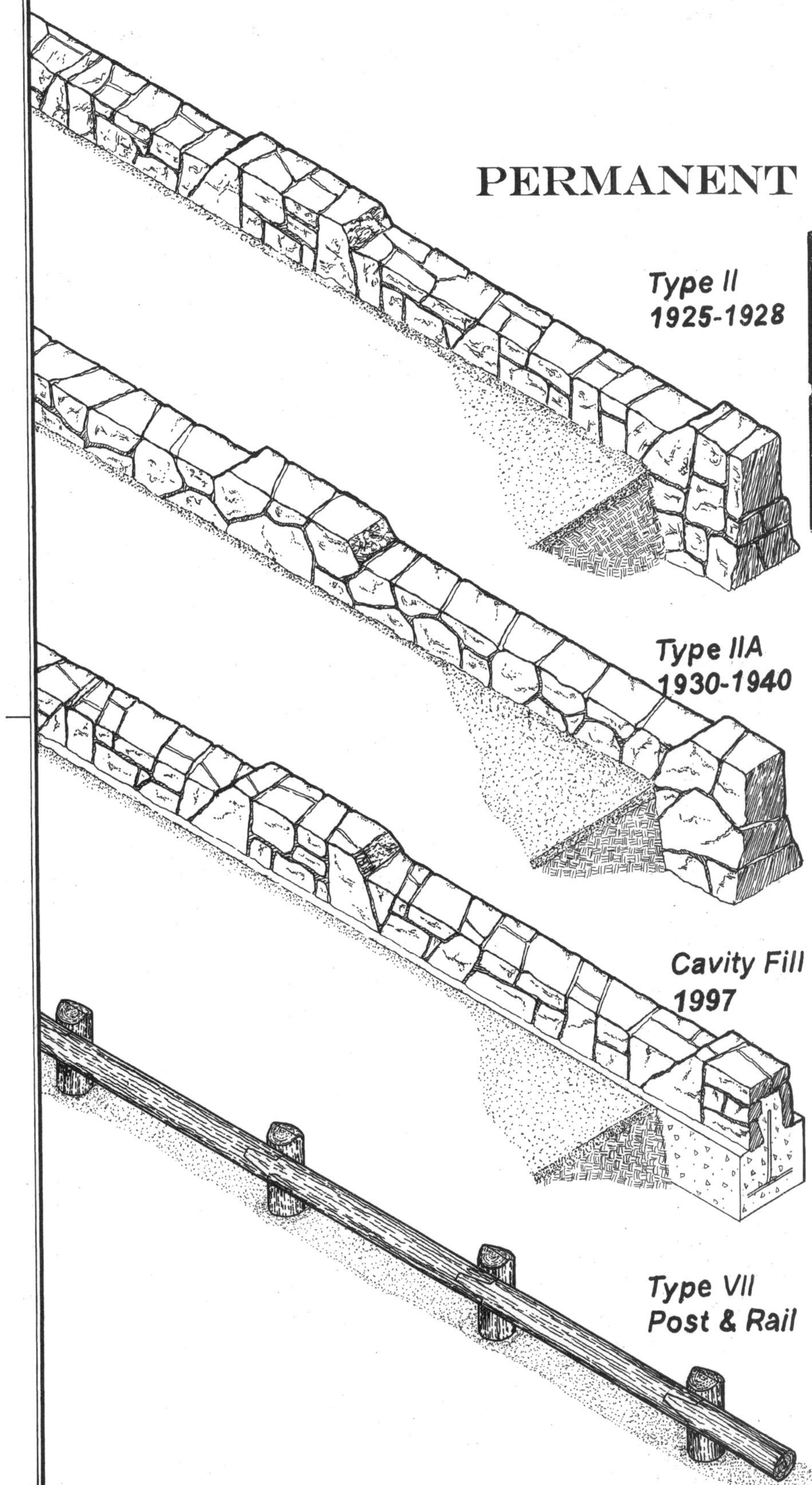

REMOVABLE

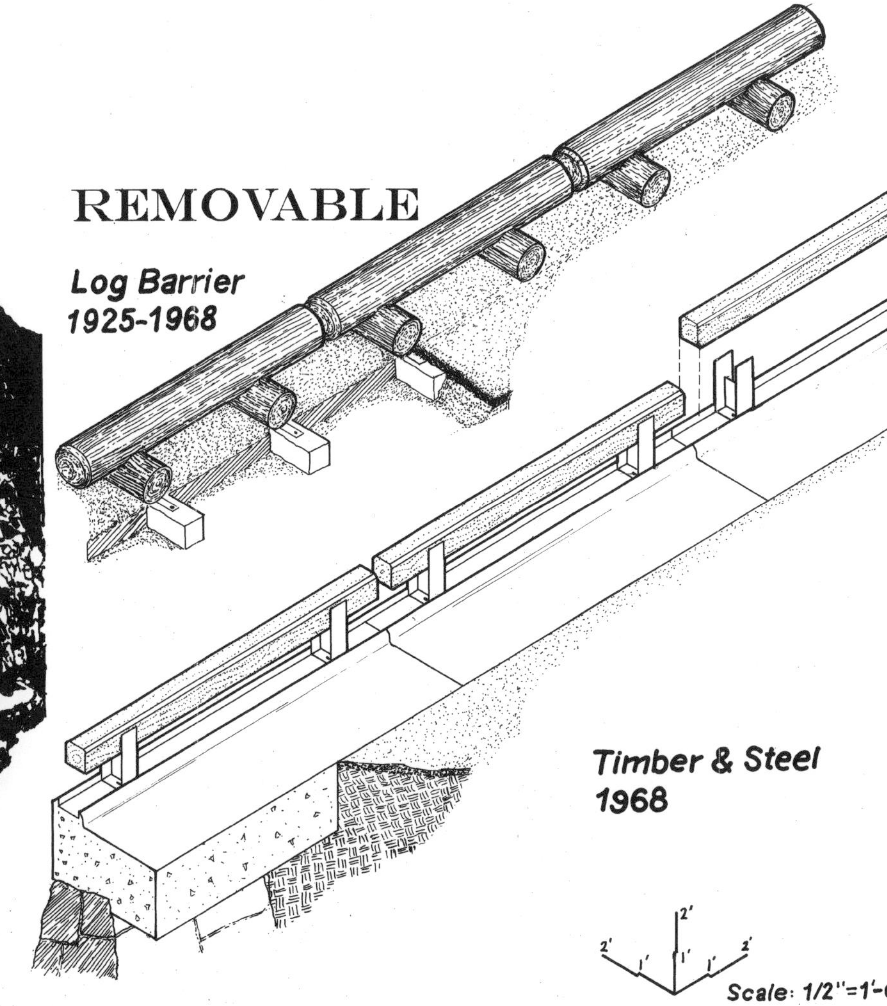

The Type II crenellated stone masonry guard wall was built exclusively on the west side of the Williams and Douglas contract of 1925-1928. The wall is characterized by a random rubble mosaic or web wall pattern with transverse and longitudinal jointing in the top course and large stones at the crenellations.

The Type IIA crenellated stone masonry guard wall built between 1930 and 1940 is characterized by an uncoursed roughly squared rubble pattern with only transverse jointing in the top course.

The concrete cavity fill guard wall is a modern solution to historic wall replacement on the road. Originally it was called "cavity filled", where stone gathered in the park was used to provide a wall exterior filled with reinforced concrete that replicated the historic Type II and Type IIA walls.

The Type VII post and rail guard rail is a standard NPS design used in Glacier primarily at waysides and to discourage vehicles from pulling off at undesignated areas along the road. It is also used to direct visitors and protect resources. Historically it was constructed as a guardrail.

Due to frequent damage to guard walls from avalanches certain portions of the Going-to-the-Sun Road are fenced with removable guard rails that are taken down each winter. Log Barrier was a standard early NPS design but was proven ineffective and was replaced by the current timber and steel rail in 1968.

RETAINING WALLS

Historic retaining walls on the Going-to-the-Sun Road are of a masonry design that, like the guard walls, incorporates stone from the immediate area of the wall in order to blend in with the landscape. Modern replacements are sometimes concrete with stone facing.

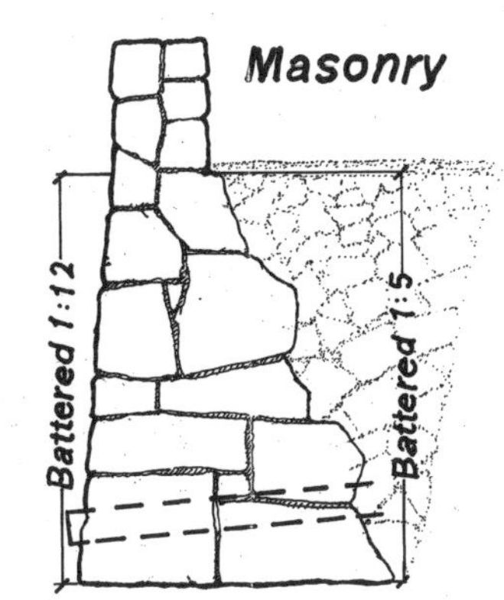

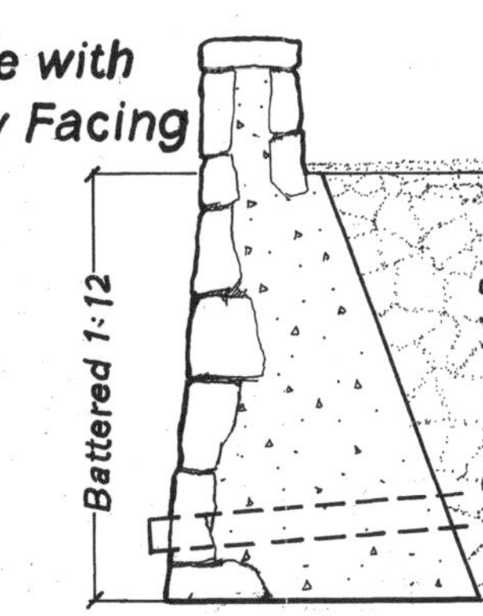

DELINEATED BY: Lucas Dupuis 2000

NPS PARK ROADS
RECORDING PROGRAM
NATIONAL PARK SERVICE
UNITED STATES DEPARTMENT OF THE INTERIOR

ADDENDUM TO GOING-TO-THE-SUN ROAD - 1933
GLACIER NATIONAL PARK
WEST GLACIER VICINITY FLATHEAD COUNTY MONTANA

SHEET 9 OF 14

HISTORIC AMERICAN ENGINEERING RECORD
MT-67

PARK ENTRANCES

WEST GLACIER

The West Glacier entrance has evolved considerably since the park was created in 1910. Visitors initially crossed the Flathead River on a timber bridge, built in 1897, near the Belton train station and followed a narrow dirt road along the riverbank. This approach was improved with the development of the park administrative village in 1924. This log cabin served as the official entrance station until 1938 when a new bridge was constructed and the main approach road rerouted. An automobile-oriented checking station was constructed for this new entrance in 1941. This structure was modified in 1963 by two additional checking cubicles in the "wings" of the original facility. These additions allowed two lanes of traffic to be processed simultaneously, and allowed for the handling of the increasing number of oversized recreational vehicles.

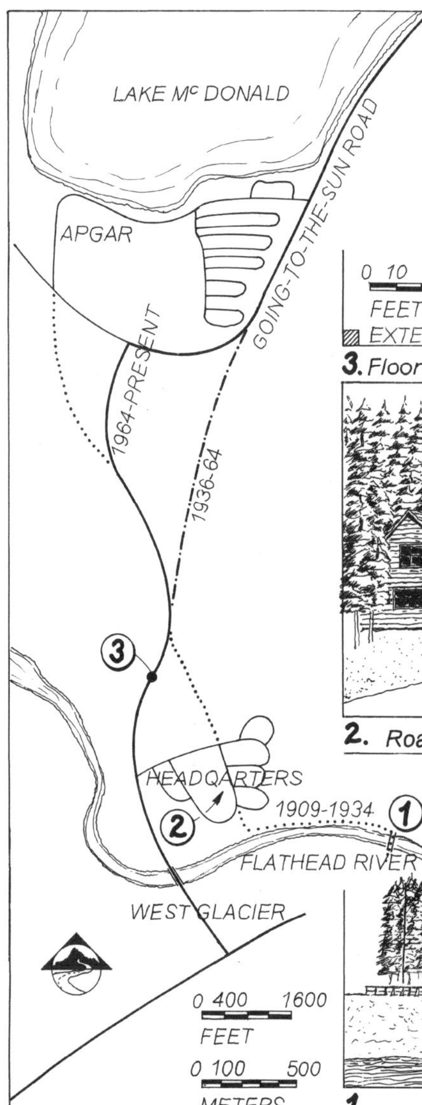

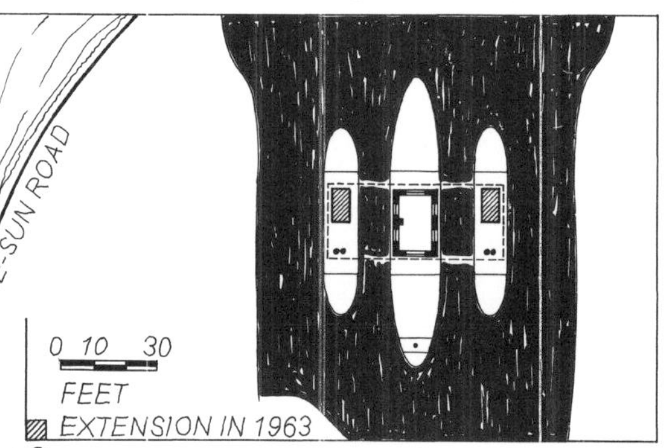

3. Floor plan of the entrance station, built in 1941

2. Road through the headquarters in the 1930s

1. Bridge at Belton in 1909

ST. MARY

The east entrance to the park has also changed dramatically over time. The first "checking" station from 1918, was a tent, located closer to the lake, near the 1913 Ranger Station, but no fees were collected. In the 1930's the course of the road was changed to its present route and an entrance station structure was built on the road, about 1/2 mile south of the present visitor center. In June of 1942, the park completed construction of a new checking station at the site of the former building, with the same design that was used in West Glacier. This served until the construction of the present-day kiosks (right next to the main Visitor Center) in 1968 - as the continuation of the Mission 66 design. The new entrance station was intended to combine traditional rustic esthetics with modern materials and stylistic concerns.

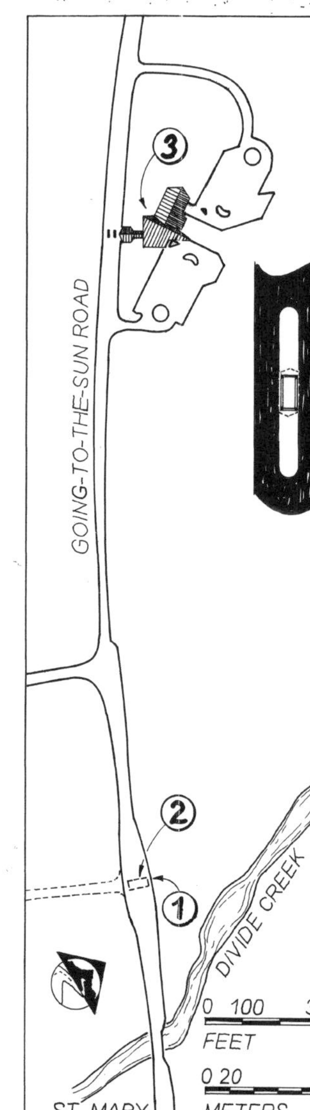

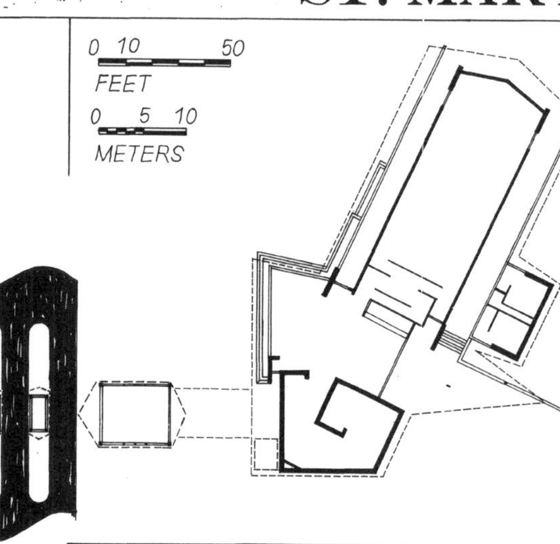

3. Floor plan of the visitor center, opened in 1964

2. Entrance station, built in 1942

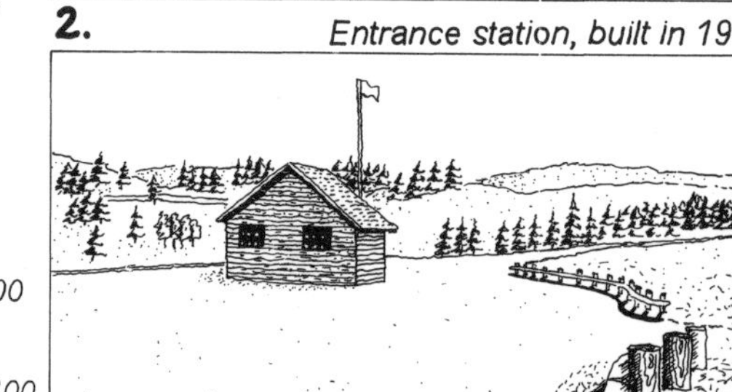

1. Entrance station, built in the 1930s

DELINEATED BY: Eszter Vogel, 2000

NPS PARK ROADS RECORDING PROGRAM
NATIONAL PARK SERVICE
UNITED STATES DEPARTMENT OF THE INTERIOR

ADDENDUM TO GOING-TO-THE-SUN ROAD - 1933
GLACIER NATIONAL PARK
FLATHEAD COUNTY

WEST GLACIER VICINITY

MONTANA

SHEET 10 OF 14

HISTORIC AMERICAN ENGINEERING RECORD
MT-67

THE LOOP

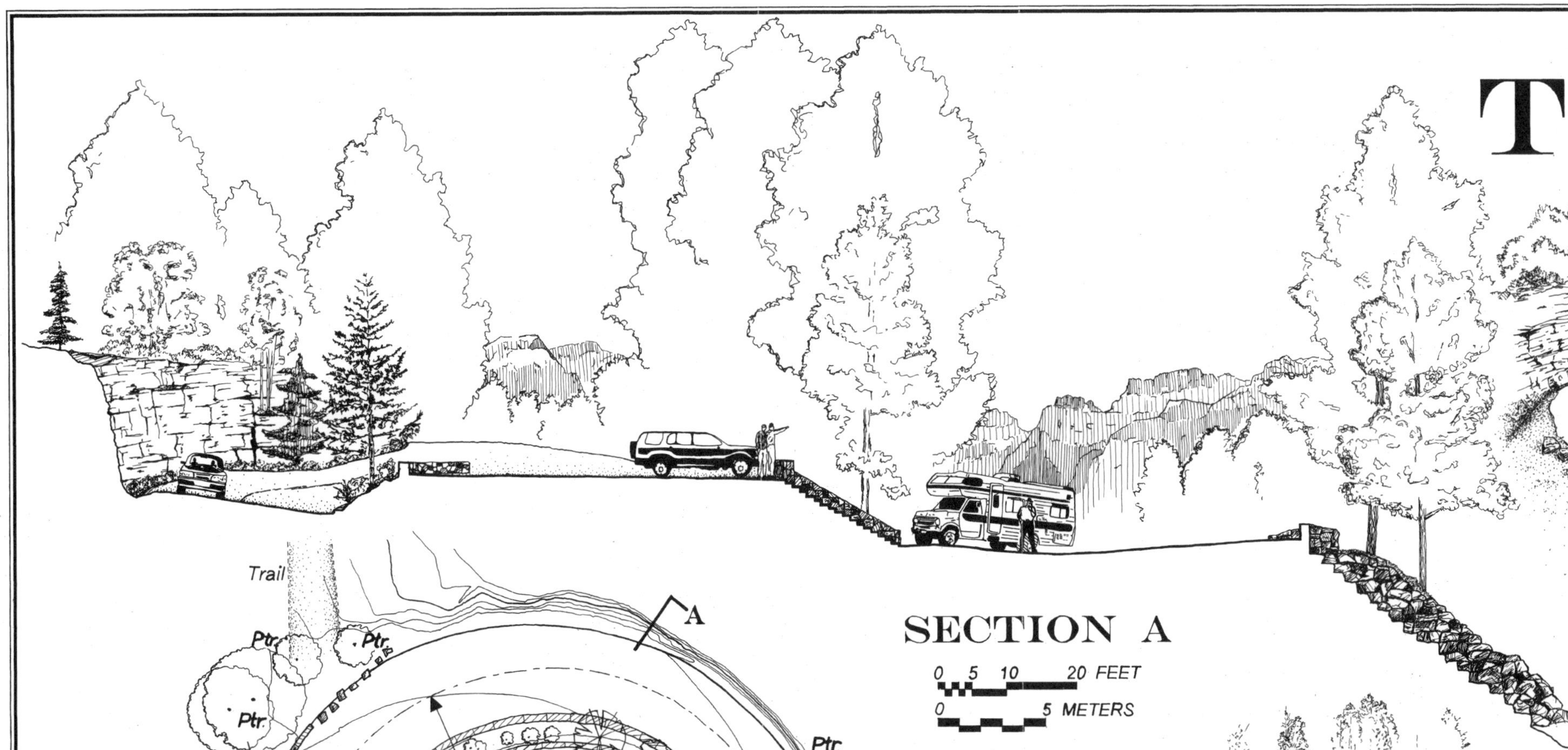

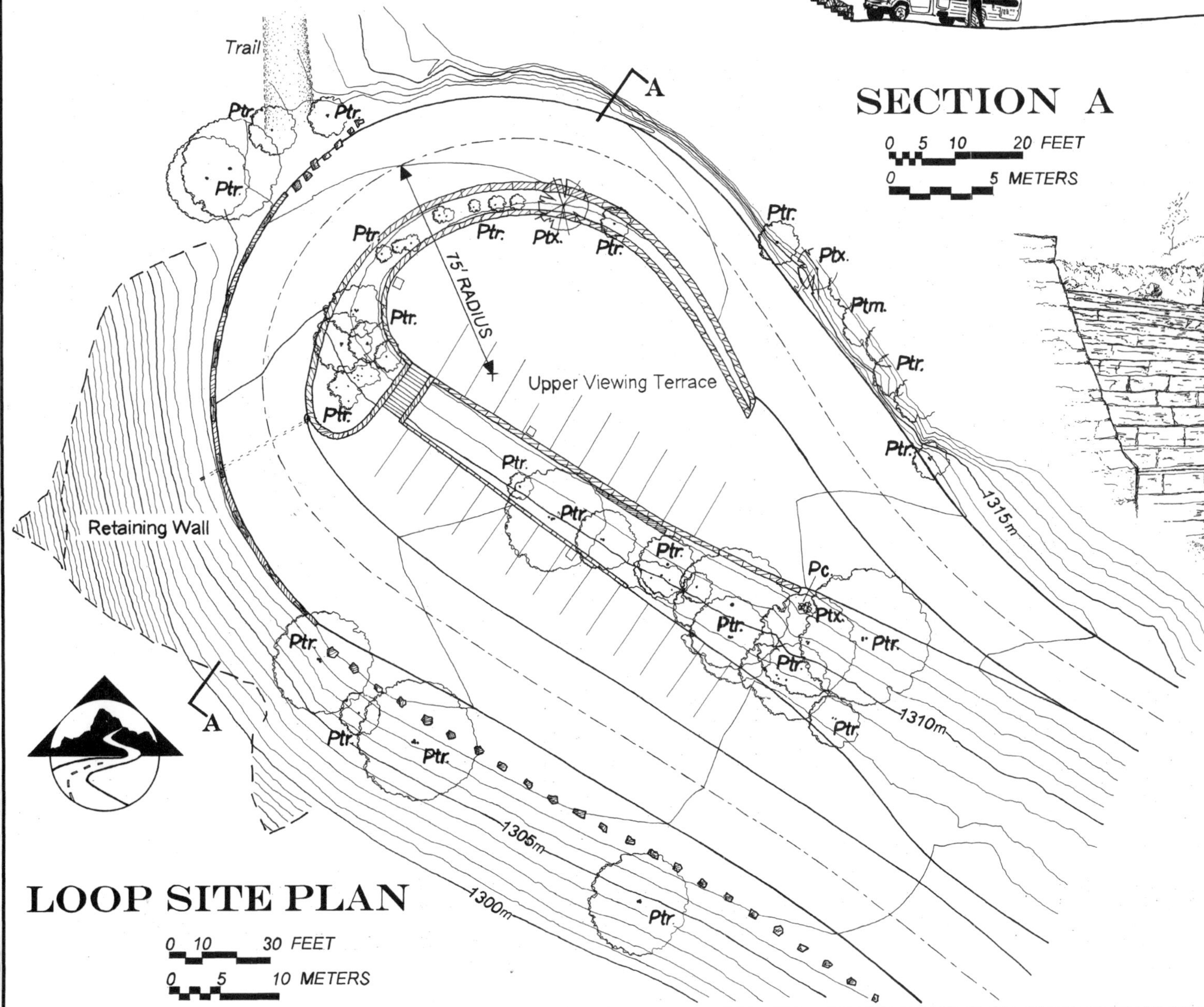

The Going-to-the-Sun Road climbs the west side of Logan Pass in two long gentle grades connected by one sharp switchback curve. The area where this change of direction takes place is known as "The Loop." Realizing that motorists would have to slow down to negotiate this turn, Going-to-the-Sun Road designers took advantage of the opportunity to develop a parking area, trailhead, and viewing platform at this highly scenic spot. The Loop marks a transformation in scenic experience as well as a change in direction. Below this spot, the road is bordered by sheltering trees. Above The Loop, the motorist ventures onto the Garden Wall, a region of sub-alpine terrain, dizzying cliffs, and sweeping mountain vistas.

Elements of the original design scheme, the rustic staircase and stone walls mark out a divided circulation pattern benched out from the steep slope. As a result of so much fill, an extensive stone retaining wall, hand-laid without mortar, supports the lower curve of The Loop. As of the turn of the year 2000, the wall was undergoing repair.

Tree Legend

(Existing 2000)

Ptr. -*Populus trichocarpa* (Black cottonwood)
Ptx. -*Pseudotsuga menziesii* (Douglas Fir)
Ptm. -*Populus tremuloides* (Quaking aspen)
Pc. -*Picea engelmannii* (Engelmann Spruce)

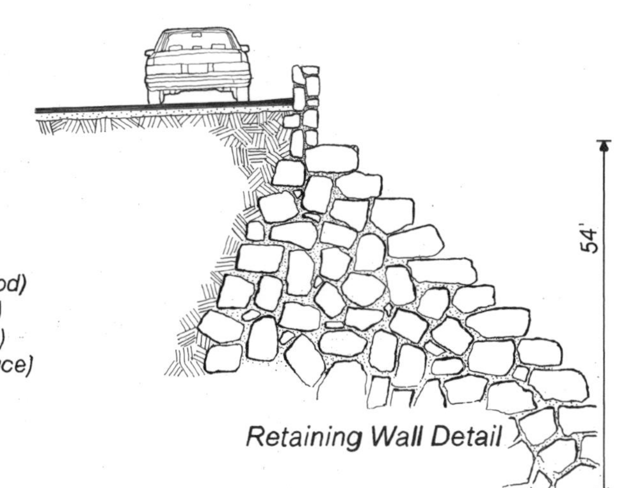

Retaining Wall Detail

DELINEATED BY: *Christopher A. Boldt, 2000*
NPS PARK ROADS RECORDING PROGRAM
NATIONAL PARK SERVICE
UNITED STATES DEPARTMENT OF THE INTERIOR

ADDENDUM TO GOING-TO-THE-SUN ROAD - 1933
GLACIER NATIONAL PARK
FLATHEAD COUNTY
WEST GLACIER VICINITY
MONTANA

SHEET 12 OF 14

HISTORIC AMERICAN ENGINEERING RECORD
MT-67

ROES CREEK AUTO CABIN CAMP - 1930

Early park visitors camped wherever they pleased. Resulting damage to vegetation prompted the park to begin planning a series of campgrounds.

Now known as Rising Sun, this campground was *not* constructed to its 1930 plan. The design was an example of the "Teepee Type" cabin arrangement. Each "Council Unit" circle would be made up of octagonal cabins surrounding a central "Council Fire." Cars would be parked either between the cabins or surrounding the fire, where their headlights could be used for evening "councils."

Swiftcurrent Auto Cabin Camp, in the Many Glacier area of the park, was built to a similar design in 1933 and 1936, but with simpler rectangular cabins and vegetation in the center replacing the fire.

Swiftcurrent Auto Cabins

SPRAGUE CREEK CAMPGROUND - 1942

The campground was designed and constructed as a chain of auto spurs arranged about naturalistically winding narrow loop roads. Tables and fireplaces, to be shared among the campers as necessary, were distributed haphazardly among the trees and clearings to give the overall effect of camping in the woods. Gravel pads for the tents were added later to cut down on mud and mitigate damage to the vegetation.

The design offered versatility, easily accommodating visitors arriving with a car and a tent, those with a camping vehicle, and backpackers with only a tent.

ST. MARY LAKE CAMPGROUND - 1964

As camping vehicles became larger, the older campgrounds proved inadequate to accommodate them. New campgrounds were constructed in which each vehicle was given its own pull-through loop. The loop roads on which these were arranged were designed with large smooth curves for easy vehicle maneuverability. Even considering their greater capacity, these campgrounds covered much larger areas of land than those built earlier.

Eventually, it was realized that even these newer campgrounds were insufficient for the largest camping vehicles being manufactured, and a total vehicle length limit for St. Mary Lake Campground was set at 35 feet.

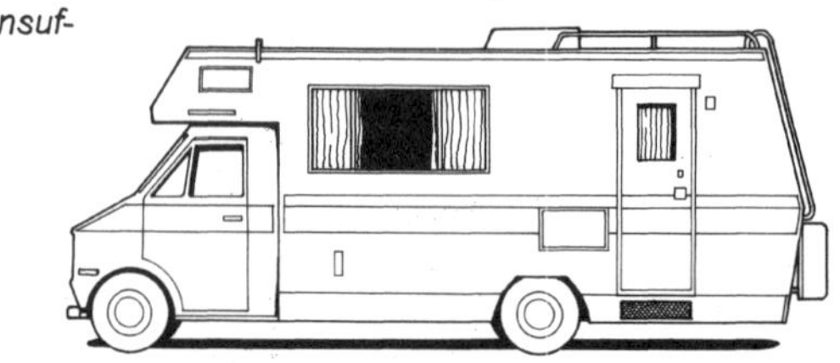

AUTO CAMP DESIGN

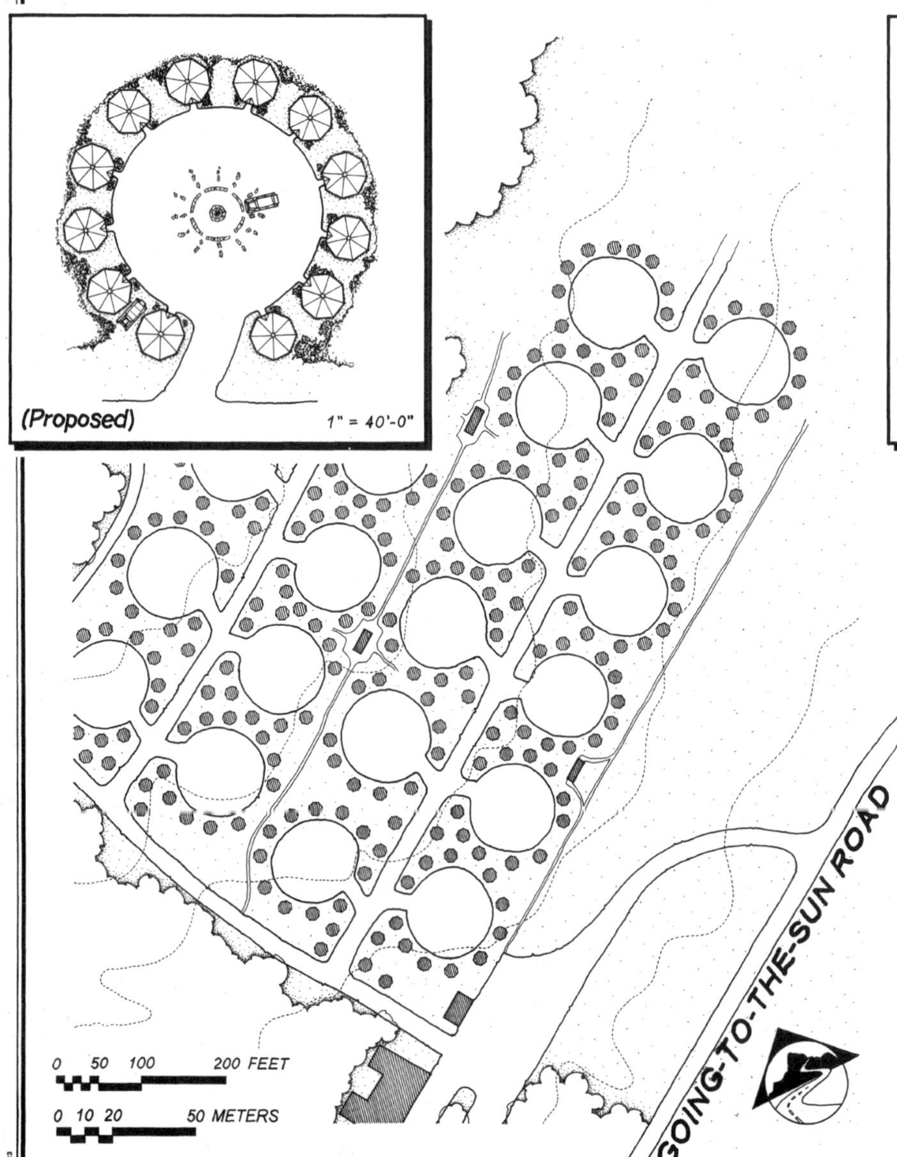

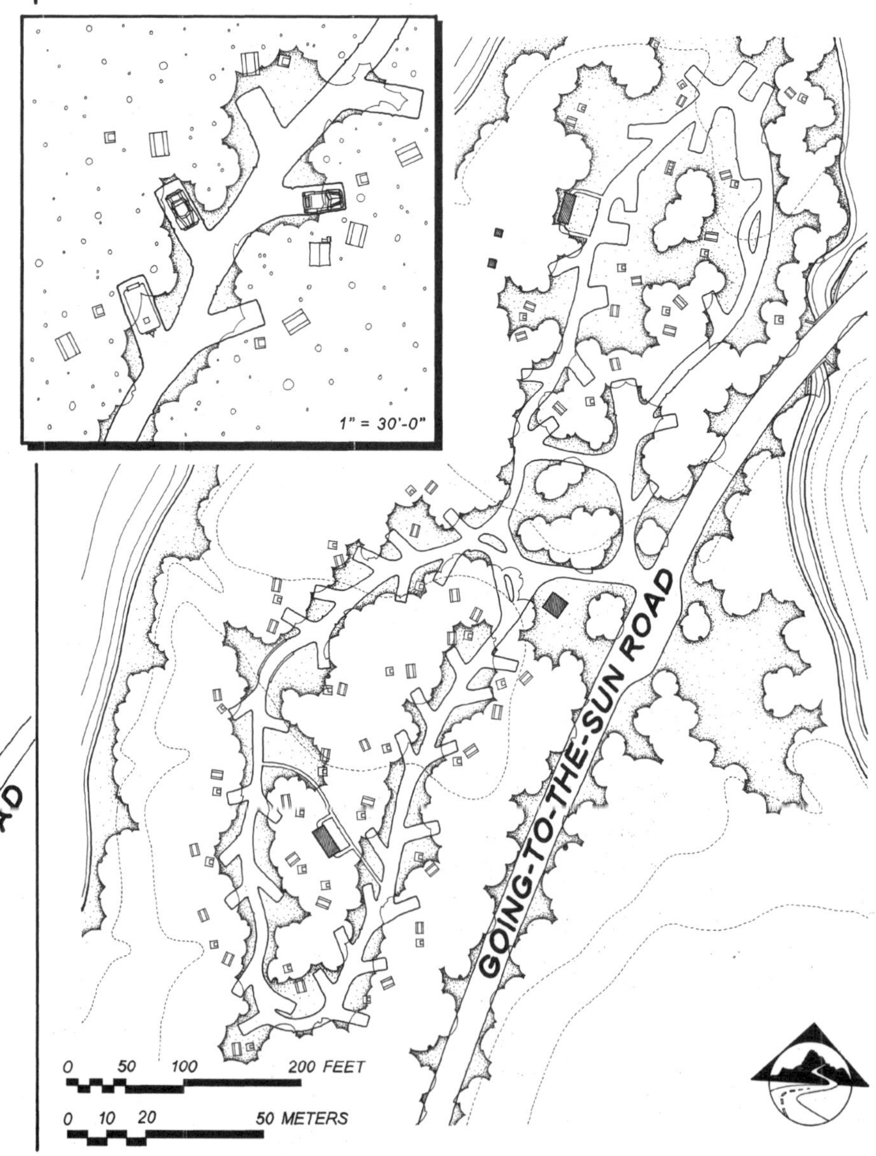

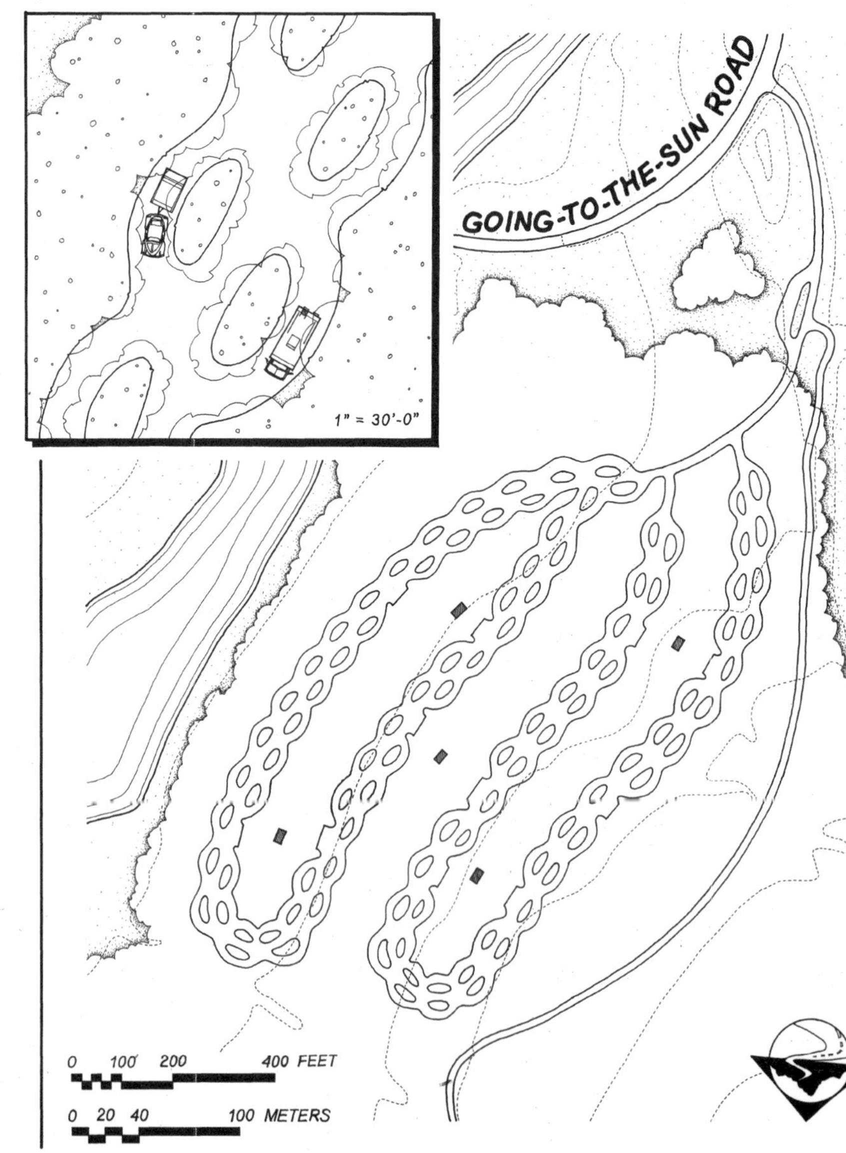

DELINEATED BY: Arin Streeter, 2000

NPS PARK ROADS RECORDING PROGRAM
NATIONAL PARK SERVICE
UNITED STATES DEPARTMENT OF THE INTERIOR

ADDENDUM TO GOING-TO-THE-SUN ROAD - 1933
GLACIER NATIONAL PARK
FLATHEAD COUNTY

WEST GLACIER VICINITY

MONTANA

SHEET 13 OF 14

HISTORIC AMERICAN ENGINEERING RECORD
MT-67

IF REPRODUCED, PLEASE CREDIT: HISTORIC AMERICAN ENGINEERING RECORD, NATIONAL PARK SERVICE, NAME OF DELINEATOR, DATE OF THE DRAWING

SNOW REMOVAL

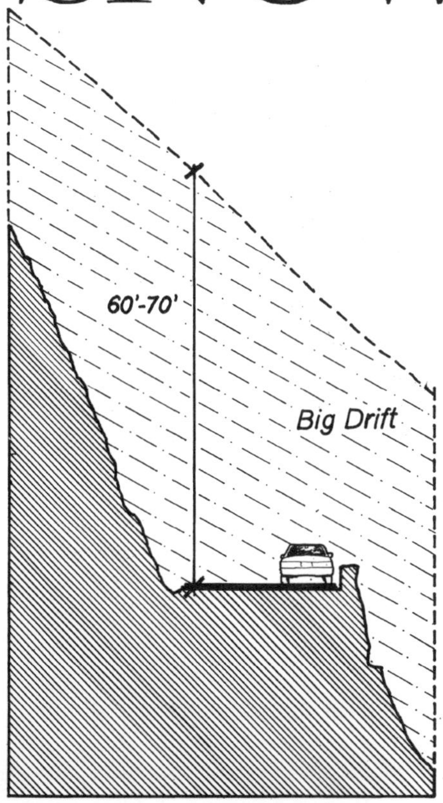

Due to Glacier National Park's high altitude, steep terrain, heavy snows, and cool climate, deep snow lingers in higher elevations well into June and July. The Going-to-the-Sun Road was carefully located to avoid shaded valleys and maximize sun exposure, but deep drifts and dense snowpacks make opening the road an annual epic of men and machines against the forces of nature. Since the road's opening in 1933, maintenance crews have developed a carefully coordinated snow removal process aimed at clearing the roadway as early and safely as possible with minimal damage to historic road structures and the fragile subalpine environment. The road crews begin clearing snow in April, but the process is long, difficult, and frequently dangerous. The safety of the road crew is always questionable while they are clearing the numerous avalanche chutes of the Garden Wall, the Big Drift and near the East Tunnel. "Spotters" are given radios and scan the mountainsides above the road for signs of avalanches. If any snow starts sliding the equipment operators are warned through the radios. Even in moderate snow years the Going-to-the-Sun Road is often not fully open until the end of May or early in June. The road crew works weeks on end to open the road, and repair the damage the snows may have caused through the winter. This process of opening the road maximizes the length of time that the road is open and allows more visitors to view what the Going-to-the-Sun Road has to offer.

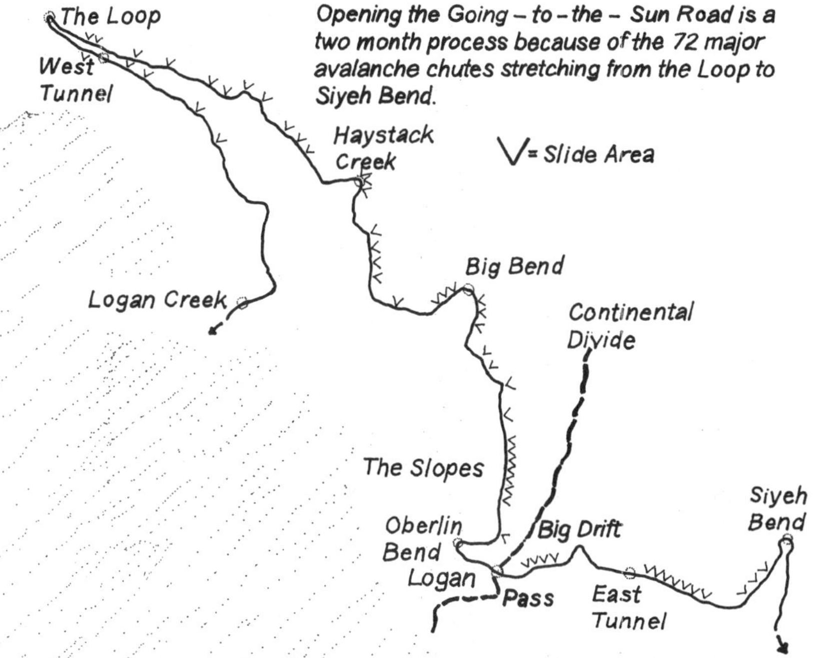

Opening the Going-to-the-Sun Road is a two month process because of the 72 major avalanche chutes stretching from the Loop to Siyeh Bend.

A - *Surveyors mark the center line of the road across large drifts to minimize equipment damage to the masonry guard walls and the asphalt surface and ensure the safety of the road crew.*

B - *A bulldozer makes initial cuts through snow drifts to provide other equipment a clearly marked and level working area.*

C - *A front end loader with a bucket specially designed for snow removal scoops snow over the sides of the deepening cut through the drift.*

D - *A rotary snow plow throws snow far away from the road cut to avoid any further snow accumulation near the roadway.*

One lane cleared near the East Tunnel. Drawing based on historical photograph.

DELINEATED BY: *Nathan Junkert 2000*

NPS PARK ROADS RECORDING PROGRAM
NATIONAL PARK SERVICE
UNITED STATES DEPARTMENT OF THE INTERIOR

WEST GLACIER VICINITY

ADDENDUM TO GOING-TO-THE-SUN ROAD - 1933
GLACIER NATIONAL PARK
FLATHEAD COUNTY

MONTANA

SHEET 14 OF 14

HISTORIC AMERICAN ENGINEERING RECORD
MT-67

IF REPRODUCED, PLEASE CREDIT: HISTORIC AMERICAN ENGINEERING RECORD, NATIONAL PARK SERVICE, NAME OF DELINEATOR, DATE OF THE DRAWING

Tennessee State Flag

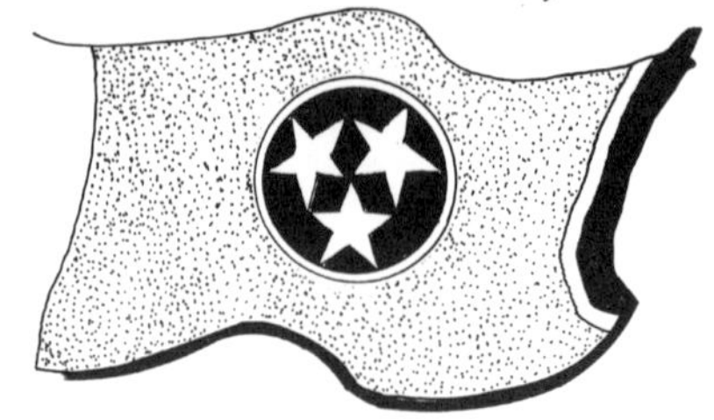

North Carolina State Flag

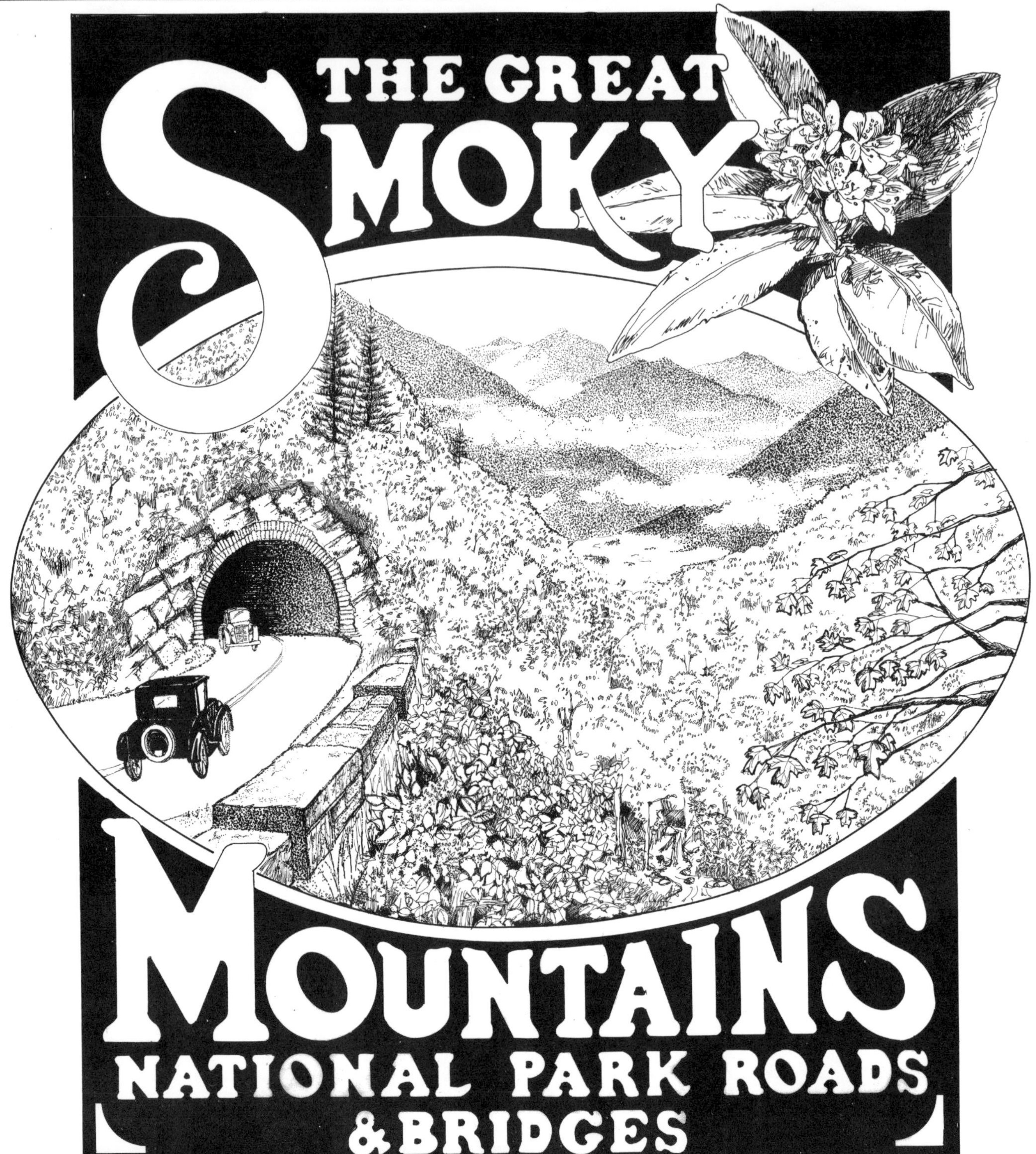

The history of Great Smoky Mountains National Park has always been intimately linked to the history of its motor roads. The movement to establish the park began when Willis Davis promoted the idea during a Knoxville Automobile Club board meeting in 1923. Alarmed at excessive logging and desirous of federally funded roads, the Automobile Club created the Smoky Mountains Conservation Association to lobby the federal government for a national park in the Smokies. When Congress finally announced authorization of the park in 1926, the Knoxville Automobile Club held a celebratory banquet.

The motor roads of Great Smoky Mountains National Park did not simply materialize with the establishment of the park. Long before tourists began motoring over Newfound Gap and causing traffic jams in Cades Cove, the Cherokee, white settlers, and loggers greatly influenced the Smokies' motor road system. After migrating to the region more than a thousand years ago from the upper Ohio River, the Cherokee blazed a sophisticated network of footpaths throughout the mountains for hunting, for socializing, and, on some occasions, for making war. Today's road from Cataloochee along the northeastern boundary of the park closely follows one of these Cherokee trails. White settlers migrated to the Smokies during the 1820s and 1830s along these same Cherokee footpaths. After widening many of them, settlers also began building new wagon roads throughout the mountains that evolved into park thoroughfares such as the Parson Branch and Rich Mountain roads near Cades Cove. Unlike these settlers, loggers sought out the more inaccessible regions of the Smokies. Timber firms including the Little River Lumber Company began logging in 1901 and by the mid-1930s had cut approximately 85 percent of what was one of the largest virgin deciduous forests on earth, leaving behind a legacy of abandoned railroad beds that were easily transformed into park motorways such as Little River Road.

Immediately after the federal government officially established Great Smoky Mountains National Park in 1934, the National Park Service began reconstructing Newfound Gap Road, a state-built road running up and over the Appalachian divide in the very heart of the Smokies. Although during the early 1930s the Park Service built several mountain roads in eastern parks including Skyline Drive in Shenandoah and Cadillac Mountain Road in Acadia, none posed the technical challenges of Newfound Gap Road in the Smokies. To eliminate sharp curves and steep grades the Park Service constructed a loop-over structure as well as a tunnel and several bridges along the almost thirty-mile highway. As important as these technical changes were aesthetic improvements. The Park Service added numerous overlooks to provide motorists with better access to the park's breathtaking scenery and made the road itself more pleasing by extensively landscaping shoulders and using native materials in the construction of bridges and tunnels whenever possible. Due to this naturalistic, or "rustic," approach, and because it passes through an incredibly diverse scenery, Newfound Gap Road is today considered one of the finest park roads in the country.

While road boosters such as Willis Davis saw the national park as a means of developing motorways through the Smokies, during the 1930s others advocating the preservation of wilderness also began influencing the motor road system of Great Smoky Mountains National Park. The notion that American wilderness was worth protecting was in a very real sense born in the Smokies, when Robert Marshall, Benton McKaye, and a local lawyer named Harvey Broome met in Knoxville, Tennessee to organize opposition to the construction of several proposed highways along the Appalachian divide. Not only did this campaign result in the founding of the Wilderness Society, one of America's most influential environmental organizations, but it also defeated plans for a "Skyway" along the crest of the Smokies. Passage of the Wilderness Act in 1964 likewise resulted in the indefinite postponement of the proposed Northshore Road along the southern boundary of the park between Bryson City and Fontana Dam.

Funding for road improvements throughout the national park system increased dramatically during Mission 66, a ten-year program aimed at expanding and rehabilitating the infrastructure of national parks. While the Park Service undertook several road projects in Great Smoky Mountains National Park, none compare to the work done on the Roaring Fork Motor Nature Trail. Due to past experiences with wilderness advocates, Park Service designers kept the road narrow and one-way for most of its length, and aligned so as to follow the twists and turns of the natural terrain. The result is a ten-mile road through a wild landscape unlike other motorways in the park and Mission 66 projects nationwide.

This project is part of the Historic American Engineering Record, administered by the Historic American Buildings Survey/Historic American Engineering Record Division (HABS/HAER) of the National Park Service, U.S. Department of the Interior (E. Blaine Cliver, Chief). and was cosponsored during the summer of 1996 by HAER and GRSMNP (Karen Wade, Superintendent). The project was funded by the Federal Lands Highway Program (Thomas Edick, Administrator) through the National Park Service Park Roads and Parkways Program (Mark Hartsoe, Manager).

NATIONAL PARK SERVICE
Department of the Interior

The field work, measured drawings, historical reports, and photographs were prepared under the direction of Program Manager Todd A. Croteau and Richard Quin, Program Historian. The recording team consisted of Edward Lupyak, Field Supervisor; Landscape Architects Matthew Regnier, Karen Young and Dorota Sikora (ICOMOS intern, Poland). Historical reports were prepared by Project Historians Michael Kelleher and Neil Maher. Formal large-format photography was done by David Haas.

Note: Drawing concept taken from Great Smoky Mountains National Park 50th anniversary poster and historic photographs.

DELINEATED BY Edward Lupyak, 1996
NATIONAL PARK SERVICE ROADS & BRIDGES RECORDING PROGRAM
NATIONAL PARK SERVICE
UNITED STATES DEPARTMENT OF THE INTERIOR
GREAT SMOKY MOUNTAINS NATIONAL PARK ROADS & BRIDGES
SEVIER COUNTY
GATLINBURG VICINITY
TENNESSEE
SHEET 1 of 11
HISTORIC AMERICAN ENGINEERING RECORD
TN-35

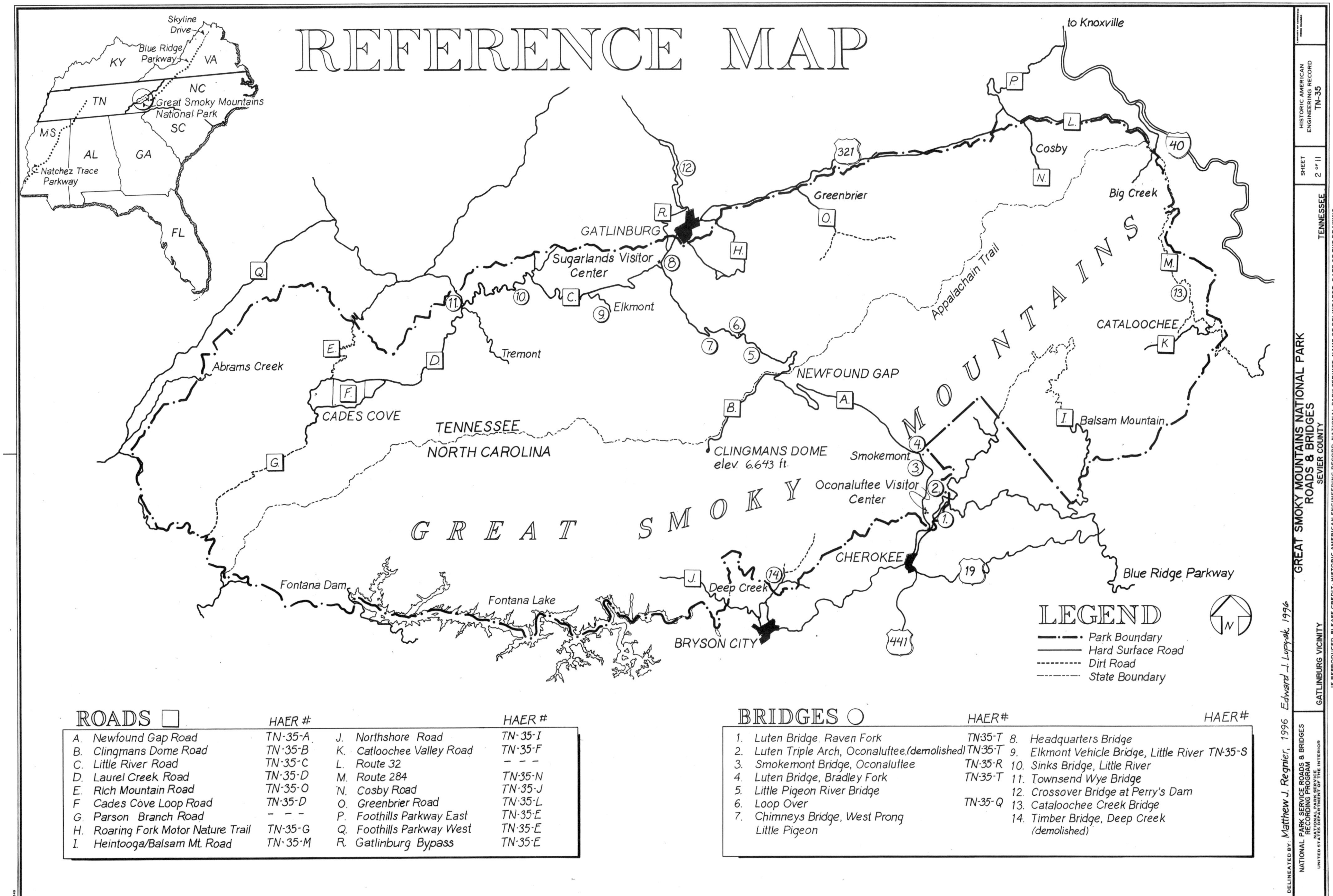
REFERENCE MAP
Skyline Drive
Blue Ridge Parkway
KY
VA
NC
TN
Great Smoky Mountains National Park
MS
AL
GA
SC
FL
Natchez Trace Parkway
to Knoxville
P.
L.
321
40
Cosby
N.
Greenbrier
O.
Big Creek
12
R.
GATLINBURG
H.
8
Sugarlands Visitor Center
C.
9 Elkmont
10
11
Q.
E.
D.
Tremont
Abrams Creek
F.
CADES COVE
6
7
5
NEWFOUND GAP
Appalachian Trail
M.
13
CATALOOCHEE
K.
GREAT SMOKY MOUNTAINS
A.
B.
TENNESSEE
NORTH CAROLINA
G.
CLINGMANS DOME elev. 6,643 ft.
Smokemont
4
3
Oconaluftee Visitor Center
2
1
I. Balsam Mountain
CHEROKEE
19
Blue Ridge Parkway
J.
Deep Creek
14
Fontana Dam
Fontana Lake
BRYSON CITY
441
LEGEND
Park Boundary
Hard Surface Road
Dirt Road
State Boundary
N
ROADS □
HAER #
A. Newfound Gap Road TN-35-A
B. Clingmans Dome Road TN-35-B
C. Little River Road TN-35-C
D. Laurel Creek Road TN-35-D
E. Rich Mountain Road TN-35-O
F. Cades Cove Loop Road TN-35-D
G. Parson Branch Road - - -
H. Roaring Fork Motor Nature Trail TN-35-G
I. Heintooga/Balsam Mt. Road TN-35-M
J. Northshore Road TN-35-I
K. Catloochee Valley Road TN-35-F
L. Route 32 - - -
M. Route 284 TN-35-N
N. Cosby Road TN-35-J
O. Greenbrier Road TN-35-L
P. Foothills Parkway East TN-35-E
Q. Foothills Parkway West TN-35-E
R. Gatlinburg Bypass TN-35-E
BRIDGES ○
HAER #
1. Luten Bridge, Raven Fork TN-35-T
2. Luten Triple Arch, Oconaluftee (demolished) TN-35-T
3. Smokemont Bridge, Oconaluftee TN-35-R
4. Luten Bridge, Bradley Fork TN-35-T
5. Little Pigeon River Bridge
6. Loop Over TN-35-Q
7. Chimneys Bridge, West Prong Little Pigeon
8. Headquarters Bridge
9. Elkmont Vehicle Bridge, Little River TN-35-S
10. Sinks Bridge, Little River
11. Townsend Wye Bridge
12. Crossover Bridge at Perry's Dam
13. Cataloochee Creek Bridge
14. Timber Bridge, Deep Creek (demolished)
DELINEATED BY: Matthew J. Regnier, 1996 Edward J. Lupyak 1996
NATIONAL PARK SERVICE ROADS & BRIDGES RECORDING PROGRAM
NATIONAL PARK SERVICE
UNITED STATES DEPARTMENT OF THE INTERIOR
GREAT SMOKY MOUNTAINS NATIONAL PARK ROADS & BRIDGES
SEVIER COUNTY
TENNESSEE
GATLINBURG VICINITY
SHEET 2 of 11
HISTORIC AMERICAN ENGINEERING RECORD TN-35
IF REPRODUCED, PLEASE CREDIT: HISTORIC AMERICAN ENGINEERING RECORD, NATIONAL PARK SERVICE, NAME OF DELINEATOR, DATE OF THE DRAWING

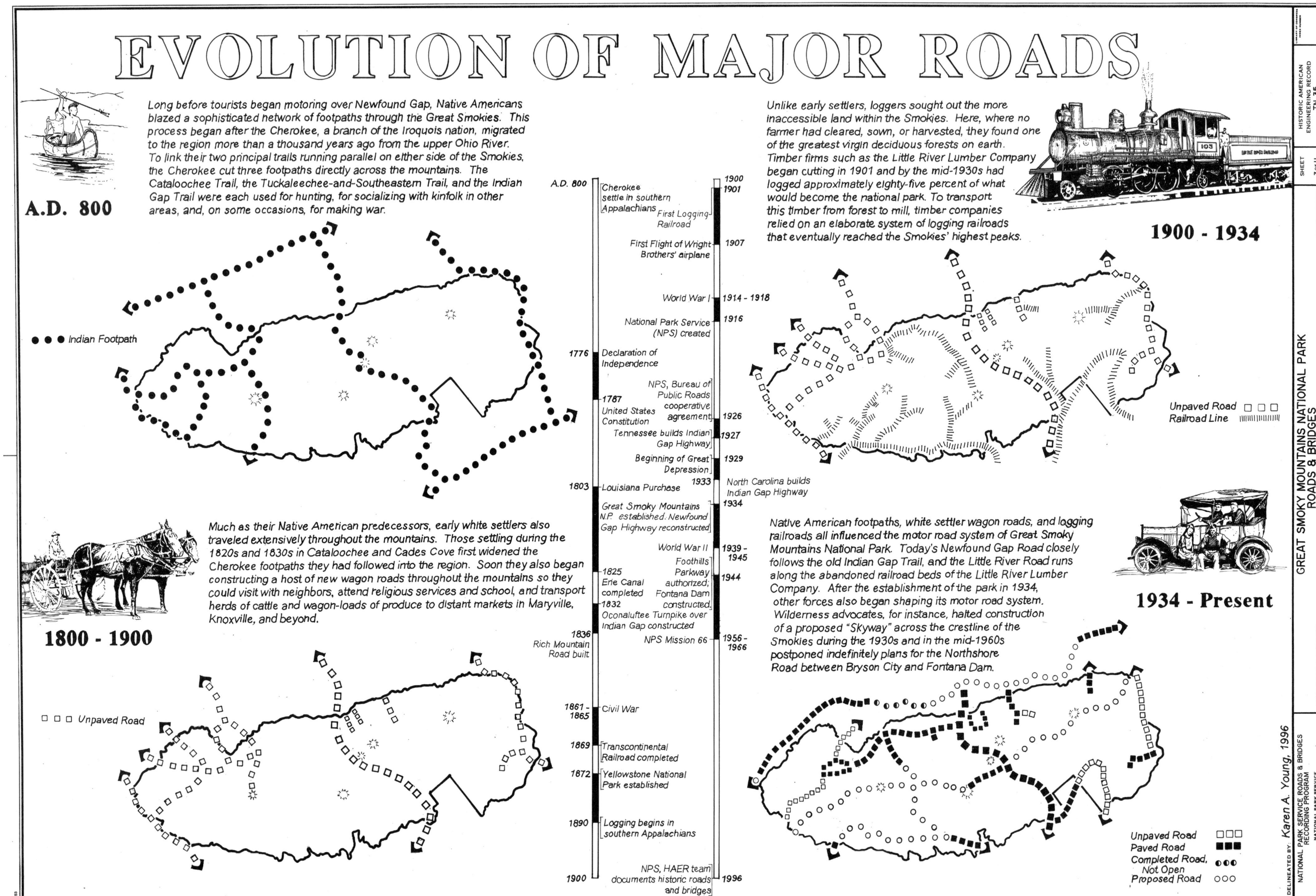
EVOLUTION OF MAJOR ROADS
A.D. 800
Long before tourists began motoring over Newfound Gap, Native Americans blazed a sophisticated network of footpaths through the Great Smokies. This process began after the Cherokee, a branch of the Iroquois nation, migrated to the region more than a thousand years ago from the upper Ohio River. To link their two principal trails running parallel on either side of the Smokies, the Cherokee cut three footpaths directly across the mountains. The Cataloochee Trail, the Tuckaleechee-and-Southeastern Trail, and the Indian Gap Trail were each used for hunting, for socializing with kinfolk in other areas, and, on some occasions, for making war.
● ● ● Indian Footpath
1800 - 1900
Much as their Native American predecessors, early white settlers also traveled extensively throughout the mountains. Those settling during the 1820s and 1830s in Cataloochee and Cades Cove first widened the Cherokee footpaths they had followed into the region. Soon they also began constructing a host of new wagon roads throughout the mountains so they could visit with neighbors, attend religious services and school, and transport herds of cattle and wagon-loads of produce to distant markets in Maryville, Knoxville, and beyond.
□ □ □ Unpaved Road
A.D. 800 Cherokee settle in southern Appalachians
1776 Declaration of Independence
1787 United States Constitution
1803 Louisiana Purchase
1825 Erie Canal completed
1832 Oconaluftee Turnpike over Indian Gap constructed
1836 Rich Mountain Road built
1861 - 1865 Civil War
1869 Transcontinental Railroad completed
1872 Yellowstone National Park established
1890 Logging begins in southern Appalachians
1900
1900
1901 First Logging Railroad
1907 First Flight of Wright Brothers' airplane
1914 - 1918 World War I
1916 National Park Service (NPS) created
1926 NPS, Bureau of Public Roads cooperative agreement
1927 Tennessee builds Indian Gap Highway
1929 Beginning of Great Depression
1933 North Carolina builds Indian Gap Highway
1934 Great Smoky Mountains N.P. established; Newfound Gap Highway reconstructed
1939 - 1945 World War II
1944 Foothills Parkway authorized; Fontana Dam constructed
1956 - 1966 NPS Mission 66
1996 NPS, HAER team documents historic roads and bridges
1900 - 1934
Unlike early settlers, loggers sought out the more inaccessible land within the Smokies. Here, where no farmer had cleared, sown, or harvested, they found one of the greatest virgin deciduous forests on earth. Timber firms such as the Little River Lumber Company began cutting in 1901 and by the mid-1930s had logged approximately eighty-five percent of what would become the national park. To transport this timber from forest to mill, timber companies relied on an elaborate system of logging railroads that eventually reached the Smokies' highest peaks.
Unpaved Road □ □ □
Railroad Line
1934 - Present
Native American footpaths, white settler wagon roads, and logging railroads all influenced the motor road system of Great Smoky Mountains National Park. Today's Newfound Gap Road closely follows the old Indian Gap Trail, and the Little River Road runs along the abandoned railroad beds of the Little River Lumber Company. After the establishment of the park in 1934, other forces also began shaping its motor road system. Wilderness advocates, for instance, halted construction of a proposed "Skyway" across the crestline of the Smokies during the 1930s and in the mid-1960s postponed indefinitely plans for the Northshore Road between Bryson City and Fontana Dam.
Unpaved Road □□□
Paved Road ■■■
Completed Road, Not Open ●●●
Proposed Road ○○○
DELINEATED BY: Karen A. Young, 1996
NATIONAL PARK SERVICE ROADS & BRIDGES RECORDING PROGRAM
NATIONAL PARK SERVICE
UNITED STATES DEPARTMENT OF THE INTERIOR
GREAT SMOKY MOUNTAINS NATIONAL PARK ROADS & BRIDGES
SEVIER COUNTY
GATLINBURG VICINITY
TENNESSEE
SHEET 3 of 11
HISTORIC AMERICAN ENGINEERING RECORD TN-35
IF REPRODUCED, PLEASE CREDIT: HISTORIC AMERICAN ENGINEERING RECORD, NATIONAL PARK SERVICE, NAME OF DELINEATOR, DATE OF THE DRAWING

DESIGN PRINCIPLES

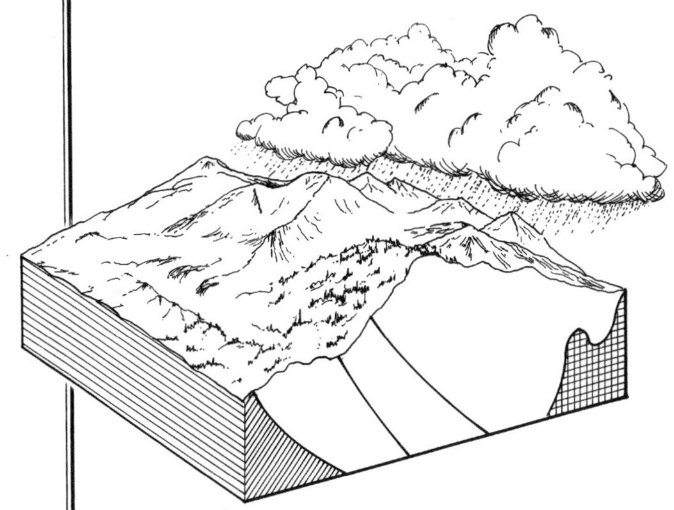

Geology and Road Construction

The geologic makeup of GSMNP is predominantly metamorphosed sedimentary rock, ranging from phyllites, schists, and quartzites, to slates, shales, sandstones, and metasiltstones. The natural instability of slates and metasiltstones caused numerous landslides during road construction, and stabilizing roadbeds built on these formations was crucial. In contrast, sandstone is stronger and erosion resistant, and excavating roadbeds through this rock required heavy machinery and took a great deal of time. However, this excavated rock provided excellent building material for stone walls and bridges.

Thousands of years of geologic change and erosion have shaped the Great Smoky Mountains, which are characterized by high mountain peaks, steep hillsides, deep river valleys, and fertile coves. This difficult terrain, and the underlying bedrock, presented numerous challenges for the designers of the roads in Great Smoky Mountains National Park (GSMNP). J. Ross Eakin, the park's first Superintendent, wrote that in comparison to the mountains of the west, where the National Park Service had a great deal of experience building roads, the geologically older mountains of the east are "cut up by many more gullies," and "to secure good [road] alignment along our mountain sides will require tremendous cuts and fills," thereby creating "a terrible scar" on the roadsides.

Attempting to avoid as much scarring as possible, road designers used tunnels, bridges, and retaining and revetment walls. Tunnels cut through ridges where, if not used, the construction of the road would create an unsightly and unstable road scar, and would present an unsafe turn radii for motorists (see fig. 2 & fig. 3). Bridges spanned the numerous streams and rivers the roads had to cross, and created safer road alignments by allowing for consistent grades. Where the routes followed streams closely, revetment walls were constructed to hold back fill and prevent the streams from undercutting the roadbeds. Techniques such as rounding cuts and revegetating roadsides (see fig. 1) controlled erosion and reduced visible road scars. Using all of these methods, designers were able to create some of the most pleasurable and scenic roads in the national park system.

Fig. 1

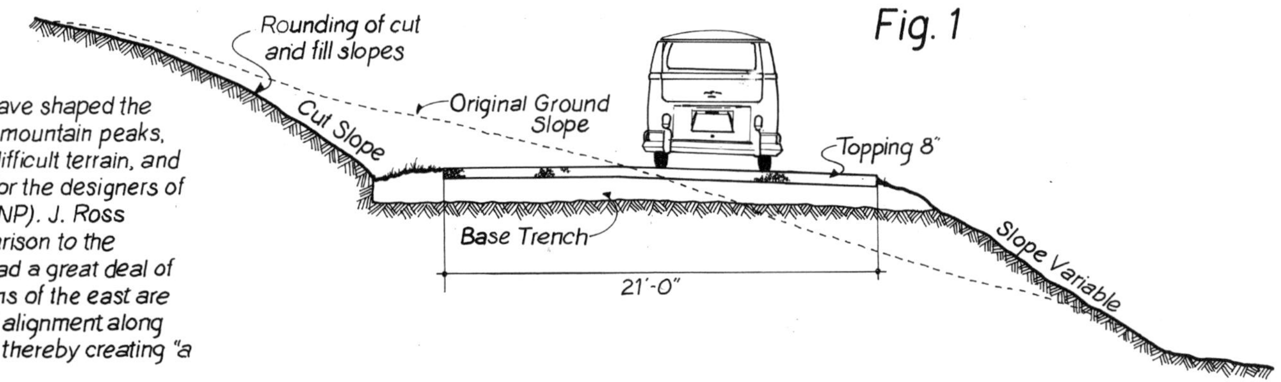

Fig. 2

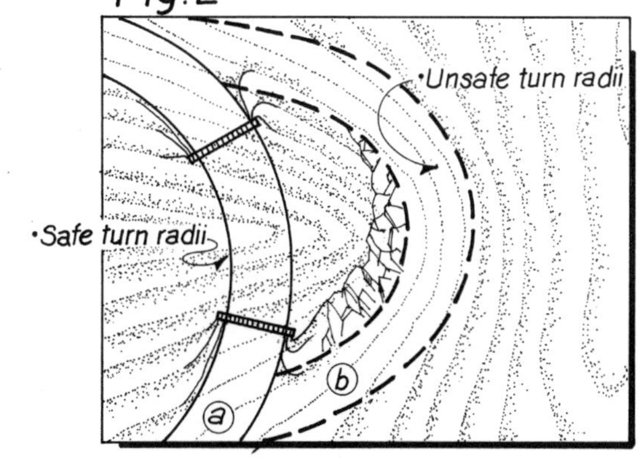

Fig. 3

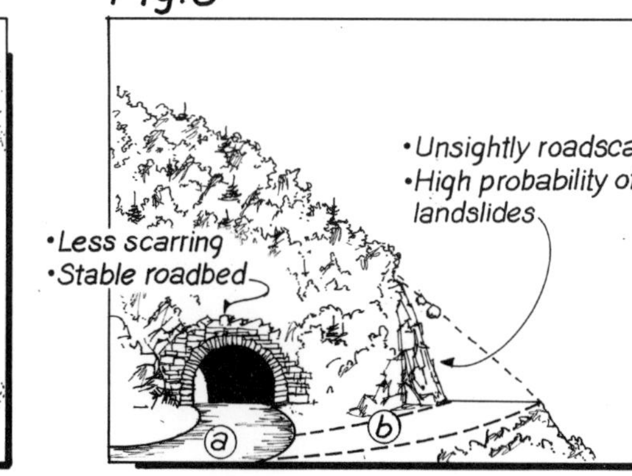

EARLY ROADS

Before the park was established in 1934, white settlers built rough wagon roads along river valleys, and through gaps in the mountains. Lacking heavy equipment, these early road builders used hand tools and man power. Rocks which were too large to move were either blown apart with blasting powder, or cracked through a combination of heating by fire followed by cooling with cold water. Oxen and other beasts of burden hauled away excavated earth and rock. These simple construction techniques made digging large road cuts, hauling material up and down steep slopes, and building bridges over streams difficult. As a result, the settlers built winding roads along the contours of the land.

1930S & 1940S

When the National Park Service created GSMNP in 1934, the existing roads had to be improved and new roads were needed in order to make the park accessible to visitors. New engineering concepts, improved technology, and better tools and equipment made the construction and improvement of roads in this period easier than it was for the early settlers in the Smokies. Men and animals still did much of the work, but new earth-moving machines such as steam shovels and diesel trucks with diesel engines made road construction more efficient. Concrete, replacing wood and stone, made bridges stronger and easier to build. These factors allowed road designers to go beyond building winding roads along the contours of the land, and use bridges and tunnels to create safer and more pleasurable road alignments for motorists in the park.

1950S & 1960S

With the increasing number of visitors to the park in the years following World War II, roads were once again improved and expanded by the federal government through a program to update park infrastructure, known as Mission 66. In addition, the faster vehicles used by park motorists required that road alignments be improved. During this period of road construction, designers were able to work with the new technology of the postwar era. In place of steam shovels, gigantic earth-moving machines dug out cuts for roadbeds, placed fill, and removed excavated material at faster rates. Mobile cranes were now available to lift huge steel beams into place in the construction of bridges. Combined with the advanced engineering techniques of the period, this improved technology let road designers overcome the constricting topography of the Great Smoky Mountains.

DELINEATED BY: Edward Lupyak, 1996.
NATIONAL PARK SERVICE ROADS & BRIDGES RECORDING PROGRAM
NATIONAL PARK SERVICE
UNITED STATES DEPARTMENT OF THE INTERIOR

GREAT SMOKY MOUNTAINS NATIONAL PARK
ROADS & BRIDGES
GATLINBURG VICINITY SEVIER COUNTY TENNESSEE

SHEET 5 OF 11

HISTORIC AMERICAN ENGINEERING RECORD
TN-35

IF REPRODUCED, PLEASE CREDIT: HISTORIC AMERICAN ENGINEERING RECORD, NATIONAL PARK SERVICE, NAME OF DELINEATOR, DATE OF THE DRAWING

MOTORIST EXPERIENCE

The roads in Great Smoky Mountains National Park offer a wide range of experiences for the motorist. Scenic overlooks and planned vistas were designed throughout the park to direct the visitor's eye. The roads travel over a variety of topography and provide differing views ranging from; gentle farmlands, deep river valleys, climbing mountain roads, steep gravel switchbacks, sweeping ridgetops, historical and natural points of interest.

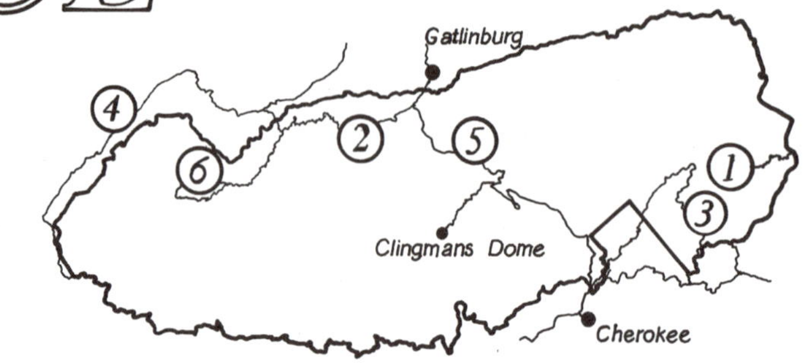

1.

Points of Interest - Several roads are designed to showcase specific natural and historical points of interest. Roads such as Cataloochee provide visitors with access to historic structures and intimate natural scenery from their vehicles.

2.

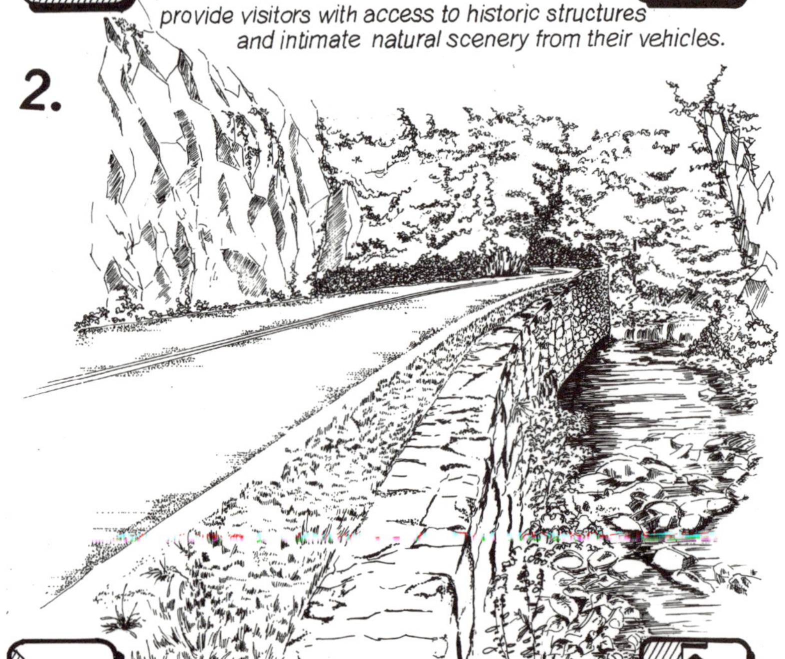

Deep River Valleys - The Little River and Laurel Creek roads twist and turn between narrow gorge walls and offer only limited views of the scenic river and walls above. The road, carried by a revetment wall, is dimly lit as the tree canopy is dense and the gorge walls are tall.

3.

Steep Switchbacks - Rich Mountain Road and other similar gravel routes maintain the character of historical pioneer travel through the Smoky Mountains. The single-lane roads follow the topography of the land up, over and around ridges in a series of switchbacks.

4.

Sweeping Ridgetops - Roads such as the Foothills Parkway follow ridges which offer gentle curving alignments and wider roadways. Broad views of distant mountains and valleys are presented through areas cleared of trees.

5.

Mountain Climbs - The Newfound Gap Road clings to the side of mountains through most of its ascent with dramatic views of nearby peaks, ridges and valleys. The roadway is benched into the steep slopes using retaining walls and many pullouts are provided to enjoy the vistas.

6.

Gentle Farmlands - Areas such as Cades Cove and Cataloochee present the historical farmsteads of pioneers and the open landscapes which they cleared. The Cades Cove Loop meanders along the edge of pastureland and the forest.

DELINEATED BY: DOROTA SIKORA, 1996

NATIONAL PARK SERVICE ROADS & BRIDGES RECORDING PROGRAM
NATIONAL PARK SERVICE
UNITED STATES DEPARTMENT OF THE INTERIOR

GREAT SMOKY MOUNTAINS NATIONAL PARK ROADS & BRIDGES
SEVIER COUNTY
GATLINBURG VICINITY
TENNESSEE

SHEET 6 OF 11

HISTORIC AMERICAN ENGINEERING RECORD
TN-35

STONEWORK

The roads of Great Smoky Mountains National Park (GSMNP) were carefully designed to harmonize with their surroundings. Using design standards based upon the principles of naturalism and scenery preservation, the roads and their accompanying features were constructed so as to have as little impact on the natural features and topography of the park as possible. The stonemasonry used in the construction of road structures such as bridges, culverts, and retaining walls were essential parts of this concept.

The stonework in national parks was built to be naturalistic, or "rustic", in appearance, which required use of stone of the same color and characteristics as the surrounding rock, and avoidance of right angles and straight lines in the cut and arrangement of the stone. Stonework in GSMNP did not adhere to the "rustic" style as closely as in other national parks, as there were more rectangular stones laid horizontally and mortar joints that roughly followed horizontal and vertical lines. These factors, combined with the use of ornamental details such as buttressing piers and string courses in bridge construction gave a more formal appearance to the road structures in the park. Although not as rustic in appearance as the stonework in other parks, the bridges, retaining walls and other masonrywork in GSMNP blend in with their surroundings extremely well.

MASONRY WALLS

Construction of retaining walls was necessary to hold the fill on which roads were constructed. In order to compensate for the stress on the walls created by the pressure of the fill, retaining walls were built according to certain engineering specifications for height, the thickness of the base, and the amount of masonry material needed for each linear foot of wall (see chart). The footing of the retaining walls differed depending on whether they were built on rock or earth, and whether or not the ground underneath the wall was sloped.

HEIGHT Feet	BASE Feet	Masonry per Lin. Ft. of Wall Cubic Yards
3	2'-9"	0.27
4	3'-0"	0.37
5	3'-4 1/2"	0.50
6	3'-11"	0.64
7	4'-5 1/2"	0.79
8	5'-0"	0.96
9	5'-7 1/2"	1.16
10	6'-3"	1.38
11	6'-10 1/2"	1.62
12	7'-6"	1.93
13	8'-1 1/2"	2.14
14	8'-9"	2.49
15	9'-4 1/2"	2.83
16	10'-0"	3.19
17	10'-7 1/2"	3.57
18	11'-3"	3.97
19	11'-10 1/2"	4.40
20	12'-6"	4.85

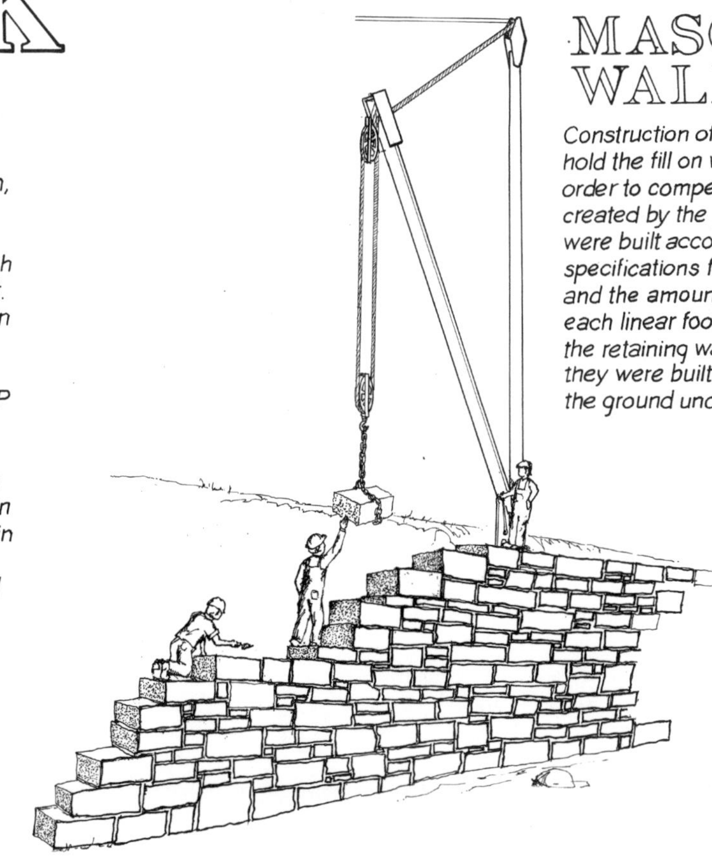

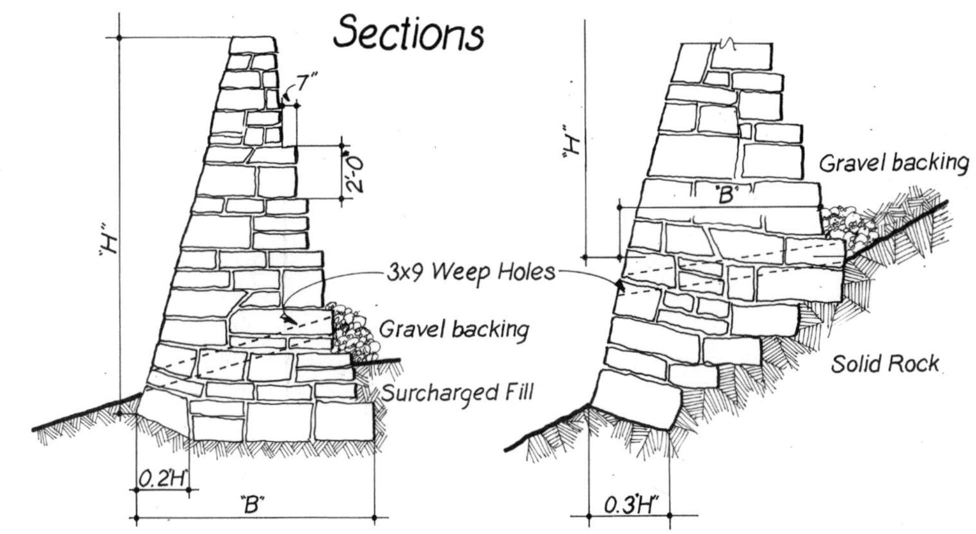

Revetment and retaining walls were built where fill was needed to construct roadbeds on steep mountain slopes and in river valleys. The large stones were hauled into place by cranes or winches, where they were dry or wet laid by masons.

TUNNEL PORTALS

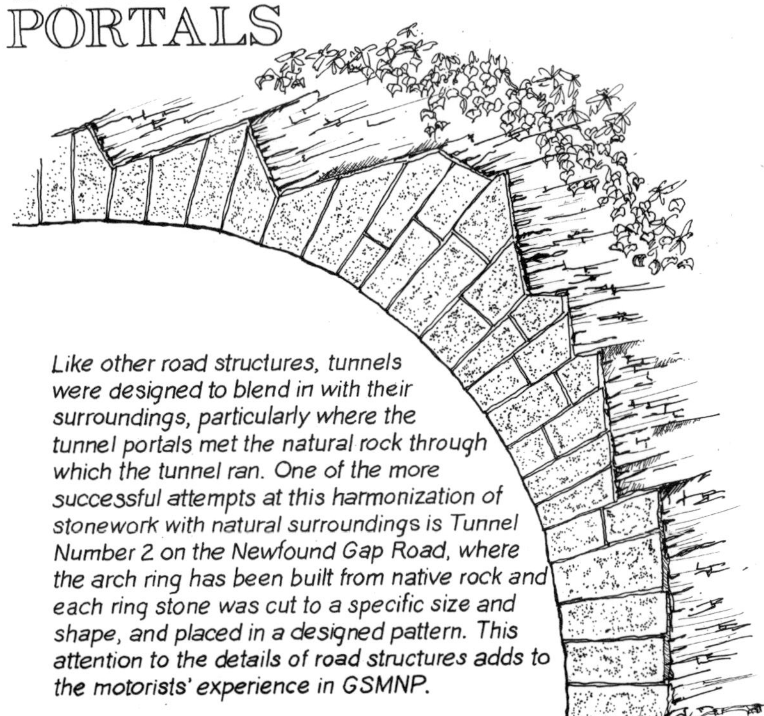

Like other road structures, tunnels were designed to blend in with their surroundings, particularly where the tunnel portals met the natural rock through which the tunnel ran. One of the more successful attempts at this harmonization of stonework with natural surroundings is Tunnel Number 2 on the Newfound Gap Road, where the arch ring has been built from native rock and each ring stone was cut to a specific size and shape, and placed in a designed pattern. This attention to the details of road structures adds to the motorists' experience in GSMNP.

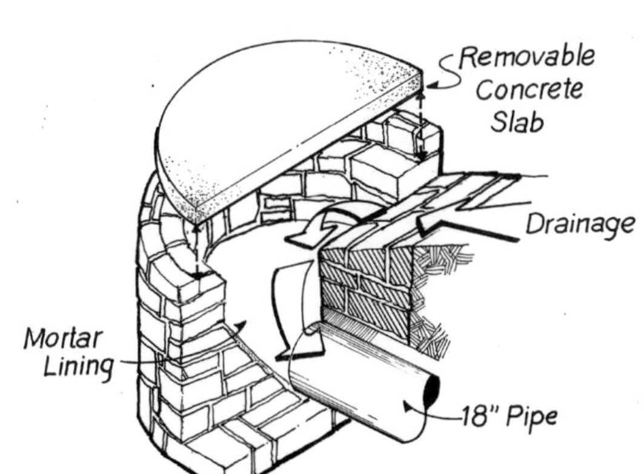

Drop culverts were constructed of stone, not for appearance, but for the durable property of this material.

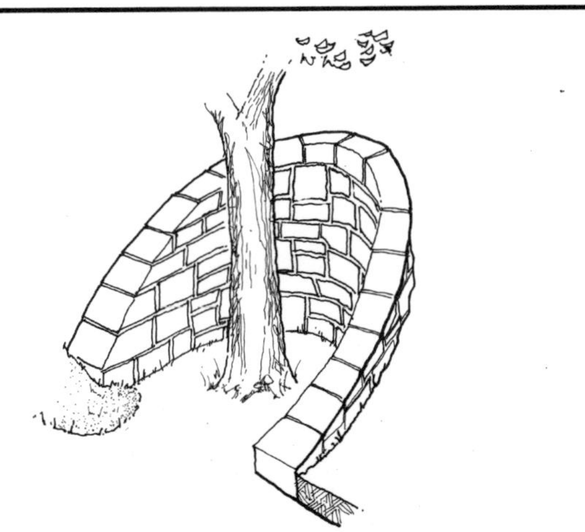

In order to save some of the larger trees along newly constructed roads, stone wells were built to hold back fill which would otherwise overwhelm them.

Curbs and sidewalks at bridges and pullouts prevented automobiles from hitting the stone guardwalls.

After years of repaving, the sidewalks and curbs have been covered with asphalt, leaving the guardwalls susceptible to damage.

DELINEATED BY: Edward Lupyak, 1996

NATIONAL PARK SERVICE ROADS & BRIDGES RECORDING PROGRAM
NATIONAL PARK SERVICE
UNITED STATES DEPARTMENT OF THE INTERIOR

GREAT SMOKY MOUNTAINS NATIONAL PARK
ROADS & BRIDGES
GATLINBURG VICINITY SEVIER COUNTY TENNESSEE

SHEET 7 OF 11

HISTORIC AMERICAN ENGINEERING RECORD TN-35

LIBRARY OF CONGRESS INDEX NUMBER

ROADSIDE BARRIERS

Roadside barriers in Great Smoky Mountains National Park provide for the safety of both motorists and park resources. For the motorist, the barriers define the edge of the road and may prevent a car from leaving the roadway. For park resources, the barriers may prevent a motorist from pulling to the side of the road and damaging a natural or historic area. GSMNP uses standard park architectural plans for guardwalls and guardrails, which require that the barriers be constructed in the rustic style. The most common roadside barriers are a series of 18" stonemasonry parapet walls carefully designed by the National Park Service. Within the typical proportions used, diversity and variation were encouraged to meet design standards ensuring irregularity, informality, and harmonization with the surrounding landscape. To break the monotony of long linear expanses of guardwalls, various crenulation patterns were introduced. Functionally, the guardwalls are freestanding or extended from retaining walls, culverts, and bridges. Other types of barriers include large boulders placed on the shoulder to deter motorists from traveling off the pavement. Guardrails are typically horizontal wood timbers attached to vertical wood posts, more recently with steel plates. An earlier type, now mostly replaced, consisting of roughly hewn posts, was typically used in forested landscapes.

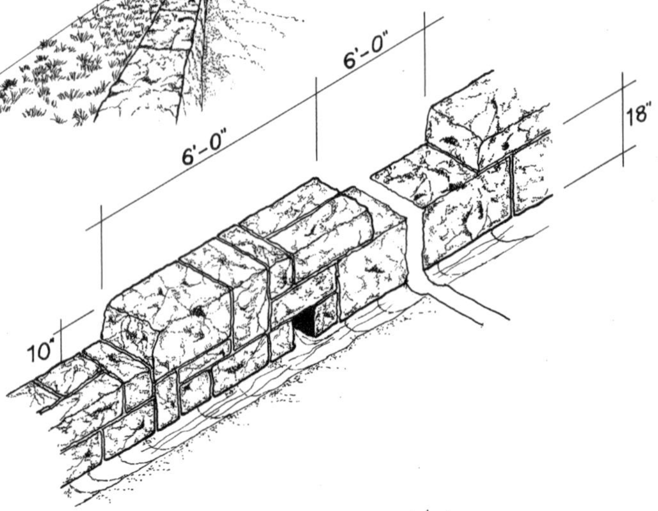

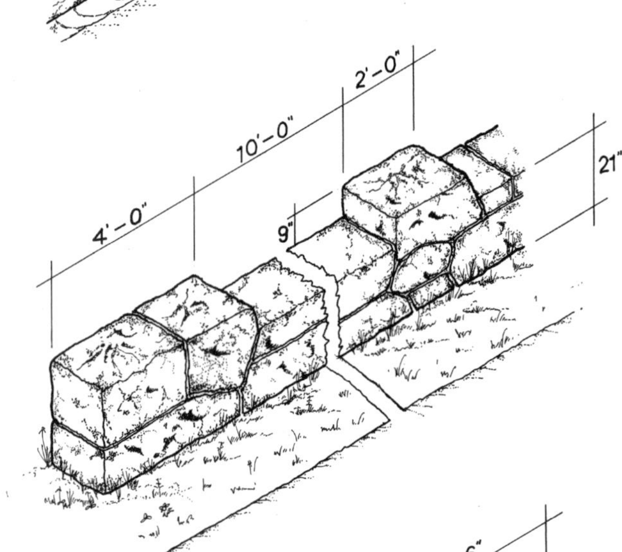

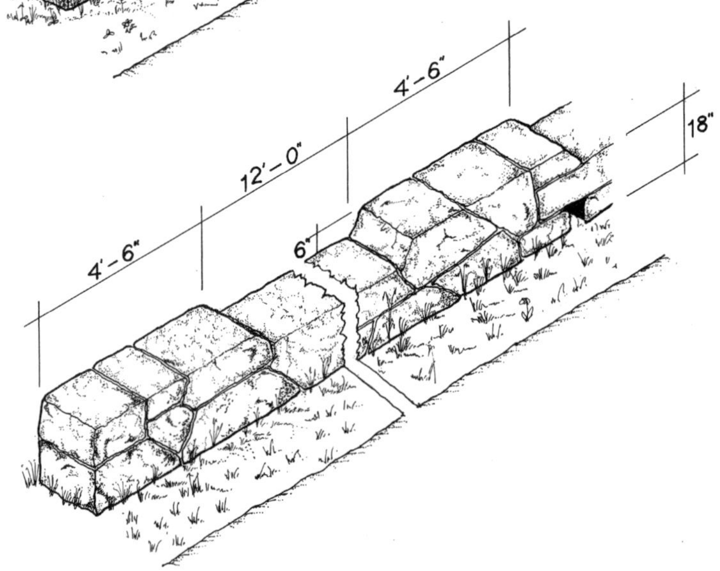

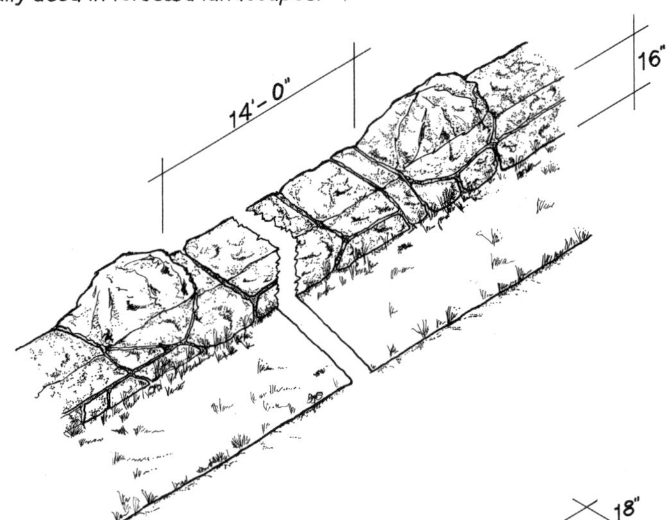

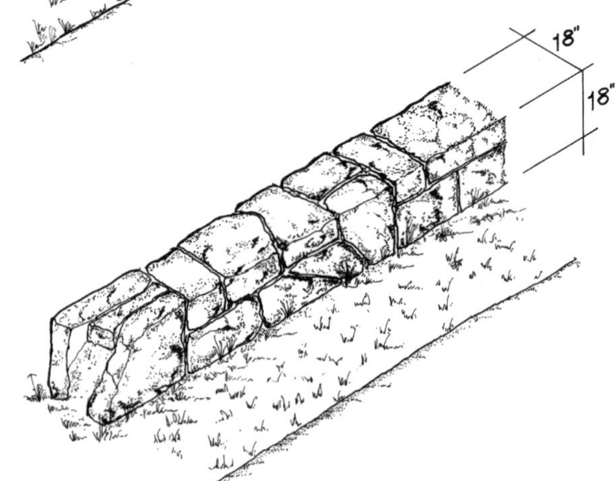

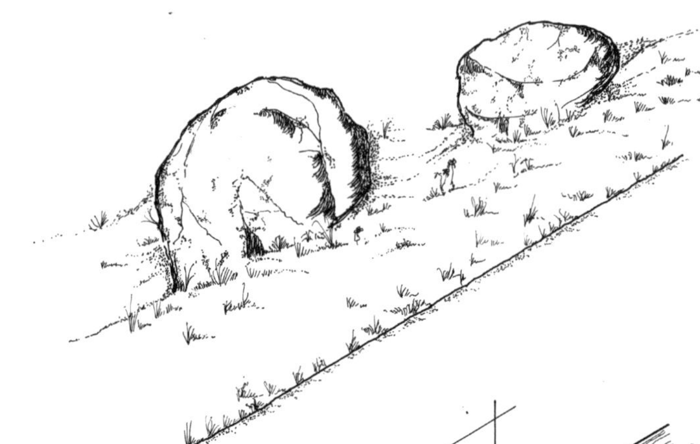

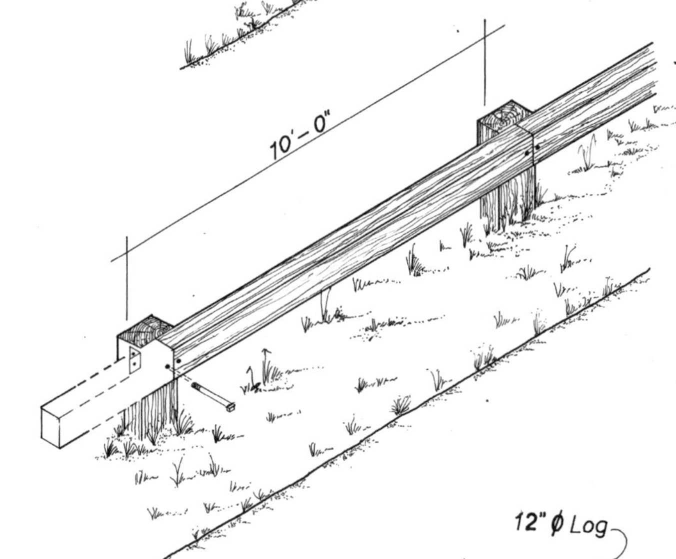

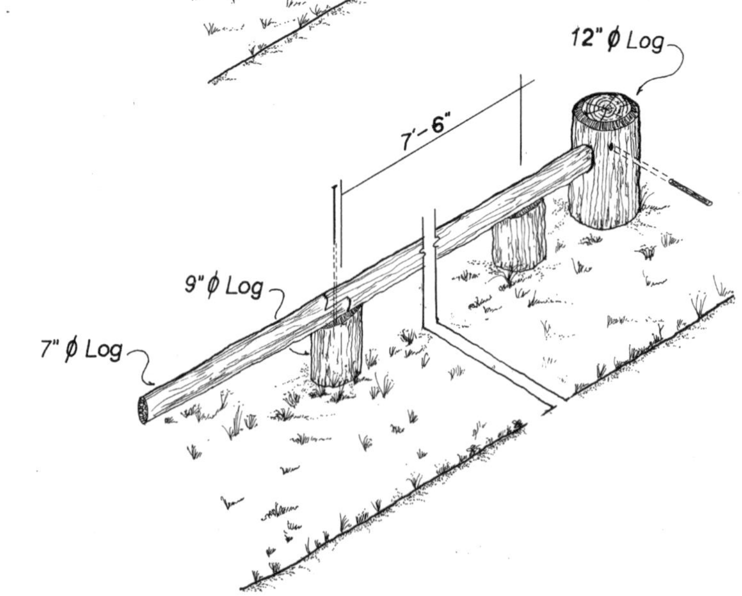

Scale: 1/2"=1'-0"

Note: Drawing based on Historic Stonemasonry of Great Smoky Mountains National Park GPS Survey 1993, and field photographs.

DELINEATED BY: Matthew J. Regnier, 1996

NATIONAL PARK SERVICE ROADS & BRIDGES RECORDING PROGRAM
NATIONAL PARK SERVICE
UNITED STATES DEPARTMENT OF THE INTERIOR

GREAT SMOKY MOUNTAINS NATIONAL PARK
ROADS & BRIDGES
SEVIER COUNTY
GATLINBURG VICINITY · TENNESSEE

SHEET 8 OF 11

HISTORIC AMERICAN ENGINEERING RECORD
TN-35

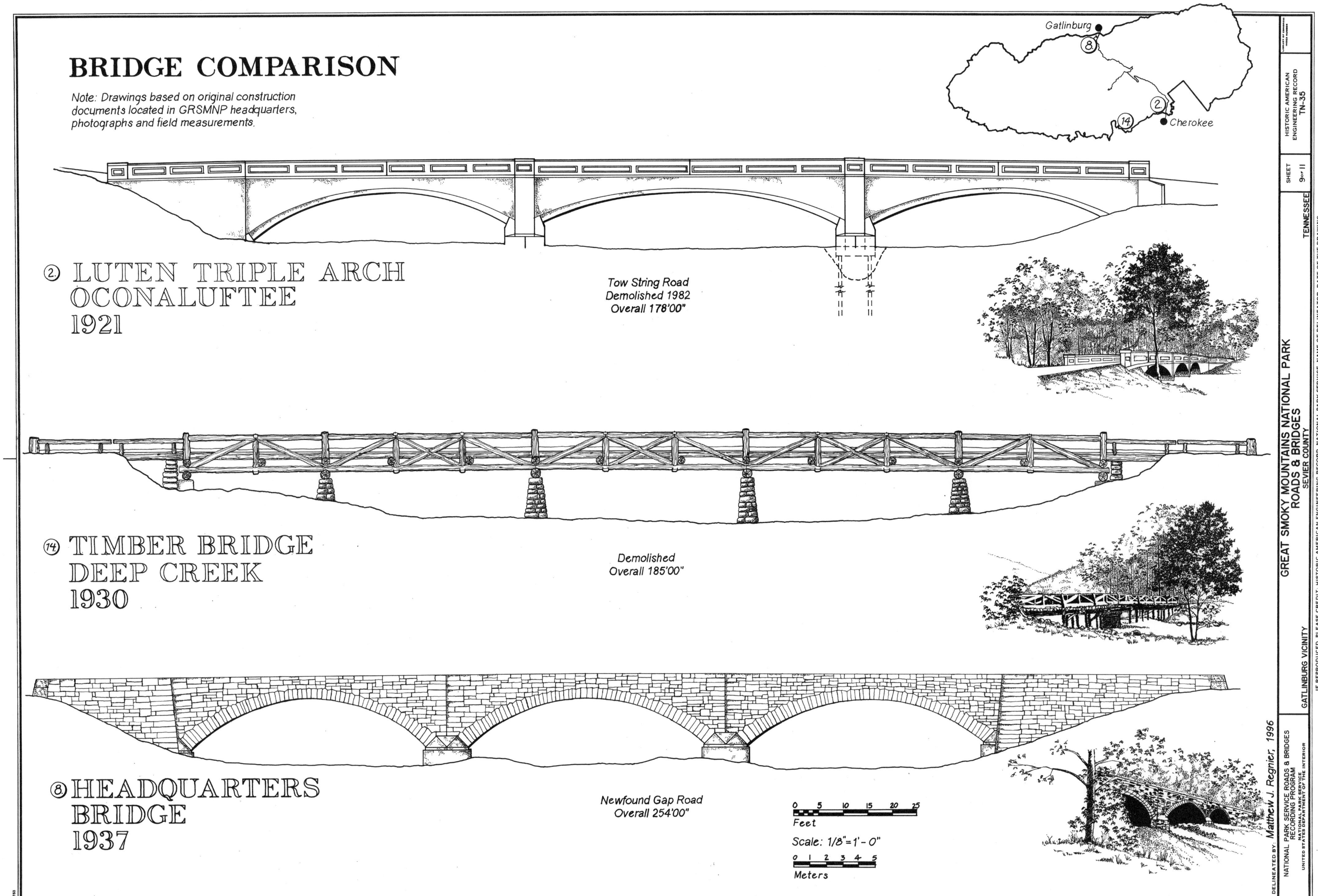
BRIDGE COMPARISON
Note: Drawings based on original construction documents located in GRSMNP headquarters, photographs and field measurements.
Gatlinburg
8
2
14
Cherokee
② LUTEN TRIPLE ARCH
OCONALUFTEE
1921
Tow String Road
Demolished 1982
Overall 178'00"
⑭ TIMBER BRIDGE
DEEP CREEK
1930
Demolished
Overall 185'00"
⑧ HEADQUARTERS
BRIDGE
1937
Newfound Gap Road
Overall 254'00"
0 5 10 15 20 25
Feet
Scale: 1/8"=1'-0"
0 1 2 3 4 5
Meters
DELINEATED BY: Matthew J. Regnier, 1996
NATIONAL PARK SERVICE ROADS & BRIDGES RECORDING PROGRAM
GREAT SMOKY MOUNTAINS NATIONAL PARK
ROADS & BRIDGES
GATLINBURG VICINITY
SEVIER COUNTY
TENNESSEE
SHEET 9 OF 11
HISTORIC AMERICAN ENGINEERING RECORD
TN-35
IF REPRODUCED, PLEASE CREDIT: HISTORIC AMERICAN ENGINEERING RECORD, NATIONAL PARK SERVICE, NAME OF DELINEATOR, DATE OF THE DRAWING

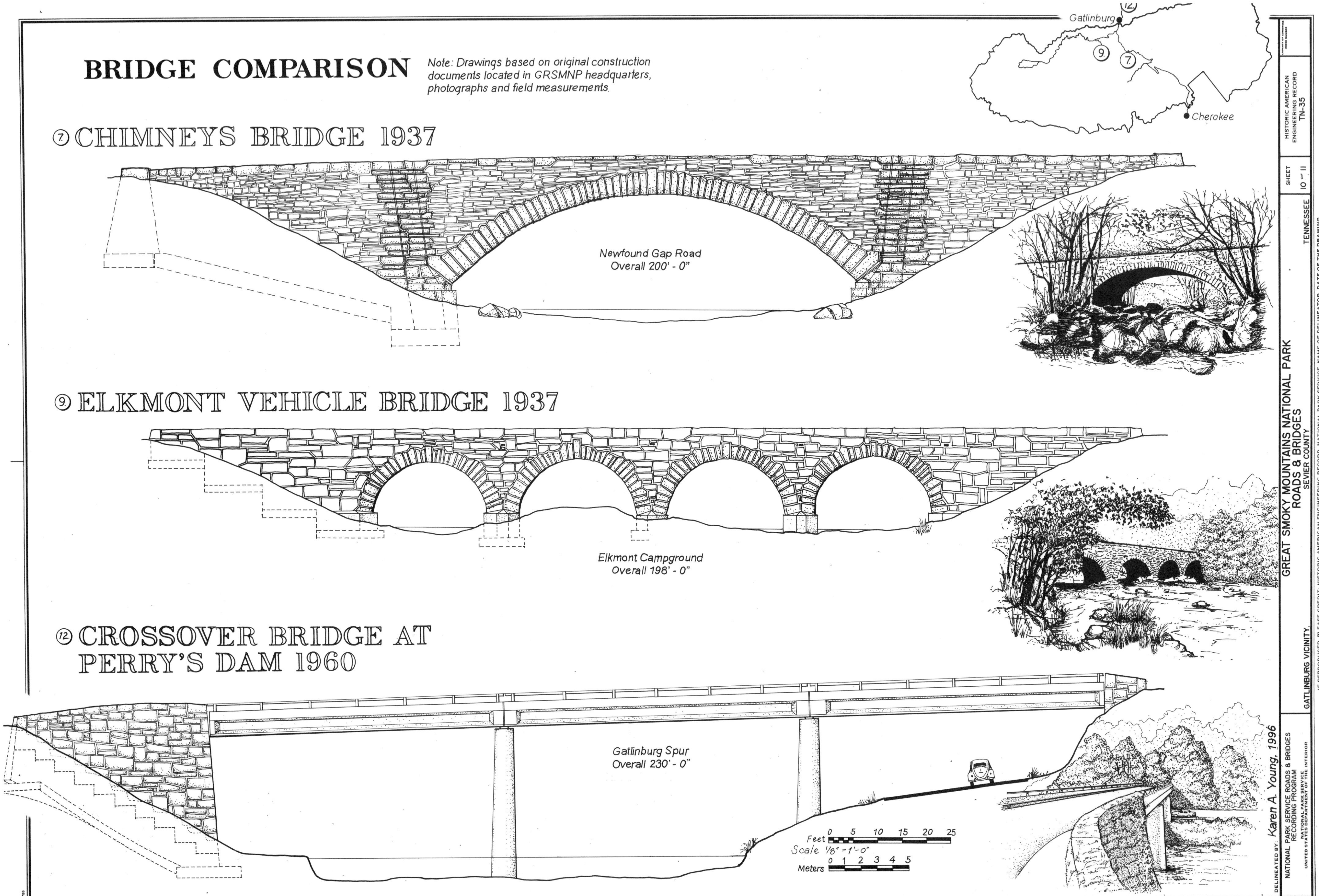

BRIDGE COMPARISON
Note: Drawings based on original construction documents located in GRSMNP headquarters, photographs and field measurements.
Gatlinburg
12
9
7
Cherokee
7 CHIMNEYS BRIDGE 1937
Newfound Gap Road
Overall 200' - 0"
9 ELKMONT VEHICLE BRIDGE 1937
Elkmont Campground
Overall 198' - 0"
12 CROSSOVER BRIDGE AT PERRY'S DAM 1960
Gatlinburg Spur
Overall 230' - 0"
Feet 0 5 10 15 20 25
Scale 1/8" = 1'-0"
Meters 0 1 2 3 4 5
DELINEATED BY: Karen A. Young, 1996
NATIONAL PARK SERVICE ROADS & BRIDGES RECORDING PROGRAM
NATIONAL PARK SERVICE
UNITED STATES DEPARTMENT OF THE INTERIOR
GREAT SMOKY MOUNTAINS NATIONAL PARK
ROADS & BRIDGES
SEVIER COUNTY
GATLINBURG VICINITY.
TENNESSEE
SHEET
10 OF 11
HISTORIC AMERICAN ENGINEERING RECORD
TN-35
IF REPRODUCED, PLEASE CREDIT: HISTORIC AMERICAN ENGINEERING RECORD, NATIONAL PARK SERVICE, NAME OF DELINEATOR, DATE OF THE DRAWING

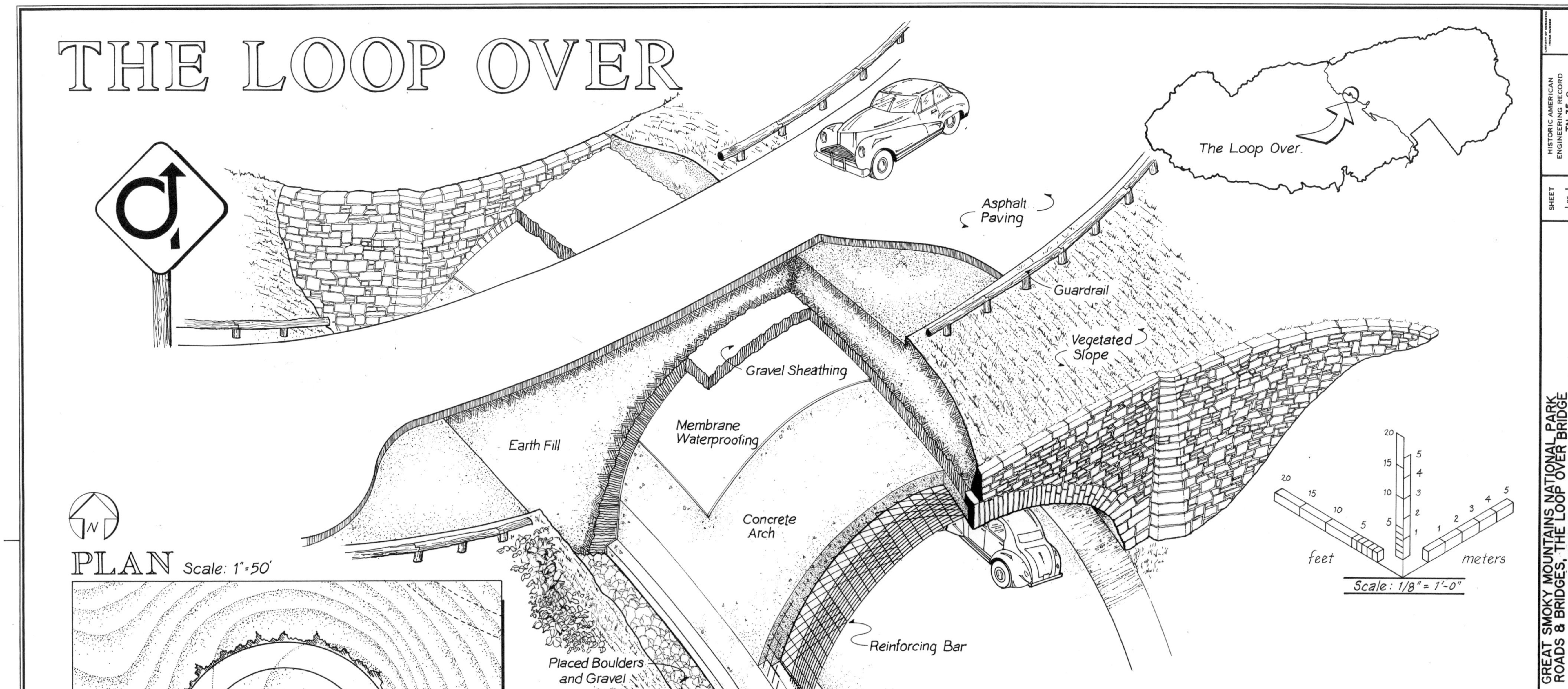

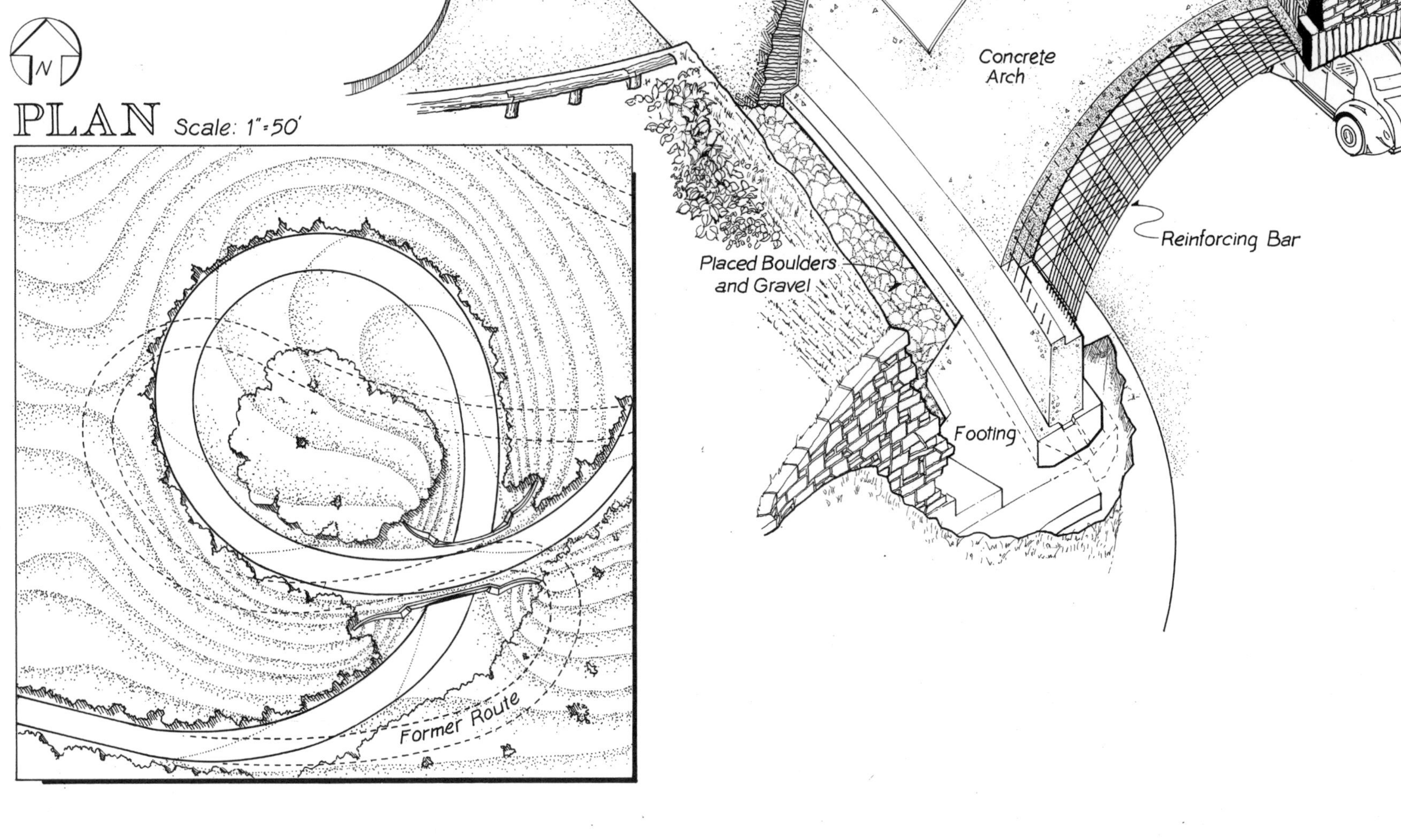

Prior to the establishment of Great Smoky Mountains National Park, the State of Tennessee built the Indian Gap Highway from Gatlinburg to Newfound Gap. This road did not meet federal standards both in terms of road design principles regarding grade and alignment, as well as scenic qualities important to a road through a national park. As a result, between 1933 and 1938, the NPS and U.S. Bureau of Public Roads rebuilt the State road by widening the right-of-way, reducing the grades, removing dangerous switchbacks, and revegetating construction scars.

To replace two tight switchbacks designers incorporated a loopover. The structure carries the road over and under itself allowing the new Newfound Gap Road to continue at a safe grade and alignment. The resulting road forms a helix, or corkscrew-like shape. The bridge is a reinforced concrete arch with stone facing walls. Large amounts of earth fill were placed for the roadbed in order to provide superelevation throughout the curve for safer travel

Following the completion of the Loopover in 1935, Great Smoky Mountains National Park Superintendent Ross Eakin remarked that the unique design of this feature of the Newfound Gap Road "attracted almost nation-wide attention." The Loopover had become one of the major attractions for motorists in the park, appearing as a centerpiece on postcards and souvenirs.

ELKMONT VEHICLE BRIDGE

CONSTRUCTION PEEL-AWAY

Construction of the Elkmont Vehicle Bridge was begun in June 1936 by the Civilian Conservation Corp (C.C.C.). The bridge is comprised of four multi-plate corrugated steel arches and stone masonry arch rings and spandrel walls. Concrete piers were constructed first to provide foundations for the arches. The arches were then assembled by bolting the steel plates together. After completing the steel arches, the spandrel walls were laid. When the walls reached a suitable height the arches were covered with fill which would support the roadbed. The roadbed was originally a crushed stone surface and was later paved with asphalt.

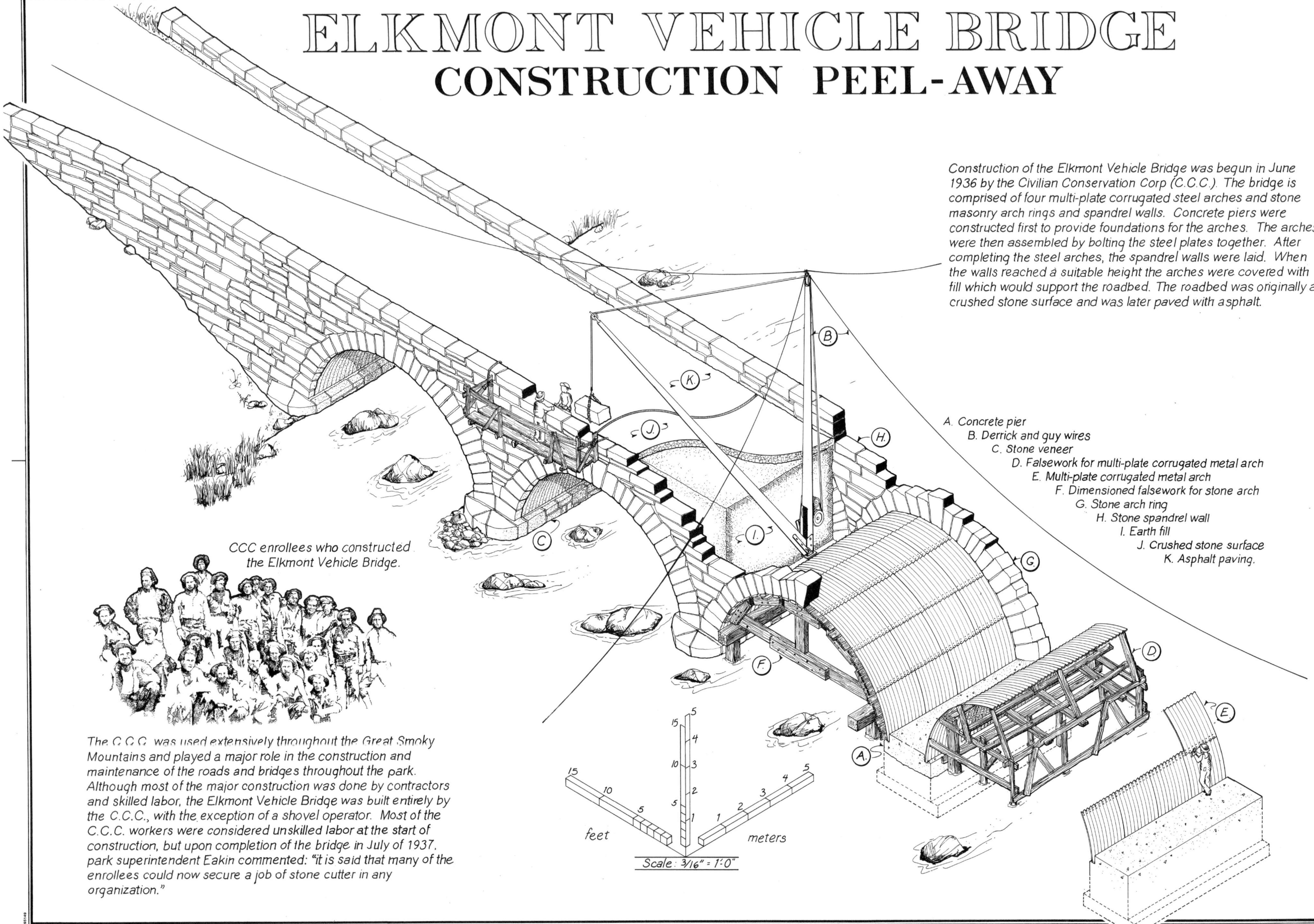

CCC enrollees who constructed the Elkmont Vehicle Bridge.

The C.C.C. was used extensively throughout the Great Smoky Mountains and played a major role in the construction and maintenance of the roads and bridges throughout the park. Although most of the major construction was done by contractors and skilled labor, the Elkmont Vehicle Bridge was built entirely by the C.C.C., with the exception of a shovel operator. Most of the C.C.C. workers were considered unskilled labor at the start of construction, but upon completion of the bridge in July of 1937, park superintendent Eakin commented: "it is said that many of the enrollees could now secure a job of stone cutter in any organization."

DELINEATED BY Edward J. Lupyak, 1996

NATIONAL PARK SERVICE ROADS & BRIDGES RECORDING PROGRAM
NATIONAL PARK SERVICE
UNITED STATES DEPARTMENT OF THE INTERIOR

GREAT SMOKY MOUNTAINS NATIONAL PARK
ROADS & BRIDGES, ELKMONT VEHICLE BRIDGE
SEVIER COUNTY
GATLINBURG VICINITY
TENNESSEE

SHEET 1 of 1

HISTORIC AMERICAN ENGINEERING RECORD
TN-35-S

IF REPRODUCED, PLEASE CREDIT: HISTORIC AMERICAN ENGINEERING RECORD, NATIONAL PARK SERVICE, NAME OF DELINEATOR, DATE OF THE DRAWING

CADES COVE

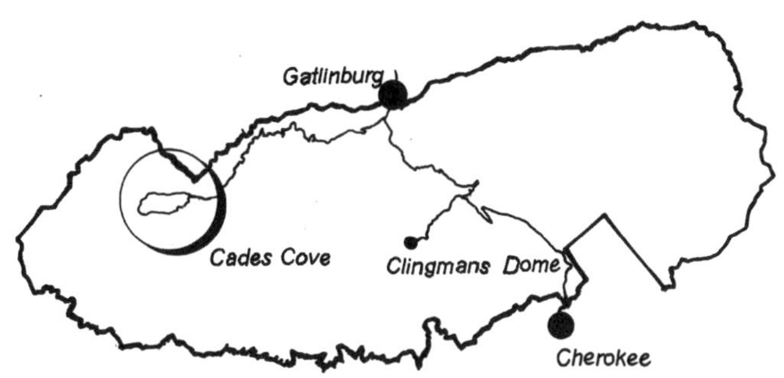

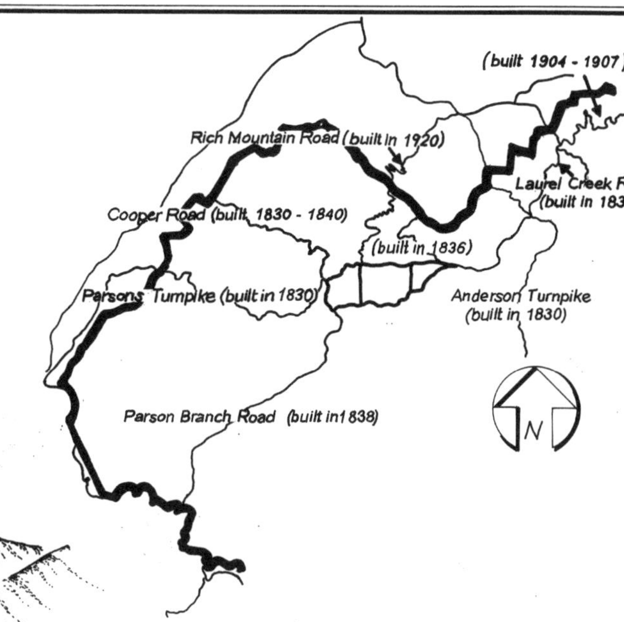

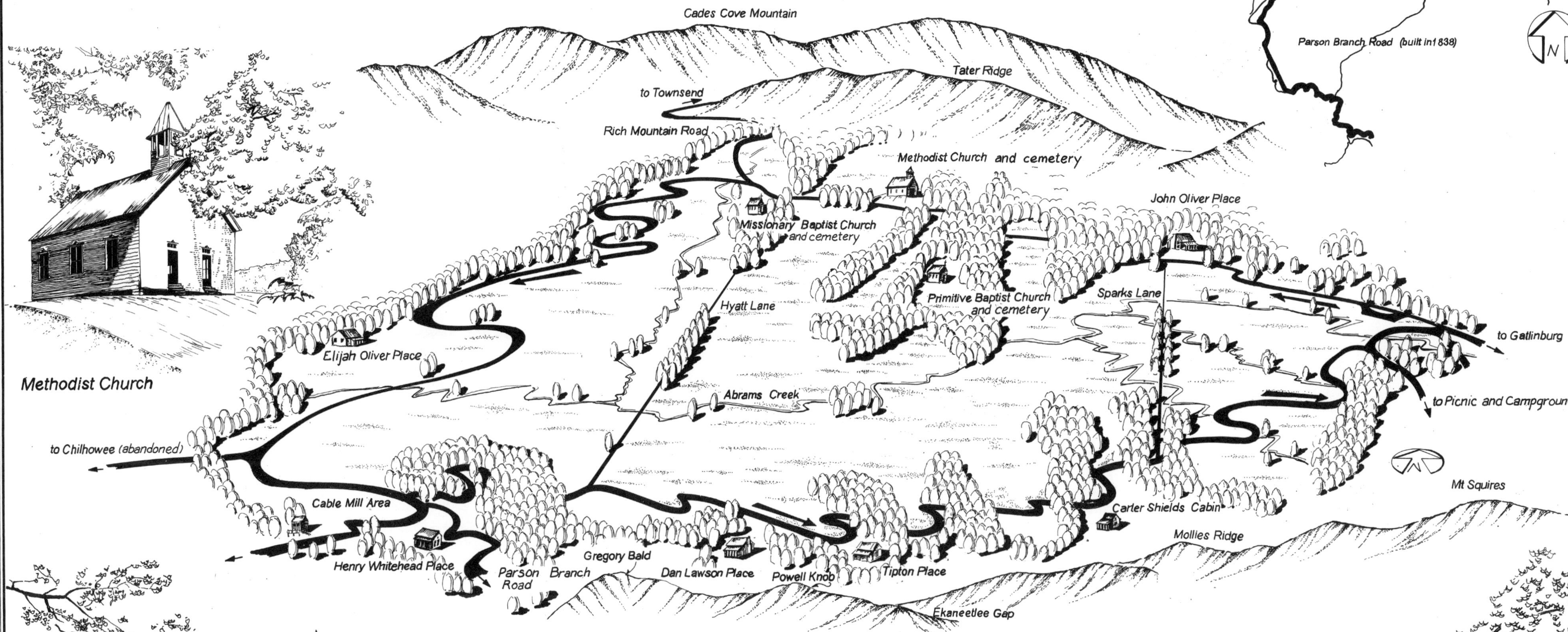

Roads in Cades Cove are worn down by years of travel and erosion.

Cades Cove, a 2 by 5-mile expanse of bottom land amidst the Great Smoky Mountains, offered white settlers coming to the region in the early nineteenth century an opportunity to establish a homestead and raise crops and livestock. By 1850, Cades Cove was a community of 685 people in 137 households. Today, Great Smoky Mountain National Park's 11-mile Cades Cove Loop Road follows the settlers' circular road system, which skirted the edges of the cove, connecting the homesteads but leaving the valuable bottom land undisturbed and open for agriculture. After the establishment of the park in 1934, the National Park Service removed many of the buildings in Cades Cove, which their owners had been forced to vacate by the coming of the park. However, several homes, barns, churches and a mill were left in the cove in order to interpret the lives of the former inhabitants of the area for park visitors. Today, the Cades Cove Loop Road allows visitors to visit these historic places and see the beauty of the opening in the mountains which drew these people into the Smokies.

Fords are frequently used in Cades Cove to cross streams. Water is allowed to flow underneath the ford, acting like a culvert, or above it when the stream is high.

DELINEATED BY: DOROTA SIKORA, 1996

NATIONAL PARK SERVICE ROADS & BRIDGES RECORDING PROGRAM

NATIONAL PARK SERVICE
UNITED STATES DEPARTMENT OF THE INTERIOR

GREAT SMOKY MOUNTAINS NATIONAL PARK
ROADS & BRIDGES, CADES COVE AND LAUREL CREEK ROAD

GATLINBURG VICINITY | SEVIER COUNTY | TENNESSEE

HISTORIC AMERICAN ENGINEERING RECORD
TN-35-D

SHEET 1 of 1

IF REPRODUCED, PLEASE CREDIT: HISTORIC AMERICAN ENGINEERING RECORD, NATIONAL PARK SERVICE, NAME OF DELINEATOR, DATE OF THE DRAWING

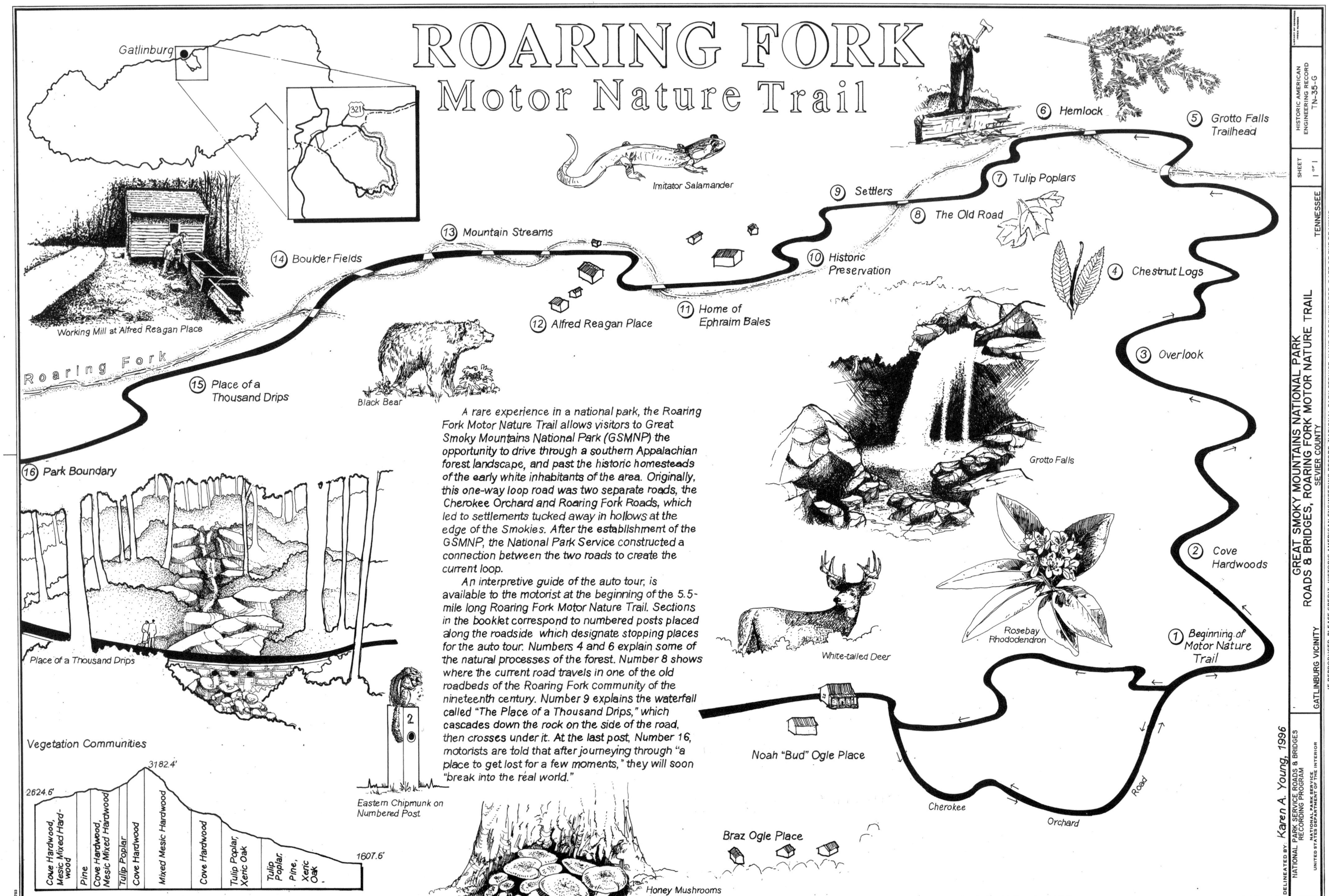
ROARING FORK
Motor Nature Trail
Gatlinburg
321
Imitator Salamander
Working Mill at Alfred Reagan Place
Roaring Fork
(1) Beginning of Motor Nature Trail
(2) Cove Hardwoods
(3) Overlook
(4) Chestnut Logs
(5) Grotto Falls Trailhead
(6) Hemlock
(7) Tulip Poplars
(8) The Old Road
(9) Settlers
(10) Historic Preservation
(11) Home of Ephraim Bales
(12) Alfred Reagan Place
(13) Mountain Streams
(14) Boulder Fields
(15) Place of a Thousand Drips
(16) Park Boundary
Black Bear
Grotto Falls
Rosebay Rhododendron
White-tailed Deer
Place of a Thousand Drips
Eastern Chipmunk on Numbered Post
Noah "Bud" Ogle Place
Braz Ogle Place
Cherokee Orchard Road
Honey Mushrooms
A rare experience in a national park, the Roaring Fork Motor Nature Trail allows visitors to Great Smoky Mountains National Park (GSMNP) the opportunity to drive through a southern Appalachian forest landscape, and past the historic homesteads of the early white inhabitants of the area. Originally, this one-way loop road was two separate roads, the Cherokee Orchard and Roaring Fork Roads, which led to settlements tucked away in hollows at the edge of the Smokies. After the establishment of the GSMNP, the National Park Service constructed a connection between the two roads to create the current loop.
An interpretive guide of the auto tour, is available to the motorist at the beginning of the 5.5-mile long Roaring Fork Motor Nature Trail. Sections in the booklet correspond to numbered posts placed along the roadside which designate stopping places for the auto tour. Numbers 4 and 6 explain some of the natural processes of the forest. Number 8 shows where the current road travels in one of the old roadbeds of the Roaring Fork community of the nineteenth century. Number 9 explains the waterfall called "The Place of a Thousand Drips," which cascades down the rock on the side of the road, then crosses under it. At the last post, Number 16, motorists are told that after journeying through "a place to get lost for a few moments," they will soon "break into the real world."
Vegetation Communities
2624.6'
3182.4'
1607.6'
Cove Hardwood, Mesic Mixed Hard-wood
Pine
Cove Hardwood, Mesic Mixed Hardwood
Tulip Poplar
Cove Hardwood
Mixed Mesic Hardwood
Cove Hardwood
Tulip Poplar, Xeric Oak
Tulip Poplar, Pine, Xeric Oak
DELINEATED BY: Karen A. Young, 1996
NATIONAL PARK SERVICE ROADS & BRIDGES RECORDING PROGRAM
NATIONAL PARK SERVICE
UNITED STATES DEPARTMENT OF THE INTERIOR
GATLINBURG VICINITY
GREAT SMOKY MOUNTAINS NATIONAL PARK
ROADS & BRIDGES, ROARING FORK MOTOR NATURE TRAIL
SEVIER COUNTY
TENNESSEE
SHEET 1 OF 1
HISTORIC AMERICAN ENGINEERING RECORD
TN-35-G
IF REPRODUCED, PLEASE CREDIT: HISTORIC AMERICAN ENGINEERING RECORD, NATIONAL PARK SERVICE, NAME OF DELINEATOR, DATE OF THE DRAWING

MARSH BILLINGS ROCKEFELLER
NATIONAL HISTORICAL PARK

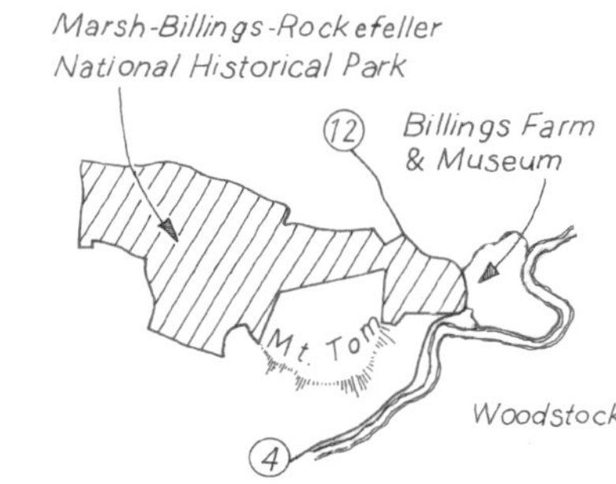

Carriage Roads

The Marsh-Billings-Rockefeller National Historical Park is a 550-acre site encompassing most of Mount Tom, a small peak rising 1359 feet above the village of Woodstock, Vermont. The park is named for the property's three most prominent former residents: pioneering environmentalist George Perkins Marsh, businessman and scientific forestry supporter Frederick Billings, and conservationist Laurence Rockefeller. The carriage road system consists of approximately 10 miles of dirt and gravel roadways that provide access to various portions of the park and lead to the summit of Mount Tom, which lies just beyond the park boundary. The majority of the carriage roads were constructed at Billings's behest between 1880 and 1895.

The Mount Tom carriage roads exemplify the marriage of form and function. When Frederick Billings purchased the boyhood home of George Perkins Marsh in 1869, he set about creating a carriage road system that would be as useful as it was attractive. Having made a fortune in mining and railroads in the American West, Billings was determined to show that profitable land management could coincide with environmental stewardship and aesthetic improvement. His model farm and forestry operations were intended to demonstrate how Vermonters could improve their land to restore its appearance, environmental stability, and profitability. The carriage road system was an essential component of this effort. Billings hoped that visitors attracted by the roadways' scenic views would be influenced by his carefully planned forest improvement projects, which were among the earliest scientific forestry efforts in the nation.

Soon after Billings purchased the Marsh property, he began expanding the existing collection of utilitarian farm roads to create a comprehensive transportation network that would accommodate day-to-day management activities while serving as pleasure routes for excursionists on foot or in carriages. Billings had engaged one of America's most prominent landscape gardeners, Robert Morris Copeland, to embellish the mansion grounds, but there is no record that Copeland made specific suggestions for the broader carriage drive system.

The Marsh-Billings-Rockefeller National Historical Park Carriage Road Recording Project was undertaken during the summer of 2001 and is part of the Historic American Engineering Record (HAER), a long-range program to document historically significant engineering and industrial works in the United States. HAER (Eric DeLony, Chief) is administered by the Historic American Buildings Survey/Historic American Engineering Record (E. Blaine Cliver, Chief), a division of the National Park Service, U.S. Department of the Interior. The project was funded by the U.S. Department of Transportation's Federal Lands Highway Program (Art Hamilton, Administrator) through the NPS Park Roads and Parkways Program (Todd A. Croteau, Manager) and co-sponsored by Marsh-Billings-Rockefeller National Historical Park (Rolf Diamant, Superintendent) in cooperation with the Woodstock Foundation, Inc. (David Donath, Executive Director) and Billings Farm and Museum (Marian Koetsier, Executive Director).

The field work, measured drawings, and historical report were completed under the direction of Christopher Marston, Project Leader and Tim Davis, Program Historian. The recording team consisted of Field Supervisor and Landscape Architect Chris Gray (Texas A & M University); Landscape Architect Aaron Feldman-Grosse (Penn State University); Botanical Illustrator Amy Mark (Southern Oregon State College); and Historian Katie Wollan (UC, Santa Barbara). Large format photography was completed by David Haas.

Nevertheless, the carriage roads exemplify the picturesque landscape design principles that were fashionable at this time. The drives display a striking variety of scenic effects, from shady tunnels of Norway Spruce to brilliant green meadows and sun-dappled open woodlands composed of towering mixed hardwoods. A tumbling waterfall enlivens the main roadway, which was embellished with rustic bridges and culverts, stone watering troughs, and an exhilarating stone causeway that provided sweeping vistas over the surrounding landscape until the surrounding trees grew to obstruct the view. One of the many engaging side roads loops around the artfully reconfigured pond known as The Pogue. The culmination of the trip is the summit of Mount Tom, which offers panoramic vistas of Woodstock and its environs.

Following Billings's death in 1890, his widow Julia and their daughters continued to improve and maintain the estate and its roads. Billings's grand-daughter, Mary French Rockefeller, and her husband, Laurance Rockefeller, inherited the bulk of the property in 1953; the south peak of Mount Tom was deeded to the Town of Woodstock. The Rockefellers donated their portion of the estate to the National Park Service in 1992. The park was formally opened in 1998.

The Mount Tom carriage road system continues to lead visitors through a rejuvenated landscape that reflects some of the earliest conservation efforts in Vermont. The roads wind in and out of majestic stands of hundred-year old trees, run along summer pastures and hayfields lined with ancient sugar maples, and climb to provide entrancing views of Woodstock from the south peak of Mount Tom. As the roads wind through the park's fields and forests, they connect present-day visitors with the history of forestry and conservation in Vermont and with the legacies of the park's three namesakes: Marsh, Billings, and Rockefeller.

DELINEATED BY: Chris Gray, 2001

NPS ROADS & BRIDGES RECORDING PROGRAM
NATIONAL PARK SERVICE
UNITED STATES DEPARTMENT OF THE INTERIOR

MARSH-BILLINGS-ROCKEFELLER CARRIAGE ROADS
MARSH-BILLINGS-ROCKEFELLER NATIONAL HISTORICAL PARK
WOODSTOCK WINDSOR COUNTY VERMONT

SHEET 1 OF 10

HISTORIC AMERICAN ENGINEERING RECORD
VT-27

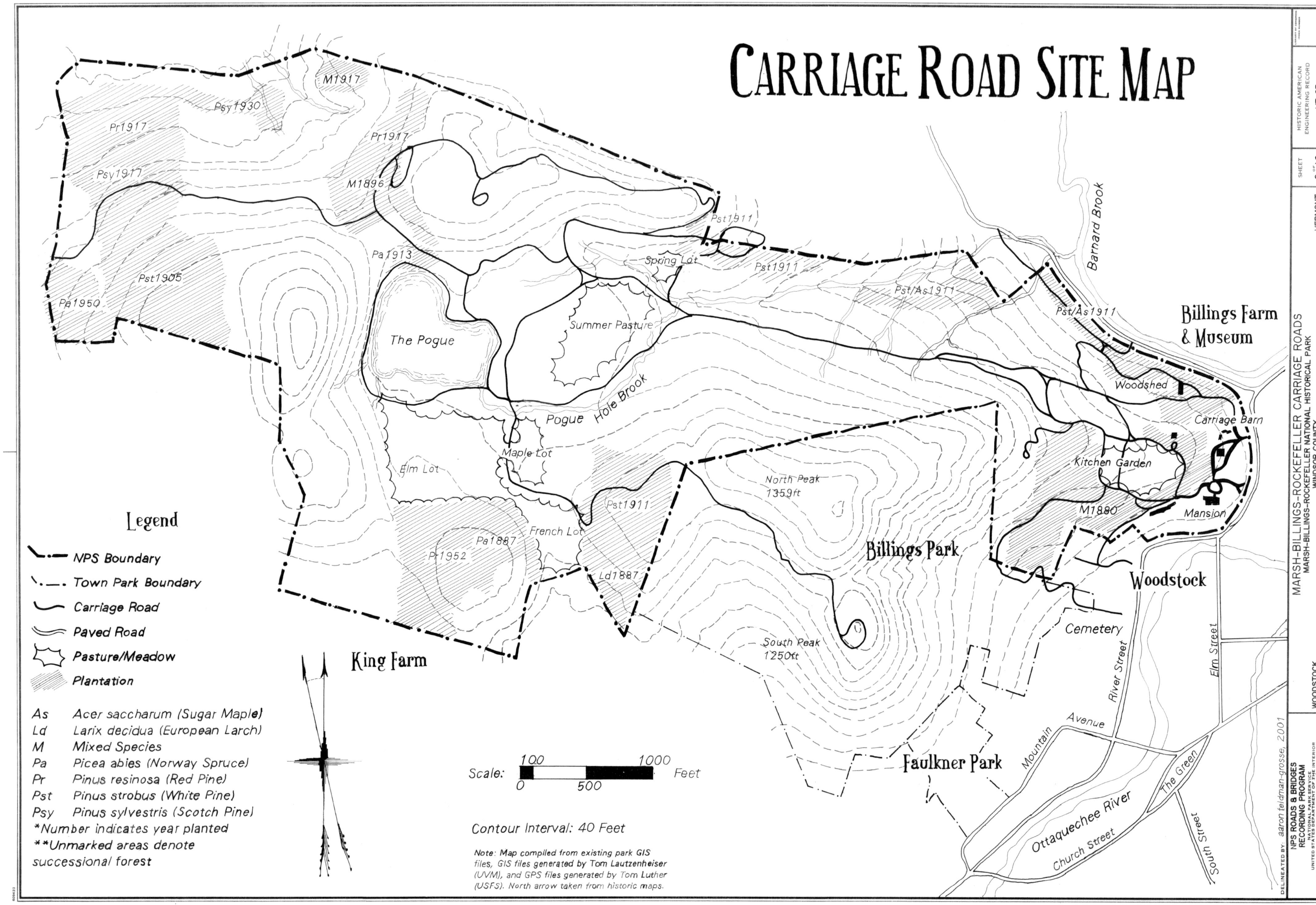

Carriage Road Site Map
Billings Farm & Museum
Woodstock
Billings Park
King Farm
Faulkner Park
Barnard Brook
Pogue Hole Brook
The Pogue
Summer Pasture
Spring Lot
Maple Lot
Elm Lot
French Lot
North Peak 1359ft
South Peak 1250ft
Woodshed
Carriage Barn
Kitchen Garden
Mansion
Cemetery
River Street
Elm Street
Mountain Avenue
Ottauquechee River
Church Street
The Green
South Street
M1917
Psy1930
Pr1917
Psy1917
M1896
Pst1911
Pa1913
Pst1905
Pa1950
Pst/As1911
M1880
Pa1887
Pr1952
Ld1887
Legend
NPS Boundary
Town Park Boundary
Carriage Road
Paved Road
Pasture/Meadow
Plantation
As Acer saccharum (Sugar Maple)
Ld Larix decidua (European Larch)
M Mixed Species
Pa Picea abies (Norway Spruce)
Pr Pinus resinosa (Red Pine)
Pst Pinus strobus (White Pine)
Psy Pinus sylvestris (Scotch Pine)
*Number indicates year planted
**Unmarked areas denote successional forest
Scale: 0 100 500 1000 Feet
Contour Interval: 40 Feet
Note: Map compiled from existing park GIS files, GIS files generated by Tom Lautzenheiser (UVM), and GPS files generated by Tom Luther (USFS). North arrow taken from historic maps.
DELINEATED BY: aaron feldman-grosse, 2001
NPS ROADS & BRIDGES RECORDING PROGRAM
MARSH-BILLINGS-ROCKEFELLER CARRIAGE ROADS
MARSH-BILLINGS-ROCKEFELLER NATIONAL HISTORICAL PARK
WINDSOR COUNTY
WOODSTOCK
VERMONT
SHEET 2 OF 10
HISTORIC AMERICAN ENGINEERING RECORD
VT-27

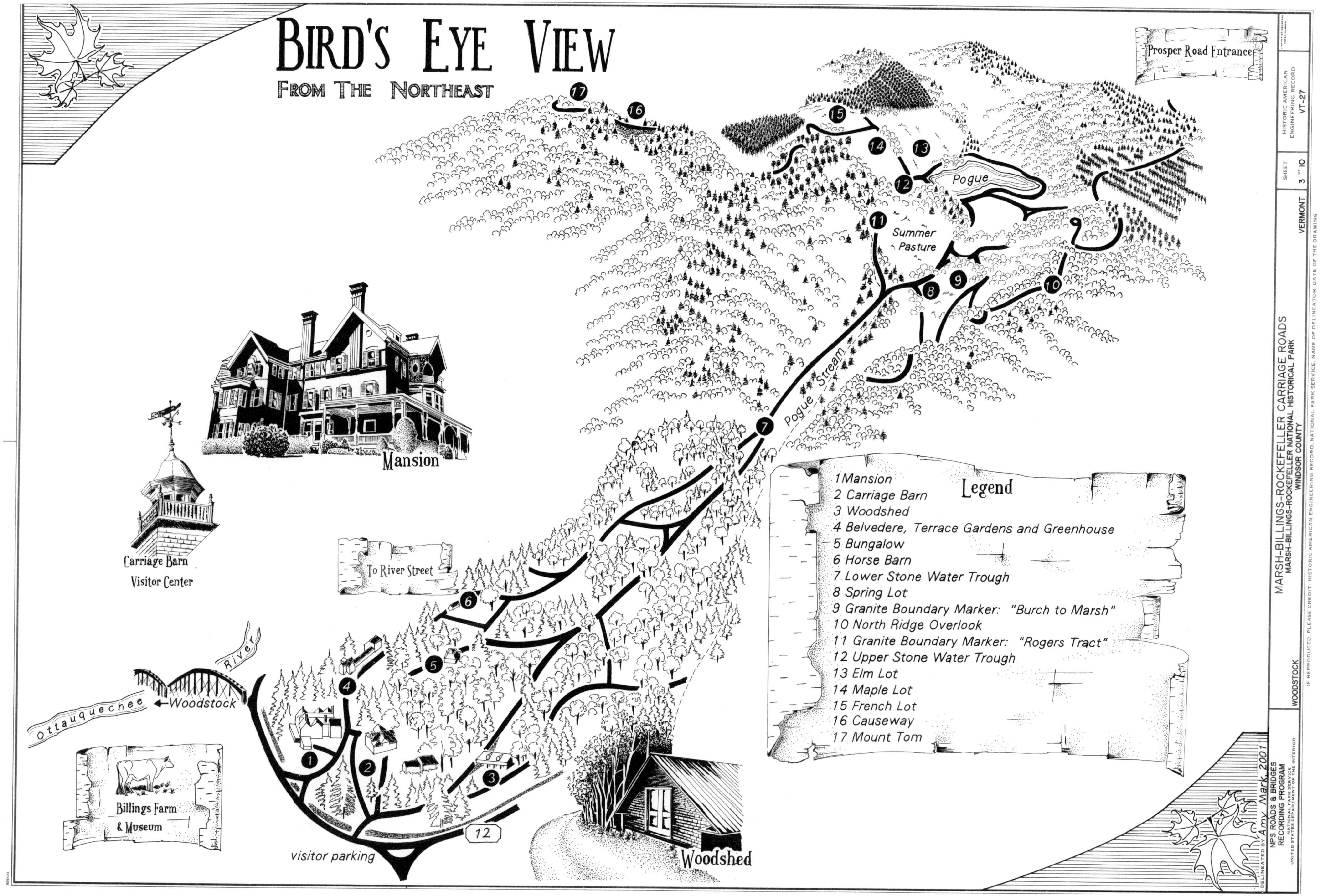
BIRD'S EYE VIEW
FROM THE NORTHEAST
Prosper Road Entrance
Pogue
Summer Pasture
Pogue Stream
Mansion
Carriage Barn
Visitor Center
To River Street
River
Ottauquechee
← Woodstock
Billings Farm & Museum
visitor parking
12
Woodshed
Legend
1 Mansion
2 Carriage Barn
3 Woodshed
4 Belvedere, Terrace Gardens and Greenhouse
5 Bungalow
6 Horse Barn
7 Lower Stone Water Trough
8 Spring Lot
9 Granite Boundary Marker: "Burch to Marsh"
10 North Ridge Overlook
11 Granite Boundary Marker: "Rogers Tract"
12 Upper Stone Water Trough
13 Elm Lot
14 Maple Lot
15 French Lot
16 Causeway
17 Mount Tom
DELINEATED BY: Amy Mark, 2001
NPS ROADS & BRIDGES RECORDING PROGRAM
NATIONAL PARK SERVICE
UNITED STATES DEPARTMENT OF THE INTERIOR
MARSH-BILLINGS-ROCKEFELLER CARRIAGE ROADS
MARSH-BILLINGS-ROCKEFELLER NATIONAL HISTORICAL PARK
WINDSOR COUNTY
WOODSTOCK
VERMONT
SHEET 3 OF 10
HISTORIC AMERICAN ENGINEERING RECORD
VT-27
IF REPRODUCED, PLEASE CREDIT: HISTORIC AMERICAN ENGINEERING RECORD, NATIONAL PARK SERVICE, NAME OF DELINEATOR, DATE OF THE DRAWING

THE RECOVERING LANDSCAPE

Before European settlement, the Vermont landscape was dominated by mature forests of mixed hardwoods and evergreens that were similar in appearance to the forests of today. Native Americans maintained small cornfields but their impact was minimal. By the mid-eighteenth century settlers were beginning to migrate to this undeveloped frontier. With them came agricultural practices that would result in the clearing of the virgin forest.

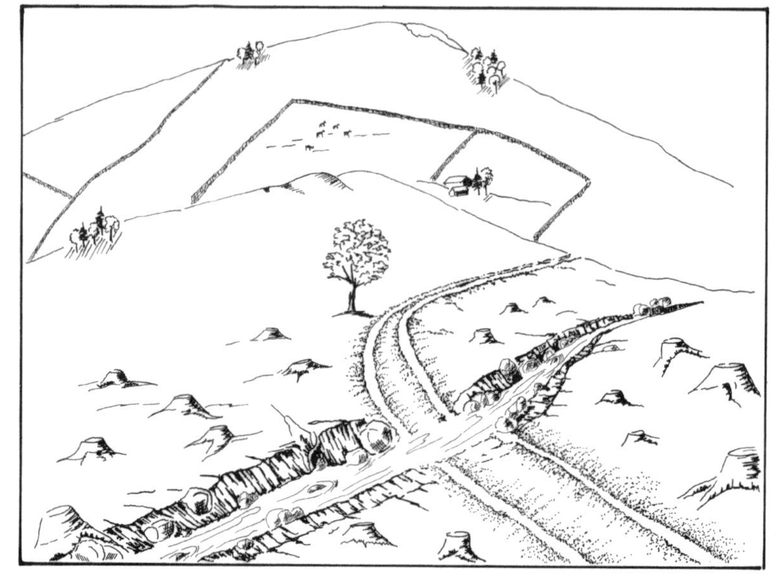

Forests throughout Vermont were cleared for tillable land, sheep pasture, potash, and timber. Small farms and villages dotted the landscape and were connected by a sprawling network of primitive or modestly improved roads. By the mid-nineteenth century, nearly 70% of the state's land had been cleared of its forests, significantly altering ecological systems, but presenting an image of agricultural tranquility.

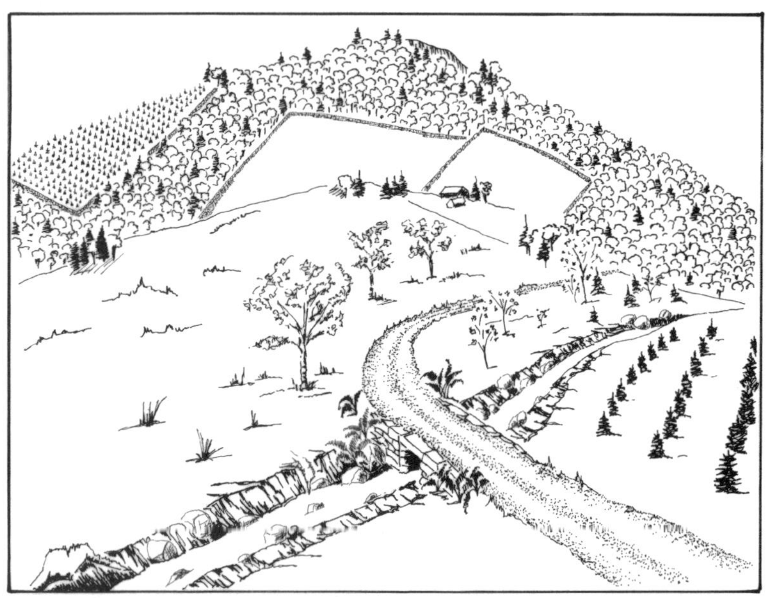

By 1890 Vermont was a landscape in transition. The economy was in decline, farms were being abandoned at an alarming rate, and villages were losing population to cities and more productive agricultural regions. Natural forest succession began to fill old fields and pastures with new growth. Frederick Billings and others assisted this revegetation process by initiating progressive forestry and land stewardship programs.

With farm abandonment continuing throughout the twentieth century and stewardship efforts expanding, Vermont's forests have returned to a semblance of their original condition. The state is now almost 80% forested. The reforestation of Vermont provides a compelling story of ecological recovery, but the disappearance of farmland may eventually compromise the state's long-cherished image as a timeless repository of traditional New England scenery and rural life.

DELINEATED BY: Amy Mark, 2001

NPS ROADS & BRIDGES RECORDING PROGRAM
NATIONAL PARK SERVICE
UNITED STATES DEPARTMENT OF THE INTERIOR

MARSH-BILLINGS-ROCKEFELLER CARRIAGE ROADS
MARSH-BILLINGS-ROCKEFELLER NATIONAL HISTORICAL PARK
WINDSOR COUNTY
WOODSTOCK VERMONT

SHEET 5 OF 10

HISTORIC AMERICAN ENGINEERING RECORD
VT-27

Forest Experience

Marsh-Billings-Rockefeller National Historical Park contains a diverse array of managed and natural landscapes that surround the carriage roads with seasonally varying scenic delights. The constantly changing views reveal a range of historic practices and ecological associations. Stone walls and old fields bordered by Sugar Maples recall the area's agricultural heritage. The legacy of Frederick Billings's pioneering scientific forestry efforts can be seen in the large stands of carefully selected trees that have grown to maturity on the flanks of Mount Tom. Natural forest succession has filled in much of the land between Billings' plantations with mixed hardwoods and other native species.

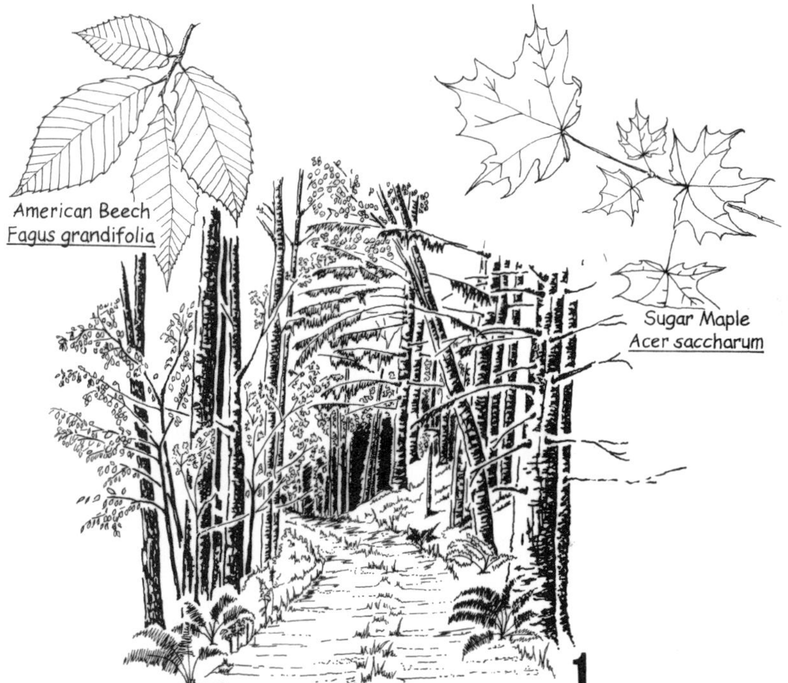

Northern Hardwood Forest

Much of the carriage road system passes through Northern hardwood forest, with its characteristic mix of Eastern Hemlock and deciduous trees including Yellow Birch, American Beech, Sugar Maple and White Ash. During the summer a vibrant understory of ferns and wood nettle create a lush green effect.

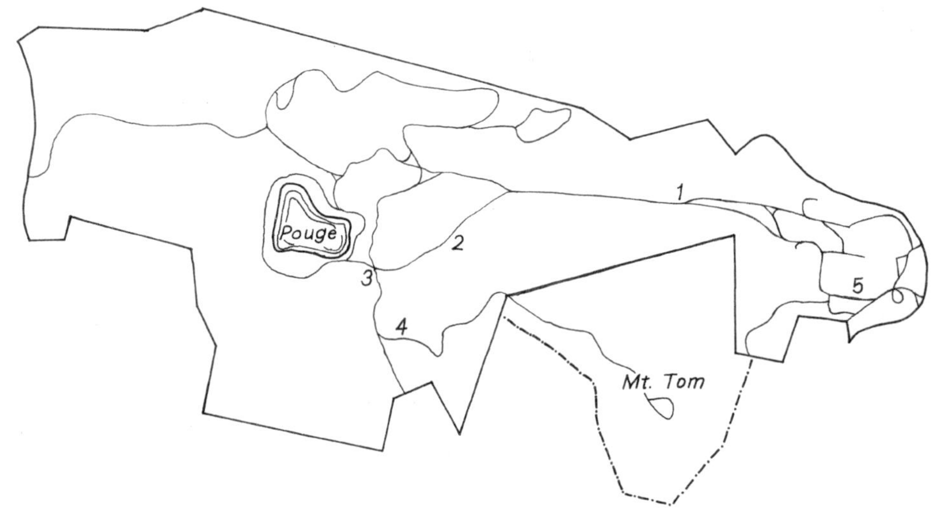

Early Norway Spruce Plantings

Leaving the mansion and traveling past the Belvedere, the carriage road passes through ornamental plantings of Norway Spruce and Eastern Hemlock. Landscape gardener Robert Morris Copland recommended these picturesque conifers as ideal choices for creating a romantic landscape effect.

Hemlock Forest

As the carriage road approaches the Pogue Stream, moisture-loving Eastern Hemlock dominates the landscape. The thick forest canopy allows little light to break through to the forest floor, limiting the growth of ground cover and creating a more open understory.

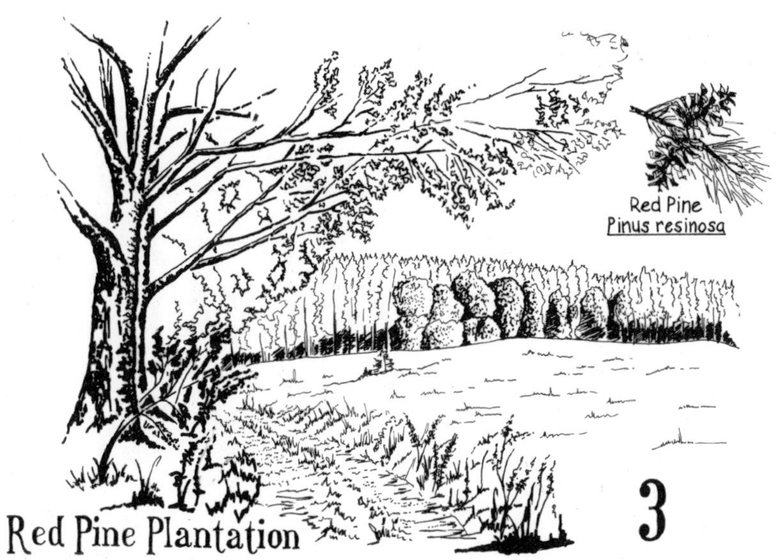

Red Pine Plantation

Entering the Elm Lot on the road to Mount Tom, the carriage road provides views of a large Red Pine plantation to the southwest. The thick growth of pine offers a stark contrast to the open hayfields. In autumn, their dark greenery complements the blazing fall foliage. In winter, these evergreens stand out against the leafless hardwoods.

The French Lot

Stately Sugar Maples parallel the carriage road as it runs along the edge of old fields, serving as reminders of earlier farming activities. Plantings of European Larch border the hayfield on the east, while a stand of Norway Spruce borders it on the west, framing the vista beyond.

DELINEATED BY Amy Mark, 2001

NPS ROADS & BRIDGES RECORDING PROGRAM
NATIONAL PARK SERVICE
UNITED STATES DEPARTMENT OF THE INTERIOR

WOODSTOCK

MARSH-BILLINGS-ROCKEFELLER CARRIAGE ROADS
MARSH-BILLINGS-ROCKEFELLER NATIONAL HISTORICAL PARK
WINDSOR COUNTY

VERMONT

SHEET 6 OF 10

HISTORIC AMERICAN ENGINEERING RECORD
VT-27

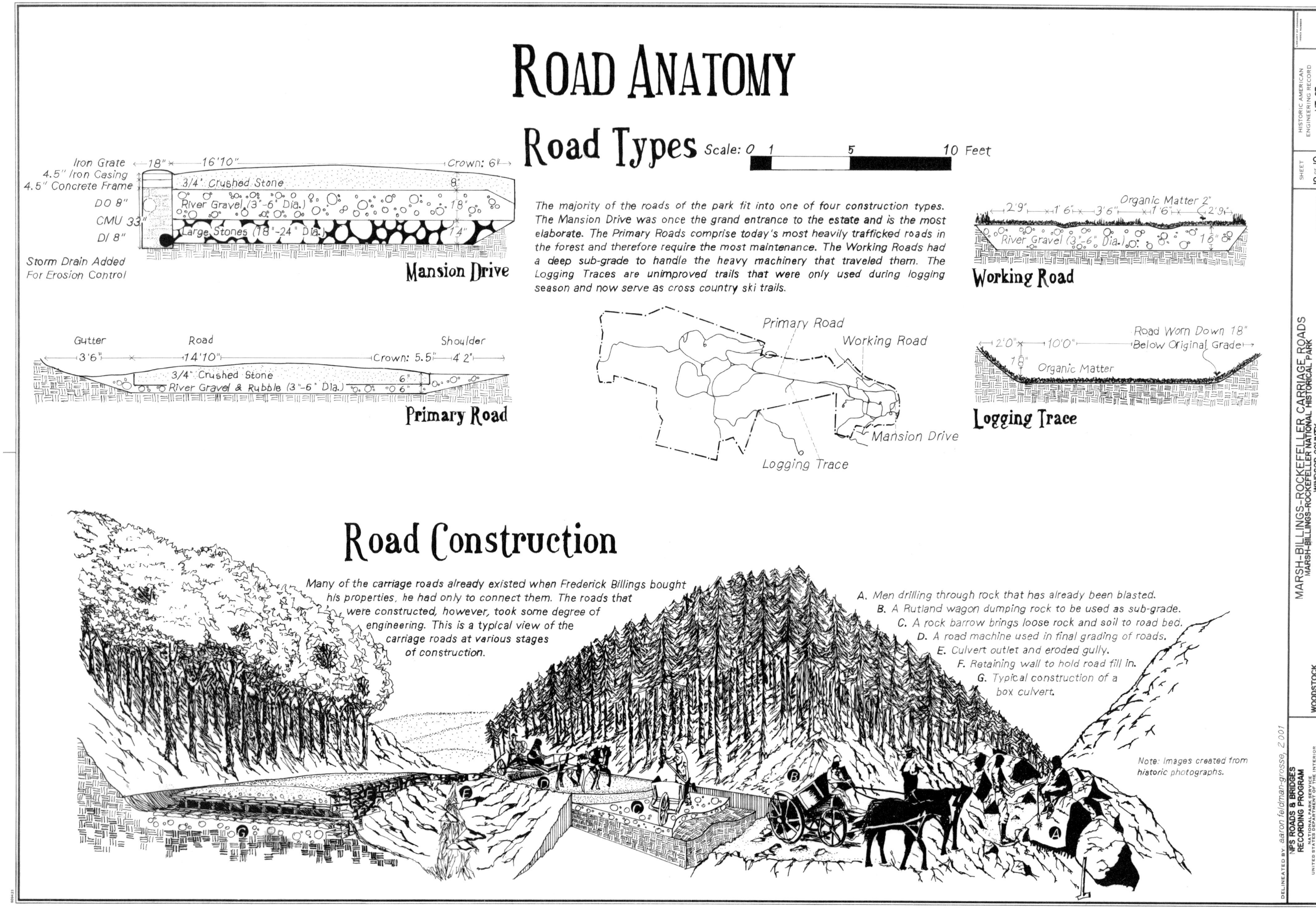

ROAD ANATOMY
Road Types
Scale: 0 1 5 10 Feet
Iron Grate
4.5" Iron Casing
4.5" Concrete Frame
DO 8"
CMU 33"
DI 8"
Storm Drain Added For Erosion Control
18"
16'10"
Crown: 6"
3/4" Crushed Stone
8"
River Gravel (3"-6" Dia.)
18"
Large Stones (18"-24" Dia.)
14"
Mansion Drive
The majority of the roads of the park fit into one of four construction types. The Mansion Drive was once the grand entrance to the estate and is the most elaborate. The Primary Roads comprise today's most heavily trafficked roads in the forest and therefore require the most maintenance. The Working Roads had a deep sub-grade to handle the heavy machinery that traveled them. The Logging Traces are unimproved trails that were only used during logging season and now serve as cross country ski trails.
Organic Matter 2"
2'9"
1'6"
3'6"
1'6"
2'9"
River Gravel (3"-6" Dia.)
16"
Working Road
Gutter
Road
Shoulder
3'6"
14'10"
Crown: 5.5"
4'2"
3/4" Crushed Stone
River Gravel & Rubble (3"-6" Dia.)
6"
6"
Primary Road
Primary Road
Working Road
Mansion Drive
Logging Trace
2'0"
10'0"
Road Worn Down 18" Below Original Grade
18"
Organic Matter
Logging Trace
Road Construction
Many of the carriage roads already existed when Frederick Billings bought his properties, he had only to connect them. The roads that were constructed, however, took some degree of engineering. This is a typical view of the carriage roads at various stages of construction.
A. Men drilling through rock that has already been blasted.
B. A Rutland wagon dumping rock to be used as sub-grade.
C. A rock barrow brings loose rock and soil to road bed.
D. A road machine used in final grading of roads.
E. Culvert outlet and eroded gully.
F. Retaining wall to hold road fill in.
G. Typical construction of a box culvert.
Note: Images created from historic photographs.
MARSH-BILLINGS-ROCKEFELLER CARRIAGE ROADS
MARSH-BILLINGS-ROCKEFELLER NATIONAL HISTORICAL PARK
WINDSOR COUNTY
VERMONT
WOODSTOCK
DELINEATED BY: aaron feldman-grosse, 2001
NPS ROADS & BRIDGES RECORDING PROGRAM
NATIONAL PARK SERVICE
UNITED STATES DEPARTMENT OF THE INTERIOR
HISTORIC AMERICAN ENGINEERING RECORD
VT-27
SHEET 10 OF 10
IF REPRODUCED, PLEASE CREDIT: HISTORIC AMERICAN ENGINEERING RECORD, NATIONAL PARK SERVICE, NAME OF DELINEATOR, DATE OF THE DRAWING

THE POGUE LOOP

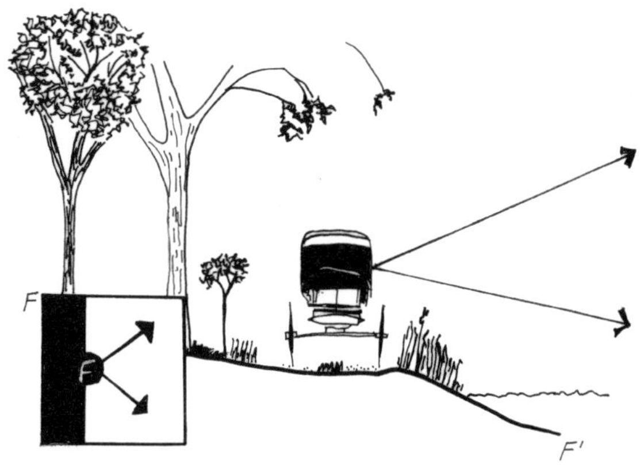

Cross Sections

Standing at point **e** *(cross-section, below), there are more than 100 feet of trees and brush between the road and the water's edge affording only filtered views of the Pogue through Sugar Maple (Acer saccharum) and Yellow Birch (Betula alleghaniensis). This view is characteristic of the views along the east side of the carriage road. After passing the intersection at the northwest corner of the Pogue, the carriage loop comes to within five feet of the water, offering expansive views of the entire lake similar to the one at point* **f** *(cross-section, above).*

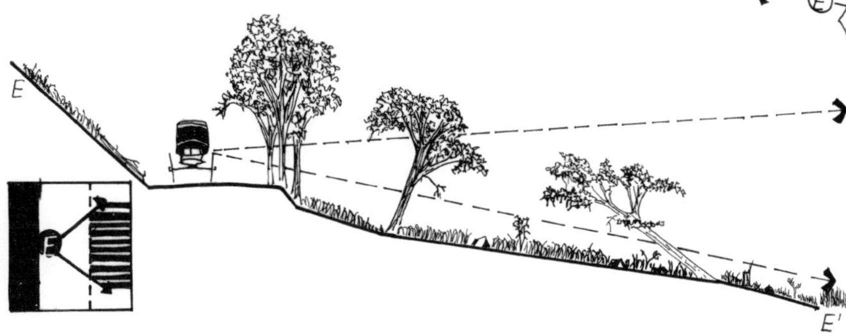

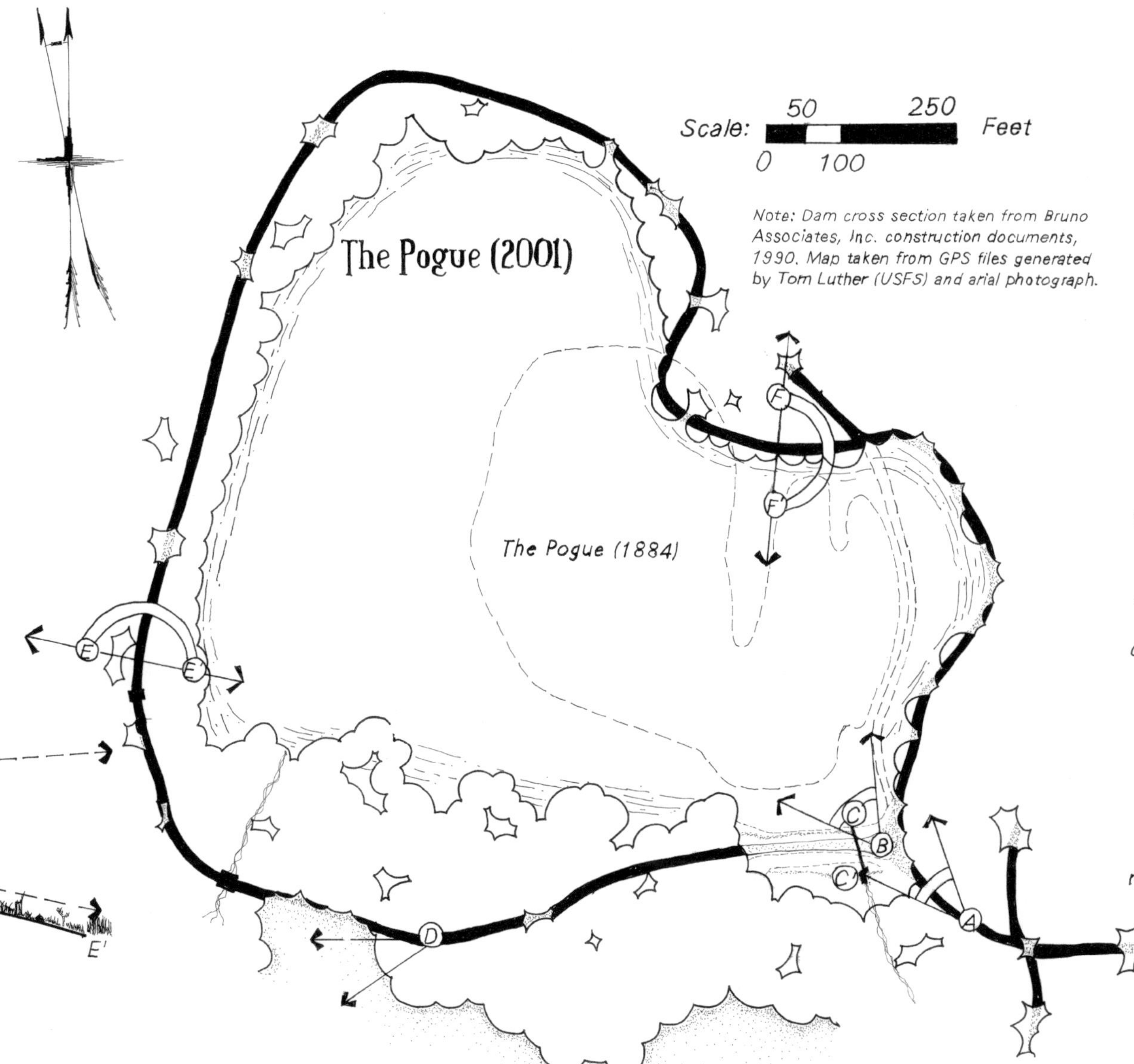

Note: Dam cross section taken from Bruno Associates, Inc. construction documents, 1990. Map taken from GPS files generated by Tom Luther (USFS) and arial photograph.

Evolution of the Pogue

Since construction was completed in 1895, the Pogue has been a popular destination for its scenic, recreational and productive qualities. The Pogue lies in the southwest corner of what was once the Dana farm, a property acquired by Frederick Billings in 1884. In 1890, Billings began construction on a carriage loop around the Pogue. It was not until 1895, however, that the lake was enlarged to fit into the 14 acre plot defined by the road. This order of construction suggests the Pogue is a designed feature in the landscape. The sequence of views the design created can still be seen while travelling clockwise around the pond.

Approach to the Pogue

Approaching the Pogue along the main carriage road creates a sense of anticipation and surprise. As one reaches point **a** *(below), an opening in the forest canopy can be seen ahead, but nothing more. It is not until reaching point* **b** *(above) that one reaches the opening to see the water of the Pogue.*

Elm Lot

Heading clockwise around the Pogue from the dam, the road veers away from the water's edge, eventually masking the view to the right entirely. At point **d** *(left), the road emerges out of the forest and into the Elm Lot. This transition provides a stark visual contrast between the dark cover of the forest and the bright open space created by the pasture.*

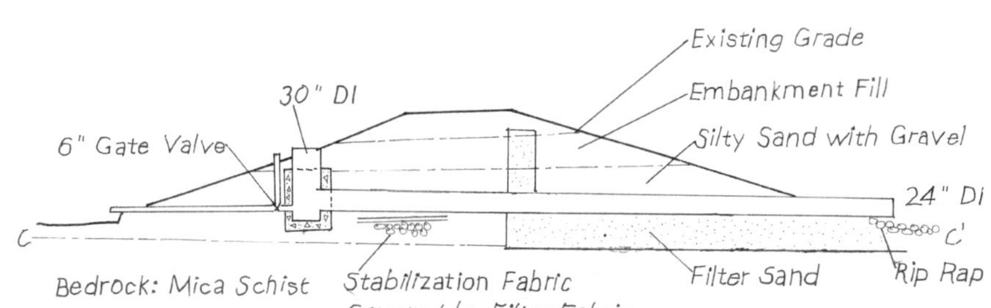

Pogue Dam

*In 1895, the Pogue was enlarged with the construction of a dam (***c***, left) under the southwest section of the carriage loop. Since construction, however, various recurring problems have appeared, including annual flooding in the winter and excessive erosion along the banks of Pogue Hole Brook year-round. In 1991, the dam was rebuilt (shown above) to address these problems.*

DELINEATED BY: aaron feldman-grosse, 2001

NPS ROADS & BRIDGES RECORDING PROGRAM
NATIONAL PARK SERVICE
UNITED STATES DEPARTMENT OF THE INTERIOR

WOODSTOCK

MARSH-BILLINGS-ROCKEFELLER CARRIAGE ROADS
MARSH-BILLINGS-ROCKEFELLER NATIONAL HISTORICAL PARK
WINDSOR COUNTY

VERMONT

SHEET 9 OF 10

HISTORIC AMERICAN ENGINEERING RECORD
VT-27

LIBRARY OF CONGRESS INDEX NUMBER

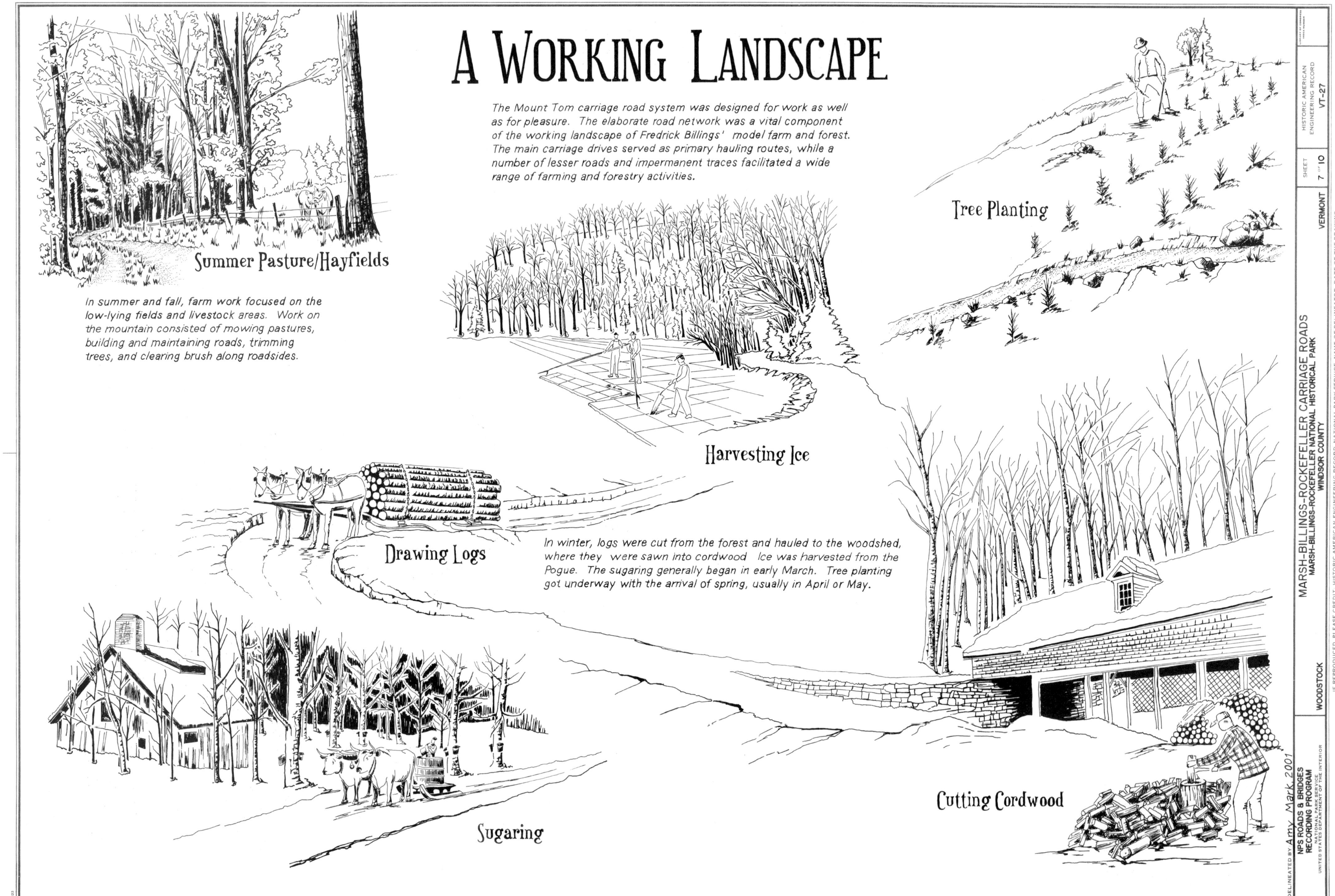

A WORKING LANDSCAPE
The Mount Tom carriage road system was designed for work as well as for pleasure. The elaborate road network was a vital component of the working landscape of Fredrick Billings' model farm and forest. The main carriage drives served as primary hauling routes, while a number of lesser roads and impermanent traces facilitated a wide range of farming and forestry activities.
Summer Pasture/Hayfields
In summer and fall, farm work focused on the low-lying fields and livestock areas. Work on the mountain consisted of mowing pastures, building and maintaining roads, trimming trees, and clearing brush along roadsides.
Tree Planting
Harvesting Ice
Drawing Logs
In winter, logs were cut from the forest and hauled to the woodshed, where they were sawn into cordwood Ice was harvested from the Pogue. The sugaring generally began in early March. Tree planting got underway with the arrival of spring, usually in April or May.
Sugaring
Cutting Cordwood
DELINEATED BY: Amy Mark 2001
NPS ROADS & BRIDGES RECORDING PROGRAM
NATIONAL PARK SERVICE
UNITED STATES DEPARTMENT OF THE INTERIOR
WOODSTOCK
MARSH-BILLINGS-ROCKEFELLER CARRIAGE ROADS
MARSH-BILLINGS-ROCKEFELLER NATIONAL HISTORICAL PARK
WINDSOR COUNTY
VERMONT
SHEET 7 OF 10
HISTORIC AMERICAN ENGINEERING RECORD
VT-27
IF REPRODUCED, PLEASE CREDIT: HISTORIC AMERICAN ENGINEERING RECORD, NATIONAL PARK SERVICE, NAME OF DELINEATOR, DATE OF THE DRAWING

MOUNT RAINIER ROADS & BRIDGES
MOUNT RAINIER NATIONAL PARK, WASHINGTON

Design with the Land

Mount Rainier is a dormant volcano rising to 14,410 feet above sea level, the top of which is covered with perpetual snow. Besides the mountain itself, the national park contains many features of scenic and scientific interest; including living glaciers, waterfalls, lakes, canyons, old-growth forests and subalpine meadows.

Mount Rainier National Park was established on March 2, 1899, contemporanious with the rise of the automobile and the desire to experience America's natural wonders. During the early national park years, pre-1916, the U.S. Army Corps of Engineers designed the park roads and bridges to harmonize with the superb scenery. Following the creation of the National Park Service in 1916, NPS engineers and landscape architects worked together to continue this tradition. Design and construction of park roads was transferred to the Bureau of Public Roads in 1925.

Between 1920 and 1940 the "Rustic Style" of architecture became the standard of design for park construction. Native stone, logs and other indigenous materials were commonly used for buildings, bridges, retaining walls, and signage. During this period, natural planting practices became the standard for revegetation work. Some of this work was carried out by the Emergency Conservation Works program, the parent organization of the Civilian Conservation Corps, a massive public works project that provided relief for many unemployed men during the Depression.

Nisqually Entrance Gate, based on historic photograph

This project is part of the Historic American Engineering Record (HAER), a long-range program to document historically significant engineering and industrial works in the United States. The HAER program is administered by the Historic American Buildings Survey/Historic American Engineering Record Division (HABS/HAER) of the National Park Service, U.S. Department of the Interior. The Mount Rainier Roads and Bridges Recording Project was cosponsored during the summer of 1992 by HAER under the general direction of Dr. Robert J. Kapsch, Chief; the National Park Service Roads and Bridges Program, John Gingles, Manager; and by Mount Rainier National Park, William Briggle, Superintendent.

The field work, measured drawings, historical reports, and photographs were prepared under the direction of Project Leader Eric N. DeLony, Chief of HAER. The recording team consisted of Todd A. Croteau, team supervisor; architectural technicians Bryan D. Fish and Daniela Trettel (U.S./ICOMOS, Argentina); and landscape architect Julie Ann Dickson Historical reports were prepared by project historian Richard H. Quin. Formal large-format photography was done by Jet Lowe, HAER staff photographer.

The landscape work related to the roads and bridges of Mount Rainier National Park maintains a sensitivity to the natural features of the land. Most built components were designed so they integrated into the system as a whole, in harmony with the natural landscapes of the mountain region. Planned to preserve the natural beauty of the park, Mount Rainier's roads and bridges provide safe and convenient access for millions of visitors. Without them, the national park experience would be markedly different from today

Bridges of various construction have spanned Mount Rainier's major streams and lesser tributaries, many of which have been washed away by glacial mud flows or replaced due to decay in the harsh environment. The first bridges were constructed of logs or rough-hewn timber; only the Nisqually River Suspension Bridge remains. Many of the park bridges appear to be constructed of stone masonry. Actually, these arch bridges are built of concrete and merely faced in stone, characteristic of the NPS Rustic Style. Several exposed-concrete structures exist and an unusual three-hinged steel web-arch spans Fryingpan Creek. Even smaller bridges and culverts were designed to blend with the rugged mountain scenery. Most of the park bridges are constructed on tangents or curves, frequently superelevated or banked, offering a smooth transition from road to span. Such designs help the park roads sweep gracefully through the spectacular terrain.

DELINEATED BY: Daniela Trettel, Todd A. Croteau, 1992

MOUNT RAINIER ROADS AND BRIDGES RECORDING PROJECT
NATIONAL PARK SERVICE
UNITED STATES DEPARTMENT OF THE INTERIOR

MOUNT RAINIER ROADS AND BRIDGES
MOUNT RAINIER NATIONAL PARK | PIERCE/LEWIS COUNTIES | WASHINGTON

SHEET 1 OF 2

HISTORIC AMERICAN ENGINEERING RECORD
WA - 35

LIBRARY OF CONGRESS INDEX NUMBER

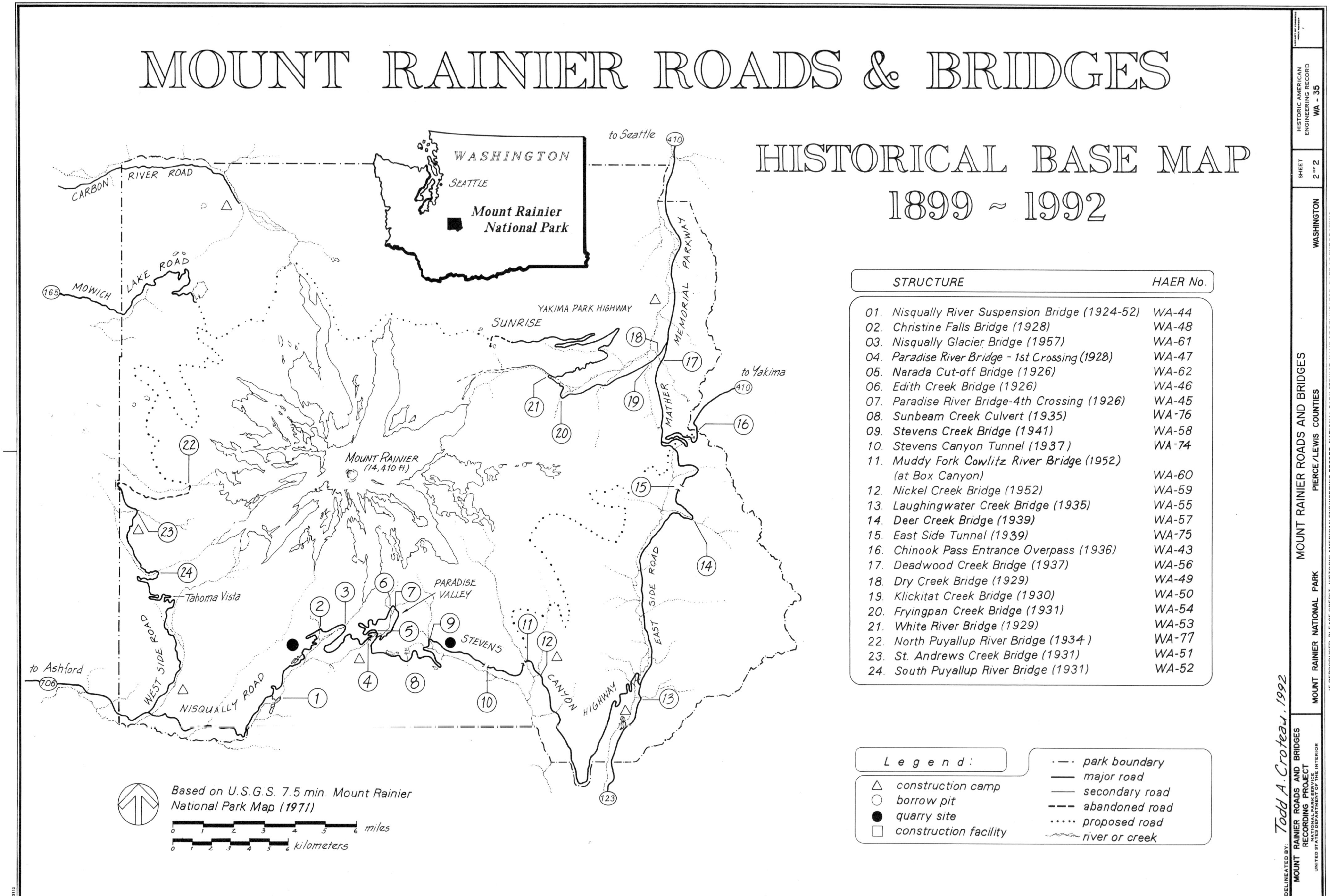
MOUNT RAINIER ROADS & BRIDGES
HISTORICAL BASE MAP
1899 ~ 1992
WASHINGTON
SEATTLE
Mount Rainier National Park
to Seattle
to Yakima
to Ashford
CARBON RIVER ROAD
MOWICH LAKE ROAD
YAKIMA PARK HIGHWAY
SUNRISE
MATHER MEMORIAL PARKWAY
MOUNT RAINIER (14,410 ft)
PARADISE VALLEY
STEVENS CANYON HIGHWAY
EAST SIDE ROAD
WEST SIDE ROAD
NISQUALLY ROAD
Tahoma Vista
STRUCTURE
HAER No.
01. Nisqually River Suspension Bridge (1924-52) WA-44
02. Christine Falls Bridge (1928) WA-48
03. Nisqually Glacier Bridge (1957) WA-61
04. Paradise River Bridge - 1st Crossing (1928) WA-47
05. Narada Cut-off Bridge (1926) WA-62
06. Edith Creek Bridge (1926) WA-46
07. Paradise River Bridge-4th Crossing (1926) WA-45
08. Sunbeam Creek Culvert (1935) WA-76
09. Stevens Creek Bridge (1941) WA-58
10. Stevens Canyon Tunnel (1937) WA-74
11. Muddy Fork Cowlitz River Bridge (1952) (at Box Canyon) WA-60
12. Nickel Creek Bridge (1952) WA-59
13. Laughingwater Creek Bridge (1935) WA-55
14. Deer Creek Bridge (1939) WA-57
15. East Side Tunnel (1939) WA-75
16. Chinook Pass Entrance Overpass (1936) WA-43
17. Deadwood Creek Bridge (1937) WA-56
18. Dry Creek Bridge (1929) WA-49
19. Klickitat Creek Bridge (1930) WA-50
20. Fryingpan Creek Bridge (1931) WA-54
21. White River Bridge (1929) WA-53
22. North Puyallup River Bridge (1934) WA-77
23. St. Andrews Creek Bridge (1931) WA-51
24. South Puyallup River Bridge (1931) WA-52
Legend:
construction camp
borrow pit
quarry site
construction facility
park boundary
major road
secondary road
abandoned road
proposed road
river or creek
Based on U.S.G.S. 7.5 min. Mount Rainier National Park Map (1971)
miles
kilometers
Todd A. Croteau, 1992
DELINEATED BY:
MOUNT RAINIER ROADS AND BRIDGES RECORDING PROJECT
NATIONAL PARK SERVICE
UNITED STATES DEPARTMENT OF THE INTERIOR
MOUNT RAINIER NATIONAL PARK
MOUNT RAINIER ROADS AND BRIDGES
PIERCE/LEWIS COUNTIES
WASHINGTON
SHEET 2 OF 2
HISTORIC AMERICAN ENGINEERING RECORD
WA - 35
IF REPRODUCED, PLEASE CREDIT: HISTORIC AMERICAN ENGINEERING RECORD, NATIONAL PARK SERVICE, NAME OF DELINEATOR, DATE OF THE DRAWING

NISQUALLY RIVER SUSPENSION BRIDGE

1924 / 1952

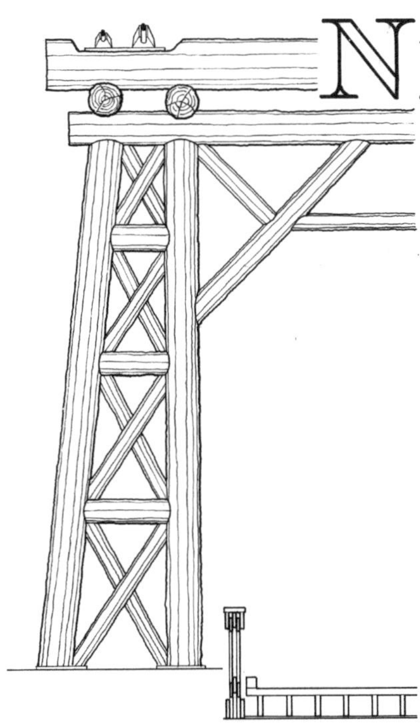

ORIGINAL TOWER

The only surviving timber-frame vehicular suspension bridge designed by the National Park Service, the Nisqually River Suspension Bridge at Longmire was built in 1923 and 1924 to provide access to the Longmire Public Auto Campground (recently closed). The original structure, which had log towers supporting a Town lattice stiffening truss, was built by the Monson-Trierweiler Company of Portland, Oregon at a cost of roughly $15,000. The bridge was reconstructed in 1951 and 1952 by contractor Thomas M. Mock of Tacoma, Washington. At this point, the original log towers were replaced with new towers of heavy sawn posts and milled lumber, and the lattice stiffening truss was replaced by a modified Howe truss. Only the original metal cables, cable saddles and suspenders were re-used in the new structure. Nevertheless, the reconstructed bridge continues to reflect the National Park Service's "rustic style" of architecture through its use of native material in an attempt to harmonize with the rugged park landscape The bridge is listed in the National Register of Historic Places as part of the Longmire Historic District.

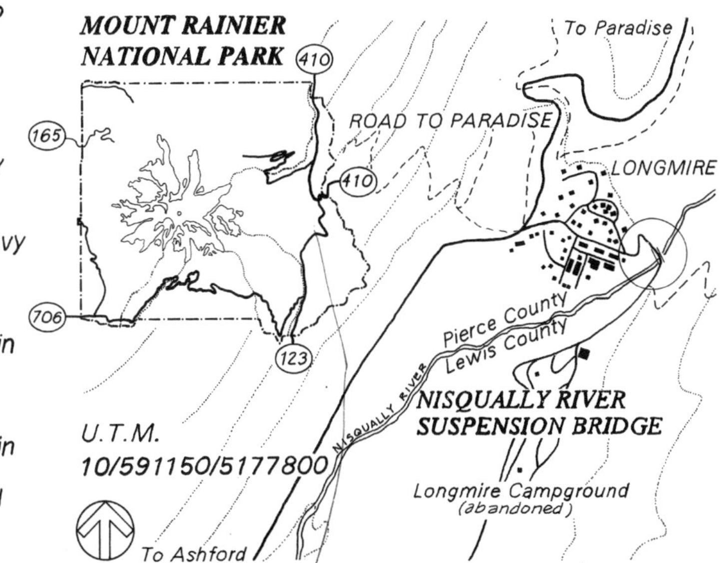

This project is part of the Historic American Engineering Record (HAER), a long-range program to document historically significant engineering and industrial works in the United States. The HAER program is administered by the Historic American Buildings Survey/Historic American Engineering Record Division (HABS/HAER) of the National Park Service, U.S. Department of the Interior. The Mount Rainier Roads and Bridges Recording Project was cosponsored during the summer of 1992 by HAER under the general direction of Dr. Robert J. Kapsch, Chief; the National Park Service Roads and Bridges Program, John Gingles, Manager; and by Mount Rainier National Park, William Briggle, Superintendent.

The field work, measured drawings, historical reports, and photographs were prepared under the direction of Project Leader Eric N. DeLony, Chief of HAER. The recording team consisted of Todd A. Croteau, team supervisor; architectural technicians Bryan D. Fish and Daniela Trettel (U.S./ICOMOS, Argentina); and landscape architect Julie Ann Dickson Historical reports were prepared by project historian Richard H. Quin. Formal large-format photography was done by Jet Lowe, HAER staff photographer.

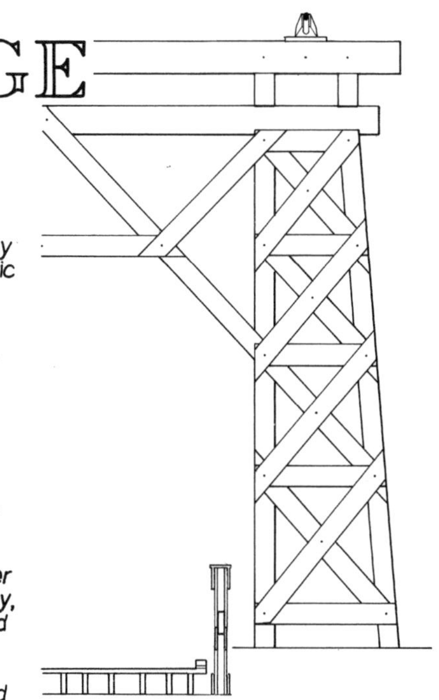

PRESENT TOWER

Drawing based on field measurements historical photographs and original design documents...See Federal Highway Administration Structure no. 9450-002P

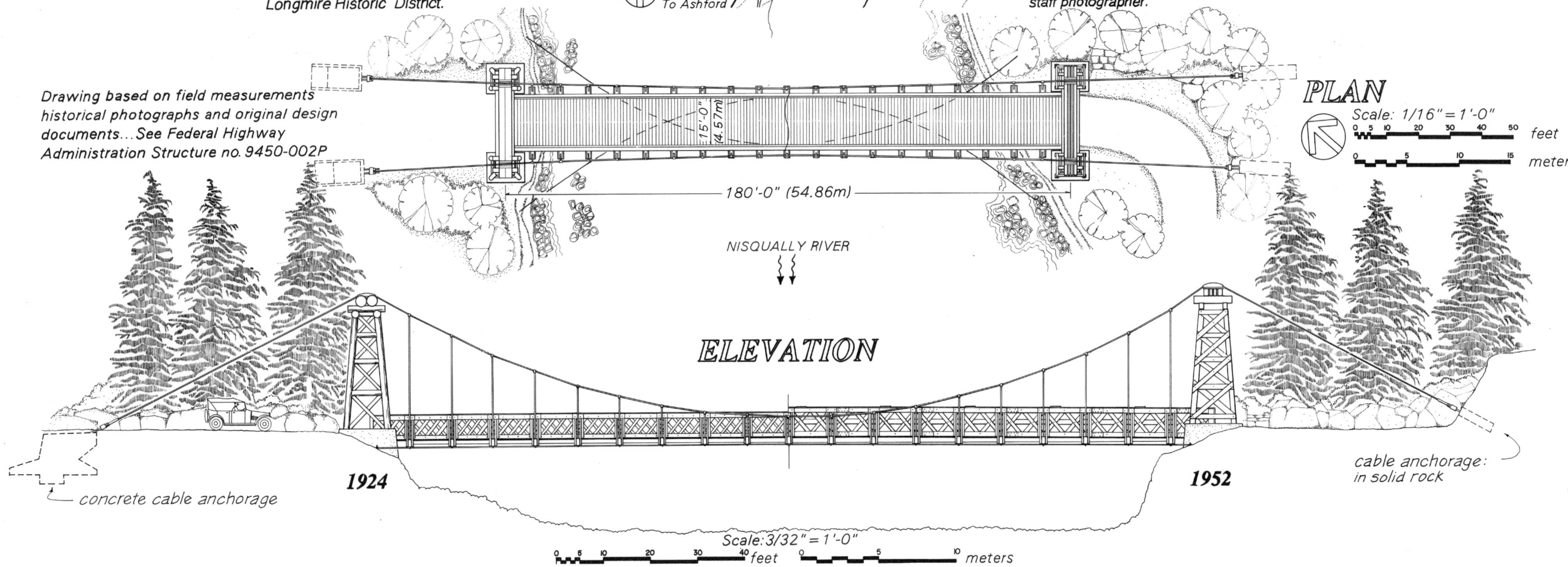

DELINEATED BY: Bryan D. Fish 1992

MOUNT RAINIER ROADS AND BRIDGES RECORDING PROJECT
NATIONAL PARK SERVICE
UNITED STATES DEPARTMENT OF THE INTERIOR

NISQUALLY RIVER SUSPENSION BRIDGE - 1924 / 1952
LONGMIRE SERVICE ROAD
MOUNT RAINIER NATIONAL PARK PIERCE / LEWIS COUNTY WASHINGTON

SHEET 1 of 3

HISTORIC AMERICAN ENGINEERING RECORD WA-44

IF REPRODUCED, PLEASE CREDIT: HISTORIC AMERICAN ENGINEERING RECORD, NATIONAL PARK SERVICE, NAME OF DELINEATOR, DATE OF THE DRAWING

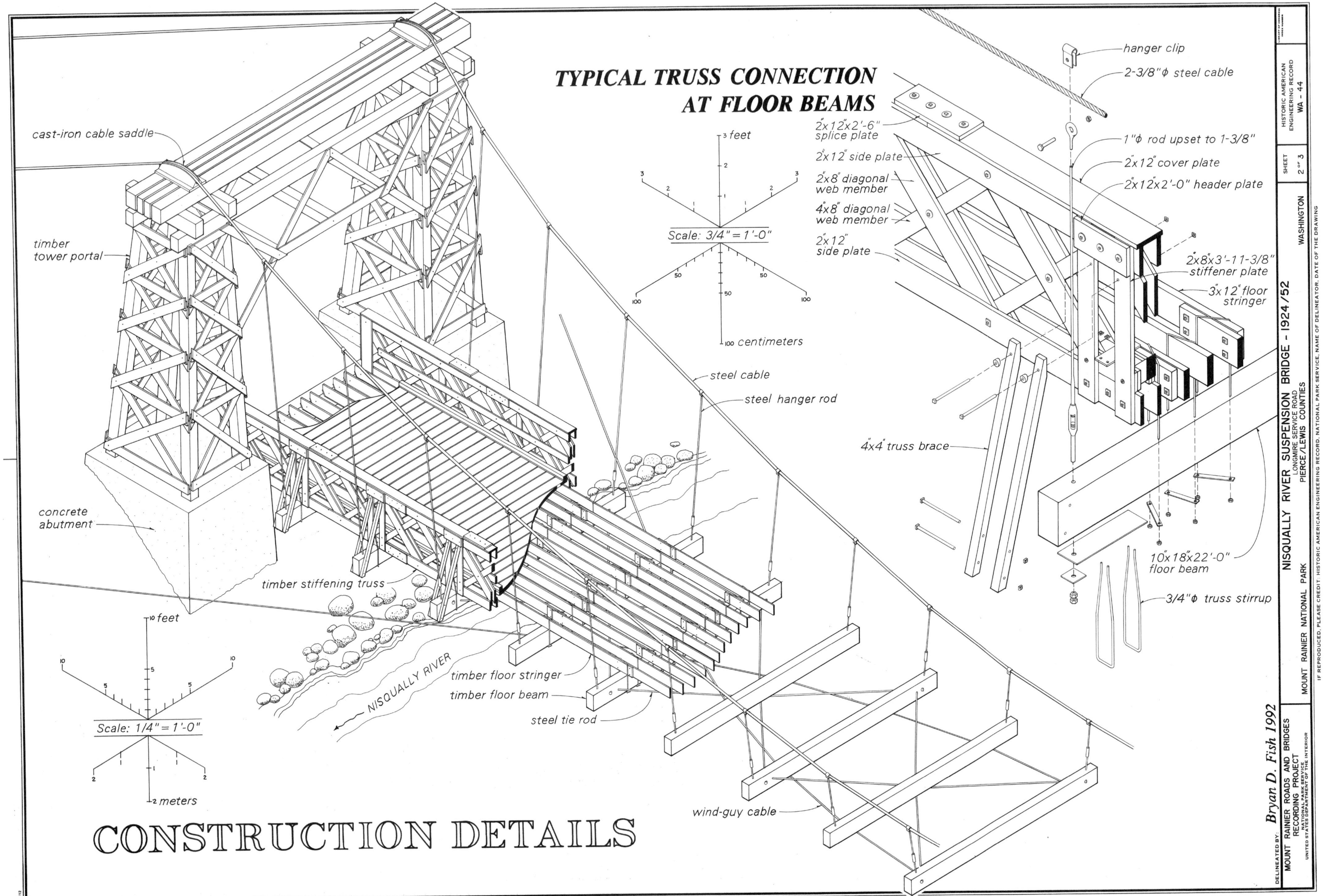
TYPICAL TRUSS CONNECTION AT FLOOR BEAMS
cast-iron cable saddle
timber tower portal
concrete abutment
timber stiffening truss
steel cable
steel hanger rod
timber floor stringer
timber floor beam
steel tie rod
wind-guy cable
NISQUALLY RIVER
Scale: 3/4" = 1'-0"
3 feet
100 centimeters
Scale: 1/4" = 1'-0"
10 feet
2 meters
hanger clip
2-3/8"ϕ steel cable
2"x12"x2'-6" splice plate
2"x12" side plate
2"x8" diagonal web member
4"x8" diagonal web member
2"x12" side plate
1"ϕ rod upset to 1-3/8"
2"x12" cover plate
2"x12"x2'-0" header plate
2"x8"x3'-11-3/8" stiffener plate
3"x12" floor stringer
4"x4" truss brace
10"x18"x22'-0" floor beam
3/4"ϕ truss stirrup
CONSTRUCTION DETAILS
NISQUALLY RIVER SUSPENSION BRIDGE - 1924/52
LONGMIRE SERVICE ROAD
PIERCE/LEWIS COUNTIES
WASHINGTON
MOUNT RAINIER NATIONAL PARK
HISTORIC AMERICAN ENGINEERING RECORD
WA-44
SHEET 2 OF 3
DELINEATED BY: Bryan D. Fish 1992
MOUNT RAINIER ROADS AND BRIDGES RECORDING PROJECT
NATIONAL PARK SERVICE
UNITED STATES DEPARTMENT OF THE INTERIOR
IF REPRODUCED, PLEASE CREDIT: HISTORIC AMERICAN ENGINEERING RECORD, NATIONAL PARK SERVICE, NAME OF DELINEATOR, DATE OF THE DRAWING

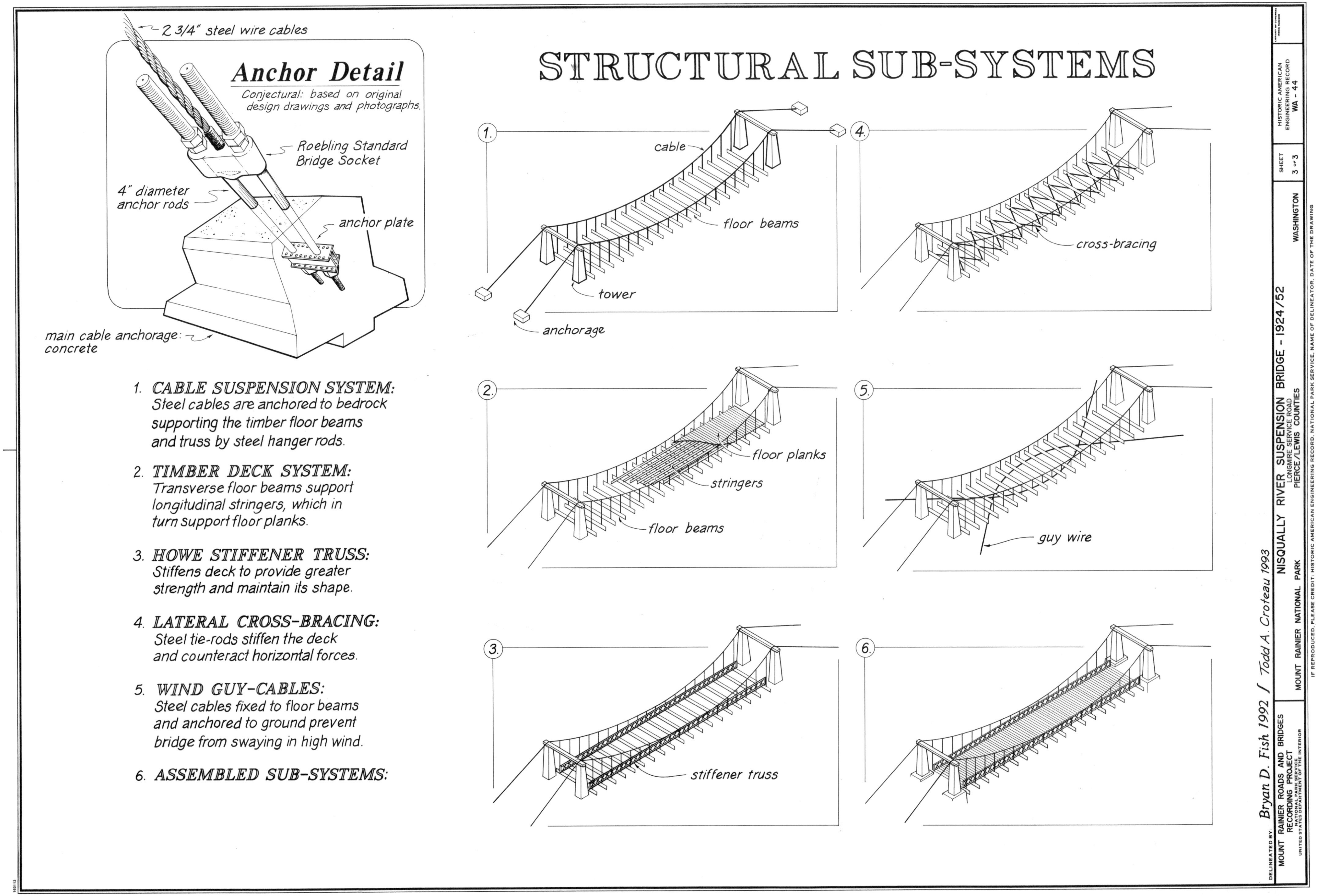
STRUCTURAL SUB-SYSTEMS
Anchor Detail
Conjectural: based on original design drawings and photographs.
2 3/4" steel wire cables
Roebling Standard Bridge Socket
4" diameter anchor rods
anchor plate
main cable anchorage: concrete
1. CABLE SUSPENSION SYSTEM: Steel cables are anchored to bedrock supporting the timber floor beams and truss by steel hanger rods.
2. TIMBER DECK SYSTEM: Transverse floor beams support longitudinal stringers, which in turn support floor planks.
3. HOWE STIFFENER TRUSS: Stiffens deck to provide greater strength and maintain its shape.
4. LATERAL CROSS-BRACING: Steel tie-rods stiffen the deck and counteract horizontal forces.
5. WIND GUY-CABLES: Steel cables fixed to floor beams and anchored to ground prevent bridge from swaying in high wind.
6. ASSEMBLED SUB-SYSTEMS:
1.
cable
floor beams
tower
anchorage
2.
floor planks
stringers
floor beams
3.
stiffener truss
4.
cross-bracing
5.
guy wire
6.
DELINEATED BY: Bryan D. Fish 1992 / Todd A. Croteau 1993
MOUNT RAINIER ROADS AND BRIDGES RECORDING PROJECT
NATIONAL PARK SERVICE
UNITED STATES DEPARTMENT OF THE INTERIOR
NISQUALLY RIVER SUSPENSION BRIDGE - 1924/52
LONGMIRE SERVICE ROAD
PIERCE/LEWIS COUNTIES
MOUNT RAINIER NATIONAL PARK
WASHINGTON
SHEET 3 of 3
HISTORIC AMERICAN ENGINEERING RECORD
WA-44
IF REPRODUCED, PLEASE CREDIT: HISTORIC AMERICAN ENGINEERING RECORD, NATIONAL PARK SERVICE, NAME OF DELINEATOR, DATE OF THE DRAWING

CHRISTINE FALLS BRIDGE
1928

A curving timber deck truss spanned Van Trump Creek at Christine Falls as early as 1908. The first bridge, built by the Army Corps of Engineers, was designed with an "intention to follow the graceful curves of the natural surface, being more pleasing and far less distructive than those laid with mathematical precision." This philosophy was continued on through the NPS years.

The present Christine Falls Bridge was designed by the Bureau of Public Roads in cooperation with the NPS Branch of Plans and Designs, Western Division, which recommended the use of stone facing. Constructed in 1927-28 as part of the Nisqually Road improvements, this bridge is an excellent example of rustic styling employed to harmonize a structure with its natural setting.

The three-centered arch of reinforced concrete has a clear span of 56 feet. Care was taken to face the filled arch with native stone and construct parapet walls similarly. NPS Landscape Architect, E.A. Davidson supervised the contractor's stonework.

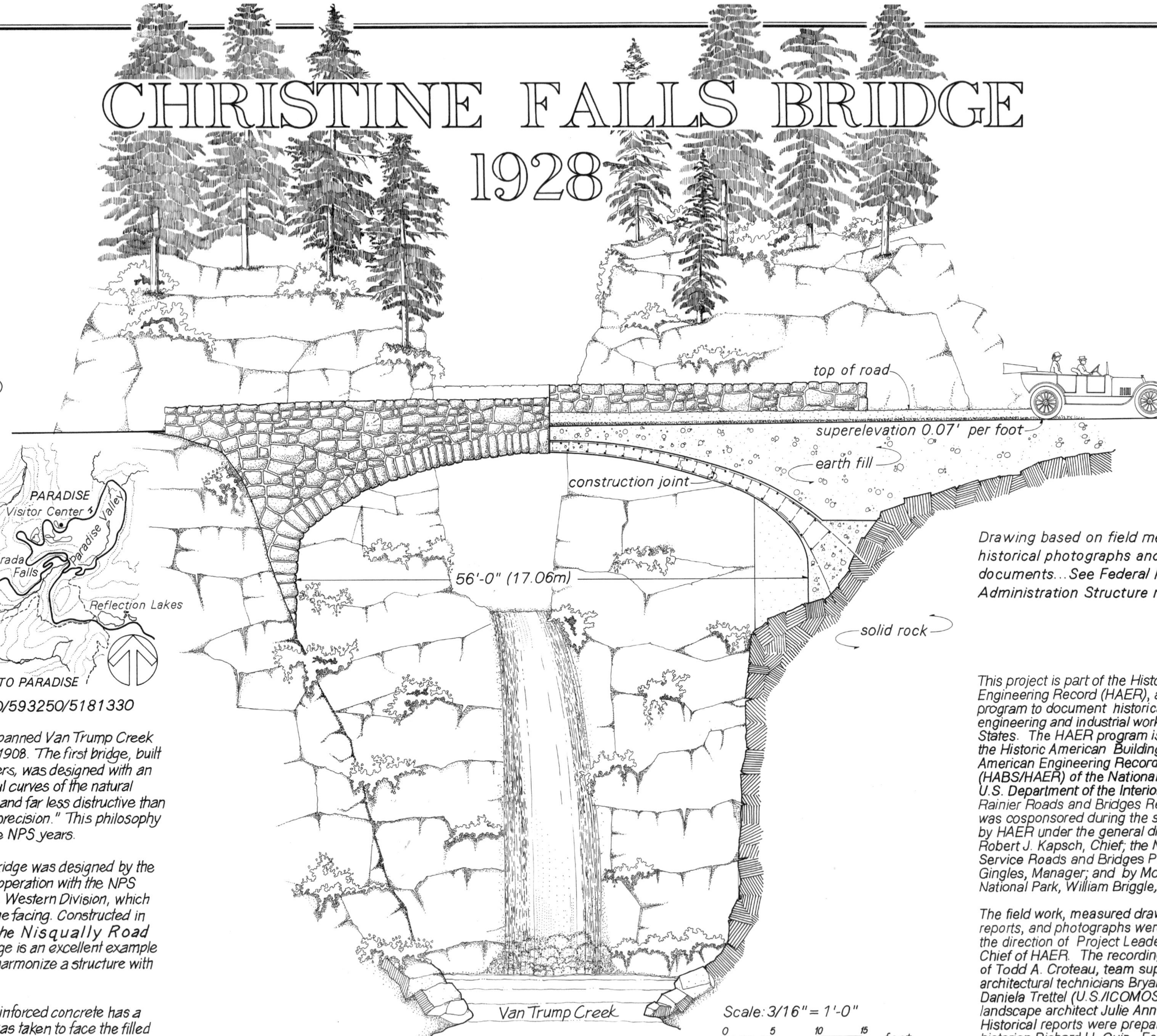

ELEVATION/SECTION

Drawing based on field measurements historical photographs and original design documents... See Federal Highway Administration Structure no. 9450-003P

This project is part of the Historic American Engineering Record (HAER), a long-range program to document historically significant engineering and industrial works in the United States. The HAER program is administered by the Historic American Buildings Survey/Historic American Engineering Record Division (HABS/HAER) of the National Park Service, U.S. Department of the Interior. The Mount Rainier Roads and Bridges Recording Project was cosponsored during the summer of 1992 by HAER under the general direction of Dr. Robert J. Kapsch, Chief; the National Park Service Roads and Bridges Program, John Gingles, Manager; and by Mount Rainier National Park, William Briggle, Superintendent.

The field work, measured drawings, historical reports, and photographs were prepared under the direction of Project Leader Eric N. DeLony, Chief of HAER. The recording team consisted of Todd A. Croteau, team supervisor; architectural technicians Bryan D. Fish and Daniela Trettel (U.S./ICOMOS, Argentina); and landscape architect Julie Ann Dickson. Historical reports were prepared by project historian Richard H. Quin. Formal large-format photography was done by John T. "Jet" Lowe, HAER staff photographer.

DELINEATED BY: Daniela Trettel, 1992

MOUNT RAINIER ROADS AND BRIDGES RECORDING PROJECT
NATIONAL PARK SERVICE
UNITED STATES DEPARTMENT OF THE INTERIOR

MOUNT RAINIER NATIONAL PARK

CHRISTINE FALLS BRIDGE - 1928
NISQUALLY ROAD
PIERCE COUNTY
WASHINGTON

SHEET 1 OF 2

HISTORIC AMERICAN ENGINEERING RECORD
WA-48

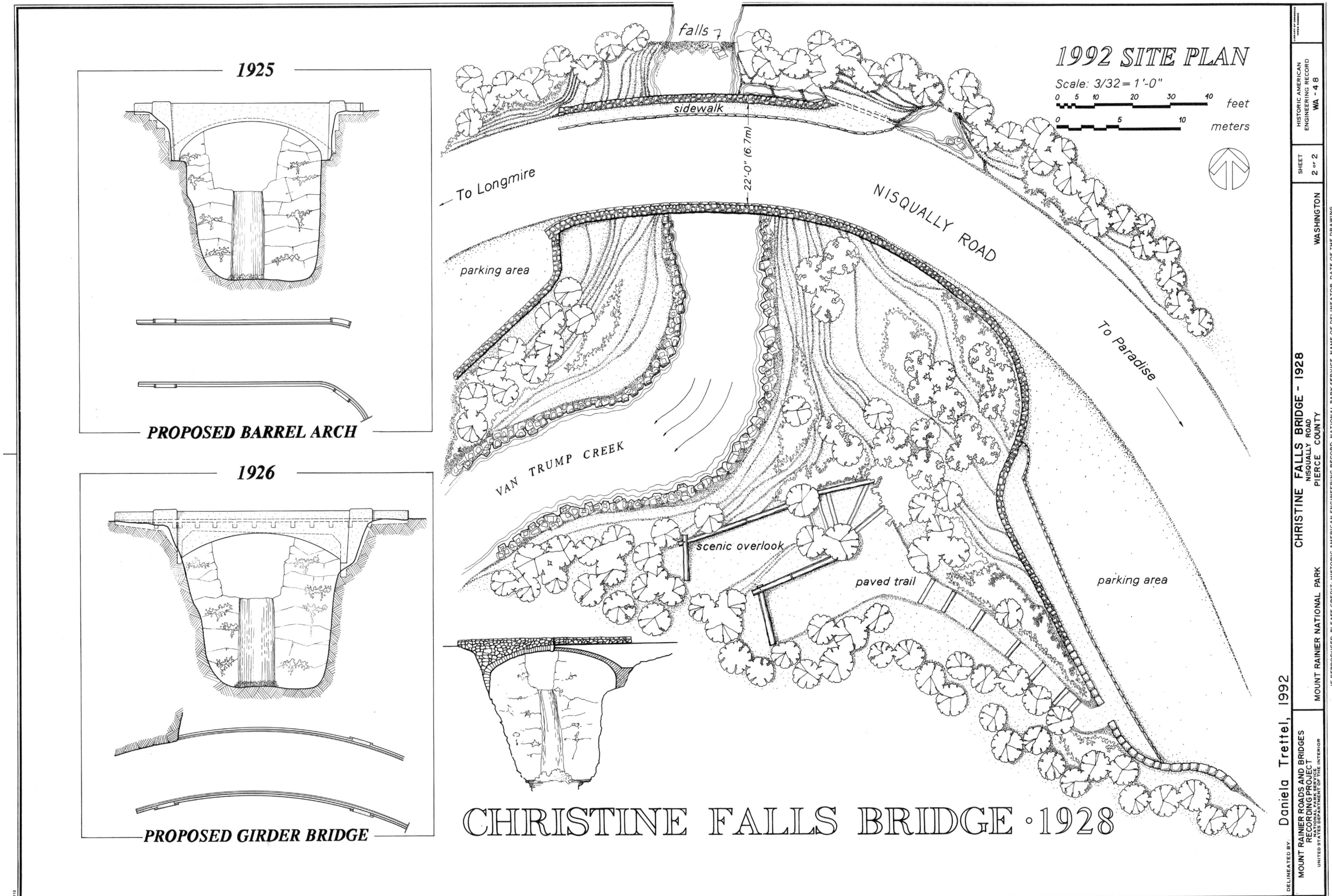
1925
PROPOSED BARREL ARCH
1926
PROPOSED GIRDER BRIDGE
falls
sidewalk
22'-0" (6.7m)
To Longmire
NISQUALLY ROAD
To Paradise
parking area
VAN TRUMP CREEK
scenic overlook
paved trail
parking area
1992 SITE PLAN
Scale: 3/32 = 1'-0"
0 5 10 20 30 40 feet
0 5 10 meters
CHRISTINE FALLS BRIDGE · 1928
DELINEATED BY: Daniela Trettel, 1992
MOUNT RAINIER ROADS AND BRIDGES RECORDING PROJECT
NATIONAL PARK SERVICE
UNITED STATES DEPARTMENT OF THE INTERIOR
CHRISTINE FALLS BRIDGE - 1928
NISQUALLY ROAD
MOUNT RAINIER NATIONAL PARK
PIERCE COUNTY
WASHINGTON
SHEET
2 of 2
HISTORIC AMERICAN ENGINEERING RECORD
WA - 48
IF REPRODUCED, PLEASE CREDIT: HISTORIC AMERICAN ENGINEERING RECORD, NATIONAL PARK SERVICE, NAME OF DELINEATOR, DATE OF THE DRAWING

ST. ANDREWS CREEK BRIDGE
1931

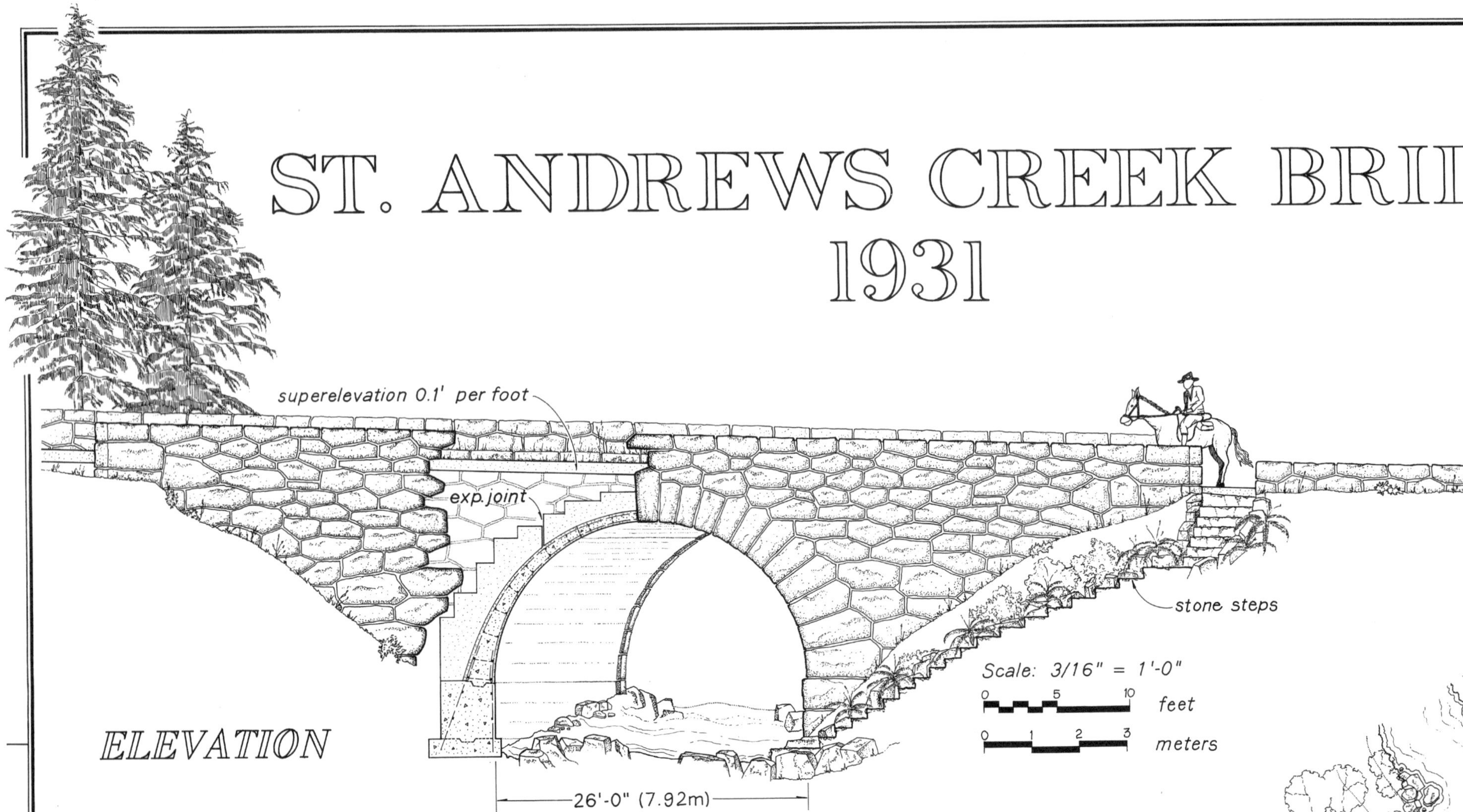

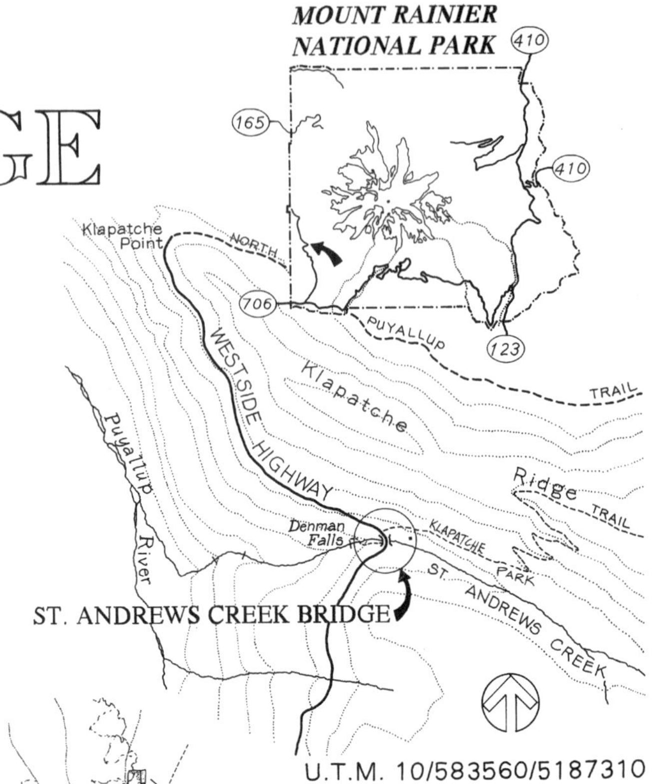

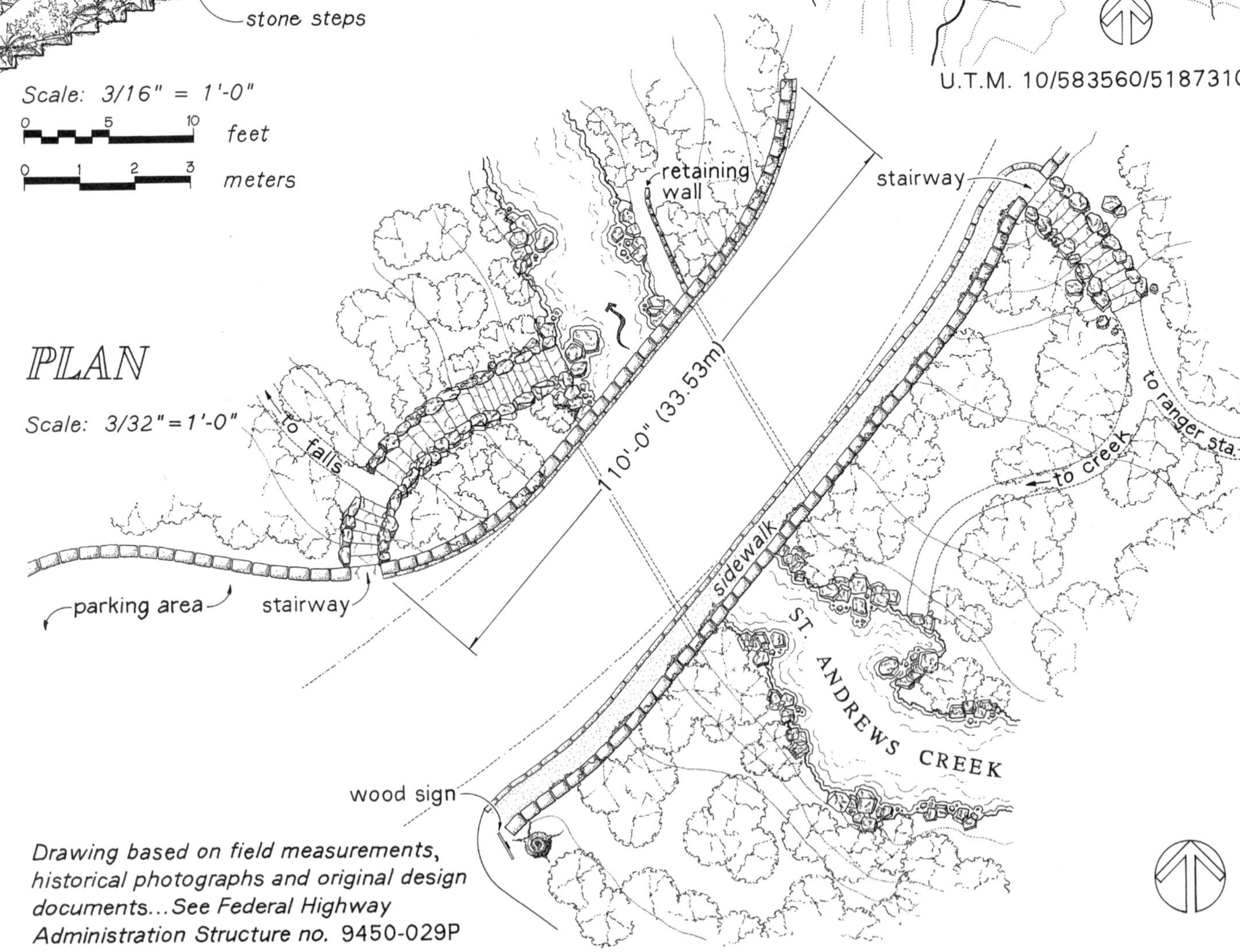

Construction of the West Side Road was the most ambitious work yet undertaken by the Bureau of Public Roads at Mount Rainier. Although original plans called for a two-lane gravel road linking the Nisqually Road with the Carbon River Entrance, the project was never fully realized. A ten-mile section to the North Puyallup River was completed in a laborious eight years. This work entailed a bridge across St. Andrews Creek. The contract for this bridge and a nearby span across the South Puyallup River was awarded in July 1930 to the W.T. Butler Company of Seattle. The two spans were completed 17 August 1931 at a combined cost of $41,985.94.

St. Andrews Creek Bridge is a reinforced concrete filled spandrel arch structure faced in native stone. The bridge spans the creek on a 26-foot circular arch with massive stone voussoirs and follows the graceful curve of the road. Park Service Landscape Engineer E.A. Davidson oversaw the bridge's masonry work and landscape design. At his suggestion stone steps were later constructed at opposite corners of the span to provide access to the creek, Denman Falls and a nearby ranger station. The bridge was designed to harmonize with the landscape through its use of rough-cut stone and native species in landscape planting.

This project is part of the Historic American Engineering Record (HAER), a long-range program to document historically significant engineering and industrial works in the United States. The HAER program is administered by the Historic American Buildings Survey/Historic American Engineering Record Division (HABS/HAER) of the National Park Service, U.S. Department of the Interior. The Mount Rainier Roads and Bridges Recording Project was cosponsored during the summer of 1992 by HAER under the general direction of Dr. Robert J. Kapsch, Chief; the National Park Service Roads and Bridges Program, John Gingles, Manager; and by Mount Rainier National Park, William Briggle, Superintendent.

The field work, measured drawings, historical reports, and photographs were prepared under the direction of Project Leader Eric N. DeLony, Chief of HAER. The recording team consisted of Todd A. Croteau, team supervisor; architectural technicians Bryan D. Fish and Daniela Trettel (U.S./ICOMOS, Argentina); and landscape architect Julie Ann Dickson. Historical reports were prepared by project historian Richard H. Quin. Formal large-format photography was done by Jet Lowe, HAER staff photographer.

Drawing based on field measurements, historical photographs and original design documents... See Federal Highway Administration Structure no. 9450-029P

DELINEATED BY *Julie Ann Dickson, 1992*

MOUNT RAINIER ROADS AND BRIDGES RECORDING PROJECT
NATIONAL PARK SERVICE
UNITED STATES DEPARTMENT OF THE INTERIOR

MOUNT RAINIER NATIONAL PARK

SAINT ANDREWS CREEK BRIDGE - 1931
WEST SIDE ROAD
PIERCE COUNTY

WASHINGTON

SHEET 1 of 1

HISTORIC AMERICAN ENGINEERING RECORD
WA - 51

IF REPRODUCED, PLEASE CREDIT: HISTORIC AMERICAN ENGINEERING RECORD, NATIONAL PARK SERVICE, NAME OF DELINEATOR, DATE OF THE DRAWING

LAUGHINGWATER CREEK BRIDGE · 1935

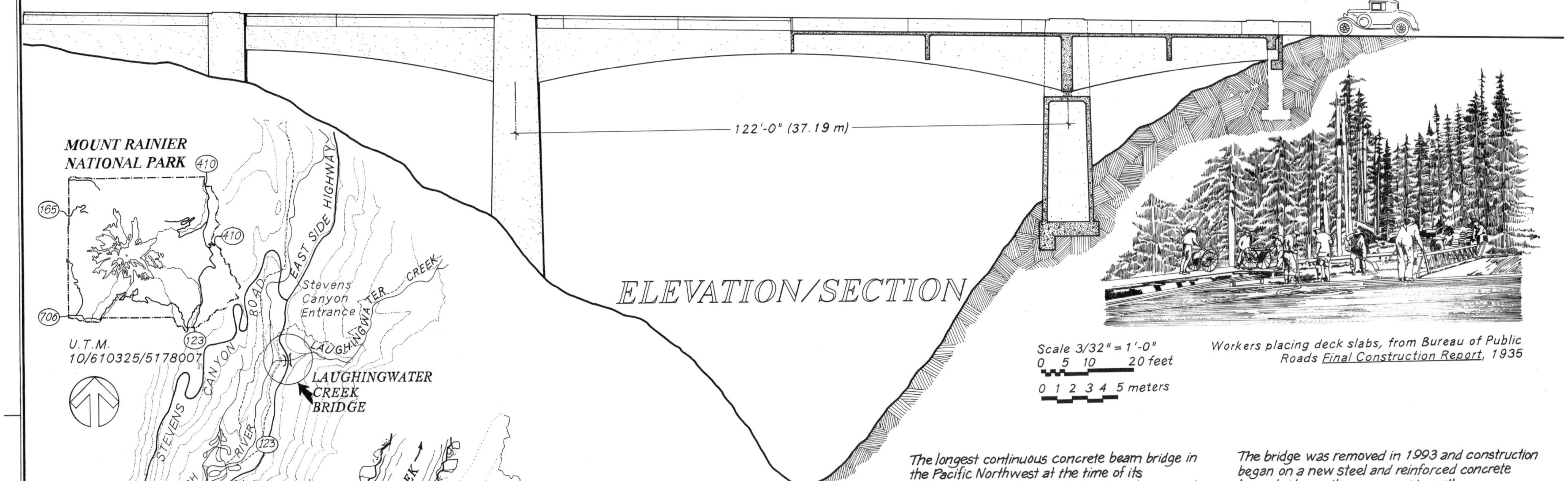

Workers placing deck slabs, from Bureau of Public Roads Final Construction Report, 1935

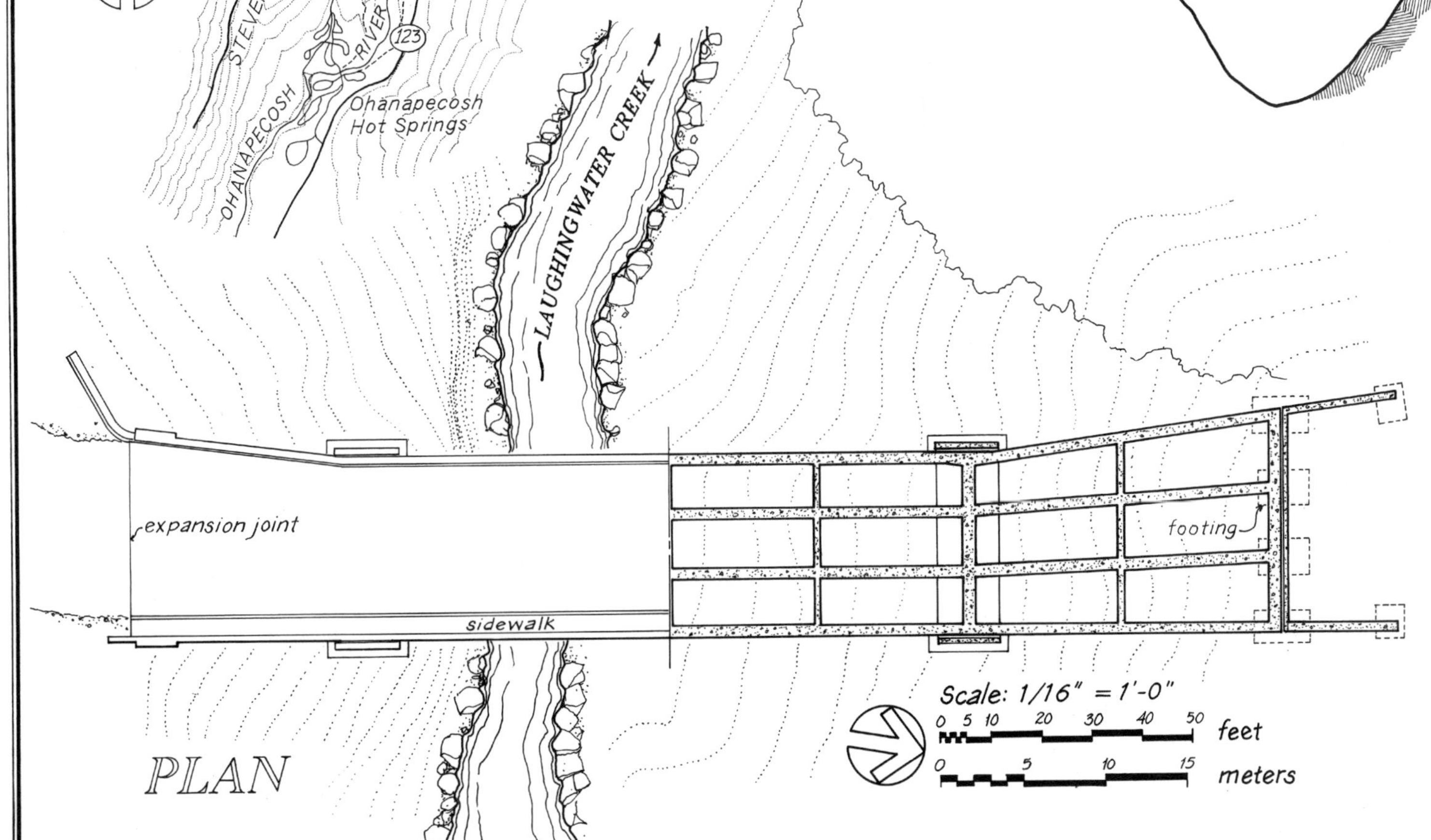

The longest continuous concrete beam bridge in the Pacific Northwest at the time of its completion, Laughingwater Creek Bridge carried the East Side Highway over its namesake creek 1-1/2 miles north of Ohanapecosh Hot Springs. Construction of the permanent bridge began in May 1935. The work was carried out by the Portland, Oregon firm of Joplin and Eldon who also worked on a number of other road projects at Mount Rainier. The firm completed the structure in October 1935 at a total cost of $54,542.93.

The bridge's 122' central span was the largest continuous concrete beam constructed by the Bureau of Public Roads in the region. The length of the span was made possible by the unusually fine quality of concrete aggregate available, which provided for a very high working stress. In order to lessen the glare from the concrete and to allow the bridge to blend in with its surroundings, the concrete finish was sandblasted, then washed with acid and stained with two coats of a copperas solution.

By the mid-1950s, the bridge began to deteriorate due to a silica-alkali reaction which caused excessive leaching and exfoliation. In 1986, Federal Highway Administration officials became concerned that the structure might fail due to debonding of the reinforcing steel.

The bridge was removed in 1993 and construction began on a new steel and reinforced concrete beam bridge with masonry side walls.

This project is part of the Historic American Engineering Record (HAER), a long-range program to document historically significant engineering and industrial works in the United States. The HAER program is administered by the Historic American Buildings Survey/Historic American Engineering Record (HABS/HAER) Division of the National Park Service, U.S. Department of the Interior. The Mount Rainier Roads and Bridges Project was cosponsored during the summer of 1992 by the NPS Roads and Bridges Program, John Gingles, Manager; HABS/HAER, Robert J. Kapsch, Chief; and by Mount Rainier National Park, William Briggle, Superintendent.

The fieldwork, measured drawings, histories and photographs were prepared under the direction of Eric N. DeLony, Chief and Principal Architect, HAER. The recording team consisted of Todd A. Croteau, field supervisor; architectural technicians Bryan D. Fish and Daniela Trettel (U.S./ICOMOS, Argentina); and landscape architect Julie Ann Dickson. Historical reports were prepared by project historian Richard H. Quin. Formal on-site photography was done by Jet Lowe, HAER staff photographer.

DELINEATED BY: Todd Croteau, Christopher Marston, 1992

MOUNT RAINIER ROADS AND BRIDGES RECORDING PROJECT
NATIONAL PARK SERVICE
UNITED STATES DEPARTMENT OF THE INTERIOR

MOUNT RAINIER NATIONAL PARK

LAUGHINGWATER CREEK BRIDGE - 1935
EAST SIDE HIGHWAY
PIERCE COUNTY
WASHINGTON

SHEET 1 OF 1

HISTORIC AMERICAN ENGINEERING RECORD
WA - 55

CHINOOK PASS ENTRANCE 1936

Drawing based on field measurements historical photographs and original design documents...drawing #R-3049 (MORA Archives)
*original signage depicted

ELEVATION

Scale: 1/4" = 1'-0"

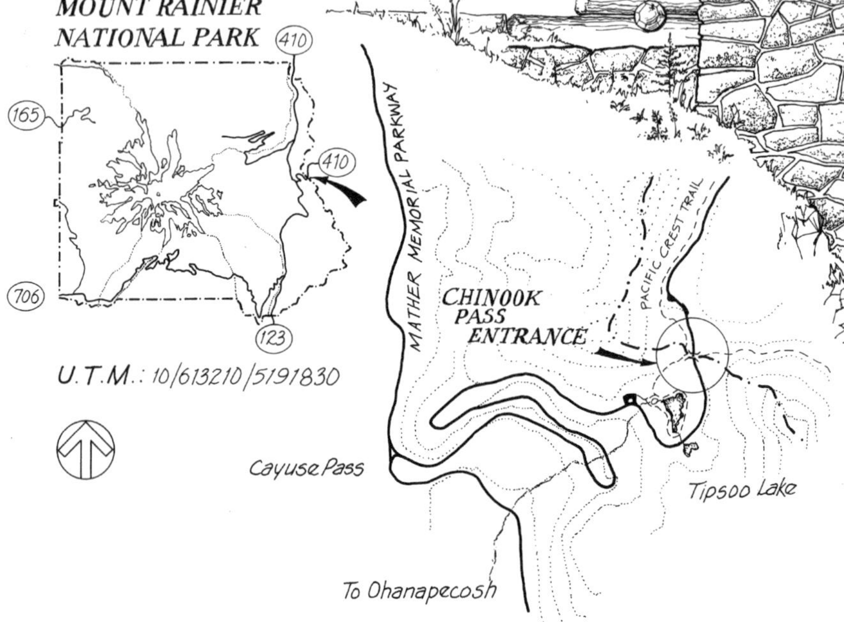

U.T.M.: 10/613210/5191830

A state highway up to Chinook Pass, and on to Yakima Valley, was opened to traffic in 1932. This road resulted in a new entrance to Mount Rainier National Park at the crest of the Cascade range. Interested in developing visitor facilities in this area, Superintendent Tomlinson requested the Branch of Plans and Designs, supervised by T.C. Vint, to prepare plans for a ranger station, picnic area, comfort station and entrance arch. Although Tomlinson's ambitious plans were never fully realized, two comfort stations and an entrance arch were constructed. In 1933, plans were drawn for the Chinook Pass Entrance and in 1936 the arch was constructed by the Civilian Conservation Corps (CCC).

Unlike other entrances, this structure is both a portal and a bridge. Carrying a horse trail over Mather Memorial Parkway (Hwy. 410), this span assures the safe crossing of trail riders while providing vehicular traffic with an uninterrupted passage into the park. Stone masonry piers support the heavy beveled cedar log stringers and rails across the top, while flagstone approaches lead to the roughly hewn cedar puncheons of the deck. Altered very little since its construction, this span carries the Pacific Crest National Scenic Trail.

PLAN

Scale: 1/8" = 1'-0"

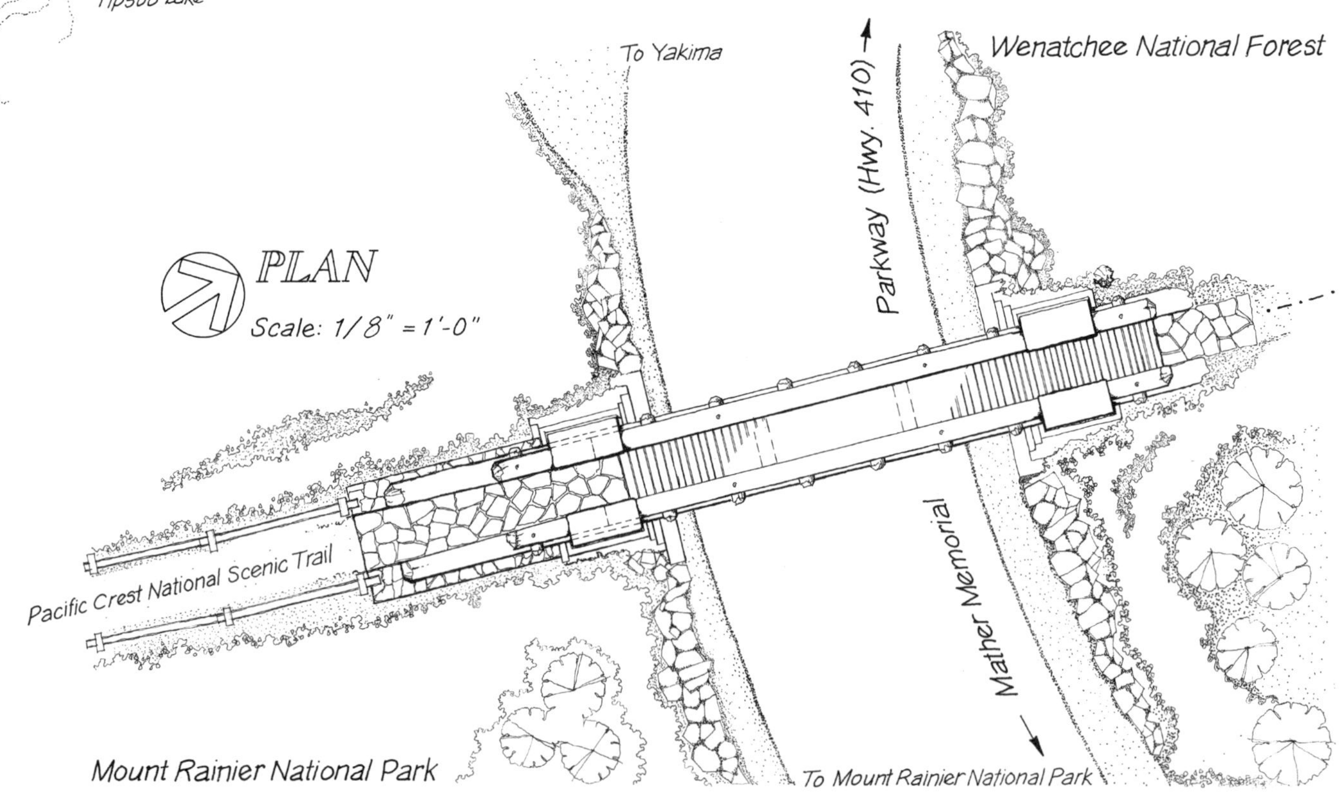

This project is part of the Historic American Engineering Record (HAER), a long-range program to document historically significant engineering and industrial works in the United States. The HAER program is administered by the Historic American Buildings Survey/Historic American Engineering Record Division (HABS/HAER) of the National Park Service, U.S. Department of the Interior. The Mount Rainier Roads and Bridges Recording Project was cosponsored during the summer of 1992 by HAER under the general direction of Dr. Robert J. Kapsch, Chief; the National Park Service Roads and Bridges Program, John Gingles, Manager; and by Mount Rainier National Park, William Briggle, Superintendent.

The field work, measured drawings, historical reports, and photographs were prepared under the direction of Project Leader Eric N. DeLony, Chief of HAER. The recording team consisted of Todd A. Croteau, team supervisor; architectural technicians Bryan D. Fish and Daniela Trettel (U.S./ICOMOS, Argentina); and landscape architect Julie Ann Dickson. Historical reports were prepared by project historian Richard H. Quin. Formal large-format photography was done by Jet Lowe, HAER staff photographer.

DELINEATED BY: Todd A. Croteau, 1992

MOUNT RAINIER ROADS AND BRIDGES RECORDING PROJECT
NATIONAL PARK SERVICE
UNITED STATES DEPARTMENT OF THE INTERIOR

MOUNT RAINIER NATIONAL PARK

CHINOOK PASS ENTRANCE - 1936
MATHER MEMORIAL PARKWAY (HWY 410)
PIERCE COUNTY

WASHINGTON

SHEET 1 OF 2

HISTORIC AMERICAN ENGINEERING RECORD
WA - 43

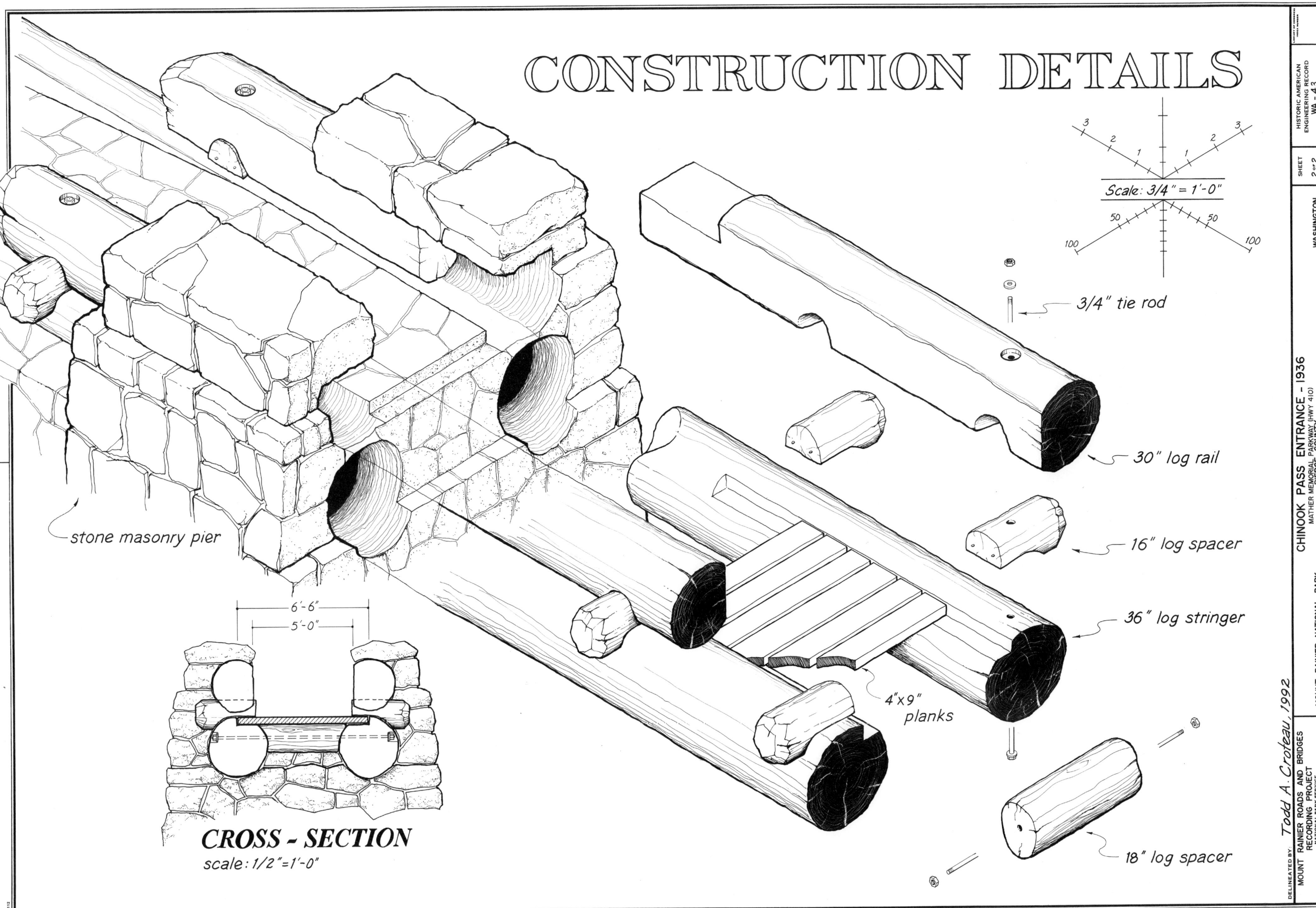

CONSTRUCTION DETAILS
Scale: 3/4" = 1'-0"
3/4" tie rod
30" log rail
16" log spacer
36" log stringer
4"x9" planks
18" log spacer
stone masonry pier
6'-6"
5'-0"
CROSS - SECTION
scale: 1/2"=1'-0"
DELINEATED BY: Todd A. Croteau, 1992
MOUNT RAINIER ROADS AND BRIDGES RECORDING PROJECT
NATIONAL PARK SERVICE
UNITED STATES DEPARTMENT OF THE INTERIOR
CHINOOK PASS ENTRANCE - 1936
MATHER MEMORIAL PARKWAY (HWY 410)
MOUNT RAINIER NATIONAL PARK
PIERCE COUNTY
WASHINGTON
SHEET 2 OF 2
HISTORIC AMERICAN ENGINEERING RECORD
WA-43
IF REPRODUCED, PLEASE CREDIT: HISTORIC AMERICAN ENGINEERING RECORD, NATIONAL PARK SERVICE, NAME OF DELINEATOR, DATE OF THE DRAWING

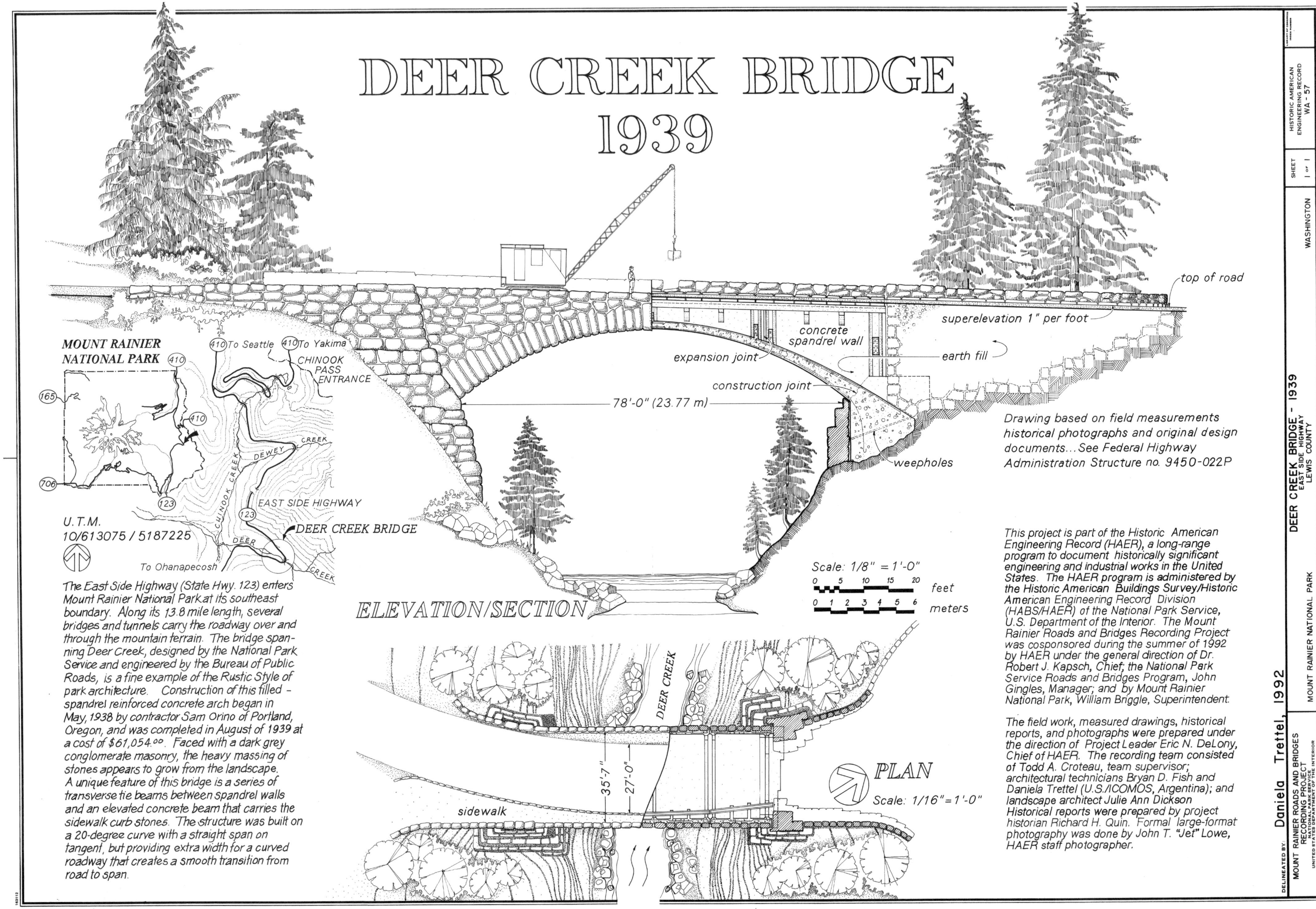
DEER CREEK BRIDGE
1939
top of road
superelevation 1" per foot
concrete spandrel wall
expansion joint
earth fill
construction joint
78'-0" (23.77 m)
weepholes
Drawing based on field measurements historical photographs and original design documents... See Federal Highway Administration Structure no. 9450-022P
MOUNT RAINIER NATIONAL PARK
410 To Seattle
410 To Yakima
CHINOOK PASS ENTRANCE
165
410
706
123
DEWEY CREEK
CHINOOK CREEK
EAST SIDE HIGHWAY
DEER CREEK BRIDGE
DEER CREEK
To Ohanapecosh
U.T.M.
10/613075 / 5187225
ELEVATION/SECTION
Scale: 1/8" = 1'-0"
0 5 10 15 20 feet
0 1 2 3 4 5 6 meters
The East Side Highway (State Hwy. 123) enters Mount Rainier National Park at its southeast boundary. Along its 13.8 mile length, several bridges and tunnels carry the roadway over and through the mountain terrain. The bridge spanning Deer Creek, designed by the National Park Service and engineered by the Bureau of Public Roads, is a fine example of the Rustic Style of park architecture. Construction of this filled-spandrel reinforced concrete arch began in May, 1938 by contractor Sam Orino of Portland, Oregon, and was completed in August of 1939 at a cost of $61,054.00. Faced with a dark grey conglomerate masonry, the heavy massing of stones appears to grow from the landscape. A unique feature of this bridge is a series of transverse tie beams between spandrel walls and an elevated concrete beam that carries the sidewalk curb stones. The structure was built on a 20-degree curve with a straight span on tangent, but providing extra width for a curved roadway that creates a smooth transition from road to span.
DEER CREEK
35'-7"
27'-0"
sidewalk
PLAN
Scale: 1/16" = 1'-0"
This project is part of the Historic American Engineering Record (HAER), a long-range program to document historically significant engineering and industrial works in the United States. The HAER program is administered by the Historic American Buildings Survey/Historic American Engineering Record Division (HABS/HAER) of the National Park Service, U.S. Department of the Interior. The Mount Rainier Roads and Bridges Recording Project was cosponsored during the summer of 1992 by HAER under the general direction of Dr. Robert J. Kapsch, Chief; the National Park Service Roads and Bridges Program, John Gingles, Manager; and by Mount Rainier National Park, William Briggle, Superintendent.
The field work, measured drawings, historical reports, and photographs were prepared under the direction of Project Leader Eric N. DeLony, Chief of HAER. The recording team consisted of Todd A. Croteau, team supervisor; architectural technicians Bryan D. Fish and Daniela Trettel (U.S./ICOMOS, Argentina); and landscape architect Julie Ann Dickson. Historical reports were prepared by project historian Richard H. Quin. Formal large-format photography was done by John T. "Jet" Lowe, HAER staff photographer.
DELINEATED BY: Daniela Trettel, 1992
MOUNT RAINIER ROADS AND BRIDGES RECORDING PROJECT
NATIONAL PARK SERVICE
UNITED STATES DEPARTMENT OF THE INTERIOR
MOUNT RAINIER NATIONAL PARK
DEER CREEK BRIDGE - 1939
EAST SIDE HIGHWAY
LEWIS COUNTY
WASHINGTON
SHEET 1 OF 1
HISTORIC AMERICAN ENGINEERING RECORD
WA-57
IF REPRODUCED, PLEASE CREDIT: HISTORIC AMERICAN ENGINEERING RECORD, NATIONAL PARK SERVICE, NAME OF DELINEATOR, DATE OF THE DRAWING

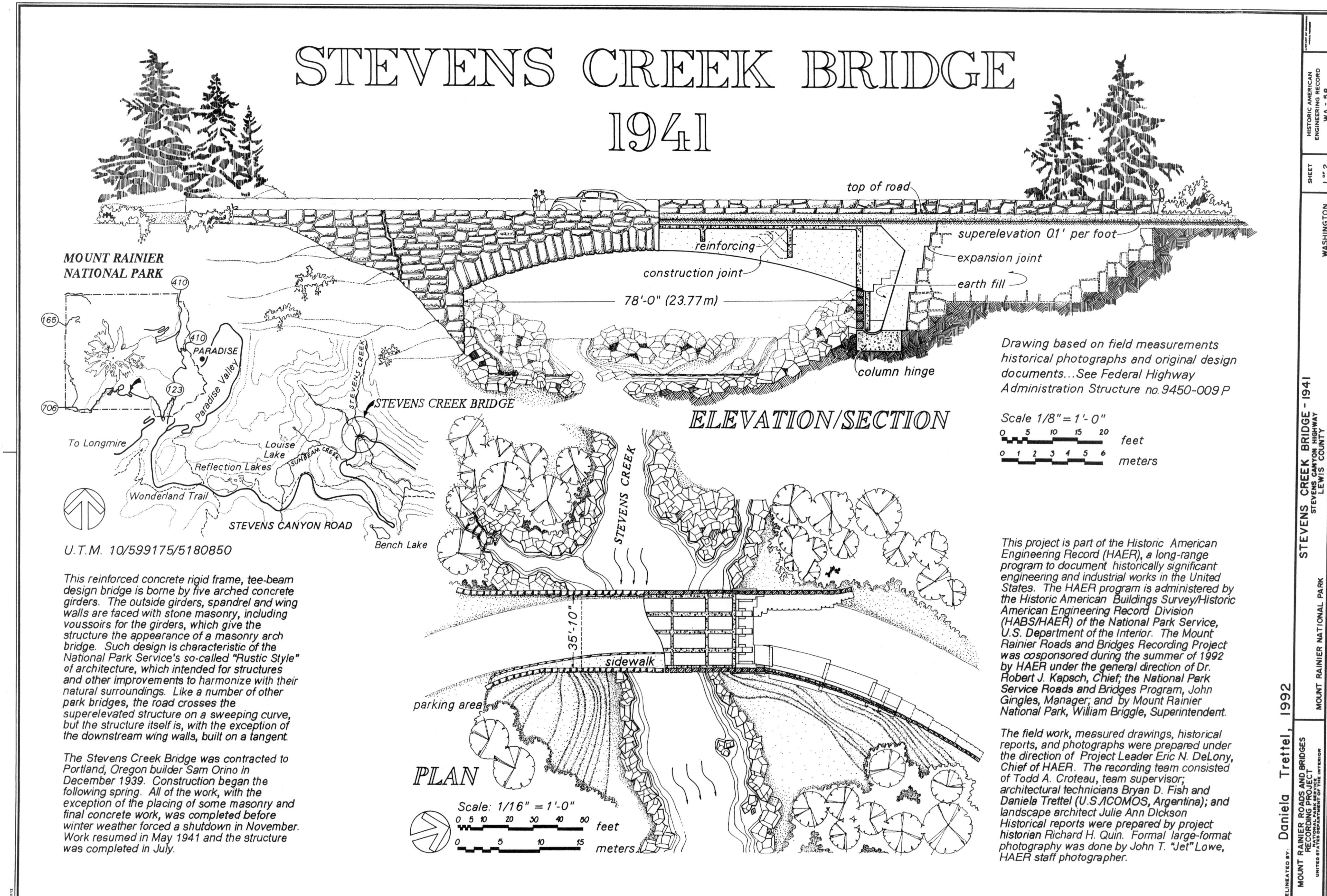

STEVENS CREEK BRIDGE
1941
top of road
superelevation 01' per foot
reinforcing
expansion joint
construction joint
earth fill
78'-0" (23.77m)
column hinge
ELEVATION/SECTION
MOUNT RAINIER
NATIONAL PARK
410
165
410
PARADISE
123
706
Paradise Valley
STEVENS CREEK
STEVENS CREEK BRIDGE
To Longmire
Louise Lake
Reflection Lakes
SUNBEAM CREEK
Wonderland Trail
STEVENS CANYON ROAD
Bench Lake
U.T.M. 10/599175/5180850
Drawing based on field measurements historical photographs and original design documents...See Federal Highway Administration Structure no.9450-009P
Scale 1/8" = 1'-0"
0 5 10 15 20 feet
0 1 2 3 4 5 6 meters
STEVENS CREEK
35'-10"
sidewalk
parking area
PLAN
Scale: 1/16" = 1'-0"
0 5 10 20 30 40 50 feet
0 5 10 15 meters
This reinforced concrete rigid frame, tee-beam design bridge is borne by five arched concrete girders. The outside girders, spandrel and wing walls are faced with stone masonry, including voussoirs for the girders, which give the structure the appearance of a masonry arch bridge. Such design is characteristic of the National Park Service's so-called "Rustic Style" of architecture, which intended for structures and other improvements to harmonize with their natural surroundings. Like a number of other park bridges, the road crosses the superelevated structure on a sweeping curve, but the structure itself is, with the exception of the downstream wing walls, built on a tangent.
The Stevens Creek Bridge was contracted to Portland, Oregon builder Sam Orino in December 1939. Construction began the following spring. All of the work, with the exception of the placing of some masonry and final concrete work, was completed before winter weather forced a shutdown in November. Work resumed in May 1941 and the structure was completed in July.
This project is part of the Historic American Engineering Record (HAER), a long-range program to document historically significant engineering and industrial works in the United States. The HAER program is administered by the Historic American Buildings Survey/Historic American Engineering Record Division (HABS/HAER) of the National Park Service, U.S. Department of the Interior. The Mount Rainier Roads and Bridges Recording Project was cosponsored during the summer of 1992 by HAER under the general direction of Dr. Robert J. Kapsch, Chief; the National Park Service Roads and Bridges Program, John Gingles, Manager; and by Mount Rainier National Park, William Briggle, Superintendent.
The field work, measured drawings, historical reports, and photographs were prepared under the direction of Project Leader Eric N. DeLony, Chief of HAER. The recording team consisted of Todd A. Croteau, team supervisor; architectural technicians Bryan D. Fish and Daniela Trettel (U.S./ICOMOS, Argentina); and landscape architect Julie Ann Dickson. Historical reports were prepared by project historian Richard H. Quin. Formal large-format photography was done by John T. "Jet" Lowe, HAER staff photographer.
DELINEATED BY: Daniela Trettel, 1992
MOUNT RAINIER ROADS AND BRIDGES RECORDING PROJECT
NATIONAL PARK SERVICE
UNITED STATES DEPARTMENT OF THE INTERIOR
STEVENS CREEK BRIDGE - 1941
STEVENS CANYON HIGHWAY
LEWIS COUNTY
MOUNT RAINIER NATIONAL PARK
WASHINGTON
SHEET 1 OF 2
HISTORIC AMERICAN ENGINEERING RECORD
WA - 58
IF REPRODUCED, PLEASE CREDIT: HISTORIC AMERICAN ENGINEERING RECORD, NATIONAL PARK SERVICE, NAME OF DELINEATOR, DATE OF THE DRAWING

MUDDY FORK RIVER BRIDGE
at BOX CANYON
1952

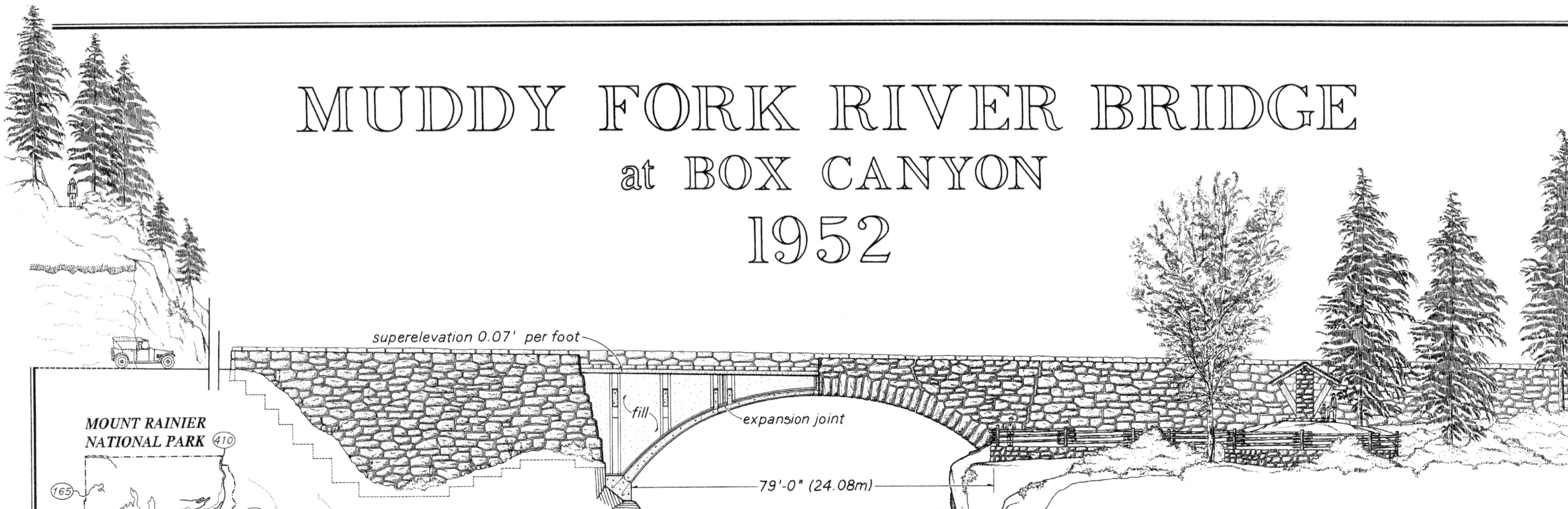

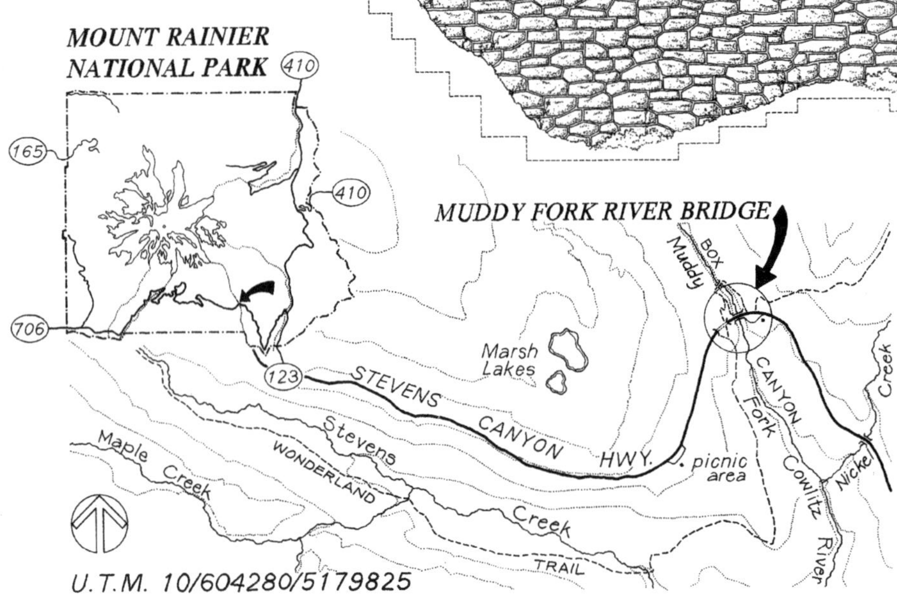

In May 1950, contractors Hawkins and Armstrong set up camp near Nickel Creek on the Stevens Canyon Highway, and began to quarry the face stone for the Muddy Fork of the Cowlitz River Bridge. By November the ring stones had been set in place and the concrete barrel arch poured. Work continued, weather permitting, until completion on October 23, 1952, 10 years after inception. Designed by the Bureau of Public Roads in cooperation with the NPS, this bridge is one of the finest examples of park service architecture integrating with the landscape.

The 79-foot reinforced-concrete filled spandrel arch is so skillfully crafted that it appears to have grown from the picturesque gorge of Box Canyon. Careful selection of stones ensured that the structure would blend with the unique glacial polished rock of the canyon. Special attention was also given to the design of the tunnel, due to its proximity to the bridge. Of the various masonry designs considered, an exposed rock portal was chosen, in order to minimize the intrusions in the landscape and preserve the natural beauty of this unusual site.

SECTION / ELEVATION

Drawing based on field measurements, historical photographs and original design documents.

This project is part of the Historic American Engineering Record (HAER), a long-range program to document historically significant engineering and industrial works in the United States. The HAER program is administered by the Historic American Buildings Survey/Historic American Engineering Record Division (HABS/HAER) of the National Park Service, U.S. Department of the Interior. The Mount Rainier Roads and Bridges Recording Project was cosponsored during the summer of 1992 by HAER under the general direction of Dr. Robert J. Kapsch, Chief; the National Park Service Roads and Bridges Program, John Gingles, Manager; and by Mount Rainier National Park, William Briggle, Superintendent.

The field work, measured drawings, historical reports, and photographs were prepared under the direction of Project Leader Eric N. DeLony, Chief of HAER. The recording team consisted of Todd A. Croteau, team supervisor; architectural technicians Bryan D. Fish and Daniela Trettel (U.S./ICOMOS, Argentina); and landscape architect Julie Ann Dickson. Historical reports were prepared by project historian Richard H. Quin, and formal large-format photography was done by Jet Lowe, HAER staff photographer.

DELINEATED BY: Julie Ann Dickson, 1992

MOUNT RAINIER ROADS AND BRIDGES RECORDING PROJECT
NATIONAL PARK SERVICE
UNITED STATES DEPARTMENT OF THE INTERIOR

MUDDY FORK COWLITZ RIVER BRIDGE (1952)
STEVENS CANYON HIGHWAY
MOUNT RAINIER NATIONAL PARK — LEWIS COUNTY — WASHINGTON

SHEET 1 OF 2

HISTORIC AMERICAN ENGINEERING RECORD
WA - 60

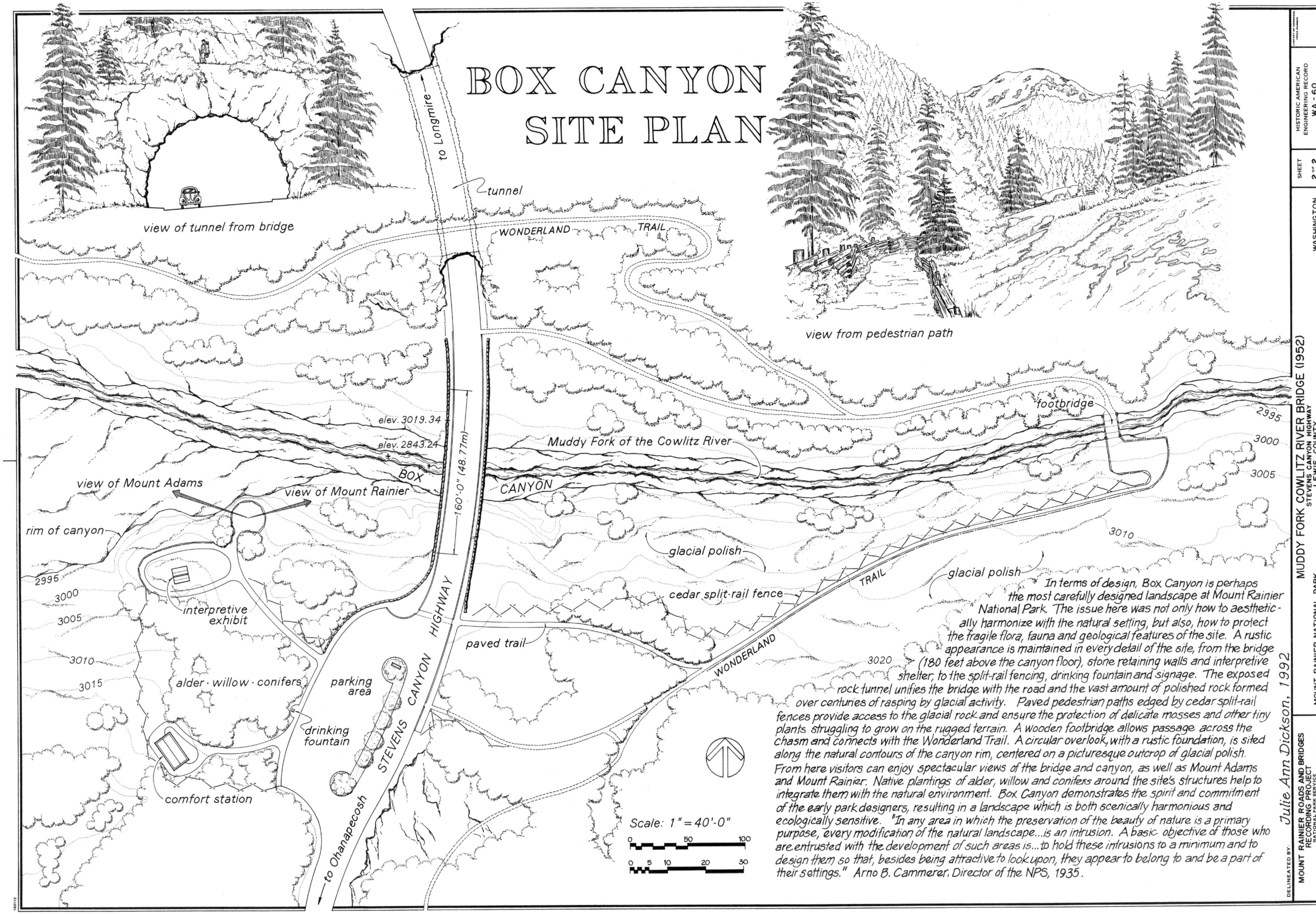
BOX CANYON SITE PLAN
view of tunnel from bridge
to Longmire
tunnel
WONDERLAND TRAIL
view from pedestrian path
footbridge
elev. 3019.34
elev. 2843.24
160'-0" (48.77m)
Muddy Fork of the Cowlitz River
BOX CANYON
view of Mount Adams
view of Mount Rainier
rim of canyon
glacial polish
glacial polish
cedar split-rail fence
interpretive exhibit
paved trail
STEVENS CANYON HIGHWAY
WONDERLAND TRAIL
alder · willow · conifers
parking area
drinking fountain
comfort station
to Ohanapecosh
2995
3000
3005
3010
3015
3020
Scale: 1" = 40'-0"
0 50 100
0 5 10 20 30
In terms of design, Box Canyon is perhaps the most carefully designed landscape at Mount Rainier National Park. The issue here was not only how to aesthetically harmonize with the natural setting, but also, how to protect the fragile flora, fauna and geological features of the site. A rustic appearance is maintained in every detail of the site, from the bridge (180 feet above the canyon floor), stone retaining walls and interpretive shelter, to the split-rail fencing, drinking fountain and signage. The exposed rock tunnel unifies the bridge with the road and the vast amount of polished rock formed over centuries of rasping by glacial activity. Paved pedestrian paths edged by cedar split-rail fences provide access to the glacial rock and ensure the protection of delicate mosses and other tiny plants struggling to grow on the rugged terrain. A wooden footbridge allows passage across the chasm and connects with the Wonderland Trail. A circular overlook, with a rustic foundation, is sited along the natural contours of the canyon rim, centered on a picturesque outcrop of glacial polish. From here visitors can enjoy spectacular views of the bridge and canyon, as well as Mount Adams and Mount Rainier. Native plantings of alder, willow and conifers around the site's structures help to integrate them with the natural environment. Box Canyon demonstrates the spirit and commitment of the early park designers, resulting in a landscape which is both scenically harmonious and ecologically sensitive. "In any area in which the preservation of the beauty of nature is a primary purpose, every modification of the natural landscape...is an intrusion. A basic objective of those who are entrusted with the development of such areas is...to hold these intrusions to a minimum and to design them so that, besides being attractive to look upon, they appear to belong to and be a part of their settings." Arno B. Cammerer, Director of the NPS, 1935.
DELINEATED BY: Julie Ann Dickson, 1992
MOUNT RAINIER ROADS AND BRIDGES RECORDING PROJECT
NATIONAL PARK SERVICE
UNITED STATES DEPARTMENT OF THE INTERIOR
MUDDY FORK COWLITZ RIVER BRIDGE (1952)
STEVENS CANYON HIGHWAY
LEWIS COUNTY
MOUNT RAINIER NATIONAL PARK
WASHINGTON
SHEET 2 of 2
HISTORIC AMERICAN ENGINEERING RECORD
WA-60
IF REPRODUCED, PLEASE CREDIT: HISTORIC AMERICAN ENGINEERING RECORD, NATIONAL PARK SERVICE, NAME OF DELINEATOR, DATE OF THE DRAWING

ROCKY MOUNTAIN NATIONAL PARK ROADS

The road system of Rocky Mountain National Park offers visitors access to diverse ecosystems characterizing the higher regions of the central Rocky Mountains. The roads take visitors through lowland meadows and aspen groves, along swift-flowing rivers and up through subalpine forest to more than 12,000' in elevation. No other national park roads offers the dramatic experience of a long drive across the tundra region, and few offer such a wide variety of experiences. The careful relationship of the park roads to the landscape results in a road system that generally harmonizes with the environment. Roadway alignments were chosen to highlight natural features, and scenic vistas and overlooks were provided to allow visitors to take in the magnificent terrain. The stone parapet walls and road-related structures, constructed in the National Park Service's distinctive "rustic style" of architecture, relate well to their natural surroundings and help evoke a distinctive "national park road" experience. Not surprisingly, the park roads are the principal attraction for most of the nearly three million visitors who flock to Rocky Mountain National Park each year.

American explorers had considered the Rocky Mountains impenetrable, but native peoples of the Ute and Arapahoe tribes had long passed back and forth over the range; the two park roads now crossing the mountains follow their general routes.

The first to cross the mountains was the Fall River Road, constructed by the State of Colorado and Larimer and Grand counties to encourage tourism. Built between 1913 and 1920, this narrow unpaved single-lane road climbed up the deep Fall River Valley to Fall River Pass, then dropped down a series of sharp switchbacks to the Colorado River in the Kawuneechee Valley. This road proved difficult for early automobiles to traverse, and clearing the shaded route of snow each year was a difficult and dangerous undertaking. Soon after it was completed, the park began planning a replacement.

The Rocky Mountain National Park Roads Recording Project was undertaken during the summer of 2000 and is part of the Historic American Engineering Record (HAER), a long-range program to document historically significant engineering and industrial works in the United States. HAER (Eric DeLony, Chief) is administered by the Historic American Buildings Survey/Historic American Engineering Record (E. Blaine Cliver, Chief), a division of the National Park Service, U.S. Department of the Interior. The project was funded by the U.S. Department of Transportation's Federal Lands Highway Program (Art Hamilton, Administrator) through the NPS Park Roads and Parkways Program (Lou DeLorme, Manager) and cosponsored by Rocky Mountain National Park (Randy Jones, Superintendent) and the NPS Cooperative Program at Montana State University (Barry Sulam, Manager).

The field work, measured drawings, and historical reports were completed under the direction of Todd A. Croteau, Project Leader, and Tim Davis, Program Historian. The recording team consisted of Field Supervisor Brandy Dubs (Montana State University); Architects Arin Streeter (University of Tennessee), Eszter Vogel (US/ICOMOS, Hungary), Lucas Dupuis and Nathan Junkert (Montana State University), Christopher Boldt (University of Washington); Landscape Architect Magdalena M. Lisowska (US/ICOMOS, Poland); and Historian Richard Quin

The new Trail Ridge Road, constructed between 1926 and 1932, climbed nearly a thousand feet higher but crossed the more open terrain of Trail Ridge. This two-lane roadway was carefully designed to avoid damage to the fragile alpine scenery it crossed. Reaching 12,183' on Trail Ridge, it is the highest continuous highway in the United States.

The roads in the Bear Lake, Moraine Park, Lily Lake and Wild Basin areas were built as county or private roads to small holdings predating the establishment of the park in 1915; today, all are under park maintenance.

The road system of Rocky Mountain National Park continues to provide visitors with access to most majestic scenery. The roads wind through deep forest glades and across the open treeless tundra, providing glimpses of boldly colored wildflowers and magnificent wild animals. Even today, decades after they were built, excursions along these remarkable roads provide memorable experiences.

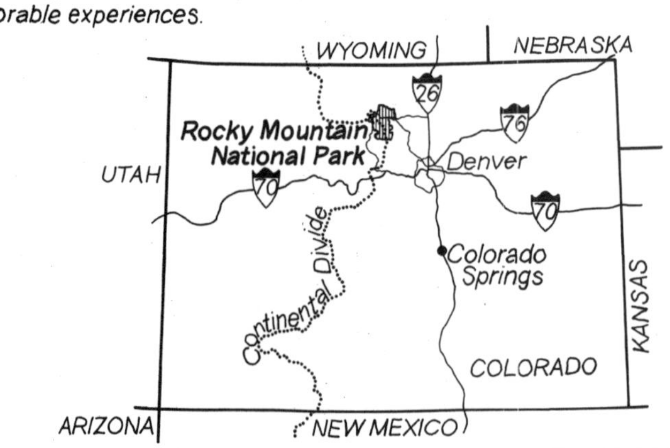

DELINEATED BY Magdalena M. Lisowska, 2000

NPS PARK ROADS RECORDING PROGRAM
NATIONAL PARK SERVICE
UNITED STATES DEPARTMENT OF THE INTERIOR

ESTES PARK VICINITY
LARIMER COUNTY
COLORADO

ROCKY MOUNTAIN NATIONAL PARK ROADS - 1920/1932

SHEET 1 OF 8

HISTORIC AMERICAN ENGINEERING RECORD
CO-78

FALL RIVER ROAD - 1920

GRADE

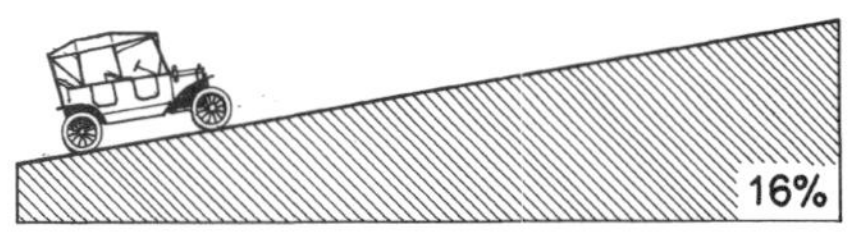

The road climbed to Fall River Pass on steep grades sometimes reaching 16%. Some early automobiles had to climb in reverse due to their weak engines and gravity-fed fuel systems. Surfacing materials washed off quickly.

TURN RADIUS

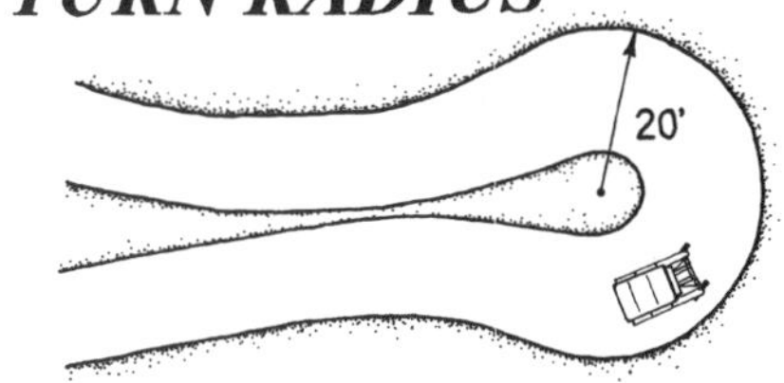

Motorists had to negotiate sixteen switchbacks with radii as tight as 20'. Some vehicles had to turn back and forth repeatedly to make the curves.

ROAD WIDTH

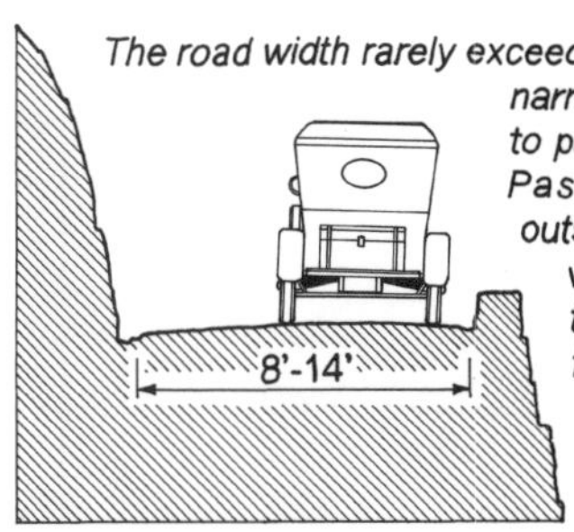

The road width rarely exceeded 14', too narrow for cars to pass safely. Passing turnouts were provided, but these were few and far between.

ELEVATION GAIN

The roadway was largely built out from the hillside. In the steepest places, multiple switchbacks were stacked one above the other.

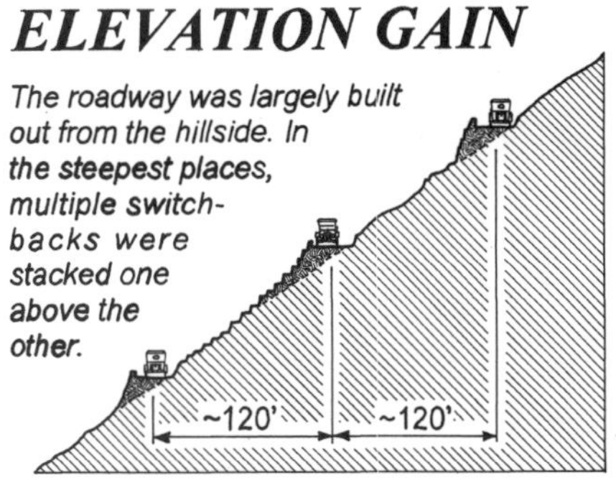

PULLOUTS

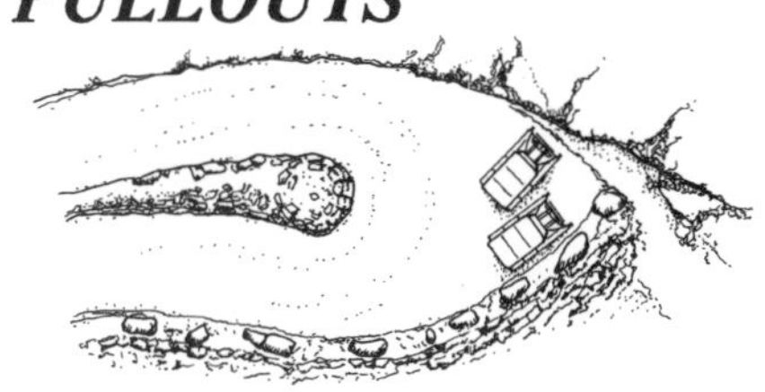

Few pullouts were provided to allow motorists to stop; some were located on switchbacks, making the curves even more difficult.

The Fall River Road was the first road to cross the Rocky Mountains in northern Colorado. Started by Larimer and Grant counties in 1913 before the establishment of Rocky Mountain National Park, the early construction was carried out by state prison inmates. The convicts made little progress and contractors completed the road in 1920. The unpaved mountain road climbed from Horseshoe Park up the steep Fall River Valley over a series of tight switchbacks to reach Fall River Pass at 11,796'; from there, the road continued down the west side of the mountains to Grand Lake. Despite the popularity of the route, the road proved difficult to traverse and maintain. Automobiles had trouble mastering the steep grade and tight curves, and avalanches often buried the road forty feet deep. Following the completion of the Trail Ridge Road in 1932, the eastern half of the Fall River Road became a one-way scenic drive from Endovalley up to Fall River Pass; the western half was abandoned or overbuilt by the Trail Ridge Road.

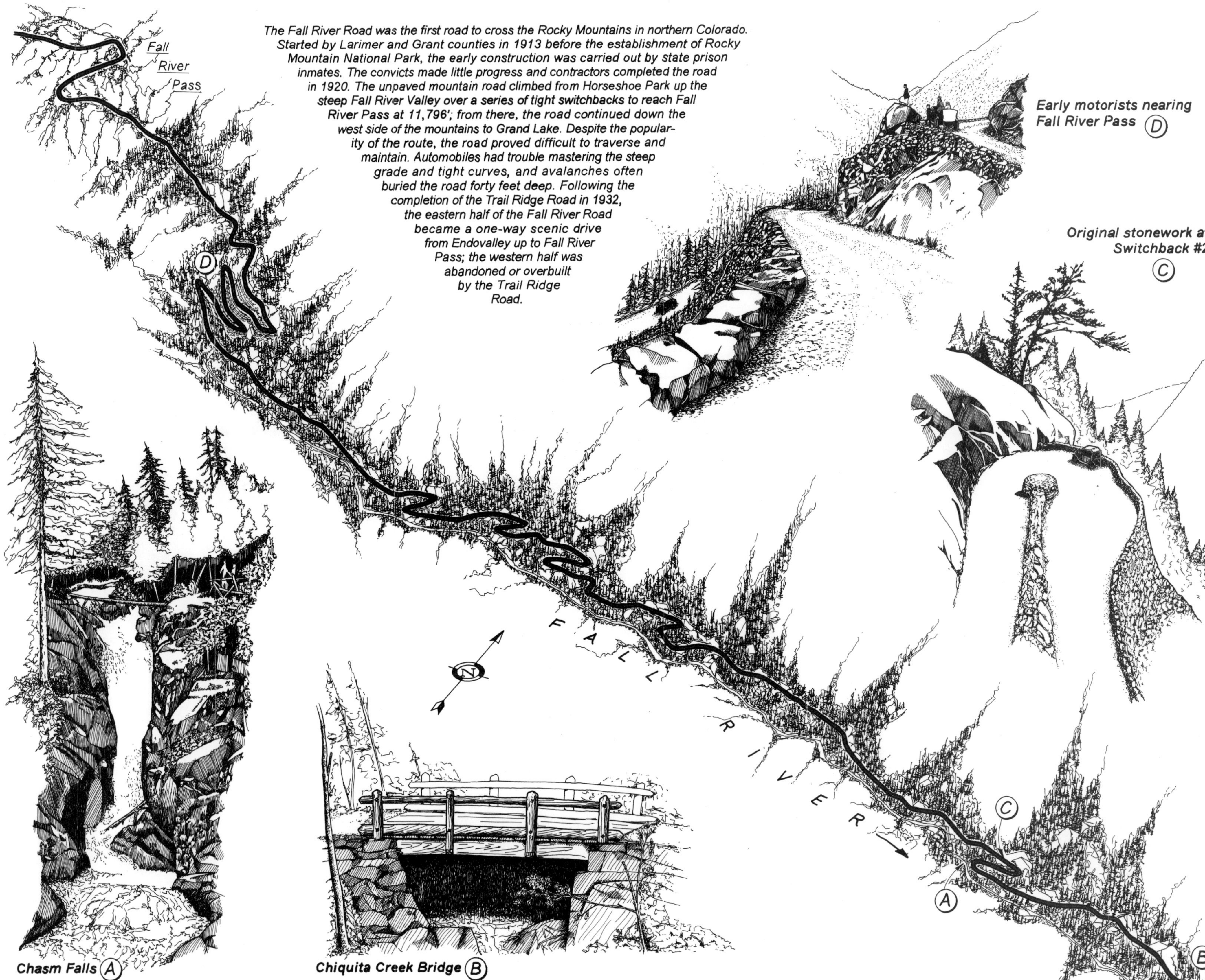

DELINEATED BY: Arin Streeter, 2000

NPS PARK ROADS RECORDING PROGRAM
NATIONAL PARK SERVICE
UNITED STATES DEPARTMENT OF THE INTERIOR

ROCKY MOUNTAIN NATIONAL PARK ROADS - 1920/1932
ESTES PARK VICINITY
LARIMER COUNTY
COLORADO

SHEET 4 OF 8

HISTORIC AMERICAN ENGINEERING RECORD
CO-78

TRAIL RIDGE ROAD - 1932

Trail Ridge Road provides spectacular views of the majestic scenery of Rocky Mountain National Park. It is the highest continuous motorway in the United States, with more than eight miles lying above 11,000' and a maximum elevation of 12,183'. The name "Trail Ridge Road" derives from its proximity to historic pathways used by native peoples to cross the Rocky Mountains.

Trail Ridge Road was designed to replace the Fall River Road, which proved inadequate for modern motor travel as soon as it opened in 1920. Trail Ridge Road was designed to have more gentle grades, broader curves, and a greater variety of scenic experiences. The sunny, ridge-top location would also reduce snow accumulation and allow Trail Ridge Road to open earlier than its shady, snow-laden predecessor.

Trail Ridge Road was constructed between 1926 and 1932 through the collaborative efforts of the National Park Service and the Bureau of Public Roads (now the Federal Highway Administration). Construction crews had to contend with imposing terrain, harsh weather, short working seasons, and stringent design criteria, which were intended to ensure that the road would "lie lightly on the land," displaying the region's rich scenic diversity with minimal impact on the natural environment. Trail Ridge Road opened in July 1932, providing motorists with access to impressive views, memorable wildlife viewing opportunities, and spectacular high mountain terrain.

GRADE

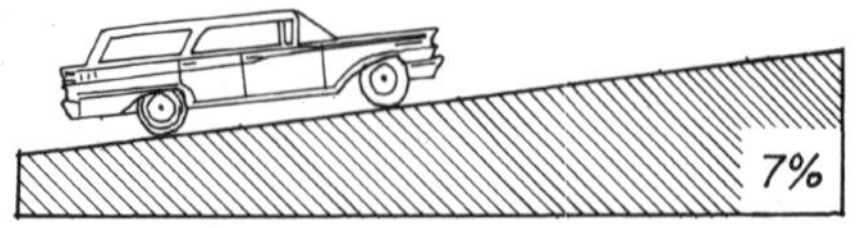

The road was designed with a ruling grade generally less than 5% and never exceeding 7%, less than half as steep as the Fall River Road.

TURN RADIUS

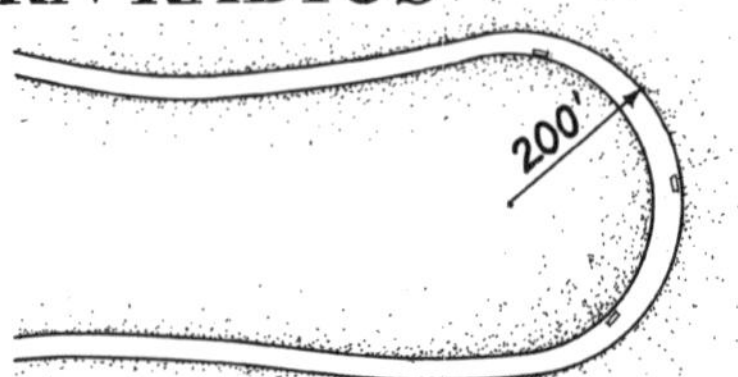

Minimum radii for open curves was 100', and 200' on blind curves. Many curves were designed to sweep across but not dominate, the landscape.

ROAD WIDTH

Unlike the single-track Fall River Road, Trail Ridge Road was designed as a two-lane with a 22' roadbed and 3' ditches in cut sections.

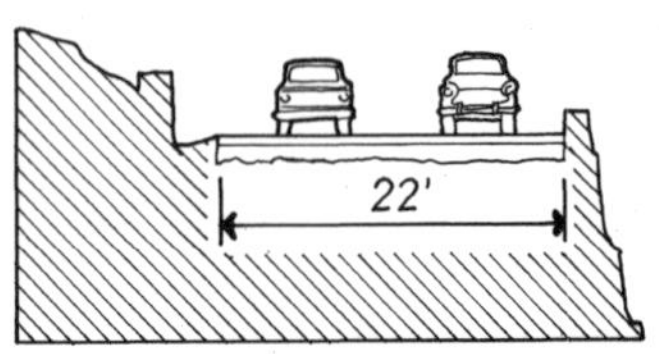

ELEVATION GAIN

The roadway was largely built into the hillside, elevated on rock fill once it reached the tundra. Long continuous curves were used to gain elevation.

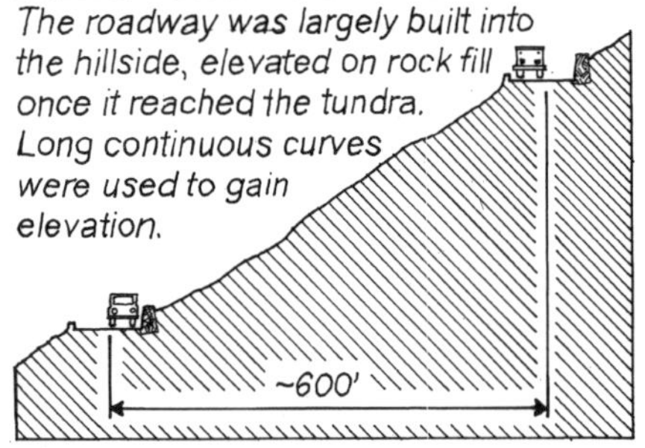

PULLOUTS

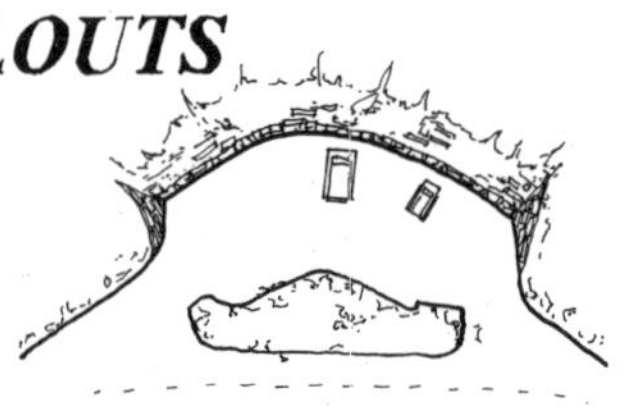

The commodious stone-walled turnouts, often located on major curves, provided ample space for visitors to take in the views.

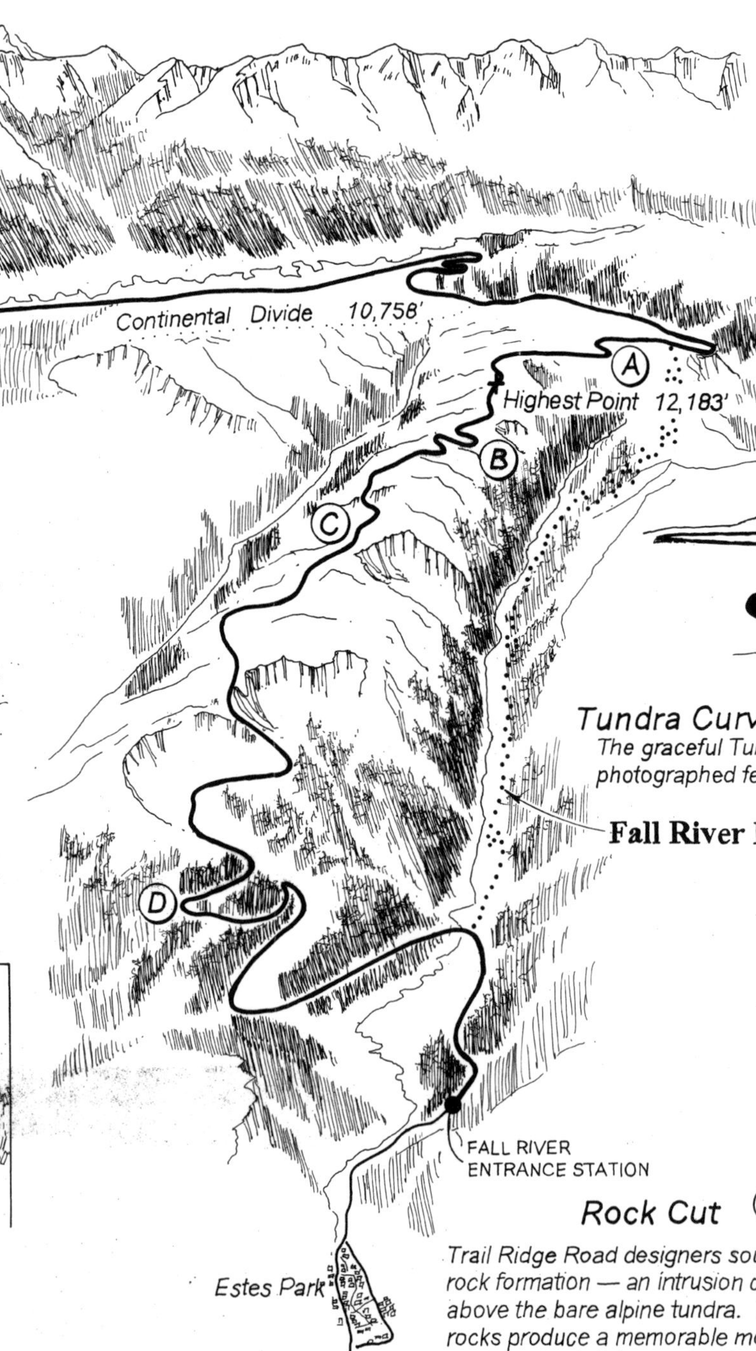

Kawuneeche Valley (E)

On the west side of the park, Trail Ridge Road passes through the Kawuneeche Valley, a region of dense forests, lush meadows, and winding streams. Several small creeks are bridged with attractive masonry culverts.

Many Parks Curve (D)

Trail Ridge Road was one of the first places the National Park Service developed a coordinated system of scenic pullouts designed to allow motorists to stop and safely view the surrounding scenery. The pedestrian walkway at Many Parks Curve was added in the 1960s to improve safety and provide even more impressive views.

Alpine Visitors Center (A)

The heavy logs atop the Alpine Visitors Center are designed to withstand heavy snows and keep the roof from blowing away in fierce winter storms where winds can exceed 200 mph.

Tundra Curves and Lava Cliffs (B)

The graceful Tundra Curves are one of the most photographed features of Trail Ridge Road.

Rock Cut (C)

Trail Ridge Road designers sought to showcase this unusual rock formation — an intrusion of igneous rock that rises above the bare alpine tundra. The looming rocks produce a memorable motoring experience that contrasts with the open tundra environment.

DELINEATED BY: Brandy Dubs, 2000

NPS PARK ROADS RECORDING PROGRAM
NATIONAL PARK SERVICE
UNITED STATES DEPARTMENT OF THE INTERIOR

ROCKY MOUNTAIN NATIONAL PARK ROADS - 1920/1932
ESTES PARK VICINITY | LARIMER COUNTY | COLORADO

SHEET 5 OF 8

HISTORIC AMERICAN ENGINEERING RECORD CO-78

IF REPRODUCED, PLEASE CREDIT: HISTORIC AMERICAN ENGINEERING RECORD, NATIONAL PARK SERVICE, NAME OF DELINEATOR, DATE OF THE DRAWING

ROADSIDE EDGING

A variety of roadside treatments were employed along the edge of the Trail Ridge Road. The hand-laid rustic-style stone walls are the most extensively used form of roadside barrier. The original walls were "Type 3" crenellated parapet walls constructed from native stone chosen to harmonize with roadside outcrops. Over the years, some of these walls were partially buried as the road was repaved. Some of these sections are now being replaced with concrete core walls faced with native stone, much of it salvaged from the older walls. The park also has installed segments of precast concrete walls faced in artificial stone. These are used to stabilize roadsides and to prevent uphill slopes from eroding, reducing the money and labor spent in ditch clearing and boulder removal.

Wood poles resting on small cross logs are generally employed to discourage parking on narrow road sections; asphalt curbs flank other sections. Heavier log rails set into vertical posts formed roadside parapets on older road segments. Light A-frame pole fences are employed to delineate a separation between the roads and the wilderness beyond. On steep unstable sections prone to slides, gabions, rock-filled wire baskets, were used to stabilize the slopes.

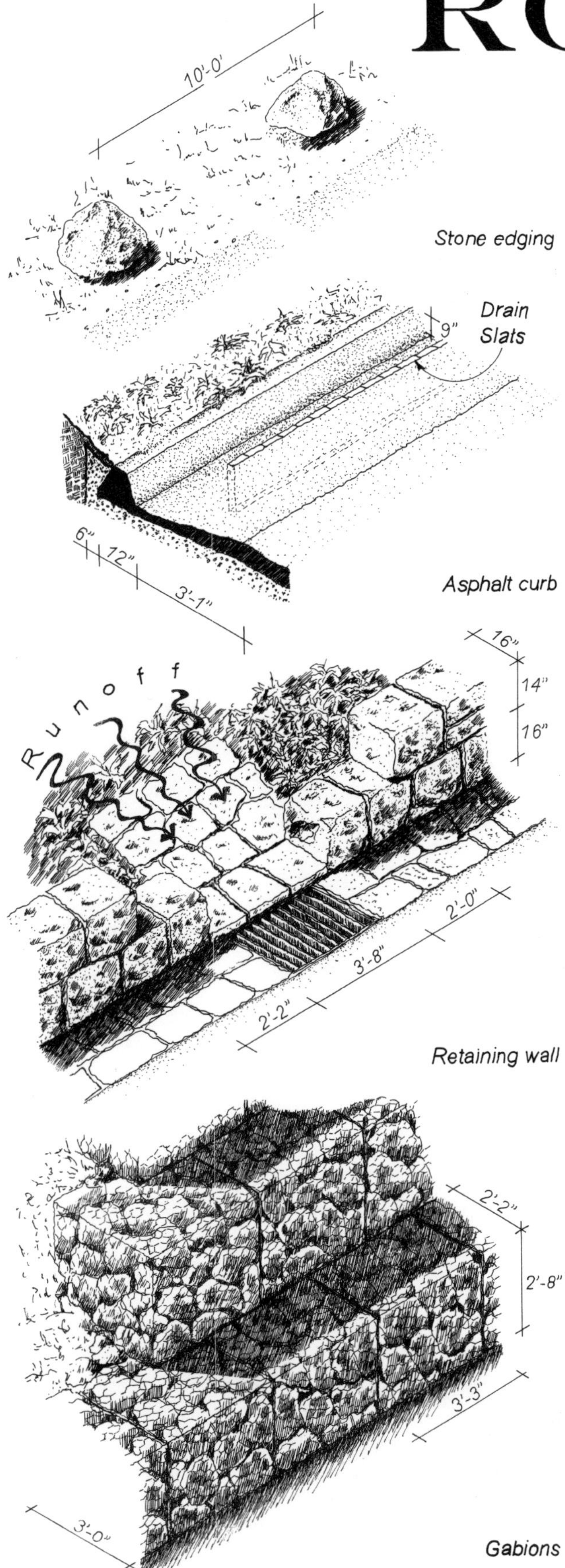

WALL RECONSTRUCTION

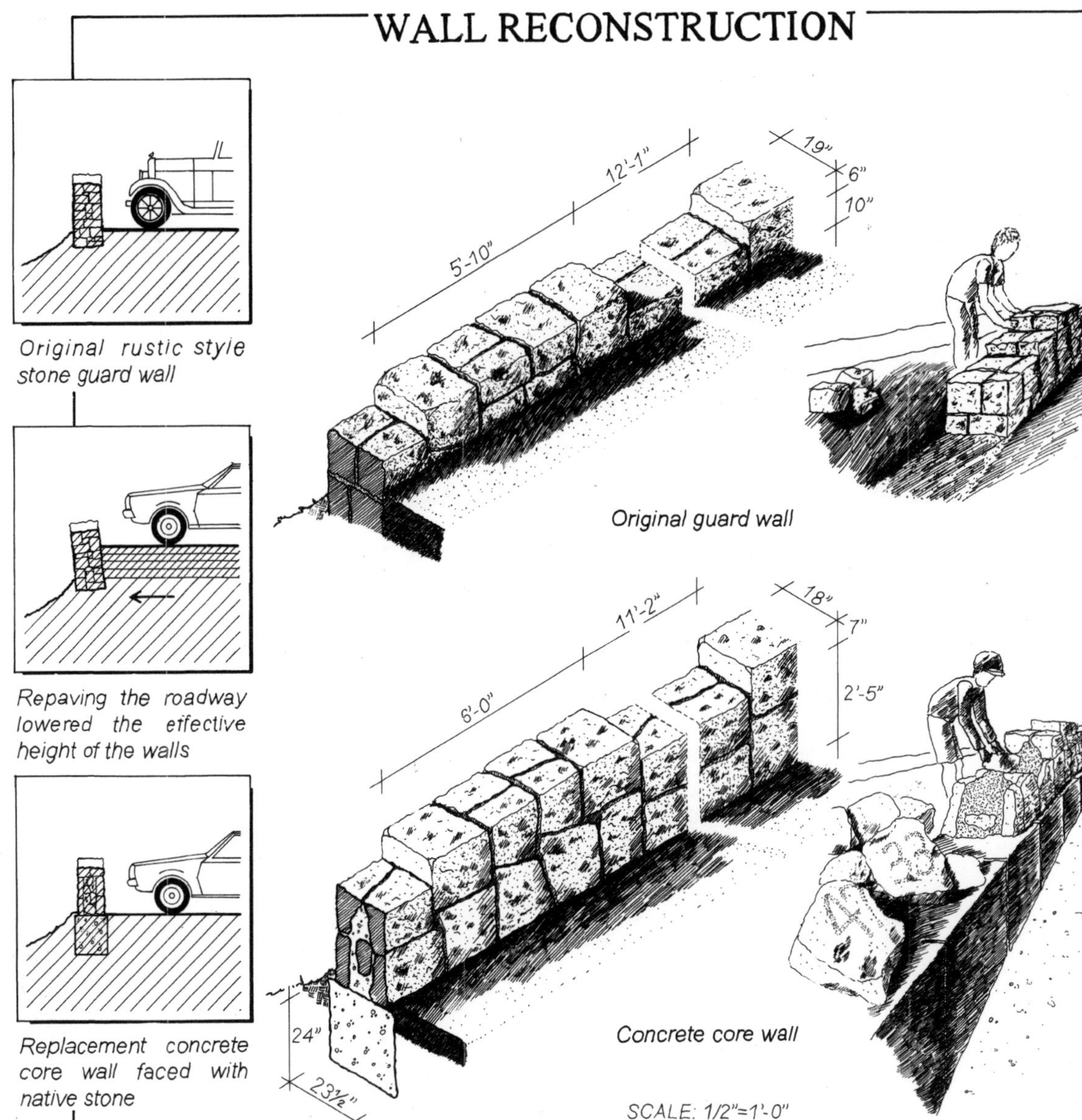

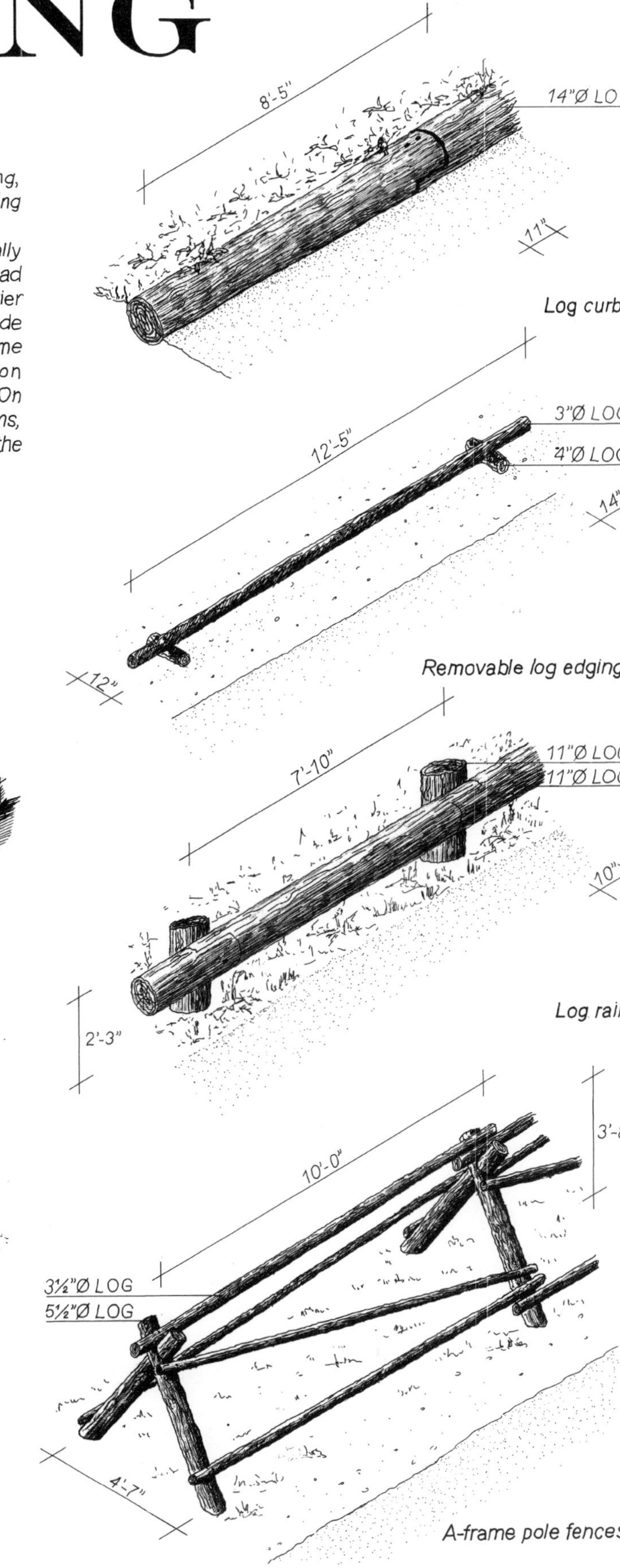

DELINEATED BY: Eszter Vogel, 2000

NPS PARK ROADS RECORDING PROGRAM
NATIONAL PARK SERVICE
UNITED STATES DEPARTMENT OF THE INTERIOR

ESTES PARK VICINITY | LARIMER COUNTY

ROCKY MOUNTAIN NATIONAL PARK ROADS - 1920/1932

COLORADO

SHEET 7 OF 8

HISTORIC AMERICAN ENGINEERING RECORD CO-78

SNOW REMOVAL

Reaching heights of more than 12,000' above sea level, the roads in Rocky Mountain National Park are frequently buried deep in snow. The winter snowpack is often more than twenty feet deep, and twice that in areas buried by avalanches. Removing snow from the roads is a difficult and dangerous maintenance task.

Immediately after the Fall River Road was completed in 1920, park administrators realized that clearing snow would be a daunting task. The road climbed to Fall River Pass in a deep shaded valley, where avalanche chutes dumped deep accumulations of snow. The road was initially cleared with hand labor. Working conditions were severe. Cold temperatures combined with winds ranging up to more than 200 mph created extreme wind chill conditions. The high altitude sun caused snowblindness and severe sunburn. New avalanches would often cover cleared sections, forcing crews to repeat their work. Experiments were made with dynamiting and the use of blow torches, kerosene flares and carbide lamps, but it soon became clear that heavy equipment would be required. In 1925, the park acquired a specially designed steam shovel. In 1931 a "Sno-Go" rotary plow was placed in operation. Still, the tremendous snowpacks along the road were a major factor in deciding to construct the Trail Ridge Road on more open terrain.

While the Trail Ridge Road was designed to pass through more open country with fewer places obstructed by heavy snow accumulation, the road climbed nearly a thousand feet higher than the Fall River Road and was also buried under deep snowpacks most years. By the time the road was completed in 1932, the park was using its new rotary plow. Even so, all old snow deeper than 45" had to be loosened with dynamite before it could be removed with machines. While heavier plows are now employed, opening the two park roads remains a tremendous and expensive undertaking. Despite periodic calls from tourist interests in nearby communities to keep the roads open year-round, crews face major challenges opening Trail Ridge Road by its target date of Memorial Day and Fall River Road by the Fourth of July.

FALL RIVER ROAD

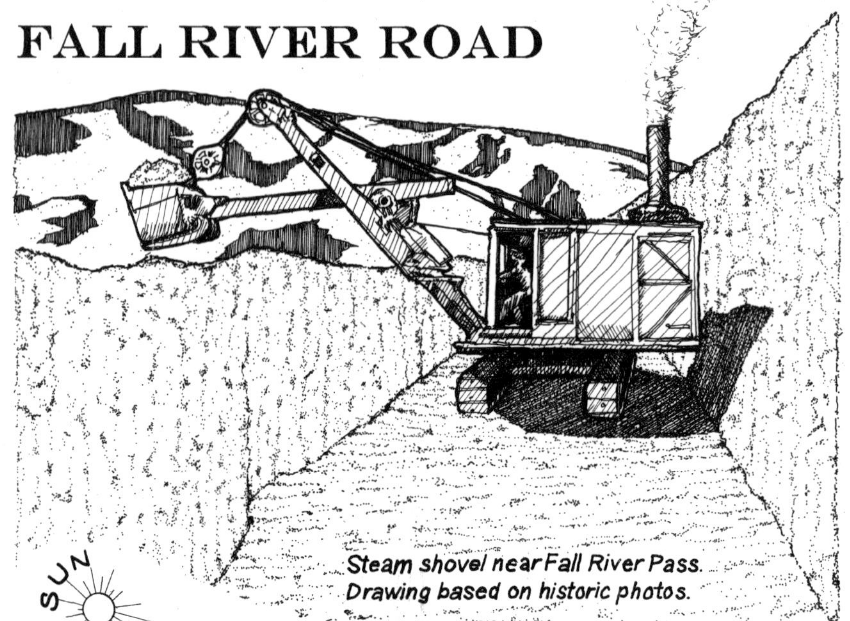

Steam shovel near Fall River Pass. Drawing based on historic photos.

Clearing the road with hand labor. Drawing based on historic photo.

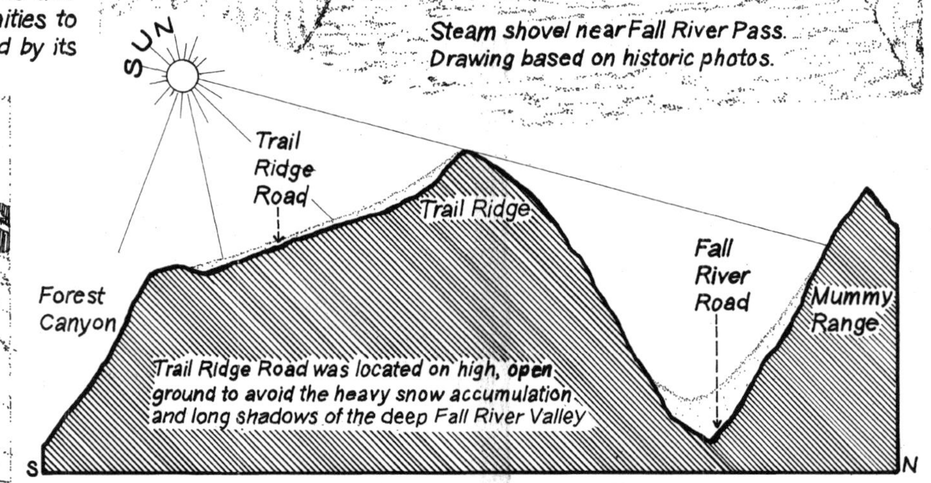

Tour group driving over the recently opened Fall River Road. Drawing based on historic photo.

TRAIL RIDGE ROAD

Snowblast rotary plow at work. Drawing based on Park Service photo.

Rotary plow or "Sno-Go" and bulldozer on Trail Ridge Road. Drawing based on historic photos.

HISTORIC AMERICAN ENGINEERING RECORD CO-78

SHEET 8 of 8

ROCKY MOUNTAIN NATIONAL PARK ROADS - 1920/1932

COLORADO · LARIMER COUNTY · ESTES PARK VICINITY

NPS PARK ROADS RECORDING PROGRAM
NATIONAL PARK SERVICE
UNITED STATES DEPARTMENT OF THE INTERIOR

DELINEATED BY: Nathan Junkert 2000

SCOTTS BLUFF NATIONAL MONUMENT

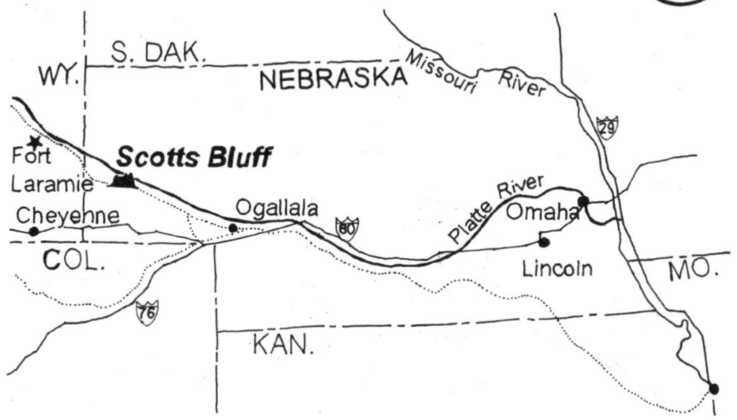

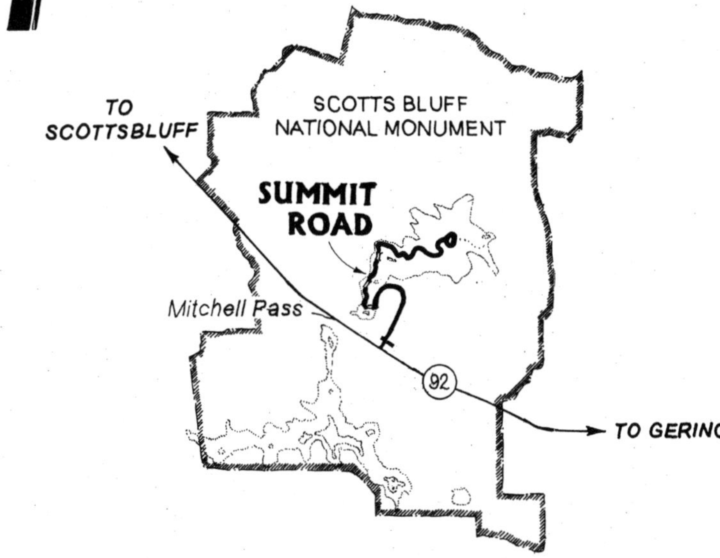

SUMMIT ROAD

Rising high above the grasslands of western Nebraska, Scotts Bluff was one of the most widely recognized landmarks on the Oregon Trail. From approximately 1841 through 1869, this historic wagon route, stretching from the Missouri River to the Pacific Ocean, was followed by thousands of emigrants including missionaries, gold seekers, and families in search of better prospects. Scotts Bluff National Monument was created in 1919 to preserve and commemorate this important landmark and to educate the public about the historic process of westward migration. Early monument boosters believed that the panoramic views from the summit would inspire visitors with an appreciation for the struggles of the early pioneers and reinforce their faith in traditional American values and institutions.

Early access to the summit was limited to a steep foot trail that stretched straight up the eastern side of the bluff. In 1927, the National Park Service (NPS), backed by a combination of federal and local funding, designed a less treacherous "zigzag" or switchback trail at the same location. With the rising popularity of motor travel and auto tourism in the 1920s, local government officials and community leaders began to petition NPS officials in Washington for an automobile road to the summit. Their calls were answered with the advent of Franklin D. Roosevelt's New Deal relief programs in 1933. Construction of the Summit Road began that year, providing needed relief work for unemployed sugar beet farmers and other local workers hit hard by the Great Depression. Over the next six years, the pace of construction was largely dictated by the availability of federal funding, through programs including the Civil Works Administration (CWA), Public Works Administration (PWA) and Civilian Conservation Corps (CCC).

Summit Road construction also benefited from the 1925 interbureau agreement between the National Park Service and the Bureau of Public Roads (BPR), an arrangement that facilitated the survey, construction, and improvement of roads throughout the National Park System. In planning the Summit Road, NPS landscape architects and BPR engineers were primarily concerned with preserving the beauty and historical integrity of the natural landscape. Maintaining the original appearance of the bluff was considered integral to the monument's commemorative goals. The western side of Scotts Bluff was chosen as the most inconspicuous site for the road, which featured three tunnels and continuous reinforced-concrete slab paving. The highly erosive nature of the sandstone formation as well as the relative inaccessibility of the site posed a number of challenges throughout the construction years.

Rough grading for the first section of road was completed in the spring of 1933. Work continued the following November with more than 200 men working on the summit parking area. Excavation of the first of three tunnels began in December of 1933, and the third tunnel was finally completed in 1936. Paving of the road commenced the following June, with the Summit Road opening to the public on September 19, 1937. Portals for three of the tunnel entrances were constructed in 1939. Over the next 50 years, improvements were minimal, confined primarily to periodic rockslide cleanup, and road patching. In 1989, the Federal Highway Administration constructed new portals on both entrances of Tunnel No. 1 and the upper entrance of Tunnel No. 2, excavated the overhanging sandstone cliffs between Tunnels No. 2 and No. 3, and stabilized the resulting slopes with shotcrete. Also at this time, the road was repaired by patching rather than complete resurfacing, in recognition of its historic significance as the oldest known concrete-paved road in Nebraska. Totaling 1.582 miles in length, the Summit Road constitutes the primary motorist experience in Scotts Bluff National Monument.

The Scotts Bluff Summit Road Recording Project was undertaken during the summer of 2000 and is part of the Historic American Engineering Record (HAER), a long-range program to document historically significant engineering and industrial works in the United States. HAER (Eric DeLony, Chief) is administered by the Historic American Buildings Survey/Historic American Engineering Record (E. Blaine Cliver, Chief) a division of the National Park Service, U.S. Department of the Interior. The project was funded by the U.S. Department of Transportation's Federal Lands Highway Program (Art Hamilton, Administrator) through the NPS Park Roads and Parkways Program (Lou DeLorme, Manager) and cosponsored by Scotts Bluff National Monument (Valerie Naylor, Superintendent) and the NPS Cooperative Program at Montana State University (Barry Sulam, Manager). The field work, measured drawings, historical reports, and photography were completed under the direction of Christopher Marston, Project Leader; Tim Davis, Program Historian; and Todd A. Croteau, Program Manager. The recording team consisted of Field Supervisor Chris Gray (Texas A & M University); Architects Todd Delyea (University of Idaho), Tim Grinstead and Stacey Heckaman (Montana State University), and Roger Michel (US/ICOMOS, Haiti); Landscape Architect Christine Magdalenos (Ball State University); and Historian Alicia Barber (University of Texas at Austin). Large-format photography was produced by David Haas.

NATIONAL PARK SERVICE
DEPARTMENT OF THE INTERIOR

HISTORIC AMERICAN ENGINEERING RECORD
NE-11
SHEET 1 OF 10
NEBRASKA
SCOTTS BLUFF SUMMIT ROAD
SCOTTS BLUFF NATIONAL MONUMENT
SCOTTS BLUFF COUNTY
GERING
DELINEATED BY: Tim Grinstead
NPS ROADS & BRIDGES RECORDING PROGRAM
NATIONAL PARK SERVICE
UNITED STATES DEPARTMENT OF THE INTERIOR

HISTORIC MIGRATION CORRIDOR

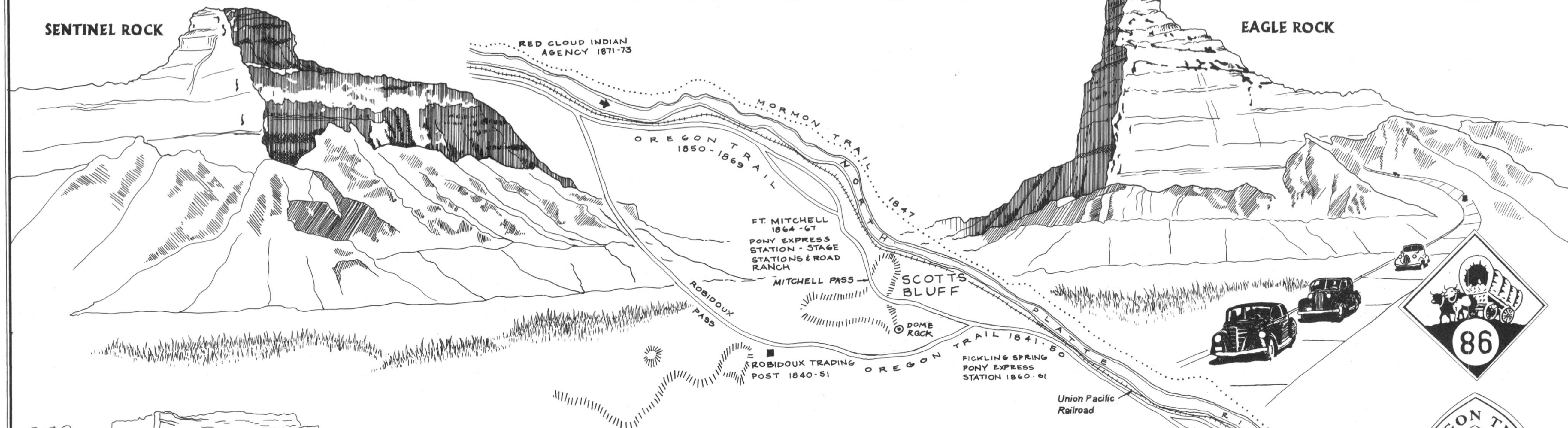

Fur Traders, and Native Americans use the North Platte River as a route for migration to the Buffalo grounds, and beaver territory.

The area now comprising Scotts Bluff National Monument has served as a migration corridor for centuries. Native American tribes including the Sioux, Pawnee, and Cheyenne followed the route of the North Platte River, as did early fur traders of various nationalities. From the 1840s through the 1860s, thousands of emigrants passed by Scotts Bluff on their way to Oregon Territory, the California gold fields and other western destinations. Following approximately the same route, the Pony Express, established in 1860, carried mail overland until being rendered obsolete less than two years later by the installation of the transcontinental telegraph line. Railroads arrived in the North Platte Valley in the first decade of the 20th century, with the competing Chicago, Burlington and Quincy, and Union Pacific lines. Soon afterward, the old Mitchell Pass road was upgraded and designated State Highway 86 (now 92). Completion of the Summit Road in 1937 encouraged the influx of a new population of transients: tourists. At the end of the 20th century, Scotts Bluff National Monument was welcoming nearly 150,000 visitors per year.

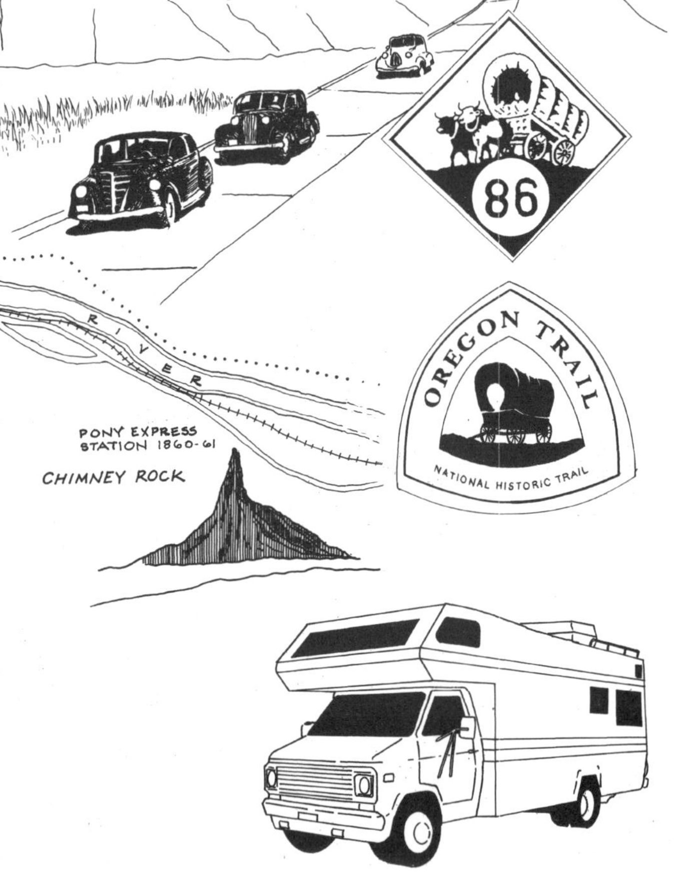

Tourism has flourished at Scotts Bluff since the completion of the summit road in 1937.

From 1846-69 Mormon pioneers traveled the north bank of the Platte on a route parallel to the Oregon trail.

Scotts Bluff was a celebrated landmark of the Oregon Trail. The traditional route of overland migration to Oregon, California, and Utah.

The advancing telegraph ended the short-lived Pony Express, (1860-61).

In 1900 the Burlington Railroad was uilt up the north side of the Platte, and he town of Scottsbluff was born.

DELINEATED BY: Todd Delyea 2000

NPS ROADS & BRIDGES RECORDING PROGRAM
NATIONAL PARK SERVICE
UNITED STATES DEPARTMENT OF THE INTERIOR

GERING

SCOTTS BLUFF SUMMIT ROAD
SCOTTS BLUFF NATIONAL MONUMENT
SCOTTS BLUFF COUNTY

NEBRASKA

SHEET 3 OF 10

HISTORIC AMERICAN ENGINEERING RECORD NE-11

HIDING THE ROAD

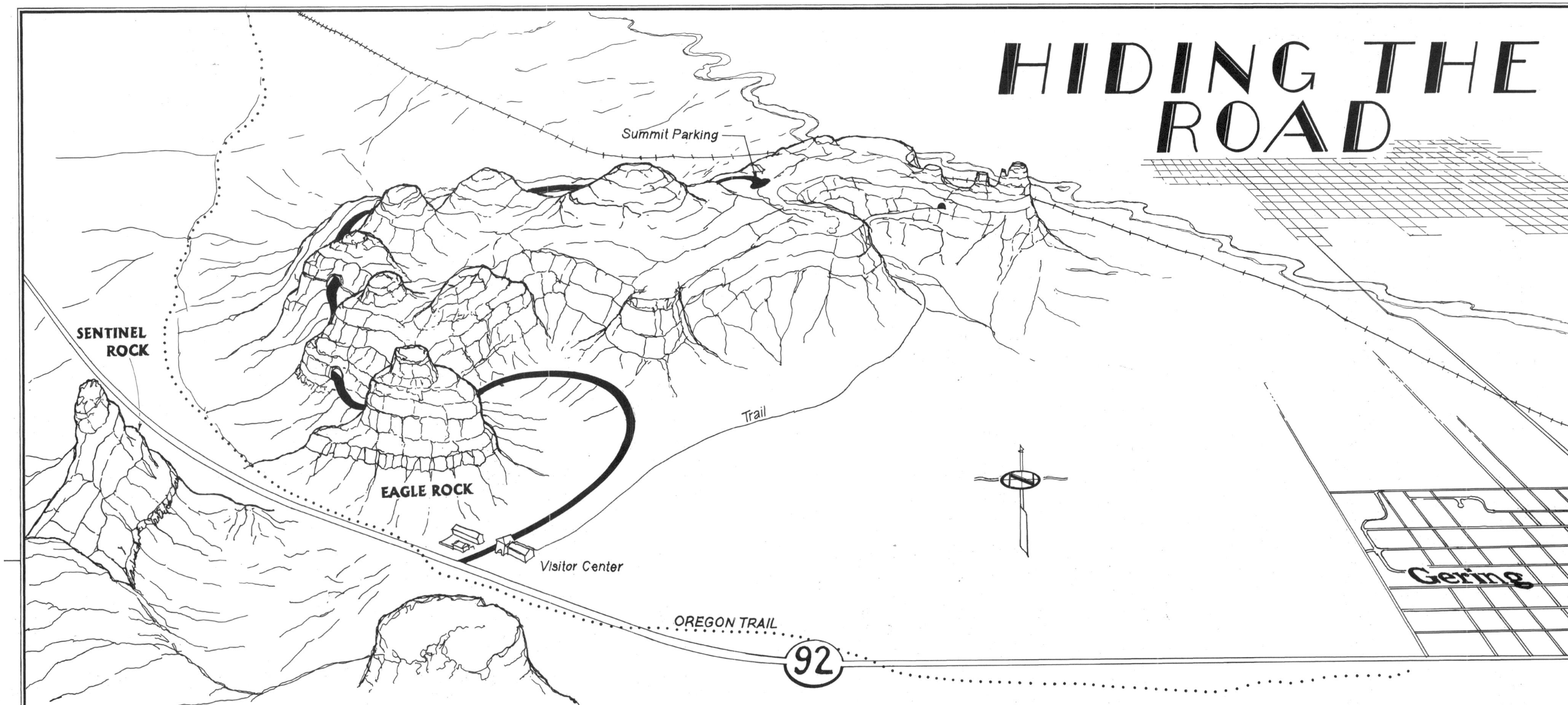

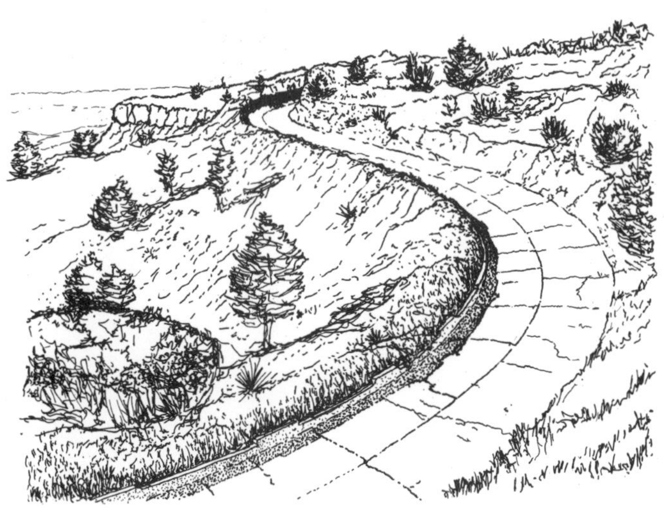

A continuous berm along the outer edge of the Summit Road made it hard to detect from below.

"This highway is being built in such a manner that it does not disfigure the hill and will make it possible for countless people to drive to the top of that hill and get one of the finest views…to be had anywhere in the west." –Harold Cook, Custodian, Scotts Bluff National Monument, 1934.

Planners intended to make the road as inconspicuous as possible, thereby providing access to the summit without marring the natural beauty and historical integrity of the landscape. Located on the western side of Scotts Bluff, the majority of the Summit Road is not visible from the Visitor Center or the eastern entrance, the primary approach to the monument. The roadbed was nestled into the rock in order to ensure safe motorist travel while further hiding the road from view. The original gunite used to line and support the tunnel portals and rock face was carefully tinted to match the natural color and rough texture of the bluff.

The Summit Road is inconspicuous as visitors approach Scotts Bluff from the east along U.S. highway 92, allowing the splendor of this magnificent earthen formation to be seen in all of its historical glory.

HISTORIC AMERICAN ENGINEERING RECORD NE-11
SHEET 4 OF 10
NEBRASKA
SCOTTS BLUFF SUMMIT ROAD
SCOTTS BLUFF NATIONAL MONUMENT
SCOTTS BLUFF COUNTY
GERING
NPS ROADS & BRIDGES RECORDING PROGRAM
NATIONAL PARK SERVICE
UNITED STATES DEPARTMENT OF THE INTERIOR
DELINEATED BY: Tim Grinstead

BUILDING A ROAD UP A ROCK

The steep and highly erosive sandstone bluff presented significant challenges for road builders. The severe terrain limited the amount of equipment and labor that could be used at one time and made it difficult to employ conventional power machinery. The steep slopes and constant threat of rock slides necessitated a variety of stabilizing techniques, ranging from gunite patches to prevent cave-ins and rockfall to extensive plantings and other measures designed to minimize erosion.

Hand-labor was employed for grading and excavation at the top of the bluff, which was initially inaccessible to power machinery. A hand-operated mining-style railway was used to haul rocks and fill. Long chutes were constructed to dump excess material out of sight well below the road.

A major rockslide in the spring of 1934 sent 100 cubic yards of rock crashing into the roadway, crushing an unoccupied truck. gunite was sprayed on the steepest slopes to stabilize them.

Newly graded areas were gently sloped and then sodded to resist erosion. Planting seedlings for trees and bushes helped anchor the sod, retain moisture, and promote soil stabilization.

DELINEATED BY: Brandy Dubs, 2000

NPS ROADS & BRIDGES RECORDING PROGRAM
NATIONAL PARK SERVICE
UNITED STATES DEPARTMENT OF THE INTERIOR

GERING

SCOTTS BLUFF SUMMIT ROAD
SCOTTS BLUFF NATIONAL MONUMENT
SCOTTS BLUFF COUNTY

NEBRASKA

SHEET 5 OF 10

HISTORIC AMERICAN ENGINEERING RECORD NE-11

LIBRARY OF CONGRESS INDEX NUMBER

PUTTING MEN TO WORK

Shift change at summit parking area.

Workers carried culverts and all other materials and tools by hand to the summit.

Men loading dump trucks by hand during the excavation of Tunnel No.1.

A hand-operated railway was used to haul excavated material to the summit for use in building fill.

Putting Men to Work

The Great Depression devastated western Nebraska's agricultural economy, creating a desperate need for relief employment. Local politicians and park administrators found mutual benefit in the federal relief projects that funded road construction while providing needed jobs for local workers. Beginning in 1933, construction of the Summit Road employed hundreds of men, with federal funding provided by New Deal programs including the Civil Works Administration (CWA), Public Works Administration (PWA), and Civilian Conservation Corps (CCC). In order to employ as many workers as possible, labor-intensive processes supplanted the use of heavy equipment wherever practicable. This extensive use of hand labor enabled those unskilled in construction work to participate in the project. For both the Park Service and public works agencies, the objective was not simply to construct a roadway but to rebuild the economic and physical health of the community by giving work to the unemployed and teaching them new skills.

"CCC work...achieves the double purpose of building parks and building men." -1937 CCC Handbook

CCC enrollees used picks and shovels to slope the banks beneath the roadway.

C.W.A. workers receiving relief checks at the administrative office in Gering.

DELINEATED BY: *Christine Magdalenos, 2000*

NPS ROADS & BRIDGES RECORDING PROGRAM
NATIONAL PARK SERVICE
UNITED STATES DEPARTMENT OF THE INTERIOR

GERING

SCOTTS BLUFF SUMMIT ROAD
SCOTTS BLUFF NATIONAL MONUMENT
SCOTTS BLUFF COUNTY

NEBRASKA

SHEET 6 OF 10

HISTORIC AMERICAN ENGINEERING RECORD
NE-11

LIBRARY OF CONGRESS INDEX NUMBER

TUNNEL CONSTRUCTION

(1) Tunnel No. 1

EXCAVATION

1. Excavation began in the fall of 1933 with a four-man crew working on the lower entrance to Tunnel No. 1. Workers initially used only manual tools such as picks and shovels, but the Brule clay of the first two tunnels proved stable enough to accommodate careful dynamite blasting.

2. The ashy composition of Tunnel No. 3 was found to be less stable than that of the others. In order to support the roof of this tunnel during the excavation process, engineers initially bored two smaller tunnels into the rock, leaving a center wall between them. The unique appearance of this double-barreled structure led planners to consider retaining this configuration, but the tunnel was later excavated to match the others.

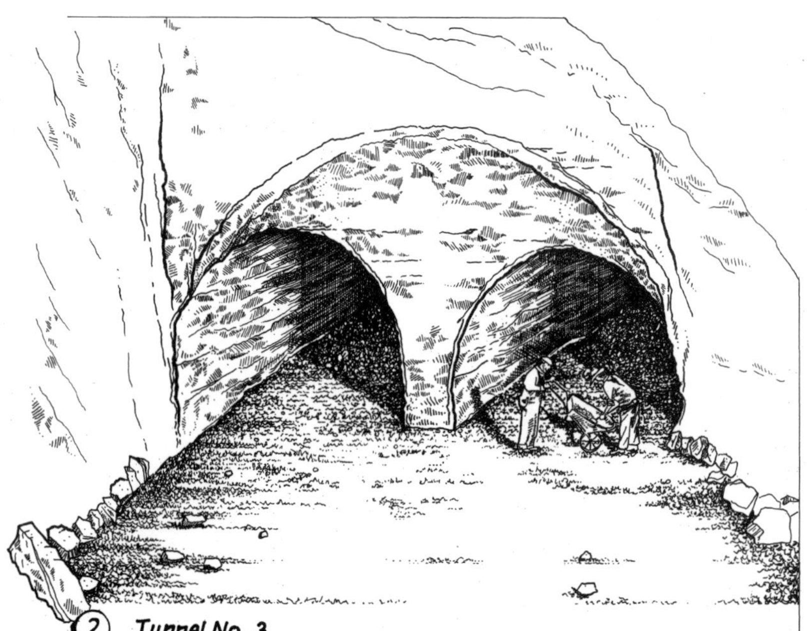

(2) Tunnel No. 3

(3) Tunnel No. 2

LINING

3. In the spring of 1934, concern was expressed that the constant high prairie wind would rapidly erode the newly exposed rock surfaces. In June 1935, the first two tunnels were lined with a metal mesh attached with galvanized spikes to the tunnel walls. Gunite, a highly liquid form of concrete, was then applied to a thickness of about 4 inches.

4. The instability of the rock at the third tunnel site necessitated a 12" reinforced concrete lining. Concrete was poured in 12'6" sections using movable forms. Test beams were obtained from each pour to ascertain when the steel forms could be moved.

(4) Tunnel No. 3

(5) Tunnel No. 3

PORTALS

5. In March 1939, portals were constructed at the lower entrances to Tunnels No. 2 and 3. These reinforced concrete portals stabilized the ends of the bore, acted as rockfall catches, and provided pleasing entrances for motorists. Once construction was finished, the portal walls were sandblasted and then coated with a colored gunite wash in an attempt to blend the construction features with the bluff. The shade was carefully matched to the rock surface as well as to the stucco of the headquarters buildings below.

6. Concerned about the continuing threat of rockslides, the Federal Highway Administration designed and built additional portals at both entrances to Tunnel No. 1 and the upper entrance to Tunnel No. 2. On the upper entrance to Tunnel No. 3, concrete was pumped to fill the area behind the existing tunnel liner, which had originally extended several feet from the rock face. This work was undertaken in 1989.

(6) Tunnel No. 1

DELINEATED BY: Todd Delyea 2000

NPS ROADS & BRIDGES RECORDING PROGRAM
NATIONAL PARK SERVICE
UNITED STATES DEPARTMENT OF THE INTERIOR

GERING

SCOTTS BLUFF SUMMIT ROAD
SCOTTS BLUFF NATIONAL MONUMENT
SCOTTS BLUFF COUNTY

NEBRASKA

SHEET 7 OF 10

HISTORIC AMERICAN ENGINEERING RECORD
NE-11

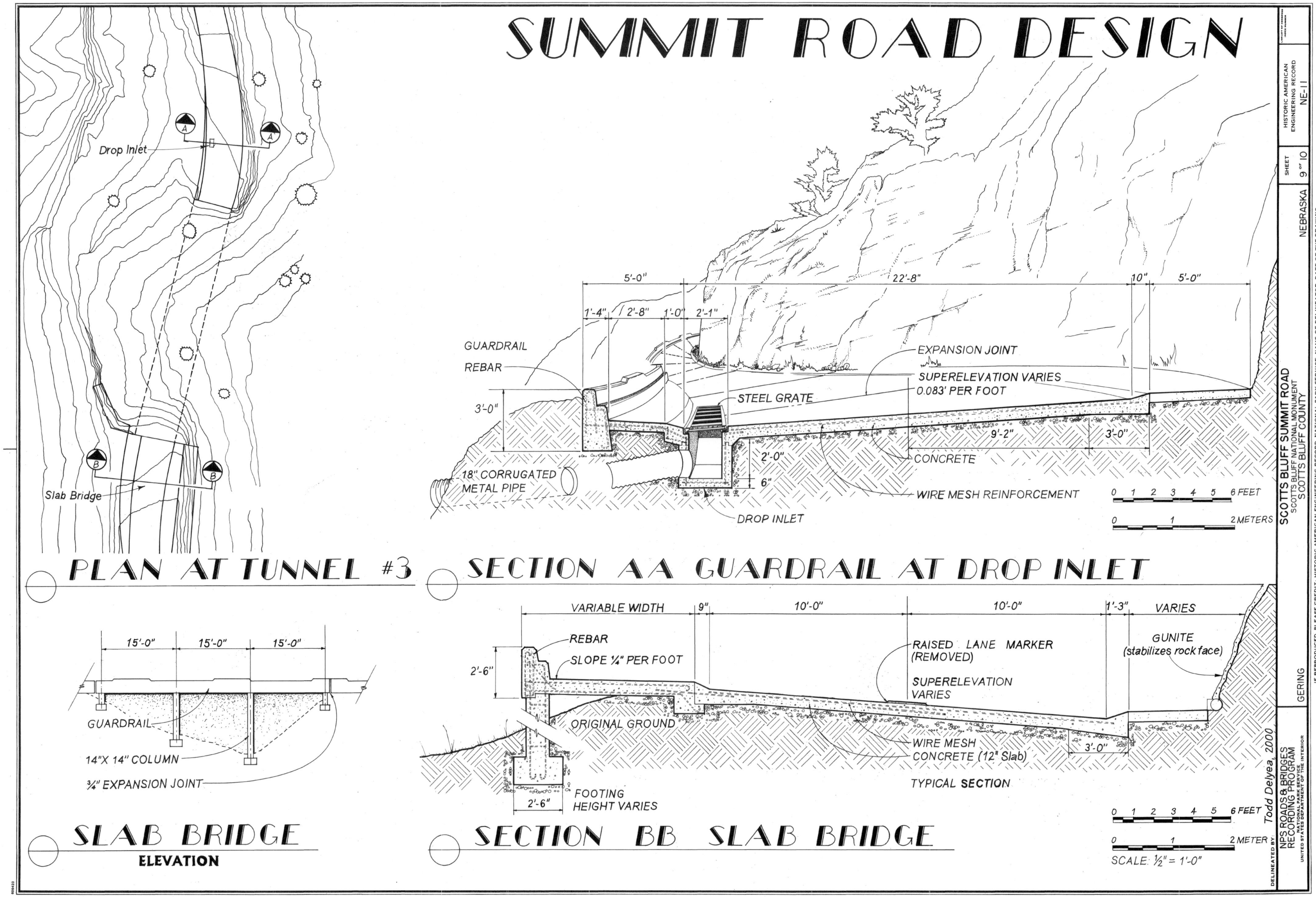
SUMMIT ROAD DESIGN
Drop Inlet
Slab Bridge
PLAN AT TUNNEL #3
GUARDRAIL
REBAR
3'-0"
5'-0"
22'-8"
10"
5'-0"
1'-4"
2'-8"
1'-0"
2'-1"
EXPANSION JOINT
SUPERELEVATION VARIES
0.083' PER FOOT
STEEL GRATE
9'-2"
3'-0"
2'-0"
6"
CONCRETE
18" CORRUGATED METAL PIPE
WIRE MESH REINFORCEMENT
DROP INLET
0 1 2 3 4 5 6 FEET
0 1 2 METERS
SECTION AA GUARDRAIL AT DROP INLET
15'-0"
15'-0"
15'-0"
GUARDRAIL
14"X 14" COLUMN
¾" EXPANSION JOINT
SLAB BRIDGE
ELEVATION
VARIABLE WIDTH
9"
10'-0"
10'-0"
1'-3"
VARIES
REBAR
SLOPE ¼" PER FOOT
2'-6"
RAISED LANE MARKER (REMOVED)
GUNITE (stabilizes rock face)
SUPERELEVATION VARIES
ORIGINAL GROUND
WIRE MESH
CONCRETE (12" Slab)
3'-0"
TYPICAL SECTION
FOOTING HEIGHT VARIES
2'-6"
0 1 2 3 4 5 6 FEET
0 1 2 METER
SCALE: ½" = 1'-0"
SECTION BB SLAB BRIDGE
HISTORIC AMERICAN ENGINEERING RECORD
NE-11
SHEET 9 OF 10
NEBRASKA
SCOTTS BLUFF SUMMIT ROAD
SCOTTS BLUFF NATIONAL MONUMENT
SCOTTS BLUFF COUNTY
GERING
DELINEATED BY Todd Delyea, 2000
NPS ROADS & BRIDGES RECORDING PROGRAM
NATIONAL PARK SERVICE
UNITED STATES DEPARTMENT OF THE INTERIOR
IF REPRODUCED, PLEASE CREDIT: HISTORIC AMERICAN ENGINEERING RECORD, NATIONAL PARK SERVICE, NAME OF DELINEATOR, DATE OF THE DRAWING

SUMMIT DEVELOPMENT
View of Summit Parking Area
N
Drawing based on field sketches and photographs and historical photographs.
Summit Map adapted from USGS 7.5 Minute Series, Scottsbluff South Quad., Nebraska, 1963 and original design documents, 1948.
4690
4680
4670
4660
4650
4640
4630
4620
4610
4600
Parking Area
South Overlook
Hiram Scotts' Memorial, 1939
North Overlook
Erosion Benchmark
To Visitor Center
LEGEND
0 160 320 480 640 FT.
0 96 192 m.
Rock Ledges
Coniferous Trees: Ponderosa Pine Rocky Mountain Juniper
Summit Trails: 1939/1953
South Overlook Trail: 1958
Saddle Rock Trail: 1933/1966
The primary motivation for constructing the Summit Road was to provide motorists with access to the panoramic views visible from the top of the bluff. It was believed that viewing the sites of the Oregon Trail and other important landmarks could inspire visitors with feelings of patriotism and pride in their American heritage. In order to experience the various views and sites on the summit, visitors are encouraged to leave their vehicles behind to explore the summit on foot. From the parking lot, two short foot trails lead to overlooks at which the surrounding historical and natural landmarks are located and described. A number of small monuments and plaques along the trails explain different aspects of the site.
Saddle Rock Trail Cave
This benchmark, placed flush with the apex of the bluff in 1933, stood 16.5" above the surface in July 2000, demonstrating the rapid rate of erosion at this exposed sandstone feature.
View from South Overlook:
Opening stretch of Summit Road and Saddle Rock Trail; Tunnel No. 1; Visitor Center; and Mitchell Pass, framed by Eagle Rock and Sentinel Rock.
View Westward:
North Platte River; Oregon Trail Route; Nebraska farm and ranch land.
View from North Overlook:
North Platte River; Union Pacific Railroad; badlands, and the towns of Gering and Scottsbluff.
DELINEATED BY: Stacey Heckaman, 2000
NPS ROADS & BRIDGES RECORDING PROGRAM
NATIONAL PARK SERVICE
UNITED STATES DEPARTMENT OF THE INTERIOR
SCOTTS BLUFF SUMMIT ROAD
SCOTTS BLUFF NATIONAL MONUMENT
SCOTTS BLUFF COUNTY
GERING
NEBRASKA
SHEET 10 OF 10
HISTORIC AMERICAN ENGINEERING RECORD
NE-11
IF REPRODUCED, PLEASE CREDIT: HISTORIC AMERICAN ENGINEERING RECORD, NATIONAL PARK SERVICE, NAME OF DELINEATOR, DATE OF THE DRAWING

SEQUOIA NATIONAL PARK, CALIFORNIA

GENERALS HIGHWAY

BRIDGES, DETAILS, AND LANDSCAPE

ASH MOUNTAIN ENTRANCE SIGN

Sequoia was designated a national park on September 25, 1890. The early military superintendents of the park soon discovered that only two wagon roads entered the park: the Mineral King and Colony Mill Roads. These two roads, although steep, prone to winter washout, and narrow, would remain the primary roads into the Park until the opening of the Generals Highway in 1926. The Mineral King Wagon and Toll Road, built in 1879, provided an access route into the new Park as it connected the foothill community of Three Rivers with the mining community of Mineral King. The Colony Mill Road, also known as the Giant Forest Road, was begun by the socialist Kaweah Colony in 1886 as a link to their land claims in the Giant Forest area and the nearby mill. The colonists' road was only completed to their mill site, 8.7 miles below the Giant Forest.

The need for improved roads in Sequoia was apparent to the park's early managers, as a road was desperately needed to reach the highly desirable Giant Forest area of the Big Trees. For many years a lack of funding prohibited road improvements within the Park, but finally in 1900 the Park received its first appropriation and the extension of the Colony Mill Road to the Giant Forest was completed in 1903.

The Mount Whitney Power Co. Road or Elk Park Road, was the third road to enter the Park prior to the construction of the Generals Highway. The power company was granted a right of way, by contract with the Department of the Interior, allowing for the construction of flumes, ditches, etc. within the park. In this agreement the power company constructed a $25,000 wagon road along the right of way up the Middle Fork Canyon to Hospital Rock. The Park in anticipation began the construction of a road down from Giant Forest to meet with the road to Hospital Rock. The "Smith Grade," begun in 1909, was only completed from Camp Sequoia at Commissary Curve to Eleven Range Point, never connecting with the power company road at Hospital Rock.

The wagon roads, constantly in disrepair, were anticipated to be inadequate for automobile traffic, and under the recommendation of National Park Service Director Stephen T. Mather, a new road was surveyed to access the Giant Forest. The Peters survey, completed in 1919, was to become the route of the Generals Highway. The new road would connect the existing Mt. Whitney Power Co. route to Hospital Rock, with the Giant Forest by way of a switchback section through Deer Ridge. The original design laid out a one-way road into the park, with the Colony Mill route serving as the exit road for outgoing traffic. Construction of the one-way highway began in 1921, but by 1923 the road design was widened to a two-way road. The Colony Mill Road was abandoned with the completion of the Generals Highway to Giant Forest in 1926.

The highway's extension from Giant Forest Village to the northern park boundary and on to General Grant National Park was begun by the Bureau of Public Roads in 1926. The National Park Service and the Bureau of Public Roads signed a joint agreement, in January of 1926, for the survey, construction, and improvement of roads and trails within the National Parks and Monuments. Under this agreement the Bureau of Public Roads supervised the extension of the highway from Giant Forest Village to the Park's northern boundary at Lost Grove, the construction of bridges and culverts, the reconstruction of the road below Hospital Rock, and the paving of the highway. The Generals Highway was dedicated on June 23, 1935 as a one-day scenic loop connecting Sequoia and General Grant national parks.

This project is part of the Historic American Engineering Record (HAER), a long range program to document historically significant engineering and industrial works in the United States. The HAER program is administered by the Historic American Buildings Survey/Historic American Engineering Record Division (HABS/HAER) of the National Park Service, U.S. Department of the Interior. The Generals Highway, Sequoia National Park recording project was cosponsored during the summer of 1993 by HAER under the general direction of Dr. Robert J. Kapsch, Chief; the National Park Service Roads and Bridges Recording Project; Tom Mulhern, Cultural Resource Division of the Western Regional Office; and by Sequoia National Park, J. Thomas Ritter, Superintendent.

The field work, measured drawings, historical reports, and photographs were prepared under the direction of Eric N. Delony, Chief of HAER, and project leader Todd A. Croteau. The recording team consisted of Carolyn J. Kiernat, team supervisor and architectural technician, Arizona State University; B. Devon Perkins, architectural technician, Yale University; Renata Stachanczyk, landscape architect, ICOMOS, Poland; Christina Slattery, historian, Ball State University; and large format photography by Brian C. Grogan, photographer.

DELINEATED BY: RENATA STACHANCZYK, 1993
GENERALS HIGHWAY RECORDING PROJECT
NATIONAL PARK SERVICE
UNITED STATES DEPARTMENT OF THE INTERIOR
GENERALS HIGHWAY
SEQUOIA NATIONAL PARK — TULARE COUNTY — CALIFORNIA
SHEET 1 OF 10
HISTORIC AMERICAN ENGINEERING RECORD CA-140
LIBRARY OF CONGRESS INDEX NUMBER

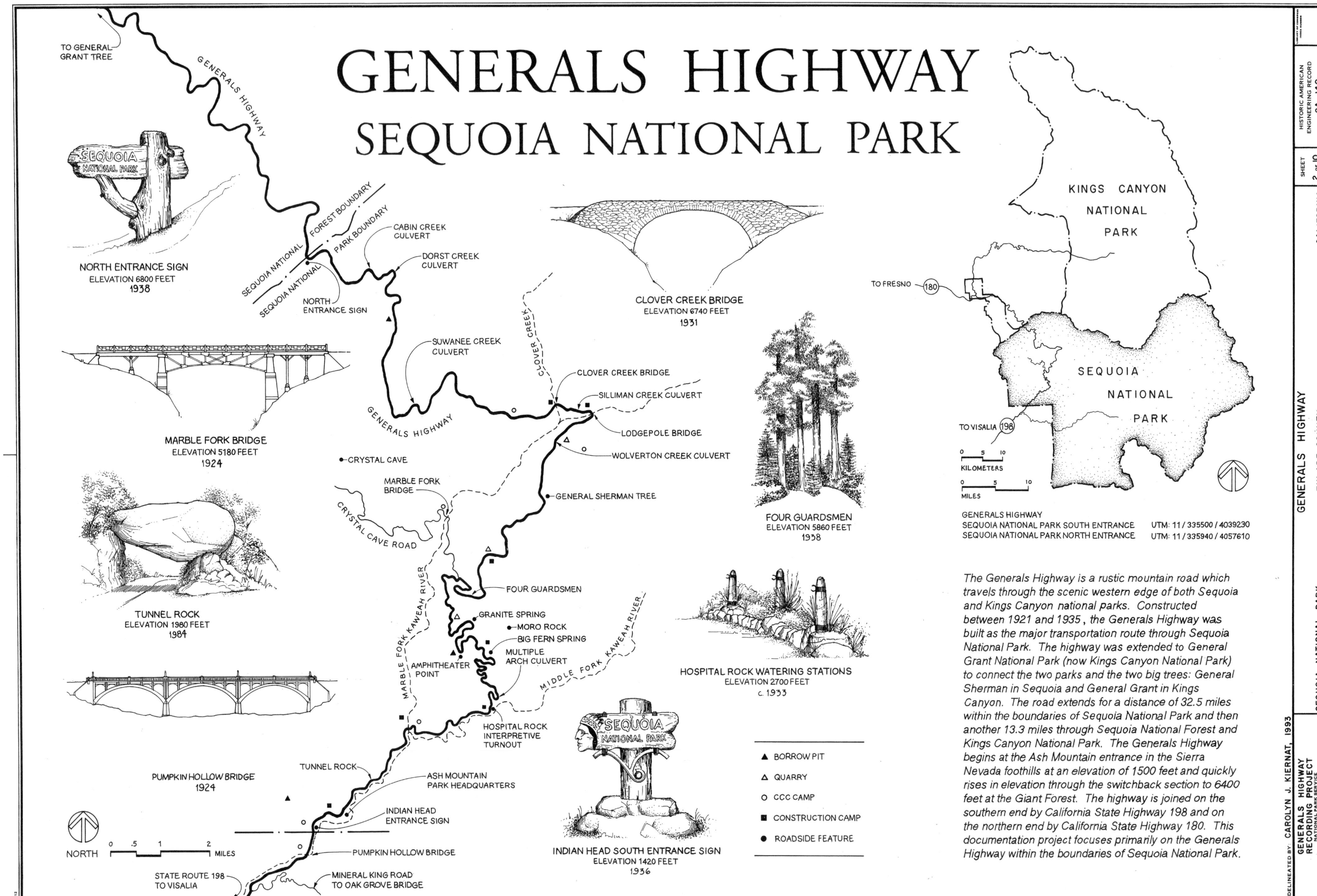
GENERALS HIGHWAY
SEQUOIA NATIONAL PARK
TO GENERAL GRANT TREE
GENERALS HIGHWAY
SEQUOIA NATIONAL PARK
NORTH ENTRANCE SIGN
ELEVATION 6800 FEET
1938
SEQUOIA NATIONAL FOREST BOUNDARY
SEQUOIA NATIONAL PARK BOUNDARY
CABIN CREEK CULVERT
DORST CREEK CULVERT
NORTH ENTRANCE SIGN
CLOVER CREEK BRIDGE
ELEVATION 6740 FEET
1931
MARBLE FORK BRIDGE
ELEVATION 5180 FEET
1924
SUWANEE CREEK CULVERT
CLOVER CREEK
CLOVER CREEK BRIDGE
SILLIMAN CREEK CULVERT
LODGEPOLE BRIDGE
WOLVERTON CREEK CULVERT
GENERALS HIGHWAY
CRYSTAL CAVE
MARBLE FORK BRIDGE
CRYSTAL CAVE ROAD
GENERAL SHERMAN TREE
FOUR GUARDSMEN
ELEVATION 5860 FEET
1938
TUNNEL ROCK
ELEVATION 1980 FEET
1984
FOUR GUARDSMEN
GRANITE SPRING
MORO ROCK
BIG FERN SPRING
MULTIPLE ARCH CULVERT
AMPHITHEATER POINT
MARBLE FORK KAWEAH RIVER
MIDDLE FORK KAWEAH RIVER
HOSPITAL ROCK WATERING STATIONS
ELEVATION 2700 FEET
c. 1933
HOSPITAL ROCK INTERPRETIVE TURNOUT
PUMPKIN HOLLOW BRIDGE
1924
TUNNEL ROCK
ASH MOUNTAIN PARK HEADQUARTERS
INDIAN HEAD ENTRANCE SIGN
SEQUOIA NATIONAL PARK
INDIAN HEAD SOUTH ENTRANCE SIGN
ELEVATION 1420 FEET
1936
NORTH
0 .5 1 2 MILES
PUMPKIN HOLLOW BRIDGE
STATE ROUTE 198 TO VISALIA
MINERAL KING ROAD TO OAK GROVE BRIDGE
BORROW PIT
QUARRY
CCC CAMP
CONSTRUCTION CAMP
ROADSIDE FEATURE
KINGS CANYON NATIONAL PARK
TO FRESNO 180
SEQUOIA NATIONAL PARK
TO VISALIA 198
0 5 10 KILOMETERS
0 5 10 MILES
GENERALS HIGHWAY
SEQUOIA NATIONAL PARK SOUTH ENTRANCE UTM: 11 / 335500 / 4039230
SEQUOIA NATIONAL PARK NORTH ENTRANCE UTM: 11 / 335940 / 4057610
The Generals Highway is a rustic mountain road which travels through the scenic western edge of both Sequoia and Kings Canyon national parks. Constructed between 1921 and 1935, the Generals Highway was built as the major transportation route through Sequoia National Park. The highway was extended to General Grant National Park (now Kings Canyon National Park) to connect the two parks and the two big trees: General Sherman in Sequoia and General Grant in Kings Canyon. The road extends for a distance of 32.5 miles within the boundaries of Sequoia National Park and then another 13.3 miles through Sequoia National Forest and Kings Canyon National Park. The Generals Highway begins at the Ash Mountain entrance in the Sierra Nevada foothills at an elevation of 1500 feet and quickly rises in elevation through the switchback section to 6400 feet at the Giant Forest. The highway is joined on the southern end by California State Highway 198 and on the northern end by California State Highway 180. This documentation project focuses primarily on the Generals Highway within the boundaries of Sequoia National Park.
HISTORIC AMERICAN ENGINEERING RECORD CA-140
SHEET 2 OF 10
CALIFORNIA
GENERALS HIGHWAY
TULARE COUNTY
SEQUOIA NATIONAL PARK
DELINEATED BY CAROLYN J. KIERNAT, 1993
GENERALS HIGHWAY RECORDING PROJECT
NATIONAL PARK SERVICE
UNITED STATES DEPARTMENT OF THE INTERIOR
IF REPRODUCED, PLEASE CREDIT: HISTORIC AMERICAN ENGINEERING RECORD, NATIONAL PARK SERVICE, NAME OF DELINEATOR, DATE OF THE DRAWING

TUNNEL ROCK
GENERALS HIGHWAY

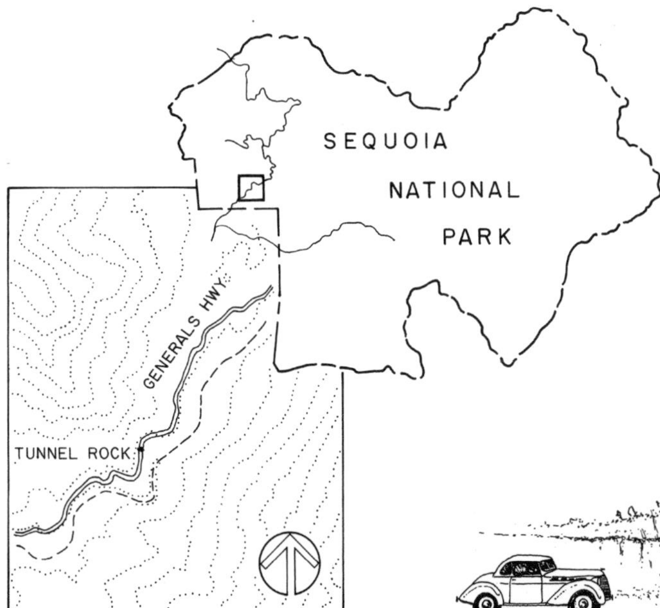

UTM: 11 / 337780 / 4040970
BASED ON 1987 USGS 7.5 MINUTE MAP
GIANT FOREST QUADRANGLE

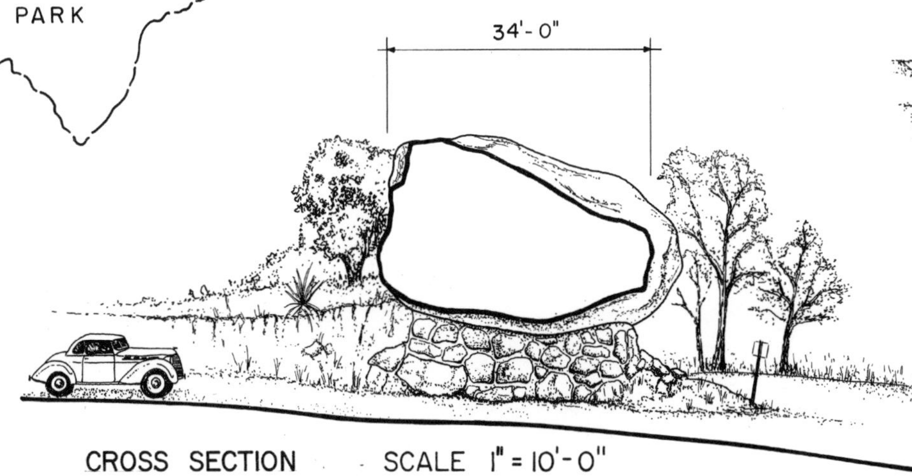

CROSS SECTION SCALE 1" = 10'-0"

Adapted from historic photo c.1934

SITE PLAN
WITH EXISTING PLANTING

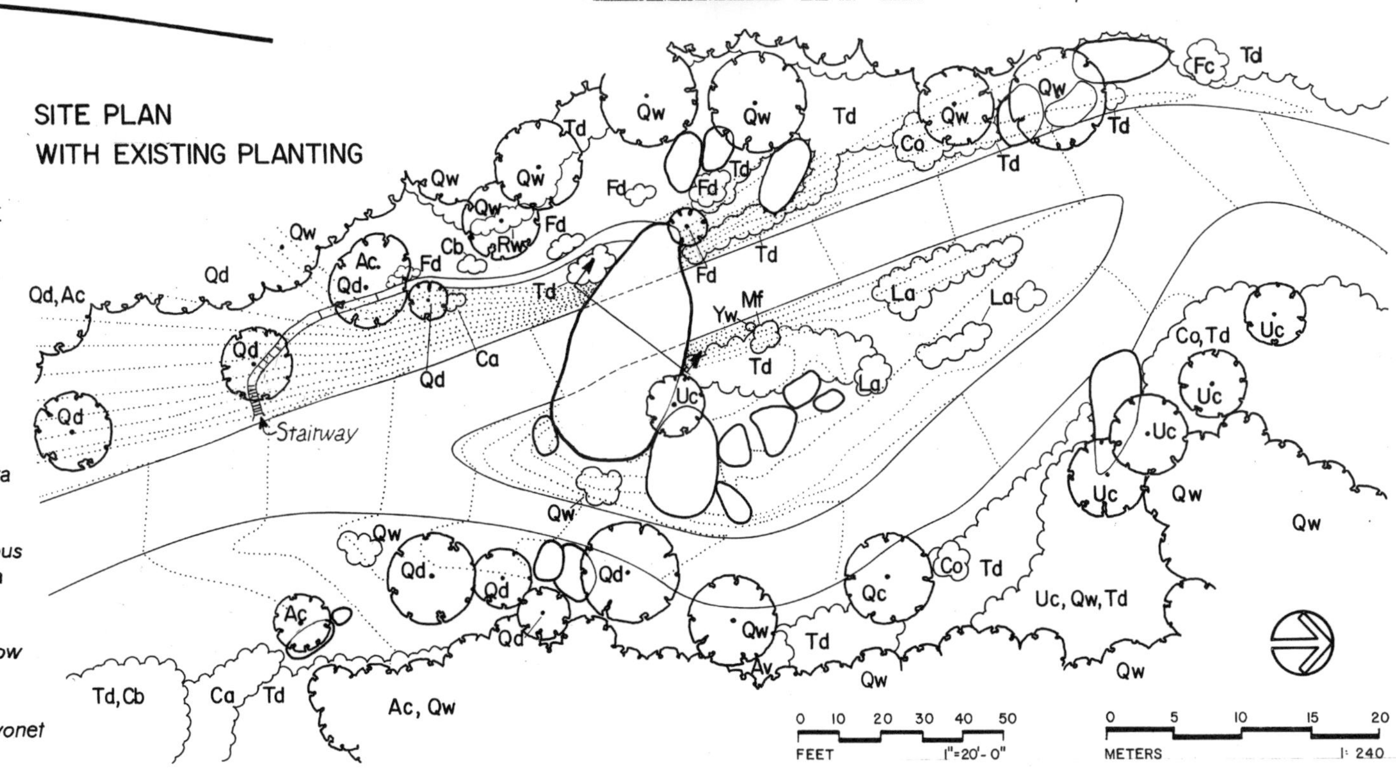

In 1934 a tunnel was excavated through this massive granite rock, located on the Generals Highway about one mile from the Ash Mountain Headquarters. The excavation by enlistees of the Civilian Conservation Corps (CCC) secured proper road alignment of the highway. The original road, constructed by the Mount Whitney Power Company, traveled around the rock to the east, similar to the current by-pass. In 1938, the CCC constructed masonry support walls on both sides of the tunnel to stabilize the earth walls. The CCC also built a stone stairway climbing to the top of the rock. The drawing on this page was adapted from an historic photograph and shows both the rock excavated from the boulder and the tunnel prior to the construction of the masonry support walls.

Drawings based on field measurements, historical photographs, and original design documents.

PLANT LIST

Key	*Botanical Name*	*Common Name*
Trees		
Ac	Aesculus californica	California Buckeye
Fd	Fraxinus dipetala	Foothill Ash
Uc	Umbellularia californica	California Laurel
Qc	Quercus chrysolepis	Canyon Live Oak
Qd	Quercus douglasii	Blue Oak
Qw	Quercus wislizenii	Interior Live Oak
Shrubs		
Av	Arctostaphyllos viscida	Whiteleaf manzanita
Ca	Ceanothus cuneatus	Buck Brush
Co	Cercis occidentalis	Redbud
Cb	Cercocarpus betuloides	Birchleaf cercocarpus
Fc	Fremontodendron californicum	California fremontia
La	Lupinus arboreus	Tree Lupine
Mf	Malvastrum fremonti	Fremont Falsemallow
Td	Toxicodendron diversilobum	Poison Oak
Yw	Yucca whipplei var. caespitosa	Yucca Spanish Bayonet

HISTORIC AMERICAN ENGINEERING RECORD CA-140
SHEET 3 OF 10
GENERALS HIGHWAY
CALIFORNIA
TULARE COUNTY
SEQUOIA NATIONAL PARK
IF REPRODUCED, PLEASE CREDIT: HISTORIC AMERICAN ENGINEERING RECORD, NATIONAL PARK SERVICE, NAME OF DELINEATOR, DATE OF THE DRAWING
DELINEATED BY: B. DEVON PERKINS, 1993
GENERALS HIGHWAY RECORDING PROJECT
NATIONAL PARK SERVICE
UNITED STATES DEPARTMENT OF THE INTERIOR

HOSPITAL ROCK
GENERALS HIGHWAY

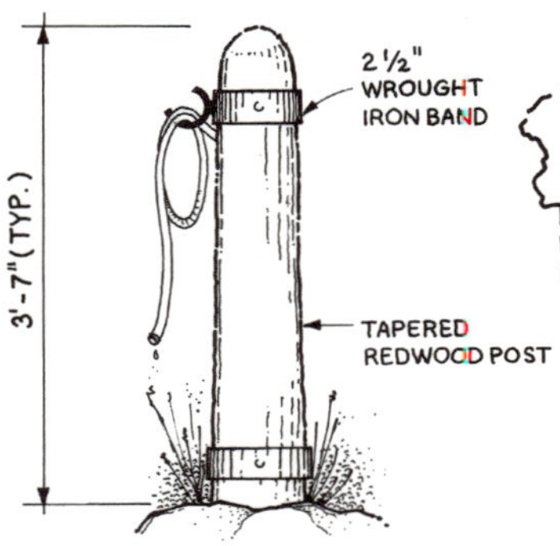

ELEVATION 1" = 1'-0"

U.T.M. 11 / 341490 / 4043030

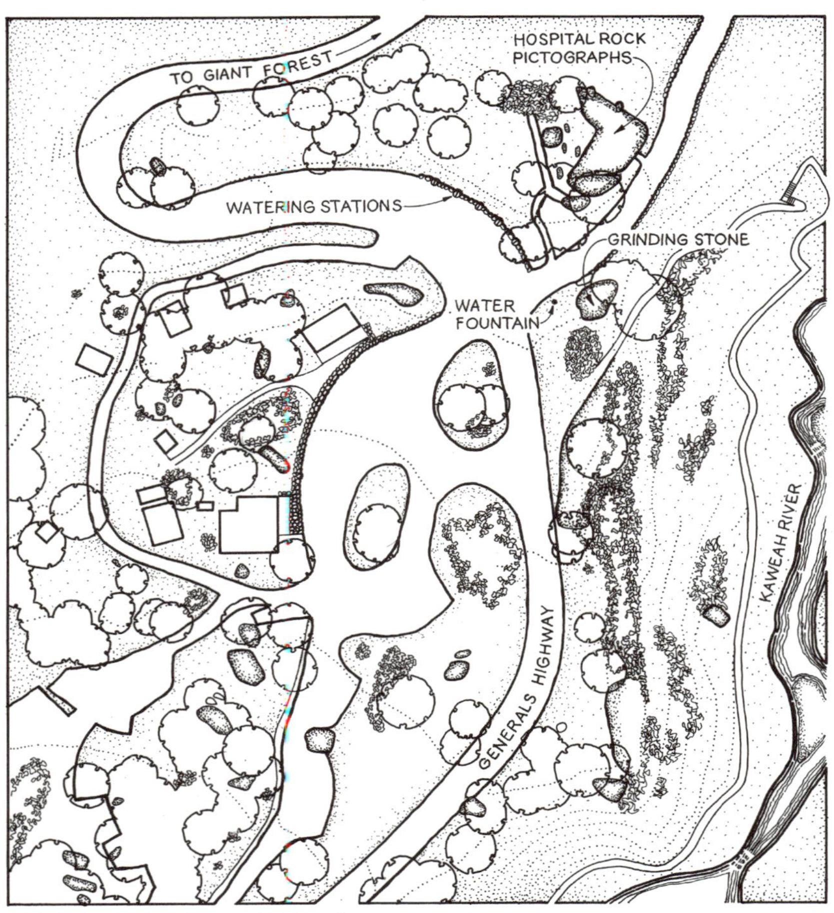

SITE PLAN BASED ON 1938 NATIONAL PARK SERVICE MAP — SEKI ARCHIVES

WATERING STATIONS

NATIVE AMERICAN GRINDING HOLES

GRANITE WATER FOUNTAIN

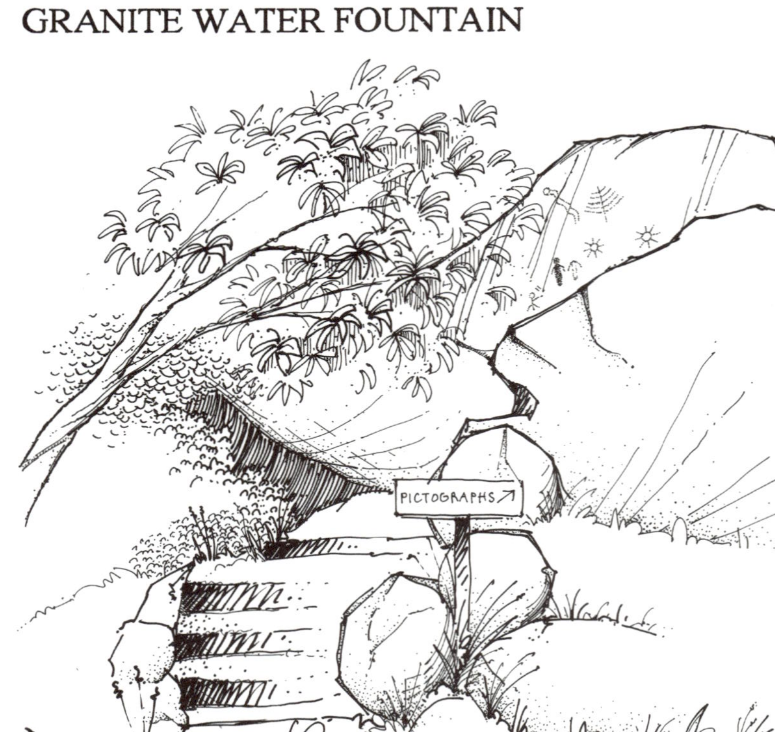

NATIVE AMERICAN PICTOGRAPHS

Hospital Rock interpretive turnout is located on Generals Highway 6.1 miles from the Ash Mountain Park entrance. Hospital Rock was the site of a Native American village of the Potwisha tribe. A large granite rock 60' long and 20' thick overhangs to create a spacious room which was used by the tribe as a shelter for the sick and as a gathering and ceremonial space for the village. Located adjacent to the rock are grinding holes which were used by the tribe to grind acorns to create flour for food preparation. Evidence of habitation prior to the Potwisha can be seen in the collection of pictographs painted on the rock face. The meanings of the painted designs are not known, but similar designs among other tribes have often had religious, ritual or magical significance.

The site earned the name "Hospital Rock" in 1873, ten years after the Potwisha tribe had abandoned the village, when it was used by some of the first explorers of the area to shelter and aid an injured man.

The development of the site as an interpretive turnout was largely completed by enrollees of the Civilian Conservation Corps (CCC) from November 1933 to April 1934. The men of the CCC laid out a parking area and campground, built a natural stone stairway to replace the wooden stairs reaching the pictographs, constructed automobile watering stations, completed roadside planting in the area and constructed a stone drinking fountain to mimic the character of the Native American grinding holes. Today the site serves as a reminder of the early inhabitants of the area prior to its creation as a national park.

DELINEATED BY: CAROLYN J. KIERNAT, 1993

GENERALS HIGHWAY RECORDING PROJECT NATIONAL PARK SERVICE UNITED STATES DEPARTMENT OF THE INTERIOR	GENERALS HIGHWAY SEQUOIA NATIONAL PARK	TULARE COUNTY	CALIFORNIA	SHEET 4 of 10	HISTORIC AMERICAN ENGINEERING RECORD CA-140

FOUR GUARDSMEN
GENERALS HIGHWAY

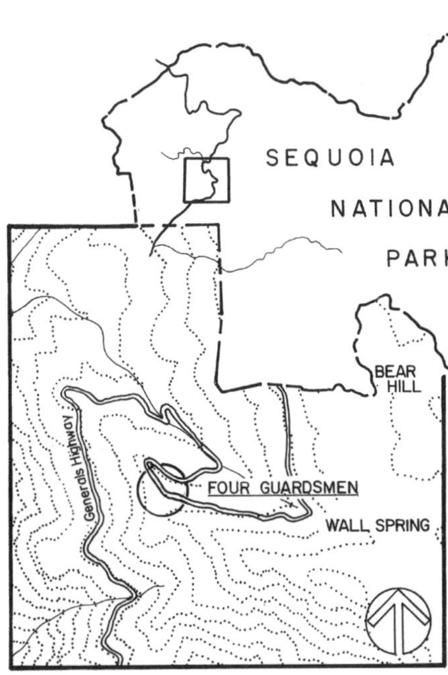

UTM: 11 / 340520 / 4046740
BASED ON 1987 USGS 7.5 MINUTE MAP
GIANT FOREST QUADRANGLE

ORIGINAL
ALIGNMENT - 1937

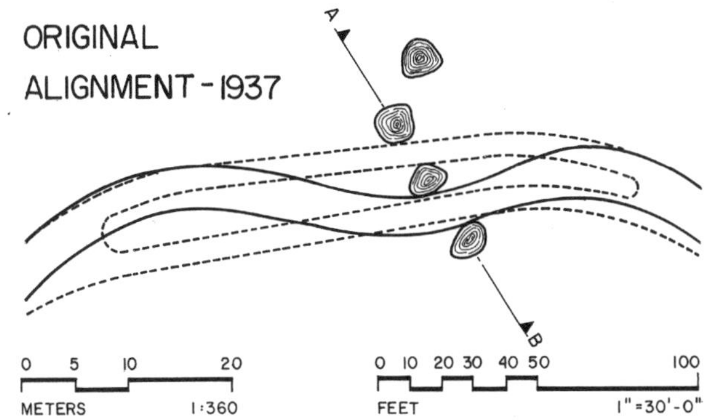

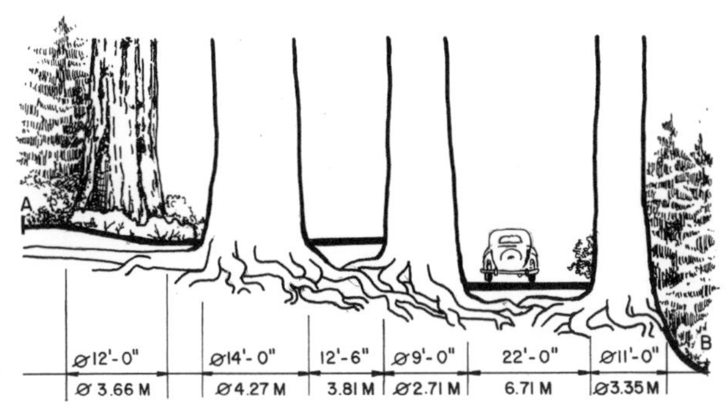

CROSS SECTION - 1993

VIEW
LOOKING NORTH, NO SCALE

The FOUR GUARDSMEN, also called the Four Guardians and the Gateway Group, are four large sequoia trees growing closely together which have been intersected by the Generals Highway.
The Four Guardsmen, located about 14.9 miles above the Ash Mountain entrance and 2 miles below the Giant Forest Village, form a natural gateway to Giant Forest. The group of four trees were named by George Welsh, surveyor of the first road through the area.
The original road design was a two-way single lane road between two of the trees. The narrow width of the road led to a bottleneck of traffic at the point and in 1939 a second lane was constructed cutting between the four trees.
This area is a good example of road design which used the natural landscape elements to create a unique experience for park visitors.

The Four Guardsmen are located among the other big trees in the forested (middle) elevation zone.

SOURCES: SEKI Archives, NPS Construction Documents
HABS/HAER Field Survey, 1993

SURVEY OF EXISTING PLANTING - 1993

PLAN LEGEND

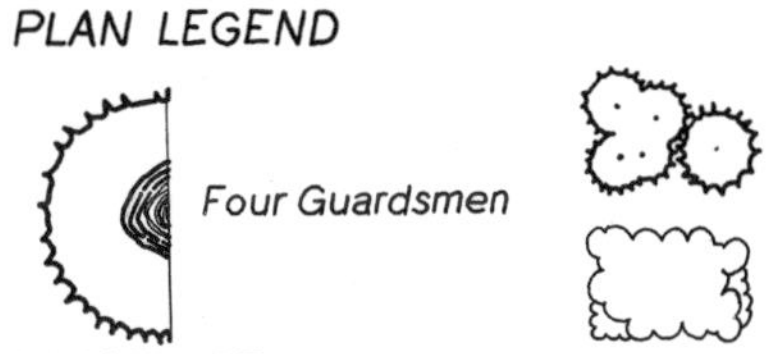

PLANT LIST

Key	Botanical Name	Common Name
Trees		
Ac	*Abies concolor*	*White Fir*
Cd	*Calocedrus deccurens*	*Incense Cedar*
Pl	*Pinus lambertiana*	*Sugar Pine*
Sg	*Sequoiadendron giganteum*	*Giant Sequoia*
Shrubs and Ferns		
Ci	*Ceanothus integerrimus*	*Deer Brush*
Cn	*Cornus nuttallii*	*Pacific Dogwood*
Cc	*Corylus cornuta var. calif.*	*California Hazelnut*
Pa	*Pteridium aquilinum*	*Bracken Fern*
Rp	*Rubus parviflorus*	*Thimbleberry*
Ss	*Salix scoulerana*	*Scouler Willow*

EXISTING
PLANTING PLAN - 1993

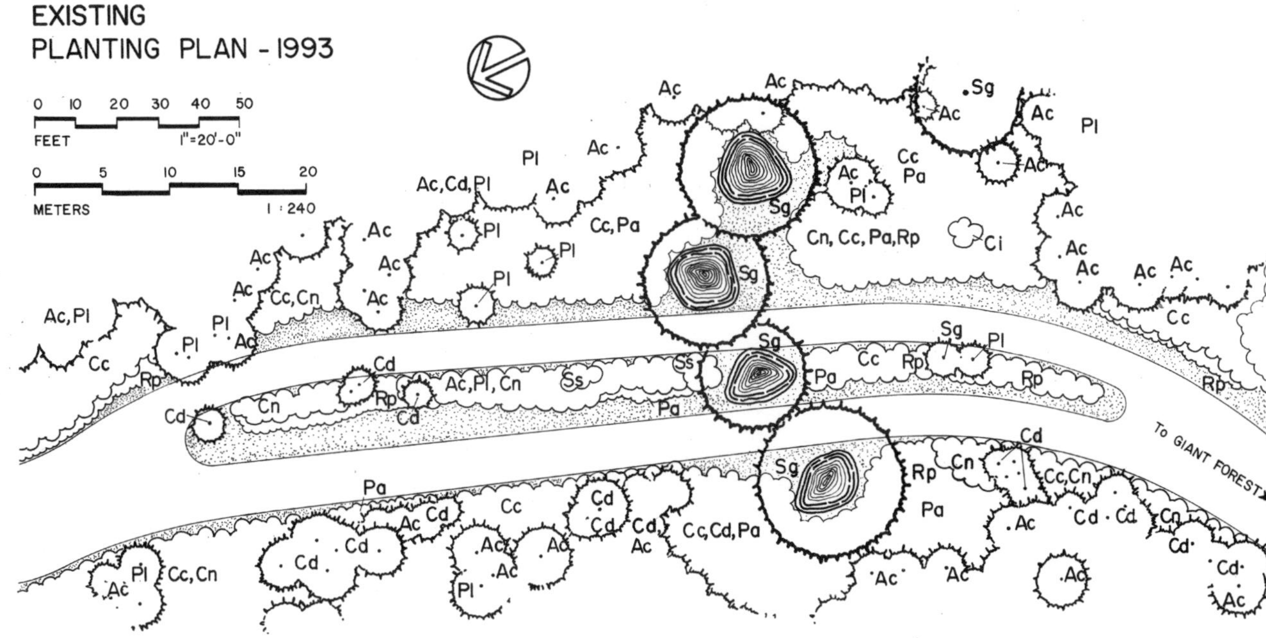

HISTORIC AMERICAN ENGINEERING RECORD
CA-140
SHEET 5 OF 10
GENERALS HIGHWAY
SEQUOIA NATIONAL PARK
TULARE COUNTY
CALIFORNIA
DELINEATED BY: RENATA STACHAŃCZYK, 1993
GENERALS HIGHWAY RECORDING PROJECT
NATIONAL PARK SERVICE
UNITED STATES DEPARTMENT OF THE INTERIOR

STONEWORK
GENERALS HIGHWAY

BIG FERN SPRING

The Generals Highway contains a variety of stonework which contributes to the rustic character of the road. The stonework within the park reflects its surroundings through the use of local schist and granite stone. Much of the stonework was constructed by the Civilian Conservation Corps (CCC) between 1933 and 1942. As a New Deal program the CCC was established to provide jobs for unemployed young men through work with the government. The contributions of the ten CCC camps within Sequoia were numerous and included extensive stonework along the highway. The CCC worked to improve road drainage through the construction of schist gutters and accompanying drop culverts in 1936 from the southern park boundary to just beyond the park headquarters. The CCC also improved the springs along the highway, including the construction of a stone wall and watering station at Big Fern Spring in 1934.

The Suwanee Creek Culvert is located on the highway between the Clover Creek Bridge and Dorst Creek. Designed in 1930 by the Bureau of Public Roads, this culvert is a true stone masonry arch with an entire length of over 150'. The opening is 6' wide and 7' high with an arch radius of 3'.

The Silliman Creek Culvert is located on the highway between the Lodgepole and Clover Creek Bridges. It was built during 1930 and 1931 under the same Bureau of Public Roads contract as the bridges. The structure is a reinforced concrete slab which spans 16' between rubble masonry abutments. The interior walls are 9'-6" high at the upstream elevation and are also of stone.

Along the Generals Highway a variety of stone guardwalls have been constructed using materials corresponding to the stone of the area. Mostly low smooth profile stone walls are used to outline parking areas and road edges, but one location has a crenellated wall. The Kaweah Canyon and switchback areas contain mostly schist walls but also some granite, while the big tree area uses large granite blocks which relate better to the scale of the sequoia trees. In the ridgetop area, granite is also the more common type of guardwall.

SUWANEE CREEK CULVERT

SILLIMAN CREEK CULVERT

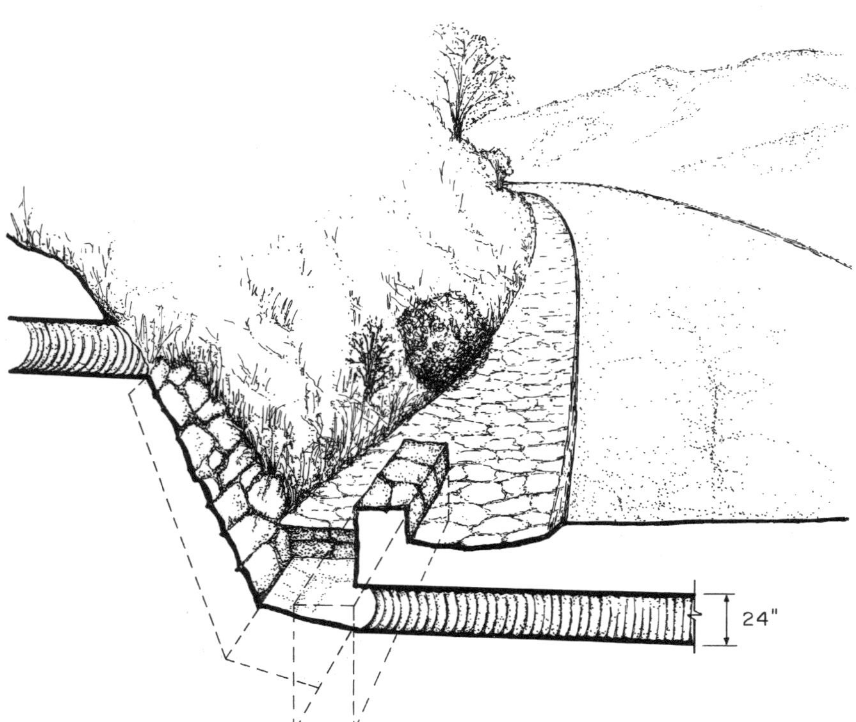

DROP CULVERT & SCHIST GUTTER

SCHIST GUARDWALL & TURNOUT

DELINEATED BY: B. DEVON PERKINS, 1993

GENERALS HIGHWAY RECORDING PROJECT NATIONAL PARK SERVICE UNITED STATES DEPARTMENT OF THE INTERIOR	GENERALS HIGHWAY SEQUOIA NATIONAL PARK TULARE COUNTY CALIFORNIA	SHEET 8 of 10	HISTORIC AMERICAN ENGINEERING RECORD CA-140	

APPROACH BRIDGES
SEQUOIA NATIONAL PARK

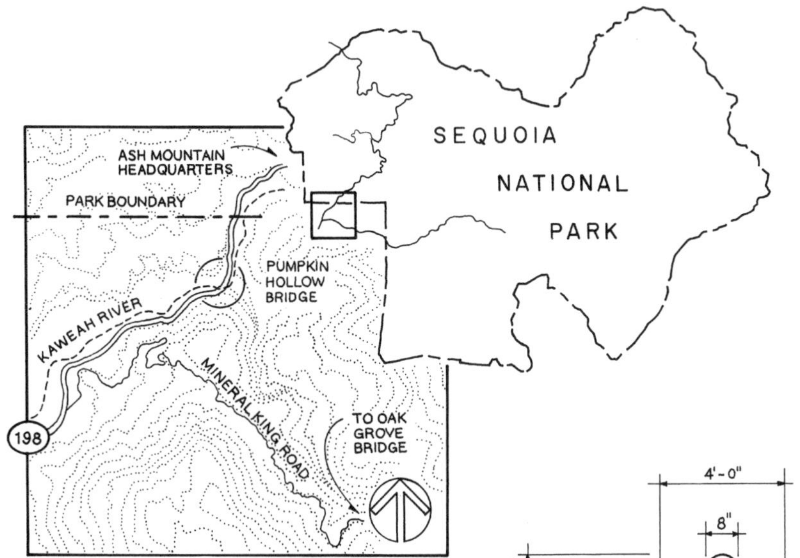

PUMPKIN HOLLOW UTM: 11 / 335300 / 4038460
OAK GROVE UTM: 11 / 339390 / 4035080
BASED ON 1987 USGS 7.5 MINUTE MAP
GIANT FOREST QUADRANGLE

The Oak Grove and Pumpkin Hollow bridges, built as reinforced concrete arch structures by Tulare County and local contractor, Nate Lovelace, serve as approach bridges to the park. The Oak Grove Bridge, completed in 1923, spans the East Fork of the Kaweah River and is located on Mineral King Road. The Pumpkin Hollow Bridge was completed the following year, 1924, and spans the Middle Fork of the Kaweah River on Highway 198. The bridges exhibit a modern design aesthetic, which is in contrast to the "rustic" style typically found within the park.

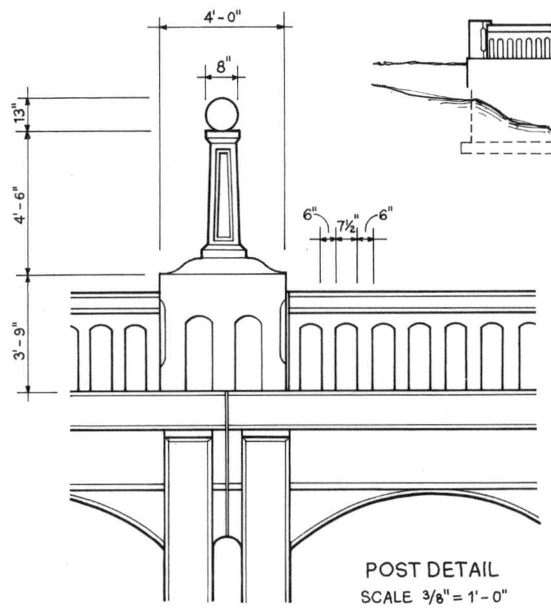

POST DETAIL
SCALE 3/8" = 1'-0"

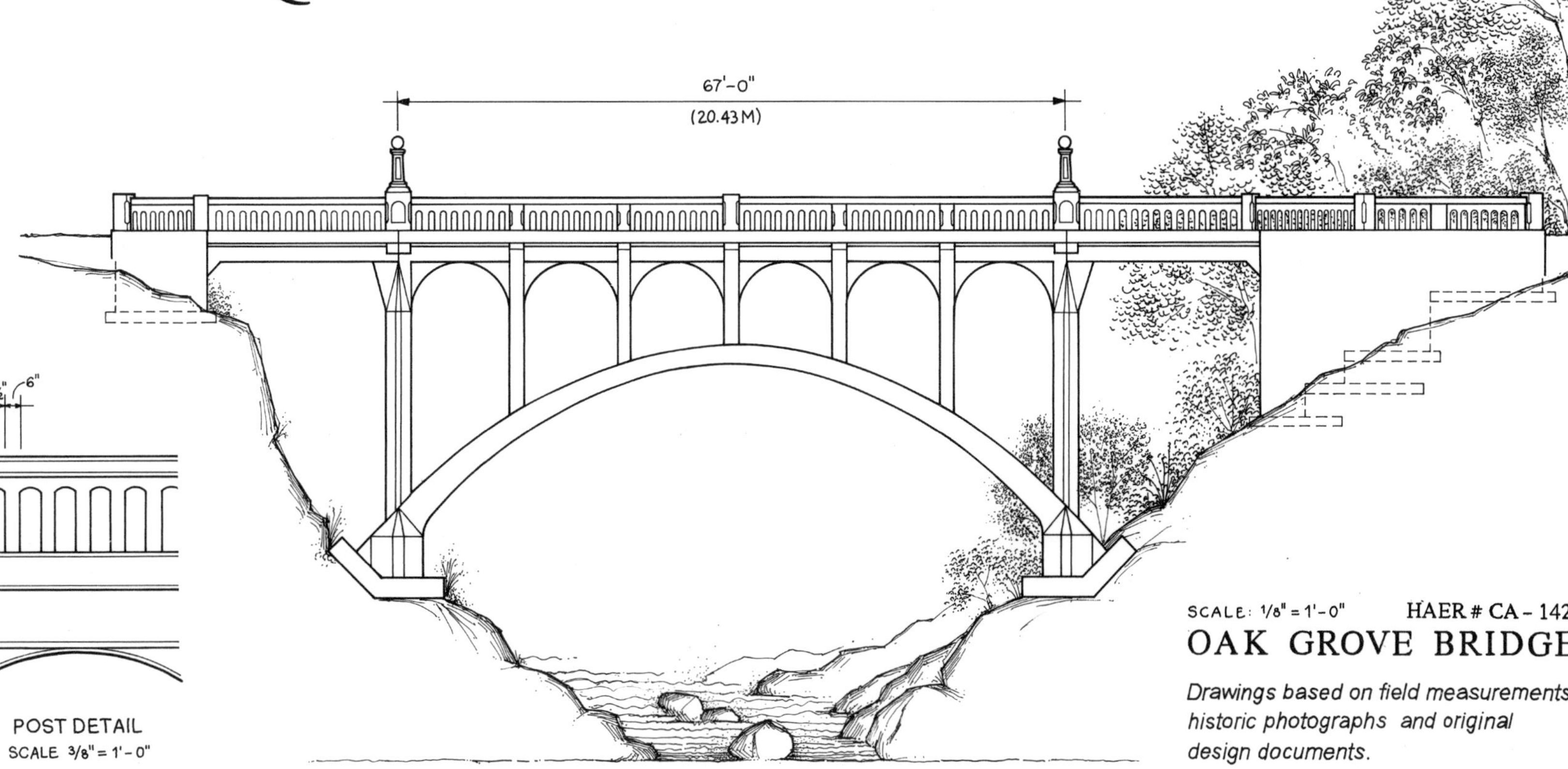

SCALE: 1/8" = 1'-0" HAER # CA - 142
OAK GROVE BRIDGE

Drawings based on field measurements, historic photographs and original design documents.

SCALE 1/8" = 1'-0" HAER # CA - 143
PUMPKIN HOLLOW BRIDGE

DELINEATED BY: CAROLYN J. KIERNAT, 1993
GENERALS HIGHWAY RECORDING PROJECT
NATIONAL PARK SERVICE
UNITED STATES DEPARTMENT OF THE INTERIOR

GENERALS HIGHWAY
THREE RIVERS TULARE COUNTY CALIFORNIA

SHEET 9 OF 10

HISTORIC AMERICAN ENGINEERING RECORD
CA-140

IF REPRODUCED, PLEASE CREDIT: HISTORIC AMERICAN ENGINEERING RECORD, NATIONAL PARK SERVICE, NAME OF DELINEATOR, DATE OF THE DRAWING

HISTORIC BRIDGES
GENERALS HIGHWAY

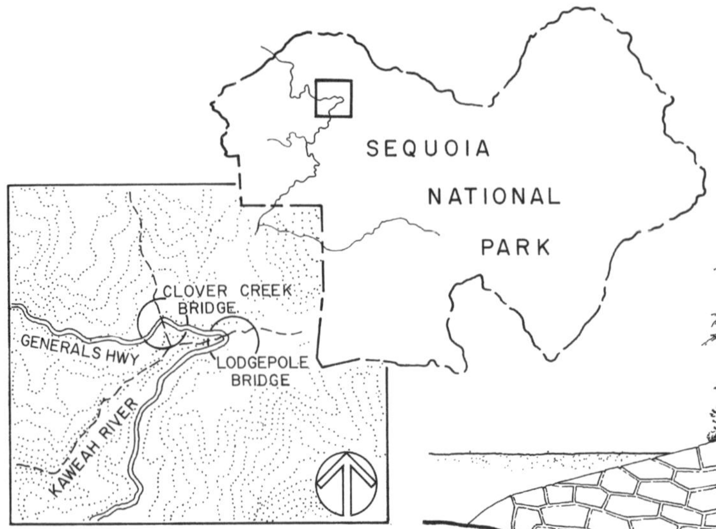

LODGEPOLE UTM: 11 / 344960 / 4052200
CLOVER CREEK UTM: 11 / 343820 / 4052530
BASED ON 1987 USGS 7.5 MINUTE MAP
LODGEPOLE QUADRANGLE

The Clover Creek and Marble Fork (Lodgepole) bridges, constructed between 1930 and 1931, have the appearance of traditional stone bridges, but are actually reinforced concrete arch designs with a stone veneer. The two bridges are located only .8 miles apart on the Generals Highway, but vary dramatically in size and setting. The Clover Creek Bridge spans 90' across a barren granite bedrock canyon while the Marble Fork Bridge, with a span half the size, 45', is set in a wooded forest. Both bridges were designed by NPS landscape architect, John Wosky, and were constructed by contract under the supervision of the Bureau of Public Roads.

This project is part of the Historic American Engineering Record (HAER), a long range program to document historically significant engineering and industrial works in the United States. The HAER program is administered by the Historic American Buildings Survey/Historic American Engineering Record Division (HABS/HAER) of the National Park Service, U.S. Department of the Interior. The Generals Highway, Sequioa National Park recording project was cosponsored during the summer of 1993 by HAER under the general direction of Dr. Robert J. Kapsch, Chief; the National Park Service Roads and Bridges Recording Project; Tom Mulhern, Cultural Resource Division of the Western Regional Office; and by Sequoia National Park, J. Thomas Ritter, Superintendent.

The field work, measured drawings, historical reports, and photographs were prepared under the direction of Eric N. Delony, Chief of HAER, and project leader Todd A. Croteau. The recording team consisted of Carolyn J.Kiernat, team supervisor and architectural technician, Arizona State University; B.Devon Perkins, architectural technician, Yale University; Renata Stachanczyk, landscape architect, ICOMOS, Poland; Christina Slattery, historian, Ball State University; and large format photography by Brian C. Grogan, photographer.

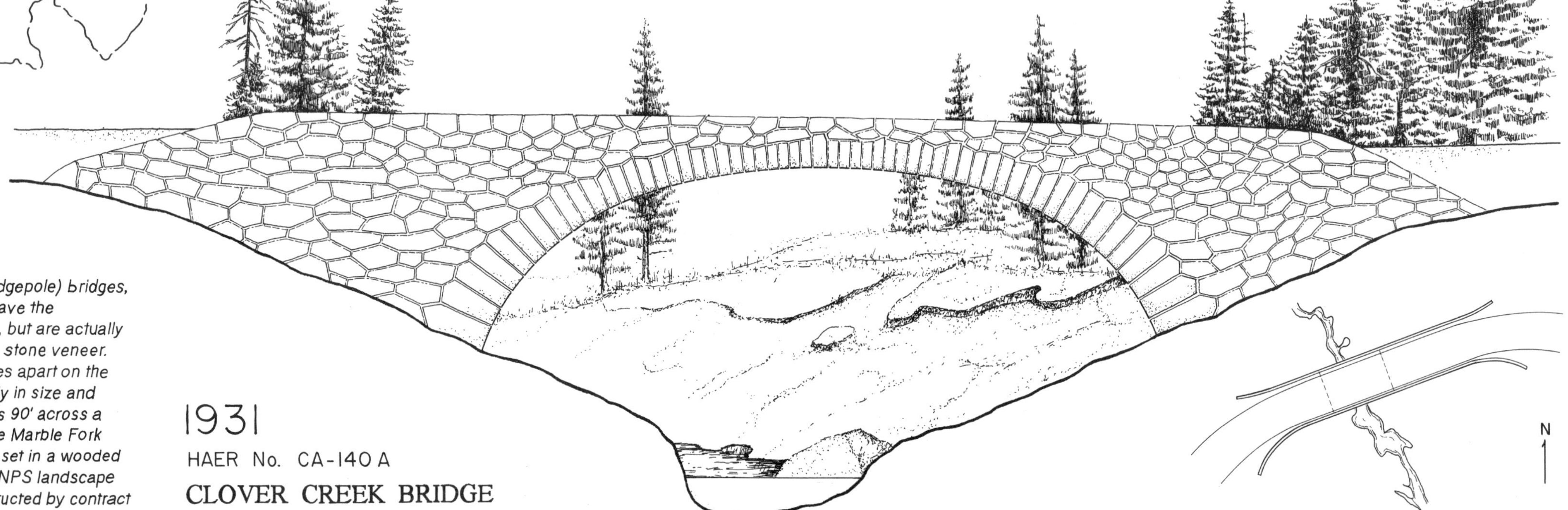

1931
HAER No. CA-140 A
CLOVER CREEK BRIDGE

25 20 15 10 5 0 10 20 30 60 90
METERS 1:96 1/8" = 1'-0" FEET

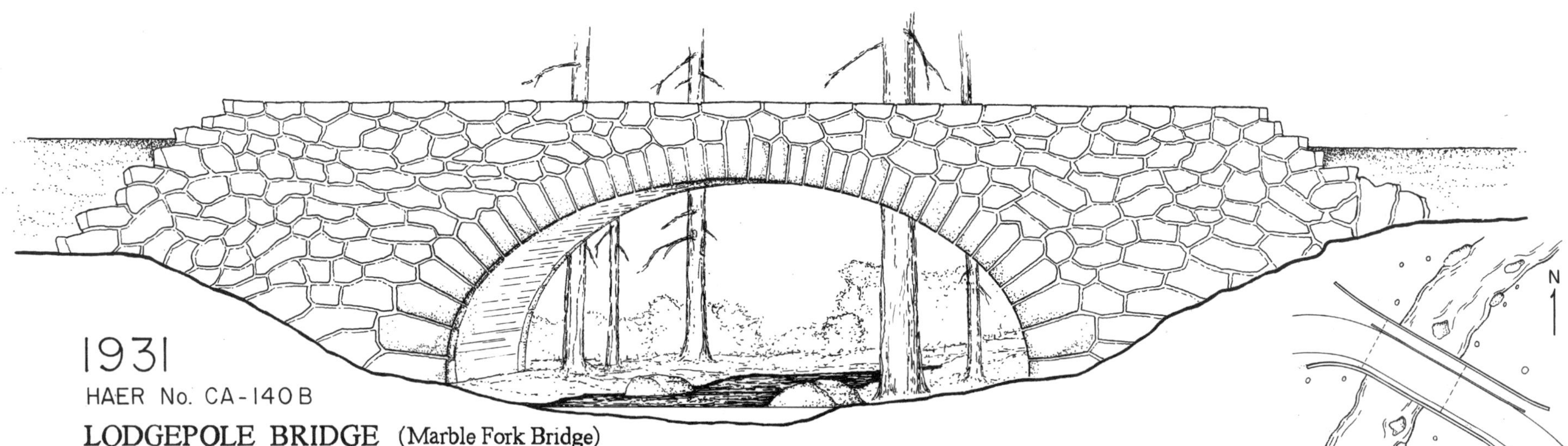

1931
HAER No. CA-140 B
LODGEPOLE BRIDGE (Marble Fork Bridge)

15 10 5 0 5 10 20 30 40 50
METERS 1:64 3/16" = 1'-0" FEET

DELINEATED BY: B. DEVON PERKINS, 1993
GENERALS HIGHWAY RECORDING PROJECT
NATIONAL PARK SERVICE
UNITED STATES DEPARTMENT OF THE INTERIOR

SEQUOIA NATIONAL PARK
GENERALS HIGHWAY
LODGEPOLE VICINITY
TULARE COUNTY
CALIFORNIA

SHEET 10 of 10
HISTORIC AMERICAN ENGINEERING RECORD
CA-140

IF REPRODUCED, PLEASE CREDIT: HISTORIC AMERICAN ENGINEERING RECORD, NATIONAL PARK SERVICE, NAME OF DELINEATOR, DATE OF THE DRAWING

SKYLINE DRIVE

SHENANDOAH NATIONAL PARK, VIRGINIA

DEDICATION JULY 3, 1936

Over seventy miles long and varying in width from less than one mile to just over 13 miles, Shenandoah National Park straddles the Blue Ridge Mountains of Virginia. Here the ancient and narrow front range of the Appalachian Mountains, shrouded in a perpetual blue haze of humidity, rises abruptly from the eastern Piedmont plateau before descending sharply into the Shenandoah Valley on the west. Winding along the ridge crest, Skyline Drive provides physical and scenic access along the park's entire length, offering views over the western valley and eastern plateau some 2,500 to 3,500 feet below.

Emerging from the same social, economic, and technological movements during the first quarter of the twentieth century, Shenandoah National Park and Skyline Drive have parallel but separate histories. The creation of Shenandoah represents the culmination of a movement to establish national parks in the eastern United States. Reporting to Interior Secretary Hubert Work in 1924, the Southern Appalachian National Park Commission recommended the Virginia Blue Ridge between Front Royal and Waynesboro as its primary choice for a park, noting that the site lay "within a 3-hour ride" of the nation's capital "and within a day's ride" of 40 million Americans. Acting on the Commission's recommendation, Congress authorized the park in 1925 (along with another in the Great Smoky Mountains) if sufficient land became available.

Unlike western national parks which were carved out of the public domain, Shenandoah was assembled from numerous tracts in private or corporate ownership. Since Congress did not appropriate funds for the acquisition of national park land, the Commonwealth of Virginia spent a decade acquiring the park's original 176,429 acres through donations, purchase and "blanket condemnation" under the right of eminent domain. Virginia presented the property to the U.S. Government which approved the park in December 1935.

Just as Shenandoah represented one of the National Park Service's (NPS) first ventures in eastern park establishment, Skyline Drive was one of the first mountain road building projects undertaken by the NPS in the East. The concept of a scenic ridge drive had been vocalized by the Southern Appalachian National Park Commission (1924), when it noted:

> *The greatest single feature [of the proposed park] ... is a possible sky-line drive along the mountain top, following a continuous ridge and looking down westerly on the Shenandoah Valley ... and also commanding a view of the Piedmont Plain stretching easterly to the Washington Monument....*

The Commission's statement was a tacit recognition of the advent and proliferation of the automobile as a recreational vehicle. By 1929, over 23 million passenger cars were registered in the U.S., making the automobile the most significant social and technological change of the period, reaching even into the ways Americans spent their leisure time. Federal, state and local officials realized that the new national park would, by necessity, be a park for the motorist.

SKYLINE DRIVE FROM STONY MAN

"... these mountains were just made for a highway," observed President Herbert Hoover to National Park Service Director Horace Albright in the fall of 1930, on the summit of the Blue Ridge Mountains near Camp Rapidan. From the same heights half a decade later, Hoover's successor Franklin D. Roosevelt dedicated Shenandoah, touting the park's scenic mountaintop road, then two-thirds complete: "You who are here know of the great usefulness to humanity which this Skyline Drive achieves." Direct involvement by these two presidents made Skyline Drive and Shenandoah National Park possible.

Skyline Drive, Shenandoah National Park Roads and Bridges Recording Project was undertaken by the Historic American Engineering Record, Eric N. DeLony, Chief, a long-range program to document historically significant engineering and industrial works in the United States. The program is part of the Historic American Buildings Survey/Historic American Engineering Record (HABS/HAER) Division of the National Park Service, U.S. Department of the Interior, E. Blaine Cliver, Chief.

Documentation was sponsored by the National Park Service and Iowa State University of Science and Technology through a cooperative agreement. Cooperators included Shenandoah National Park, Bill Wade, Superintendent; and Department of Landscape Architecture, ISU, J. Timothy Keller, Chair.

SHENANDOAH NATIONAL PARK

Funding was provided by the Federal Lands Highway Office, Thomas Edick, Administrator, through the NPS Park Roads and Parkways Program, Mark Hartsoe, Manager.

Fieldwork, measured drawings, historical reports, and photographs were prepared under the overall supervision and direction of project leader Christopher H. Marston, HAER Architect, and Richard H. Quin, HAER Historian. The Iowa State recording team consisted of Robert R. Harvey, ASLA, Field Supervisor; harlen d. Groe, Landscape Architect; Michael P. Lanning, and Christopher J. Seeger, Landscape Architectural Technicians; Shane P. Wirth, Architectural Technician; James R. Hill III, Historian. Large format photography was prepared by William Faust, Charlottesville, Va.

Skyline Drive was tailored specifically for the motoring tourist. As Charles E. Peterson, head of the I IPS's Eastern Division, Branch of Plans and Design in the 1930s, recalled: "The overall idea was that motorists should be able to drive out of Washington for a Sunday's mountain experience and get back home by night."

Construction of Skyline Drive actually predated the dedication of Shenandoah National Park by five years. In 1931, President Hoover (a proponent of the road since 1929) authorized drought relief funds to finance the work, provided that much of the labor be by locals using traditional hand tools and farm implements. A 100 foot right-of-way was acquired through the park area, and work commenced in mid-1931 under a cooperative agreement between the Bureau of Public Roads (BPR) and the NPS. This partnership, formalized in 1926, had been forged and perfected on western park road building projects. BPR engineers provided the technical expertise, while NPS landscape architects provided naturalistic design knowledge and exercised final review authority over the project. NPS designers and engineers adapted mountain drive standards from western parks to the gentler topography of the east, coupling this with the design principles standardized in New York's Westchester County urban parkways.

Skyline Drive was built in three, roughly equal, sections. The central section (from Thornton Gap to Swift Run Gap) opened to long lines of motorists in 1934. In 1936, the north section (Thornton Gap to Front Royal) opened, followed by the south section in 1939.

During the Roosevelt Administration, the drive and park again received the direct attention of the president. Under the New Deal, the Public Works Administration, Works Progress Administration, and Civilian Conservation Corps allocated funds and labor for drive construction, beautification, and development.

As the first park road in the eastern mountains, the drive was an instant success. Hardly had the project begun before it was expanded into a Shenandoah-Great Smoky Mountains Parkway (ultimately the 469 mile long Blue Ridge Parkway, commenced in 1934). One year after Shenandoah's dedication and two years before completion of the drive, the park attracted one million visitors via its scenic ridge road, making it the first national park to do so.

While technically not a parkway, Skyline Drive displays elements standard to that road type (limited access, noncommercial recreational traffic, no commercial development along the right-of-way, and access to adjoining recreational facilities). The roadway's geometric design preserves the natural topography whilc presenting the landscape's scenic values. As a result, the drive influenced later national parkway design. The 105.5 mile, paved, two-lane Skyline Drive, with its 35 mile-an-hour speed limit, serves as Shenandoah National Park's principal transportation corridor. It includes the park's four entrance stations, 67 scenic overlooks, and 40 plus roadside and trail head parking areas, in addition to supporting seven picnic areas, four campgrounds, two visitor centers, and six developed areas providing wayside, dining, and/or lodging facilities. The Appalachian Trail runs beside Skyline Drive through the park, crossing the drive at a number of points.

DELINEATED BY: Robert R. Harvey, 1996

NATIONAL PARK SERVICE ROADS & BRIDGES RECORDING PROJECT

NATIONAL PARK SERVICE, UNITED STATES DEPARTMENT OF THE INTERIOR

SKYLINE DRIVE

SHENANDOAH NATIONAL PARK

PAGE COUNTY

LURAY VICINITY

VIRGINIA

SHEET 1 OF 18

HISTORIC AMERICAN ENGINEERING RECORD VA-119

ENTRANCE STATIONS

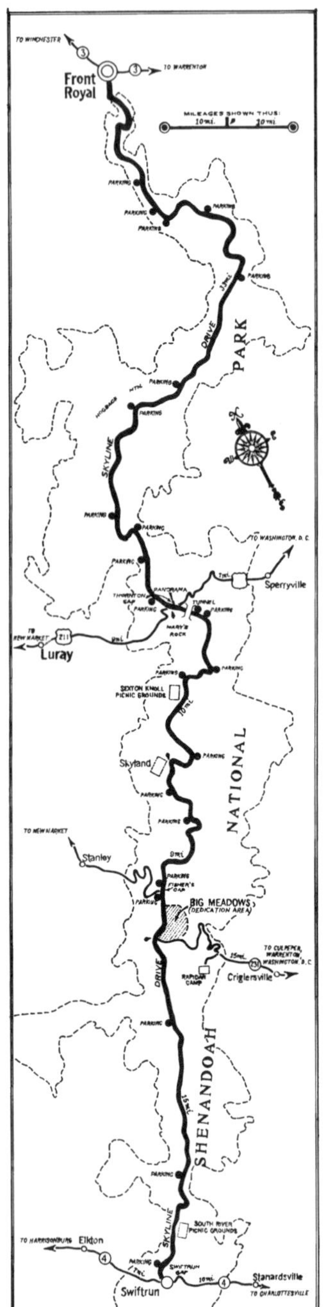

Commemorative Map
July 3, 1936 from Dedication
Ceremony showing extent of
Skyline Drive

Skyline Drive & Panorama
crossing Lee Highway & Panorama, 1935

Shenandoah National Park's Skyline Drive is accessed through four entrance stations located (from north to south) at Front Royal, Thornton Gap, Swift Run Gap, and Rockfish Gap. The Thornton Gap and Swift Run Gap entrances are located at the drive's intersection with two historic transmontane routes: the Lee Highway (U.S. 211) and the Spotswood Trail (U.S. 33). These highways are the only two lateral roads through the park which remain open to public motor vehicle travel.

Park entrance facilities have evolved from temporary, small one-room log "information and checking stations" (c. 1936) at the roadside to the sprawling grade separations and entrance stations of Mission 66 and more recent design. The Rockfish Gap entrance at the drive's southern end features the park's oldest extant entrance station (c. 1939). Constructed on a central traffic island, the building is a one-room board-and-batten structure roofed with concrete tile shingles simulating wood shakes. Two similar buildings formerly served the Thornton Gap entrance. In addition to regulating access to the drive, checking traffic and collecting fees, rangers at the entrances provide information on current road and park conditions.

Thornton Gap — Entrance Station
modest entrance station — mile post 31.5

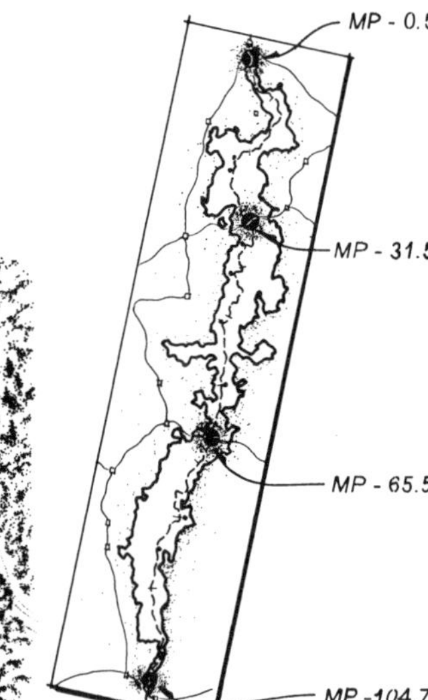

Locator Map

Skyline Drive crossing Lee Highway
Current View from Marys Rock — Mission 66 era

Front Royal — Entrance Station
1974 - 75 Post Mission 66 — mile post 0.5

Swift Run — Entrance Station
1974 - 75 Post Mission 66 — mile post 65.5

Rockfish Gap — Entrance Station
1938 - 39 — mile post 104.7

DELINEATED BY: harlen d. Groe, 1996

NATIONAL PARK SERVICE
ROADS & BRIDGES RECORDING PROJECT
NATIONAL PARK SERVICE
UNITED STATES DEPARTMENT OF THE INTERIOR

LURAY VICINITY

SKYLINE DRIVE
SHENANDOAH NATIONAL PARK
PAGE COUNTY

VIRGINIA

SHEET
6 of 18

HISTORIC AMERICAN
ENGINEERING RECORD
VA-119

DEVELOPMENT OF THE DRIVE
ROAD CONSTRUCTION

EXCAVATION ~ GRADING

After the route had been surveyed, plans prepared, and a 100' right-of-way acquired, construction crews began excavating and grading the roadbed. This 1934 photograph taken north of Gooney Run Overlook (mile post 6.8) shows a typical grading operation. As the drilling crew works in the foreground, the powder man (seated among boxes of explosives) prepares a charge. The young driller's helpers standing on the drill tongue counterweight the rig against overturning. Behind them, a power shovel works in loose rock loading dump trucks which empty the fill in the distance. A pipe crew has placed the culvert on the slope in advance of the fill. Controlled blasting techniques minimized unsightly landscape scars and conserved stone for other uses in construction.

ROAD SUB BASE

Once rough grading operations were completed, preparation of the road subgrade or foundation layer began. Since the subgrade would eventually support the weight of all traffic traveling on the pavement, weak spots in this foundation had to be eliminated. Subgrade soil maintaining a high degree of density, uniformity, and stability was brought in prior to the placement of the road base or surface layers. Grading crews cut and shaped the compacted layer of soil to the proper road cross-section using motor patrol or tractor-drawn blade graders and power scrapers. A crew could prepare 1,000 to 2,500 feet of subgrade in two eight hour shifts on a typical work day. Left: Construction of Hughes River Gap/Stony Man Overlook (mile post 38.6).

CHESTNUT CRIBBING

Constructed by a method of side-hill cut and fill, the roadway lay on a shelf carved out of or built onto the mountainside. This technique was perfected in western mountain parks like Yellowstone, and became a standard procedure for building park roads. Rock and soil excavated in cuts were used to build up fill areas; careful road alignment achieved a balance between the two. Extensive use of fills in the construction of overlooks provided adequate bases for parking areas. Some filled areas in the Central District utilized a cribbing of American Chestnut logs (like the one built near Panorama in 1933) to support the road shoulder and masonry guard walls. The deteriorated cribbing was removed under the drive's 1983 rehabilitation program.

FINAL GRADING

Skyline Drive's road base consisted of six to eight inches of crushed stone. Much of the stone came from roadway excavation and quarry sites near the right-of-way. Portable crushing mills pulverized the unearthed rock into an aggregate 1-1/2 inches in size or smaller, which was stockpiled prior to it's application to the road in layers. Loose stone was compacted by constant blading and dragging, and the day to day traffic of construction vehicles, produced a traffic-bound macadam base. Although, the Bureau of Public Roads' construction contractors were responsible for finishing operations such as trimming road shoulders, side, and ditch slopes, much of this work was completed by the Civilian Conservation Corps in order to finish the roadway to National Park Service standards.

DRY LAID WALLS

Dry laid rubble retaining walls or hand laid rock embankments (such as this one depicted in 1935, near the park's north entrance, mile post 1.5) help stabilize the slopes of several of the drive's steepest fills. After a slope face was thoroughly compacted and an adequate footing prepared at its base, stone rubble taken from adjacent roadway excavations was hand-placed without mortar up the embankment. One such wall in the South District rose to a height of 60 feet with a slope ratio of 3/4:1. A typical crew consisted of a foreman, two masons and about 20 laborers. These hand laid rock embankments (along with cement masonry retaining walls) reduced excavation quantities, minimized unsightly landscape scars, and protected newly planted vegetation.

ROAD SURFACING

Paving operations began in the late spring or summer following the completion of the road base the preceeding fall. Loose stone was bladed into windrows along the roadside or in one lane. After the road base was repaired, swept, and a prime coat of road tar applied, the windrowed stone was spread in a uniform layer over the road surface. Liquid asphalt was "shot" over the aggregate, which was mixed and blended back into windrows. Then, the process was repeated. Once blended satisfactorily and spread a final time, this mixed-in-place bituminous surface was rolled. A seal coat, waterproofed the pavement and gave it the proper surface texture. Finally, the completed section of the drive was opened to traffic. Parking overlooks and entrance roads to the park's developed areas were treated in similar fashion.

HISTORIC AMERICAN ENGINEERING RECORD
VA-119

SHEET 7 of 18

VIRGINIA

SKYLINE DRIVE
SHENANDOAH NATIONAL PARK
PAGE COUNTY

LURAY VICINITY

DELINEATED BY:
NATIONAL PARK SERVICE
ROADS & BRIDGES RECORDING PROJECT
NATIONAL PARK SERVICE
UNITED STATES DEPARTMENT OF THE INTERIOR

DEVELOPMENT OF THE DRIVE

PARAPET CONSTRUCTION

HUB POST FABRICATION

Roadside guard rail barriers protect motorists from Skyline Drive's sharp curves and steep shoulder drop-offs. In 1932, a year after construction began, National Park Service (NPS) and Bureau of Public Roads (BPR) officials agreed to build stone guard walls along the drive and discouraged the use of wooden rail barriers for safety and aesthetic reasons. Meanwhile, large chestnut logs served as curbs at dangerous places. Construction of the stone walls proceeded slowly, forcing the approval of temporary timber guard rails the summer of 1935. That winter, Civilian Conservation Corps (CCC) crews began selecting chestnut logs during forest cleanup for use as rails and hub posts the following spring. Prior to installation, hub posts were treated with a preservative coat of creosote.

STONE REMOVAL

Discussing the construction of Skyline Drive's guard walls in September 1933, NPS engineer in charge James Lassiter observed: "The getting out of the stone is a larger job than laying the wall." NPS tractors and trucks outfitted with cranes or winches were used to load rock at quarries and stockpiles, and to hoist stones into place on the walls. While the BPR insisted on building the guard walls, the NPS provided the unskilled labor and heavy equipment necessary to accomplish the task. CCC enrollees assigned to Shenandoah, gathered stones as part of their roadside cleanup and park beautification work. Rocks ranging in size from 6" x 8" x 12" and up to nine feet in length were delivered to roadside construction sites or to stockpiles. Bureau masons were assisted by other CCC crews in laying of the walls.

GUARD RAIL PLACEMENT

The temporary chestnut log guard rail employed, along the drive was a slightly modified version of a design on BPR plans used at Sexton Knoll picnic (Pinnacles) area. CCC crews constructed the log guard rails as Emergency Conservation Work projects. The rail, 20" high consisted of 16' long logs pinned atop 4' to 5' long hub posts, equally spaced and set into the ground. The ends of each guard rail section tapered down to the road shoulder. By 1944, most of the timber guard rail used on Skyline Drive was located in the North District, where it constituted one half of the section's existing guard barriers (compared to six percent in the Central District, and less than five percent in the South District). After WWII, the railing was systematically removed or replaced by stone guard walls.

STONE GUARDWALL

This 1934 photograph (mile post 33.0) of Hazel Mountain Overlook illustrates the interrelation between guard wall, overlook and drive construction. Here as elswhere, power shovels (foreground) are used to finishe the road cut, flattening the slope and improving sight distances around the curve. Trucks from CCC Camp #10, relocate stone to the overlook (background) and guard wal site, across the drive. Wood frames and string lines, set by NPS and CCC landscape foremen or engineers, indicate the wall's location and assist masons in maintaining proper alignment of height and batter. In what will become a planting island, a bulldozer (center, behind shovel) sits. This island is designed to separate and screen the overlook from the drive. In the distance, grading has begun on the viewing terrace, with its distinctive rock outcropping.

QUARRYING STONE

Native stone was used in all masonry work, such as guard walls, retaining walls, culvert headwalls, tree wells, hand laid rock embankments, and gutters, as well as for much of the crushed stone base. Rock was obtained from excavations along the drive or from quarries within the park or just outside its boundaries. NPS officials selected sites where few or no landscape scars would be visible to passing traffic on the drive, adjacent trails, or from distant view points within the park. Some quarries were located off the right-of-way, screened by vegetation and topography. Others were actually situated on the road alignment. The quarry on Big Run Creek, supplied rock for work between Swift Run and Browns gap (mile 65.5 to 82.9).

STONE RETAINING WALL

Stone retaining walls were erected at a number of points along the drive. Cement stone masonry walls stabilized steep roadway embankments, and were typically located out of motorists' sight. Massive dry-laid retaining walls, such as this one under construction at Crescent Rock (mile post 44.4) in 1934, held back the earth fill which formed artificial parking and viewing terraces at many overlooks. The stone parapets capping these overlook walls were similar to the roadside guard walls, except that they were built using small of amounts of cement mortar to bind the stones together for safety. Generally, overlook parapets stood three feet from the edge of the parking area to provide room for a sidewalk and stone curb.

DELINEATED BY *harlen d. Groe, 1996*

NATIONAL PARK SERVICE
ROADS & BRIDGES/RECORDING PROJECT
NATIONAL PARK SERVICE
UNITED STATES DEPARTMENT OF THE INTERIOR

LURAY VICINITY

SKYLINE DRIVE
SHENANDOAH NATIONAL PARK
PAGE COUNTY

VIRGINIA

SHEET 8 OF 18

HISTORIC AMERICAN ENGINEERING RECORD
VA-119

DEVELOPMENT OF THE DRIVE
LANDSCAPE TECHNIQUES

SLOPE STABILIZATION

NPS landscape architects drew upon the naturalistic landscape gardening practices of the 1920s, and their own previous experience in western parks as they designed Skyline Drive to lie lightly along the crest of the Blue Ridge Mountains. An extensive program of road bank naturalization accomplished this by blending the drive into its surroundings and preventing roadside soil erosion. Slope stabilization of the steeper cut and fill areas was a crucial step in this process. This necessitated the placement of various types of temporary log and brush cribbing, or rows of boulders across slope faces. The slopes were then revegetated with native plants in naturalistic plantings.

TREE COLLECTION

Evergreens such as eastern white pines and Blue Ridge firs and some deciduous trees over 10 feet high were removed from the road right-of-way for transplanting elsewhere along Skyline Drive. CCC enrollees hand balled and burlapped trees, wenched them onto trucks and moved them to park nurseries or developed areas like Skyland, Big Meadows, and Elkwallow. In addition to transplanting established trees, the CCC planted hundreds of thousands of seedlings for park beautification and erosion control. Prior to the establishment of the nursery at Big Meadows (1936), seedlings were supplied by outside sources. In November 1934, 2,525 hickory and walnut trees were received from Bixby Nurseries on Long Island, New York.

SLOPE BLENDING

By the time construction began on Skyline Drive, the NPS had adopted a routine program of road bank blending based on curvilinear grading of earth slopes. Bank and cut crests were rounded, slopes flattened, and the ends of cuts and fills warped to thoroughly integrate the new surfaces with the old. Specifications for Skyline Drive issued in 1936 by the Bureau of Public Roads (BPR) and approved by the NPS stipulated maximum slope ratios ranging from 4:1 (on vertical heights measuring 3 feet or less) to 1 1/2:1 (on slopes exceeding 15 feet in height). The Civilian Conservation Corps (CCC) performed much of the road bank blending along the drive using power shovels to cut back the slopes then rounding them with hand labor.

PLANT NURSERY

Nurseries operated by the CCC at Big Meadows and Front Royal supplied native plant material for revegetation. Trees and shrubs salvaged from the drive and other construction right-of-ways, abandoned house sites, and farmsteads were transplanted to nursery hot beds. Small plants were transferred to cans, which were then placed in shaded beds prior to final transplanting. Other stock was propagated from seed or seedlings. Trees included hemlock, oak, black locust, black walnut, alternate leaved dogwood, red spruce, and various species of pine and fir. Shrubs included strawberry bush, mountain laurel, arrowwood viburnum, Canada yew, rhododendron and azalea.

BANK EROSION CONTROL

NPS slope blending techniques used on Skyline Drive produced erosion resistant road banks able to be revegetated quickly. After the completion of preliminary bank stabilization and flattening, topsoil that had been stripped away and stored during road construction was reapplied to the shoulders and cut and fill slopes. Plantings of small trees, shrubs, and vines were interspersed behind erosion control brush cribbing on steep slopes. Flattened banks (particularly those in open meadow lands) were sodded, trenched, or broadcast seeded with grasses, annual and perennial forbs. Then the slopes were mulched with leaves, forest duff, or a clay-leaf mixture to further prevent erosion and to provide an organic base for understory growth.

MTN. LAUREL PLANTING

The roadside and landscape naturalization programs for Skyline Drive and Shenandoah began in late Spring 1933. A resident staff of NPS landscape architects supervised CCC projects propagating and planting of hundreds of thousands of trees and shrubs beside the drive and throughout the park. Vines, such as Virginia creeper, wild grape, American bittersweet, and clematis were rooted in rock cuts and along guard walls. Road banks and overlooks were planted in mountain laurel and other small stock. In developed areas trees and shrubs provided proper visual aesthetics. Open bays in the planting scheme varied the forest outline, while trimmed vistas opened spectacular views from the drive and overlooks.

DELINEATED BY harlen d. Groe, ROBERT R. HARVEY, 1996

NATIONAL PARK SERVICE
ROADS & BRIDGES RECORDING PROJECT
UNITED STATES DEPARTMENT OF THE INTERIOR

SKYLINE DRIVE
SHENANDOAH NATIONAL PARK
PAGE COUNTY

LURAY VICINITY

VIRGINIA

SHEET 9 of 18

HISTORIC AMERICAN ENGINEERING RECORD
VA-119

IF REPRODUCED, PLEASE CREDIT: HISTORIC AMERICAN ENGINEERING RECORD, NATIONAL PARK SERVICE, NAME OF DELINEATOR, DATE OF THE DRAWING

ROAD DETAILS

In 1983, the Federal Highway Administration enforced an extensive rehabilitation program along Skyline Drive to meet modern safety standards. As part of this program, the original stone guard walls and drainage features were refurbished or replaced, and in some instances new features are being added.

Old Guard Wall

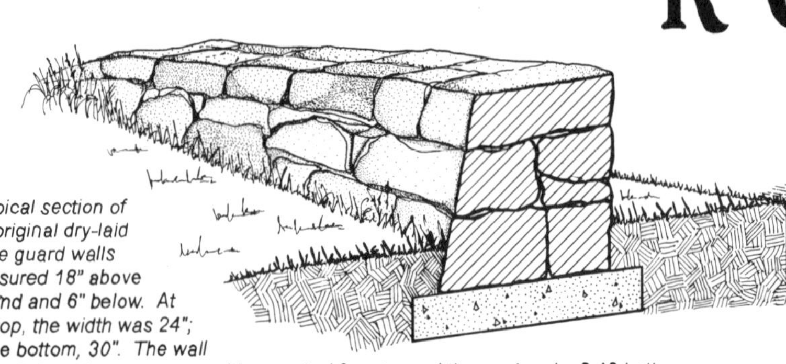

A typical section of the original dry-laid stone guard walls measured 18" above ground and 6" below. At the top, the width was 24"; at the bottom, 30". The wall stood 5' from the pavement, with a vertical face toward the road and a 3:12 batter in back. Laid horizontally, no four stones were to have contiguous corners. Bunching of similar-sized stone and excessive use of spalls (stone fragments) was avoided. By 1938, the top course was set in mortar with deeply raked joints.

New Guard Wall

The stone veneer utilizes stones from the old walls. Intended to resemble the old walls in appearance, the new masonry guard walls are characterized by smaller stones with wide mortar joints.

The new masonry guard walls consist of a concrete core wall 39" high (27" exposed above ground) with stone facing.

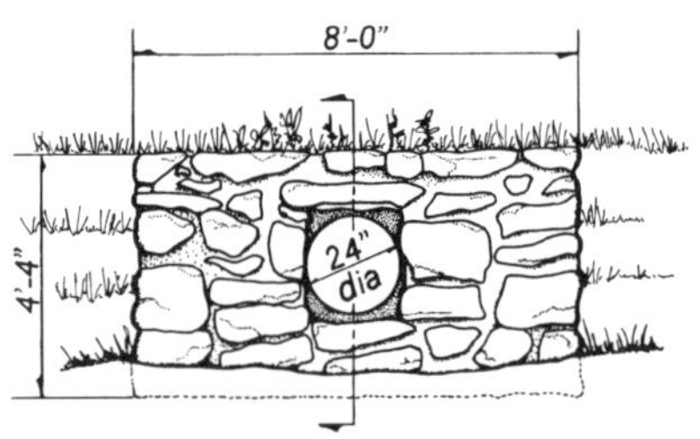

Headwall Elevation
scale: 3/8" = 1'-0"

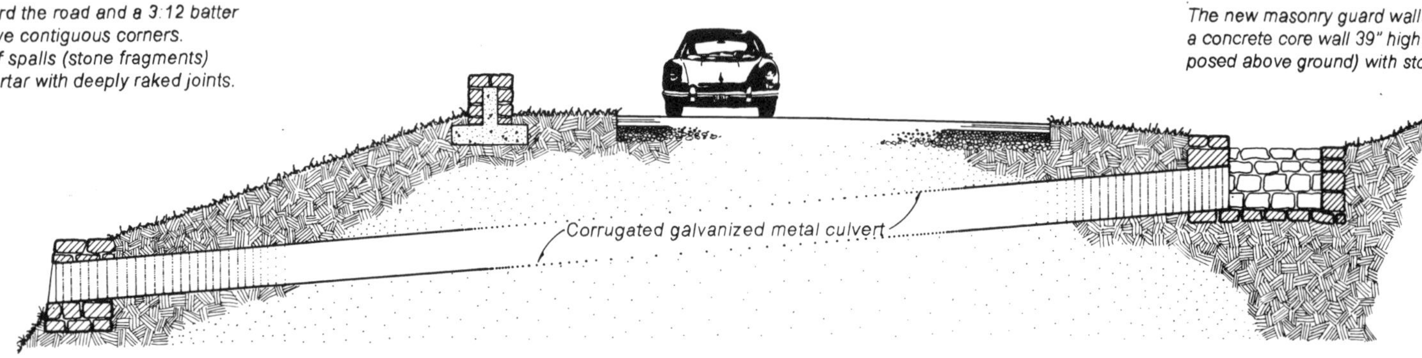

Road Cross Section scale: 1/4" = 1'-0"

Feet 0 1 2 3 4 5 10
Meters 1:48 0 1 2 3

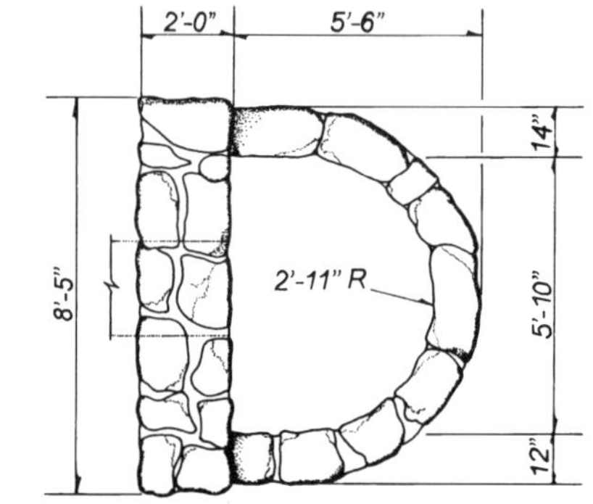

Drop Inlet Plan
scale: 3/8" = 1'-0"

Feet 0 1 2 3 4 5
Meters 1:32 0 1 2

The Bureau of Public Roads (BPR), under a cooperative agreement with the National Park Service (NPS), supervised the construction of Skyline Drive from 1931 to 1939. Strict design and engineering specifications ensured the uniform construction of the scenic drive. A typical roadway section is 30' wide, composed of a 20' road surface with 5' shoulders in fill, and 3' shoulders plus 2' ditches in cut. (Roadway width in the North District is 34', with 5' shoulders and 2' ditches throughout.) Curves are spiraled, providing a wide radius easily negotiated by vehicles, and superelevated or banked for safety.

The road surface consists of a two-layer "traffic-bound" crushed stone base covered with a bituminous "road-mix" asphalt surface which has been repaved and resealed several times since 1950. Corrugated galvanized metal pipe culverts provide cross drainage under the road subbase. Drop inlets or stone masonry headwalls serve as pipe termini. Vitrified clay tile underdrains laid beneath roadside ditches further direct water away from the subbase.

Construction of Stone Waterway

Grouted rubble or rock paved gutters carry surface runoff from the drive through roadside ditches to drop inlets which empty into cross culverts. Placed on a 4" gravel bed, stones 6" or larger with flat top surfaces were laid with their long dimension parallel to the gutter's center line. A gutter crew consisted of two assistant masons and six laborers. Originally, Skyline Drive also featured dry rubble and bituminous gutters. Replacement gutters built since 1983 are similar to the original rock paved gutters, but feature wide mortar joints and modified drop inlets.

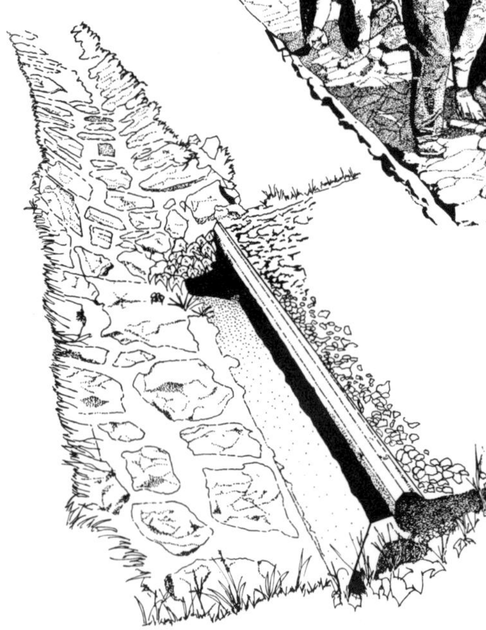

Modern Drop Inlet and Grouted Rubble Gutter

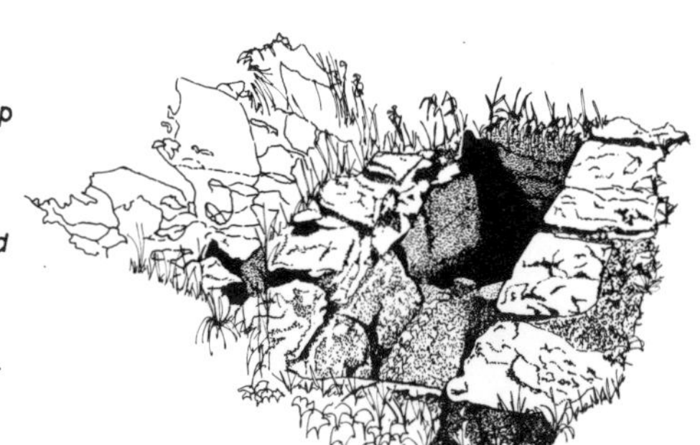

Rectangular Drop Inlet

Double Drainage Outlet

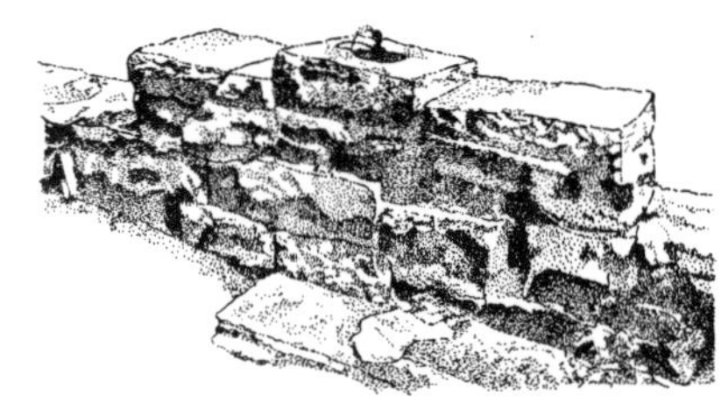

Three types of water fountains or "bubblers" are found in the park and along the drive: small boulder bubblers, fountains incorporated into masonry walls, and stacked rock bubblers (which replaced earlier chestnut log fountains in the 1950s).

Springs, like the one at Hemlock Spring Overlook, supply water to the roadside fountains. Water from the spring box is piped under the drive and overlook parking area to the fountain. This spring also supplies water to the fountain and comfort station at Stony Man Overlook.

Lock Concrete Spring Box, Hemlock Spring
(Typical arrangement of spring box and disinfectant facility at a gravity-fed roadside water source.)

HISTORIC AMERICAN ENGINEERING RECORD VA-119

SHEET 10 OF 18

VIRGINIA

SKYLINE DRIVE
SHENANDOAH NATIONAL PARK
PAGE COUNTY

LURAY VICINITY

NATIONAL PARK SERVICE
ROADS & BRIDGES RECORDING PROJECT
UNITED STATES DEPARTMENT OF THE INTERIOR

DELINEATED BY: Shane P. Wirth, 1996

IF REPRODUCED, PLEASE CREDIT: HISTORIC AMERICAN ENGINEERING RECORD, NATIONAL PARK SERVICE, NAME OF DELINEATOR, DATE OF THE DRAWING

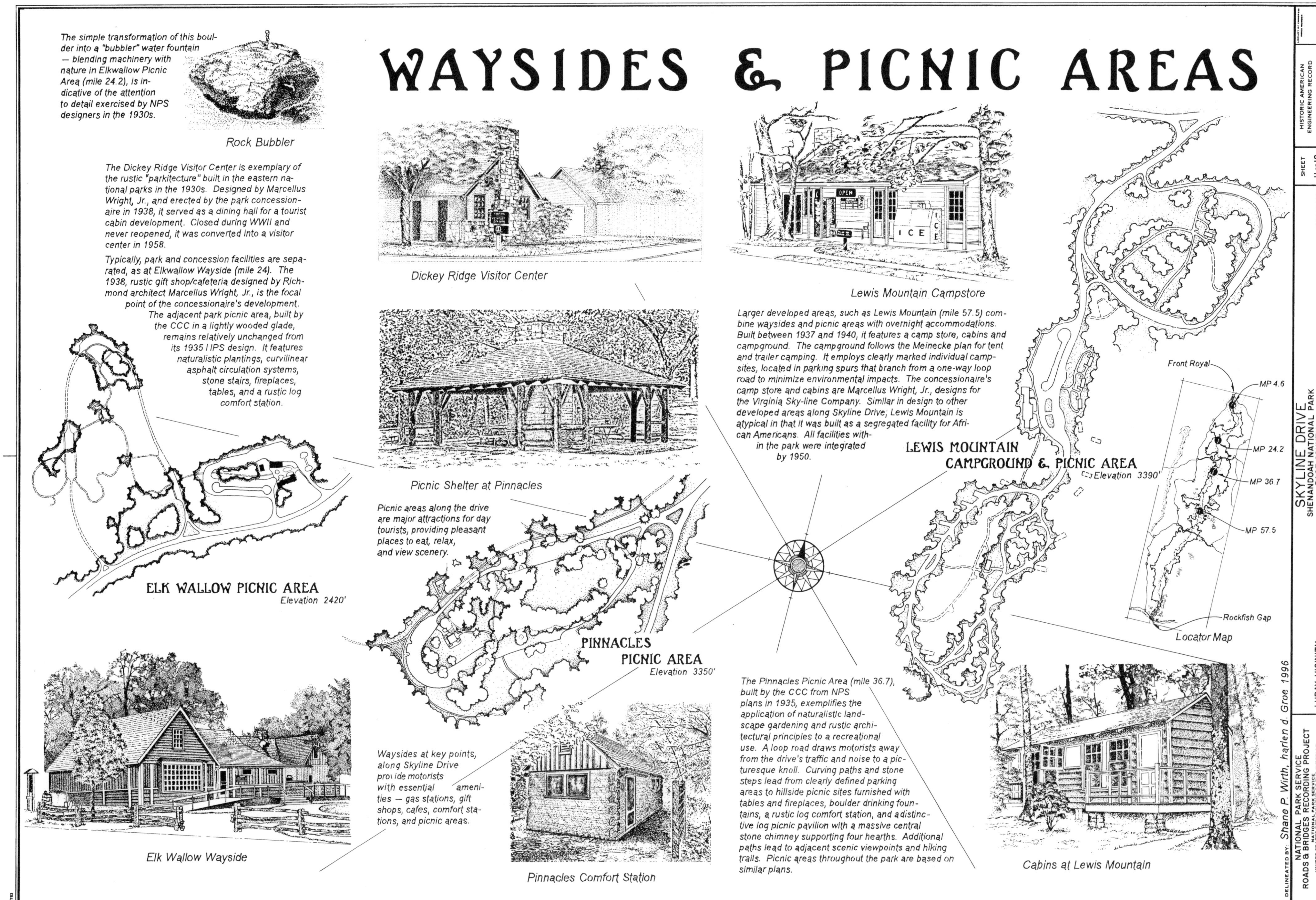
WAYSIDES & PICNIC AREAS
The simple transformation of this boulder into a "bubbler" water fountain — blending machinery with nature in Elkwallow Picnic Area (mile 24.2), is indicative of the attention to detail exercised by NPS designers in the 1930s.
Rock Bubbler
The Dickey Ridge Visitor Center is exemplary of the rustic "parkitecture" built in the eastern national parks in the 1930s. Designed by Marcellus Wright, Jr., and erected by the park concessionaire in 1938, it served as a dining hall for a tourist cabin development. Closed during WWII and never reopened, it was converted into a visitor center in 1958.
Typically, park and concession facilities are separated, as at Elkwallow Wayside (mile 24). The 1938, rustic gift shop/cafeteria designed by Richmond architect Marcellus Wright, Jr., is the focal point of the concessionaire's development. The adjacent park picnic area, built by the CCC in a lightly wooded glade, remains relatively unchanged from its 1935 IPS design. It features naturalistic plantings, curvilinear asphalt circulation systems, stone stairs, fireplaces, tables, and a rustic log comfort station.
Dickey Ridge Visitor Center
Lewis Mountain Campstore
OPEN
ICE
Larger developed areas, such as Lewis Mountain (mile 57.5) combine waysides and picnic areas with overnight accommodations. Built between 1937 and 1940, it features a camp store, cabins and campground. The campground follows the Meinecke plan for tent and trailer camping. It employs clearly marked individual campsites, located in parking spurs that branch from a one-way loop road to minimize environmental impacts. The concessionaire's camp store and cabins are Marcellus Wright, Jr., designs for the Virginia Sky-line Company. Similar in design to other developed areas along Skyline Drive, Lewis Mountain is atypical in that it was built as a segregated facility for African Americans. All facilities within the park were integrated by 1950.
Picnic Shelter at Pinnacles
Picnic areas along the drive are major attractions for day tourists, providing pleasant places to eat, relax, and view scenery.
ELK WALLOW PICNIC AREA
Elevation 2420'
LEWIS MOUNTAIN CAMPGROUND & PICNIC AREA
Elevation 3390'
Front Royal
MP 4.6
MP 24.2
MP 36.7
MP 57.5
Rockfish Gap
Locator Map
PINNACLES PICNIC AREA
Elevation 3350'
Elk Wallow Wayside
Waysides at key points along Skyline Drive provide motorists with essential amenities — gas stations, gift shops, cafes, comfort stations, and picnic areas.
Pinnacles Comfort Station
The Pinnacles Picnic Area (mile 36.7), built by the CCC from NPS plans in 1935, exemplifies the application of naturalistic landscape gardening and rustic architectural principles to a recreational use. A loop road draws motorists away from the drive's traffic and noise to a picturesque knoll. Curving paths and stone steps lead from clearly defined parking areas to hillside picnic sites furnished with tables and fireplaces, boulder drinking fountains, a rustic log comfort station, and a distinctive log picnic pavilion with a massive central stone chimney supporting four hearths. Additional paths lead to adjacent scenic viewpoints and hiking trails. Picnic areas throughout the park are based on similar plans.
Cabins at Lewis Mountain
DELINEATED BY: Shane P. Wirth, harlen d. Groe 1996
NATIONAL PARK SERVICE
ROADS & BRIDGES RECORDING PROJECT
LURAY VICINITY
SKYLINE DRIVE
SHENANDOAH NATIONAL PARK
PAGE COUNTY
VIRGINIA
SHEET 11 of 18
HISTORIC AMERICAN ENGINEERING RECORD
VA-119
IF REPRODUCED, PLEASE CREDIT: HISTORIC AMERICAN ENGINEERING RECORD, NATIONAL PARK SERVICE, NAME OF DELINEATOR, DATE OF THE DRAWING

OVERLOOKS ~ NORTH DISTRICT

VIEW FROM RANGE VIEW OVERLOOK
Elevation 2810'

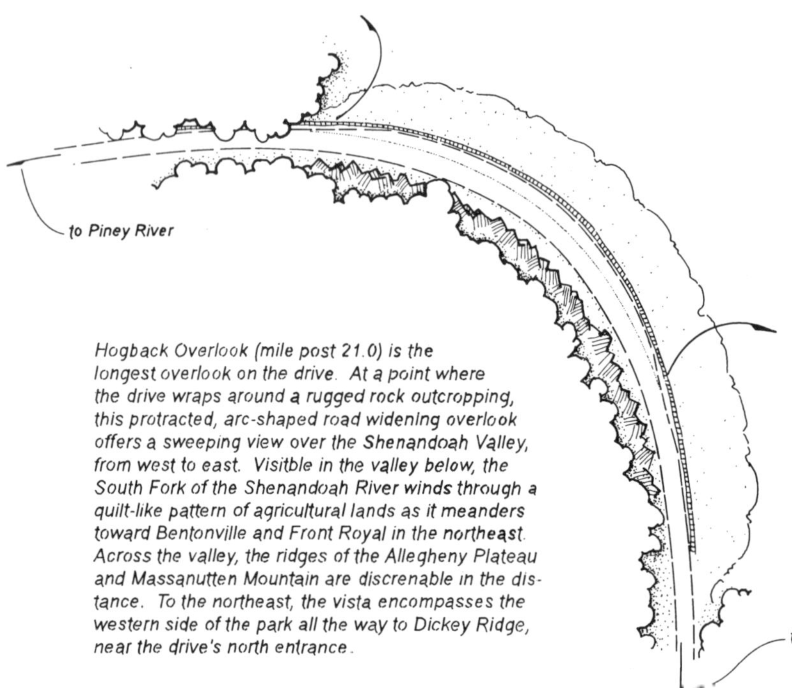

Hogback Overlook (mile post 21.0) is the longest overlook on the drive. At a point where the drive wraps around a rugged rock outcropping, this protracted, arc-shaped road widening overlook offers a sweeping view over the Shenandoah Valley, from west to east. Visitble in the valley below, the South Fork of the Shenandoah River winds through a quilt-like pattern of agricultural lands as it meanders toward Bentonville and Front Royal in the northeast. Across the valley, the ridges of the Allegheny Plateau and Massanutten Mountain are discrenable in the distance. To the northeast, the vista encompasses the western side of the park all the way to Dickey Ridge, near the drive's north entrance.

Scenic Overlooks

The 69 scenic overlooks lining Skyline Drive in Shenandoah National Park are integral to the park visitor's experience. These frequent vantage points offer motorists a succession of panoramic views of the Blue Ridge peaks and hollows, and of the patchwork landscape of the valley and piedmont below. Also, they provide visitors opportunities to leave their automobiles for short hikes along park trails and the Appalachian Trail to scenic or natural points of interest, to read interpretive signs, or take photographs.

NPS landscape architects and engineers employed two types of overlooks along the drive: road widenings and parking overlooks. Road widenings, simple extensions of the road surface, range from basic "drive throughs" with or without guard walls and interpretive signs (like No Name Overlook, mile post 5.4) to larger pull-offs featuring limited parallel parking and sidewalks. Parking overlooks are generally larger turnouts separated and often screened from the drive by planting islands. In addition to head-in parking, they may include landscaped elements such as, paved walkways, stone walls and curbs, spring-fed water fountains, picnic tables, and an occasional comfort station.

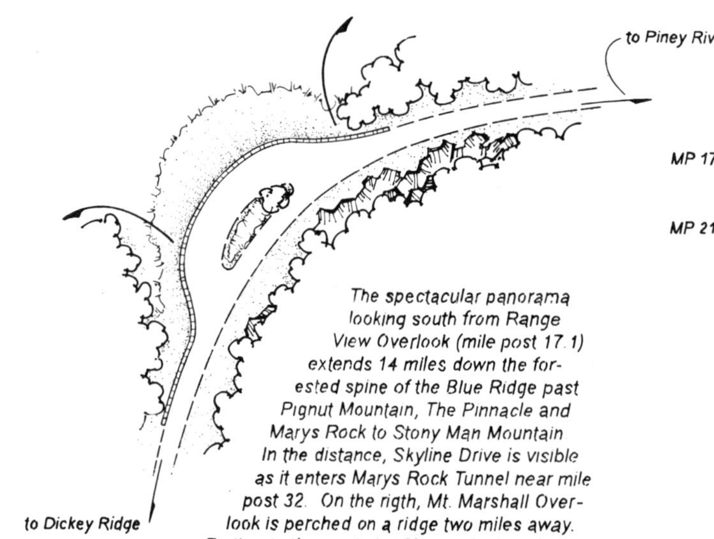

The spectacular panorama looking south from Range View Overlook (mile post 17.1) extends 14 miles down the forested spine of the Blue Ridge past Pignut Mountain, The Pinnacle and Marys Rock to Stony Man Mountain. In the distance, Skyline Drive is visible as it enters Marys Rock Tunnel near mile post 32. On the rigth, Mt. Marshall Overlook is perched on a ridge two miles away. Farther to the west, the Shenandoah Valley, Massanutten, and Allegheny Mountains disappear in the distance. Formerly known as Harris Hollow Overlook, this parking turnout features a grassy, planting island with a single specimen locust tree, stone-veneer concrete core replacement guard wall, and interpretive signage.

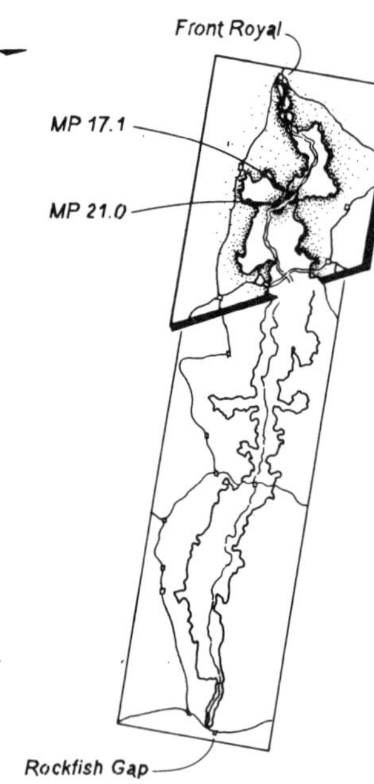

Locator Map

VIEW FROM HOGBACK OVERLOOK
Elevation 3385'

DELINEATED BY: *Shane P. Wirth, Michael P. Lanning, harlen d. Groe 1996*

NATIONAL PARK SERVICE
ROADS & BRIDGES RECORDING PROJECT
NATIONAL PARK SERVICE
UNITED STATES DEPARTMENT OF THE INTERIOR

LURAY VICINITY

SKYLINE DRIVE
SHENANDOAH NATIONAL PARK
PAGE COUNTY

VIRGINIA

SHEET 12 OF 18

HISTORIC AMERICAN ENGINEERING RECORD
VA-119

OVERLOOKS ~ CENTRAL DISTRICT

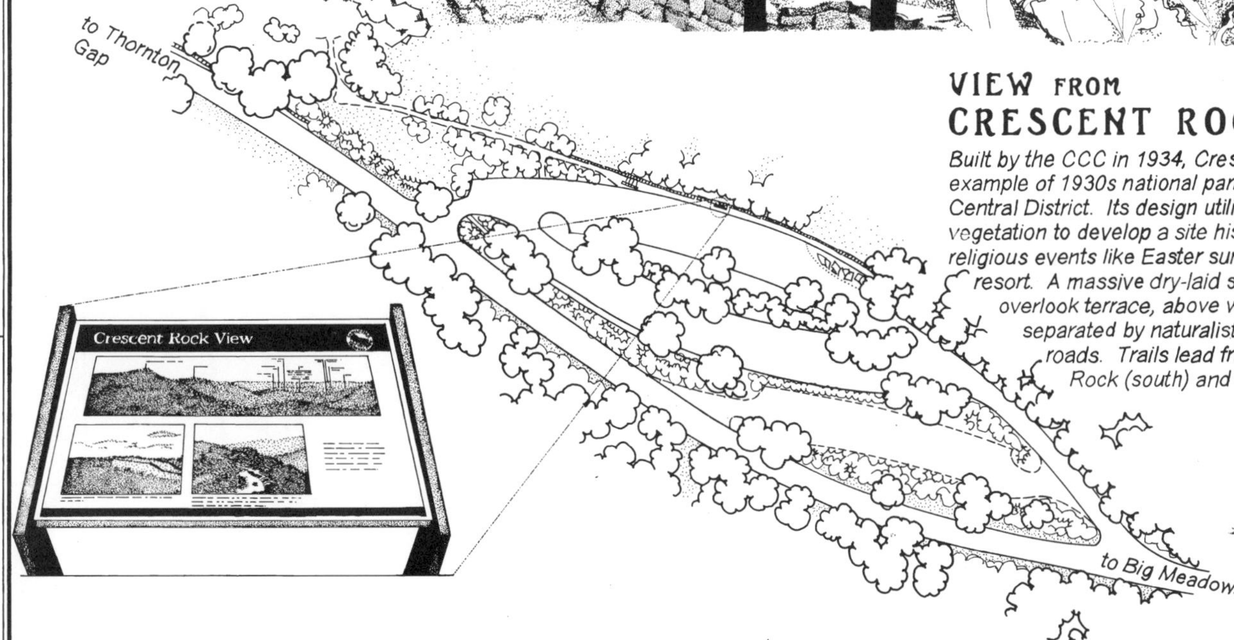

VIEW FROM CRESCENT ROCK OVERLOOK Milepost 44.4

Built by the CCC in 1934, Crescent Rock Overlook (mile post 44.4) is a classic example of 1930s national park road design, as are most of the overlooks in the Central District. Its design utilized existing surface contours, rock formations, and vegetation to develop a site historically frequented by large numbers of visitors for religious events like Easter sunrise services or on day hikes from the Skyland resort. A massive dry-laid stone retaining wall and guard parapet form the overlook terrace, above which is a multi-tiered parking lot (the tiers separated by naturalistic planting islands) with one way access roads. Trails lead from the overlook to the cliffs of Crescent Rock (south) and Bettys Rock (north).

Locator Map

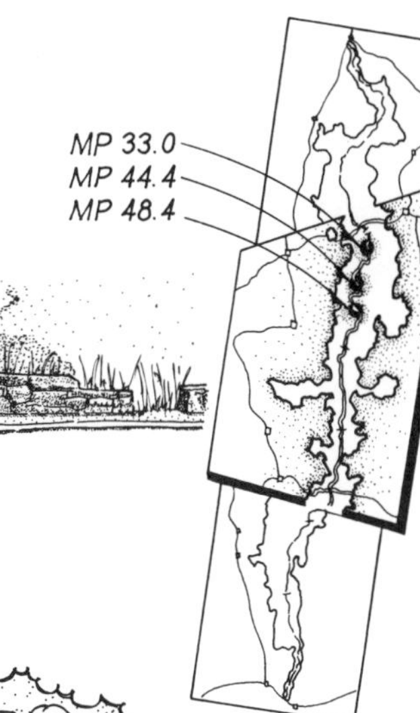

Hazel Mountain Overlook (mile post 33.0), built in 1935, was designed around a picturesque granite outcropping. The base of the outcropping was excavated and additional rock added to embellish this striking feature, which forms the overlook's viewing terrace. Curvilinear stone walls emerge from either side of the boulders, stepping down to blend with the drive's guard wall and demarking the parking area from the terrace. Incorporated into the left hand wall are a water fountain and stone steps which lead to the top of the outcropping and the view. Screening the overlook from the drive, and further blending it into the natural surroundings, is a stone-edged planting island filled with native pines, oaks, and mountain laurel.

Tree branches form natural arbors over the road between mile 48.4 and 49.2, and again between mile 58.8 and 59, creating "Green Tunnels" in summer. These tree-lined corridors represent a small portion of the hundreds of thousands of plantings along the drive made by the CCC under the direction of NPS landscape architects. Through these plantings, the drive's designers re-created a "natural" setting from a largely disturbed mountaintop landscape. As the vegetation along the drive matured, some distant views were replaced by intimate woodland scenes. This planned juxtaposition between dense forest and spectacular highland vista gives Skyline Drive much of its unique character.

THE "GREEN TUNNEL"

Milepost 48.4 -49 2 and 58.8 -59.

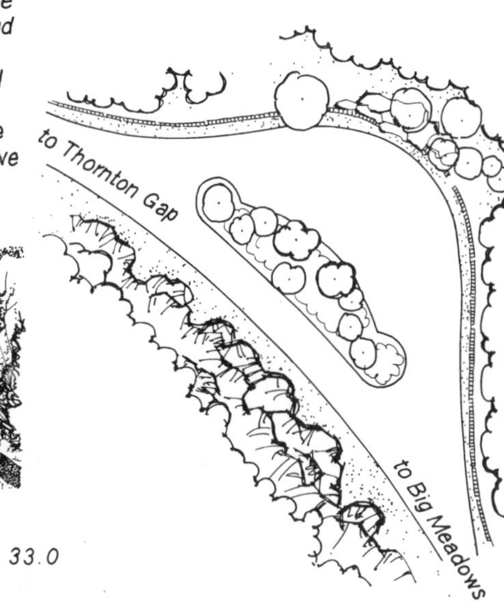

VIEW FROM HAZEL MTN OVERLOOK Milepost 33.0

DELINEATED BY: harlen d. Groe, Shane P. Wirth, 1996 Ryan Lundquist, 2001

NATIONAL PARK SERVICE
ROADS & BRIDGES RECORDING PROJECT
NATIONAL PARK SERVICE
UNITED STATES DEPARTMENT OF THE INTERIOR

SKYLINE DRIVE
SHENANDOAH NATIONAL PARK
PAGE COUNTY
LURAY VICINITY
VIRGINIA

SHEET 13 OF 18

HISTORIC AMERICAN ENGINEERING RECORD
VA-119

MARYS ROCK TUNNEL

Marys Rock Tunnel, located at mile 32.4, is the only engineering structure of its kind on the Skyline Drive. According to Bureau of Public Roads engineers and NPS landscape architects, tunneling was necessary to eliminate an extensive cut and fill around a sharp ridge below Marys Rock, thus preventing unsightly scarring to the mountain side. Averaging fifteen feet per day, three fifteen-man crews worked eight-hour shifts around the clock, six days a week, to drill, blast, and clear 10,799 cubic yards of granite from the ridge opening. In order to excavate the tunnel consistently at its full height and width, crews drilled 40 twelve-foot deep holes, three or four feet apart, across the rock face. These were loaded with dynamite (500 lbs. of explosives per blast); the tunnel cleared of men and machines; and the charge electrically detonated. Several hours were then required to remove the rubble. This process was repeated twice every 24 hours. The 610' long tunnel was completed in January 1932, a little more than three months after construction began, at a cost of $32,397.00.

In 1958 and 1959, the NPS added a pneumatic concrete liner to Marys Rock Tunnel in an attempt to remedy the problem of spring runoff and enormous icicles forming over the road in winter. The cost for repairs were six times the original construction expenditures.

—North Portal

Construction of retaining walls above North Portal prior to excavation.

The entrances above Marys Rock Tunnel crumbled during initial excavation, but once the surface material slid away solid rock was exposed, thus obviating the need for formal portals designed by the NPS Branch of Plans and Design in 1933. Some slope stabilization and landscape design work was performed around the tunnel entrances, largely in the form of a 47 foot dry-rubble wall on the ridge above the north entrance.

Marys Rock Tunnel provided the NPS and the BPR a technological showcase in which to demonstrate, within an eastern national park just 80 miles from Washington, one of the more spectacular aspects of their mountain road building expertise perfected in the western parks a decade earlier. The first tunnel that eastern national park visitors encountered, it was extremely popular – as a 1935 photograph of the tunnel overlook attests. Plans for a much longer tunnel under Black Rock Mountain, on the southern section of the Skyline Drive, proved too costly and were subsequently dropped.

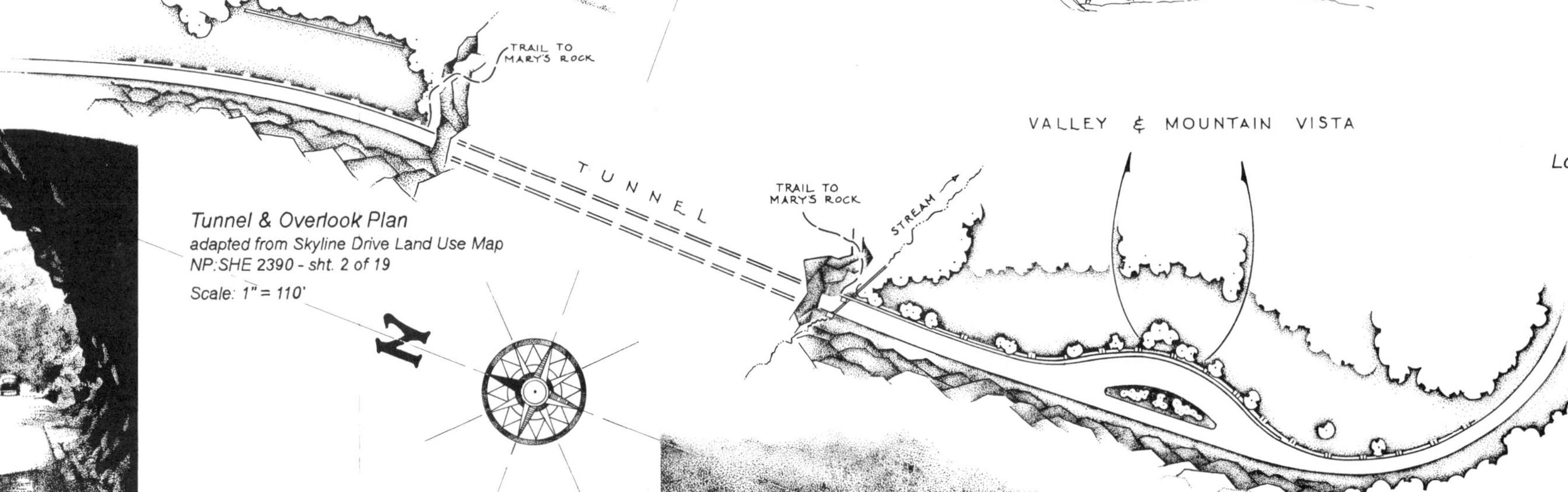

Tunnel & Overlook Plan
adapted from Skyline Drive Land Use Map
NP:SHE 2390 - sht. 2 of 19
Scale: 1" = 110'

Locator Map

Cut Rock — Original Interior

Interior View with Icicle

Removal of Icicles

South Portal —

Overlook, 1934

Overlook, 1996

DELINEATED BY: Michael P. Lanning, 1996

NATIONAL PARK SERVICE
ROADS & BRIDGES RECORDING PROJECT

LURAY VICINITY

SKYLINE DRIVE
SHENANDOAH NATIONAL PARK
PAGE COUNTY

VIRGINIA

SHEET 14 OF 18

HISTORIC AMERICAN ENGINEERING RECORD
VA-119

SKYLAND

PERSPECTIVE OF SKYLAND
From the 1937 Master Plan — N.P. SHE. 1155

Stony Man Camp
Adapted from Catalogue Cover 1897–8

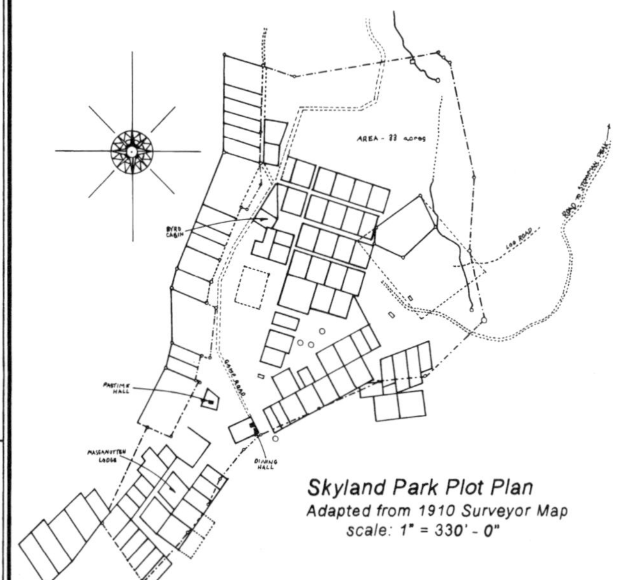

Skyland Park Plot Plan
Adapted from 1910 Surveyor Map
scale: 1" = 330' - 0"

Skyland has accommodated tourists since the late 1880s, when the young George Freeman Pollock created a Blue Ridge mountain resort on mining lands in which his father held part interest. Located on a small plateau (elev. 3560 feet) slightly west of Stony Man peak, Pollock's Skyland Park (originally known as Stony Man Camp in the 1890s) attracted wealthy vacationers from Washington, D.C. and other cities. Pollock charmed his guests, mountaineers, and rattlesnakes alike with his eccentric showmanship, summoning campers with a bugle and staging grand pageants, bonfires, dances, and other chautauqua-like entertainments.

Skyland's early development contained a central recreational space dividing two cabin areas; one on the plateau's western rim overlooking the Shenandoah Valley, the other tucked against the cliff face to the east. Byrds Nest cottage, built in 1906 for Richard E. Byrd, Speaker of the Virginia House of Delegates, exemplifies the resort's modest, rustic architectural style. Architect Victor E. Mindeleff designed the distinctive Adirondack style Massanutten Lodge (1911) for Pollock's bride Addie Nairn Hunter. By the mid 1920s, Skyland cottage design had evolved into standardized plans, typified by Boulder Cabin. The resort's landscaped spaces bore the stamp of individual cabin owners; best known were the gardens of Mr. and Mrs. George H. Judd, begun in 1911.

A lifelong advocate of Northern Virginia's scenic Blue Ridge Mountains, Pollock hosted the National Conference on State Parks at Skyland in 1925. His promotional efforts helped to galvanize critical support for the establishment of Shenandoah National Park. In 1930, the Commonwealth of Virginia purchased the Skyland property, deeding it for inclusion within the national park five years later. The Pollocks continued to operate Skyland until 1937, when they reluctantly turned the resort over to the new park's concessioner. However, they retained life tenancy of Massanutten Lodge until George Pollock's death in 1949.

Between 1930 and 1939, Skyland underwent changes that were influenced by the construction of the Skyline Drive (begun in 1931) and the creation of Shenandoah National Park (dedicated in 1936). Upgrading the resort to park standards, National Park Service designers constructed a loop access road, removed a number of cottages and outbuildings, and rehabilitated existing buildings and structures. To accommodate increasing postwar visitation, the park concessioner erected the modernistic lodge and dining room (1952) and adjacent guest registration building (1956), during the Mission 66 period. Today, Skyland is operated by ARAMARK, Virginia Sky-Line Company, Inc.

MASSANUTTEN LODGE ca.1911

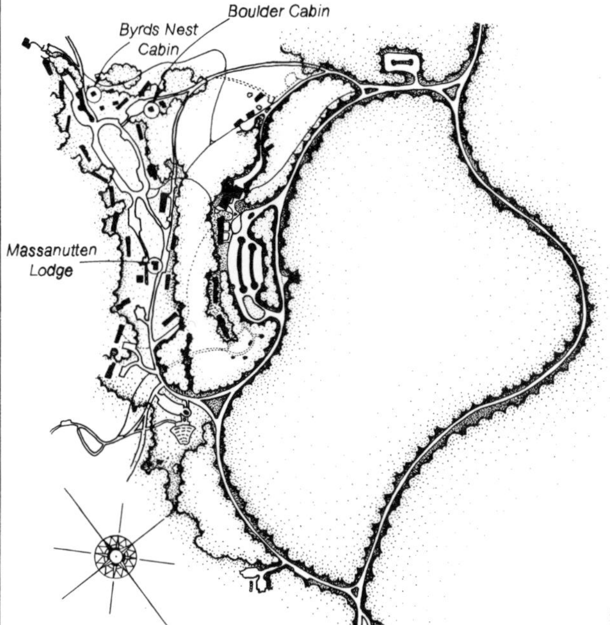

Skyland Development Plan
scale: 1" = 510' - 0" Mile Mk 51.2

BYRDS NEST CABIN ca.1906

BOULDER CABIN ca.1925

Locator Map

DELINEATED BY: Shane P. Wirth, 1996
NATIONAL PARK SERVICE
ROADS & BRIDGES RECORDING PROJECT
NATIONAL PARK SERVICE
UNITED STATES DEPARTMENT OF THE INTERIOR

SKYLINE DRIVE
SHENANDOAH NATIONAL PARK
PAGE COUNTY
LURAY VICINITY
VIRGINIA

SHEET 15 OF 18

HISTORIC AMERICAN ENGINEERING RECORD
VA-119

IF REPRODUCED, PLEASE CREDIT: HISTORIC AMERICAN ENGINEERING RECORD, NATIONAL PARK SERVICE, NAME OF DELINEATOR, DATE OF THE DRAWING

BIG MEADOWS

VIEW FROM BYRD VISITOR CENTER

Big Meadows, bisected by the Skyline Drive between mileposts 51 and 52, is located in the park's Central District. Archeological evidence suggests that the meadows may have been cleared by burning since precolonial times. Civilian Conservation Corps Camp NP-2 (Camp Fechner, so-named for the Director of the CCC) operated here from 1933-42, a reminder of President Franklin D. Roosevelt's commitment to the Skyline Drive and Shenandoah National Park. FDR dedicated the park from this site in 1936.

Big Meadows featured the park's first campground in 1937. A picnic area was added two years later. The concessioner built wayside (1938) and lodge (1939) were designed by Richmond architect Marcellus Wright, Jr. Today, this developed area includes park maintenance and interpretive facilities; picnic area and campground; concessioner operated wayside, dining and lodging services. The 300 acre mowed area has no trails and visitors may wander at will.

RANGER/CHECKING STATION

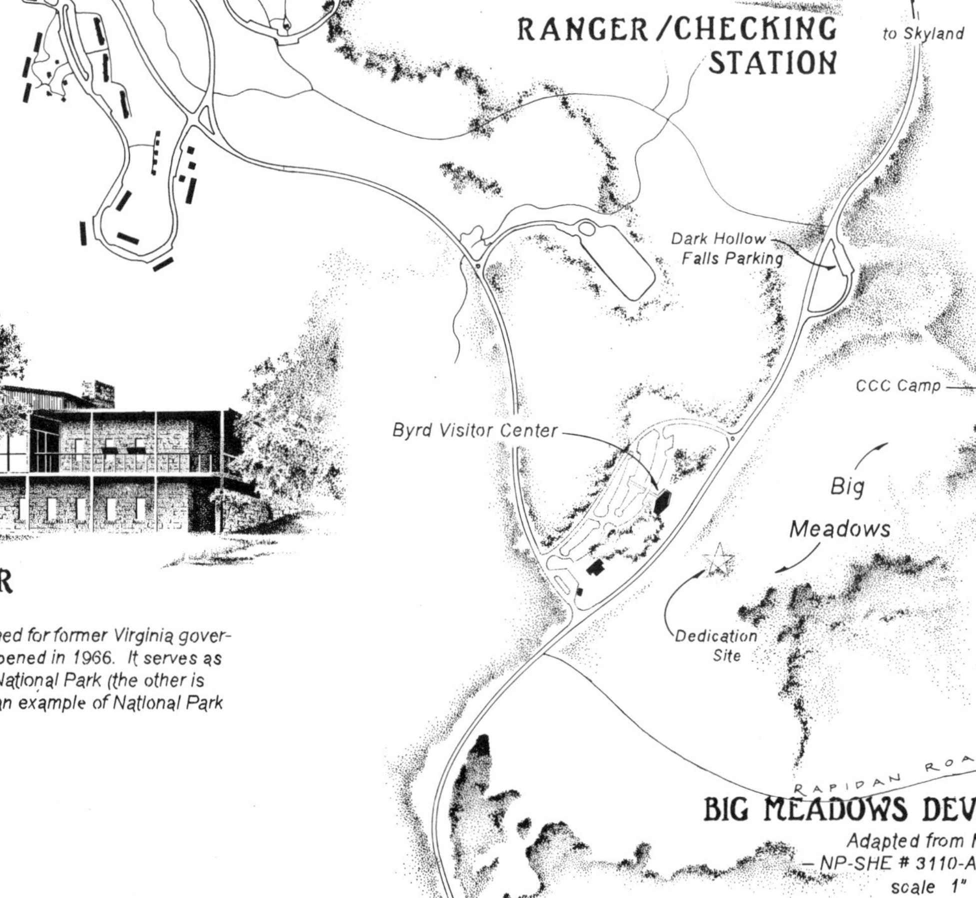

BIG MEADOWS DEVELOPMENT PLAN

Adapted from Master Plan
– NP-SHE # 3110-A and Aerial Photo
scale 1" = 500'

BYRD VISITOR CENTER

The Harry Flood Byrd Visitor Center – named for former Virginia governor, U.S. Senator, and park enthusiast – opened in 1966. It serves as one of two visitor centers in Shenandoah National Park (the other is located at Dickey Ridge). The building is an example of National Park Service Mission 66 design.

President Herbert Hoover established a trout fishing camp/retreat (1929) on property that he had purchased below Big Meadows. Situated at the confluence of two mountain streams forming the Rapidan River, this "Summer White House" offered the President an escape from the formality and heat of Washington, DC. While horseback riding through the area, Hoover encouraged both state and NPS officials to begin construction of the Skyline Drive.

When Hoover left office, Rapidan Camp (renamed Camp Hoover) served the Boy Scouts of America for several years. Later, as maintenance proved too costly for the scouting organization, the camp and the surrounding Hoover property were deeded to the National Park Service. Under the terms of this transfer, the facility remains available for use by government officials; otherwise use of Camp Hoover is at the park's disposal.

F. Roosevelt

H. Hoover

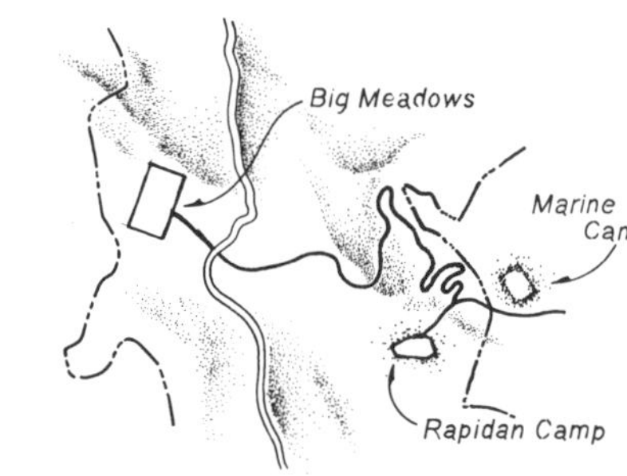

Locator map

PRESIDENT'S CABIN

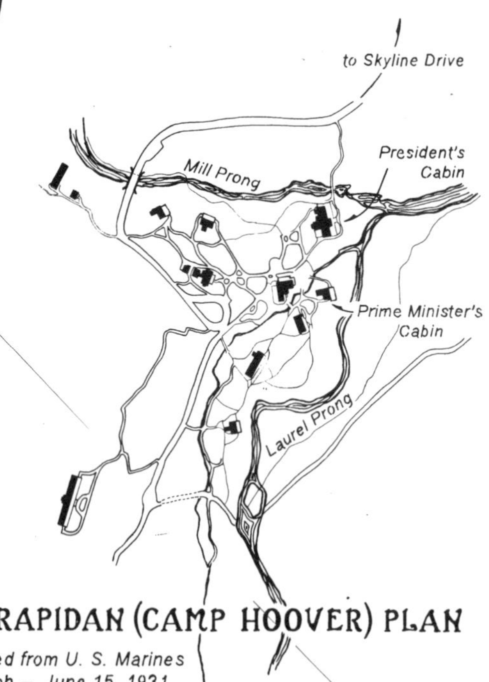

CAMP RAPIDAN (CAMP HOOVER) PLAN

Adapted from U. S. Marines
sketch – June 15, 1931
scale 1" = 200'

HISTORIC AMERICAN ENGINEERING RECORD
VA-119

SHEET 16 OF 18

VIRGINIA

SKYLINE DRIVE
SHENANDOAH NATIONAL PARK
PAGE COUNTY

LURAY VICINITY

NATIONAL PARK SERVICE
ROADS & BRIDGES RECORDING PROJECT
NATIONAL PARK SERVICE
UNITED STATES DEPARTMENT OF THE INTERIOR

DELINEATED BY *Michael P. Lanning, harlen d. Groe, 1996*

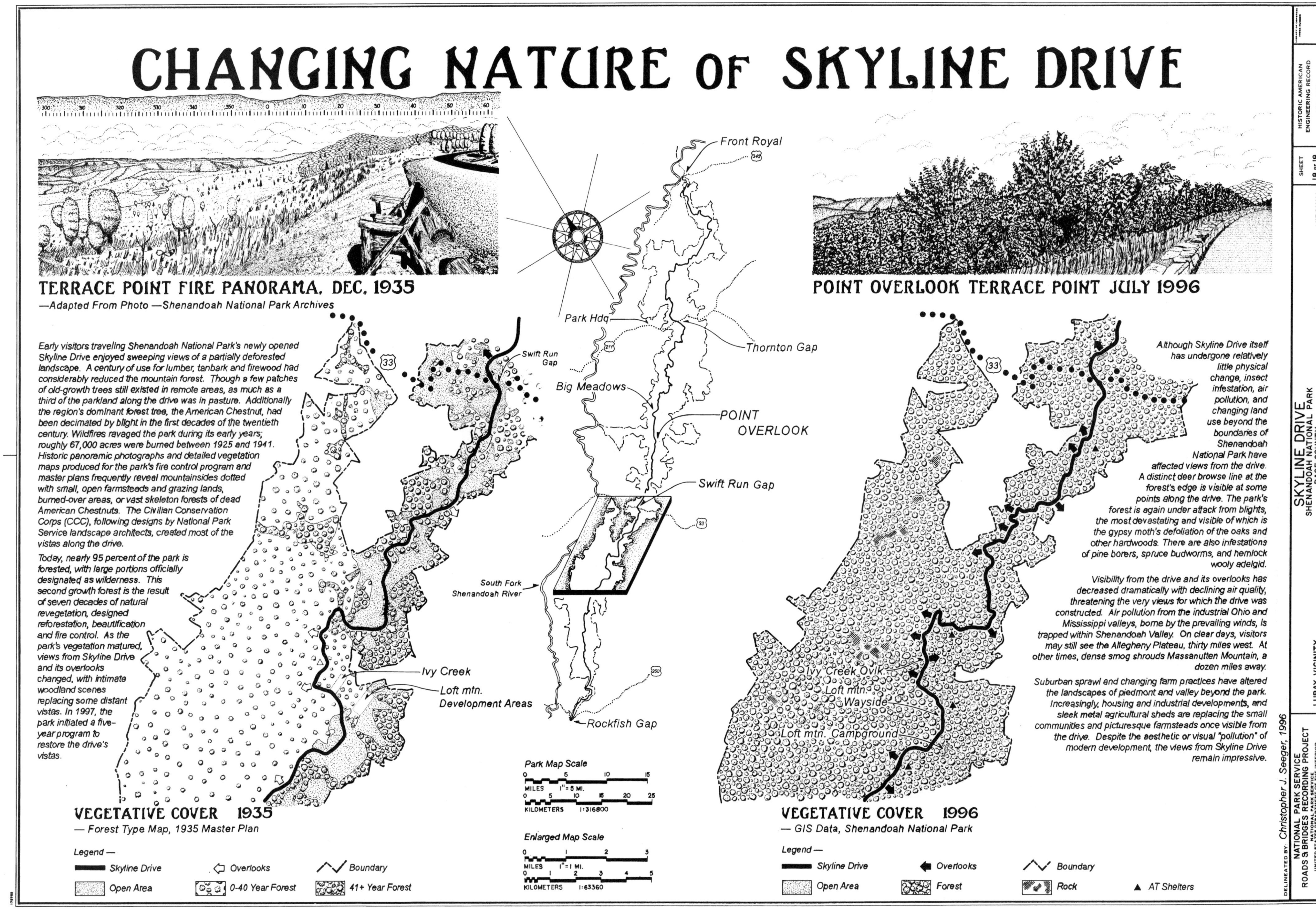

CHANGING NATURE OF SKYLINE DRIVE
TERRACE POINT FIRE PANORAMA, DEC, 1935
—Adapted From Photo —Shenandoah National Park Archives
POINT OVERLOOK TERRACE POINT JULY 1996
Front Royal
Park Hdq
Thornton Gap
Big Meadows
POINT OVERLOOK
Swift Run Gap
South Fork Shenandoah River
Rockfish Gap
Swift Run Gap
Ivy Creek
Loft mtn. Development Areas
Ivy Creek Ovlk.
Loft mtn. Wayside
Loft mtn. Campground
Early visitors traveling Shenandoah National Park's newly opened Skyline Drive enjoyed sweeping views of a partially deforested landscape. A century of use for lumber, tanbark and firewood had considerably reduced the mountain forest. Though a few patches of old-growth trees still existed in remote areas, as much as a third of the parkland along the drive was in pasture. Additionally the region's dominant forest tree, the American Chestnut, had been decimated by blight in the first decades of the twentieth century. Wildfires ravaged the park during its early years; roughly 67,000 acres were burned between 1925 and 1941. Historic panoramic photographs and detailed vegetation maps produced for the park's fire control program and master plans frequently reveal mountainsides dotted with small, open farmsteads and grazing lands, burned-over areas, or vast skeleton forests of dead American Chestnuts. The Civilian Conservation Corps (CCC), following designs by National Park Service landscape architects, created most of the vistas along the drive.
Today, nearly 95 percent of the park is forested, with large portions officially designated as wilderness. This second growth forest is the result of seven decades of natural revegetation, designed reforestation, beautification and fire control. As the park's vegetation matured, views from Skyline Drive and its overlooks changed, with intimate woodland scenes replacing some distant vistas. In 1997, the park initiated a five-year program to restore the drive's vistas.
Although Skyline Drive itself has undergone relatively little physical change, insect infestation, air pollution, and changing land use beyond the boundaries of Shenandoah National Park have affected views from the drive. A distinct deer browse line at the forest's edge is visible at some points along the drive. The park's forest is again under attack from blights, the most devastating and visible of which is the gypsy moth's defoliation of the oaks and other hardwoods. There are also infestations of pine borers, spruce budworms, and hemlock wooly adelgid.
Visibility from the drive and its overlooks has decreased dramatically with declining air quality, threatening the very views for which the drive was constructed. Air pollution from the industrial Ohio and Mississippi valleys, borne by the prevailing winds, is trapped within Shenandoah Valley. On clear days, visitors may still see the Allegheny Plateau, thirty miles west. At other times, dense smog shrouds Massanutten Mountain, a dozen miles away.
Suburban sprawl and changing farm practices have altered the landscapes of piedmont and valley beyond the park. Increasingly, housing and industrial developments, and sleek metal agricultural sheds are replacing the small communities and picturesque farmsteads once visible from the drive. Despite the aesthetic or visual "pollution" of modern development, the views from Skyline Drive remain impressive.
VEGETATIVE COVER 1935
— Forest Type Map, 1935 Master Plan
Legend —
Skyline Drive
Overlooks
Boundary
Open Area
0-40 Year Forest
41+ Year Forest
Park Map Scale
MILES 1"=5 MI.
KILOMETERS 1:316800
Enlarged Map Scale
MILES 1"=1 MI.
KILOMETERS 1:63360
VEGETATIVE COVER 1996
— GIS Data, Shenandoah National Park
Legend —
Skyline Drive
Overlooks
Boundary
Open Area
Forest
Rock
AT Shelters
DELINEATED BY: Christopher J. Seeger, 1996
NATIONAL PARK SERVICE
ROADS & BRIDGES RECORDING PROJECT
NATIONAL PARK SERVICE
UNITED STATES DEPARTMENT OF THE INTERIOR
LURAY VICINITY
SKYLINE DRIVE
SHENANDOAH NATIONAL PARK
PAGE COUNTY
VIRGINIA
SHEET 18 OF 18
HISTORIC AMERICAN ENGINEERING RECORD
VA-119
IF REPRODUCED, PLEASE CREDIT: HISTORIC AMERICAN ENGINEERING RECORD, NATIONAL PARK SERVICE, NAME OF DELINEATOR, DATE OF THE DRAWING

YELLOWSTONE ROADS & BRIDGES

THE ROAD SYSTEM PAST & PRESENT

YELLOWSTONE NATIONAL PARK

The historic bridges of Yellowstone National Park represent the important role of landscape architecture within the National Park Service. The philosophy of harmonizing structures and buildings with the natural scenery was first expressed by Lt. Dan Kingman of the U.S. Army Corps of Engineers in 1883. The Army engineers, particularly Dan Kingman and Hiram Chittenden, were responsible for road and bridge construction from 1883-1918. The National Park Service, created in 1916, assumed responsibility for the road and bridge construction in 1918, at which time official policy reiterated the earlier philosophy and called for the employment of trained "engineers who either possess a knowledge of landscape architecture or have a proper appreciation of the aesthetic value of park lands." In 1926, the direction was turned over to the Bureau of Public Roads. The Cub Creek Bridge (1928) and all of the bridges designed and/or built in the 1930s exhibit the planning and collaboration between Bureau of Public Roads engineers and National Park Service landscape architects.

Evidence of the military's role is expressed by the two steel bridges _Obsidian Creek Bridge (1910) and Firehole River Bridge at the end of the Fountain Freight Road (1911).

The road system in Yellowstone National Park is a near mirror image of the plan that the first superintendent, Nathanial P. Langford, imagined in the early 1870s. Throughout the decades, all road and bridge builders experienced the same difficulties - bad weather, unusual geologic and thermal features, and a short construction season.

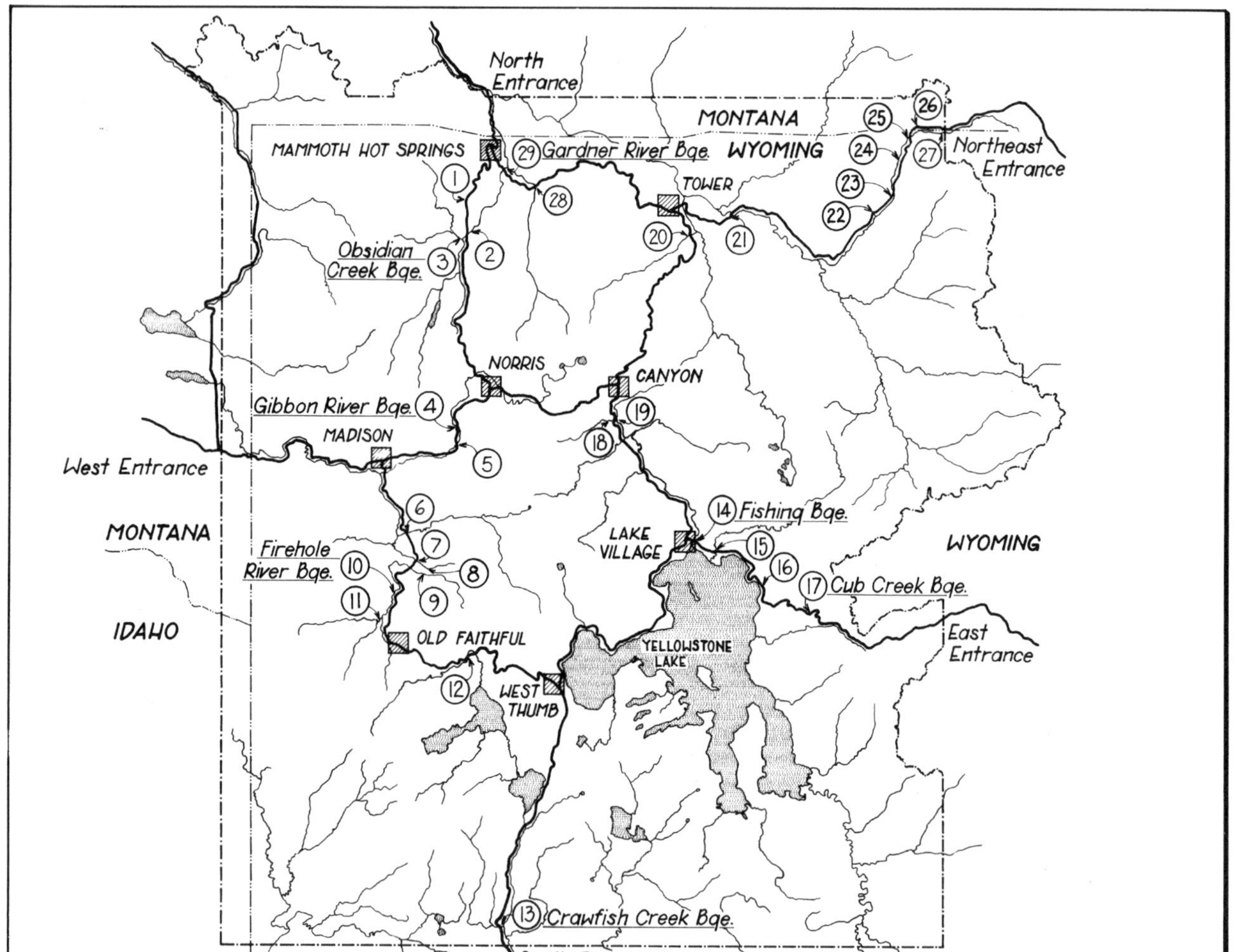

REFERENCE:

1. GOLDEN GATE VIADUCT (HAER N° WY-46)
2. SEVEN MILE BRIDGE (HAER N° WY-28)
3. OBSIDIAN CREEK BRIDGE (HAER N° WY-10)
4. GIBBON RIVER BRIDGE # 2 MP 2569 (HAER N° WY-29)
5. GIBBON RIVER BRIDGE # 3 MP 2691 (HAER N° WY-30)
6. NEZ PERCE CREEK BRIDGE (HAER N° WY-48)
7. FIREHOLE LAKE DRIVE _ 4TH BRIDGE (HAER N° WY-42)
8. FIREHOLE LAKE DRIVE _ 5TH BRIDGE (HAER N° WY-43)
9. FIREHOLE LAKE DRIVE _ 2ND BRIDGE (HAER N° WY-41)
10. FIREHOLE RIVER BRIDGE (ARMY) (HAER N° WY-27)
11. FIREHOLE RIVER BRIDGE (NEAR MORNING GLORY POOL) (HAER N° WY-44)
12. ISA LAKE BRIDGE (HAER N° WY-31)
13. CRAWFISH CREEK BRIDGE (HAER N° WY-26)
14. FISHING BRIDGE (HAER N° WY-9)
15. PELICAN CREEK BRIDGE (HAER N° WY-49)
16. SEDGE CREEK BRIDGE (HAER N° WY-35)
17. CUB CREEK BRIDGE (HAER N° WY-25)
18. OTTER CREEK BRIDGE (HAER N° WY-32)
19. OTTER CREEK BRIDGE (CAMPGRD. RD.) (HAER N° WY-47)
20. TOWER CREEK BRIDGE (HAER N° WY-33)
21. LAMAR RIVER BRIDGE (HAER N° WY-12)
22. PEBBLE CREEK BRIDGE (HAER N° WY-40)
23. SODA BUTTE CREEK # 1 (HAER N° WY-39)
24. FHWA CREEK (STA. 268 + 57) (HAER N° WY-38)
25. RWC CREEK (STA. 243 + 51) (HAER N° WY-37)
26. TLF CREEK (STA. 198 + 85) (HAER N° WY-36)
27. SODA BUTTE CREEK # 2 (HAER N° WY-50)
28. LAVA CREEK BRIDGE (HAER N° WY-34)
29. GARDNER RIVER BRIDGE (HAER N° WY-7)

Yellowstone National Park (location of documented bridges)

In 1988, the Federal Highway Administration began a 20-year reconstruction program in Yellowstone National Park. This documentation is being used as a part of the planning effort of the National Park Service's "A Road System Rehabilitation Plan and Its Accompanying Environmental Documents."

This recording project is part of the Historic American Engineering Record (HAER), a long range program to document historically significant engineering and industrial works in the United States. HAER is an agency of the National Park Service, U. S. Dept. of the Interior. The Yellowstone Roads and Bridges Recording Project was cosponsored by Yellowstone National Park, Robert Barbee, Superintendent; the NPS Roads and Bridges Program, John Gingles, Manager; Lorraine Mintzmyer, Regional Director, and Rodd Wheaton, Division Chief, Cultural Resources, Rocky Mountain Region, NPS, and the Historic American Buildings Survey/Historic American Engineering Record (HABS/HAER) under the general direction of Robert J. Kapsch, Chief.

The fieldwork, measured drawings, historical reports, and photography were completed under the direction of Eric DeLony, Chief and Principal Architect, HAER. The recording team consisted of Mary Shivers Culpin, Project Historian, Rocky Mountain Region, NPS, Julie E. Pearson, Supervising Architect, Texas Tech University; Steven M. Varner, Project Engineer, Virginia Polytechnic Institute; Elizabeth A. Harvey, Project Illustrator, Iowa State University; and Architect Technicians Laura E. Salarano, U.S./ICOMOS, Universidad Nacional de Rosario, Argentina, and Gerald J. Hansen, North Dakota State University. Formal on site photography was done by Jet Lowe, HAER staff photographer.

DELINEATED BY: ELIZABETH A. HARVEY, LAURA E. SALARANO · 1989

YELLOWSTONE ROADS AND BRIDGES
HISTORIC AMERICAN ENGINEERING RECORD
NATIONAL PARK SERVICE
UNITED STATES DEPARTMENT OF THE INTERIOR

YELLOWSTONE N.P.

YELLOWSTONE ROADS & BRIDGES
PARK, TETON COUNTIES
TETON COUNTY

WYOMING

SHEET 1 OF 1

HISTORIC AMERICAN ENGINEERING RECORD
WY - 24

OBSIDIAN CREEK BRIDGE 1910

INDIAN CREEK CAMPGROUND

YELLOWSTONE NATIONAL PARK

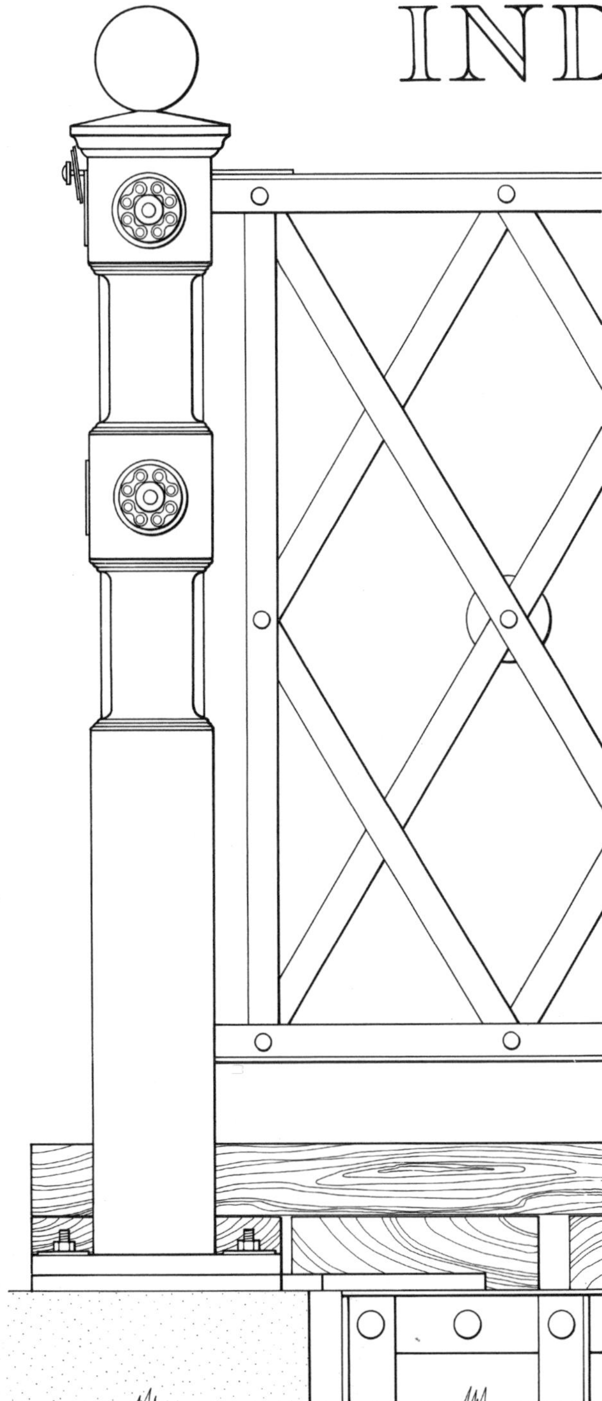

Railing Detail
Scale: 3" = 1'-0" (Metric: 1:4)

The Obsidian Creek Bridge, completed in 1910, carries the Indian Creek Campground Road over Obsidian Creek just off the Grand Loop Road between Mammoth Hot Springs and Norris Junction. The bridge was constructed by the U.S. Army Corps of Engineers in their second phase of bridge construction. The Army Engineers began the construction of roads and bridges in Yellowstone National Park in 1883 and continued the work until 1918 when the responsibility was assumed by the National Park Service. In 1909, an Army bridge inspection stated "... all of these wooden bridges were built in the earlier days [1880s 1890s] of road construction in the Park, the material being pine in all cases. The life of the pine in that altitude is not to exceed 12 to 15 years if clear from earth and moisture, and considerably less if in contact with earth and moisture. Therefore, most of the structures in the Park have reached the limit of safety for use as bridges.

The first four steel bridges in the Park were constructed in 1901, followed by nine during 1903-1904, four in 1910, and three in 1911.

The 1909 inspection report recommended that a 36-foot plate girder span on concrete abutments replace the large culvert pipe covered with rock filling, which had proved inadequate for water flow. Prior to the culvert pipe, an A-truss wooden span crossed Obsidian Creek.

The bridge is important to Yellowstone National Park for its association with the U.S. Army Corps of Engineers.

This recording project is part of the Historic American Engineering Record (HAER), a long range program to document historically significant engineering and industrial works in the United States. HAER is an agency of the National Park Service, U.S. Department of the Interior. The Yellowstone Roads and Bridges Recording Project was cosponsored by Yellowstone National Park, Robert Barbee, Superintendent; the NPS Roads and Bridges Program, John Gingles, Manager; Rocky Mountain Region, NPS, Lorraine Mintzmyer, Regional Director, and Rodd Wheaton, Division Chief, Cultural Resources; and the Historic American Buildings Survey/Historic American Engineering Record (HABS/HAER) under the general direction of Robert J. Kapsch, Chief.

The fieldwork, measured drawings, historical reports, and photography were completed under the direction of Eric DeLony, Chief and Principal Architect, HAER. The recording team consisted of Mary Shivers Culpin, Project Historian, Rocky Mountain Region, NPS; Julie E. Pearson, Supervising Architect, Texas Tech University; Steven M. Varner, Project Engineer, Virginia Polytechnic Institute; Elizabeth A. Harvey, Project Illustrator, Iowa State University; and Architect Technicians Laura E. Salarano, U.S. ICOMOS, Universidad Nacional de Rosario, Argentina, and Gerald J. Hansen, North Dakota State University. Formal on site photography was done by Jet Lowe, HAER staff photographer.

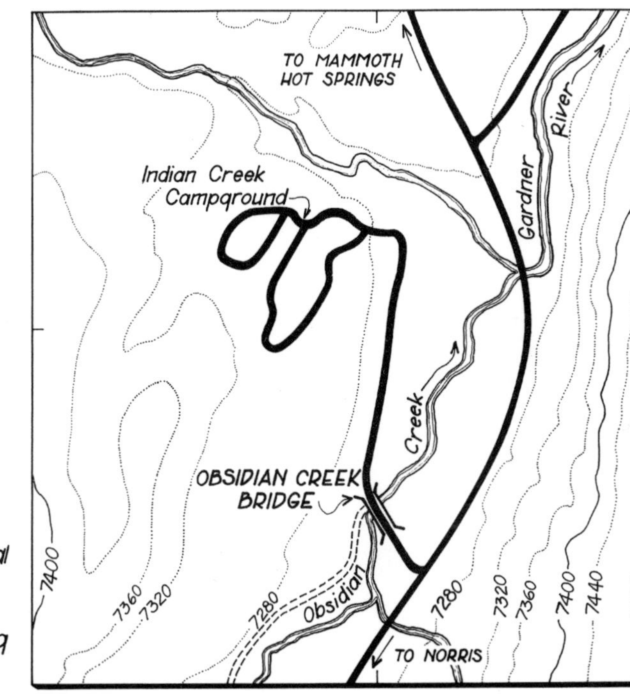

Based on U.S.G.S. 7.5 Minute Series
Mammoth Quadrangle (1986)
U.T.M.: 12.525280.4978260

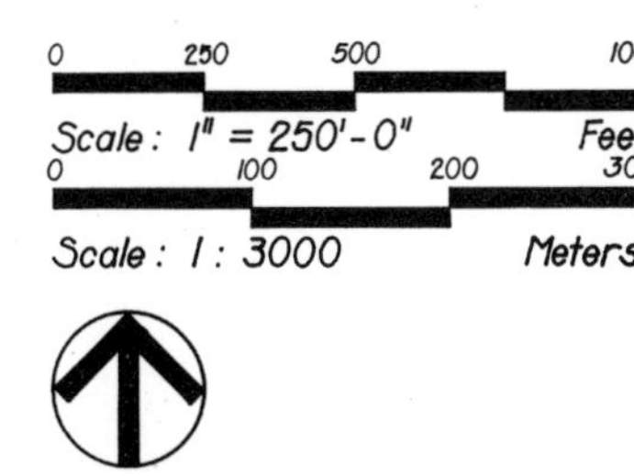

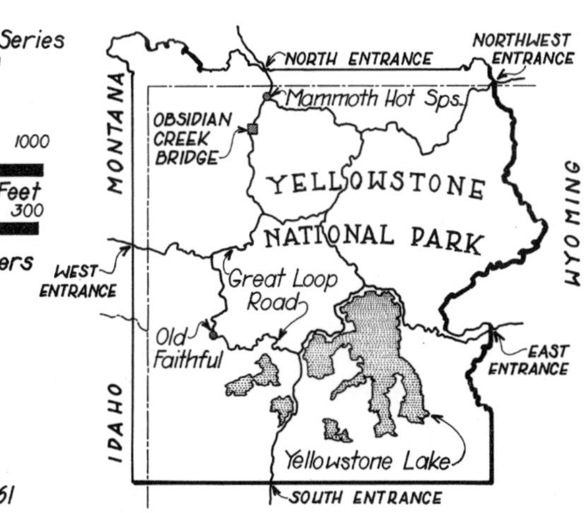

Based on U.S.G.S. Yellowstone National Park
Wyoming-Montana-Idaho 1961

DELINEATED BY: ELIZABETH A. HARVEY · LAURA E. SALARANO, 1989

YELLOWSTONE ROADS AND BRIDGES
HISTORIC AMERICAN ENGINEERING RECORD
NATIONAL PARK SERVICE
UNITED STATES DEPARTMENT OF THE INTERIOR

OBSIDIAN CREEK BRIDGE 1910
SPANNING OBSIDIAN CREEK ON INDIAN CREEK CAMPGROUND ROUND
YELLOWSTONE NATIONAL PARK PARK COUNTY WYOMING

SHEET 1 OF 2

HISTORIC AMERICAN ENGINEERING RECORD
WY-10

IF REPRODUCED, PLEASE CREDIT: HISTORIC AMERICAN ENGINEERING RECORD, NATIONAL PARK SERVICE, NAME OF DELINEATOR, DATE OF THE DRAWING

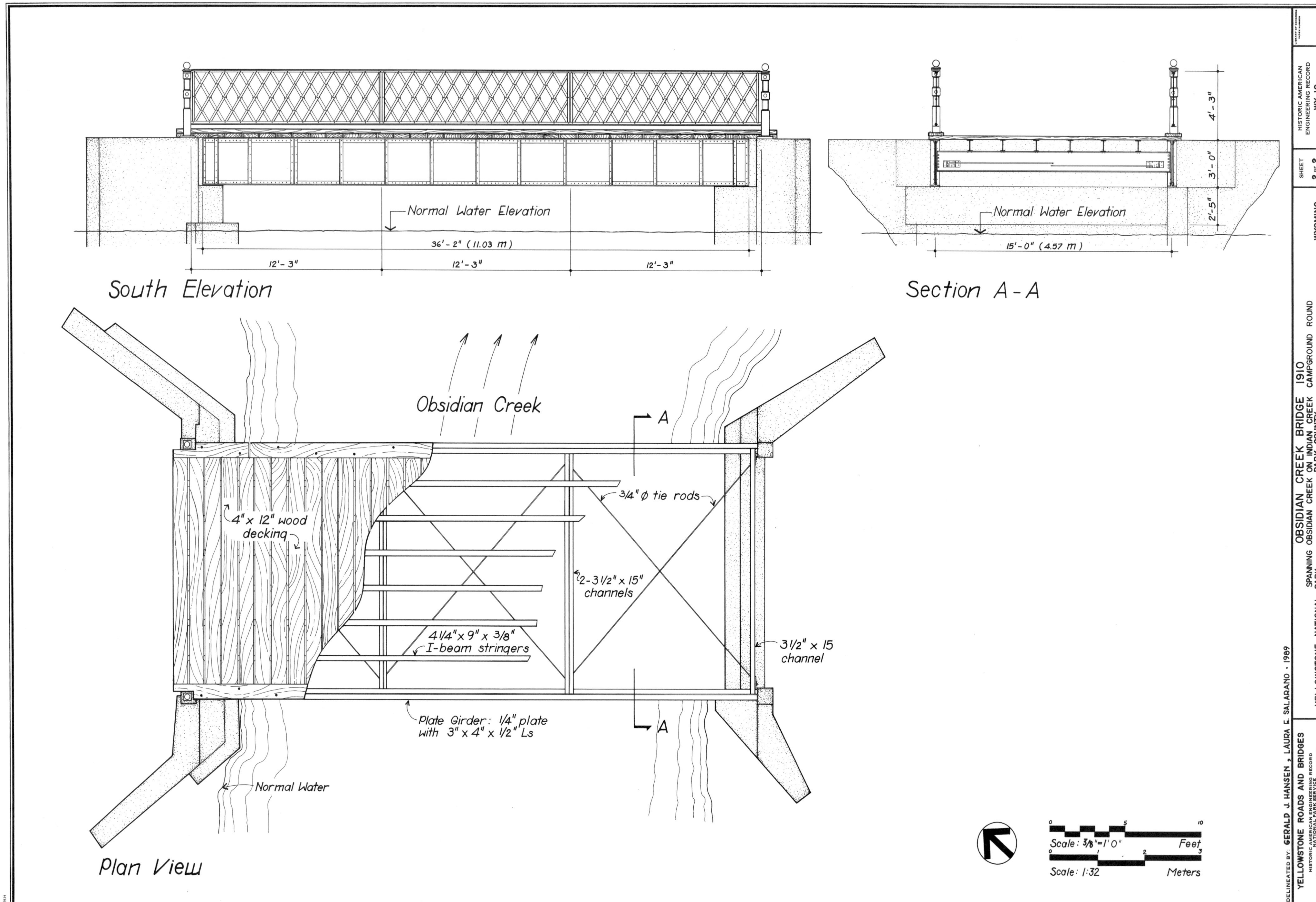

Normal Water Elevation
36'-2" (11.03 m)
12'-3"
12'-3"
12'-3"
South Elevation
4'-3"
3'-0"
2'-5"
Normal Water Elevation
15'-0" (4.57 m)
Section A-A
Obsidian Creek
A
A
3/4" ϕ tie rods
4" x 12" wood decking
2-3 1/2" x 15" channels
4 1/4" x 9" x 3/8" I-beam stringers
3 1/2" x 15 channel
Plate Girder: 1/4" plate with 3" x 4" x 1/2" Ls
Normal Water
Plan View
Scale: 3/8"=1'0"
Feet
Scale: 1:32
Meters
DELINEATED BY: GERALD J. HANSEN, LAURA E. SALARANO · 1989
YELLOWSTONE ROADS AND BRIDGES
OBSIDIAN CREEK BRIDGE 1910
SPANNING OBSIDIAN CREEK ON INDIAN CREEK CAMPGROUND ROUND
YELLOWSTONE NATIONAL PARK
PARK COUNTY
WYOMING
SHEET 2 of 2
HISTORIC AMERICAN ENGINEERING RECORD
WY-10

FISHING BRIDGE · 1936
YELLOWSTONE RIVER NEAR LAKE
YELLOWSTONE NATIONAL PARK

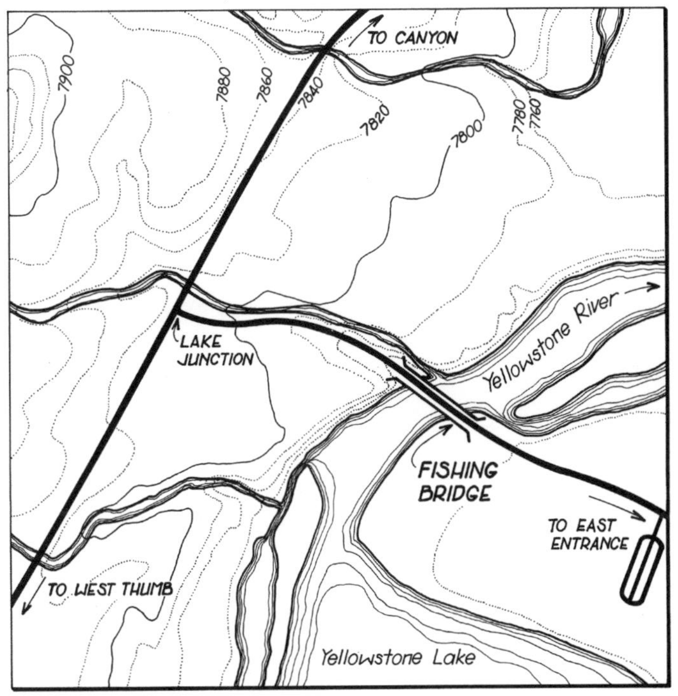

Based on U.S.G.S. 7.5 Minute Series
Lake, Wyoming Quadrangle (1986)
U.T.M.: 12.549080.4931970

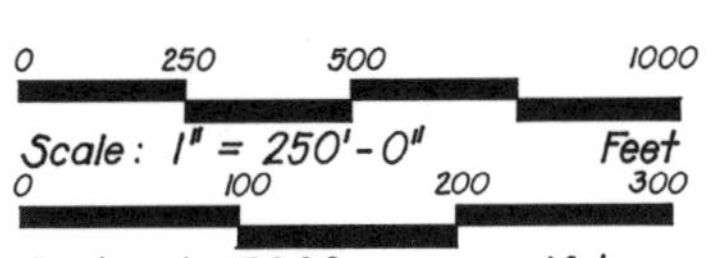

Based on U.S.G.S. Yellowstone
National Park
Wyoming-Montana-Idaho 1961

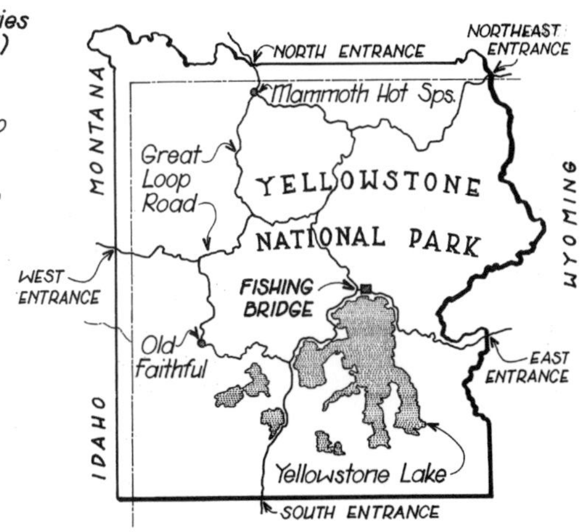

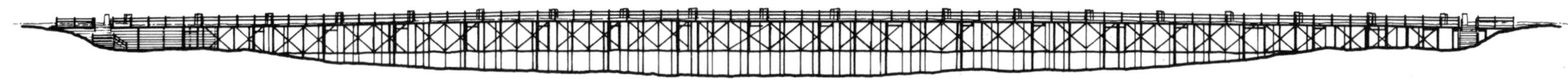

North Elevation
no scale

The Fishing Bridge, on the East Entrance Road, crosses the Yellowstone River just above its outlet from Yellowstone Lake. The bridge replaced a narrow, wooden structure which was built as part of the U.S. Army Corps of Engineer's 1907 East Entrance Road project. The popularity of fishing from the old bridge, which was designed for horse-drawn traffic, led the National Park Service to request a structure which could safely accommodate this type of public usage. The National Park Service Division of Plans and Design also requested that the structure be of rustic design.

The location survey was made in 1931. The selected crossing, which was about 100 feet below the old bridge on the south or east side of the river and on a 45 degree skew and intersecting the old bridge on the north or west side of the river, provided several advantages. Ice damage would be lessened as it had more room to break up and the farther distance from the lake permitted boating sheltered from the frequent and unpredictable storms on the lake.

Plans for Fishing Bridge were developed in the Regional Office of the Bureau of Public Roads, with the architectural plans being completed by the Division of Plans and Design of the National Park Service. Strong and Grant, of Springville, Utah were awarded the bridge contract November 27, 1935. Work began on the bridge May 26, 1936. Material for the log spans came from the Pacific Coast. Port Orford cedar was used for the bents. They were obtained near Marshfield, Oregon and were treated by the Mineralized Cell Wood Preserving Company of Portland, Oregon. The Douglas fir caps and stringers were obtained near Wauna, Oregon and were treated with Wolman Salts of the Crossett Western Company in Wauna, Oregon. The structure was stained a brownish color. The bridge has nineteen 28 foot spans for a total length of 532 feet. Two 5 foot sidewalks flank the 24 foot curb to curb roadway with stairways to the ground at the reinforced concrete abutments. The cost of the Fishing Bridge was $92,408.09, exclusive of the 68% of the total contract amount for engineering services. With timber cribbing at abutments and stairs, the bridge is a very good example of rustic design.

This recording project is part of the Historic American Engineering Record (HAER), a long range program to document historically significant engineering and industrial works in the United States. The HAER program is an agency of the National Park Service, U.S. Department of the Interior. The Yellowstone Roads and Bridges Recording Project was cosponsored by Yellowstone National Park, Robert Barbee, Superintendent; the NPS Roads and Bridges Program, John Gingles, Manager; Rocky Mountain Region, NPS, Lorraine Mintzmyer, Regional Director, and Rodd Wheaton, Division Chief, Cultural Resources; and the Historic American Buildings Survey/Historic American Engineering Record (HABS/HAER) under the general direction of Robert J. Kapsch, Chief.

The fieldwork, measured drawings, historical reports and photography were completed under the direction of Eric DeLony, Chief and Principal Architect, HAER. The recording team consisted of Mary Shivers Culpin, Project Historian, Rocky Mountain Region, NPS; Julie E. Pearson, Supervising Architect, Texas Tech University; Steven M. Varner, Project Engineer, Virginia Polytechnic Institute; Elizabeth A. Harvey, Project Illustrator, Iowa State University; and Architect Technicians Laura E. Salarano, U.S./ICOMOS, Universidad Nacional de Rosario, Argentina, and Gerald J. Hansen, North Dakota State University. Formal on site photography was done by Jet Lowe, HAER staff photographer.

DELINEATED BY: JULIE E. PEARSON · LAURA E. SALARANO · 1989

YELLOWSTONE ROADS AND BRIDGES
HISTORIC AMERICAN ENGINEERING RECORD
NATIONAL PARK SERVICE
UNITED STATES DEPARTMENT OF THE INTERIOR

FISHING BRIDGE 1936
SPANNING YELLOWSTONE RIVER ON EAST ENTRANCE ROAD
YELLOWSTONE NATIONAL PARK PARK COUNTY WYOMING

SHEET 1 of 3

HISTORIC AMERICAN ENGINEERING RECORD
WY-9

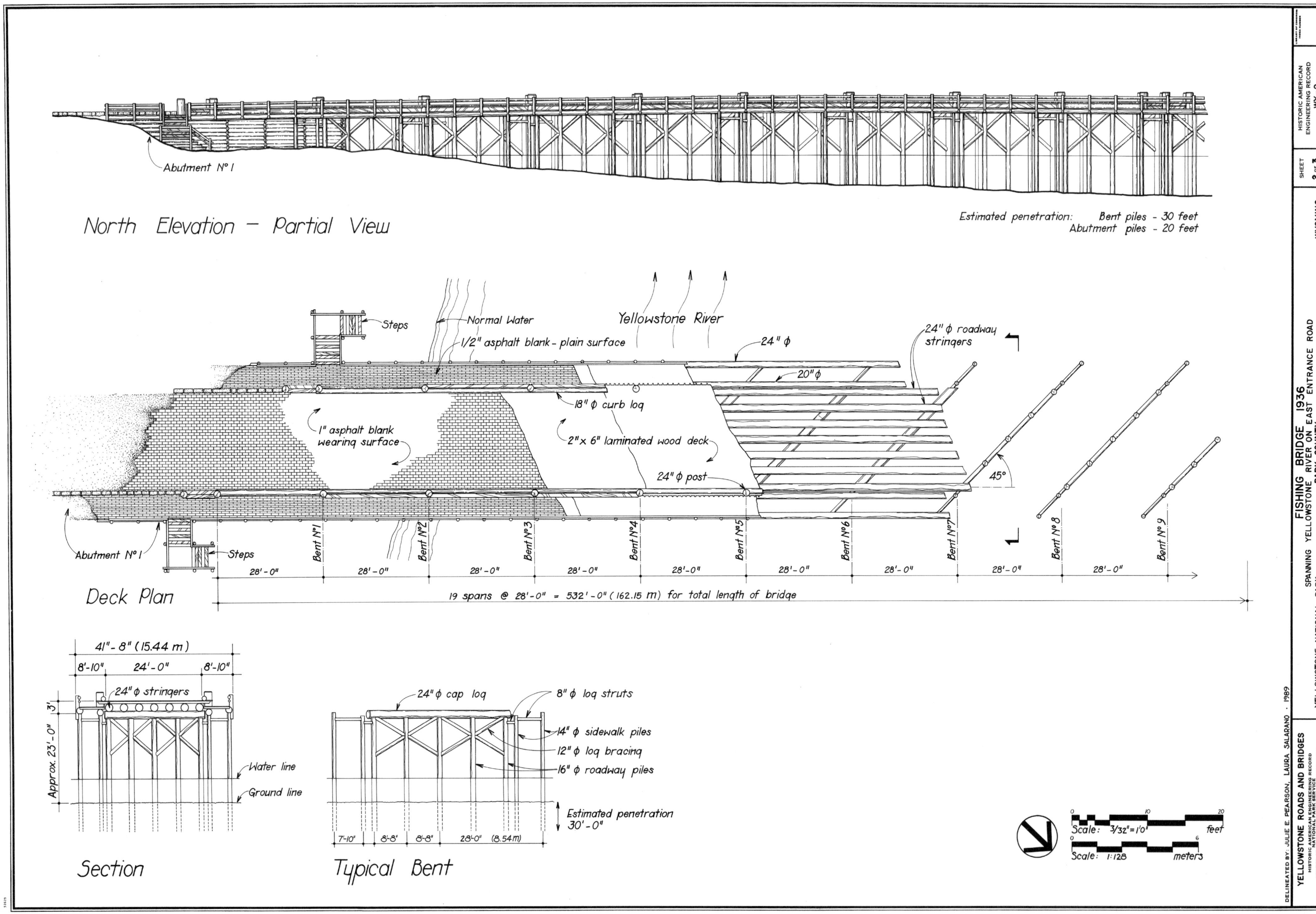
Abutment N° I
North Elevation – Partial View
Estimated penetration: Bent piles - 30 feet
Abutment piles - 20 feet
Steps
Normal Water
Yellowstone River
1/2" asphalt blank - plain surface
24" ϕ
24" ϕ roadway stringers
20" ϕ
18" ϕ curb log
1" asphalt blank wearing surface
2" x 6" laminated wood deck
24" ϕ post
45°
Abutment N° I
Steps
Bent N° 1
Bent N° 2
Bent N° 3
Bent N° 4
Bent N° 5
Bent N° 6
Bent N° 7
Bent N° 8
Bent N° 9
28'-0"
Deck Plan
19 spans @ 28'-0" = 532'-0" (162.15 m) for total length of bridge
41'-8" (15.44 m)
8'-10"
24'-0"
8'-10"
24" ϕ stringers
3'
Approx. 23'-0"
Water line
Ground line
Section
24" ϕ cap log
8" ϕ log struts
14" ϕ sidewalk piles
12" ϕ log bracing
16" ϕ roadway piles
Estimated penetration 30'-0"
7'-10"
8'-8"
8'-8"
28'-0" (8.54 m)
Typical Bent
Scale: 3/32" = 1'0"
feet
Scale: 1:128
meters
DELINEATED BY: JULIE E. PEARSON, LAURA SALARANO · 1989
YELLOWSTONE ROADS AND BRIDGES
HISTORIC AMERICAN ENGINEERING RECORD
NATIONAL PARK SERVICE
UNITED STATES DEPARTMENT OF THE INTERIOR
FISHING BRIDGE 1936
SPANNING YELLOWSTONE RIVER ON EAST ENTRANCE ROAD
YELLOWSTONE NATIONAL PARK
PARK COUNTY
WYOMING
SHEET 2 of 3
HISTORIC AMERICAN ENGINEERING RECORD
WY-9
IF REPRODUCED, PLEASE CREDIT: HISTORIC AMERICAN ENGINEERING RECORD, NATIONAL PARK SERVICE, NAME OF DELINEATOR, DATE OF THE DRAWING

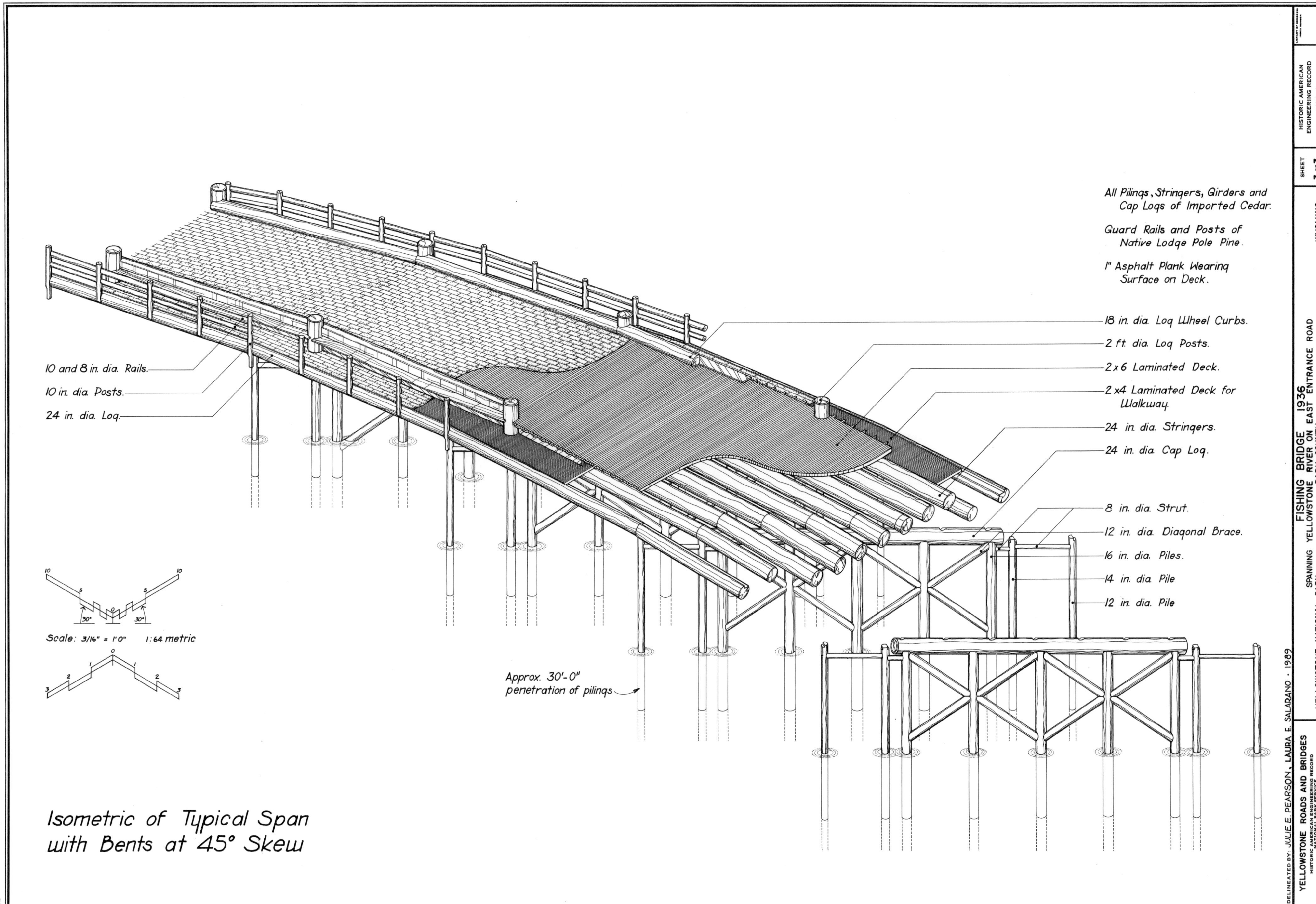

All Pilings, Stringers, Girders and Cap Logs of Imported Cedar.
Guard Rails and Posts of Native Lodge Pole Pine.
1" Asphalt Plank Wearing Surface on Deck.
18 in. dia. Log Wheel Curbs.
2 ft. dia. Log Posts.
2x6 Laminated Deck.
2x4 Laminated Deck for Walkway.
24 in. dia. Stringers.
24 in. dia. Cap Log.
8 in. dia. Strut.
12 in. dia. Diagonal Brace.
16 in. dia. Piles.
14 in. dia. Pile
12 in. dia. Pile
10 and 8 in. dia. Rails.
10 in. dia. Posts.
24 in. dia. Log.
Approx. 30'-0" penetration of pilings
Scale: 3/16" = 1'0" 1:64 metric
Isometric of Typical Span with Bents at 45° Skew
FISHING BRIDGE 1936
SPANNING YELLOWSTONE RIVER ON EAST ENTRANCE ROAD
YELLOWSTONE NATIONAL PARK
PARK COUNTY
WYOMING
SHEET 3 of 3
HISTORIC AMERICAN ENGINEERING RECORD
WY-9
DELINEATED BY: JULIE E. PEARSON, LAURA E. SALARANO · 1989
YELLOWSTONE ROADS AND BRIDGES
HISTORIC AMERICAN ENGINEERING RECORD
NATIONAL PARK SERVICE
UNITED STATES DEPARTMENT OF THE INTERIOR
IF REPRODUCED, PLEASE CREDIT: HISTORIC AMERICAN ENGINEERING RECORD, NATIONAL PARK SERVICE, NAME OF DELINEATOR, DATE OF THE DRAWING

CRAWFISH CREEK BRIDGE · 1936

SOUTH ENTRANCE HIGHWAY

YELLOWSTONE NATIONAL PARK

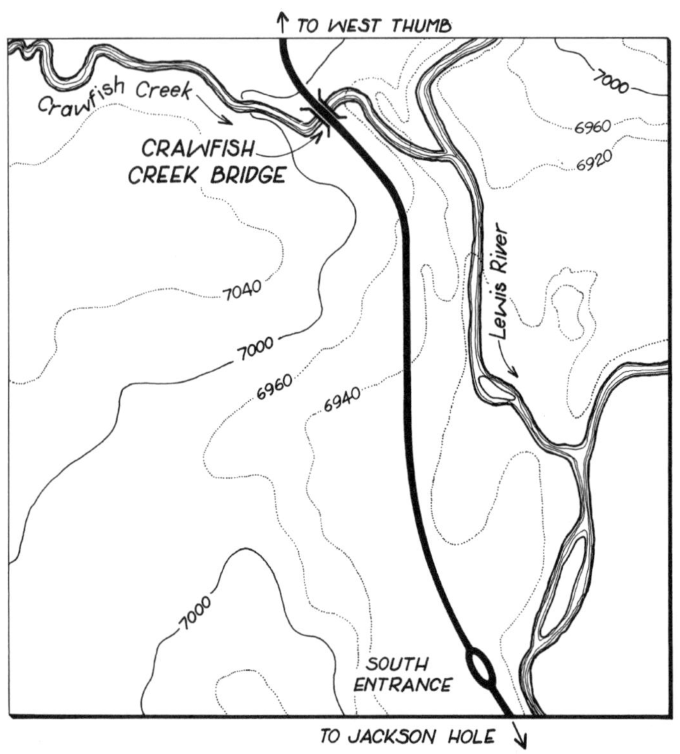

Based on U.S.G.S. 15 Minute Series
Huckleberry Mountain Quadrangle (1986)
U.T.M.: 12.526004888550

0 250 500 1000
Scale: 1" = 250'-0" Feet
0 100 200 300
Scale: 1 : 3000 Meters

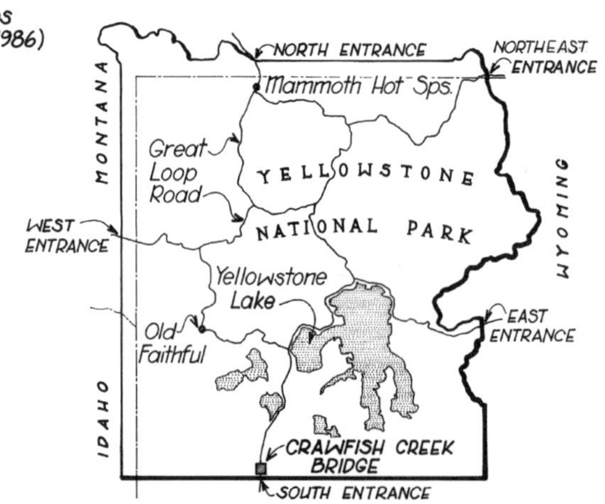

Based on U.S.G.S. Yellowstone National Park
Wyoming - Montana - Idaho 1961

The Crawfish Bridge,located 1.5 miles north of the South Entrance of Yellowstone National Park, was completed in 1936. The Bureau of Public Roads completed the bridge survey in 1932. After extensive investigation and discussion, the site and design of the bridge was approved during the winter of 1933-34. The Yellowstone National Park landscape architects selected the design. Olaf Nelson was awarded the construction contract on September 19, 1934, and Carl Jensen of Denver, Colorado, received a sub-contract for the construction of all of the masonry work on the proyect.

Considerable field engineering work was needed during the bridge's construction because of its architectural type and the use of local building materials. An outcropping of volcanic stone, partly of obsidian composition, on the west bank of the Lewis River 1/4 miles south of the bridge site, was quarried for use. During the 1935 construction season all of the concrete had been poured, a major portion of the masonry and masonry facing had been finished, and the water proofing of the arch barrel and spandrel walls started.

The new Crawfish Creek Bridge, which replaced a very narrow steel truss bridge constructed by the U.S. Army Corps of Engineers, is a 72-foot span concrete arch barrel with masonry facing and wing walls. One detail requested by the National Park Service was that (random ashlar pattern) rectangular stone be used instead of five sided stone and that weathered stone faces not be used. At the time of its completion, the resident landscape architect, Sanford Hill, felt that it was one of the best bridges in the Park.

This recording project is part of the Historic American Engineering Record (HAER), a long-range program to document historically significant engineering and industrial works in the United States. The HAER program is an agency of the National Park Service, U.S. Department of the Interior. The Yellowstone Roads and Bridges Recording Project was co-sponsored by Yellowstone National Park, Robert Barbee, Superintendent; the NPS Roads and Bridges Program, John Gingles, Manager, Rocky Mountain Region, NPS, Lorraine Mintzmyer, Regional Director, and Rodd Wheaton, Division Chief, Cultural Resources, and the Historic American Buildings Survey/Historic American Engineering Record (HABS/HAER) under the general direction of Robert J. Kapsch, Chief.

The fieldwork, measured drawings, historical reports and photography were completed under the direction of Eric De Lony, Chief and Principal Architect, HAER. The recording team consisted of Mary Shivers Culpin, Project Historian, Rocky Mountain Region, NPS; Julie E. Pearson, Supervising Architect, Texas Tech University; Steven M. Varner, Project Engineer, Virginia Polytechnic Institute; Elizabeth A. Harvey, Project Illustrator, Iowa State University; and Architect Technicians Laura E. Salarano, US/ICOMOS, Universidad Nacional de Rosario, Argentina, and Gerald J. Hansen, North Dakota State University. Formal on-site photography was done by Jet Lowe, HAER staff photographer.

HISTORIC AMERICAN ENGINEERING RECORD WY-26
SHEET 1 of 3
CRAWFISH CREEK BRIDGE 1936
SPANNING CRAWFISH CREEK ON SOUTH ENTRANCE ROAD
YELLOWSTONE NATIONAL PARK TETON COUNTY WYOMING
DELINEATED BY: LAURA E. SALARANO, 1989
YELLOWSTONE ROADS AND BRIDGES
HISTORIC AMERICAN ENGINEERING RECORD
NATIONAL PARK SERVICE
UNITED STATES DEPARTMENT OF THE INTERIOR

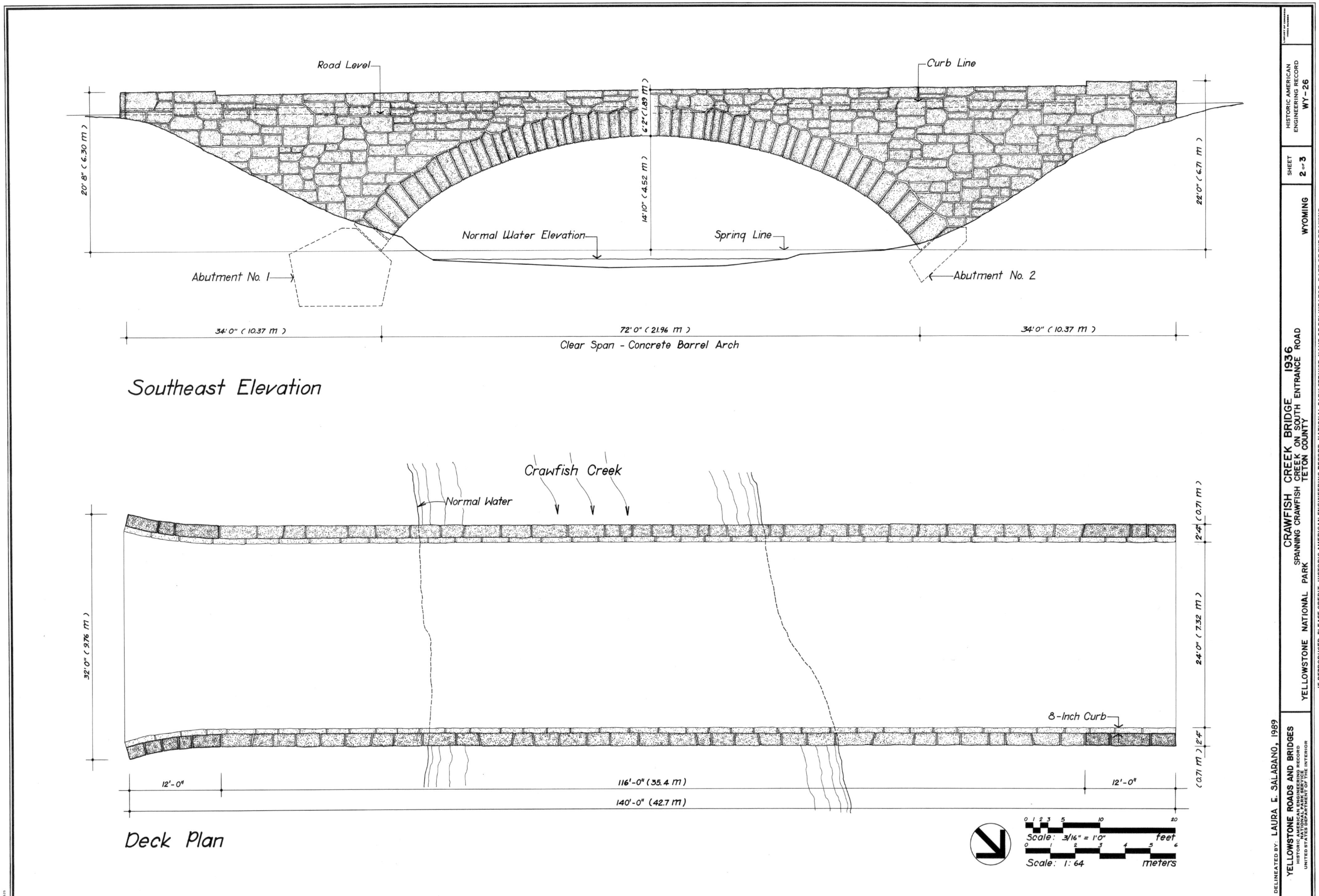

Road Level
Curb Line
6'-2" (1.89 m)
14'-10" (4.52 m)
20'-8" (6.30 m)
22'-0" (6.71 m)
Normal Water Elevation
Spring Line
Abutment No. 1
Abutment No. 2
34'-0" (10.37 m)
72'-0" (21.96 m)
34'-0" (10.37 m)
Clear Span - Concrete Barrel Arch
Southeast Elevation
Crawfish Creek
Normal Water
32'-0" (9.76 m)
2'-4" (0.71 m)
24'-0" (7.32 m)
8-Inch Curb
12'-0"
116'-0" (35.4 m)
140'-0" (42.7 m)
Deck Plan
Scale: 3/16" = 1'0"
feet
Scale: 1:64
meters
DELINEATED BY: LAURA E. SALARANO, 1989
YELLOWSTONE ROADS AND BRIDGES
HISTORIC AMERICAN ENGINEERING RECORD
NATIONAL PARK SERVICE
UNITED STATES DEPARTMENT OF THE INTERIOR
CRAWFISH CREEK BRIDGE 1936
SPANNING CRAWFISH CREEK ON SOUTH ENTRANCE ROAD
YELLOWSTONE NATIONAL PARK TETON COUNTY
WYOMING
SHEET 2 of 3
HISTORIC AMERICAN ENGINEERING RECORD
WY-26
IF REPRODUCED, PLEASE CREDIT: HISTORIC AMERICAN ENGINEERING RECORD, NATIONAL PARK SERVICE, NAME OF DELINEATOR, DATE OF THE DRAWING

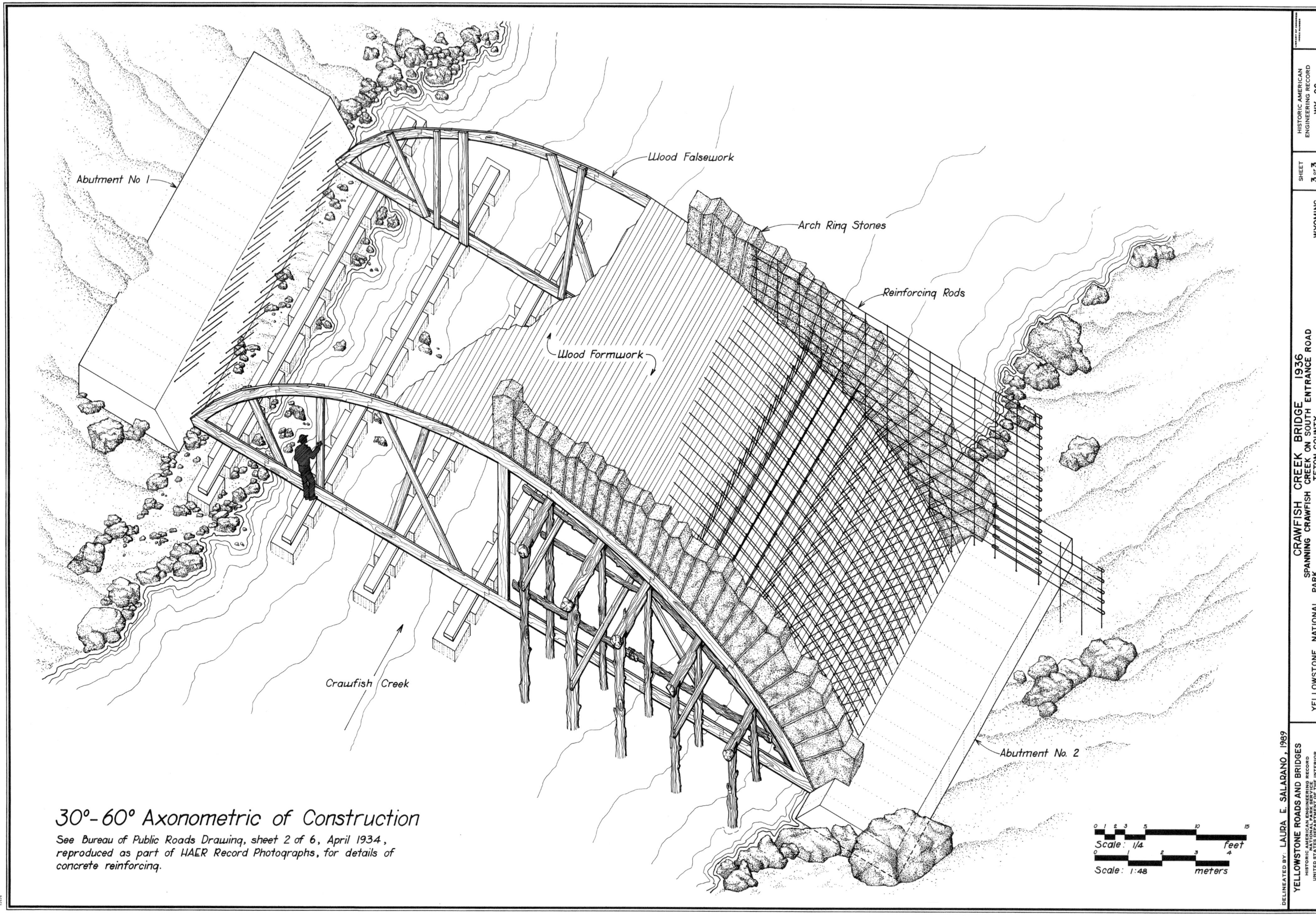

Abutment No 1
Wood Falsework
Arch Ring Stones
Reinforcing Rods
Wood Formwork
Crawfish Creek
Abutment No. 2
30°-60° Axonometric of Construction
See Bureau of Public Roads Drawing, sheet 2 of 6, April 1934, reproduced as part of HAER Record Photographs, for details of concrete reinforcing.
Scale: 1/4 feet
Scale: 1:48 meters
DELINEATED BY: LAURA E. SALARANO, 1989
YELLOWSTONE ROADS AND BRIDGES
HISTORIC AMERICAN ENGINEERING RECORD
NATIONAL PARK SERVICE
UNITED STATES DEPARTMENT OF THE INTERIOR
CRAWFISH CREEK BRIDGE 1936
SPANNING CRAWFISH CREEK ON SOUTH ENTRANCE ROAD
YELLOWSTONE NATIONAL PARK
TETON COUNTY
WYOMING
SHEET 3 OF 3
HISTORIC AMERICAN ENGINEERING RECORD
WY-26
IF REPRODUCED, PLEASE CREDIT: HISTORIC AMERICAN ENGINEERING RECORD, NATIONAL PARK SERVICE, NAME OF DELINEATOR, DATE OF THE DRAWING

GARDNER RIVER BRIDGE · 1939

EAST OF MAMMOTH HOT SPRINGS

YELLOWSTONE NATIONAL PARK

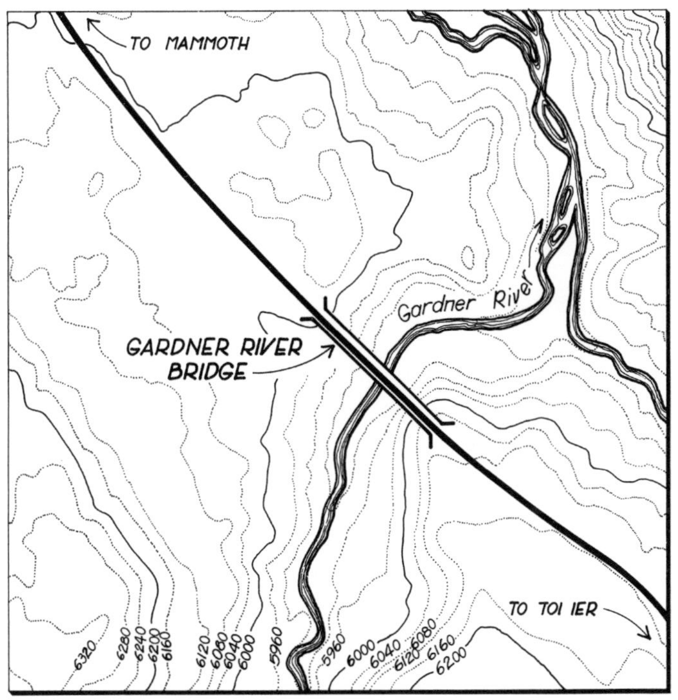

Based on U.S.G.S. 7.5 Minute Series
Mammoth Quadrangle (1986)
U.T.M.: 12.525280.4978260

0 250 500 1000
Scale: 1" = 250'-0" Feet
0 100 200 300
Scale: 1 : 3000 Meters

Based on U.S.G.S. Yellowstone
National Park
Wyoming-Montana-Idaho 1961

The Gardner River Bridge on the road between Mammoth Hot Springs and Tower Junction was completed November 14, 1939. The first location survey was completed in 1930, but the question of whether to build a "high bridge" or choose a lower elevation for the bridge site took the National Park Service several years to decide. By 1937, the "high bridge" site was chosen. The final design was completed during 1937–1938 under the direction of Senior Highway Bridge Engineer of the Western Region headquarter's office of the Public Roads Administration (formerly the Bureau of Public Roads). Mr. Guy H. James of Tulsa, Oklahoma, received the contract for construction on January 5, 1939.

The bridge consists of four 184 foot steel deck truss spans supported on reinforced concrete U-type abutments and 4-column steel towers resting on reinforced concrete pedestals. The bridge has a 24 foot clear concrete roadway with railings of structural steel. The abutments have a solid concrete railing. The overall length of the structure is 942 feet end to end of abutment slabs. The steel portion of the bridge is 805 feet between abutments. From the water line to the roadway grade, the maximum height is 201 feet. The alignment of the bridge is on a tangent on a uniform downgrade from east to west of 2.106%. The reinforcing and structural steel was fabricated by J. B. Klein Iron and Foundry Co. of Oklahoma City, Oklahoma.

Because of the prominent location and the topography of the bridge site, the Gardner River Bridge received considerable attention. As early as 1929, the well-known Landscape Architect Gilmore Clark of New York's Westchester Parks County Commission noted in his "Mammoth Plan" that a high bridge "was suitable from a landscape standpoint.

This recording project is part of the Historic American Engineering Record (HAER) a long range program to document historically significant engineering and industrial works in the United States. The HAER program is an agency of the National Park Service, U.S. Dept. of the Interior. The Yellowstone Roads and Bridges Recording Project was cosponsored by Yellowstone National Park, Robert Barbee, Superintendent; the NPS Roads and Bridges Program, John Gingles, Manager; Rocky Mountain Region, NPS, Lorraine Mintzmyer, Regional Director, and Rodd Wheaton, Division Chief, Cultural Resources; and the Historic American Buildings Survey/Historic American Engineering Record (HABS/HAER) under the general direction of Robert J. Kapsch, Chief.

The fieldwork, measured drawings, historical reports and photography were completed under the direction of Eric DeLony, Chief and Principal Architect, HAER. The recording team consisted of Mary Shivers Culpin, Project Historian, Rocky Mountain Region, NPS; Julie E. Pearson, Supervising Architect, Texas Tech University; Steven M. Varner, Project Engineer, Virginia Polytechnic Institute; Elizabeth A. Harvey, Project Illustrator, Iowa State University; and Architect Technicians Laura E. Salarano, U.S./ICOMOS, Universidad Nacional de Rosario, Argentina, and Gerald J. Hansen, North Dakota State University. Formal on site photography was done by Jet Lowe HAER staff photographer.

DELINEATED BY: ELIZABETH A. HARVEY · LAURA E. SALARANO, 1989

YELLOWSTONE ROADS AND BRIDGES
HISTORIC AMERICAN ENGINEERING RECORD
NATIONAL PARK SERVICE
UNITED STATES DEPARTMENT OF THE INTERIOR

GARDNER RIVER BRIDGE 1939
SPANNING GARDNER RIVER ON NORTH ENTRANCE ROAD
YELLOWSTONE NATIONAL PARK PARK COUNTY WYOMING

SHEET 1 OF 3

HISTORIC AMERICAN ENGINEERING RECORD
WY-7

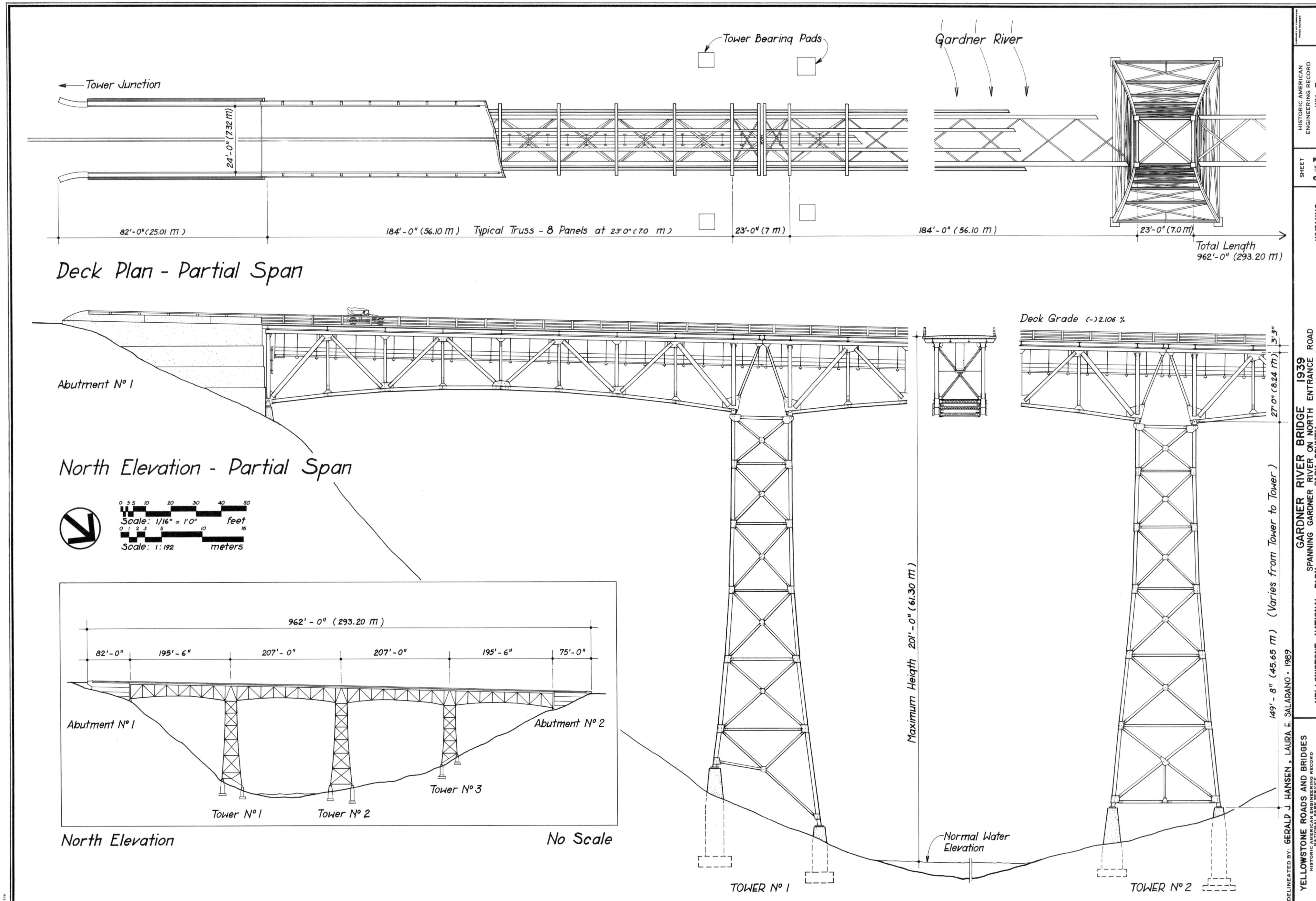
Tower Junction
Tower Bearing Pads
Gardner River
24'-0" (7.32 m)
82'-0" (25.01 m)
184'-0" (56.10 m) Typical Truss - 8 Panels at 23'0" (7.0 m)
23'-0" (7 m)
184'-0" (56.10 m)
23'-0" (7.0 m)
Total Length 962'-0" (293.20 m)
Deck Plan - Partial Span
Deck Grade (-) 2.106 %
3'3"
27'0" (8.24 m)
Abutment N° 1
North Elevation - Partial Span
Scale: 1/16" = 1'0" feet
Scale: 1:192 meters
Maximum Heigth 201'-0" (61.30 m)
149'-8" (45.65 m) (Varies from Tower to Tower)
Normal Water Elevation
TOWER N° 1
TOWER N° 2
962'-0" (293.20 m)
82'-0"
195'-6"
207'-0"
207'-0"
195'-6"
75'-0"
Abutment N° 1
Abutment N° 2
Tower N° 1
Tower N° 2
Tower N° 3
North Elevation
No Scale
DELINEATED BY: GERALD J. HANSEN, LAURA E. SALARANO · 1989
YELLOWSTONE ROADS AND BRIDGES
GARDNER RIVER BRIDGE 1939
SPANNING GARDNER RIVER ON NORTH ENTRANCE ROAD
YELLOWSTONE NATIONAL PARK
PARK COUNTY
WYOMING
SHEET 2 of 3
HISTORIC AMERICAN ENGINEERING RECORD
WY-7
IF REPRODUCED, PLEASE CREDIT: HISTORIC AMERICAN ENGINEERING RECORD, NATIONAL PARK SERVICE, NAME OF DELINEATOR, DATE OF THE DRAWING

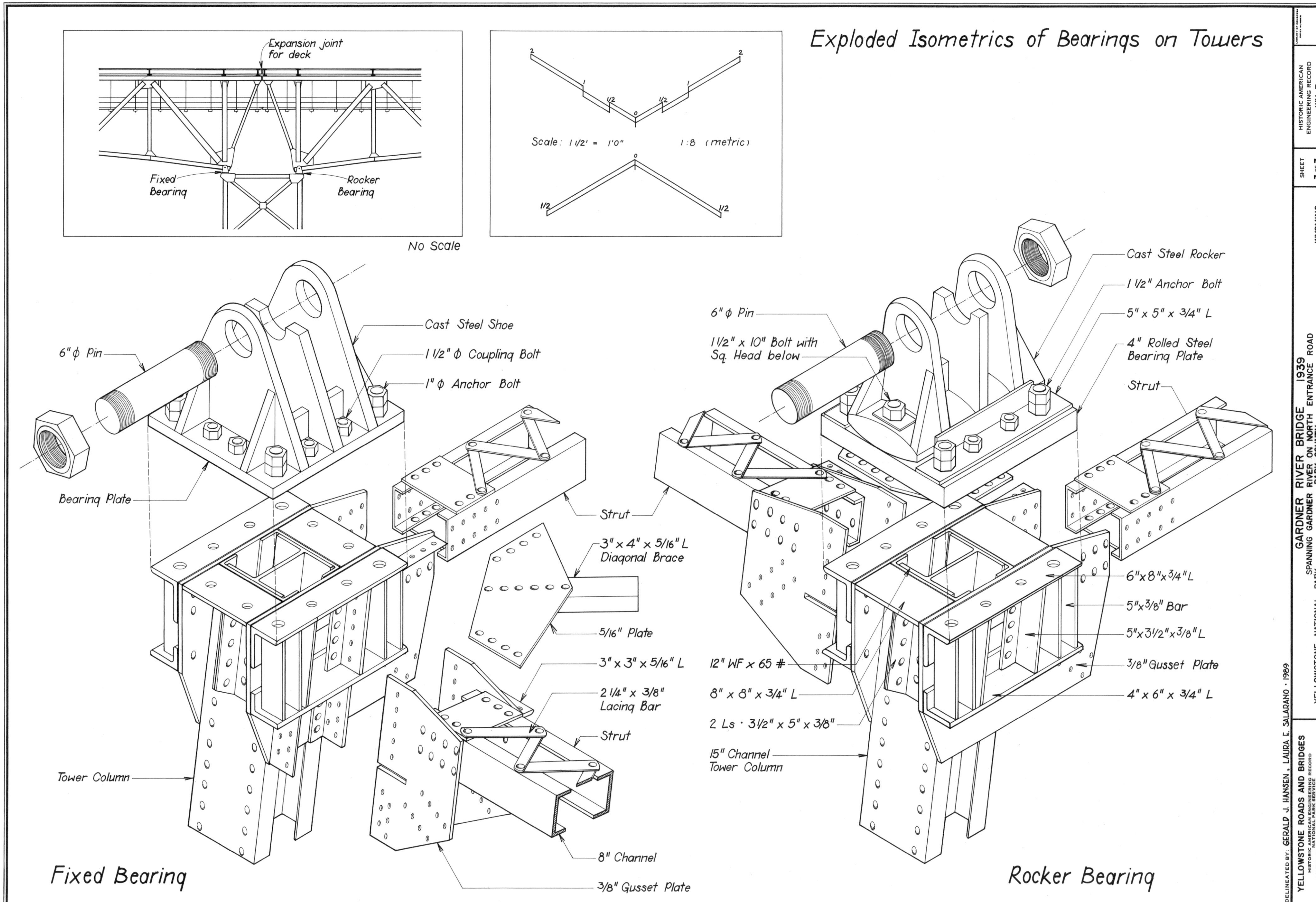

Exploded Isometrics of Bearings on Towers
Expansion joint for deck
Fixed Bearing
Rocker Bearing
No Scale
Scale: 1 1/2" = 1'0"
1:8 (metric)
2
1
1/2
0
Cast Steel Shoe
1 1/2" ϕ Coupling Bolt
1" ϕ Anchor Bolt
6" ϕ Pin
Bearing Plate
Strut
3" x 4" x 5/16" L Diagonal Brace
5/16" Plate
3" x 3" x 5/16" L
2 1/4" x 3/8" Lacing Bar
Strut
Tower Column
8" Channel
3/8" Gusset Plate
Fixed Bearing
Cast Steel Rocker
1 1/2" Anchor Bolt
5" x 5" x 3/4" L
4" Rolled Steel Bearing Plate
Strut
6" ϕ Pin
1 1/2" x 10" Bolt with Sq. Head below
6"x8"x3/4"L
5"x3/8" Bar
5"x3 1/2"x3/8"L
3/8" Gusset Plate
4" x 6" x 3/4" L
12" WF x 65 #
8" x 8" x 3/4" L
2 Ls · 3 1/2" x 5" x 3/8"
15" Channel Tower Column
Rocker Bearing
HISTORIC AMERICAN ENGINEERING RECORD
WY-7
SHEET 3 of 3
GARDNER RIVER BRIDGE 1939
SPANNING GARDNER RIVER ON NORTH ENTRANCE ROAD
YELLOWSTONE NATIONAL PARK
PARK COUNTY
WYOMING
IF REPRODUCED, PLEASE CREDIT: HISTORIC AMERICAN ENGINEERING RECORD, NATIONAL PARK SERVICE, NAME OF DELINEATOR, DATE OF THE DRAWING
DELINEATED BY: GERALD J. HANSEN, LAURA E. SALARANO · 1989
YELLOWSTONE ROADS AND BRIDGES
HISTORIC AMERICAN ENGINEERING RECORD
NATIONAL PARK SERVICE
UNITED STATES DEPARTMENT OF THE INTERIOR

YELLOWSTONE ROADS

A CULTURAL LANDSCAPE

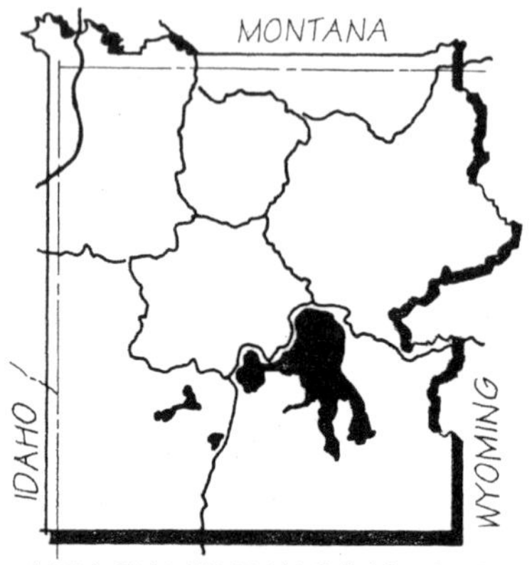

YELLOWSTONE NATIONAL PARK

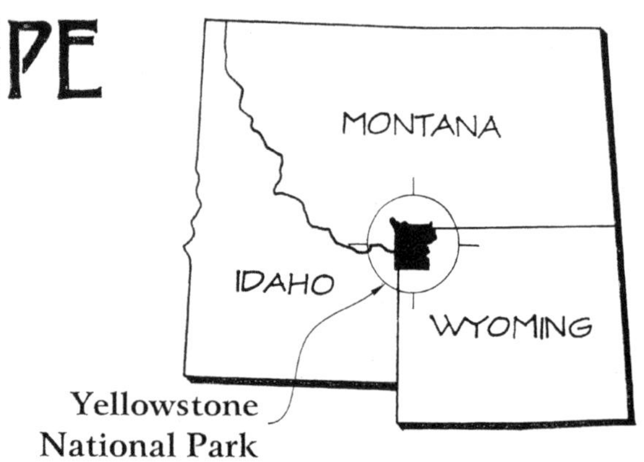

YELLOWSTONE NATIONAL PARK

Roosevelt Arch at Gardiner, Montana, built 1903

In 1872, the United States Congress set aside land in the territories of Wyoming and Montana as Yellowstone National Park, to be retained in its natural condition and made accessible for the enjoyment of the public. The nation's first national park, Yellowstone encompassed a vast range of natural and scenic features, including mountains and valleys, rivers and lakes, canyons and waterfalls, diverse wildlife, and the world's largest concentration of hot springs and geysers.

Providing public access to these natural wonders proved a challenge to early park administrators, who focused on connecting the major features of the Park with a "Grand Loop Road". The park's first superintendents and the officers of the U.S. Army Corps of Engineers, placed in charge of road construction in 1883, developed a park roads philosophy that called for roads to blend harmoniously with the landscape. When the National Park Service assumed responsibility for park administration in 1918, NPS landscape architects involved in road design adopted a similar approach.

The Yellowstone National Park road system continues to provide access to the scenic features the park was created to protect for the enjoyment and for the benefit of future generations.

The Yellowstone Roads and Bridges Recording Project is part of the Historic American Engineering Record (HAER), a long range program to document historically significant engineering and industrial works in the United States. The HAER program is administered by the Historic American Buildings Survey/Historic American Engineering Record Division (HABS/HAER) of the National Park Service, U. S. Department of the Interior. The Yellowstone Roads and Bridges Recording Project was co-sponsored during the summer of 1999 by HABS/HAER under the general direction of E. Blaine Cliver, Chief, by the Federal Lands Highway Program of the U. S. Department of Transportation, by the National Park Service Park Roads and Parkways Program, and by Yellowstone National Park, Michael Finley, Superintendent.

The fieldwork, measured drawings, historical reports, and photography for this recording project were prepared under the direction of Project Leader Todd A. Croteau, HAER Staff Architect, and Tim Davis, HAER Historian. The recording team included Jill Patricia Caouette, field supervisor and architect from the University of New Mexico; Nancy M. McClure, historian from Oklahoma State University; Forrest P. Huisman, architect from Montana State University; and Meredith Mitchell, landscape architect from the University of Manitoba, Canada, through the U. S. ICOMOS program. Formal large-format photography was done by Jet Lowe, HAER Staff Photographer.

DELINEATED BY: Jill Patricia Caouette, 1999, 2000

YELLOWSTONE ROADS RECORDING PROJECT
NATIONAL PARK SERVICE
UNITED STATES DEPARTMENT OF THE INTERIOR

ADDENDUM TO YELLOWSTONE NATIONAL PARK ROADS & BRIDGES
YELLOWSTONE NATIONAL PARK PARK & TETON COUNTIES WYOMING, MONTANA, IDAHO

SHEET 1 OF 10

HISTORIC AMERICAN ENGINEERING RECORD
WY-24

LIBRARY OF CONGRESS INDEX NUMBER

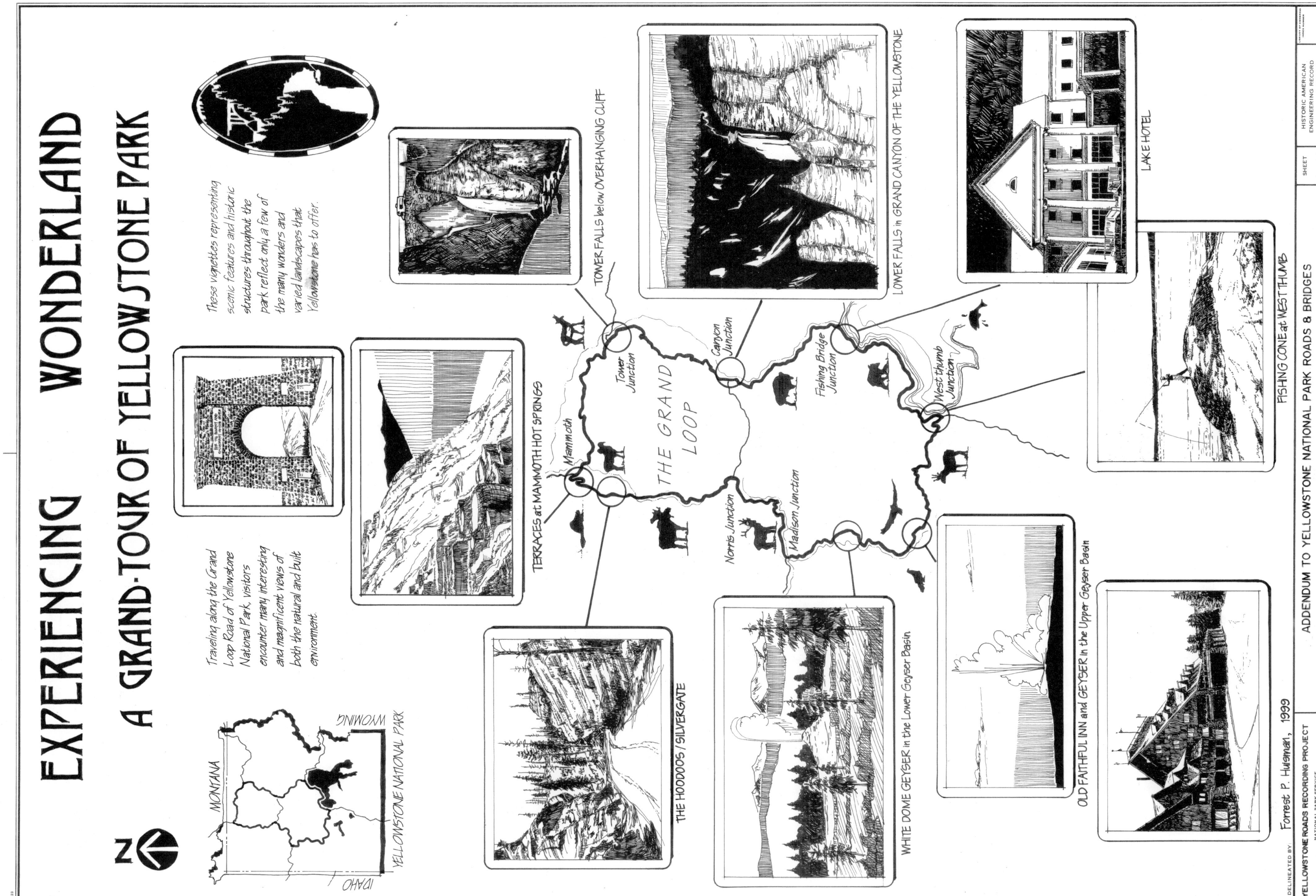
EXPERIENCING WONDERLAND
A GRAND-TOUR OF YELLOWSTONE PARK
These vignettes representing scenic features and historic structures throughout the park reflect only a few of the many wonders and varied landscapes that Yellowstone has to offer.
Traveling along the Grand Loop Road of Yellowstone National Park, visitors encounter many interesting and magnificent views of both the natural and built environment.
MONTANA
WYOMING
IDAHO
YELLOWSTONE NATIONAL PARK
N
THE GRAND LOOP
Mammoth
Tower Junction
Canyon Junction
Fishing Bridge Junction
West thumb Junction
Madison Junction
Norris Junction
TERRACES at MAMMOTH HOT SPRINGS
TOWER FALLS below OVERHANGING CLIFF
LOWER FALLS in GRAND CANYON OF THE YELLOWSTONE
LAKE HOTEL
FISHING CONE at WEST THUMB
OLD FAITHFUL INN and GEYSER in the Upper Geyser Basin
WHITE DOME GEYSER in the Lower Geyser Basin
THE HOODOOS / SILVERGATE
DELINEATED BY
Forrest P. Hulsman, 1999
YELLOWSTONE ROADS RECORDING PROJECT
NATIONAL PARK SERVICE
UNITED STATES DEPARTMENT OF THE INTERIOR
ADDENDUM TO YELLOWSTONE NATIONAL PARK ROADS & BRIDGES
YELLOWSTONE NATIONAL PARK
PARK & TETON COUNTIES
WYOMING, MONTANA, IDAHO
SHEET 2 OF 10
HISTORIC AMERICAN ENGINEERING RECORD
WY-24
IF REPRODUCED, PLEASE CREDIT: HISTORIC AMERICAN ENGINEERING RECORD, NATIONAL PARK SERVICE, NAME OF DELINEATOR, DATE OF THE DRAWING

EVOLUTION OF YELLOWSTONE NATIONAL PARK ROADWAYS

LEGEND

- Abandoned Route
- Existing Road
- New Road
- Scenic Route or Bypass
- ◆ Entrance Station
- ● Major Attraction
- ○ Town
- Wagon Trail

Roadways in Yellowstone National Park are designed to connect major attractions and ease visitors' travels to them. The current road system grew from early wagon trails that followed river valleys and lakeshores. By 1880 some trails had been improved, linking attractions in the northwest portion of the park. By 1890 the U.S. Army Corps of Engineers had built new roads connecting the east and west sides of the park. During the first 30 years of development, the Yellowstone road system was in constant change, but by 1905 the system of roads connecting attractions, hotels and entrance roads–styled the "Grand Loop"–had crystallized. Although some small sections have been abandoned or transformed into scenic routes, the Grand Loop remains in 2000 much as it was in 1905.

NOTE: This drawing is based on historic maps of Yellowstone National Park located in the Library of Congress, especially: "Map of the Yellowstone National Park" (1881); "Yellowstone National Park. Northern Pacific Railroad" (1895)"; & "Northern Pacific, Yellowstone National Park" (1910). For more information, see HAER Field Notebooks.

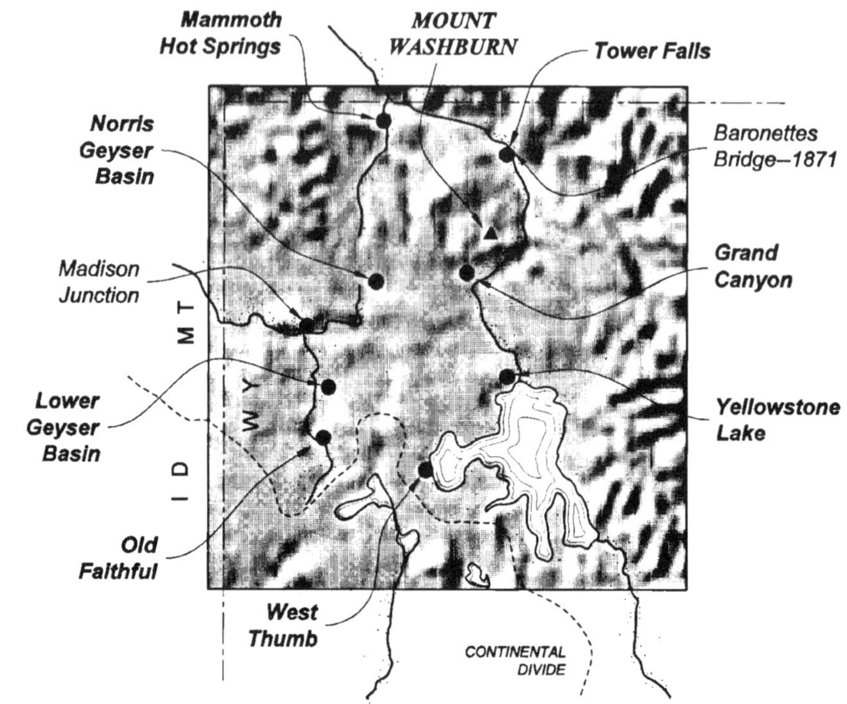

1872 - Park Founded

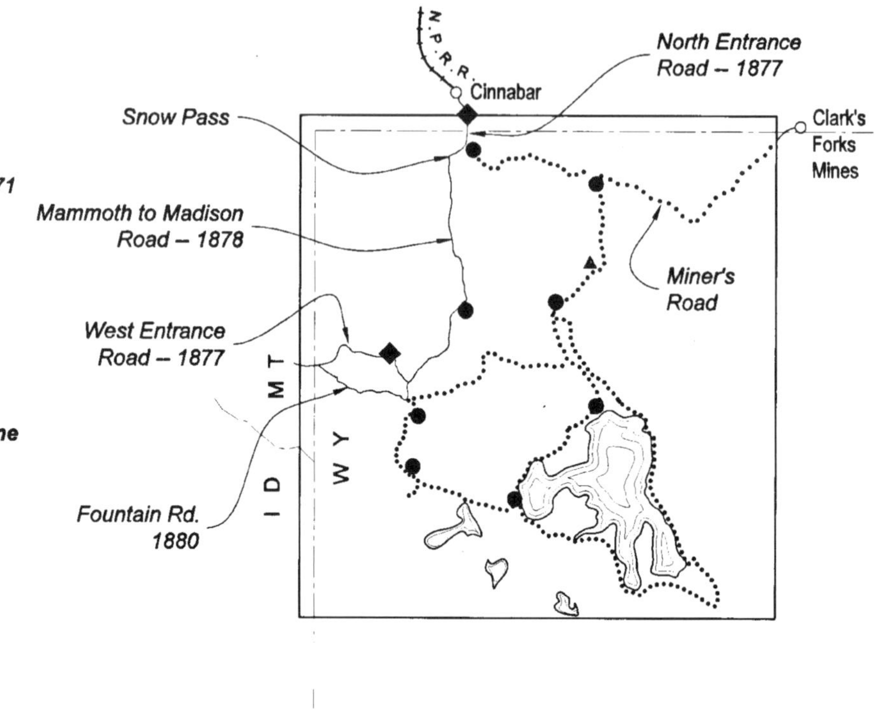

1872 - 1880

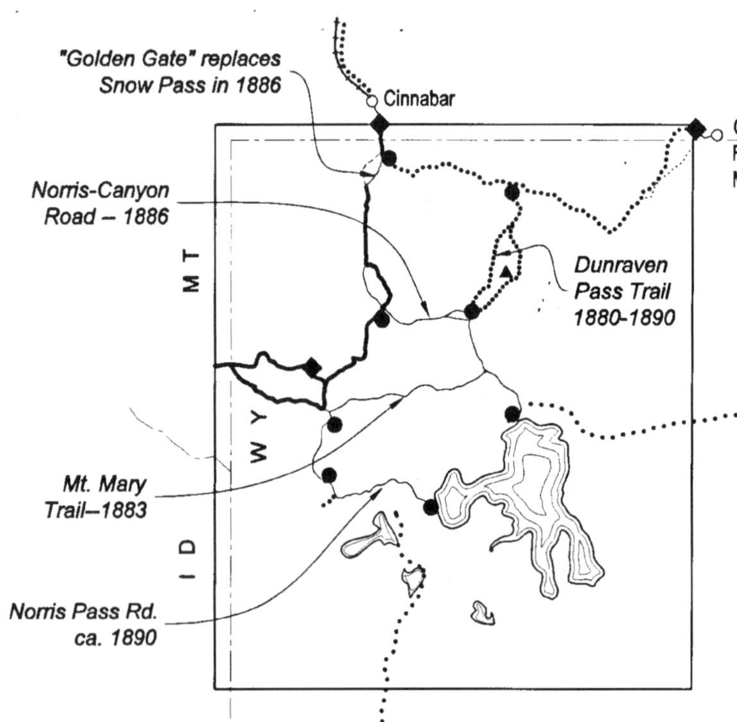

1881 - 1890

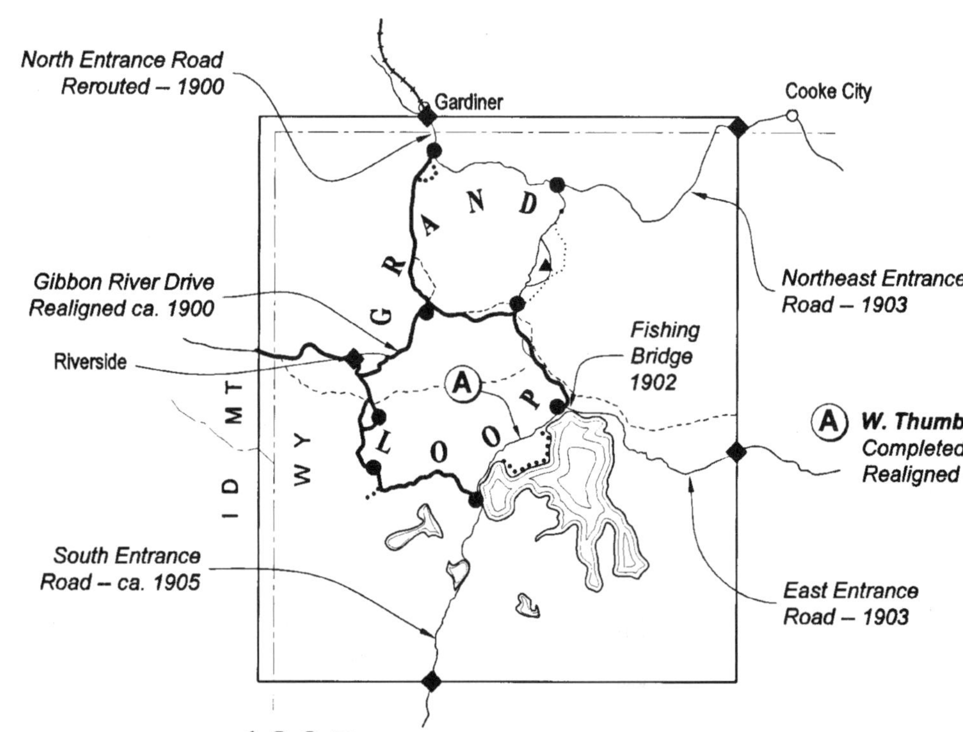

1905 - Grand Loop Completed

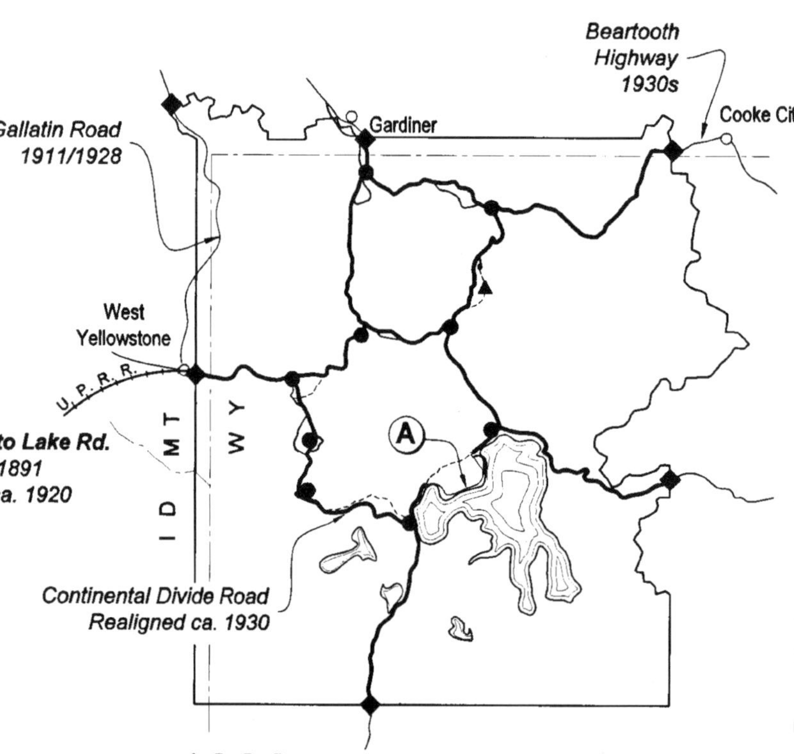

1999 - Paved Roadways

DELINEATED BY: Forrest P. Huisman, 1999

YELLOWSTONE ROADS RECORDING PROJECT

ADDENDUM TO YELLOWSTONE NATIONAL PARK ROADS & BRIDGES

YELLOWSTONE NATIONAL PARK — PARK & TETON COUNTIES — WYOMING, MONTANA, IDAHO

HISTORIC AMERICAN ENGINEERING RECORD WY-24

SHEET 3 OF 10

ROADWAY CONSTRUCTION

METHODS AND EVOLUTION

Archaeological evidence indicates that, throughout history, travelers used similar corridors through what became Yellowstone National Park. For centuries, American Indians followed game trails to hunting grounds and other destinations. These same corridors were used by fur trappers, and later by official explorations that led to the creation of the first national park in 1872. Early park superintendents struggled to improve trails and construct roads with limited funding. As visitation increased, more effort was devoted to improving visitor access. In 1883, the U. S. Army Corps of Engineers took over responsibility for road construction. Under the Corps, road standards and the basic tourist route through the park took shape. The narrow roads accommodated stage coach traffic and carried visitors between the major park attractions. By the turn of the twentieth century, water tanks stationed along the roadway allowed sprinkling of the road surface to keep down the dust. By 1905, the Corps had completed the "Grand Loop," the basic route still traveled by visitors today.

Following completion of the road system, engineers concentrated on improving existing roads by widening, straightening, and macadamizing the surface. After the National Park Service took over the administration of Yellowstone in 1918, crews continued work to control the dust by oiling and upgrading roads for automobile traffic. Paving with a bituminous, or asphalt surface began in the 1920s. Minor reroutes and further widening continued throughout the twentieth century. In 1999, Yellowstone was in the middle of a twenty year road rehabilitation project to bring the roads up to modern standards. Wider traffic lanes, broader shoulders, and expansive recovery zones have in many areas substantially altered the visitors' experience.

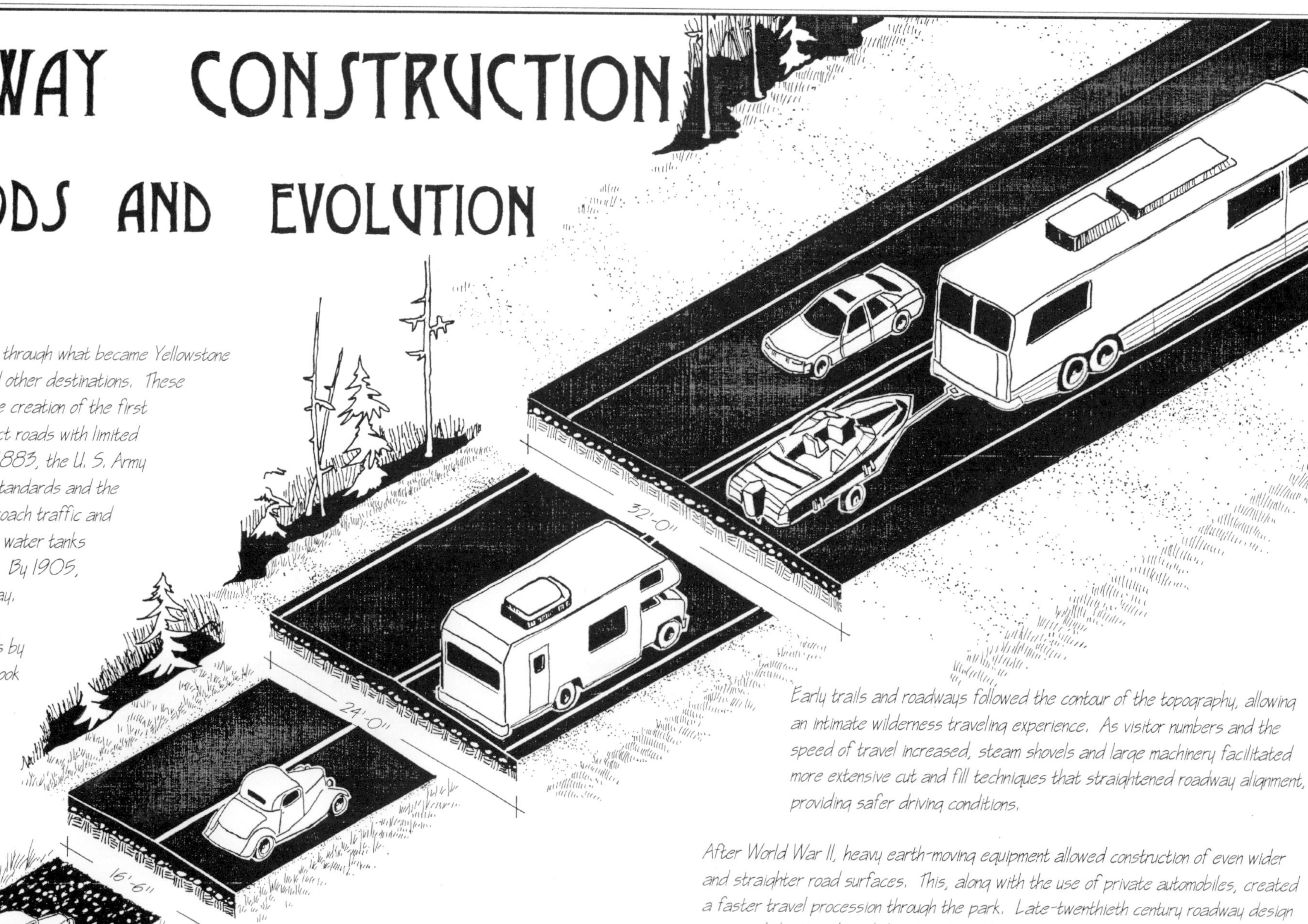

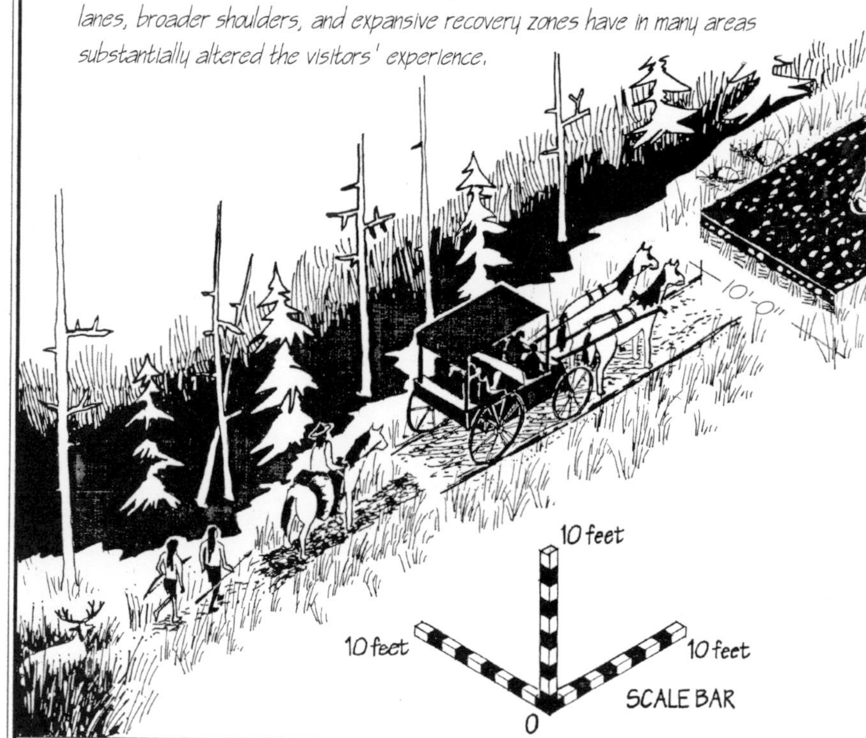

Early trails and roadways followed the contour of the topography, allowing an intimate wilderness traveling experience. As visitor numbers and the speed of travel increased, steam shovels and large machinery facilitated more extensive cut and fill techniques that straightened roadway alignment, providing safer driving conditions.

After World War II, heavy earth-moving equipment allowed construction of even wider and straighter road surfaces. This, along with the use of private automobiles, created a faster travel procession through the park. Late-twentieth century roadway design continued this trend, and the intimate wilderness experience that greeted earlier visitors to Yellowstone has been diminished for travelers through the park.

EARLY ROADS

1930s & 1940s

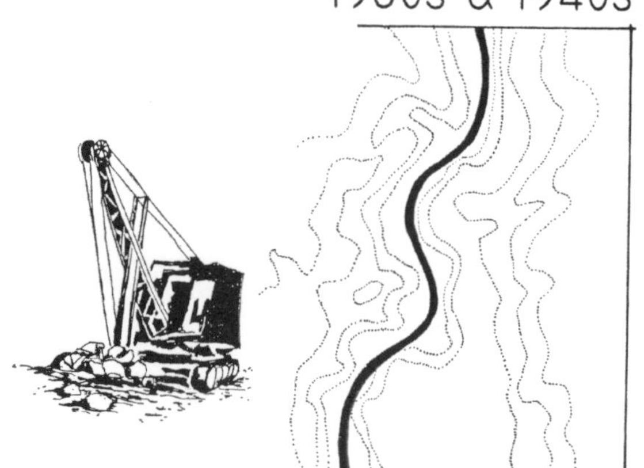

1950s & PRESENT

DELINEATED BY: Forrest P. Huisman, 1999

YELLOWSTONE ROADS RECORDING PROJECT
NATIONAL PARK SERVICE
UNITED STATES DEPARTMENT OF THE INTERIOR

ADDENDUM TO YELLOWSTONE NATIONAL PARK ROADS & BRIDGES

YELLOWSTONE NATIONAL PARK | PARK & TETON COUNTIES | WYOMING, MONTANA, IDAHO

SHEET 5 OF 10

HISTORIC AMERICAN ENGINEERING RECORD WY-24

LIBRARY OF CONGRESS INDEX NUMBER

ENTRANCE CHECKING STATIONS

NORTH ENTRANCE STATION photo ca. 1923

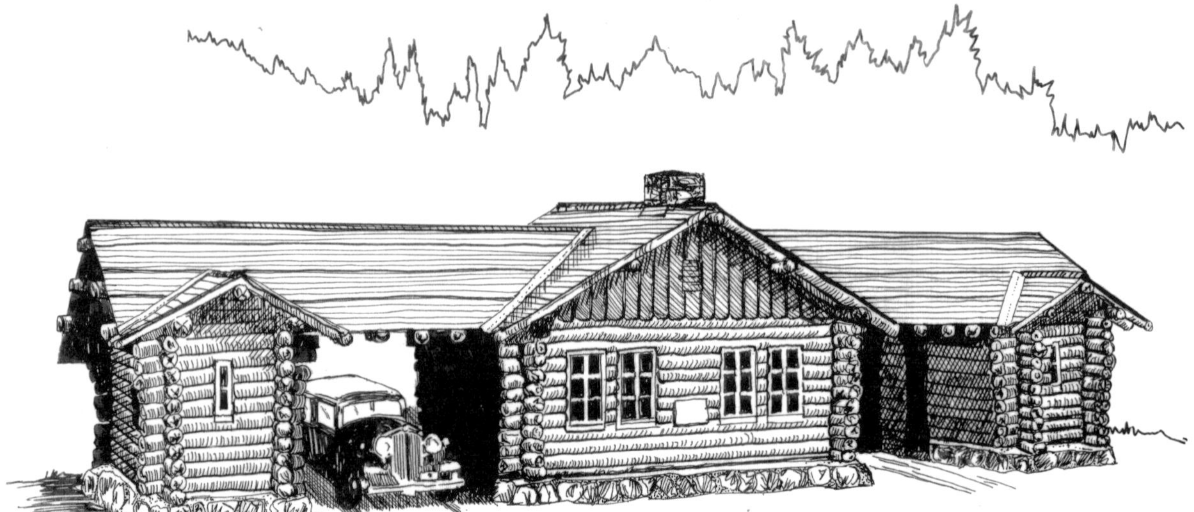

NORTHEAST ENTRANCE STATION constructed and photo 1935

WEST ENTRANCE STATION photo ca. 1924

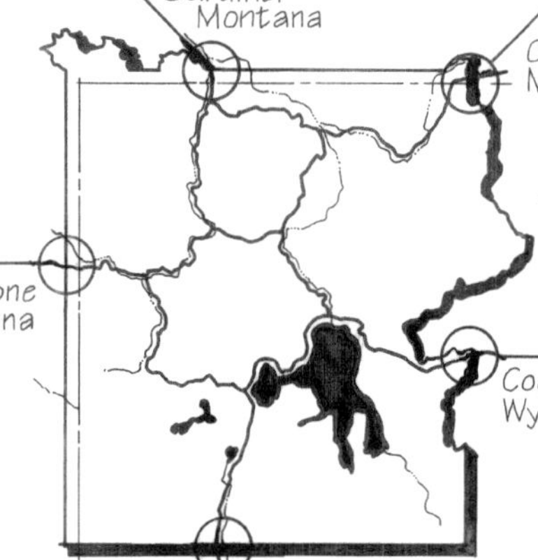

EAST ENTRANCE STATION constructed 1966 photo ca. 1992

Yellowstone National Park has five entrance roads that provide access to the park from Montana and Wyoming. The North, Northeast, East, South, and West entrances have each had a succession of structures that served as checking stations. Shown are stations from various eras of the park's history. The Northeast Entrance Station, constructed in 1935, was declared a National Historic Landmark in 1987, in recognition of its national significance as an outstanding example of the rustic design style employed by the National Park Service.

Developed during the Mission 66 Program, the East Entrance Station combines a park aesthetic of modern forms and materials. Function serves a greater role in its design with attendants booth set into the ground for ergonmic considerations.

SOUTH ENTRANCE STATION photo ca. 1935

Besides providing a physical "gate" to the park, and a place to collect entrance fees and orient visitors, the entrance station serves a less tangible purpose. Through its architecture, the entrance station signals that the visitor has stepped into unique territory. Its design offers a glimpse of what is to come as travelers cross into our national parks.

"The building was not only the physical boundary, but the psychological boundary between the rest of the world and what was set aside as a permanently wild place."
–Laura Soulliere Harrison, 1986

Drawing based on historical photographs.

DELINEATED BY: Jill Patricia Caouette, 1999

YELLOWSTONE ROADS RECORDING PROJECT
NATIONAL PARK SERVICE
UNITED STATES DEPARTMENT OF THE INTERIOR

ADDENDUM TO YELLOWSTONE NATIONAL PARK ROADS & BRIDGES
YELLOWSTONE NATIONAL PARK — PARK & TETON COUNTIES — WYOMING, MONTANA, IDAHO

SHEET 6 OF 10

HISTORIC AMERICAN ENGINEERING RECORD
WY-24

LIBRARY OF CONGRESS INDEX NUMBER

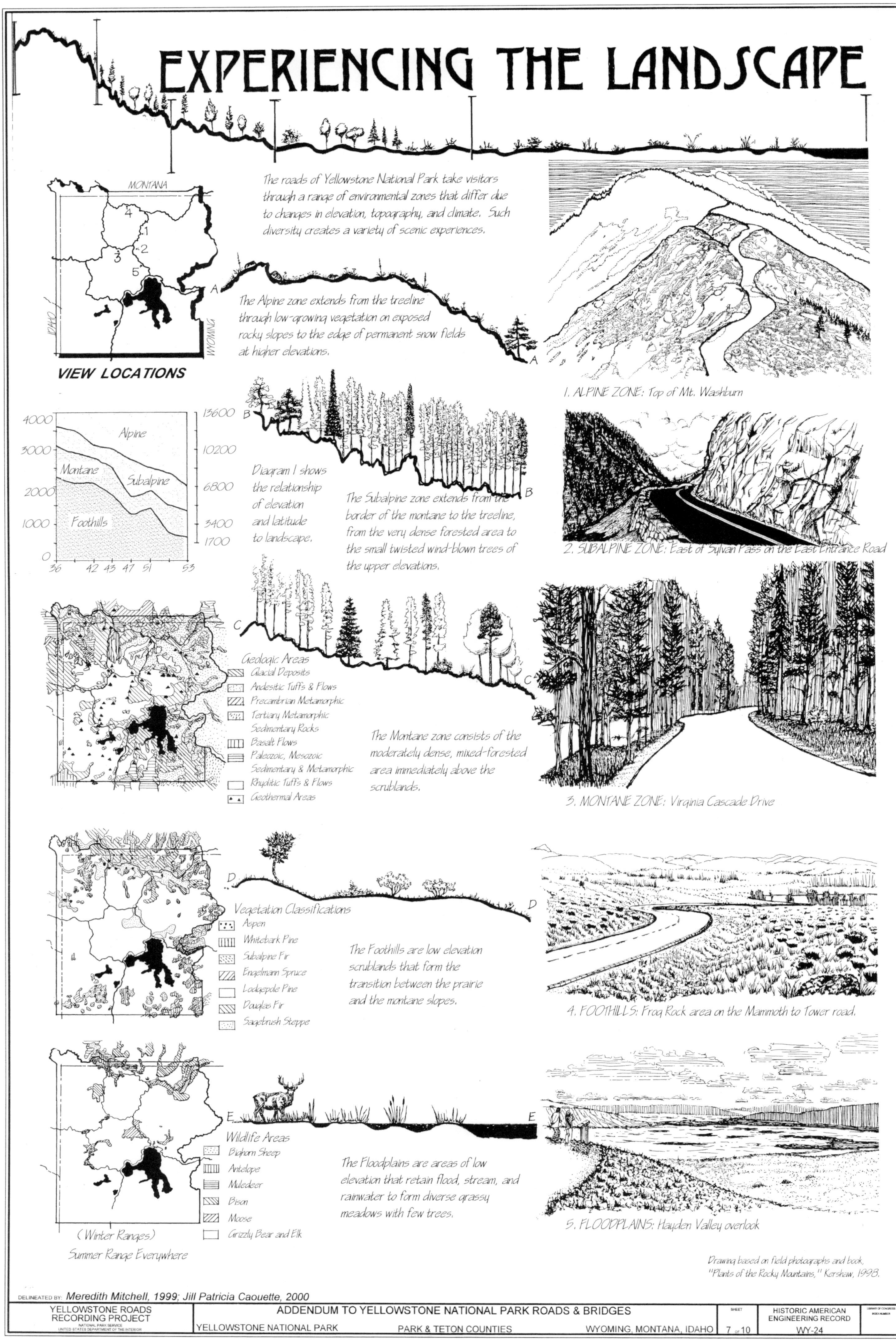
EXPERIENCING THE LANDSCAPE
The roads of Yellowstone National Park take visitors through a range of environmental zones that differ due to changes in elevation, topography, and climate. Such diversity creates a variety of scenic experiences.
MONTANA
IDAHO
WYOMING
VIEW LOCATIONS
The Alpine zone extends from the treeline through low-growing vegetation on exposed rocky slopes to the edge of permanent snow fields at higher elevations.
1. ALPINE ZONE: Top of Mt. Washburn
Alpine
Montane
Subalpine
Foothills
Diagram 1 shows the relationship of elevation and latitude to landscape.
The Subalpine zone extends from the border of the montane to the treeline, from the very dense forested area to the small twisted wind-blown trees of the upper elevations.
2. SUBALPINE ZONE: East of Sylvan Pass on the East Entrance Road
Geologic Areas
Glacial Deposits
Andesitic Tuffs & Flows
Precambrian Metamorphic
Tertiary Metamorphic Sedimentary Rocks
Basalt Flows
Paleozoic, Mesozoic Sedimentary & Metamorphic
Rhyolitic Tuffs & Flows
Geothermal Areas
The Montane zone consists of the moderately dense, mixed-forested area immediately above the scrublands.
3. MONTANE ZONE: Virginia Cascade Drive
Vegetation Classifications
Aspen
Whitebark Pine
Subalpine Fir
Engelmann Spruce
Lodgepole Pine
Douglas Fir
Sagebrush Steppe
The Foothills are low elevation scrublands that form the transition between the prairie and the montane slopes.
4. FOOTHILLS: Frog Rock area on the Mammoth to Tower road.
Wildlife Areas
Bighorn Sheep
Antelope
Muledeer
Bison
Moose
Grizzly Bear and Elk
(Winter Ranges)
Summer Range Everywhere
The Floodplains are areas of low elevation that retain flood, stream, and rainwater to form diverse grassy meadows with few trees.
5. FLOODPLAINS: Hayden Valley overlook
Drawing based on field photographs and book, "Plants of the Rocky Mountains," Kershaw, 1998.
DELINEATED BY: Meredith Mitchell, 1999; Jill Patricia Caouette, 2000
YELLOWSTONE ROADS RECORDING PROJECT
NATIONAL PARK SERVICE
UNITED STATES DEPARTMENT OF THE INTERIOR
ADDENDUM TO YELLOWSTONE NATIONAL PARK ROADS & BRIDGES
YELLOWSTONE NATIONAL PARK
PARK & TETON COUNTIES
WYOMING, MONTANA, IDAHO
SHEET 7 of 10
HISTORIC AMERICAN ENGINEERING RECORD
WY-24

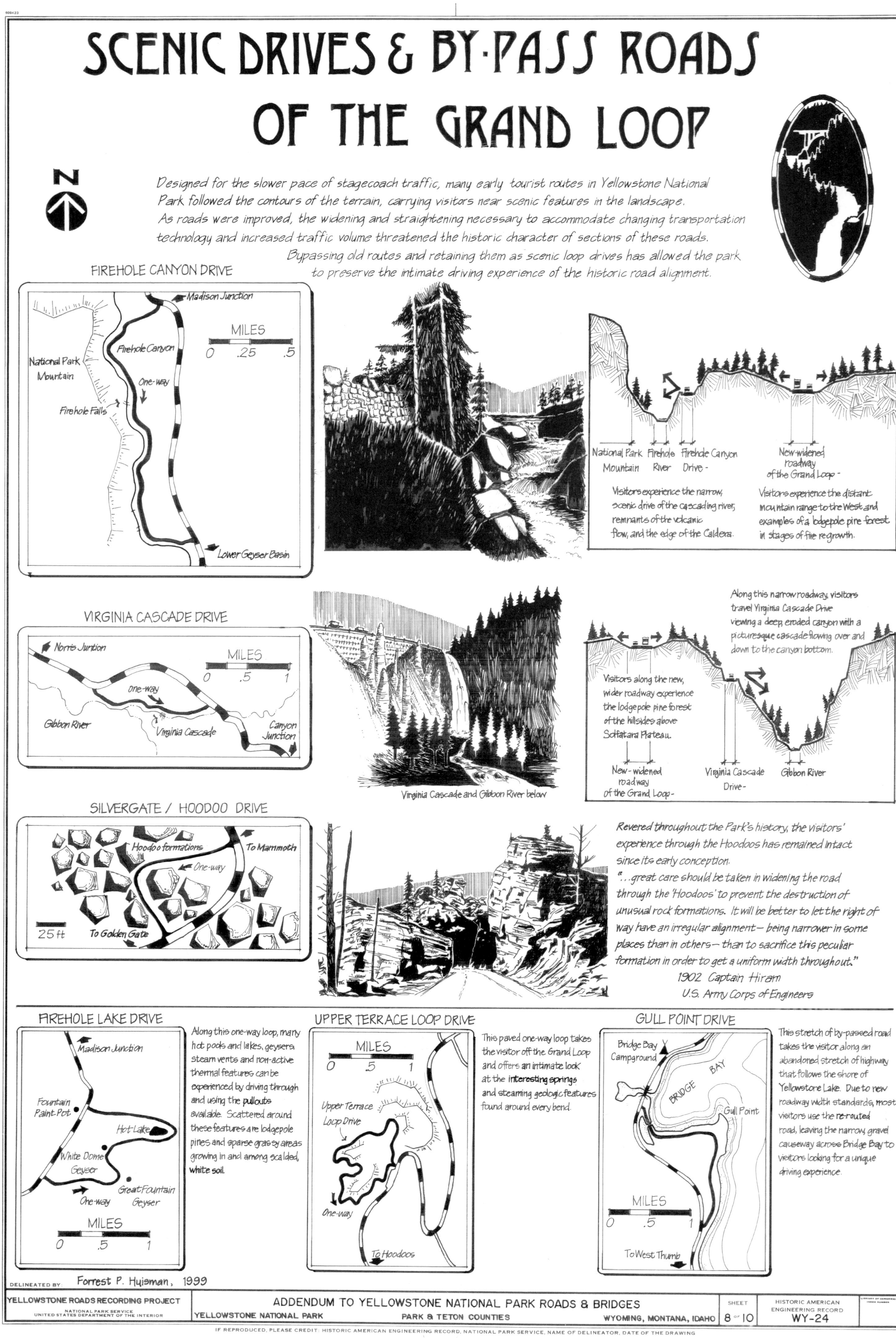
SCENIC DRIVES & BY-PASS ROADS OF THE GRAND LOOP
N
Designed for the slower pace of stagecoach traffic, many early tourist routes in Yellowstone National Park followed the contours of the terrain, carrying visitors near scenic features in the landscape.
As roads were improved, the widening and straightening necessary to accommodate changing transportation technology and increased traffic volume threatened the historic character of sections of these roads.
Bypassing old routes and retaining them as scenic loop drives has allowed the park to preserve the intimate driving experience of the historic road alignment.
FIREHOLE CANYON DRIVE
Madison Junction
MILES
0 .25 .5
National Park Mountain
Firehole Canyon
One-way
Firehole Falls
Lower Geyser Basin
National Park Mountain
Firehole River
Firehole Canyon Drive -
New-widened roadway of the Grand Loop -
Visitors experience the narrow, scenic drive of the cascading river, remnants of the volcanic flow, and the edge of the Caldera.
Visitors experience the distant mountain range to the West and examples of a lodgepole pine forest in stages of fire regrowth.
VIRGINIA CASCADE DRIVE
Norris Juntion
MILES
0 .5 1
one-way
Gibbon River
Virginia Cascade
Canyon Junction
Virginia Cascade and Gibbon River below
Along this narrow roadway, visitors travel Virginia Cascade Drive viewing a deep, eroded canyon with a picturesque cascade flowing over and down to the canyon bottom.
Visitors along the new, wider roadway experience the lodgepole pine forest of the hillsides above Solfatara Plateau.
New-widened roadway of the Grand Loop -
Virginia Cascade Drive -
Gibbon River
SILVERGATE / HOODOO DRIVE
Hoodoo formations
To Mammoth
One-way
25 ft
To Golden Gate
Revered throughout the Park's history, the visitors' experience through the Hoodoos has remained intact since its early conception.
"...great care should be taken in widening the road through the 'Hoodoos' to prevent the destruction of unusual rock formations. It will be better to let the right of way have an irregular alignment— being narrower in some places than in others— than to sacrifice this peculiar formation in order to get a uniform width throughout."
1902 Captain Hiram
U.S. Army Corps of Engineers
FIREHOLE LAKE DRIVE
Madison Junction
Fountain Paint Pot
Hot Lake
White Dome Geyser
Great Fountain Geyser
One-way
MILES
0 .5 1
Along this one-way loop, many hot pools and lakes, geysers, steam vents and non-active thermal features can be experienced by driving through and using the pullouts available. Scattered around these features are lodgepole pines and sparse grassy areas growing in and among scalded, white soil.
UPPER TERRACE LOOP DRIVE
MILES
0 .5 1
Upper Terrace Loop Drive
One-way
To Hoodoos
This paved one-way loop takes the visitor off the Grand Loop and offers an intimate look at the interesting springs and steaming geologic features found around every bend.
GULL POINT DRIVE
Bridge Bay Campground
BRIDGE BAY
Gull Point
MILES
0 .5 1
To West Thumb
This stretch of by-passed road takes the visitor along an abandoned stretch of highway that follows the shore of Yellowstone Lake. Due to new roadway width standards, most visitors use the re-routed road, leaving the narrow, gravel causeway across Bridge Bay to visitors looking for a unique driving experience.
DELINEATED BY: Forrest P. Huisman, 1999
YELLOWSTONE ROADS RECORDING PROJECT
NATIONAL PARK SERVICE
UNITED STATES DEPARTMENT OF THE INTERIOR
ADDENDUM TO YELLOWSTONE NATIONAL PARK ROADS & BRIDGES
YELLOWSTONE NATIONAL PARK
PARK & TETON COUNTIES
WYOMING, MONTANA, IDAHO
SHEET 8 OF 10
HISTORIC AMERICAN ENGINEERING RECORD
WY-24
IF REPRODUCED, PLEASE CREDIT: HISTORIC AMERICAN ENGINEERING RECORD, NATIONAL PARK SERVICE, NAME OF DELINEATOR, DATE OF THE DRAWING

WAYSIDES AND PULLOUTS

Yellowstone National Park roads offer visitors many opportunities to stop along the way to enjoy the park. Waysides and pullouts provide visitor facilities, informational and interpretive signage, and access to nearby natural, historic, and scenic features.

"Signs are miniature symbols that National Parks can use to show off their individuality as well as have a specific purpose. They can by directional, designative, regulatory, cautionary, or informative."
-Albert Good, Park and Recreation Structures

Plans and illustrations based on 1933 YNP Master Plans. Specifications from Design Standards of YNP Landscape Architects Divison.

Moose-bog Wayside, photo ca. 1950

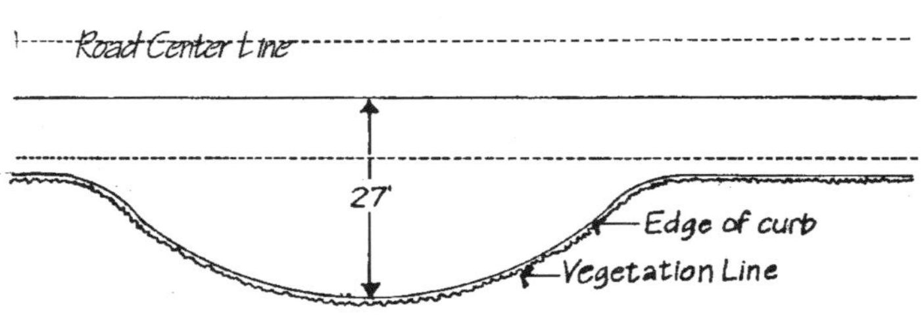

General Vehicular Pullout
Scale 3¼"= 50'

Drive-by Pullout

- 2-5 minute stop
- No seating
- Usually no designed views
- New pullouts generally paved
- No curbing
- Room for 1-2 cars

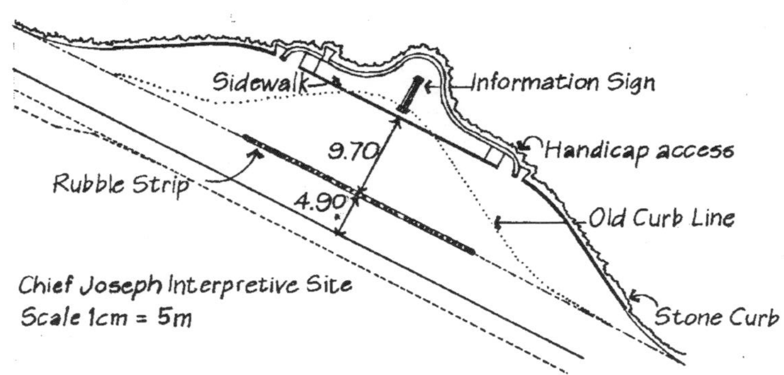

Chief Joseph Interpretive Site
Scale 1cm = 5m

Information Pullout

- 5-8 minute stop
- No seating
- Information sign about specific view or subject to site
- Curbing
- Room for 2-3 cars

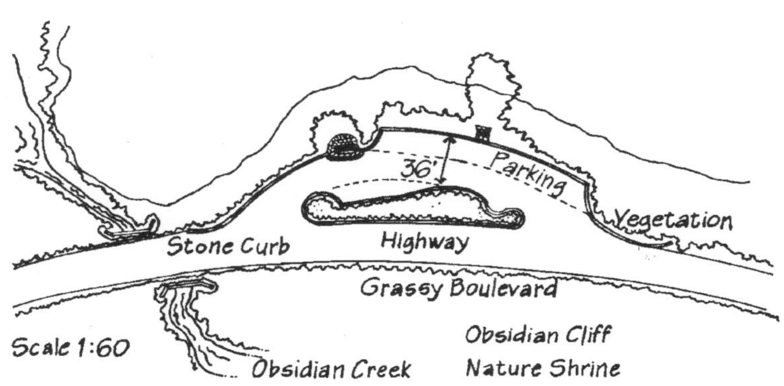

Scale 1:60

Interpretative Wayside

- 10-15 minute stop
- No or some seating
- Directed views or vistas
- Larger information sign, nature shrine or kiosk
- Landscaped or naturalized

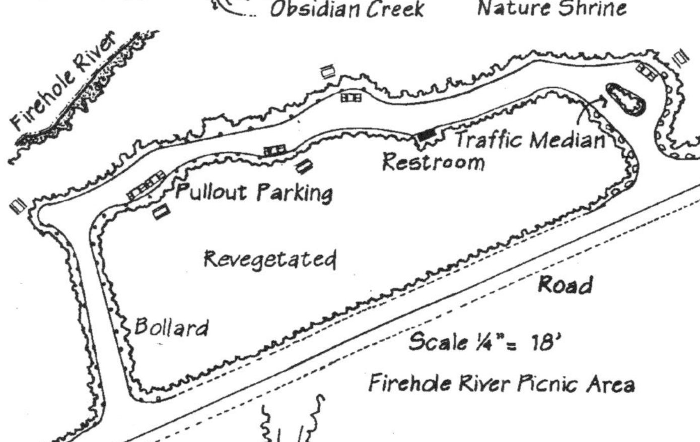

Scale ¼"= 18'
Firehole River Picnic Area

Picnic Wayside

- 15 minute to 1 hour stop
- Seating, restrooms, facilities
- Picnic tables
- Notable view or vista
- Information sign
- Limited activity areas
- Landscaped or naturalized
- One way
- Paved and lined parking

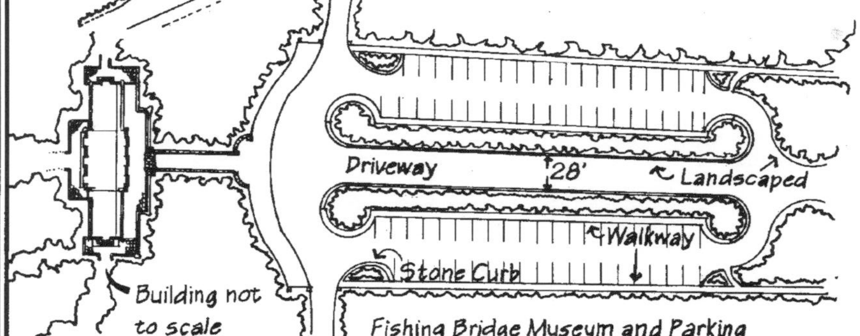

Fishing Bridge Museum and Parking

Interactive Wayside

- More than one hour stop
- Seating and check-in areas
- Amenities and facilities
- Specific directed view and destinations
- Large kiosk, structure or museum
- Extended activities and trailheads
- Landscaped and/ or naturalized
- Paved parking

Fishing Bridge Museum

DELINEATED BY: Meredith Mitchell, 1999; Jill Patricia Caouette, 2000

YELLOWSTONE ROADS RECORDING PROJECT NATIONAL PARK SERVICE UNITED STATES DEPARTMENT OF THE INTERIOR	ADDENDUM TO YELLOWSTONE NATIONAL PARK ROADS & BRIDGES YELLOWSTONE NATIONAL PARK — PARK & TETON COUNTIES — WYOMING, MONTANA, IDAHO	SHEET 9 of 10	HISTORIC AMERICAN ENGINEERING RECORD WY-24

IF REPRODUCED, PLEASE CREDIT THE HISTORIC AMERICAN ENGINEERING RECORD, NATIONAL PARK SERVICE, NAME OF DELINEATOR, DATE OF DRAWING

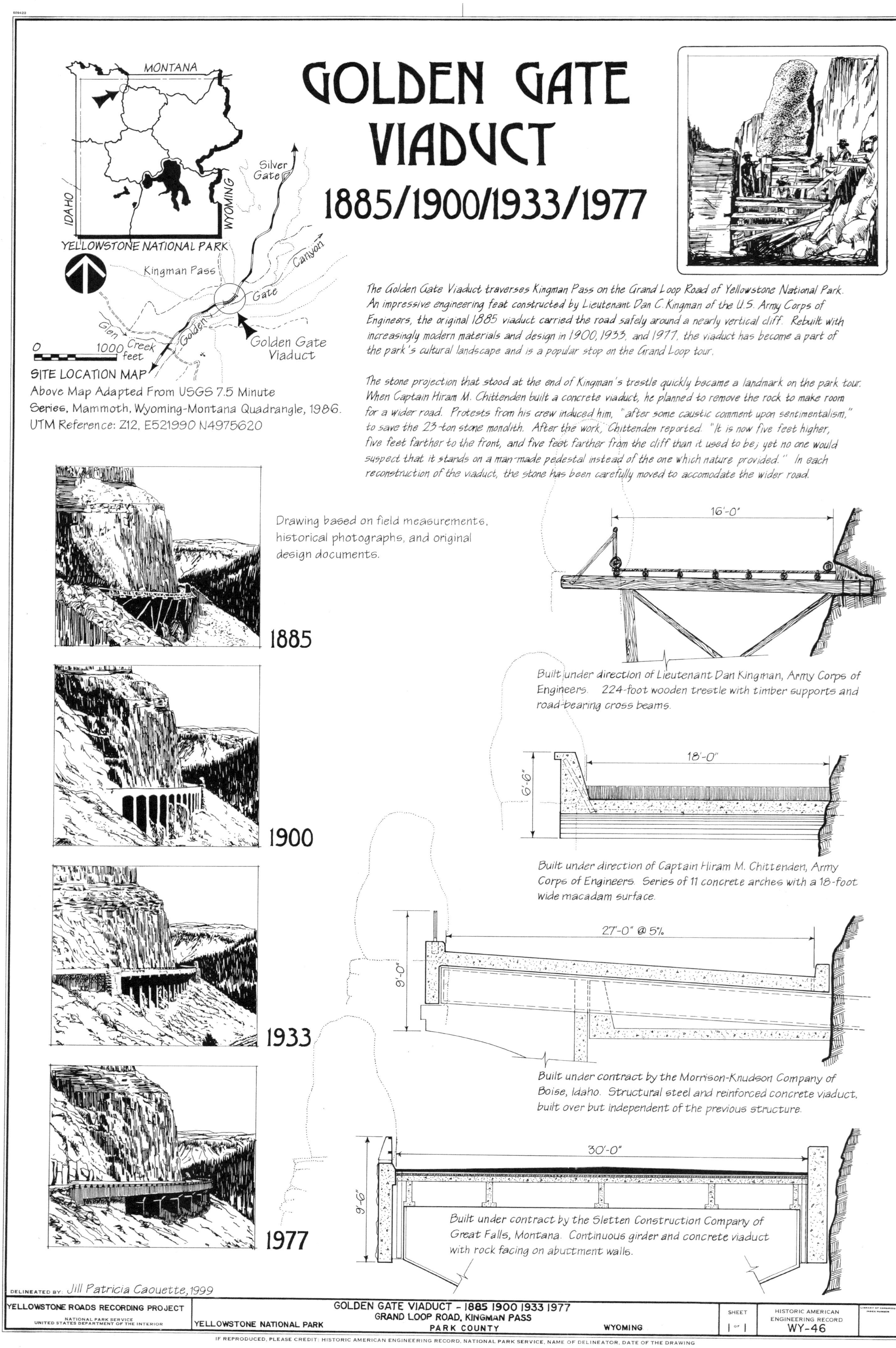

GOLDEN GATE VIADUCT
1885/1900/1933/1977
MONTANA
IDAHO
WYOMING
YELLOWSTONE NATIONAL PARK
Silver Gate
Kingman Pass
Canyon
Gate
Glen Creek
Golden
Golden Gate Viaduct
0 1000 feet
SITE LOCATION MAP
Above Map Adapted From USGS 7.5 Minute Series, Mammoth, Wyoming-Montana Quadrangle, 1986.
UTM Reference: Z12, E521990 N4975620
The Golden Gate Viaduct traverses Kingman Pass on the Grand Loop Road of Yellowstone National Park. An impressive engineering feat constructed by Lieutenant Dan C. Kingman of the U.S. Army Corps of Engineers, the original 1885 viaduct carried the road safely around a nearly vertical cliff. Rebuilt with increasingly modern materials and design in 1900, 1933, and 1977, the viaduct has become a part of the park's cultural landscape and is a popular stop on the Grand Loop tour.
The stone projection that stood at the end of Kingman's trestle quickly became a landmark on the park tour. When Captain Hiram M. Chittenden built a concrete viaduct, he planned to remove the rock to make room for a wider road. Protests from his crew induced him, "after some caustic comment upon sentimentalism," to save the 23-ton stone monolith. After the work, Chittenden reported: "It is now five feet higher, five feet farther to the front, and five feet farther from the cliff than it used to be; yet no one would suspect that it stands on a man-made pedestal instead of the one which nature provided." In each reconstruction of the viaduct, the stone has been carefully moved to accomodate the wider road.
Drawing based on field measurements, historical photographs, and original design documents.
1885
16'-0"
Built under direction of Lieutenant Dan Kingman, Army Corps of Engineers. 224-foot wooden trestle with timber supports and road-bearing cross beams.
1900
18'-0"
6'-6"
Built under direction of Captain Hiram M. Chittenden, Army Corps of Engineers. Series of 11 concrete arches with a 18-foot wide macadam surface.
1933
27'-0" @ 5%
9'-0"
Built under contract by the Morrison-Knudson Company of Boise, Idaho. Structural steel and reinforced concrete viaduct, built over but independent of the previous structure.
1977
30'-0"
9'-6"
Built under contract by the Sletten Construction Company of Great Falls, Montana. Continuous girder and concrete viaduct with rock facing on abutment walls.
DELINEATED BY: Jill Patricia Caouette, 1999
YELLOWSTONE ROADS RECORDING PROJECT
NATIONAL PARK SERVICE
UNITED STATES DEPARTMENT OF THE INTERIOR
GOLDEN GATE VIADUCT - 1885 1900 1933 1977
GRAND LOOP ROAD, KINGMAN PASS
YELLOWSTONE NATIONAL PARK
PARK COUNTY
WYOMING
SHEET 1 of 1
HISTORIC AMERICAN ENGINEERING RECORD
WY-46
IF REPRODUCED, PLEASE CREDIT: HISTORIC AMERICAN ENGINEERING RECORD, NATIONAL PARK SERVICE, NAME OF DELINEATOR, DATE OF THE DRAWING

CORKSCREW BRIDGE 1904/1919

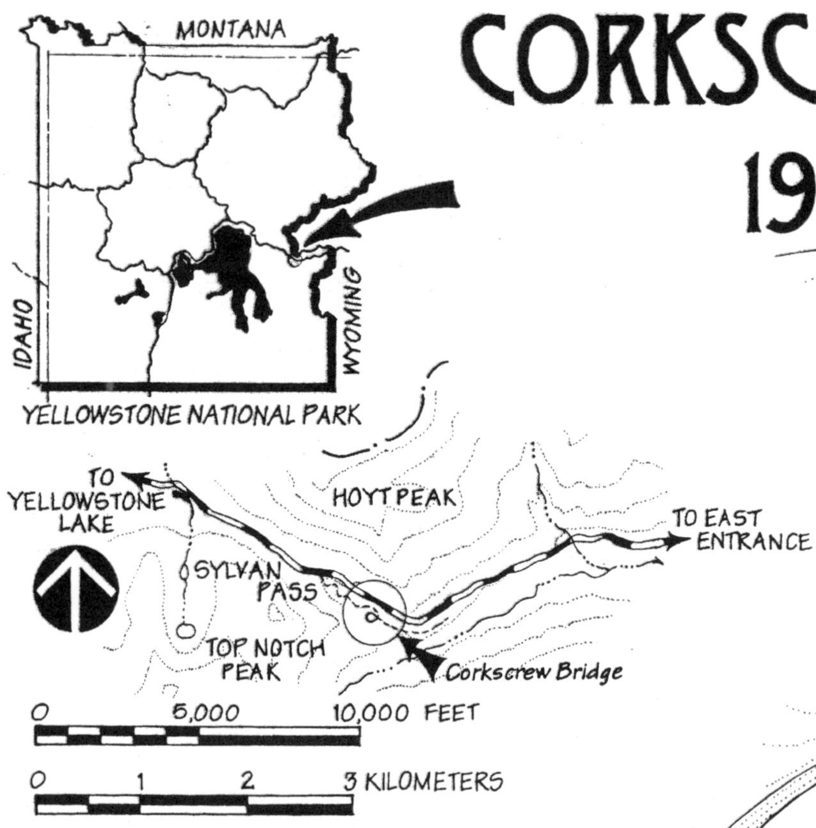

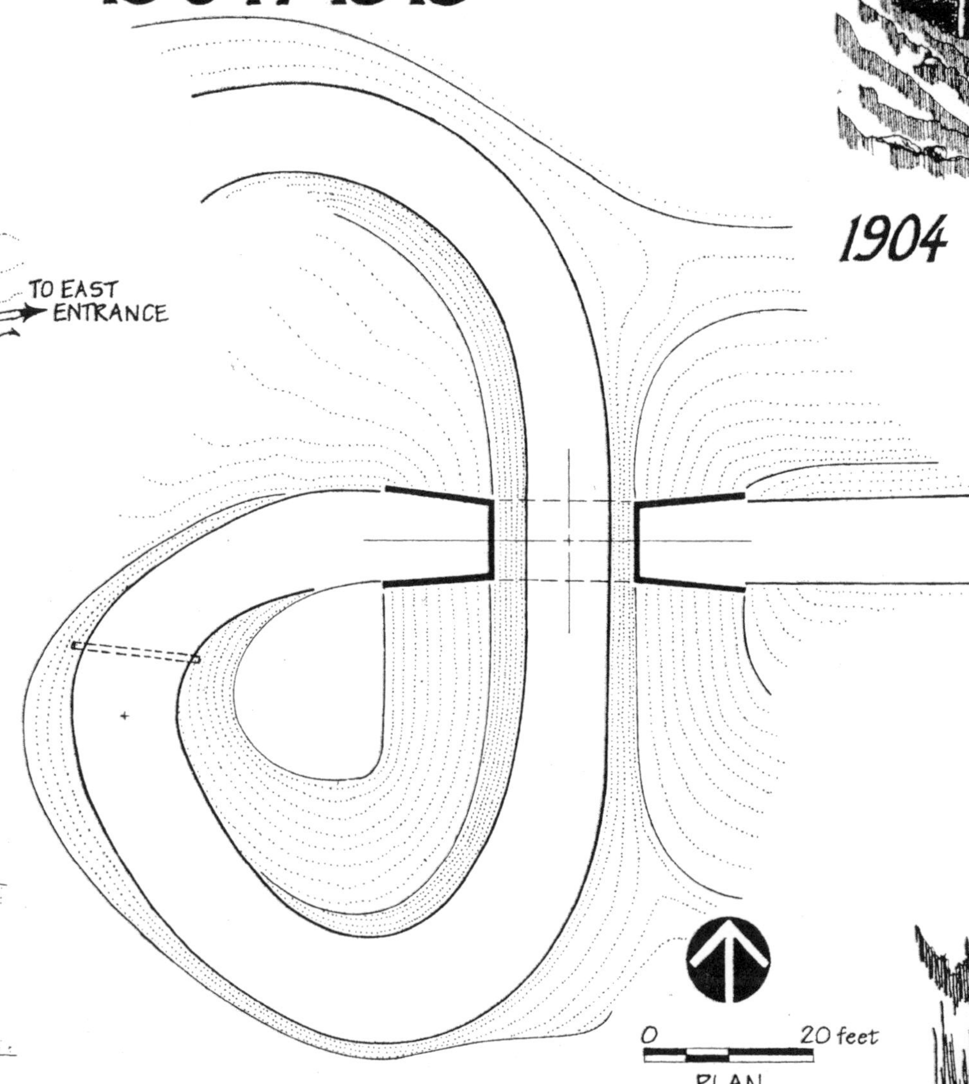

Building a road through Sylvan Pass presented a formidable challenge. The steep descent on the east side of the Absaroka Range delayed the construction of the East Entrance Road until after the turn of the twentieth century. To carry the road safely down the slope at a maximum grade of 10 percent, U.S. Army Corps of Engineers Captain Hiram M. Chittenden introduced a spiral to quickly change elevation in a short distance. Completed in 1904, the timber trestle supported the road that then looped around to pass beneath it. The trestle was replaced by a shorter span in 1916. In 1919 Superintendent Horace M. Albright approved specifications for a reinforced concrete "underpass" bridge in the same location. By 1929, advances in road construction technology and the demands of automobile traffic rendered the narrow bridge obsolete. A new road cut into the hillside above gradually made the descent from Sylvan Pass over eight miles, and the Corkscrew Bridge was bypassed. To observant motorists, it is still visible in the valley below.

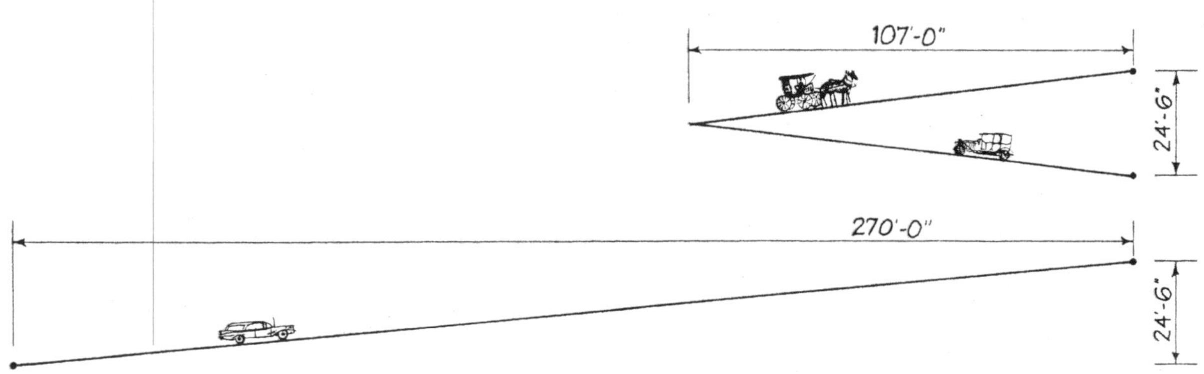

DELINEATED BY: Jill Patricia Caouette, 1999

YELLOWSTONE ROADS RECORDING PROJECT
NATIONAL PARK SERVICE
UNITED STATES DEPARTMENT OF THE INTERIOR

YELLOWSTONE NATIONAL PARK

CORKSCREW BRIDGE (LOOP BRIDGE) - 1904/1919
OLD EAST ENTRANCE ROAD, SYLVAN PASS
PARK COUNTY

WYOMING

SHEET 1 OF 1

HISTORIC AMERICAN ENGINEERING RECORD
WY-86

LIBRARY OF CONGRESS INDEX NUMBER

IF REPRODUCED, PLEASE CREDIT THE HISTORIC AMERICAN ENGINEERING RECORD, NATIONAL PARK SERVICE, NAME OF DELINEATOR, DATE OF DRAWING

CANYON BRIDGE OVER JAY CREEK
1895/1915

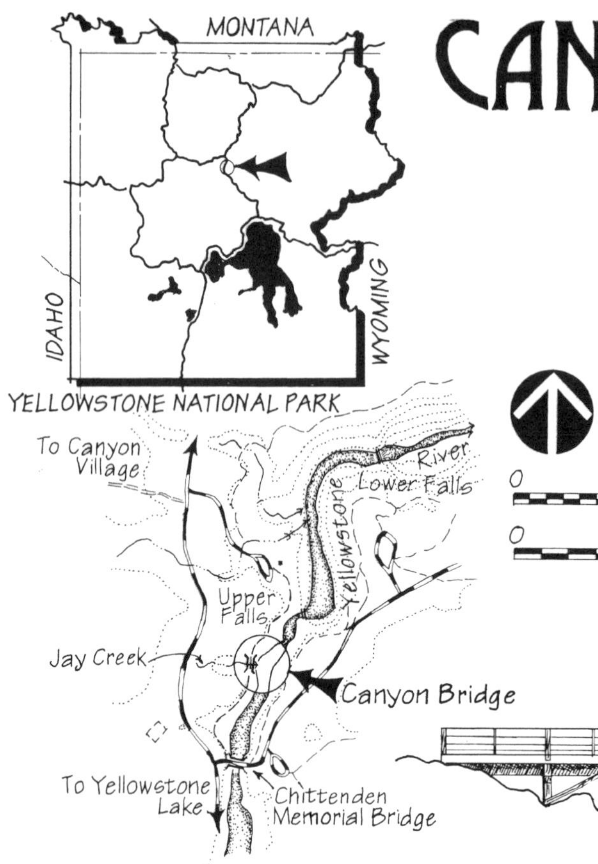

SITE LOCATION MAP
Above Map Adapted From USGS 7.5 Minute Series, Crystal Falls, Wyoming Quadrangle, 1986.
UTM Reference: Z12, E539505 N4950790

Drawing based on field measurements, historic photographs, and original design documents.

1895 ELEVATION 1915

The Canyon Bridge spans the often dry ravine of Jay Creek on the north side of the Yellowstone River near the Upper Falls. Built by contract for the U.S. Army Corps of Engineers in 1915, the existing structure is a reinforced concrete rib arch with open spandrels. It replaced a wooden arch truss bridge completed on the same site in 1895. By bridging the ravine, the structure improved an existing foot and bridge trail to allow for its use by stagecoaches, and later by park buses and automobiles. By providing park visitors access to views of the Grand Canyon of the Yellowstone, the Canyon Bridge reflects the design philosophy developed by the Corps - that the road system should provide access to the scenic features without overwhelming the beauty of nature that the park was created to protect.

No longer used for vehicular traffic, the Canyon Bridge provides pedestrian access to Lookout Point, Grand View, and Inspiration Point. As part of the North Rim Trail, the historic bridge and road alignment offer hikers a peaceful walk near the river, away from traffic. The bridge stands as a historic remnant of Yellowstone's early road system.

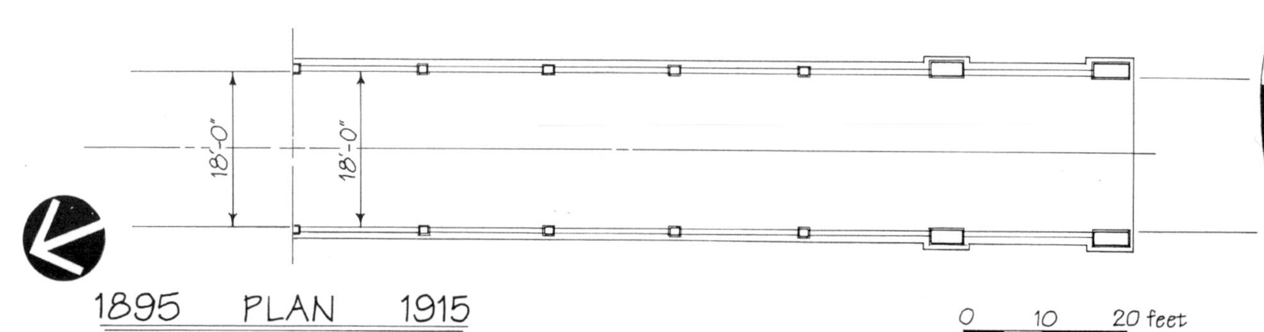

1895 PLAN 1915

DELINEATED BY: Jill Patricia Caouette, 1999
YELLOWSTONE ROADS RECORDING PROJECT
NATIONAL PARK SERVICE
UNITED STATES DEPARTMENT OF THE INTERIOR
CANYON BRIDGE - 1895 1915
NORTH RIM TRAIL SPANNING JAY CREEK
YELLOWSTONE NATIONAL PARK
PARK COUNTY
WYOMING
SHEET 1 OF 1
HISTORIC AMERICAN ENGINEERING RECORD
WY-87

CHITTENDEN MEMORIAL BRIDGE

1903/1962

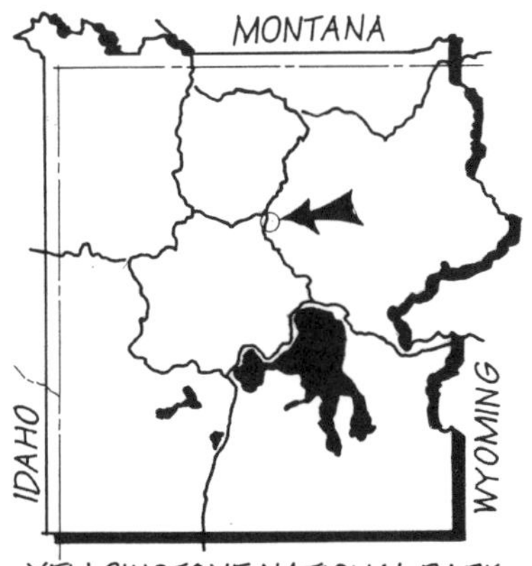

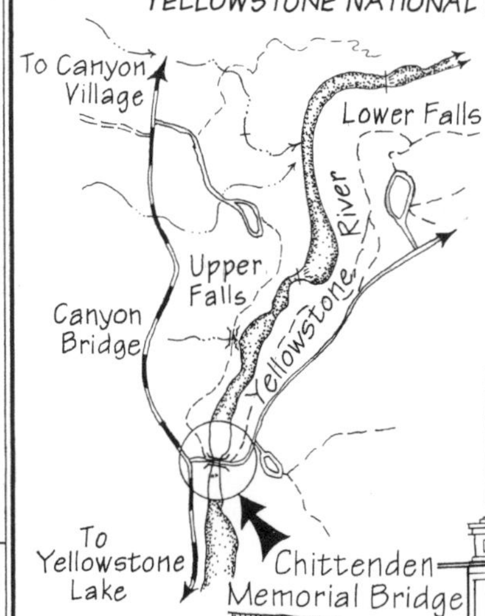

SITE LOCATION MAP
Map Adapted From USGS 7.5 Minute Series, Crystal Falls, Wyoming Quadrangle, 1986.
UTM Reference: Z12, E539450 N4950435

0 3000 feet
0 500 1000 meters

"This section of the river, immediately above the Upper Falls, presented numerous sites on which an ordinary bridge could be built at no great cost; but the remarkable scenic attractions of the rapids made any ordinary structure seem out of place. Accordingly it was decided to build something worthy of the situation and a single arch of slender profile was selected."

- Hiram M. Chittenden

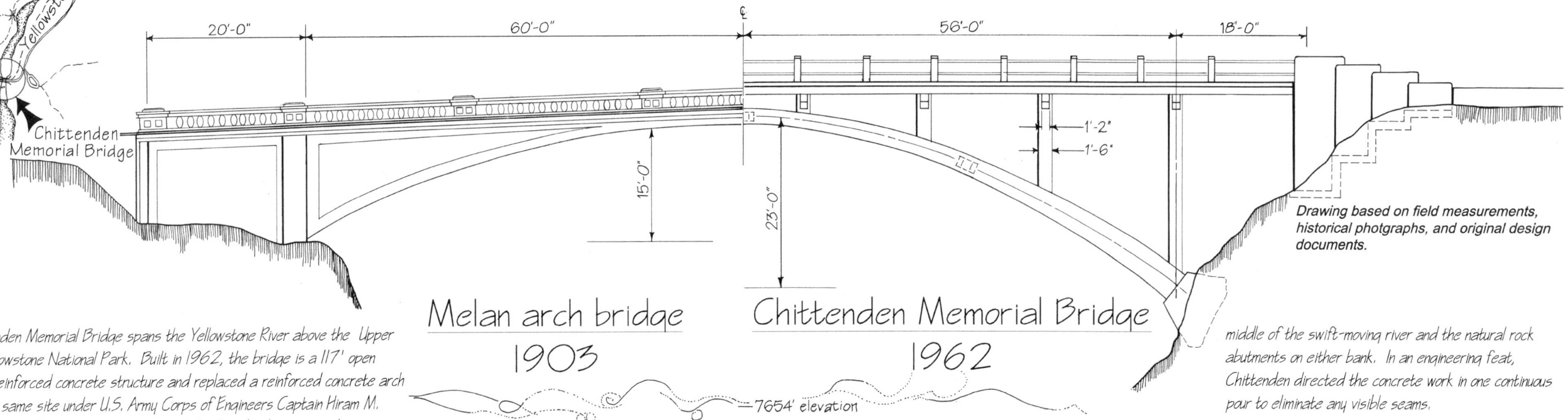

The Chittenden Memorial Bridge spans the Yellowstone River above the Upper Falls in Yellowstone National Park. Built in 1962, the bridge is a 117' open spandrel, reinforced concrete structure and replaced a reinforced concrete arch built on the same site under U.S. Army Corps of Engineers Captain Hiram M. Chittenden in 1903. In the construction of the first bridge to provide vehicular access to Artist Point on the south rim of the canyon, Chittenden employed a design developed by Austrian engineer Joseph Melan. Chittenden believed in finding the proper balance between development of the park to provide access to it's scenery and the need to make such development as unobtrusive on the natural setting as possible. He saw the landscape around the bridge site as some of the most magnificent in the park, and favored the graceful Melan arch to blend with the surroundings.

The bridge's location over the rapids of the river made the erection of the formwork for the pouring of the concrete the most dangerous part of construction. The formwork was supported by a temporary pier built in the middle of the swift-moving river and the natural rock abutments on either bank. In an engineering feat, Chittenden directed the concrete work in one continuous pour to eliminate any visible seams.

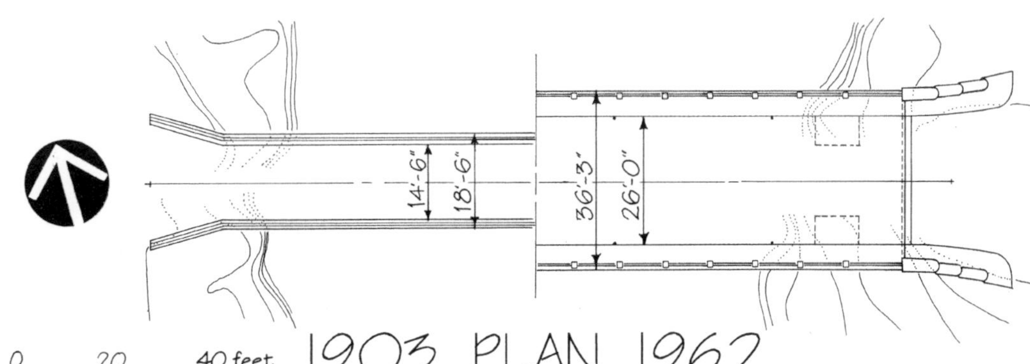

1903 PLAN 1962

By 1960, the deteriorated condition and narrow width of the Melan arch bridge caused serious concern for the safety of visitors. After considerable debate, National Park Service officials decided to replace it with a wider, stronger bridge. By this time, Melan's design was considered obsolete. The new span was a more modern reinforced concrete open-spandrel arch. Demolition of the old and construction of the new bridge on the same site began in 1961. Yellowstone National Park dedicated the new bridge, named in Chittenden's honor, on August 8, 1963. The Chittenden Memorial Bridge provides unparalleled views of the Upper and Lower Falls in the Grand Canyon of the Yellowstone, still one of the most scenic and most visited areas on the Grand Loop Road.

DELINEATED BY: Jill Patricia Cauette, 2000

YELLOWSTONE ROADS RECORDING PROJECT
NATIONAL PARK SERVICE
UNITED STATES DEPARTMENT OF THE INTERIOR

CHITTENDEN BRIDGE/CHITTENDEN MEMORIAL BRIDGE - 1903/1962
ARTISTS POINT ROAD SPANNING YELLOWSTONE RIVER
YELLOWSTONE NATIONAL PARK PARK COUNTY WYOMING

SHEET 1 OF 1

HISTORIC AMERICAN ENGINEERING RECORD
WY-88

YOSEMITE ROADS & BRIDGES

YOSEMITE NATIONAL PARK, CALIFORNIA

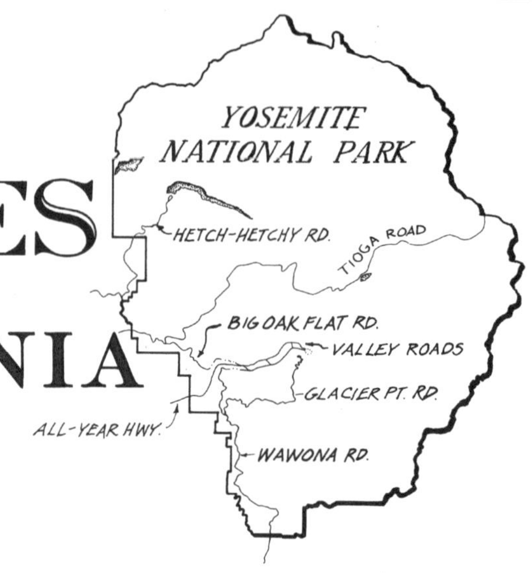

This drawing was adapted from several historic photographs located in the Yosemite Research Library.

For more than two decades following the discovery of the Yosemite Valley in 1851, the only access to the Valley was by rough pack and foot trails. In 1864, the Yosemite Valley and the Mariposa Grove of giant sequoias were set aside by Congress as a protective reserve, but it would be another ten years before the first road reached the area.

Hoping to capitalize on tourism, citizens of Mariposa and Tuolumne counties sought to entice Yosemite-bound visitors by constructing toll roads into the Valley. By the early 1870s, private toll roads were under construction from Coulterville, Big Oak Flat and Mariposa. A great road-building race ensued. Each community hoped to finish its road first and thereby capture the tourist trade. Dr. John Taylor McLean completed his Coulterville Road before the others on 17 June 1874. His hopes for a transportation monopoly were dashed only twenty-nine days later when the "Chinese Camp and Yo Semite Turnpike," forerunner of the Big Oak Flat Road, reached the Valley. From the south, a cartel headed by Albert Henry Washburn pushed another toll road from Mariposa to Big Tree Station (now Wawona) and on to the Valley in 1875. North of Yosemite Valley, the "Great Sierra Wagon Road" was completed through the high country to reach gold and silver mines near Tioga Pass in 1833. Though not built for the tourist trade, this route would evolve into the present Tioga Road, which at a height of nearly ten thousand feet, is the highest paved mountain highway in California. Construction of the early roads was a major engineering feat of the period: all of these early roads were built across the rough, mountainous terrain using hand labor, aided by picks, shovels and "Giant" blasting powder. On the Valley floor, the Board of Commissioners of the Yosemite Grant sporadically developed a system of carriage drives, connected by wood and iron truss bridges.

Yosemite was designated a national park in 1890. The United States Cavalry assumed command of the park's administration and managed affairs until 1914. During this period, they oversaw the construction of three small bridges at the base of Bridalveil Fall in 1913.

In 1900, the first automobile struggled over steep, unimproved and inadequate roads to reach the Valley. Although they were banned by the park's acting military administration in 1907, the age of the automobile had nearly arrived, and in 1913, Secretary of the Interior, Franklin K. Lane opened park gates to the new and soon-to-be dominant form of transportation.

The National Park Service was created in 1916 to protect the Nation's most important scenic and cultural resources. Its first Director, Stephen T. Mather, was a native Californian deeply interested in the Yosemite Valley. In 1915, when he was still Assistant Secretary of the Interior, Mather and some business associates purchased the old Tioga Road and deeded it over to the park. As Director, he signed a Memorandum of Agreement with the Bureau of Public Roads (BPR) in 1925, under which the BPR would take charge of road construction in the national parks. Following the policies described by the Organic Act, the NPS worked with the Bureau of Public Roads to ensure that the new Yosemite road system was designed to blend with the natural setting. In the 1930s, the Wawona, Big Oak Flat and parts of the Tioga Road were constructed under BPR supervision. To preserve the park landscape, miles of stone retaining walls were designed to conceal the roads from principal views, and four tunnels were constructed to avoid scarring of granite cliffs. One of these, the Wawona Tunnel, was the longest in the West at the time of its construction in 1933. In Yosemite Valley, the NPS landscape architecture division and the BPR collaborated to design a series of "rustic-style" bridges, made of concrete and steel, but faced with granite or large redwood logs so as to integrate with the natural setting.

After World War II, the Federal Highway Administration and the National Park Service completed the reconstruction of the old Tioga Road as a modern park highway, to the delight of motorists but to the dismay of some conservationists. In the 1980s, annual visitation passed the three million mark and park roads are now approaching their carrying capacity. Today, on-going road improvement plans and new park environmental policies pose a threat to road structures throughout Yosemite. The Historic American Engineering Record, a division of the National Park Service, documented significant features of the road system in 1991. Histories, measured drawings and photographs produced by HAER will be used to evaluate the features as historic structures, significant in portraying the character of the park. Should they be destroyed, a historic record exists to guide the integrity of future construction endeavors.

This recording project is part of the Historic American Engineering Record (HAER), a long-range program to document historically significant engineering and industrial works in the United States. HAER is an agency of the National Park Service, U.S. Department of the Interior. The Yosemite Roads and Bridges Project was co-sponsored by Yosemite National Park, Michael Finley, Superintendent; Kevin Cann, Chief of Maintenance and Engineering; the NPS Roads and Bridges Program, John Gingles, Manager; and the Historic American Buildings Survey/Historic American Engineering Record (HABS/HAER) under the general direction of Robert J. Kapsch, Chief.

The fieldwork, measured drawings, histories and photographs were completed under the direction of Eric DeLony, Chief and Principal Architect, HAER. The recording team consisted of Industrial Designer Todd A. Croteau, field supervisor; Architectural Technicians Dione DeMartelaere, David R. Fleming and Marie-Claude LeSauteur; Project Historian, Richard H. Quin. Formal on-site photography was done by Brian C. Grogan

DELINEATED BY: DIONE M. DE MARTELAERE 1991

YOSEMITE ROADS AND BRIDGES RECORDING PROJECT
UNITED STATES DEPARTMENT OF THE INTERIOR

YOSEMITE ROADS AND BRIDGES
YOSEMITE NATIONAL PARK
MARIPOSA COUNTY
CALIFORNIA

HISTORIC AMERICAN ENGINEERING RECORD
CA-117

SHEET 1 OF 2

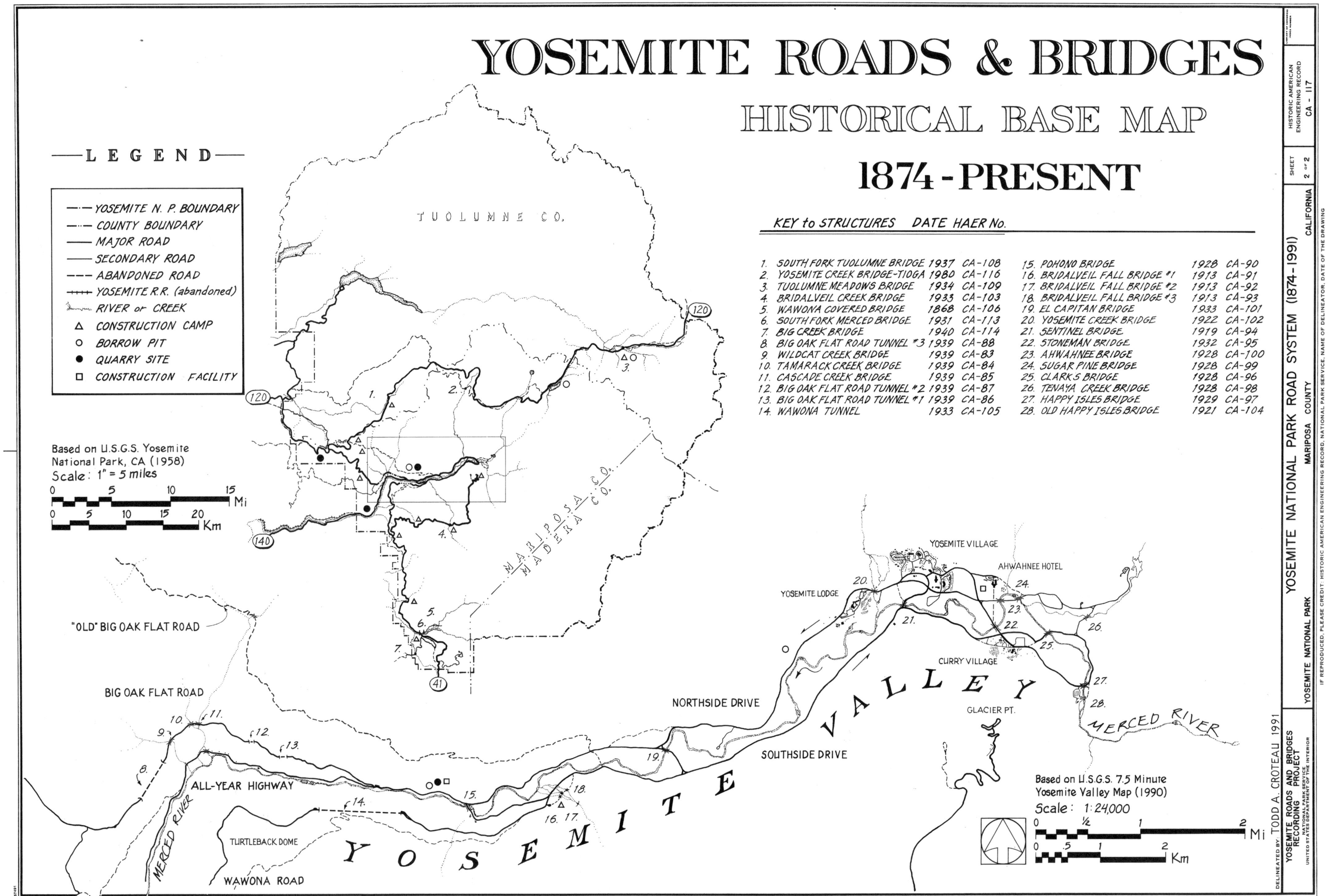

YOSEMITE ROADS & BRIDGES
HISTORICAL BASE MAP
1874 - PRESENT
KEY to STRUCTURES DATE HAER No.
1. SOUTH FORK TUOLUMNE BRIDGE 1937 CA-108
2. YOSEMITE CREEK BRIDGE-TIOGA 1980 CA-116
3. TUOLUMNE MEADOWS BRIDGE 1934 CA-109
4. BRIDALVEIL CREEK BRIDGE 1933 CA-103
5. WAWONA COVERED BRIDGE 1868 CA-106
6. SOUTH FORK MERCED BRIDGE 1931 CA-113
7. BIG CREEK BRIDGE 1940 CA-114
8. BIG OAK FLAT ROAD TUNNEL #3 1939 CA-88
9. WILDCAT CREEK BRIDGE 1939 CA-83
10. TAMARACK CREEK BRIDGE 1939 CA-84
11. CASCADE CREEK BRIDGE 1939 CA-85
12. BIG OAK FLAT ROAD TUNNEL #2 1939 CA-87
13. BIG OAK FLAT ROAD TUNNEL #1 1939 CA-86
14. WAWONA TUNNEL 1933 CA-105
15. POHONO BRIDGE 1928 CA-90
16. BRIDALVEIL FALL BRIDGE #1 1913 CA-91
17. BRIDALVEIL FALL BRIDGE #2 1913 CA-92
18. BRIDALVEIL FALL BRIDGE #3 1913 CA-93
19. EL CAPITAN BRIDGE 1933 CA-101
20. YOSEMITE CREEK BRIDGE 1922 CA-102
21. SENTINEL BRIDGE 1919 CA-94
22. STONEMAN BRIDGE 1932 CA-95
23. AHWAHNEE BRIDGE 1928 CA-100
24. SUGAR PINE BRIDGE 1928 CA-99
25. CLARKS BRIDGE 1928 CA-96
26. TENAYA CREEK BRIDGE 1928 CA-98
27. HAPPY ISLES BRIDGE 1929 CA-97
28. OLD HAPPY ISLES BRIDGE 1921 CA-104
LEGEND
YOSEMITE N. P. BOUNDARY
COUNTY BOUNDARY
MAJOR ROAD
SECONDARY ROAD
ABANDONED ROAD
YOSEMITE R.R. (abandoned)
RIVER or CREEK
CONSTRUCTION CAMP
BORROW PIT
QUARRY SITE
CONSTRUCTION FACILITY
Based on U.S.G.S. Yosemite National Park, CA (1958)
Scale: 1" = 5 miles
0 5 10 15 Mi
0 5 10 15 20 Km
TUOLUMNE CO.
MARIPOSA CO.
MADERA CO.
120
140
41
"OLD" BIG OAK FLAT ROAD
BIG OAK FLAT ROAD
ALL-YEAR HIGHWAY
MERCED RIVER
TURTLEBACK DOME
WAWONA ROAD
NORTHSIDE DRIVE
SOUTHSIDE DRIVE
YOSEMITE VALLEY
YOSEMITE VILLAGE
AHWAHNEE HOTEL
YOSEMITE LODGE
CURRY VILLAGE
GLACIER PT.
MERCED RIVER
Based on U.S.G.S. 7.5 Minute Yosemite Valley Map (1990)
Scale: 1:24,000
0 ½ 1 2 Mi
0 .5 1 2 Km
DELINEATED BY: TODD A. CROTEAU 1991
YOSEMITE ROADS AND BRIDGES RECORDING PROJECT
NATIONAL PARK SERVICE
UNITED STATES DEPARTMENT OF THE INTERIOR
YOSEMITE NATIONAL PARK
YOSEMITE NATIONAL PARK ROAD SYSTEM (1874-1991)
MARIPOSA COUNTY
CALIFORNIA
SHEET 2 of 2
HISTORIC AMERICAN ENGINEERING RECORD
CA - 117
IF REPRODUCED, PLEASE CREDIT: HISTORIC AMERICAN ENGINEERING RECORD, NATIONAL PARK SERVICE, NAME OF DELINEATOR, DATE OF THE DRAWING

WAWONA COVERED BRIDGE

The main truss timbers of the bridge are hand-hewn ponderosa pine. When reconstructed in 1956, historic techniques and tools were employed.

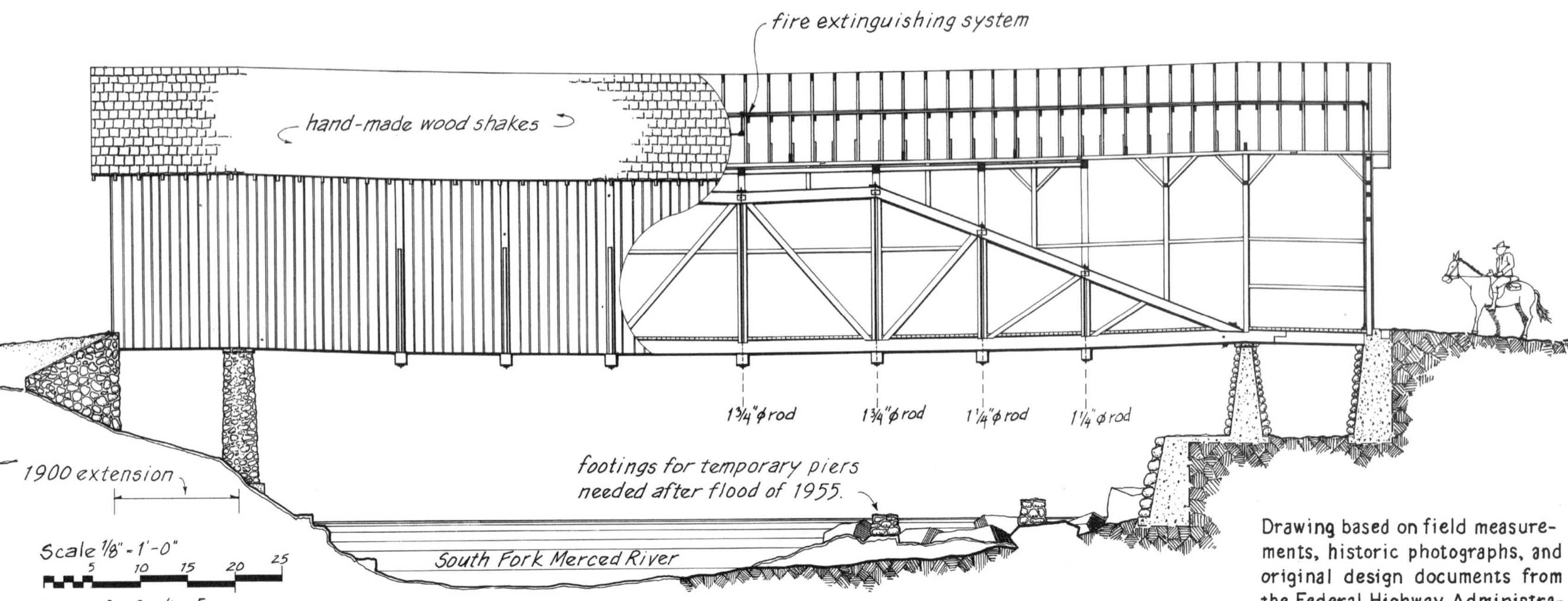

ELEVATION / SECTION

Drawing based on field measurements, historic photographs, and original design documents from the Federal Highway Administration, Drwgs. # RG 2420-A & B, August 1983.

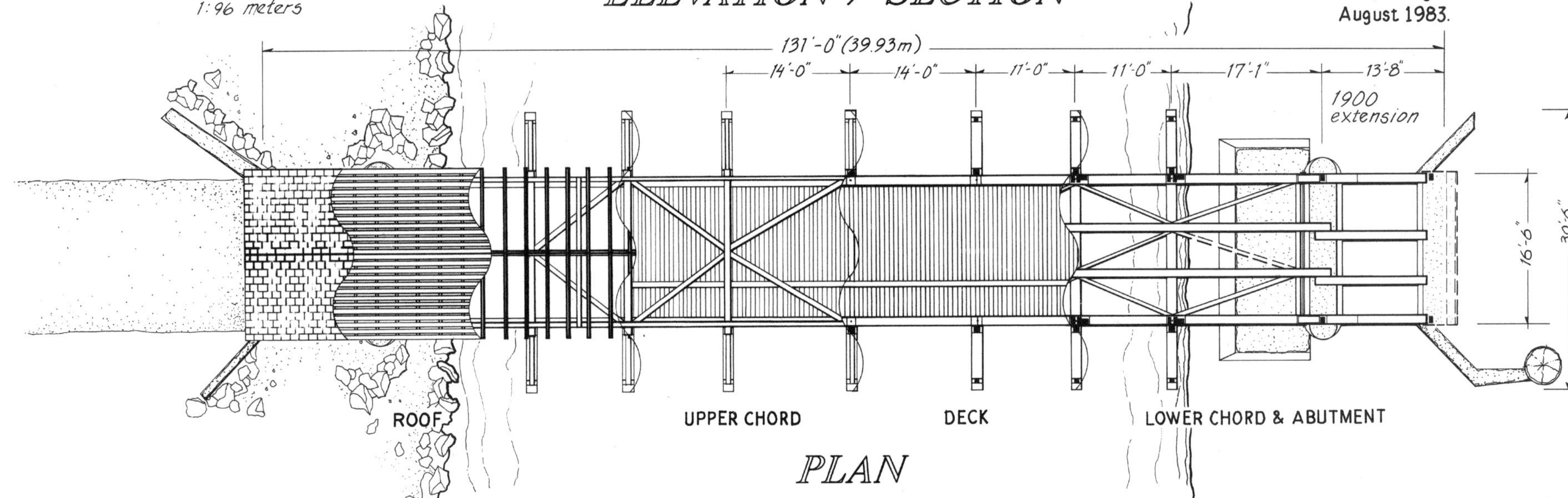

PLAN

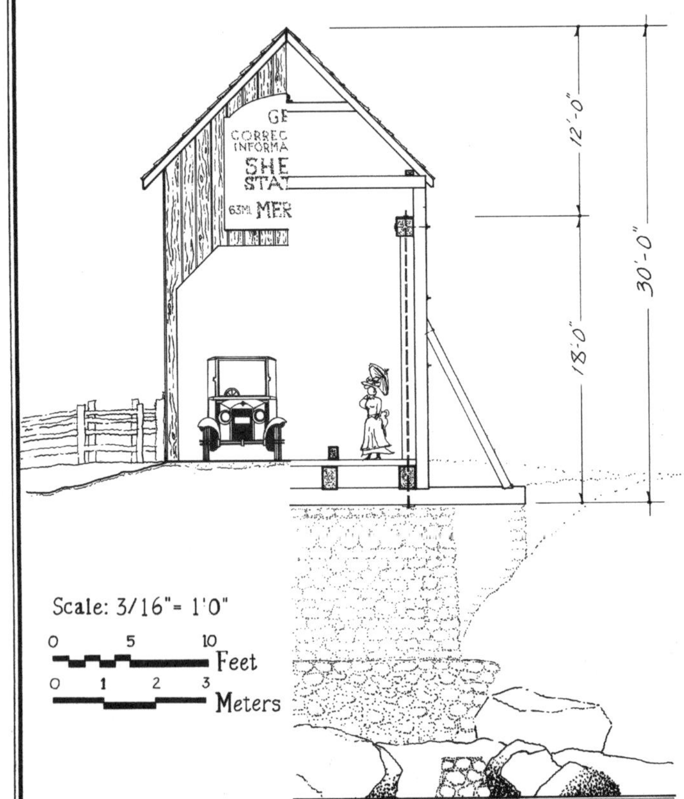

ELEVATION / SECTION

Spanning the South Fork of the Merced River in the southern part of Yosemite National Park, the Wawona Covered Bridge was originally constructed as an open-deck structure. It is supported by modified Queen-Post trusses of native woods, braced by steel (originally iron) tie rods sandwiched between vertical posts. The main chord members and floor beams are hand-hewn Ponderosa pine, while vertical and diagonal members are Douglas fir. The housing, added in 1878, consists of an internal frame of Douglas fir and external knee braces extending from the transverse floor beams. The bridge originally rested on wooden cribs filled with stone rubble; these were later replaced in 1957 by reinforced concrete piers and abutments faced with stone.

The Wawona Bridge was built in 1868 as an open pony-truss bridge by Galen Clark, the first settler at Wawona and the first Guardian of the Yosemite Grant. In 1878, the bridge was acquired by Washburn, Coffman & Chapman and converted to a covered bridge, supposedly ordered by Washburn to remind him of his New England home. John Conway, builder of the original Wawona Road, was contracted to cover the bridge and used wood from Washburn's sawmill. The bridge was extended in 1900 by short additions at each end. The structure has been repaired numerous times, and following a major flood in 1955, was dismantled and reconstructed the following year. The Wawona Covered Bridge is one of the oldest surviving bridges in California. Now a part of the Pioneer Yosemite History Center, the bridge continues to carry horse, stage and pedestrian traffic.

This recording project is part of the Historic American Engineering Record (HAER), a long-range program to document historically significant engineering and industrial works in the United States. HAER is an agency of the National Park Service, U.S. Department of the Interior. The Yosemite Roads and Bridges Project was co-sponsored by Yosemite National Park, Michael Finley, Superintendent; Kevin Cann, Chief of Maintenance and Engineering; the NPS Roads and Bridges Program, John Gingles, Manager; and the Historic American Buildings Survey/Historic American Engineering Record (HABS/HAER) under the general direction of Robert J. Kapsch, Chief.

The Fieldwork, measured drawings, histories and photographs were completed under the direction of Eric DeLony, Chief and Principal Architect, HAER. The recording team consisted of Industrial Designer Todd A. Croteau, field supervisor; Architectural Technicians Dione DeMartelaere, David R. Fleming and Marie-Claude LeSauteur; Project Historian, Richard H. Quin. Formal on-site photography was done by Brian C. Grogan.

DELINEATED BY: DIONE M. DE MARTELAERE 1991

YOSEMITE ROADS AND BRIDGES RECORDING PROJECT
NATIONAL PARK SERVICE
UNITED STATES DEPARTMENT OF THE INTERIOR

YOSEMITE NATIONAL PARK

WAWONA COVERED BRIDGE - 1868
PIONEER YOSEMITE HISTORY CENTER SPANNING THE SOUTH FORK MERCED RIVER
MARIPOSA COUNTY

CALIFORNIA

SHEET 1 OF 2

HISTORIC AMERICAN ENGINEERING RECORD
CA - 106

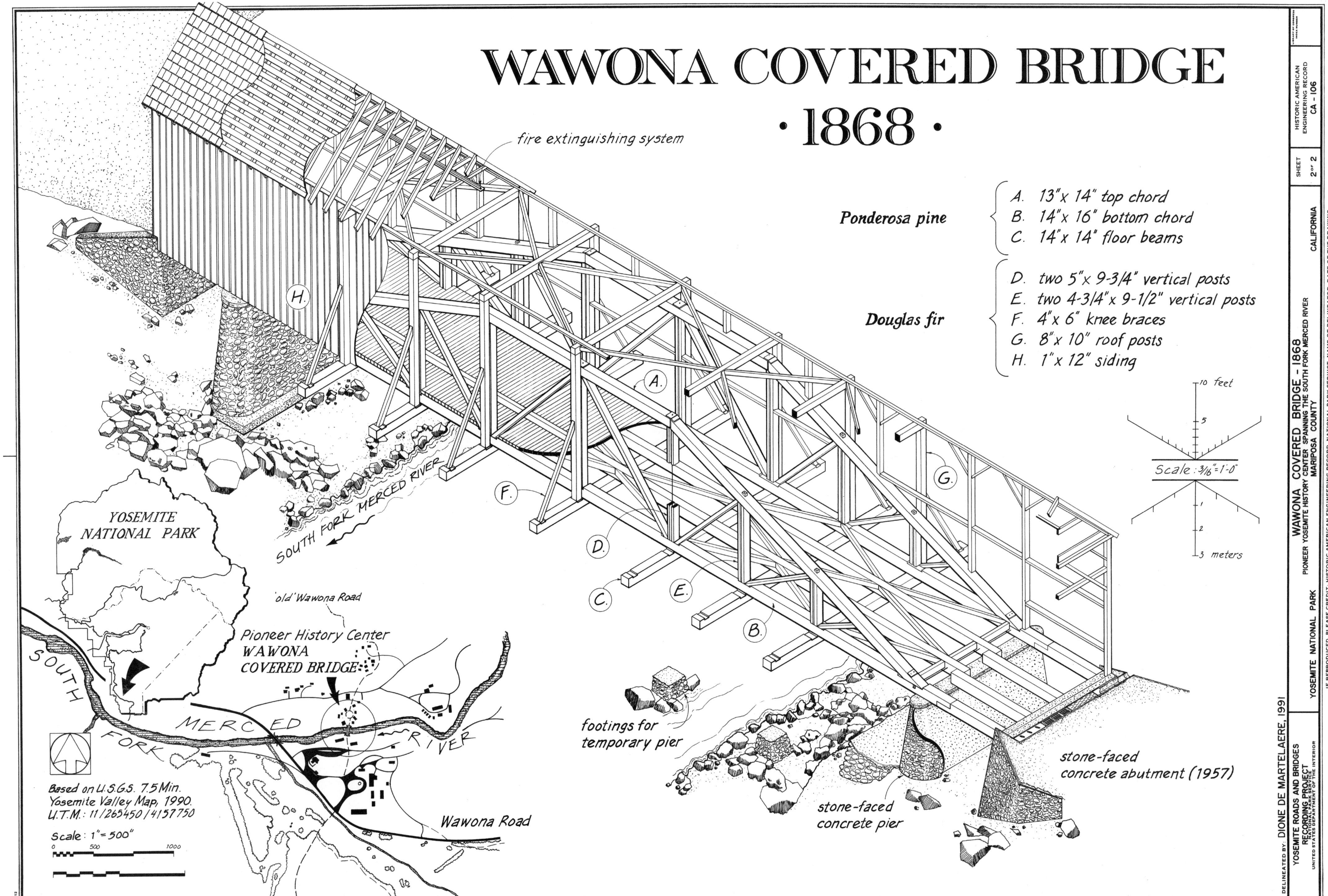
WAWONA COVERED BRIDGE
· 1868 ·
fire extinguishing system
Ponderosa pine
A. 13"x 14" top chord
B. 14"x 16" bottom chord
C. 14"x 14" floor beams
Douglas fir
D. two 5"x 9-3/4" vertical posts
E. two 4-3/4"x 9-1/2" vertical posts
F. 4"x 6" knee braces
G. 8"x 10" roof posts
H. 1"x 12" siding
10 feet
5
Scale: 3/16"=1'-0"
1
2
3 meters
SOUTH FORK MERCED RIVER
YOSEMITE NATIONAL PARK
'old' Wawona Road
Pioneer History Center
WAWONA COVERED BRIDGE
SOUTH FORK MERCED RIVER
Wawona Road
Based on U.S.G.S. 7.5 Min. Yosemite Valley Map, 1990.
U.T.M.: 11/265450/4157750
Scale: 1"=500"
0 500 1000
footings for temporary pier
stone-faced concrete pier
stone-faced concrete abutment (1957)
DELINEATED BY: DIONE DE MARTELAERE, 1991
YOSEMITE ROADS AND BRIDGES RECORDING PROJECT
NATIONAL PARK SERVICE
UNITED STATES DEPARTMENT OF THE INTERIOR
WAWONA COVERED BRIDGE - 1868
PIONEER YOSEMITE HISTORY CENTER SPANNING THE SOUTH FORK MERCED RIVER
YOSEMITE NATIONAL PARK
MARIPOSA COUNTY
CALIFORNIA
SHEET 2 OF 2
HISTORIC AMERICAN ENGINEERING RECORD
CA - 106
IF REPRODUCED, PLEASE CREDIT: HISTORIC AMERICAN ENGINEERING RECORD, NATIONAL PARK SERVICE, NAME OF DELINEATOR, DATE OF THE DRAWING

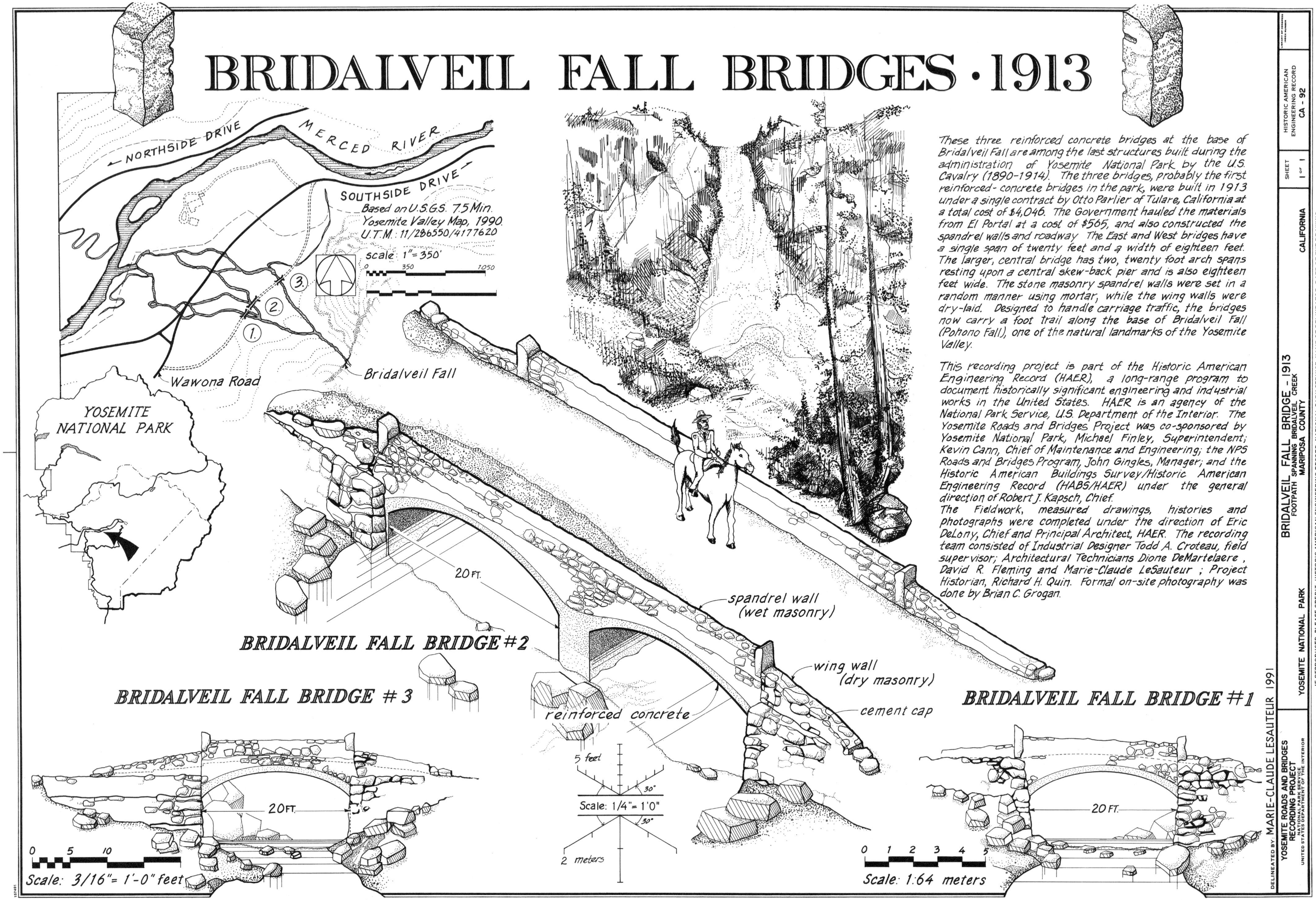

BRIDALVEIL FALL BRIDGES · 1913
← NORTHSIDE DRIVE
MERCED RIVER
SOUTHSIDE DRIVE →
Based on U.S.G.S. 7.5 Min. Yosemite Valley Map, 1990. U.T.M.: 11/286550/4177620
scale: 1" = 350'
0
350
1050
1.
2.
3.
Wawona Road
Bridalveil Fall
YOSEMITE NATIONAL PARK
These three reinforced concrete bridges at the base of Bridalveil Fall are among the last structures built during the administration of Yosemite National Park by the U.S. Cavalry (1890-1914). The three bridges, probably the first reinforced-concrete bridges in the park, were built in 1913 under a single contract by Otto Parlier of Tulare, California at a total cost of $4,046. The Government hauled the materials from El Portal at a cost of $565, and also constructed the spandrel walls and roadway The East and West bridges have a single span of twenty feet and a width of eighteen feet. The larger, central bridge has two, twenty foot arch spans resting upon a central skew-back pier and is also eighteen feet wide. The stone masonry spandrel walls were set in a random manner using mortar, while the wing walls were dry-laid. Designed to handle carriage traffic, the bridges now carry a foot trail along the base of Bridalveil Fall (Pohono Fall), one of the natural landmarks of the Yosemite Valley.
This recording project is part of the Historic American Engineering Record (HAER), a long-range program to document historically significant engineering and industrial works in the United States. HAER is an agency of the National Park Service, U.S. Department of the Interior. The Yosemite Roads and Bridges Project was co-sponsored by Yosemite National Park, Michael Finley, Superintendent; Kevin Cann, Chief of Maintenance and Engineering; the NPS Roads and Bridges Program, John Gingles, Manager; and the Historic American Buildings Survey/Historic American Engineering Record (HABS/HAER) under the general direction of Robert J. Kapsch, Chief.
The Fieldwork, measured drawings, histories and photographs were completed under the direction of Eric DeLony, Chief and Principal Architect, HAER. The recording team consisted of Industrial Designer Todd A. Croteau, field supervisor; Architectural Technicians Dione DeMartelaere, David R. Fleming and Marie-Claude LeSauteur; Project Historian, Richard H. Quin. Formal on-site photography was done by Brian C. Grogan.
20 FT.
spandrel wall (wet masonry)
BRIDALVEIL FALL BRIDGE #2
wing wall (dry masonry)
cement cap
reinforced concrete
BRIDALVEIL FALL BRIDGE # 3
BRIDALVEIL FALL BRIDGE #1
5 feet
30°
Scale: 1/4" = 1'0"
2 meters
20 FT.
0 5 10
Scale: 3/16" = 1'-0" feet
0 1 2 3 4 5
Scale: 1:64 meters
DELINEATED BY: MARIE-CLAUDE LESAUTEUR 1991
YOSEMITE ROADS AND BRIDGES RECORDING PROJECT
NATIONAL PARK SERVICE
UNITED STATES DEPARTMENT OF THE INTERIOR
YOSEMITE NATIONAL PARK
BRIDALVEIL FALL BRIDGE - 1913
FOOTPATH SPANNING BRIDALVEIL CREEK
MARIPOSA COUNTY
CALIFORNIA
SHEET 1 OF 1
HISTORIC AMERICAN ENGINEERING RECORD CA - 92
IF REPRODUCED, PLEASE CREDIT: HISTORIC AMERICAN ENGINEERING RECORD, NATIONAL PARK SERVICE, NAME OF DELINEATOR, DATE OF THE DRAWING

SENTINEL BRIDGE · 1919

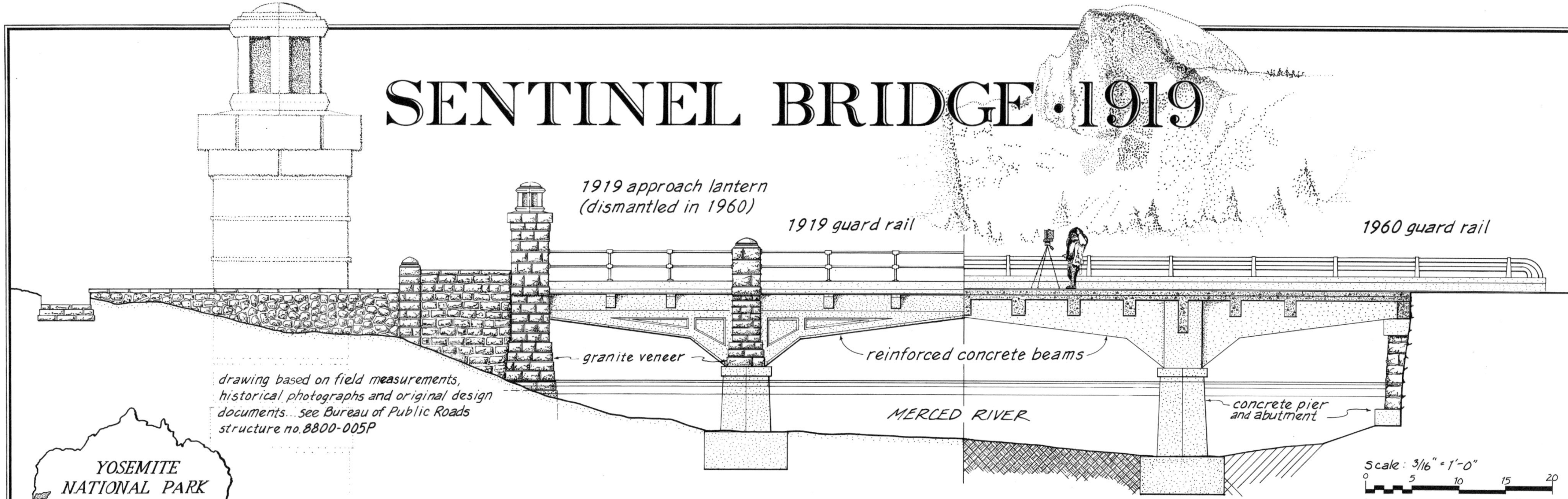

1919 ◦ ELEVATION / SECTION ◦ 1960

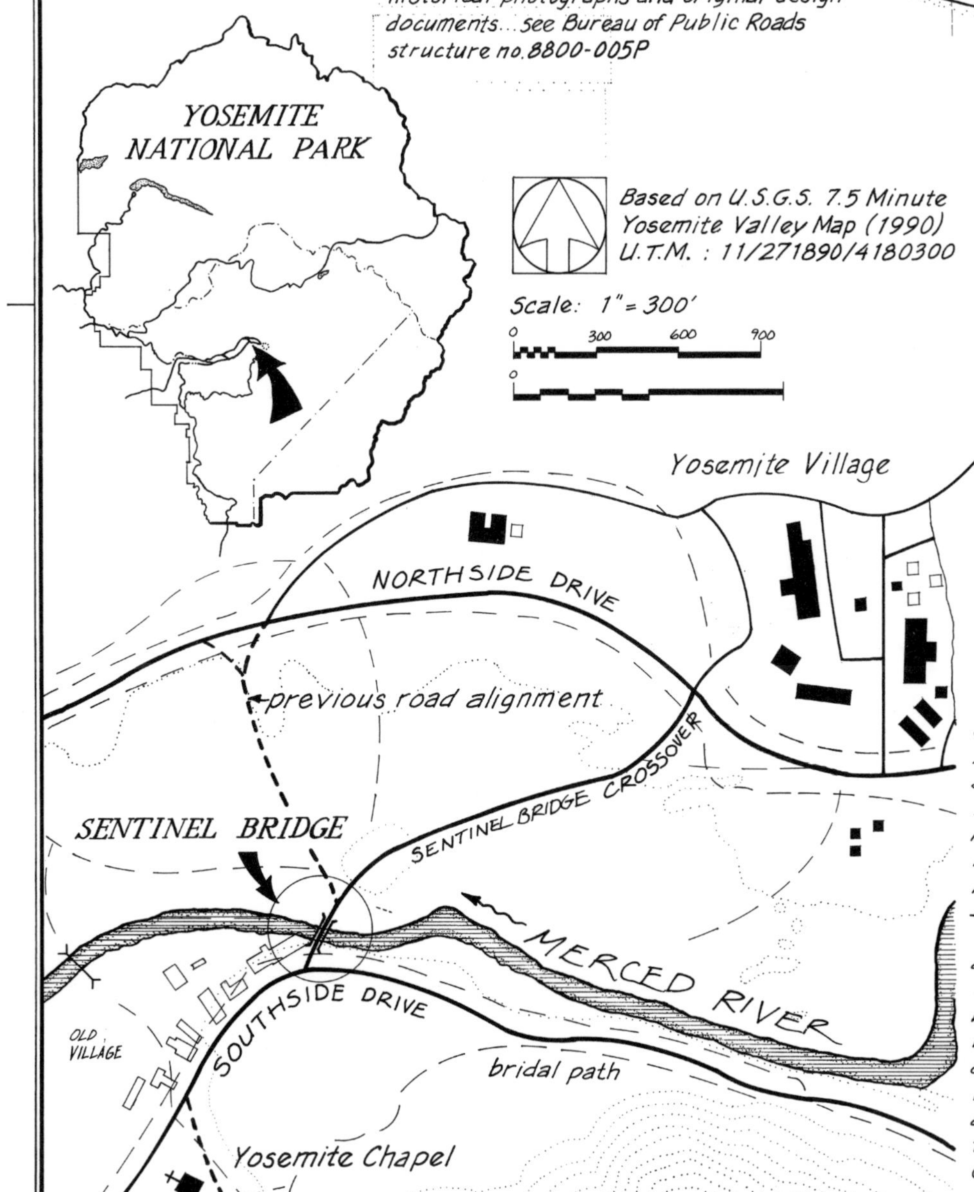

PLAN

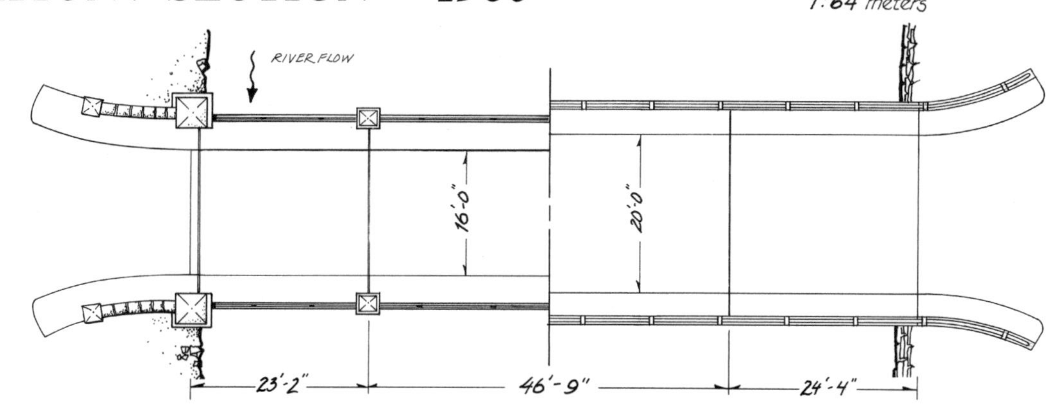

The oldest surviving Merced River bridge in Yosemite Valley, the Sentinel Bridge also marks the site of the first bridge, a crude log span erected about 1859 by hotelier Gustavus Adolphus Hite. Some time before 1865, Hite's structure was replaced with another log bridge by Yosemite pioneer and entrepreneur James Mason Huthchings. This span was washed away by a flood on December 23, 1867. Hutchings had a new truss bridge built of finished lumber. In 1878, the Board of Commissioners of the Yosemite Grant replaced Hutchings' bridge with a 96' iron bowstring truss bridge with a wooden deck. It was first known as "Upper Iron Bridge" and later called the "Sentinel Hotel Bridge" or Sentinel Bridge, after another adjacent hotel. The iron bridge was rehabilitated in 1898 and continued in use for another two decades, carrying the heaviest traffic loads of the Valley in the vicinity of the original Yosemite Village.

In 1916 and 1917, a new park maintenance area was developed on the north side of the Valley near the base of Yosemite Falls. Over the next few years, employee residences and operation facilities were relocated to this site from the Old Village on the south side of the Merced River. To hasten traffic flow between the two centers, the old Sentinel Bridge was replaced in 1918-19 with a new reinforced concrete span designed by the National Park Service's landscape architecture division. The new bridge was 97' long, 23' wide and carried two lanes of traffic. Its abutments and piers are of native granite. It originally featured decorative lanterns atop the approach pylons. Ansel Adams' famous photograph, "Moon Over HALF DOME" was taken from this bridge. The Sentinel Bridge was widened and heightened in 1960 and the lanterns were removed, changing its appearance.

This recording project is part of the Historic American Engineering Record (HAER), a long-range program to document historically significant engineering and industrial works in the United States. HAER is an agency of the National Park Service, U.S. Department of the Interior. The Yosemite Roads and Bridges Project was co-sponsored by Yosemite National Park, Michael Finley, Superintendent; Kevin Cann, Chief of Maintenance and Engineering; the NPS Roads and Bridges Program, John Gingles, Manager; and the Historic American Buildings Survey/Historic American Engineering Record (HABS/HAER) under the general direction of Robert J. Kapsch, Chief.

The Fieldwork, measured drawings, histories and photographs were completed under the direction of Eric DeLony, Chief and Principal Architect, HAER. The recording team consisted of Industrial Designer Todd A. Croteau, field supervisor; Architectural Technicians Dione DeMartelaere, David R. Fleming and Marie-Claude LeSauteur. Project Historian, Richard H. Quin. Formal on-site photography was done by Brian C. Grogan.

DELINEATED BY: MARIE-CLAUDE LE SAUTEUR, 1991

YOSEMITE ROADS AND BRIDGES RECORDING PROJECT
NATIONAL PARK SERVICE
UNITED STATES DEPARTMENT OF THE INTERIOR

SENTINEL BRIDGE - 1919
SENTINEL BRIDGE CROSSOVER SPANNING THE MERCED RIVER
MARIPOSA COUNTY
YOSEMITE NATIONAL PARK
CALIFORNIA

HISTORIC AMERICAN ENGINEERING RECORD CA-94

SHEET 1 OF 2

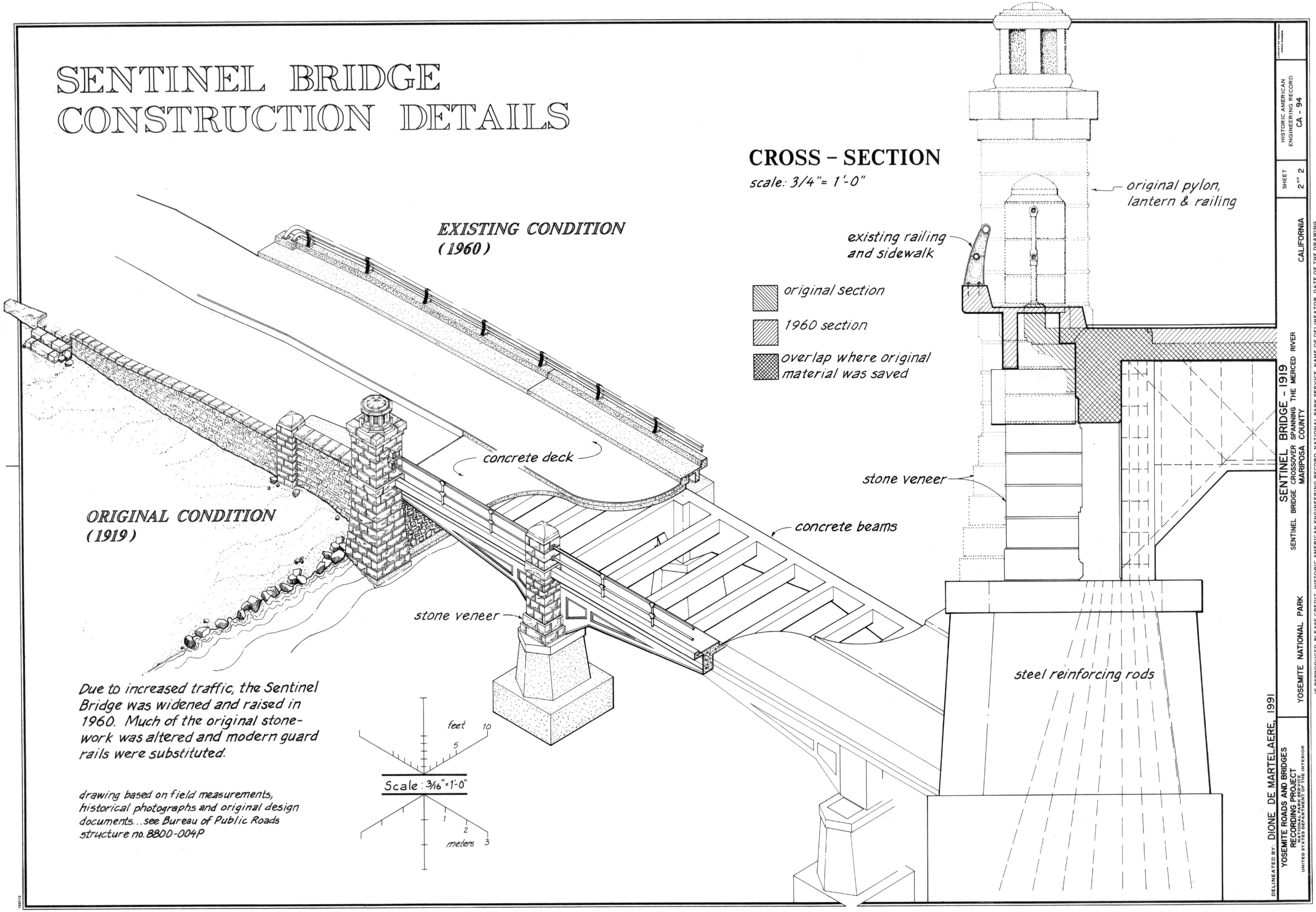
SENTINEL BRIDGE
CONSTRUCTION DETAILS
EXISTING CONDITION
(1960)
ORIGINAL CONDITION
(1919)
concrete deck
stone veneer
concrete beams
CROSS - SECTION
scale: 3/4" = 1'-0"
original pylon,
lantern & railing
existing railing
and sidewalk
original section
1960 section
overlap where original
material was saved
stone veneer
steel reinforcing rods
Due to increased traffic, the Sentinel Bridge was widened and raised in 1960. Much of the original stonework was altered and modern guard rails were substituted.
drawing based on field measurements, historical photographs and original design documents... see Bureau of Public Roads structure no. 8800-004P
feet
5
10
Scale: 3/16" = 1'-0"
1
2
3
meters
HISTORIC AMERICAN ENGINEERING RECORD
CA-94
SHEET
2 OF 2
SENTINEL BRIDGE - 1919
SENTINEL BRIDGE CROSSOVER SPANNING THE MERCED RIVER
MARIPOSA COUNTY
YOSEMITE NATIONAL PARK
CALIFORNIA
DELINEATED BY: DIONE DE MARTELAERE, 1991
YOSEMITE ROADS AND BRIDGES RECORDING PROJECT
NATIONAL PARK SERVICE
UNITED STATES DEPARTMENT OF THE INTERIOR
IF REPRODUCED, PLEASE CREDIT: HISTORIC AMERICAN ENGINEERING RECORD, NATIONAL PARK SERVICE, NAME OF DELINEATOR, DATE OF THE DRAWING

YOSEMITE CREEK BRIDGE · 1921

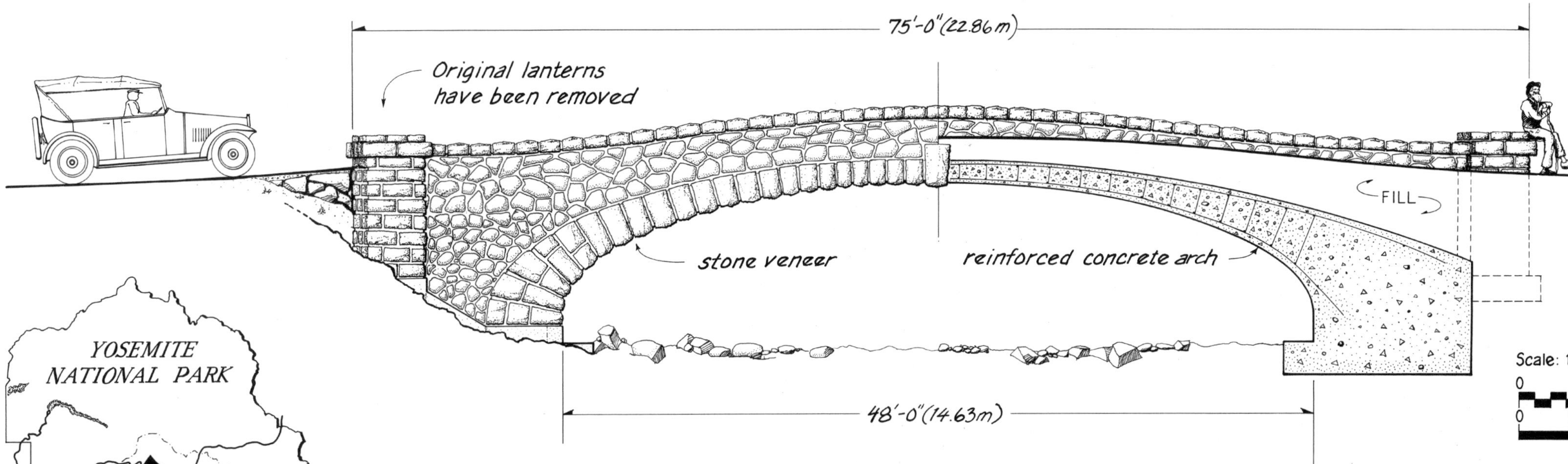

Drawing based on field measurements, historic photographs and original design documents..... See NPS drawing file NP-YOS-50

Scale: 1/4" = 1'0"
0 5 10 20 Feet
0 1 2 3 4 5 Meters

ELEVATION / SECTION

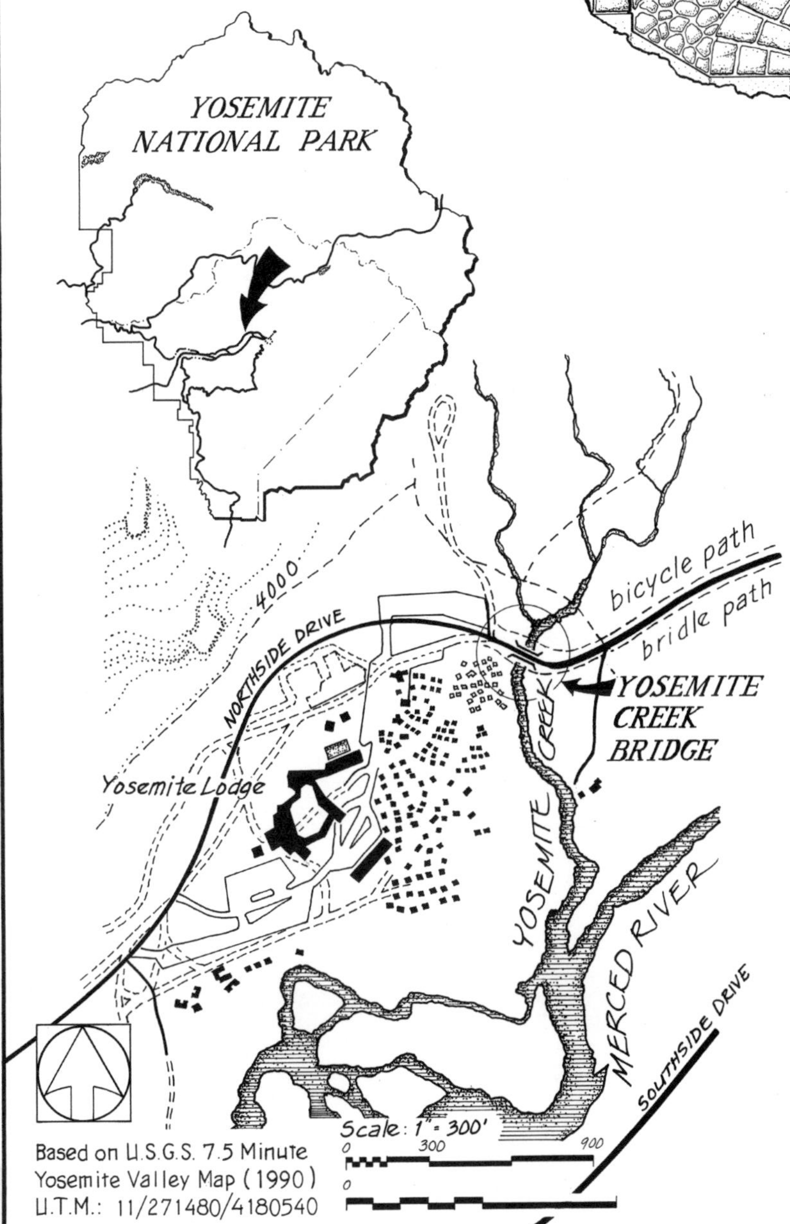

Based on U.S.G.S. 7.5 Minute Yosemite Valley Map (1990)
U.T.M.: 11/271480/4180540

Scale: 1" = 300'
0 300 900

James Mason Hutchings, Yosemite pioneer and entrepreneur, built the first bridge over Yosemite Creek before 1864. It was a small wood structure, providing access for guests at Hutchings' hotel to the base of Yosemite Falls, the tallest waterfall in North America and one of the chief attractions of the Yosemite Valley. Hutchings' bridge was destroyed by a flood in December, 1867. Apparently, Hutchings did not replace the span. The second bridge on the site was probably built by the Yosemite Board of Commissioners, which established a circuit road around the Valley in the 1880's.

The Acting Superintendent of Yosemite National Park noted that a bridge on this location was replaced in the fall of 1912. The new bridge lasted only about a decade. In 1920, Park Superintendent Washington B. Lewis described it as a "prehistoric bridge" in a "very marked state of decay," and called for its replacement. Funds for a new structure were quickly appropriated. Lewis reported in 1921 that a new masonry-faced concrete arch bridge had been completed over Yosemite Creek at a total cost of $32,000. The Yosemite Creek Bridge is the oldest of the stone-faced masonry arch bridges in Yosemite Valley.

This recording project is part of the Historic American Engineering Record (HAER), a long-range program to document historically significant engineering and industrial works in the United States. HAER is an agency of the National Park Service, U.S. Department of the Interior. The Yosemite Roads and Bridges Project was co-sponsored by Yosemite National Park, Michael Finley, Superintendent; Kevin Cann, Chief of Maintenance and Engineering; the NPS Roads and Bridges Program, John Gingles, Manager; and the Historic American Building Survey/Historic American Engineering Record (HABS/HAER) under the general direction of Robert J. Kapsch, Chief.

The Fieldwork, measured drawings, histories and photographs were completed under the direction of Eric DeLony, Chief and Principal Architect, HAER. The recording team consisted of Industrial Designer Todd A. Croteau, field supervisor; Architectural Technicians Dione DeMartelaere, David R. Fleming and Marie-Claude LeSauteur; Project Historian, Richard H. Quin. Formal on-site photography was done by Brian C. Grogan.

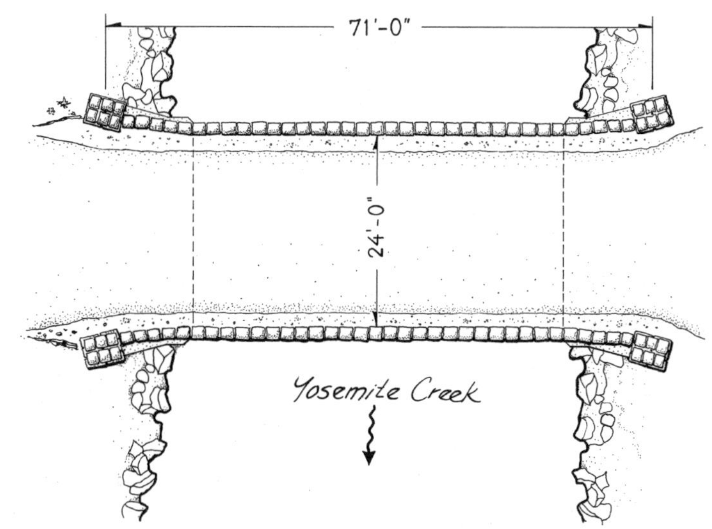

Scale: 3/32" = 1'0"
0 10 20 30 40 Feet
0 5 10 Meters

PLAN

DELINEATED BY: TODD A. CROTEAU 1991

YOSEMITE ROADS AND BRIDGES RECORDING PROJECT
NATIONAL PARK SERVICE
UNITED STATES DEPARTMENT OF THE INTERIOR

YOSEMITE CREEK BRIDGE - 1921
NORTHSIDE DRIVE SPANNING YOSEMITE CREEK
YOSEMITE NATIONAL PARK
MARIPOSA COUNTY
CALIFORNIA

SHEET 1 of 1

HISTORIC AMERICAN ENGINEERING RECORD
CA-102

IF REPRODUCED, PLEASE CREDIT: HISTORIC AMERICAN ENGINEERING RECORD, NATIONAL PARK SERVICE, NAME OF DELINEATOR, DATE OF THE DRAWING

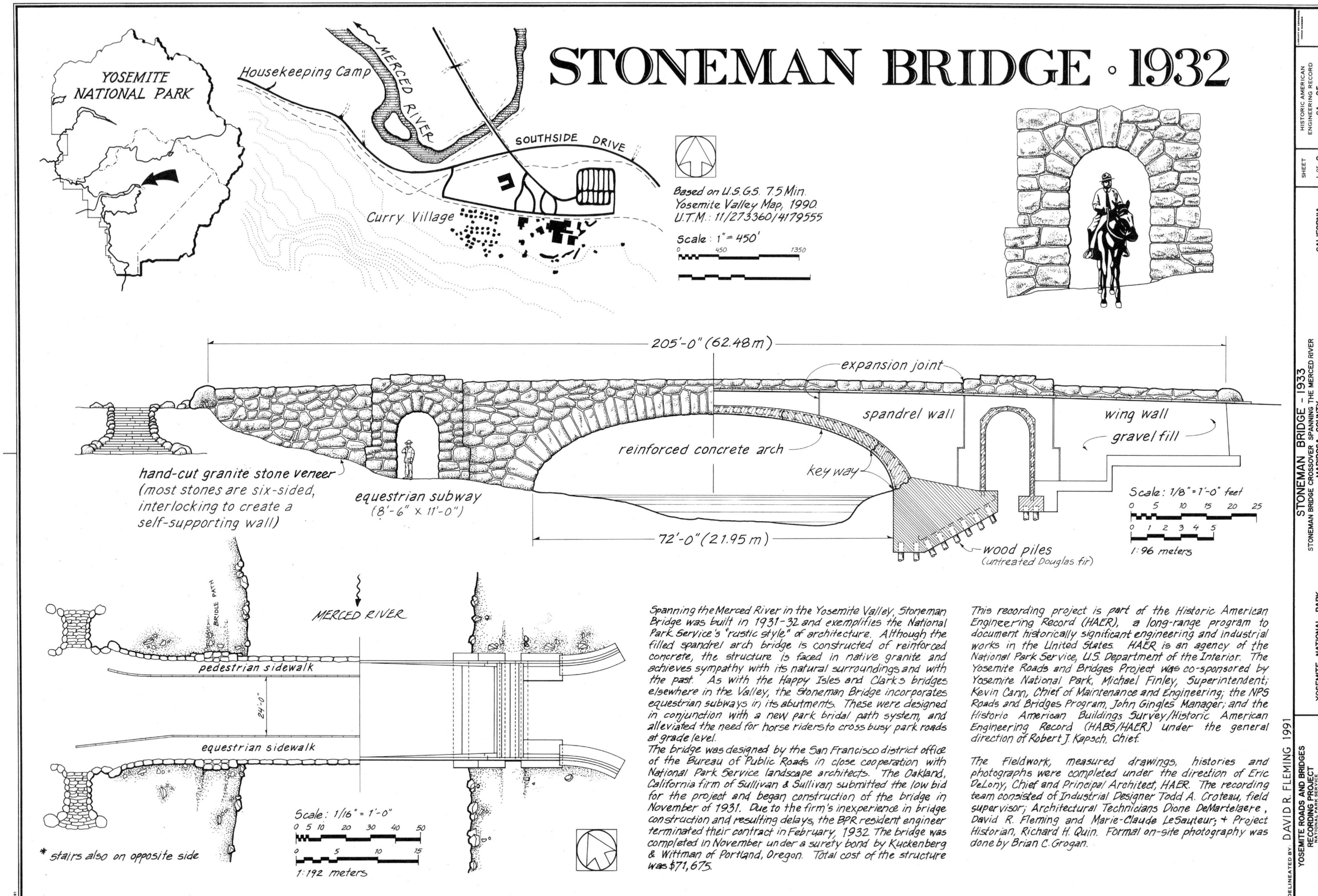

STONEMAN BRIDGE · 1932
YOSEMITE NATIONAL PARK
Housekeeping Camp
MERCED RIVER
SOUTHSIDE DRIVE
Curry Village
Based on U.S.G.S. 7.5 Min. Yosemite Valley Map, 1990 U.T.M.: 11/273360/4179555
Scale: 1" = 450'
0 450 1350
205'-0" (62.48 m)
expansion joint
spandrel wall
wing wall
gravel fill
reinforced concrete arch
key way
hand-cut granite stone veneer (most stones are six-sided, interlocking to create a self-supporting wall)
equestrian subway (8'-6" x 11'-0")
72'-0" (21.95 m)
wood piles (untreated Douglas fir)
Scale: 1/8" = 1'-0" feet
0 5 10 15 20 25
0 1 2 3 4 5
1:96 meters
BRIDLE PATH
MERCED RIVER
pedestrian sidewalk
24'-0"
equestrian sidewalk
Scale: 1/16" = 1'-0"
0 5 10 20 30 40 50
0 5 10 15
1:192 meters
* stairs also on opposite side
Spanning the Merced River in the Yosemite Valley, Stoneman Bridge was built in 1931-32 and exemplifies the National Park Service's "rustic style" of architecture. Although the filled spandrel arch bridge is constructed of reinforced concrete, the structure is faced in native granite and achieves sympathy with its natural surroundings and with the past. As with the Happy Isles and Clarks bridges elsewhere in the Valley, the Stoneman Bridge incorporates equestrian subways in its abutments. These were designed in conjunction with a new park bridal path system, and alleviated the need for horse riders to cross busy park roads at grade level.
The bridge was designed by the San Francisco district office of the Bureau of Public Roads in close cooperation with National Park Service landscape architects. The Oakland, California firm of Sullivan & Sullivan submitted the low bid for the project and began construction of the bridge in November of 1931. Due to the firm's inexperience in bridge construction and resulting delays, the BPR resident engineer terminated their contract in February, 1932. The bridge was completed in November under a surety bond by Kuckenberg & Wittman of Portland, Oregon. Total cost of the structure was $71,675.
This recording project is part of the Historic American Engineering Record (HAER), a long-range program to document historically significant engineering and industrial works in the United States. HAER is an agency of the National Park Service, U.S. Department of the Interior. The Yosemite Roads and Bridges Project was co-sponsored by Yosemite National Park, Michael Finley, Superintendent; Kevin Cann, Chief of Maintenance and Engineering; the NPS Roads and Bridges Program, John Gingles Manager; and the Historic American Buildings Survey/Historic American Engineering Record (HABS/HAER) under the general direction of Robert J. Kapsch, Chief.
The Fieldwork, measured drawings, histories and photographs were completed under the direction of Eric DeLony, Chief and Principal Architect, HAER. The recording team consisted of Industrial Designer Todd A. Croteau, field supervisor; Architectural Technicians Dione DeMartelaere, David R. Fleming and Marie-Claude LeSauteur; + Project Historian, Richard H. Quin. Formal on-site photography was done by Brian C. Grogan.
DELINEATED BY: DAVID R. FLEMING 1991
YOSEMITE ROADS AND BRIDGES RECORDING PROJECT
NATIONAL PARK SERVICE
UNITED STATES DEPARTMENT OF THE INTERIOR
YOSEMITE NATIONAL PARK
STONEMAN BRIDGE - 1933
STONEMAN BRIDGE CROSSOVER SPANNING THE MERCED RIVER
MARIPOSA COUNTY
CALIFORNIA
SHEET 1 OF 2
HISTORIC AMERICAN ENGINEERING RECORD
CA-95
IF REPRODUCED, PLEASE CREDIT HISTORIC AMERICAN ENGINEERING RECORD, NATIONAL PARK SERVICE, NAME OF DELINEATOR, DATE OF THE DRAWING

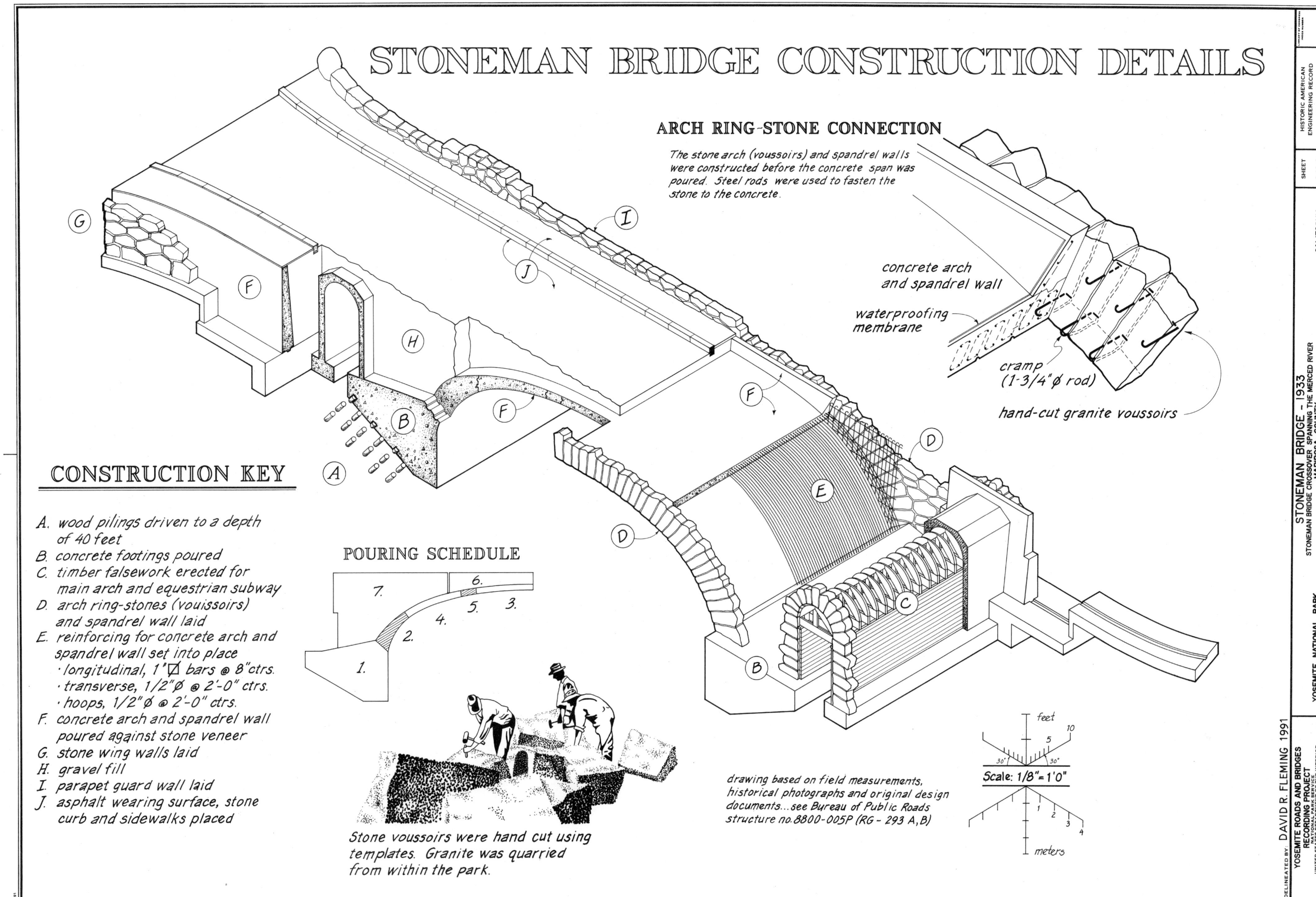
STONEMAN BRIDGE CONSTRUCTION DETAILS
ARCH RING-STONE CONNECTION
The stone arch (voussoirs) and spandrel walls were constructed before the concrete span was poured. Steel rods were used to fasten the stone to the concrete.
concrete arch and spandrel wall
waterproofing membrane
cramp (1-3/4"ø rod)
hand-cut granite voussoirs
G
F
H
B
A
F
I
J
F
D
E
D
C
B
CONSTRUCTION KEY
A. wood pilings driven to a depth of 40 feet
B. concrete footings poured
C. timber falsework erected for main arch and equestrian subway
D. arch ring-stones (voussoirs) and spandrel wall laid
E. reinforcing for concrete arch and spandrel wall set into place
· longitudinal, 1"☐ bars @ 8" ctrs.
· transverse, 1/2"ø @ 2'-0" ctrs.
· hoops, 1/2"ø @ 2'-0" ctrs.
F. concrete arch and spandrel wall poured against stone veneer
G. stone wing walls laid
H. gravel fill
I. parapet guard wall laid
J. asphalt wearing surface, stone curb and sidewalks placed
POURING SCHEDULE
7.
6.
5.
3.
4.
2.
1.
Stone voussoirs were hand cut using templates. Granite was quarried from within the park.
drawing based on field measurements, historical photographs and original design documents... see Bureau of Public Roads structure no.8800-005P (RG - 293 A,B)
feet
5
10
30°
30°
Scale: 1/8" = 1'0"
1
2
3
4
meters
HISTORIC AMERICAN ENGINEERING RECORD
CA - 95
SHEET 2 OF 2
CALIFORNIA
STONEMAN BRIDGE - 1933
STONEMAN BRIDGE CROSSOVER SPANNING THE MERCED RIVER
MARIPOSA COUNTY
YOSEMITE NATIONAL PARK
DELINEATED BY: DAVID R. FLEMING 1991
YOSEMITE ROADS AND BRIDGES RECORDING PROJECT
NATIONAL PARK SERVICE
UNITED STATES DEPARTMENT OF THE INTERIOR
IF REPRODUCED, PLEASE CREDIT: HISTORIC AMERICAN ENGINEERING RECORD, NATIONAL PARK SERVICE, NAME OF DELINEATOR, DATE OF THE DRAWING

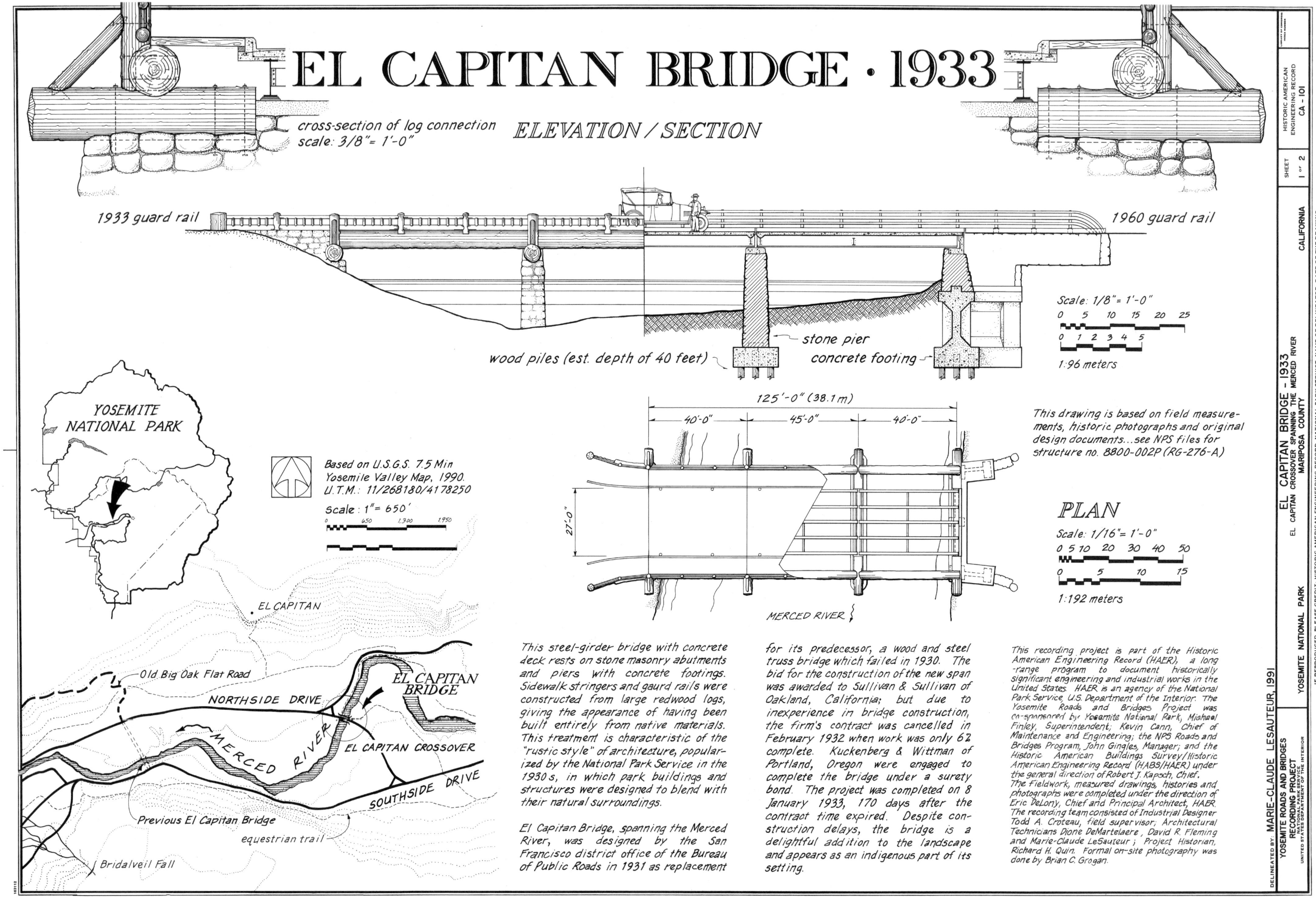
EL CAPITAN BRIDGE · 1933
cross-section of log connection
scale: 3/8" = 1'-0"
ELEVATION / SECTION
1933 guard rail
1960 guard rail
Scale: 1/8" = 1'-0"
0 5 10 15 20 25
0 1 2 3 4 5
1:96 meters
stone pier
wood piles (est. depth of 40 feet)
concrete footing
125'-0" (38.1 m)
40'-0"
45'-0"
40'-0"
27'-0"
MERCED RIVER
This drawing is based on field measurements, historic photographs and original design documents...see NPS files for structure no. 8800-002P (RG-276-A)
PLAN
Scale: 1/16" = 1'-0"
0 5 10 20 30 40 50
0 5 10 15
1:192 meters
YOSEMITE NATIONAL PARK
Based on U.S.G.S. 7.5 Min Yosemite Valley Map, 1990.
U.T.M.: 11/268180/4178250
Scale: 1" = 650'
0 650 1,300 1,950
EL CAPITAN
Old Big Oak Flat Road
NORTHSIDE DRIVE
EL CAPITAN BRIDGE
MERCED RIVER
EL CAPITAN CROSSOVER
SOUTHSIDE DRIVE
Previous El Capitan Bridge
equestrian trail
Bridalveil Fall
This steel-girder bridge with concrete deck rests on stone masonry abutments and piers with concrete footings. Sidewalk stringers and gaurd rails were constructed from large redwood logs, giving the appearance of having been built entirely from native materials. This treatment is characteristic of the "rustic style" of architecture, popularized by the National Park Service in the 1930s, in which park buildings and structures were designed to blend with their natural surroundings.
El Capitan Bridge, spanning the Merced River, was designed by the San Francisco district office of the Bureau of Public Roads in 1931 as replacement for its predecessor, a wood and steel truss bridge which failed in 1930. The bid for the construction of the new span was awarded to Sullivan & Sullivan of Oakland, California; but due to inexperience in bridge construction, the firm's contract was cancelled in February 1932 when work was only 62 complete. Kuckenberg & Wittman of Portland, Oregon were engaged to complete the bridge under a surety bond. The project was completed on 8 January 1933, 170 days after the contract time expired. Despite construction delays, the bridge is a delightful addition to the landscape and appears as an indigenous part of its setting.
This recording project is part of the Historic American Engineering Record (HAER), a long-range program to document historically significant engineering and industrial works in the United States. HAER is an agency of the National Park Service, U.S. Department of the Interior. The Yosemite Roads and Bridges Project was co-sponsored by Yosemite National Park, Michael Finley, Superintendent; Kevin Cann, Chief of Maintenance and Engineering; the NPS Roads and Bridges Program, John Gingles, Manager; and the Historic American Buildings Survey/Historic American Engineering Record (HABS/HAER) under the general direction of Robert J. Kapsch, Chief.
The Fieldwork, measured drawings, histories and photographs were completed under the direction of Eric DeLony, Chief and Principal Architect, HAER. The recording team consisted of Industrial Designer Todd A. Croteau, field supervisor; Architectural Technicians Dione DeMartelaere, David R. Fleming and Marie-Claude LeSauteur; Project Historian, Richard H. Quin. Formal on-site photography was done by Brian C. Grogan.
DELINEATED BY: MARIE-CLAUDE LESAUTEUR, 1991
YOSEMITE ROADS AND BRIDGES RECORDING PROJECT
NATIONAL PARK SERVICE
UNITED STATES DEPARTMENT OF THE INTERIOR
YOSEMITE NATIONAL PARK
EL CAPITAN BRIDGE - 1933
EL CAPITAN CROSSOVER SPANNING THE MERCED RIVER
MARIPOSA COUNTY
CALIFORNIA
SHEET
1 OF 2
HISTORIC AMERICAN ENGINEERING RECORD
CA-101
IF REPRODUCED, PLEASE CREDIT: HISTORIC AMERICAN ENGINEERING RECORD, NATIONAL PARK SERVICE, NAME OF DELINEATOR, DATE OF THE DRAWING

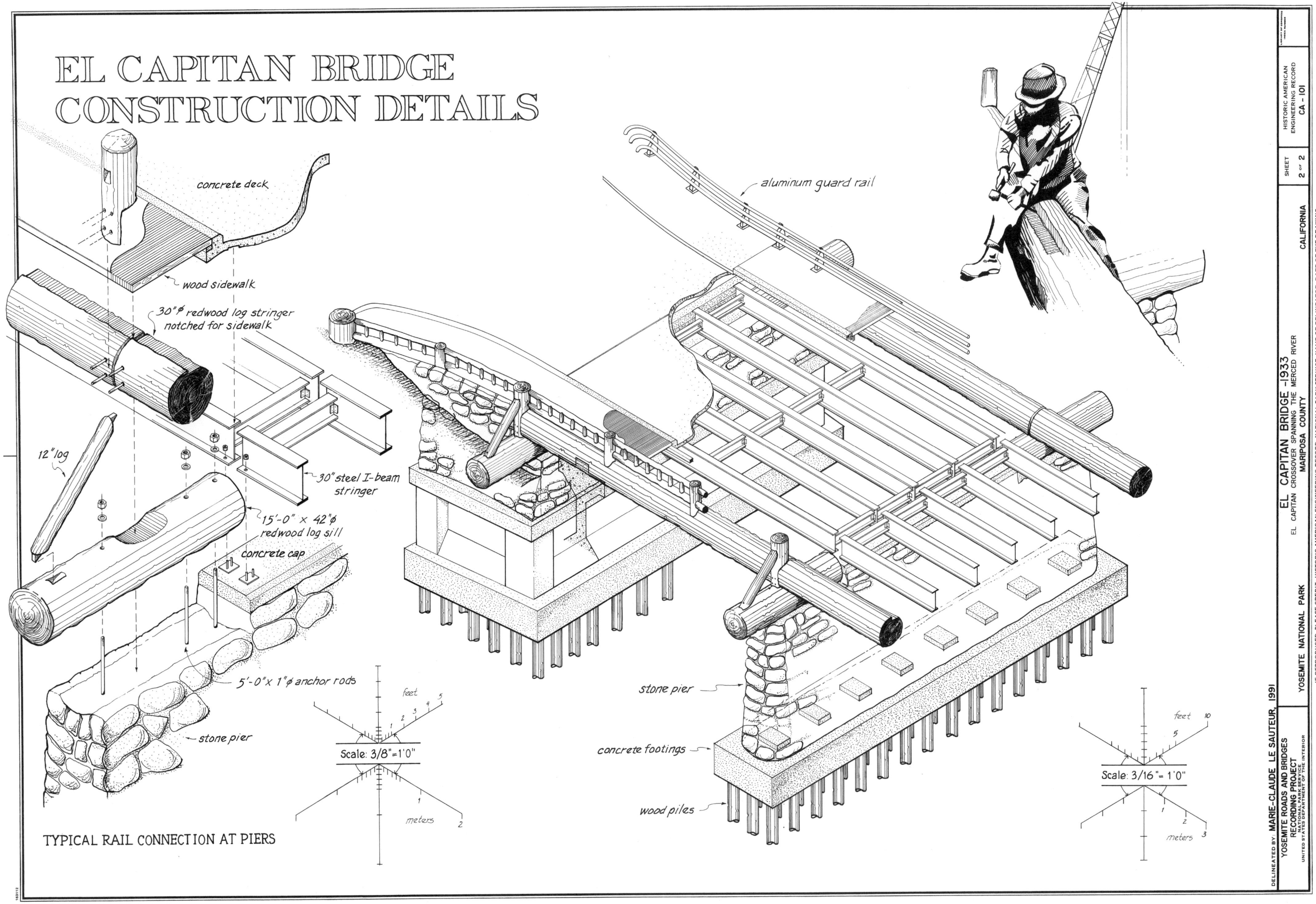
EL CAPITAN BRIDGE
CONSTRUCTION DETAILS
concrete deck
wood sidewalk
30"ϕ redwood log stringer notched for sidewalk
12" log
30" steel I-beam stringer
15'-0" × 42"ϕ redwood log sill
concrete cap
5'-0" x 1"ϕ anchor rods
stone pier
Scale: 3/8"=1'0"
feet
meters
TYPICAL RAIL CONNECTION AT PIERS
aluminum guard rail
stone pier
concrete footings
wood piles
Scale: 3/16"= 1'0"
DELINEATED BY: MARIE-CLAUDE LE SAUTEUR, 1991
YOSEMITE ROADS AND BRIDGES RECORDING PROJECT
NATIONAL PARK SERVICE
UNITED STATES DEPARTMENT OF THE INTERIOR
EL CAPITAN BRIDGE -1933
EL CAPITAN CROSSOVER SPANNING THE MERCED RIVER
YOSEMITE NATIONAL PARK
MARIPOSA COUNTY
CALIFORNIA
SHEET 2 OF 2
HISTORIC AMERICAN ENGINEERING RECORD
CA - 101
IF REPRODUCED, PLEASE CREDIT: HISTORIC AMERICAN ENGINEERING RECORD, NATIONAL PARK SERVICE, NAME OF DELINEATOR, DATE OF THE DRAWING

WAWONA TUNNEL · 1933

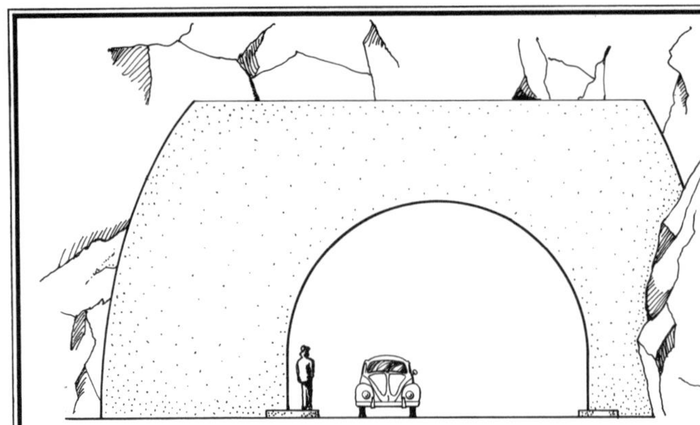

WEST PORTAL

EAST PORTAL

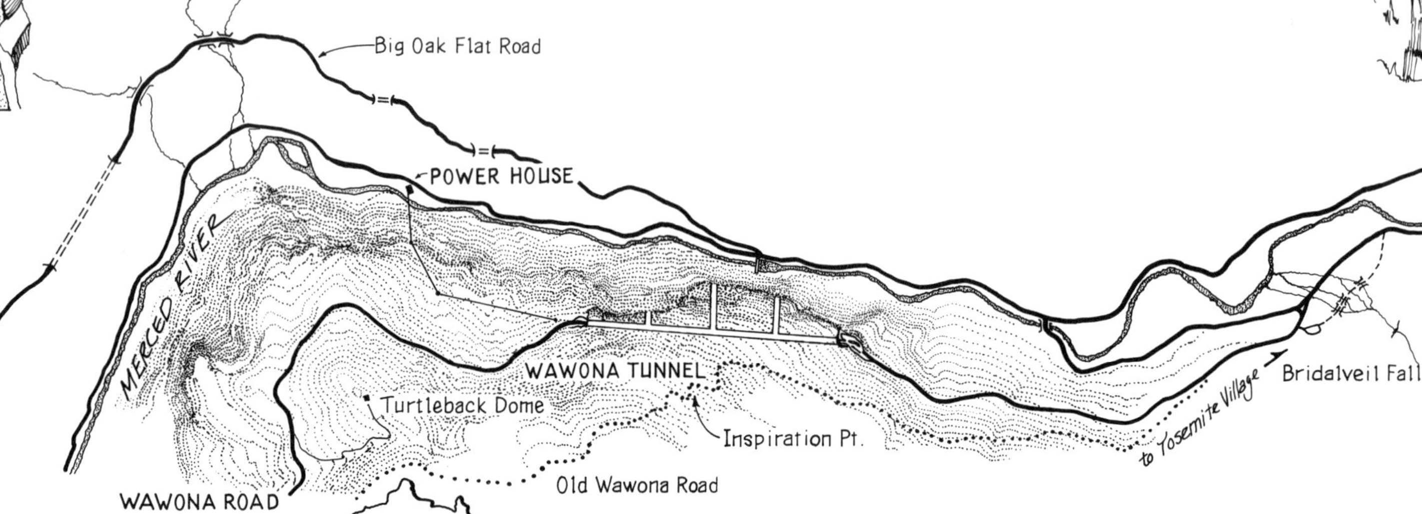

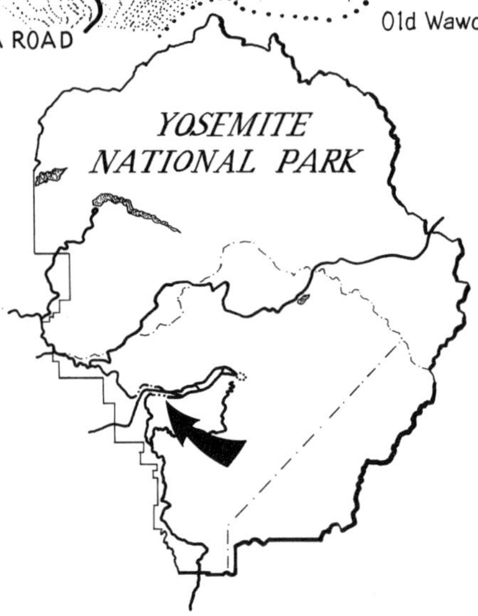

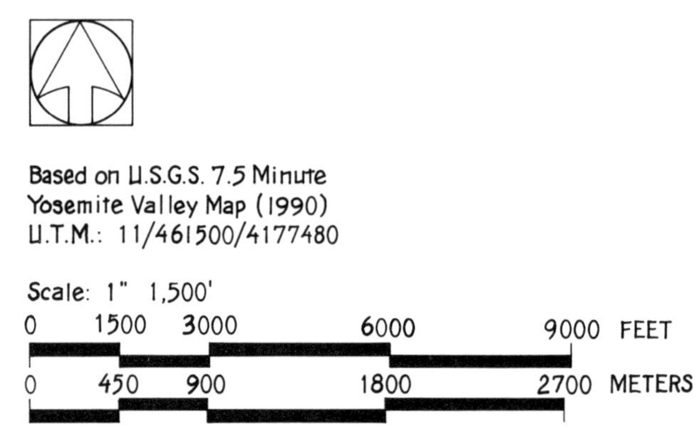

Completed in 1933, the Wawona Tunnel was the longest vehicular tunnel in the western United States at the time of its construction. It is significant not only for its engineering advancements, but also for minimizing the negative effects of road construction on the Yosemite landscape by avoiding disfigurement of the Valley walls. The 4,230 foot long tunnel carries the Wawona Road through solid granite east of Turtleback Dome. Relocation of the old route via a tunnel guaranteed a minimal impact of the new road on the Yosemite Valley landscape and created a major new vista for visitors entering the park from the south.

The Bureau of Public Roads designed and supervised the construction, which was done by contractor Goerig & Dahlberg of Seattle, Washington. Work began on 30 January 1931. The tunnel was bored through solid granite and was originally unlined except for sections where rock failures necessitated a concrete lining. The tunnel was bored from the east, or lower end only. More than 275 tons of blasting powder was used in shots averaging 2,000 pounds each and 85 tons of drill steel was worn out. Two battery-powered locomotives and twenty dump cars were employed to move the excavated rock. The tunnel bore was completed on 6 January 1932. Three fresh air adits or openings were provided for natural ventilation. In addition, three 9-foot diameter ventilating fans, activated by carbon monoxide detectors, were located in the central adit to supplement natural flow. When carbon monoxide built up beyond the capacity of the ventilation system, semaphores at the tunnel ends would drop and warning klaxons were sounded to halt traffic from entering.

The tunnel grade is five percent, and that of the approaches six percent, with a transverse grade of two inches down per 24 feet of road to the north. The maximum depth of rock above the tunnel is 550 feet; maximum distance in from the cliff face is 503 feet. The east portal was left unfinished so as not to call attention to the structure; the west portal is a concrete vault.

Work was completed without one serious injury or death, on 13 April, 1933 by Goerig & Dahlberg (A.C. Goerig Company after April 1932), at a total cost of $690,000. Formal dedication ceremonies for the tunnel were held on site 10 June 1933.

Two turnouts constructed outside the east portal provide visitors with dramatic new vistas of the Yosemite Valley. These parking areas, built on fill blasted from the tunnel bore, are now one of the most popular panoramas in Yosemite National Park. Called "Discovery View," the turnout presents the visitor with stunning views of the Valley dominated by Half Dome, Bridalveil Fall and El Capitan.

This recording project is part of the Historic American Engineering Record (HAER), a long-range program to document historically significant engineering and industrial works in the United States. HAER is an agency of the National Park Service, U.S. Department of the Interior. The Yosemite Roads and Bridges Project was co-sponsored by Yosemite National Park, Michael Finley, Superintendent; Kevin Cann, Chief of Maintenance and Engineering; the NPS Roads and Bridges Program, John Gingles, Manager; and the Historic American Buildings Survey/Historic American Engineering Record (HABS/HAER) under the general direction of Robert J. Kapsch, Chief.

The Fieldwork, measured drawings, histories and photographs were completed under the direction of Eric DeLony, Chief and Principal Architect, HAER. The recording team consisted of Industrial Designer Todd A. Croteau, field supervisor; Architectural Technicians Dione DeMartelaere, David R. Fleming and Marie-Claude LeSauteur; Project Historian, Richard H. Quin. Formal on-site photography was done by Brian C. Grogan.

LENGTH: 4,230ft.(1,289.3m) WIDTH: 28ft.(8.4m) HEIGHT: 19ft.(5.8m) GRADE: 5%

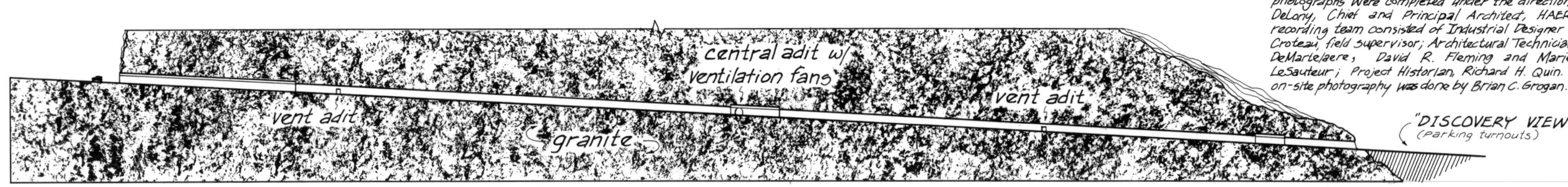

SECTION

Scale: 1"= 200'

DELINEATED BY: TODD A. CROTEAU 1991

YOSEMITE ROADS AND BRIDGES RECORDING PROJECT
NATIONAL PARK SERVICE
UNITED STATES DEPARTMENT OF THE INTERIOR

WAWONA TUNNEL - 1933
CALIFORNIA HIGHWAY 41 EAST OF TURTLEBACK DOME
YOSEMITE NATIONAL PARK — MARIPOSA COUNTY — CALIFORNIA

SHEET 1 OF 2

HISTORIC AMERICAN ENGINEERING RECORD CA - 105

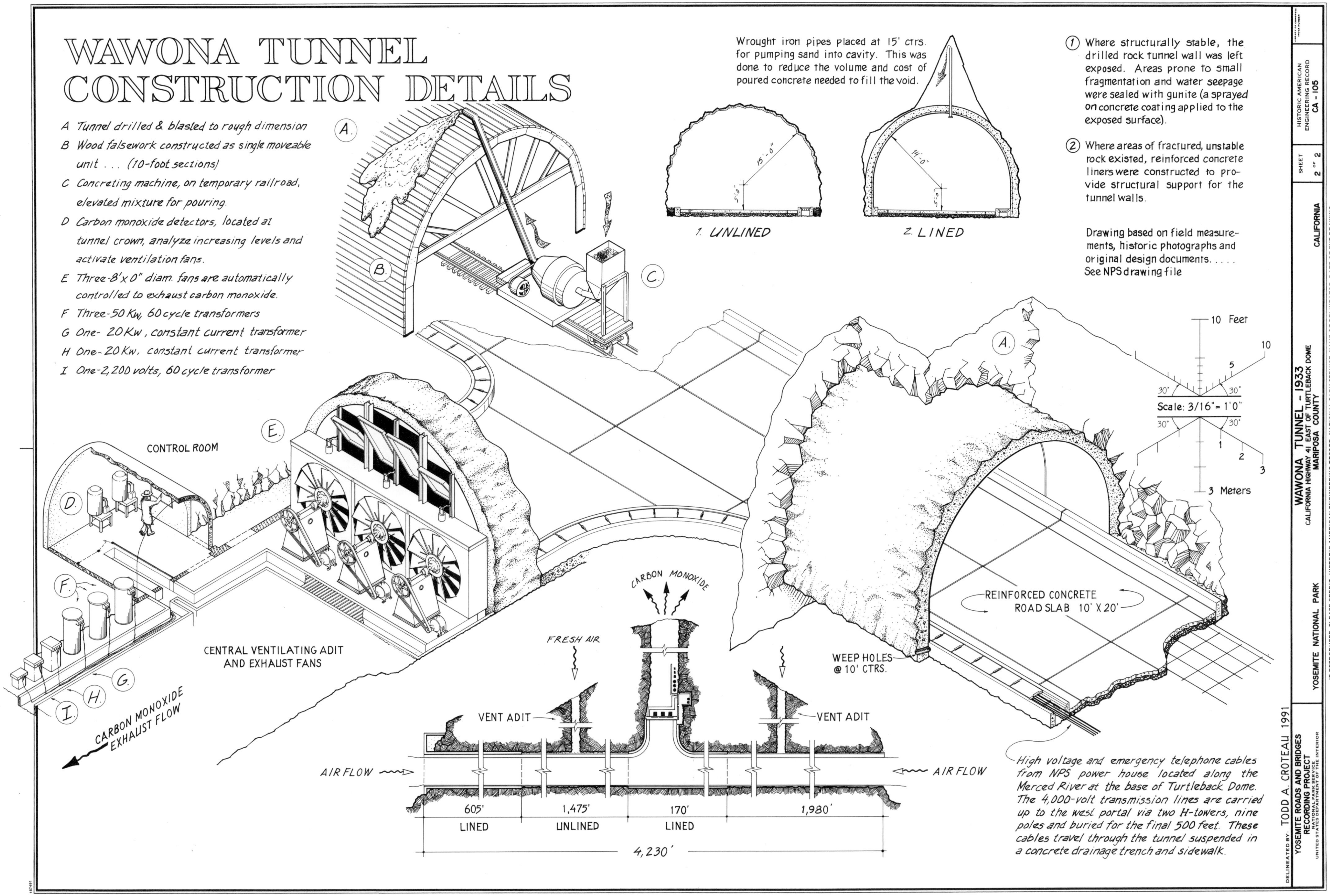

WAWONA TUNNEL
CONSTRUCTION DETAILS
A Tunnel drilled & blasted to rough dimension
B Wood falsework constructed as single moveable unit . . . (10-foot sections)
C Concreting machine, on temporary railroad, elevated mixture for pouring.
D Carbon monoxide detectors, located at tunnel crown, analyze increasing levels and activate ventilation fans.
E Three-8'x 0" diam. fans are automatically controlled to exhaust carbon monoxide.
F Three-50 Kw, 60 cycle transformers
G One- 20 Kw, constant current transformer
H One-20 Kw, constant current transformer
I One-2,200 volts, 60 cycle transformer
Wrought iron pipes placed at 15' ctrs. for pumping sand into cavity. This was done to reduce the volume and cost of poured concrete needed to fill the void.
15'-0"
5'-0"
1. UNLINED
14'-0"
5'-0"
2. LINED
(1) Where structurally stable, the drilled rock tunnel wall was left exposed. Areas prone to small fragmentation and water seepage were sealed with gunite (a sprayed on concrete coating applied to the exposed surface).
(2) Where areas of fractured, unstable rock existed, reinforced concrete liners were constructed to provide structural support for the tunnel walls.
Drawing based on field measurements, historic photographs and original design documents. . . . . See NPS drawing file
10 Feet
10
5
30°
30°
Scale: 3/16" = 1'0"
30°
30°
1
2
3
3 Meters
A.
B.
C.
A.
E.
D.
F.
G.
H.
I.
CONTROL ROOM
CENTRAL VENTILATING ADIT AND EXHAUST FANS
CARBON MONOXIDE EXHAUST FLOW
CARBON MONOXIDE
FRESH AIR
REINFORCED CONCRETE ROAD SLAB 10' X 20'
WEEP HOLES @ 10' CTRS.
VENT ADIT
VENT ADIT
AIR FLOW
AIR FLOW
605'
LINED
1,475'
UNLINED
170'
LINED
1,980'
4,230'
High voltage and emergency telephone cables from NPS power house located along the Merced River at the base of Turtleback Dome. The 4,000-volt transmission lines are carried up to the west portal via two H-towers, nine poles and buried for the final 500 feet. These cables travel through the tunnel suspended in a concrete drainage trench and sidewalk.
DELINEATED BY: TODD A. CROTEAU 1991
YOSEMITE ROADS AND BRIDGES RECORDING PROJECT
WAWONA TUNNEL - 1933
CALIFORNIA HIGHWAY 41 EAST OF TURTLEBACK DOME
MARIPOSA COUNTY
YOSEMITE NATIONAL PARK
CALIFORNIA
IF REPRODUCED, PLEASE CREDIT: HISTORIC AMERICAN ENGINEERING RECORD, NATIONAL PARK SERVICE, NAME OF DELINEATOR, DATE OF THE DRAWING
SHEET 2 of 2
HISTORIC AMERICAN ENGINEERING RECORD
CA-105

CASCADE CREEK BRIDGE ·1939·

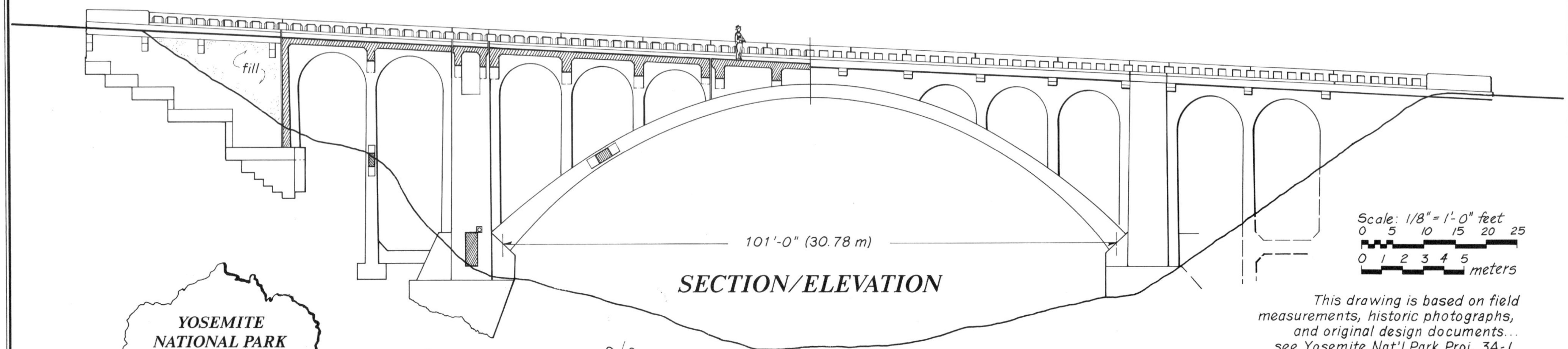

SECTION/ELEVATION

This drawing is based on field measurements, historic photographs, and original design documents... see Yosemite Nat'l Park Proj. 3A-1

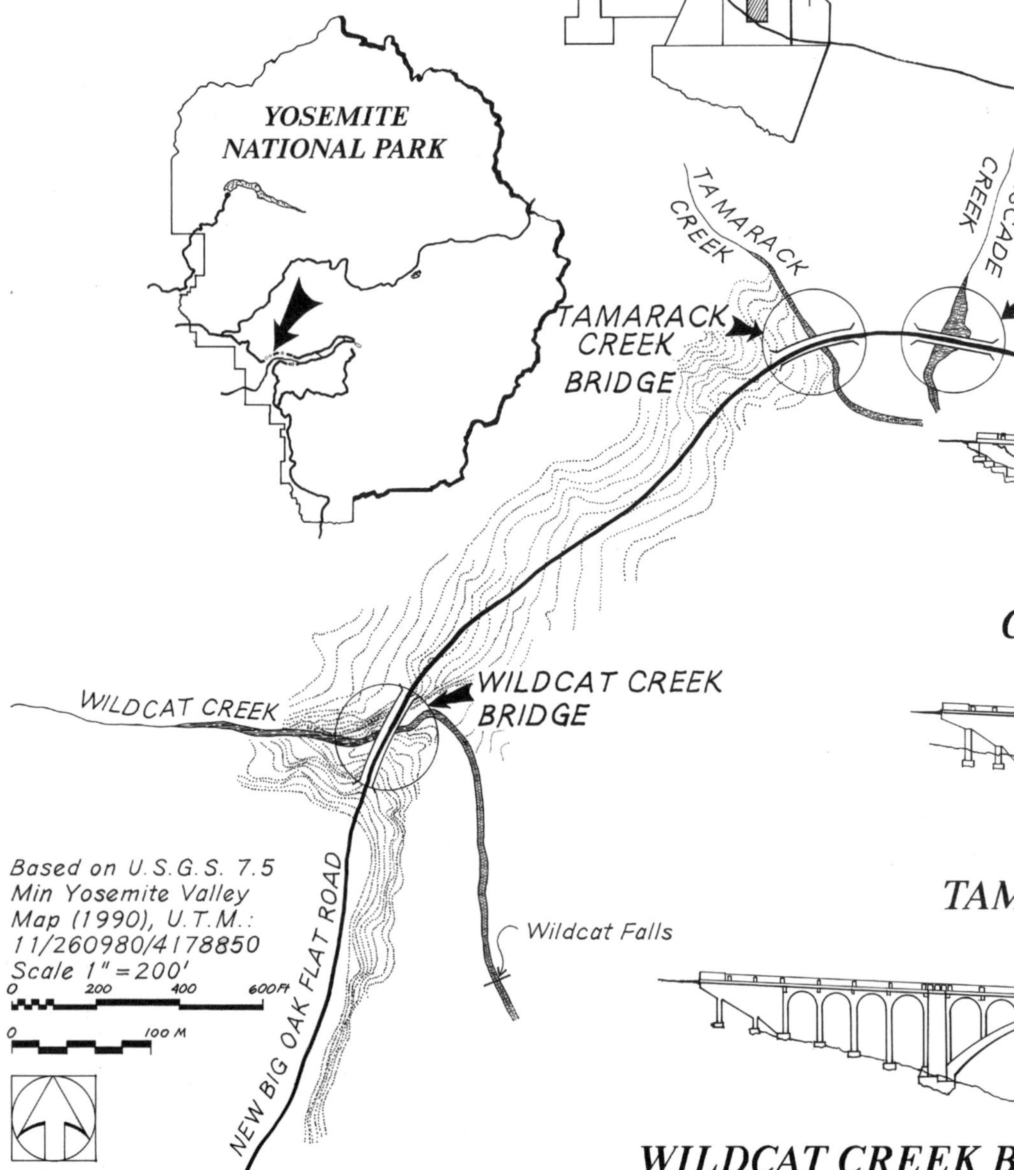

CASCADE CREEK BRIDGE
HAER NO. CA-85

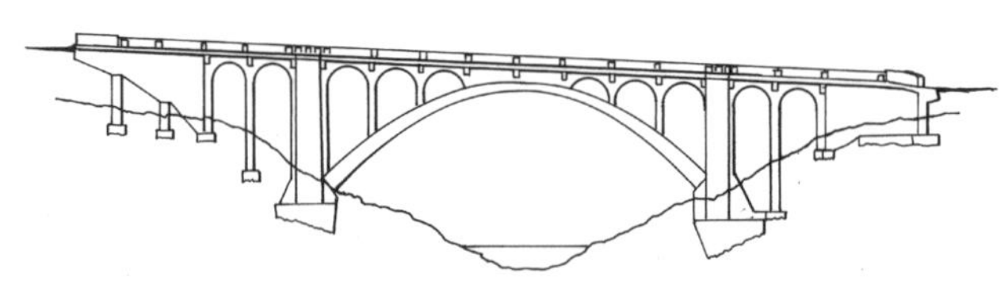

TAMARACK CREEK BRIDGE
HAER NO. CA-84

WILDCAT CREEK BRIDGE HAER NO. CA-83

The 1939 bridge over Cascade Creek on the New Big Oak Flat Road, like adjacent structures over Wildcat and Tamarack creeks, marks a departure from earlier park bridges constructed with facings of native materials in the "rustic style" popularized by the National Park Service. The new reinforced concrete bridges are characterized by their graceful open spandrel arches and reflect the streamline aesthetic of the 1930s.

The Cascade Creek and neighboring bridges were designed in 1937 and 1938 by the San Francisco regional office of the Public Roads Administration, a federal relief program which was the Depression-era successor to the Bureau of Public Roads. Contractor for the construction was John Rocca of San Rafael, California. Construction began in October 1938. The three bridges were built under a combined contract and were completed 21 October 1939 at a cost of $161,823,69.

The three bridges, along with three tunnels and a series of stone retaining and parapet walls, are the major features of the New Big Oak Flat Road, constructed between 1935 and 1940. The new highway was part of a major road rebuilding project in Yosemite Park which began in 1932 with the Wawona Road and concluded with the reconstruction of parts of the Tioga Road in the late 1930s. The New Big Oak Flat Road remains the chief access road to Yosemite Valley from northern California.

This recording project is part of the Historic American Engineering Record (HAER), a long-range program to document historically significant engineering and industrial works in the United States. HAER is an agency of the National Park Service, U.S. Department of the Interior. The Yosemite Roads and Bridges Project was cosponsored by Yosemite National Park, Michael Finley, Superintendent; Kevin Cann, Chief of Engineering; the NPS Roads and Bridges Program, John Gingles, Manager; and the Historic American Buildings Survey/Historic American Engineering Record (HABS/HAER) under the general direction of Robert J. Kapsch, Chief.

The fieldwork, measured drawings, histories and photography were completed under the direction of Eric N. DeLony, Chief and Principal Architect, HAER. The recording team consisted of Industrial Designer Todd A. Croteau, field supervisor (Rhode Island School of Design); Architectural Technicians Dione de Martelaere (North Dakota State University), David R. Fleming (the Catholic University of America), and Marie-Claude Lesauteur (U.S. ICOMOS, McGill University); and Project Historian Richard H. Quin (Middle Tennessee State University). Formal on-site photography was done by Brian C. Grogan (Hampden-Sydney College).

DELINEATED BY: Todd Croteau, Christopher Marston, 1991

YOSEMITE ROADS AND BRIDGES RECORDING PROJECT
NATIONAL PARK SERVICE
UNITED STATES DEPARTMENT OF THE INTERIOR

CASCADE CREEK BRIDGE - 1939
NEW BIG OAK FLAT ROAD SPANNING CASADE CREEK
YOSEMITE NATIONAL PARK MARIPOSA COUNTY CALIFORNIA

SHEET 1 OF 1

HISTORIC AMERICAN ENGINEERING RECORD
CA-85

YOSEMITE

National Park Roads and Bridges

View from Inspiration Point along Old Wawona Road

Tourists began visiting Yosemite Valley in the 1850s to see its "scenes of wonder and curiosity." The journey was extremely difficult, as no roads yet existed. Early visitors endured miles of arduous travel over primitive horse trails. They often arrived too exhausted to fully enjoy the glorious landscape awaiting them.

Following the designation of the Yosemite Grant by the federal government in 1864, citizens in Tuolumne and Mariposa counties sought to entice Yosemite-bound visitors by constructing toll roads to the valley. In the early 1870s, companies constructing toll roads from the communities of Coulterville, Mariposa and Big Oak Flat raced to be the first to complete a wagon route and profit from the resulting traffic. The Coulterville & Yosemite Turnpike won, opening on 17 June 1874. The Chinese Camp & Yo Semite Turnpike reached the valley from Big Oak Flat less than a month later. A year later, Albert Henry Washburn and his partners pushed a road from Mariposa to Big Tree Station (Wawona) and on to Yosemite Valley. These roads were difficult to construct, as only hand tools and blasting powder were available. The three toll road companies competed for a limited number of visitors. The Big Oak Flat Road managed to break even, but the Coulterville Road was a financial disaster. Only the Washburn group realized any substantial return on their investment.

The construction of toll roads allowed stagecoaches to replace saddle horses as the primary means of transportation to Yosemite. With the extension of railway service to nearby communities, stagecoach companies began regular service to and from the valley. The trip remained difficult, as the primitive roads were steep, narrow, and dusty.

North of Yosemite Valley, The Great Sierra Consolidated Silver Mining Company constructed a wagon road to serve its mines on the Sierra Crest in 1882. The Great Sierra Wagon Road, commonly called the Tioga Road, remained in use for only a year before the mines played out and the road was abandoned.
In 1907, the Yosemite Valley Rail Road completed a line to El Portal on the western edge of the park and built a connecting stage road to the valley. Most visitors took this route and the longer stage roads declined. By the early twentieth century, the roads came under government control.

The first automobile entered Yosemite Valley in 1900. In response to increasing automobile traffic, park authorities banned their use in 1907. Outraged motorists aided by the California Automobile Association convinced the Secretary of the Interior to overturn the ban in 1913. The automobile soon became the favored mode of transportation. The State of California built the improved "All-Year Highway" up the Merced River to El Portal, and motorists began using this route in conjunction with the former stage road to reach the valley in all seasons.

In the 1930s the federal government reconstructed the principal park roads to facilitate automobile travel. The National Park Service and the Bureau of Public Roads rebuilt the Wawona Road and relocated much of the Big Oak Flat Road to provide for a safer and faster approach to the Valley. Improvements included the construction of four tunnels to avoid massive scars to the valley walls and the provision of overlooks to allow motorists to take in the striking views. The long-neglected Tioga Road was repaired and put back into service. The State of California built a connecting road up to Tioga Pass, and the road became an important route across the mountains. Much of this road was rerouted in the 1950s during the National Park Service's Mission 66 program. The newly widened and straightened roadway was reopened in 1961.

The Yosemite roads, both old and new, are significant for their engineering achievements, for the role they played in the development and enjoyment of Yosemite, and for the way in which they were constructed to harmonize with their natural settings. Natural materials were often used in the construction of road-related structures, and rustic-style amenities complemented the landscape. Travel on the park roads today is far easier than it was in the nineteenth century, but the experience remains a compelling adventure for visitors to one of America's most popular national parks.

NATIONAL PARK SERVICE
Department of the Interior

DELINEATED BY: Team, 2001
NPS ROADS & BRIDGES RECORDING PROGRAM
NATIONAL PARK SERVICE
UNITED STATES DEPARTMENT OF THE INTERIOR

YOSEMITE NATIONAL PARK ROADS
YOSEMITE VICINITY | MARIPOSA & TUOLUMNE COUNTIES | CALIFORNIA
SHEET 1 of 19
HISTORIC AMERICAN ENGINEERING RECORD
CA-117

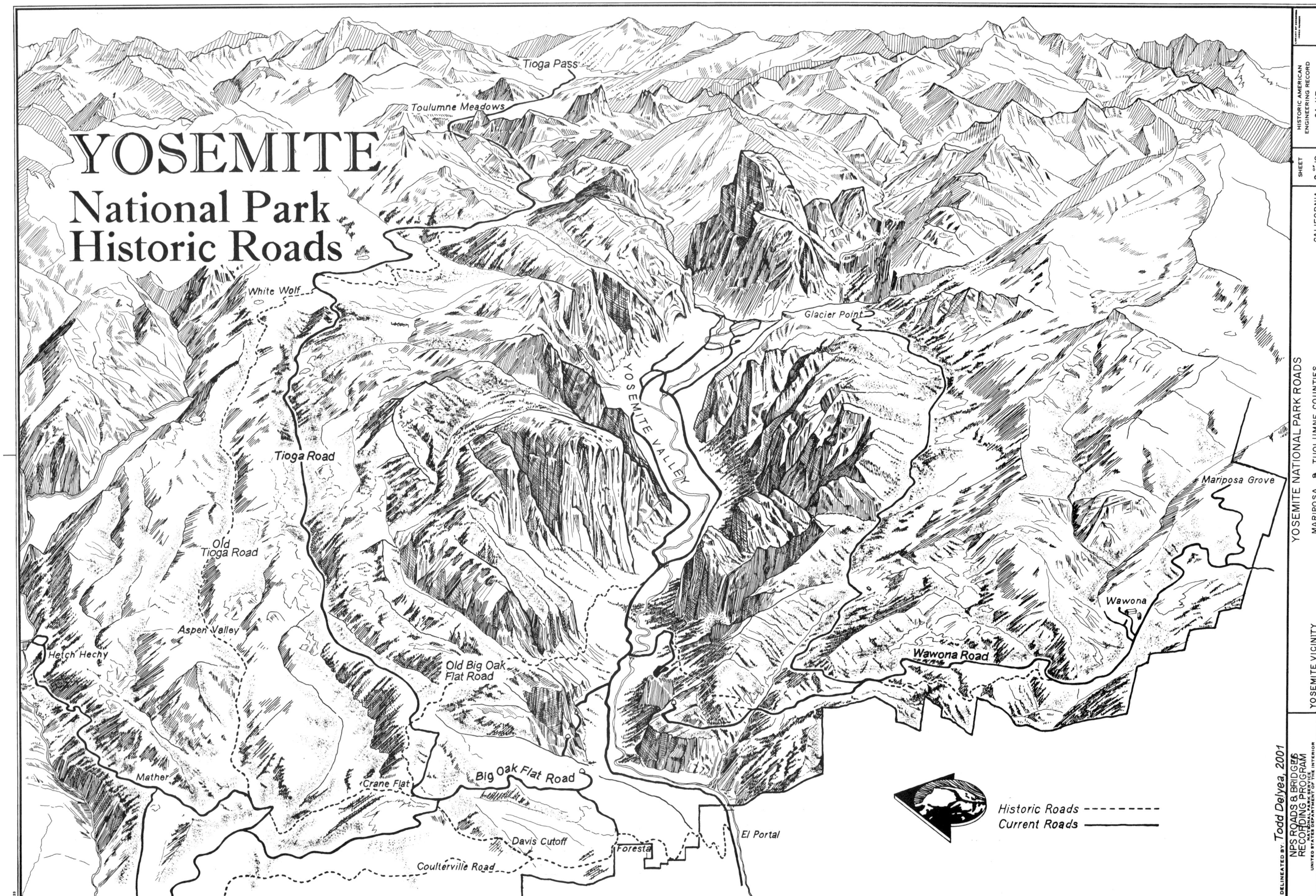
YOSEMITE
National Park
Historic Roads
Tioga Pass
Toulumne Meadows
White Wolf
Glacier Point
YOSEMITE VALLEY
Tioga Road
Old Tioga Road
Aspen Valley
Hetch Hechy
Old Big Oak Flat Road
Big Oak Flat Road
Crane Flat
Mather
Davis Cutoff
Foresta
Coulterville Road
El Portal
Wawona Road
Wawona
Mariposa Grove
Historic Roads
Current Roads
DELINEATED BY: Todd Delyea, 2001
NPS ROADS & BRIDGES RECORDING PROGRAM
NATIONAL PARK SERVICE
UNITED STATES DEPARTMENT OF THE INTERIOR
YOSEMITE NATIONAL PARK ROADS
YOSEMITE VICINITY
MARIPOSA & TUOLUMNE COUNTIES
CALIFORNIA
SHEET 2 OF 19
HISTORIC AMERICAN ENGINEERING RECORD
CA-117
IF REPRODUCED, PLEASE CREDIT: HISTORIC AMERICAN ENGINEERING RECORD, NATIONAL PARK SERVICE, NAME OF DELINEATOR, DATE OF THE DRAWING

RACE TO THE VALLEY

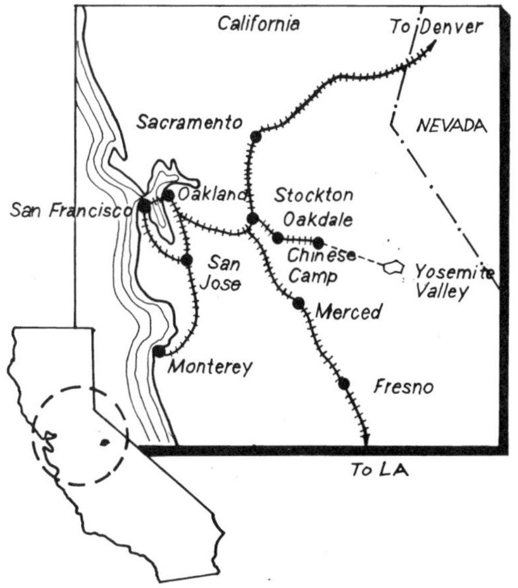

As interest in the wonders of Yosemite Valley increased, three rival companies sought to construct wagon roads to the Valley in order to capture tolls from the visitors. The Coulterville & Yosemite Turnpike was completed in June 1874, and the rival Chinese Camp & Yosemite Turnpike (Big Oak Flat Road) less than a month later. From the south, the Mariposa Big Tree Grove and Yo Semite Turnpike (Wawona Road) reached the Valley the following year. With competition from the north and west divided, only the Wawona Road proved profitable. All of the roads were purchased by the government by the 1910s.

Coulter and Murphy's Hotel was decorated with signs, flags, and banners for the formal opening celebration of the Wawona Road on July 22, 1875

Descending the steep trails Into Yosemite Valley from the Rim was a precarious nerve-racking Experience for visitors.

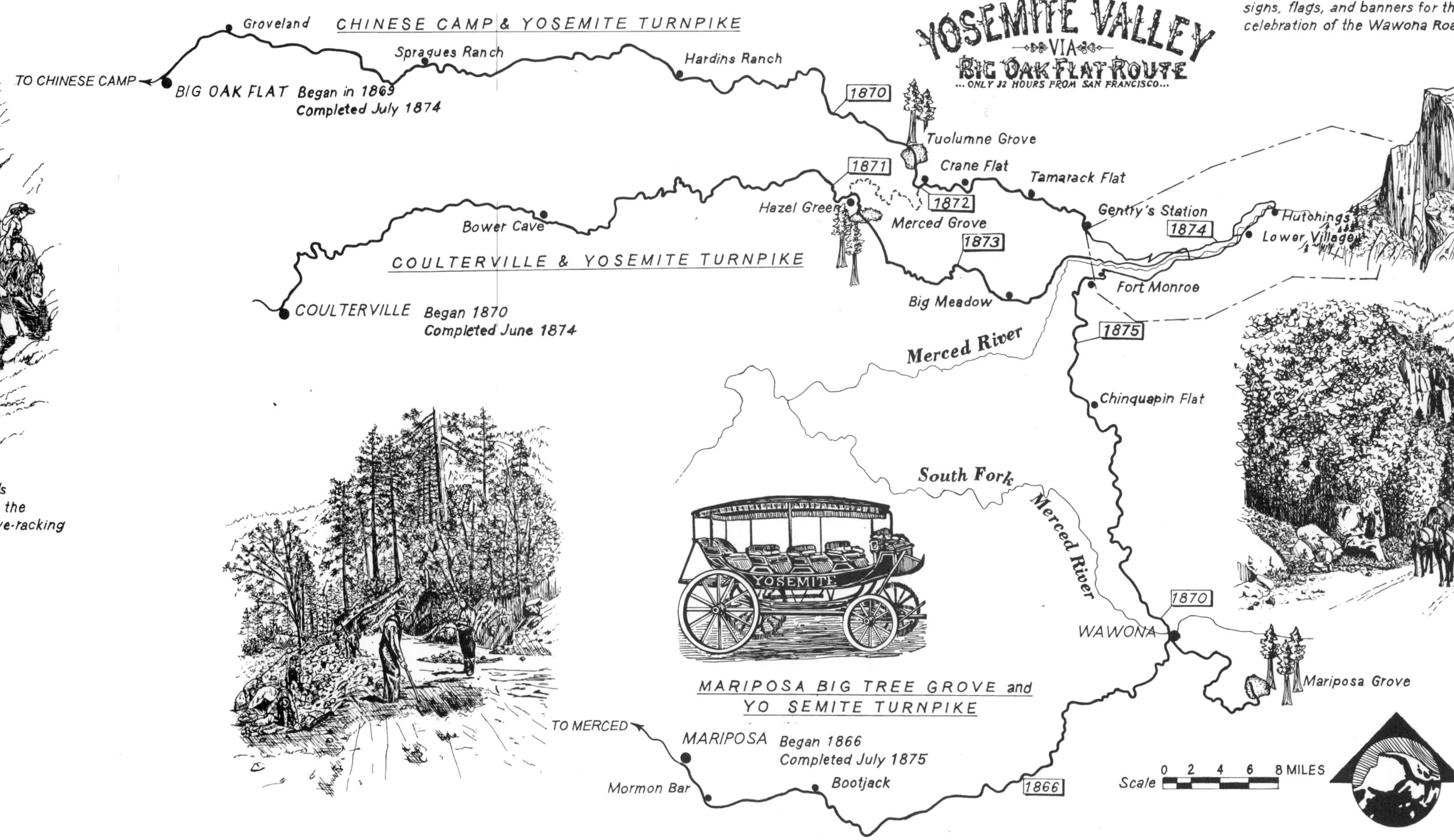

HISTORIC AMERICAN ENGINEERING RECORD CA-117
SHEET 3 of 19
CALIFORNIA
YOSEMITE NATIONAL PARK ROADS
MARIPOSA & TUOLUMNE COUNTIES
YOSEMITE VICINITY
NPS ROADS & BRIDGES RECORDING PROGRAM
NATIONAL PARK SERVICE
UNITED STATES DEPARTMENT OF THE INTERIOR
DELINEATED BY: Ann Kero 2001

COULTERVILLE ROAD 1874

The first road to reach the Valley, the Coulterville and Yosemite Turnpike was begun by investors in the small mining town of Coulterville but completed in 1874 by Dr. John Taylor McClean. Running from Bower Cave on the Merced to the Valley floor at Cascade Creek, its monopoly was short-lived and the road never carried the expected loads of tourists. In 1913, it became the first road in the park over which automobiles could legally travel. Most of the road was closed in 1982 following the massive "Cookie" rockslide; it is now a hiking trail.

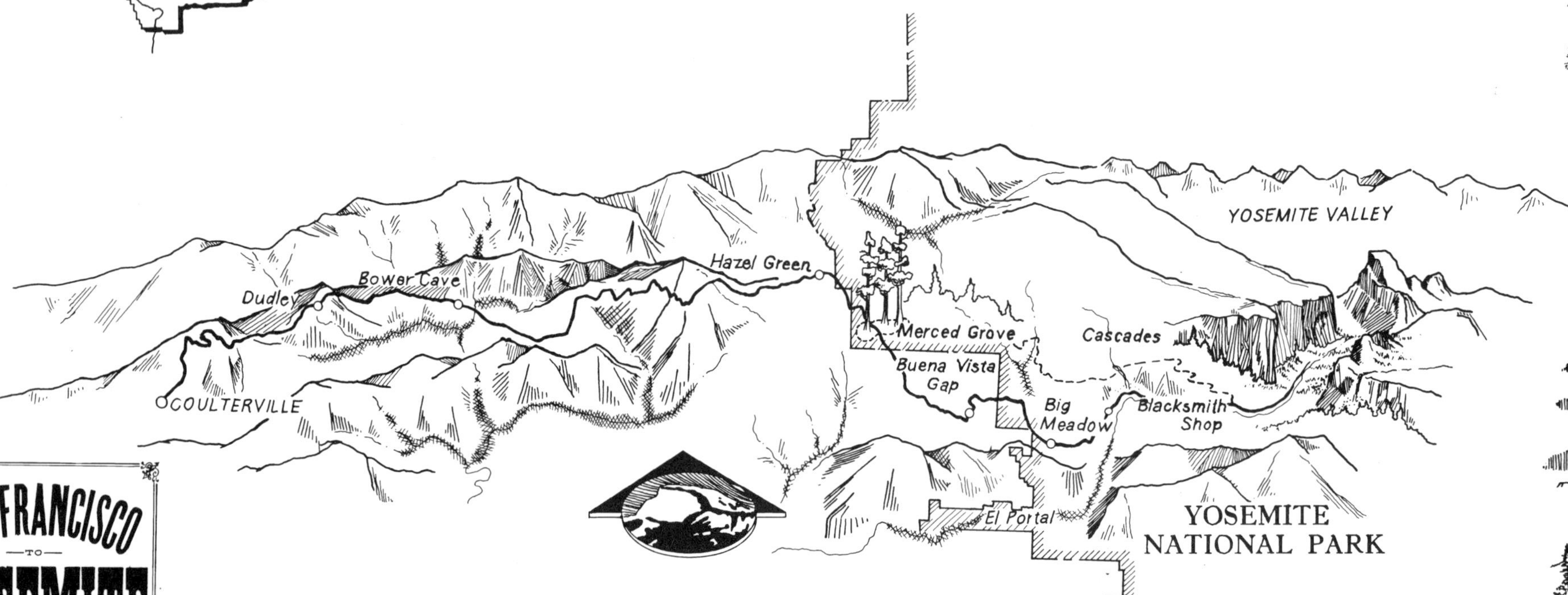

The Coulterville Road passed directly through the Merced Grove of Big Trees, a primary attraction for early visitors to Yosemite.

Hazel Green acted as the lunch stop on the Coulterville Road for many years. From Hazel Green the road headed south through the Merced Grove then southwest past Buena Vista gap for the first views of Yosemite Valley.

An early automobile heads down the steepest section of the Coulterville Road towards its junction with the Merced River.

The Coulterville Roads "Collision with the geological history of the park was inevidible" claimed Yosemite's superintendent, After a rockslide closed it on April 3, 1982.

DELINEATED BY: *Todd Delyea, 2001*

NPS ROADS & BRIDGES RECORDING PROGRAM
NATIONAL PARK SERVICE
UNITED STATES DEPARTMENT OF THE INTERIOR

YOSEMITE NATIONAL PARK ROADS
YOSEMITE VICINITY — MARIPOSA & TUOLUMNE COUNTIES — CALIFORNIA

SHEET 4 OF 19
HISTORIC AMERICAN ENGINEERING RECORD
CA-117

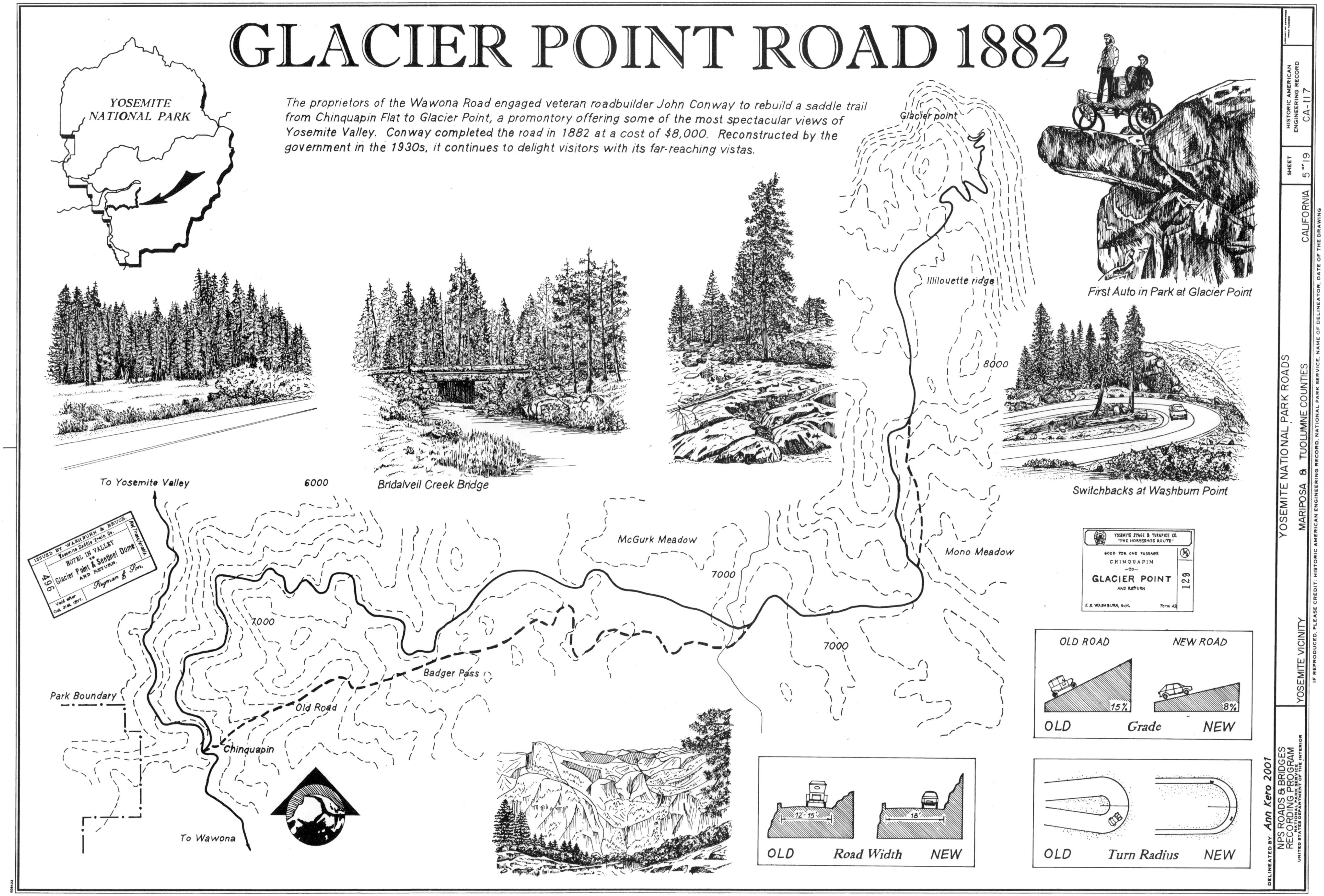

GLACIER POINT ROAD 1882
The proprietors of the Wawona Road engaged veteran roadbuilder John Conway to rebuild a saddle trail from Chinquapin Flat to Glacier Point, a promontory offering some of the most spectacular views of Yosemite Valley. Conway completed the road in 1882 at a cost of $8,000. Reconstructed by the government in the 1930s, it continues to delight visitors with its far-reaching vistas.
YOSEMITE NATIONAL PARK
Bridalveil Creek Bridge
First Auto in Park at Glacier Point
Switchbacks at Washburn Point
Glacier point
Illilouette ridge
8000
To Yosemite Valley
6000
McGurk Meadow
Mono Meadow
7000
7000
7000
Badger Pass
Old Road
Park Boundary
Chinquapin
To Wawona
ISSUED BY WASHBURN & BRUCE.
Yosemite Saddle Train Co.
HOTEL IN VALLEY
TO
Glacier Point & Sentinel Dome
AND RETURN.
496
YOSEMITE STAGE & TURNPIKE CO.
"THE HORSESHOE ROUTE"
GOOD FOR ONE PASSAGE
CHINQUAPIN
—TO—
GLACIER POINT
AND RETURN
129
OLD ROAD
NEW ROAD
15%
8%
OLD
Grade
NEW
12'-15'
18'
OLD
Road Width
NEW
OLD
Turn Radius
NEW
DELINEATED BY: Ann Kero 2001
NPS ROADS & BRIDGES RECORDING PROGRAM
NATIONAL PARK SERVICE
UNITED STATES DEPARTMENT OF THE INTERIOR
YOSEMITE NATIONAL PARK ROADS
YOSEMITE VICINITY
MARIPOSA & TUOLUMNE COUNTIES
CALIFORNIA
SHEET 5 OF 19
HISTORIC AMERICAN ENGINEERING RECORD
CA-117
IF REPRODUCED, PLEASE CREDIT: HISTORIC AMERICAN ENGINEERING RECORD, NATIONAL PARK SERVICE, NAME OF DELINEATOR, DATE OF THE DRAWING

BIG OAK FLAT ROAD 1874/ 1940

The "Chinese Camp & Yo Semite Turnpike" was constructed in the early 1870s as a toll route to the park from the northwest. After showcasing the Valley from Rainbow Point, It picked its way down the cliffs on a series of remarkable switchbacks called the "Zig Zag." To accommodate modern motorists, it was replaced by a new modern highway with easier grades in the 1930s. The New Big Oak Flat Road utilizes sweeping curves and three tunnels to eliminate the need for the earlier switchbacks.

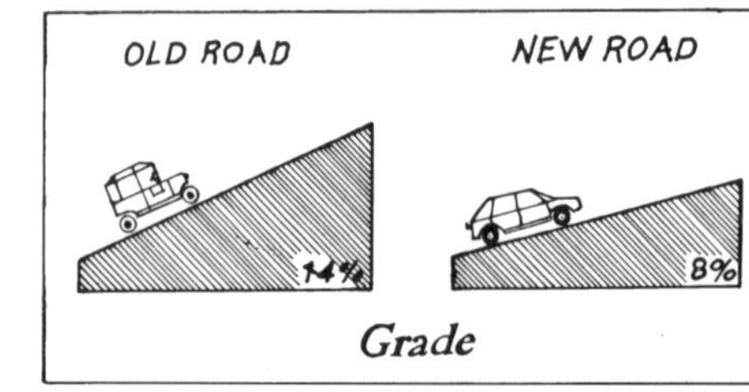

The tunnel through the Dead Giant in the Tuolumne Grove was cut in 1878 to entice Tourists to travel the Big Oak Flat Road.

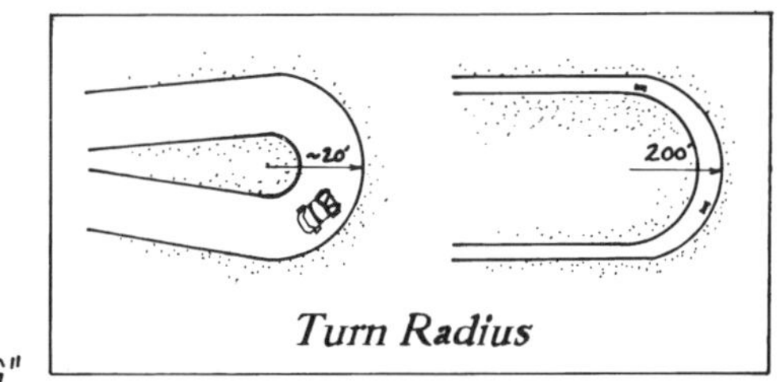

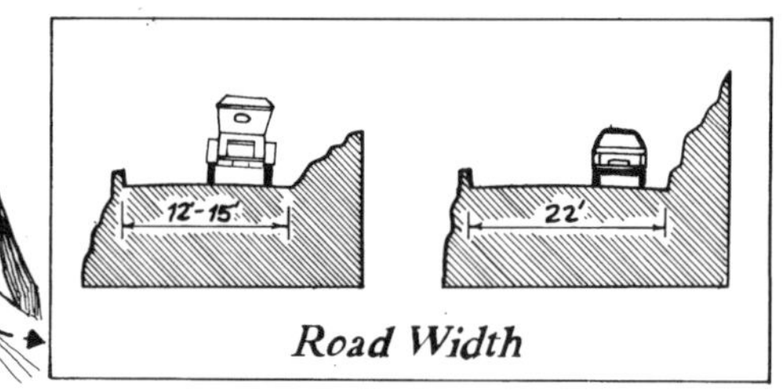

Horse stages ascend the "Zig Zag" Switchback on the Old Big Oak Flat Road.

An early type motor bus descends the Old Big Oak Flat Road just above its junction with the Merced River.

Tunnel Construction on the Big Oak Flat road Allowed The Park Service to provide a steady grade, and minimize scarring.

The streamlined bridges on the Big Oak Flat Road Eliminated the numerous switchbacks that were on The old road.

DELINEATED BY: Ann Kero 2001

NPS ROADS & BRIDGES RECORDING PROGRAM
NATIONAL PARK SERVICE
UNITED STATES DEPARTMENT OF THE INTERIOR

YOSEMITE NATIONAL PARK ROADS
YOSEMITE VICINITY
MARIPOSA & TUOLUMNE COUNTIES
CALIFORNIA

SHEET 6 of 19

HISTORIC AMERICAN ENGINEERING RECORD CA-117

IF REPRODUCED, PLEASE CREDIT: HISTORIC AMERICAN ENGINEERING RECORD, NATIONAL PARK SERVICE, NAME OF DELINEATOR, DATE OF THE DRAWING

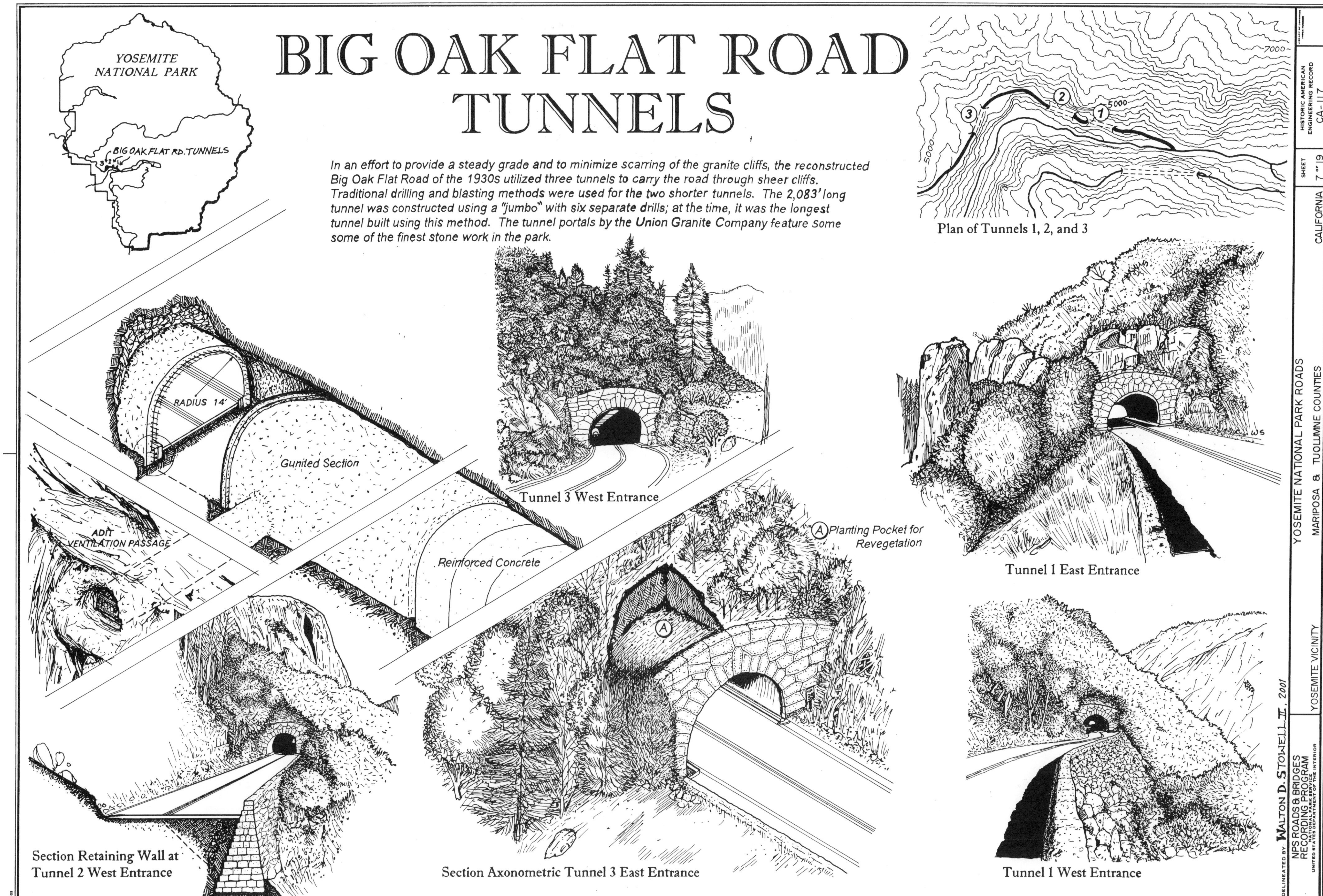
BIG OAK FLAT ROAD TUNNELS
In an effort to provide a steady grade and to minimize scarring of the granite cliffs, the reconstructed Big Oak Flat Road of the 1930s utilized three tunnels to carry the road through sheer cliffs. Traditional drilling and blasting methods were used for the two shorter tunnels. The 2,083' long tunnel was constructed using a "jumbo" with six separate drills; at the time, it was the longest tunnel built using this method. The tunnel portals by the Union Granite Company feature some some of the finest stone work in the park.
YOSEMITE NATIONAL PARK
BIG OAK FLAT RD. TUNNELS
Plan of Tunnels 1, 2, and 3
RADIUS 14'
Gunited Section
Reinforced Concrete
ADIT VENTILATION PASSAGE
Tunnel 3 West Entrance
Tunnel 1 East Entrance
Ⓐ Planting Pocket for Revegetation
Section Retaining Wall at Tunnel 2 West Entrance
Section Axonometric Tunnel 3 East Entrance
Tunnel 1 West Entrance
DELINEATED BY: WALTON D. STOWELL II, 2001
NPS ROADS & BRIDGES RECORDING PROGRAM
NATIONAL PARK SERVICE
UNITED STATES DEPARTMENT OF THE INTERIOR
YOSEMITE VICINITY
YOSEMITE NATIONAL PARK ROADS
MARIPOSA & TUOLUMNE COUNTIES
CALIFORNIA
SHEET 7 OF 19
HISTORIC AMERICAN ENGINEERING RECORD
CA-117
IF REPRODUCED, PLEASE CREDIT: HISTORIC AMERICAN ENGINEERING RECORD, NATIONAL PARK SERVICE, NAME OF DELINEATOR, DATE OF THE DRAWING

WAWONA ROAD 1875/1933

The most successful of the early toll roads, the "Mariposa Big Trees and Yo Semite Turnpike" was started in the 1860s by Yosemite guardian Galen Clark but was taken over by Mariposa County investors in the 1870s. Reaching the Valley floor in 1875, it was the most popular route to the park for decades. Reconstructed by the National Park Service in the 1930s, it remains the principal approach to the park from the south. After passing through forests most of the way, travelers emerge through the Wawona Tunnel to one of the most exciting views of Yosemite Valley.

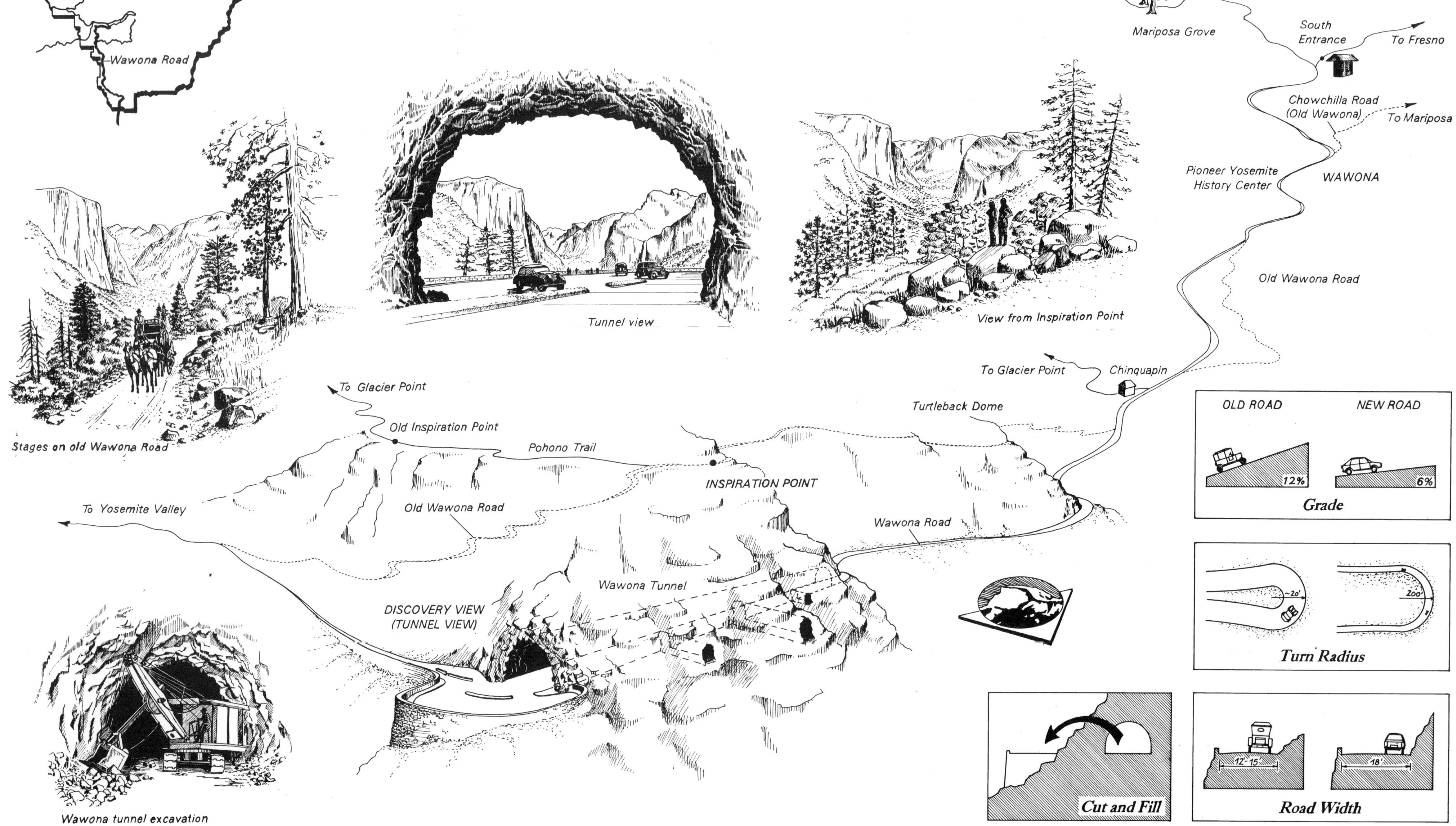

DELINEATED BY: Anne Teresiak 2001
NPS ROADS & BRIDGES RECORDING PROGRAM
NATIONAL PARK SERVICE
UNITED STATES DEPARTMENT OF THE INTERIOR
ADDENDUM TO: YOSEMITE NATIONAL PARK ROADS & BRIDGES
YOSEMITE VICINITY
MARIPOSA & TUOLUMNE COUNTIES
CALIFORNIA
SHEET 8 of 19
HISTORIC AMERICAN ENGINEERING RECORD
CA-117
IF REPRODUCED, PLEASE CREDIT: HISTORIC AMERICAN ENGINEERING RECORD, NATIONAL PARK SERVICE, NAME OF DELINEATOR, DATE OF THE DRAWING

EL PORTAL ROAD

"The All-Year Highway"

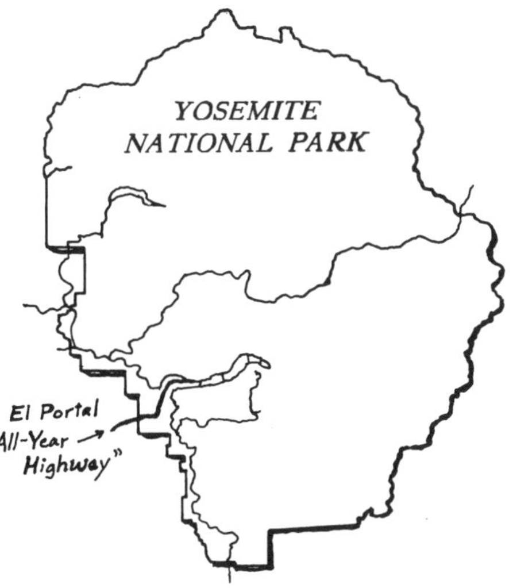

Long thought impassible, the steep and flood-prone Merced River Valley was first conquered by the Yosemite Valley Rail Road, which built a line to El Portal and a connecting road to the Valley in 1908. In the 1930s, the State of California, looking to provide reliable year-round access to the park, built a modern highway to El Portal, and the National Park Service reconstructed the old wagon road into the Valley. Following disastrous floods of the Merced, the road has been realigned and heavily reconstructed.

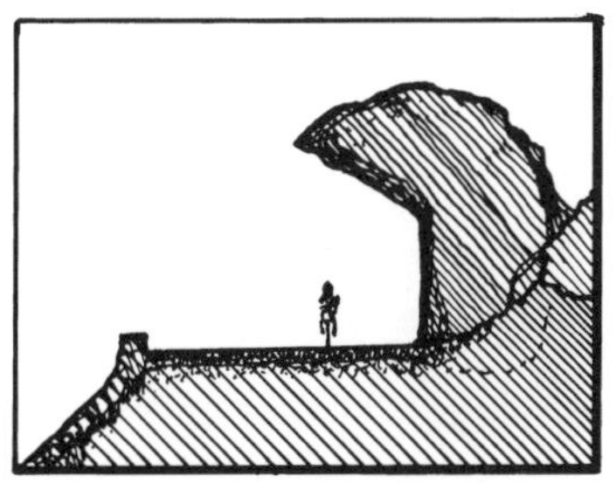

Half-Arch Rock

Arch Rock

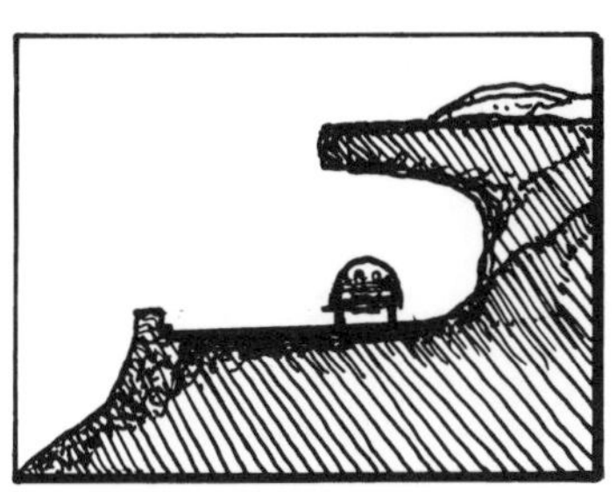

Dog Rock

1927 ¼ mile below Arch Rock, traffic awaiting entry permits

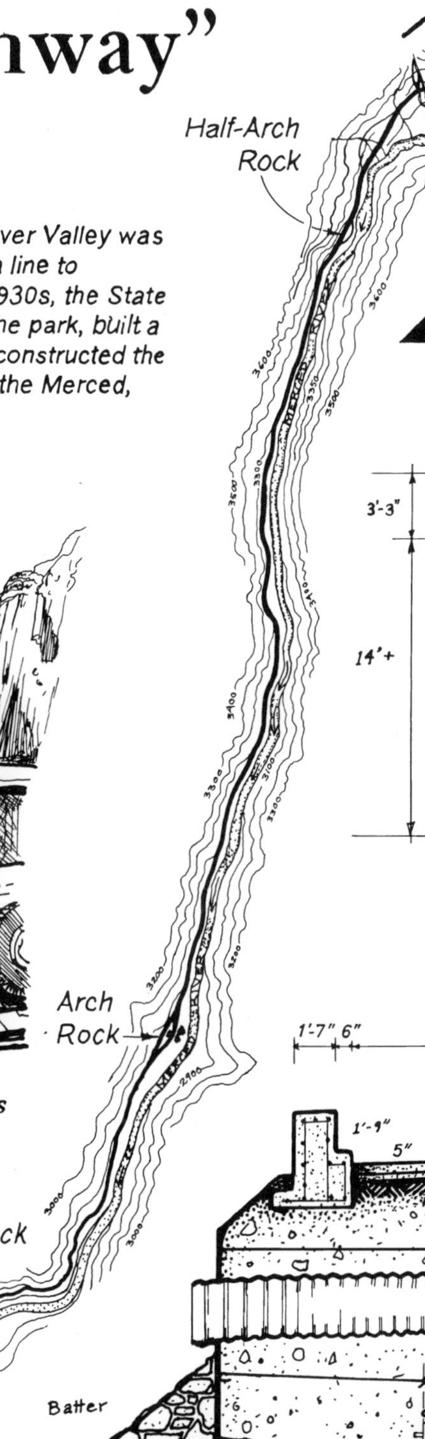

View to El Portal
from Yosemite Valley

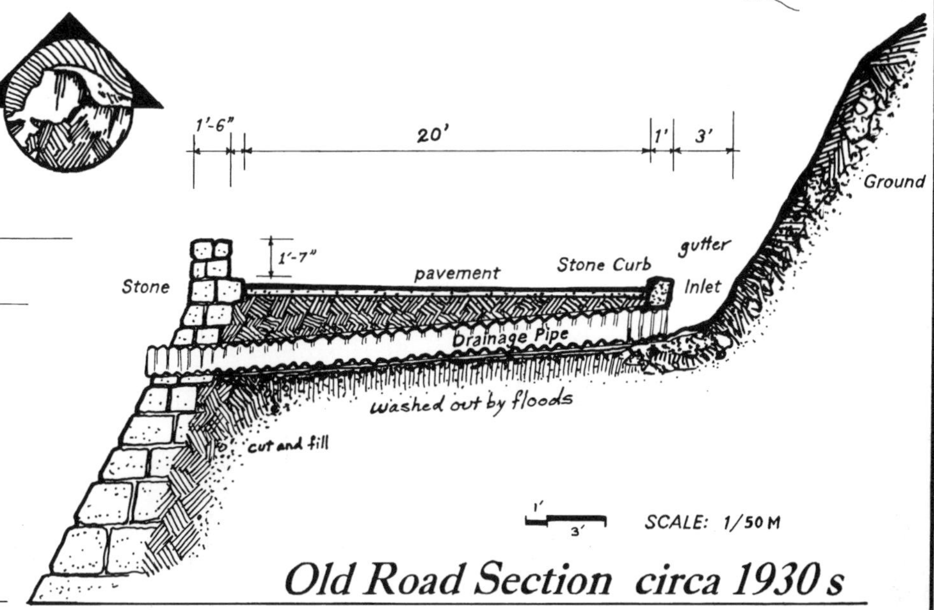

Old Road Section circa 1930s

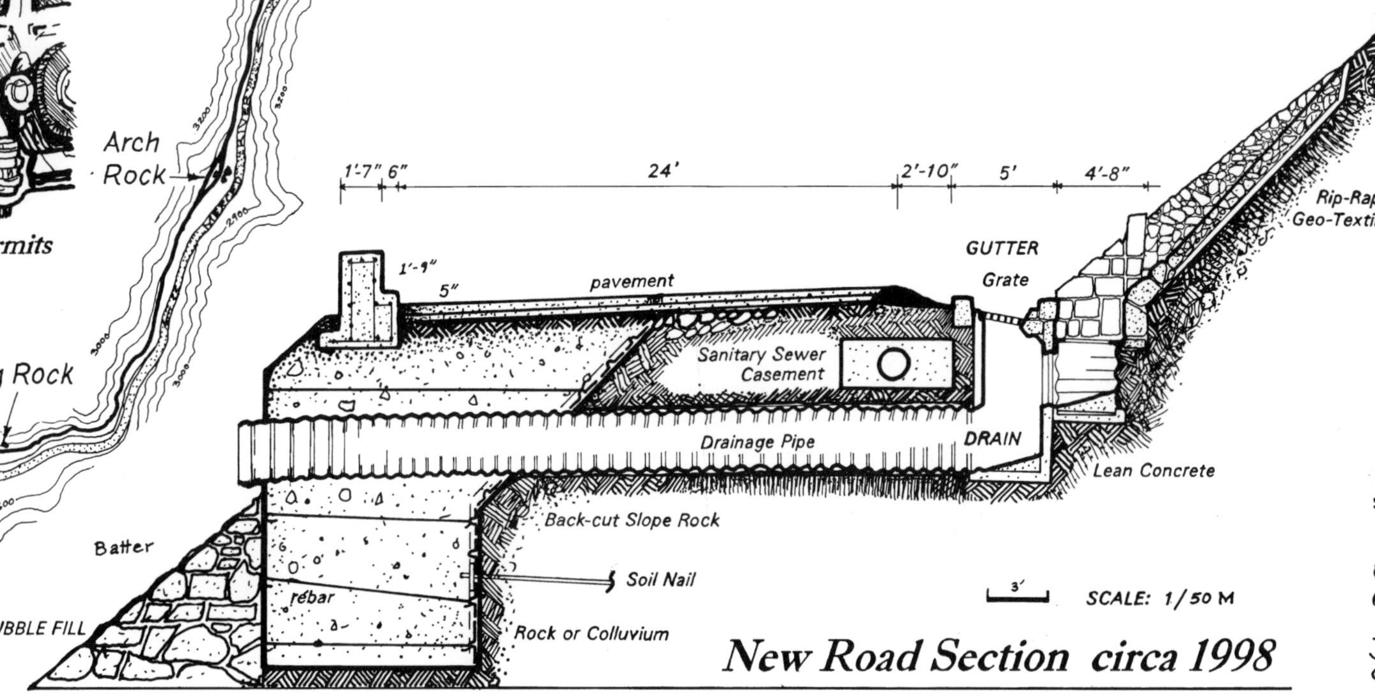

New Road Section circa 1998

DELINEATED BY: Walton D. Stowell II, 2001

NPS ROADS & BRIDGES RECORDING PROGRAM
UNITED STATES DEPARTMENT OF THE INTERIOR

YOSEMITE NATIONAL PARK ROADS
YOSEMITE VICINITY
MARIPOSA & TUOLUMNE COUNTIES
CALIFORNIA

SHEET 9 of 19

HISTORIC AMERICAN ENGINEERING RECORD
CA-117

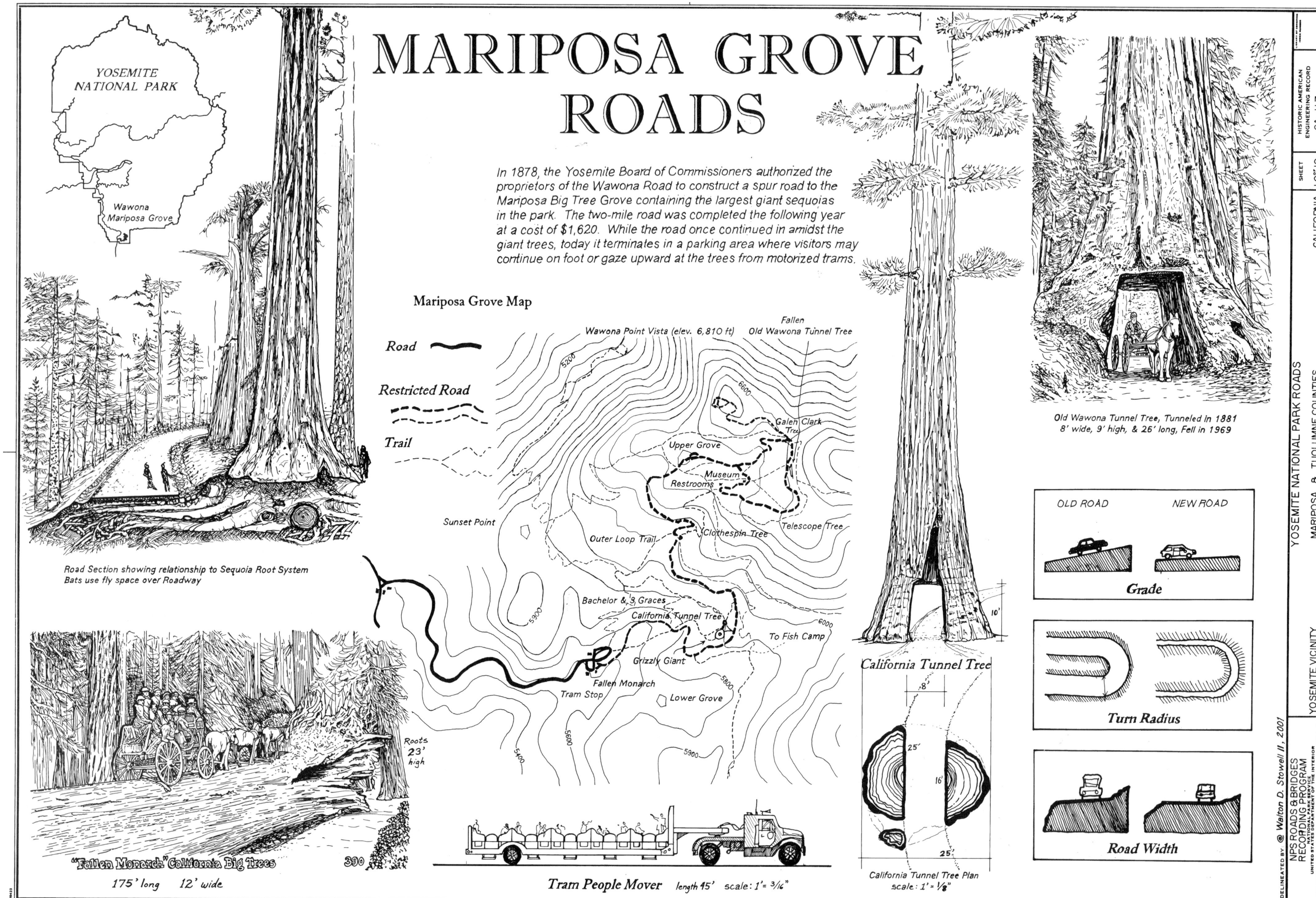
MARIPOSA GROVE ROADS
In 1878, the Yosemite Board of Commissioners authorized the proprietors of the Wawona Road to construct a spur road to the Mariposa Big Tree Grove containing the largest giant sequoias in the park. The two-mile road was completed the following year at a cost of $1,620. While the road once continued in amidst the giant trees, today it terminates in a parking area where visitors may continue on foot or gaze upward at the trees from motorized trams.
YOSEMITE NATIONAL PARK
Wawona
Mariposa Grove
Road Section showing relationship to Sequoia Root System
Bats use fly space over Roadway
Mariposa Grove Map
Road
Restricted Road
Trail
Wawona Point Vista (elev. 6,810 ft)
Fallen Old Wawona Tunnel Tree
Galen Clark Tree
Upper Grove
Museum
Restrooms
Sunset Point
Outer Loop Trail
Clothespin Tree
Telescope Tree
Bachelor & 3 Graces
California Tunnel Tree
To Fish Camp
Grizzly Giant
Fallen Monarch
Tram Stop
Lower Grove
5200
6600
5900
6000
5800
5600
5400
"Fallen Monarch" California Big Trees
390
Roots 23' high
175' long 12' wide
Tram People Mover length 45' scale: 1' = 3/16"
California Tunnel Tree
8'
10'
25'
16'
California Tunnel Tree Plan
scale: 1' = 1/8"
Old Wawona Tunnel Tree, Tunneled in 1881
8' wide, 9' high, & 26' long, Fell in 1969
OLD ROAD
NEW ROAD
Grade
Turn Radius
Road Width
DELINEATED BY © Walton D. Stowell II, 2001
NPS ROADS & BRIDGES RECORDING PROGRAM
NATIONAL PARK SERVICE
UNITED STATES DEPARTMENT OF THE INTERIOR
YOSEMITE NATIONAL PARK ROADS
YOSEMITE VICINITY
MARIPOSA & TUOLUMNE COUNTIES
CALIFORNIA
SHEET 10 OF 19
HISTORIC AMERICAN ENGINEERING RECORD
CA-117
IF REPRODUCED, PLEASE CREDIT: HISTORIC AMERICAN ENGINEERING RECORD, NATIONAL PARK SERVICE, NAME OF DELINEATOR, DATE OF THE DRAWING

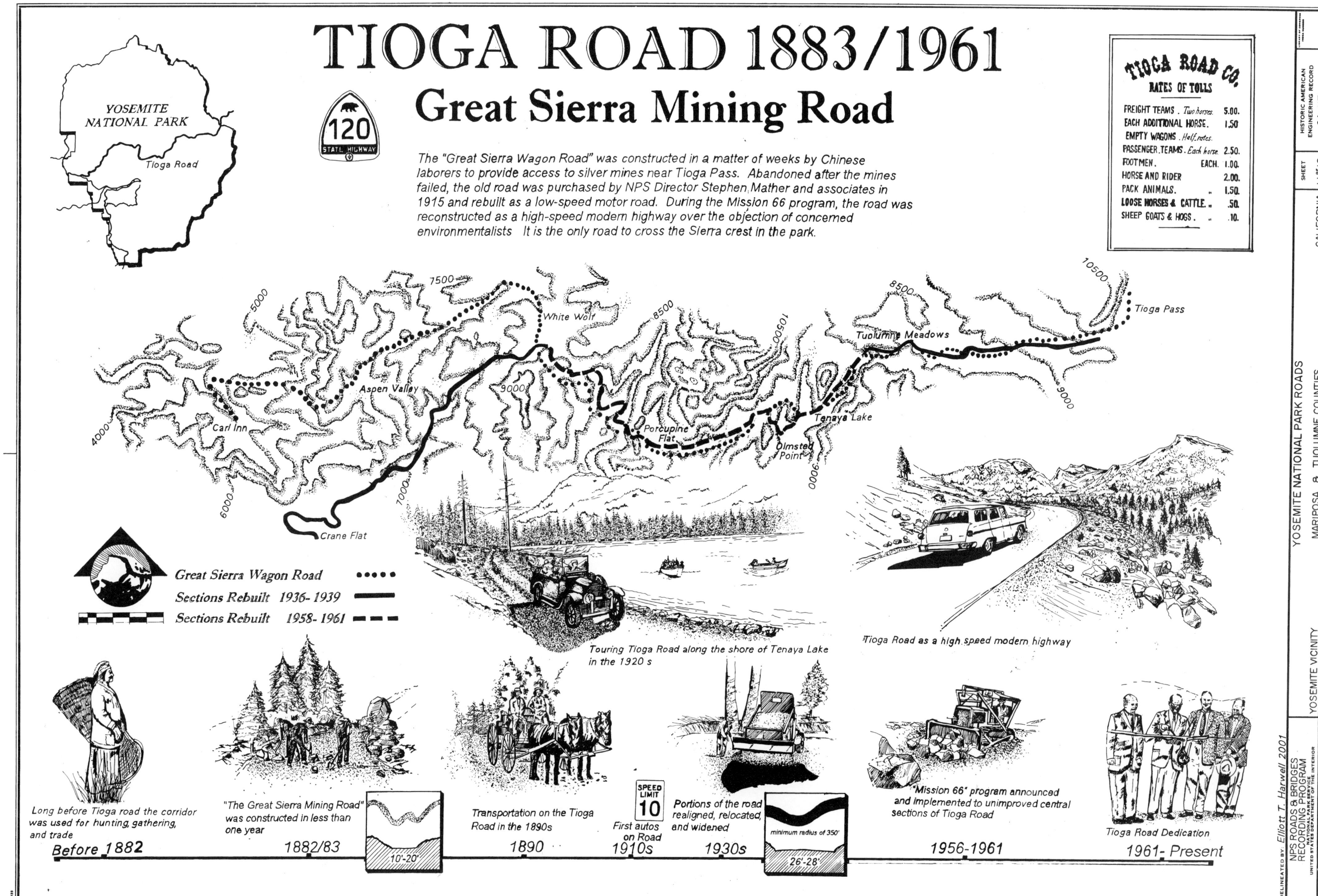

TIOGA ROAD 1883/1961
Great Sierra Mining Road
120
STATE HIGHWAY
The "Great Sierra Wagon Road" was constructed in a matter of weeks by Chinese laborers to provide access to silver mines near Tioga Pass. Abandoned after the mines failed, the old road was purchased by NPS Director Stephen Mather and associates in 1915 and rebuilt as a low-speed motor road. During the Mission 66 program, the road was reconstructed as a high-speed modern highway over the objection of concerned environmentalists It is the only road to cross the Sierra crest in the park.
YOSEMITE NATIONAL PARK
Tioga Road
TIOGA ROAD CO.
RATES OF TOLLS
FREIGHT TEAMS . Two horses. 5.00.
EACH ADDITIONAL HORSE. 1.50
EMPTY WAGONS . Half rates.
PASSENGER TEAMS . Each horse 2.50.
FOOTMEN. EACH. 1.00.
HORSE AND RIDER 2.00.
PACK ANIMALS. " 1.50.
LOOSE HORSES & CATTLE. " .50.
SHEEP GOATS & HOGS. " .10.
7500
5000
4000
6000
7000
8500
9000
10500
White Wolf
Aspen Valley
Carl Inn
Crane Flat
Porcupine Flat
Olmsted Point
Tenaya Lake
Tuolumne Meadows
Tioga Pass
Great Sierra Wagon Road
Sections Rebuilt 1936-1939
Sections Rebuilt 1958-1961
Touring Tioga Road along the shore of Tenaya Lake in the 1920 s
Tioga Road as a high speed modern highway
Long before Tioga road the corridor was used for hunting, gathering, and trade
"The Great Sierra Mining Road" was constructed in less than one year
10'-20'
Transportation on the Tioga Road in the 1890s
SPEED LIMIT 10
First autos on Road
Portions of the road realigned, relocated, and widened
minimum radius of 350'
26'-28'
"Mission 66" program announced and implemented to unimproved central sections of Tioga Road
Tioga Road Dedication
Before 1882
1882/83
1890
1910s
1930s
1956-1961
1961- Present
DELINEATED BY: Elliott T. Harwell 2001
NPS ROADS & BRIDGES RECORDING PROGRAM
NATIONAL PARK SERVICE
UNITED STATES DEPARTMENT OF THE INTERIOR
YOSEMITE NATIONAL PARK ROADS
YOSEMITE VICINITY
MARIPOSA & TUOLUMNE COUNTIES
CALIFORNIA
SHEET
11 of 19
HISTORIC AMERICAN ENGINEERING RECORD
CA-117
IF REPRODUCED, PLEASE CREDIT: HISTORIC AMERICAN ENGINEERING RECORD, NATIONAL PARK SERVICE, NAME OF DELINEATOR, DATE OF THE DRAWING

VALLEY ROADS

The first roads in Yosemite Valley were constructed in the 1860s, years before connecting roads made their way in from the park boundaries. In 1871, guardian Galen Clark had a wagon packed in over the saddle trails and began offering tours. In the early 1900s, a carriage road system was constructed by the U.S. Army, which then administered the park. Over the years, much of the system has been reconstructed. In 1970, the roads at the eastern end of the Valley were closed to automobiles, offering hikers and cyclists a refuge from the now heavily congested road system.

1930's Parking Traffic in the Valley

Yosemite Alternative Fuel Shuttle Bus Project

Old Roads Restricted now for walking or bicycles

21st Century Public Transit

1883

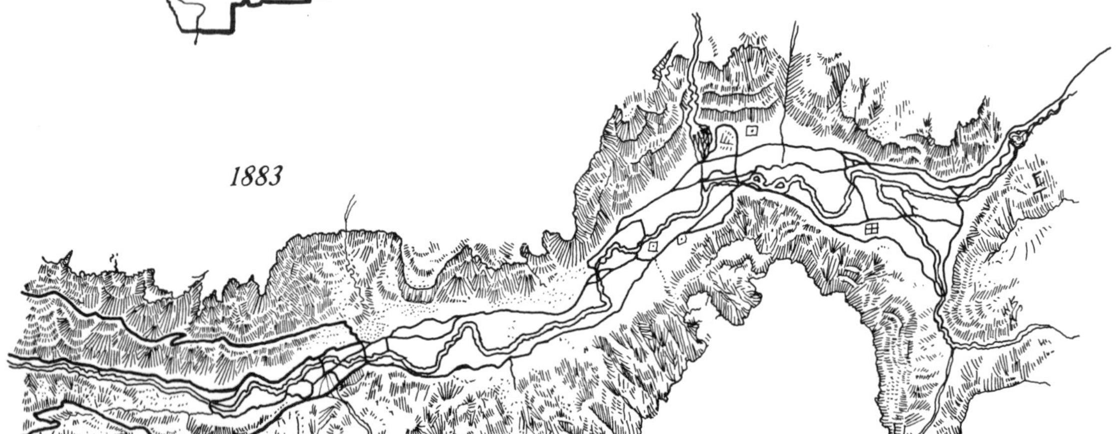

1926 Automobiles in the Valley

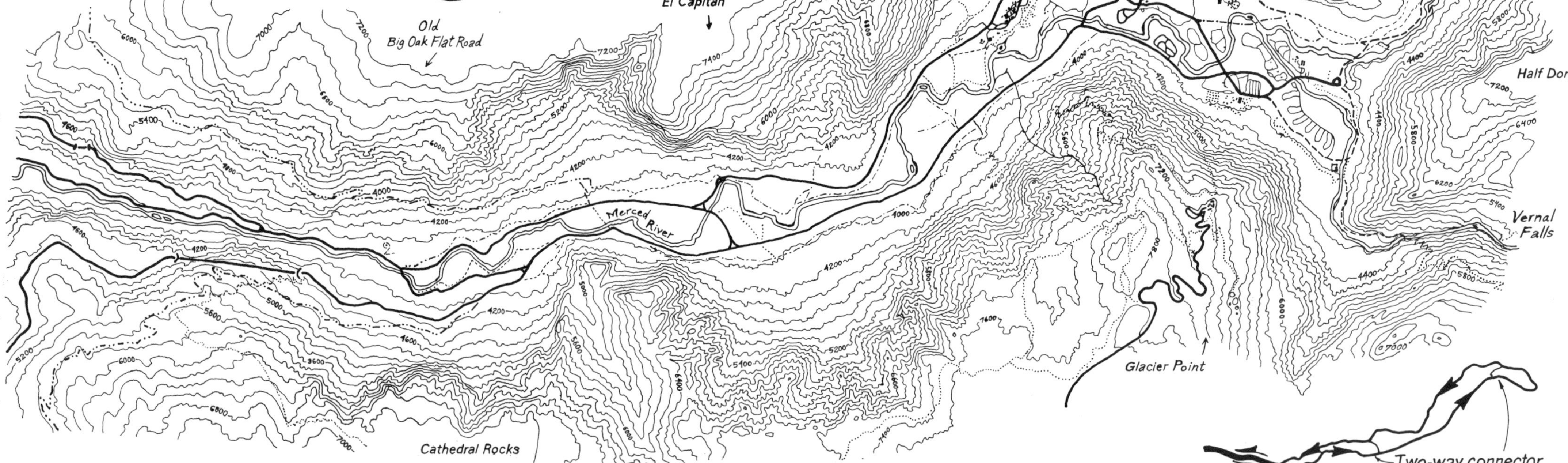

Road —— Restricted Road or Trail -·-·-·-·-·-

2001 Valley Roads Plan

VALLEY TRAFFIC FLOW

DELINEATED BY: Walton D. Stowell II, 2001

NPS ROADS & BRIDGES RECORDING PROGRAM
NATIONAL PARK SERVICE
UNITED STATES DEPARTMENT OF THE INTERIOR

YOSEMITE VICINITY MARIPOSA & TUOLUMNE COUNTIES CALIFORNIA

SHEET 13 OF 19

HISTORIC AMERICAN ENGINEERING RECORD CA-117

IF REPRODUCED, PLEASE CREDIT: HISTORIC AMERICAN ENGINEERING RECORD, NATIONAL PARK SERVICE, NAME OF DELINEATOR, DATE OF THE DRAWING

ENTRANCE STATIONS

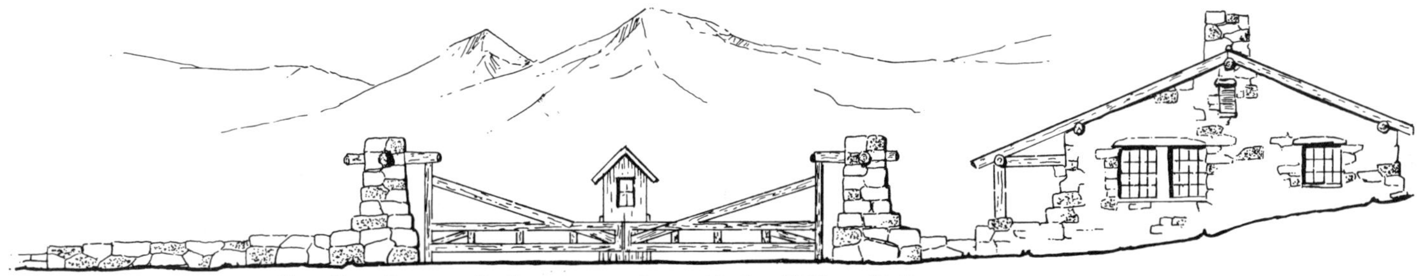

Tioga Pass Entrance Station was constructed in the 1930 s with the "Rustic" style used by the National Park Service

Sign located at Big Oak Flat Entrance in 1941

Even before automobiles were admitted to Yosemite, an entrance fee was charged for admission. With the opening of the park roads to motorists, entrance stations were established to collect the fees and to provide information for visitors. The earliest entrance stations were built of logs and boulders in the prevailing rustic style. Later fee booths sometimes incorporated native materials but were strictly functional structures of simpler and less expensive construction.

"Rustic" Arch Rock Entrance in the 1930 s

"Mission 66" sign located at Big Oak Flat Entrance

"Mission 66" South Entrance Station erected in the late 1950 s

Coulterville Entrance in 1913

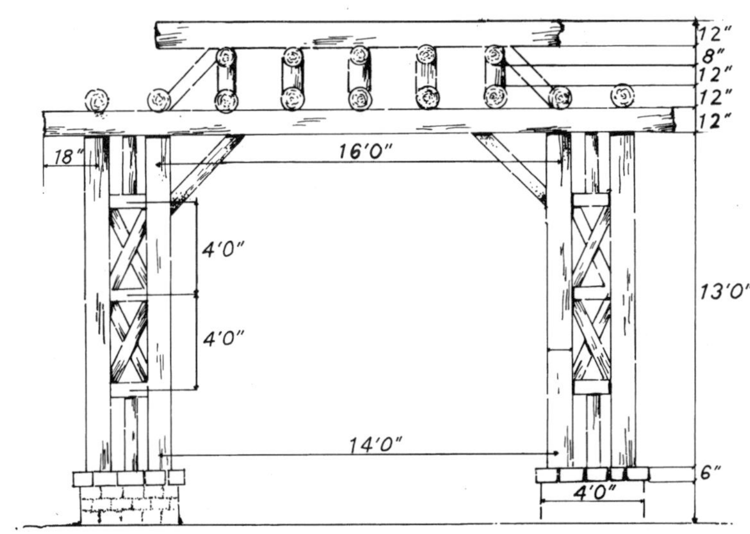

Wawona Entrance on the Chunchilla Mountain Road constructed in the 1920 s

Aspen Valley Entrance in the 1910 s

HISTORIC AMERICAN ENGINEERING RECORD
CA-117

SHEET 14 OF 19

CALIFORNIA

YOSEMITE NATIONAL PARK ROADS

MARIPOSA & TUOLUMNE COUNTIES

YOSEMITE VICINITY

DELINEATED BY: Elliott T. Harwell 2001

NPS ROADS & BRIDGES RECORDING PROGRAM
NATIONAL PARK SERVICE
UNITED STATES DEPARTMENT OF THE INTERIOR

IF REPRODUCED, PLEASE CREDIT: HISTORIC AMERICAN ENGINEERING RECORD, NATIONAL PARK SERVICE, NAME OF DELINEATOR, DATE OF THE DRAWING

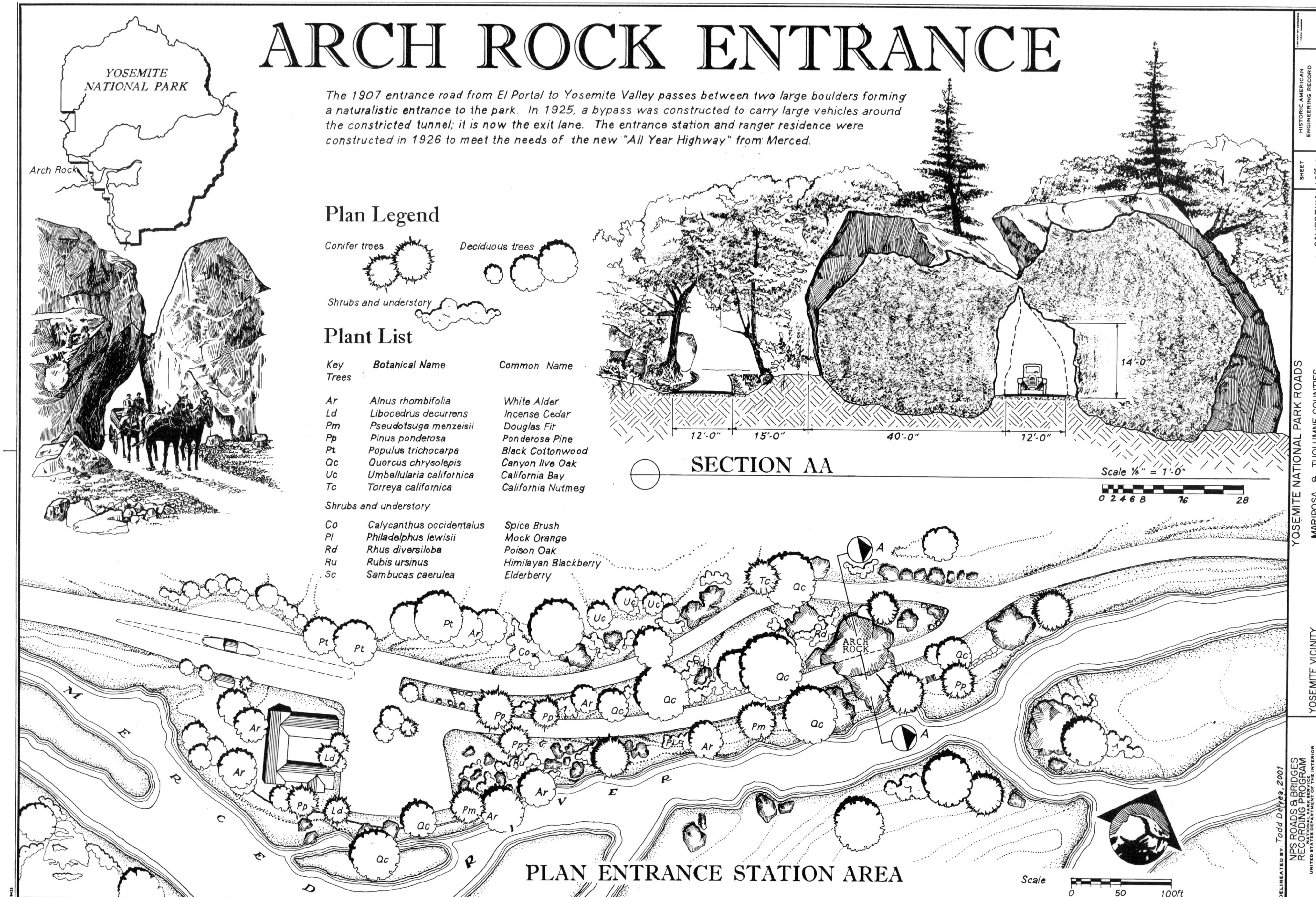
ARCH ROCK ENTRANCE
YOSEMITE NATIONAL PARK
Arch Rock
The 1907 entrance road from El Portal to Yosemite Valley passes between two large boulders forming a naturalistic entrance to the park. In 1925, a bypass was constructed to carry large vehicles around the constricted tunnel; it is now the exit lane. The entrance station and ranger residence were constructed in 1926 to meet the needs of the new "All Year Highway" from Merced.
Plan Legend
Conifer trees
Deciduous trees
Shrubs and understory
Plant List
Key Trees
Botanical Name
Common Name
Ar Alnus rhombifolia White Alder
Ld Libocedrus decurrens Incense Cedar
Pm Pseudotsuga menzeisii Douglas Fir
Pp Pinus ponderosa Ponderosa Pine
Pt Populus trichocarpa Black Cottonwood
Qc Quercus chrysolepis Canyon live Oak
Uc Umbellularia californica California Bay
Tc Torreya californica California Nutmeg
Shrubs and understory
Co Calycanthus occidentalus Spice Brush
Pl Philadelphus lewisii Mock Orange
Rd Rhus diversiloba Poison Oak
Ru Rubis ursinus Himilayan Blackberry
Sc Sambucas caerulea Elderberry
14'-0"
12'-0"
15'-0"
40'-0"
12'-0"
SECTION AA
Scale ⅛" = 1'-0"
0 2 4 6 8 16 28
A
ARCH ROCK
MERCED RIVER
PLAN ENTRANCE STATION AREA
Scale
0 50 100ft
HISTORIC AMERICAN ENGINEERING RECORD
CA-117
SHEET 15 OF 19
CALIFORNIA
YOSEMITE NATIONAL PARK ROADS
MARIPOSA & TUOLUMNE COUNTIES
YOSEMITE VICINITY
NPS ROADS & BRIDGES RECORDING PROGRAM
NATIONAL PARK SERVICE
UNITED STATES DEPARTMENT OF THE INTERIOR
DELINEATED BY: Todd Delyea, 2001
IF REPRODUCED, PLEASE CREDIT HISTORIC AMERICAN ENGINEERING RECORD, NATIONAL PARK SERVICE, NAME OF DELINEATOR, DATE OF THE DRAWING

GUARD WALLS

A variety of stone guard walls or parapet was employed to prevent vehicles from plunging off the park roads. Most of the modern motor roads from the 1930s have their own unique styles. The early walls were constructed of native stone, cut and mortared to fit in varying patterns. Some of these guard walls are now being replaced with concrete core walls to meet increasingly stringent safety standards.

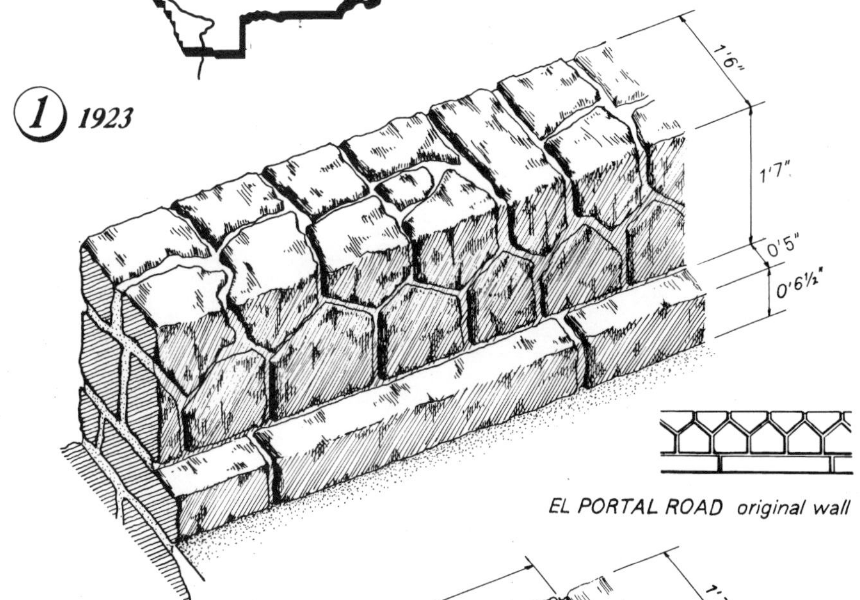

EL PORTAL ROAD original wall

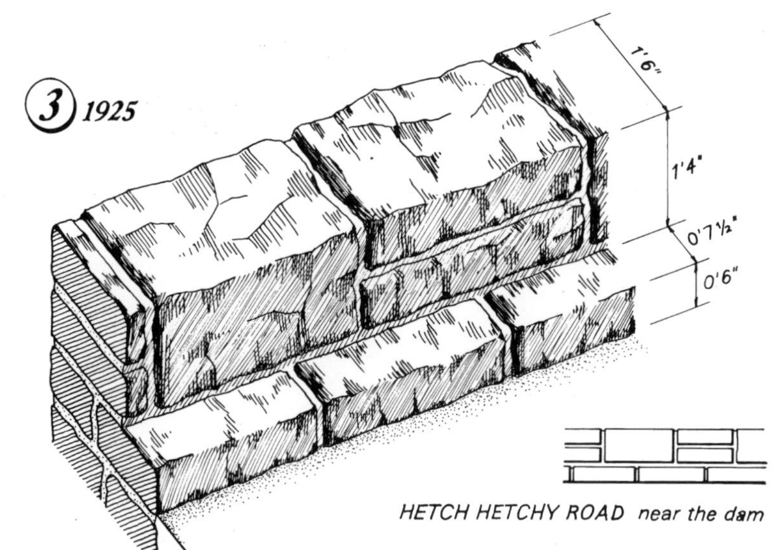

HETCH HETCHY ROAD near the dam

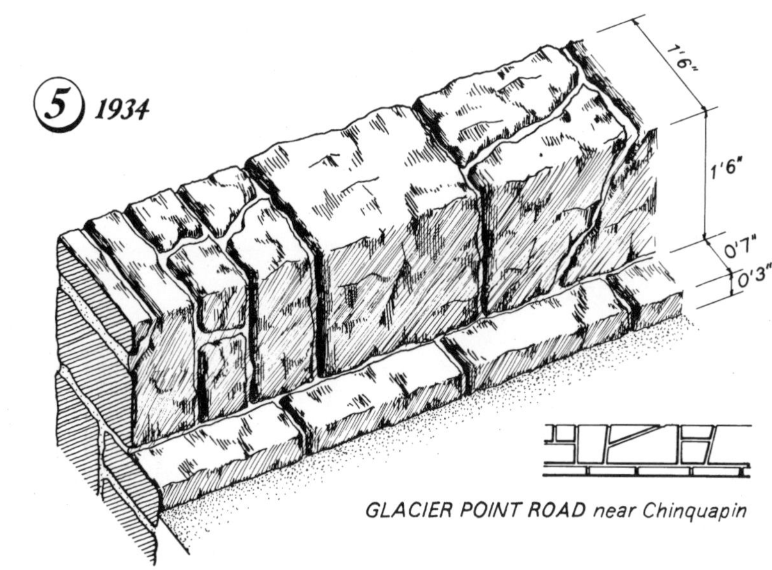

GLACIER POINT ROAD near Chinquapin

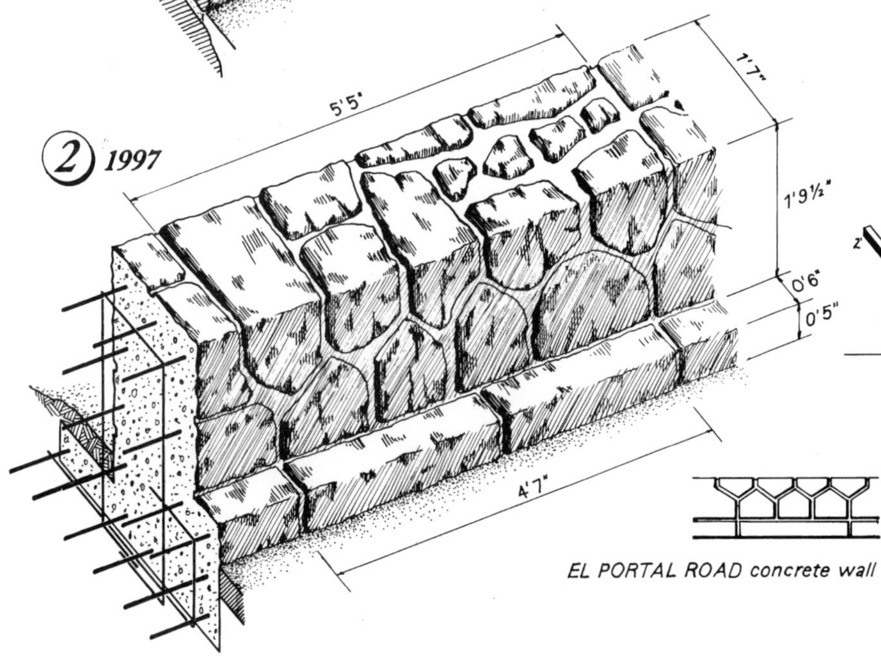

EL PORTAL ROAD concrete wall

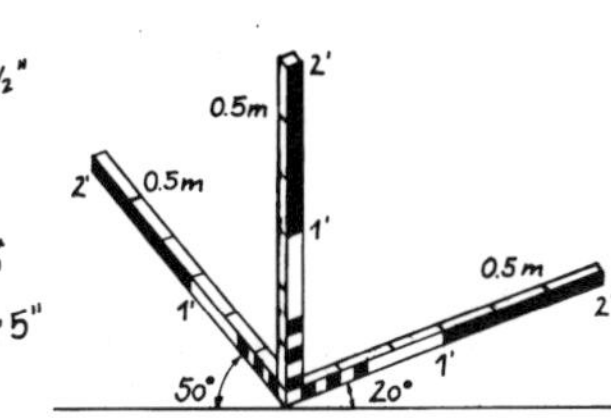

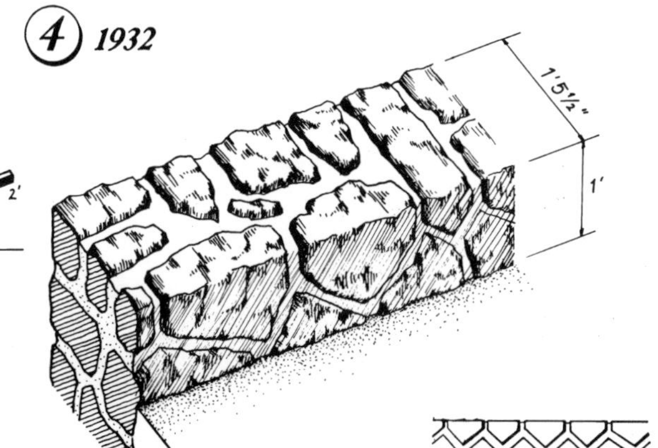

WAWONA ROAD near Bishop Creek

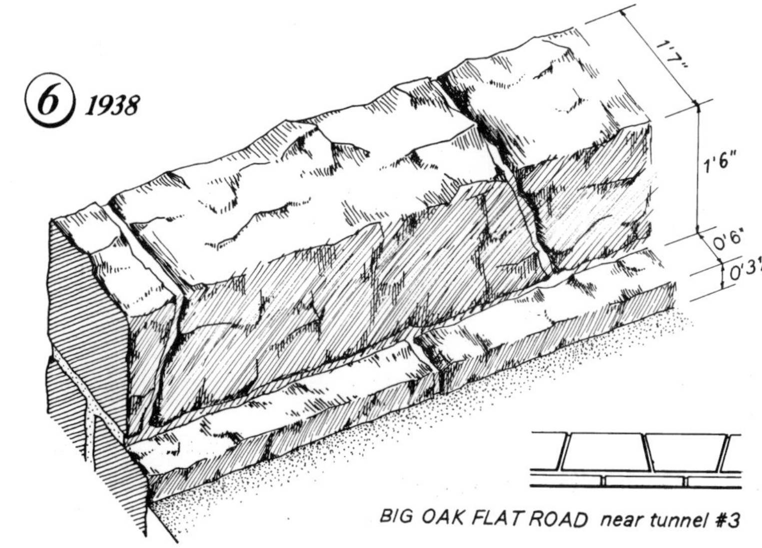

BIG OAK FLAT ROAD near tunnel #3

DELINEATED BY: Anne Teresiak 2001

NPS ROADS & BRIDGES RECORDING PROGRAM
NATIONAL PARK SERVICE
UNITED STATES DEPARTMENT OF THE INTERIOR

YOSEMITE NATIONAL PARK ROADS
YOSEMITE VICINITY
MARIPOSA & TUOLUMNE COUNTIES
CALIFORNIA

SHEET 16 OF 19

HISTORIC AMERICAN ENGINEERING RECORD
CA-117

IF REPRODUCED, PLEASE CREDIT: HISTORIC AMERICAN ENGINEERING RECORD, NATIONAL PARK SERVICE, NAME OF DELINEATOR, DATE OF THE DRAWING

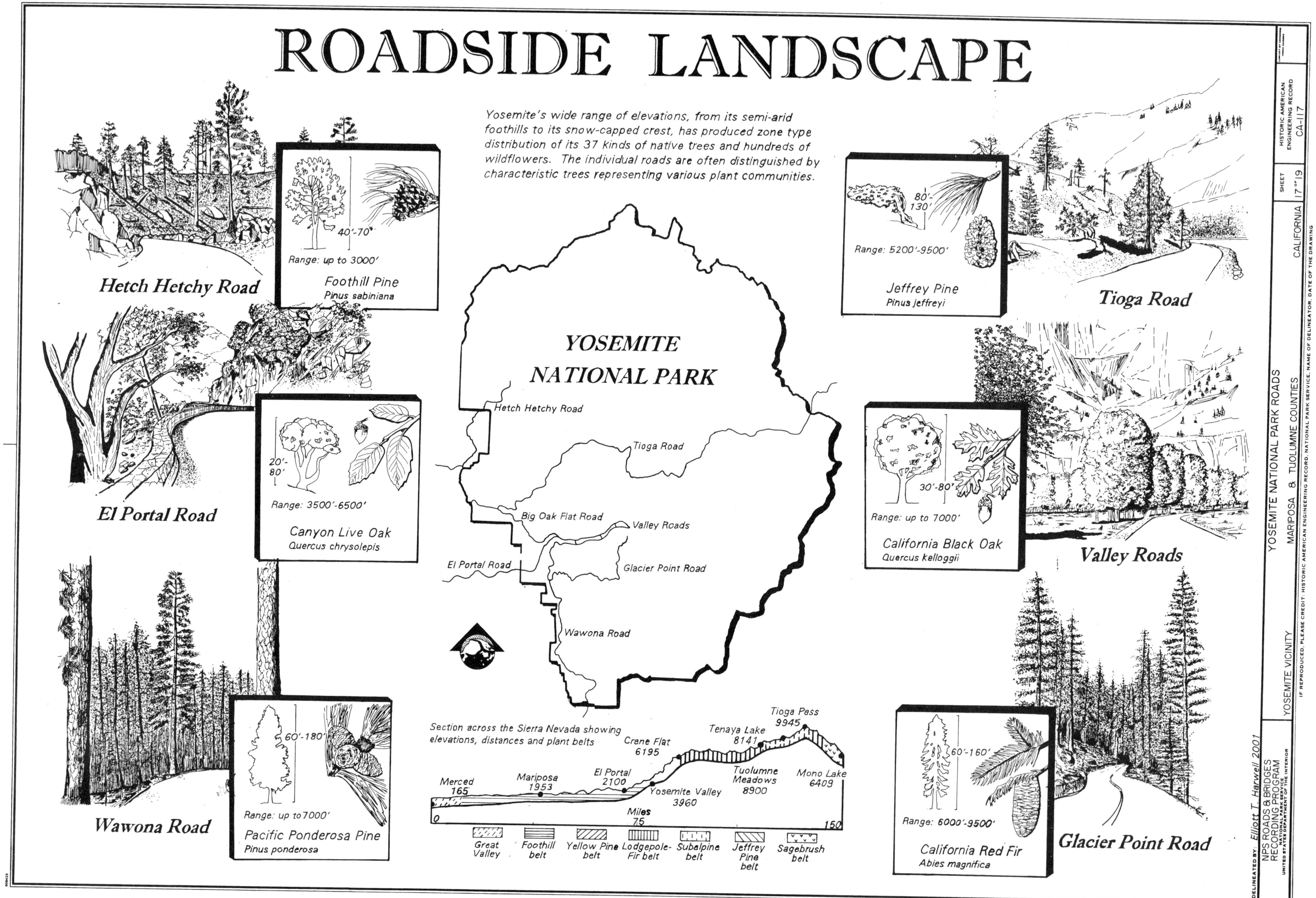
ROADSIDE LANDSCAPE
Yosemite's wide range of elevations, from its semi-arid foothills to its snow-capped crest, has produced zone type distribution of its 37 kinds of native trees and hundreds of wildflowers. The individual roads are often distinguished by characteristic trees representing various plant communities.
Hetch Hetchy Road
40'-70'
Range: up to 3000'
Foothill Pine
Pinus sabiniana
El Portal Road
20'-80'
Range: 3500'-6500'
Canyon Live Oak
Quercus chrysolepis
Wawona Road
60'-180'
Range: up to 7000'
Pacific Ponderosa Pine
Pinus ponderosa
YOSEMITE NATIONAL PARK
Hetch Hetchy Road
Tioga Road
Big Oak Flat Road
Valley Roads
El Portal Road
Glacier Point Road
Wawona Road
80'-130'
Range: 5200'-9500'
Jeffrey Pine
Pinus jeffreyi
Tioga Road
30'-80'
Range: up to 7000'
California Black Oak
Quercus kelloggii
Valley Roads
60'-160'
Range: 6000'-9500'
California Red Fir
Abies magnifica
Glacier Point Road
Section across the Sierra Nevada showing elevations, distances and plant belts
Merced 165
Mariposa 1953
El Portal 2100
Crane Flat 6195
Yosemite Valley 3960
Tenaya Lake 8141
Tuolumne Meadows 8900
Tioga Pass 9945
Mono Lake 6409
Miles
0
75
150
Great Valley
Foothill belt
Yellow Pine belt
Lodgepole-Fir belt
Subalpine belt
Jeffrey Pine belt
Sagebrush belt
DELINEATED BY: Elliott T. Harwell 2001
NPS ROADS & BRIDGES RECORDING PROGRAM
UNITED STATES DEPARTMENT OF THE INTERIOR
YOSEMITE NATIONAL PARK ROADS
YOSEMITE VICINITY
MARIPOSA & TUOLUMNE COUNTIES
CALIFORNIA
SHEET 17 OF 19
HISTORIC AMERICAN ENGINEERING RECORD
CA-117
IF REPRODUCED, PLEASE CREDIT: HISTORIC AMERICAN ENGINEERING RECORD, NATIONAL PARK SERVICE, NAME OF DELINEATOR, DATE OF THE DRAWING

BRIDGE EVOLUTION
19th Century - 1920 s

Yosemite's earliest bridges were simple wooden structures, easily erected from locally available materials. Over time, more complex trusses were erected by the military, followed by the first reinforced concrete spans. The first rustic style bridges were erected in 1913.

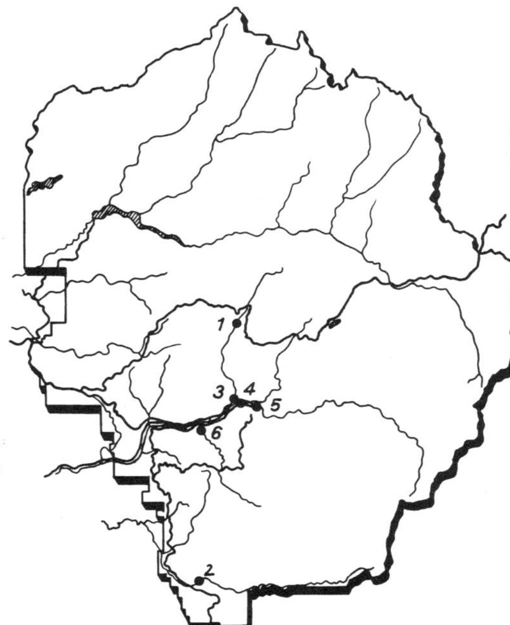

1 Log Bridges

The earliest bridges were built of large logs and primarily used as footbridges. Simple queen post trusses were used for crossing short spans with heavier loads.
This 3-panel bridge over Yosemite Creek was a modification of the king post truss with a horizontal chord for reinforcement.

Early bridge over Yosemite Creek

Simple log bridge in Yosemite Valley

2 Wawona Covered Bridge

A covered bridge protects the interior wooden truss system from decay. The 1868 open deck bridge at Wawona was covered in 1878 and remains the only covered bridge in Yosemite. Outside the park but close to the boundary there was another covered bridge across the South Fork Tuolumne River.

Wawona Covered Bridge 1878

3 Timber Truss Bridges

As spans grew longer and vehicle weights increased, more complex truss systems were employed. Such bridges spanned across Yosemite Creek and the Merced River in Yosemite Valley. Also the old Cascade Creek Bridge built in 1907 was timber truss construction.

Old bridge over Yosemite Creek 1884

Old Cascade Creek Bridge 1907

4 Steel Truss Bridges

More expensive than wooden trusses, steel bridges were much more durable and could carry heavier loads. The second Sentinel Bridge - an early representative of this group - was erected in 1879 as an iron truss. Built in 1915, the old El Capitan Bridge was a combined timber and steel truss system.

Sentinel Bridge 1879

5 Reinforced Concrete Bridges

By 1920, concrete bridges had made their appearance on the Valley floor. Over time, most of them were replaced, like the Old Stoneman Bridge & the Old Sentinel Bridge. One of the earliest concrete bridges was erected in 1921 at Happy Isles. The original concrete railing was later replaced by a Mission 66 railing.

Old Happy Isles Bridge 1921

present state

6 Early Rustic Concrete Bridges

By facing three simple reinforced concrete bridges at Bridalveil Falls with stone, the military set the precedent for rustic bridge design in 1913.

Bridalveil Fall Bridge #1 1913

DELINEATED BY: *Anne Teresiak 2001*

NPS ROADS & BRIDGES RECORDING PROGRAM NATIONAL PARK SERVICE UNITED STATES DEPARTMENT OF THE INTERIOR	YOSEMITE NATIONAL PARK ROADS YOSEMITE VICINITY MARIPOSA & TUOLUMNE COUNTIES CALIFORNIA	SHEET 18 of 19	HISTORIC AMERICAN ENGINEERING RECORD CA-117

BRIDGE EVOLUTION
1920 s - Present

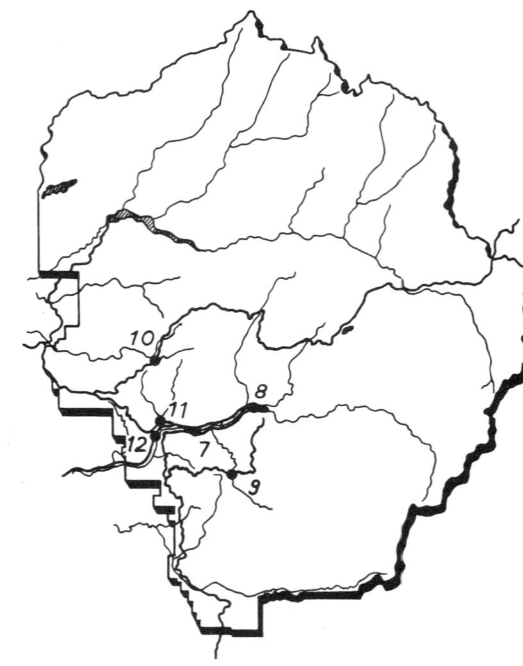

In the 1920s, the National Park Service embraced the rustic style, characterized by the use of native materials, but would employ more modern designs where rustic spans were inappropriate. Today's bridges reflect a revival of earlier rustic design.

7 Rustic Stone Arched Bridges

Most of the Valley bridges were graceful arches faced in native stone and with large ring stones defining the arch. The first one was Yosemite Creek Bridge built in 1921. It was followed by Pohono Bridge, Tenaya Creek Bridge, Ahwanee Bridge and Sugar Pine Bridge, which were designed in a plan for five bridges in 1927.

Pohono Bridge 1927

8 Rustic Arches with Subways

Several of the Valley bridges incorporated tunnels or subways to accommodate passage of foot and bridle trails.
The first of this style was Clark's Bridge built in 1928 as one of the five bridges contract of 1927. It was followed by the erection of Happy Isles Bridge in 1929 and Stoneman Bridge in 1932.

Clark's Bridge 1928

9 Rustic Steel Girder Bridges

In an interesting rustic variant, massive logs were bolted to bridges to conceal their interior steel girder construction. Both Upper Bridalveil Creek Bridge and El Capitan Bridge are still in use, while Upper Yosemite Creek Bridge and South Fork Merced River Bridge are condemned. Over time, all of them lost their original railing.

Upper Bridalveil Creek Bridge 1933

Section present railing

10 Concrete Girder Deck Bridges

Less expensive than arched bridges, concrete girder spans still employed native stone for abutments.
There are only two bridges of this style in Yosemite, and both are found on Tioga Pass Road:
Tuolumne River Bridge, built in 1934, and South Fork Tuolumne River Bridge, erected in 1937.

South Fork Tuolumne River Bridge 1937

11 Reinforced Concrete Open Spandrel Arches

The three soaring open spandrel arch bridges on the Big Oak Flat Road reflect a streamlined modern style.
They were built across Tamarack Creek, Wildcat Creek and Cascade Creek in 1939.

Cascade Creek Bridge 1939

12 Contemporary Rustic Bridges

Recent bridges in the park reflect a return to the rustic style in their use of native stone facing. The Cascade Bridge on El Portal Road was built in 1985. The section shows the interior concrete construction. Another example is the new Sentinel Bridge in Yosemite Valley, built in 1994.

Cascade Bridge 1985

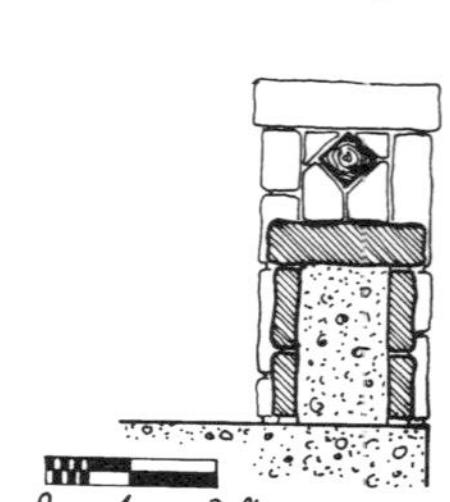

Section railing

DELINEATED BY: Anne Teresiak 2001

NPS ROADS & BRIDGES RECORDING PROGRAM NATIONAL PARK SERVICE UNITED STATES DEPARTMENT OF THE INTERIOR	YOSEMITE NATIONAL PARK ROADS YOSEMITE VICINITY · MARIPOSA & TUOLUMNE COUNTIES · CALIFORNIA	SHEET 19 of 19	HISTORIC AMERICAN ENGINEERING RECORD CA-117

IF REPRODUCED, PLEASE CREDIT: HISTORIC AMERICAN ENGINEERING RECORD, NATIONAL PARK SERVICE, NAME OF DELINEATOR, DATE OF THE DRAWING

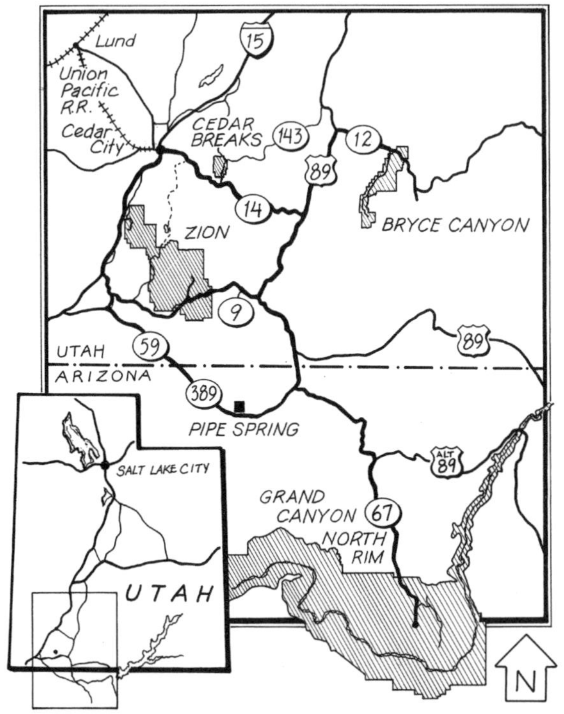

ZION

ROADS & BRIDGES

ZION NATIONAL PARK, UTAH

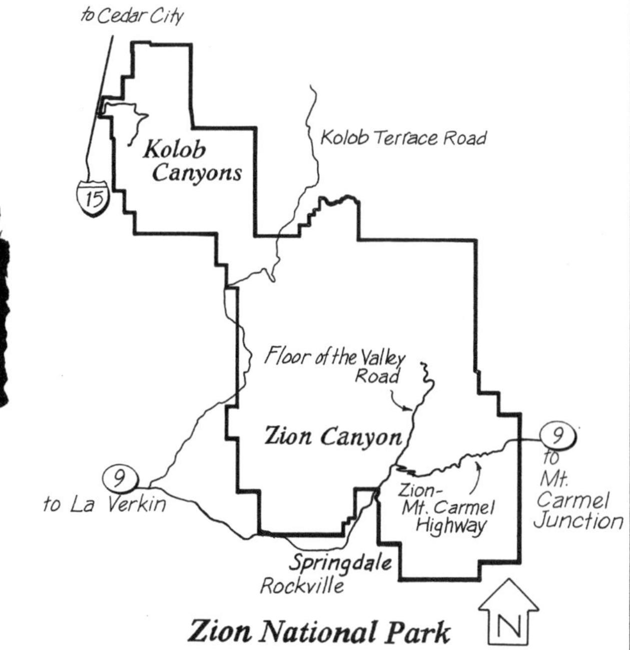

Zion National Park, encompassing 229 square miles of southern Utah wilderness, contains many natural wonders found only in this region of the country. Deep canyons twist between high plateaus, testifying to the patient strength of erosion. Petrified forests, rivers, waterfalls, and wildlife are surrounded by varied and spectacular rock formations, including one of the world's largest natural arches--Kolob Arch, spanning 310 feet.

The area was relatively isolated until the first settlers entered the area in 1858, founding Virgin City, Utah, located south of Zion Canyon. Mormon pioneers developed rough wagon roads between small settlements, frequently following trails first established by Native American residents and Spanish explorers. The uniqueness of the park area was soon recognized, leading to the creation of Mukuntuweap National Monument in 1909 (later changed to Zion National Park in 1919).

The first automobiles to travel into Zion National Park followed a road constructed in 1917 with a $15,000 special appropriation, extending four miles from the south entrance to Weeping Rock. In 1925, a longer "Government Road" replaced it. The gravel-surfaced highway ran 7.5 miles from the south entrance to the Temple of Sinawava and cost $70,000. This road was replaced in 1932 with today's Floor of the Valley Road, which now carries nearly a million vehicles per year through Zion Canyon.

Further refinements of the existing road systems into and surrounding Zion were not begun until 1920, when an aggressive road building campaign was initiated, spearheaded by the Union Pacific Railroad in conjunction with the National Park Service, Utah Road Commission, and Bureau of Public Roads. The Union Pacific sought to connect Zion, Bryce Canyon, Cedar Breaks, and the North Rim of the Grand Canyon with a series of roads creating a "circle" or loop tour. Visitors could access the area by train into Cedar City, Utah, board a large touring limousine, visit the parks and stay at lodges run by a subsidiary of the Union Pacific. Completion of an adequate road system was the key to the long-range plans developed by the corporation.

Illustration based on historic photographs of a Union Pacific Railroad touring limousine ca. 1930s.

This project is part of the Historic American Engineering Record (HAER), a long-range program to document historically significant engineering and industrial works in the United States. The HAER program is administered by the Historic American Buildings Survey/Historic American Engineering Record Division (HABS/HAER) of the National Park Service, U.S. Department of the Interior. The Zion Roads and Bridges Recording Project was cosponsored during the summer of 1993 by HAER under the general direction of Dr. Robert J. Kapsch, Chief; the National Park Service Roads Program, John Gingles, Manager; and by Zion National Park, Donald Falvey, Superintendent.

The field work, measured drawings, historical reports, and photographs were prepared under the direction of Project Leaders Eric DeLony, Chief of HAER, and Todd A. Croteau, HAER Staff Architect. The recording team consisted of Todd A. Croteau, field supervisor; Architectural technicians Laura J. Culberson and Christopher Payne. Historical reports were prepared by project historians Michael Anderson, Christine Madrid, and Richard H. Quin. Formal large-format photography was done by Brian Grogan.

During this time, State engineer Howard Means and Bureau of Public Roads Engineers B.J. Finch and R.R. Mitchell located and surveyed a 25-mile long route which would travel through Zion National Park, meeting with Highway 89 near Mount Carmel. Completion of this route was essential to the Union Pacific's overall plan to increase tourism in the area. The NPS allotted $1,500,000 to construct the initial 8.5-mile segment of the road from the North Fork Virgin River, up the talus slopes of Pine Creek canyon through seven switchbacks, into the cliffs themselves with a 5,613-foot long tunnel and on to the park's east boundary. Two bridges were included in the construction, North Fork Virgin River and Pine Creek bridges, both constructed in a rustic style harmonious with the natural environment of each structures individual setting.

The remainder of the road, east of the tunnel, was finished by the State of Utah as a federal aid project at a cost of over $400,000, including the Co-op and Clear Creek bridges (replaced in 1993). With the completion of the Zion National Park road system, tourists (with the Union Pacific or otherwise) were able to visit parks within the circle tour loop on an interconnected, modern system of highways. Although the Union Pacific plans for the area were successful, they became short-lived, as more and more Americans sought to explore the wilderness regions of Utah and Arizona with the flexibility and privacy of their own personal vehicles. Today, nearly 3,000,000 visitors travel through Zion National Park yearly to stand in wonder beneath its massive sandstone monoliths and meander among the valley's lush canopy of cottonwoods and ashes.

DELINEATED BY: Todd A. Croteau, 1993
ZION NATIONAL PARK ROADS AND BRIDGES RECORDING PROJECT
NATIONAL PARK SERVICE
UNITED STATES DEPARTMENT OF THE INTERIOR

ZION NATIONAL PARK ROADS AND BRIDGES
SPRINGDALE VICINITY
WASHINGTON COUNTY
UTAH

SHEET 1 of 2

HISTORIC AMERICAN ENGINEERING RECORD
UT-72

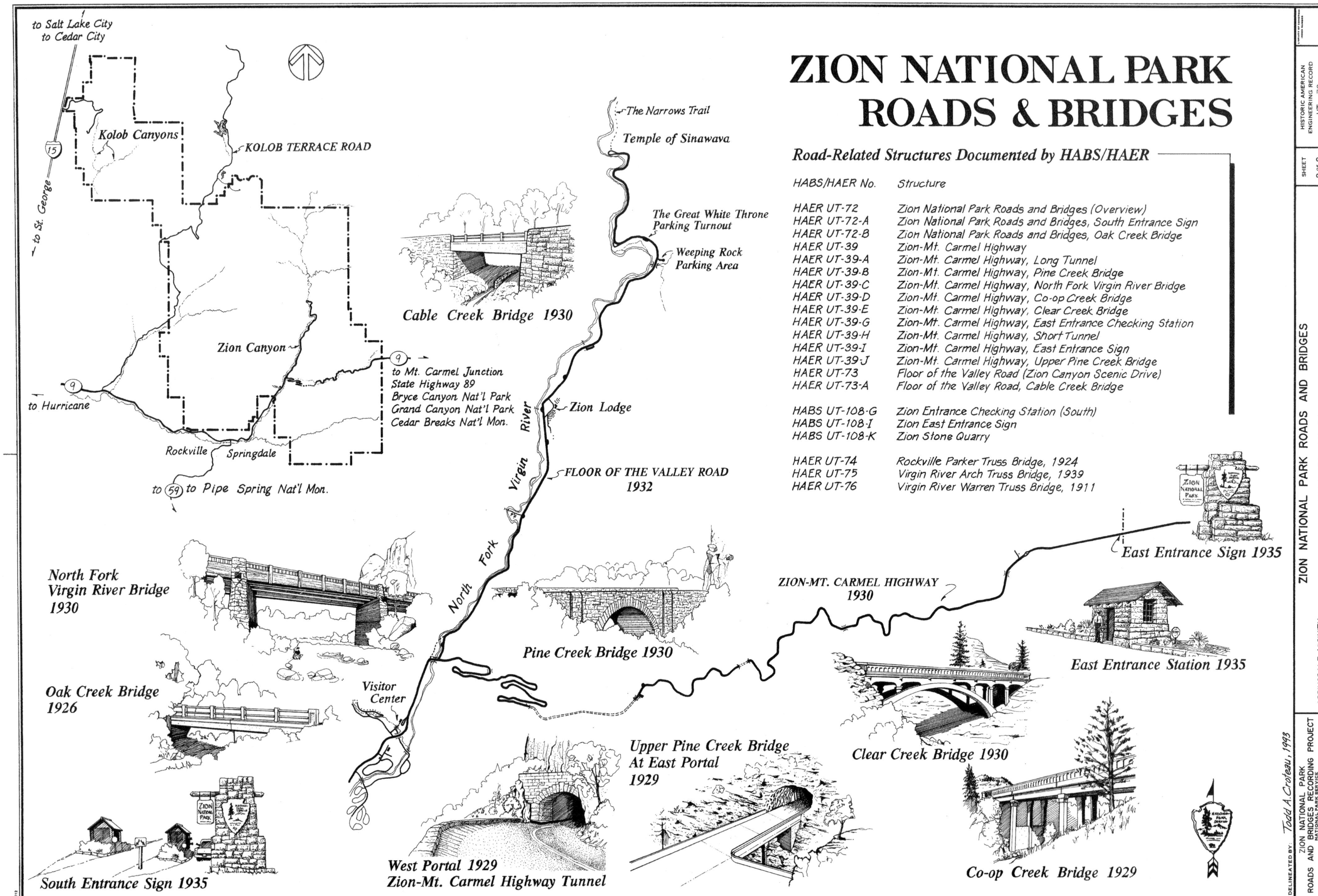
ZION NATIONAL PARK
ROADS & BRIDGES
Road-Related Structures Documented by HABS/HAER
HABS/HAER No. Structure
HAER UT-72 Zion National Park Roads and Bridges (Overview)
HAER UT-72-A Zion National Park Roads and Bridges, South Entrance Sign
HAER UT-72-B Zion National Park Roads and Bridges, Oak Creek Bridge
HAER UT-39 Zion-Mt. Carmel Highway
HAER UT-39-A Zion-Mt. Carmel Highway, Long Tunnel
HAER UT-39-B Zion-Mt. Carmel Highway, Pine Creek Bridge
HAER UT-39-C Zion-Mt. Carmel Highway, North Fork Virgin River Bridge
HAER UT-39-D Zion-Mt. Carmel Highway, Co-op Creek Bridge
HAER UT-39-E Zion-Mt. Carmel Highway, Clear Creek Bridge
HAER UT-39-G Zion-Mt. Carmel Highway, East Entrance Checking Station
HAER UT-39-H Zion-Mt. Carmel Highway, Short Tunnel
HAER UT-39-I Zion-Mt. Carmel Highway, East Entrance Sign
HAER UT-39-J Zion-Mt. Carmel Highway, Upper Pine Creek Bridge
HAER UT-73 Floor of the Valley Road (Zion Canyon Scenic Drive)
HAER UT-73-A Floor of the Valley Road, Cable Creek Bridge
HABS UT-108-G Zion Entrance Checking Station (South)
HABS UT-108-I Zion East Entrance Sign
HABS UT-108-K Zion Stone Quarry
HAER UT-74 Rockville Parker Truss Bridge, 1924
HAER UT-75 Virgin River Arch Truss Bridge, 1939
HAER UT-76 Virgin River Warren Truss Bridge, 1911
to Salt Lake City
to Cedar City
15
to St. George
Kolob Canyons
KOLOB TERRACE ROAD
Zion Canyon
9
to Hurricane
9
to Mt. Carmel Junction
State Highway 89
Bryce Canyon Nat'l Park
Grand Canyon Nat'l Park
Cedar Breaks Nat'l Mon.
Rockville
Springdale
to 59 to Pipe Spring Nat'l Mon.
The Narrows Trail
Temple of Sinawava
The Great White Throne Parking Turnout
Weeping Rock Parking Area
Cable Creek Bridge 1930
Zion Lodge
North Fork Virgin River
FLOOR OF THE VALLEY ROAD 1932
North Fork Virgin River Bridge 1930
Oak Creek Bridge 1926
Visitor Center
Pine Creek Bridge 1930
ZION-MT. CARMEL HIGHWAY 1930
East Entrance Sign 1935
ZION NATIONAL PARK
East Entrance Station 1935
Clear Creek Bridge 1930
Upper Pine Creek Bridge At East Portal 1929
Co-op Creek Bridge 1929
West Portal 1929
Zion-Mt. Carmel Highway Tunnel
South Entrance Sign 1935
ZION NATIONAL PARK
DELINEATED BY: Todd A. Croteau, 1993
ZION NATIONAL PARK ROADS AND BRIDGES RECORDING PROJECT
ZION NATIONAL PARK ROADS AND BRIDGES
SPRINGDALE VICINITY
WASHINGTON COUNTY
UTAH
SHEET 2 OF 2
HISTORIC AMERICAN ENGINEERING RECORD UT-72
IF REPRODUCED, PLEASE CREDIT: HISTORIC AMERICAN ENGINEERING RECORD, NATIONAL PARK SERVICE, NAME OF DELINEATOR, DATE OF THE DRAWING

ZION-MOUNT CARMEL HIGHWAY TUNNEL
1930

WEST PORTAL *(masonry-faced)*

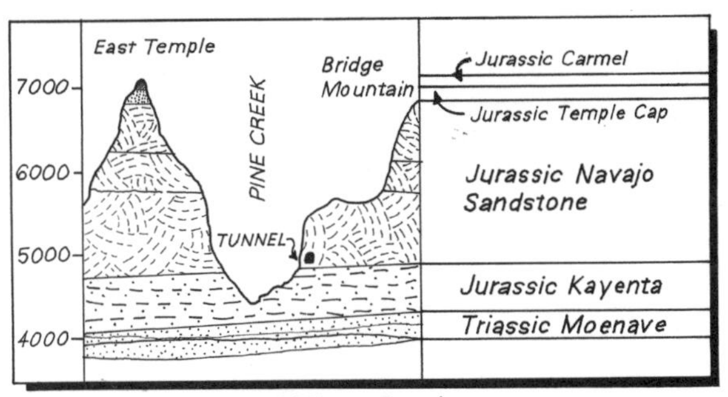

Cross Section of Zion Geology

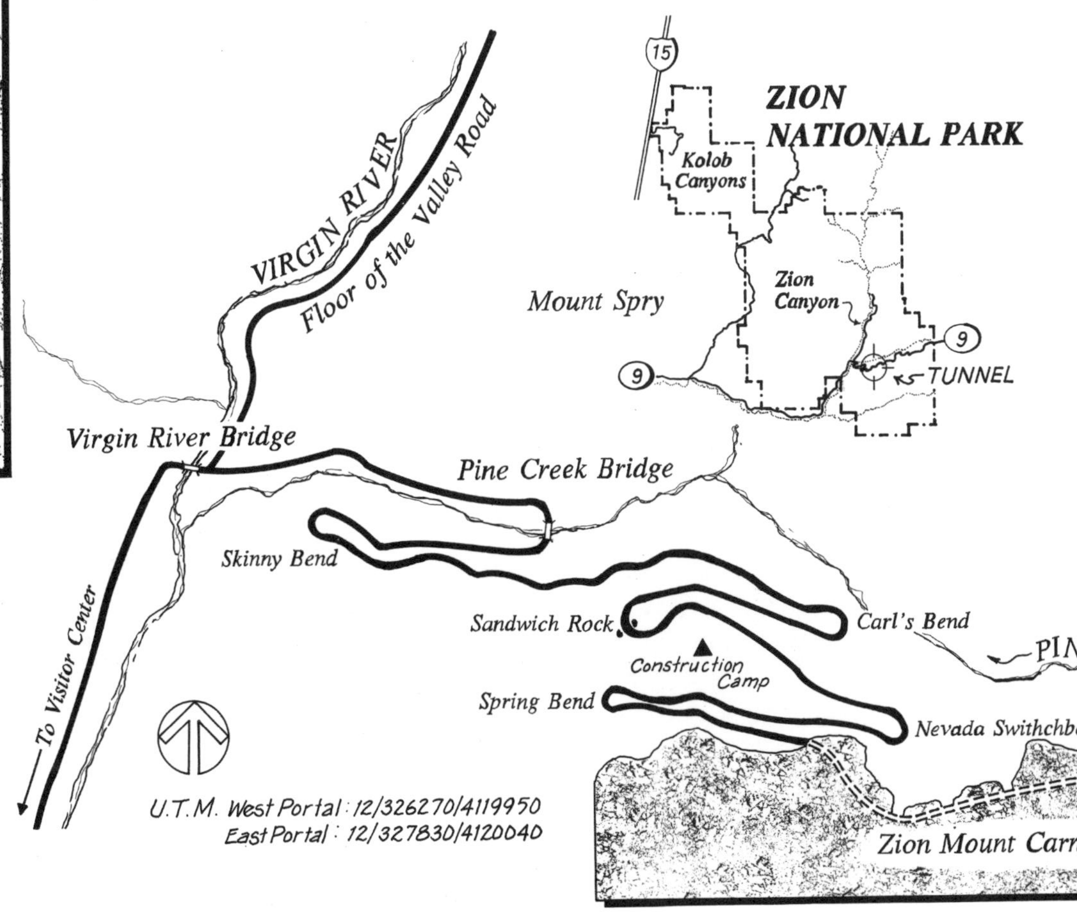

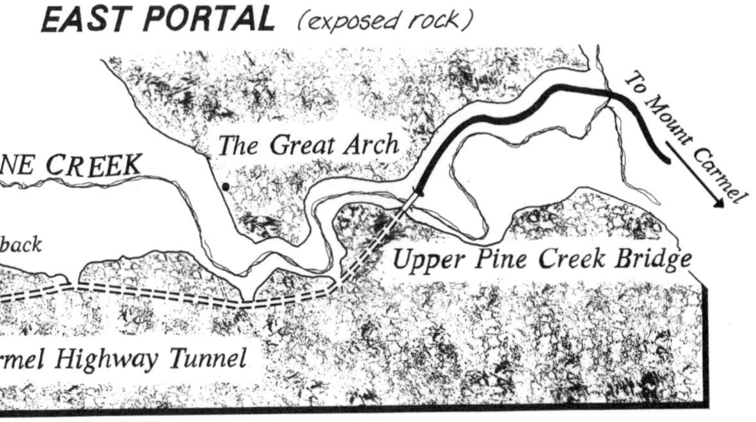

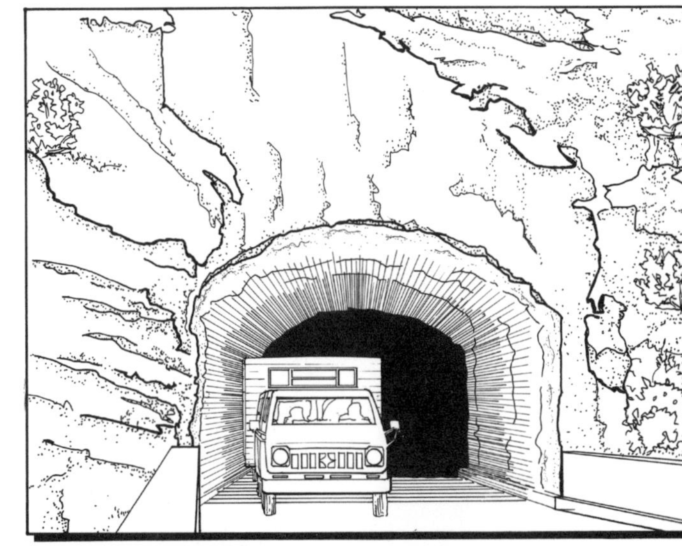

EAST PORTAL *(exposed rock)*

Drawing based on field measurements historical photographs and original design documents. Map based on U.S.G.S.

At 5,613 feet, the Zion-Mount Carmel Highway Tunnel is the longest vehicular tunnel constructed in a National Park. Designed co-operatively by the Bureau of Public Roads and the National Park Service in 1927, the tunnel was completed and dedicated by July 4, 1930 at a cost of $763,527.

Consistent with Park Service aesthetic ideals of the time, the tunnel was designed to impact the visible landscape as little as possible. Five (later six) ventilation galleries were plotted along the length of the proposed tunnel, connected by linear tangents. Curves with a minimum radius of 225 feet at the galleries created a smooth uninterrupted passage through the rock. A consistent distance of 21 feet was maintained between the centerline of the tunnel and the face of the cliff.

The west portal of the tunnel is framed by a masonry facade and symmetrical arch constructed of 2' thick cut sandstone laid by Shea and Shea in 1929. Due to differing conditions within the sandstone tunnel, a variety of linings were used. Masonry lining, continuous concrete lining, intermittent concrete ribs, and timber lining are all found protecting the interior walls. A masonry guardwall, originally 36" high and raised to 42" in the late 1950s, encloses the edge of each gallery, now closed to visitor access.

Difficulties presented by the inaccessible and unstable site chosen for the tunnel required the builder, Nevada Contracting Company, to utilize excavating techniques typically used in mining operations. The company bored an exterior stope, or stepped vertical shaft, to reach roadway grade (at Gallery 4), built the tunnel from the inside out through the galleries (rather than through the entrance portals), and used ring drilling equipment to complete this imaginative project.

This project is part of the Historic American Engineering Record (HAER), a long-range program to document historically significant engineering and industrial works in the United States. The HAER program is administered by the Historic American Buildings Survey/Historic American Engineering Record Division (HABS/HAER) of the National Park Service, U.S. Department of the Interior. The Zion Roads and Bridges Recording Project was cosponsored during the summer of 1993 by HAER under the general direction of Dr. Robert J. Kapsch, Chief; the National Park Service Roads Program, John Gingles, Manager; and by Zion National Park, Donald Falvey, Superintendent.

The field work, measured drawings, historical reports, and photographs were prepared under the direction of Project Leaders Eric DeLony, Chief of HAER, and Todd A. Croteau, HAER Staff Architect. The recording team consisted of Todd A. Croteau, field supervisor; Architectural technicians Laura J. Culberson and Christopher Payne. Historical reports were prepared by project historians Michael Anderson, Christine Madrid, and Richard H. Quin. Formal large-format photography was done by Brian Grogan.

DELINEATED BY: Laura J. Culberson 1993, Todd A. Croteau 1994

ZION NATIONAL PARK ROADS AND BRIDGES RECORDING PROJECT
NATIONAL PARK SERVICE, UNITED STATES DEPARTMENT OF THE INTERIOR

ZION-MT. CARMEL HIGHWAY, ZION-MT. CARMEL HIGHWAY TUNNEL - 1930
BETWEEN MILE POST AND ZION-MT. CARMEL HIGHWAY (STATE HIGHWAY 9)
SPRINGDALE VICINITY WASHINGTON COUNTY UTAH

SHEET 1 of 3

HISTORIC AMERICAN ENGINEERING RECORD UT-39-A

IF REPRODUCED, PLEASE CREDIT: HISTORIC AMERICAN ENGINEERING RECORD, NATIONAL PARK SERVICE, NAME OF DELINEATOR, DATE OF THE DRAWING

STAGES OF TUNNEL CONSTRUCTION

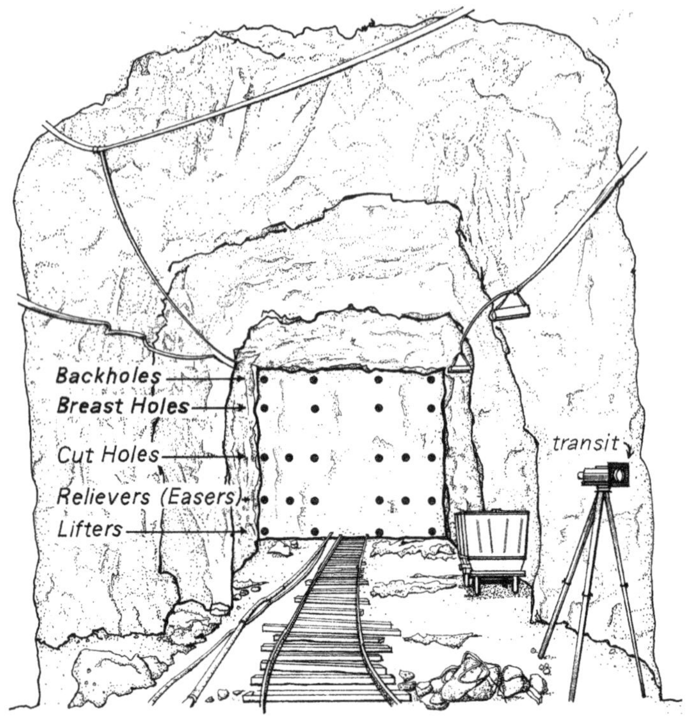

PILOT TUNNEL

Pilot tunnel crews began their eight-hour shift by loading spoil and muck into mining cars (A) using Sullivan and Butler mucking machines, then running the cars to the nearest gallery with a Sullivan compressed-air tugger and dumping the debris into Pine Creek Canyon. Survey engineers determined center line and drift (B) while other crew members placed drilling equipment (C). Men spent 3-4 hours drilling five rows of eight-foot-deep holes in the tunnel face. From the top, these were back holes, breast holes cut holes, relievers (easers), and lifters at the bottom (D). After exploding the rows in two-minute intervals, crews moved back in and scaled loose rubble from the ceiling and walls before completing their shift and returning to camp. In this manner, pilot tunnel crews worked 1,275 consecutive eight-hour shifts to blast an 8' x 9' tunnel more than one mile through solid sandstone.

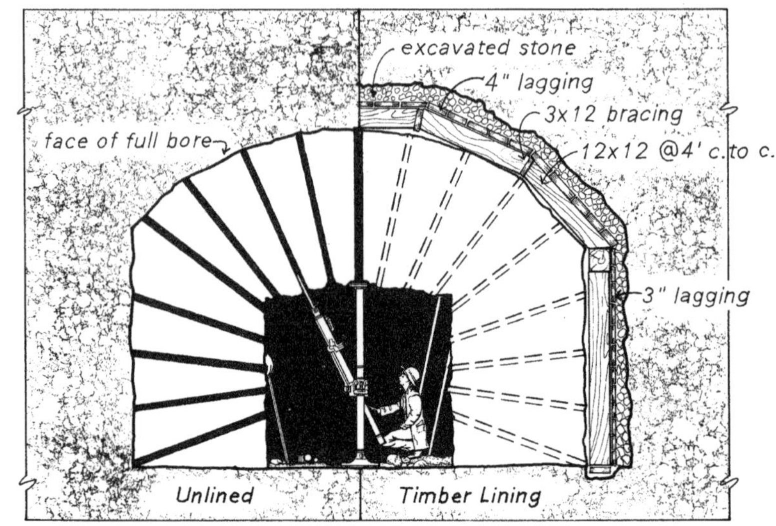

FULL BORE

Drawing based on field measurements historical photographs and original design documents.

Following pilot tunnel crews at a safe distance, full bore crews used a process known as "ring-drilling" to excavate the tunnel to its main dimensions of 16' x 22'. They setup several self-radiating, reverse feed cylinder, stoping drills at three-foot intervals within the pilot bore and drilled columns or "rings" consisting of seventeen holes each (E). When 10-12 rings were completed, each ring was packed with fifty pounds of dynamite and all were exploded at once (F). Crews then used an Erie 1/2-yard compressed air shovel to load the spoil into Mack two-yard trucks (G), and dumped the debris into Pine Creek Valley. Full bore crews moved forward from 30 to 36 feet each eight-hour shift and needed eight months to drill and blast the entire one-mile of tunnel.

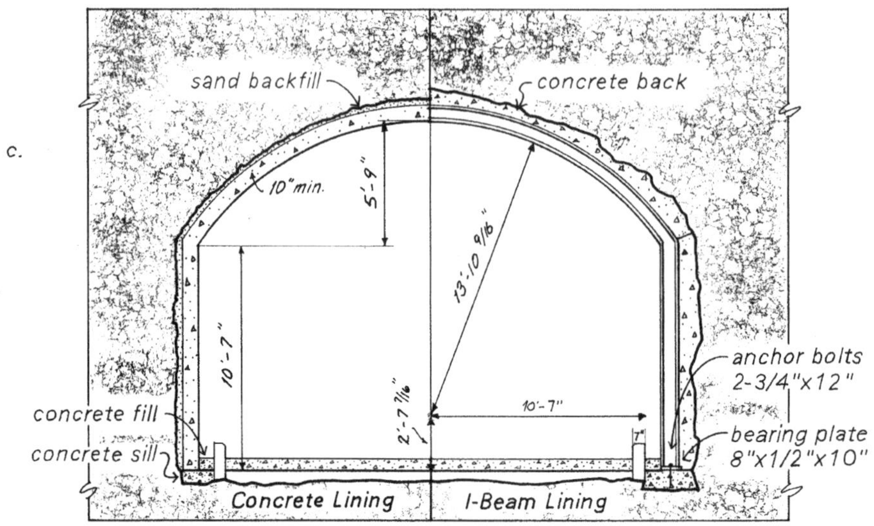

LINING

Scale: 1/4" = 1'-0"

0 1 2 3 4 5 10 15 feet

0 1 2 3 4 meters

Engineers did not expect to line the tunnel, but excavation revealed unstable sandstone at various points. In the initial project, 580 feet of tunnel was lined with 8" x 8" and 12" x 12" hewn timbers (H). Later inspections revealed cracks which required another 400 feet of 12" x 12" timber lining, concrete lining at Galleries 6 and 2, and one-inch-thick gunite lining on the entire tunnel ceiling, spring line to spring, where timber was not in place. Further signs of instability led to an additional 164 feet of 12" x 12" timber and 726 feet of continuous and intermittent (ribbed) concrete lining in 1932-33. In 1937, the remainder of the tunnel was lined with continuous or intermittent concrete.

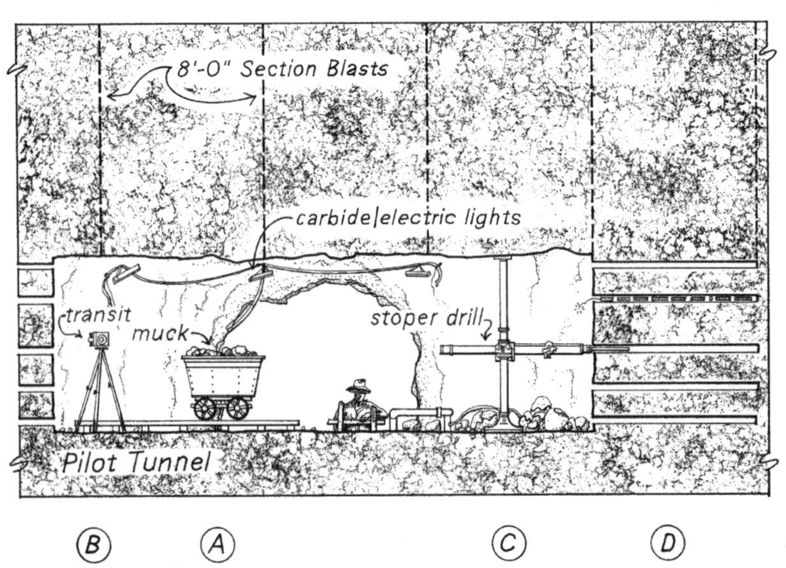

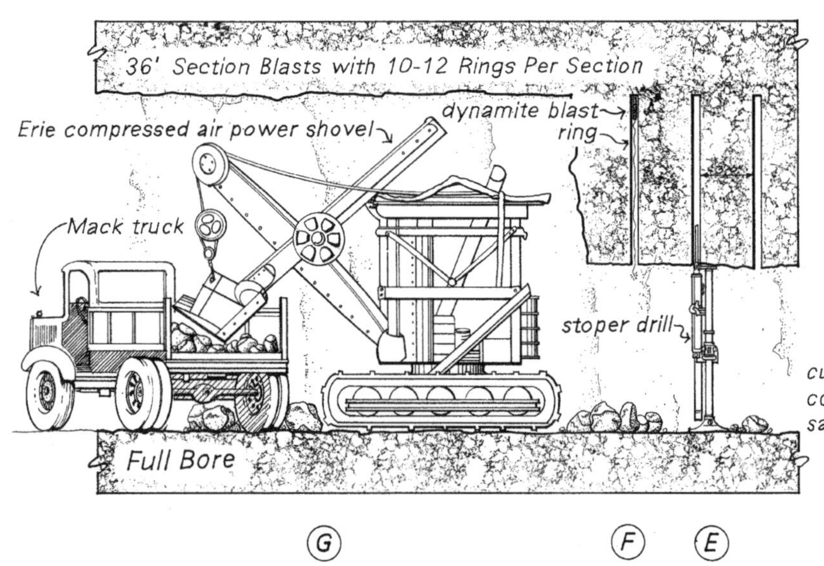

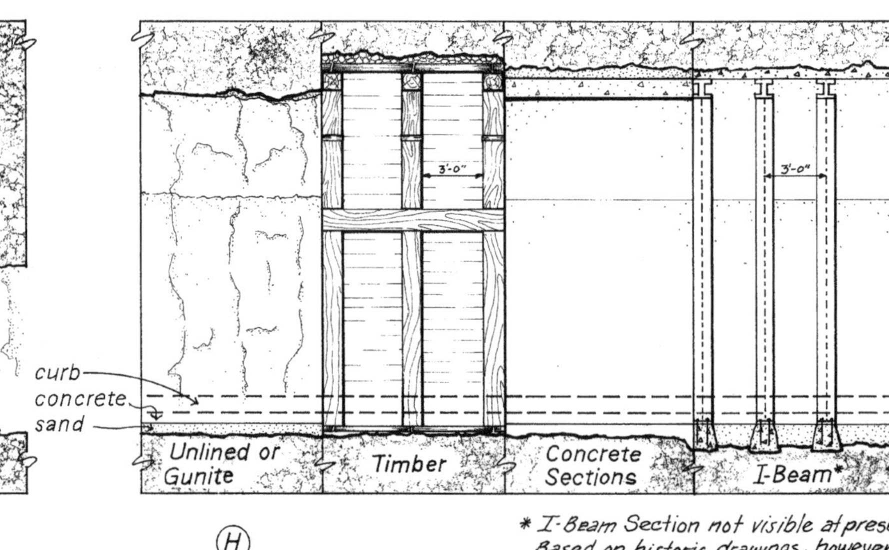

** I-Beam Section not visible at present. Based on historic drawings, however, research provided no evidence of construction.*

DELINEATED BY: Laura J. Culberson, Todd A. Croteau, 1993

ZION NATIONAL PARK
ROADS AND BRIDGES RECORDING PROJECT
NATIONAL PARK SERVICE
UNITED STATES DEPARTMENT OF THE INTERIOR

ZION-MT. CARMEL HIGHWAY, ZION-MT. CARMEL HIGHWAY TUNNEL - 1930
BETWEEN MILE POST AND , ZION-MT. CARMEL HIGHWAY (STATE HIGHWAY 9)
SPRINGDALE VICINITY WASHINGTON COUNTY UTAH

SHEET 2 OF 3

HISTORIC AMERICAN ENGINEERING RECORD UT-39-A

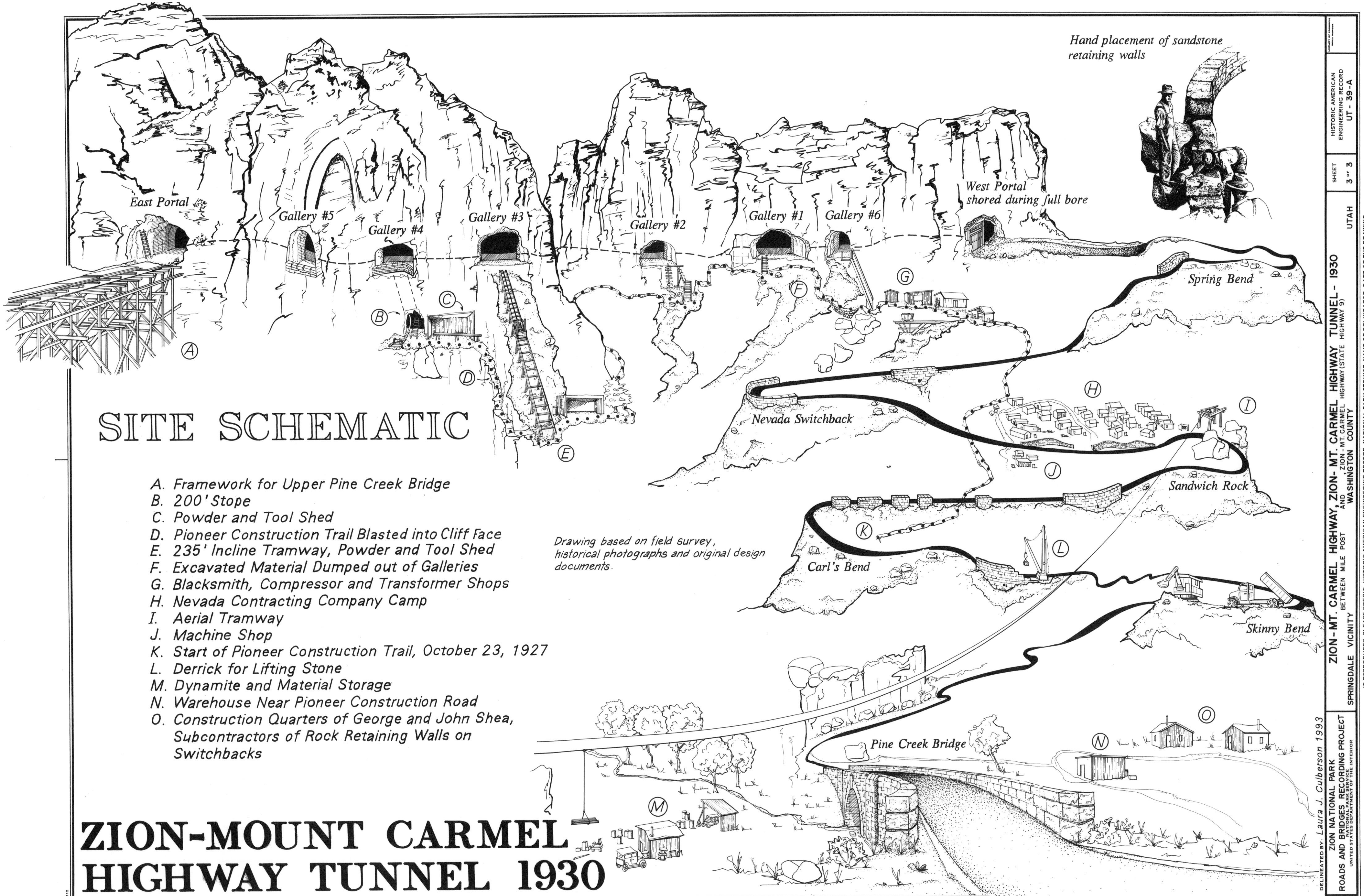

Hand placement of sandstone retaining walls
East Portal
Gallery #5
Gallery #4
Gallery #3
Gallery #2
Gallery #1
Gallery #6
West Portal shored during full bore
Spring Bend
Nevada Switchback
Sandwich Rock
Carl's Bend
Skinny Bend
Pine Creek Bridge
A
B
C
D
E
F
G
H
I
J
K
L
M
N
O
SITE SCHEMATIC
A. Framework for Upper Pine Creek Bridge
B. 200' Stope
C. Powder and Tool Shed
D. Pioneer Construction Trail Blasted into Cliff Face
E. 235' Incline Tramway, Powder and Tool Shed
F. Excavated Material Dumped out of Galleries
G. Blacksmith, Compressor and Transformer Shops
H. Nevada Contracting Company Camp
I. Aerial Tramway
J. Machine Shop
K. Start of Pioneer Construction Trail, October 23, 1927
L. Derrick for Lifting Stone
M. Dynamite and Material Storage
N. Warehouse Near Pioneer Construction Road
O. Construction Quarters of George and John Shea, Subcontractors of Rock Retaining Walls on Switchbacks
Drawing based on field survey, historical photographs and original design documents.
ZION-MOUNT CARMEL HIGHWAY TUNNEL 1930
DELINEATED BY: Laura J. Culberson 1993
ZION NATIONAL PARK
ROADS AND BRIDGES RECORDING PROJECT
NATIONAL PARK SERVICE
UNITED STATES DEPARTMENT OF THE INTERIOR
ZION-MT. CARMEL HIGHWAY, ZION-MT. CARMEL HIGHWAY TUNNEL - 1930
BETWEEN MILE POST AND ZION-MT. CARMEL HIGHWAY (STATE HIGHWAY 9)
SPRINGDALE VICINITY WASHINGTON COUNTY UTAH
SHEET 3 OF 3
HISTORIC AMERICAN ENGINEERING RECORD
UT-39-A
IF REPRODUCED, PLEASE CREDIT: HISTORIC AMERICAN ENGINEERING RECORD, NATIONAL PARK SERVICE, NAME OF DELINEATOR, DATE OF THE DRAWING

PINE CREEK BRIDGE
1930

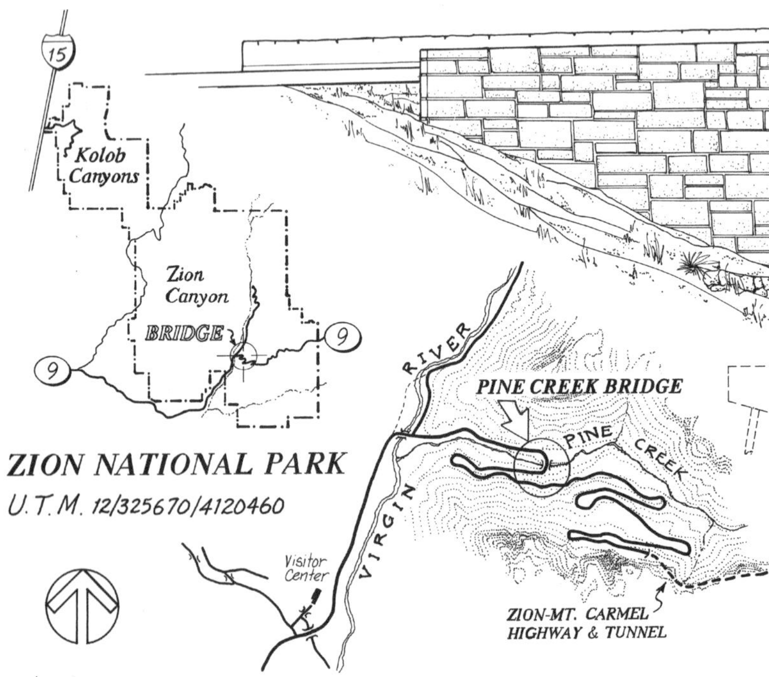

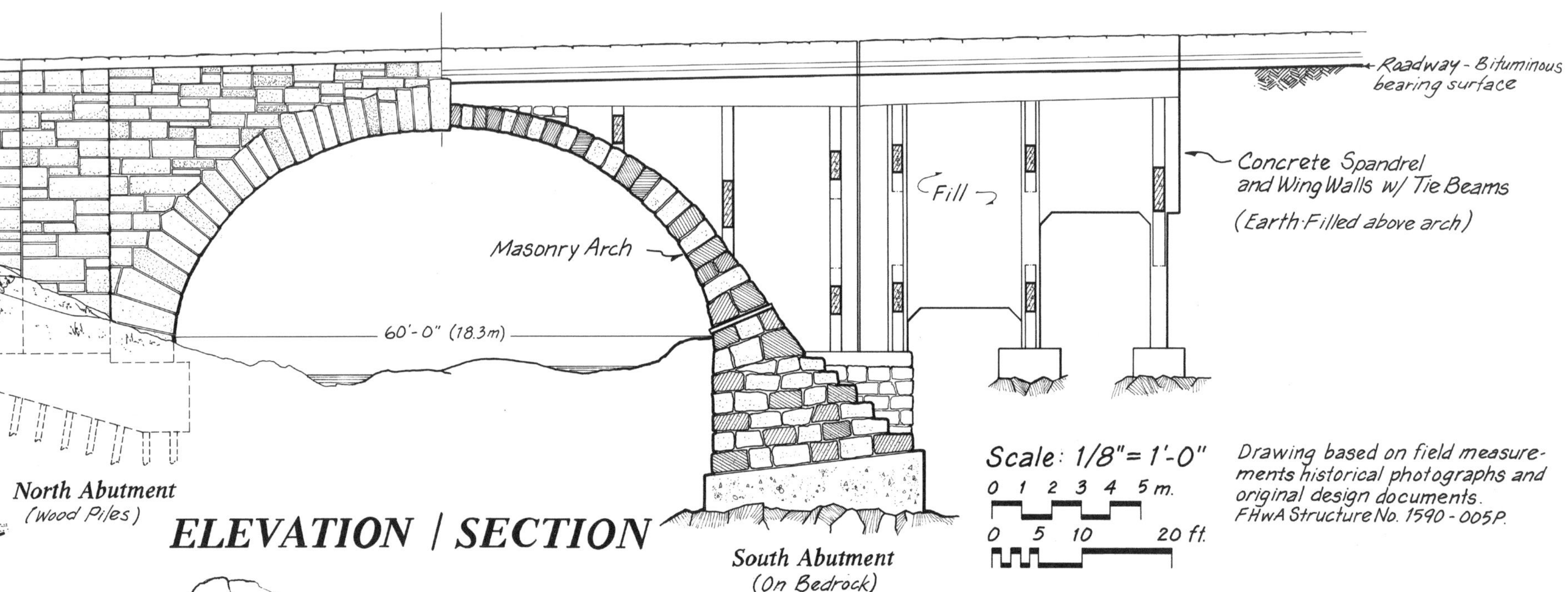

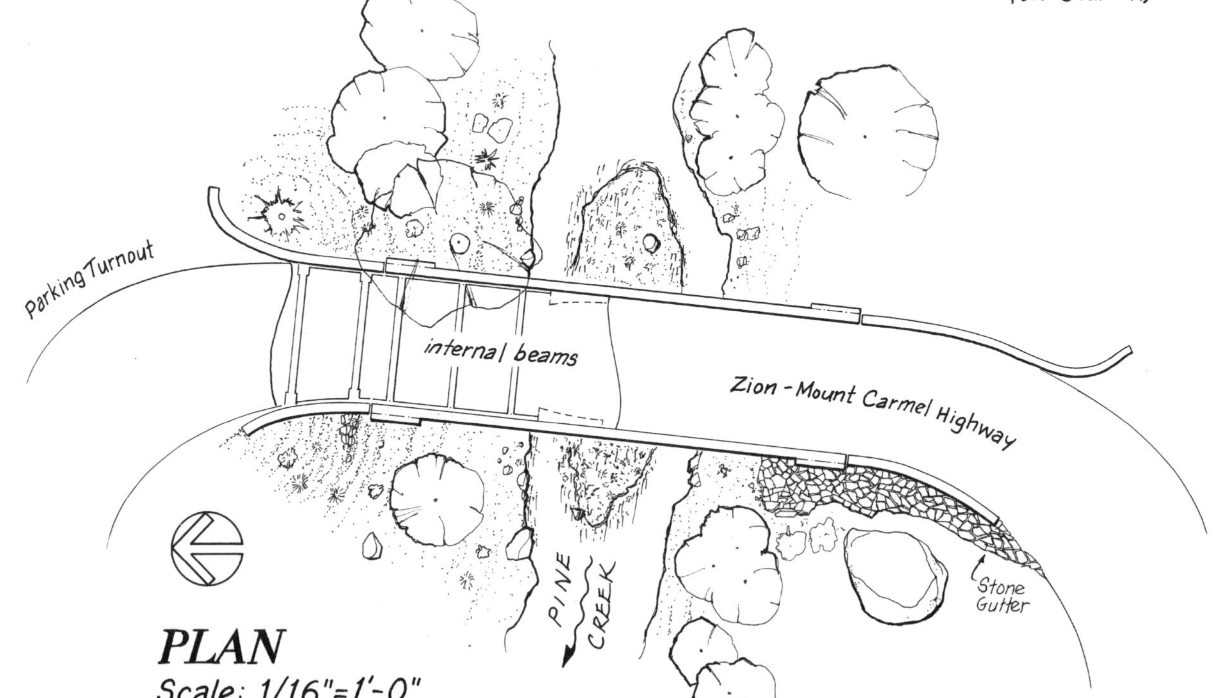

PLAN
Scale: 1/16" = 1'-0"

Pine Creek Bridge, designed in a rustic style consistent with the aesthetic goals of the National Park Service, was constructed in 1929-30 as part of the 25-mile long route of the Zion-Mount Carmel Highway. The bridge blends with the landscape and reflects the general shape and color of The Great Arch of Zion rock formation located a short distance up the canyon. In addition, the structure was deliberately planned to account for the unique needs of automobiles, being constructed with a wider, superelevated roadway to accomodate the curved alignment, thus allowing for easier negotiation of the curve. Original plans for the structure specified that each abutment be supported by concrete piles and footings. During construction, however, it was necessary to modify the plans, resulting in a cellular concrete and masonry abutment on bed-rock supporting the south end while the north abutment required a concrete footing with wood piles driven 35-46 feet to the underlying bedrock. When completed, the total cost for the bridge (including revisions) was $72,947.95.

The 120-foot long bridge has a 23-foot high solid masonry keystone arch of multi-hued Navajo sandstone with a clear span of 60 feet. Concrete spandrel and wing walls with concrete tie beams are faced with a similar ashlar stone. Pine Creek Bridge is believed to be one of few masonry arch auto bridges designed for the National Park Service by the Bureau of Public Roads.

This project is part of the Historic American Engineering Record (HAER), a long-range program to document historically significant engineering and industrial works in the United States. The HAER program is administered by the Historic American Buildings Survey/Historic American Engineering Record Division (HABS/HAER) of the National Park Service, U.S. Department of the Interior. The Zion Roads and Bridges Recording Project was cosponsored during the summer of 1993 by HAER under the general direction of Dr. Robert J. Kapsch, Chief; the National Park Service Roads and Bridges Program, John Gingles, Manager; and by Zion National Park, Donald Falvey, Superintendent.

The field work, measured drawings, historical reports, and photographs were prepared under the direction of Project Leaders Eric N. DeLony, Chief of HAER and Todd A. Croteau, HAER Staff Architect. The recording team consisted of Todd A. Croteau, field supervisor; architectural technicians Laura J. Culberson and Christopher Payne. Historical reports were prepared by project historians Michael Anderson, Christine Madrid, and Richard H. Quin. Formal large-format photography was done by Brian Grogan.

DELINEATED BY: Todd A. Croteau, 1993

ZION NATIONAL PARK
ROADS AND BRIDGES RECORDING PROJECT
NATIONAL PARK SERVICE
UNITED STATES DEPARTMENT OF THE INTERIOR

ZION - MT. CARMEL HIGHWAY, PINE CREEK BRIDGE - 1930
SPANNING PINE CREEK ON ZION - MT. CARMEL HIGHWAY (STATE HIGHWAY 9)
SPRINGDALE VICINITY WASHINGTON COUNTY UTAH

SHEET 1 of 2

HISTORIC AMERICAN ENGINEERING RECORD
UT-39-B

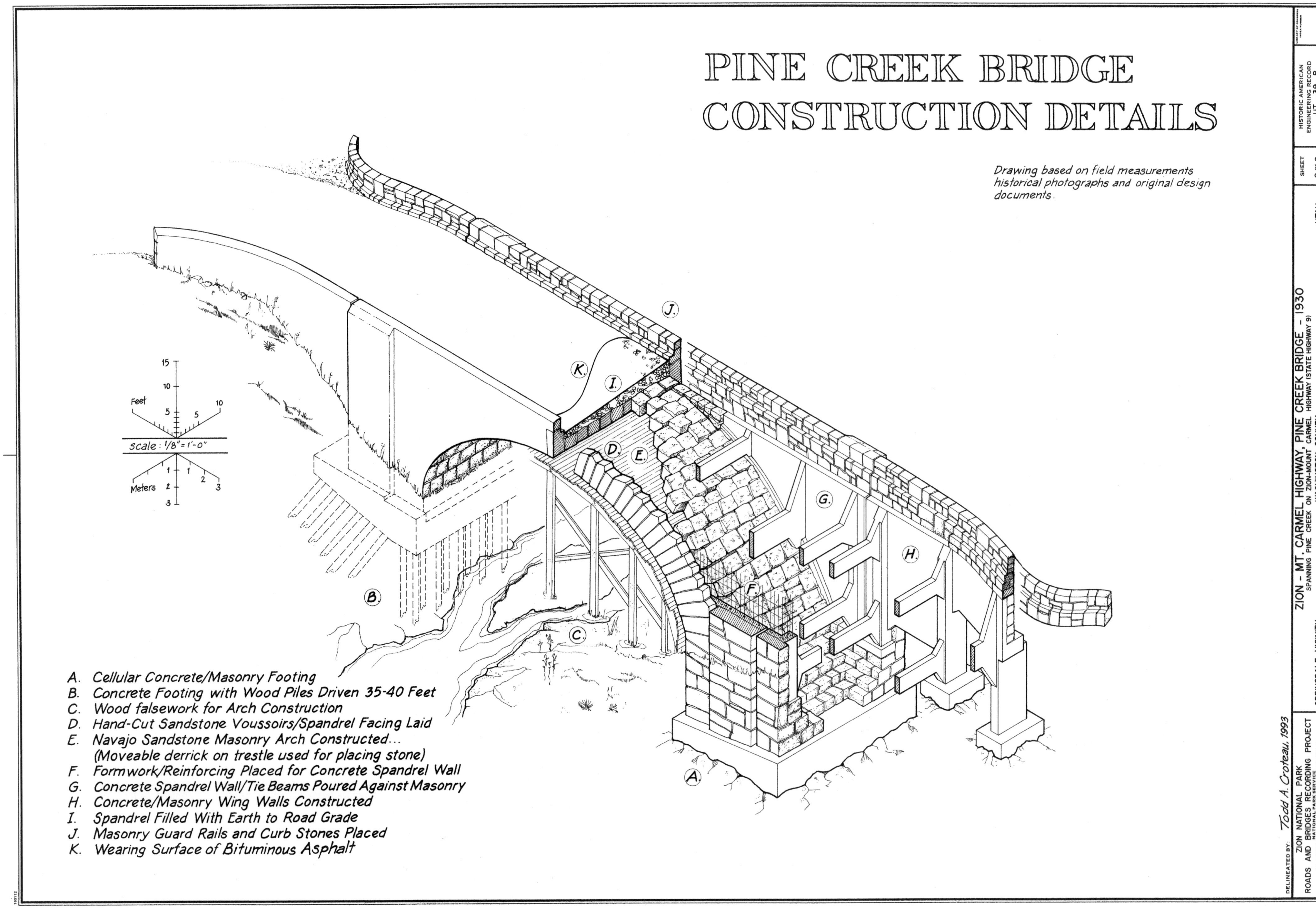
PINE CREEK BRIDGE
CONSTRUCTION DETAILS
Drawing based on field measurements historical photographs and original design documents.
J.
K.
I.
D.
E.
G.
H.
F.
B.
C.
A.
15
10
Feet
5
5
10
Scale: 1/8" = 1'-0"
1
1
2
Meters
2
3
3
A. Cellular Concrete/Masonry Footing
B. Concrete Footing with Wood Piles Driven 35-40 Feet
C. Wood falsework for Arch Construction
D. Hand-Cut Sandstone Voussoirs/Spandrel Facing Laid
E. Navajo Sandstone Masonry Arch Constructed...
(Moveable derrick on trestle used for placing stone)
F. Formwork/Reinforcing Placed for Concrete Spandrel Wall
G. Concrete Spandrel Wall/Tie Beams Poured Against Masonry
H. Concrete/Masonry Wing Walls Constructed
I. Spandrel Filled With Earth to Road Grade
J. Masonry Guard Rails and Curb Stones Placed
K. Wearing Surface of Bituminous Asphalt
DELINEATED BY Todd A. Croteau, 1993
ZION NATIONAL PARK
ROADS AND BRIDGES RECORDING PROJECT
NATIONAL PARK SERVICE
UNITED STATES DEPARTMENT OF THE INTERIOR
ZION - MT. CARMEL HIGHWAY, PINE CREEK BRIDGE - 1930
SPANNING PINE CREEK ON ZION-MOUNT CARMEL HIGHWAY (STATE HIGHWAY 9)
SPRINGDALE VICINITY
WASHINGTON COUNTY
UTAH
SHEET 2 OF 2
HISTORIC AMERICAN ENGINEERING RECORD
UT-39-B
IF REPRODUCED, PLEASE CREDIT: HISTORIC AMERICAN ENGINEERING RECORD, NATIONAL PARK SERVICE, NAME OF DELINEATOR, DATE OF THE DRAWING

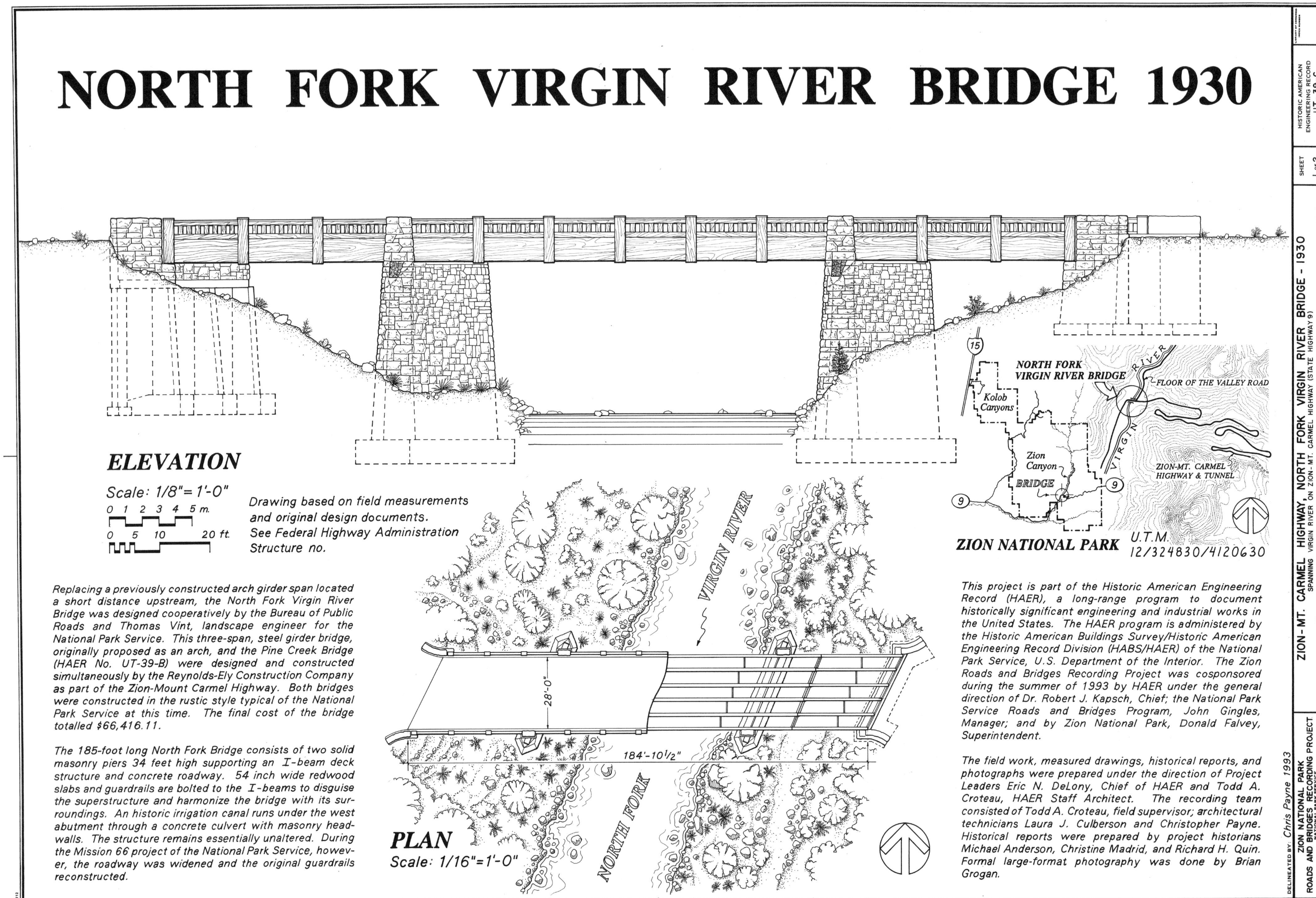

NORTH FORK VIRGIN RIVER BRIDGE 1930
ELEVATION
Scale: 1/8"=1'-0"
0 1 2 3 4 5 m.
0 5 10 20 ft.
Drawing based on field measurements and original design documents. See Federal Highway Administration Structure no.
NORTH FORK VIRGIN RIVER BRIDGE
FLOOR OF THE VALLEY ROAD
Kolob Canyons
Zion Canyon
BRIDGE
ZION-MT. CARMEL HIGHWAY & TUNNEL
VIRGIN RIVER
15
9
ZION NATIONAL PARK
U.T.M. 12/324830/4120630
VIRGIN RIVER
28'-0"
184'-10 1/2"
NORTH FORK
PLAN
Scale: 1/16"=1'-0"
Replacing a previously constructed arch girder span located a short distance upstream, the North Fork Virgin River Bridge was designed cooperatively by the Bureau of Public Roads and Thomas Vint, landscape engineer for the National Park Service. This three-span, steel girder bridge, originally proposed as an arch, and the Pine Creek Bridge (HAER No. UT-39-B) were designed and constructed simultaneously by the Reynolds-Ely Construction Company as part of the Zion-Mount Carmel Highway. Both bridges were constructed in the rustic style typical of the National Park Service at this time. The final cost of the bridge totalled $66,416.11.
The 185-foot long North Fork Bridge consists of two solid masonry piers 34 feet high supporting an I-beam deck structure and concrete roadway. 54 inch wide redwood slabs and guardrails are bolted to the I-beams to disguise the superstructure and harmonize the bridge with its surroundings. An historic irrigation canal runs under the west abutment through a concrete culvert with masonry headwalls. The structure remains essentially unaltered. During the Mission 66 project of the National Park Service, however, the roadway was widened and the original guardrails reconstructed.
This project is part of the Historic American Engineering Record (HAER), a long-range program to document historically significant engineering and industrial works in the United States. The HAER program is administered by the Historic American Buildings Survey/Historic American Engineering Record Division (HABS/HAER) of the National Park Service, U.S. Department of the Interior. The Zion Roads and Bridges Recording Project was cosponsored during the summer of 1993 by HAER under the general direction of Dr. Robert J. Kapsch, Chief; the National Park Service Roads and Bridges Program, John Gingles, Manager; and by Zion National Park, Donald Falvey, Superintendent.
The field work, measured drawings, historical reports, and photographs were prepared under the direction of Project Leaders Eric N. DeLony, Chief of HAER and Todd A. Croteau, HAER Staff Architect. The recording team consisted of Todd A. Croteau, field supervisor; architectural technicians Laura J. Culberson and Christopher Payne. Historical reports were prepared by project historians Michael Anderson, Christine Madrid, and Richard H. Quin. Formal large-format photography was done by Brian Grogan.
DELINEATED BY: Chris Payne 1993
ZION NATIONAL PARK
ROADS AND BRIDGES RECORDING PROJECT
NATIONAL PARK SERVICE
UNITED STATES DEPARTMENT OF THE INTERIOR
ZION-MT. CARMEL HIGHWAY, NORTH FORK VIRGIN RIVER BRIDGE - 1930
SPANNING VIRGIN RIVER ON ZION-MT. CARMEL HIGHWAY (STATE HIGHWAY 9)
SPRINGDALE VICINITY WASHINGTON COUNTY UTAH
SHEET 1 OF 2
HISTORIC AMERICAN ENGINEERING RECORD
UT-39-C
IF REPRODUCED, PLEASE CREDIT: HISTORIC AMERICAN ENGINEERING RECORD, NATIONAL PARK SERVICE, NAME OF DELINEATOR, DATE OF THE DRAWING

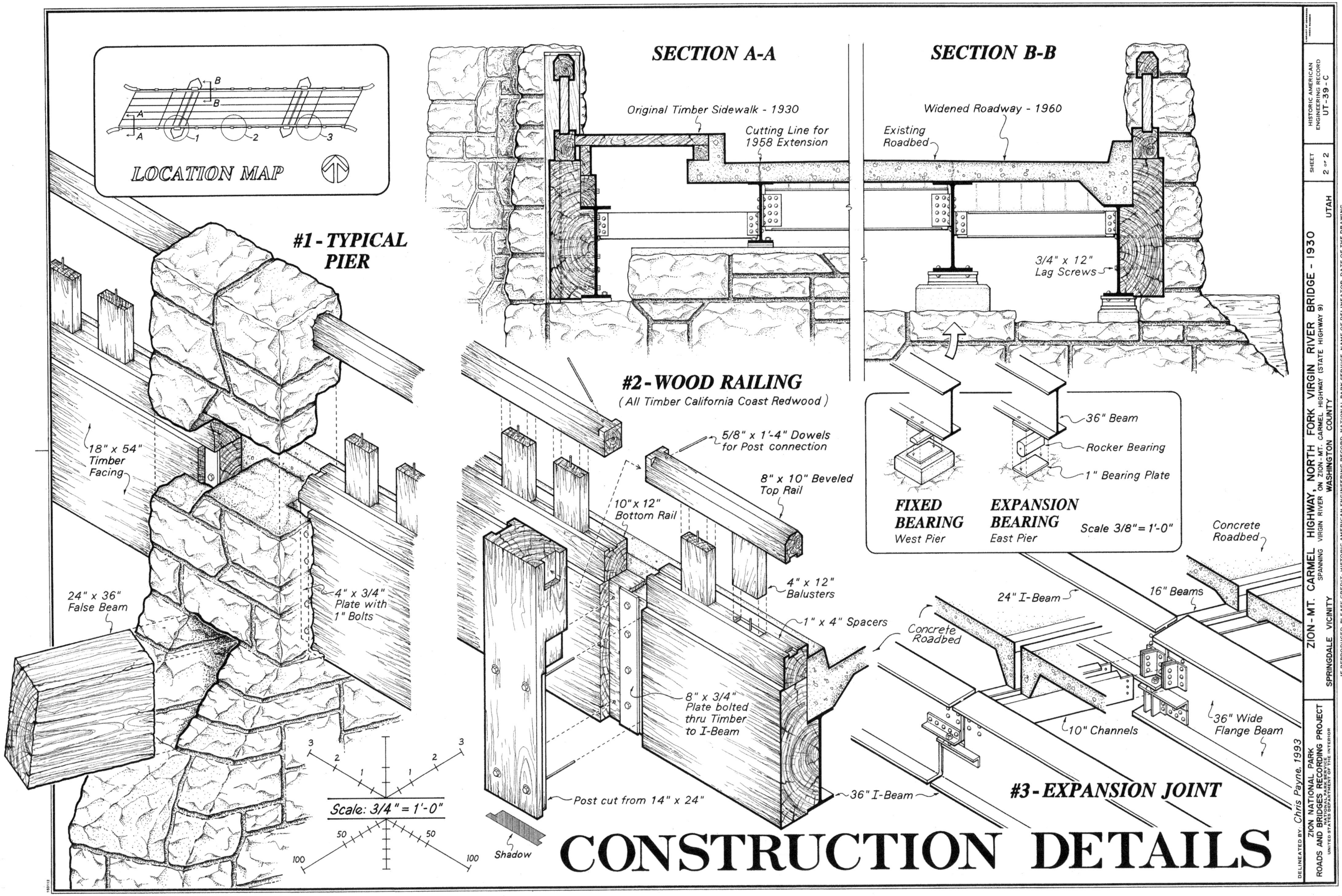
LOCATION MAP
B
B
A
A
1
2
3
#1 - TYPICAL PIER
18" x 54" Timber Facing
24" x 36" False Beam
4" x 3/4" Plate with 1" Bolts
Scale: 3/4" = 1'-0"
SECTION A-A
Original Timber Sidewalk - 1930
Cutting Line for 1958 Extension
SECTION B-B
Widened Roadway - 1960
Existing Roadbed
3/4" x 12" Lag Screws
#2 - WOOD RAILING
(All Timber California Coast Redwood)
5/8" x 1'-4" Dowels for Post connection
8" x 10" Beveled Top Rail
10" x 12" Bottom Rail
4" x 12" Balusters
1" x 4" Spacers
8" x 3/4" Plate bolted thru Timber to I-Beam
Post cut from 14" x 24"
36" I-Beam
Shadow
36" Beam
Rocker Bearing
1" Bearing Plate
FIXED BEARING
West Pier
EXPANSION BEARING
East Pier
Scale 3/8" = 1'-0"
Concrete Roadbed
24" I-Beam
16" Beams
Concrete Roadbed
10" Channels
36" Wide Flange Beam
#3 - EXPANSION JOINT
CONSTRUCTION DETAILS
ZION - MT. CARMEL HIGHWAY, NORTH FORK VIRGIN RIVER BRIDGE - 1930
SPANNING VIRGIN RIVER ON ZION-MT. CARMEL HIGHWAY (STATE HIGHWAY 9)
SPRINGDALE VICINITY
WASHINGTON COUNTY
UTAH
HISTORIC AMERICAN ENGINEERING RECORD
UT-39-C
SHEET 2 of 2
DELINEATED BY: Chris Payne, 1993
ZION NATIONAL PARK
ROADS AND BRIDGES RECORDING PROJECT
IF REPRODUCED, PLEASE CREDIT: HISTORIC AMERICAN ENGINEERING RECORD, NATIONAL PARK SERVICE, NAME OF DELINEATOR, DATE OF THE DRAWING

Parkways

The National Park Service's parkways epitomize the harmonious integration of highway engineering and landscape design. Stretching for miles through scenic backcountry, bucolic farmland, urban woodlands, and hallowed historic ground, national parkways preserve invaluable natural and cultural resources while affording a wide variety of attractive motoring experiences.

HABS/HAER's parkway documentation began with a detailed mapping of Rock Creek and Potomac Parkway, a 2½-mile-long urban oasis in the heart of the nation's capital. The George Washington Memorial Parkway project took a more interpretive approach, delineating design concepts and construction processes while providing a representative overview of key sites and structures along the Potomac River near Washington, D.C. Virginia's Colonial Parkway was portrayed in similar fashion, with greater emphasis placed on the precise rendition of landscape features and construction details.

The extraordinary length and diversity of the Blue Ridge Parkway (469 miles) generated a correspondingly ambitious and wide-ranging analysis. Artistically rendered views captured the distinguishing characteristics of representative parkway landscapes. Additional drawings portrayed visitor facilities, signs, bridges, and key management practices. A similar approach was taken in the documentation of Natchez Trace Parkway and Baltimore-Washington Parkway, with additional drawings tracing the historical evolution of these historic transportation corridors.

Belcher's Curve, Blue Ridge Parkway, 1997
Photographer: David Haas, HAER

National Parkways

Baltimore-Washington Parkway

Cover Sheet	214
Places of Interest	215
Evolving Roadway Types	216
Road Details	217
Vegetation	218

Blue Ridge Parkway

Cover Sheet	219
Map	220
The Ridge	221
The Plateau	222
The Highlands	223
The Black Mountains	224
The Balsam Mountains	225
Humpback Rocks	226
Otter Creek	227
Peaks of Otter	228
Rocky Knob	229
Mabry Mill	230
Cumberland Knob	231
Doughton Park	232
Julian Price Memorial Park	233
Linn Cove Viaduct	234
Craggy Gardens	235
Mount Pisgah	236
Agricultural Leases, Scenic Easements	237
Management Zones	238
Reinforced Concrete Slab and Arch Bridges	239
Bridges	240
Viaducts	241
Tunnels	242
Signs	243
Fences	244
Overlooks	245
Vegetation Management	246

Colonial Parkway

Cover Sheet	247
Development of the Parkway	248
Building the Parkway	249
Roadway Design	250
Parkway Views, James/York Rivers	251
Parkway Views, Jones Mill Pond/Halfway Creek	252
Brick Veneer Bridges	253
C&O Railroad/Lafayette Street Bridge, 1937	254
C&O Railroad/Lafayette Street Bridge Isometric	255
Williamsburg Tunnel	256
Williamsburg Tunnel Construction	257
Jamestown Island Loop Road	258

George Washington Memorial Parkway

Cover Sheet	259
A Highway of History and Commemoration	260
Evolution of Transportation Routes, 1776–1970	261
Parkway Proposals, 1890–1930	262
Planned Views	263
Looking Down the Road	264
Parkway Construction Sequence	265
Intersections	266
Mount Vernon Terminus	267
Roadside Details	268
Bridges of the GWMP	269
Boundary Channel Bridge	271

Natchez Trace Parkway

Cover Sheet	272
Evolution of the Trace	273
Cultural Resources	274
Natural Resources	275
Interpreting Agricultural Landscapes	276
Motorist Experience	277
Motorist Waysides	278
Road Alignment and Construction	279
Double Arch Bridge	280
Bridge Construction	281
Bridge Comparison	282

Rock Creek and Potomac Parkway

Cover Sheet	283
Evolution of the Design, 1908	284
Evolution of the Design, 1916–33	285
Index to Plans	286
Map (Lincoln Memorial)	287
Map (P Street Bridge)	288
Map (Waterside Drive Overpass)	289
Map (William Taft Memorial Bridge)	290
Map (Calvert Street Bridge)	291
Parkway Sections	292
Bridges of Rock Creek and Potomac Parkway	293
Boulder Bridge, 1902	295
Boulder Bridge Cutaway	296
Ross Drive Bridge, 1907	297
Ross Drive Bridge Details	298

BALTIMORE-WASHINGTON PARKWAY

1954-1999

View looking north at Good Luck Road Bridge as the parkway passes through Greenbelt Park. From an NPS photograph by Abbie Rowe, 1954.

The Baltimore-Washington Parkway was designed to be a safe and attractive entrance to the nation's capital. The first limited-access, divided roadway in Maryland, it offered a much needed alternative to the dangerous, commercialized U.S. Route 1. The parkway also served as a vital link between the nation's capital and several important federal agencies. These destinations included defense installations, research centers, and recreation areas such as Greenbelt Park. First proposed in the 1920s, the parkway was not completed until 1954.

The National Park Service and the U.S. Bureau of Public Roads designed and constructed the 19-mile long parkway, which extends from the junction of Route 50 at the Maryland/D.C. line to Fort Meade near the Route 175 interchange. The parkway is part of the Route 295 right-of-way, which continues along the Anacostia River through Washington, D.C. to the Woodrow Wilson Bridge to the south. From the north, the parkway becomes the "Baltimore-Washington Expressway" (Rt 295), extending past BWI Airport into Baltimore City under the jurisdiction of the state of Maryland.

The parkway represents an intermediate stage in road design between the relatively narrow, winding, undivided parkways of the pre-World War II era to the wider, straighter, less visually appealing freeways and interstate highways. The parkway incorporates design measures made necessary by post-war demands for roadways that allowed for moderately high-speed, efficient and safe movement of traffic. These features included long, gradual curves linked with smooth spiral transitions, mountable curbs, grade-separated interchanges, and merge lanes at exits and entrances. The visual character included border plantings to screen out undesirable views and improve the parkway's appearance; wide, variable-width medians with landscape treatments that range from open meadows to dense woodlands; and an ever-changing succession of attractively composed views of native forest, meadows, and picturesquely located specimen trees.

The parkway has continued to maintain its general character despite pressures from increased traffic volume and suburban development. A major renovation project, undertaken from 1985 to 2001, has attempted to respect the parkway's scenic qualities while enabling it to better accommodate modern traffic demands.

This recording project is part of the Historic American Engineering Record (HAER), administered by the Historic American Buildings Survey/Historic American Engineering Record (HABS/HAER) Division of the National Park Service, U.S. Department of the Interior (E. Blaine Cliver, Chief).

The Baltimore-Washington Parkway Recording Project was cosponsored in 1999 by HABS/HAER, the NPS National Capital Region (Terry Carlstrom, Director), and NPS-NCR National Capital Parks-East (John Hale, Superintendent). The project was funded by the Federal Lands Highway Program (Arthur Hamilton, Administrator) through the National Park Service Park Roads and Bridges Program (Lou DeLorme, Manager).

The measured drawings, historical reports and photographs were prepared under the direction of Christopher Marston, HAER Project Leader, Tim Davis, HAER Historian, and Todd Croteau, HAER Roads and Bridges Program Manager. The recording team consisted of Francesca Cassara, field supervisor and landscape architect (University of Michigan); Lee Albritton (The George Washington University) and Ana Cabeza (US-ICOMOS, Venezuela), architectural technicians; and Tim Simandl (NPS-NCR National Capital Parks – Central), historian. Ana Dymek (US-ICOMS, Poland) completed the landscape sheet. Large format photography was done by Jet Lowe. Michelle Matulac and Katalina Fernandez (Catholic University of America) edited the drawings.

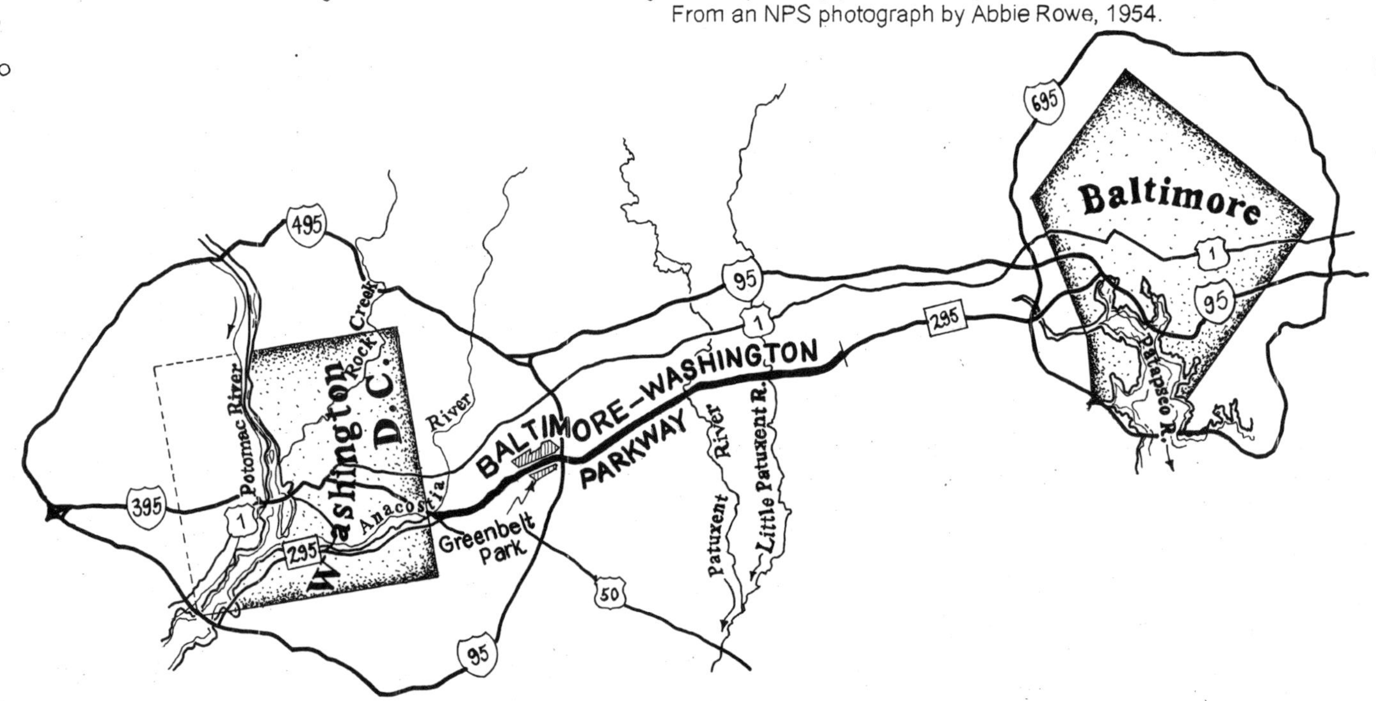

HISTORIC AMERICAN ENGINEERING RECORD MD-129

SHEET 1 OF 10

MARYLAND

BALTIMORE-WASHINGTON PARKWAY

PRINCE GEORGE'S COUNTY

GREENBELT VICINITY

NPS ROADS & BRIDGES RECORDING PROJECT

NATIONAL PARK SERVICE

UNITED STATES DEPARTMENT OF THE INTERIOR

DELINEATED BY: Michelle Matulac, 1999

IF REPRODUCED, PLEASE CREDIT: HISTORIC AMERICAN ENGINEERING RECORD, NATIONAL PARK SERVICE, NAME OF DELINEATOR, DATE OF THE DRAWING

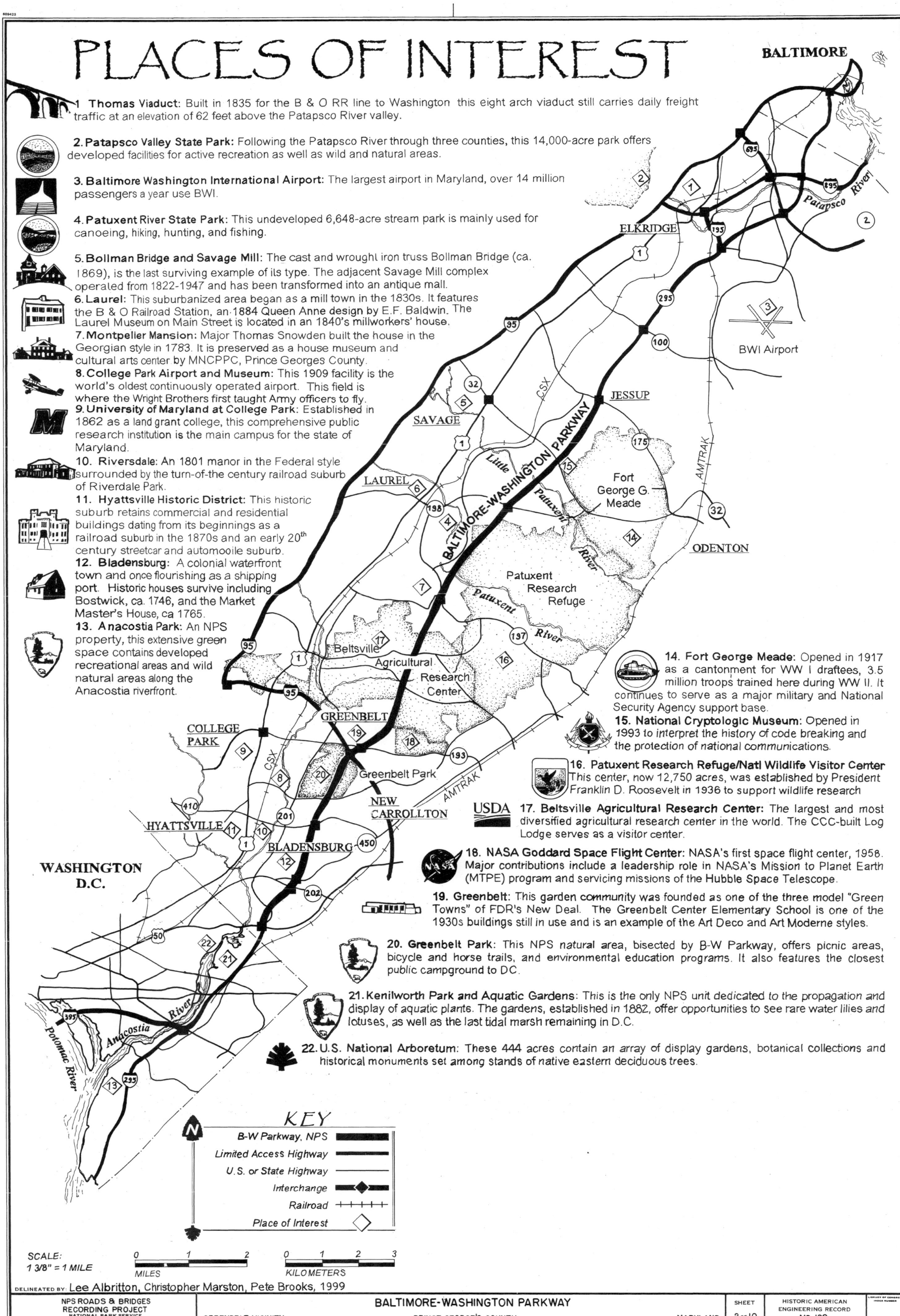
PLACES OF INTEREST
BALTIMORE
1 Thomas Viaduct: Built in 1835 for the B & O RR line to Washington this eight arch viaduct still carries daily freight traffic at an elevation of 62 feet above the Patapsco River valley.
2. Patapsco Valley State Park: Following the Patapsco River through three counties, this 14,000-acre park offers developed facilities for active recreation as well as wild and natural areas.
3. Baltimore Washington International Airport: The largest airport in Maryland, over 14 million passengers a year use BWI.
4. Patuxent River State Park: This undeveloped 6,648-acre stream park is mainly used for canoeing, hiking, hunting, and fishing.
5. Bollman Bridge and Savage Mill: The cast and wrought iron truss Bollman Bridge (ca. 1869), is the last surviving example of its type. The adjacent Savage Mill complex operated from 1822-1947 and has been transformed into an antique mall.
6. Laurel: This suburbanized area began as a mill town in the 1830s. It features the B & O Railroad Station, an 1884 Queen Anne design by E.F. Baldwin. The Laurel Museum on Main Street is located in an 1840's millworkers' house.
7. Montpelier Mansion: Major Thomas Snowden built the house in the Georgian style in 1783. It is preserved as a house museum and cultural arts center by MNCPPC, Prince Georges County.
8. College Park Airport and Museum: This 1909 facility is the world's oldest continuously operated airport. This field is where the Wright Brothers first taught Army officers to fly.
9. University of Maryland at College Park: Established in 1862 as a land grant college, this comprehensive public research institution is the main campus for the state of Maryland.
10. Riversdale: An 1801 manor in the Federal style surrounded by the turn-of-the century railroad suburb of Riverdale Park.
11. Hyattsville Historic District: This historic suburb retains commercial and residential buildings dating from its beginnings as a railroad suburb in the 1870s and an early 20th century streetcar and automobile suburb.
12. Bladensburg: A colonial waterfront town and once flourishing as a shipping port. Historic houses survive including Bostwick, ca. 1746, and the Market Master's House, ca 1765.
13. Anacostia Park: An NPS property, this extensive green space contains developed recreational areas and wild natural areas along the Anacostia riverfront.
14. Fort George Meade: Opened in 1917 as a cantonment for WW I draftees, 3.5 million troops trained here during WW II. It continues to serve as a major military and National Security Agency support base.
15. National Cryptologic Museum: Opened in 1993 to interpret the history of code breaking and the protection of national communications.
16. Patuxent Research Refuge/Natl Wildlife Visitor Center This center, now 12,750 acres, was established by President Franklin D. Roosevelt in 1936 to support wildlife research
USDA
17. Beltsville Agricultural Research Center: The largest and most diversified agricultural research center in the world. The CCC-built Log Lodge serves as a visitor center.
18. NASA Goddard Space Flight Center: NASA's first space flight center, 1958. Major contributions include a leadership role in NASA's Mission to Planet Earth (MTPE) program and servicing missions of the Hubble Space Telescope.
19. Greenbelt: This garden community was founded as one of the three model "Green Towns" of FDR's New Deal. The Greenbelt Center Elementary School is one of the 1930s buildings still in use and is an example of the Art Deco and Art Moderne styles.
20. Greenbelt Park: This NPS natural area, bisected by B-W Parkway, offers picnic areas, bicycle and horse trails, and environmental education programs. It also features the closest public campground to DC.
21. Kenilworth Park and Aquatic Gardens: This is the only NPS unit dedicated to the propagation and display of aquatic plants. The gardens, established in 1882, offer opportunities to see rare water lilies and lotuses, as well as the last tidal marsh remaining in D.C.
22. U.S. National Arboretum: These 444 acres contain an array of display gardens, botanical collections and historical monuments set among stands of native eastern deciduous trees.
BALTIMORE-WASHINGTON PARKWAY
ELKRIDGE
BWI Airport
JESSUP
SAVAGE
LAUREL
Fort George G. Meade
ODENTON
Patuxent Research Refuge
Little Patuxent River
Patuxent River
Beltsville
Agricultural Research Center
GREENBELT
Greenbelt Park
NEW CARROLLTON
COLLEGE PARK
HYATTSVILLE
BLADENSBURG
WASHINGTON D.C.
Anacostia River
Potomac River
Patapsco River
CSX
AMTRAK
KEY
B-W Parkway, NPS
Limited Access Highway
U.S. or State Highway
Interchange
Railroad
Place of Interest
SCALE:
1 3/8" = 1 MILE
MILES
KILOMETERS
DELINEATED BY: Lee Albritton, Christopher Marston, Pete Brooks, 1999
NPS ROADS & BRIDGES RECORDING PROJECT
NATIONAL PARK SERVICE
UNITED STATES DEPARTMENT OF THE INTERIOR
BALTIMORE-WASHINGTON PARKWAY
GREENBELT VICINITY
PRINCE GEORGE'S COUNTY
MARYLAND
SHEET 2 of 10
HISTORIC AMERICAN ENGINEERING RECORD
MD-129
IF REPRODUCED, PLEASE CREDIT: HISTORIC AMERICAN ENGINEERING RECORD, NATIONAL PARK SERVICE, NAME OF DELINEATOR, DATE OF THE DRAWING

EVOLVING ROADWAY TYPES

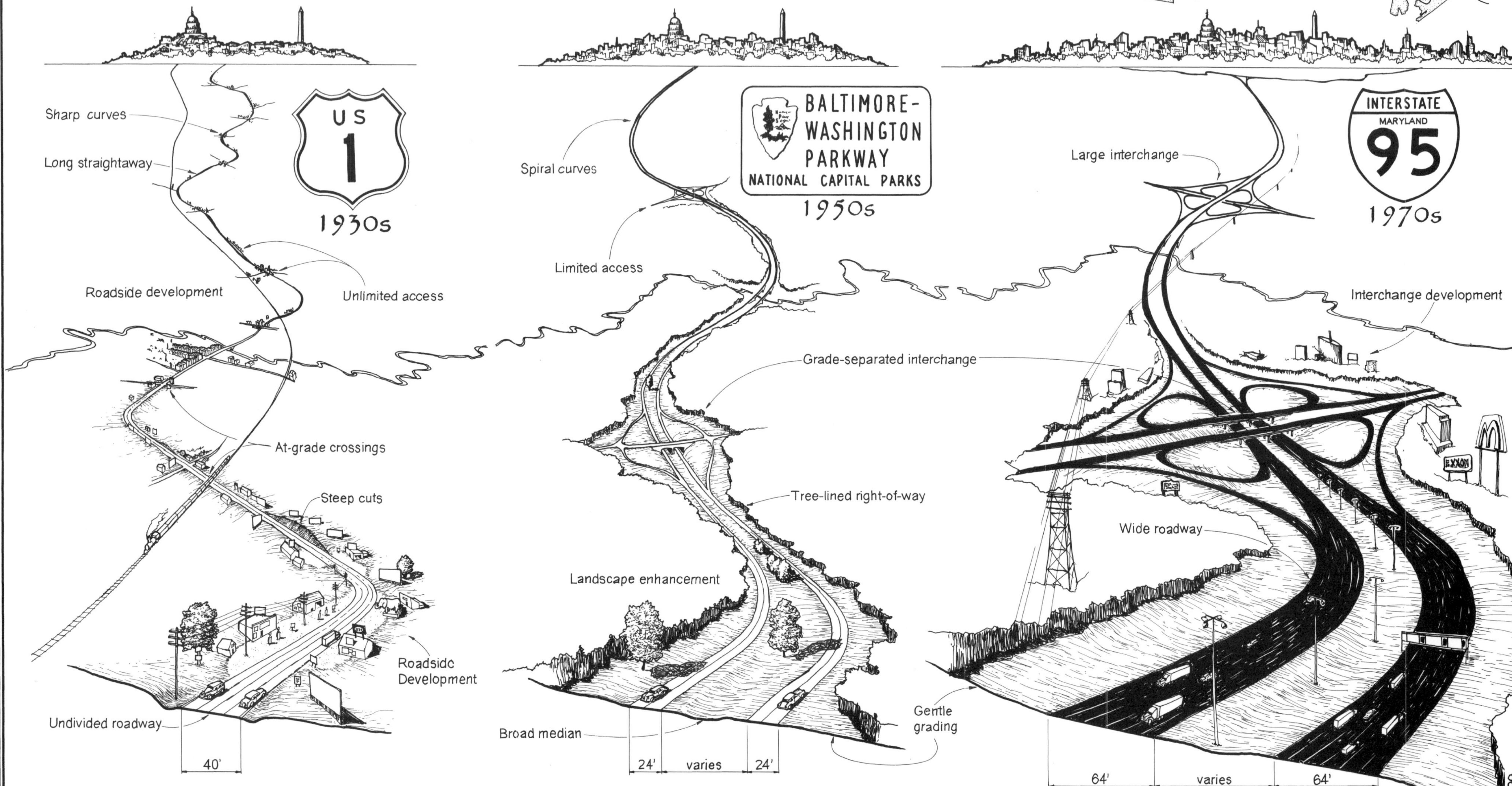

U.S. Route 1 typified the problems associated with ordinary highways. While it was paved with modern asphalt and concrete, most design features were unchanged from horse and buggy days. The road itself was a haphazard mix of sharp curves, long straightaways and uneven grades. Multiple at-grade intersections, unrestricted access from roadside properties and the lack of median dividers produced an unsafe and inefficient transportation corridor. Billboards, gas stations, roadside eateries and telephone poles lined the narrow right-of-way, distracting motorists and obscuring roadside scenery.

Parkways provided an attractive and efficient alternative to conventional highway construction. Carefully designed roadways with gentle grades and sweeping spiral curves were safer and more attractive than traditional alignments. Broad medians, grade-separated interchanges, and strict limitations on access from cross-streets and abutting properties greatly enhanced safety and efficiency. A wide, tree-lined right-of-way screened out unsightly roadside development and provided opportunities for landscape enhancement. Prohibitions on trucks and other commercial traffic made driving safer and more comfortable.

Interstate highways employed many of the basic design features pioneered by motor parkways but placed less emphasis on scenic values. Wider, straighter roadways with additional lanes and longer merging zones accommodated higher speeds and traffic volumes but took up a larger portion of the right-of-way. Access was tightly controlled, but viewshed protection was often limited, especially at interchanges. Landscape enhancements were minimal and the roadway dominated the forward view. Grade-separation structures were larger and less artistically designed. Interstates accommodated all types of modern motor traffic, including large trucks and buses.

DELINEATED BY: Pete Brooks, 1999

NPS ROADS & BRIDGES RECORDING PROJECT
NATIONAL PARK SERVICE
UNITED STATES DEPARTMENT OF THE INTERIOR

GREENBELT VICINITY

BALTIMORE-WASHINGTON PARKWAY

PRINCE GEORGE'S COUNTY

MARYLAND

SHEET 5 of 10

HISTORIC AMERICAN ENGINEERING RECORD MD-129

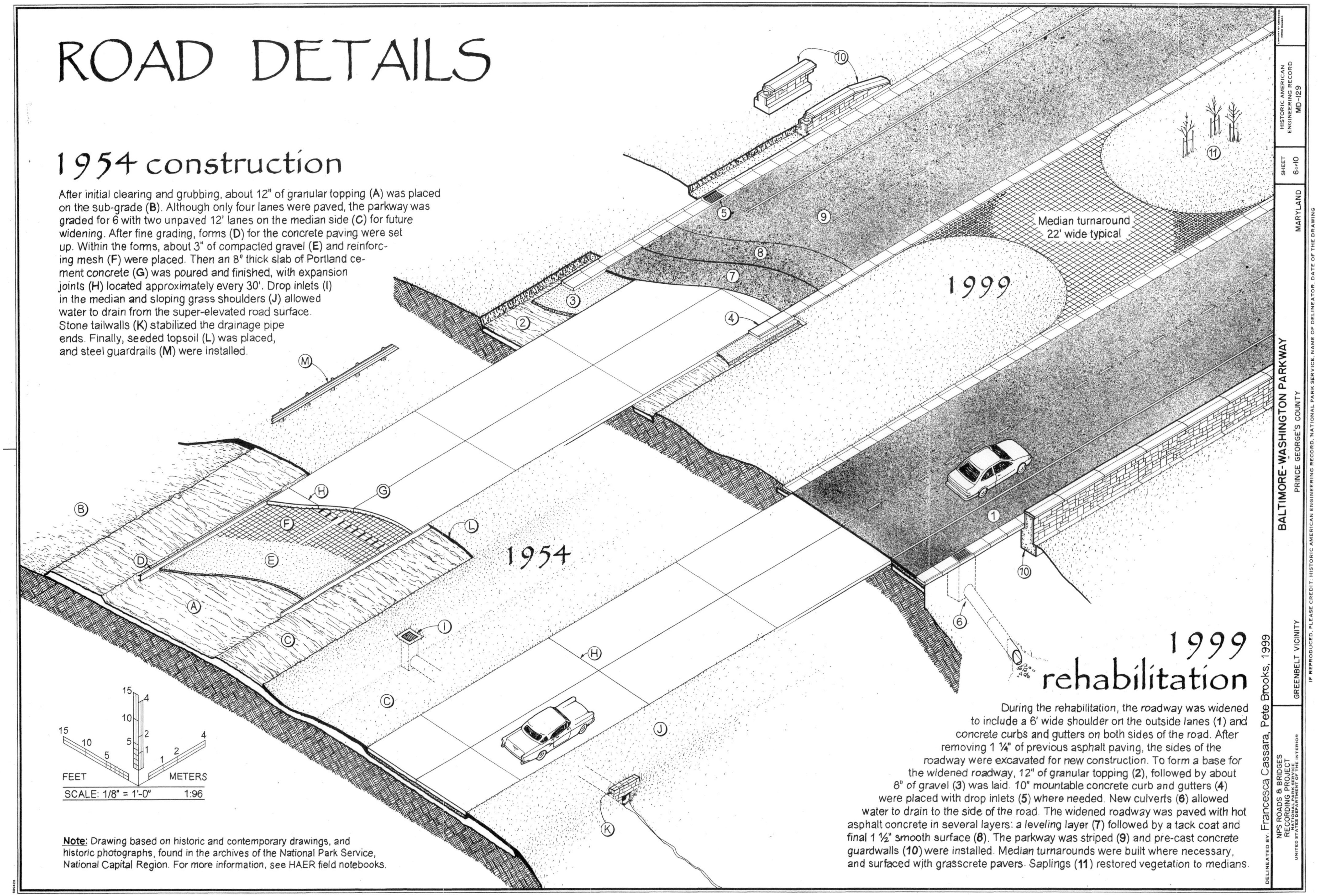
ROAD DETAILS
1954 construction
After initial clearing and grubbing, about 12" of granular topping (A) was placed on the sub-grade (B). Although only four lanes were paved, the parkway was graded for 6 with two unpaved 12' lanes on the median side (C) for future widening. After fine grading, forms (D) for the concrete paving were set up. Within the forms, about 3" of compacted gravel (E) and reinforcing mesh (F) were placed. Then an 8" thick slab of Portland cement concrete (G) was poured and finished, with expansion joints (H) located approximately every 30'. Drop inlets (I) in the median and sloping grass shoulders (J) allowed water to drain from the super-elevated road surface. Stone tailwalls (K) stabilized the drainage pipe ends. Finally, seeded topsoil (L) was placed, and steel guardrails (M) were installed.
Median turnaround
22' wide typical
1999
1954
1999
rehabilitation
During the rehabilitation, the roadway was widened to include a 6' wide shoulder on the outside lanes (1) and concrete curbs and gutters on both sides of the road. After removing 1 ¼" of previous asphalt paving, the sides of the roadway were excavated for new construction. To form a base for the widened roadway, 12" of granular topping (2), followed by about 8" of gravel (3) was laid. 10" mountable concrete curb and gutters (4) were placed with drop inlets (5) where needed. New culverts (6) allowed water to drain to the side of the road. The widened roadway was paved with hot asphalt concrete in several layers: a leveling layer (7) followed by a tack coat and final 1 ½" smooth surface (8). The parkway was striped (9) and pre-cast concrete guardwalls (10) were installed. Median turnarounds were built where necessary, and surfaced with grasscrete pavers. Saplings (11) restored vegetation to medians.
FEET
METERS
SCALE: 1/8" = 1'-0" 1:96
Note: Drawing based on historic and contemporary drawings, and historic photographs, found in the archives of the National Park Service, National Capital Region. For more information, see HAER field notebooks.
DELINEATED BY: Francesca Cassara, Pete Brooks, 1999
NPS ROADS & BRIDGES RECORDING PROJECT
NATIONAL PARK SERVICE
UNITED STATES DEPARTMENT OF THE INTERIOR
BALTIMORE-WASHINGTON PARKWAY
GREENBELT VICINITY
PRINCE GEORGE'S COUNTY
MARYLAND
SHEET 6 OF 10
HISTORIC AMERICAN ENGINEERING RECORD
MD-129
IF REPRODUCED, PLEASE CREDIT: HISTORIC AMERICAN ENGINEERING RECORD, NATIONAL PARK SERVICE, NAME OF DELINEATOR, DATE OF THE DRAWING

VEGETATION

Tulip Tree *Liriodendron tulipifera*

Virginia Pine *Inus virginiana*

Bald cypress *Taxodium distinchum*

Skunk Cabbage *Symplocarpus foetidus*

1. **Native vegetation** on the Baltimore-Washington Parkway consists largely of oak-hickory and coniferous forest with an understory of shrubs such as pawpaw and mountain laurel. Native and planted grasses are maintained in the median and along the edges through sculpting of mown bays.

2. **Planting and vegetation management** are important aspects of the overall landscape plan. Trees are planted in select areas and wildflowers are sown alongside tall grasses in the median. These provide not only a colorful view for the motorist but also a habitat for wildlife.

3. **Natural wetlands** are found in low-lying areas where they are not always obvious to motorists. Although NPS naturalists strive to protect these ecosystems within the parkway boundary, wetlands are impacted by neighboring development along the parkway corridor.

4. **Interchange drainage** areas have been designed to simulate natural wetlands. These water retention basins not only provide an area to filter water but also form a habitat for wildlife and diverse plants.

Sweet Gum *Liquidambar styraciflua*

Pin Oak *Querus plaustris*

Lance-leaved Coreposis *Coreposis lanceolata*

Blue Vervain *Vervain hastata*

Cattail *Typha latifolia*

Sedge *Juncus effusus*

DELINEATED BY: Anna Dymek, 1999

NPS ROADS & BRIDGES RECORDING PROJECT
NATIONAL PARK SERVICE
UNITED STATES DEPARTMENT OF THE INTERIOR

BALTIMORE-WASHINGTON PARKWAY
GREENBELT VICINITY PRINCE GEORGE'S COUNTY MARYLAND

SHEET 7 OF 10

HISTORIC AMERICAN ENGINEERING RECORD MD-129

IF REPRODUCED, PLEASE CREDIT: HISTORIC AMERICAN ENGINEERING RECORD, NATIONAL PARK SERVICE, NAME OF DELINEATOR, DATE OF THE DRAWING

BLUE RIDGE PARKWAY

VIRGINIA TO NORTH CAROLINA
1935 TO 1987

The legislated purpose of the Blue Ridge Parkway was to link Shenandoah and Great Smoky Mountains national parks with a road designed for pleasant motoring as well as to conserve and interpret the unique natural and cultural resources of the Southern Highlands. Rather than just a park-to-park connection, the parkway was conceived and planned as an elongated national park providing recreational and park activities. Traveling the parkway was intended to be a "ride-a-while, stop-a-while" experience complete with scenic pullouts, recreation areas, and visitor contact stations.

The impetus behind the construction of the 469-mile Blue Ridge Parkway was firmly rooted in the New Deal. It was proposed as a public works project to relieve the unemployment found in the Appalachians during the Great Depression. In November 1933, the Blue Ridge Parkway was approved with $4 million allotted to begin construction. The implementation of this project developed into a four-way partnership. The states of Virginia and North Carolina acquired the needed right-of-way to construct the road and the Bureau of Public Roads provided the technical assistance and expertise in roadway construction. The fourth partner was the National Park Service (NPS). Under the direction of NPS landscape architects, these partners worked together to plan, design, and construct the parkway. Companies engaged on the project were required to hire as many unemployed local men as possible for parkway construction, often under quotas set in the contract. The first section was contracted in September of 1935, but it would take another 52 years before the road was completed with the completion around Grandfather Mountain in 1987. The parkway was now a unified drive between Shenandoah and Great Smoky Mountain national parks.

The parkway was the longest road ever to be planned as a single unit up to that time in America. Following existing practice, the road was designed and constructed in sections, twenty in Virginia and twenty-four in North Carolina. Unlike earlier parkways, it is located away from cities and is in the mountains. In contrast to earlier roads in national parks, it runs through settled countryside as well as wild mountain landscapes. It follows the Blue Ridge for the first 355 of its 469 miles. In the remaining 115 miles it crosses some of the highest and most rugged Southern Appalachians, including the Black Mountains, Great Craggies, Pisgah Ledge, Great Balsam, and Plott Balsam ranges.

In contrast to Shenandoah's Skyline Drive, the Blue Ridge Parkway was not planned as a ridge-line route, although extensive segments follow the crest of the mountains. Instead, variety was introduced by routing the road along mountain sides, plateaus, streams, and through broad river valleys, providing visitors with one of the most diversified motoring experiences in the world. Likewise, the designers provided the road with a high standard of grade and curvature so that motorists could safely devote their attention to the scenery.

The Blue Ridge Parkway Roads and Bridges Recording Project was undertaken in 1996-97 by the Historic American Engineering Record (HAER), Eric N. DeLony, Chief, a long-range program to document historically significant engineering, industrial, and maritime works in the United States. The HAER program is part of the Historic American Buildings Survey/ Historic American Engineering Record (HABS/HAER) Division of the National Park Service, Department of the Interior, E. Blaine Cliver, Chief. Funding was provided by the Federal Lands Highway Office, Thomas Edick, Administrator, through the NPS Park Roads and Parkways Program. This recording project was cosponsored by HAER and the Blue Ridge Parkway, Gary Everhardt, Superintendent; Gary Johnson, Chief of Resource Planning and Professional Services; Allen Hess, Cultural Resources Specialist; and Will Orr, Landscape Architect.

The documentation was prepared under the direction of NPS historian Richard Quin and HAER architect Christopher Marston. The recording team consisted of field supervisor Lia Dikigoropoulou; architects Natascha Weiner (foreman), Matthew Stormont, and Carlos Jimenez Rosa (ICOMOS, Spain); landscape architects Ian Shanklin, Cheria Yost, and Lidia Klupsz (ICOMOS, Poland); and illustrator Jennifer K. Cuthbertson. The overview history was prepared by Richard Quin, and bridge reports by field project historian Brian Cleven. Large format photography was produced by David Haas.

***DOUGHTON PARK** milepost 244*

DELINEATED BY: *Jennifer K. Cuthbertson, 1997; edited by Elisabeth Dubin, 1997*

NATIONAL PARK SERVICE
ROADS & BRIDGES RECORDING PROJECT

ASHEVILLE VICINITY | BLUE RIDGE PARKWAY | BUNCOMBE COUNTY | NORTH CAROLINA

SHEET 1 OF 28

HISTORIC AMERICAN ENGINEERING RECORD NC-42

BLUE RIDGE PARKWAY

The Blue Ridge Parkway runs between Shenandoah National Park and the Great Smoky Mountains National Park - a distance of 469 miles. When construction began in 1935, it was the longest federally planned roadway in the country. The Parkway was designed and constructed in sections: as land was purchased by the states, rights-of-way were approved and contracts secured through the Bureau of Public Roads. Groundbreaking was celebrated in September 1935, upon awarding of the first contract for section 2-A extending 12.5 miles from the Virginia-North Carolina state line to Cumberland Knob, North Carolina. Subsequent work continued on sections 2-B through 2-E. Construction in Virginia began in Feb. 1936 on sections 1-P and 1-A. Section 1-A, from Rockfish Gap to Jarmans Gap, eventually became part of Skyline Drive in 1961.

As work progressed, a fifty mile section near Roanoke opened to the public in April, 1939. Later that year, the parkway opened to traffic between the state line and Blowing Rock. When work halted due to the outbreak of World War II, some 170 miles of parkway were complete, with another 160 miles under construction. Construction resumed in the 1950s and by 1968 the parkway was complete except for a 7.7 mile stretch around Grandfather Mountain. It was not until the section including the Linn Cove Viaduct was completed in 1987 that the Blue Ridge Parkway fully opened, 52 years after the 469-mile project began.

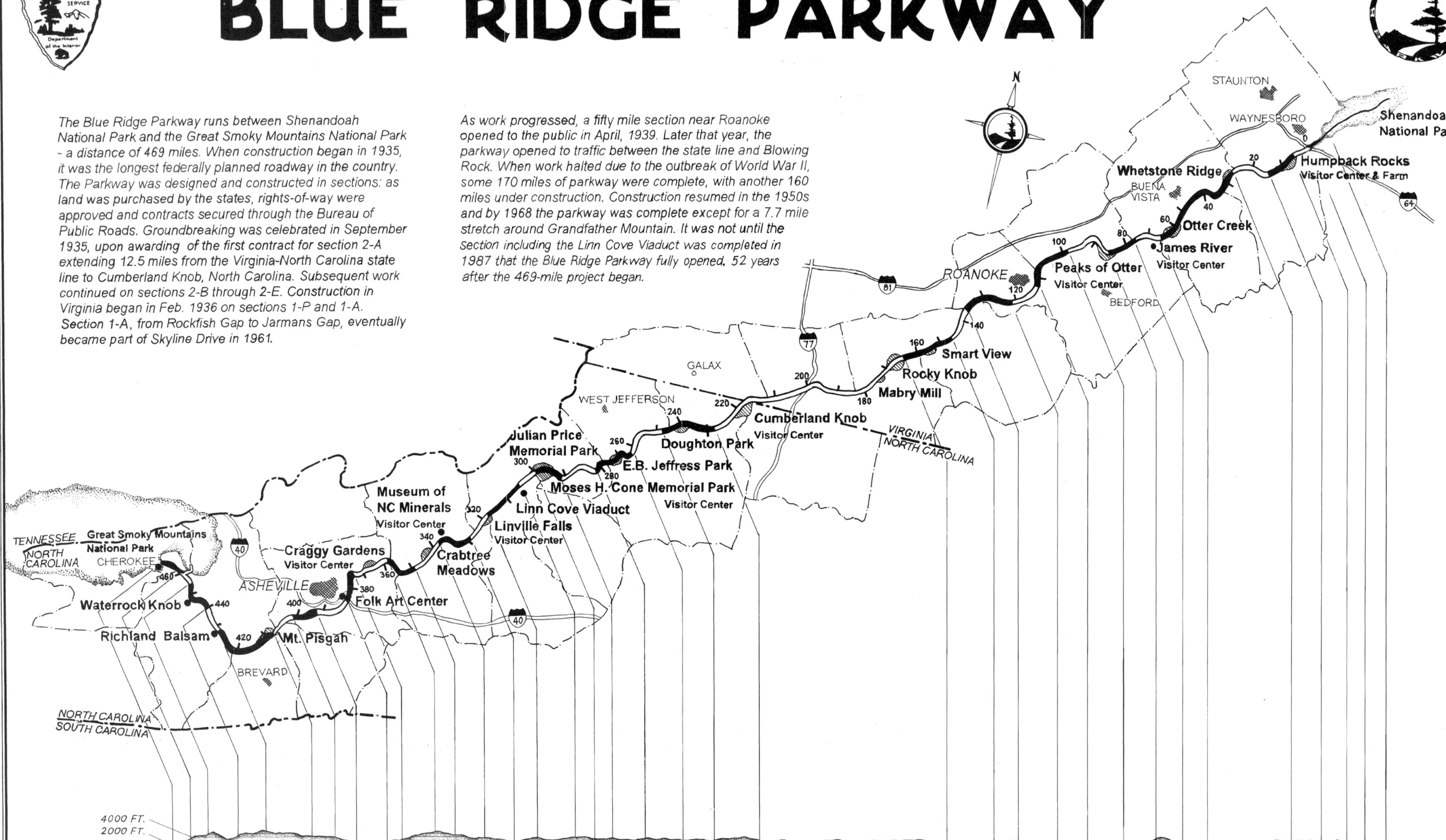

COMPLETED	'57	'42	1960	1962	1955		'65	'63	1967	'52	1950	'41	1940	1939	1941	1940	'87	'60	1958	1937	1938	1937		
SECTIONS	2-Z	2-Y	2-X	2-W	2-V	2-U	2-T	2-S	2-R	2-Q	2-P	2-N	2-M	2-L	2-K	2-J	2-H	2-G	2-F	2-E	2-D	2-C	2-B	2-A
SHEET	BALSAM MOUNTAINS										BLACK MOUNTAINS								HIGHLANDS					
STATE	NORTH CAROLINA																							

COMPLETED	1939					1938		1936	1960	1965	1950	1941	1940	1941	1942			1940	1941	'40	
SECTIONS	1-W	1-V	1-U	1-T	1-S	1-R	1-Q	1-P	1-N	1-M	1-L	1-K	1-J	1-H	1-G	1-F	1-E	1-D	1-C	1-B	1-A
SHEET	PLATEAU										RIDGE										
STATE	VIRGINIA																				

DELINEATED BY: *Lia M. Dikigoropoulou, 1997; edited by Elisabeth Dubin, 1997*

NATIONAL PARK SERVICE
ROADS & BRIDGES RECORDING PROJECT
NATIONAL PARK SERVICE
UNITED STATES DEPARTMENT OF THE INTERIOR

BLUE RIDGE PARKWAY
ASHEVILLE VICINITY
BUNCOMBE COUNTY
NORTH CAROLINA

SHEET 2 of 28

HISTORIC AMERICAN ENGINEERING RECORD
NC-42

PINE TREE OVERLOOK *milepost 95.1*

APPALACHIAN TRAIL CROSSING *milepost 96.0*

WHETSTONE RIDGE RESTAURANT and CRAFT SHOP *milepost 29.0*

THE RIDGE

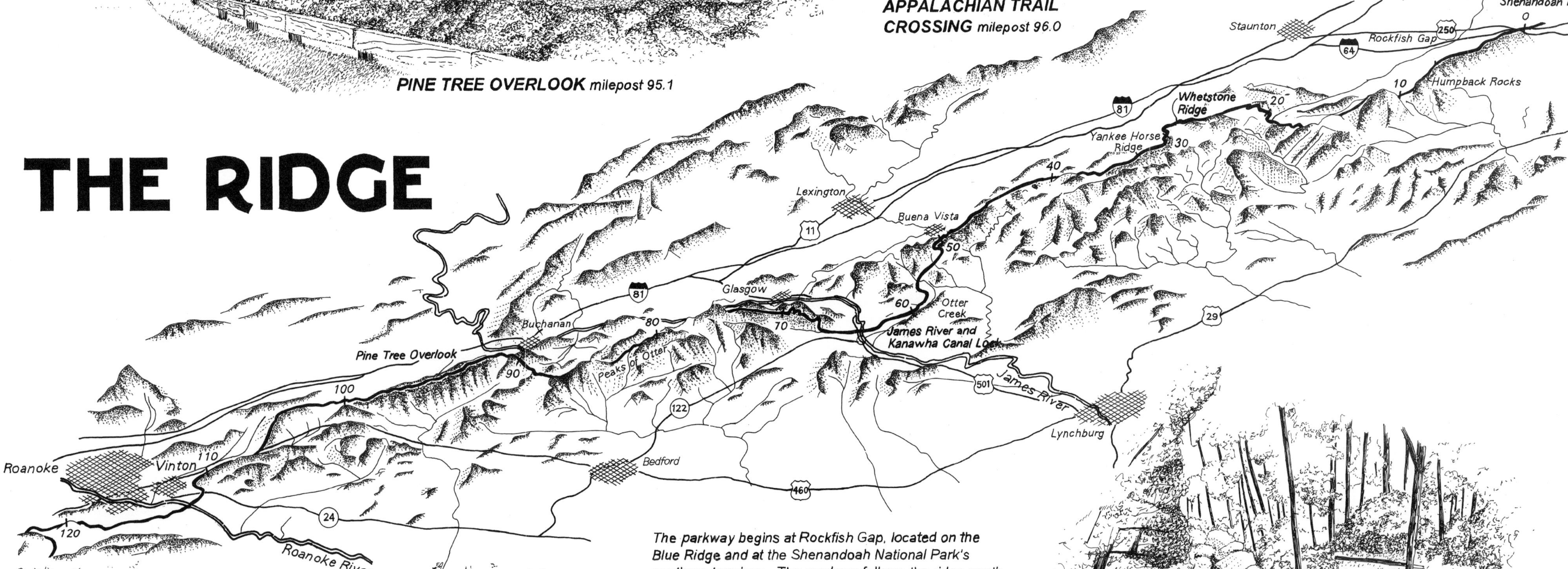

The parkway begins at Rockfish Gap, located on the Blue Ridge and at the Shenandoah National Park's southern terminus. The roadway follows the ridge south to Roanoke except when crossing the James River Water Gap, the lowest elevation on the parkway. This section of roadway is the oldest with Sections 1B through 1K completed prior to 1942. The only exception is Section 1L which was finished in 1950. Most of the parkway's best examples of Appalachian culture can be found in this region – Humpback Rocks Farm, Johnson Farm, Howardsville Turnpike, Irish Creek Railway, and the James River and Kanawha Canal Locks. In an area where the lands were once cleared for agriculture and trade routes,and where living off the land was the norm, nature has reasserted itself. This section , with the majority of the parkway running through the Washington and Jefferson national forests, can now be described as a mountain forest experience. It is here the section's major interpretative emphasis, the Blue Ridge's flora and fauna, is focused.

JAMES RIVER AND KANAWHA CANAL LOCK *milepost 63.6*

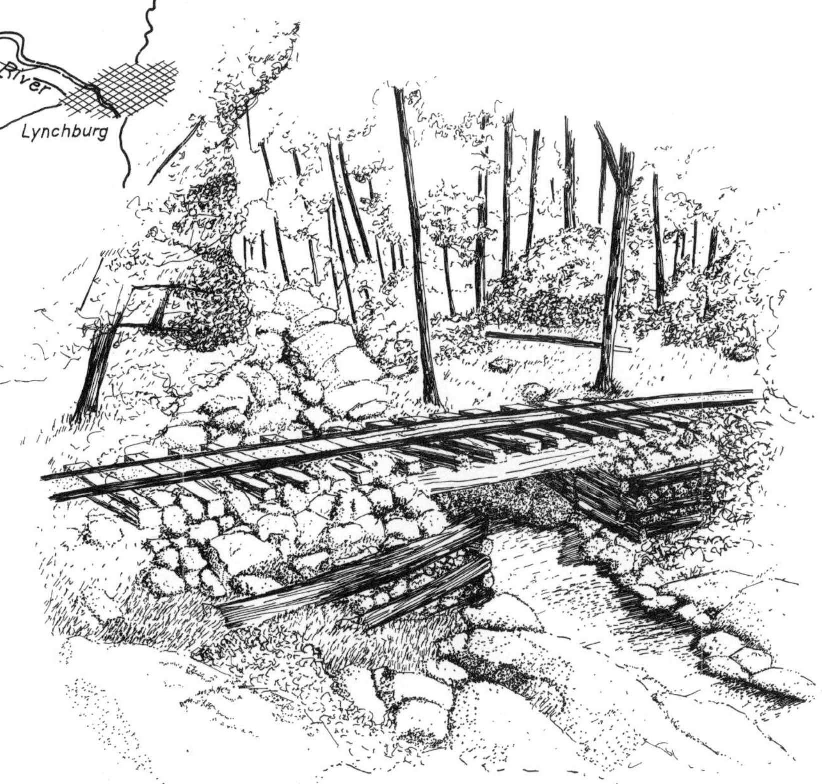

IRISH CREEK RAILROAD BRIDGE *milepost 34.4*

DELINEATED BY: *Natascha V. Wiener and Cherla Yost, 1997*

NATIONAL PARK SERVICE
ROADS & BRIDGES RECORDING PROJECT
NATIONAL PARK SERVICE
UNITED STATES DEPARTMENT OF THE INTERIOR

BLUE RIDGE PARKWAY
ASHEVILLE VICINITY BUNCOMBE COUNTY NORTH CAROLINA

SHEET 3 OF 28

HISTORIC AMERICAN ENGINEERING RECORD
NC-42

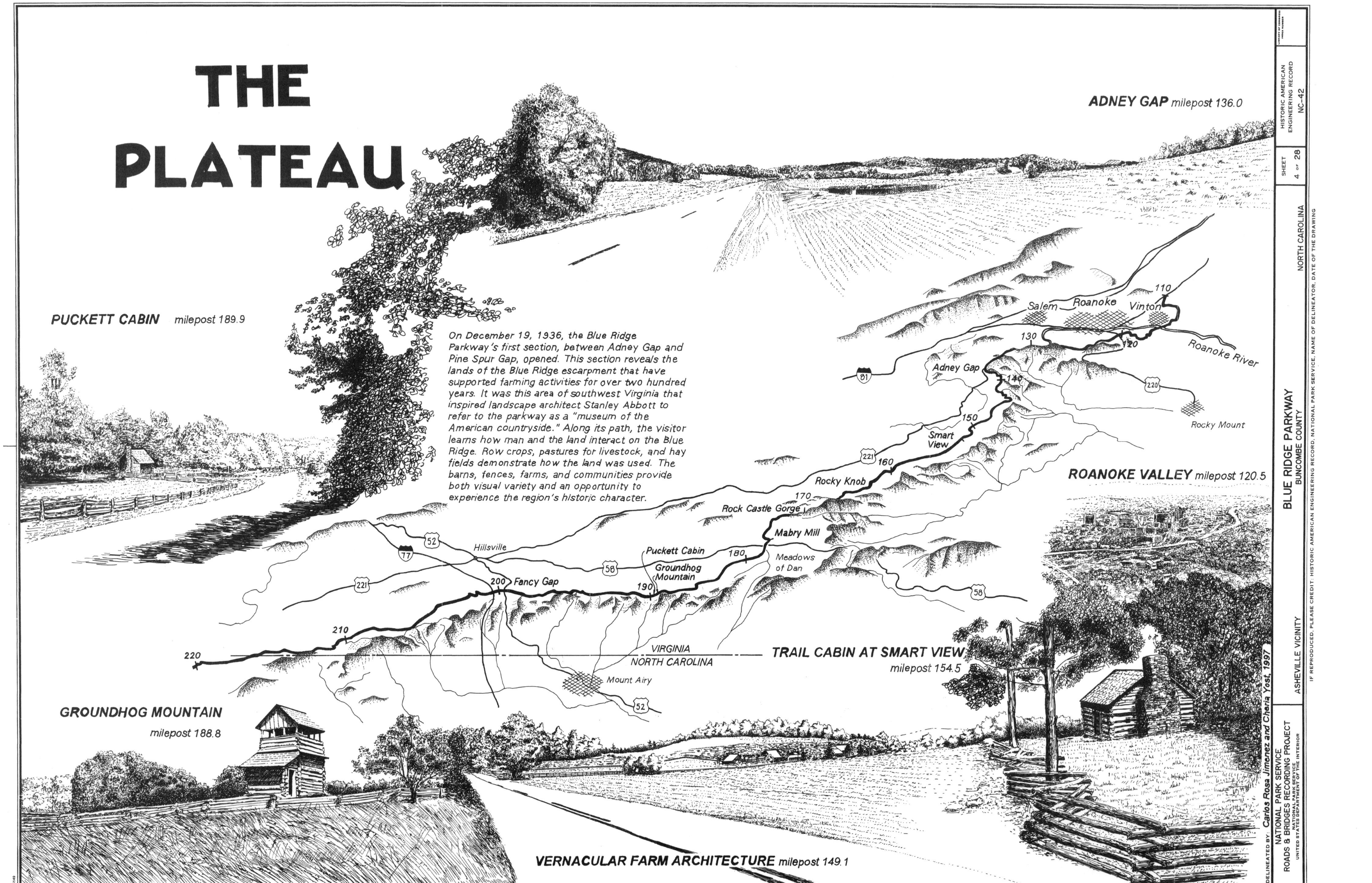

THE PLATEAU
ADNEY GAP milepost 136.0
PUCKETT CABIN milepost 189.9
On December 19, 1936, the Blue Ridge Parkway's first section, between Adney Gap and Pine Spur Gap, opened. This section reveals the lands of the Blue Ridge escarpment that have supported farming activities for over two hundred years. It was this area of southwest Virginia that inspired landscape architect Stanley Abbott to refer to the parkway as a "museum of the American countryside." Along its path, the visitor learns how man and the land interact on the Blue Ridge. Row crops, pastures for livestock, and hay fields demonstrate how the land was used. The barns, fences, farms, and communities provide both visual variety and an opportunity to experience the region's historic character.
Salem
Roanoke
Vinton
110
130
120
Roanoke River
81
Adney Gap
140
220
Rocky Mount
150
Smart View
221
160
Rocky Knob
ROANOKE VALLEY milepost 120.5
170
Rock Castle Gorge
Mabry Mill
52
77
Hillsville
Puckett Cabin
180
Meadows of Dan
58
Groundhog Mountain
221
200
Fancy Gap
190
58
210
220
VIRGINIA
NORTH CAROLINA
TRAIL CABIN AT SMART VIEW milepost 154.5
Mount Airy
52
GROUNDHOG MOUNTAIN milepost 188.8
VERNACULAR FARM ARCHITECTURE milepost 149.1
DELINEATED BY: Carlos Rosa Jimenez and Cheria Yost, 1997
NATIONAL PARK SERVICE
ROADS & BRIDGES RECORDING PROJECT
NATIONAL PARK SERVICE
UNITED STATES DEPARTMENT OF THE INTERIOR
ASHEVILLE VICINITY
BLUE RIDGE PARKWAY
BUNCOMBE COUNTY
NORTH CAROLINA
SHEET 4 OF 28
HISTORIC AMERICAN ENGINEERING RECORD
NC-42
IF REPRODUCED, PLEASE CREDIT: HISTORIC AMERICAN ENGINEERING RECORD, NATIONAL PARK SERVICE, NAME OF DELINEATOR, DATE OF THE DRAWING

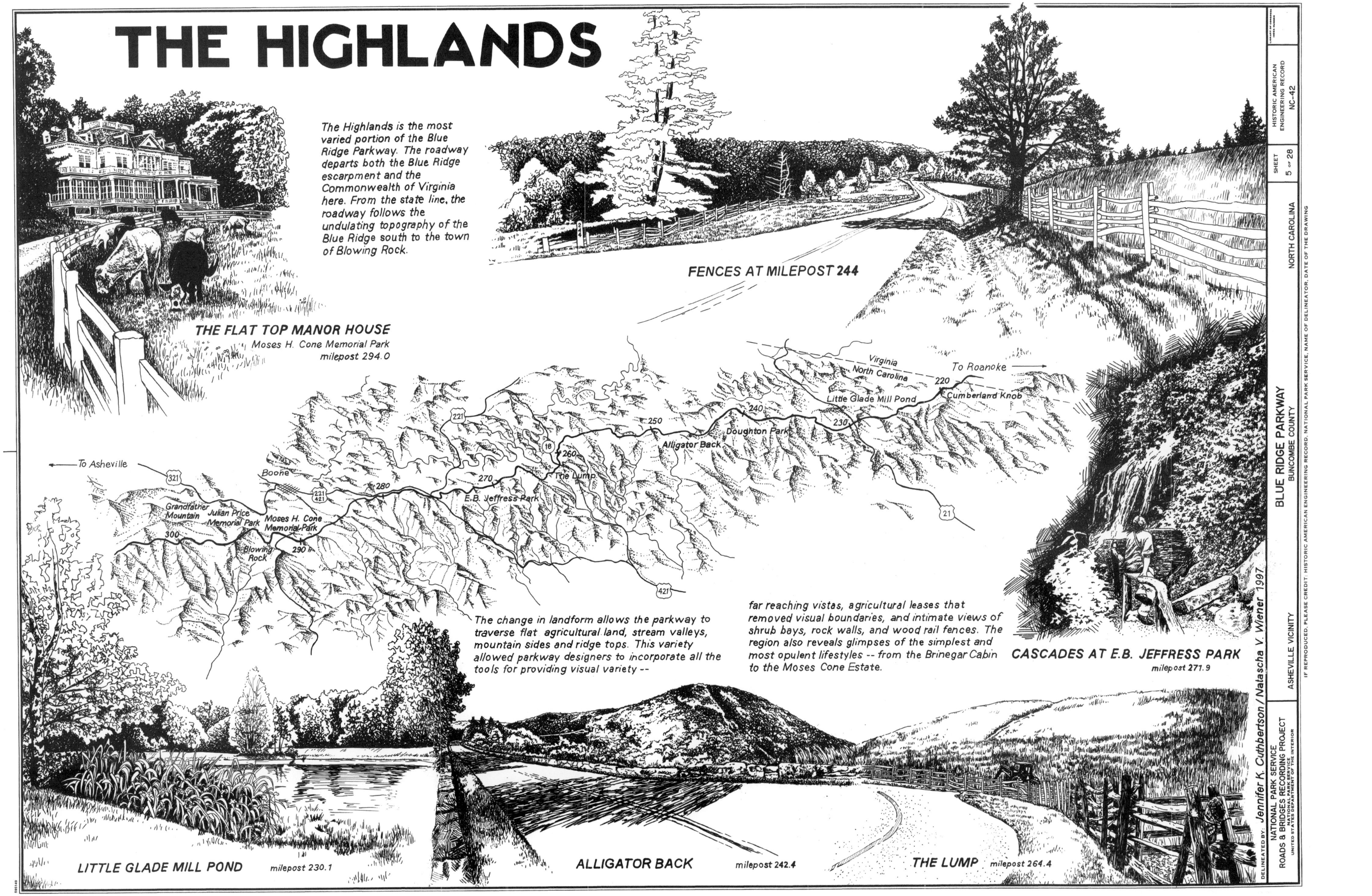

THE HIGHLANDS
The Highlands is the most varied portion of the Blue Ridge Parkway. The roadway departs both the Blue Ridge escarpment and the Commonwealth of Virginia here. From the state line, the roadway follows the undulating topography of the Blue Ridge south to the town of Blowing Rock.
THE FLAT TOP MANOR HOUSE
Moses H. Cone Memorial Park
milepost 294.0
FENCES AT MILEPOST 244
To Asheville
To Roanoke
Virginia
North Carolina
Cumberland Knob
Little Glade Mill Pond
Doughton Park
Alligator Back
The Lump
E.B. Jeffress Park
Boone
Grandfather Mountain
Julian Price Memorial Park
Moses H. Cone Memorial Park
Blowing Rock
321
221
421
18
21
220
230
240
250
260
270
280
290
300
The change in landform allows the parkway to traverse flat agricultural land, stream valleys, mountain sides and ridge tops. This variety allowed parkway designers to incorporate all the tools for providing visual variety --
far reaching vistas, agricultural leases that removed visual boundaries, and intimate views of shrub bays, rock walls, and wood rail fences. The region also reveals glimpses of the simplest and most opulent lifestyles -- from the Brinegar Cabin to the Moses Cone Estate.
CASCADES AT E.B. JEFFRESS PARK
milepost 271.9
LITTLE GLADE MILL POND
milepost 230.1
ALLIGATOR BACK
milepost 242.4
THE LUMP
milepost 264.4
HISTORIC AMERICAN ENGINEERING RECORD
NC-42
SHEET
5 of 28
NORTH CAROLINA
BLUE RIDGE PARKWAY
BUNCOMBE COUNTY
ASHEVILLE VICINITY
DELINEATED BY: Jennifer K. Cuthbertson / Natascha V. Wiener 1997
NATIONAL PARK SERVICE
ROADS & BRIDGES RECORDING PROJECT
IF REPRODUCED, PLEASE CREDIT: HISTORIC AMERICAN ENGINEERING RECORD, NATIONAL PARK SERVICE, NAME OF DELINEATOR, DATE OF THE DRAWING

THE BLACK MOUNTAINS

BLACK MOUNTAINS OVERLOOKING THE ASHEVILLE WATERSHED *milepost 358*

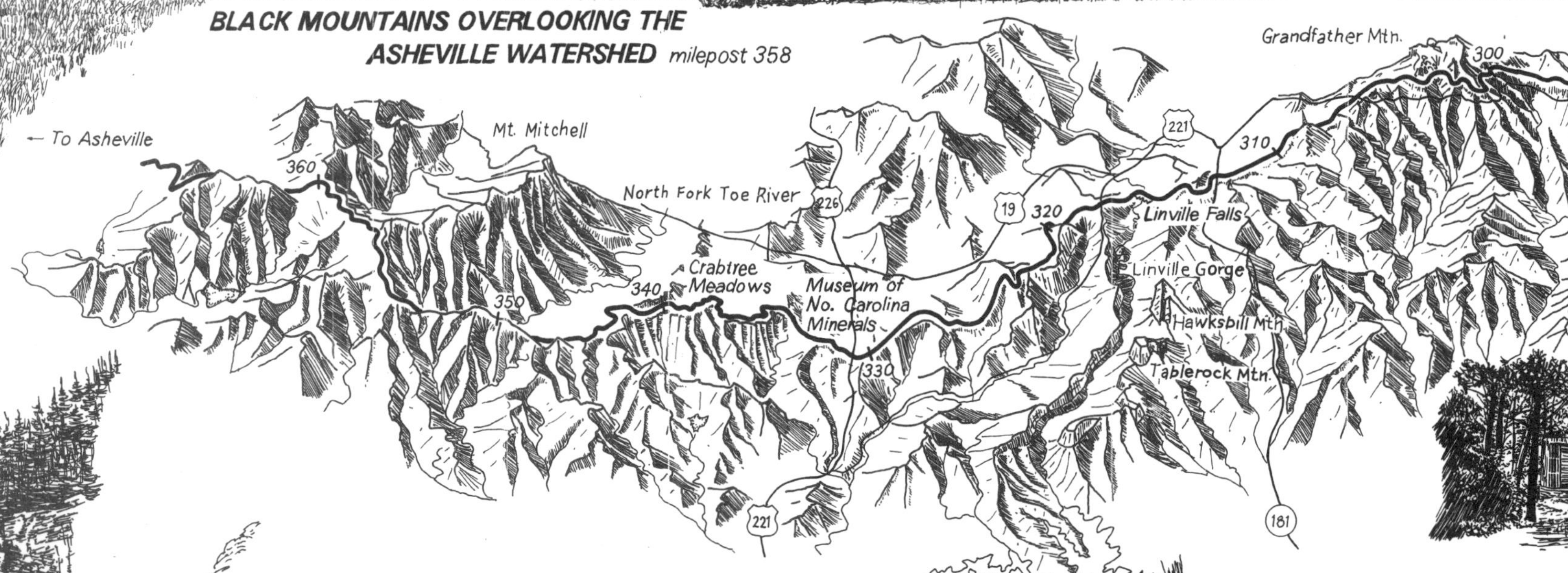

AMPHITHEATER AT CRABTREE MEADOWS *milepost 340*

LINVILLE FALLS & OVERLOOK *milepost 316.4*

LINVILLE RIVER BRIDGE *milepost 316.6*

Soon after the roadway crosses Grandfather Mountain, the highest point on the Blue Ridge, it passes the 1,500 foot deep chasm, Linville Gorge. Geology is one of the main features of this region. The Museum of North Carolina Minerals was constructed to showcase the mineral wealth of the region. Past the Linville area, the Parkway leaves the Blue Ridge and skirts the Black Mountains and the highest peak east of the Mississippi, Mount Mitchell at 6,684 feet. The departure from the Blue Ridge is necessary to provide the connection to the Great Smoky Mountains National Park. From Mount Mitchell, the Parkway negotiates the Great Craggy Mountains and its open heath balds on its way to the city of Asheville.

TANAWHA TRAIL BRIDGING BOONE FORK *milepost 300*

LINN COVE VIADUCT, ROUGH RIDGE, & "THE GREAT WALL" AT GRANDFATHER MOUNTAIN *milepost 304.0*

DELINEATED BY: *Lidia Klupsz, Ian Shanklin & Elisabeth Dubin, 1997*

NATIONAL PARK SERVICE
ROADS & BRIDGES RECORDING PROJECT
NATIONAL PARK SERVICE
UNITED STATES DEPARTMENT OF THE INTERIOR

BLUE RIDGE PARKWAY
ASHEVILLE VICINITY
BUNCOMBE COUNTY
NORTH CAROLINA

SHEET 6 OF 28

HISTORIC AMERICAN ENGINEERING RECORD
NC-42

IF REPRODUCED, PLEASE CREDIT: HISTORIC AMERICAN ENGINEERING RECORD, NATIONAL PARK SERVICE, NAME OF DELINEATOR, DATE OF THE DRAWING

THE BALSAM MOUNTAINS
GRAVEYARD FIELDS FROM SHINING ROCK SPUR ROAD milepost 420.2
CRAGGY FLATS TUNNEL milepost 365.5
TERMINUS BRIDGE OVER OCCONALUFTEE RIVER milepost 469.0
HIGHEST POINT ON BLUE RIDGE PARKWAY (AT 6047 FEET) milepost 431.4
This southernmost region is the most rugged section of the Blue Ridge Parkway, crossing the Pisgah Ledge and Great Balsam and Plott Balsam mountains before entering the Great Smoky Mountains National Park. It could also be named "the tunnel district," since twenty of the parkway's twenty-six tunnels are located in this mountainous region.
Great Smoky Mountains National Park
Heintooga Overlook
441
Soco Gap
460
Cherokee
19
450
Balsam Gap
440
Waynesville
40
23 74
Sylva
430
420
Devils Courthouse
Mt. Pisgah 410
Looking Glass Rock
Brevard
French Broad River
400
26
390
Asheville
23
Folk Art Center
380
370
360
Mt. Mitchell
To Roanoke
The parkway reaches its southernmost point at Tennessee Bald, at the junction of the Pisgah Ledge and Great Balsam ranges, and its highest elevation of 6,047 feet on the shoulder of Richland Mountain of the Balsams.
DEVILS COURTHOUSE milepost 422.4
CRAGGY GARDENS FROM CRAGGY PINNACLE OVERLOOK milepost 364.0
LOOKING GLASS ROCK milepost 417
DELINEATED BY Ian Shanklin & Elisabeth Dubin, 1997
NATIONAL PARK SERVICE
ROADS & BRIDGES RECORDING PROJECT
ASHEVILLE VICINITY
BLUE RIDGE PARKWAY
BUNCOMBE COUNTY
NORTH CAROLINA
SHEET 7 of 28
HISTORIC AMERICAN ENGINEERING RECORD
NC-42
IF REPRODUCED, PLEASE CREDIT: HISTORIC AMERICAN ENGINEERING RECORD, NATIONAL PARK SERVICE, NAME OF DELINEATOR, DATE OF THE DRAWING

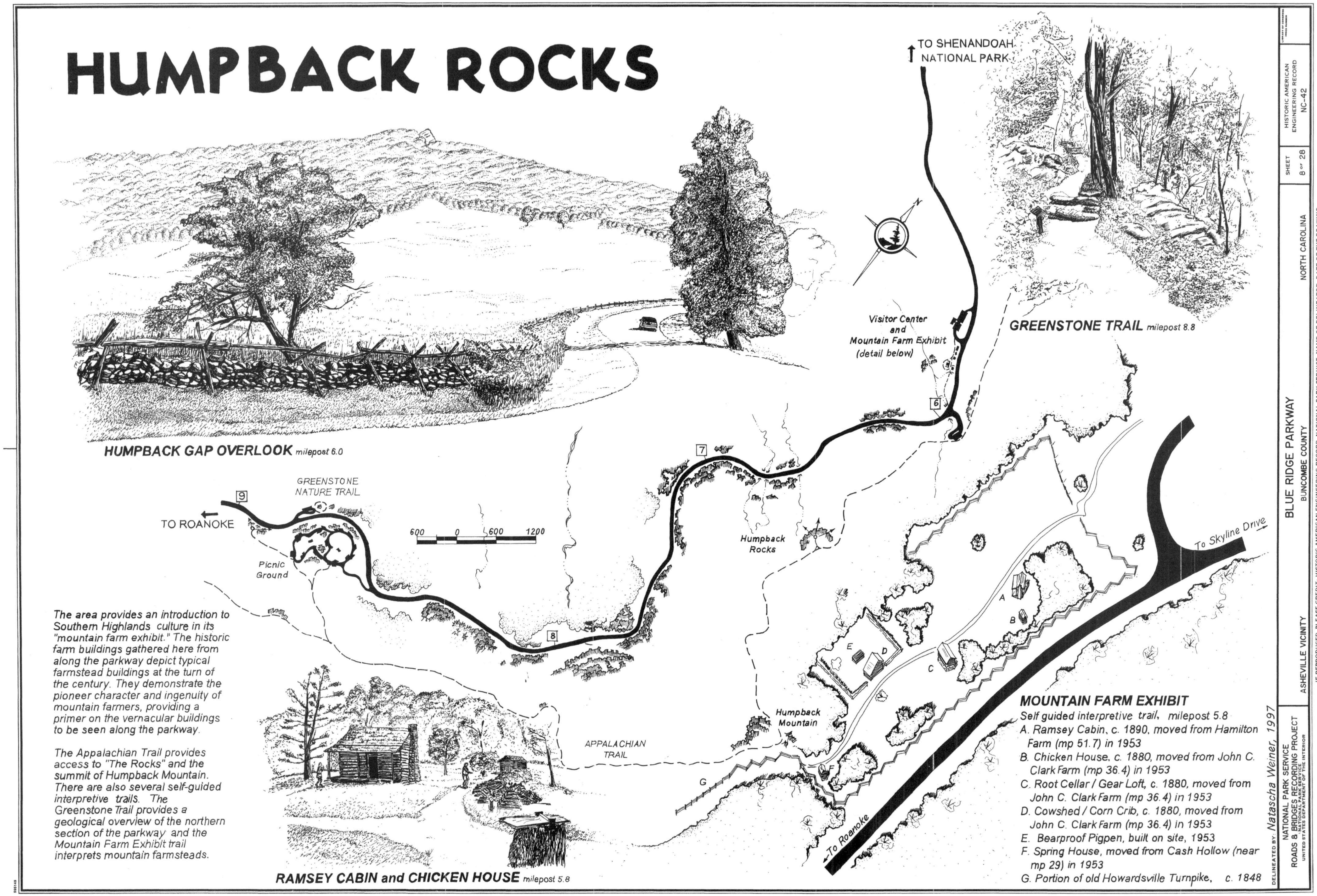
HUMPBACK ROCKS
TO SHENANDOAH NATIONAL PARK
N
Visitor Center and Mountain Farm Exhibit (detail below)
GREENSTONE TRAIL milepost 8.8
HUMPBACK GAP OVERLOOK milepost 6.0
6
7
8
9
GREENSTONE NATURE TRAIL
TO ROANOKE
Picnic Ground
600 0 600 1200
Humpback Rocks
To Skyline Drive
A
B
C
D
E
F
G
Humpback Mountain
APPALACHIAN TRAIL
To Roanoke
The area provides an introduction to Southern Highlands culture in its "mountain farm exhibit." The historic farm buildings gathered here from along the parkway depict typical farmstead buildings at the turn of the century. They demonstrate the pioneer character and ingenuity of mountain farmers, providing a primer on the vernacular buildings to be seen along the parkway.
The Appalachian Trail provides access to "The Rocks" and the summit of Humpback Mountain. There are also several self-guided interpretive trails. The Greenstone Trail provides a geological overview of the northern section of the parkway and the Mountain Farm Exhibit trail interprets mountain farmsteads.
RAMSEY CABIN and CHICKEN HOUSE milepost 5.8
MOUNTAIN FARM EXHIBIT
Self guided interpretive trail, milepost 5.8
A. Ramsey Cabin, c. 1890, moved from Hamilton Farm (mp 51.7) in 1953
B. Chicken House, c. 1880, moved from John C. Clark Farm (mp 36.4) in 1953
C. Root Cellar / Gear Loft, c. 1880, moved from John C. Clark Farm (mp 36.4) in 1953
D. Cowshed / Corn Crib, c. 1880, moved from John C. Clark Farm (mp 36.4) in 1953
E. Bearproof Pigpen, built on site, 1953
F. Spring House, moved from Cash Hollow (near mp 29) in 1953
G. Portion of old Howardsville Turnpike, c. 1848
DELINEATED BY: Natascha Weiner, 1997
NATIONAL PARK SERVICE
ROADS & BRIDGES RECORDING PROJECT
UNITED STATES DEPARTMENT OF THE INTERIOR
ASHEVILLE VICINITY
BLUE RIDGE PARKWAY
BUNCOMBE COUNTY
NORTH CAROLINA
SHEET 8 OF 28
HISTORIC AMERICAN ENGINEERING RECORD
NC-42
IF REPRODUCED, PLEASE CREDIT: HISTORIC AMERICAN ENGINEERING RECORD, NATIONAL PARK SERVICE, NAME OF DELINEATOR, DATE OF THE DRAWING

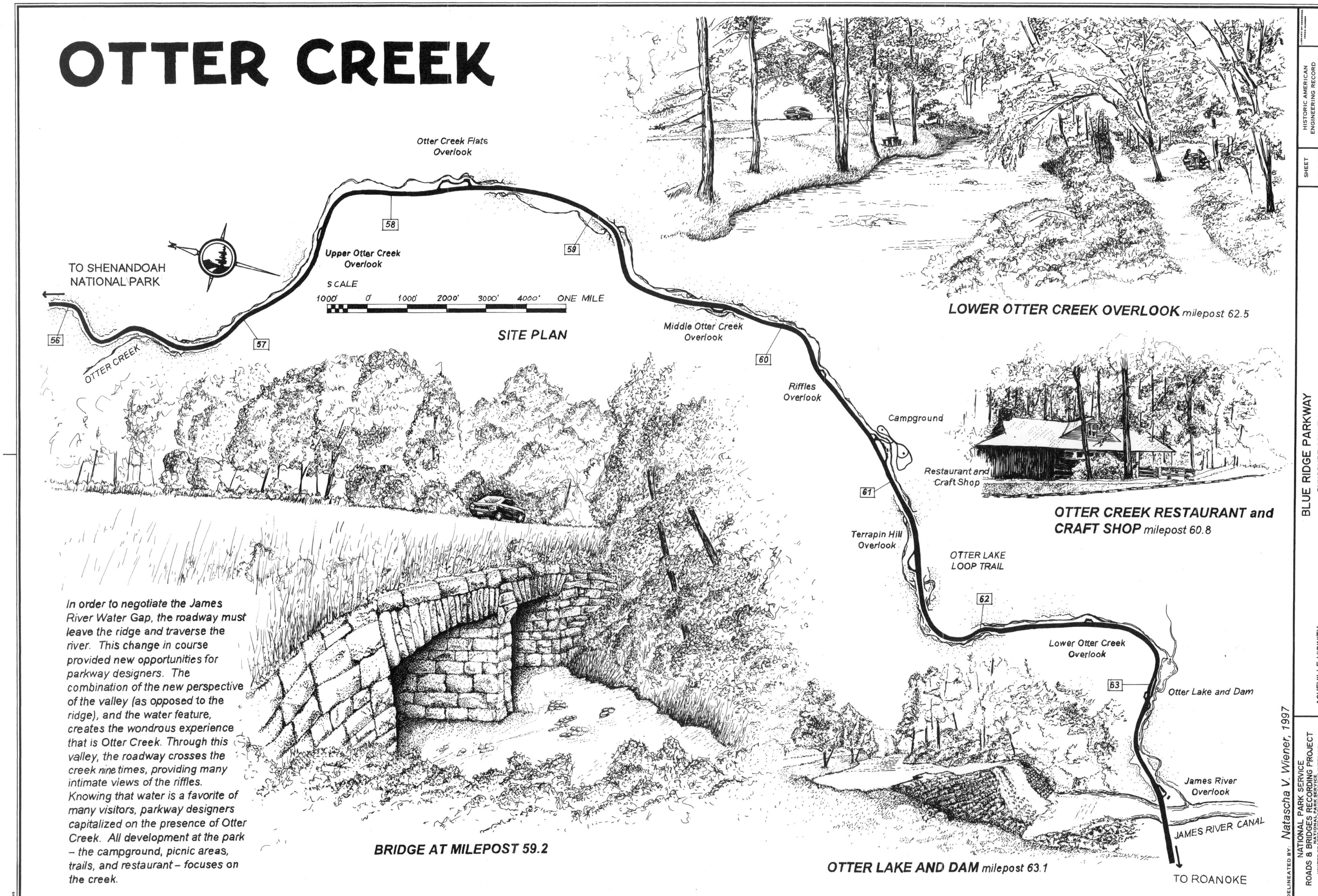
OTTER CREEK
Otter Creek Flats Overlook
58
59
Upper Otter Creek Overlook
TO SHENANDOAH NATIONAL PARK
SCALE
1000' 0' 1000' 2000' 3000' 4000' ONE MILE
SITE PLAN
56
57
OTTER CREEK
Middle Otter Creek Overlook
60
Riffles Overlook
Campground
Restaurant and Craft Shop
61
Terrapin Hill Overlook
OTTER LAKE LOOP TRAIL
62
Lower Otter Creek Overlook
63
Otter Lake and Dam
James River Overlook
JAMES RIVER CANAL
TO ROANOKE
LOWER OTTER CREEK OVERLOOK milepost 62.5
OTTER CREEK RESTAURANT and CRAFT SHOP milepost 60.8
BRIDGE AT MILEPOST 59.2
OTTER LAKE AND DAM milepost 63.1
In order to negotiate the James River Water Gap, the roadway must leave the ridge and traverse the river. This change in course provided new opportunities for parkway designers. The combination of the new perspective of the valley (as opposed to the ridge), and the water feature, creates the wondrous experience that is Otter Creek. Through this valley, the roadway crosses the creek nine times, providing many intimate views of the riffles. Knowing that water is a favorite of many visitors, parkway designers capitalized on the presence of Otter Creek. All development at the park – the campground, picnic areas, trails, and restaurant – focuses on the creek.
DELINEATED BY: Natascha V. Wiener, 1997
NATIONAL PARK SERVICE
ROADS & BRIDGES RECORDING PROJECT
ASHEVILLE VICINITY
BLUE RIDGE PARKWAY
BUNCOMBE COUNTY
NORTH CAROLINA
SHEET 9 OF 28
HISTORIC AMERICAN ENGINEERING RECORD
NC-42
IF REPRODUCED, PLEASE CREDIT: HISTORIC AMERICAN ENGINEERING RECORD, NATIONAL PARK SERVICE, NAME OF DELINEATOR, DATE OF THE DRAWING

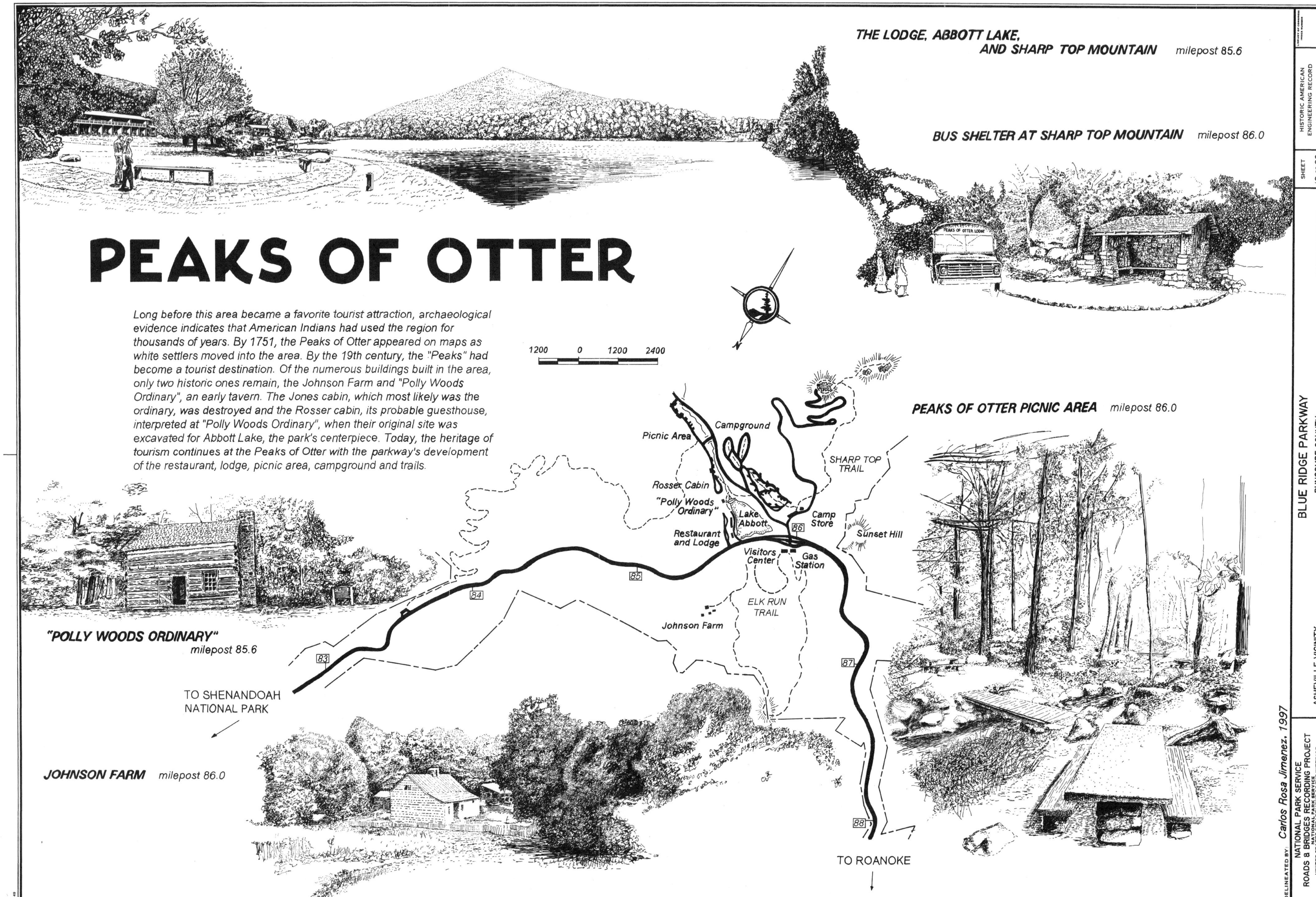
THE LODGE, ABBOTT LAKE,
AND SHARP TOP MOUNTAIN milepost 85.6
BUS SHELTER AT SHARP TOP MOUNTAIN milepost 86.0
PEAKS OF OTTER LODGE
PEAKS OF OTTER
Long before this area became a favorite tourist attraction, archaeological evidence indicates that American Indians had used the region for thousands of years. By 1751, the Peaks of Otter appeared on maps as white settlers moved into the area. By the 19th century, the "Peaks" had become a tourist destination. Of the numerous buildings built in the area, only two historic ones remain, the Johnson Farm and "Polly Woods Ordinary", an early tavern. The Jones cabin, which most likely was the ordinary, was destroyed and the Rosser cabin, its probable guesthouse, interpreted at "Polly Woods Ordinary", when their original site was excavated for Abbott Lake, the park's centerpiece. Today, the heritage of tourism continues at the Peaks of Otter with the parkway's development of the restaurant, lodge, picnic area, campground and trails.
1200 0 1200 2400
N
PEAKS OF OTTER PICNIC AREA milepost 86.0
Campground
Picnic Area
SHARP TOP TRAIL
Rosser Cabin
"Polly Woods Ordinary"
Lake Abbott
Camp Store
Restaurant and Lodge
Sunset Hill
Visitors Center
Gas Station
86
85
84
83
87
88
ELK RUN TRAIL
Johnson Farm
"POLLY WOODS ORDINARY"
milepost 85.6
TO SHENANDOAH NATIONAL PARK
JOHNSON FARM milepost 86.0
TO ROANOKE
HISTORIC AMERICAN ENGINEERING RECORD
NC-42
SHEET 10 OF 28
NORTH CAROLINA
BLUE RIDGE PARKWAY
BUNCOMBE COUNTY
ASHEVILLE VICINITY
DELINEATED BY: Carlos Rosa Jimenez, 1997
NATIONAL PARK SERVICE
ROADS & BRIDGES RECORDING PROJECT
IF REPRODUCED, PLEASE CREDIT: HISTORIC AMERICAN ENGINEERING RECORD, NATIONAL PARK SERVICE, NAME OF DELINEATOR, DATE OF THE DRAWING

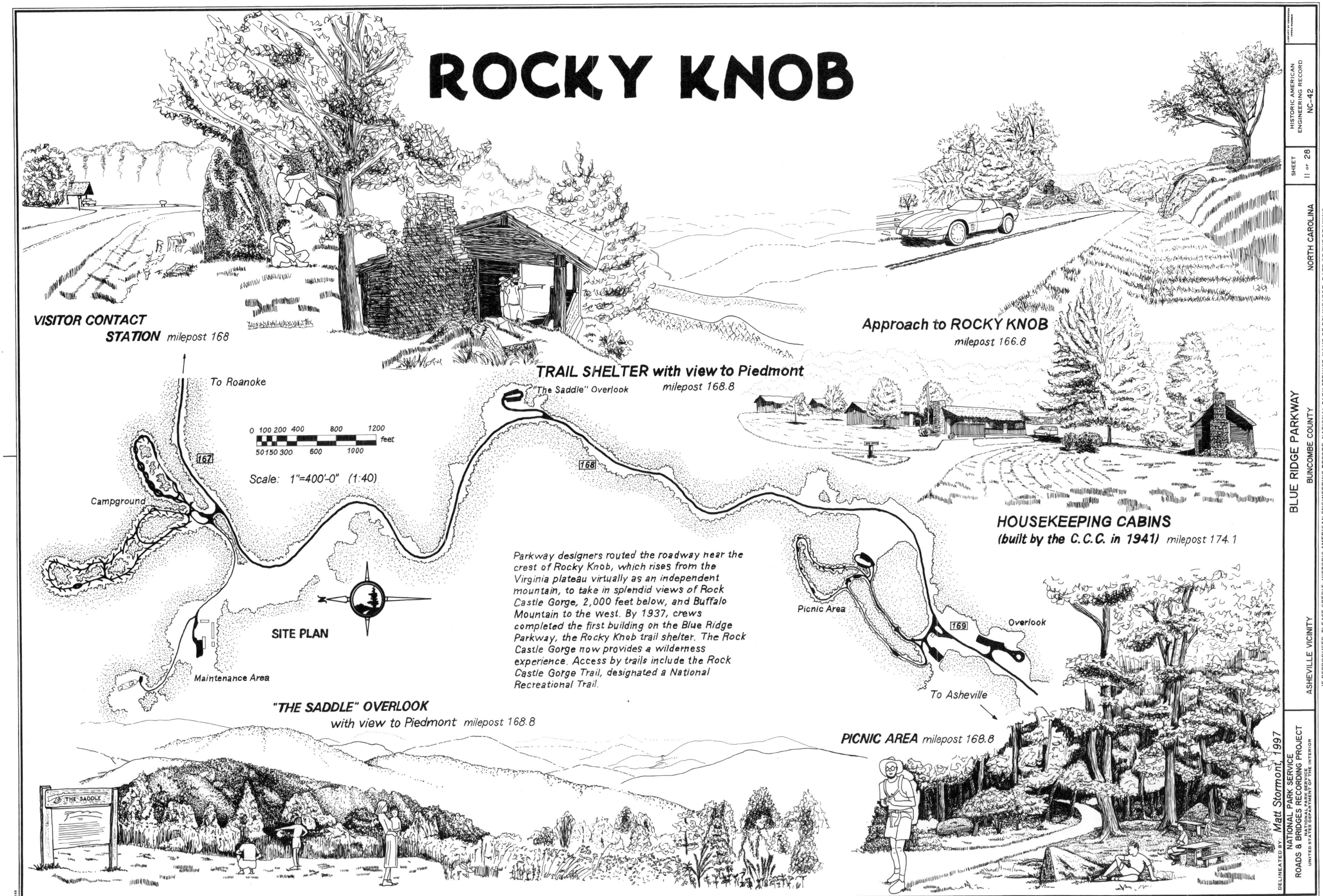

ROCKY KNOB
VISITOR CONTACT STATION milepost 168
TRAIL SHELTER with view to Piedmont milepost 168.8
Approach to ROCKY KNOB milepost 166.8
HOUSEKEEPING CABINS (built by the C.C.C. in 1941) milepost 174.1
To Roanoke
"The Saddle" Overlook
0 100 200 400 800 1200 feet
50 150 300 600 1000
Scale: 1"=400'-0" (1:40)
167
168
169
Campground
SITE PLAN
Maintenance Area
Picnic Area
Overlook
To Asheville
Parkway designers routed the roadway near the crest of Rocky Knob, which rises from the Virginia plateau virtually as an independent mountain, to take in splendid views of Rock Castle Gorge, 2,000 feet below, and Buffalo Mountain to the west. By 1937, crews completed the first building on the Blue Ridge Parkway, the Rocky Knob trail shelter. The Rock Castle Gorge now provides a wilderness experience. Access by trails include the Rock Castle Gorge Trail, designated a National Recreational Trail.
"THE SADDLE" OVERLOOK with view to Piedmont milepost 168.8
THE SADDLE
PICNIC AREA milepost 168.8
DELINEATED BY: Matt Stormont, 1997
NATIONAL PARK SERVICE
ROADS & BRIDGES RECORDING PROJECT
NATIONAL PARK SERVICE
UNITED STATES DEPARTMENT OF THE INTERIOR
ASHEVILLE VICINITY
BUNCOMBE COUNTY
BLUE RIDGE PARKWAY
NORTH CAROLINA
SHEET 11 of 28
HISTORIC AMERICAN ENGINEERING RECORD
NC-42
IF REPRODUCED, PLEASE CREDIT: HISTORIC AMERICAN ENGINEERING RECORD, NATIONAL PARK SERVICE, NAME OF DELINEATOR, DATE OF THE DRAWING

MABRY MILL
This scenic setting created by the parkway's landscape architects is one of the most photographed views in the National Park Service. Although the mill is not as old as other structures interpreted on the parkway, planners found the mill's texture and three separate roof-lines attractive. Contributing to its popularity, an easily accessible path guides visitors through the mountain industry museum setting - past buildings, exhibits, and the restored mill. The dredged pond provides a dramatic reflection while its edge directs visitors to the "perfect view."
FLUME AND TRAIL
MABRY MILL AND POND milepost 176.2
SITE PLAN
To Asheville
To Roanoke
176
A
B
C
D
E
F
G
H
I
J
K
SITE PLAN
A. Restaurant & Gift Shop
B. Mabry Mill & Pond, c. 1903-1914, restored 1942, pond excavated 1942
C. Lumber Drying Rack, moved from Rocky Knob in 1957
D. Log Cart, origins unclear
E. Matthew's Cabin, c. 1896, moved from Carroll County in 1956 (former site of original Mabry home, c. 1914, demolished in 1942 by the Blue Ridge Parkway)
F. Moonshine Still, c. pre-prohibition, moved from Great Smoky Mountains National Park in 1957
G. Sorghum Mill, moved from Rocky Knob in 1957
H. Blacksmith Shop, c. 1898, moved from west of the Mabry Mill in the line of the Parkway in 1942
I. Farming Tools
J. Wash House, origins unclear
K. Flume
PANORAMA OF MABRY MILL TRAIL
DELINEATED BY: Jennifer K. Cuthbertson / Natascha V. Wiener, 1997
NATIONAL PARK SERVICE
ROADS & BRIDGES RECORDING PROJECT
ASHEVILLE VICINITY
BLUE RIDGE PARKWAY
BUNCOMBE COUNTY
NORTH CAROLINA
SHEET 12 of 28
HISTORIC AMERICAN ENGINEERING RECORD
NC-42
IF REPRODUCED, PLEASE CREDIT: HISTORIC AMERICAN ENGINEERING RECORD, NATIONAL PARK SERVICE, NAME OF DELINEATOR, DATE OF THE DRAWING

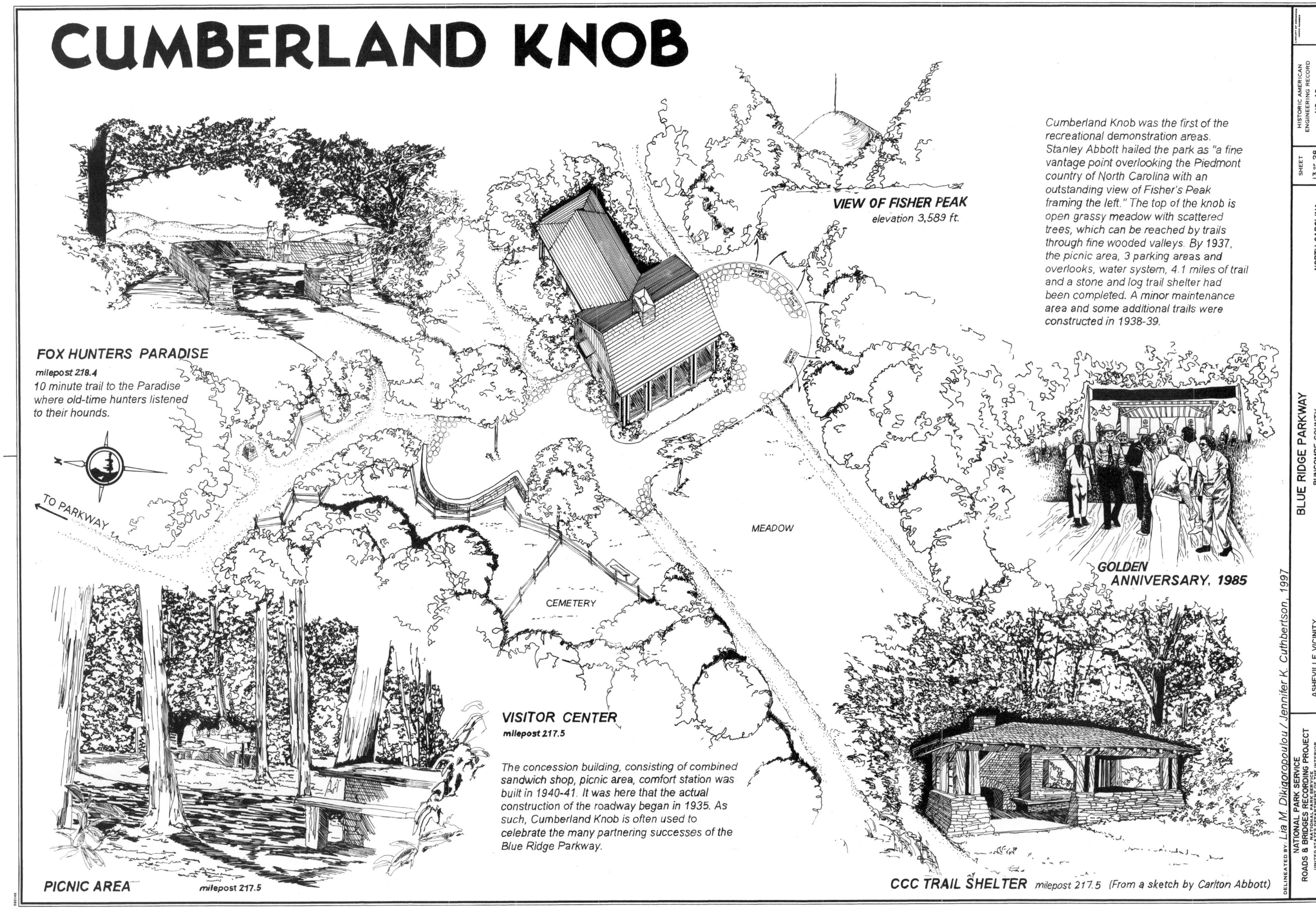
CUMBERLAND KNOB
VIEW OF FISHER PEAK
elevation 3,589 ft.
Cumberland Knob was the first of the recreational demonstration areas. Stanley Abbott hailed the park as "a fine vantage point overlooking the Piedmont country of North Carolina with an outstanding view of Fisher's Peak framing the left." The top of the knob is open grassy meadow with scattered trees, which can be reached by trails through fine wooded valleys. By 1937, the picnic area, 3 parking areas and overlooks, water system, 4.1 miles of trail and a stone and log trail shelter had been completed. A minor maintenance area and some additional trails were constructed in 1938-39.
FOX HUNTERS PARADISE
milepost 218.4
10 minute trail to the Paradise where old-time hunters listened to their hounds.
TO PARKWAY
MEADOW
CEMETERY
GOLDEN ANNIVERSARY, 1985
VISITOR CENTER
milepost 217.5
The concession building, consisting of combined sandwich shop, picnic area, comfort station was built in 1940-41. It was here that the actual construction of the roadway began in 1935. As such, Cumberland Knob is often used to celebrate the many partnering successes of the Blue Ridge Parkway.
PICNIC AREA
milepost 217.5
CCC TRAIL SHELTER milepost 217.5 (From a sketch by Carlton Abbott)
DELINEATED BY: Lia M. Dikigoropoulou / Jennifer K. Cuthbertson, 1997
NATIONAL PARK SERVICE
ROADS & BRIDGES RECORDING PROJECT
NATIONAL PARK SERVICE
UNITED STATES DEPARTMENT OF THE INTERIOR
ASHEVILLE VICINITY
BLUE RIDGE PARKWAY
BUNCOMBE COUNTY
NORTH CAROLINA
SHEET
13 OF 28
HISTORIC AMERICAN ENGINEERING RECORD
NC-42
LIBRARY OF CONGRESS INDEX NUMBER
IF REPRODUCED, PLEASE CREDIT: HISTORIC AMERICAN ENGINEERING RECORD, NATIONAL PARK SERVICE, NAME OF DELINEATOR, DATE OF THE DRAWING

DOUGHTON PARK
ALLIGATOR BACK
milepost 242.4
FENCES AT MILEPOST 244
BRINEGAR CABIN
milepost 238.5
The old Martin Brinegar cabin was restored for interpretive purposes starting in 1941, but with the grounds providing an interpretive landscape, not a historic restoration.
Blue Ridge Parkway
246
TO ASHEVILLE
244
243
Bluff Mountain
242
Picnic Area
241
Restaurant & Gas
Lodge Area
240
Overlook Shelter
Wildcat Rocks
Campgrounds
239
Basin Coves Overlook
Caudill Cabin
Brinegar Cabin
238
TO ROANOKE
Cove Creek
Basin Creek
This high mountain meadow park features broad open fields broken by forested groves and terminating in a 1000 foot cliff which gave the area its original name, the Bluffs. Most of the park is forested, but its wide open meadows are the most memorable feature to parkway motorists. Crews of Works Progress Administration (WPA) laborers constructed the park's picnic grounds, parking areas, campgrounds, fences, and the most extensive trail system along the parkway while CCC workers concentrated on landscape work. Concession facilities were built after WWII.
BLUFFS LODGE
milepost 241.4
CCC TRAIL SHELTER
milepost 241.1
DELINEATED BY: Jennifer K. Cuthbertson / Ian Shanklin, 1997
NATIONAL PARK SERVICE
ROADS & BRIDGES RECORDING PROJECT
ASHEVILLE VICINITY
BLUE RIDGE PARKWAY
BUNCOMBE COUNTY
NORTH CAROLINA
SHEET
14 OF 28
HISTORIC AMERICAN ENGINEERING RECORD
NC-42
IF REPRODUCED, PLEASE CREDIT: HISTORIC AMERICAN ENGINEERING RECORD, NATIONAL PARK SERVICE, NAME OF DELINEATOR, DATE OF THE DRAWING

JULIAN PRICE MEMORIAL PARK

PRICE LAKE *milepost 296.8*

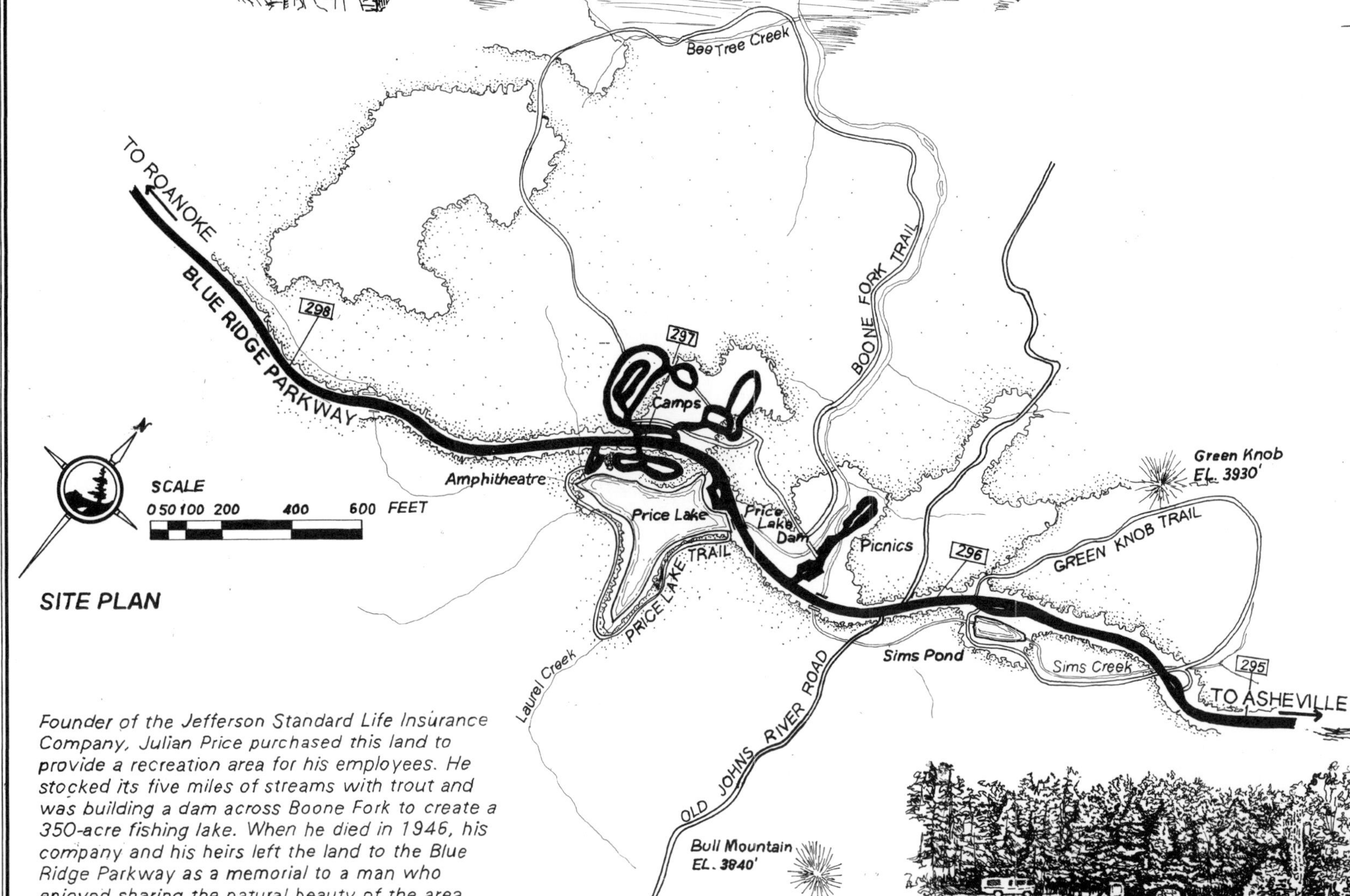

SITE PLAN

Founder of the Jefferson Standard Life Insurance Company, Julian Price purchased this land to provide a recreation area for his employees. He stocked its five miles of streams with trout and was building a dam across Boone Fork to create a 350-acre fishing lake. When he died in 1946, his company and his heirs left the land to the Blue Ridge Parkway as a memorial to a man who enjoyed sharing the natural beauty of the area with others. The Park Service completed the lake begun by Price, though it was reduced in scale to 47 acres. Price Lake, the chief scenic feature of the park, is used for fishing and boating. The area's campground and picnic areas are among the most popular on the parkway.

AMPHITHEATER *milepost 296.9*

SIMS POND *milepost 295.9*

PRICE LAKE DAM *milepost 296.8*

DELINEATED BY *Lia M. Dikigoropoulou, 1997*

NATIONAL PARK SERVICE
ROADS & BRIDGES RECORDING PROJECT

ASHEVILLE VICINITY

BLUE RIDGE PARKWAY
BUNCOMBE COUNTY

NORTH CAROLINA

SHEET 15 OF 28

HISTORIC AMERICAN ENGINEERING RECORD
NC-42

IF REPRODUCED, PLEASE CREDIT: HISTORIC AMERICAN ENGINEERING RECORD, NATIONAL PARK SERVICE, NAME OF DELINEATOR, DATE OF THE DRAWING

LINN COVE VIADUCT

THE FIRST SEGMENTAL STRUCTURE IN NORTH AMERICA BUILT USING PROGRESSIVE PLACEMENT

PEDESTRIAN TRAIL BELOW VIADUCT

Due to the environmental sensitivity of Grandfather Mountain, the exact location of the final section of the Blue Ridge Parkway created a lengthy and heated controversy between private individuals and the National Park Service. North Carolina Governor Dan K. Moore finally negotiated a compromise location. In order to eliminate massive cuts and fills, it was agreed the road should be elevated, or bridged, wherever possible. Figg and Muller Engineers, Inc., developed the bridge design and construction method. The result was the most complicated concrete bridge ever built at the time. It includes nearly every kind of alignment geometry used in highway construction. The 1,243' viaduct was constructed from the top down to minimize disturbance to the natural environment. The only construction that occurred at ground level was the drilling of foundations for the seven piers upon which the viaduct rests. The viaduct itself was the only access road for construction. Each of the 153 unique, 50 ton, pre-cast segments was lowered by stiff-leg crane and epoxied into position against the proceeding segment. Steel cables threaded through the segments secured the entire bridge deck. Construction was begun in 1979 and completed in 1983 at a cost of almost 10 million.

VIEW OF VIADUCT ON GRANDFATHER MOUNTAIN
milepost 304

SEGMENTAL CONSTRUCTION PROCESS:

CASTING THE SEGMENTS

TRANSPORTING THE SEGMENTS

Contractors employed a short-line match-casting system which permitted the segments' horizontal and vertical curves to be cast economically and with a high degree of accuracy. A low-boy tractor trailer then carried the segments to the end of the cantilever. Because of the limited access, the truck had to be backed up the access road on to the completed portion of the viaduct.

ERECTING THE SEGMENTS

A stiff-leg crane lifted the precast superstructure and pier segments from the truck around to the end of the cantilever where they were placed into position. The precast segmental pier design enabled workers to quickly construct and immediately load a pier upon completion, thereby minimizing delays in the erection of the superstructure. A temporary wood bridge allowed workers to build the pier foundations ahead of the superstructure in order to speed up construction with minimum of disturbance to the site.

DELINEATED BY: *Carlos Rosa Jimenez & Ian Shanklin, 1997*

NATIONAL PARK SERVICE
ROADS & BRIDGES RECORDING PROJECT
NATIONAL PARK SERVICE
UNITED STATES DEPARTMENT OF THE INTERIOR

BLUE RIDGE PARKWAY
ASHEVILLE VICINITY — BUNCOMBE COUNTY — NORTH CAROLINA

SHEET 16 OF 28

HISTORIC AMERICAN ENGINEERING RECORD
NC-42

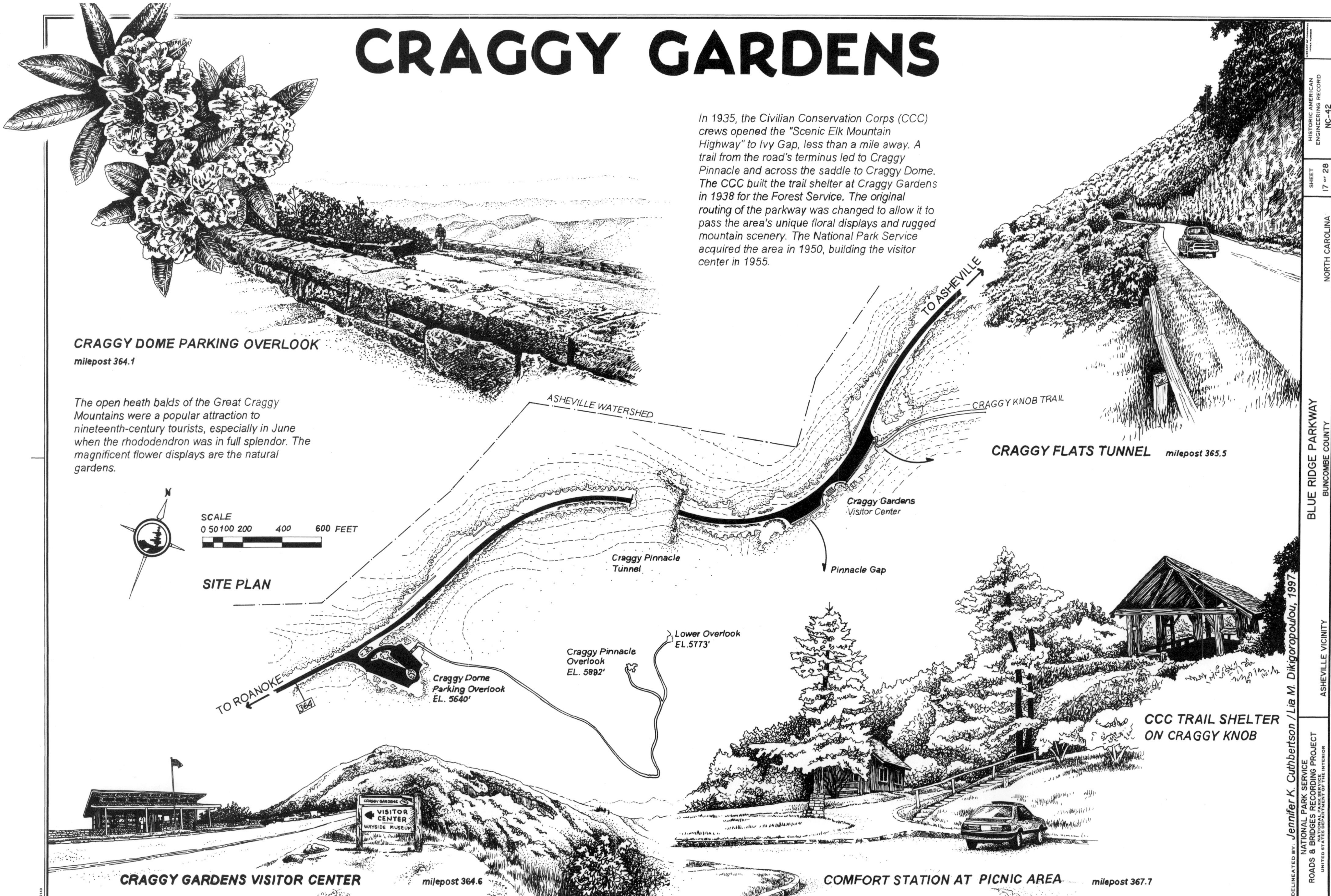
CRAGGY GARDENS
In 1935, the Civilian Conservation Corps (CCC) crews opened the "Scenic Elk Mountain Highway" to Ivy Gap, less than a mile away. A trail from the road's terminus led to Craggy Pinnacle and across the saddle to Craggy Dome. The CCC built the trail shelter at Craggy Gardens in 1938 for the Forest Service. The original routing of the parkway was changed to allow it to pass the area's unique floral displays and rugged mountain scenery. The National Park Service acquired the area in 1950, building the visitor center in 1955.
CRAGGY DOME PARKING OVERLOOK
milepost 364.1
The open heath balds of the Great Craggy Mountains were a popular attraction to nineteenth-century tourists, especially in June when the rhododendron was in full splendor. The magnificent flower displays are the natural gardens.
N
SCALE
0 50 100 200 400 600 FEET
SITE PLAN
ASHEVILLE WATERSHED
TO ASHEVILLE
CRAGGY KNOB TRAIL
CRAGGY FLATS TUNNEL
milepost 365.5
Craggy Gardens Visitor Center
Pinnacle Gap
Craggy Pinnacle Tunnel
Lower Overlook EL.5773'
Craggy Pinnacle Overlook EL. 5892'
Craggy Dome Parking Overlook EL. 5640'
TO ROANOKE
364
CCC TRAIL SHELTER ON CRAGGY KNOB
CRAGGY GARDENS
VISITOR CENTER
WAYSIDE MUSEUM
CRAGGY GARDENS VISITOR CENTER
milepost 364.6
COMFORT STATION AT PICNIC AREA
milepost 367.7
DELINEATED BY: Jennifer K. Cuthbertson / Lia M. Dikigoropoulou, 1997
NATIONAL PARK SERVICE
ROADS & BRIDGES RECORDING PROJECT
NATIONAL PARK SERVICE
UNITED STATES DEPARTMENT OF THE INTERIOR
ASHEVILLE VICINITY
BLUE RIDGE PARKWAY
BUNCOMBE COUNTY
NORTH CAROLINA
SHEET
17 OF 28
HISTORIC AMERICAN ENGINEERING RECORD
NC-42
IF REPRODUCED, PLEASE CREDIT: HISTORIC AMERICAN ENGINEERING RECORD, NATIONAL PARK SERVICE, NAME OF DELINEATOR, DATE OF THE DRAWING

MOUNT PISGAH

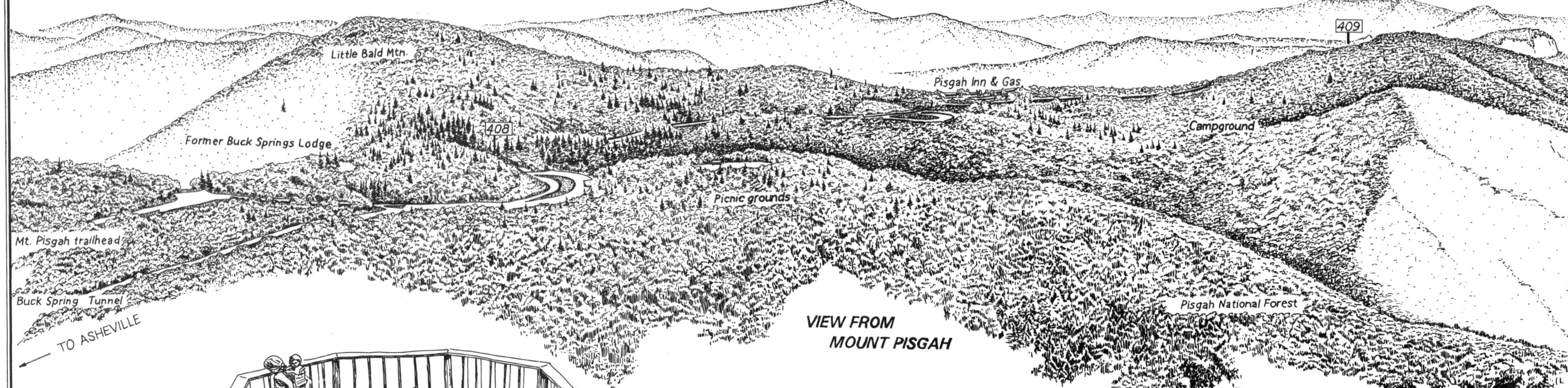

VIEW FROM MOUNT PISGAH

Once part of Biltmore owner George W. Vanderbilt's Pisgah Forest tract, this area was acquired by the United States Forest Service and became the nucleus of the Pisgah National Forest. Vanderbilt's farm superintendent, George Weston and his wife, Mary, established the Pisgah National Forest Inn on the site adjacent to the present Pisgah Inn in 1919. Concessionaires have operated here ever since. The campground was established by the forest service in the 1930s as the "Frying Pan Gap Camping Area." Due to years of deterioration, Vanderbilt's Buck Spring Lodge was removed prior to the transfer of the land to the NPS. The Pisgah National Forest Inn received only limited use after 1968, and was demolished in 1990. Remnants of the Buck Spring Lodge can still be seen along the Buck Spring Trail.

OBSERVATION PLATFORM ON MOUNT PISGAH *milepost 407.8*

CAMP SCENE *milepost 408.6*

BUCK SPRING TRAILHEAD *milepost 407.8*

CAMPSITE NESTLED IN THE RHODODENDRON *milepost 408.6*

VIEW FROM RESTAURANT AT PISGAH INN *milepost 408.5*

Ian Shanklin 1997

DELINEATED BY:
NATIONAL PARK SERVICE
ROADS & BRIDGES RECORDING PROJECT
NATIONAL PARK SERVICE
UNITED STATES DEPARTMENT OF THE INTERIOR

ASHEVILLE VICINITY
BLUE RIDGE PARKWAY
BUNCOMBE COUNTY
NORTH CAROLINA

HISTORIC AMERICAN ENGINEERING RECORD
NC-42
SHEET 18 OF 28

IF REPRODUCED, PLEASE CREDIT: HISTORIC AMERICAN ENGINEERING RECORD, NATIONAL PARK SERVICE, NAME OF DELINEATOR, DATE OF THE DRAWING

AGRICULTURAL LEASES

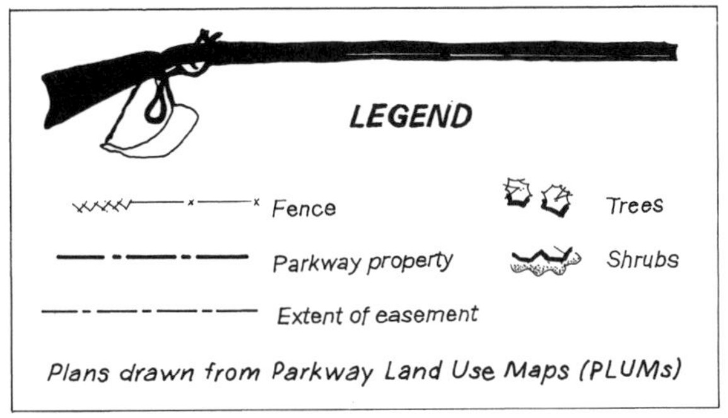

1" = 100'

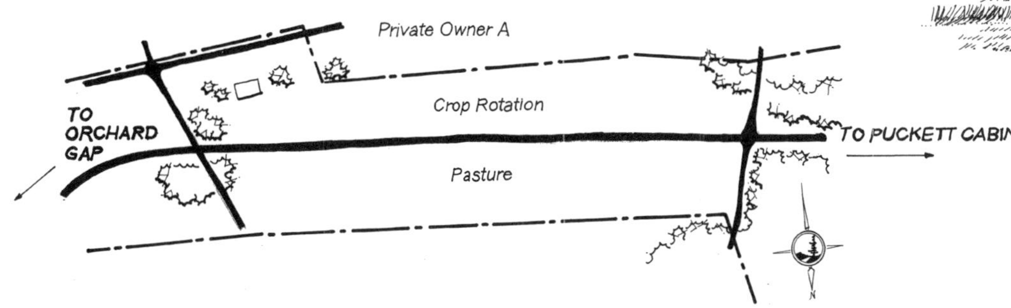

VOLUNTEER GAP milepost 192

With the narrow confines of the parkway, some unique solutions were needed to maintain the pastoral setting through which the roadway passed and to control future development. The parkway achieves this through agricultural leases and scenic easements. Along many portions, land is leased to individuals who use the land for pasturing animals, for hay fields, and for growing row crops. These provide both an agricultural setting along the road and openings to see out from the roadway. In certain sections, additional land is covered by scenic easements, which restrict the rights of the owner to develop their land or remove scenic landscape components. The successful implementation of these programs allows the view of the pastoral countryside to remain and the boundary lines to disappear.

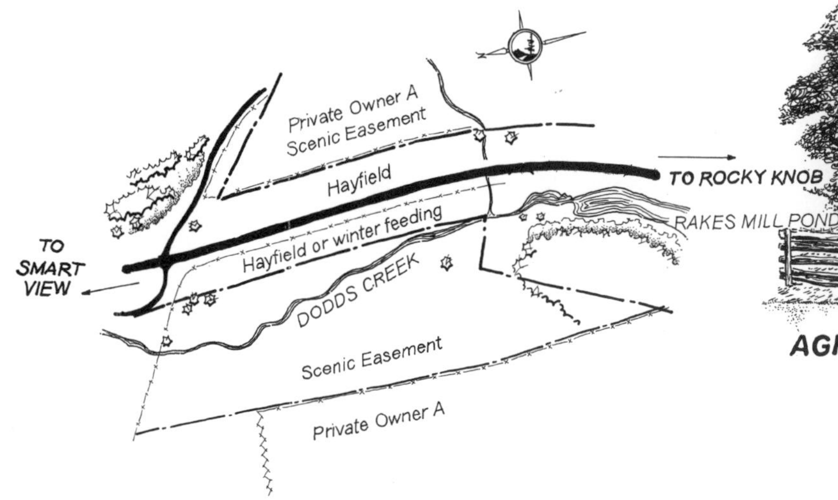

AGRICULTURE AT MILEPOST 161.5

BELCHER CURVE milepost 173

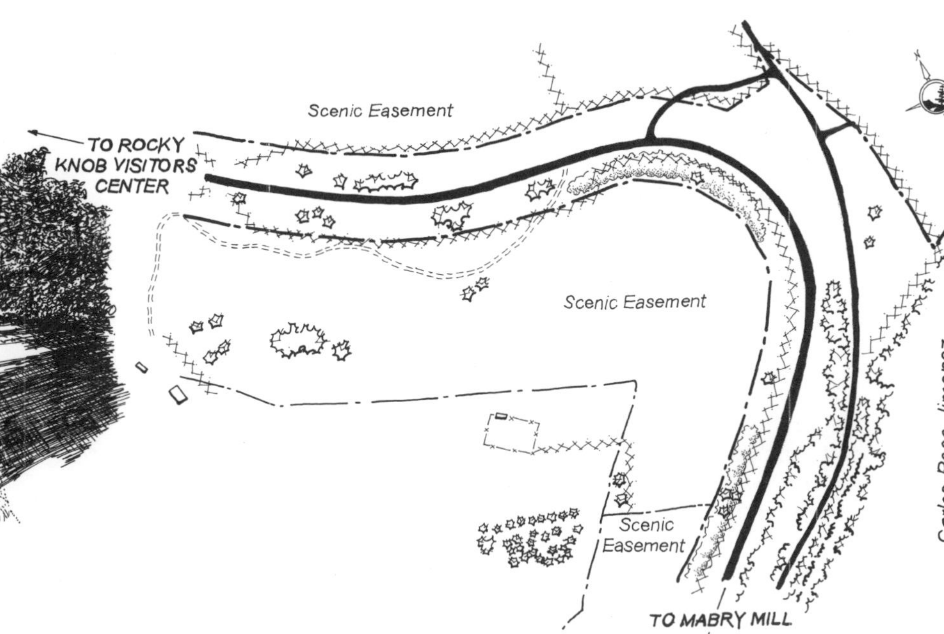

SCENIC EASEMENTS

HISTORIC AMERICAN ENGINEERING RECORD
NC-42

SHEET 19 of 28

BLUE RIDGE PARKWAY
BUNCOMBE COUNTY
NORTH CAROLINA

ASHEVILLE VICINITY

DELINEATED BY Carlos Rose Jimenez & Cheria Yost, 1997

NATIONAL PARK SERVICE
ROADS & BRIDGES RECORDING PROJECT
NATIONAL PARK SERVICE
UNITED STATES DEPARTMENT OF THE INTERIOR

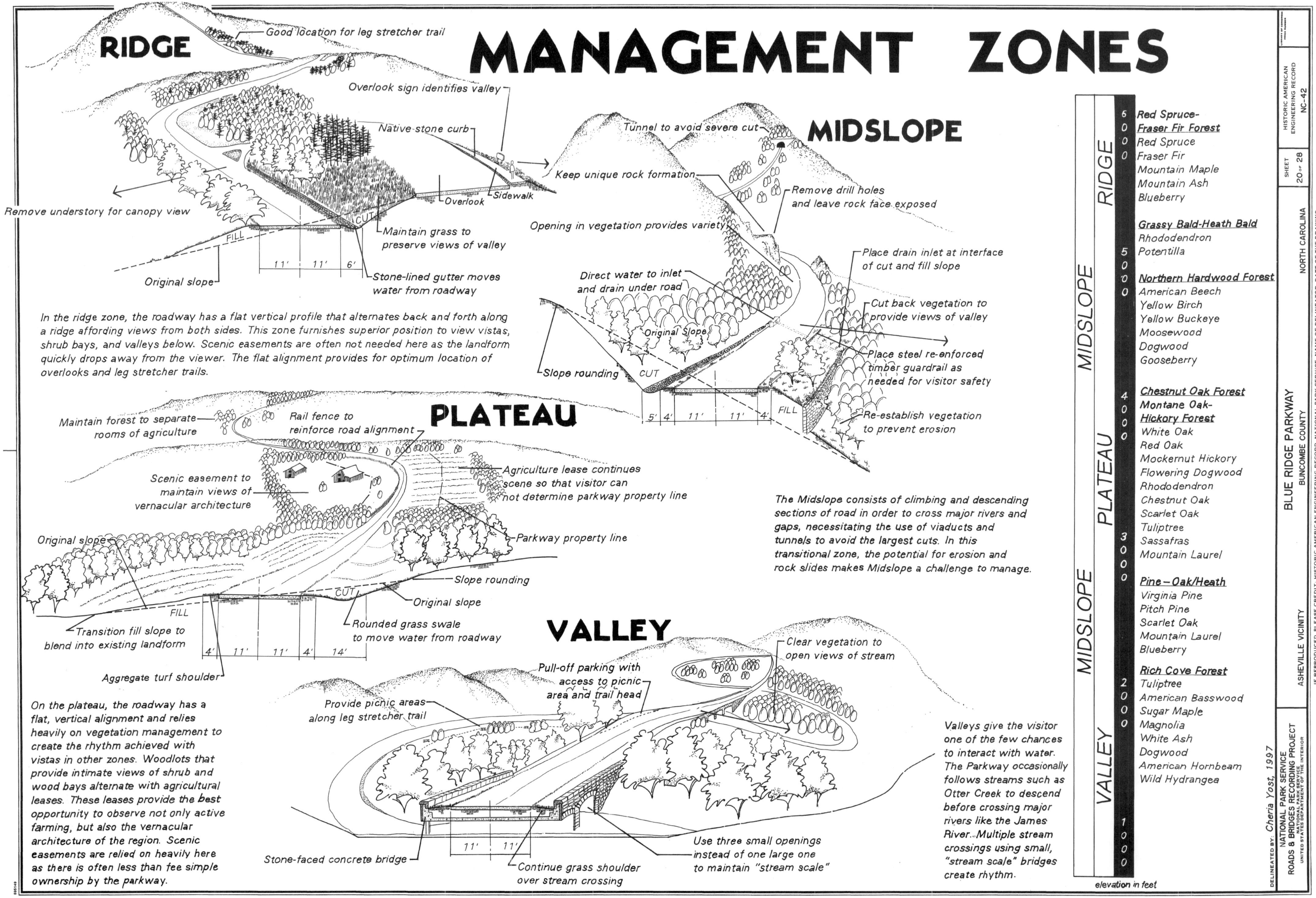

MANAGEMENT ZONES
RIDGE
Good location for leg stretcher trail
Overlook sign identifies valley
Native stone curb
Overlook
Sidewalk
Remove understory for canopy view
CUT
FILL
Maintain grass to preserve views of valley
11′ 11′ 6′
Original slope
Stone-lined gutter moves water from roadway
In the ridge zone, the roadway has a flat vertical profile that alternates back and forth along a ridge affording views from both sides. This zone furnishes superior position to view vistas, shrub bays, and valleys below. Scenic easements are often not needed here as the landform quickly drops away from the viewer. The flat alignment provides for optimum location of overlooks and leg stretcher trails.
MIDSLOPE
Tunnel to avoid severe cut
Keep unique rock formation
Remove drill holes and leave rock face exposed
Opening in vegetation provides variety
Place drain inlet at interface of cut and fill slope
Direct water to inlet and drain under road
Original Slope
Cut back vegetation to provide views of valley
Place steel re-enforced timber guardrail as needed for visitor safety
Slope rounding
CUT
FILL
5′ 4′ 11′ 11′ 4′
Re-establish vegetation to prevent erosion
The Midslope consists of climbing and descending sections of road in order to cross major rivers and gaps, necessitating the use of viaducts and tunnels to avoid the largest cuts. In this transitional zone, the potential for erosion and rock slides makes Midslope a challenge to manage.
PLATEAU
Maintain forest to separate rooms of agriculture
Rail fence to reinforce road alignment
Scenic easement to maintain views of vernacular architecture
Agriculture lease continues scene so that visitor can not determine parkway property line
Parkway property line
Original slope
Slope rounding
CUT
Original slope
FILL
Transition fill slope to blend into existing landform
Rounded grass swale to move water from roadway
4′ 11′ 11′ 4′ 14′
Aggregate turf shoulder
On the plateau, the roadway has a flat, vertical alignment and relies heavily on vegetation management to create the rhythm achieved with vistas in other zones. Woodlots that provide intimate views of shrub and wood bays alternate with agricultural leases. These leases provide the best opportunity to observe not only active farming, but also the vernacular architecture of the region. Scenic easements are relied on heavily here as there is often less than fee simple ownership by the parkway.
VALLEY
Pull-off parking with access to picnic area and trail head
Clear vegetation to open views of stream
Provide picnic areas along leg stretcher trail
11′ 11′
Stone-faced concrete bridge
Continue grass shoulder over stream crossing
Use three small openings instead of one large one to maintain "stream scale"
Valleys give the visitor one of the few chances to interact with water. The Parkway occasionally follows streams such as Otter Creek to descend before crossing major rivers like the James River. Multiple stream crossings using small, "stream scale" bridges create rhythm.
RIDGE
MIDSLOPE
PLATEAU
MIDSLOPE
VALLEY
6000
5000
4000
3000
2000
1000
elevation in feet
Red Spruce-Fraser Fir Forest
Red Spruce
Fraser Fir
Mountain Maple
Mountain Ash
Blueberry
Grassy Bald-Heath Bald
Rhododendron
Potentilla
Northern Hardwood Forest
American Beech
Yellow Birch
Yellow Buckeye
Moosewood
Dogwood
Gooseberry
Chestnut Oak Forest
Montane Oak-Hickory Forest
White Oak
Red Oak
Mockernut Hickory
Flowering Dogwood
Rhododendron
Chestnut Oak
Scarlet Oak
Tuliptree
Sassafras
Mountain Laurel
Pine–Oak/Heath
Virginia Pine
Pitch Pine
Scarlet Oak
Mountain Laurel
Blueberry
Rich Cove Forest
Tuliptree
American Basswood
Sugar Maple
Magnolia
White Ash
Dogwood
American Hornbeam
Wild Hydrangea
HISTORIC AMERICAN ENGINEERING RECORD
NC-42
SHEET 20 OF 28
BLUE RIDGE PARKWAY
NORTH CAROLINA
BUNCOMBE COUNTY
ASHEVILLE VICINITY
DELINEATED BY: Cheria Yost, 1997
NATIONAL PARK SERVICE
ROADS & BRIDGES RECORDING PROJECT
NATIONAL PARK SERVICE
UNITED STATES DEPARTMENT OF THE INTERIOR
IF REPRODUCED, PLEASE CREDIT: HISTORIC AMERICAN ENGINEERING RECORD, NATIONAL PARK SERVICE, NAME OF DELINEATOR, DATE OF THE DRAWING

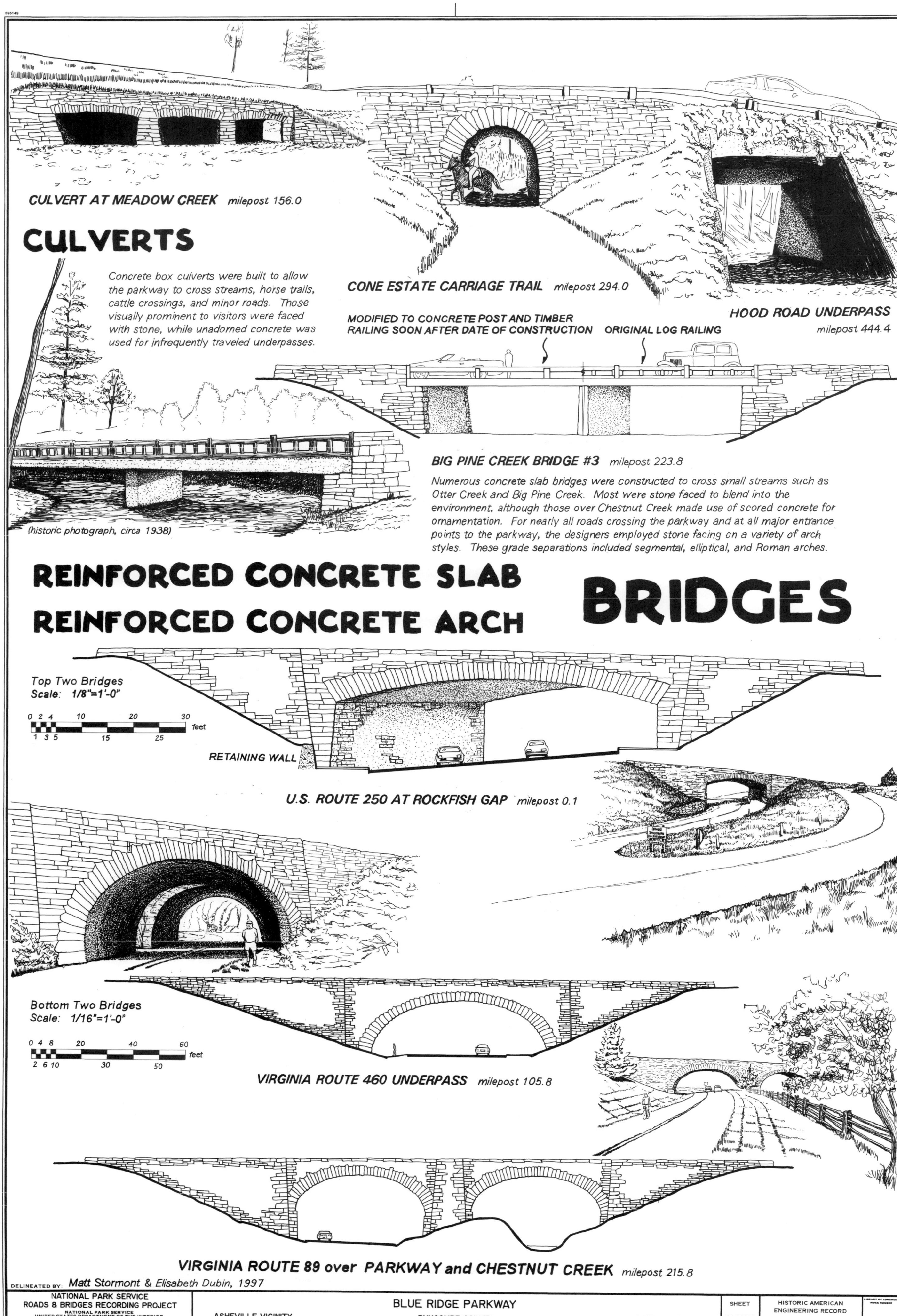

CULVERT AT MEADOW CREEK milepost 156.0
CULVERTS
Concrete box culverts were built to allow the parkway to cross streams, horse trails, cattle crossings, and minor roads. Those visually prominent to visitors were faced with stone, while unadorned concrete was used for infrequently traveled underpasses.
CONE ESTATE CARRIAGE TRAIL milepost 294.0
HOOD ROAD UNDERPASS milepost 444.4
MODIFIED TO CONCRETE POST AND TIMBER RAILING SOON AFTER DATE OF CONSTRUCTION
ORIGINAL LOG RAILING
(historic photograph, circa 1938)
BIG PINE CREEK BRIDGE #3 milepost 223.8
Numerous concrete slab bridges were constructed to cross small streams such as Otter Creek and Big Pine Creek. Most were stone faced to blend into the environment, although those over Chestnut Creek made use of scored concrete for ornamentation. For nearly all roads crossing the parkway and at all major entrance points to the parkway, the designers employed stone facing on a variety of arch styles. These grade separations included segmental, elliptical, and Roman arches.
REINFORCED CONCRETE SLAB
REINFORCED CONCRETE ARCH
BRIDGES
Top Two Bridges
Scale: 1/8"=1'-0"
0 2 4 10 20 30 feet
1 3 5 15 25
RETAINING WALL
U.S. ROUTE 250 AT ROCKFISH GAP milepost 0.1
Bottom Two Bridges
Scale: 1/16"=1'-0"
0 4 8 20 40 60 feet
2 6 10 30 50
VIRGINIA ROUTE 460 UNDERPASS milepost 105.8
VIRGINIA ROUTE 89 over PARKWAY and CHESTNUT CREEK milepost 215.8
DELINEATED BY: Matt Stormont & Elisabeth Dubin, 1997
NATIONAL PARK SERVICE
ROADS & BRIDGES RECORDING PROJECT
NATIONAL PARK SERVICE
UNITED STATES DEPARTMENT OF THE INTERIOR
BLUE RIDGE PARKWAY
ASHEVILLE VICINITY
BUNCOMBE COUNTY
NORTH CAROLINA
SHEET 21 OF 28
HISTORIC AMERICAN ENGINEERING RECORD
NC-42
IF REPRODUCED, PLEASE CREDIT: HISTORIC AMERICAN ENGINEERING RECORD, NATIONAL PARK SERVICE, NAME OF DELINEATOR, DATE OF THE DRAWING

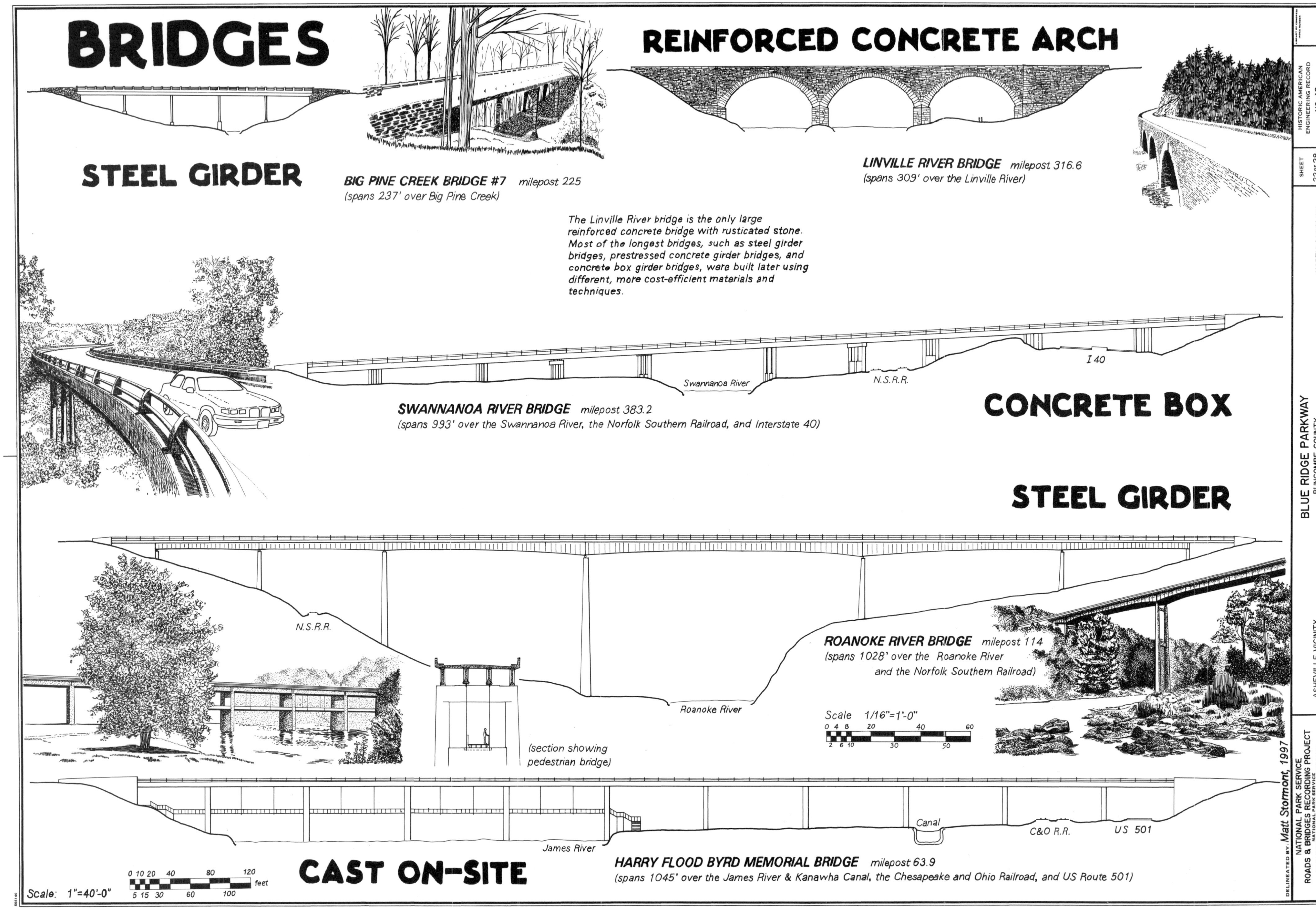

BRIDGES
STEEL GIRDER
BIG PINE CREEK BRIDGE #7 milepost 225
(spans 237' over Big Pine Creek)
REINFORCED CONCRETE ARCH
LINVILLE RIVER BRIDGE milepost 316.6
(spans 309' over the Linville River)
The Linville River bridge is the only large reinforced concrete bridge with rusticated stone. Most of the longest bridges, such as steel girder bridges, prestressed concrete girder bridges, and concrete box girder bridges, were built later using different, more cost-efficient materials and techniques.
Swannanoa River
N.S.R.R.
I 40
SWANNANOA RIVER BRIDGE milepost 383.2
(spans 993' over the Swannanoa River, the Norfolk Southern Railroad, and Interstate 40)
CONCRETE BOX
STEEL GIRDER
N.S.R.R.
Roanoke River
ROANOKE RIVER BRIDGE milepost 114
(spans 1028' over the Roanoke River and the Norfolk Southern Railroad)
Scale 1/16"=1'-0"
0 4 8 20 40 60
2 6 10 30 50
(section showing pedestrian bridge)
Canal
C&O R.R.
US 501
James River
CAST ON-SITE
HARRY FLOOD BYRD MEMORIAL BRIDGE milepost 63.9
(spans 1045' over the James River & Kanawha Canal, the Chesapeake and Ohio Railroad, and US Route 501)
Scale: 1"=40'-0"
0 10 20 40 80 120
5 15 30 60 100
feet
HISTORIC AMERICAN ENGINEERING RECORD
NC-42
SHEET 22 OF 28
NORTH CAROLINA
BLUE RIDGE PARKWAY
BUNCOMBE COUNTY
ASHEVILLE VICINITY
NATIONAL PARK SERVICE
ROADS & BRIDGES RECORDING PROJECT
UNITED STATES DEPARTMENT OF THE INTERIOR
DELINEATED BY: Matt Stormont, 1997
IF REPRODUCED, PLEASE CREDIT: HISTORIC AMERICAN ENGINEERING RECORD, NATIONAL PARK SERVICE, NAME OF DELINEATOR, DATE OF THE DRAWING

VIADUCTS

Masonry pier arch and steel girder

Concrete box and concrete pier

WILSON CREEK

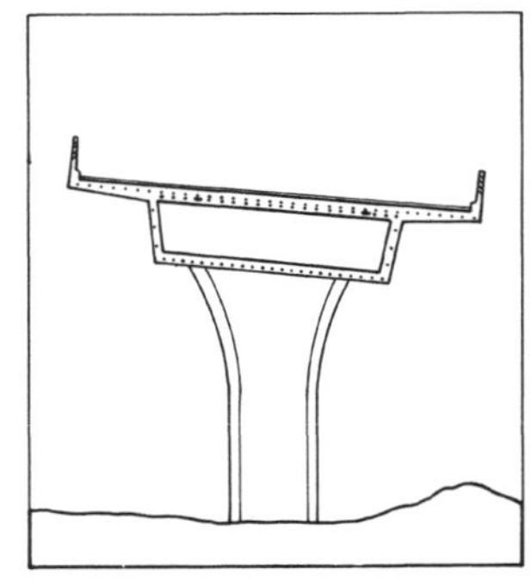

Viaducts are bridges which cross ravines with little or no water or are used to eliminate switch back and steep grades. Their construction eliminates massive rock cuts and fills which would be necessary on steep slopes, helping to protect the natural landscape. A variety of forms have been used depending on a site's conditions and its visibility from other sections of the Blue Ridge Parkway.

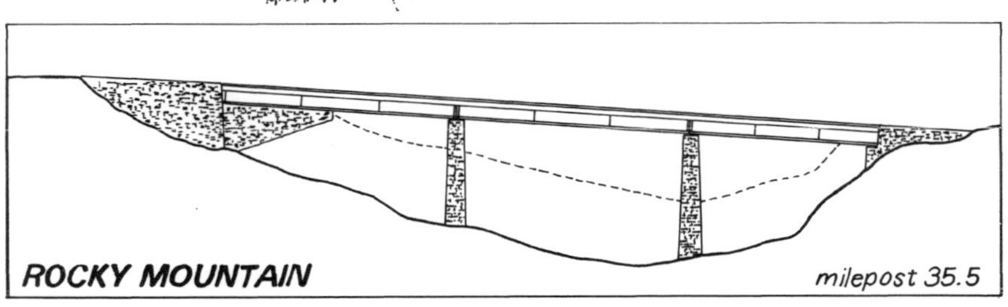

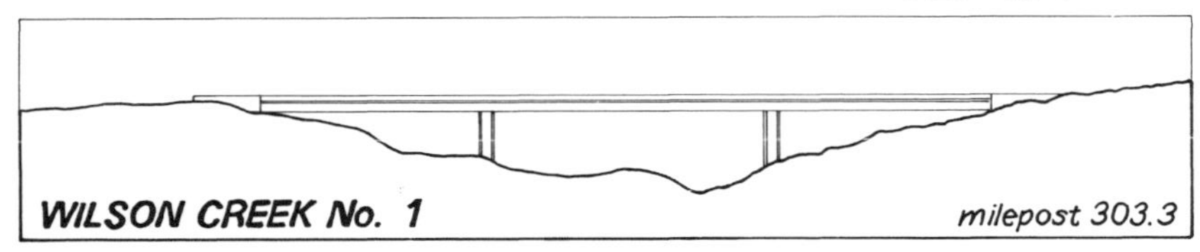

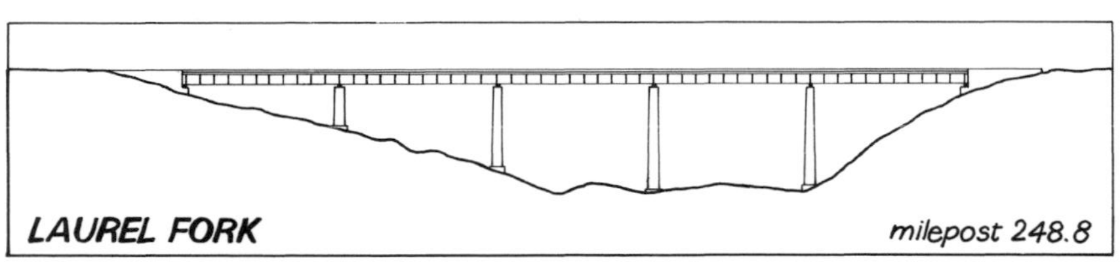

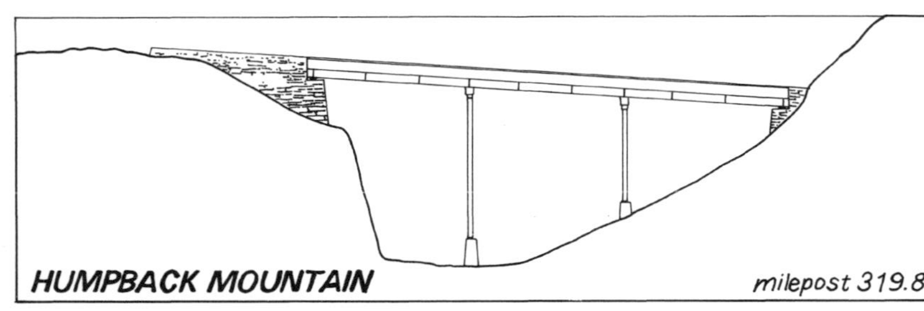

Concrete arch pier and steel girder

Steel girder and steel pier

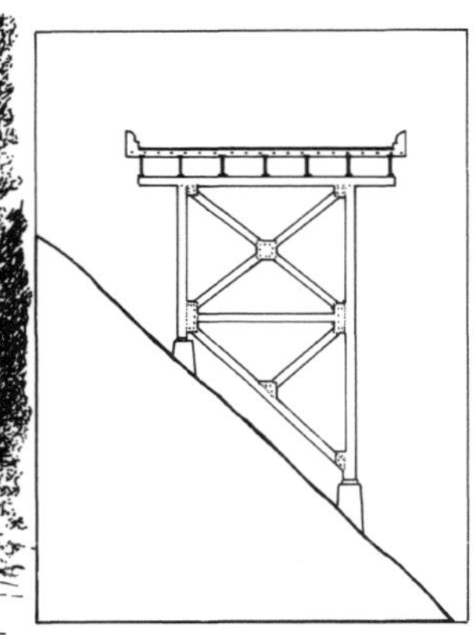

Steel girder and stone faced concrete pier

Steel girder and concrete pier

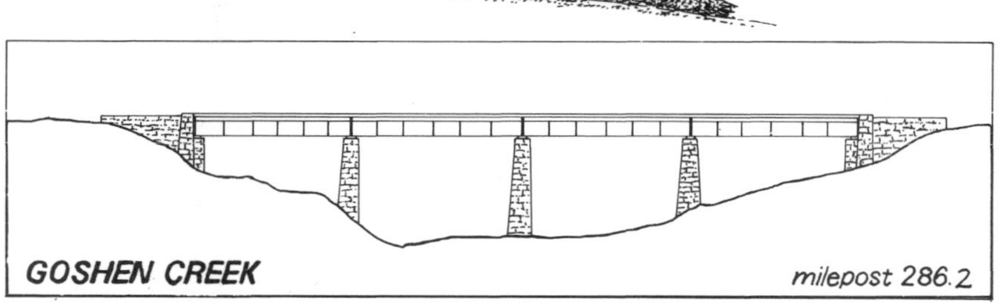

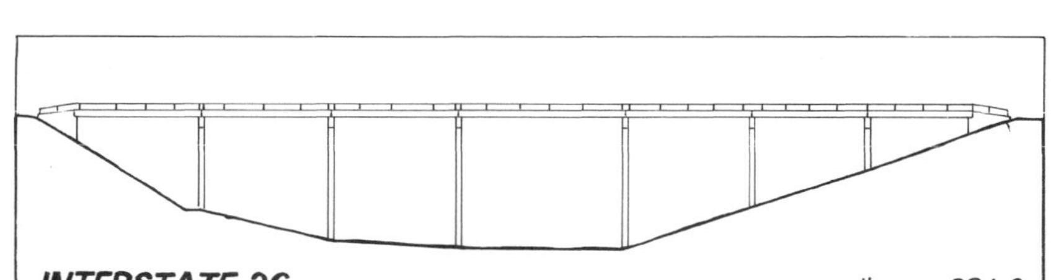

DELINEATED BY: Carlos Rosa Jimenez, 1997

NATIONAL PARK SERVICE ROADS & BRIDGES RECORDING PROJECT NATIONAL PARK SERVICE UNITED STATES DEPARTMENT OF THE INTERIOR	BLUE RIDGE PARKWAY ASHEVILLE VICINITY — BUNCOMBE COUNTY — NORTH CAROLINA	SHEET 23 OF 28	HISTORIC AMERICAN ENGINEERING RECORD NC-42

TUNNELS

In keeping with the design standard of minimizing impacts to the land, tunnels are used in areas of steep terrain with ridges that run perpendicular to the roadway alignment. Tunnels are used more extensively in North Carolina (twenty five times) due to more rugged terrain versus Virginia which only has one tunnel.
The construction of the tunnels was a costly and potentially hazardous part of building the road. Tunnels were excavated in the prewar period using compressed air drills and dynamite. Later, the use of a jumbo rig made it possible to drill and blast an entire tunnel face in one operation. Tunnel lengths range from 150 feet at Rough Ridge to1434 feet at Pine Mountain and all but three have a minimum height of 13 feet. Lining was first used in the Devil's Courthouse Tunnel due to tunnel cave-ins during construction. Approximately one quarter of the structure was subsequently lined with concrete. Acting Superintendent Abbuehl observed a noticeable improvement in lighting on the lined section, a point he felt favored lining of other structures. An additional benefit was the elimination of much of the moisture entering the tunnels which caused ice problems in the winter.

NORTH PORTAL OF THE SOUTH TUNNEL
milepost 344.7

LITTLE SWITZERLAND TUNNEL
milepost 333.4

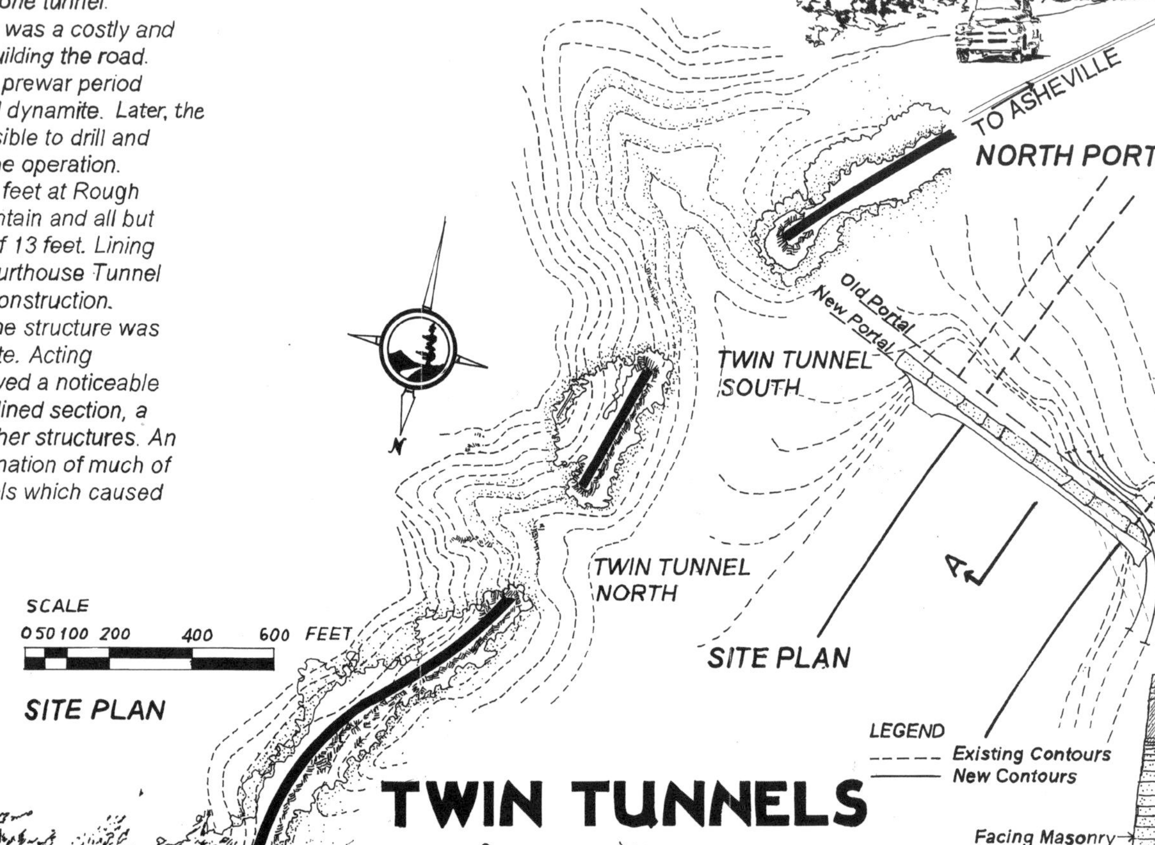

SITE PLAN

SITE PLAN

TWIN TUNNELS

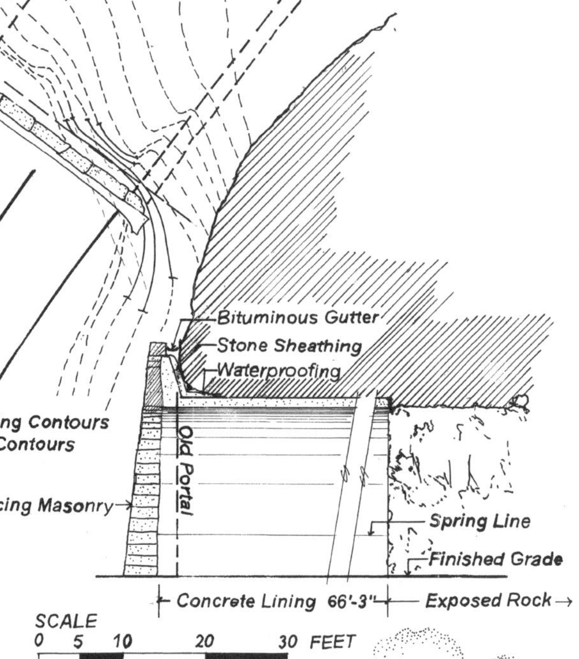

SECTION A-A

NORTH PORTAL OF THE NORTH TUNNEL
milepost 344.5

Masonry portals were added in the 1950s to eliminate rock and ice falls at the tunnel openings.

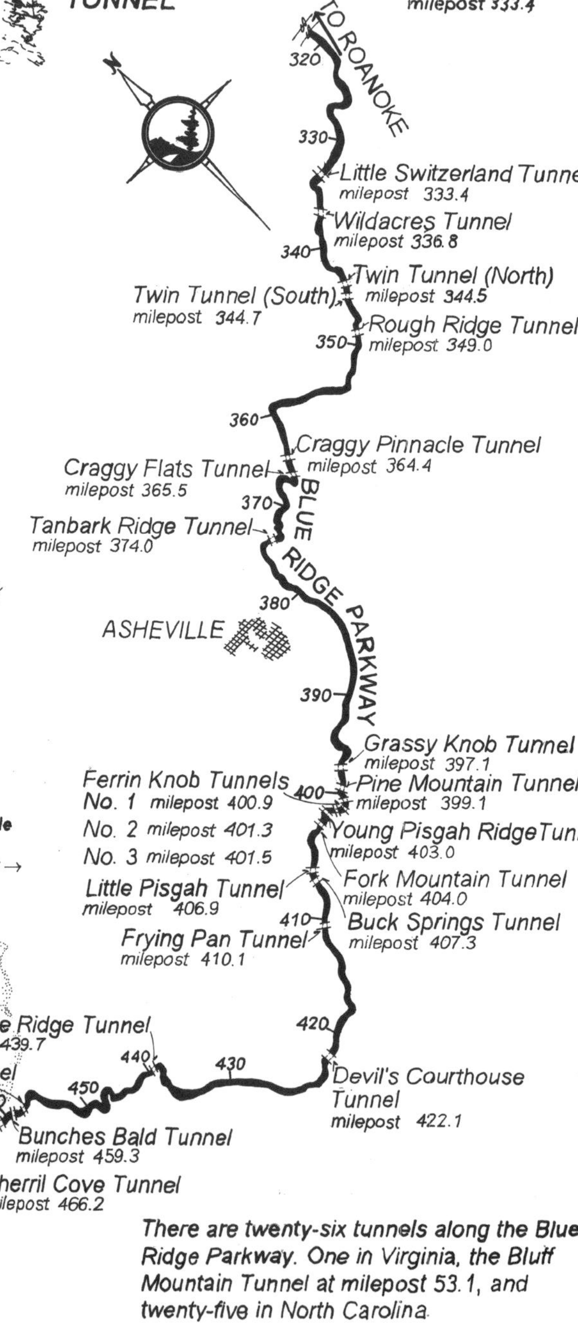

LOCATOR MAP

There are twenty-six tunnels along the Blue Ridge Parkway. One in Virginia, the Bluff Mountain Tunnel at milepost 53.1, and twenty-five in North Carolina.

DELINEATED BY: Lia M. Dikigoropoulou, 1997

NATIONAL PARK SERVICE
ROADS & BRIDGES RECORDING PROJECT
UNITED STATES DEPARTMENT OF THE INTERIOR

BLUE RIDGE PARKWAY
BUNCOMBE COUNTY
ASHEVILLE VICINITY
NORTH CAROLINA

SHEET 24 OF 28

HISTORIC AMERICAN ENGINEERING RECORD
NC-42

SIGNS

REGULATORY

INTERPRETIVE

Blue Ridge Mountains
Parkway
Lone Virginia Pine
Parkway Blue "to simulate the deep blue of the mountains in the late evening"
Driftwood Gray Stain

INFORMATION

ENTRY SIGN, 1948
(undated BLRI Archives photo)

Signs, an integral design element of the Blue Ridge Parkway, are divided into a hierarchy by the information provided. A number of design changes have occurred over time. Originally, all signs were routed of wood. Today, several materials are used to provide needed information.

Regulatory *signs are the most standardized signs providing information regarding safety and directions to connecting routes. They all meet the standards as set forth in the Manual on Uniform Traffic Control Devices.*

Information *signs provide primary direction to services and locations. They have evolved from wood routed signs to painted plywood with attached letters and symbols to match NPS guidelines. However, there are still many wood routed signs to keep the original design character of the parkway.*

Interpretive *signs provide the "story behind the scenery." By conveying details about what is seen, they provide a better understanding of the cultural resources found along the parkway. There are many sign types from wood routed "gunboards" to modern day "modulite" panels.*

GUN BOARD, 1941

CURRENT ENTRY SIGN, *from 1949*

LIFT-TOP EASEL, 1950s
(undated BLRI Archives photo)

TAVERN-STYLE SIGN, 1941
(Photo 1952)

EARLY STOP SIGN
(Photo 1949)

SELF-GUIDED TRAIL SIGN, 1970s

MILE POST MARKER, 1947
(Photo 1953)

MODERN STOP SIGN

MODULITE MARKER, 1984
FABRICATED AT HARPER'S FERRY CENTER

TRAIL SIGN, 1940s

EARLY ACCESS SIGN
(Photo 1953)

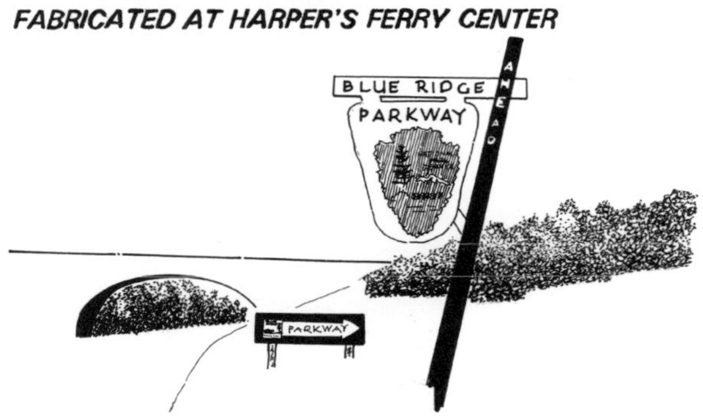

1960s ACCESS SIGN *(undated BLRI Archives photo)*

MODERN ACCESS SIGN

DELINEATED BY: *Cheria Yost and Matt Stormont, 1997*

NATIONAL PARK SERVICE ROADS & BRIDGES RECORDING PROJECT NATIONAL PARK SERVICE UNITED STATES DEPARTMENT OF THE INTERIOR	ASHEVILLE VICINITY	BLUE RIDGE PARKWAY BUNCOMBE COUNTY	NORTH CAROLINA	SHEET 25 OF 28	HISTORIC AMERICAN ENGINEERING RECORD NC-42

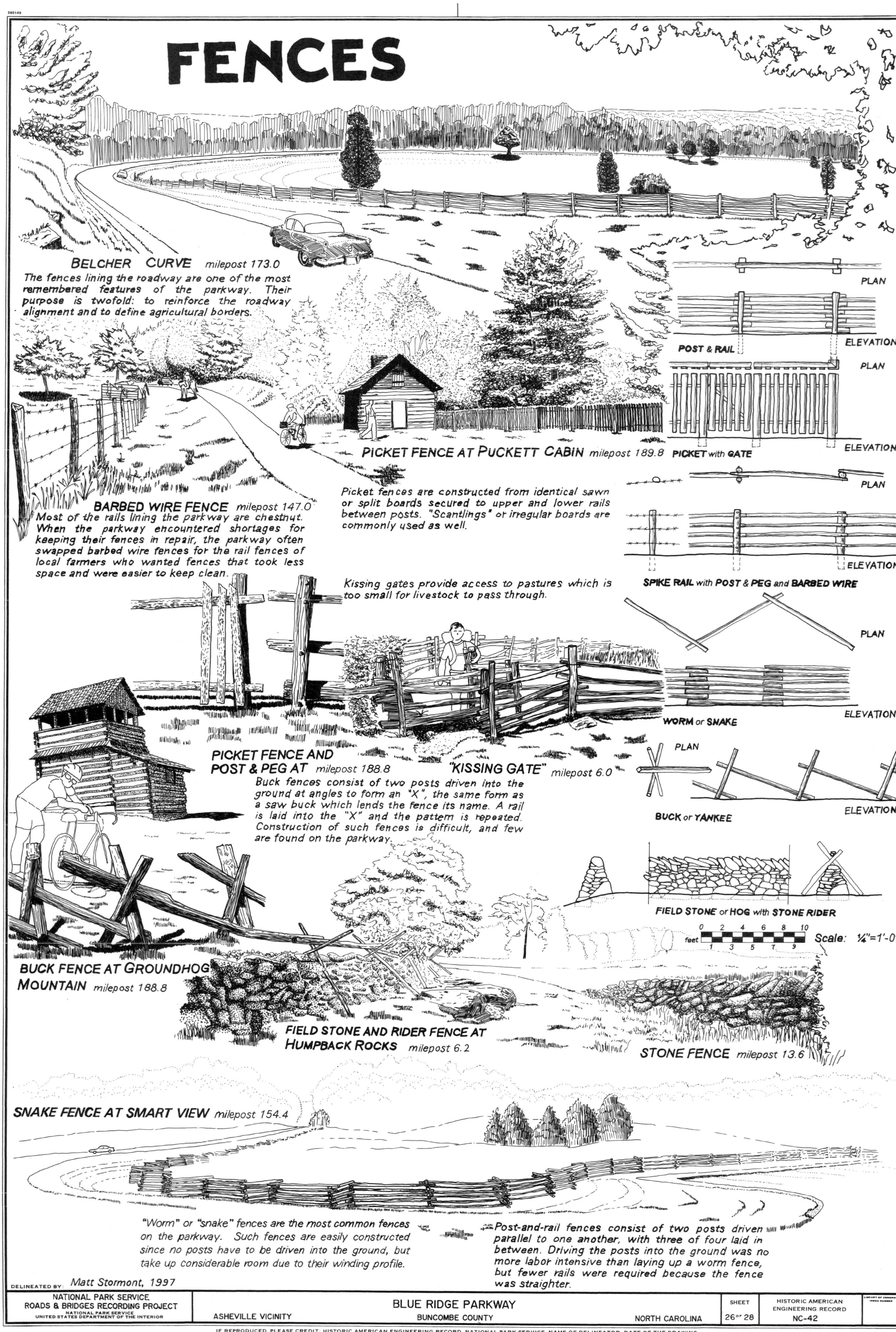
FENCES
BELCHER CURVE milepost 173.0
The fences lining the roadway are one of the most remembered features of the parkway. Their purpose is twofold: to reinforce the roadway alignment and to define agricultural borders.
PLAN
ELEVATION
POST & RAIL
PLAN
ELEVATION
PICKET with GATE
PICKET FENCE AT PUCKETT CABIN milepost 189.8
PLAN
ELEVATION
SPIKE RAIL with POST & PEG and BARBED WIRE
BARBED WIRE FENCE milepost 147.0
Most of the rails lining the parkway are chestnut. When the parkway encountered shortages for keeping their fences in repair, the parkway often swapped barbed wire fences for the rail fences of local farmers who wanted fences that took less space and were easier to keep clean.
Picket fences are constructed from identical sawn or split boards secured to upper and lower rails between posts. "Scantlings" or irregular boards are commonly used as well.
Kissing gates provide access to pastures which is too small for livestock to pass through.
PLAN
ELEVATION
WORM or SNAKE
PICKET FENCE AND POST & PEG AT milepost 188.8
"KISSING GATE" milepost 6.0
PLAN
ELEVATION
BUCK or YANKEE
Buck fences consist of two posts driven into the ground at angles to form an "X", the same form as a saw buck which lends the fence its name. A rail is laid into the "X" and the pattern is repeated. Construction of such fences is difficult, and few are found on the parkway.
FIELD STONE or HOG with STONE RIDER
0 2 4 6 8 10
feet
1 3 5 7 9
Scale: ¼"=1'-0"
BUCK FENCE AT GROUNDHOG MOUNTAIN milepost 188.8
FIELD STONE AND RIDER FENCE AT HUMPBACK ROCKS milepost 6.2
STONE FENCE milepost 13.6
SNAKE FENCE AT SMART VIEW milepost 154.4
"Worm" or "snake" fences are the most common fences on the parkway. Such fences are easily constructed since no posts have to be driven into the ground, but take up considerable room due to their winding profile.
Post-and-rail fences consist of two posts driven parallel to one another, with three of four laid in between. Driving the posts into the ground was no more labor intensive than laying up a worm fence, but fewer rails were required because the fence was straighter.
DELINEATED BY: Matt Stormont, 1997
NATIONAL PARK SERVICE
ROADS & BRIDGES RECORDING PROJECT
NATIONAL PARK SERVICE
UNITED STATES DEPARTMENT OF THE INTERIOR
BLUE RIDGE PARKWAY
ASHEVILLE VICINITY
BUNCOMBE COUNTY
NORTH CAROLINA
SHEET
26 of 28
HISTORIC AMERICAN ENGINEERING RECORD
NC-42
IF REPRODUCED, PLEASE CREDIT: HISTORIC AMERICAN ENGINEERING RECORD, NATIONAL PARK SERVICE, NAME OF DELINEATOR, DATE OF THE DRAWING

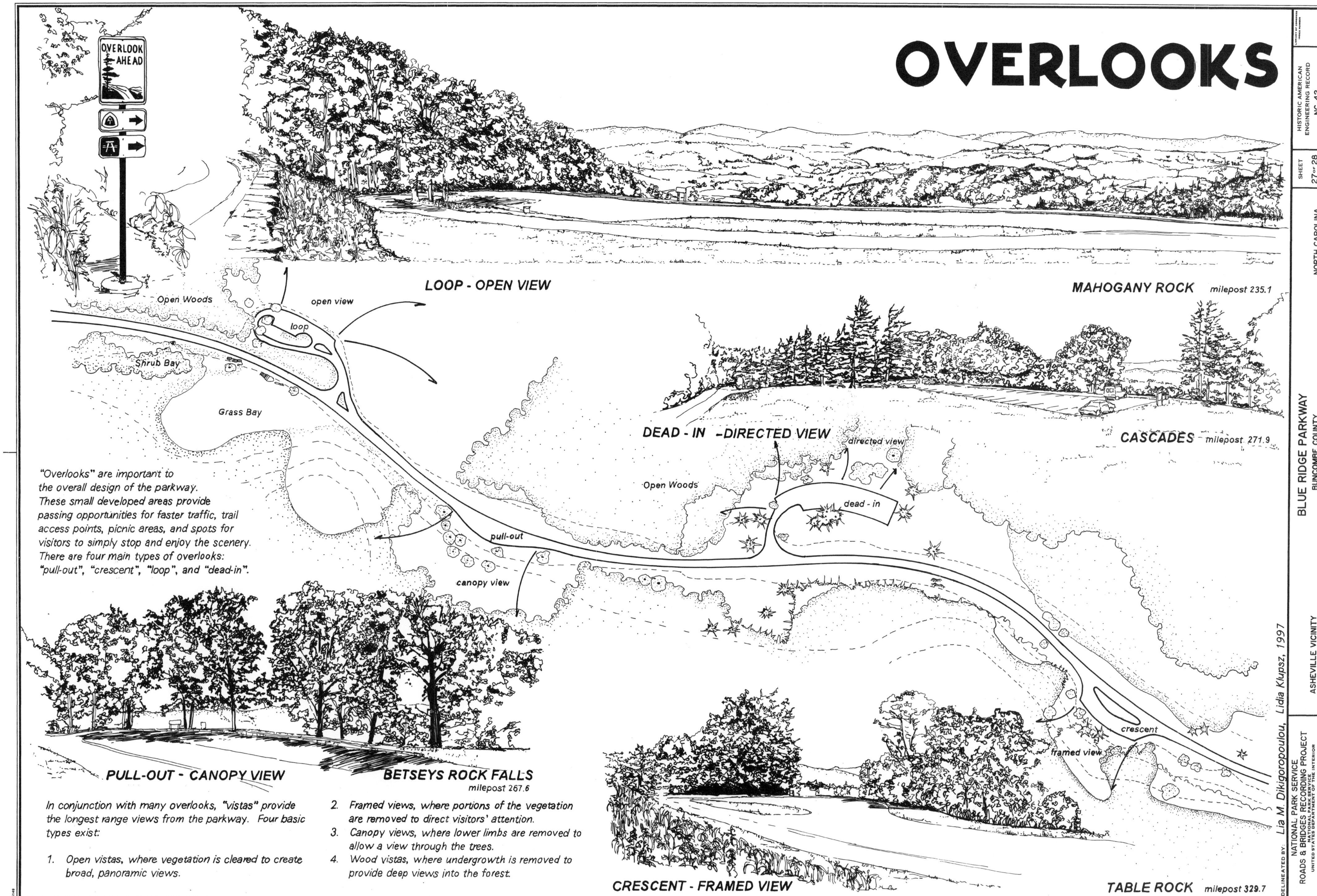
OVERLOOKS
OVERLOOK AHEAD
LOOP - OPEN VIEW
MAHOGANY ROCK milepost 235.1
Open Woods
open view
loop
Shrub Bay
Grass Bay
DEAD - IN -DIRECTED VIEW
directed view
Open Woods
dead - in
CASCADES milepost 271.9
pull-out
canopy view
"Overlooks" are important to the overall design of the parkway. These small developed areas provide passing opportunities for faster traffic, trail access points, picnic areas, and spots for visitors to simply stop and enjoy the scenery. There are four main types of overlooks: "pull-out", "crescent", "loop", and "dead-in".
PULL-OUT - CANOPY VIEW
BETSEYS ROCK FALLS milepost 267.6
crescent
framed view
In conjunction with many overlooks, "vistas" provide the longest range views from the parkway. Four basic types exist:
1. Open vistas, where vegetation is cleared to create broad, panoramic views.
2. Framed views, where portions of the vegetation are removed to direct visitors' attention.
3. Canopy views, where lower limbs are removed to allow a view through the trees.
4. Wood vistas, where undergrowth is removed to provide deep views into the forest.
CRESCENT - FRAMED VIEW
TABLE ROCK milepost 329.7
DELINEATED BY: Lia M. Dikigoropoulou, Lidia Klupsz, 1997
NATIONAL PARK SERVICE
ROADS & BRIDGES RECORDING PROJECT
NATIONAL PARK SERVICE
UNITED STATES DEPARTMENT OF THE INTERIOR
ASHEVILLE VICINITY
BLUE RIDGE PARKWAY
BUNCOMBE COUNTY
NORTH CAROLINA
SHEET 27 OF 28
HISTORIC AMERICAN ENGINEERING RECORD
NC-42
IF REPRODUCED, PLEASE CREDIT: HISTORIC AMERICAN ENGINEERING RECORD, NATIONAL PARK SERVICE, NAME OF DELINEATOR, DATE OF THE DRAWING

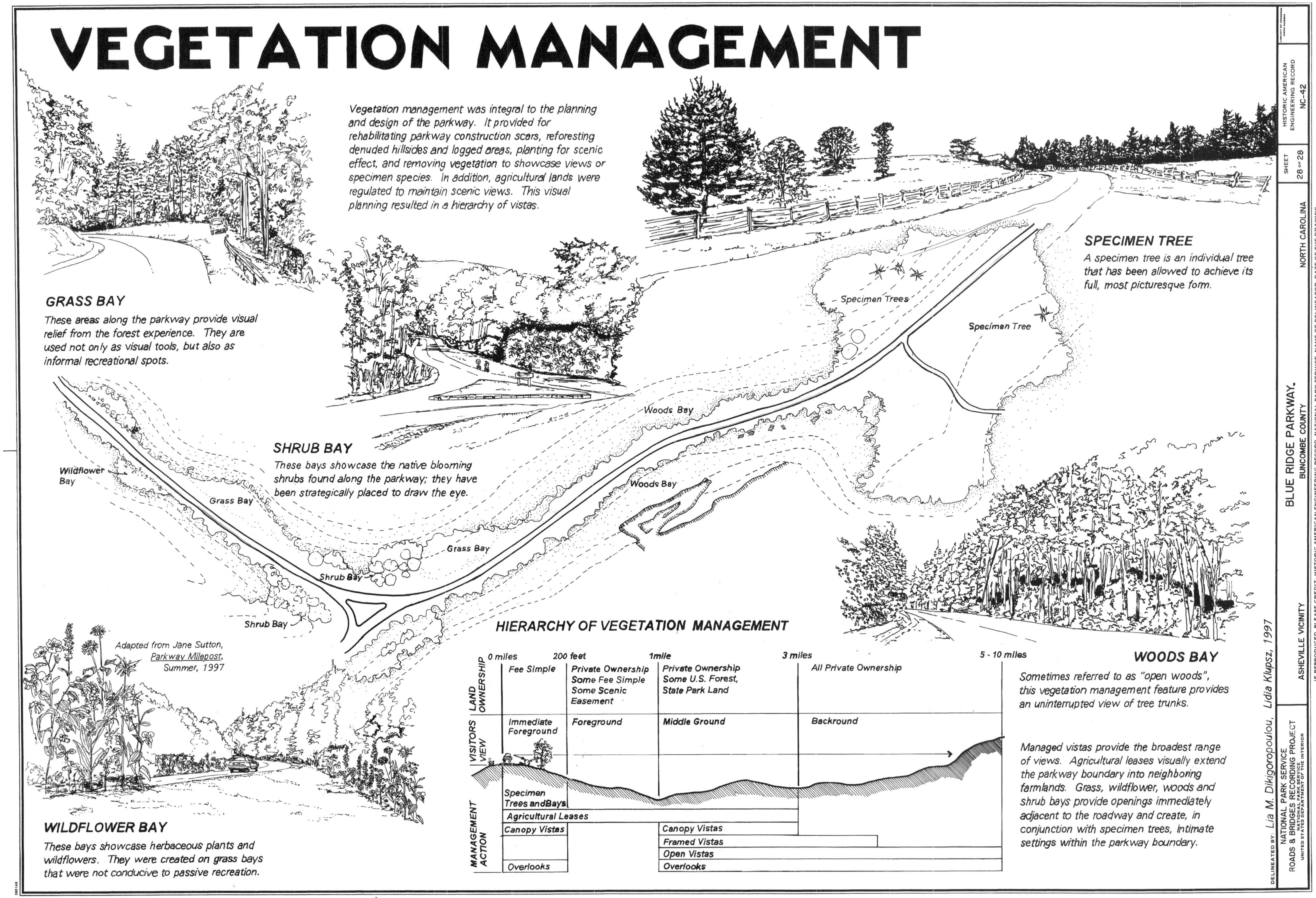
VEGETATION MANAGEMENT
Vegetation management was integral to the planning and design of the parkway. It provided for rehabilitating parkway construction scars, reforesting denuded hillsides and logged areas, planting for scenic effect, and removing vegetation to showcase views or specimen species. In addition, agricultural lands were regulated to maintain scenic views. This visual planning resulted in a hierarchy of vistas.
GRASS BAY
These areas along the parkway provide visual relief from the forest experience. They are used not only as visual tools, but also as informal recreational spots.
SHRUB BAY
These bays showcase the native blooming shrubs found along the parkway; they have been strategically placed to draw the eye.
SPECIMEN TREE
A specimen tree is an individual tree that has been allowed to achieve its full, most picturesque form.
Specimen Trees
Specimen Tree
Woods Bay
Woods Bay
Wildflower Bay
Grass Bay
Grass Bay
Shrub Bay
Shrub Bay
Adapted from Jane Sutton, Parkway Milepost, Summer, 1997
WILDFLOWER BAY
These bays showcase herbaceous plants and wildflowers. They were created on grass bays that were not conducive to passive recreation.
HIERARCHY OF VEGETATION MANAGEMENT
0 miles
200 feet
1mile
3 miles
5 - 10 miles
LAND OWNERSHIP
Fee Simple
Private Ownership Some Fee Simple Some Scenic Easement
Private Ownership Some U.S. Forest, State Park Land
All Private Ownership
VISITORS VIEW
Immediate Foreground
Foreground
Middle Ground
Backround
MANAGEMENT ACTION
Specimen Trees and Bays
Agricultural Leases
Canopy Vistas
Canopy Vistas
Framed Vistas
Open Vistas
Overlooks
Overlooks
WOODS BAY
Sometimes referred to as "open woods", this vegetation management feature provides an uninterrupted view of tree trunks.
Managed vistas provide the broadest range of views. Agricultural leases visually extend the parkway boundary into neighboring farmlands. Grass, wildflower, woods and shrub bays provide openings immediately adjacent to the roadway and create, in conjunction with specimen trees, intimate settings within the parkway boundary.
DELINEATED BY: Lia M. Dikigoropoulou, Lidia Klupsz, 1997
NATIONAL PARK SERVICE
ROADS & BRIDGES RECORDING PROJECT
NATIONAL PARK SERVICE
UNITED STATES DEPARTMENT OF THE INTERIOR
ASHEVILLE VICINITY
BLUE RIDGE PARKWAY.
BUNCOMBE COUNTY
NORTH CAROLINA
SHEET 28 OF 28
HISTORIC AMERICAN ENGINEERING RECORD
NC-42
IF REPRODUCED, PLEASE CREDIT: HISTORIC AMERICAN ENGINEERING RECORD, NATIONAL PARK SERVICE, NAME OF DELINEATOR, DATE OF THE DRAWING

COLONIAL PARKWAY

COLONIAL NATIONAL HISTORICAL PARK, VIRGINIA

1931

Established in 1930, Colonial National Historical Park is a 10,221 acre unit of the National Park Service located on the "Lower Neck" of Virginia between the York and the James rivers. The region is characterized by marine and freshwater wetlands, although there are also a variety of pine and hardwood forests, agricultural fields, rivers, ponds, and coastal bluffs. But it is not nature that attracts most visitors to the park; it is the history and cultural landscape of Tidewater Virginia. Within an area of about twenty miles a visitor can see the sites of Jamestown, the first permanent English settlement in North America founded in 1607; Williamsburg, Virginia's colonial capital between 1699 and 1780; and Yorktown, an important colonial tobacco port and site of the surrender of General Cornwallis' forces in 1781.

The historical relationship and geographic proximity of this "historic triangle" has attracted the attention of historians and antiquarians since the late nineteenth century. In the language of Victorian America, Jamestown, Williamsburg, and Yorktown were "sacred shrines" of the American experience. Since the 1880s and 1890s, civic groups such as the Association for the Preservation of Virginia Antiquities (APVA) actively pursued the conservation of the colonial landscape as a symbolic reminder of Virginia's traditional society. By the 1920s, however, communities and state agencies like the Virginia Conservation and Development Commission under William Carson began to promote heritage as a way to foster economic growth.

The 1920s marked a period of expansion for the National Park Service under Stephen Mather and Horace Albright. In an attempt to gain broader public and congressional support, NPS administrators increased activities in the east. In 1930, Congress authorized the establishment of a historical park to include Jamestown Island and the Yorktown battlefield, with ties to Williamsburg which was being restored with funds from John D. Rockefeller, Jr. Originally designated as a national monument (later becoming a national historical park in 1936), the park represents one of the National Park Service's first ventures into the field of historic preservation where many of the standards that guide programs today were developed.

COLONIAL PARKWAY CROSSING UNDER PARKWAY DRIVE BRIDGE

The original bill, written by Congressman Louis Cramton of Michigan, included the construction of a "Colonial Parkway" to link the sites and provide a scenic corridor for the increasing numbers of motoring visitors to America's national parks. By the 1920s, automobile access was a primary concern for Park Service administrators and landscape architects who had to develop plans which balanced use and conservation. In 1926, an interbureau agreement between the NPS and the Bureau of Public Roads, an agency of the Department of Agriculture, provided parks with the most modern engineering practices available through their direct supervision of construction projects. From this point on, the development of park roads and bridges was a collaborative effort between NPS landscape architects and BPR highway engineers.

Along with other NPS parkways constructed in the 1930s, the Colonial Parkway marks an important development in Park Service road-building programs. The Eastern Division of the Branch of Plans and Design under Charles E. Peterson, and the engineers of the BPR integrated parkway design principles standardized in Westchester County, New York, with the Park Service's own traditions of landscape architecture. The Colonial Parkway is a continuous strip of concrete with gentle sweeping curves set in a broad, tree-lined right-of-way devoid of commercial development. Primary regard was given to landscape features along the parkway through the creation of vistas, extensive planting, and slope beautification. Taking their form from the region's material culture, most bridges and drainage structures were clad with hand-made "Virginia style" brick. The roadway's pavement was limited to 30' width, and specially "broomed" to expose the aggregate to recreate the feel of a country road. Together, these design features were incorporated to provide continuity to the historical experience as visitors motor from one period of American colonial history to another.

Completed in 1957, the Colonial Parkway is the unifying component of the Park's road system that includes interpretive tour roads on Jamestown Island and the Yorktown Battlefield. The two tour roads provide visitors with a more specialized and personal experience of the island and the battlefield through roadside exhibits and historical markers.

Documentation of the Colonial National Historical Park Roads and Bridge was undertaken by the Historic American Engineering Record in the Summer of 1995 under the direction of project leader Christopher H. Marston. The HAER recording team included field supervisor Robert R. Harvey, Iowa State University; ICOMOS landscape architect Magdalena Bielecka, The Center for the Preservation of Historic Landscapes, Warsaw, Poland; architect Catherine Lee Doar, Varina, Virginia; landscape architect Kevin Doniere, Cincinnati, Ohio; and HAER historian Michael G. Bennett. Large format photography was prepared by Jack Boucher, HABS staff photographer.

Historic site symbols are based on publication, "Colonial Parkway," U.S. Government Printing Office, 1965 of-789-013.

Note: See also HAER No. VA-116, Jamestown Island Loop Road.
HAER No. VA-117, Yorktown Battlefield Tour Roads.

DELINEATED BY: ROBERT R. HARVEY, 1995

COLONIAL NHP
ROADS & BRIDGES RECORDING PROJECT
NATIONAL PARK SERVICE
UNITED STATES DEPARTMENT OF THE INTERIOR

YORKTOWN VICINITY

COLONIAL PARKWAY
YORKTOWN TO JAMESTOWN ISLAND
YORK COUNTY

VIRGINIA

SHEET 1 of 9

HISTORIC AMERICAN ENGINEERING RECORD
VA-48

DEVELOPMENT OF THE PARKWAY

COLONIAL PARKWAY ENTRANCE MARKER

KEY TO STRUCTURES	DATE	HAER No.
1. Ballard Creek Culvert	1931	VA-48-E
2. Bracken's Pond Culvert	1931	VA-48-F
3. Navy Mine Depot Overpass	1931, 1966, 1980	VA-48-A
4. Jones Mill Pond Dam	1932	VA-48-G
5. Indian Field Creek Bridge	1933, 1980	VA-48-H
6. Felgates Creek Bridge	1933, 1980	VA-48-I
7. King Creek Bridge	1933, 1980	VA-48-J
8. Capital Landing Underpass	1936	VA-48-B
9. C & O Railroad Underpass	1937	VA-48-C
10. Williamsburg Tunnel	1942	VA-48-D
11. Halfway Creek Bridge	1942	VA-48-K
12. U.S. Rt. 143 Bridge	1948	VA-48-L
13. College Creek Bridge	1956	VA-48-M
14. Mill Creek Bridge	1956	VA-48-N
15. Powhatan Creek Bridge	1956	VA-48-O
16. Isthmus Bridge	1956	VA-48-P
17. Yorktown Creek Bridge	1956	VA-48-Q
18. U.S. Route 17 Bridge	1956	VA-48-R
19. VA Route. 238 Bridge	1956	VA-48-S
20. Glebe Gut Bridge	1956	VA-48-T
21. Newport Avenue Bridge	1957	VA-48-U
22. North Pier Bridge	1962	VA-48-V
23. VA Rt. 641 Bridge	1964	VA-48-W
24. Hubbard Lane Bridge	1964	VA-48-X
25. I-64 Bridge	1965	VA-48-Y
26. VA Rt. 199 Bridge	1966, 1973	VA-48-Z
27. Parkway Drive Bridge	1972	VA-48-AA

The development of the Colonial Parkway encompassed a twenty-six year construction history marked by routing conflicts, war, limited funding, and land acquisition problems. Between 1931 and 1937, the road was completed from Ballard Creek, near Yorktown, to Williamsburg. The only major construction between 1937 and 1955 was a tunnel under the restored capital, and a bridge over Halfway Creek. The parkway was completed during an extremely active period between 1955 and 1957 in preparation for Jamestown's 350th anniversary.

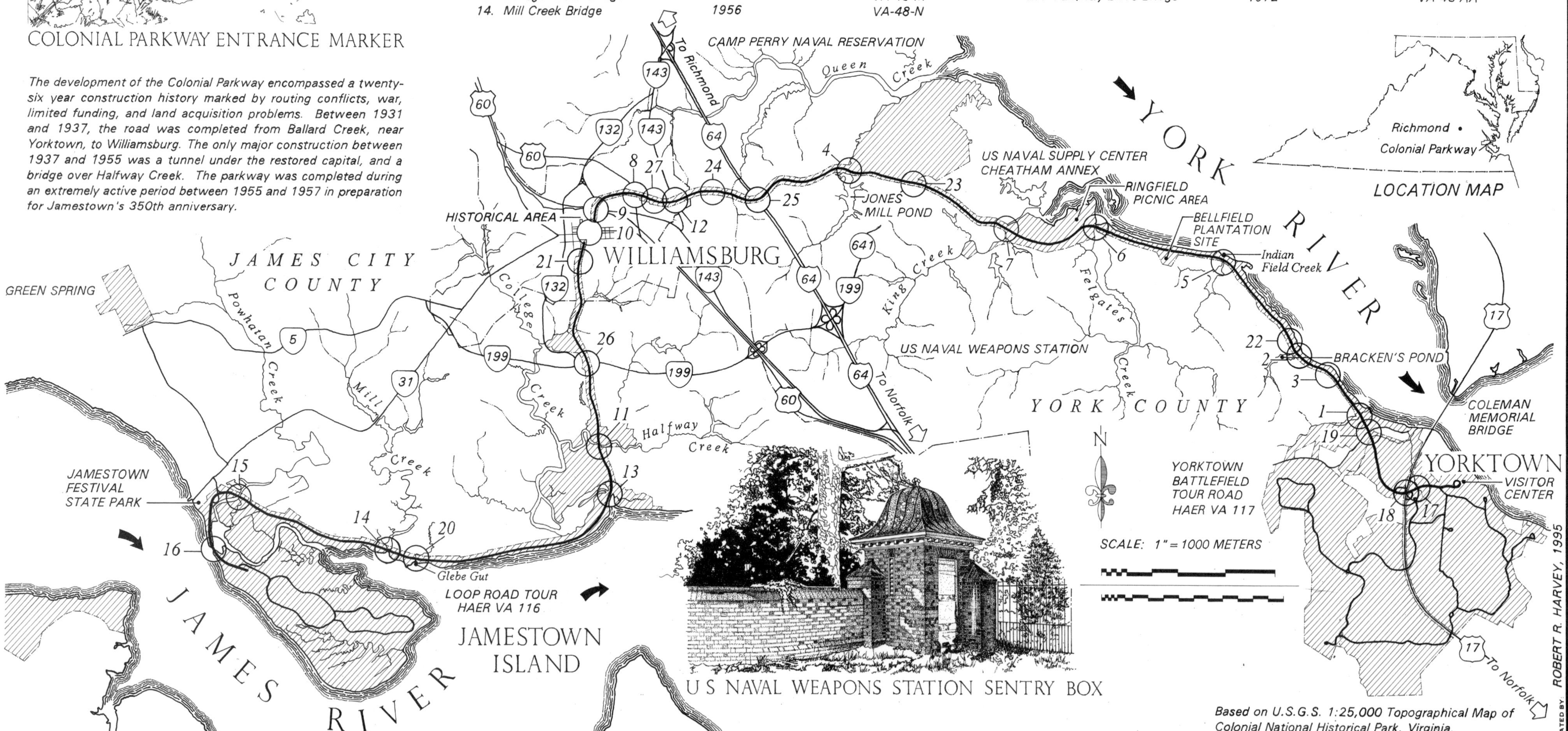

LOCATION MAP

U S NAVAL WEAPONS STATION SENTRY BOX

Based on U.S.G.S. 1:25,000 Topographical Map of Colonial National Historical Park, Virginia. 37076-B5-PF-025, 1981

DELINEATED BY: *ROBERT R. HARVEY, 1995*

COLONIAL NHP
ROADS & BRIDGES RECORDING PROJECT
NATIONAL PARK SERVICE
UNITED STATES DEPARTMENT OF THE INTERIOR

YORKTOWN VICINITY

COLONIAL PARKWAY
YORKTOWN TO JAMESTOWN ISLAND
YORK COUNTY

VIRGINIA

SHEET 2 of 9

HISTORIC AMERICAN ENGINEERING RECORD
VA-48

BUILDING THE PARKWAY

CLEARING & GRUBBING

After the road alignment was approved, Bureau of Public Roads survey teams set stakes along the location of the road's slopes, and all ground cover was cleared to a distance of 5' beyond the stakes. Directed by the landscape architects, the supervisory engineers were responsible for marking certain trees for preservation. All locust and cedar trees with a minimum diameter of 6" were cut to board length and saved. All other material cleared was burned at night in specified areas.

EXCAVATION & GRADING

With the right-of-way cleared and grubbed, the bedding for drainage structures was excavated, and the roadway, overlooks, and parking areas were graded in accordance with the specifications on the plans. Topsoil was removed from the area to be paved, and stored for replacement upon slopes for planting. Drainage structures were constructed and backfilled, and a suitable subgrade was attained. Excavated earth was used to create embankments, and to top areas of hydraulic fill.

SLOPE DEVELOPMENT

Landscape architects gave special attention to the development of vegetated slopes along the Colonial Parkway to set the road into the landscape. Tree stumps along slope areas were left below the surface to help stabilize the embankments. A varying grade between 2:1 and 5:1 was used, but the engineer was responsible for on-site supervision to adjust the slopes to local conditions. Following detailed plans from the landscape division, the proper grade was attained, and crews seeded the slopes to stabilize the earth before beautification.

HYDRAULIC FILL

The decision to align the parkway along the York and James rivers necessitated the extensive use of hydraulic fill to create a suitable road grade at a minimum elevation of 11' above sea level. Timber bulkheads were constructed to stabilize the fill material dredged from specified areas along the river bed. Utilizing the bleeding process of hydraulic fill, a discharge pipe was placed at the centerline of the embankment through which dredged material was allowed to run out and settle naturally.

PAVING

A two course subgrade between 9" and 12" thick underlies most of the pavement along the parkway. Form work and reinforcement mesh were constructed for the concrete sections, which were 10' wide and between 40' and 60' long. Concrete was mixed with a local aggregate (primarily marl) and poured first in the center 10' wide crowned section. Special treatments, including hosing, brooming, and acid washes were used to expose the aggregate and create the feel of a "country road." (see drawing No. 6 for detail).

PLANTING

The primary planting season along the Colonial Parkway usually extended from the fall to early winter. New trees, such as pines, dogwoods, and cedars, were planted along the road, and many existing trees were moved to create balance and variety to the roadscape. While funds for purchasing plant material were allocated to the CCC camps through Emergency Conservation Works programs, deals were often made with local landowners to donate trees in exchange for the grading and surfacing of roads on their property.

DELINEATED BY: *Kevin Doniere, 1995*

COLONIAL NHP
ROADS & BRIDGES RECORDING PROJECT
NATIONAL PARK SERVICE
UNITED STATES DEPARTMENT OF THE INTERIOR

YORKTOWN VICINITY

COLONIAL PARKWAY
YORKTOWN TO JAMESTOWN ISLAND
YORK COUNTY

VIRGINIA

SHEET 3 OF 9

HISTORIC AMERICAN ENGINEERING RECORD
VA-48

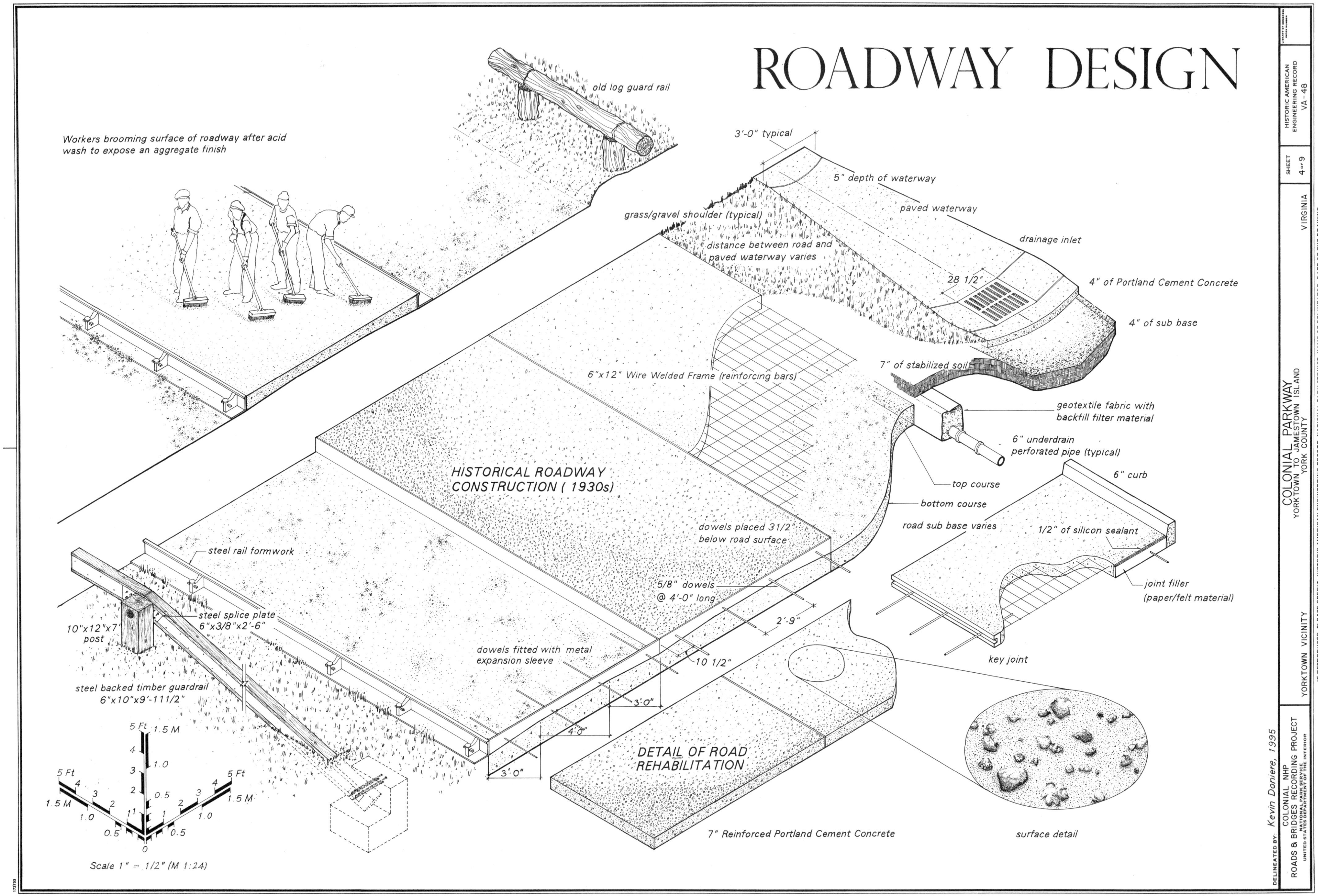

ROADWAY DESIGN
old log guard rail
Workers brooming surface of roadway after acid wash to expose an aggregate finish
3'-0" typical
5" depth of waterway
paved waterway
grass/gravel shoulder (typical)
distance between road and paved waterway varies
drainage inlet
28 1/2"
4" of Portland Cement Concrete
4" of sub base
7" of stabilized soil
6"x12" Wire Welded Frame (reinforcing bars)
geotextile fabric with backfill filter material
6" underdrain perforated pipe (typical)
HISTORICAL ROADWAY CONSTRUCTION (1930s)
top course
bottom course
6" curb
road sub base varies
1/2" of silicon sealant
dowels placed 3 1/2" below road surface
steel rail formwork
joint filler (paper/felt material)
5/8" dowels @ 4'-0" long
2'-9"
steel splice plate 6"x3/8"x2'-6"
10"x12"x7' post
key joint
dowels fitted with metal expansion sleeve
10 1/2"
steel backed timber guardrail 6"x10"x9'-11 1/2"
3'0"
4'0"
3'0"
DETAIL OF ROAD REHABILITATION
7" Reinforced Portland Cement Concrete
surface detail
Scale 1" = 1/2" (M 1:24)
HISTORIC AMERICAN ENGINEERING RECORD VA-48
SHEET 4 of 9
VIRGINIA
COLONIAL PARKWAY
YORKTOWN TO JAMESTOWN ISLAND
YORK COUNTY
YORKTOWN VICINITY
DELINEATED BY: Kevin Doniere, 1995
COLONIAL NHP
ROADS & BRIDGES RECORDING PROJECT
NATIONAL PARK SERVICE
UNITED STATES DEPARTMENT OF THE INTERIOR
IF REPRODUCED, PLEASE CREDIT: HISTORIC AMERICAN ENGINEERING RECORD, NATIONAL PARK SERVICE, NAME OF DELINEATOR, DATE OF THE DRAWING

LOOKING EAST INTO SANDY BAY

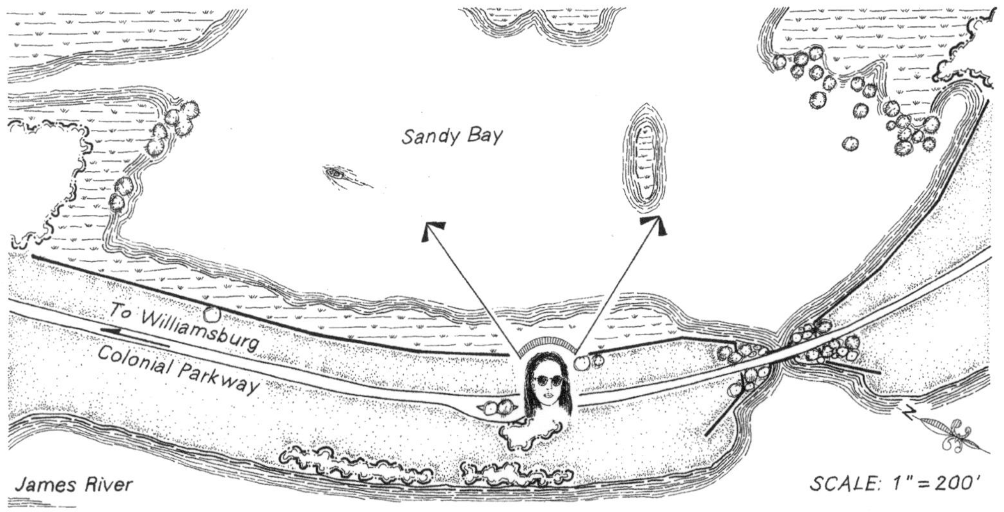

LIST OF PLANT MATERIAL

Botanical Name	*Common Name*
JAMES RIVER	
Juniperus virginiana	*RED CEDAR*
Lonicera japonica	*HONEYSUCKLE*
Myrica cerifera	*SOUTHERN BAYBERRY*
Pinus taeda	*LOBLOLLY PINE*
Platanus occidentalis	*SYCAMORE*
Quercus alba	*WHITE OAK*
Taxodium distichum	*BALDCYPRESS*
YORK RIVER	
Celtis occidentalis	*COMMON HACKBERRY*
Cornus florida	*FLOWERING DOGWOOD*
Juglans nigra	*WALNUT*
Juniperus virginiana	*RED CEDAR*
Liriodendron tulipifera	*TULIP TREE*
Platanus occidentalis	*SYCAMORE*
Robinia pseudoacacia	*LOCUST*
Sassafras albidum	*SASSAFRAS*

JAMES RIVER

PARKWAY VIEWS

YORK RIVER

Integral to the natural and historical development of Tidewater Virginia, the York and the James rivers broaden the interpretive and scenic qualities of the Colonial Parkway. The decision to align the parkway along the river banks provided landscape architects the opportunity to develop dramatic open views of the rivers. These vistas, or unified landscape compositions framed by natural features, were also developed for inward views of marshes and ponds. Land management plans guide maintenance crews in the selective cutting of vegetation to open viewsheds and highlight exotic and flowering trees along the parkway. Just as significant, however, is the use of vegetated screens to hide views not related to the overall feel or intent of the parkway, such as commercial or industrial sites.

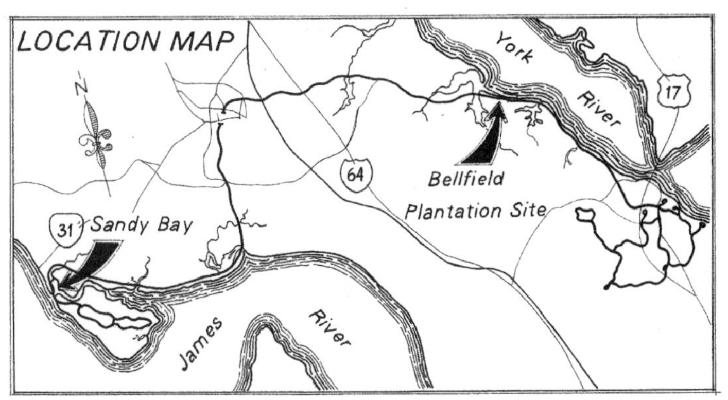

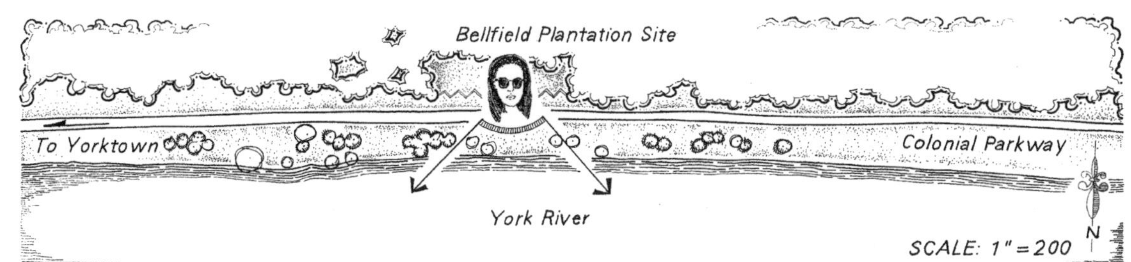

LOOKING NORTHEAST OVER THE YORK RIVER

DELINEATED BY: *Magdalena Bielecka, 1995*

COLONIAL NHP
ROADS & BRIDGES RECORDING PROJECT
NATIONAL PARK SERVICE
UNITED STATES DEPARTMENT OF THE INTERIOR

YORKTOWN VICINITY

COLONIAL PARKWAY
YORKTOWN TO JAMESTOWN ISLAND
YORK COUNTY

VIRGINIA

SHEET 5 OF 9

HISTORIC AMERICAN ENGINEERING RECORD
VA-48

LOOKING WEST INTO JONES MILL POND

PARKWAY VIEWS

JONES MILL POND

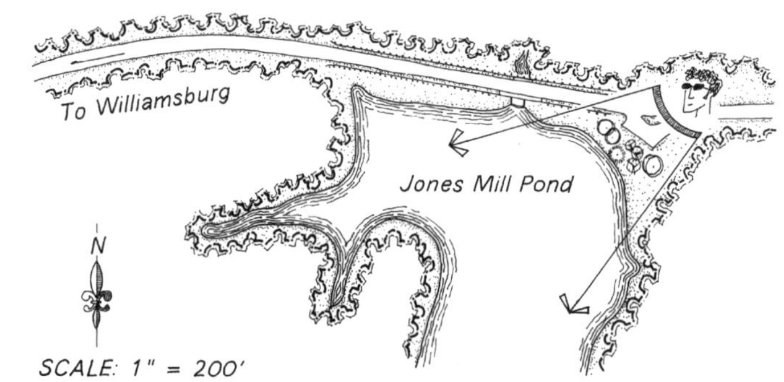

LIST OF PLANT MATERIAL
JONES MILL POND

Botanical Name	Common Name
Acer saccharum	SUGAR MAPLE
Cercis canadensis	REDBUD
Fagus grandifolia	BEECH
Juniperus virginiana	RED CEDAR
Kalmia latifolia	MOUNTAIN LAUREL
Liquidambar styraciflua	SWEETGUM
Myrica cerifera	SOUTHERN BAYBERRY
Pinus taeda	LOBLOLLY PINE
Platanus occidentalis	SYCAMORE
Quercus sp.	OAKS

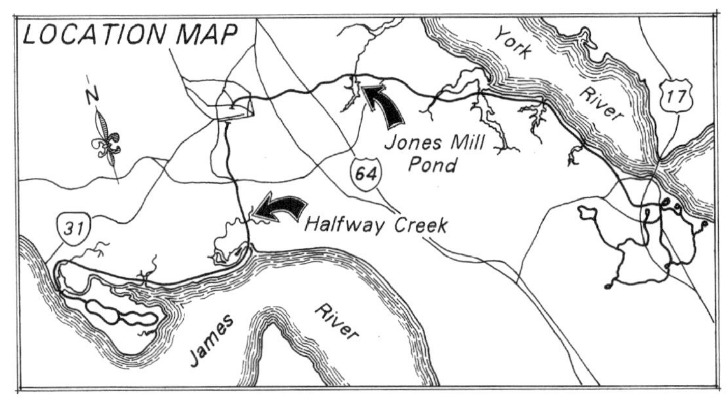

HALFWAY CREEK

LIST OF PLANT MATERIAL
HALFWAY CREEK

Botanical Name	Common Name	Botanical Name	Common Name
Carya ovata	HICKORY	Pinus sylvestris	SCOTCH PINE
Castanea dentata	CHESTNUT	Prunus serotina	BLACK CHERRY
Cornus florida	DOGWOOD	Quercus sp.	OAKS
Juglans nigra	WALNUT	Robinia sp.	LOCUST
Liriodendron tulipifera	TULIP TREE	Sassafras albidum	SASSAFRAS

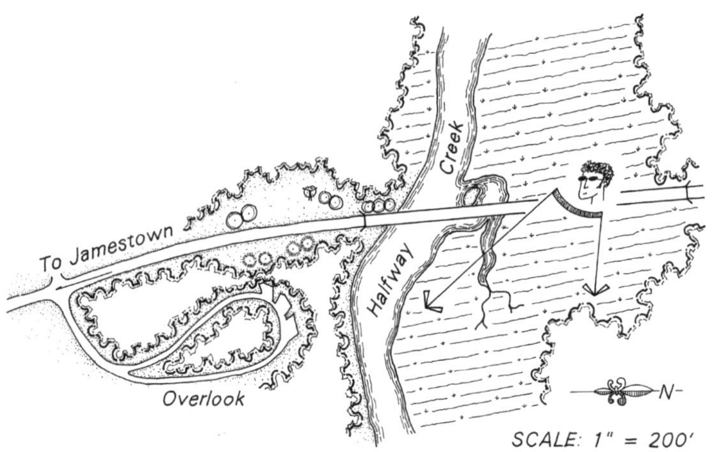

LOOKING EAST OVER HALFWAY CREEK

DELINEATED BY: Kevin Doniere, 1995

COLONIAL NHP
ROADS & BRIDGES RECORDING PROJECT
NATIONAL PARK SERVICE
UNITED STATES DEPARTMENT OF THE INTERIOR

YORKTOWN VICINITY

COLONIAL PARKWAY
YORKTOWN TO JAMESTOWN ISLAND
YORK COUNTY

VIRGINIA

SHEET 6 of 9

HISTORIC AMERICAN ENGINEERING RECORD
VA-48

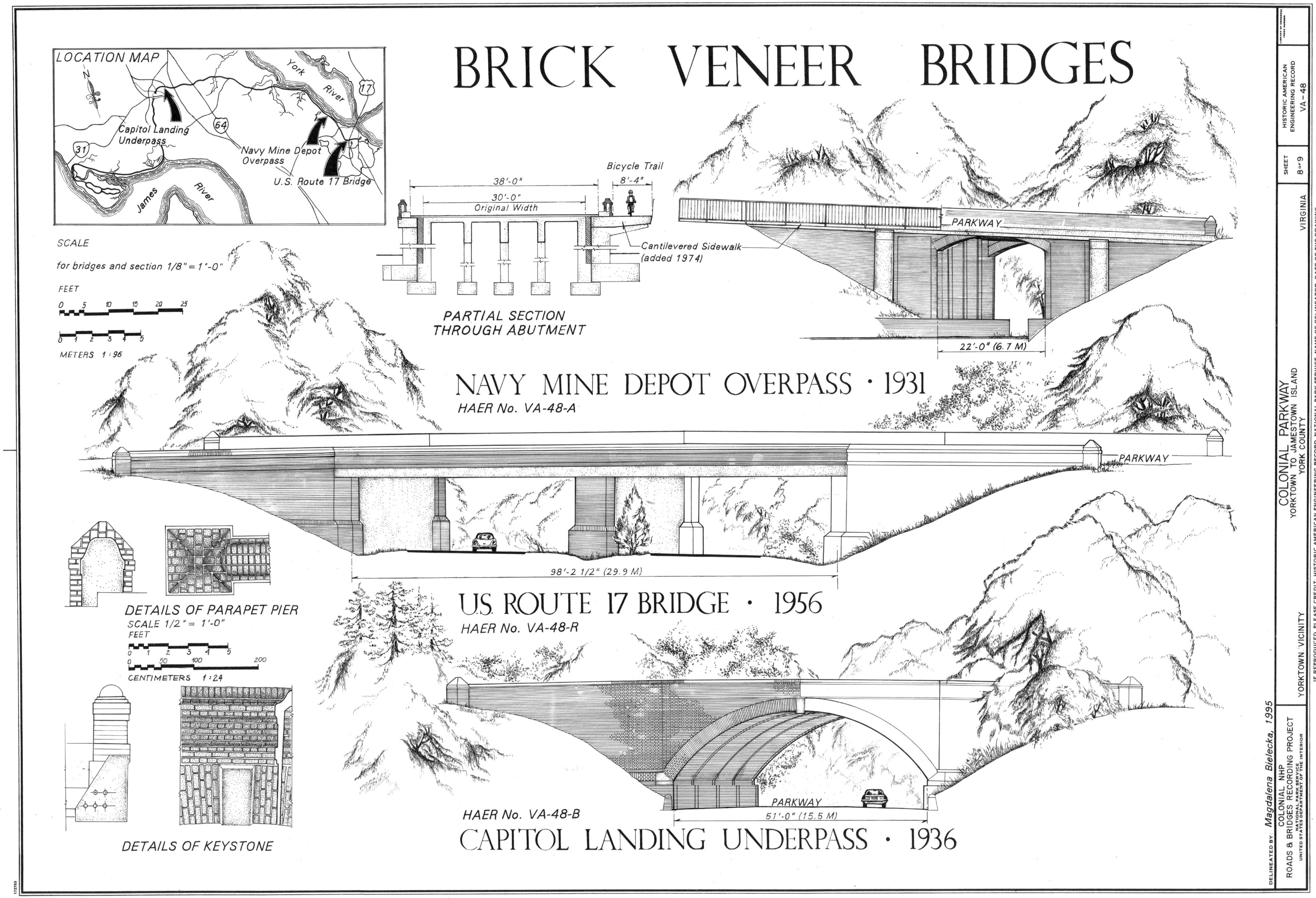
BRICK VENEER BRIDGES
LOCATION MAP
N
York River
17
64
31
Capitol Landing Underpass
Navy Mine Depot Overpass
U.S. Route 17 Bridge
James River
SCALE
for bridges and section 1/8" = 1'-0"
FEET
0 5 10 15 20 25
0 1 2 3 4 5
METERS 1:96
38'-0"
30'-0"
Original Width
Bicycle Trail
8'-4"
Cantilevered Sidewalk
(added 1974)
PARTIAL SECTION
THROUGH ABUTMENT
PARKWAY
22'-0" (6.7 M)
NAVY MINE DEPOT OVERPASS · 1931
HAER No. VA-48-A
PARKWAY
98'-2 1/2" (29.9 M)
U.S. ROUTE 17 BRIDGE · 1956
HAER No. VA-48-R
DETAILS OF PARAPET PIER
SCALE 1/2" = 1'-0"
FEET
0 1 2 3 4 5
0 50 100 200
CENTIMETERS 1:24
DETAILS OF KEYSTONE
PARKWAY
51'-0" (15.5 M)
HAER No. VA-48-B
CAPITOL LANDING UNDERPASS · 1936
DELINEATED BY Magdalena Bielecka, 1995
COLONIAL NHP
ROADS & BRIDGES RECORDING PROJECT
NATIONAL PARK SERVICE
UNITED STATES DEPARTMENT OF THE INTERIOR
YORKTOWN VICINITY
COLONIAL PARKWAY
YORKTOWN TO JAMESTOWN ISLAND
YORK COUNTY
VIRGINIA
SHEET 8 of 9
HISTORIC AMERICAN ENGINEERING RECORD
VA-48
IF REPRODUCED, PLEASE CREDIT: HISTORIC AMERICAN ENGINEERING RECORD, NATIONAL PARK SERVICE, NAME OF DELINEATOR, DATE OF THE DRAWING

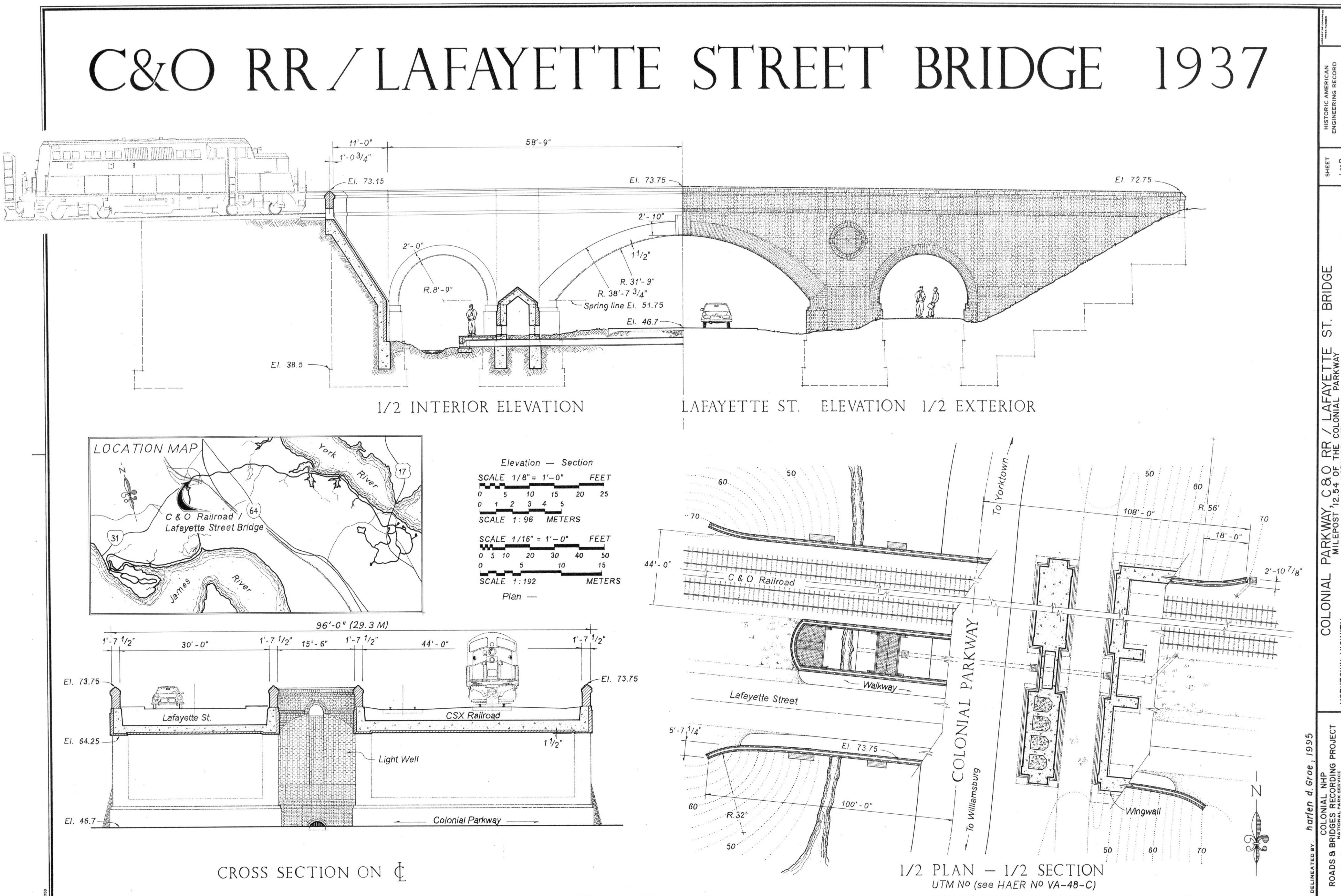
C&O RR / LAFAYETTE STREET BRIDGE 1937
1/2 INTERIOR ELEVATION
LAFAYETTE ST. ELEVATION 1/2 EXTERIOR
11'-0"
58'-9"
1'-0 3/4"
El. 73.15
El. 73.75
El. 72.75
2'-10"
2'-0"
1 1/2"
R. 8'-9"
R. 31'-9"
R. 38'-7 3/4"
Spring line El. 51.75
El. 46.7
El. 38.5
LOCATION MAP
York River
C & O Railroad / Lafayette Street Bridge
James River
Elevation — Section
SCALE 1/8" = 1'-0" FEET
SCALE 1:96 METERS
SCALE 1/16" = 1'-0" FEET
SCALE 1:192 METERS
Plan —
96'-0" (29.3 M)
1'-7 1/2"
30'-0"
15'-6"
44'-0"
El. 73.75
El. 64.25
El. 46.7
Lafayette St.
CSX Railroad
Light Well
Colonial Parkway
CROSS SECTION ON ℄
To Yorktown
108'-0"
R. 56'
18'-0"
2'-10 7/8"
44'-0"
C & O Railroad
Walkway
COLONIAL PARKWAY
Lafayette Street
5'-7 1/4"
El. 73.75
100'-0"
R. 32'
To Williamsburg
Wingwall
N
1/2 PLAN — 1/2 SECTION
UTM No (see HAER No VA-48-C)
DELINEATED BY: harlen d. Groe, 1995
COLONIAL NHP
ROADS & BRIDGES RECORDING PROJECT
NATIONAL PARK SERVICE
UNITED STATES DEPARTMENT OF THE INTERIOR
COLONIAL PARKWAY, C&O RR / LAFAYETTE ST. BRIDGE
MILEPOST 12.54 OF THE COLONIAL PARKWAY
YORKTOWN VICINITY
YORK COUNTY
VIRGINIA
SHEET 1 of 2
HISTORIC AMERICAN ENGINEERING RECORD
VA-48-C
IF REPRODUCED, PLEASE CREDIT: HISTORIC AMERICAN ENGINEERING RECORD, NATIONAL PARK SERVICE, NAME OF DELINEATOR, DATE OF THE DRAWING

CONSTRUCTION ISOMETRIC

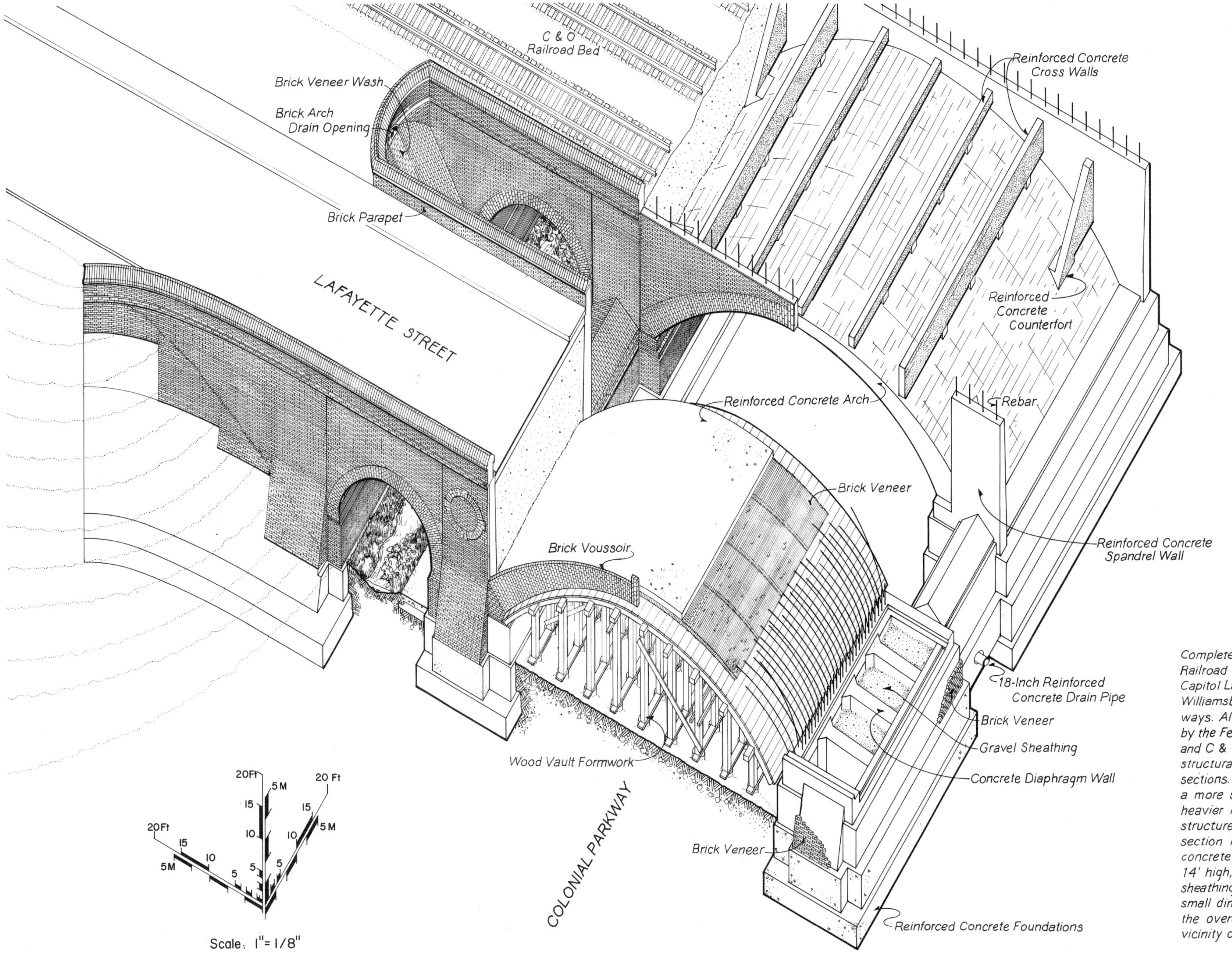

Completed in 1937, the Lafayette Street and C & O Railroad underpass was constructed along with the Capitol Landing underpass to provide an approach to Williamsburg without intersecting existing right-of-ways. Although considered two separate structures by the Federal Highway Administration, the Lafayette and C & O bridge is a single structure with different structural elements for the street and railroad sections. The northern section carrying the railroad is a more substantial structure to accommodate the heavier loads of railroad traffic. While the entire structure was built upon a 3'-6" base, the railroad section has an additional 12' of solid reinforced concrete to form its foundation, compared with the 14' high, 2' wide diaphragm walls filled with gravel sheathing for the road section. The bridge's relatively small dimensions and brick veneer harmonize with the overall design of the Colonial Parkway in the vicinity of Williamsburg. (See HAER No. VA-48-C)

HISTORIC AMERICAN ENGINEERING RECORD VA-48-C

SHEET 2 of 2

COLONIAL PARKWAY, C&O RR / LAFAYETTE ST. BRIDGE
MILEPOST 12.54 OF THE COLONIAL PARKWAY
YORK COUNTY

YORKTOWN VICINITY

VIRGINIA

COLONIAL NHP
ROADS & BRIDGES RECORDING PROJECT
NATIONAL PARK SERVICE
UNITED STATES DEPARTMENT OF THE INTERIOR

DELINEATED BY: Catherine Lee Dear, 1995 + Edward J. Lupyak 1997

WILLIAMSBURG TUNNEL 1936-1949

In a letter to National Park Service director Horace Albright, superintendent B. Floyd Flickenger stated, "I feel that the tunnel may prove to be one of our worst stumbling blocks and will probably cause us more trouble than any project we have yet attempted in this Park."

The issue of routing the parkway either around or through Williamsburg created considerable conflict between the Park Service, the Williamsburg Foundation and the citizens of Williamsburg. By 1935 a few possible alignments were under consideration. They were all a variation of two alignments: a route north and west of the city proposed by Arthur Shurcliff and the Williamsburg Foundation to keep the parkway away from the estate of John D. Rockefeller, Jr. at Bassett Hall; and a route south and east of the city put forth by Charles Peterson in 1930 to balance the York River section of the road with the James River.

By early 1936, another route through the city by way of a tunnel entered into the debate. In May, 1936, St. Louis planning consultant Harland Bartholomew studied the three routes and recommended the construction of a tunnel to provide one main entrance to the city and a direct connection to the business district by way of Henry Street. After extensive study, the Park Service's proposal to build a tunnel received the endorsement of the Williamsburg Foundation and the city council in 1939.

Employing "cut and cover" technology, the J. G. Attaway Construction Co. of Statesboro, Georgia, constructed the tunnel between 1940 and 1942. Numerous problems plagued the project, including both material and labor shortages, and cave-ins which resulted in considerable damage to structures along the tunnel's route. World War II curtailed finishing work on the tunnel, and the portals were barricaded until 1948 when a lighting system was installed and the tunnel was paved. An official ceremony on May 10, 1949, opened the tunnel to traffic. While it took thirteen years to by-pass Williamsburg, the needed section was completed to continue the parkway to Jamestown Island.

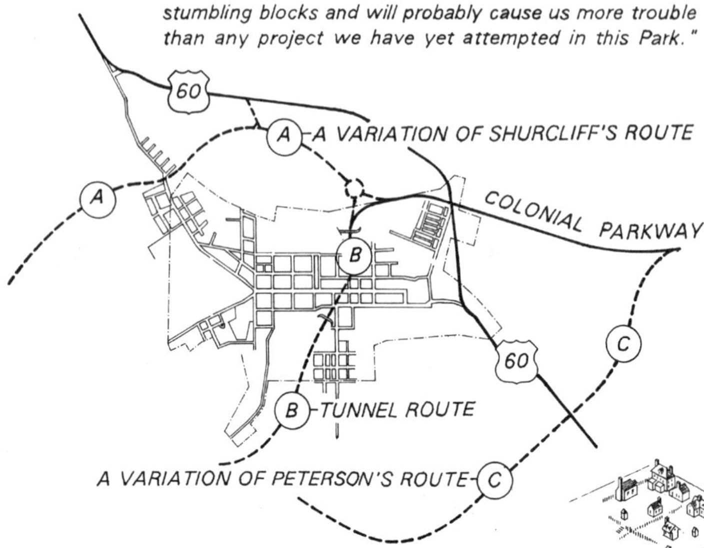

EVOLUTION OF THE COLONIAL PARKWAY ROUTES AT WILLIAMSBURG

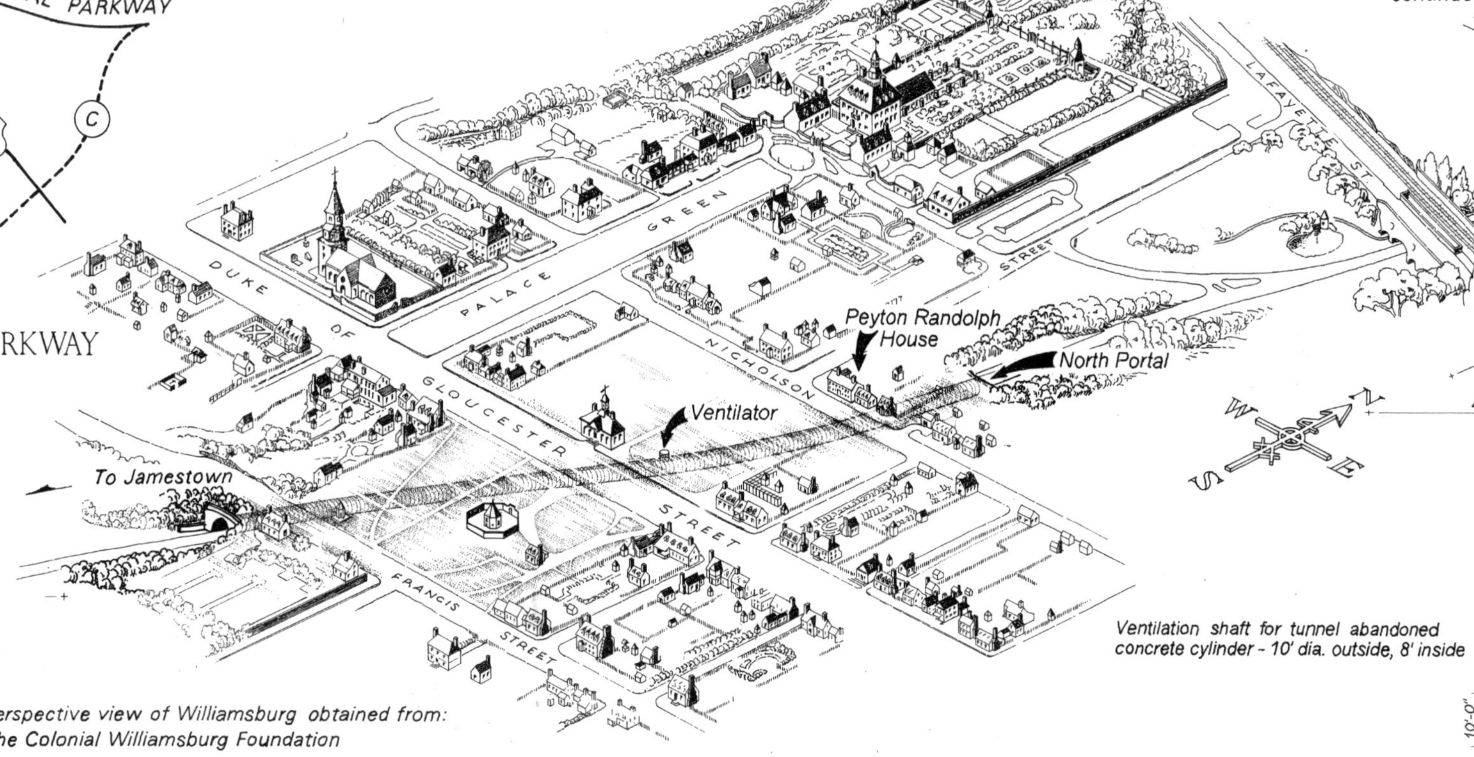

LOCATION OF TUNNEL PASSING BENEATH THE HISTORIC DISTRICT

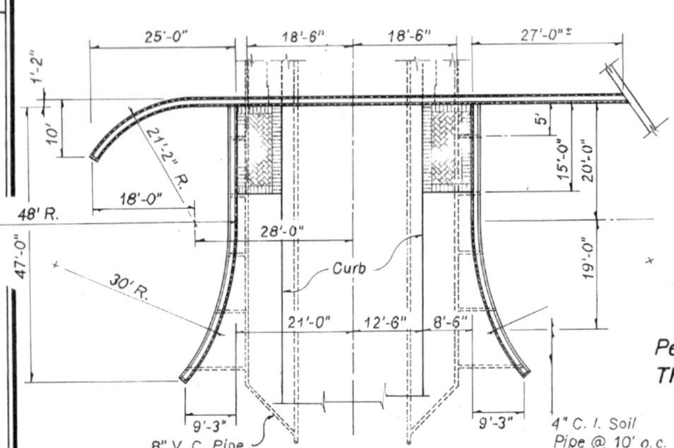

PLAN

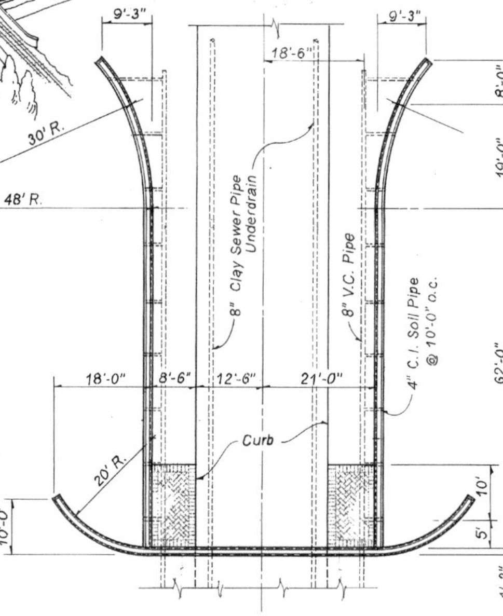

PLAN

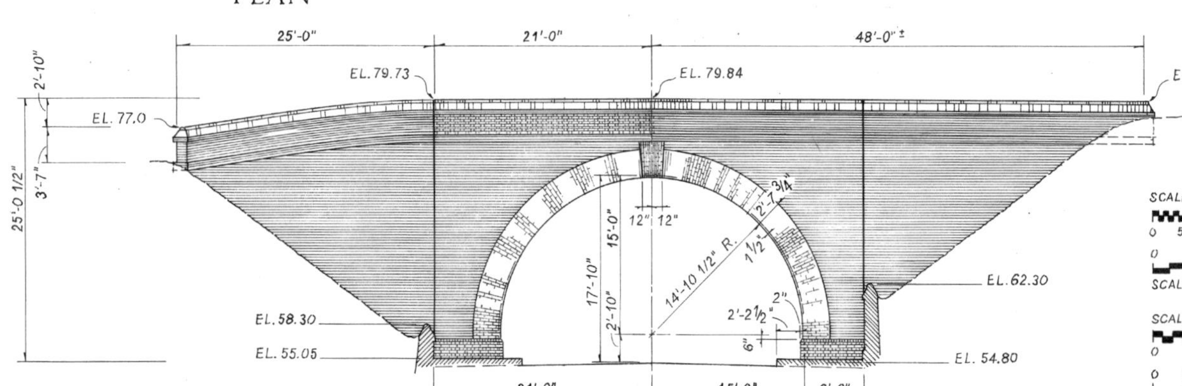

SOUTH PORTAL

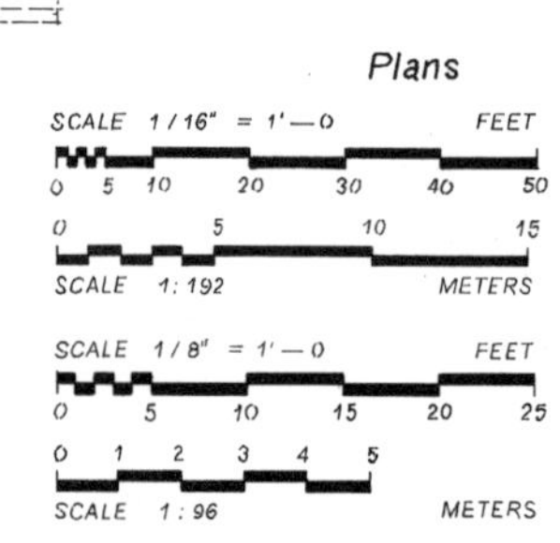

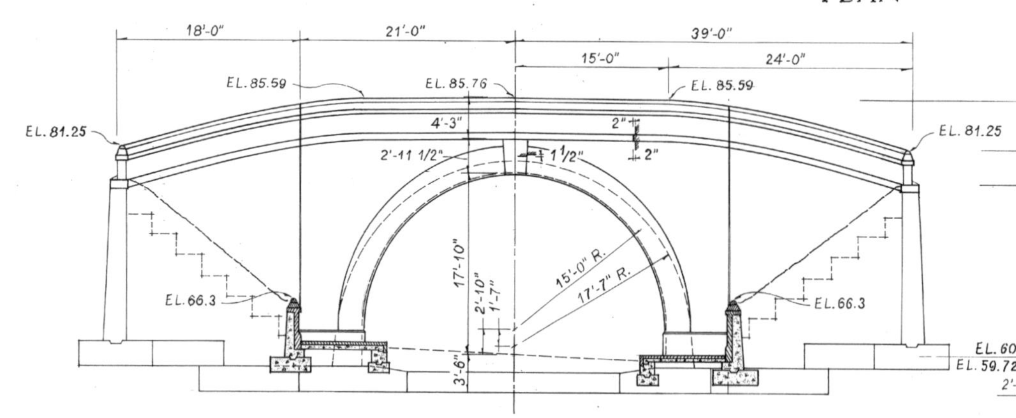

NORTH PORTAL

HISTORIC AMERICAN ENGINEERING RECORD VA-48-D

SHEET 1 of 2

COLONIAL PARKWAY, WILLIAMSBURG TUNNEL
.3 MILES SOUTH OF C&O RR/LAFAYETTE ST. BRIDGE
YORK COUNTY

YORKTOWN VICINITY — VIRGINIA

DELINEATED BY: ROBERT R. HARVEY, harlen d. Groe, 1995

COLONIAL NHP
ROADS & BRIDGES RECORDING PROJECT
NATIONAL PARK SERVICE
UNITED STATES DEPARTMENT OF THE INTERIOR

WILLIAMSBURG TUNNEL CONSTRUCTION

SHORING OF TUNNEL EXCAVATION

Due to numerous cave-ins during excavation of the 50' wide trench for the tunnel, extensive shoring was utilized to stabilize the trench walls and the foundations of nearby structures.

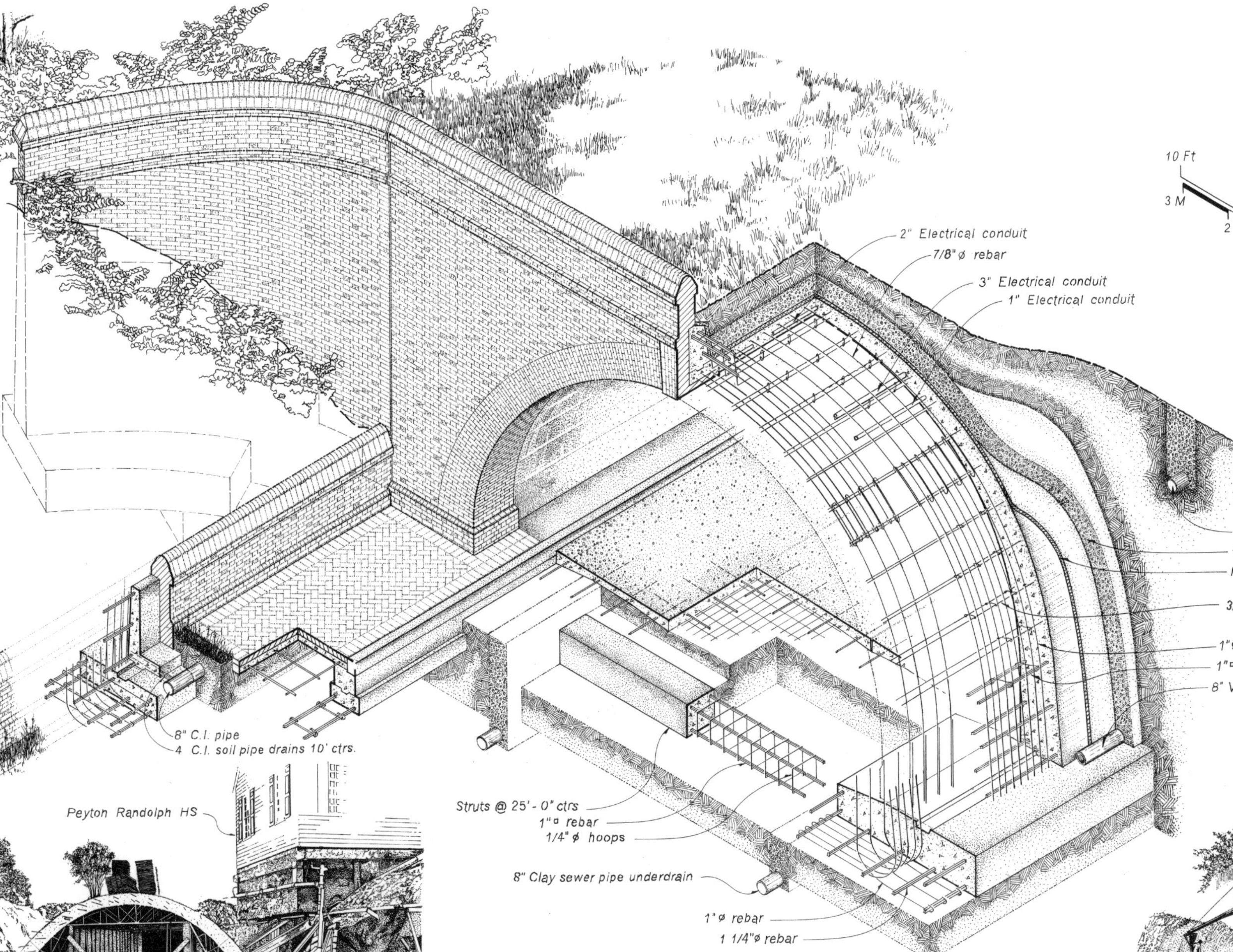

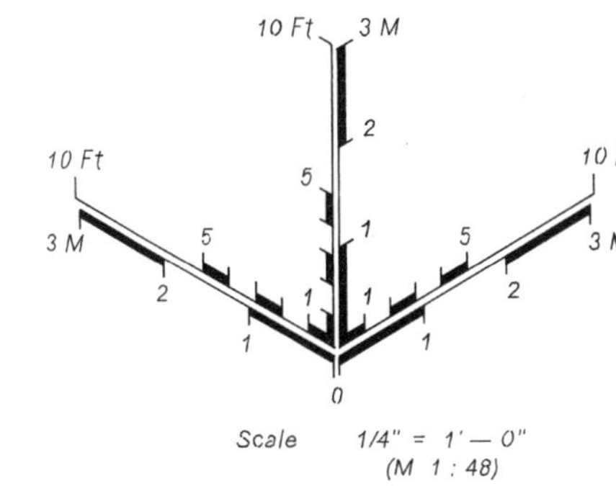

NORTH PORTAL ISOMETRIC

The steel formwork for the 30' wide tunnel arch was comprised of arch rib trusses spaced 30" center-to-center with 30" x 30" steel plates on both the inside and outside of the trusswork, anchored with flat steel tie bars.

CONSTRUCTION OF TUNNEL ADJACENT TO HISTORIC STRUCTURE

Drawings of tunnel construction based on photographs obtained from: The Colonial Williamsburg Foundation.

A concrete mixing plant was constructed adjacent to the trench, and was fitted with an elevated swing-shaft capable of pouring concrete directly into shutes attached to the tunnel's formwork. Concrete was poured in 30' sections. Once set, the concrete was waterproofed with hot tar, fabric and asphalt plank, and encased with 12" gravel sheathing. In order to re-open roads and resume landscaping, backfilling operations began as each section was completed.

PLACEMENT OF CONCRETE

DELINEATED BY: ROBERT R. HARVEY, harlen d Groe, 1995

COLONIAL NHP
ROADS & BRIDGES RECORDING PROJECT
NATIONAL PARK SERVICE
UNITED STATES DEPARTMENT OF THE INTERIOR

COLONIAL PARKWAY, WILLIAMSBURG TUNNEL
.3 MILES SOUTH OF C&O RR/LAFAYETTE ST. BRIDGE
YORK COUNTY

YORKTOWN VICINITY

VIRGINIA

SHEET 2 of 2

HISTORIC AMERICAN ENGINEERING RECORD
VA-48-D

JAMESTOWN ISLAND LOOP ROAD

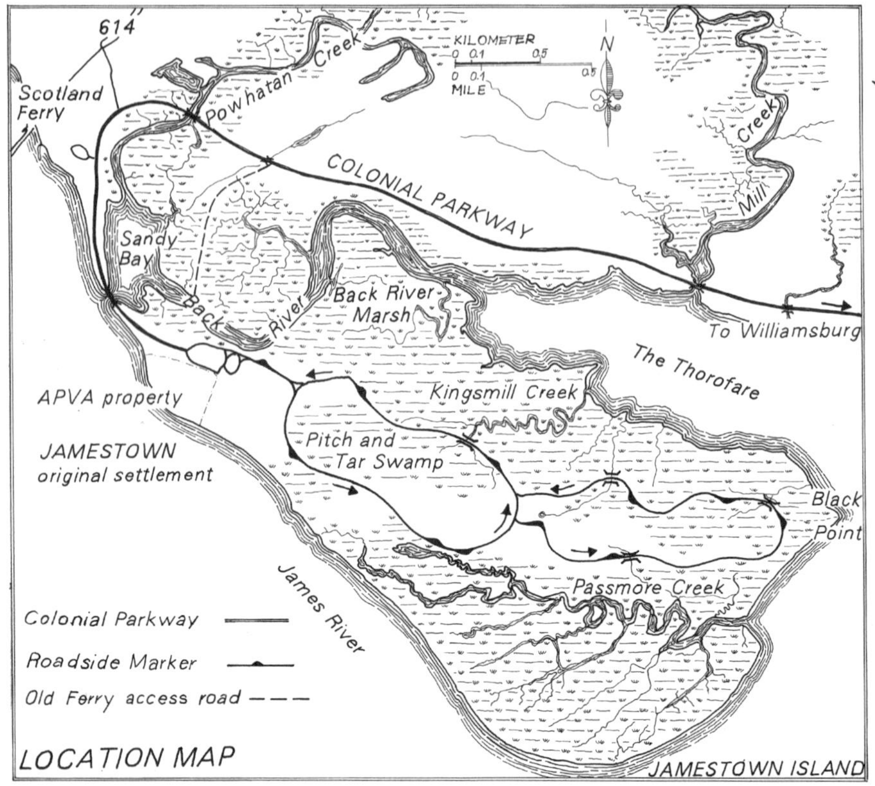

LOCATION MAP

Note: See also HAER No VA-48, Colonial Parkway
HAER No VA-117, Yorktown Battlefield Tour Roads

ARCHEOLOGY ON JAMESTOWN ISLAND

TIMBER TRESTLE CONSTRUCTION DETAILS

SCALE: 1/2" = 1'-0"

- A - piles, 3 per bent 12" butt
- B - 3"x8" sway bracing fastening
- C - beam
- D - 6"x12"x12" block over pile caps
- E - 3/4" bolts
- F - 6"x12" stringers
- G - 6"x4"x12" block
- H - 6"x6" wheel guard
- I - 2"x4" bridging span ctrs 2' - 6" spikes each end
- J - 3"x8"x14' plank flooring

Scale 1" = 1/2" (M 1:24)

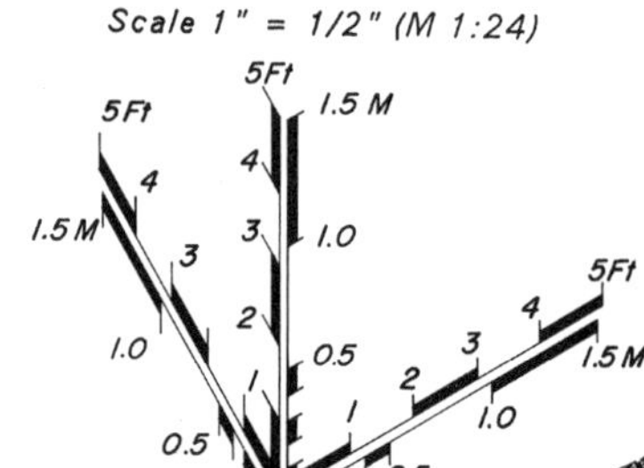

GALLERY IN THE WOODS

"An experience in history, leading the most interested visitor deeper into the throughtful process of understanding the colony."
1961 Master Plan

VIEW OF TIMBER TRESTLE

Jamestown Island is comprised of 1,561 acres of wetlands and pine-oak forests. Jamestown, the first permanent English settlement in North America, was founded along the island's shore in 1607. The only aboveground structure remaining from the seventeenth century is the old church tower located on twenty-two acres of land owned by the Association for the Preservation of Virginia Antiquities (APVA). Archeological excavations provide the bulk of the park's knowledge of the settlement and development of Jamestown.

As part of an overall development plan for Jamestown's 350th anniversary, single-lane "wilderness roads" to the interior of the island were constructed in the 1950s to provide the visitor with a feeling of the primitive isolation of the 17th century frontier. Along the roads a series of oil paintings by Sidney King depict the early settler's use of the land from the perspective of 1950s interpretive programs. Ongoing research continues to expand the Park Service's understanding of the island, from which new interpretive programs will be developed.

DELINEATED BY: Magdalena Bielecka, 1995

COLONIAL NHP
ROADS & BRIDGES RECORDING PROJECT
NATIONAL PARK SERVICE
UNITED STATES DEPARTMENT OF THE INTERIOR

JAMESTOWN VICINITY

JAMESTOWN ISLAND LOOP ROAD
JAMESTOWN ISLAND
JAMES CITY COUNTY

VIRGINIA

SHEET 1 OF 1

HISTORIC AMERICAN ENGINEERING RECORD
VA-116

GEORGE WASHINGTON MEMORIAL PARKWAY

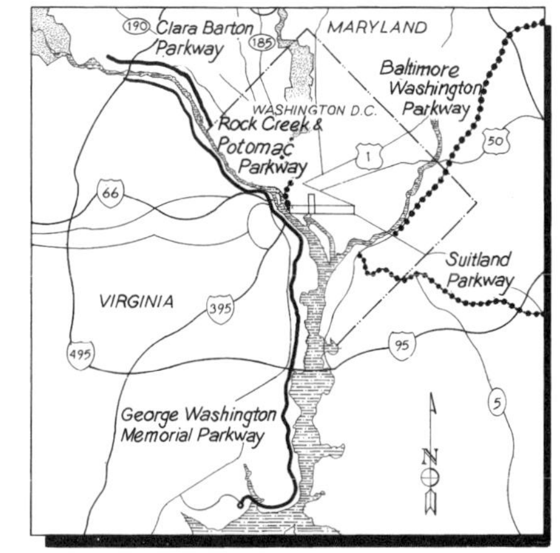

George Washington Memorial Parkway occupies an important position in the history of American landscape architecture, highway construction, and regional planning. Constructed between 1929-1970, the parkway provides an indispensable link in the transportation system of the national capital region, serves as a memorial to the nation's first president, and preserves invaluable historic, recreational, and natural resources along the Potomac River. As the first modern motor parkway built by the federal government, it embodied state-of-the-art design principles and strongly influenced parkway and highway development throughout the country.

The original Mount Vernon Memorial Highway (MVMH) section of George Washington Memorial Parkway (GWMP) was authorized by Congress in 1928 and constructed between 1929-32. It connects Washington, D.C. and George Washington's home at Mount Vernon with an attractively landscaped motor road that serves both tourists and daily commuters. MVMH also contains a wide variety of historic sites, memorials, recreation areas, and wildlife habitats.

The Capper-Cramton Act of 1930 established the George Washington Memorial Parkway, which was intended to incorporate MVMH and provide similar development on both sides of the Potomac from Mount Vernon to Great Falls. Funding problems delayed significant additional construction until after World War II. The original scheme was eventually scaled back, eliminating the southeast leg between Washington and Fort Washington in Prince George's County, Maryland, and stopping development on the northern end at the Capital Beltway in Virginia and MacArthur Boulevard on the Maryland side. Post-World War II portions of the parkway were designed for higher speeds and increased traffic volume, but continued to exemplify the potential for landscape architects, traffic engineers, and planners to work together to produce attractive and efficient transportation facilities that also provide for resource protection and recreational development.

The history of George Washington Memorial Parkway long predates the invention of the automobile. Mount Vernon's status as the patriotic tourist's "American Mecca" was established during Washington's lifetime. Visitors flocked to the site in increasing numbers with every transportation advance from carriage to steamboat, electric railway, and automobile. Alexandria businessmen first proposed connecting Washington and Mount Vernon with a monumental national road in the 1880s. The Mount Vernon Avenue Association was formed in 1888 to promote this goal. The proposed boulevard would improve the region's notoriously poor road system, address the needs of Mount Vernon-bound tourists, and provide a boost to local commerce and property values. The association advocated a formal boulevard lined with elaborate exhibition buildings, statues of former presidents and vice-presidents, and memorials honoring civil and military heroes from every state in the union.

George Washington

In 1890 the U.S. Army Corps of Engineers surveyed and reported on three alternative routes, but did not initiate any further action. The 1901 Senate Park Commission endorsed the concept of a national road to Mount Vernon and advocated preserving the banks of the Potomac between Washington and Great Falls through parkway construction as well. These early suggestions produced few immediate results, but they established the general concept that led to the eventual creation of MVMH and GWMP.

Drawing based on 1932 American Motorist magazine cover

Documentation of George Washington Memorial Parkway was undertaken as a two-year project by the Washington office of the Historic American Buildings Survey/Historic American Engineering Record (HABS/HAER), a division of the National Park Service, Robert Kapsch, Chief of HABS/HAER; Paul Dolinsky, Chief of HABS; Eric Delony, Chief of HAER. The project was cosponsored by the the National Park Service Roads and Bridges Program, John Gingles, Manager. The project supervisor was Sara Amy Leach, HABS Historian. HAER staff architect Todd Croteau provided design support. Large-format photographs were produced by HABS photographer Jack Boucher and HAER photographer Jet Lowe.

The summer 1993 documentation team consisted of architect technicians Gary McCloud (Catholic University) and Peter Ratcliffe (Catholic University). The summer 1994 team consisted of architects Robert Dawson (University of Arizona) and Michael Gala (Catholic University), and landscape architects Ed Lupyak (Pennsylvania State University) and Anna Maria Marconi-Betka (ICOMOS/Poland). The 1994 team leader was landscape architect Tim Mackey (Harvard University). The historic overview was written by Timothy Davis (University of Texas). Bridge reports were prepared by Michael Kucher (University of Delaware) and Jennifer Wentzien (University of Washington).

The growing popularity of the automobile during the 1920s produced widespread dissatisfaction with existing routes to Mount Vernon and helped parkway promoters secure congressional authorization for a roadway that would incorporate the latest advances in parkway design. Generous funding ensured that the road would be monumental in character and that it would be completed in time to accommodate the surge in tourist traffic expected for the celebration of the bicentennial of George Washington's birth in 1932. The parkway was also touted as an historic corridor and as an aesthetic and engineering masterpiece that would itself serve as the consummate means of honoring George Washington's memory. The inclusion of historic sites and the installation of numerous memorial plantings, tablets, and monuments contributed a didactic emphasis on cultural features that distinguishes GWMP from most other parkways and National Park Service roads.

MVMH was designed and built by the U.S. Bureau of Public Roads (BPR). Since New York's Westchester County Park Commission was the acknowledged leader in motor parkway design, the BPR engaged Westchester veterans Gilmore Clarke and Jay Downer to serve as design consultants. Clarke also designed the bridge elevations for the original MVMH section. The BPR's R. E. Toms was the principal highway engineer in charge of the project. BPR District Engineer J. W. Johnson was in charge of construction, while Westchester-trained BPR landscape architect Wilbur Simonson supervised the development of the parkway landscape. BPR engineer J. V. McNary oversaw construction of bridges, overpasses, and other engineered structures.

MVMH was heralded as a paragon of parkway development and as an example of successful cooperation between landscape architects, planners, and highway engineers. The popular and professional press heralded it as "America's Most Modern Motorway" and used it to demonstrate the value of broad right-of-ways, limited access, careful attention to landscape design, continuously-curving roadway alignment, and sophisticated traffic circulation features. Many of these elements later became standard aspects of highway design practice.

The northern portions of GWMP were largely completed during the 1950s-60s. The post-World War II sections of GWMP continued to employ advanced parkway design and construction techniques as they wound along the rugged terrain of the Potomac Palisades. The later parkway sections were distinguished by longer, more sweeping curves, continuous medians, independent road alignments for north- and south-bound traffic lanes, and soaring concrete bridges spanning the steep ravines along the upper Potomac. The last stretch between Chain Bridge and the District of Columbia line was completed in 1970. The northern sections of GWMP were designed and built by the BPR in cooperation with National Park Service staff architects and landscape architects.

During the 1930s, parkway construction was seen as a desirable means of combining recreational development and landscape preservation. By the time the project wound toward a close in the 1960s, high-speed motorways were generally regarded as incompatible with natural resource protection. While political and economic concerns were largely responsible for killing the southeast leg of the parkway, preservationists played an important role in stopping GWMP's northern extension well south of the proposed terminus at Great Falls. The Maryland segment was redesignated Clara Barton Parkway in November 1989.

Clara Barton

DRAWN BY: Ed Lupyak, 1994

GEORGE WASHINGTON MEMORIAL PARKWAY DOCUMENTATION PROJECT 1994
NATIONAL PARK SERVICE
UNITED STATES DEPARTMENT OF THE INTERIOR

NAME AND LOCATION OF STRUCTURE: GEORGE WASHINGTON MEMORIAL PARKWAY
MOUNT VERNON, FAIRFAX COUNTY, VIRGINIA

SHEET 1 OF 21 SHEETS

HISTORIC AMERICAN ENGINEERING RECORD VA-69

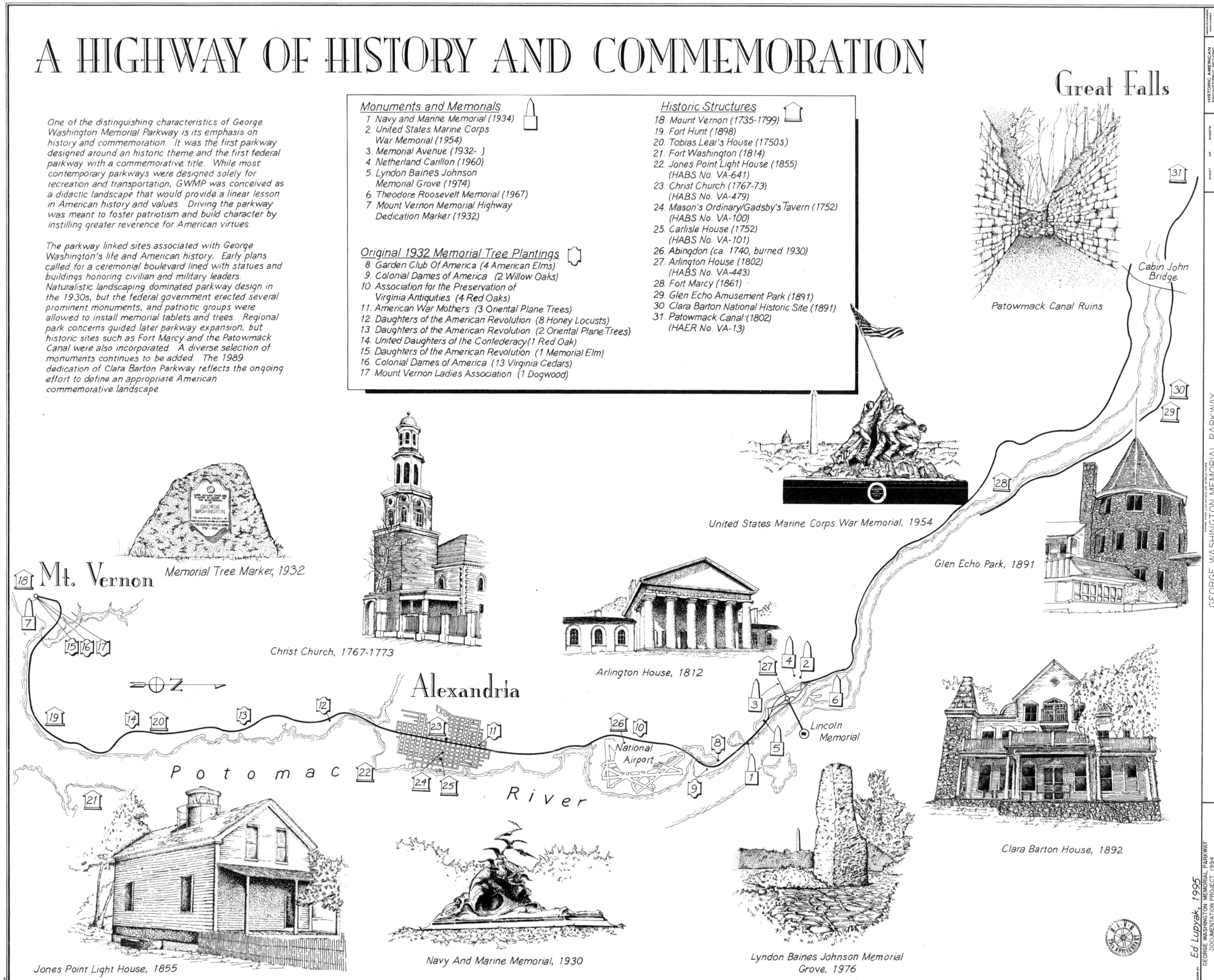

A HIGHWAY OF HISTORY AND COMMEMORATION
One of the distinguishing characteristics of George Washington Memorial Parkway is its emphasis on history and commemoration. It was the first parkway designed around an historic theme and the first federal parkway with a commemorative title. While most contemporary parkways were designed solely for recreation and transportation, GWMP was conceived as a didactic landscape that would provide a linear lesson in American history and values. Driving the parkway was meant to foster patriotism and build character by instilling greater reverence for American virtues.
The parkway linked sites associated with George Washington's life and American history. Early plans called for a ceremonial boulevard lined with statues and buildings honoring civilian and military leaders. Naturalistic landscaping dominated parkway design in the 1930s, but the federal government erected several prominent monuments, and patriotic groups were allowed to install memorial tablets and trees. Regional park concerns guided later parkway expansion, but historic sites such as Fort Marcy and the Patowmack Canal were also incorporated. A diverse selection of monuments continues to be added. The 1989 dedication of Clara Barton Parkway reflects the ongoing effort to define an appropriate American commemorative landscape.
Monuments and Memorials
1. Navy and Marine Memorial (1934)
2. United States Marine Corps War Memorial (1954)
3. Memorial Avenue (1932-)
4. Netherland Carillon (1960)
5. Lyndon Baines Johnson Memorial Grove (1974)
6. Theodore Roosevelt Memorial (1967)
7. Mount Vernon Memorial Highway Dedication Marker (1932)
Original 1932 Memorial Tree Plantings
8. Garden Club Of America (4 American Elms)
9. Colonial Dames of America (2 Willow Oaks)
10. Association for the Preservation of Virginia Antiquities (4 Red Oaks)
11. American War Mothers (3 Oriental Plane Trees)
12. Daughters of the American Revolution (8 Honey Locusts)
13. Daughters of the American Revolution (2 Oriental Plane Trees)
14. United Daughters of the Confederacy (1 Red Oak)
15. Daughters of the American Revolution (1 Memorial Elm)
16. Colonial Dames of America (13 Virginia Cedars)
17. Mount Vernon Ladies Association (1 Dogwood)
Historic Structures
18. Mount Vernon (1735-1799)
19. Fort Hunt (1898)
20. Tobias Lear's House (1750s)
21. Fort Washington (1814)
22. Jones Point Light House (1855) (HABS No. VA-641)
23. Christ Church (1767-73) (HABS No. VA-479)
24. Mason's Ordinary/Gadsby's Tavern (1752) (HABS No. VA-100)
25. Carlisle House (1752) (HABS No. VA-101)
26. Abingdon (ca. 1740, burned 1930)
27. Arlington House (1802) (HABS No. VA-443)
28. Fort Marcy (1861)
29. Glen Echo Amusement Park (1891)
30. Clara Barton National Historic Site (1891)
31. Patowmack Canal (1802) (HAER No. VA-13)
Great Falls
Patowmack Canal Ruins
Cabin John Bridge
United States Marine Corps War Memorial, 1954
Glen Echo Park, 1891
Mt. Vernon
Memorial Tree Marker, 1932
Christ Church, 1767-1773
Alexandria
Arlington House, 1812
Lincoln Memorial
National Airport
Potomac River
Clara Barton House, 1892
Jones Point Light House, 1855
Navy And Marine Memorial, 1930
Lyndon Baines Johnson Memorial Grove, 1976
Ed Lupyak, 1995
GEORGE WASHINGTON MEMORIAL PARKWAY
MOUNT VERNON, FAIRFAX COUNTY, VIRGINIA
HISTORIC AMERICAN ENGINEERING RECORD VA-69

GEORGE WASHINGTON MEMORIAL PARKWAY

EVOLUTION OF TRANSPORTATION ROUTES 1776 – 1970

0 1 3 5 10 MILES
0 1 3 5 10 KILOMETERS

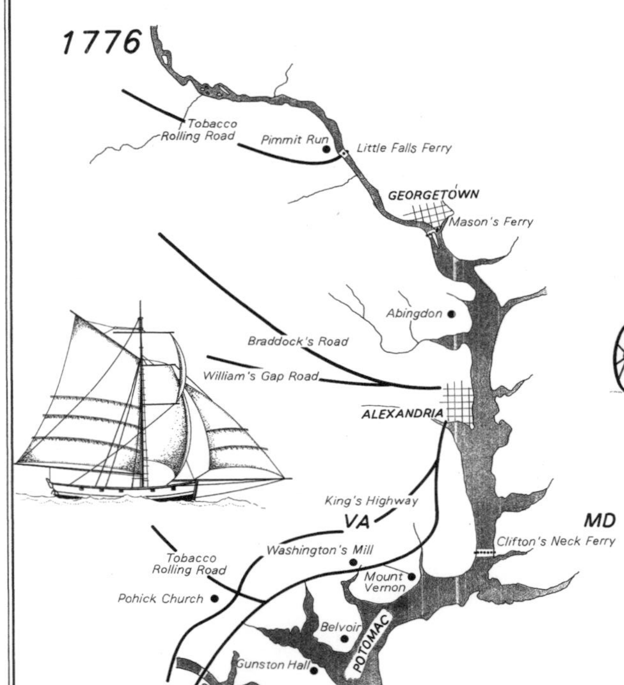

Colonial Era

During the colonial and early federal periods, the Potomac River provided the easiest and most efficient means of transportation. Small sailboats, barges, and large ocean-going vessels plied the Potomac, delivering goods and picking up tobacco at plantation wharves and growing port cities like Georgetown and Alexandria. A rough, unimproved road known as the King's Highway stretched from Fredericksburg to Alexandria, keeping well inland from the Potomac to avoid crossing its marshes and estuaries. The opening of the Occoquan ferry in 1691 created a more easterly route that passed closer to Mount Vernon. By the late eighteenth century, two primitive trade roads linked Alexandria with the Shenandoah valley.

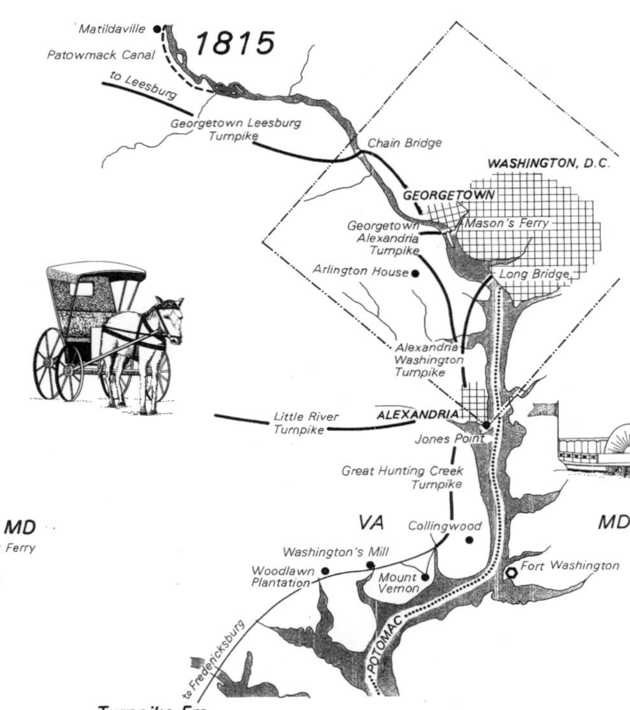

Turnpike Era

The rapid spread of toll roads at the beginning of the nineteenth century created a marked improvement in the regional road network. The first bridges across the Potomac River were also built at this time, serving as focal points for the new turnpikes. By 1815, Alexandria was connected to Washington by turnpikes leading to the new Long Bridge and to the ferry at Mason's Island. A new bridge over Great Hunting Creek led to the virtual abandonment of the old King's Highway in favor of a turnpike being built toward Accotink. Turnpikes linked Georgetown and Alexandria with the Shenandoah Valley, but still kept away from the rugged Potomac palisades.

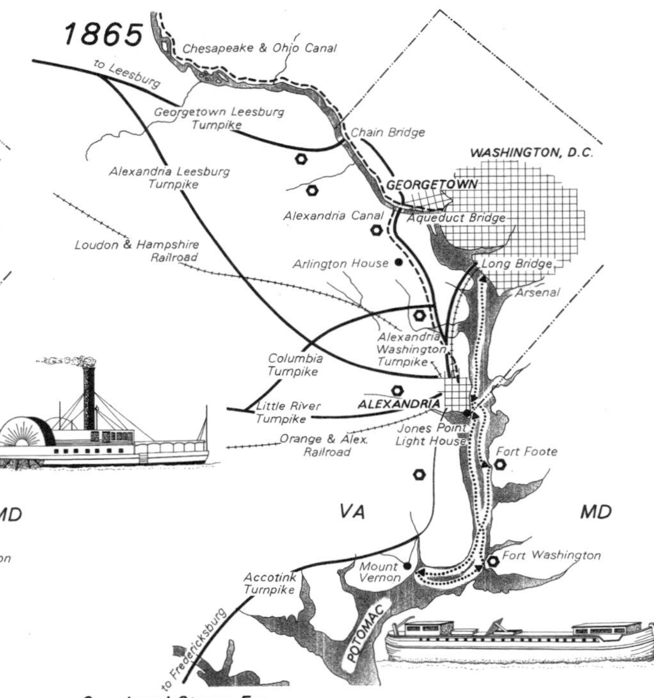

Canal and Steam Era

The mid nineteenth century witnessed the construction of an extensive canal network along the Potomac, along with the widespread application of steam engines to move goods and people throughout the region. Steamboats operated along the Potomac by the 1820s, becoming the preferred means of long distance travel and serving as the primary tourist conveyance to Mount Vernon. The Chesapeake and Ohio and Alexandria canals effectively transported heavy, non-perishable goods. Several steam railroads also provided service to the Washington/Alexandria area at the outbreak of the Civil War. Unfortunately, these new forms of transportation caused the region's marginal road network to fall into greater decline.

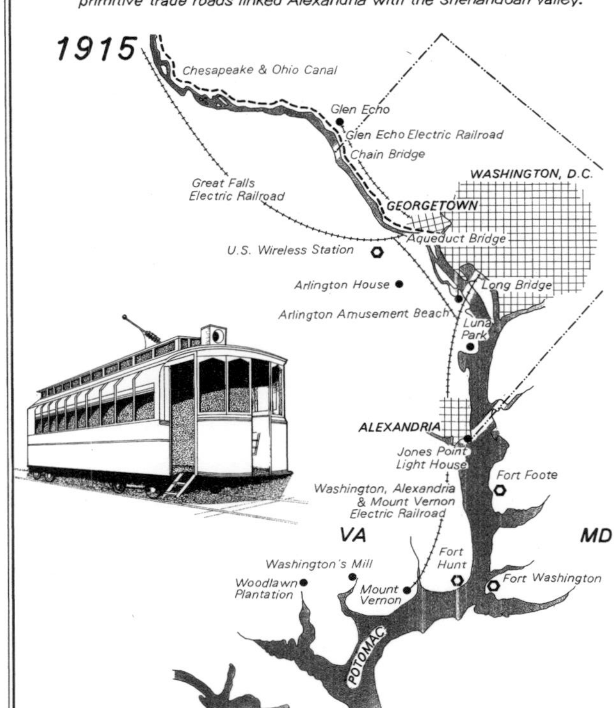

Electric Railway Era

The construction of an electric railway line from Alexandria to Mount Vernon in 1892 dramatically increased the number of visitors and helped initiate the transformation of northern Virginia into a suburb of Washington. The trolley line was extended from Alexandria to Washington in 1896. The electric railway's lower cost and frequent service helped democratize the Mount Vernon tourist experience by offering access to a broader segment of the American public. Electric railways soon extended north of the city to Great Falls and Glen Echo, and several amusement parks were built along the trolley lines. Local citizens also used the electric railroad to commute to work and to seek out picnic areas along the Potomac shoreline.

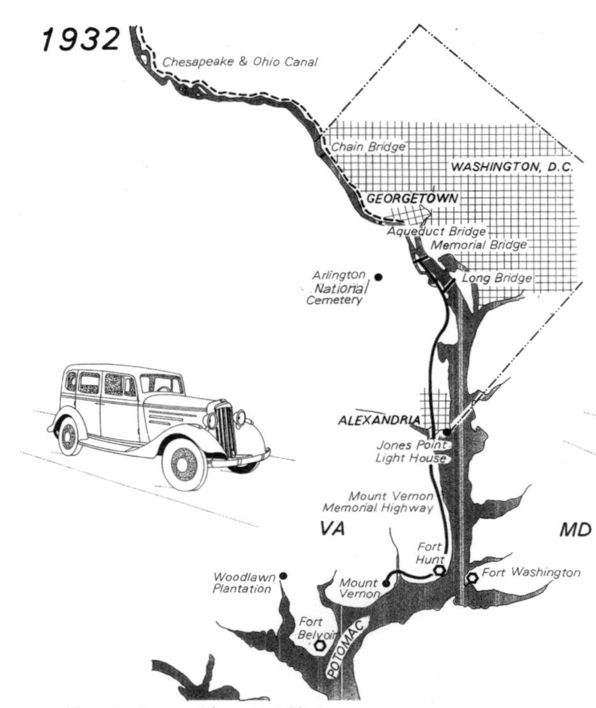

Mount Vernon Memorial Highway

By the 1920s it was obvious that existing roadways were inadequate for automobile traffic. The primary road from Washington to Mount Vernon was congested, poorly laid out, and lined with billboards and commercial establishments. In 1928 the Bureau of Public Roads was authorized to build a state-of-the-art motor parkway along the Potomac shoreline. The swamps and creek mouths that had long thwarted potential road builders were conquered with handsome bridges and extensive landfilling operations. By 1932, motorists could enjoy a paved, four-lane motorway that wound from Washington to Mount Vernon through attractive scenery with no roadside development and no major at-grade intersections.

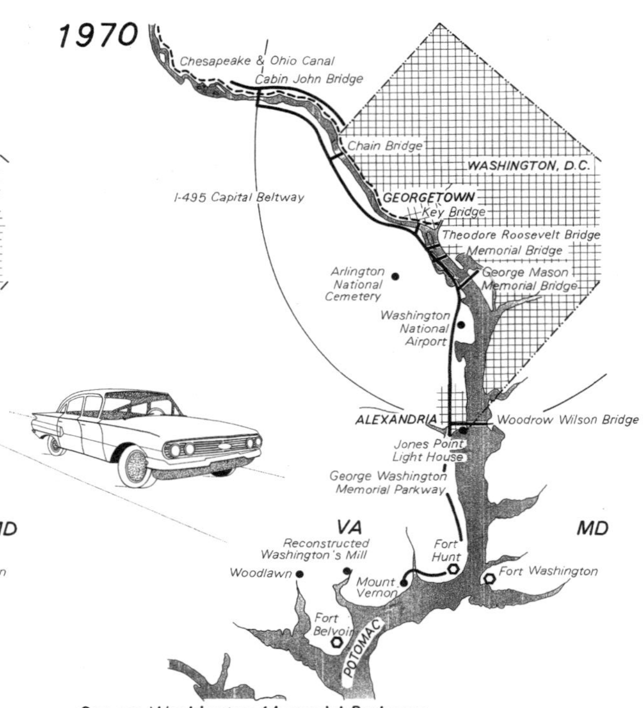

George Washington Memorial Parkway

The 1930 Capper-Cramton Act authorized the creation of a parkway along both sides of the Potomac from Mount Vernon to Great Falls. The parkway would preserve the Potomac palisades while aiding the suburban development of northern Virginia and Montgomery County, Maryland. Funding problems and logistical difficulties delayed construction for many years and scaled down the ambitious scope of the original project. The parkway was extended to the CIA headquarters near Langley in 1959, and reached its present northern terminus at the Capital Beltway in 1962. The Maryland portion was limited to a 6.6 mile stretch from Chain Bridge to MacArthur Boulevard, which was officially completed in 1970.

DRAWN BY: Anna Marconi-Betka, 1994

GEORGE WASHINGTON MEMORIAL PARKWAY DOCUMENTATION PROJECT 1994 NATIONAL PARK SERVICE UNITED STATES DEPARTMENT OF THE INTERIOR	NAME AND LOCATION OF STRUCTURE: GEORGE WASHINGTON MEMORIAL PARKWAY MOUNT VERNON, FAIRFAX COUNTY, VIRGINIA	SHEET 4 OF SHEETS 21	HISTORIC AMERICAN ENGINEERING RECORD VA-69

IF REPRODUCED, PLEASE CREDIT: HISTORIC AMERICAN BUILDINGS SURVEY, NATIONAL PARK SERVICE, NAME OF DELINEATOR, DATE OF THE DRAWING

PARKWAY PROPOSALS 1890-1930

1890

In 1890, the U.S. Army Corps of Engineers examined three possible routes for a National Road from Washington to Mount Vernon via the Aqueduct Bridge. The Mount Vernon Avenue Association preferred the middle route because it had many historical associations and provided the most extensive views of Washington and the Potomac River valley. The Senate Park Commission endorsed this route in 1901.

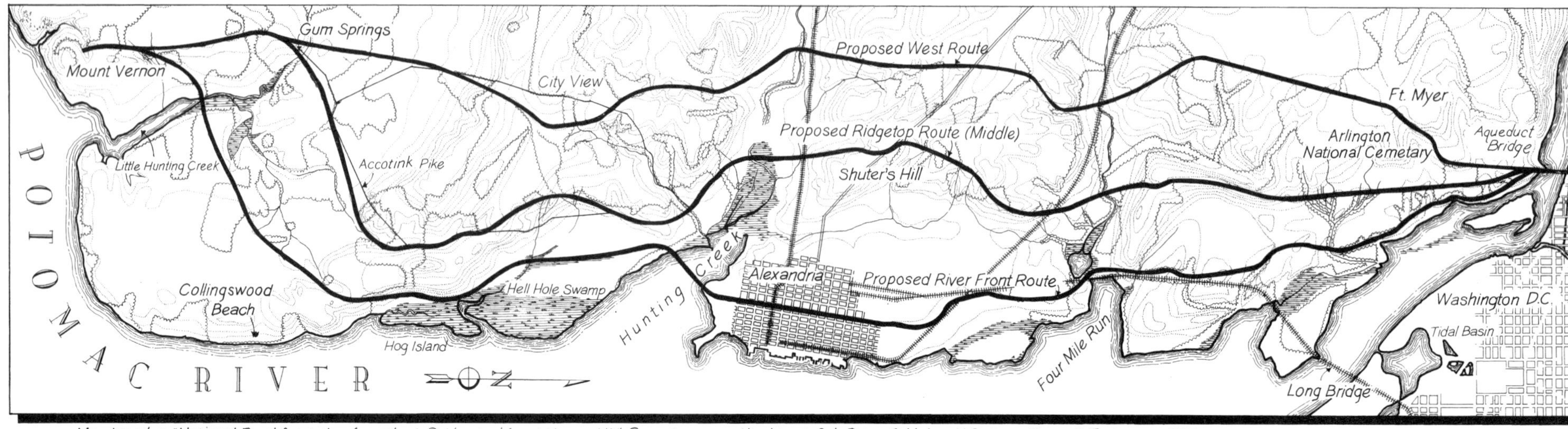

Map based on "National Road from the Aqueduct Bridge to Mount Vernon, VA" Report prepared by Lieut. Col. Peter C Hains, U.S. Army Corp of Engineers, 1890

1930

Two alignments were considered when the project was revived in the 1920s. The riverfront route was chosen on the grounds that it offered better possibilities for general park development and provided access to more historic sites. The shoreline drive would require extensive filling operations, and be more expensive to construct than an inland route, but it would be more suitable for automobile traffic than the hilly inland route with its frequent grade crossings.

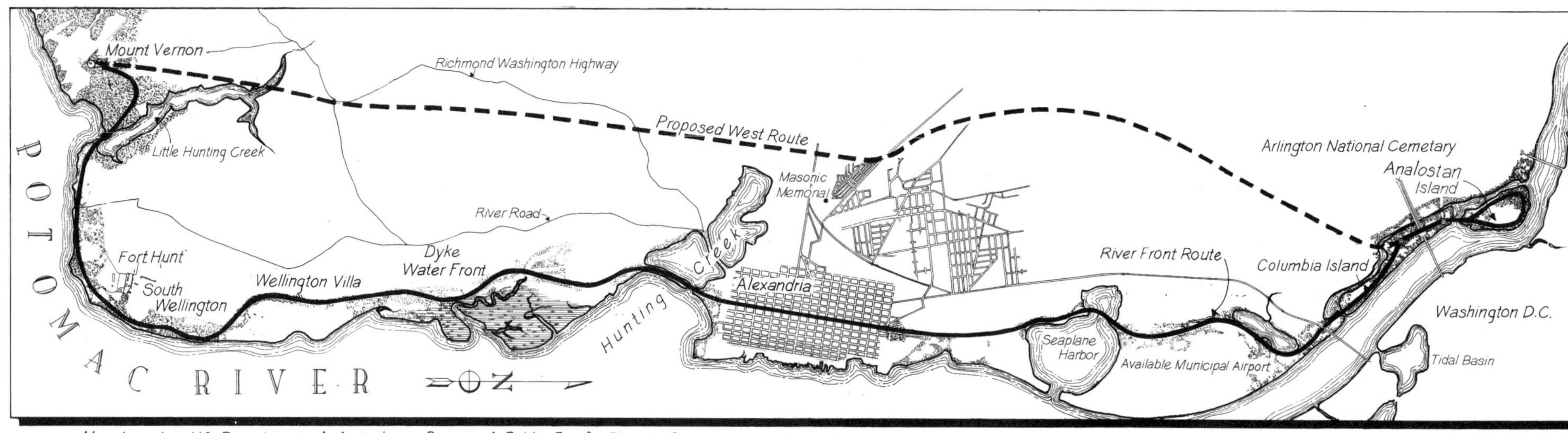

Map based on U.S. Department of Agriculture, Bureau of Public Roads, Plan for Development, Mount Vernon Memorial Highway, VA., January, 1930

The Evolution of George Washington Memorial Parkway

Today's George Washington Memorial Parkway bears little resemblance to the grand memorial boulevard promoted by the Mount Vernon Avenue Association in the late nineteenth century. It also differs significantly from the ambitious regional park plan proposed in the 1920s-30s.

Early proposals concentrated on linking historic sites and providing panoramic views that could be enjoyed by slow-moving travelers in carriages or on horseback. The original route generally followed existing roadways to keep down costs, avoiding the marshes and estuaries along the Potomac shoreline to eliminate expensive fills and bridges. While ordinary country lanes would be used at first, the avenue promoters called for an imposing formal boulevard, which would be well-paved and drained and lined with statues, memorials, and stately rows of regularly spaced trees.

When the Bureau of Public Roads took over the project in the 1920s, it ruled in favor of the river front route, which eliminated most grade crossings and took advantage of cheap wetlands and existing government holdings such as Fort Hunt. Technical advances and generous funding reduced the problems posed by the marshy, low-lying terrain.

BPR designers used aerial photography to help devise an alignement that would be safe at automotive speeds while providing a dynamic experience of constantly changing views and shielding motorists from unsightly development. Unlike the earlier avenue promoters, they also sought to preserve and restore the surrounding natural landscape to create a regional park for general recreational purposes.

1930

The original intention behind the creation of George Washington Memorial Parkway was to extend MVMH on both sides of the river from Mount Vernon to Great Falls. Bridges or ferries would enable motorists to make a complete loop. A southern continuation to Woodlawn, Gunston Hall, and Wakefield was contemplated but never begun. A variety of economic and political factors combined to eliminate the southeast leg of the parkway and prevent the northern segments from reaching Great Falls.

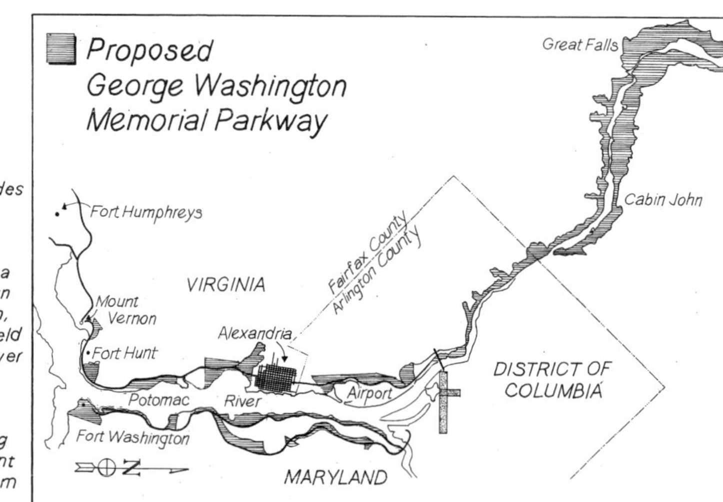

Map based on National Park and Planning Commission, 1931

HISTORIC AMERICAN ENGINEERING RECORD VA-69

SHEET 5 OF 21

GEORGE WASHINGTON MEMORIAL PARKWAY
MOUNT VERNON, FAIRFAX COUNTY, VIRGINIA

Robert Dawson and Ed Lupyak, 1994

GEORGE WASHINGTON MEMORIAL PARKWAY DOCUMENTATION PROJECT 1994

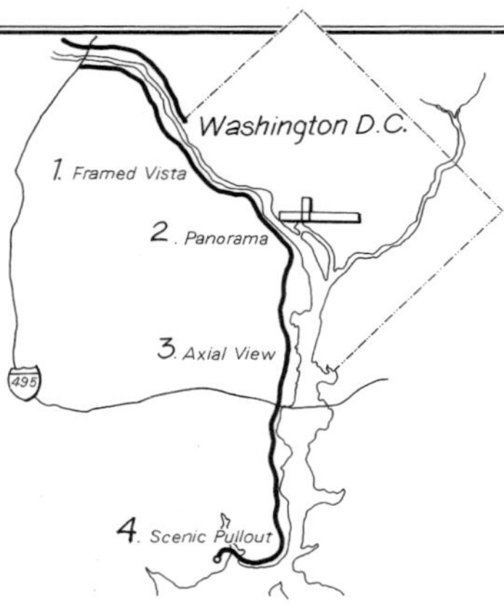

PLANNED VIEWS

Early motor parkway designers faced the challenge of adapting traditional landscape architecture methods to the new speeds and scale produced by automobiles. The influence of romantic landscape painting and picturesque park design remained strong, but landscape compositions were simplified to be appreciated at higher speeds. The ability of automobiles to easily cover distances and climb hills gave parkway designers greater ability to seek out attractive scenery and dramatic viewpoints. GWMP's designers combined this new freedom with traditional design techniques to provide access to spectacular scenery and focus attention on dramatic views and symbolic vistas.

"The automobile has . . . made necessary a broader treatment of views. Intimate and confined views cannot be appreciated from a fast moving vehicle; simplicity and breadth are required. Different views must not follow too closely upon one another, and the openings and vistas through bordering woods or shrubbery must be of far greater width. Every accent of the prospect and planting must be stronger because of the brief time in which it is seen."

-- Charles W. Eliot II, Landscape Architecture Quarterly, 1922

Scale: 1" = 40', 1:480

0 40 80 120 FEET 0 10 20 40 METERS

FRAMED VISTA -1-

While pedestrians or carriage occupants could easily enjoy lateral views, motorists had to continually watch where they were going. As speeds increased, the emphasis on forward views became increasingly important. Parkway designers frequently combined a bend in the road with a break in the bordering vegetation to frame scenes off the main axis of the parkway. These "windows" were deliberately limited in width and number to avoid creating prolonged distractions.

GWMP designers employed this technique in dramatic fashion along the Potomac Palisades, where southbound motorists are treated to striking views of Washington, D.C.

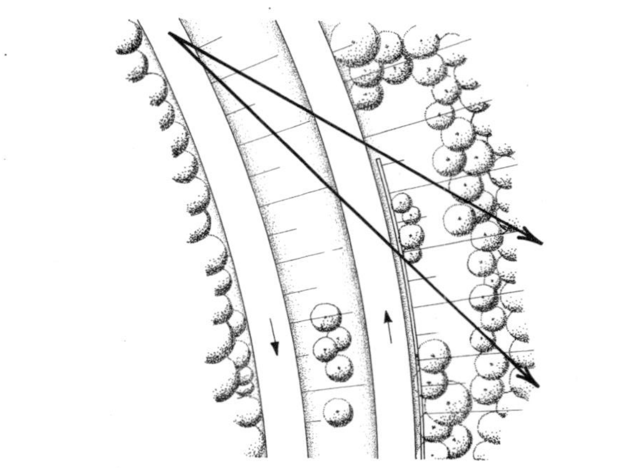

Potomac Palisades

PANORAMA -2-

Parkway designers considered the relationship between the road and Washington's monumental core to be a matter of great aesthetic and symbolic significance. The approach to Washington was designed to provide a simple yet dignified transition between the informal parkway landscape and the grand spaces and neoclassical monumentality of the national capital.

Border plantings were kept to a minimum in order to provide expansive views across the Potomac River. The circulation system of Columbia Island was designed in part to slow down motorists so they would appreciate these views at a more dignified pace. When the parkway was originally built, the heights near National Airport provided another panoramic view of the city, but the roadway was moved to lower ground when the airport was constructed.

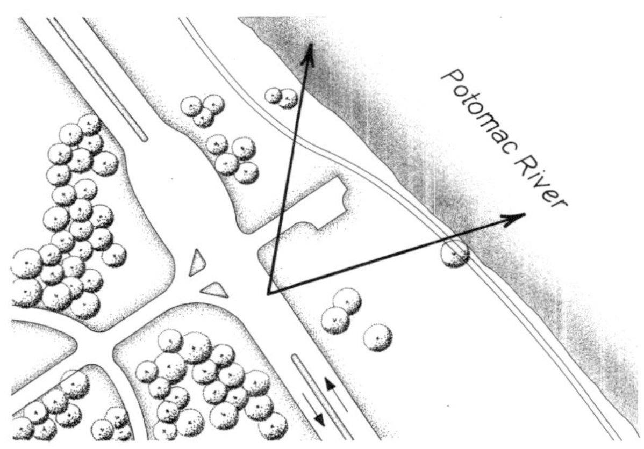

Columbia Island

AXIAL VIEW -3-

The use of long straight avenues to direct attention to objects of interest was another classic design technique employed by parkway builders. This tactic was used sparingly, since the parkway was designed primarily as an informal landscape with continuous sweeping curvature and irregular naturalistic planting.

The most striking use of the classic axial view occurs just north of Alexandria, where one of the parkway's two long straight stretches points directly at the distant Washington Monument. Tall rows of trees on either side of the parkway help focus the motorist's gaze while screening out surrounding development. This "Monument Vista" provides the first suggestion of formal Washington. It was strongly emphasized in the original parkway plans.

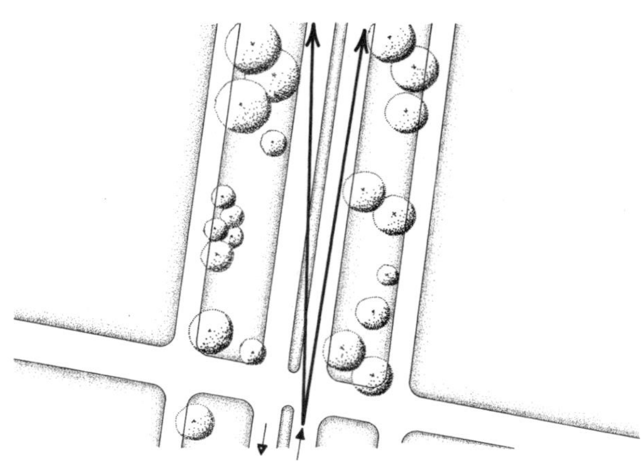

Washington Monument Vista, Alexandria

SCENIC PULLOUT -4-

Small parking areas were provided at particularly scenic areas to provide motorists with an opportunity to safely pull off the roadway and enjoy the view. These scenic pullouts range in size from minor pavement widenings to extensively developed picnic areas complete with toilet facilities, tables, fireplaces, and interpretive signs explaining the adjacent historic and natural features.

The Hillcrest overlook provides an excellent view across the Potomac to Fort Washington, an imposing edifice that guarded the southern approach to Washington from 1808-1922.

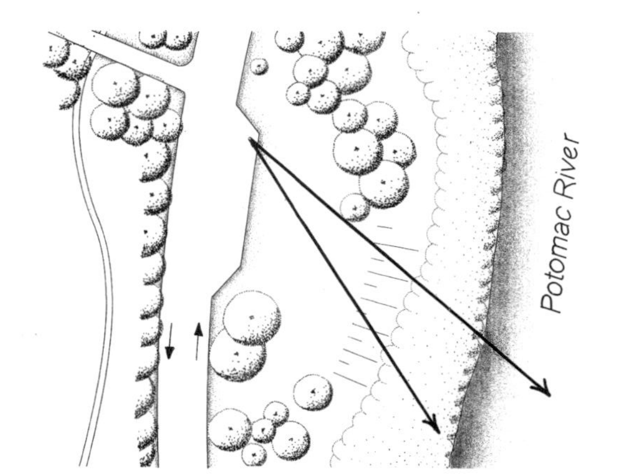

Hillcrest Overlook

DRAWN BY: Anna Marconi-Betka, 1994

GEORGE WASHINGTON MEMORIAL PARKWAY DOCUMENTATION PROJECT 1994
NATIONAL PARK SERVICE
UNITED STATES DEPARTMENT OF THE INTERIOR

NAME AND LOCATION OF STRUCTURE
GEORGE WASHINGTON MEMORIAL PARKWAY
MOUNT VERNON, FAIRFAX COUNTY, VIRGINIA

SHEET 7 OF SHEETS 21

HISTORIC AMERICAN ENGINEERING RECORD VA-69

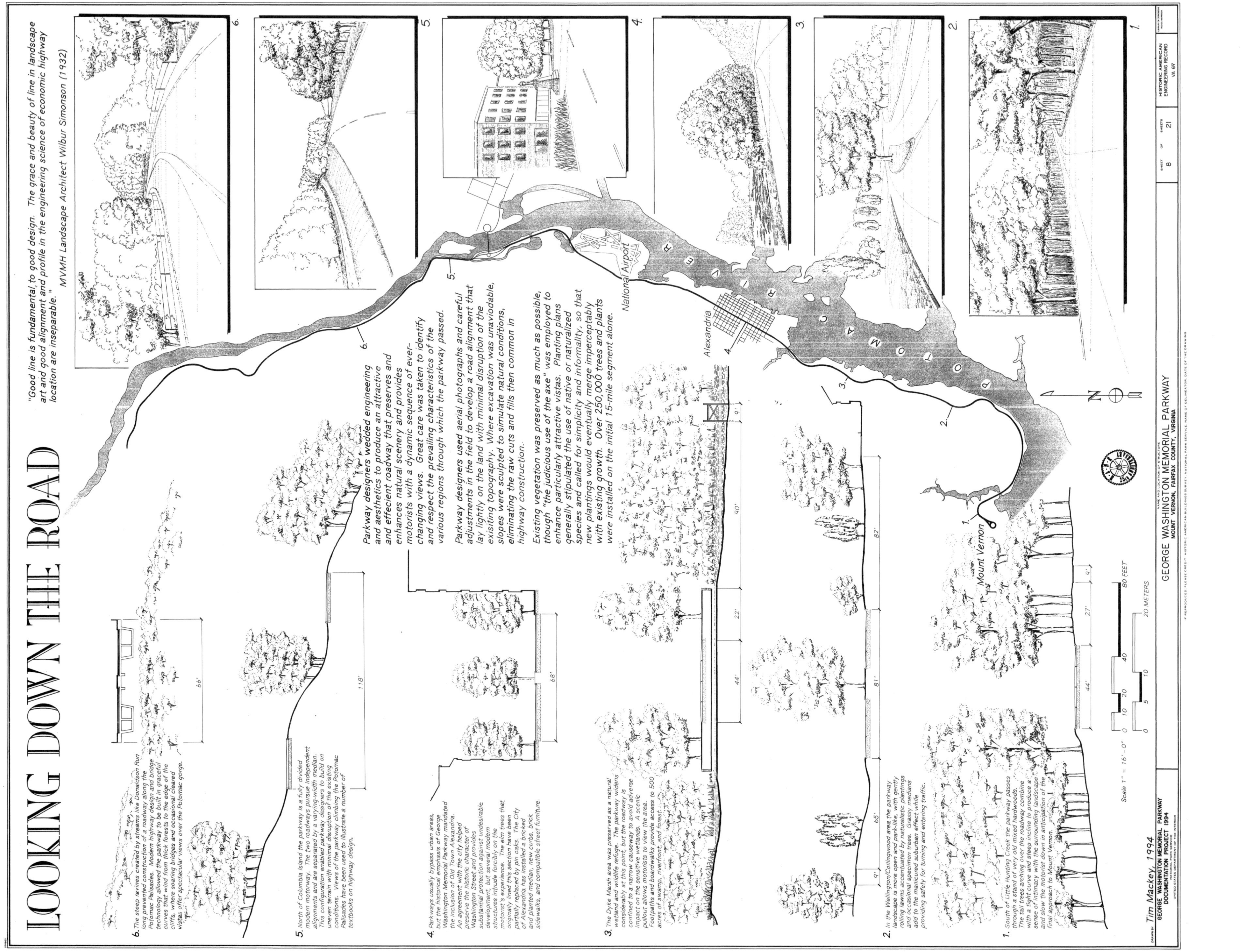

LOOKING DOWN THE ROAD
"Good line is fundamental to good design. The grace and beauty of line in landscape art and good alignment and profile in the engineering science of economic highway location are inseparable."
MVMH Landscape Architect Wilbur Simonson (1932)
6. The steep ravines created by streams like Donaldson Run long prevented construction of a roadway along the Potomac Palisades. Modern highway design and bridge technology allowed the parkway to be built in graceful curves that wind from thick forests to the edge of the cliffs, where soaring bridges and occasional cleared vistas offer spectacular views over the Potomac gorge.
66'
5. North of Columbia Island the parkway is a fully divided modern motorway. The two roadways pursue independent alignments and are separated by a varying-width median. This configuration enabled parkway designers to build on uneven terrain with minimal disruption of the existing conditions. Views of the parkway climbing the Potomac Palisades have been used to illustrate a number of textbooks on highway design.
118'
Parkway designers wedded engineering and aesthetics to produce an attractive and effecient roadway that preserves and enhances natural scenery and provides motorists with a dynamic sequence of ever-changing views. Great care was taken to identify and respect the prevailing characteristics of the various regions through which the parkway passed.
Parkway designers used aerial photographs and careful adjustments in the field to develop a road alignment that lay lightly on the land with minimal disruption of the exisiting topography. Where excavation was unavoidable, slopes were sculpted to simulate natural conditions, eliminating the raw cuts and fills then common in highway construction.
Existing vegetation was preserved as much as possible, though "the judicious use of the axe" was employed to enhance particularly attractive vistas. Planting plans generally stipulated the use of native or naturalized species and called for simplicity and informality, so that new plantings would eventually merge imperceptably with existing growth. Over 250,000 trees and plants were installed on the initial 15-mile segment alone.
4. Parkways usually bypass urban areas, but the historical emphasis of George Washington Memorial Parkway mandated the inclusion of Old Town Alexandria. An agreement with the city helped preserve the historic character of Washington Street and provides substantial protection against undesirable development, but several modern structures intrude forcibly on the motorist's experience. The elm trees that originally lined this section have been partially replaced by pin oaks. The City of Alexandria has installed a bricked and planted median, new curbs, brick sidewalks, and compatible street furniture.
68'
3. The Dyke Marsh area was preserved as a natural wetland and wildlife refuge. The parkway widens considerably at this point, but the roadway is confined to a narrow causeway to avoid adverse impact on the sensitive wetlands. A scenic pullout allows motorists to view the area. Footpaths and boardwalks provide access to 500 acres of swamp, riverfront, and forest.
44'
22'
90'
9'
2. In the Wellington/Collingwood area the parkway landscape is more open and park-like, with gently rolling lawns accentuated by naturalistic plantings and occasional specimen trees. Grassy medians add to the manicured suburban effect while providing safety for turning and entering traffic.
9'
65'
81'
82'
1. South of Little Hunting Creek the parkway passes through a stand of very old mixed hardwoods. The tall trees arching over the roadway combine with a tight curve and steep incline to produce a sense of intimacy with the surrounding landscape and slow the motorist down in anticipation of the final approach to Mount Vernon.
44'
27'
9'
National Airport
Alexandria
POTOMAC RIVER
Mount Vernon
1.
2.
3.
4.
5.
6.
N
Scale 1" = 16'-0"
0 10 20 40 80 FEET
0 5 10 20 METERS
DRAWN BY: Tim Mackey, 1994
GEORGE WASHINGTON MEMORIAL PARKWAY DOCUMENTATION PROJECT 1994
NATIONAL PARK SERVICE
UNITED STATES DEPARTMENT OF THE INTERIOR
NAME AND LOCATION OF STRUCTURE
GEORGE WASHINGTON MEMORIAL PARKWAY
MOUNT VERNON, FAIRFAX COUNTY, VIRGINIA
SHEET 8 OF SHEETS 21
HISTORIC AMERICAN ENGINEERING RECORD
VA 69
IF REPRODUCED, PLEASE CREDIT: HISTORIC AMERICAN BUILDINGS SURVEY, NATIONAL PARK SERVICE, NAME OF DELINEATOR, DATE OF THE DRAWING

PARKWAY CONSTRUCTION SEQUENCE

Parkway construction can be viewed as a sequence of highly specialized operations that accumulate in a series of layers to produce a final product that is both aesthetically pleasing and structurally sound.

First, a satisfactory route must be determined. Economic, political, aesthetic, and engineering concerns influence this decision. After the alignment is chosen, the existing landscape is shaped by cutting away high spots and filling in depressions and unstable ground in order to produce a solid and relatively level foundation. Parkway designers try to pick a route that minimizes the need for extensive grading, but MVMH's river front location required an extensive filling operation.

After the preliminary grading is complete, drainage structures are put in place and the road bed is prepared through the addition of a gravel sub-base and further grading. In modern road construction, a concrete or asphalt pavement is then laid down. Most of MVMH's pavement is reinforced concrete slab, but a flexible asphalt pavement was used over long stretches of filled land. Paving is usually followed by the construction of incidental features such as lighting fixtures, signs, and guard rails. The final step is to plant trees and shrubs, restore topsoil, complete final landscape grading, and add sod and grass seed.

MVMH's tight construction schedule made it imperative that these stages be carefully planned and coordinated so that the parkway could be finished in time for the George Washington Bicentennial Celebration in February 1932.

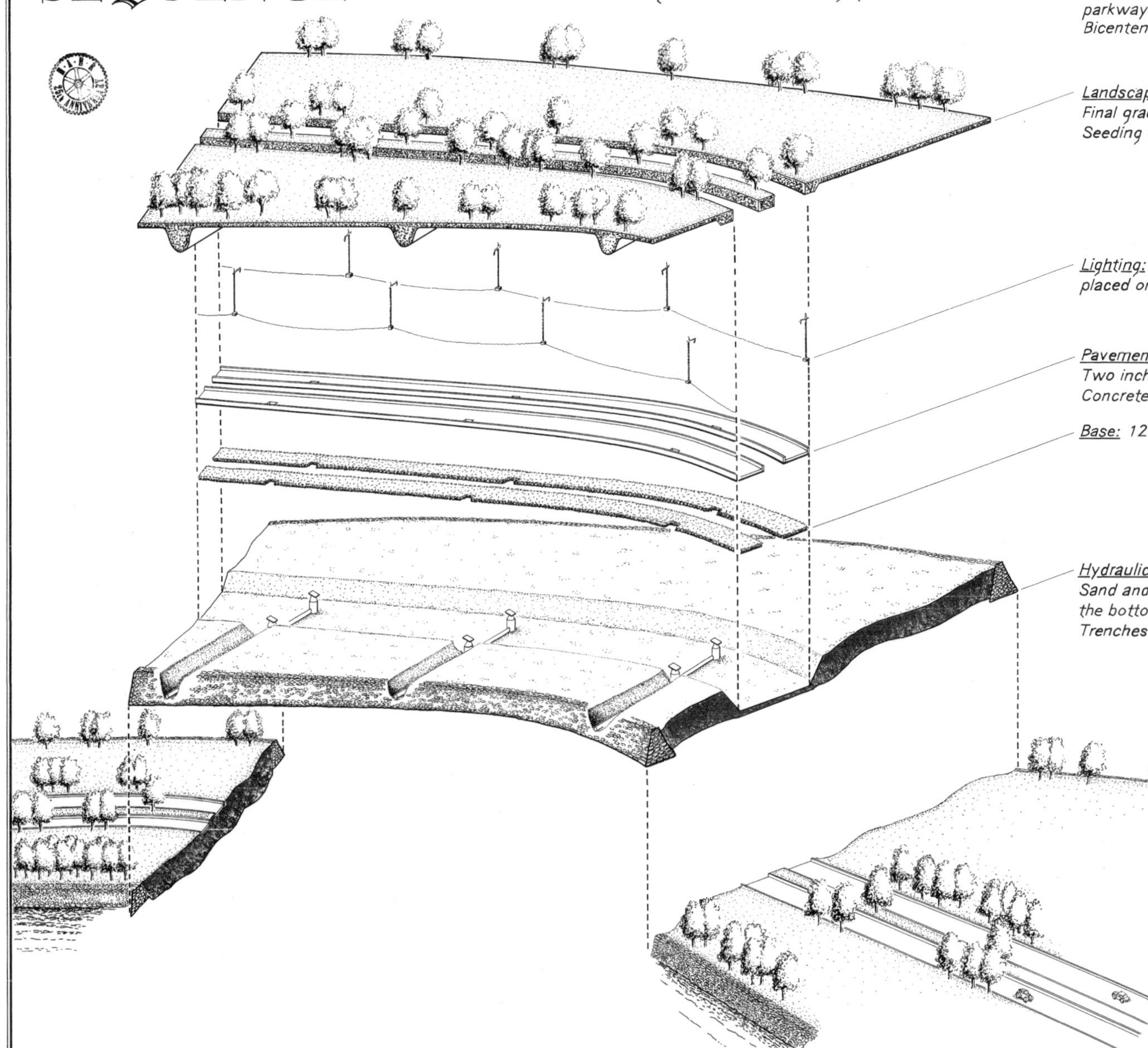

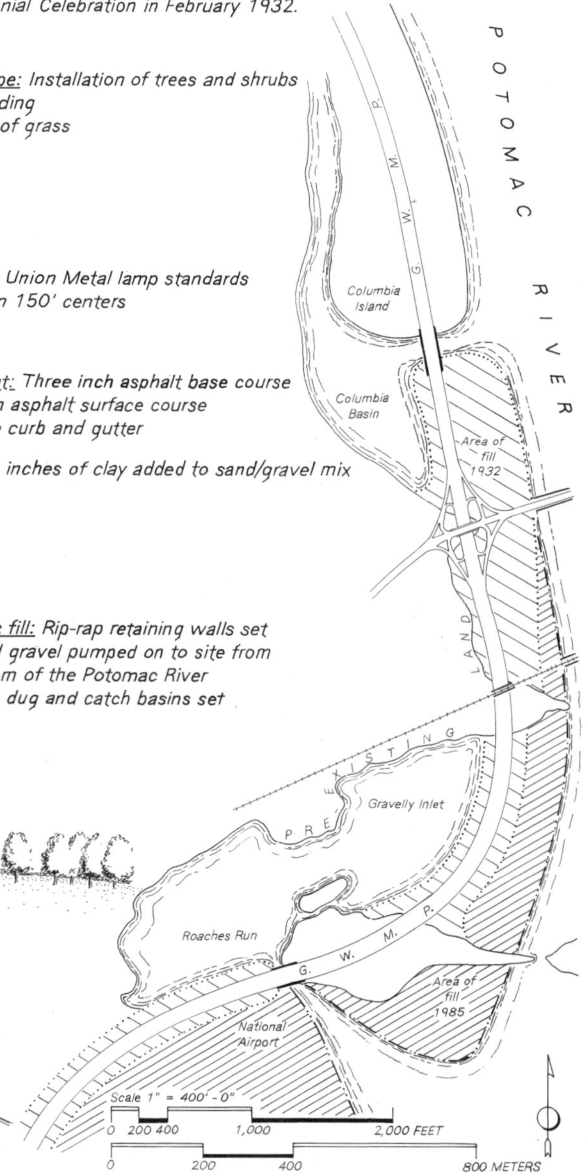

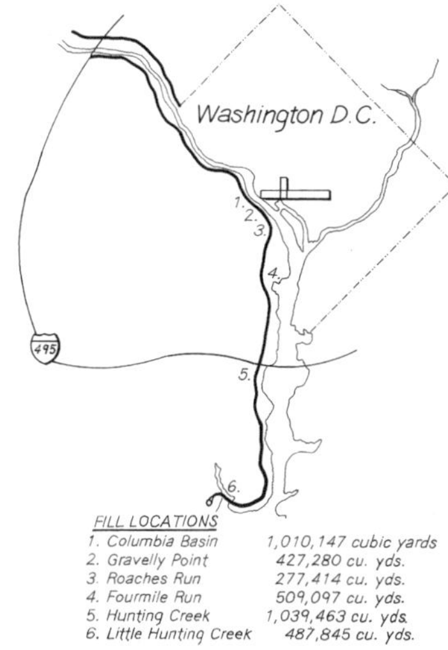

HYDRAULIC FILL

The greatest technical challenge faced by the parkway builders was constructing a stable roadway over the marshes, estuaries, and uneven shoreline of the Potomac waterfront. Approximately 2.75 miles of the original MVMH was built on newly placed sand and gravel fill, and the Potomac shoreline was extensively reconfigured to provide a more regular appearance better suited for recreational development.

Most of this material was "hydraulic fill" taken from the bottom of the Potomac River and pumped to the construction site by two powerful dredges operated by the U.S. Army Corps of Engineers. The pipeline from the dredges to the site was supported by pontoons and temporary trestles. A diesel booster pump provided additional power when the length of the pipeline exceeded the capacity of the dredges. Nearly 5 million cubic yards of fill was placed in this manner during the initial construction process. The creation of National Airport required another major hydraulic filling operation. Smaller amounts of fill have been added over the year to facilitate road relocations, provide additional park land, and further regularize the shoreline.

The "Welatka" was the most powerful 20-inch hydraulic dredge operated by the U.S Army Corps of Engineers in 1929. It was 214' long, 40' wide, and had quarters for a crew of 60. The dredge was driven by a 1500 hp triple expansion reciprocating engine powered by steam generated by four oil-fired water-tube boilers. It could dig to a depth of 45'. In combination with the diesel booster, it could pump 240 cubic yards of fill per hour over a distance of 5400'

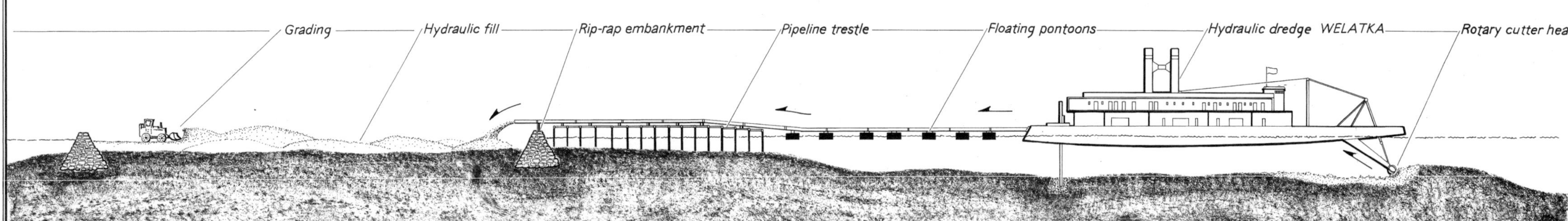

Tim Mackey, 1994

GEORGE WASHINGTON MEMORIAL PARKWAY
DOCUMENTATION PROJECT 1994

GEORGE WASHINGTON MEMORIAL PARKWAY
MOUNT VERNON, FAIRFAX COUNTY, VIRGINIA

HISTORIC AMERICAN ENGINEERING RECORD
VA-69

SHEET 9 OF 21 SHEETS

INTERSECTIONS

Parkway designers helped pioneer the development of free-flowing, limited-access motor roads, introducing numerous traffic circulation features that became standard components of modern highway design. Innovative traffic control techniques developed during the 1920s-30s for special parkway projects like MVMH eventually spread to general expressway and interstate highway construction.

Parkway designers sought to improve the safety and efficiency of motor roads by restricting the number of intersecting roadways and reducing the dangers posed by turning and entering traffic. The acquisition of broad right-of-ways enabled parkway builders to control access, eliminate commercial roadside development, and screen out distracting or undesirable views. Wide buffer strips, grade separations, traffic islands, and extensive medians added to the expense of parkway construction, but these features helped increase traffic capacity and provided a safer and more enjoyable driving experience for motorists traveling at modern speeds.

The Bureau of Public Roads' designers employed a variety of techniques to improve the safety and efficiency of MVMH. Locating the parkway along the riverfront automatically reduced the number of potentially troublesome cross roads. Several types of grade separation structures and safety intersections were used to eliminate conventional road crossings from the parkway except through the Washington Street section of downtown Alexandria.

By the time the northern sections of GWMP were constructed, continuous medians and fully-integrated, grade-separated interchanges were established parkway- and highway-building practice. The northern GWMP has some unusual traffic circulation features, including the u-turn overpass at Turkey Run, the cantilevered segment of Clara Barton Parkway, and the complex intersections with I-495 and U.S. Route 1.

Original 1932 Signage

2. Memorial Circle (1932)

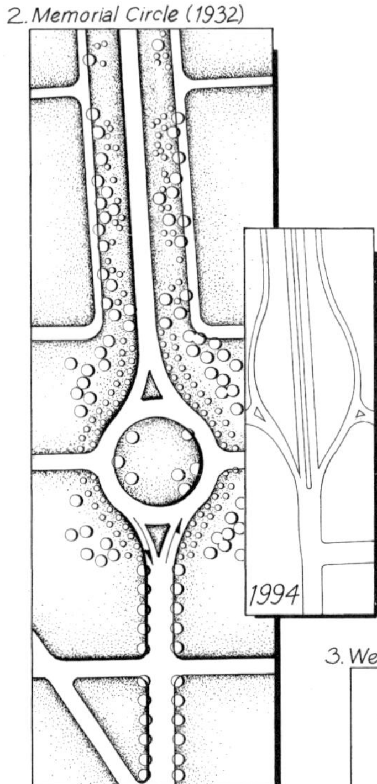

1. The First Federal Cloverleaf

The Fourteenth Street Highway Bridge interchange was the first cloverleaf built by the federal government. Cloverleafs required considerable space and were expensive to construct, but they were rapidly acknowledged as the safest type of intersection for high-speed traffic. A grade separation structure allowed through-traffic on both roadways to proceed unimpeded, while spiral connecting ramps enabled motorists to change direction without stopping or making dangerous left hand turns across traffic.

The cloverleaf was such a novel concept in 1930 that the BPR constructed a large scale model to convince the public that the design was both safe and attractive. Washington newspapers prepared motorists with detailed descriptions of how it functioned, and an extended median strip and extra signs were installed to prevent confusion. BPR landscape architect Wilbur Simonson proclaimed, "When completed it will be without question the most attractive grade-separation traffic unit ever designed."

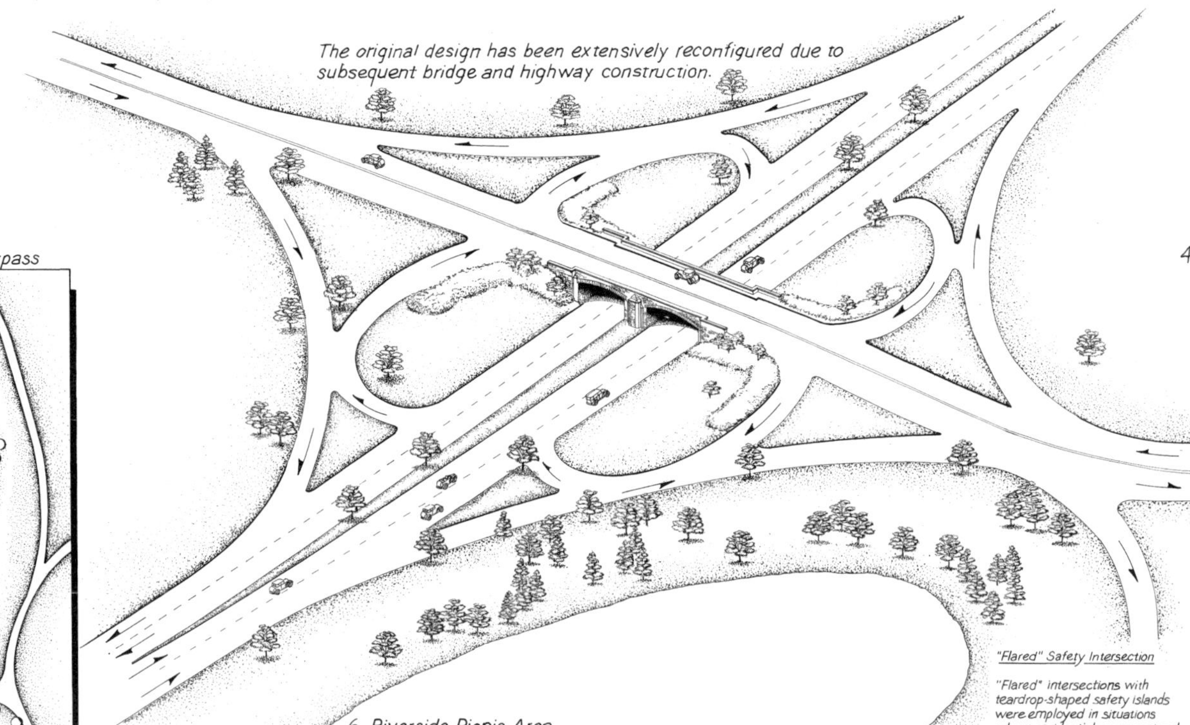

5. Fort Hunt Overpass (1932)

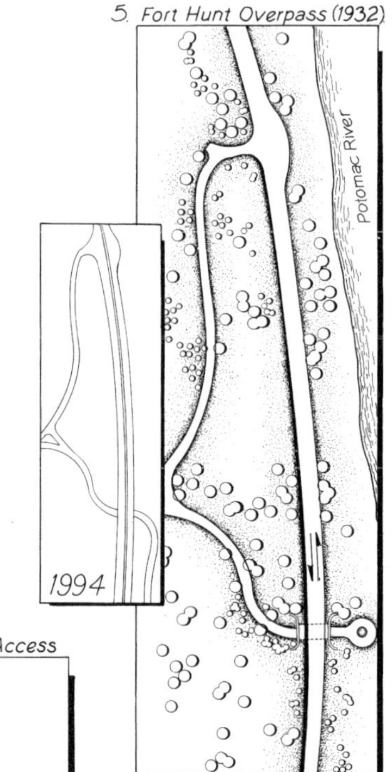

3. Wellington Underpass

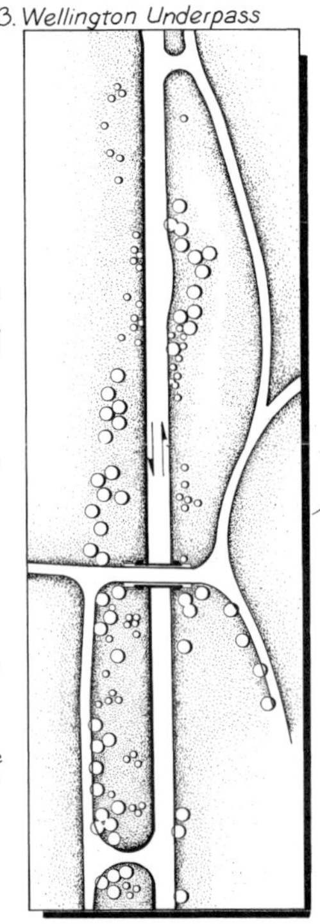

4. Wellington Lane Access

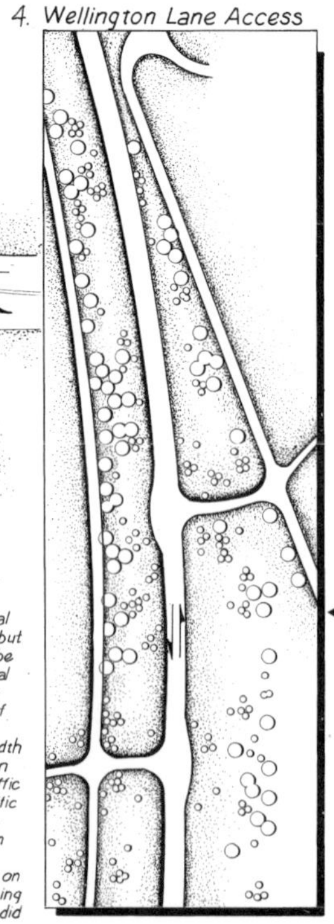

Traffic Circle/Rotary ▲

Traffic circles or rotaries were a traditional means of combining streams of traffic and marking transitions between different types of space. The Memorial Circle north of Alexandria was designed to absorb traffic from side roads and serve as a functional and symbolic transition between the free-flowing parkway and the business district of Alexandria. Traffic engineers soon recognized that rotaries were not suited for high-speed automobile traffic. The Memorial Circle proved to be the most accident-prone segment of the parkway. It was removed between 1957-62 and replaced with a full-stop intersection and traffic lights.

Underpass ▶

Intersecting roads carrying moderate traffic volumes could be effectively handled with simple grade separation structures. The Wellington Underpass allowed local traffic in this pre-existing suburb to cross MVMH without interfering with through traffic. Border roads collected traffic from the side streets and provided access to the main parkway at widely separated entranceways. The handsome structure also visually broke up the parkway's only long straight section and made an attractive frame for axial views in both directions.

6. Riverside Picnic Area

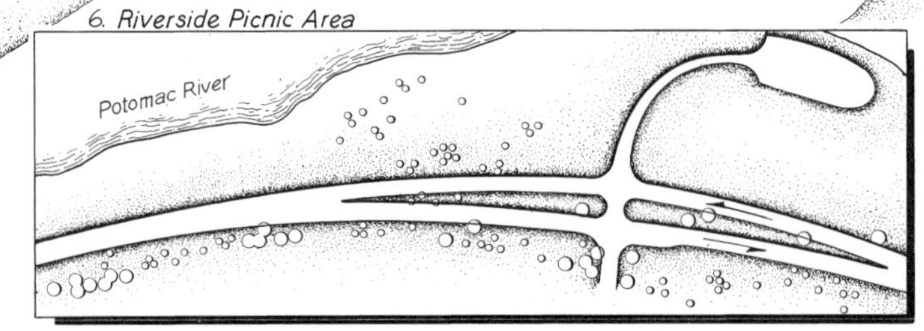

"Flared" Safety Intersection

"Flared" intersections with teardrop-shaped safety islands were employed in situations where residential or recreational development required access, but cross-traffic was expected to be light. Since most of the original MVMH carried undivided, two-way traffic, the construction of short median strips that expanded to a car-length in width at the center of the intersection enabled turning or entering traffic to safely cross one lane of traffic at a time. An extra lane-width was provided in either direction for additional safety. Visually, this configuration worked best on slight curves, where the widening required by the central islands did not appear to disrupt the continuous curvature of the highway alignment.

Overpass ▲

Like underpasses, overpasses enabled local traffic to cross the parkway without interfering with through traffic. Carrying the main parkway drive over intersecting roads generally made it easier to screen cross-traffic from view. This was particularly desirable in the case of the Fort Hunt Overpass, since the cross road was originally intended primarily for service traffic between Fort Hunt and a wharf on the Potomac shore. Tentative plans were also made for a ferry crossing to Fort Washington at this point. After Fort Hunt was converted to park land, the overpass provided an opportunity to construct a grade-separated access road enabling northbound traffic to safely enter the new picnic area.

◀ Staggered Intersection

Where the parkway was relatively straight, or a narrow right-of-way prevented the introduction of safety islands, staggered intersections provided a measure of relief. Reducing entry points with border roads and staggering crossings by approximately 400' helped reduce the danger of motorists shooting directly across the main parkway drive. Crossing traffic would enter the parkway, merge into the right line, and then exit

Drawn by: Ed Lupyak, 1994

GEORGE WASHINGTON MEMORIAL PARKWAY DOCUMENTATION PROJECT 1994
NATIONAL PARK SERVICE
UNITED STATES DEPARTMENT OF THE INTERIOR

NAME AND LOCATION OF STRUCTURE
GEORGE WASHINGTON MEMORIAL PARKWAY
MOUNT VERNON, FAIRFAX COUNTY, VIRGINIA

SHEET 10 OF 21 SHEETS

HISTORIC AMERICAN ENGINEERING RECORD
VA-69

LIBRARY OF CONGRESS INDEX NUMBER

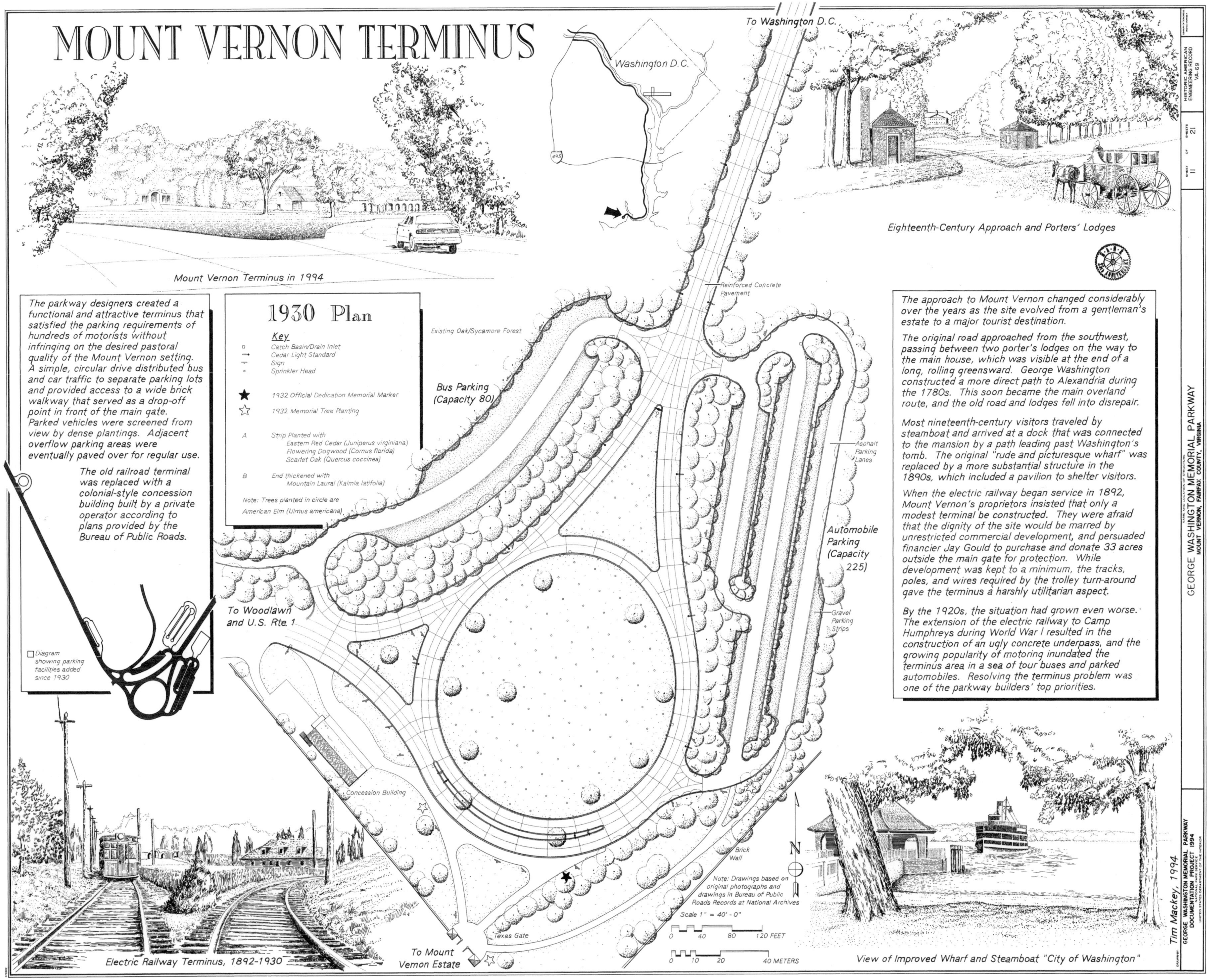
MOUNT VERNON TERMINUS
To Washington D.C.
Washington D.C.
Mount Vernon Terminus in 1994
Eighteenth-Century Approach and Porters' Lodges
The parkway designers created a functional and attractive terminus that satisfied the parking requirements of hundreds of motorists without infringing on the desired pastoral quality of the Mount Vernon setting. A simple, circular drive distributed bus and car traffic to separate parking lots and provided access to a wide brick walkway that served as a drop-off point in front of the main gate. Parked vehicles were screened from view by dense plantings. Adjacent overflow parking areas were eventually paved over for regular use.
The old railroad terminal was replaced with a colonial-style concession building built by a private operator according to plans provided by the Bureau of Public Roads.
Diagram showing parking facilities added since 1930
1930 Plan
Key
Catch Basin/Drain Inlet
Cedar Light Standard
Sign
Sprinkler Head
1932 Official Dedication Memorial Marker
1932 Memorial Tree Planting
A Strip Planted with
Eastern Red Cedar (Juniperus virginiana)
Flowering Dogwood (Cornus florida)
Scarlet Oak (Quercus coccinea)
B End thickened with
Mountain Laural (Kalmia latifolia)
Note: Trees planted in circle are American Elm (Ulmus americana)
Existing Oak/Sycamore Forest
Bus Parking (Capacity 80)
Reinforced Concrete Pavement
Asphalt Parking Lanes
Automobile Parking (Capacity 225)
Gravel Parking Strips
To Woodlawn and U.S. Rte. 1
Concession Building
Brick Wall
Texas Gate
To Mount Vernon Estate
N
Note: Drawings based on original photographs and drawings in Bureau of Public Roads Records at National Archives
Scale 1" = 40'-0"
0 40 80 120 FEET
0 10 20 40 METERS
The approach to Mount Vernon changed considerably over the years as the site evolved from a gentleman's estate to a major tourist destination.
The original road approached from the southwest, passing between two porter's lodges on the way to the main house, which was visible at the end of a long, rolling greensward. George Washington constructed a more direct path to Alexandria during the 1780s. This soon became the main overland route, and the old road and lodges fell into disrepair.
Most nineteenth-century visitors traveled by steamboat and arrived at a dock that was connected to the mansion by a path leading past Washington's tomb. The original "rude and picturesque wharf" was replaced by a more substantial structure in the 1890s, which included a pavilion to shelter visitors.
When the electric railway began service in 1892, Mount Vernon's proprietors insisted that only a modest terminal be constructed. They were afraid that the dignity of the site would be marred by unrestricted commercial development, and persuaded financier Jay Gould to purchase and donate 33 acres outside the main gate for protection. While development was kept to a minimum, the tracks, poles, and wires required by the trolley turn-around gave the terminus a harshly utilitarian aspect.
By the 1920s, the situation had grown even worse. The extension of the electric railway to Camp Humphreys during World War I resulted in the construction of an ugly concrete underpass, and the growing popularity of motoring inundated the terminus area in a sea of tour buses and parked automobiles. Resolving the terminus problem was one of the parkway builders' top priorities.
Electric Railway Terminus, 1892-1930
View of Improved Wharf and Steamboat "City of Washington"
Tim Mackey, 1994
GEORGE WASHINGTON MEMORIAL PARKWAY DOCUMENTATION PROJECT 1994
NATIONAL PARK SERVICE
UNITED STATES DEPARTMENT OF THE INTERIOR
GEORGE WASHINGTON MEMORIAL PARKWAY
MOUNT VERNON, FAIRFAX COUNTY, VIRGINIA
SHEET 11 OF 21 SHEETS
HISTORIC AMERICAN ENGINEERING RECORD
VA-69

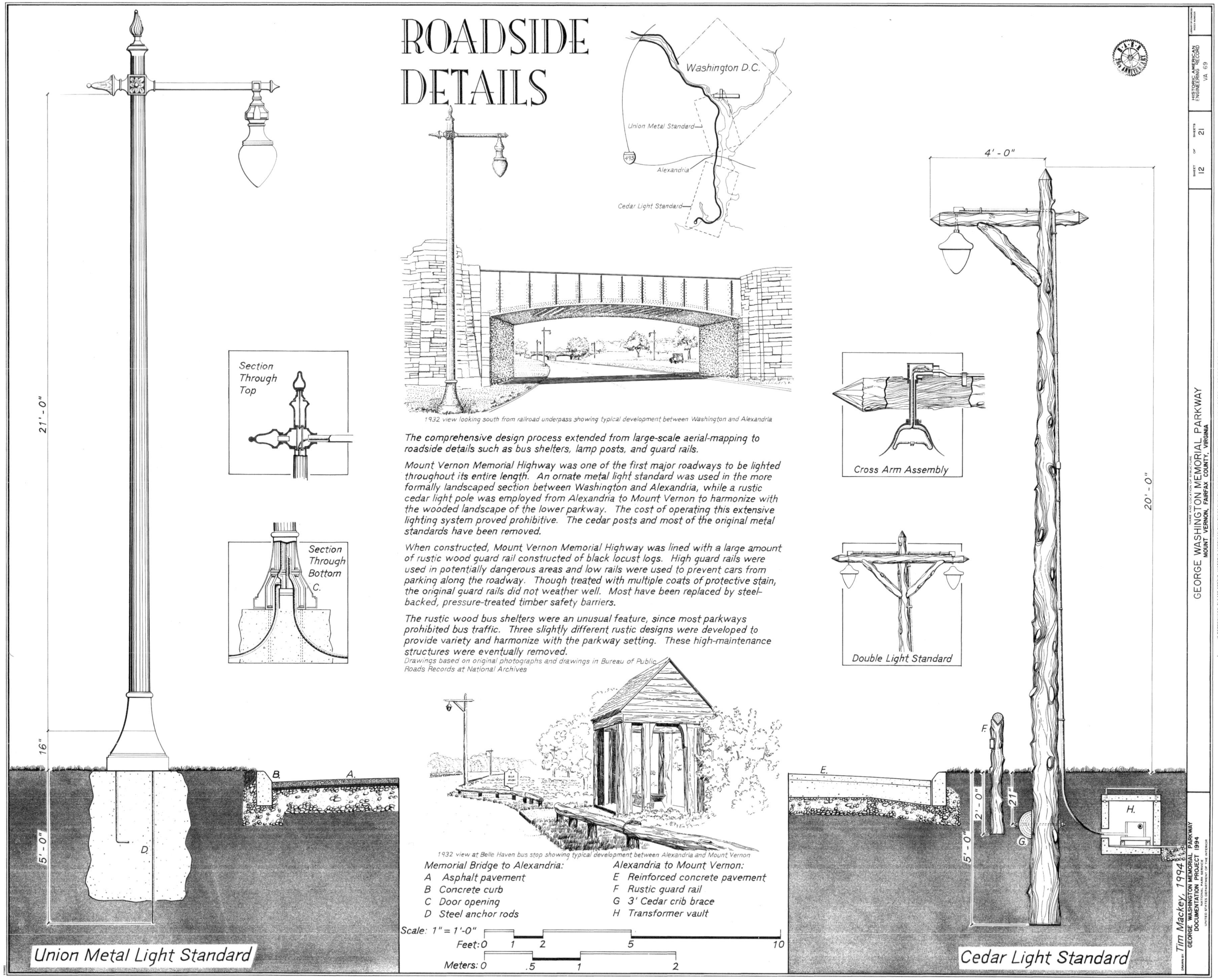
ROADSIDE DETAILS
Washington D.C.
Union Metal Standard
Alexandria
Cedar Light Standard
1932 view looking south from railroad underpass showing typical development between Washington and Alexandria
The comprehensive design process extended from large-scale aerial-mapping to roadside details such as bus shelters, lamp posts, and guard rails.
Mount Vernon Memorial Highway was one of the first major roadways to be lighted throughout its entire length. An ornate metal light standard was used in the more formally landscaped section between Washington and Alexandria, while a rustic cedar light pole was employed from Alexandria to Mount Vernon to harmonize with the wooded landscape of the lower parkway. The cost of operating this extensive lighting system proved prohibitive. The cedar posts and most of the original metal standards have been removed.
When constructed, Mount Vernon Memorial Highway was lined with a large amount of rustic wood guard rail constructed of black locust logs. High guard rails were used in potentially dangerous areas and low rails were used to prevent cars from parking along the roadway. Though treated with multiple coats of protective stain, the original guard rails did not weather well. Most have been replaced by steel-backed, pressure-treated timber safety barriers.
The rustic wood bus shelters were an unusual feature, since most parkways prohibited bus traffic. Three slightly different rustic designs were developed to provide variety and harmonize with the parkway setting. These high-maintenance structures were eventually removed.
Drawings based on original photographs and drawings in Bureau of Public Roads Records at National Archives
1932 view at Belle Haven bus stop showing typical development between Alexandria and Mount Vernon
Memorial Bridge to Alexandria:
A Asphalt pavement
B Concrete curb
C Door opening
D Steel anchor rods
Alexandria to Mount Vernon:
E Reinforced concrete pavement
F Rustic guard rail
G 3' Cedar crib brace
H Transformer vault
Scale: 1" = 1'-0"
Feet: 0 1 2 5 10
Meters: 0 .5 1 2
Section Through Top
Section Through Bottom
Cross Arm Assembly
Double Light Standard
21'-0"
16"
5'-0"
4'-0"
20'-0"
2'-0"
21"
Union Metal Light Standard
Cedar Light Standard
Tim Mackey, 1994
GEORGE WASHINGTON MEMORIAL PARKWAY DOCUMENTATION PROJECT 1994
NATIONAL PARK SERVICE
UNITED STATES DEPARTMENT OF THE INTERIOR
GEORGE WASHINGTON MEMORIAL PARKWAY
MOUNT VERNON, FAIRFAX COUNTY, VIRGINIA
SHEET 12 OF 21 SHEETS
HISTORIC AMERICAN ENGINEERING RECORD
VA 69

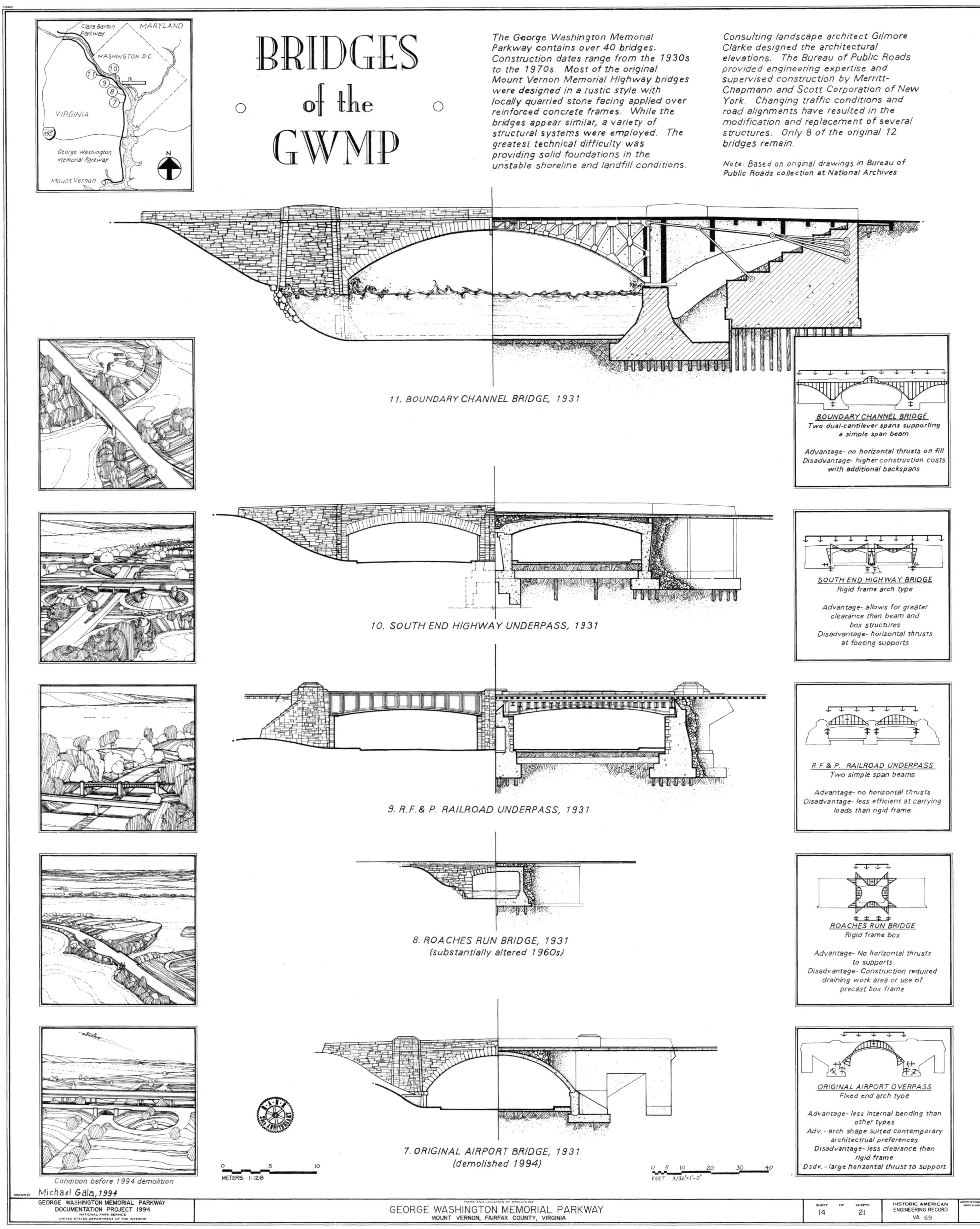
BRIDGES of the GWMP
The George Washington Memorial Parkway contains over 40 bridges. Construction dates range from the 1930s to the 1970s. Most of the original Mount Vernon Memorial Highway bridges were designed in a rustic style with locally quarried stone facing applied over reinforced concrete frames. While the bridges appear similar, a variety of structural systems were employed. The greatest technical difficulty was providing solid foundations in the unstable shoreline and landfill conditions.
Consulting landscape architect Gilmore Clarke designed the architectural elevations. The Bureau of Public Roads provided engineering expertise and supervised construction by Merritt-Chapmann and Scott Corporation of New York. Changing traffic conditions and road alignments have resulted in the modification and replacement of several structures. Only 8 of the original 12 bridges remain.
Note: Based on original drawings in Bureau of Public Roads collection at National Archives
Clara Barton Parkway
MARYLAND
WASHINGTON D.C.
VIRGINIA
George Washington Memorial Parkway
Mount Vernon
11. BOUNDARY CHANNEL BRIDGE, 1931
BOUNDARY CHANNEL BRIDGE
Two dual-cantilever spans supporting a simple span beam
Advantage- no horizontal thrusts on fill
Disadvantage- higher construction costs with additional backspans
10. SOUTH END HIGHWAY UNDERPASS, 1931
SOUTH END HIGHWAY BRIDGE
Rigid frame arch type
Advantage- allows for greater clearance than beam and box structures
Disadvantage- horizontal thrusts at footing supports
9. R.F.& P. RAILROAD UNDERPASS, 1931
R.F.& P. RAILROAD UNDERPASS
Two simple span beams
Advantage- no horizontal thrusts
Disadvantage- less efficient at carrying loads than rigid frame
8. ROACHES RUN BRIDGE, 1931
(substantially altered 1960s)
ROACHES RUN BRIDGE
Rigid frame box
Advantage- No horizontal thrusts to supports
Disadvantage- Construction required draining work area or use of precast box frame
7. ORIGINAL AIRPORT BRIDGE, 1931
(demolished 1994)
ORIGINAL AIRPORT OVERPASS
Fixed end arch type
Advantage- less internal bending than other types
Adv.- arch shape suited contemporary architectural preferences
Disadvantage- less clearance than rigid frame
Dsdv.- large horizontal thrust to support
Condition before 1994 demolition
METERS 1:128
FEET 3/32"=1'-0"
Michael Gala, 1994
GEORGE WASHINGTON MEMORIAL PARKWAY DOCUMENTATION PROJECT 1994
NATIONAL PARK SERVICE
UNITED STATES DEPARTMENT OF THE INTERIOR
GEORGE WASHINGTON MEMORIAL PARKWAY
MOUNT VERNON, FAIRFAX COUNTY, VIRGINIA
SHEET 14 OF SHEETS 21
HISTORIC AMERICAN ENGINEERING RECORD
VA 69
IF REPRODUCED, PLEASE CREDIT: HISTORIC AMERICAN BUILDINGS SURVEY, NATIONAL PARK SERVICE, NAME OF DELINEATOR, DATE OF THE DRAWING

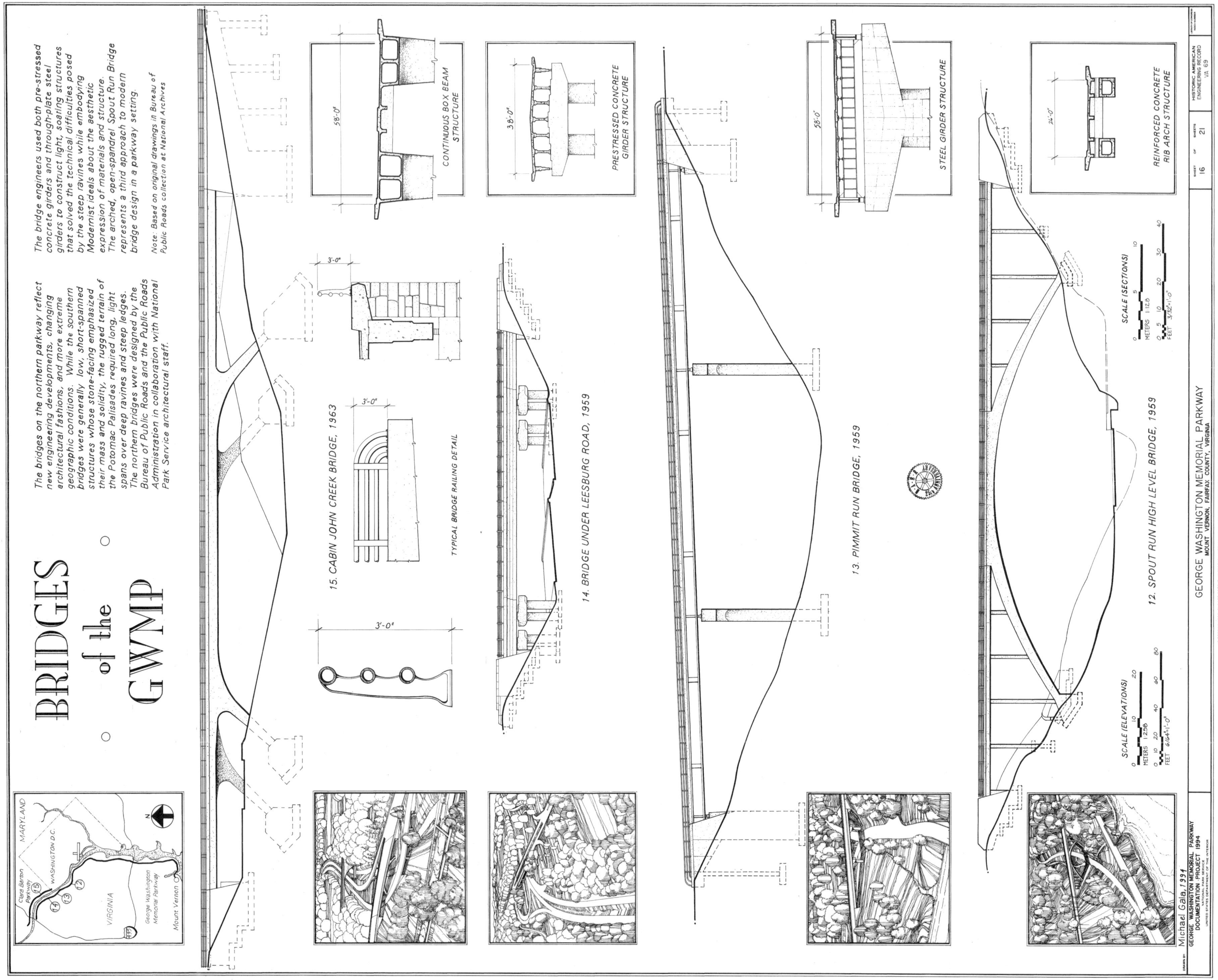

BRIDGES of the GWMP
The bridges on the northern parkway reflect new engineering developments, changing architectural fashions, and more extreme geographic conditions. While the southern bridges were generally low, short-spanned structures whose stone-facing emphasized their mass and solidity, the rugged terrain of the Potomac Palisades required long, light spans over deep ravines and steep ledges. The northern bridges were designed by the Bureau of Public Roads and the Public Roads Administration in collaboration with National Park Service architectural staff.
The bridge engineers used both pre-stressed concrete girders and through-plate steel girders to construct light, soaring structures that solved the technical difficulties posed by the steep ravines while embodying Modernist ideals about the aesthetic expression of materials and structure. The arched, open-spandrel Spout Run Bridge represents a third approach to modern bridge design in a parkway setting.
Note: Based on original drawings in Bureau of Public Roads collection at National Archives
MARYLAND
WASHINGTON D.C.
VIRGINIA
Clara Barton Parkway
George Washington Memorial Parkway
Mount Vernon
15. CABIN JOHN CREEK BRIDGE, 1963
TYPICAL BRIDGE RAILING DETAIL
CONTINUOUS BOX BEAM STRUCTURE
14. BRIDGE UNDER LEESBURG ROAD, 1959
PRESTRESSED CONCRETE GIRDER STRUCTURE
13. PIMMIT RUN BRIDGE, 1959
STEEL GIRDER STRUCTURE
12. SPOUT RUN HIGH LEVEL BRIDGE, 1959
REINFORCED CONCRETE RIB ARCH STRUCTURE
SCALE (ELEVATIONS)
SCALE (SECTIONS)
GEORGE WASHINGTON MEMORIAL PARKWAY
MOUNT VERNON, FAIRFAX COUNTY, VIRGINIA
Michael Gala, 1994
GEORGE WASHINGTON MEMORIAL PARKWAY DOCUMENTATION PROJECT 1994
HISTORIC AMERICAN ENGINEERING RECORD VA-69
SHEET 16 OF 21

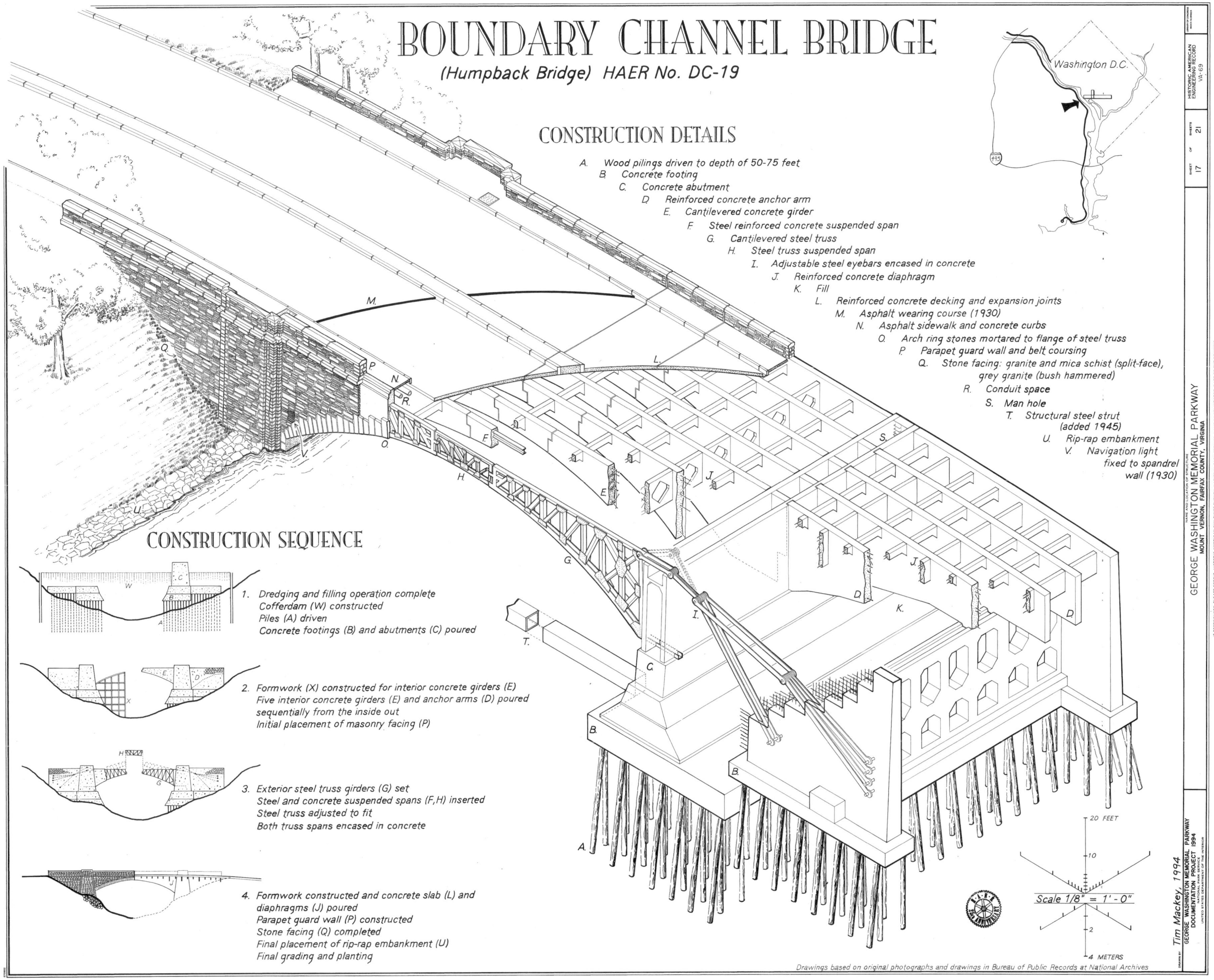

BOUNDARY CHANNEL BRIDGE
(Humpback Bridge) HAER No. DC-19
CONSTRUCTION DETAILS
A. Wood pilings driven to depth of 50-75 feet
B. Concrete footing
C. Concrete abutment
D. Reinforced concrete anchor arm
E. Cantilevered concrete girder
F. Steel reinforced concrete suspended span
G. Cantilevered steel truss
H. Steel truss suspended span
I. Adjustable steel eyebars encased in concrete
J. Reinforced concrete diaphragm
K. Fill
L. Reinforced concrete decking and expansion joints
M. Asphalt wearing course (1930)
N. Asphalt sidewalk and concrete curbs
O. Arch ring stones mortared to flange of steel truss
P. Parapet guard wall and belt coursing
Q. Stone facing: granite and mica schist (split-face), grey granite (bush hammered)
R. Conduit space
S. Man hole
T. Structural steel strut (added 1945)
U. Rip-rap embankment
V. Navigation light fixed to spandrel wall (1930)
Washington D.C.
CONSTRUCTION SEQUENCE
1. Dredging and filling operation complete
Cofferdam (W) constructed
Piles (A) driven
Concrete footings (B) and abutments (C) poured
2. Formwork (X) constructed for interior concrete girders (E)
Five interior concrete girders (E) and anchor arms (D) poured sequentially from the inside out
Initial placement of masonry facing (P)
3. Exterior steel truss girders (G) set
Steel and concrete suspended spans (F,H) inserted
Steel truss adjusted to fit
Both truss spans encased in concrete
4. Formwork constructed and concrete slab (L) and diaphragms (J) poured
Parapet guard wall (P) constructed
Stone facing (Q) completed
Final placement of rip-rap embankment (U)
Final grading and planting
20 FEET
10
Scale 1/8" = 1'-0"
2
4 METERS
Drawings based on original photographs and drawings in Bureau of Public Records at National Archives
Tim Mackey, 1994
GEORGE WASHINGTON MEMORIAL PARKWAY
DOCUMENTATION PROJECT 1994
NATIONAL PARK SERVICE
UNITED STATES DEPARTMENT OF THE INTERIOR
GEORGE WASHINGTON MEMORIAL PARKWAY
MOUNT VERNON, FAIRFAX COUNTY, VIRGINIA
SHEET 17 OF 21 SHEETS
HISTORIC AMERICAN ENGINEERING RECORD
VA-69
IF REPRODUCED, PLEASE CREDIT: HISTORIC AMERICAN BUILDINGS SURVEY, NATIONAL PARK SERVICE, NAME OF DELINEATOR, DATE OF THE DRAWING

THE NATCHEZ TRACE PARKWAY

Traversing through Mississippi, Alabama and Tennessee, the 450-mile Natchez Trace Parkway commemorates one of America's earliest transportation routes. Established in 1938 as a unit of the National Park Service (NPS), the Natchez Trace Parkway represents the culmination of federal and regional collaboration to pay tribute to an historic trail known as the Old Natchez Trace. Designed for recreational purposes, the Natchez Trace Parkway closely parallels, but does not replicate, the historic Natchez Trace. The parkway was designed to connect the cities of Natchez, Jackson, Tupelo and Nashville with a scenic byway far removed from the chaos of ordinary highway travel that would make travelling by automobile a pleasurable experience.

A regional Good Roads movement championed by the Natchez Trace Military Highway Association in 1916 continued an effort initiated by the Mississippi Daughters of the American Revolution (DAR) to commemorate the Old Natchez Trace by paving it for modern automobile use. The infusion of New Deal funding for the project brought with it federal Bureau of Public Roads and National Park Service planners. After completion of a survey of the historic trace in 1934, the NPS and the Bureau of Public Roads (BPR) initiated field surveys to stake out the centerline of the proposed roadway, which was a difficult task. The proposed design and the laying out of the parkway were not without controversy. Those expecting a modern highway were baffled by the particular restrictions, including limited access and imposed speed limits, required by a recreational parkway. Creating wide buffer zones between the roadway and unsightly land-use developments angered some adjacent landowners whose property faced condemnation.

Despite some objections, enthusiastic Natchez Trace supporters celebrated the first of many groundbreaking ceremonies on 16 September 1937. Work on the parkway slowed, however, with the advent of U.S. involvement in World War II, as shortages of equipment, materials, and labor resulted in the complete halt of all construction jobs not connected with the war effort. The post-war inflationary boom, rationing during the Korean conflict, and domestic spending cuts in deference to the Cold War military build-up further slowed construction of the Natchez Trace Parkway.

A long-awaited acceleration of the pace of construction occurred with the NPS Mission 66 program, which was a monumental effort from 1956 to 1966 to refurbish park roads, buildings, and other facilities that had suffered from several years of federal funding cuts.

The Natchez Trace Parkway today represents the culmination of decades of dedication underwritten by crucial federal funding. The parkway represents a remarkable achievement, merging a modern roadway with its surrounding rural landscape, and making scores of natural and cultural resources related to the Old Trace accessible to the public. This is due to the care that has been taken, amidst much controversy, to ensure that modern motorists feel a sense of connectedness with the road on which they travel and the scenery through which they motor.

This recording project is part of the Historic American Engineering Record (HAER), a long-range program to document historically significant engineering, industrial, and maritime works in the United States. The HAER program is administered by the National Park Service, U.S. Department of the Interior. The Natchez Trace Parkway Recording Project was cosponsored during the summer of 1998 by HAER under the general direction of the Chief of HABS/HAER, E. Blaine Cliver and the Natchez Trace Parkway's Superintendent, Wendell A. Simpson.

The field work, measured drawings, historical reports and photographs were prepared under the general direction of Eric DeLony, Chief of HAER, Todd Croteau, HAER Architect, and Tim Davis, HAER Historian.

The recording team consisted of Landscape Architect Supervisor Tim Halsey, Landscape Architect Katie Dugdill (ICOMOS), and Architect Nicholas Zydycryn. Jean Fulton produced the historical report. Large format photography was produced by William A. Faust III, HAER Contract Photographer.

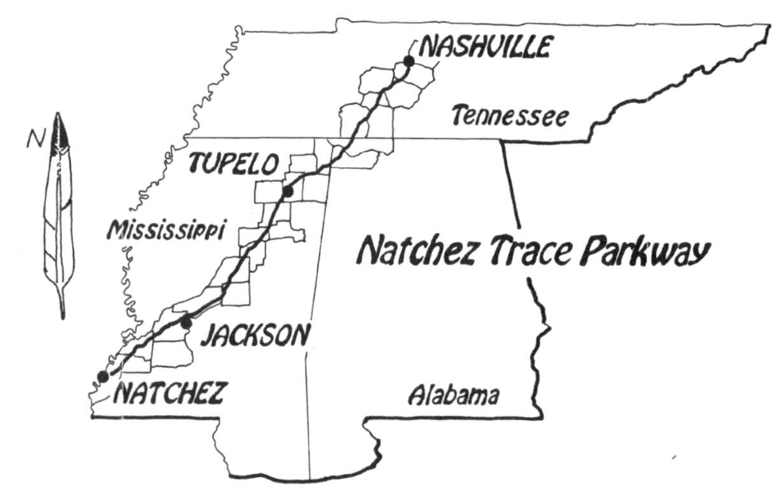

DELINEATED BY: ED LUPYAK & TODD CROTEAU, 1999

NPS ROADS AND BRIDGES RECORDING PROGRAM
NATIONAL PARK SERVICE
UNITED STATES DEPARTMENT OF THE INTERIOR

TUPELO VICINITY

NATCHEZ TRACE PARKWAY
NATCHEZ, MISSISSIPPI TO NASHVILLE, TENNESSEE
LEE COUNTY

MISSISSIPPI

SHEET 1 OF 15

HISTORIC AMERICAN ENGINEERING RECORD
MS-15

LIBRARY OF CONGRESS INDEX NUMBER

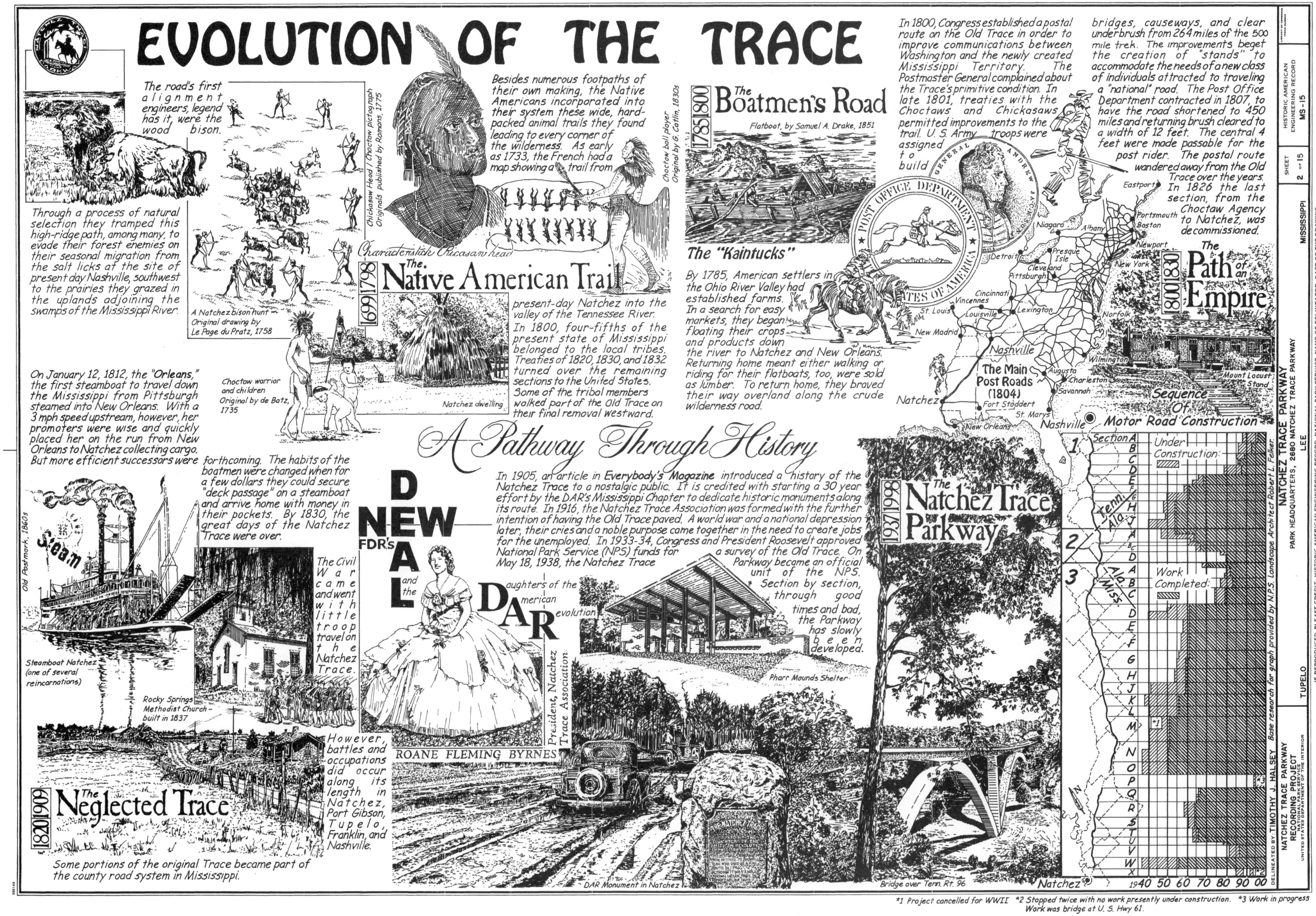

EVOLUTION OF THE TRACE
The road's first alignment engineers, legend has it, were the wood bison.
Through a process of natural selection they tramped this high-ridge path, among many, to evade their forest enemies on their seasonal migration from the salt licks at the site of present day Nashville, southwest to the prairies they grazed in the uplands adjoining the swamps of the Mississippi River.
A Natchez bison hunt
Original drawing by Le Page du Pratz, 1758
Chickasaw Head / Choctaw pictograph
Originals published by Romans, 1775
Besides numerous footpaths of their own making, the Native Americans incorporated into their system these wide, hard-packed animal trails they found leading to every corner of the wilderness. As early as 1733, the French had a map showing a trail from present-day Natchez into the valley of the Tennessee River.
In 1800, four-fifths of the present state of Mississippi belonged to the local tribes. Treaties of 1820, 1830, and 1832 turned over the remaining sections to the United States. Some of the tribal members walked part of the Old Trace on their final removal westward.
Choctaw ball player
Original by G. Catlin, 1830s
Characteristic Chicasaw head
1699 1798
The Native American Trail
Choctaw warrior and children
Original by de Batz, 1735
Natchez dwelling
1785 1800
The Boatmen's Road
Flatboat, by Samuel A. Drake, 1851
The "Kaintucks"
By 1785, American settlers in the Ohio River Valley had established farms. In a search for easy markets, they began floating their crops and products down the river to Natchez and New Orleans. Returning home meant either walking or riding for their flatboats, too, were sold as lumber. To return home, they braved their way overland along the crude wilderness road.
In 1800, Congress established a postal route on the Old Trace in order to improve communications between Washington and the newly created Mississippi Territory. The Postmaster General complained about the Trace's primitive condition. In late 1801, treaties with the Choctaws and Chickasaws permitted improvements to the trail. U. S. Army troops were assigned to build bridges, causeways, and clear underbrush from 264 miles of the 500 mile trek. The improvements beget the creation of "stands" to accommodate the needs of a new class of individuals attracted to traveling a "national" road. The Post Office Department contracted in 1807, to have the road shortened to 450 miles and returning brush cleared to a width of 12 feet. The central 4 feet were made passable for the post rider. The postal route wandered away from the Old Trace over the years. In 1826 the last section, from the Choctaw Agency to Natchez, was decommissioned.
POST OFFICE DEPARTMENT
STATES OF AMERICA
GENERAL ANDREW J
1800 1830
The Path of an Empire
The Main Post Roads (1804)
Eastport
Portsmouth
Boston
Albany
Niagara
Newport
New York
Detroit
Presque Isle
Cleveland
Pittsburgh
Cincinnati
Vincennes
St. Louis
Louisville
Lexington
New Madrid
Nashville
Norfolk
Wilmington
Augusta
Charleston
Savannah
Natchez
Fort Stoddert
St. Marys
New Orleans
Mount Locust Stand
On January 12, 1812, the "Orleans," the first steamboat to travel down the Mississippi from Pittsburgh steamed into New Orleans. With a 3 mph speed upstream, however, her promoters were wise and quickly placed her on the run from New Orleans to Natchez collecting cargo. But more efficient successors were forthcoming. The habits of the boatmen were changed when for a few dollars they could secure "deck passage" on a steamboat and arrive home with money in their pockets. By 1830, the great days of the Natchez Trace were over.
Old Postmark, 1860s
Steam
Steamboat Natchez (one of several reincarnations)
The Civil War came and went with little troop travel on the Natchez Trace.
Rocky Springs Methodist Church built in 1837
However, battles and occupations did occur along its length in Natchez, Port Gibson, Tupelo, Franklin, and Nashville.
1820 1909
The Neglected Trace
Some portions of the original Trace became part of the county road system in Mississippi.
A Pathway Through History
FDR's NEW DEAL and the DAR
Daughters of the American Revolution
ROANE FLEMING BYRNES
President, Natchez Trace Association
In 1905, an article in Everybody's Magazine introduced a history of the Natchez Trace to a nostalgic public. It is credited with starting a 30 year effort by the DAR's Mississippi Chapter to dedicate historic monuments along its route. In 1916, the Natchez Trace Association was formed with the further intention of having the Old Trace paved. A world war and a national depression later, their cries and a noble purpose came together in the need to create jobs for the unemployed. In 1933-34, Congress and President Roosevelt approved National Park Service (NPS) funds for a survey of the Old Trace. On May 18, 1938, the Natchez Trace Parkway became an official unit of the NPS. Section by section, through good times and bad, the Parkway has slowly been developed.
Pharr Mounds Shelter
DAR Monument in Natchez
1937 1998
The Natchez Trace Parkway
Bridge over Tenn. Rt. 96
Sequence Of Motor Road Construction
Nashville
Natchez
Section
1
2
3
Tenn. Ala.
Ala. Miss.
A B C D E F G H J K A B C D A B C D E F G H J K L M N O P Q R S T U V W X
Under Construction:
Work Completed:
1940 50 60 70 80 90 00
*1
*2
*3
N
Base research for graph provided by N.P.S. Landscape Architect Robert L. Felker
DELINEATED BY TIMOTHY J. HALSEY
NATCHEZ TRACE PARKWAY RECORDING PROJECT
UNITED STATES DEPARTMENT OF THE INTERIOR
NATCHEZ TRACE PARKWAY
PARK HEADQUARTERS, 2680 NATCHEZ TRACE PARKWAY
TUPELO
LEE
MISSISSIPPI
HISTORIC AMERICAN ENGINEERING RECORD
MS-15
SHEET 2 OF 15
IF REPRODUCED, PLEASE CREDIT: HISTORIC AMERICAN ENGINEERING RECORD, NATIONAL PARK SERVICE, NAME OF DELINEATOR, DATE OF THE DRAWING
*1 Project cancelled for WWII *2 Stopped twice with no work presently under construction. *3 Work in progress Work was bridge at U. S. Hwy 61.

CULTURAL RESOURCES

Closely following the original Natchez Trace, the Parkway leads contemporary travelers beside one of the most historic routes in America. The visiting motorist encounters historic sites, military battlefields, American Indian burial mounds and preserved sections of the Old Trace. Interpretive information for each site is provided by large, wooden signs and markers placed by the National Park Service at points of interest along the Parkway.

Meriwether Lewis

Meriwether Lewis, with William Clark, led the first expedition to the Pacific Northwest (1804-06). Lewis died under mysterious circumstances at Grinders Inn, on October 11, 1809. The broken column design of this monument symbolizes a life ended in its prime. (MP 385.9)

Gordon House

One of the few remaining buildings associated with the Old Trace, this brick house (ca. 1818) was the home of John Gordon. Gordon operated a ferry across the Duck River from 1801 until traffic on the Trace declined (MP 407.7).

Rocky Springs

Rocky Springs (ca. 1795) declined after the onslaught of yellow fever, drought and the Civil War. Now a ghost town, a small cemetery rests on a hill top under tall trees draped with Spanish Moss. (MP 54.8)

Mount Locust

Built in 1778 Mount Locust served as both inn or "stand," and family dwelling. This small plantation home is the only remaining stand of the more than 50 that served the Old Trace. (MP 15.5)

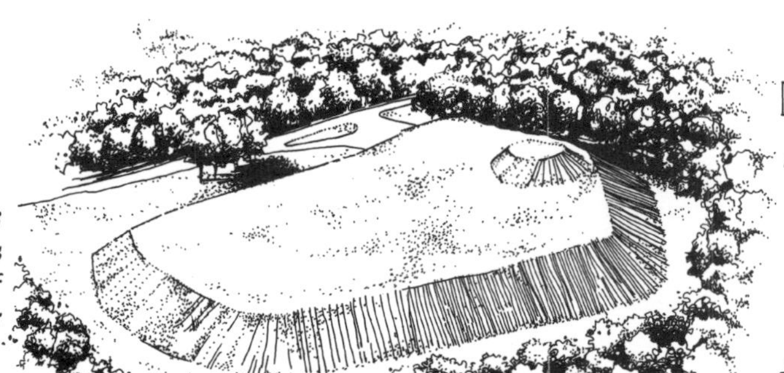

Emerald Mound

Ancestors of the Natchez Indians built this and other ceremonial mounds ca. 1400 A.D. The second largest of its type in the United States, it covers nearly eight acres. (MP 10.3.)

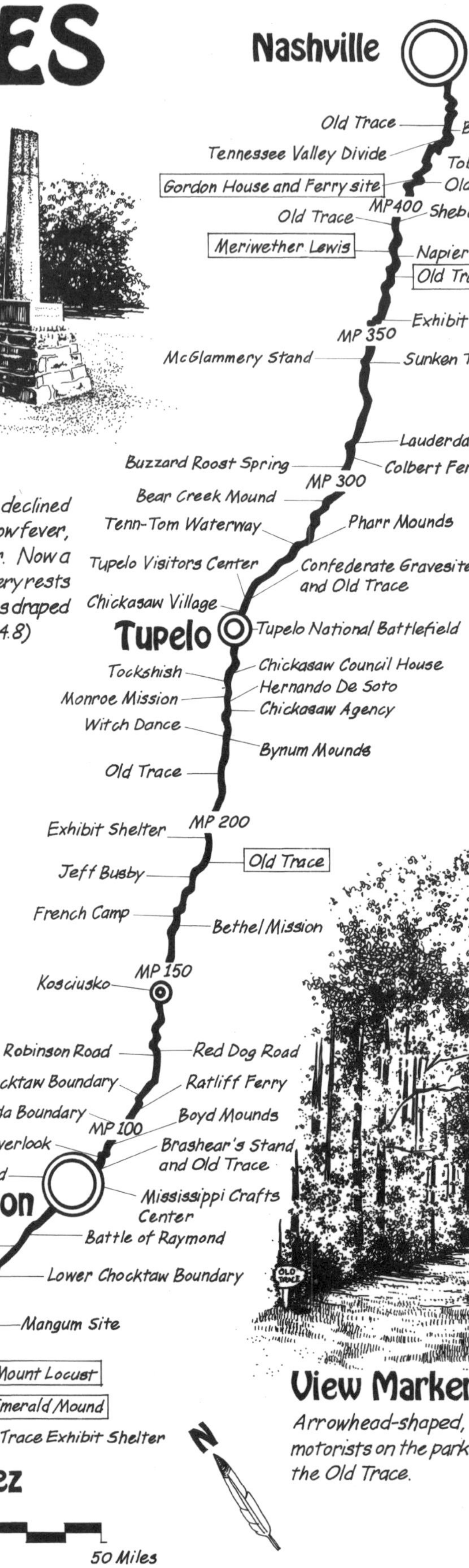

Old Trace Drive

A narrow, one-way route leads along a paved section of the Old Trace. Overlooks provide views over the surrounding countryside. (MP 375.8)

Pedestrian Exploration

Visitors are encouraged to explore Old Trace on foot and to try and imagine the hardships of travelling this hazardous route. (MP 198.6)

View Markers

Arrowhead-shaped, roadside markers allow motorists on the parkway to glimpse sections of the Old Trace.

Sunken Trace

The constant traffic of generations of travelers along the Trace, carved a deep sunken path in the ground. The trail was further eroded after heavy storms filled the route with water. (MP 41.5)

DELINEATED BY: *Katie Dugdill 1998*

NATCHEZ TRACE PARKWAY RECORDING PROJECT
NATIONAL PARK SERVICE
UNITED STATES DEPARTMENT OF THE INTERIOR

TUPELO

NATCHEZ TRACE PARKWAY
PARK HEADQUARTERS 2680 NATCHEZ TRACE PARKWAY
LEE

MISSISSIPPI

SHEET 3 of 15

HISTORIC AMERICAN ENGINEERING RECORD
MS-15

Leaf Identification: 1. Bald Cypress *Taxodium distichum* · 2. Tupelo *Nyssa sylvatica* · 3. Buttonbush *Cephalanthus occidentalis* · 4. Hazel Alder *Alnus Rugosa* · 5. Winterberry *Ilex verticillata* · 6. Black Walnut *Juglans nigra* · 7. Sycamore *Platanus occidentalis* · 8. Red Maple *A. Rubrum* · 9. Winged Elm *Ulmus alata* · 10. Overcup Oak *Q. lyrata* · 11. Swamp Chestnut Oak *Q. michauxii* · 12. Water Oak *Q. nigra* · 13. Slippery Elm *Ulmus fulva* · 14. American Elm *U. americana* · 15. Sassafras *Sassafras variifolium* · 16. Black Jack Oak *Q. marilandica* · 17. Post Oak *Q. stellata*

NATURAL RESOURCES

"As settlers advanced across the South, they cut holes in the forest as patches and plantations. They cut tunnels through the vast woods to form trails and roads: they scarcely had force enough to sweep the forest itself away."

-Clark and Kirwan (1967)

The Swamp Succession

1a. Tupelo-Baldcypress Swamp Community

- located in areas with water tables between ground level and 3 feet above ground level
- easy IDs for these trees are the buttress trunk bases on both and so-called root "knees" on the bald cypress
- named for the dominant species

Cypress Swamp (M.P. 122.0)

1b. Swamp Community

- located in areas with water tables at or slightly below ground level
- dominant species is Hazel Alder and associated shrubs including Buttonwood and Winterberry

2. Bottomland Forest Community

- this community may follow either of the other swamp communities
- a varied community with lots of spring forest wildflowers and early summer blooms
- Water Oak, Swamp Chestnut Oak, Overcup Oak, Sycamore, American Elm, Winged Elm, Slippery Elm, Red Maple, and Black Walnut

The Old Field Succession

1. Annual Grass -Weed Community

- first cover for abandoned fields
- 1 to 2 year lifespan

2. Perennial Grass -Weed Community

- 3rd year replacement
- 5 to 8 year lifespan

3. Shrub Community

- often does not occur farther south
- may last 10 to 20 years

Deer, by Joshuah Tolford, 1945

Additional Communities

Black Belt Prairie and the

a. Discontinuous prairie
- it and uplands found early conversion to pasture and farmland
- new characteristic identifier of the region is the Eastern Redcedar hammock which has ties to soils in abandoned fields

b. Upland forest
- most abundant species are Post Oak, Blackjack Oak, and Hickories

c. Lowland forest
- a varied bottomland forest with thick briar, shrub or cane undergrowth

4. Pine Community (Fire Subclimax Community)

- 75 to 100 year succession lifespan if not interrupted
- with cleansing fires, this community could last forever without a hardwood succession
- Predominately Shortleaf and Loblolly Pines

U. Choctaw Boundary (M.P. 128.4)

Beech - Maple Community

- this is the climax community for more northern states such as Ohio, Indiana, and Southern Michigan
- consider this to be an "island" community within the normal finds
- these trees like more moist soils than the Oak-Hickory Community
- Beech are also a secondary identifier of the Bottomland Forest and Oak - Hickory Climax Communities
- named for the dominant trees

5. Oak - Pine Community (Transition Stage)

- usually lasts 100 years
- most commonly seen in southern half of the Parkway
- Oaks, Pines, Hickory, Sweetgum, Winged Elm, American Elm, Tuliptree, Beech, and others

6. Oak - Hickory Climax Community

- the ultimate tree community for much of the entire region
- no fully mature example forests exist on the Parkway
- Southern Climax Species Group: Southern Red Oak, Scarlet Oak, perhaps White Oak, Mockernut Hickory, and perhaps Shagbark Hickory
- Farther North Climax Species Group: same group except add Chestnut Oak and replace Southern Red Oak with Northern Red Oak
- associated trees to this group: Sweetgum, Black Tupelo, Sourwood, Flowering Dogwood, Tuliptree, and others

Old Trace (M.P. 403.7)

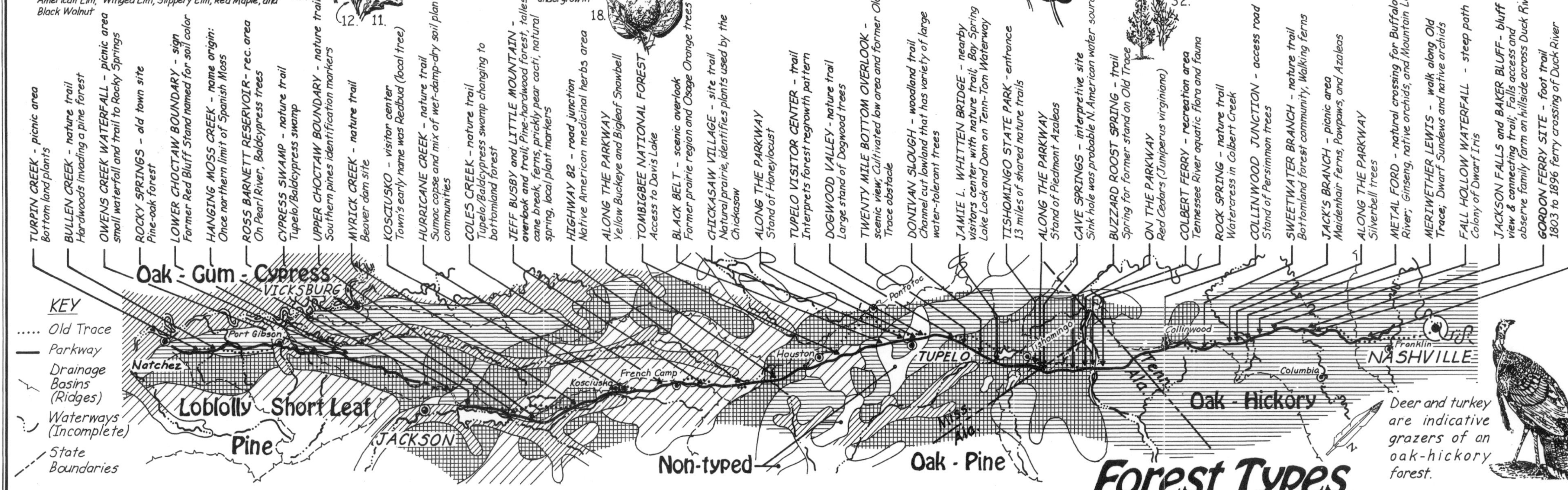

Forest Types

Deer and turkey are indicative grazers of an oak-hickory forest.

18. Osage Orange *Maclura pomifera* · 19. Shortleaf Pine *Pinus echinata* · 20. Loblolly Pine *Pinus taeda* · 21. Beech *Fagus grandifolia* · 22. Black Maple *Acer nigrum* · 23. Tulip Tree *Liriodendron tulipifera* · 24. Sweet Gum *Liquidambar styraciflua* · 25. White Oak *Q. alba* · 26. Mockernut Hickory *Carya alba or tomentosa* · 27. Flowering Dogwood *Cornus florida* · 28. Scarlet Oak *Q. coccinea* · 29. Shagbark Hickory *C. ovata* · 30. Chestnut Oak *Quercus prinus* · 31. Sourwood *Oxydendrum arboreum* · 32. Eastern Red Cedar *Juniperus virginiana*

DELINEATED BY: TIMOTHY J. HALSEY

NATCHEZ TRACE PARKWAY RECORDING PROJECT

UNITED STATES DEPARTMENT OF THE INTERIOR

NATCHEZ TRACE PARKWAY

PARK HEADQUARTERS, 2680 NATCHEZ TRACE PARKWAY

TUPELO · LEE · MISSISSIPPI

HISTORIC AMERICAN ENGINEERING RECORD MS-15

SHEET 4 OF 15

INTERPRETING AGRICULTURAL LANDSCAPES

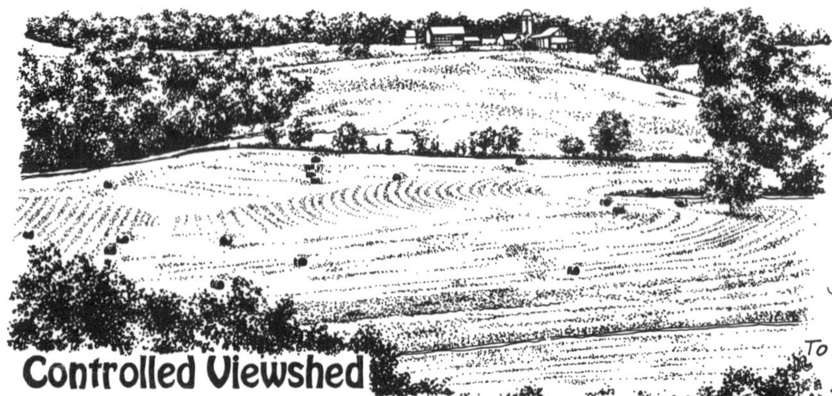

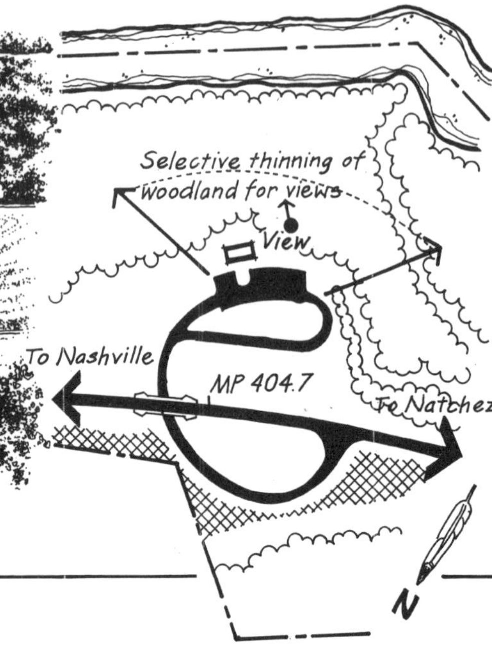

Controlled Viewshed

At Jackson Falls comfort station (MP 404.7) selective thinning of the deciduous woodland edge creates a directed view toward the Baker family farm. The once commonplace rural vista such as at Jackson Falls is now becoming increasingly rare.

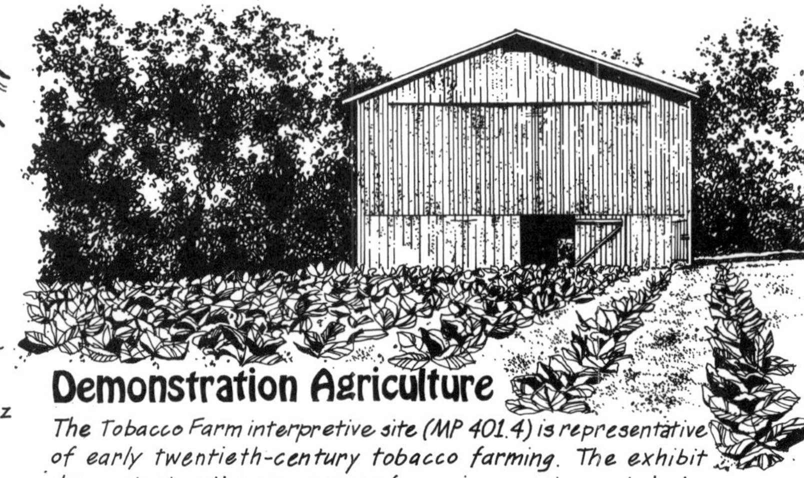

Demonstration Agriculture

The Tobacco Farm interpretive site (MP 401.4) is representative of early twentieth-century tobacco farming. The exhibit demonstrates the processes of growing, cutting and drying tobacco, an important economic crop in the South for many years.

Agricultural Leasing

Landscape plans arranged by Park Service landscape architects allow adjacent landowners to lease land within the National Park Service (NPS) right-of-way. Farmers, leasing the land, are obliged to plant their fields with characteristic Southern crops including cotton, corn, soybeans, and hay meadows, retaining a traditional rural landscape. Agricultural leasing benefits the NPS through minimizing maintenance costs in addition to providing visiting motorists a typical Southern, rural setting. During the 1970s Energy Crisis some unleased fields, owned by the Park Service, were allowed to reforest.

Scenic Easements

Scenic easements are private property in which the government has acquired a monetary interest, for the purpose of limiting its use. Easements provide visual protection for many portions of the Parkway, as a buffer and agricultural fields. This deed restriction on a property owner assures that land adjacent to the Parkway will be used for a specific compatible purpose, aimed at preventing the introduction of unsightly structures. The use of scenic easements helps the designed Parkway landscape merge with the existing countryside.

Hay fields and pasture land MP 254.9

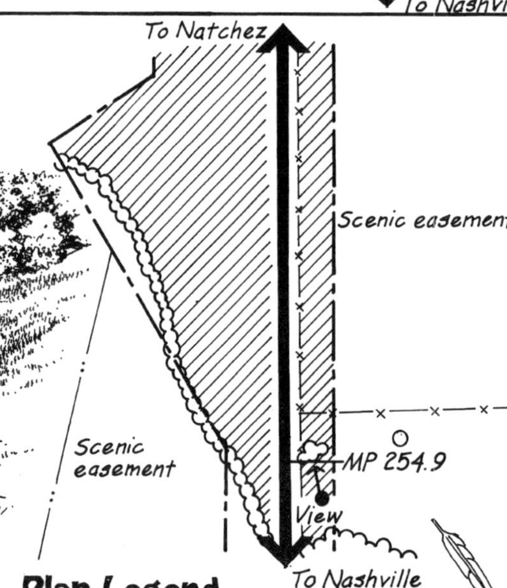

Soybean fields MP 253

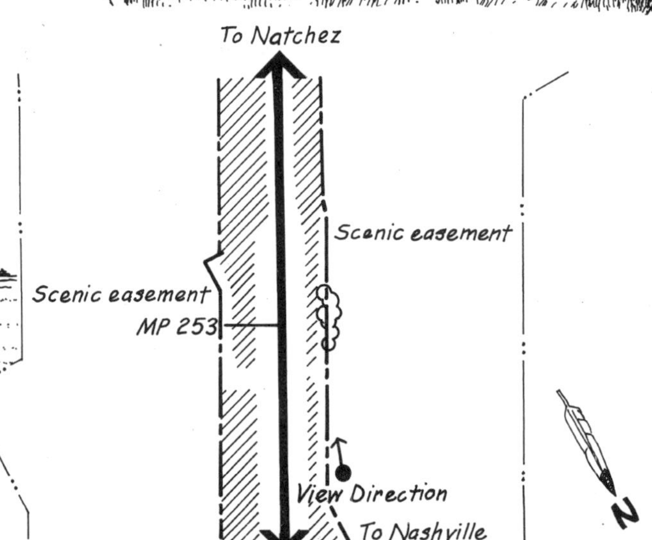

Plan Legend

Scale 1" = 200'

- Scenic Easement
- Parkway Right of Way
- Agricultural Lease
- Natural Reforestation
- Fence
- Trees

Plans drawn from "Land Use and Maintenance Maps," Natchez Trace Parkway Headquarters, Tupelo, M.S.

DELINEATED BY: Katie Dugdill 1998

NATCHEZ TRACE PARKWAY RECORDING PROJECT
NATIONAL PARK SERVICE
UNITED STATES DEPARTMENT OF THE INTERIOR

NATCHEZ TRACE PARKWAY
PARK HEADQUARTERS 2680 NATCHEZ TRACE PARKWAY
TUPELO LEE MISSISSIPPI

SHEET 6 OF 15

HISTORIC AMERICAN ENGINEERING RECORD
MS-15

MOTORIST EXPERIENCE

The Natchez Trace Parkway weaves through the scenic rural South and offers the visiting motorist a variety of experiences. The Parkway roughly follows the Old Natchez Trace, crossing and recrossing it through areas of limestone ridges, rolling farmland, deep forests, and waterlogged bottomlands.

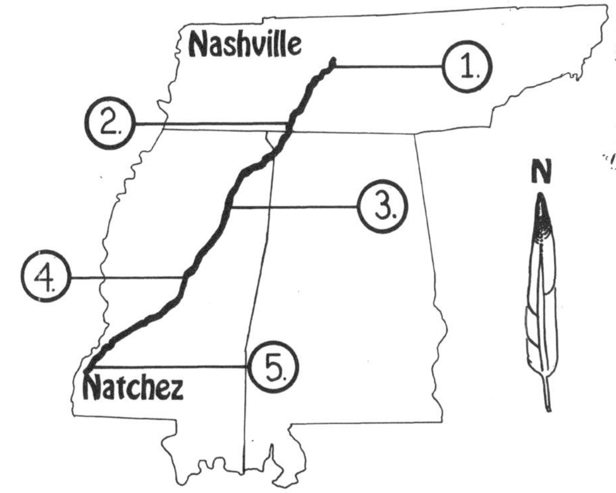

1. Sweeping in great, elegant curves through the deciduous forests of oak and hickory, the Parkway flows over and clings to steep limestone ridges. The road fits closely to the land, and blends into the native scenery. Guard walls are constructed of the indigenous gray limestone.

2. The Parkway meanders through open farmland in a gently rolling, pastoral scene of round hay bales and cattle grazing peacefully at the side of the road. The motoring visitor enjoys a leisurely drive through an ever changing rural landscape.

3. Row crops of cotton, corn and soybeans are an integral feature of this countryside dominated by large cultivated fields. The Black Belt, a once-vast prairie, is famous for its rich, black agricultural soils. The land surface is flat and the Parkway follows a long, straight course.

4. The Parkway motor road is flanked by meticulously mown shoulders and grassy bays, in addition to farmland. Other maintained features include stands of Southern pine, specimen magnolias and wildflower meadows.

5. During its southern course the Parkway travels through swamps and waterlogged bottomlands. The road winds through shady forests in low-lying terrain, and the mixed forest canopy reaches over the sun-dappled road creating an enclosed "tunnel" of vegetation. Lone trees are veiled in Spanish moss.

DELINEATED BY Katie Dugdill 1998

NATCHEZ TRACE PARKWAY RECORDING PROJECT
NATIONAL PARK SERVICE
UNITED STATES DEPARTMENT OF THE INTERIOR

NATCHEZ TRACE PARKWAY
PARK HEADQUARTERS 2680 NATCHEZ TRACE PARKWAY
LEE
TUPELO
MISSISSIPPI

SHEET 7 OF 15

HISTORIC AMERICAN ENGINEERING RECORD
MS-15

IF REPRODUCED, PLEASE CREDIT: HISTORIC AMERICAN ENGINEERING RECORD, NATIONAL PARK SERVICE, NAME OF DELINEATOR, DATE OF THE DRAWING

MOTORIST WAYSIDES

• The Natchez Trace Parkway has many waysides with various facilities and interpretive signs catering to the visiting motorist.

Post Rider Logo

The routed wooden Post Rider logo commemorates the Trace as an early postal route. Designed in the 1950s, the logo directs motorists to interpretive sites along the Parkway.

DAR Markers

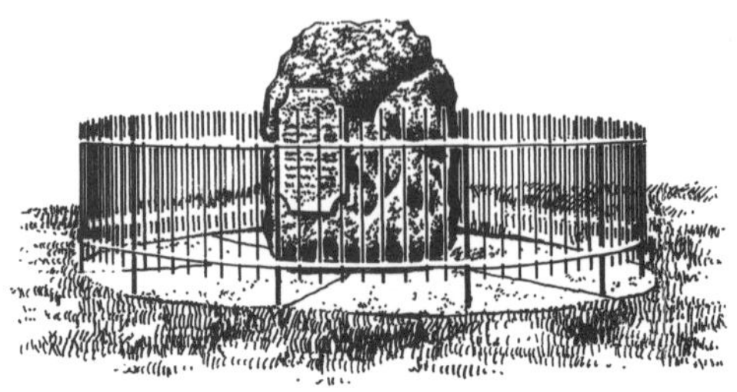

The Daughters of the American Revolution (DAR) memorialized the Trace by erecting inscribed granite monuments along its route. Since 1909, 14 markers have been placed in Mississippi, one for each county traversed by the Trace. In Tennessee the United States Daughters of the War of 1812 have marked the Trace with iron plaques.

Type 1: Roadside Pull-Off

Chickasaw Council House

Chickasaw Council House (MP 251.1) is a typical example of a basic, short stop, roadside pull-off. The motoring visitor loops off the Parkway to read the interpretive sign and rest. Both trash receptacles and interpretive panels are designed to be accessible from a vehicle.

Unique Signage

Signage forms a principal component of the design of the Natchez Trace Parkway. Signs unique to the Parkway are as illustrated.

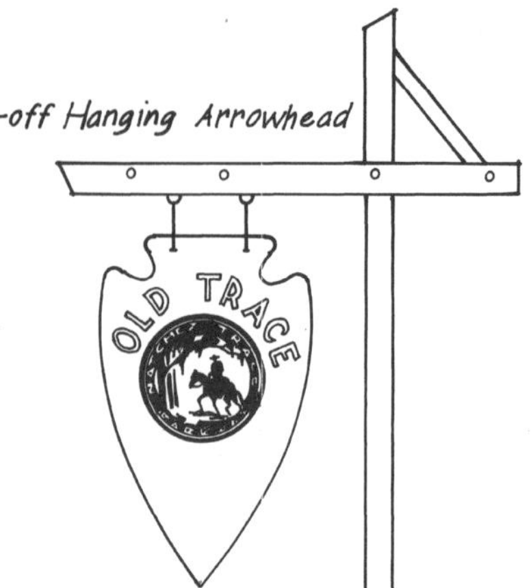

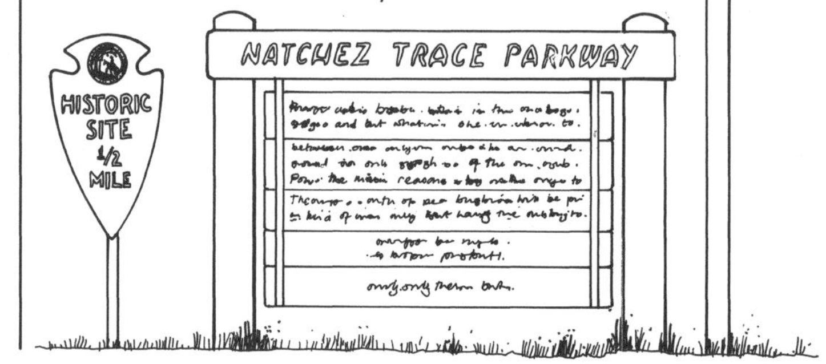

Type 2: Parking Area

Jackson Falls

Comprehensive pull-outs, including Jackson Falls (MP 404.7) offer more elaborate interpretation, restrooms, picnic tables, and trails which lead to natural and historic resources.

View of Comfort Station

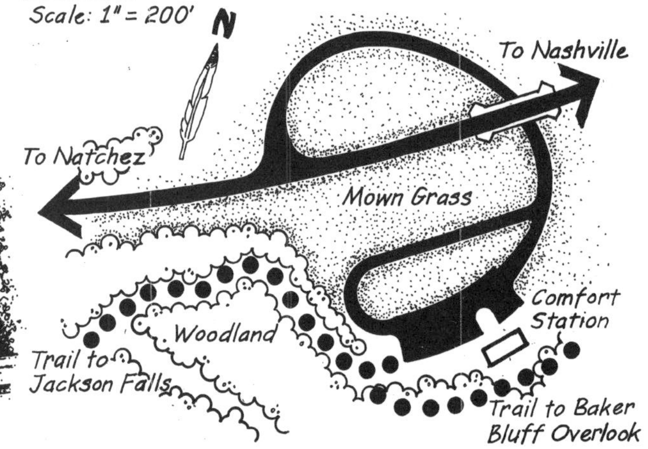

Jackson Falls Pull-Off Area

Type 3: Multi-Use Area

Jeff Busby

Several multi-use areas along the Parkway offer camping and picnic areas, nature trails, interpretive centers and restrooms. Jeff Busby (MP 193.1) is the only motorist wayside to provide a service station in addition to these other accommodations. These areas provide for stays of longer duration.

Arrowhead Old Trace Marker

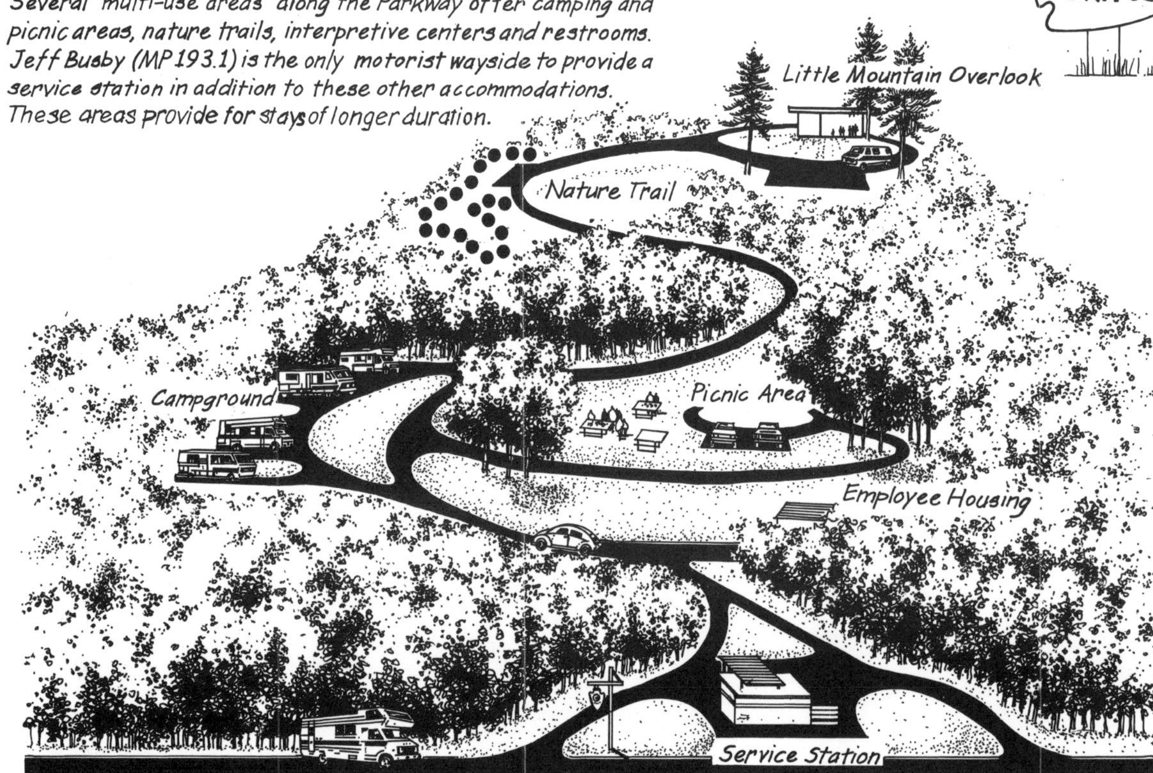

Schematic Diagram

DELINEATED BY: Katie Dugdill 1998

NATCHEZ TRACE PARKWAY RECORDING PROJECT
NATIONAL PARK SERVICE
UNITED STATES DEPARTMENT OF THE INTERIOR

NATCHEZ TRACE PARKWAY
PARK HEADQUARTERS 2680 NATCHEZ TRACE PARKWAY
LEE
TUPELO
MISSISSIPPI

SHEET 8 OF 15

HISTORIC AMERICAN ENGINEERING RECORD
MS-15

IF REPRODUCED, PLEASE CREDIT: HISTORIC AMERICAN ENGINEERING RECORD, NATIONAL PARK SERVICE, NAME OF DELINEATOR, DATE OF THE DRAWING

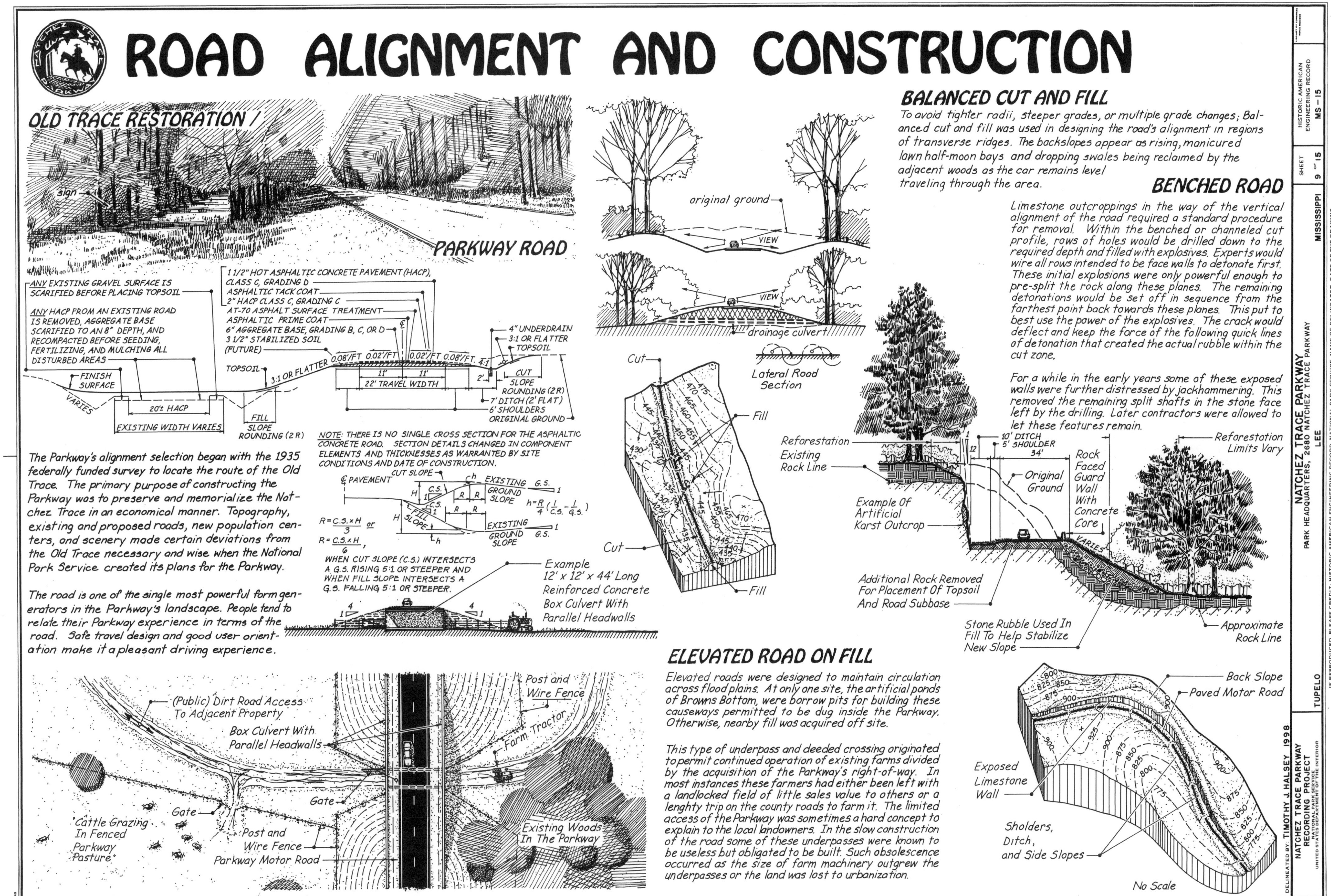
ROAD ALIGNMENT AND CONSTRUCTION
OLD TRACE RESTORATION /
sign
PARKWAY ROAD
ANY EXISTING GRAVEL SURFACE IS SCARIFIED BEFORE PLACING TOPSOIL
ANY HACP FROM AN EXISTING ROAD IS REMOVED, AGGREGATE BASE SCARIFIED TO AN 8" DEPTH, AND RECOMPACTED BEFORE SEEDING, FERTILIZING, AND MULCHING ALL DISTURBED AREAS
1 1/2" HOT ASPHALTIC CONCRETE PAVEMENT (HACP), CLASS C, GRADING D
ASPHALTIC TACK COAT
2" HACP CLASS C, GRADING C
AT-70 ASPHALT SURFACE TREATMENT
ASPHALTIC PRIME COAT
6" AGGREGATE BASE, GRADING B, C, OR D
3 1/2" STABILIZED SOIL (FUTURE)
4" UNDERDRAIN
3:1 OR FLATTER
TOPSOIL
FINISH SURFACE
VARIES
TOPSOIL
3:1 OR FLATTER
0.08'/FT 0.02'/FT 0.02'/FT 0.08'/FT 4:1
11' 11'
22' TRAVEL WIDTH
CUT SLOPE ROUNDING (2 R)
7' DITCH (2' FLAT)
6' SHOULDERS
ORIGINAL GROUND
20'± HACP
EXISTING WIDTH VARIES
FILL SLOPE ROUNDING (2 R)
NOTE: THERE IS NO SINGLE CROSS SECTION FOR THE ASPHALTIC CONCRETE ROAD. SECTION DETAILS CHANGED IN COMPONENT ELEMENTS AND THICKNESSES AS WARRANTED BY SITE CONDITIONS AND DATE OF CONSTRUCTION.
℄ PAVEMENT
CUT SLOPE
EXISTING GROUND SLOPE
G.S.
FILL SLOPE
EXISTING GROUND SLOPE
R=C.S.×H/3 or R=C.S.×H/6,
h=R/4 (1/C.S. − 1/G.S.)
WHEN CUT SLOPE (C.S.) INTERSECTS A G.S. RISING 5:1 OR STEEPER AND WHEN FILL SLOPE INTERSECTS A G.S. FALLING 5:1 OR STEEPER.
The Parkway's alignment selection began with the 1935 federally funded survey to locate the route of the Old Trace. The primary purpose of constructing the Parkway was to preserve and memorialize the Natchez Trace in an economical manner. Topography, existing and proposed roads, new population centers, and scenery made certain deviations from the Old Trace necessary and wise when the National Park Service created its plans for the Parkway.
The road is one of the single most powerful form generators in the Parkway's landscape. People tend to relate their Parkway experience in terms of the road. Safe travel design and good user orientation make it a pleasant driving experience.
Example 12' x 12' x 44' Long Reinforced Concrete Box Culvert With Parallel Headwalls
original ground
VIEW
VIEW
drainage culvert
Lateral Road Section
Cut
Fill
Cut
Fill
BALANCED CUT AND FILL
To avoid tighter radii, steeper grades, or multiple grade changes; Balanced cut and fill was used in designing the road's alignment in regions of transverse ridges. The backslopes appear as rising, manicured lawn half-moon bays and dropping swales being reclaimed by the adjacent woods as the car remains level traveling through the area.
BENCHED ROAD
Limestone outcroppings in the way of the vertical alignment of the road required a standard procedure for removal. Within the benched or channeled cut profile, rows of holes would be drilled down to the required depth and filled with explosives. Experts would wire all rows intended to be face walls to detonate first. These initial explosions were only powerful enough to pre-split the rock along these planes. The remaining detonations would be set off in sequence from the farthest point back towards these planes. This put to best use the power of the explosives. The crack would deflect and keep the force of the following quick lines of detonation that created the actual rubble within the cut zone.
For a while in the early years some of these exposed walls were further distressed by jackhammering. This removed the remaining split shafts in the stone face left by the drilling. Later contractors were allowed to let these features remain.
Reforestation Existing Rock Line
10' DITCH
5' SHOULDER
34'
Rock Faced Guard Wall With Concrete Core
Original Ground
Reforestation Limits Vary
Example Of Artificial Karst Outcrop
VARIES
Additional Rock Removed For Placement Of Topsoil And Road Subbase
Stone Rubble Used In Fill To Help Stabilize New Slope
Approximate Rock Line
(Public) Dirt Road Access To Adjacent Property
Box Culvert With Parallel Headwalls
Post and Wire Fence
Farm Tractor
Gate
Gate
Cattle Grazing In Fenced Parkway Pasture
Post and Wire Fence
Parkway Motor Road
Existing Woods In The Parkway
ELEVATED ROAD ON FILL
Elevated roads were designed to maintain circulation across flood plains. At only one site, the artificial ponds of Browns Bottom, were borrow pits for building these causeways permitted to be dug inside the Parkway. Otherwise, nearby fill was acquired off site.
This type of underpass and deeded crossing originated to permit continued operation of existing farms divided by the acquisition of the Parkway's right-of-way. In most instances these farmers had either been left with a landlocked field of little sales value to others or a lenghty trip on the county roads to farm it. The limited access of the Parkway was sometimes a hard concept to explain to the local landowners. In the slow construction of the road some of these underpasses were known to be useless but obligated to be built. Such obsolescence occurred as the size of farm machinery outgrew the underpasses or the land was lost to urbanization.
Back Slope
Paved Motor Road
Exposed Limestone Wall
Sholders, Ditch, and Side Slopes
No Scale
DELINEATED BY: TIMOTHY J. HALSEY 1998
NATCHEZ TRACE PARKWAY RECORDING PROJECT
NATIONAL PARK SERVICE
UNITED STATES DEPARTMENT OF THE INTERIOR
NATCHEZ TRACE PARKWAY
PARK HEADQUARTERS, 2680 NATCHEZ TRACE PARKWAY
TUPELO
LEE
MISSISSIPPI
HISTORIC AMERICAN ENGINEERING RECORD
MS-15
SHEET 9 OF 15
IF REPRODUCED, PLEASE CREDIT: HISTORIC AMERICAN ENGINEERING RECORD, NATIONAL PARK SERVICE, NAME OF DELINEATOR, DATE OF THE DRAWING

DOUBLE ARCH BRIDGE

=OVER TENNESSEE ROUTE 96=

The First U.S. Pre-Cast Segmental Concrete Arched Bridge

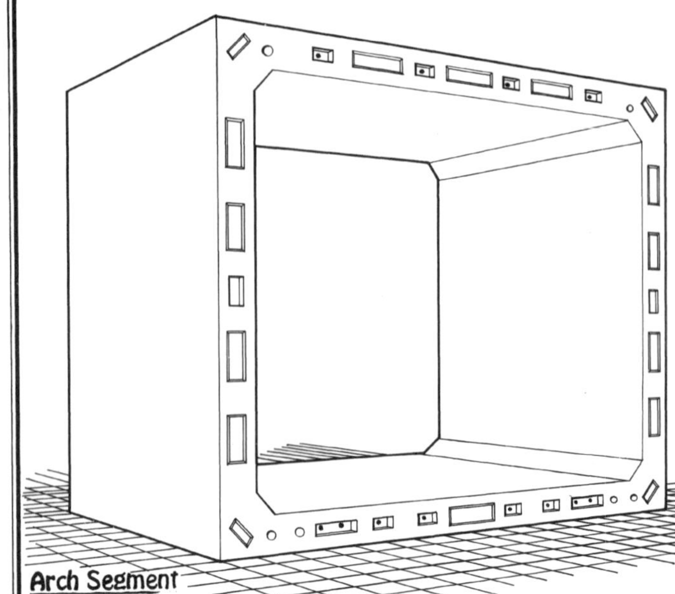

Arch Segment

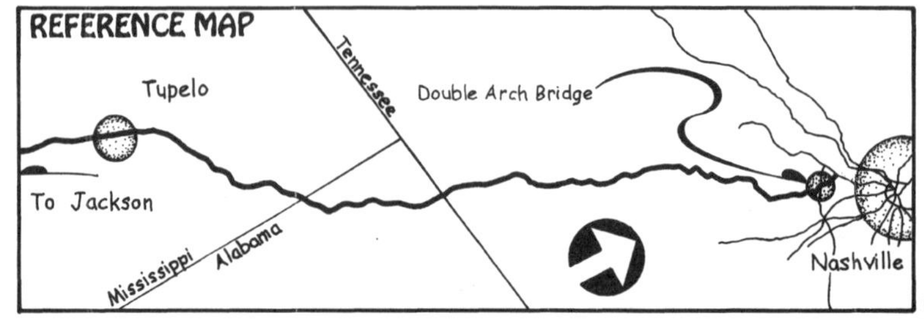

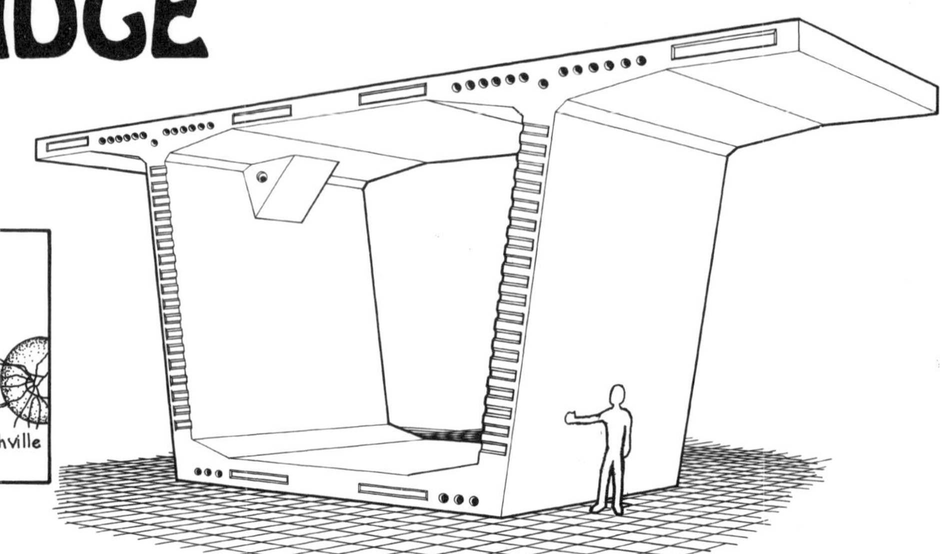

Superstructure Segment

Designed By: Figg & Mueller Engineering
Contracted By: National Park Service-Federal Highway Division

Serving as a vital link in the ultimate completion of the Natchez Trace Parkway, the Double Arch Bridge represents the most innovative engineering project in the Parkway's construction history. Completed in May 1994, and selected for a Presidential Award for Design Excellence in 1995, this bridge became the Nation's first segmentally constructed concrete arch bridge.

Spanning some 1,648 feet over State Highway 96, eight miles west of Franklin, TN Trace motorists are afforded a panoramic view 155' above the valley bottom. Open arches echo the surrounding hilly terrain, resulting in a modern bridge that complements rather than dominates its picturesque setting. Arch bridges are traditionally constructed using a number of vertical support members called spandrel columns. In an effort to create an airy structure that yields to the surrounding landscape, the spandrel columns were omitted.

A combination of concrete precast and cast-in-place segments were used to construct the bridge. The precast segments were constructed offsite at a plant under controlled conditions. Together, 196 superstructure (road deck) segments and 122 arch segments were cast and joined to complete the structure.

Both the arch and the superstructure segments were hoisted into place using a large-capacity, ground-mounted, crane. The selection of this particular type of crane substantially reduced the required amount of clearing, ground preparation and restoration. Adding to the significance of the project, the contractor erected the pre-cast concrete arch segments without relying on traditional falsework, a labor-intensive and environmentally intrusive framework constructed to support bridge erection. Instead, the cantilevered segments were held in place using temporary backstay caissons and cable stays that supported the cantilevered arch halves until cast-in-place keystone segments secured their union.

Typical road deck segments measured 8'-6" in length, and weighed between 36 and 57 tons. The segments were erected using a conventional balanced cantilever construction technique, allowing the work-in-progress to extend beyond a fulcrum point. Cables in the top and bottom of each segment joined each unit to the next. Epoxy was applied to the segment faces before the cables were tensioned in place.

A latex/concrete overlay was applied to the deck to prevent the corrosion of deck reinforcement from water seeping into the naturally porous concrete. The surface of the bridge was sprayed with a white, textured coating that gives it the visual appearance of a single unit as it covered the segment joints.

A parking area and overlook was constructed at the north end of the bridge to avoid the hazards posed by motorists using the bridge itself as a scenic vantage point. The overlook gracefully offers a breathtaking view of the Double Arch Bridge through tall trees, reflecting the care with which the landscape was preserved.

View North From South Slope

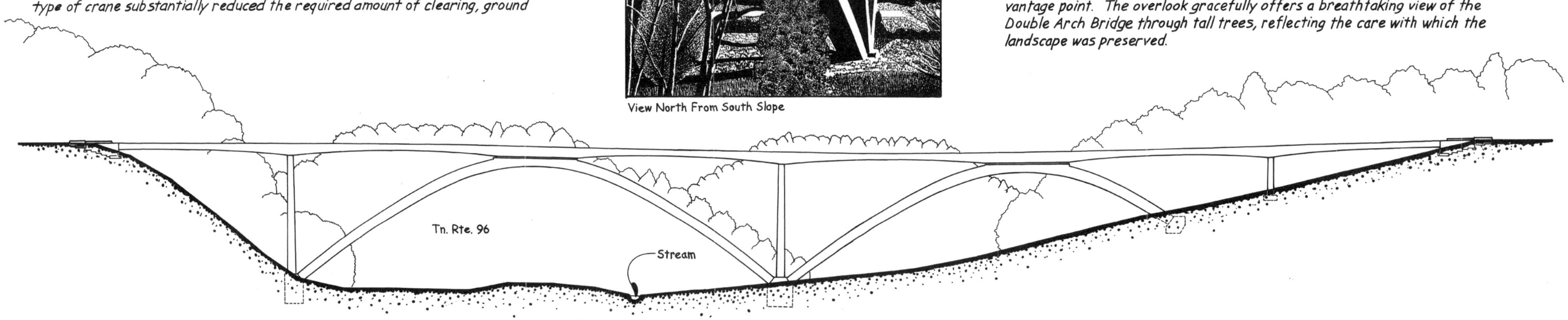

Elevation Scale: 1"=60'

DELINEATED BY NICHOLAS A. ZYDYCRYN, 1998
NATCHEZ TRACE PARKWAY RECORDING PROJECT
NATIONAL PARK SERVICE
UNITED STATES DEPARTMENT OF THE INTERIOR

NATCHEZ TRACE PARKWAY
PARK HEADQUARTERS, 2680 NATCHEZ TRACE PARKWAY
TUPELO LEE MISSISSIPPI

SHEET 10 OF 15

HISTORIC AMERICAN ENGINEERING RECORD
MS-15

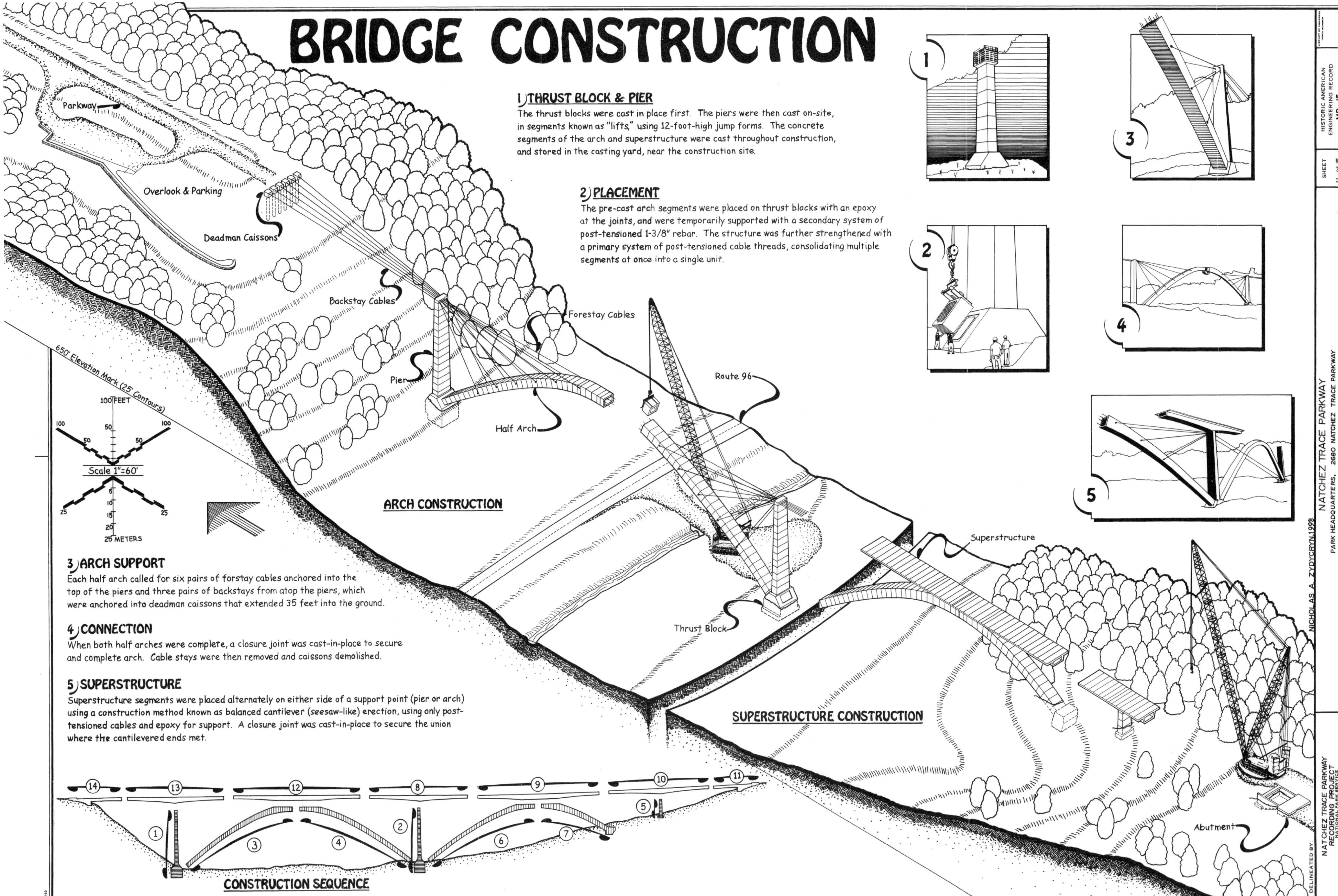
BRIDGE CONSTRUCTION
1) THRUST BLOCK & PIER
The thrust blocks were cast in place first. The piers were then cast on-site, in segments known as "lifts," using 12-foot-high jump forms. The concrete segments of the arch and superstructure were cast throughout construction, and stored in the casting yard, near the construction site.
2) PLACEMENT
The pre-cast arch segments were placed on thrust blocks with an epoxy at the joints, and were temporarily supported with a secondary system of post-tensioned 1-3/8" rebar. The structure was further strengthened with a primary system of post-tensioned cable threads, consolidating multiple segments at once into a single unit.
3) ARCH SUPPORT
Each half arch called for six pairs of forstay cables anchored into the top of the piers and three pairs of backstays from atop the piers, which were anchored into deadman caissons that extended 35 feet into the ground.
4) CONNECTION
When both half arches were complete, a closure joint was cast-in-place to secure and complete arch. Cable stays were then removed and caissons demolished.
5) SUPERSTRUCTURE
Superstructure segments were placed alternately on either side of a support point (pier or arch) using a construction method known as balanced cantilever (seesaw-like) erection, using only post-tensioned cables and epoxy for support. A closure joint was cast-in-place to secure the union where the cantilevered ends met.
Parkway
Overlook & Parking
Deadman Caissons
Backstay Cables
Forestay Cables
Pier
Half Arch
Route 96
650' Elevation Mark (25' Contours)
100 FEET
Scale 1"=60'
25 METERS
ARCH CONSTRUCTION
Thrust Block
Superstructure
SUPERSTRUCTURE CONSTRUCTION
Abutment
CONSTRUCTION SEQUENCE
HISTORIC AMERICAN ENGINEERING RECORD
MS-15
SHEET 11 OF 15
NATCHEZ TRACE PARKWAY
PARK HEADQUARTERS, 2680 NATCHEZ TRACE PARKWAY
TUPELO LEE MISSISSIPPI
NICHOLAS A. ZYDYCRYN 1998
DELINEATED BY:
NATCHEZ TRACE PARKWAY RECORDING PROJECT
NATIONAL PARK SERVICE
UNITED STATES DEPARTMENT OF THE INTERIOR
IF REPRODUCED, PLEASE CREDIT: HISTORIC AMERICAN ENGINEERING RECORD, NATIONAL PARK SERVICE, NAME OF DELINEATOR, DATE OF THE DRAWING

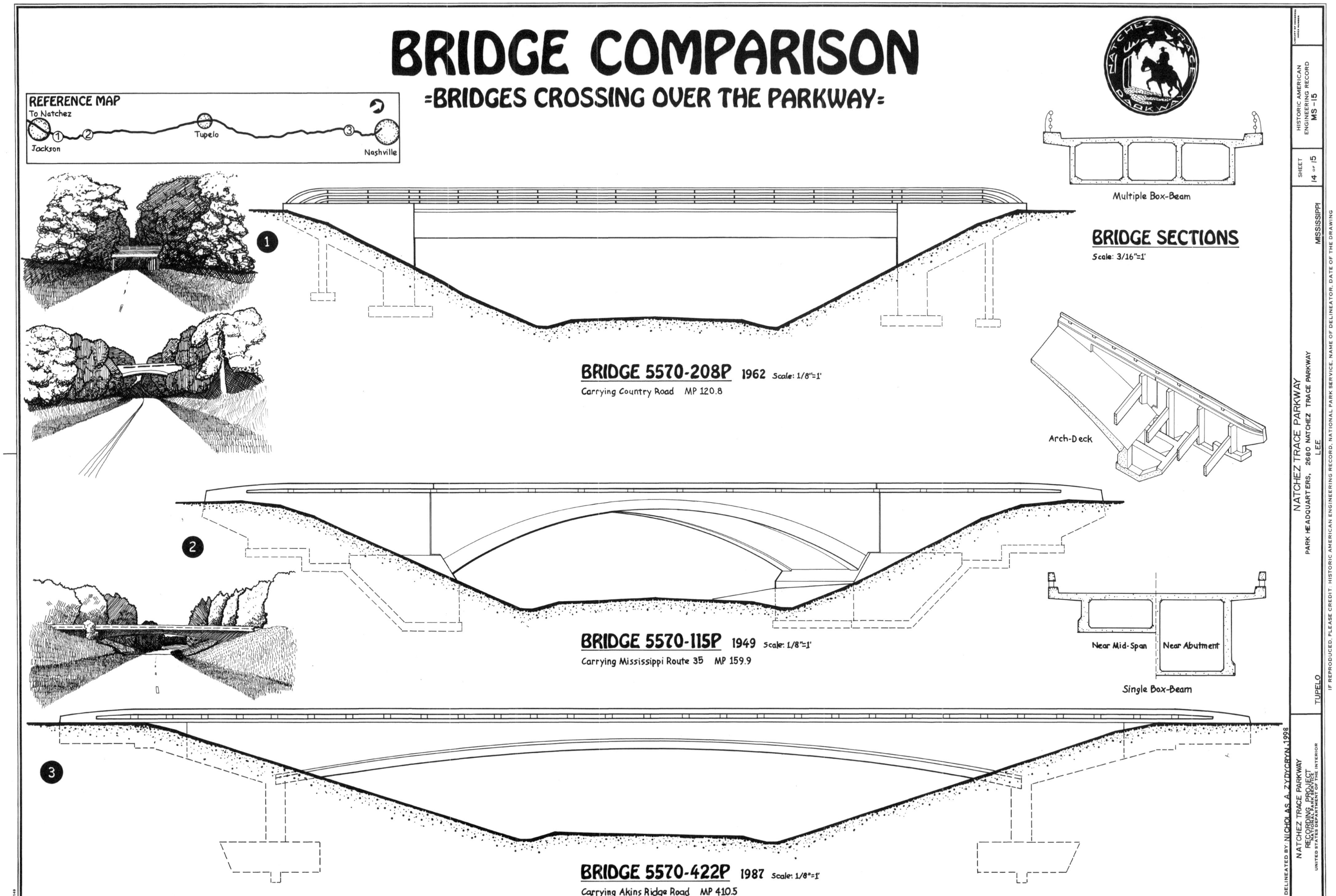

BRIDGE COMPARISON
=BRIDGES CROSSING OVER THE PARKWAY=
REFERENCE MAP
To Natchez
Jackson
Tupelo
Nashville
1
2
3
BRIDGE 5570-208P 1962 Scale: 1/8"=1'
Carrying Country Road MP 120.8
BRIDGE 5570-115P 1949 Scale: 1/8"=1'
Carrying Mississippi Route 35 MP 159.9
BRIDGE 5570-422P 1987 Scale: 1/8"=1'
Carrying Akins Ridge Road MP 410.5
Multiple Box-Beam
BRIDGE SECTIONS
Scale: 3/16"=1'
Arch-Deck
Near Mid-Span
Near Abutment
Single Box-Beam
NATCHEZ TRACE PARKWAY
HISTORIC AMERICAN ENGINEERING RECORD MS-15
SHEET 14 OF 15
NATCHEZ TRACE PARKWAY
PARK HEADQUARTERS, 2680 NATCHEZ TRACE PARKWAY
TUPELO
LEE
MISSISSIPPI
DELINEATED BY: NICHOLAS A. ZYDYCRYN 1998
NATCHEZ TRACE PARKWAY RECORDING PROJECT
NATIONAL PARK SERVICE
UNITED STATES DEPARTMENT OF THE INTERIOR
IF REPRODUCED, PLEASE CREDIT: HISTORIC AMERICAN ENGINEERING RECORD, NATIONAL PARK SERVICE, NAME OF DELINEATOR, DATE OF THE DRAWING

ROCK CREEK AND POTOMAC PARKWAY

WASHINGTON, D.C.

RELIEF PANEL BY LEON HERMANT.
CALVERT STREET BRIDGE, Ca. 1935.

BUFFALO SCULPTURE BY ALEXANDER PHIMISTER PROCTOR.
Q STREET BRIDGE, Ca. 1914.

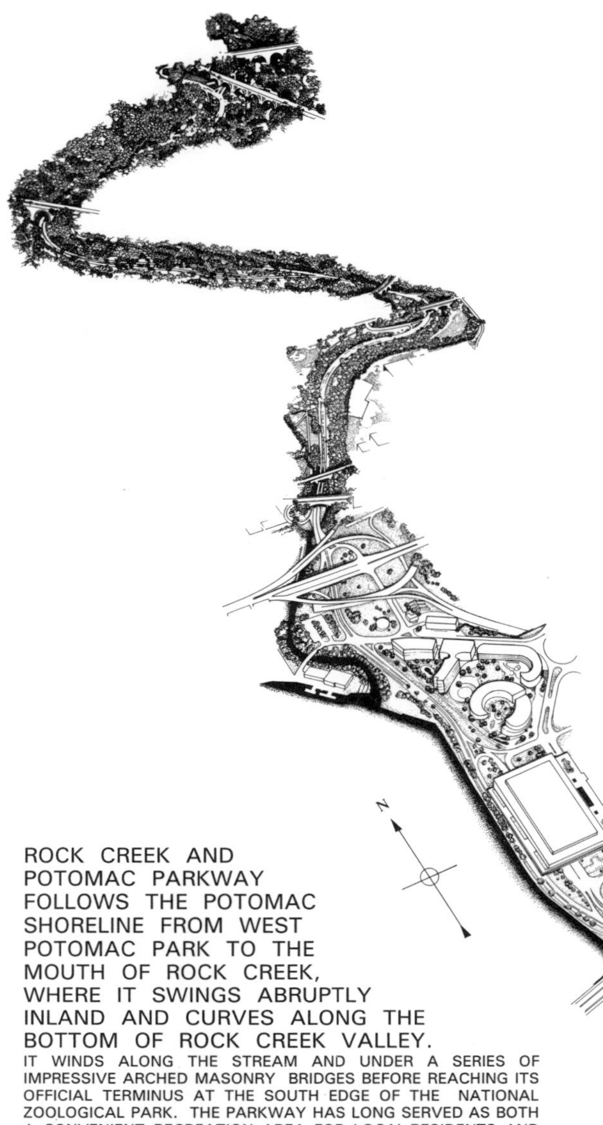

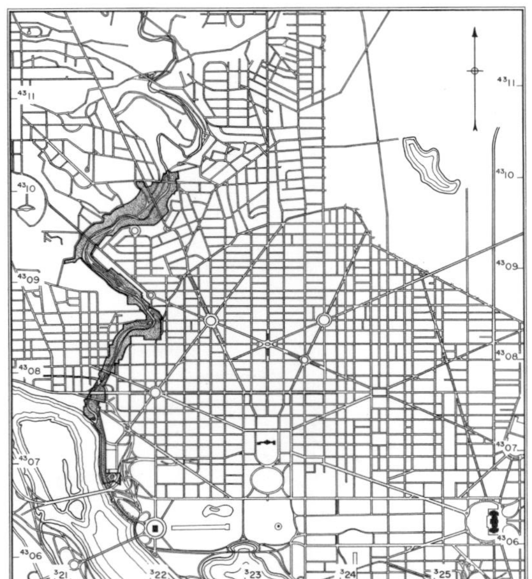

SITE LOCATION MAP
MAP ADAPTED FROM USGS WASHINGTON WEST D.C. - MD. - VA. QUADRANGLE, 1965, PHOTOREVISED 1983.
MILES: 1"= 0.38 MILES
KILOMETERS: 1 : 24,000

DOCUMENTATION OF ROCK CREEK AND POTOMAC PARKWAY WAS UNDERTAKEN AS A TWO-YEAR PILOT PROJECT TO HELP ESTABLISH STANDARDS AND GUIDELINES FOR RECORDING THE STRUCTURES AND LANDSCAPE FEATURES OF PARK ROADS AND PARKWAYS. THIS PROJECT WAS A JOINT EFFORT OF THE HISTORIC AMERICAN BUILDINGS SURVEY AND THE HISTORIC AMERICAN ENGINEERING RECORD (HABS/HAER), A COMBINED DIVISION OF THE NATIONAL PARK SERVICE, ROBERT KAPSCH, CHIEF. THE PROJECT WAS SPONSORED BY THE PARK ROADS PROGRAM OF THE NATIONAL PARK SERVICE, JOHN GINGLES, DEPUTY CHIEF, ENGINEERING AND SAFETY SERVICES DIVISION. THE PROJECT SUPERVISOR WAS SARA AMY LEACH, HABS HISTORIAN.

THE WASHINGTON-BASED SUMMER 1992 DOCUMENTATION TEAM WAS HEADED BY LANDSCAPE ARCHITECT ROBERT HARVEY (IOWA STATE UNIVERSITY-DEPARTMENT OF LANDSCAPE ARCHITECTURE) WHO SERVED AS FIELD SUPERVISOR; THE LANDSCAPE ARCHITECTS WERE DEBORAH WARSHAW (UNIVERSITY OF VIRGINIA) AND DOROTA PAPE-SILIWONCZUK (US/ICOMOS - POLAND, BOARD OF HISTORICAL PALACES AND GARDENS RESTORATION); ARCHITECTS WERE EVAN MILLER (UNIVERSITY OF COLORADO - BOULDER), STEVEN NOSE (UNIVERSITY OF MARYLAND), AND TONY ARCARO (CATHOLIC UNIVERSITY). THE HISTORIANS WERE TIMOTHY DAVIS (UNIVERSITY OF TEXAS) AND AMY ROSS (UNIVERSITY OF VIRGINIA). JACK E. BOUCHER MADE THE LARGE-FORMAT PHOTOGRAPHS; AIR SURVEY CORPORATION OF STERLING, VIRGINIA, PROVIDED AERIAL PHOTOGRAPHY AND DIGITAL MAPPING SERVICES.

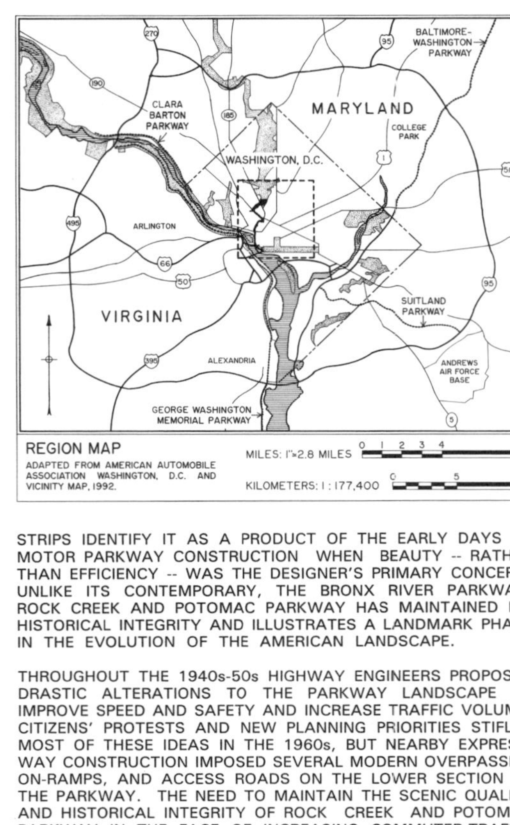

REGION MAP
ADAPTED FROM AMERICAN AUTOMOBILE ASSOCIATION WASHINGTON, D.C. AND VICINITY MAP, 1992.
MILES: 1"=2.8 MILES
KILOMETERS: 1 : 177,400

ROCK CREEK AND POTOMAC PARKWAY FOLLOWS THE POTOMAC SHORELINE FROM WEST POTOMAC PARK TO THE MOUTH OF ROCK CREEK, WHERE IT SWINGS ABRUPTLY INLAND AND CURVES ALONG THE BOTTOM OF ROCK CREEK VALLEY.

IT WINDS ALONG THE STREAM AND UNDER A SERIES OF IMPRESSIVE ARCHED MASONRY BRIDGES BEFORE REACHING ITS OFFICIAL TERMINUS AT THE SOUTH EDGE OF THE NATIONAL ZOOLOGICAL PARK. THE PARKWAY HAS LONG SERVED AS BOTH A CONVENIENT RECREATION AREA FOR LOCAL RESIDENTS AND AN ATTRACTIVE, WOODED COMMUTER ROUTE LINKING DOWNTOWN WASHINGTON WITH ITS NORTHWEST SUBURBS. IT IS A KEY CONNECTION IN THE PRECISELY PLANNED SYSTEM OF PARKS AND PARKWAYS SURROUNDING THE NATIONAL CAPITAL.

ROCK CREEK AND POTOMAC PARKWAY IS THE OLDEST OF SEVERAL SCENIC PARKWAYS IN AND AROUND WASHINGTON, D.C. ITS 2.5 MILES ENCAPSULATE A CENTURY OF AMERICAN LANDSCAPE DESIGN THEORY AND PRACTICE. FIRST PROPOSED IN 1867, THE PARKWAY WAS AUTHORIZED BY CONGRESS IN 1913 AND COMPLETED IN 1936. ITS CONCEPTION AND DESIGN ARE FIRMLY EMBEDDED IN TURN-OF-THE-CENTURY URBAN IMPROVEMENT PHILOSOPHIES AND ROMANTIC LANDSCAPE AESTHETICS, YET IT WAS COMPLETED AT A TIME WHEN NEW DIMENSIONS OF SPEED, SCALE, AND SOCIAL AND SPATIAL ORGANIZATION WERE ENTERING THE URBAN EQUATION.

THE CITY BEAUTIFUL-ERA LANDSCAPE ARCHITECTS WHO FIRST CONCEIVED ROCK CREEK AND POTOMAC PARKWAY ENVISIONED IT AS A SCENIC PLEASURE ROUTE LINKING THE DISTRICT'S TWO MAJOR PARKS. THE GRACEFUL, TREE-LINED BOULEVARDS OF CITIES SUCH AS CHICAGO, BUFFALO, AND BROOKLYN PROVIDED AMPLE PRECEDENT FOR THE SCHEME OF CARRIAGE ROADS, BRIDLE PATHS, AND PEDESTRIAN PROMENADES ORIGINALLY PROPOSED FOR ROCK CREEK VALLEY. THE OLMSTED FIRM'S RIVERWAY PROJECT IN BOSTON WAS AN EVEN MORE DIRECT INFLUENCE ON THE PARKWAY'S DESIGNERS, WHO REJECTED THE FORMAL AVENUE CONFIGURATION TYPICAL OF NINETEENTH-CENTURY BOULEVARDS IN FAVOR OF THE NATURALISTIC LANDSCAPE TREATMENT EMPLOYED ALONG PORTIONS OF THE RIVERWAY.

THE PARKWAY'S BASIC CONFIGURATION WAS ARTICULATED IN THE McMILLAN COMMISSION'S 1902 RECOMMENDATIONS FOR THE DEVELOPMENT OF WASHINGTON, D.C.'s PARK SYSTEM. LEGISLATIVE HURDLES, LOCAL POLITICS, AND FUNDING PROBLEMS DELAYED CONSTRUCTION UNTIL THE 1920s-30s. BECAUSE OF THIS PROLONGED DEVELOPMENT PHASE, ROCK CREEK AND POTOMAC PARKWAY REPRESENTS AN IMPORTANT EXAMPLE OF HOW THE AUTOMOBILE AFFECTED THE EVOLUTION OF AMERICAN LANDSCAPE DESIGN. THE SHIFTING EMPHASIS FROM GENERAL PARK IMPROVEMENTS TO THE CREATION OF A HANDSOME, LANDSCAPED TRANSPORTATION CORRIDOR REFLECTS THE TRANSFORMATION OF THE PARKWAY IDEA IN AMERICA.

WITH THE RISING POPULARITY OF AUTOMOBILE COMMUTING IN THE 1920s-30s, ROCK CREEK AND POTOMAC PARKWAY WAS INCREASINGLY PERCEIVED AS A TRANSPORTATION CORRIDOR RATHER THAN AS A PLACE FOR THE LEISURELY CONTEMPLATION OF NATURE. MULTIPLE CARRIAGE DRIVES, MEANDERING ACCESS ROADS, AND PICTURESQUE PEDESTRIAN OVERLOOKS GAVE WAY TO A SINGLE, DOMINANT AUTO ROAD WITH SIMPLIFIED LANDSCAPE TREATMENT AND LIMITED ACCESS. THE RECONFIGURED ROCK CREEK AND POTOMAC PARKWAY WAS NO MODERN EXPRESSWAY, HOWEVER. IT WAS A TRANSITIONAL LANDSCAPE, RETAINING MANY CHARACTERISTICS OF NINETEENTH-CENTURY PARK ROADS BUT EMPLOYING INNOVATIVE DESIGN FEATURES DEVELOPED TO PERFECTION IN LATER THOROUGHFARES SUCH AS CONNECTICUT'S MERRITT PARKWAY AND NEW YORK'S TACONIC PARKWAY. THE NARROW, TWISTING ROADWAY, WITH ITS ENVELOPING VEGETATION, SLOW SPEEDS, ABRUPT ENTRANCES, AND INFREQUENT MEDIAN STRIPS IDENTIFY IT AS A PRODUCT OF THE EARLY DAYS OF MOTOR PARKWAY CONSTRUCTION WHEN BEAUTY -- RATHER THAN EFFICIENCY -- WAS THE DESIGNER'S PRIMARY CONCERN. UNLIKE ITS CONTEMPORARY, THE BRONX RIVER PARKWAY, ROCK CREEK AND POTOMAC PARKWAY HAS MAINTAINED ITS HISTORICAL INTEGRITY AND ILLUSTRATES A LANDMARK PHASE IN THE EVOLUTION OF THE AMERICAN LANDSCAPE.

THROUGHOUT THE 1940s-50s HIGHWAY ENGINEERS PROPOSED DRASTIC ALTERATIONS TO THE PARKWAY LANDSCAPE TO IMPROVE SPEED AND SAFETY AND INCREASE TRAFFIC VOLUME. CITIZENS' PROTESTS AND NEW PLANNING PRIORITIES STIFLED MOST OF THESE IDEAS IN THE 1960s, BUT NEARBY EXPRESSWAY CONSTRUCTION IMPOSED SEVERAL MODERN OVERPASSES, ON-RAMPS, AND ACCESS ROADS ON THE LOWER SECTION OF THE PARKWAY. THE NEED TO MAINTAIN THE SCENIC QUALITY AND HISTORICAL INTEGRITY OF ROCK CREEK AND POTOMAC PARKWAY IN THE FACE OF INCREASING COMMUTER-TRAFFIC PRESSURE REMAINS A FORMIDABLE POLICY ISSUE FOR PARKWAY MANAGERS AND REGIONAL PLANNERS.

ROCK CREEK AND POTOMAC PARKWAY, ALSO DESIGNATED RESERVATION NO. 360, IS UNDER THE JURISDICTION OF THE NATIONAL PARK SERVICE, NATIONAL CAPITAL REGION. NATIONAL CAPITAL PARKS-CENTRAL ADMINISTERS THE SOUTH END OF THE PARKWAY ALONG THE POTOMAC WATERFRONT, AND ROCK CREEK PARK ADMINISTERS THE PARKWAY FROM VIRGINIA AVENUE NORTH TO THE NATIONAL ZOO. THE BRIDGES THAT CARRY CITY STREETS ACROSS THE PARKWAY ARE OWNED AND MAINTAINED BY THE D.C. DEPARTMENT OF PUBLIC WORKS.

DRAWN BY: EVAN MILLER, ROBERT HARVEY, AND DOUGLAS ANDERSON 1992

ROCK CREEK AND POTOMAC PARKWAY DOCUMENTATION PROJECT 1991-1992 NATIONAL PARK SERVICE UNITED STATES DEPARTMENT OF THE INTERIOR	NAME AND LOCATION OF STRUCTURE: ROCK CREEK AND POTOMAC PARKWAY, WASHINGTON, DISTRICT OF COLUMBIA	SHEET 1 OF 36 SHEETS	HISTORIC AMERICAN BUILDINGS SURVEY DC-697

IF REPRODUCED, PLEASE CREDIT: HISTORIC AMERICAN BUILDINGS SURVEY, NATIONAL PARK SERVICE, NAME OF DELINEATOR, DATE OF THE DRAWING

EVOLUTION OF THE DESIGN

IN 1867, MAJ. NATHANIEL MICHLER OF THE U.S. ARMY CORPS OF ENGINEERS RECOMMENDED TURNING THE UPPER REACHES OF ROCK CREEK VALLEY INTO A FEDERALLY PROTECTED PARK AND BUILDING AVENUES ALONG THE LOWER VALLEY TO CONNECT IT WITH DOWNTOWN WASHINGTON. CONGRESS CREATED ROCK CREEK PARK IN 1890, BUT THE LOWER VALLEY REMAINED IN PRIVATE HANDS. BY THE END OF THE NINETEENTH CENTURY, THE VALLEY BELOW MASSACHUSETTS AVENUE HAD BECOME A COMBINED DUMPING GROUND AND INDUSTRIAL ZONE. RAMSHACKLE FRAME HOUSES AND TENEMENTS LINED THE VALLEY NEAR P STREET. WASHINGTONIANS REGARDED THE AREA AS A PUBLIC HEALTH HAZARD AND CIVIC EMBARRASSMENT. THE WASHINGTON BOARD OF TRADE AND THE GEORGETOWN CITIZENS' ASSOCIATION SUGGESTED REHABILITATING THE VALLEY, BUT COULD NOT AGREE ON A DESIGN.

THE McMILLAN COMMISSION, FORMED IN 1901 TO REVITALIZE THE DISTRICT OF COLUMBIA PARK SYSTEM, EXAMINED TWO OPTIONS FOR TRANSFORMING THE POLLUTED VALLEY INTO AN ATTRACTIVE PARKWAY, LINKING ROCK CREEK PARK WITH THE POTOMAC PARKS. THE "CLOSED-VALLEY" PLAN CALLED FOR ENCLOSING ROCK CREEK IN A CONDUIT AND FILLING IN THE VALLEY ABOVE TO CREATE A HANDSOME BOULEVARD FLANKED BY SUBSTANTIAL HOUSES. THE "OPEN-VALLEY" PLAN PROPOSED RESTORING THE VALLEY TO ITS NATURAL CONDITION AND PLACING A WINDING DRIVE ALONG THE CREEK. THE McMILLAN COMMISSION STRONGLY RECOMMENDED THE SECOND ALTERNATIVE AS THE MORE ATTRACTIVE AND PRACTICAL SOLUTION. THE COMMISSION CONSIDERED ROCK CREEK AND POTOMAC PARKWAY AN ESSENTIAL COMPONENT OF ITS AMBITIOUS PLAN TO DEVELOP AN EXTENSIVE SERIES OF PARKS LINKING THE CITY'S MONUMENTAL CORE WITH ITS SURROUNDINGS.

GEORGETOWN CITIZENS CONTINUED TO ADVOCATE THE CLOSED-VALLEY APPROACH, BUT THE OFFICE OF THE D.C. ENGINEER COMMISSIONER RULED IN FAVOR OF THE OPEN-VALLEY PLAN AFTER CONDUCTING A DETAILED SURVEY IN 1908. CONGRESS AUTHORIZED THE PURCHASE OF LAND FOR THE PARKWAY IN 1913. FUNDING SHORTAGES AND LAND-ACQUISITION PROBLEMS DELAYED THE START OF CONSTRUCTION UNTIL 1926 AND PROLONGED THE PROCESS UNTIL 1936.

CHANGING CONDITIONS IN THE YEARS FOLLOWING THE PARKWAY'S CONCEPTION RESULTED IN MANY CHANGES TO THE DESIGN. MOST INVOLVED SIMPLIFYING THE LANDSCAPE AS IT EVOLVED FROM A MULTI-PURPOSE URBAN PARK TO A SCENIC TRANSPORTATION CORRIDOR DOMINATED BY AUTOMOBILE COMMUTERS. SEVERAL OF THE WINDING CARRIAGE ENTRANCES WERE DEEMED UNNECESSARY. UPGRADING THEM FOR AUTOMOBILES WOULD HAVE BEEN UNDULY DESTRUCTIVE OF NATURAL SCENERY. DEVELOPMENT ALONG THE PARKWAY'S BORDERS ELIMINATED THE FORMAL AVENUES THAT WERE TO HAVE FLANKED THE EDGES OF ROCK CREEK VALLEY.

BECAUSE THE PLANS FOR ROCK CREEK AND POTOMAC PARKWAY EVOLVED OVER SO MANY YEARS, IT IS DIFFICULT TO APPORTION CREDIT FOR ITS DESIGN. AS LANDSCAPE ARCHITECTURE EXPERT FOR THE McMILLAN COMMISSION AND THE COMMISSION OF FINE ARTS, FREDERICK LAW OLMSTED, JR., EXERTED CONSIDERABLE INFLUENCE ON THE PARKWAY'S OVERALL DESIGN AND SUPERVISED ITS GRADUAL REFINEMENT. A SUCCESSION OF LANDSCAPE ARCHITECTS AND ENGINEERS EMPLOYED BY THE ARMY CORPS OF ENGINEERS AND THE OFFICE OF PUBLIC BUILDINGS AND GROUNDS WORKED OUT THE DETAILS OF THE DESIGN AND PREPARED PLANS AND CONSTRUCTION DRAWINGS. CAPT. E. M. MARKHAM, MAJ. JAY J. MORROW, AND COL. WILLIAM S. HARTS OF THE U.S. ARMY CORPS OF ENGINEERS, AND LANDSCAPE ARCHITECTS JAMES G. LANGDON, IRVING W. PAYNE, AND THOMAS C. JEFFERS MADE SIGNIFICANT CONTRIBUTIONS TO THE PARKWAY'S DESIGN.

DEGRADED VALLEY BETWEEN O AND P STREETS
(*Rock Creek & Potomac Parkway Commission Report*, 1916)

POTOMAC WATERFRONT, ca. 1915
(Commission of Fine Arts)

DISTRICT OF COLUMBIA PARK SYSTEM PLAN
(*McMillan Commission Report*, 1902)

SECTION

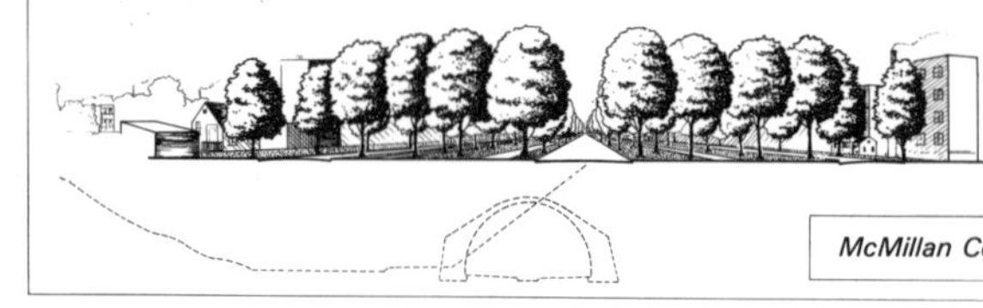

SECTION

McMillan Commission Report, 1902

CLOSED-VALLEY PLAN
1908

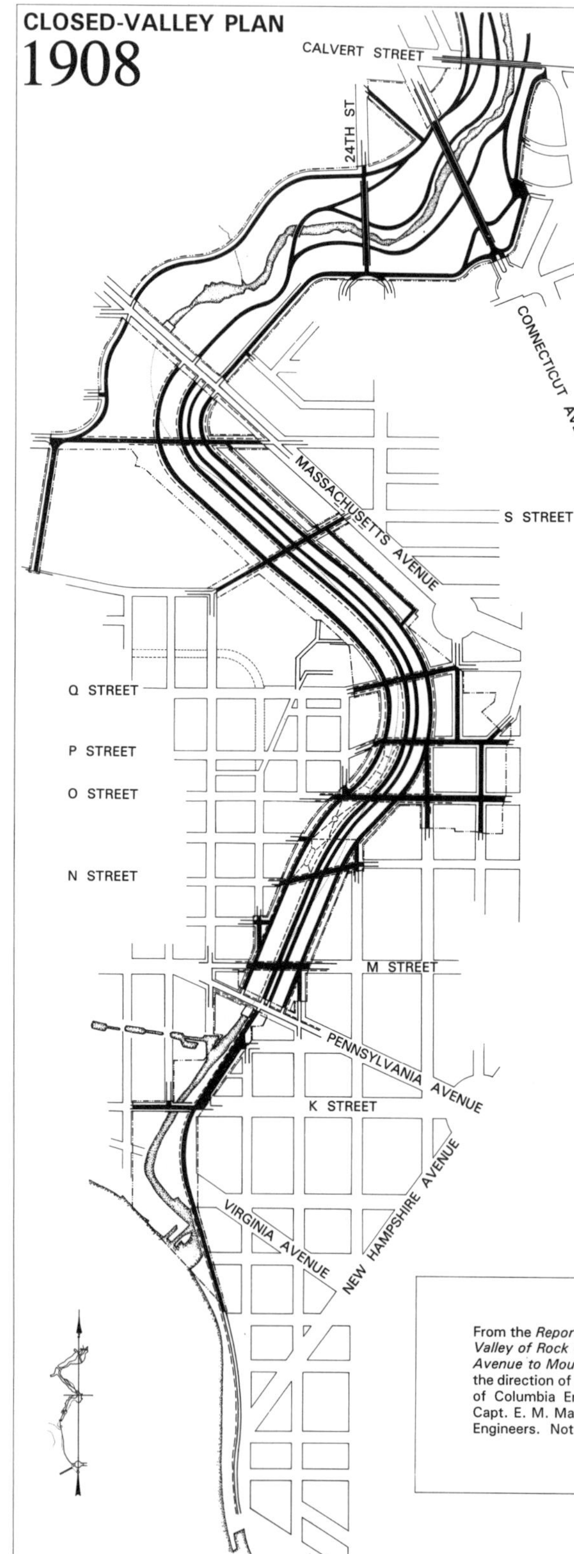

OPEN-VALLEY PLAN
1908

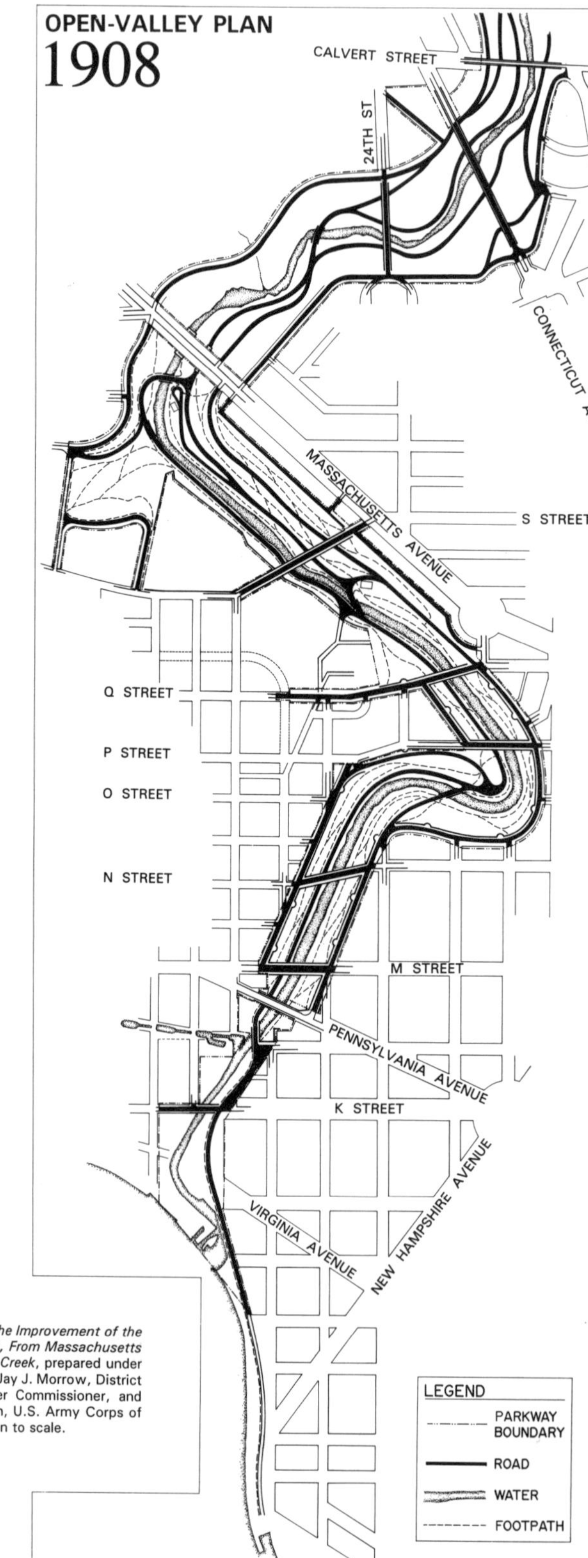

From the *Report on the Improvement of the Valley of Rock Creek, From Massachusetts Avenue to Mouth of Creek*, prepared under the direction of Maj. Jay J. Morrow, District of Columbia Engineer Commissioner, and Capt. E. M. Markham, U.S. Army Corps of Engineers. Not drawn to scale.

HISTORIC AMERICAN BUILDINGS SURVEY DC-697

SHEET 2 OF 36

ROCK CREEK AND POTOMAC PARKWAY
WASHINGTON, DISTRICT OF COLUMBIA

DOROTA PAPE-SILIWONCZUK 1992

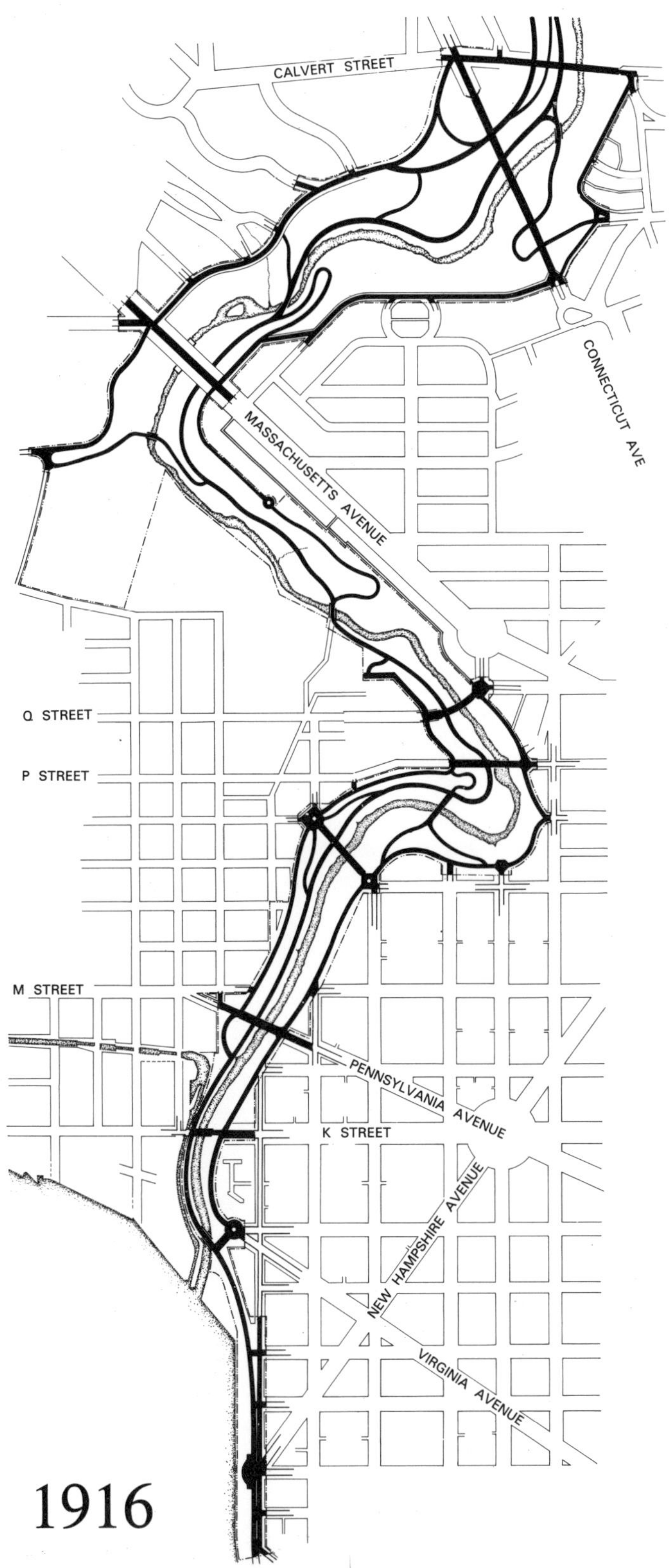

1916

THE 1916 PLAN ENCOMPASSED THE ENTIRE PARKWAY PATH FROM THE WATERFRONT TO THE ZOO. BORDER ROADS EXTENDED TO CONNECTICUT AVENUE. ADDITIONAL ENTRANCES PROVIDED ACCESS TO THE EXCLUSIVE SURROUNDING NEIGHBORHOODS. THE NEW PLAN REDUCED CONFLICTS WITH THE PARKWAY'S NEIGHBORS BY ELIMINATING THE ROUTES THROUGH OAK HILL CEMETERY AND PROPOSING TO SEPARATE THE MOUTH OF ROCK CREEK FROM THE ENTRANCE TO THE C & O CANAL. A NEW CROSSING BETWEEN N AND O STREETS WAS CALLED FOR TO REPLACE THE ANTIQUATED M STREET BRIDGE. THIS PRELIMINARY DESIGN OF ROCK CREEK AND POTOMAC PARKWAY WAS PREPARED UNDER THE DIRECTION OF COL. WILLIAM W. HARTS, OFFICER IN CHARGE OF PUBLIC BUILDINGS AND GROUNDS, BY LANDSCAPE ARCHITECT JAMES G. LANGDON.

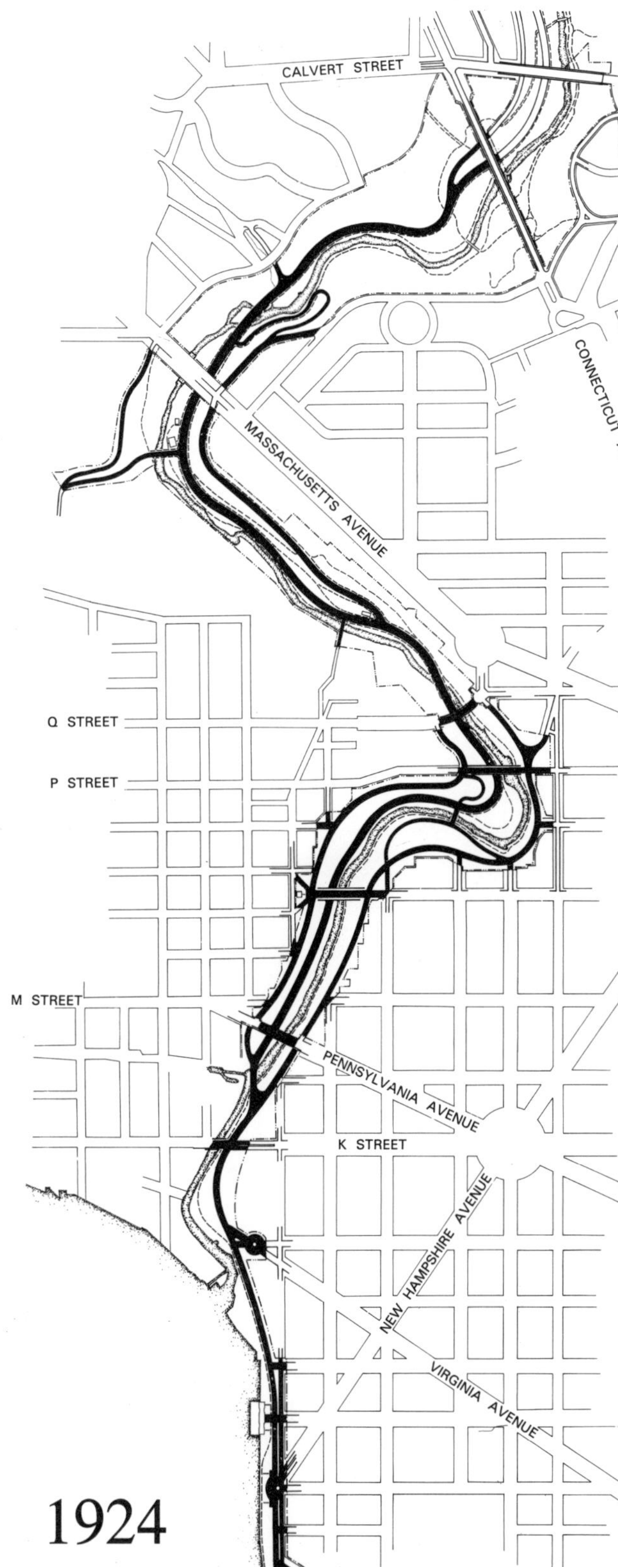

1924

THE INCREASING SIMPLIFICATION OF THE PARKWAY'S DESIGN WAS EVIDENT IN THE ELIMINATION OF THE PROPOSED CANAL SEPARATION AND REDUCTION IN THE NUMBER OF ACCESS ROADS. AN AGREEMENT WITH THE WASHINGTON GAS LIGHT COMPANY PROVIDED ROOM TO ROUTE THE PARKWAY DRIVE ON THE EAST SIDE OF ROCK CREEK. PRESSURE FROM GEORGETOWN RESIDENTS RESTORED THE M STREET BRIDGE PLACEMENT AND ELIMINATED THE PROPOSED N STREET CROSSING. THE MAIN PARKWAY DRIVE CROSSED TO THE EAST SIDE OF THE CREEK JUST ABOVE Q STREET RATHER THAN FARTHER NORTH AT LYON'S MILL ROAD. THIS PLAN FOR THE PARKWAY WAS PREPARED UNDER THE DIRECTION OF LT. COL. C. O. SHERRILL, OFFICER IN CHARGE OF PUBLIC BUILDINGS AND GROUNDS, BY OFFICE OF ENGINEERS LANDSCAPE ARCHITECTS IRVING W. PAYNE AND THOMAS C. JEFFERS.

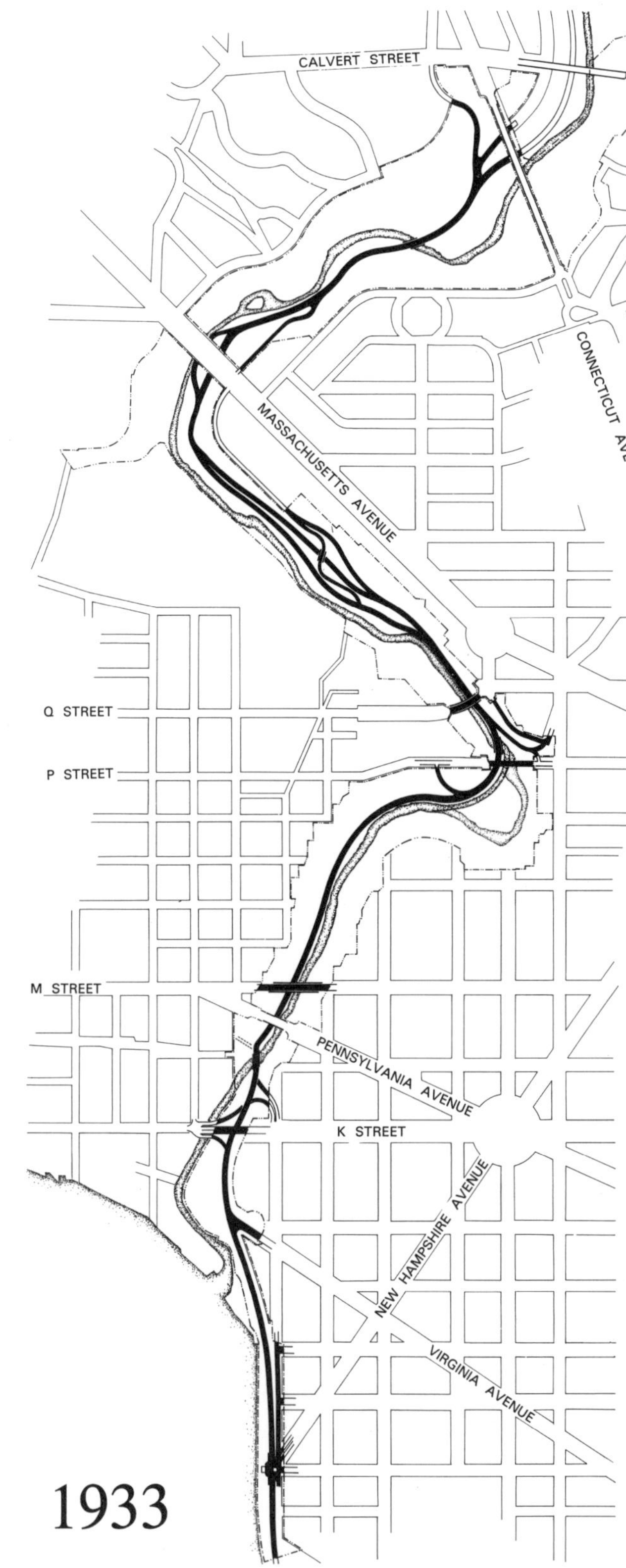

1933

DEVELOPMENT ALONG THE EDGES OF THE VALLEY, TOGETHER WITH THE WIDER, STRAIGHTER ROADS REQUIRED BY AUTOMOBILES, NECESSITATED MAJOR REVISIONS TO PRESERVE SCENERY AND SECLUDE MOTORISTS FROM THE SURROUNDING CITYSCAPE. THE MOST SIGNIFICANT CHANGES WERE THE ELIMINATION OF BORDER ROADS AND THE CREATION OF SEPARATE NORTHBOUND AND SOUTHBOUND DRIVEWAYS BETWEEN Q STREET AND MASSACHUSETTS AVENUE. LOCATION AND ALIGNMENT OF ACCESS ROADS WAS ALSO REVISED. THE COMMISSION OF FINE ARTS TRIED UNSUCCESSFULLY TO PREVENT THE DISTRICT HIGHWAY OFFICIALS FROM ROUTING PARKWAY-BOUND TRAFFIC ACROSS THE CEREMONIAL WATERGATE STEPS. THIS 1933 ROCK CREEK AND POTOMAC PARKWAY VICINITY MAP WAS PREPARED BY THE OFFICE OF PUBLIC BUILDINGS AND PUBLIC PARKS.

DRAWN BY: DOROTA PAPE-SILIWONCZUK 1992
ROCK CREEK AND POTOMAC PARKWAY DOCUMENTATION PROJECT 1991-1992
NATIONAL PARK SERVICE
UNITED STATES DEPARTMENT OF THE INTERIOR

NAME AND LOCATION OF STRUCTURE
ROCK CREEK AND POTOMAC PARKWAY
WASHINGTON, DISTRICT OF COLUMBIA

SHEET 3 OF 36 SHEETS
HISTORIC AMERICAN BUILDINGS SURVEY DC-697
LIBRARY OF CONGRESS INDEX NUMBER

INDEX TO PLANS

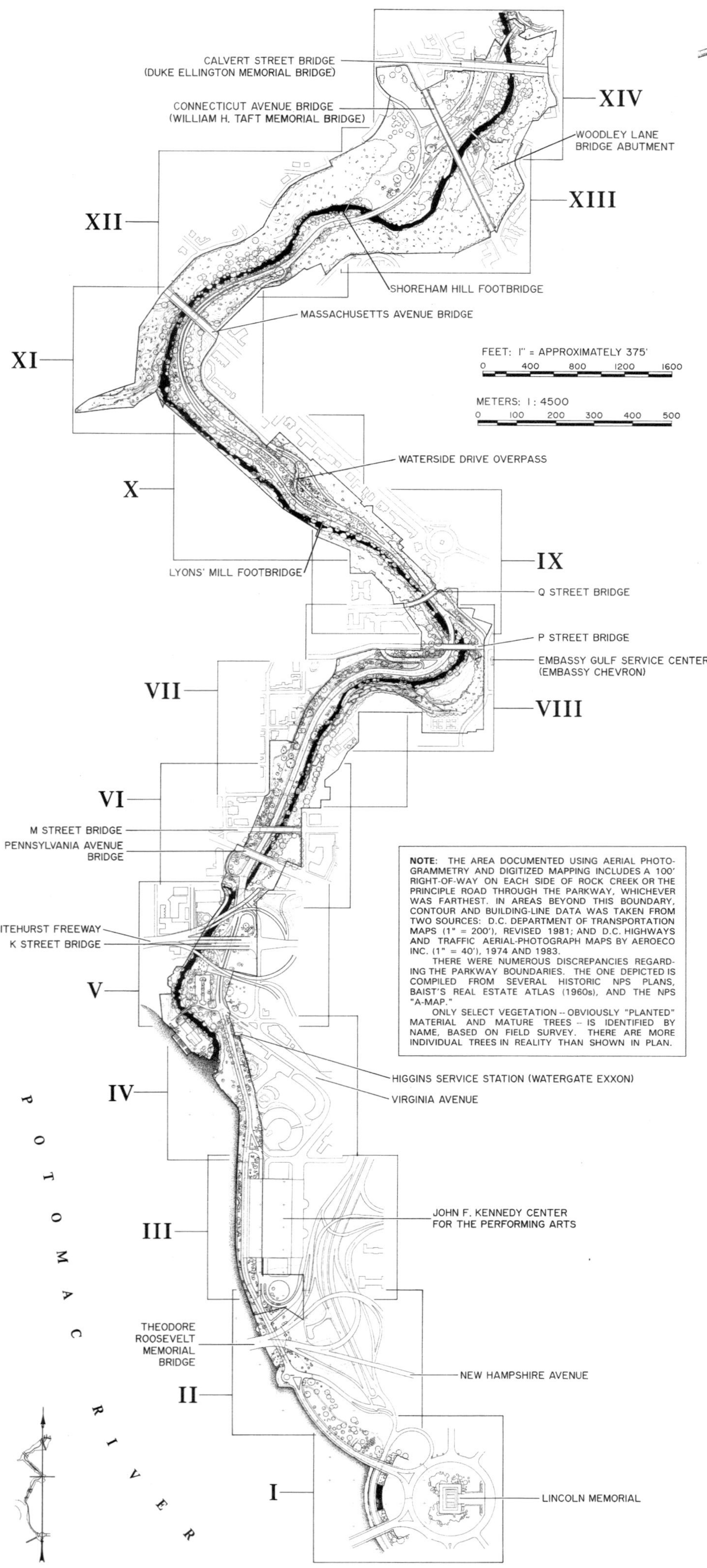

NOTE: THE AREA DOCUMENTED USING AERIAL PHOTOGRAMMETRY AND DIGITIZED MAPPING INCLUDES A 100' RIGHT-OF-WAY ON EACH SIDE OF ROCK CREEK OR THE PRINCIPLE ROAD THROUGH THE PARKWAY, WHICHEVER WAS FARTHEST. IN AREAS BEYOND THIS BOUNDARY, CONTOUR AND BUILDING-LINE DATA WAS TAKEN FROM TWO SOURCES: D.C. DEPARTMENT OF TRANSPORTATION MAPS (1" = 200'), REVISED 1981; AND D.C. HIGHWAYS AND TRAFFIC AERIAL-PHOTOGRAPH MAPS BY AEROECO INC. (1" = 40'), 1974 AND 1983.

THERE WERE NUMEROUS DISCREPANCIES REGARDING THE PARKWAY BOUNDARIES. THE ONE DEPICTED IS COMPILED FROM SEVERAL HISTORIC NPS PLANS, BAIST'S REAL ESTATE ATLAS (1960s), AND THE NPS "A-MAP."

ONLY SELECT VEGETATION -- OBVIOUSLY "PLANTED" MATERIAL AND MATURE TREES -- IS IDENTIFIED BY NAME, BASED ON FIELD SURVEY. THERE ARE MORE INDIVIDUAL TREES IN REALITY THAN SHOWN IN PLAN.

FLOWERING DOGWOOD BRANCH

DESCRIPTION

ROCK CREEK AND POTOMAC PARKWAY IS APPROXIMATELY 2.5 MILES LONG. THE WIDTH OF THE BORDERING PARKLAND RANGES FROM A NARROW GRASSY STRIP ALONG THE POTOMAC RIVER OPPOSITE THE JOHN F. KENNEDY CENTER FOR THE PERFORMING ARTS TO A MORE GENEROUS CORRIDOR AVERAGING SLIGHTLY MORE THAN 500 FEET THROUGH MOST OF ROCK CREEK VALLEY. IN THE AMPHITHEATER-LIKE SECTIONS BELOW THE CONNECTICUT AVENUE AND P STREET BRIDGES, THE PARKWAY BROADENS SIGNIFICANTLY TO PROVIDE AN EXPANSE OF OPEN LAWN BORDERED BY TREES AND SHRUBS. ROCK CREEK VALLEY BEGINS NEAR K STREET AND RAPIDLY REACHES A DEPTH OF APPROXIMATELY 50 FEET. THIS DEPTH IS MAINTAINED TO THE P STREET BEND WHERE THE VALLEY WALLS GRADUALLY STEEPEN, RISING 80 FEET ABOVE THE STREAM AT MASSACHUSETTS AVENUE, AND CLIMB TO 120 FEET NEAR THE CONNECTICUT AVENUE BRIDGE.

ROCK CREEK AND POTOMAC PARKWAY'S VALLEY LOCATION AND THICK BORDER VEGETATION ALLOW MOTORISTS TO DRIVE ALMOST TO THE HEART OF THE CITY THROUGH A PARK ENVIRONMENT THAT EFFECTIVELY CONCEALS THE SURROUNDING CITYSCAPE. SOUTHBOUND DRIVERS EXPERIENCE A GRADUAL TRANSITION FROM THE SINUOUS CURVES OF THE SECLUDED, DENSELY WOODED UPPER VALLEY TO THE SWEEPING VISTAS OF THE POTOMAC WATERFRONT. NORTHBOUND MOTORISTS RAPIDLY LEAVE THE CITY BEHIND AS THEY PASS UNDERNEATH THE CONCRETE OVERPASSES SURROUNDING THE WHITEHURST EXPRESSWAY AND ENTER THE LOWER REACHES OF ROCK CREEK VALLEY.

THE WATERFRONT SECTION OF THE PARKWAY IS FLANKED BY ROWS OF REGULARLY SPACED SYCAMORES. THE NATURAL SCENERY ON THEODORE ROOSEVELT MEMORIAL ISLAND AND THE VIRGINIA SHORE PROVIDES A COUNTERPOINT TO THE MODERNISTIC FORMS OF THE KENNEDY CENTER, WATERGATE COMPLEX, AND THEODORE ROOSEVELT MEMORIAL BRIDGE. BETWEEN THE WATERFRONT AND P STREET, THE PARKWAY ENTERS A LARGELY ARTIFICIAL LANDSCAPE THAT WAS RECLAIMED FROM CITY DUMPS AND INDUSTRIAL USE THROUGH EXTENSIVE EXCAVATION. THE LONG, OPEN BANK WEST OF THE ROAD BETWEEN M AND P STREETS IS RESPLENDENT WITH DAFFODILS IN EARLY SPRING. EXTENSIVE PLANTING PLANS WERE FOLLOWED WHEN THE PARKWAY WAS CREATED, WITH THE AIM OF RECREATING A NATURALISTIC LANDSCAPE. MOST ORIGINAL PLANTINGS HAVE BEEN LOST OVER TIME. THICK STANDS OF VOLUNTEER FLOODPLAIN SPECIES PREVAIL EXCEPT IN LIMITED AREAS KEPT CLEAR BY REGULAR MAINTENANCE. NORTH OF MASSACHUSETTS AVENUE, THE ORIGINAL FORESTED VALLEY LANDSCAPE IS GENERALLY PRESERVED. DENSE STANDS OF MATURE HARDWOODS PREDOMINATE. OCCASIONAL SPECIMEN TREES HAVE BEEN REPLANTED IN THE INTENSIVELY MANAGED AREAS BETWEEN THE MOTOR ROAD AND THE MULTI-USE TRAIL AND IN THE OPEN AREAS AT THE NORTHERN AND SOUTHERN ENDS OF THE PARKWAY.

THE PARKWAY ROAD ITSELF CONSISTS OF AN ASPHALT WEARING SURFACE LAID OVER A CRUSHED-STONE BASE, WITH PRECAST MOUNTABLE CURBS NORTH OF VIRGINIA AVENUE AND GRANITE CURBS ALONG THE POTOMAC WATERFRONT. SEVERAL AREAS, SUCH AS THE SECTION BETWEEN THE RETAINING WALLS NORTH OF Q STREET, HAVE A CONCRETE BASE.

ROAD WIDTH ON THE MAIN PARKWAY DRIVE IS GENERALLY 44 FEET, WITH SOME WIDENING AND MINOR SUPERELEVATION AT MAJOR CURVES. THROUGH MOST OF ITS LENGTH, THE PARKWAY CARRIES TWO LANES OF TRAFFIC IN EITHER DIRECTION, WITH NO MEDIAN STRIP DIVIDING ONCOMING TRAFFIC. IN THE MIDSECTION, HOWEVER, FROM JUST ABOVE DUMBARTON BRIDGE TO THE MASSACHUSETTS AVENUE BRIDGE, THE NORTHBOUND AND SOUTHBOUND LANES SEPARATE INTO TWO 24-FOOT ROADWAYS WHOSE SEPARATE PATHS HELP PRESERVE THE VALLEY'S CONTOURS AND PROVIDE FOR AN EVEN MORE RELAXED AND SCENIC DRIVING EXPERIENCE. A 4-FOOT WIDE BELGIAN BLOCK MEDIAN EXTENDS FROM THE K STREET OVERPASS TO WEST POTOMAC PARK.

RESERVATION NO. 360 CONTAINS 174.23 ACRES AND OFFICIALLY EXTENDS FROM THE SOUTH PORTAL OF THE ZOO TUNNEL TO A POINT ABOUT 200 FEET NORTH OF THE THEODORE ROOSEVELT MEMORIAL BRIDGE. THESE BOUNDARIES REFLECT THE ORIGINAL TAKING LINES MANDATED BY EARLIER LAND USES, HOWEVER, AND THE PARKWAY'S DESIGNERS AND DISTRICT PLANNERS HAVE GENERALLY CONSIDERED THE TRAFFIC CIRCLE BETWEEN THE LINCOLN MEMORIAL AND ARLINGTON MEMORIAL BRIDGE TO BE THE PARKWAY'S EFFECTIVE SOUTHERN TERMINUS. SEVERAL MINOR ALTERATIONS HAVE BEEN MADE IN THE PARKWAY'S BORDERS OVER THE YEARS, NOTABLY THE TRANSFER OF ROSE PARK TO THE DISTRICT OF COLUMBIA DEPARTMENT OF PARKS AND RECREATION IN 1972.

PLANT LIST

	BOTANICAL NAME	COMMON NAME
AA	*Ailanthus altissima*	Tree of Heaven
AC	*Amelanchier canadensis*	Serviceberry
AN	*Acer negundo*	Box Elder
AO	*Aesculus octandra*	Yellow Buckeye
AR	*Acer rubrum*	Red Maple
AS	*Acer saccharinum*	Silver Maple
ASa	*Acer saccharum*	Sugar Maple
BJ	*Berberis julianae*	Wintergreen Barberry
BN	*Betula nigra*	River Birch
BP	*Broussonetia papyrifera*	Paper Mulberry
CC	*Carpinus carolinia*	American Hornbeam
CCa	*Cercis canadensis*	Redbud
CCg	*Cotinus coggygria*	Smoketree
CD	*Cedrus deodora*	Deodar Cedar
CF	*Cornus florida*	Flowering Dogwood
CL	*Celtis laevigata*	Sugarberry
CM	*Cornus mas*	Cornelian Cherry Dogwood
CM	*Castanea mollisima*	Chinese Chestnut
CO	*Celtis occidentalis*	Hackberry
Csp	*Crataegus species*	Hawthorn
FA	*Fraxinus americana*	White Ash
FP	*Fraxinus pennsylvanica*	Green Ash
FS	*Fagus sylvatica*	Beech
GB	*Ginkgo biloba*	Ginkgo
GT	*Gleditsia tricanthos*	Honeylocust
HV	*Hamamelis virginiana*	Common Witchhazel
IO	*Ilex opaca*	American Holly
Isp	*Ilex species*	Holly
JN	*Juglans nigra*	Black Walnut
JV	*Juniperus virginiana*	Eastern Red Cedar
LI	*Lagerstroemia indica*	Crapemyrtle
LS	*Liquidambar styraciflua*	Sweet Gum
Lsp	*Lonicera species*	Honeysuckle
Lsp	*Ligustrum species*	Privet
LT	*Liriodendron tulipifera*	Tulip Poplar
Masp	*Malus species*	Crabapple
MG	*Magnolia grandiflora*	Southern Magnolia
MP	*Maclura pomifera*	Osage Orange
Msp	*Morus species*	Mulberry
NS	*Nyssa sylvatica*	Black Gum
PA	*Platanus acerifolia*	London Planetree
PD	*Populus deltoides*	Eastern Poplar
PN	*Pinus nigra*	Austrian Pine
PO	*Platanus occidentalis*	Sycamore
PS	*Pinus strobus*	White Pine
PSe	*Prunus serotina*	Black Cherry
Psp	*Prunus species*	Flowering Cherry
PSy	*Pinus sylvestris*	Scotch Pine
PT	*Paulownia tomentosa*	Royal Paulownia
PV	*Pinus virginiana*	Virginia Pine
QA	*Quercus alba*	White Oak
QB	*Quercus borealis*	Northern Red Oak
QB	*Quercus bicolor*	Swamp White Oak
QC	*Quercus coccinea*	Scarlet Oak
QF	*Quercus falcata*	Southern Red Oak
QN	*Quercus nigra*	Water Oak
QP	*Quercus phellos*	Willow Oak
QPa	*Quercus palustris*	Pin Oak
QR	*Quercus rubra*	Red Oak
QV	*Quercus velutina*	Black Oak
RC	*Rhamnus cathartica*	Common Buckthorn
RP	*Robinia pseudoacacia*	Black Locust
SJ	*Sophora japonica*	Japanese Pagodatree
TC	*Tsuga canadensis*	Canadian Hemlock
UA	*Ulmus americana*	American Elm
UP	*Ulmus pumila*	Siberian Elm
VD	*Viburnum dilatum*	Linden Viburnum
VP	*Viburnum prunifolium*	Blackhaw

DRAWN BY: EVAN E. MILLER 1992

ROCK CREEK AND POTOMAC PARKWAY DOCUMENTATION PROJECT 1991-1992
NATIONAL PARK SERVICE
UNITED STATES DEPARTMENT OF THE INTERIOR

NAME AND LOCATION OF STRUCTURE: ROCK CREEK AND POTOMAC PARKWAY
WASHINGTON, DISTRICT OF COLUMBIA

SHEET 5 OF SHEETS 36

HISTORIC AMERICAN BUILDINGS SURVEY
DC-697

IF REPRODUCED, PLEASE CREDIT: HISTORIC AMERICAN BUILDINGS SURVEY, NATIONAL PARK SERVICE, NAME OF DELINEATOR, DATE OF THE DRAWING

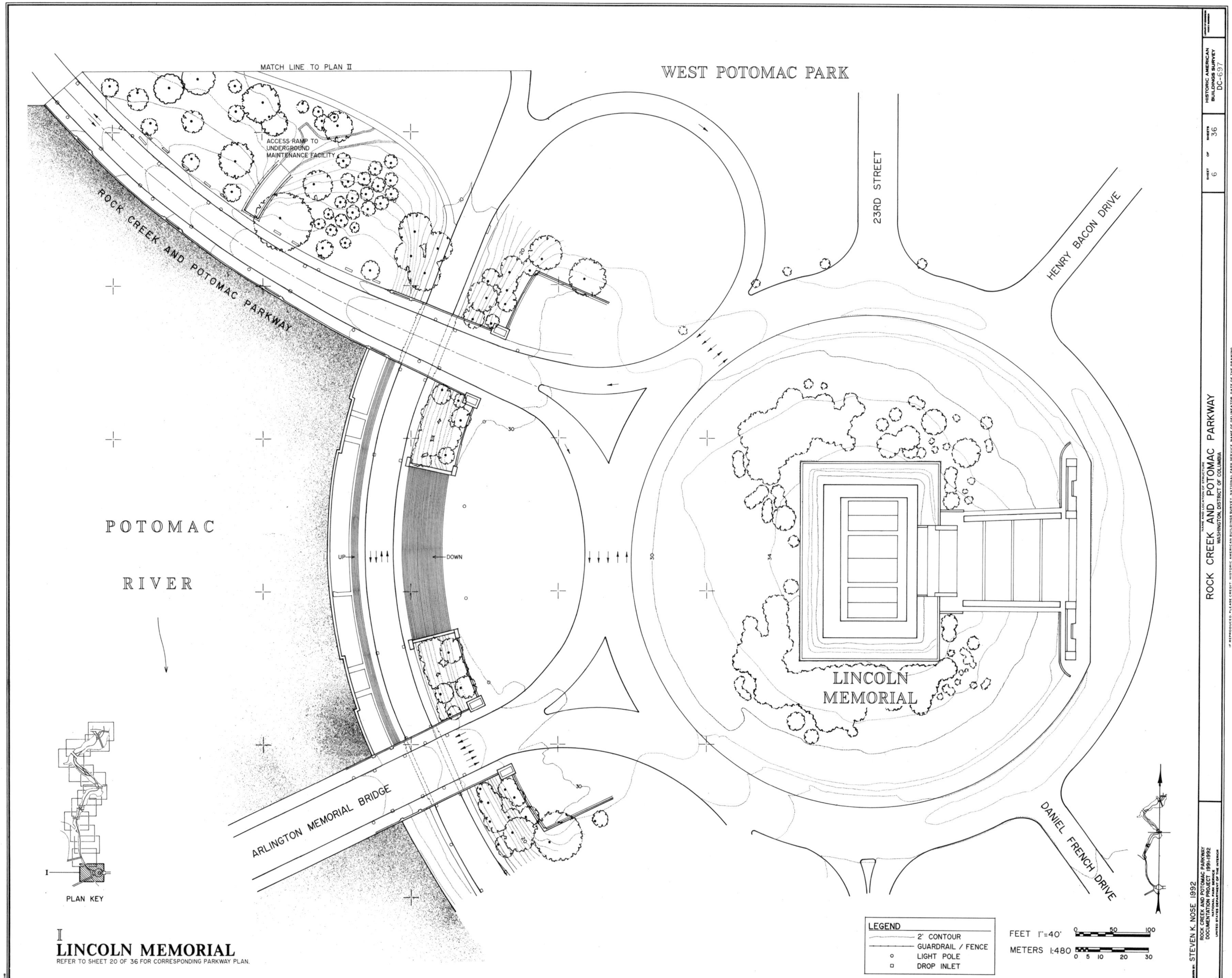

MATCH LINE TO PLAN II
WEST POTOMAC PARK
ACCESS RAMP TO UNDERGROUND MAINTENANCE FACILITY
ROCK CREEK AND POTOMAC PARKWAY
23RD STREET
HENRY BACON DRIVE
POTOMAC
RIVER
UP
DOWN
LINCOLN MEMORIAL
ARLINGTON MEMORIAL BRIDGE
DANIEL FRENCH DRIVE
PLAN KEY
I
LINCOLN MEMORIAL
REFER TO SHEET 20 OF 36 FOR CORRESPONDING PARKWAY PLAN.
LEGEND
2' CONTOUR
GUARDRAIL / FENCE
LIGHT POLE
DROP INLET
FEET 1"=40'
METERS 1:480
STEVEN K. NOSE 1992
ROCK CREEK AND POTOMAC PARKWAY
WASHINGTON, DISTRICT OF COLUMBIA
HISTORIC AMERICAN BUILDINGS SURVEY
DC-697
6
36

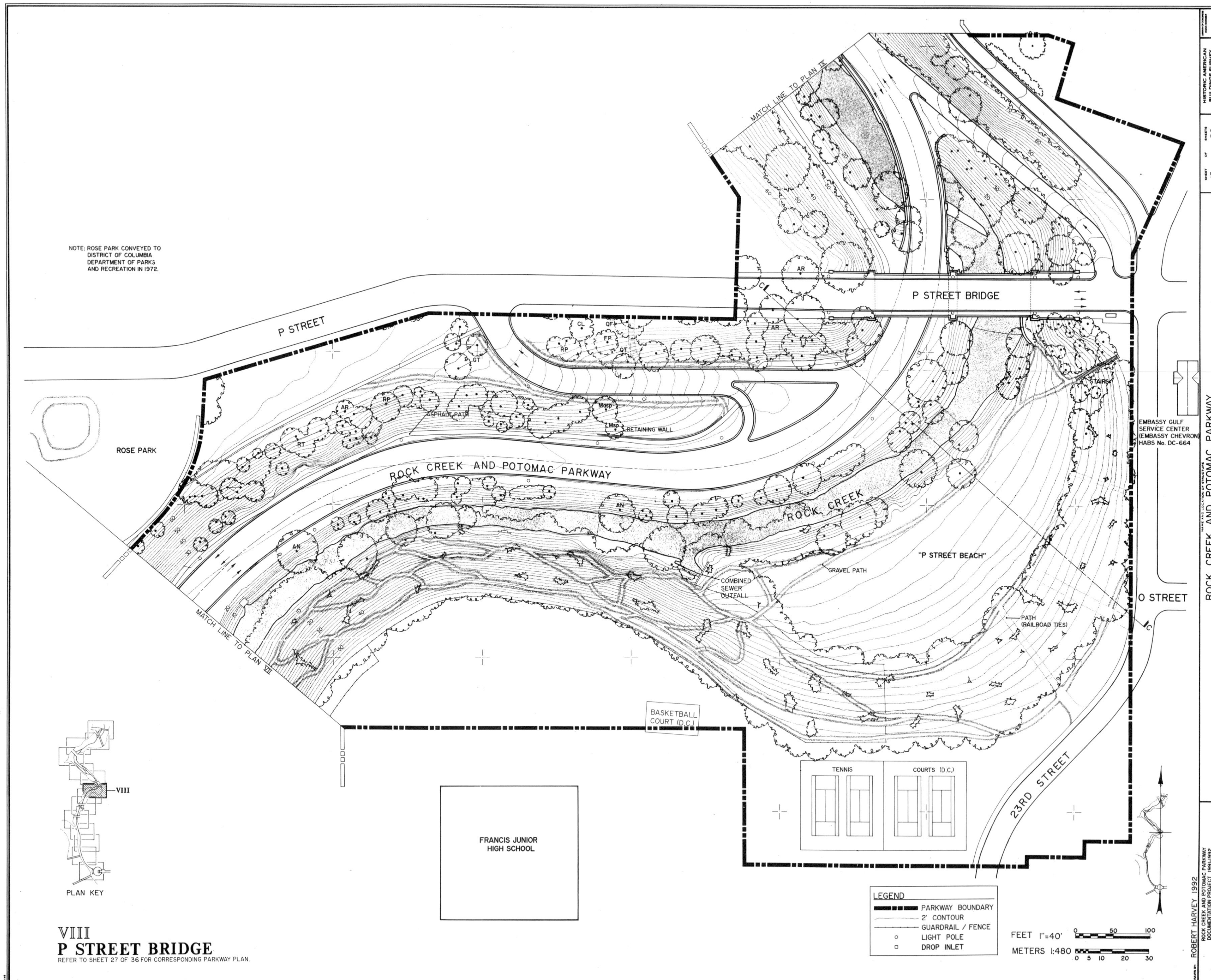
NOTE: ROSE PARK CONVEYED TO DISTRICT OF COLUMBIA DEPARTMENT OF PARKS AND RECREATION IN 1972.
P STREET
P STREET BRIDGE
ROSE PARK
ROCK CREEK AND POTOMAC PARKWAY
ROCK CREEK
ASPHALT PATH
RETAINING WALL
"P STREET BEACH"
GRAVEL PATH
COMBINED SEWER OUTFALL
PATH (RAILROAD TIES)
STAIRS
EMBASSY GULF SERVICE CENTER (EMBASSY CHEVRON) HABS No. DC-664
O STREET
23RD STREET
MATCH LINE TO PLAN IX
MATCH LINE TO PLAN VII
BASKETBALL COURT (D.C.)
TENNIS COURTS (D.C.)
FRANCIS JUNIOR HIGH SCHOOL
PLAN KEY
VIII
LEGEND
PARKWAY BOUNDARY
2' CONTOUR
GUARDRAIL / FENCE
LIGHT POLE
DROP INLET
FEET 1"=40'
METERS 1:480
VIII
P STREET BRIDGE
REFER TO SHEET 27 OF 36 FOR CORRESPONDING PARKWAY PLAN.
ROBERT HARVEY 1992
ROCK CREEK AND POTOMAC PARKWAY
WASHINGTON, DISTRICT OF COLUMBIA
SHEET 13 OF 36
HISTORIC AMERICAN BUILDINGS SURVEY DC-697

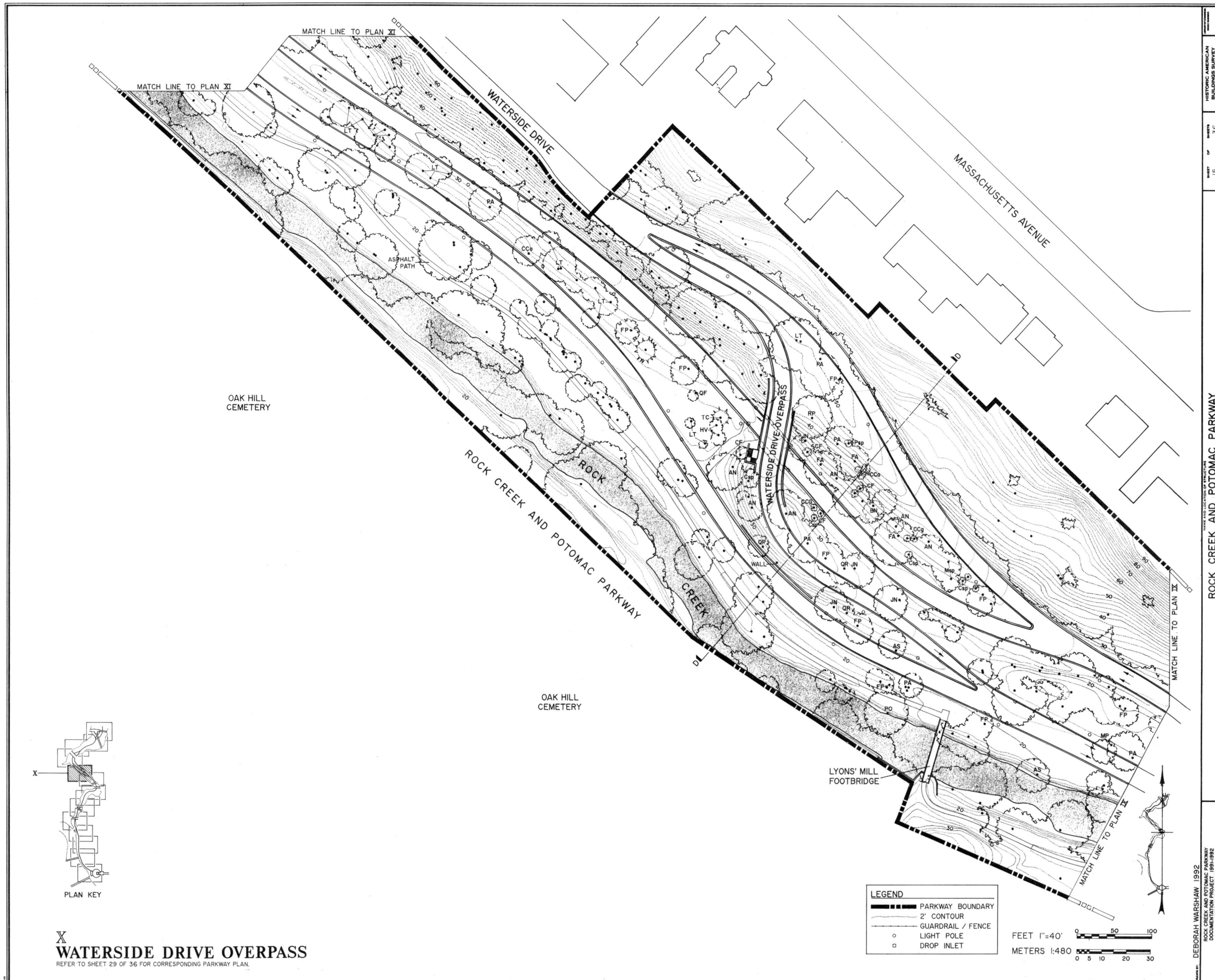
MATCH LINE TO PLAN XI
MATCH LINE TO PLAN XI
WATERSIDE DRIVE
MASSACHUSETTS AVENUE
OAK HILL CEMETERY
ASPHALT PATH
WATERSIDE DRIVE OVERPASS
ROCK CREEK AND POTOMAC PARKWAY
ROCK
CREEK
WALL
OAK HILL CEMETERY
LYONS' MILL FOOTBRIDGE
MATCH LINE TO PLAN IX
MATCH LINE TO PLAN IX
PLAN KEY
LEGEND
PARKWAY BOUNDARY
2' CONTOUR
GUARDRAIL / FENCE
LIGHT POLE
DROP INLET
FEET 1"=40'
METERS 1:480
X
WATERSIDE DRIVE OVERPASS
REFER TO SHEET 29 OF 36 FOR CORRESPONDING PARKWAY PLAN.
DEBORAH WARSHAW 1992
ROCK CREEK AND POTOMAC PARKWAY
WASHINGTON, DISTRICT OF COLUMBIA
HISTORIC AMERICAN BUILDINGS SURVEY
DC-697
15
36

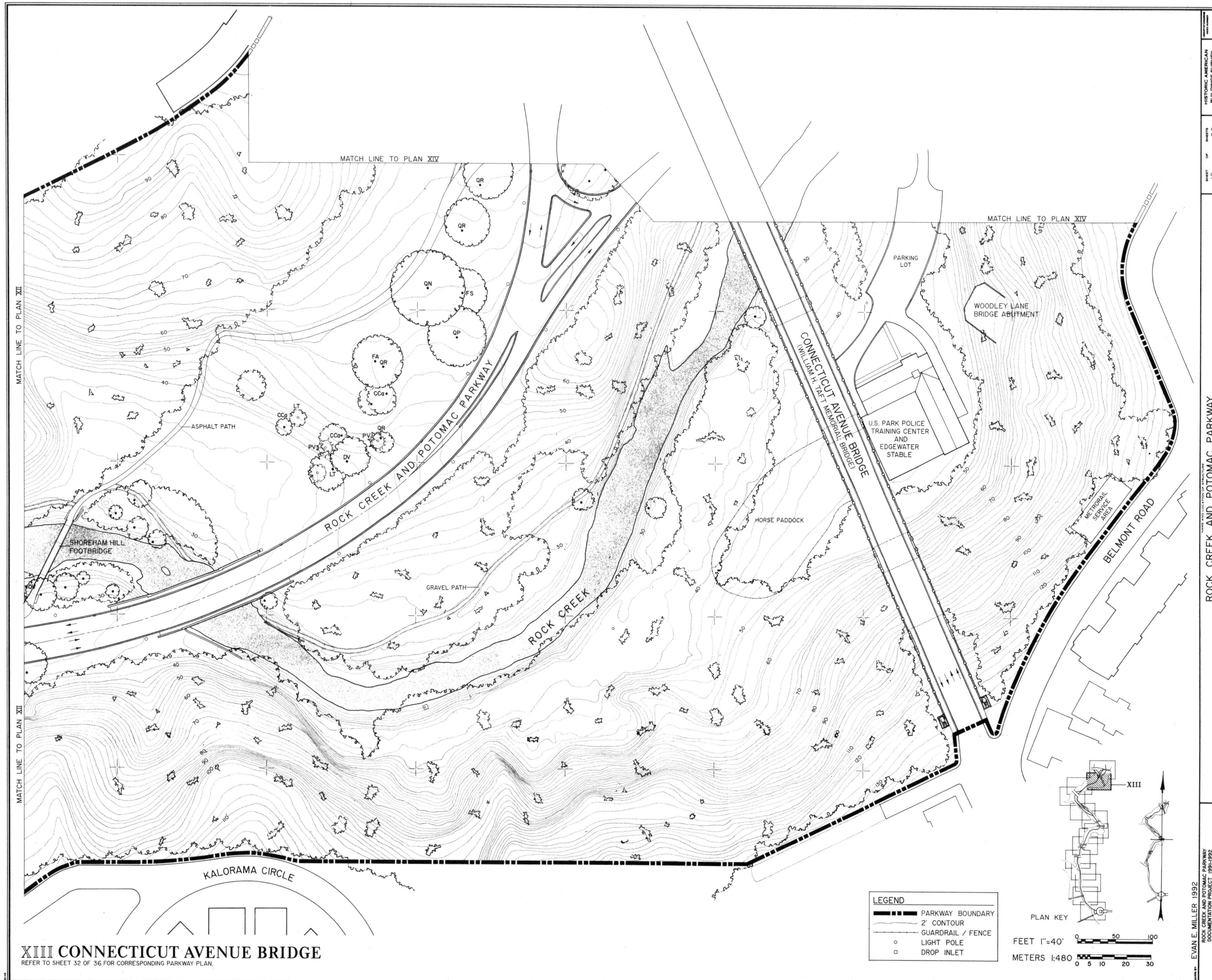
XIII CONNECTICUT AVENUE BRIDGE
REFER TO SHEET 32 OF 36 FOR CORRESPONDING PARKWAY PLAN.
MATCH LINE TO PLAN XIV
MATCH LINE TO PLAN XII
CONNECTICUT AVENUE BRIDGE
(WILLIAM H. TAFT MEMORIAL BRIDGE)
ROCK CREEK AND POTOMAC PARKWAY
ROCK CREEK
ASPHALT PATH
GRAVEL PATH
SHOREHAM HILL FOOTBRIDGE
PARKING LOT
WOODLEY LANE BRIDGE ABUTMENT
U.S. PARK POLICE TRAINING CENTER AND EDGEWATER STABLE
HORSE PADDOCK
METRORAIL SERVICE AREA
BELMONT ROAD
KALORAMA CIRCLE
LEGEND
PARKWAY BOUNDARY
2' CONTOUR
GUARDRAIL / FENCE
LIGHT POLE
DROP INLET
FEET 1"=40'
METERS 1:480
PLAN KEY
XIII
ROCK CREEK AND POTOMAC PARKWAY
WASHINGTON, DISTRICT OF COLUMBIA
EVAN E. MILLER 1992
HISTORIC AMERICAN BUILDINGS SURVEY
DC-697
18
36

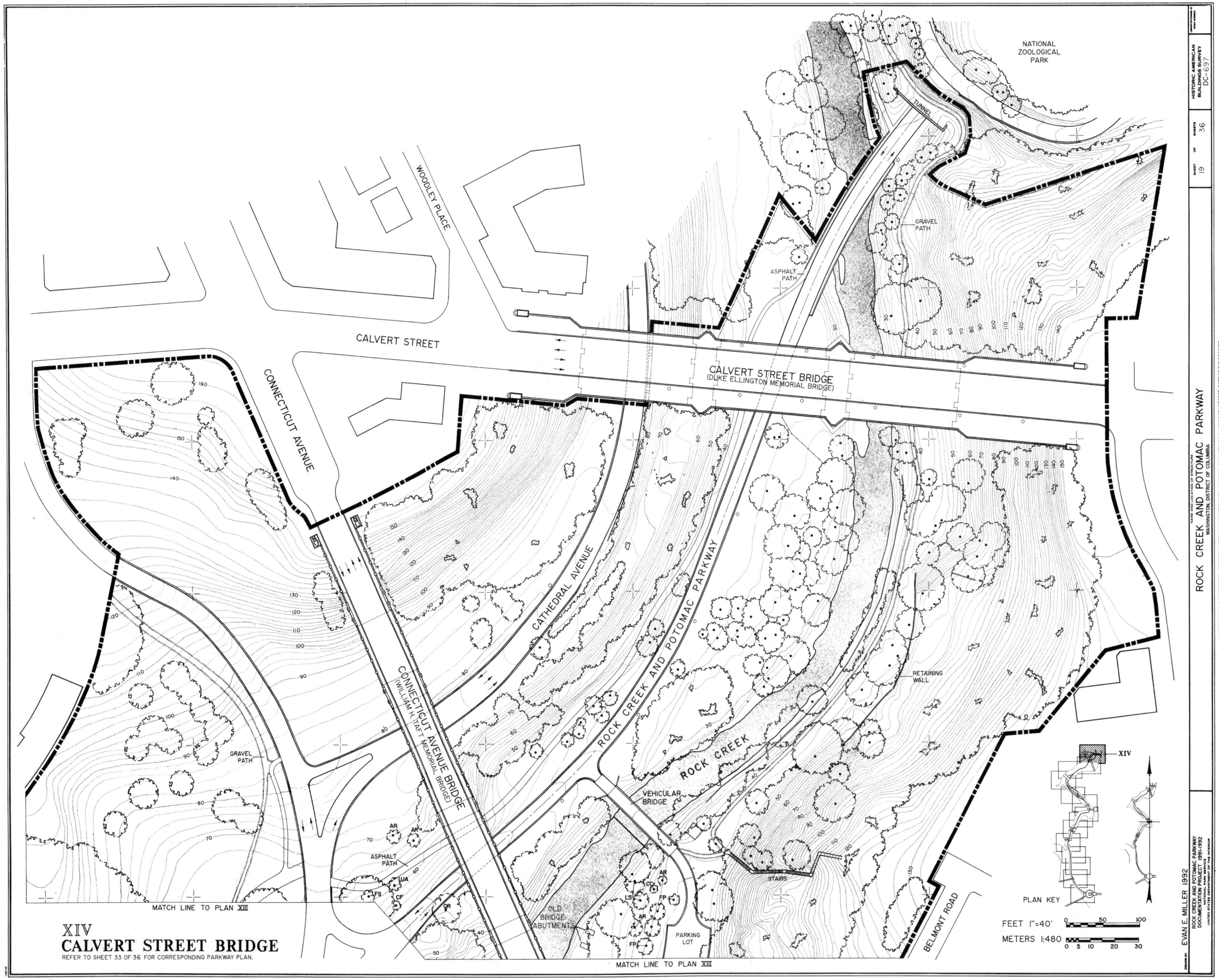
NATIONAL ZOOLOGICAL PARK
TUNNEL
GRAVEL PATH
ASPHALT PATH
WOODLEY PLACE
CALVERT STREET
CALVERT STREET BRIDGE
(DUKE ELLINGTON MEMORIAL BRIDGE)
CONNECTICUT AVENUE
CATHEDRAL AVENUE
ROCK CREEK AND POTOMAC PARKWAY
ROCK CREEK
RETAINING WALL
CONNECTICUT AVENUE BRIDGE
(WILLIAM H. TAFT MEMORIAL BRIDGE)
VEHICULAR BRIDGE
STAIRS
OLD BRIDGE ABUTMENTS
PARKING LOT
BELMONT ROAD
MATCH LINE TO PLAN XIII
PLAN KEY
XIV
FEET 1"=40'
METERS 1:480
XIV
CALVERT STREET BRIDGE
REFER TO SHEET 33 OF 36 FOR CORRESPONDING PARKWAY PLAN.
EVAN E. MILLER 1992
ROCK CREEK AND POTOMAC PARKWAY
WASHINGTON, DISTRICT OF COLUMBIA
HISTORIC AMERICAN BUILDINGS SURVEY
DC-697
19 36

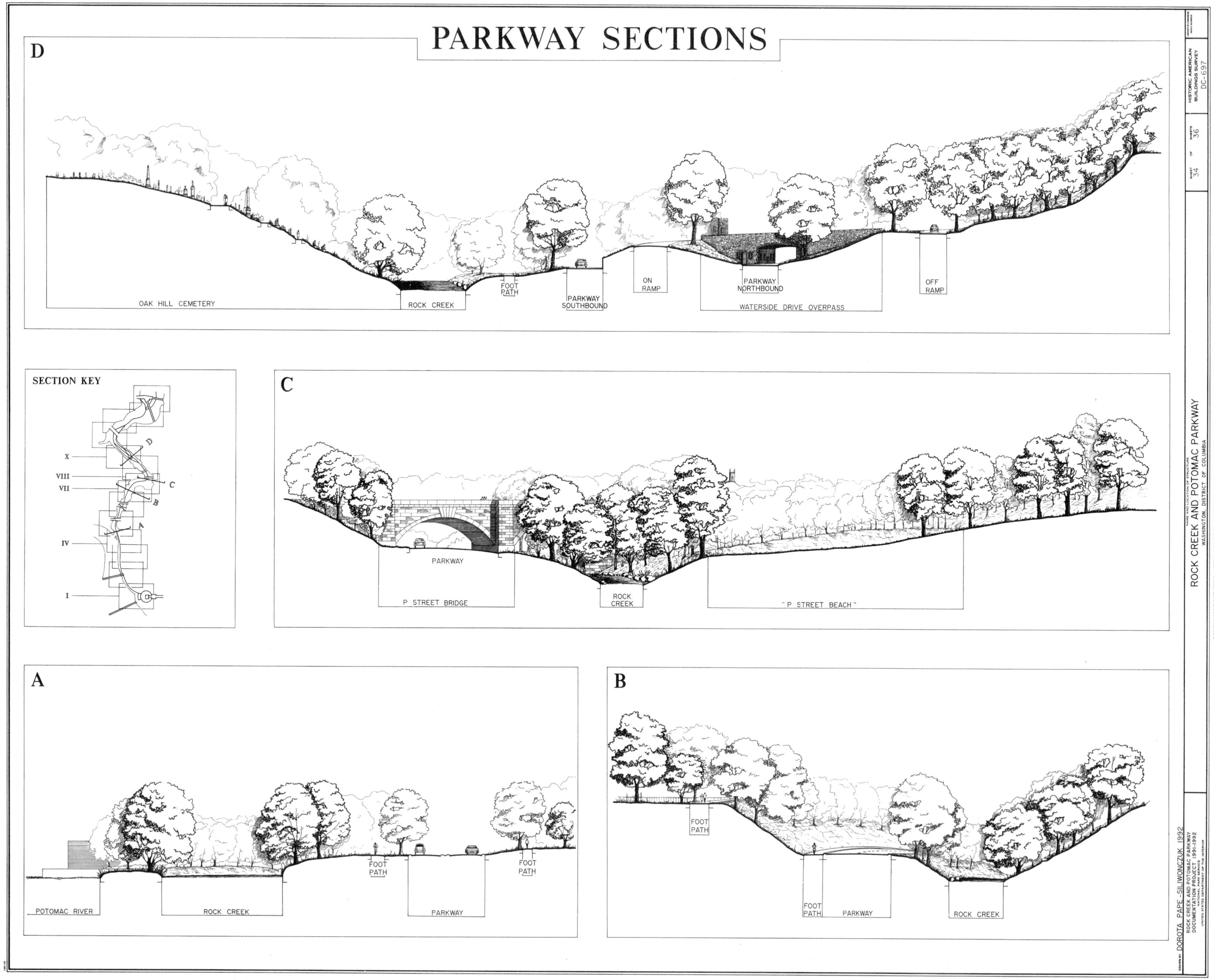
PARKWAY SECTIONS
D
OAK HILL CEMETERY
ROCK CREEK
FOOT PATH
PARKWAY SOUTHBOUND
ON RAMP
PARKWAY NORTHBOUND
WATERSIDE DRIVE OVERPASS
OFF RAMP
SECTION KEY
X
VIII
VII
IV
I
D
C
B
A
C
PARKWAY
P STREET BRIDGE
ROCK CREEK
" P STREET BEACH "
A
POTOMAC RIVER
ROCK CREEK
FOOT PATH
PARKWAY
FOOT PATH
B
FOOT PATH
FOOT PATH
PARKWAY
ROCK CREEK
HISTORIC AMERICAN BUILDINGS SURVEY
DC-697
SHEET 34 OF 36 SHEETS
ROCK CREEK AND POTOMAC PARKWAY
WASHINGTON, DISTRICT OF COLUMBIA
DRAWN BY DOROTA PAPE-SILIWONCZUK 1992
ROCK CREEK AND POTOMAC PARKWAY DOCUMENTATION PROJECT 1991-1992
NATIONAL PARK SERVICE
UNITED STATES DEPARTMENT OF THE INTERIOR

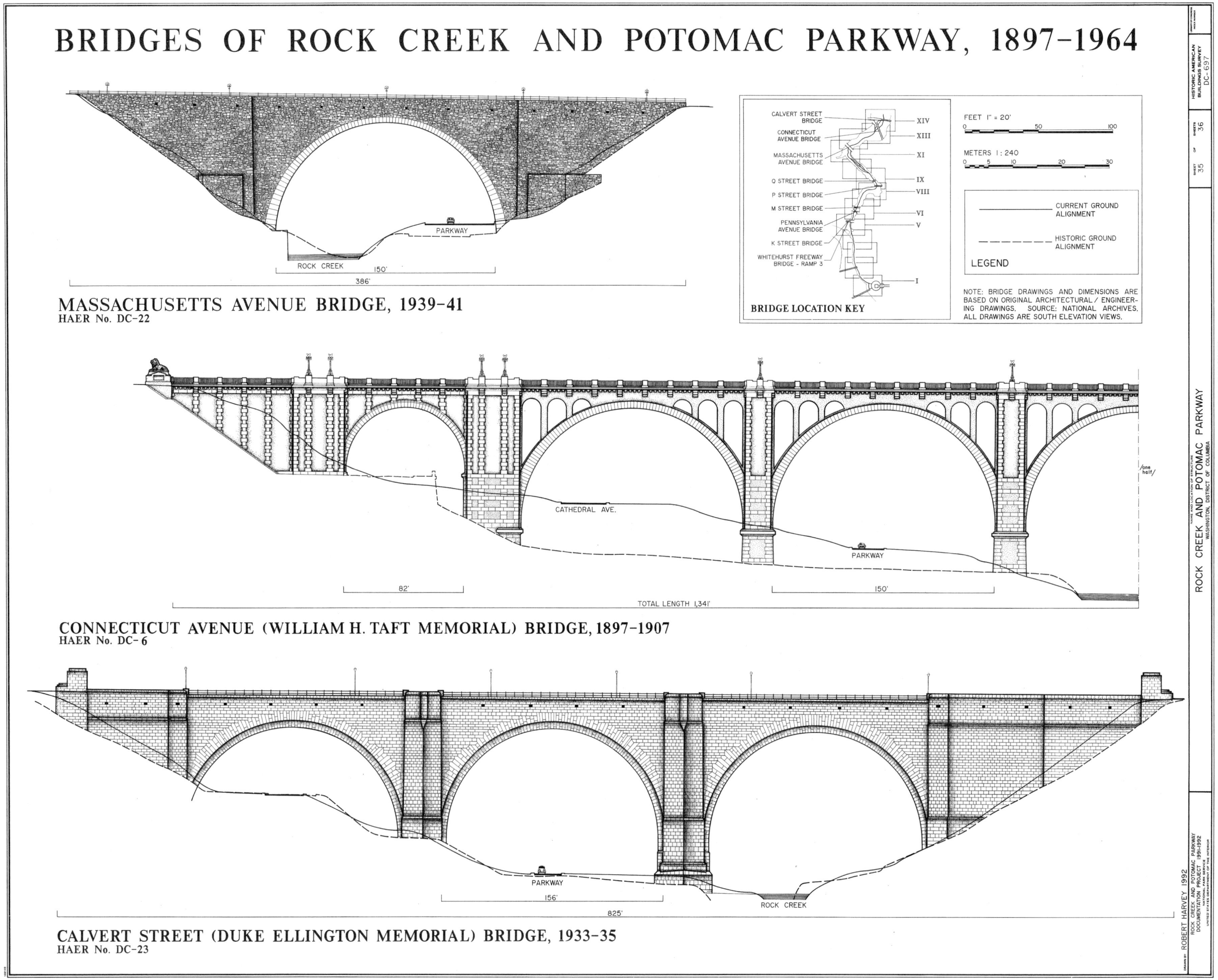
BRIDGES OF ROCK CREEK AND POTOMAC PARKWAY, 1897–1964
PARKWAY
ROCK CREEK
150'
386'
MASSACHUSETTS AVENUE BRIDGE, 1939–41
HAER No. DC-22
CALVERT STREET BRIDGE
CONNECTICUT AVENUE BRIDGE
MASSACHUSETTS AVENUE BRIDGE
Q STREET BRIDGE
P STREET BRIDGE
M STREET BRIDGE
PENNSYLVANIA AVENUE BRIDGE
K STREET BRIDGE
WHITEHURST FREEWAY BRIDGE - RAMP 3
XIV
XIII
XI
IX
VIII
VI
V
I
BRIDGE LOCATION KEY
FEET 1" = 20'
METERS 1 : 240
CURRENT GROUND ALIGNMENT
HISTORIC GROUND ALIGNMENT
LEGEND
NOTE: BRIDGE DRAWINGS AND DIMENSIONS ARE BASED ON ORIGINAL ARCHITECTURAL / ENGINEERING DRAWINGS. SOURCE: NATIONAL ARCHIVES. ALL DRAWINGS ARE SOUTH ELEVATION VIEWS.
CATHEDRAL AVE.
PARKWAY
82'
150'
TOTAL LENGTH 1,341'
CONNECTICUT AVENUE (WILLIAM H. TAFT MEMORIAL) BRIDGE, 1897–1907
HAER No. DC-6
PARKWAY
156'
825'
ROCK CREEK
CALVERT STREET (DUKE ELLINGTON MEMORIAL) BRIDGE, 1933–35
HAER No. DC-23
HISTORIC AMERICAN BUILDINGS SURVEY
DC-697
35
36
ROCK CREEK AND POTOMAC PARKWAY
WASHINGTON, DISTRICT OF COLUMBIA
ROBERT HARVEY 1992
ROCK CREEK AND POTOMAC PARKWAY DOCUMENTATION PROJECT 1991-1992

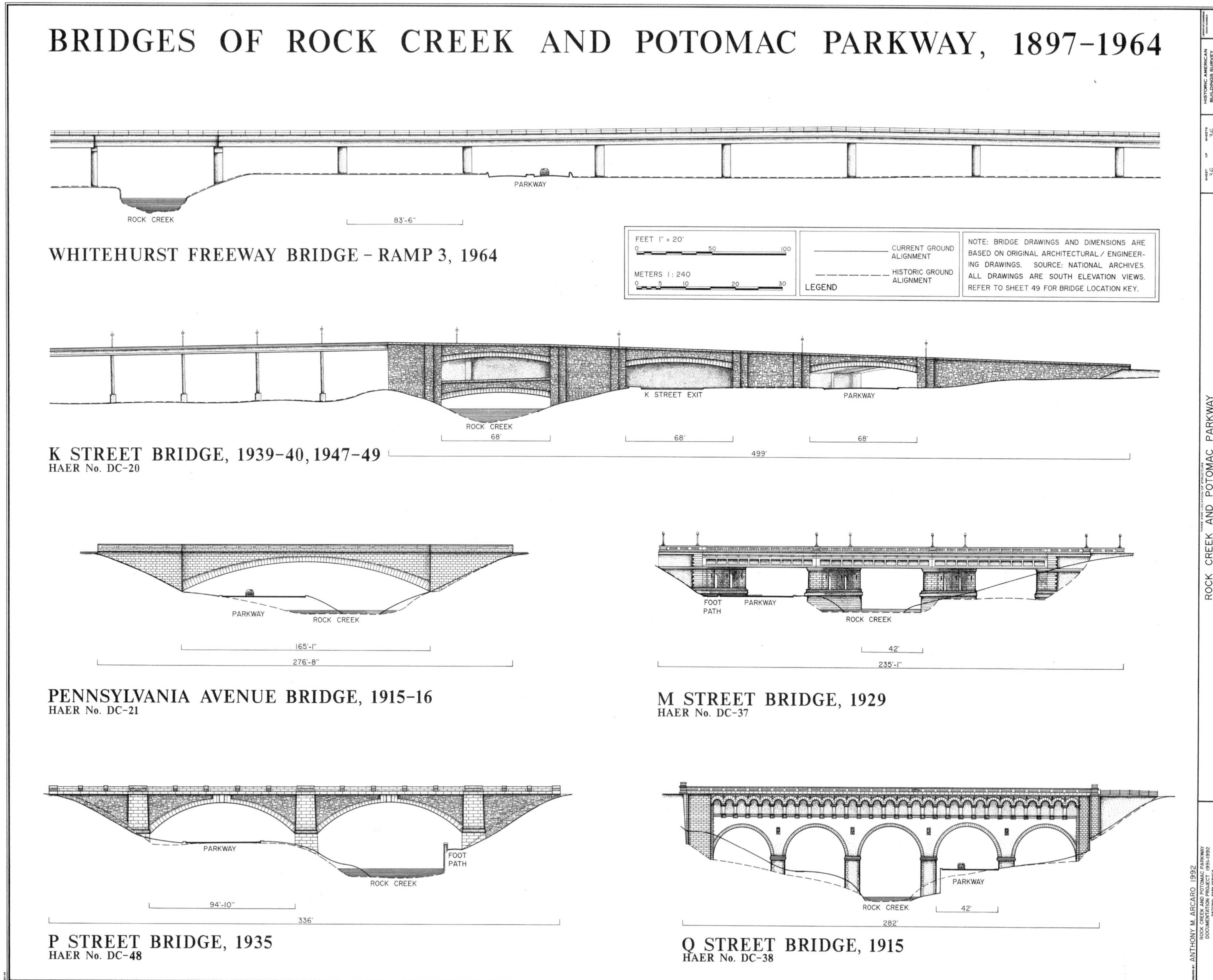
BRIDGES OF ROCK CREEK AND POTOMAC PARKWAY, 1897-1964
PARKWAY
ROCK CREEK
83'-6"
WHITEHURST FREEWAY BRIDGE - RAMP 3, 1964
FEET 1" = 20'
0 50 100
METERS 1 : 240
0 5 10 20 30
CURRENT GROUND ALIGNMENT
HISTORIC GROUND ALIGNMENT
LEGEND
NOTE: BRIDGE DRAWINGS AND DIMENSIONS ARE BASED ON ORIGINAL ARCHITECTURAL / ENGINEERING DRAWINGS. SOURCE: NATIONAL ARCHIVES. ALL DRAWINGS ARE SOUTH ELEVATION VIEWS. REFER TO SHEET 49 FOR BRIDGE LOCATION KEY.
K STREET EXIT
PARKWAY
ROCK CREEK
68'
68'
68'
499'
K STREET BRIDGE, 1939-40, 1947-49
HAER No. DC-20
PARKWAY
ROCK CREEK
165'-1"
276'-8"
PENNSYLVANIA AVENUE BRIDGE, 1915-16
HAER No. DC-21
FOOT PATH
PARKWAY
ROCK CREEK
42'
235'-1"
M STREET BRIDGE, 1929
HAER No. DC-37
PARKWAY
FOOT PATH
ROCK CREEK
94'-10"
336'
P STREET BRIDGE, 1935
HAER No. DC-48
PARKWAY
ROCK CREEK
42'
282'
Q STREET BRIDGE, 1915
HAER No. DC-38
HISTORIC AMERICAN BUILDINGS SURVEY
DC-697
36
36
ROCK CREEK AND POTOMAC PARKWAY
WASHINGTON, DISTRICT OF COLUMBIA
ANTHONY M. ARCARO 1992

BOULDER BRIDGE

·1902·

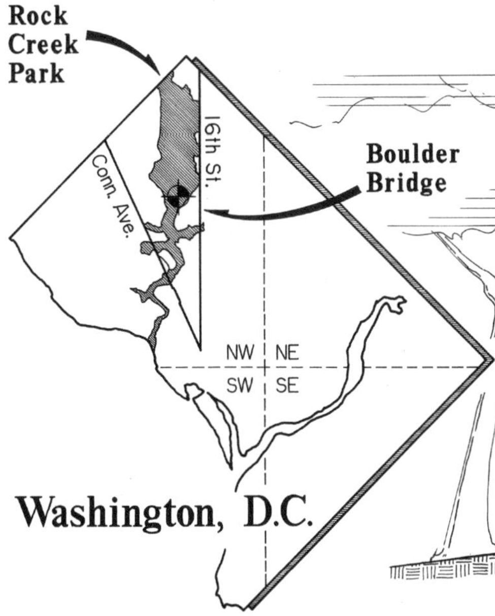

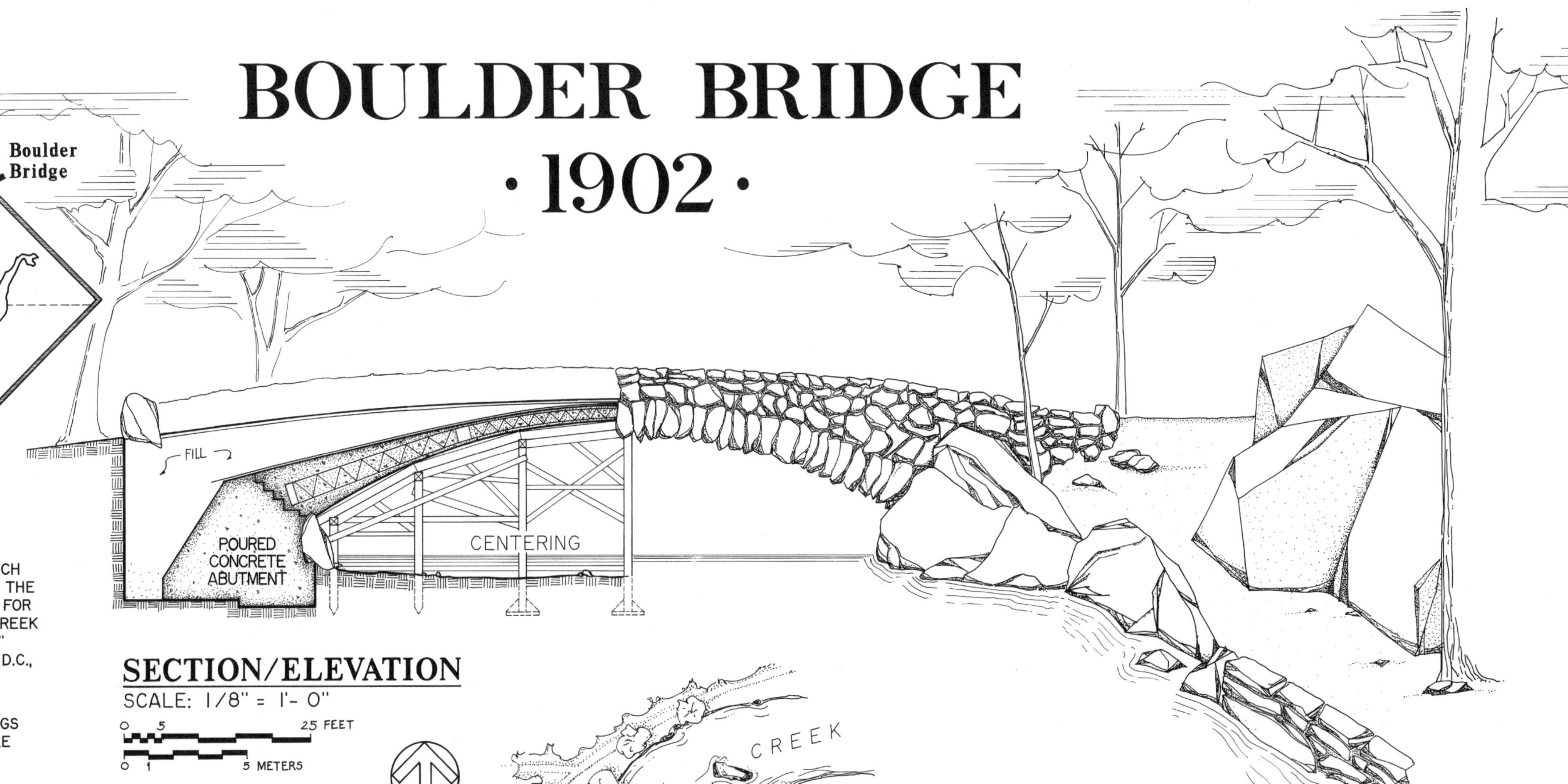

SOURCE OF DRAWINGS:

ORIGINAL PLANS AND STEEL ARCH DRAWINGS WHICH ACCOMPANIED THE AUGUST 7,1901, SPECIFICATIONS FOR "MELAN BRIDGE ACROSS ROCK CREEK ON LINE OF ROCK CREEK DRIVE," OFFICE OF THE COMMISSIONER, D.C., AT THE NATIONAL ARCHIVES.

UTM REFERENCE: Z18 E322780 N4312800 TAKEN FROM THE USGS WASHINGTON WEST QUADRANGLE 7.5 MINUTE MAP.

SECTION/ELEVATION

SCALE: 1/8" = 1'- 0"

0 5 25 FEET

0 1 5 METERS

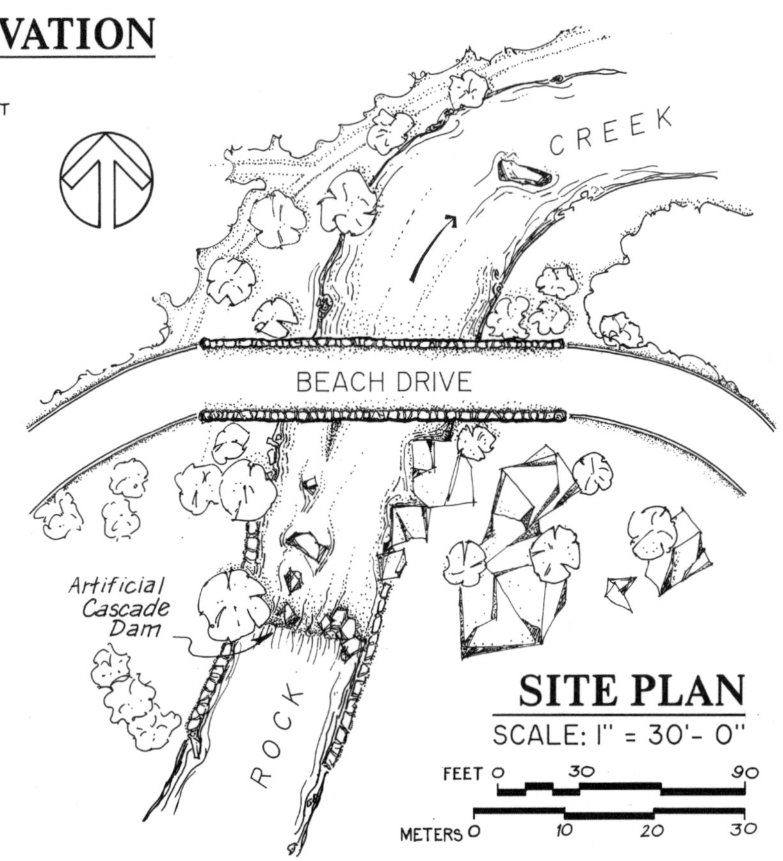

SITE PLAN

SCALE: 1" = 30'- 0"

FEET 0 30 90

METERS 0 10 20 30

In 1902 this concrete-steel arch bridge was constructed across Rock Creek on the line of Rock Creek Drive, now known as Beach Drive. The span utilized the patented Melan method of construction and was designed under the general direction of Captain Lansing H. Beach, Engineer Commissioner of the District of Columbia and member of the Board of Control in charge of Rock Creek National Park. A royalty was paid for the use of this patented technique and contractors Talty and Allen of Washington, D.C. were awarded the job. The Melan method called for the use of self-supporting steel lattice arch-trusses encased in concrete as opposed to steel rods or bars. The aesthetics of the bridge were determined by a desire to harmonize with the natural features of the site, particularly the irregular boulder-like rocks along the creek bed and the rocky ledges of the surrounding hills. Thus a boulder-faced masonry exterior was selected. Great care was taken in the selection of the boulders. Arch stones were selected by the engineer before bids were solicited. Suitable springing and newel stones were expectedly difficult to locate but the greatest trouble was finding enough through stones for the parapet as two sides would be visible. The site also features an artificial concrete dam constructed to allow water to drop over and onto large boulders placed below, thus replicating the sights and sounds of a natural cascade.

DELINEATED BY: ELAINE PIERCE, 1995

NATIONAL PARK SERVICE ROADS & BRIDGES RECORDING PROGRAM
NATIONAL PARK SERVICE
UNITED STATES DEPARTMENT OF THE INTERIOR

BOULDER BRIDGE (1907)
ROCK CREEK PARK
WASHINGTON
DISTRICT OF COLUMBIA

SHEET 1 of 2

HISTORIC AMERICAN ENGINEERING RECORD DC-12

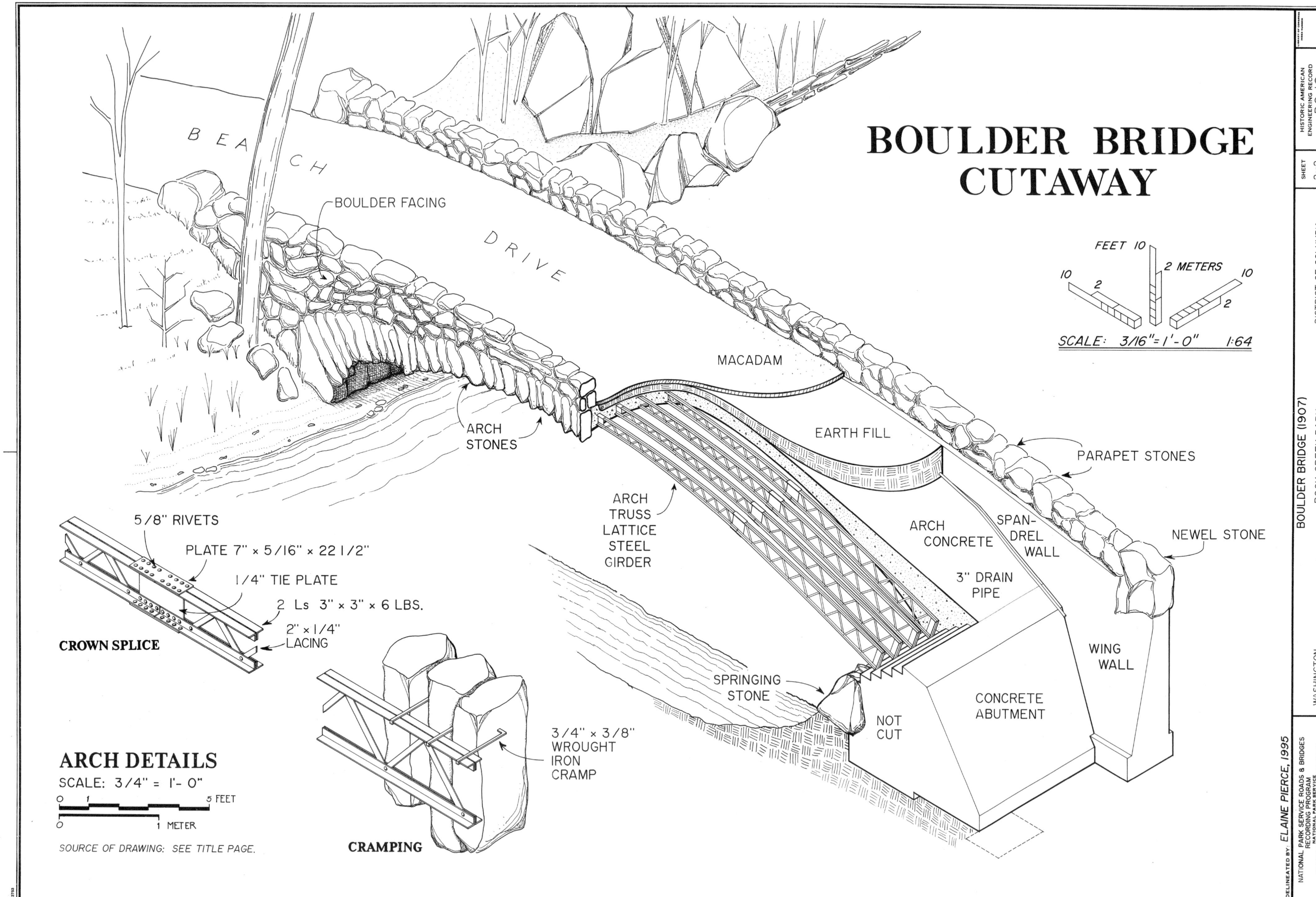

BOULDER BRIDGE
CUTAWAY
BEACH DRIVE
BOULDER FACING
MACADAM
EARTH FILL
PARAPET STONES
ARCH STONES
ARCH TRUSS LATTICE STEEL GIRDER
ARCH CONCRETE
SPAN-DREL WALL
NEWEL STONE
3" DRAIN PIPE
WING WALL
SPRINGING STONE
NOT CUT
CONCRETE ABUTMENT
FEET 10
2 METERS
10
2
SCALE: 3/16"=1'-0" 1:64
5/8" RIVETS
PLATE 7" × 5/16" × 22 1/2"
1/4" TIE PLATE
2 Ls 3" × 3" × 6 LBS.
2" × 1/4" LACING
CROWN SPLICE
3/4" × 3/8" WROUGHT IRON CRAMP
CRAMPING
ARCH DETAILS
SCALE: 3/4" = 1'- 0"
0 1 5 FEET
0 1 METER
SOURCE OF DRAWING: SEE TITLE PAGE.
DELINEATED BY: ELAINE PIERCE, 1995
NATIONAL PARK SERVICE ROADS & BRIDGES RECORDING PROGRAM
NATIONAL PARK SERVICE
UNITED STATES DEPARTMENT OF THE INTERIOR
BOULDER BRIDGE (1907)
WASHINGTON
ROCK CREEK PARK
DISTRICT OF COLUMBIA
SHEET 2 OF 2
HISTORIC AMERICAN ENGINEERING RECORD
DC-12
IF REPRODUCED, PLEASE CREDIT: HISTORIC AMERICAN ENGINEERING RECORD, NATIONAL PARK SERVICE, NAME OF DELINEATOR, DATE OF THE DRAWING

ROSS DRIVE BRIDGE · 1907

WASHINGTON, D.C.

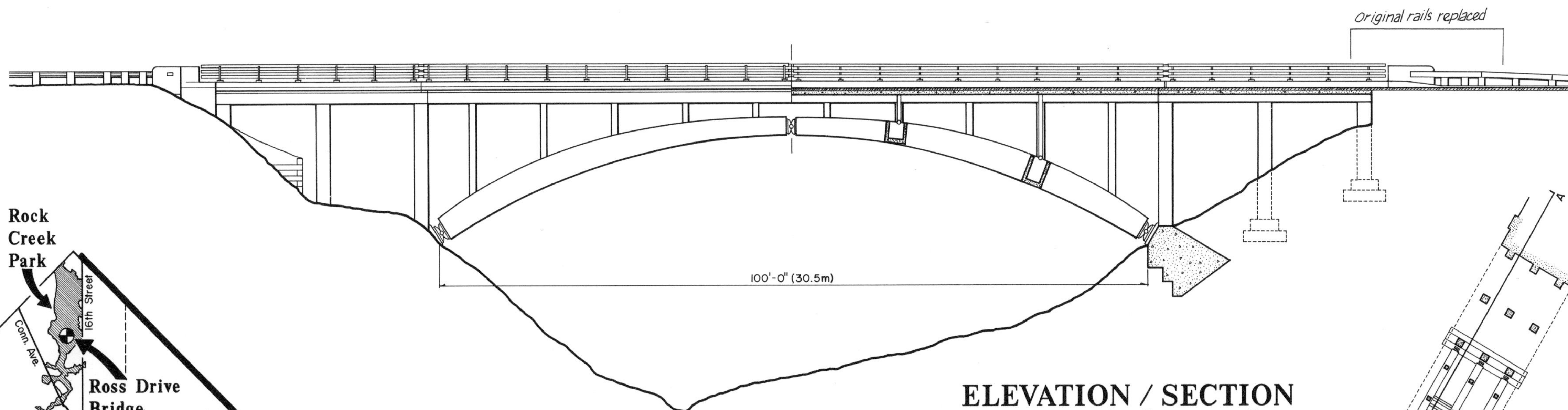

ELEVATION / SECTION

1/8" = 1'-0" FEET 0 5 20
1:96 METERS 0 1 5

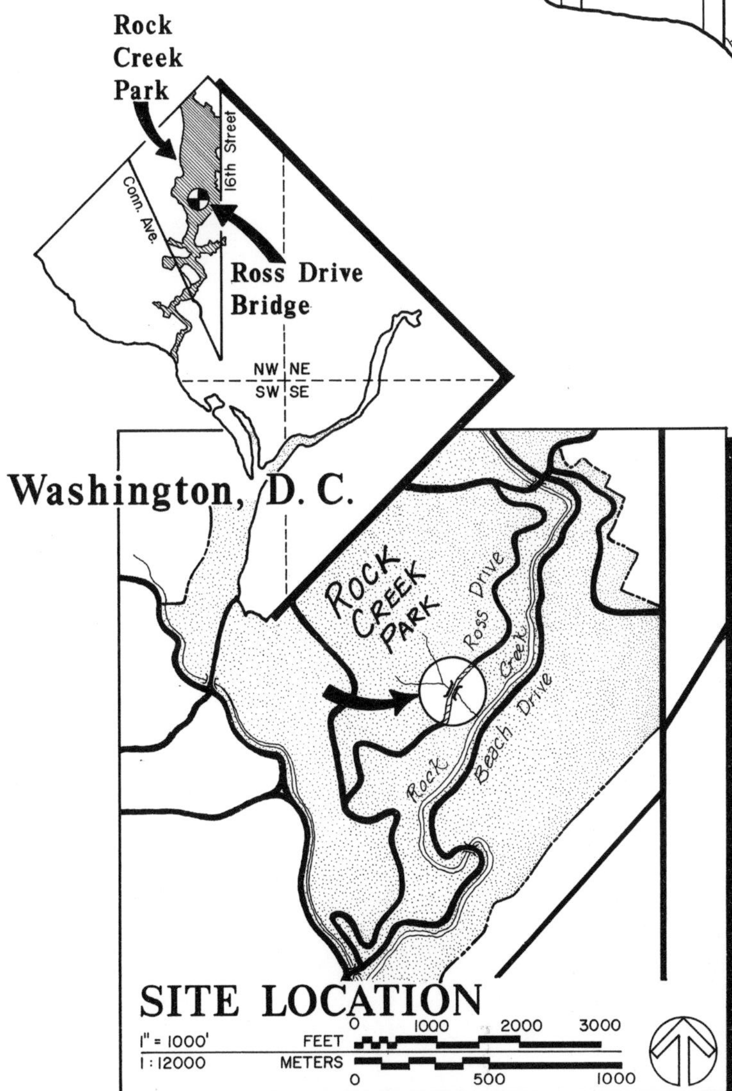

SITE LOCATION

1" = 1000' FEET 0 1000 2000 3000
1:12000 METERS 0 500 1000

U.T.M.: 18/323900/4314230

Designed by W.J. Douglas, Engineer of Bridges in the District of Columbia, to harmonize with its surroundings, Ross Drive Bridge was completed in 1907 at a cost of $8,000. So as not to obstruct views, a span consisting of three exposed three-hinged concrete arch ribs with thin spandrel columns, a concrete deck and light metal railing was constructed. Connecting struts between the arch ribs were detailed with "flower troughs" from which vines could grow and cover the structure in a veil of foliage. Continuing the integration of landscape concerns, a Macadam road surface was placed over the concrete deck with a twelve-inch vegetated shoulder on each side. The new bridge replaced an existing timber trestle which was used, with little modification, as falsework for the construction of the concrete span. Several modifications have occurred, including the application of a gunite coating to the superstructure, removal of the Macadam surface and shoulder and replacement of original railings; however, the bridge retains its original character.

Note: Drawings depict 1995 appearance and are based on field measurements, historic photographs and original design documents.

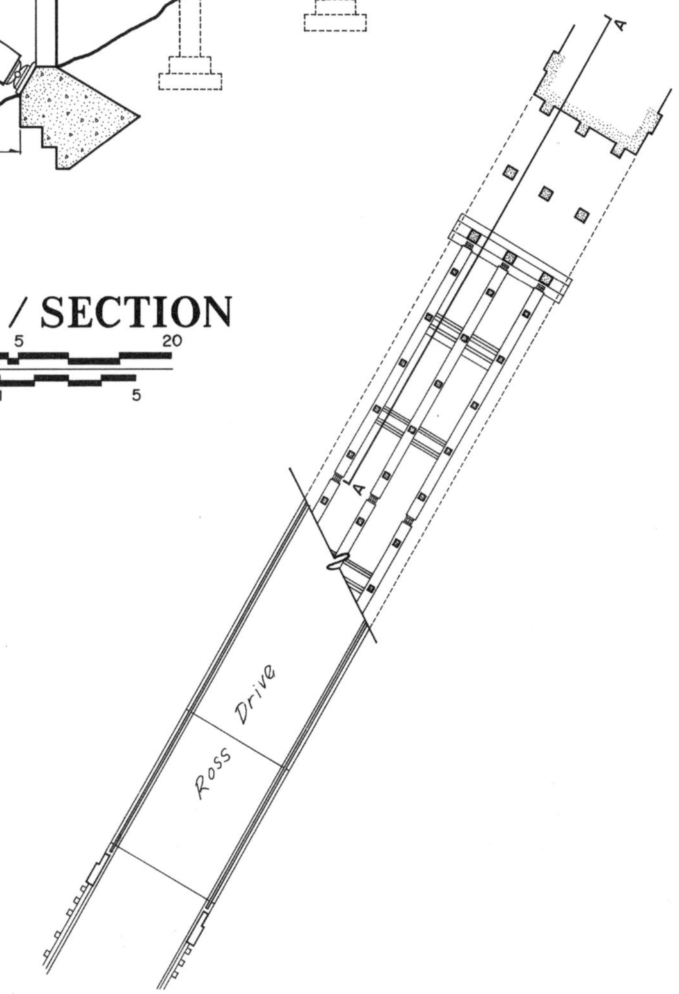

PLAN

1/16" = 1'-0" FEET 0 5 10 20 50
1:192 METERS 0 5 10 15

DELINEATED BY: DOUGLAS ANDERSON, 1995

NATIONAL PARK SERVICE ROADS & BRIDGES RECORDING PROGRAM
NATIONAL PARK SERVICE
UNITED STATES DEPARTMENT OF THE INTERIOR

ROSS DRIVE BRIDGE (1901)
ROCK CREEK PARK
WASHINGTON — DISTRICT OF COLUMBIA

SHEET 1 of 2

HISTORIC AMERICAN ENGINEERING RECORD
DC-13

IF REPRODUCED, PLEASE CREDIT: HISTORIC AMERICAN ENGINEERING RECORD, NATIONAL PARK SERVICE, NAME OF DELINEATOR, DATE OF THE DRAWING

ROSS DRIVE BRIDGE DETAILS

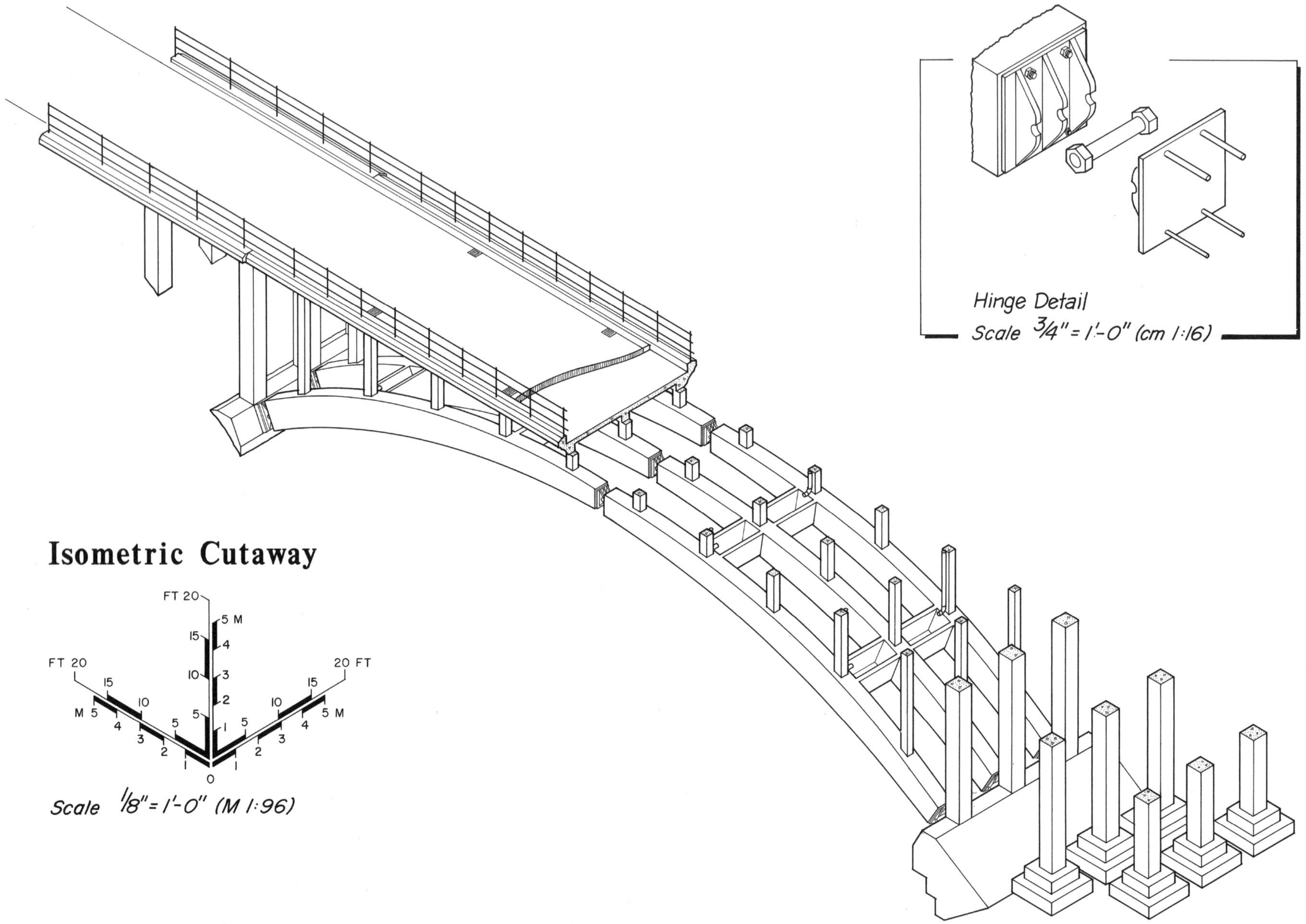

Isometric Cutaway

Hinge Detail
Scale 3/4" = 1'-0" (cm 1:16)

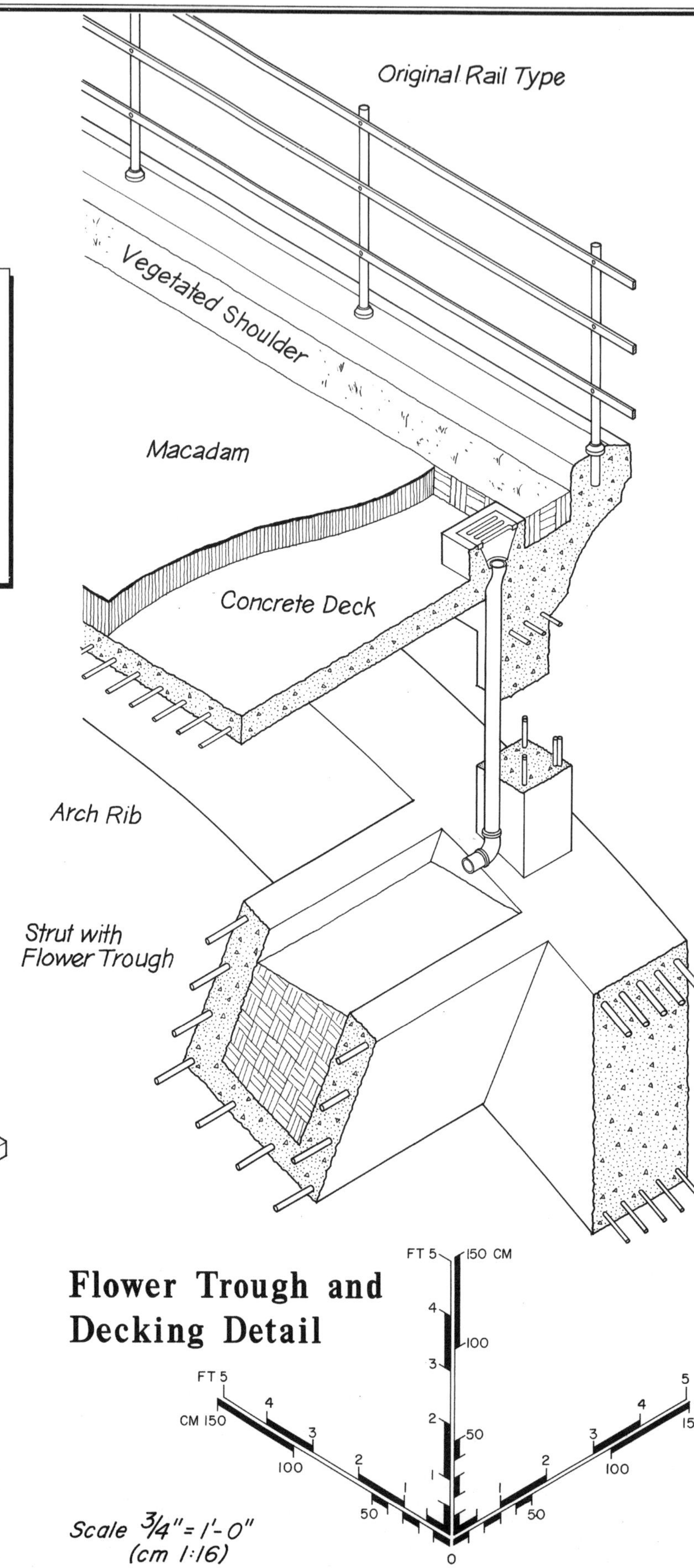

Flower Trough and Decking Detail

DELINEATED BY: DOUGLAS ANDERSON 1995

NATIONAL PARK SERVICE ROADS & BRIDGES RECORDING PROGRAM
NATIONAL PARK SERVICE
UNITED STATES DEPARTMENT OF THE INTERIOR

ROSS DRIVE BRIDGE (1901)
ROCK CREEK PARK
WASHINGTON
DISTRICT OF COLUMBIA

SHEET 2 OF 2

HISTORIC AMERICAN ENGINEERING RECORD
DC-13

National Military Parks

National military parks are a unique subcategory of national parks. They are intended to preserve and commemorate the sites of famous battles. Many were developed with the active participation of veterans groups drawn from the original combatants. While considerable emphasis was placed on preserving historic features, the desire to honor fallen comrades and narrate the course of famous events produced landscapes full of impressive monuments, various types of informational tablets, and sophisticated circulation systems.

Where historic roads were inadequate, dignified avenues and winding tour roads were constructed to provide sweeping views and access to significant sites. Bridges and other engineered structures were often built with the latest technologies and most expensive materials. Gettysburg's labor-intensive telford-paved avenues and Vicksburg's technologically advanced yet classically configured Melan arch bridges testify to the tremendous reverence accorded these hallowed sites.

The HAER documentation elucidates this complex synergy of historic, commemorative, and technological factors. Conceptual drawings detail paving techniques, dissect drainage principles, reveal design strategies, and trace long-term changes in vegetation patterns. More traditional documentation describes and categorizes a wide range of bridges, culverts, signs, fences, and historic markers. Historically focused drawings relate development patterns to battlefield precedents and suggest the ways in which park roads can provide imaginative access to the past.

Vicksburg National Military Park, 1997

Photographer: William Faust II, HAER

National Military Park Road Systems

Chickamauga and Chattanooga National Military Park Tour Roads

Cover Sheet 302
Evolution of the Road System 303
The Commemorative Landscape 304
Landscape Elements 305
The Changing Landscape 306
Motorist's Experience 307
Roadway Construction 308
Chickamauga Battlefield Bridges 309

Gettysburg National Military Park Tour Roads

Cover Sheet 310
Evolution of the Avenues, 1863–95 311
Evolution of the Avenues, 1895–1998 312
Avenue Construction 313
Avenue Cross Sections 314
Drainage 315
Bridge Comparison 316
West Confederate Avenue Bridge 317
Observation Towers 318
Signs and Markers 319
Boundary Demarcations 320
Experiencing the Battle 321

Shiloh National Military Park Tour Roads

Cover Sheet 322
Battlefield Park Road Types 323
Tour Road Interpretive Features 324
Bridges and Culverts 325

Vicksburg National Military Park Tour Roads

Cover Sheet 326
Reference Map 1 327
Reference Map 2 328
Evolution of Transportation 329
Battlefield Tour Roads, Schematic View 330
Drainage Design Principles 331
Roadway Engineering 332
Melan Arch Bridges, 1903 333
Melan Arch Bridge Construction 334
Confederate Avenue Steel Bridge, 1903 335
Connection Details 336

CHICKAMAUGA AND CHATTANOOGA

Edward E. Betts
First Park Engineer

NATIONAL MILITARY PARK TOUR ROADS

In 1888 two Union veterans touring the battlefields of Chickamauga came up with the idea to preserve and commemorate this important Civil War site. Taking as their model contemporary commemoration at Gettysburg Battlefield, the two veterans--Henry Boynton and Ferdinand Van Derveer--used their career contacts to promote the creation of a "Western Gettysburg". In May of 1890 Chickamauga and Chattanooga became the first National Military Park.

CHCHNMP was planned to combine commemoration, historical study and landscape preservation. Like the other National Military Parks that would follow it, CHCHNMP would be visited by historians, veterans, locals, and tourists. The design and construction of an effective road system was needed to provide access to the park's diverse features. In addition to clearing fields, erecting monuments, and placing markers the park commissioners would also build approach roads, construct culverts and guttering, and reconstruct roads as they had appeared during the 1863 battle. Between 1890 and its official opening in 1895 the park was home to a crew of 300 men who worked daily to finish this $750,000 project.

During this time Edward E. Betts--the first park engineer and a Chattanooga native--guided the design, construction and reconstruction of the park roads. He mapped the roads and landscape of the 1863 battle and designed elegant stone culverts to accentuate park landscapes. With the aid of Atwell Thompson, Betts reconstructed LaFayette Road (a regional highway and the crux of the 1863 Battle of Chickamauga) and purchased the steam-powered roller (photograph at right) that packed miles of loose gravel roads. They designed and built Crest Road on Missionary Ridge to connect a series of small commemorative sites and to exploit grand views of Chattanooga. And they also made plans, never fully realized, to construct a great "Military Road" stretching from Sherman Reservation north of Chattanooga to Lee and Gordon's Mill south of Chickamauga Battlefield.

The park commission's work was an astounding success. By 1900 they had built over 50 miles of new roads and restored many miles of historic roads to their approximate condition at the time of the battle. These roads not only linked the scattered pieces of the park and opened historic sites for visitors, but also formed a new transportation network that increased property values, eased local travel, and encouraged tourism in the entire area.

Work on the tour roads continued after 1900, but the aim was to refine and maintain the system: several historic bridges were replaced, new culverts were constructed and most approach roads were widened and gravelled to accomodate the ever-increasing traffic.

Steam-powered Roller, ca. 1900

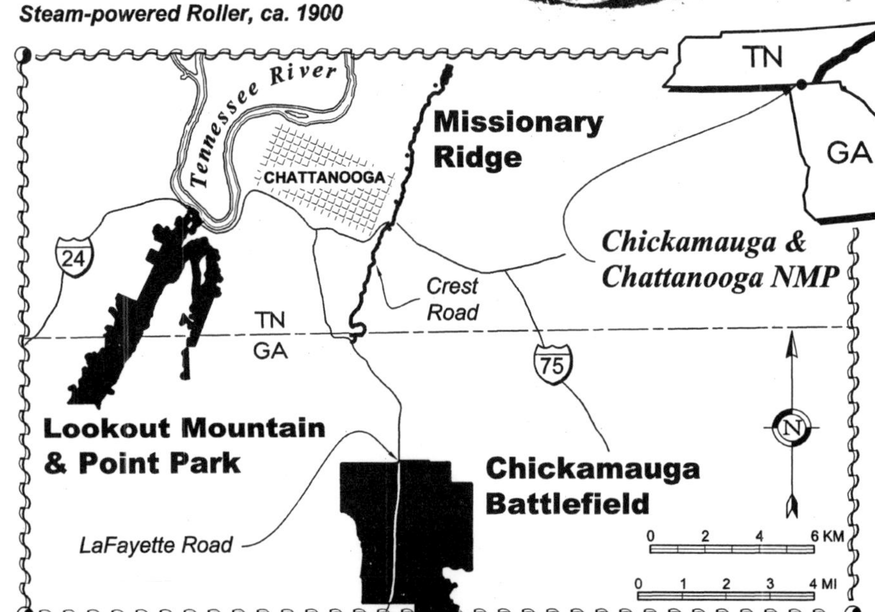

LOCATION MAP

In 1933 CHCHNMP was transferred from the War Department to the National Park Service. At this time, the Civilian Conservation Corps (CCC) brought four camps of workers to the park and initiated the first new construction in decades. The CCC and NPS widened and concreted LaFayette and other roads, built new bridges in the Park Service "rustic style," and replaced 40-year-old drainage features. The CCC work extended beyond the roads into the landscape where crews cut trails and bridle paths, planted trees and grass, and prevented erosion by sculpting roadsides.

Although the closure of the last CCC camp in 1942 brought an end to widespread improvements, tourism and traffic continued to increase. The park--bound to its roads by history and access--suffered from its own success. By the early 1960s LaFayette Road, the main route through Chickamauga Battlefield, was crowded with local, tourist and through traffic. At this time Mission 66--a program to renew America's neglected National Parks-- prompted a new era of road construction and rehabilitation at CHCHNMP. Various plans to reroute LaFayette Road began in 1964, and by 2000 a bypass route to the west of the park diverted most local traffic around the park.

The Chickamauga and Chattanooga National Military Park Tour Roads Recording Project is part of the Historic American Engineering Record (HAER), a long-range program documenting historically significant engineering sites in the United States. The HAER program is administered by the National Park Service, U.S. Department of the Interior. This project was sponsored during the summer of 1998 by the NPS Park Roads & Parkways Program through Federal Lands Highway funds; and CHCHNMP, Patrick Reed, Superintendent & Jim Ogden, park historian.

The field work, measured drawings, historical reports, and photographs were prepared under the general direction of Eric DeLony, Chief of HAER and Todd Croteau, NPS Roads & Bridges Program Manager. The recording team consisted of Pete Brooks, architect and field supervisor (Yale University); Nancy Hamburger (Georgia State University); Anna Sniegucka, (ICOMOS-Poland); and Tetyana Sprysa (ICOMOS-Ukraine), architects; David Ezell, historian (Georgia Tech); and David Haas, large format photography.

NOTE: *Map based on 1999 National Park Service Tour Map of the CHCHNMP and E. E. Betts' "Chickamauga & Chattanooga National Park & Approaches", 1910. Graphic drawn from the archives of the CHCHNMP.*

DELINEATED BY: *Pete Brooks, 2000*

NPS ROADS & BRIDGES RECORDING PROGRAM
NATIONAL PARK SERVICE
UNITED STATES DEPARTMENT OF THE INTERIOR

CHICKAMAUGA AND CHATTANOOGA NATIONAL MILITARY PARKS
TOUR ROADS
FORT OGLETHORPE VICINITY WALKER & CATOOSA GEORGIA

SHEET 1 of 15

HISTORIC AMERICAN ENGINEERING RECORD
GA-95

LIBRARY OF CONGRESS INDEX NUMBER

EVOLUTION OF THE ROAD SYSTEM

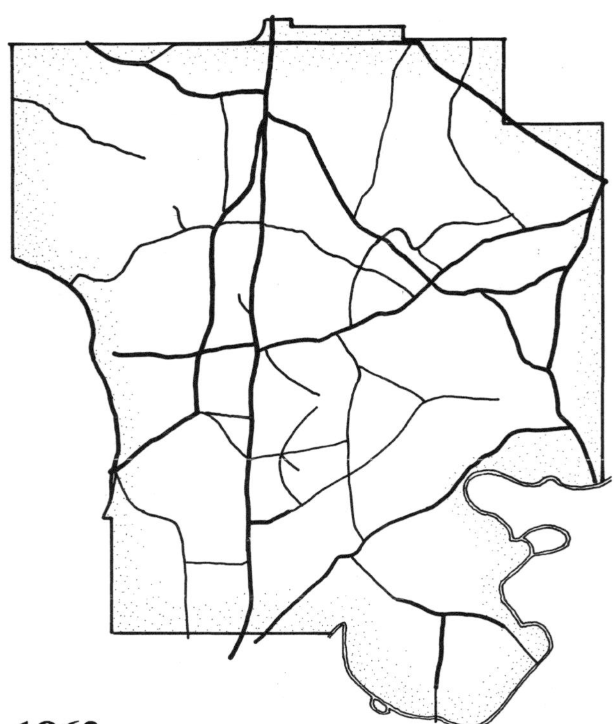

1863
Battle Conditions

A network of dirt roads ranging from LaFayette Road, a primary North-South connector, to small trails, linked homesteads which covered the area. The road system provided troop movement during the War.

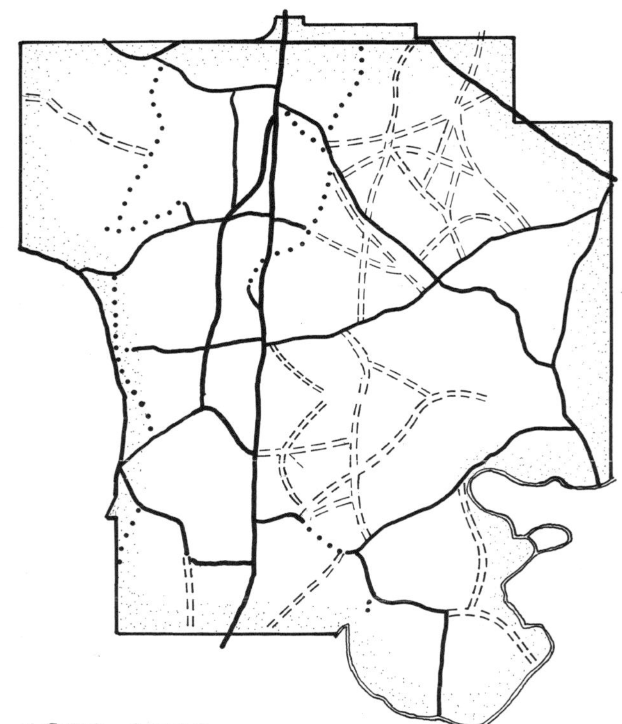

1890-1933
Park Commission Era

Park Commission efforts focused on restoring battlefield conditions. To that end, old roads in the Park were located and reopened. New roads were added providing access to monuments. During this period, roads were both added and deleted for military exercises. during the Spanish-American war and World War I.

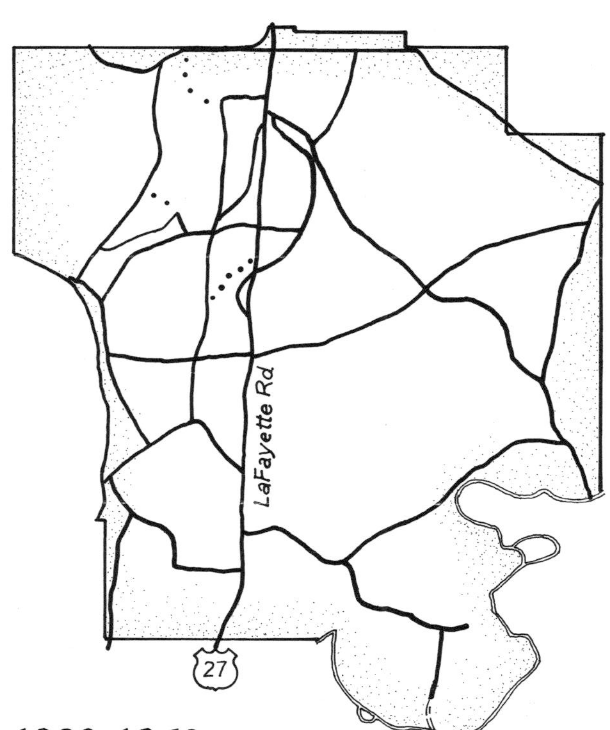

1933-1960
Early National Park Service Era

Early Park Service road construction and renovation was linked to CCC efforts. Much of their work went into road repair and forest underbrush clearing. During this period, many historic roads were paved.

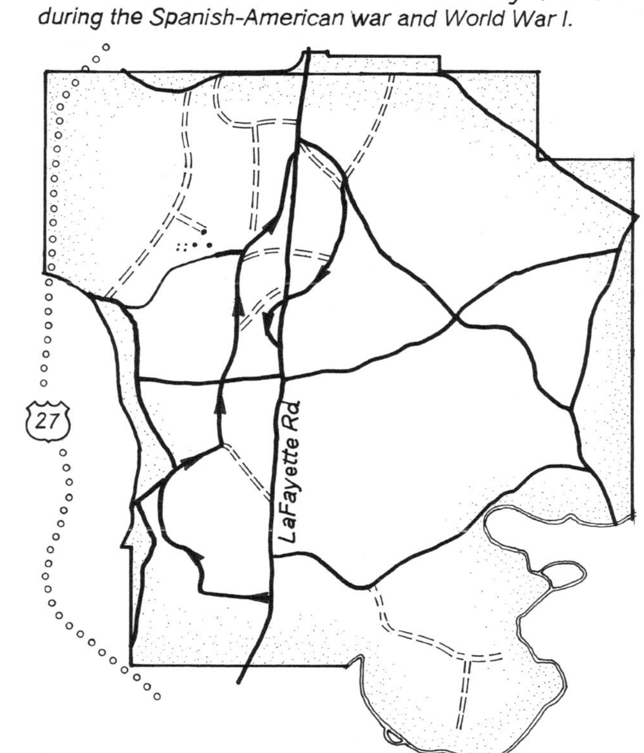

1960-Present

Since the Park's inception, roads have been shared by local commuters, transportation vehicles and designated visitors. Efforts to eliminate heavy traffic have resulted in the relocation of Hwy.27. Completion of construction will result in improved traffic conditions within the Park system.

Note: Drawing based on: 1896 Map of the Battlefield of Chickamauga, E.E. Betts, Chief Engineer; 1892 Chickamauga Battlefield Map, E.E. Betts; 1934 Chickamauga Battlefield Monument Map, Albert Pike; 1995 Park brochure; Hwy. 27 relocation brochure.

Legend:

- —— *Existing roads*
- = = = = *Obliterated roads*
- • • • • *Added roads*
- ○ ○ ○ ○ *Proposed roads*
- ◄— *One-way, tour route*

Notes:

Portions of alignment and intersections of some historic roads have changed.
Each map refers to the previous map.

DELINEATED BY: Nancy Hamburger, 1998

NPS ROADS & BRIDGES RECORDING PROGRAM NATIONAL PARK SERVICE UNITED STATES DEPARTMENT OF THE INTERIOR	CHICKAMAUGA AND CHATTANOOGA NATIONAL MILITARY PARKS TOUR ROADS FORT OGLETHORPE, VICINITY OF CATOOSA/WALKER GEORGIA	SHEET 6 OF 15	HISTORIC AMERICAN ENGINEERING RECORD GA-95

THE COMMEMORATIVE LANDSCAPE

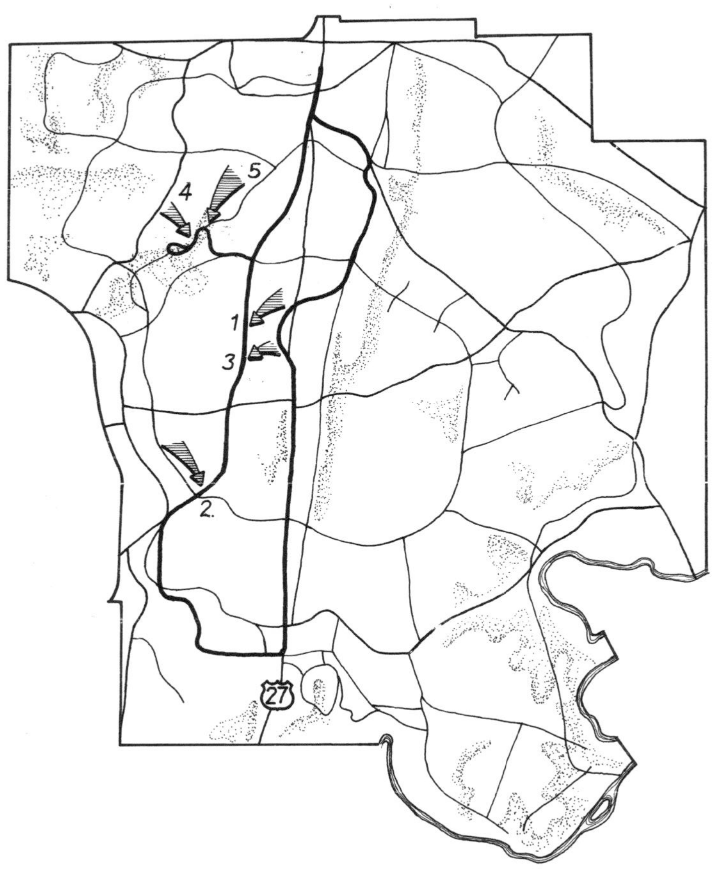

There are a variety of markers that help visitors more easily understand the commemorative landscape of the Chickamauga Battlefield. Historic structures are reconstructed at various sites, interpretive tablets and signs explain significant locations on the field, and commemorative markers indicate positions of important events in the battle These are integrated into the road network.

3.

Several cabins in the park have been reconstructed on their original foundations.

1.

These metal tablets give detailed descriptions of military actions that occurred exactly where they are placed.

4.

Stacks of cannonballs are placed throughout the park--marking headquarter locations and sites where brigade commanders died.

2.

Fingerboards are holdovers from the early days of the park-typically they direct visitors to locations or landmarks.

5.

Monuments and markers show the positions of troops in the field.

DELINEATED BY: Anna Śniegucka, 1998

NPS ROADS & BRIDGES RECORDING PROGRAM
NATIONAL PARK SERVICE
UNITED STATES DEPARTMENT OF THE INTERIOR

CHICKAMAUGA AND CHATTANOOGA NATIONAL MILITARY PARKS
TOUR ROADS
FORT OGLETHORPE, VICINITY OF CATOOSA/WALKER GEORGIA

SHEET 7 OF 15

HISTORIC AMERICAN ENGINEERING RECORD GA-95

LANDSCAPE ELEMENTS

For National Military Parks

Typically, military parks utilized similar techniques for developing interpretive landscapes, however each park created unique character for its signage and standardized monumentation.

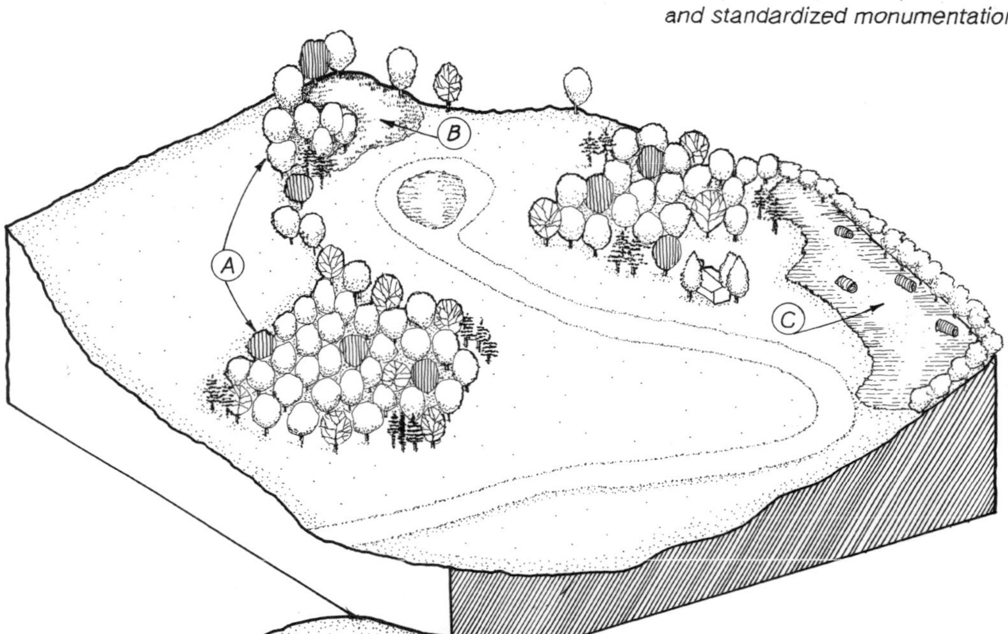

Vegetation

While forests cover much of the Park (A), a portion of the land is set aside for agricultural leases (B). An agricultural lease is an open hay field, maintained by local farmers. Mow bays arc smaller open areas maintained by park staff. A mow bay allows views of monuments and sites (C). In some instances, planting plans by the National Park Service established specific vegetation areas.

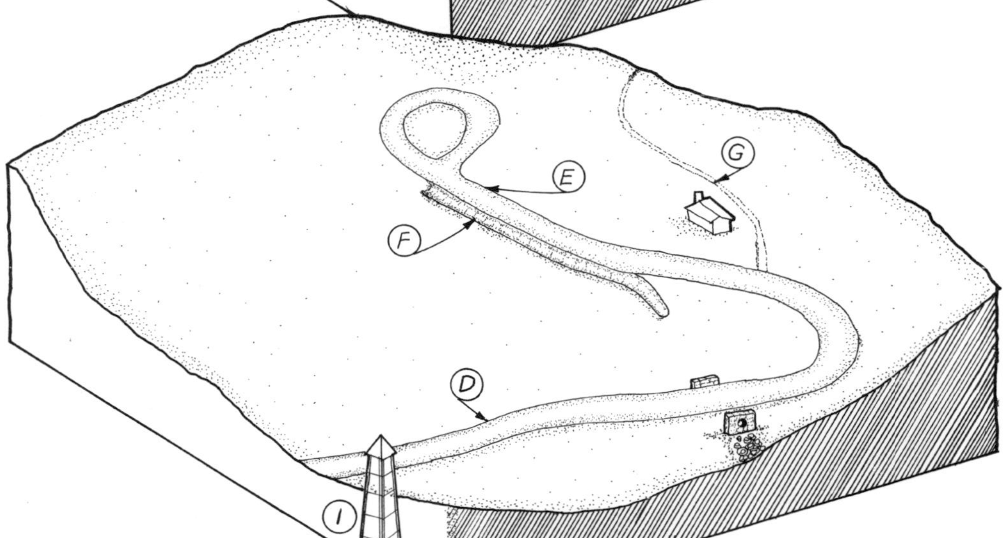

Tour Roads

Contemporary tour roads often follow historic road alignments (D). Sometimes roads or road segments were added to provide access to battle sites or monuments (E). Over time, park engineers improved the road systems; improvements included pavement, gutters, and culverts (F). Abandoned roads remain as foot and horse trails (G).

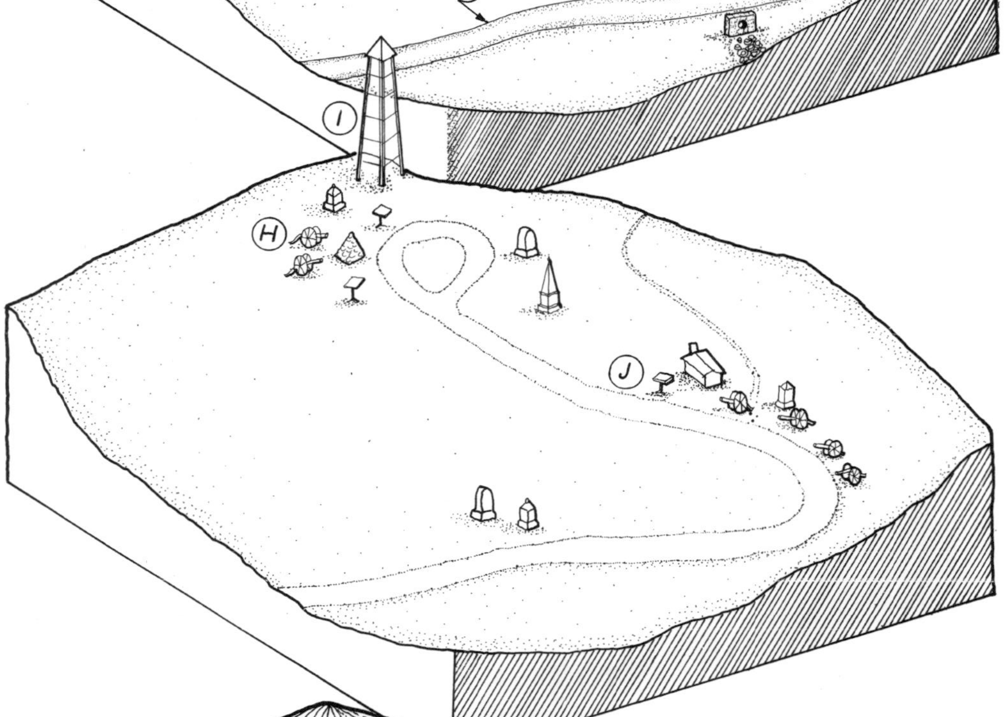

Monuments & Interpretation

Commemorative monuments and markers along the road mark the locations of units and batteries engaged in battle. They include monuments, cast iron tablets, cannons, and cannon ball pyramids (H). Towers, now removed, allowed visitors an aerial view of battle conditions (I). Interpretive devices including signs, voice recordings and markers describe actions at tour stops (J).

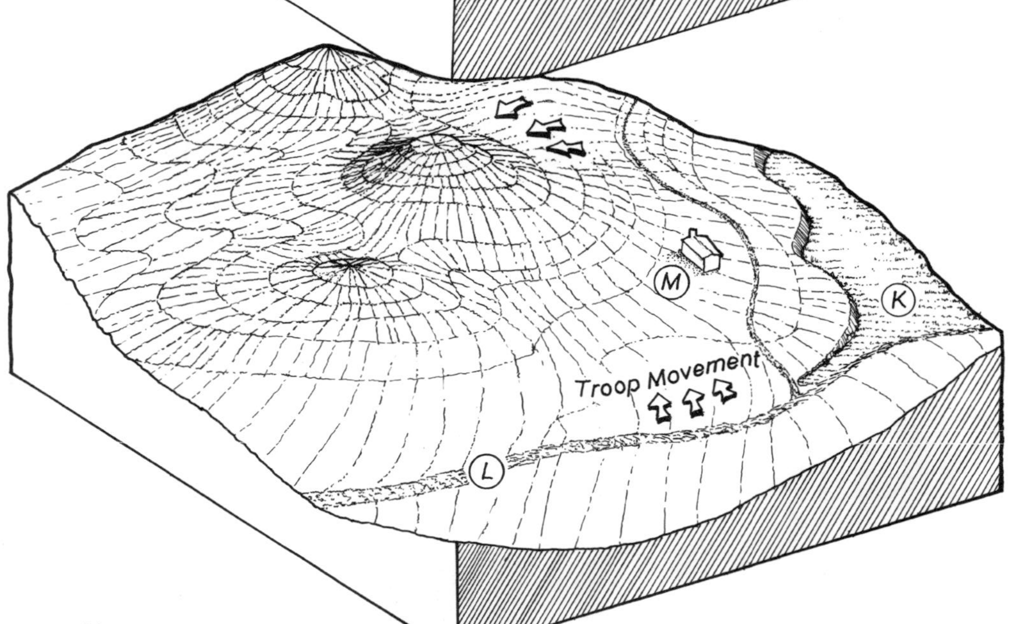

Civil War Landscape

Settled by farmers, the land was cultivated (K), logged and grazed. Fields were fenced with 5' high cedar fences while livestock was permitted to graze in the forests, clearing underbrush. Unpaved roads and trails linked scattered houses and farmsteads (L). Chickamauga's gently rolling topography combined with open fields and forests set the stage for the battle. Fighting at Chickamauga was spread over the landscape, including farms and houses (M).

Schematic: concepts based on Chickamauga battlefield.

DELINEATED BY: Nancy Hamburger, 1998

NPS ROADS & BRIDGES RECORDING PROGRAM NATIONAL PARK SERVICE UNITED STATES DEPARTMENT OF THE INTERIOR	CHICKAMAUGA AND CHATTANOOGA NATIONAL MILITARY PARKS TOUR ROADS FORT OGLETHORPE, VICINITY OF CATOOSA/WALKER GEORGIA	SHEET 8 OF 15	HISTORIC AMERICAN ENGINEERING RECORD GA-95

THE CHANGING LANDSCAPE

The landscape in 1863 was one created by farmers for their needs. This land was populated with occasional small farms, open fields and wooded areas along their periphery. Farmers considered the land to be a valuable resource, one from which they drew a variety of resources–from the food they grew to the wood used to create homes and outbuildings.

The rocky soil of the district limited the size of the farms and the density of the woodlands, keeping the region's population relatively low. The road system was a utilitarian one–dictated by the area's topography and bodies of water around them. Most of the area's residents used paths through the woods and along the fields to go from one place to another. The roads evolved naturally, with no manmade design.

The Battle of Chickamauga had a strong negative impact on the land. Entire areas were ravaged by fires that blazed out of control, charring both the woodlands and farmer's fields.

Thus, the impact of the war established the beginnings of the present landscape.

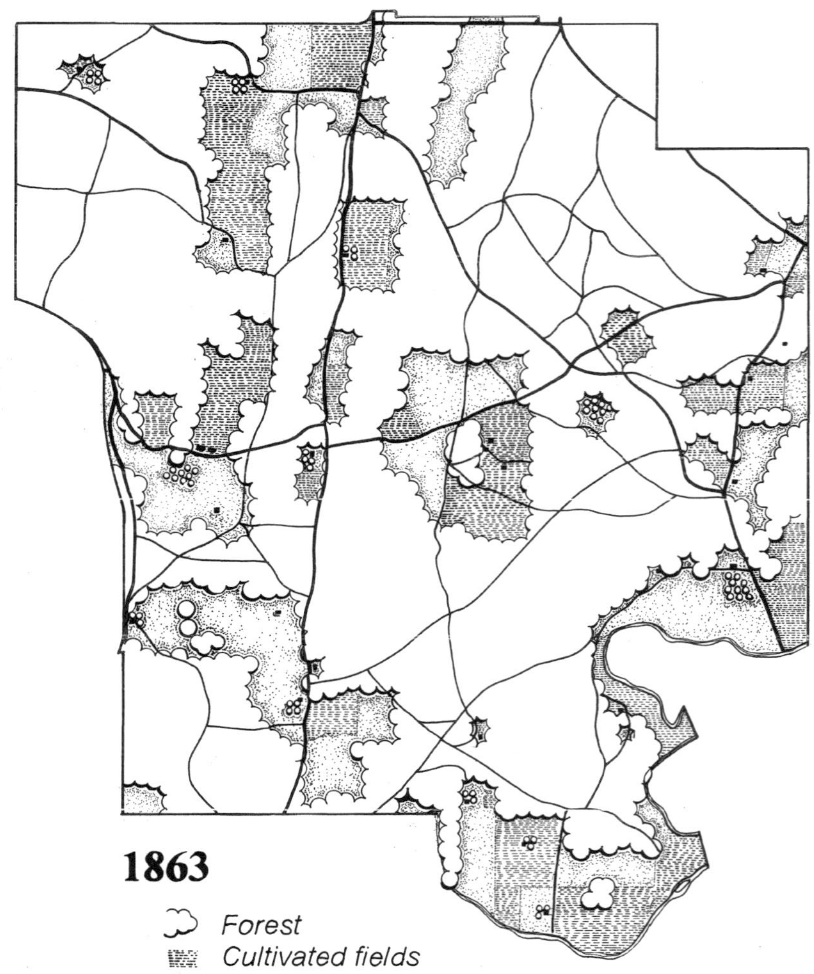

1863

Forest
Cultivated fields
Uncultivated fields
Structures
Roads

The modern landscape is one designed to commemorate the battlefield of a century before. The farms no longer exist, as the park's lands no longer allow private residences.

This change has impacted the development of the roads in the area. No longer needed to connect one farm to another, many of the roads in Chickamauga disappeared after the Civil War's end. As some roads were discontinued, other roads were developed. These roads were necessary so that visitors may be able to see the area's monumentation. The road system that exists is a basic one–the minimal amount needed for tourists to see the park's major sites.

Since the area's formation as a national park, the park's visitors have been assured that there will be no further evolution of the park land. There are a variety of scenic routes available so that tourists may see the land as it once was. However, some monuments are off of the "beaten path," placing them out of the reach of all but the heartiest enthusiasts.

Yet one element has had a detrimental impact on the area, Highway 27 cuts through the center of the park, causing problems for tourists and animals in the area. This problem will be resolved in the next few years by a rerouting of the roadway.

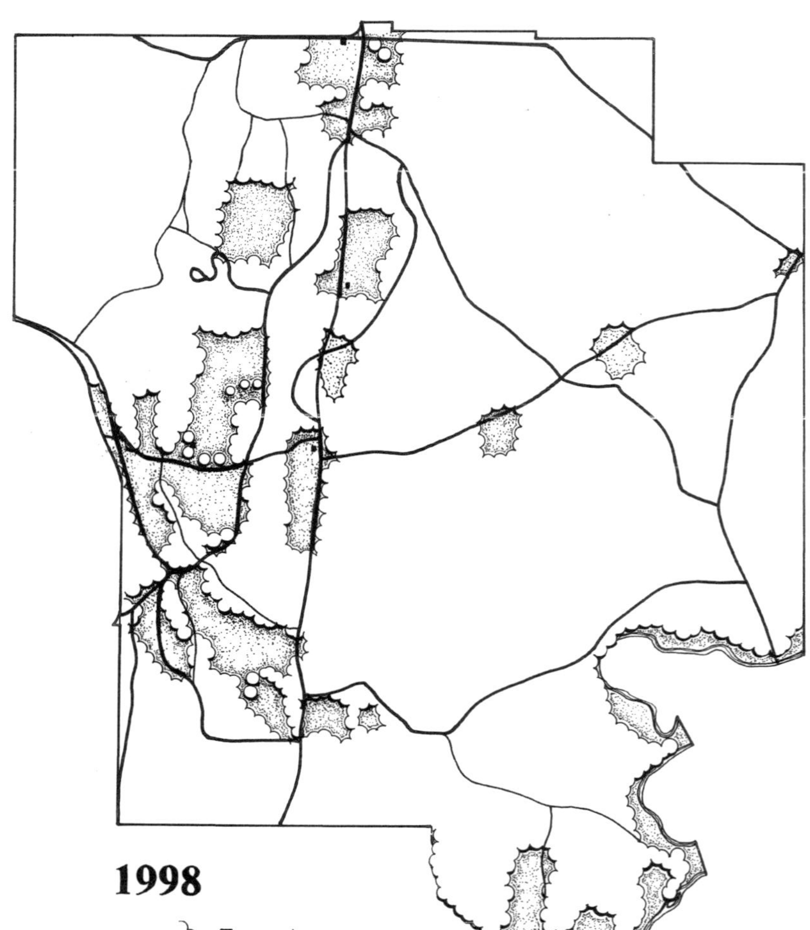

1998

Forest
Uncultivated fields
Structures
Roads

DELINEATED BY: Anna Śniegucka, 1998

NPS ROADS & BRIDGES RECORDING PROGRAM NATIONAL PARK SERVICE UNITED STATES DEPARTMENT OF THE INTERIOR	CHICKAMAUGA AND CHATTANOOGA NATIONAL MILITARY PARKS TOUR ROADS FORT OGLETHORPE, VICINITY OF CATOOSA/WALKER GEORGIA	SHEET 9 OF 15	HISTORIC AMERICAN ENGINEERING RECORD GA-95

MOTORIST'S EXPERIENCE

(Chickamauga Battlefield)

Open View

Perhaps the most lasting impression of Chickamauga Battlefield is one of open fields and broad views. The sense of space in these fields is a result of carefully maintained forest edges, although the contrast between bright fields and dark forest adds to the illusion. At Glenn Field (right) this illusion of space is magnified by monuments at the edge of the field.

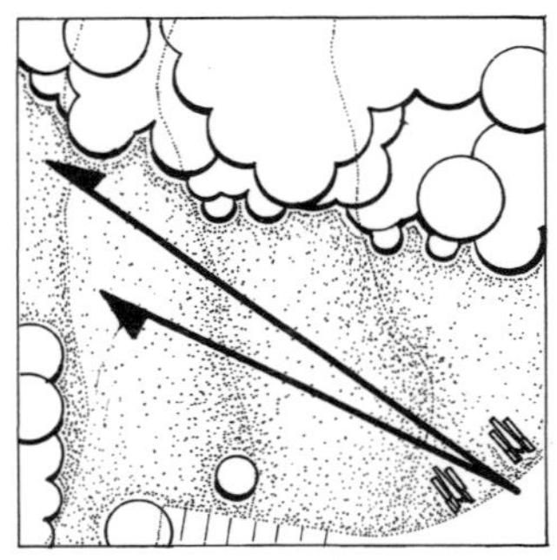

Focal Point

Sometimes monuments create focal points for the motorist's view. Backed by forest, as at the South Carolina Monument (right), these focal points often sit on small rises in the Park landscape, peaking motorist interest and contrasting with roadside interpretive tablets. Other focal points stand in the center of fields and dominate the landscape.

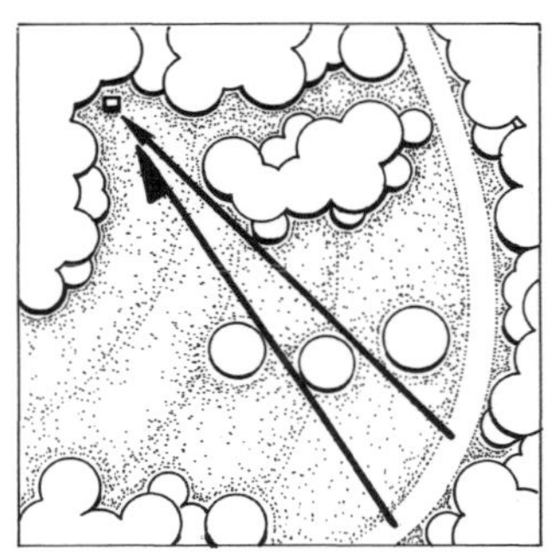

Framed View

Sometimes monuments or historic structures capture the motorist's attention to the exclusion of all other landscape features. The speed of motor travel, combined with their proximity to the road, frames these elements and pushes them into the foreground. The Brotherton Cabin, near the road and on a hill (right), is an example of a framed view.

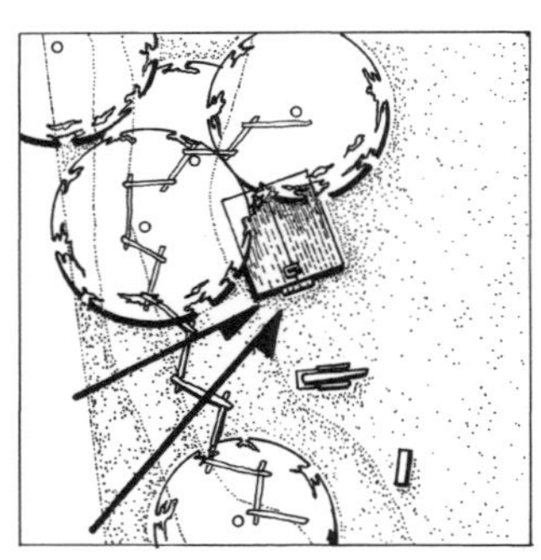

Filtered View

Cleared of underbrush, the forests at Chickamauga take on a new look, a look similar to that of the Civil War. These kinds of open woodlands along the roads form filtered views, which alternately reveal and conceal monuments, cannons, and signs. The sense of mystery created in these copses is enhanced along the Poe (right) and Battleline Roads by a density of monuments.

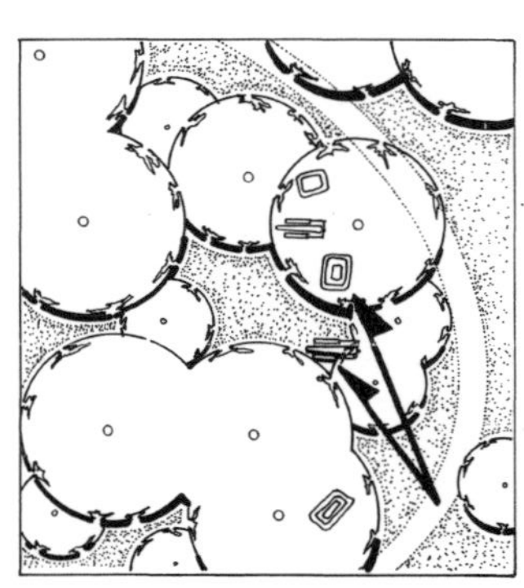

Closed View

Many of the Park roadways are closely bordered by dense forest. The winding roads make their way through these forests, leaving the visitor with a sense of narrow space–the closed view. These views heighten the expectation between open fields and historic sites. Along the Glenn-Viniard road (right), the sense that the forest has overwhelmed the land is strong.

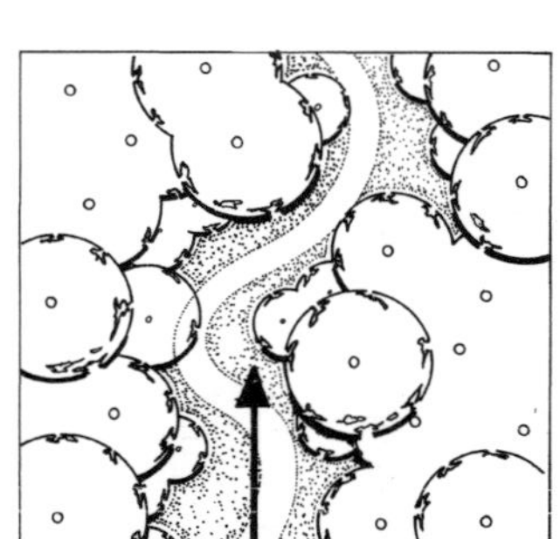

DELINEATED BY: Anna Śniegucka, 1998

NPS ROADS & BRIDGES RECORDING PROGRAM NATIONAL PARK SERVICE UNITED STATES DEPARTMENT OF THE INTERIOR	CHICKAMAUGA AND CHATTANOOGA NATIONAL MILITARY PARKS TOUR ROADS FORT OGLETHORPE, VICINITY OF CATOOSA/WALKER GEORGIA	SHEET 10 of 15	HISTORIC AMERICAN ENGINEERING RECORD GA-95

IF REPRODUCED, PLEASE CREDIT: HISTORIC AMERICAN ENGINEERING RECORD, NATIONAL PARK SERVICE, NAME OF DELINEATOR, DATE OF THE DRAWING

ROADWAY CONSTRUCTION

Steam-powered Roller, ca. 1900

Equipment

Although much of the 1890s roadwork in the Park was done by hand, early equipment for improving and constructing roads included horse-drawn gravel spreaders and rollers. A steam-powered roller (photo at left), purchased by the Park in the late 1890s, was used to compact the surfaces of macadamized and gravel roads (see "Paving" below). Also during this time, a derrick made of heavy timbers was used to place stones for culverts and bridges (photo at right). In addition, crews used the derrick to lift stone out of quarries in the Battlefield. By the 1920s, this equipment had been replaced by heavy machines.

Culvert Construction with Derrick, ca. 1900

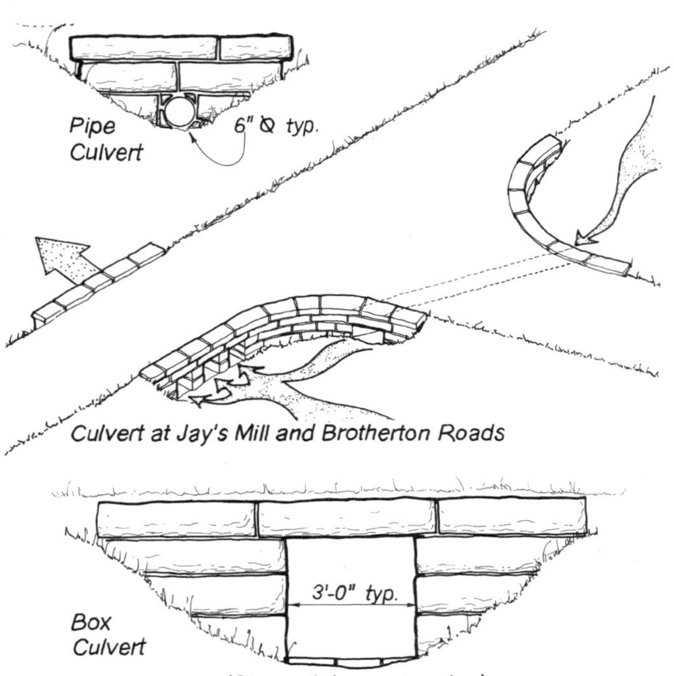

Culvert at Jay's Mill and Brotherton Roads

(Stone slab construction)

Paving & Drainage

During the 1890s, Park engineers upgraded existing dirt roads to make passage through the Battlefields more convenient. To ensure the roads would resist heavy use, many were simply graveled, while main thoroughfares, like LaFayette road, were widened and macadamized (right). Most roads were also crowned to divert runoff into hand-laid stone guttering. Water passed under roads through culverts: the most common had stone headwalls and clay piping, while larger box culverts diverted streams under the road (left). Engineers sometimes designed elaborate culverts at intersections. In low places, revetment walls raised the roadway above flood levels (see below). After 1934, the National Park Service widened and concreted main roads.

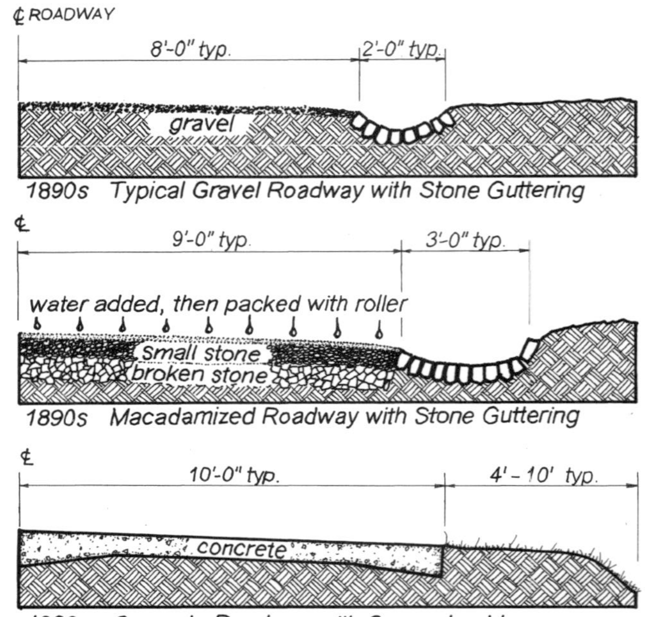

Reconstruction

1933-Current National Park Service Era
Main roads are concreted and widened (A). Tree wells (B) are constructed to preserve roadside trees. Wide grass shoulders (C) and drop culverts (D) replace stone guttering. Where necessary, existing culverts are widened (E). During the 1960s, roads are repaved with asphalt and widened (F).

1890s Park Commission Era
Roads are rebuilt on historic alignments: improvements include oiling, graveling or macadamizing surfaces (G); widening; and building gutters and culverts (see above). Revetment structures (H), made of corbeled limestone raise roads out of easily flooded areas.

1863 Civil War Era
During the Battle of Chickamauga, roads are simple dirt trails with little drainage and no paving (I)

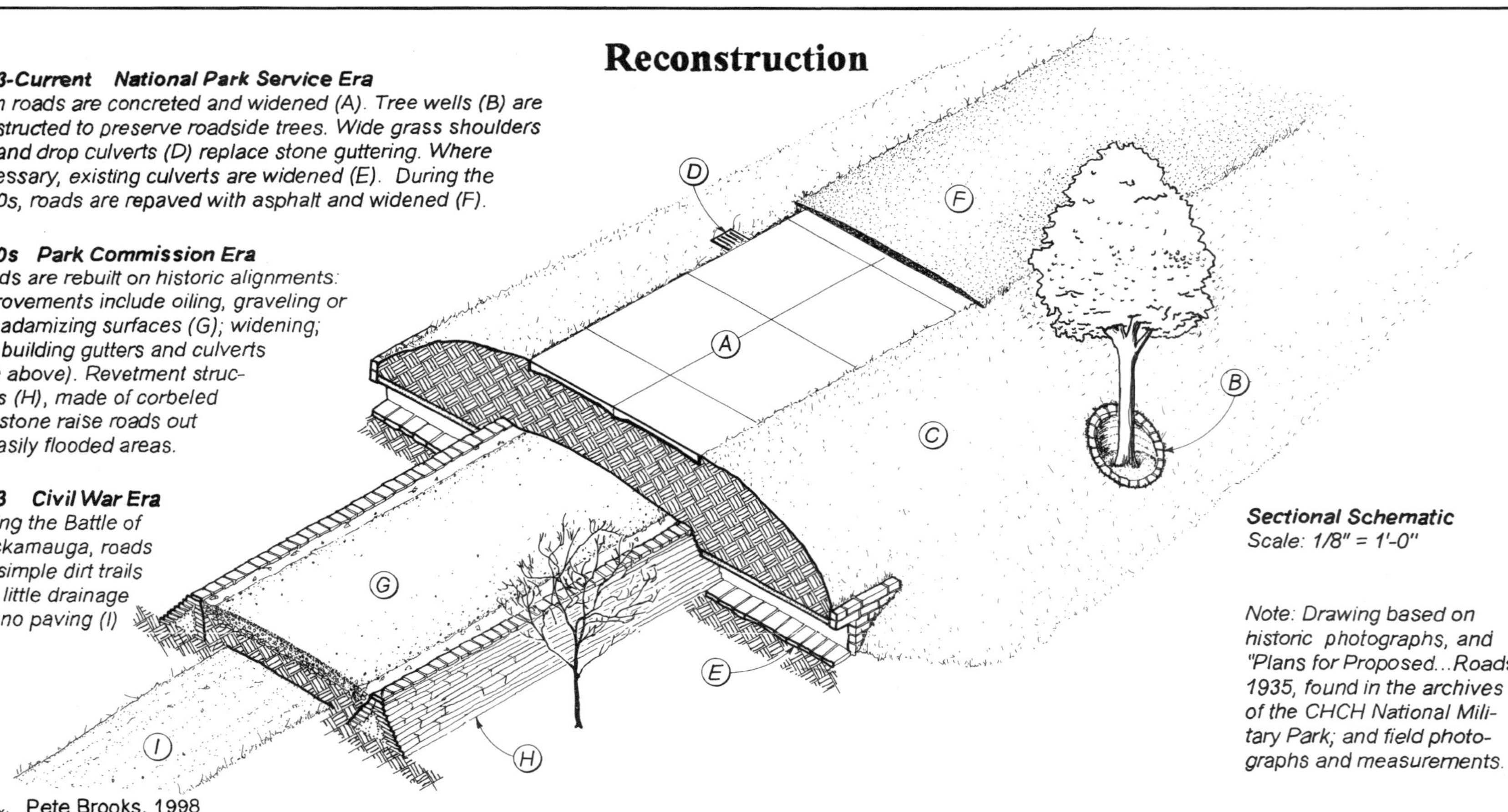

Sectional Schematic
Scale: 1/8" = 1'-0"

Note: Drawing based on historic photographs, and "Plans for Proposed...Roads" 1935, found in the archives of the CHCH National Military Park; and field photographs and measurements.

DELINEATED BY: Pete Brooks, 1998

NPS ROADS & BRIDGES RECORDING PROGRAM NATIONAL PARK SERVICE UNITED STATES DEPARTMENT OF THE INTERIOR	CHICKAMAUGA AND CHATTANOOGA NATIONAL MILITARY PARKS TOUR ROADS FORT OGLETHORPE, VICINITY OF CATOOSA/WALKER GEORGIA	SHEET 11 OF 15	HISTORIC AMERICAN ENGINEERING RECORD GA-95

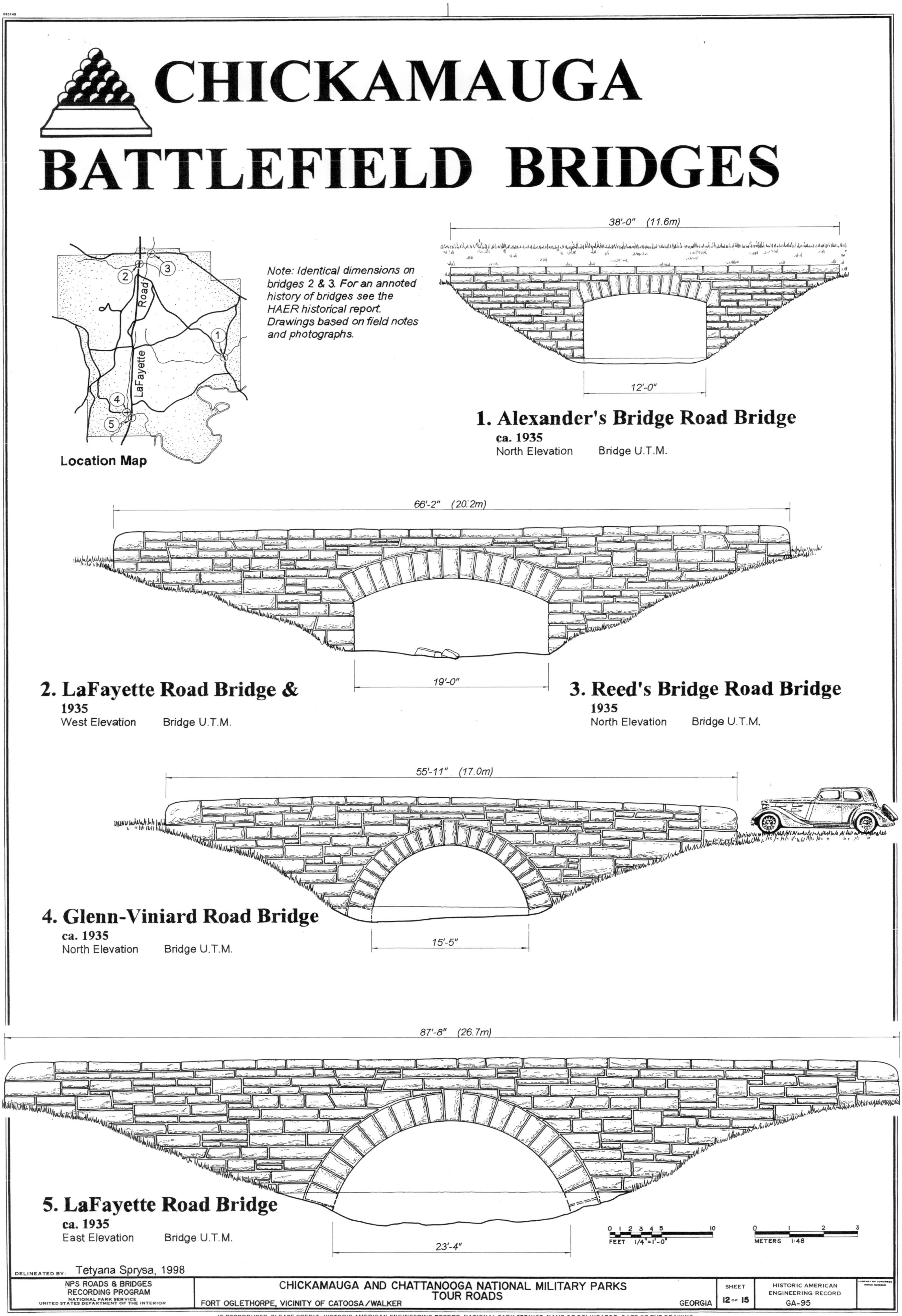

CHICKAMAUGA
BATTLEFIELD BRIDGES
Location Map
Road
LaFayette
Note: Identical dimensions on bridges 2 & 3. For an annoted history of bridges see the HAER historical report. Drawings based on field notes and photographs.
38'-0" (11.6m)
12'-0"
1. Alexander's Bridge Road Bridge
ca. 1935
North Elevation
Bridge U.T.M.
66'-2" (20.2m)
19'-0"
2. LaFayette Road Bridge &
1935
West Elevation
Bridge U.T.M.
3. Reed's Bridge Road Bridge
1935
North Elevation
Bridge U.T.M.
55'-11" (17.0m)
15'-5"
4. Glenn-Viniard Road Bridge
ca. 1935
North Elevation
Bridge U.T.M.
87'-8" (26.7m)
23'-4"
5. LaFayette Road Bridge
ca. 1935
East Elevation
Bridge U.T.M.
FEET 1/4"=1'-0"
METERS 1:48
DELINEATED BY: Tetyana Sprysa, 1998
NPS ROADS & BRIDGES RECORDING PROGRAM
NATIONAL PARK SERVICE
UNITED STATES DEPARTMENT OF THE INTERIOR
CHICKAMAUGA AND CHATTANOOGA NATIONAL MILITARY PARKS
TOUR ROADS
FORT OGLETHORPE, VICINITY OF CATOOSA/WALKER
GEORGIA
SHEET 12 OF 15
HISTORIC AMERICAN ENGINEERING RECORD
GA-95
IF REPRODUCED, PLEASE CREDIT: HISTORIC AMERICAN ENGINEERING RECORD, NATIONAL PARK SERVICE, NAME OF DELINEATOR, DATE OF THE DRAWING

GETTYSBURG

National Military Park

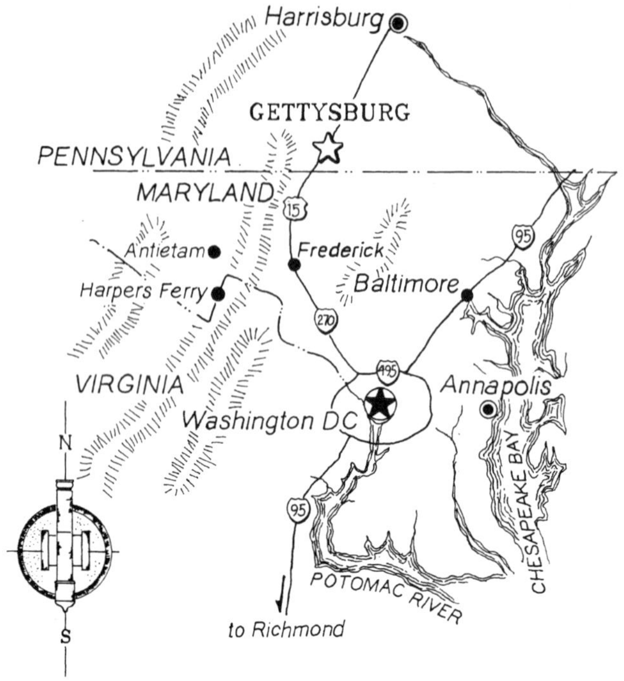

Roads & Bridges

The three days of the Battle of Gettysburg, July 1 - 3, 1863, are among the most studied moments in American history. Culminating in a dramatic infantry assault often called the "High Water Mark" of the Confederacy, the significance of this major Northern victory was immediately recognized. Efforts to preserve the battlefield began even as efforts continued to tend the wounded and bury the dead.

Just as the roads radiating from Gettysburg drew the armies together, roads have remained central to the commemoration of the battle. The battlefield has been under the successive control of three organizations: the Gettysburg Battlefield Memorial Association (1864-1895), the War Department (1895-1933), and the National Park Service (1933-present). All of these organizations created, modified, and maintained the interpretive road system which combines existing public roads, farm lanes and other historic traces in combination with avenues constructed to follow the opposing armies' lines of battle.

The famous address by President Abraham Lincoln at the Soldiers' National Cemetery offered a philosophy for the memorialization of the site. The Gettysburg Battlefield Memorial Association (GBMA), founded privately by prominent local citizens in the aftermath of the battle, followed this sentiment in attempting to preserve sections of the battlefield as a memorial to those, living and dead, who participated in the great struggle. Gettysburg became a model for Civil War battlefield parks and memorial sites. By 1880 Union veterans of the battle took an increasing role in preserving the battlefield, erecting monuments and opening avenues to make the area more accessible to visitors. While the GBMA purchased plots of land and built several Avenues, these efforts were necessarily modest; the avenues were generally unimproved and marked only Union battle lines.

This project is part of the Historic American Engineering Record (HAER), a long-range program to document historically significant engineering and industrial works in the United States. The HAER program is administered by the Historic American Buildings Survey/Historic American Engineering Record Division (HABS/HAER) of the National Park Service, U.S. Department of the Interior. Gettysburg National Military Park (GETT) Tour Roads Recording Project was cosponsored during the summer of 1998 by HAER (Eric DeLony, Chief) and GETT (John Latschar, Superintendent). The project was funded by the Federal Lands Highway Program (Allen Burden, Acting Administrator) through the National Park Service Park Roads and Parkways Program (Mark Hartsoe, Manager).

The field work, measured drawings, historical reports, and photographs were prepared under the directions of Program Manager Todd Croteau, Project Leader Christopher Marston and Program Historian Tim Davis. The recording team consisted of Edward Lupyak, field supervisor, landscape architects Nicole Steel, Nicki Yung and Christiane Weber (ICOMOS Intern, Germany). Historical reports were prepared by project historian Amanda Holmes. Formal large-format photography was done by David Haas.

On 31 May 1893, on the verge of the battle's 30th anniversary, the Secretary of War established a Gettysburg Park Commission to mark the lines of battle of both armies. The commission developed a cohesive plan for the battlefield, including more extensive monumentation, information tablets, observation towers, formal fencing, and a system of avenues both along the lines of battle and connecting the major sites. As a testament to their dedication to ensure this battlefield would inform future generations, the commission had the roads constructed using the finest method available, the Telford system, making them among the best American roads of the day. These avenues combined advanced road construction technology with great sensitivity to the landscape in which they were constructed. The battlefield park's appearance today is largely due to the efforts of this commission, conducted under the aegis of the War Department.

The National Park Service (NPS) assumed administration of the Gettysburg National Military Park in 1933, becoming steward not only of the battlefield site, but also of the commemorative landscape created by Gettysburg veterans. The NPS has had to maintain a balance between public safety and impact upon the historic resource. In its early years the NPS removed many sharp curves to accomodate automobile traffic and also created a more naturalistic park landscape. At the same time, they added different types of interpretive signage and defined tour routes as the battle moved beyond living memory. Today Gettysburg accommodates almost two million visitors per year, in a park almost ten times the size of that created by the GBMA. Nevertheless, the avenue system at Gettysburg still serves as an enduring testament to the efforts of the veterans of that conflict to commemorate their sacrifices and create meaning out of their experiences of war.

DELINEATED BY: Edward J. Lupyak and Christiane Weber, Nicki Yung, 1998

NATIONAL PARK SERVICE
ROADS & BRIDGES RECORDING PROJECT
NATIONAL PARK SERVICE
UNITED STATES DEPARTMENT OF THE INTERIOR

GETTYSBURG NATIONAL MILITARY PARK TOUR ROADS
GETTYSBURG VICINITY | ADAMS COUNTY | PENNSYLVANIA

SHEET 1 of 14

HISTORIC AMERICAN ENGINEERING RECORD
PA-485

EVOLUTION OF THE AVENUES

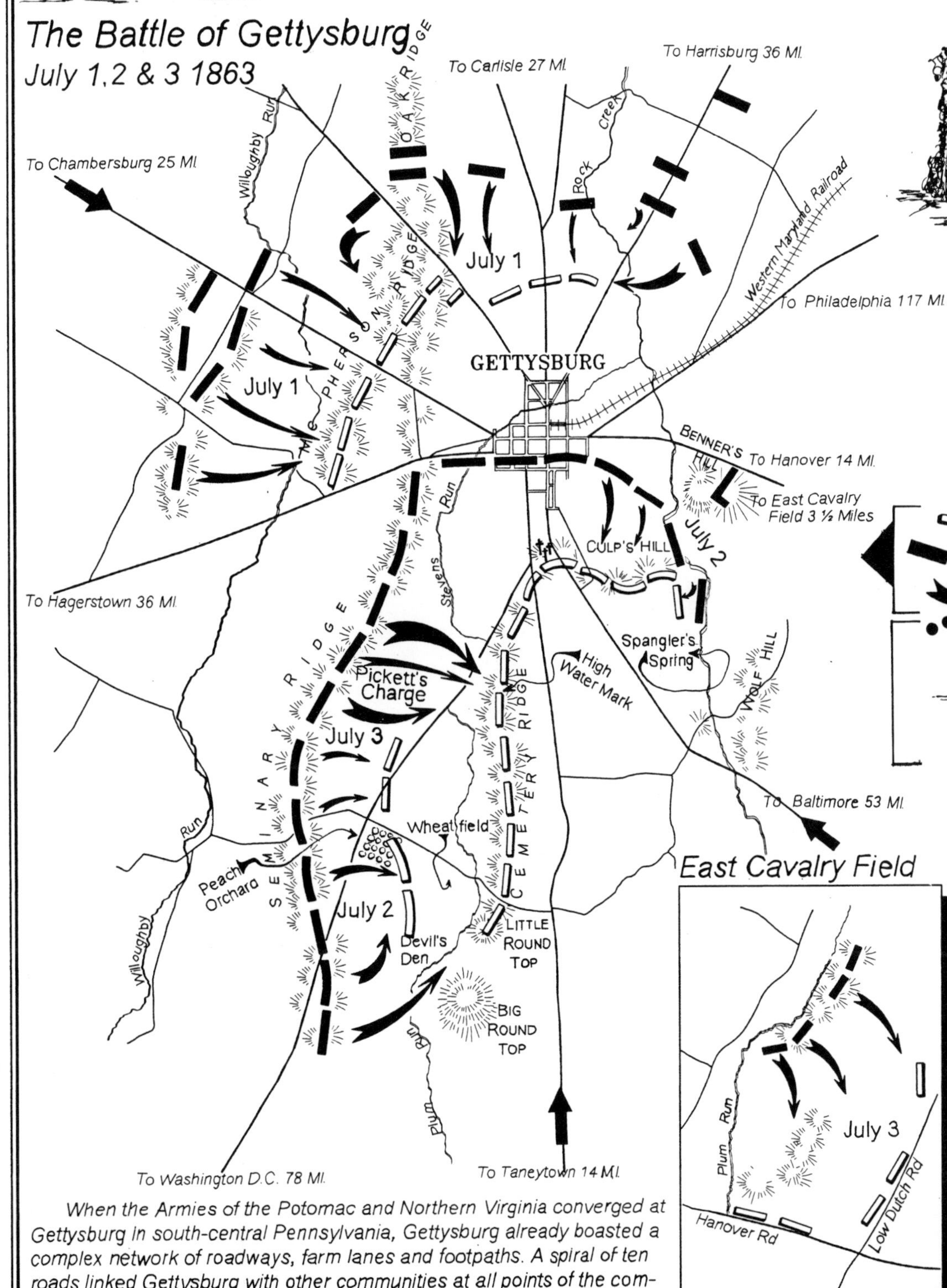

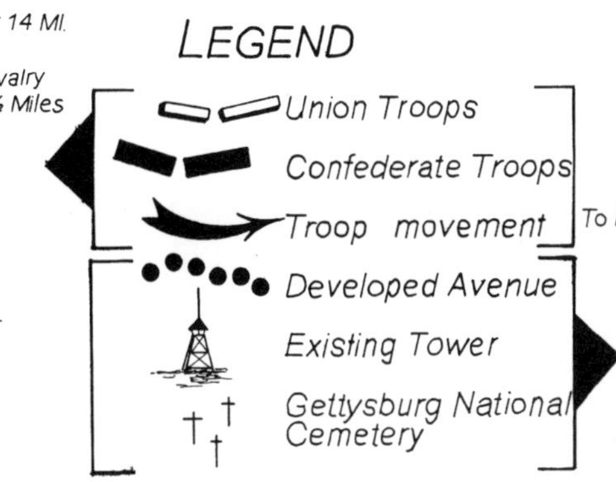

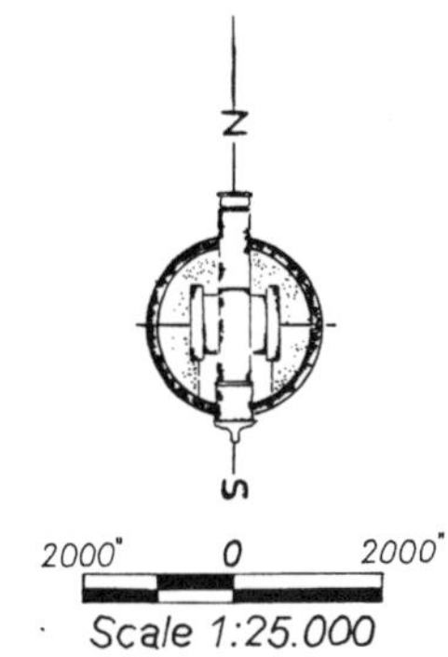

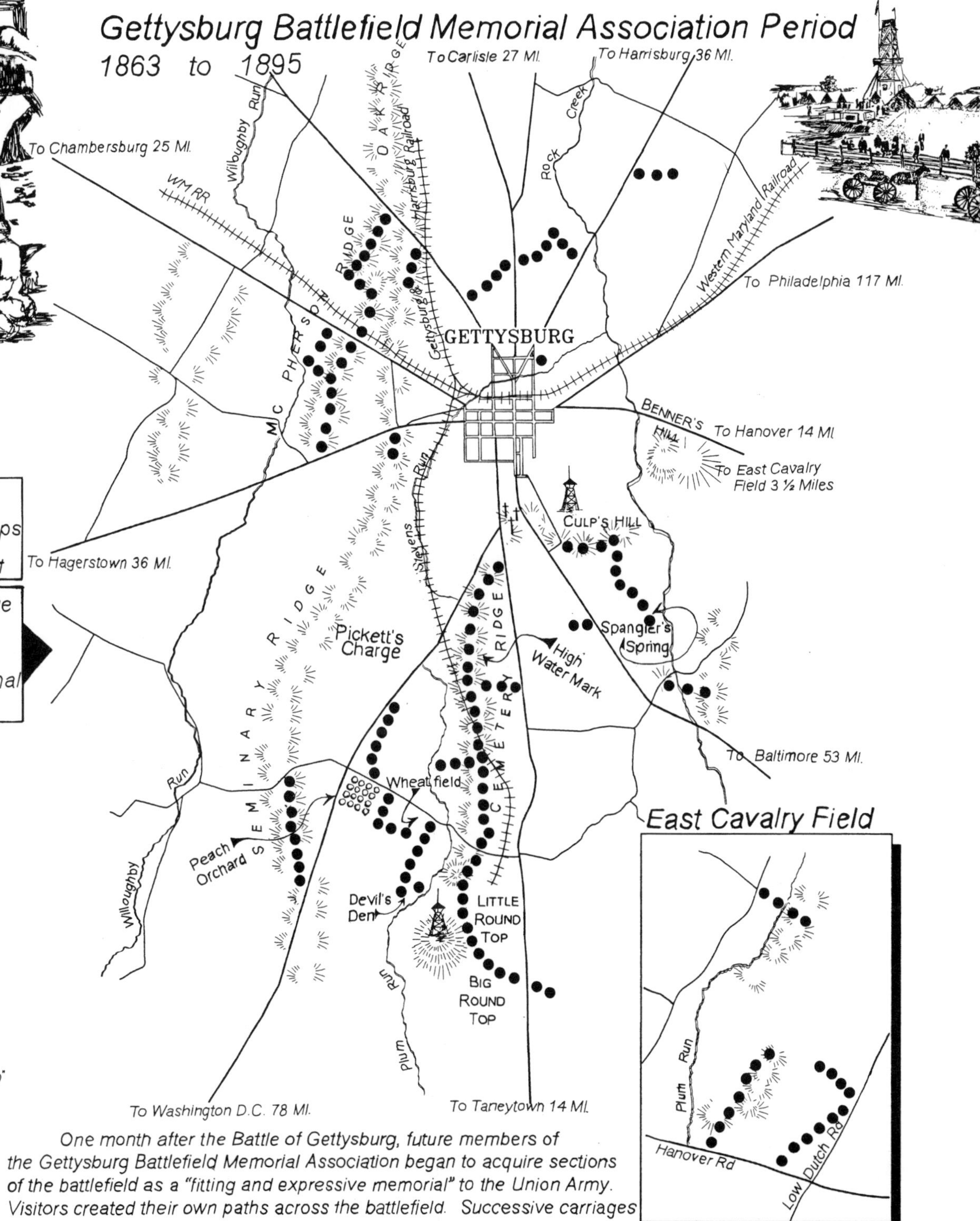

When the Armies of the Potomac and Northern Virginia converged at Gettysburg in south-central Pennsylvania, Gettysburg already boasted a complex network of roadways, farm lanes and footpaths. A spiral of ten roads linked Gettysburg with other communities at all points of the compass. The opposing forces took advantage of the access routes, but it was the general topography of the region which ultimately influenced the particulars of where the battle took place. General Robert E. Lee brought his army from the west and northwest along the Chambersburg Pike, and General George G. Meade from the south and southeast along the Taneytown Road and Baltimore Pike.

Note: Maps based on various sources located in Gettysburg National Military Park (Original Basemaps from 1863 and 1895)

One month after the Battle of Gettysburg, future members of the Gettysburg Battlefield Memorial Association began to acquire sections of the battlefield as a "fitting and expressive memorial" to the Union Army. Visitors created their own paths across the battlefield. Successive carriages carved routes up Culp's Hill, through the Wheatfield and along Cemetery Ridge to the Round Tops. The GBMA, while fully intending to expand their holdings and make all sites accessible, floundered in its early years. In the 1880's, however, Civil War veterans reorganized the GBMA and the battlefield underwent extensive monumentation and renewed avenue construction, marking and naming the Union lines of battle.

DRAWN BY: Christiane Weber, 1998

NATIONAL PARK SERVICE
ROADS & BRIDGES RECORDING PROJECT

GETTYSBURG NATIONAL MILITARY PARK TOUR ROADS
GETTYSBURG VICINITY ADAMS COUNTY PENNSYLVANIA

HISTORIC AMERICAN ENGINEERING RECORD
PA-485
SHEET 3 of 14

IF REPRODUCED, PLEASE CREDIT THE HISTORIC AMERICAN ENGINEERING RECORD, NATIONAL PARK SERVICE, NAME OF DELINEATOR, DATE OF DRAWING

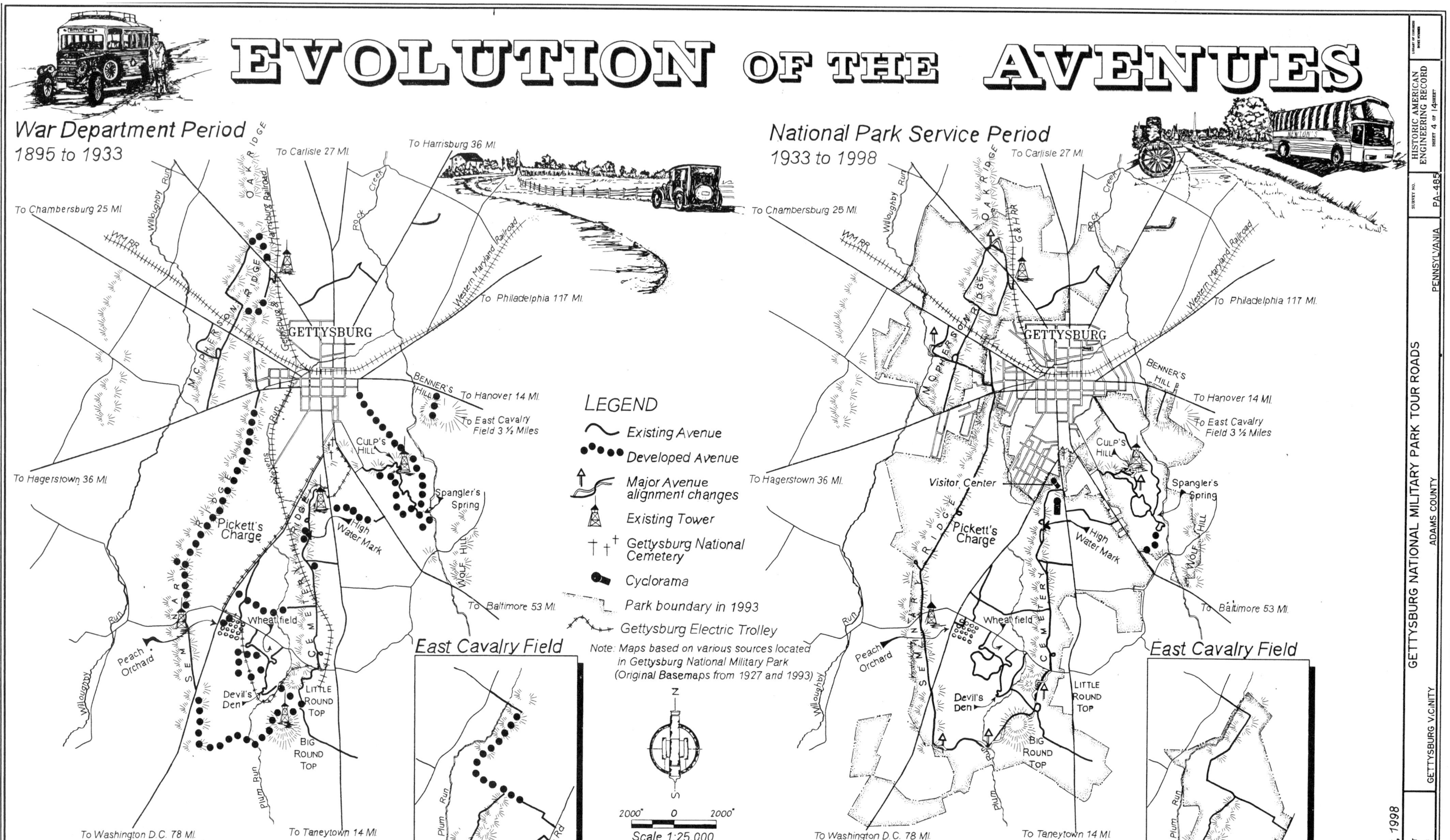

Threats imposed on the battlefield by commercial interests, such as the Gettysburg Electric Trolley, spurred the preservation-minded into action. Since the war, marking and preserving the battlefield had been the concern of local individuals and veterans' groups from states which had participated in the battle. An Act of 11 February 1895 made the Gettysburg Battlefield the concern of the Federal Government under the War Department. The Gettysburg Battlefield Commission oversaw the improvement and construction of avenues, as well as the improvement of public roads. Of paramount concern to the Commission's tenure was the marking of the lines of battle for both armies, and placing informational and non-judgmental tablets throughout the fields.

When the National Park Service assumed control of Gettysburg National Military Park in 1933, most of the Telford and Macadam avenues had been in place for over thirty years. The higher speeds and increasing use of automobiles wore out the avenues faster than they could be repaired. The NPS saw the War Department avenues as old-fashioned and dangerous, and developed a systematic plan to resurface and widen the avenues, rebuild bridges, extend the drainage system, and to realign the sharp curves. Additionally, defined tour routes were established and posted with directional signs. Visitors typically toured independently, however, official licensed tour guide services were developed along with molor tour buses.

AVENUE CONSTRUCTION

The majority of avenues in Gettysburg National Military Park originally were constructed using the Telford system, known for its solidity and durability. After surveying the land and establishing the battle lines to determine the avenues' alignment, laborers cleared and rough-graded the earthen road bed. The laborers set large shoulder stones along the edges, preparing a solid border for the 8" wedge-like foundation stones which were then hand-placed, chinked and worked for a tight fit. Next, 4" of 1 ½" stones were spread over the 8" stones, followed by a layer of clay to act as a binder, and topped with 2" of quarter-inch stone screenings. Finally, the surface was rolled with a 14 ton steam roller, both as a part of the construction and as a regular maintenance procedure.

The Telfordized system was a labor-intensive and expensive method of road building, but produced a superior road which the Gettysburg Battlefield Commission planned would last for generations as an integral part of their monumental legacy at Gettysburg. The Telford base still remains under many of the avenues today.

Telford Construction Process ca. 1895

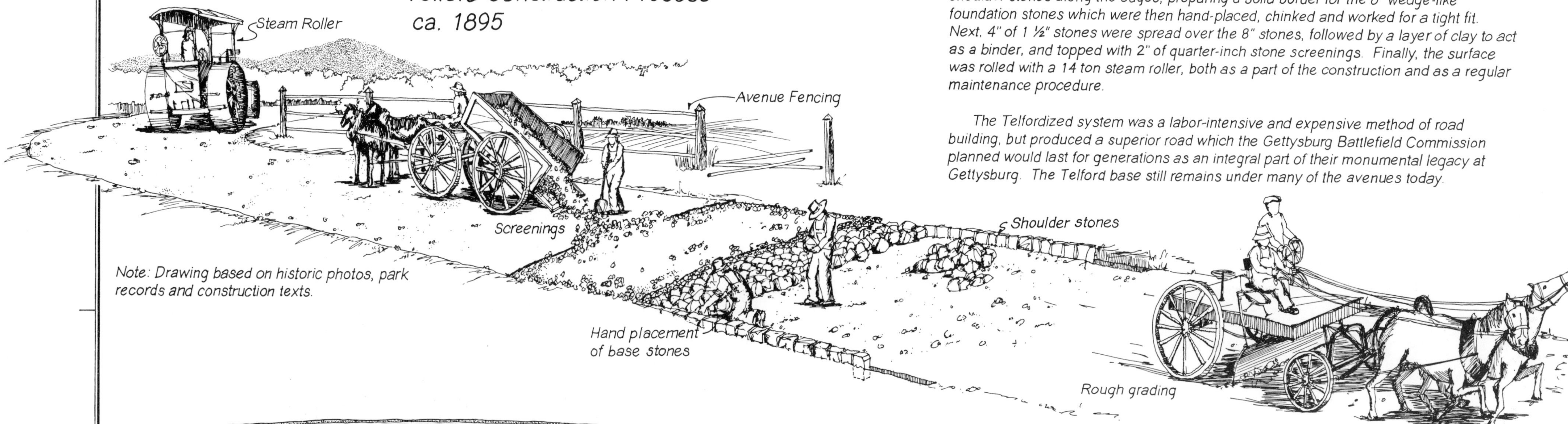

Note: Drawing based on historic photos, park records and construction texts.

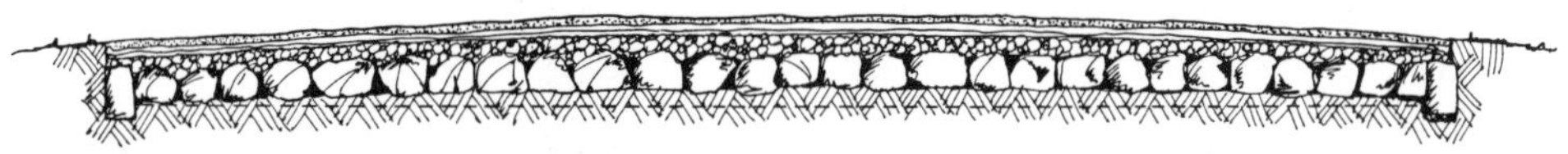

Telford Base Road

- Very durable
- Labor-intensive
- High initial cost
- Excellent in all soil conditions

The Telford method of road construction was invented in 1824 by Thomas Telford. It was first used at Gettysburg National Military Park in 1895 and approximately 22 miles of Telford Road was constructed between 1895 and 1922.

Macadam Road

- Moderately durable
- Quick construction
- Low initial cost
- Needs good drainage or will rut and erode quickly

The first Macadam road was built in 1817 in London, England by John Loudon Macadam. Of the 35 miles of avenues built by the War Department between 1895 and 1922 approximately 13 miles were Macadam.

Telford Materials

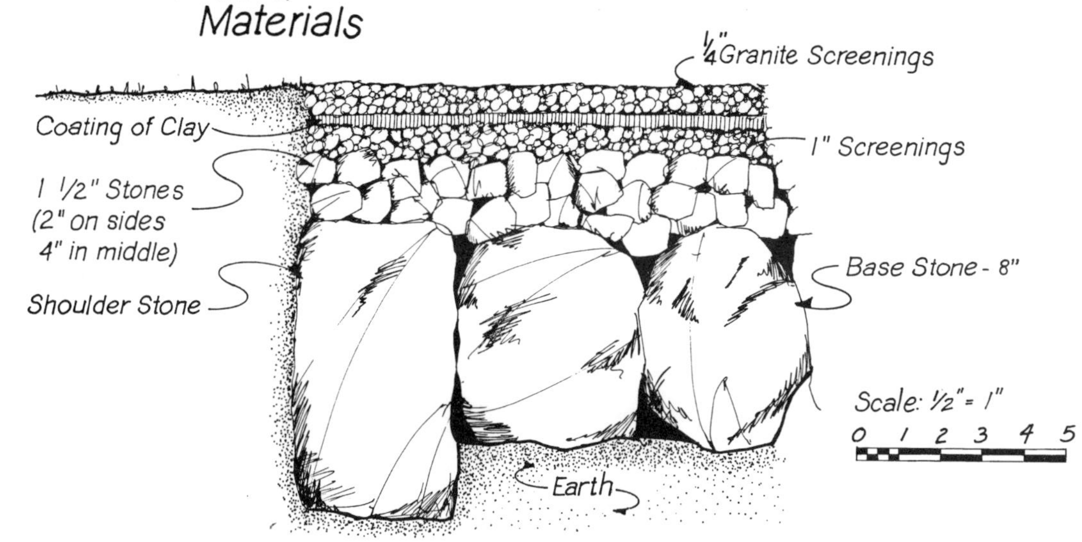

Note: final surface to be flush with shoulder.

DELINEATED BY: Edward J. Lupyak, 1998

NATIONAL PARK SERVICE
ROADS & BRIDGES RECORDING PROJECT
NATIONAL PARK SERVICE
UNITED STATES DEPARTMENT OF THE INTERIOR

GETTYSBURG NATIONAL MILITARY PARK TOUR ROADS
GETTYSBURG VICINITY ADAMS COUNTY PENNSYLVANIA

SHEET 5 of 14

HISTORIC AMERICAN ENGINEERING RECORD PA-485

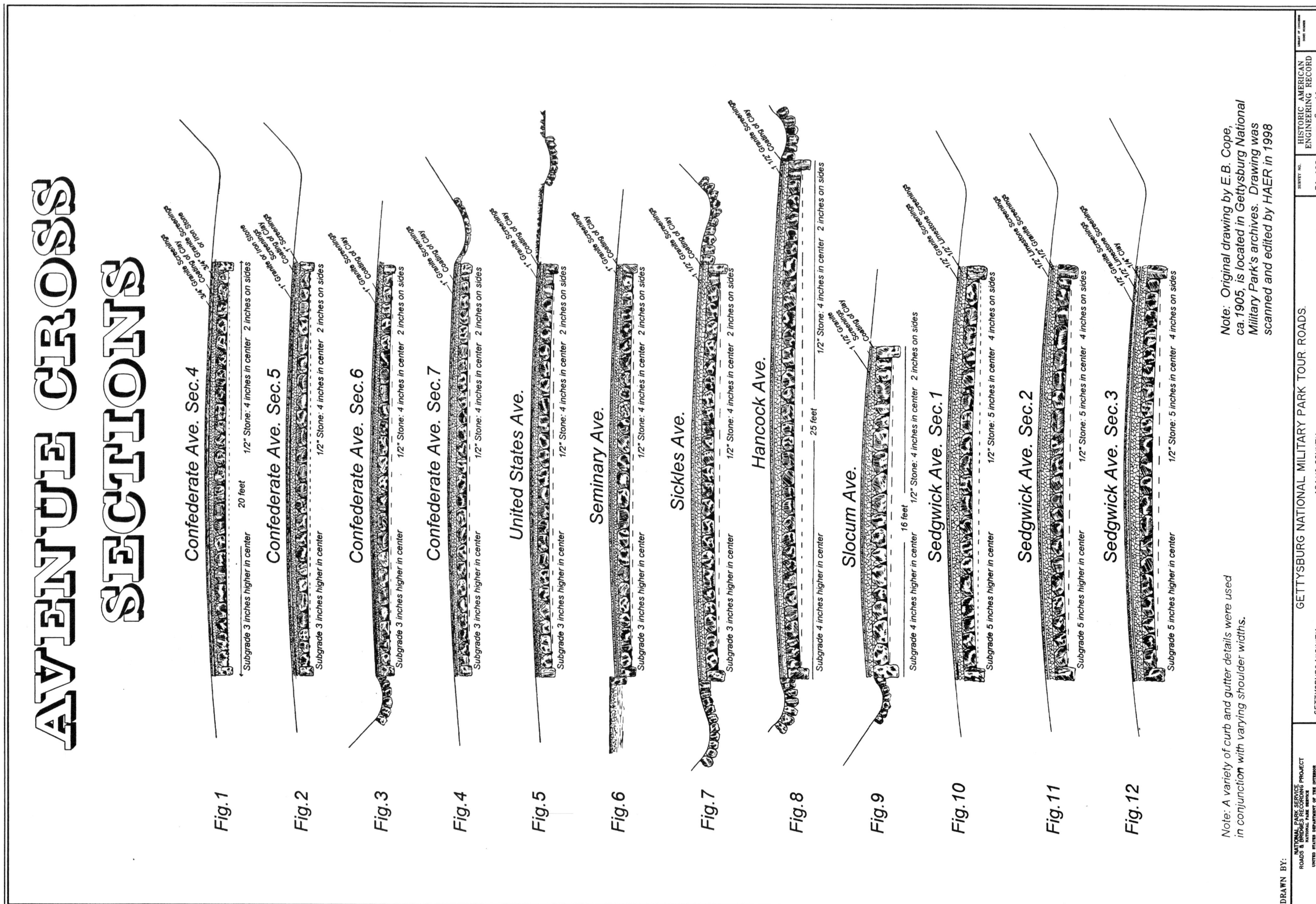
AVENUE CROSS SECTIONS
Fig.1 Confederate Ave. Sec.4
Fig.2 Confederate Ave. Sec.5
Fig.3 Confederate Ave. Sec.6
Fig.4 Confederate Ave. Sec.7
Fig.5 United States Ave.
Fig.6 Seminary Ave.
Fig.7 Sickles Ave.
Fig.8 Hancock Ave.
Fig.9 Slocum Ave.
Fig.10 Sedgwick Ave. Sec.1
Fig.11 Sedgwick Ave. Sec.2
Fig.12 Sedgwick Ave. Sec.3
Subgrade 3 inches higher in center
20 feet
1/2" Stone: 4 inches in center 2 inches on sides
3/4" Granite Screenings
Coating of Clay
Coating of Iron Stone
1" Granite Screenings
Subgrade 4 inches higher in center
25 feet
16 feet
Subgrade 5 inches higher in center
1/2" Stone: 5 inches in center 4 inches on sides
1/2" Granite Screenings
1/2" Limestone Screenings
1/4" Clay
Note: A variety of curb and gutter details were used in conjunction with varying shoulder widths.
Note: Original drawing by E.B. Cope, ca.1905, is located in Gettysburg National Military Park's archives. Drawing was scanned and edited by HAER in 1998
DRAWN BY:
NATIONAL PARK SERVICE
ROADS & BRIDGES RECORDING PROJECT
GETTYSBURG NATIONAL MILITARY PARK TOUR ROADS
GETTYSBURG VICINITY
ADAMS COUNTY
PENNSYLVANIA
PA-485
HISTORIC AMERICAN ENGINEERING RECORD
SHEET 6 OF 14

DRAINAGE

Park maintenance crew paving gutters on Webb Avenue ca. 1900 (Tipton Photo)

On 4 July 1863, after the three days of intense fighting, heavy rains began to fall washing out the already rutted local roads. Rivers and streams throughout the region flooded, impeding the Confederate Army's retreat and drowning some wounded where they lay awaiting aid.

Seldom does it rain as hard as it did that day, but successive stages of Gettysburg National Military Park's development have contended with drainage issues to protect the agricultural lands and the avenues from flooding and erosion. The Gettysburg Battlefield Commission instituted a comprehensive and carefully designed drainage system to increase the durability of their Telford roads. Contributing aesthetically to the formal commemorative landscape, nearly 13 miles of stone-paved gutters were placed along the avenues while stone catch basins and culverts diverted water under and away from the road beds.

Paved Gutters

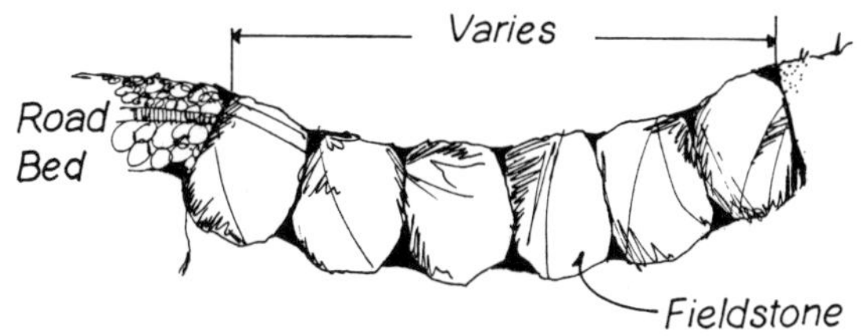

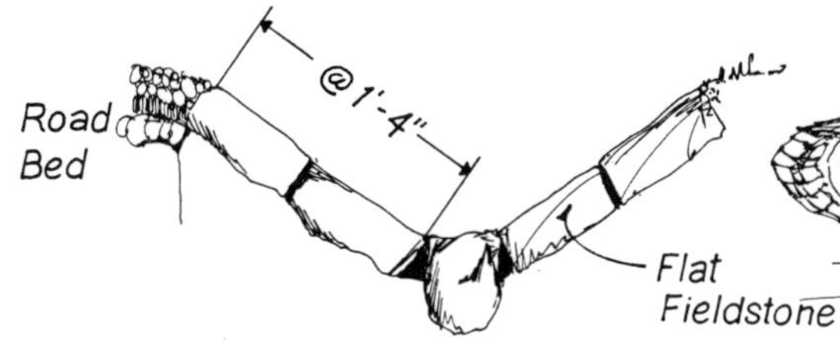

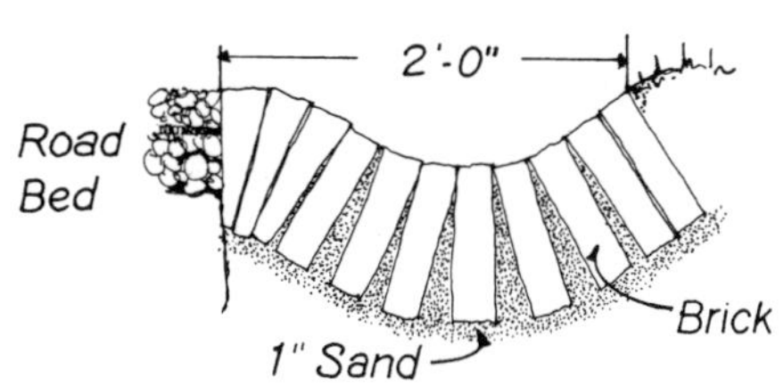

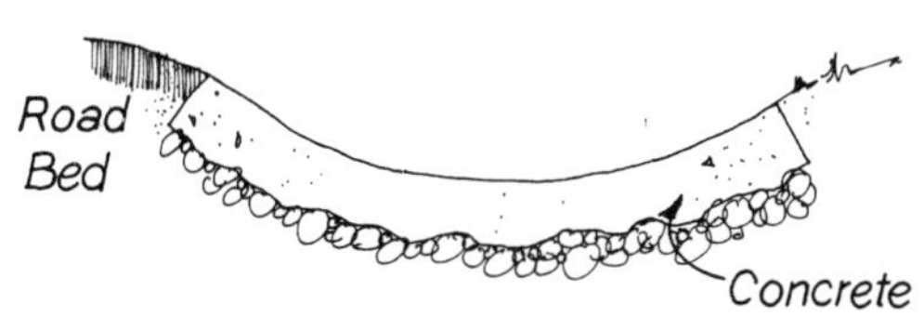

Fieldstone and brick gutters lined many of the avenues during the War Department Era of the Park. Due to high maintenance and cost all of the gutters have been removed or covered with asphalt. Others have been replaced with concrete gutters by the National Park Service.

Traces of the original system remain visible in the park. Beginning in 1933, the National Park Service extended the system and introduced new design and safety standards, contributing to Gettysburg's variety of drainage details.

Culverts

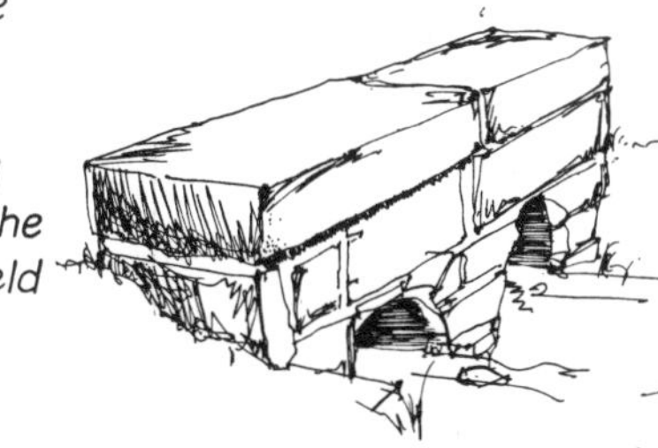

War Department Double Culvert with 12" pipes

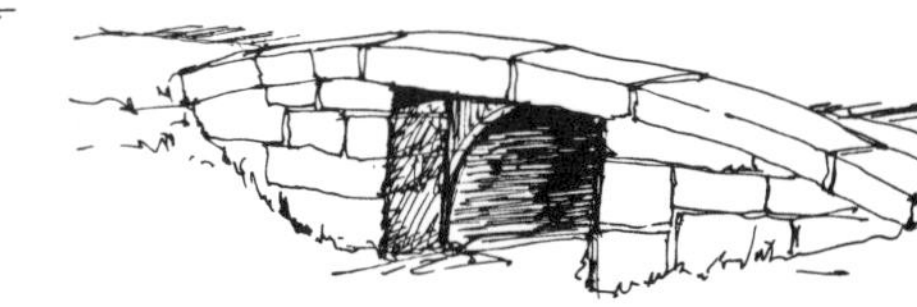

CCC Stone Culvert with 24" pipe and rounded top

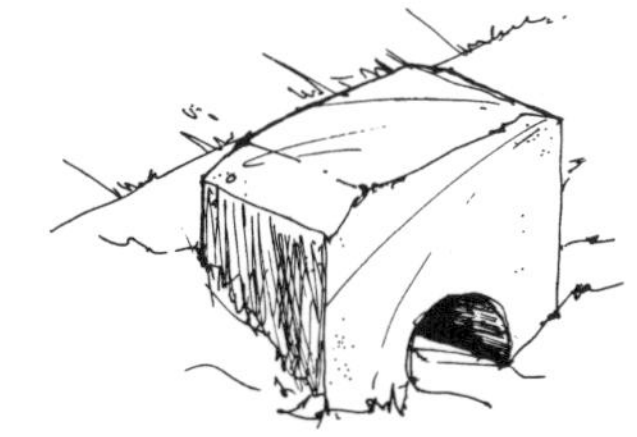

War Department Concrete head wall with 8" pipe

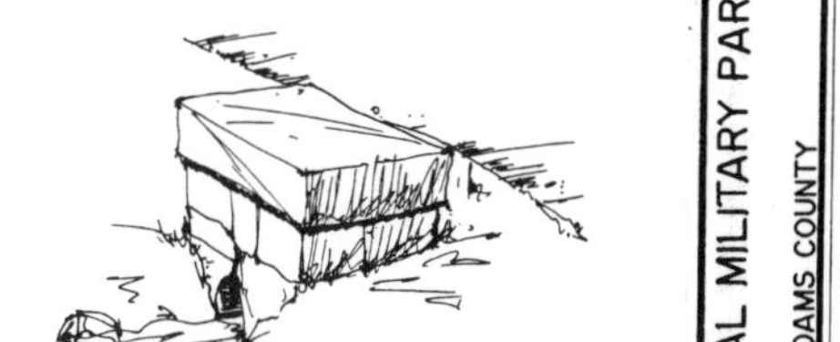

War Department Stone head wall with 8" pipe

War Department Stone Catch Basin

National Park Service Stone head wall with original brick arch from War Department era

DELINEATED BY: Edward J. Lupyak, 1998

NATIONAL PARK SERVICE
ROADS & BRIDGES RECORDING PROJECT
NATIONAL PARK SERVICE
UNITED STATES DEPARTMENT OF THE INTERIOR

GETTYSBURG NATIONAL MILITARY PARK TOUR ROADS
GETTYSBURG VICINITY ADAMS COUNTY PENNSYLVANIA

SHEET 7 of 14

HISTORIC AMERICAN ENGINEERING RECORD PA-485

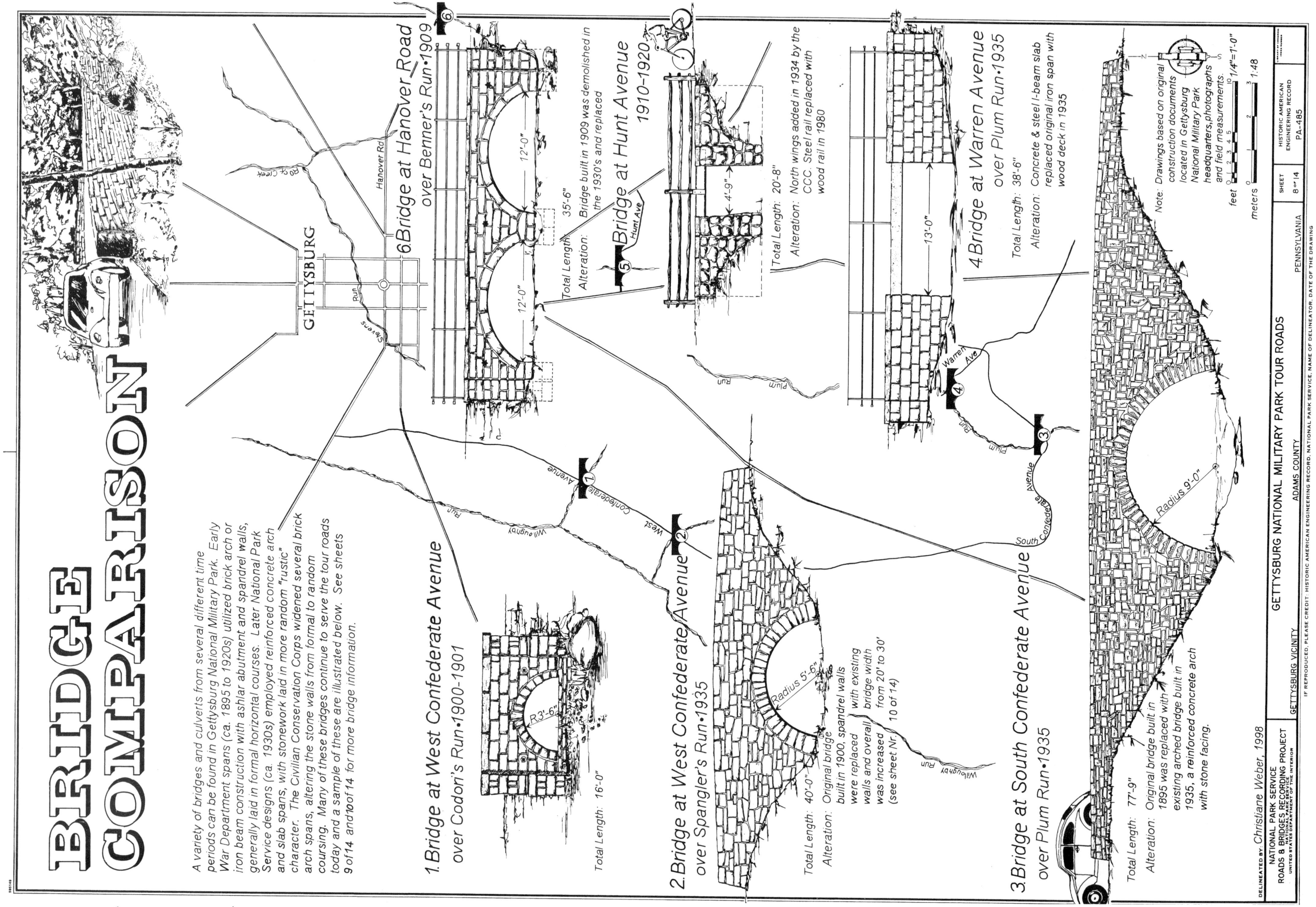
BRIDGE COMPARISON
A variety of bridges and culverts from several different time periods can be found in Gettysburg National Military Park. Early War Department spans (ca. 1895 to 1920s) utilized brick arch or iron beam construction with ashlar abutment and spandrel walls, generally laid in formal horizontal courses. Later National Park Service designs (ca. 1930s) employed reinforced concrete arch and slab spans, with stonework laid in more random "rustic" character. The Civilian Conservation Corps widened several brick arch spans, altering the stone walls from formal to random coursing. Many of these bridges continue to serve the tour roads today and a sample of these are illustrated below. See sheets 9 of 14 and 10 of 14 for more bridge information.
GETTYSBURG
Hanover Rd
Rock Creek
Stevens Run
Willoughby Run
West Confederate Avenue
South Confederate Avenue
Warren Ave
Hunt Ave
Plum Run
1. Bridge at West Confederate Avenue over Codori's Run•1900-1901
R3'-6"
Total Length: 16'-0"
2. Bridge at West Confederate Avenue over Spangler's Run•1935
Radius 5'-6"
Total Length: 40'-0"
Alteration: Original bridge built in 1900, spandrel walls were replaced with existing walls and overall bridge width was increased from 20' to 30' (see sheet Nr. 10 of 14)
3. Bridge at South Confederate Avenue over Plum Run•1935
Radius 9'-0"
Total Length: 77'-9"
Alteration: Original bridge built in 1895 was replaced with existing arched bridge built in 1935, a reinforced concrete arch with stone facing.
4. Bridge at Warren Avenue over Plum Run•1935
13'-0"
Total Length: 38'-6"
Alteration: Concrete & steel I-beam slab replaced original iron span with wood deck in 1935
5. Bridge at Hunt Avenue 1910-1920
4'-9"
Total Length: 20'-8"
Alteration: North wings added in 1934 by the CCC. Steel rail replaced with wood rail in 1980
6. Bridge at Hanover Road over Benner's Run•1909
12'-0"
12'-0"
Total Length: 35'-6"
Alteration: Bridge built in 1909 was demolished in the 1930's and replaced
Note: Drawings based on original construction documents located in Gettysburg National Military Park headquarters, photographs and field measurements.
feet
meters
1/4"=1'-0"
1:48
DELINEATED BY: Christiane Weber, 1998
NATIONAL PARK SERVICE
ROADS & BRIDGES RECORDING PROJECT
UNITED STATES DEPARTMENT OF THE INTERIOR
GETTYSBURG NATIONAL MILITARY PARK TOUR ROADS
GETTYSBURG VICINITY
ADAMS COUNTY
PENNSYLVANIA
SHEET 8 of 14
HISTORIC AMERICAN ENGINEERING RECORD
PA-485
IF REPRODUCED, PLEASE CREDIT: HISTORIC AMERICAN ENGINEERING RECORD, NATIONAL PARK SERVICE, NAME OF DELINEATOR, DATE OF THE DRAWING

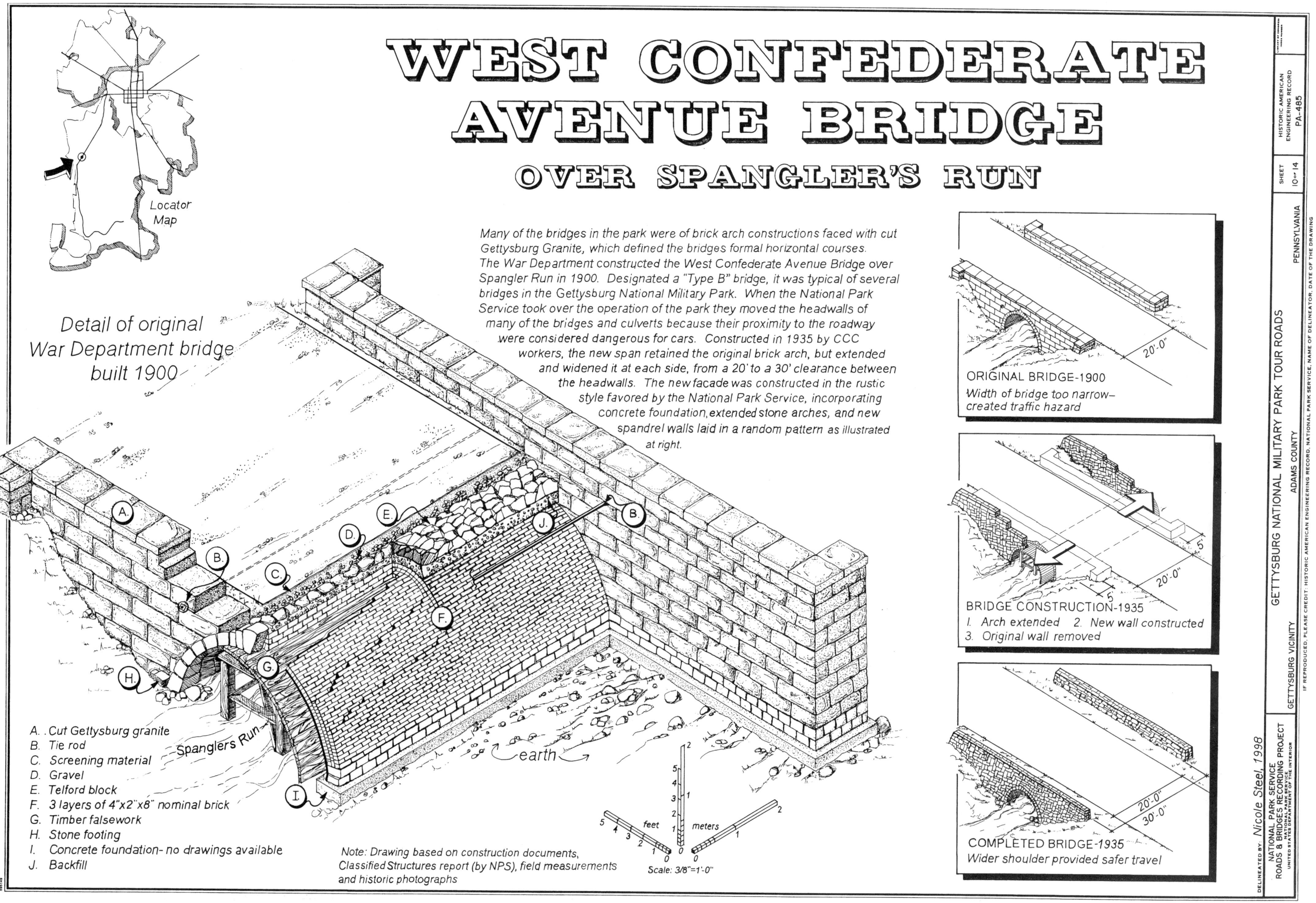

WEST CONFEDERATE AVENUE BRIDGE
OVER SPANGLER'S RUN
Locator Map
Many of the bridges in the park were of brick arch constructions faced with cut Gettysburg Granite, which defined the bridges formal horizontal courses. The War Department constructed the West Confederate Avenue Bridge over Spangler Run in 1900. Designated a "Type B" bridge, it was typical of several bridges in the Gettysburg National Military Park. When the National Park Service took over the operation of the park they moved the headwalls of many of the bridges and culverts because their proximity to the roadway were considered dangerous for cars. Constructed in 1935 by CCC workers, the new span retained the original brick arch, but extended and widened it at each side, from a 20' to a 30' clearance between the headwalls. The new facade was constructed in the rustic style favored by the National Park Service, incorporating concrete foundation, extended stone arches, and new spandrel walls laid in a random pattern as illustrated at right.
Detail of original War Department bridge built 1900
A. Cut Gettysburg granite
B. Tie rod
C. Screening material
D. Gravel
E. Telford block
F. 3 layers of 4"x2"x8" nominal brick
G. Timber falsework
H. Stone footing
I. Concrete foundation- no drawings available
J. Backfill
Spanglers Run
earth
Note: Drawing based on construction documents, Classified Structures report (by NPS), field measurements and historic photographs
feet
meters
Scale: 3/8"=1'-0"
20'-0"
ORIGINAL BRIDGE-1900
Width of bridge too narrow–created traffic hazard
5'
20'-0"
BRIDGE CONSTRUCTION-1935
1. Arch extended 2. New wall constructed 3. Original wall removed
20'-0"
30'-0"
COMPLETED BRIDGE-1935
Wider shoulder provided safer travel
DELINEATED BY: Nicole Steel, 1998
NATIONAL PARK SERVICE
ROADS & BRIDGES RECORDING PROJECT
GETTYSBURG VICINITY
GETTYSBURG NATIONAL MILITARY PARK TOUR ROADS
ADAMS COUNTY
PENNSYLVANIA
SHEET 10 of 14
HISTORIC AMERICAN ENGINEERING RECORD PA-485
IF REPRODUCED, PLEASE CREDIT: HISTORIC AMERICAN ENGINEERING RECORD, NATIONAL PARK SERVICE, NAME OF DELINEATOR, DATE OF THE DRAWING

Note: Original drawing of tower located in Gettysburg National Military Park Archives.

OBSERVATION TOWERS

No sooner had Congress and the Senate agreed to establish the Gettysburg National Military Park in 1895 when the Gettysburg Battlefield Commission approved the plans for the erection of five towers at key points in the battlefield. The five steel observation towers were part of the original design plan for the park and its avenue system. Colonel Emmor B. Cope's design for the 75' and 60' steel towers were constructed by the Variety Iron Works Company, of Cleveland, Ohio. The addition of the towers took the plan of the park beyond simply fixing the battle lines with markers and avenues to acknowledging the greater scope and significance of the battle for veterans and for the American public. Their placement in Ziegler's Grove, on Big Round Top, Warfield Ridge, Oak Ridge, and Culp's Hill provided park-wide complementary views of the battlefield for use by tourists and those visiting the park for tactical military study. In the 1960's the National Park Service removed the towers from Big Round Top and Ziegler's Grove, and truncated the Oak Ridge tower. Although encroaching vegetation now obstructs some of the viewshed, the remaining towers still offer an unsurpassed visual link between significant points on the battlefield and stand as monuments to the work of the Commission.

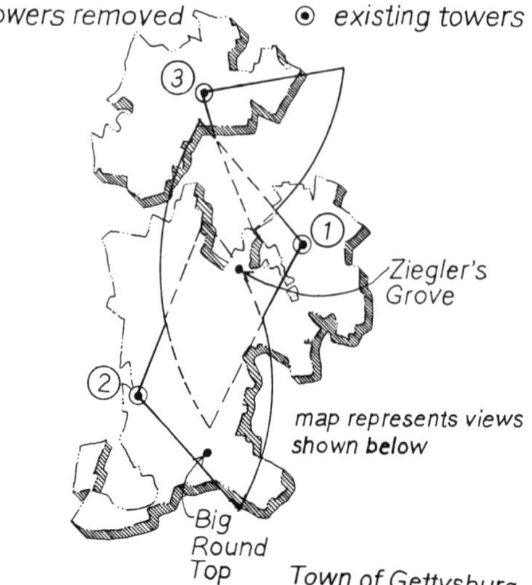

① Culp's Hill

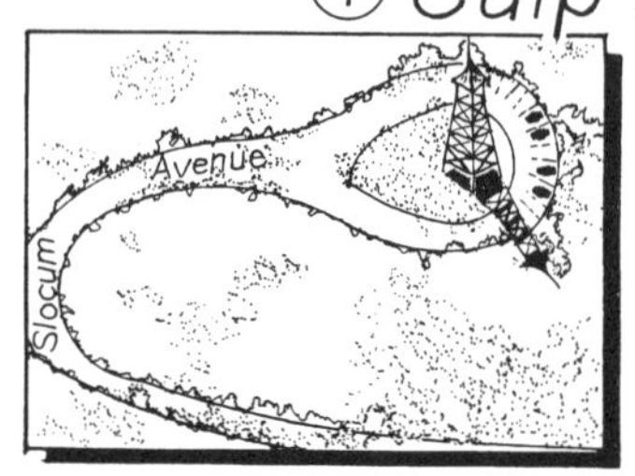

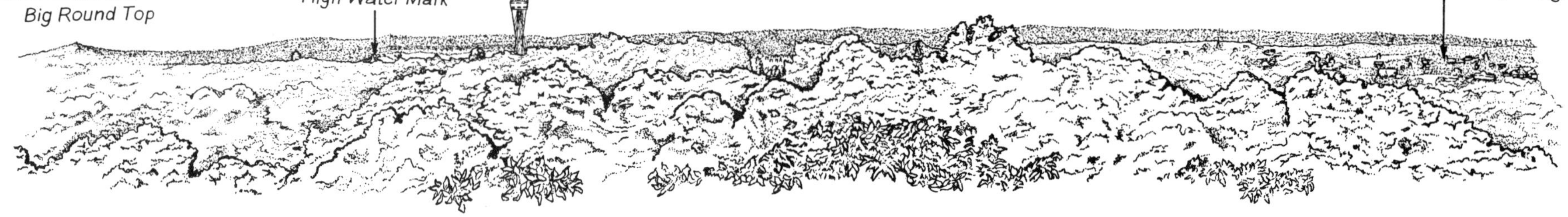

② Warfield Ridge

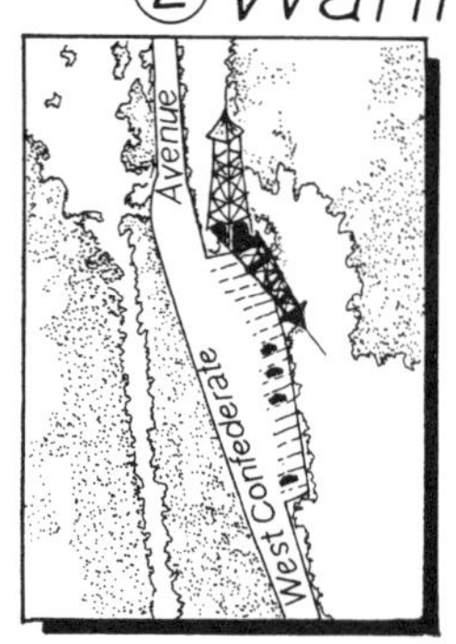

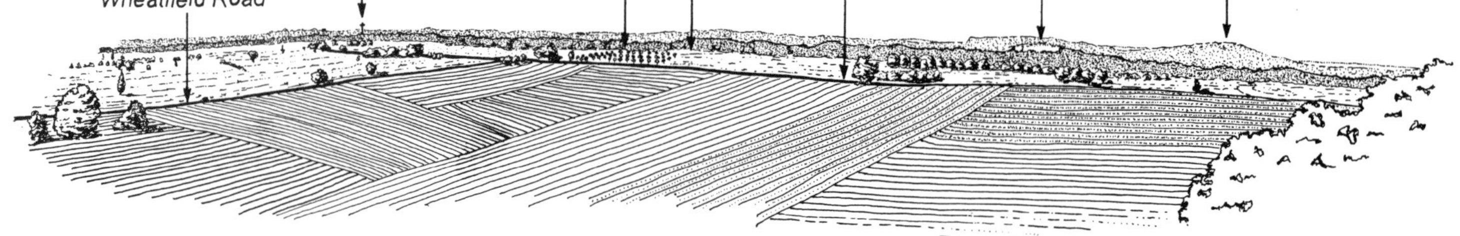

③ Oak Ridge

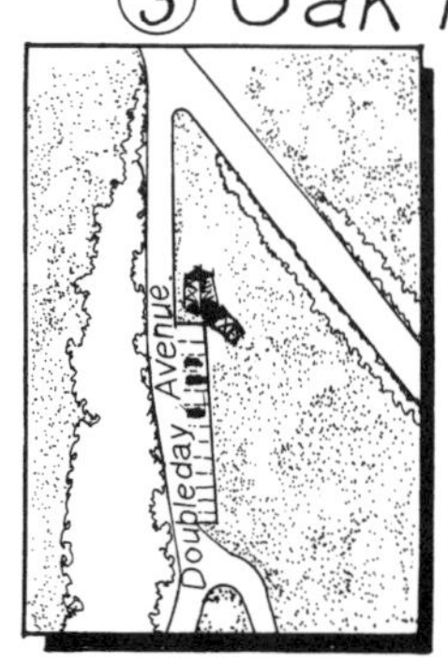

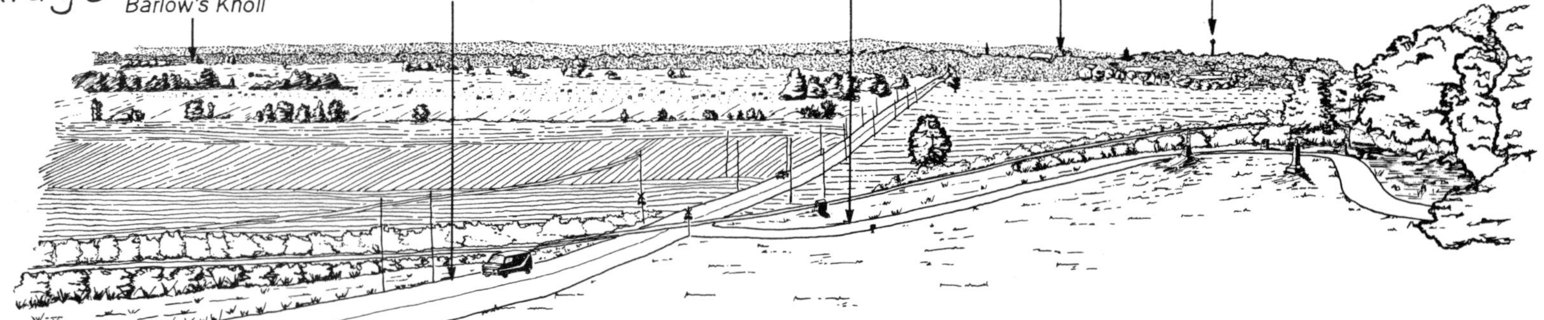

DRAWN BY: Nicole Steel, 1998

NATIONAL PARK SERVICE
ROADS & BRIDGES RECORDING PROJECT

GETTYSBURG NATIONAL MILITARY PARK TOUR ROADS
GETTYSBURG VICINITY ADAMS COUNTY PENNSYLVANIA

HISTORIC AMERICAN ENGINEERING RECORD
PA-485
SHEET 11 OF 14

SIGNS & MARKERS

Note: Drawings based on field photographs, notes, historical photographs and original drawings.

Many of the earliest visitors to the Gettysburg Battlefield were Civil War Veterans, their families, and others who were familiar with the events of the battle. Unofficial Battlefield Guides offered their services to those who wanted on-site interpretation and a tour of the battlefield. In the 1890s the War Department instituted a more comprehensive and uniform sign system. Iron tablets identified sites on the battlefield, such as strategic topographical features, farmsteads, and avenue names. Ever wary of visitor behavior, they also posted regulatory signs requesting that people respect the monuments and avenues. With the passing of the veterans and their contemporaries came a cry for more interpretation. In the 1930s the National Park Service instituted directional and speed limit signs and later added wayside exhibits and on-site audio programs. A blend of signs from nearly all of these eras can still be found along the avenues today.

By the mid-1880s many veteran organizations began to mark their units' positions on the battlefield with elaborate monuments. In 1895 the Gettysburg Battlefield Commission was authorized to mark the locations where both the Union and Confederate soldiers participated in the battle, resulting in an extensive system of markers, tablets, and further monument development throughout the battlefield. For many years this array served as the major means of interpretation of the battle, offering matter-of-factual details of troop movements and numbers. Civil War era cannon tubes, mounted upright on granite, mark the headquarters of Meade, Lee, and the corps commanders. Bronze inscription tablets mounted on rough-hewn granite designate the general positions of the corps and divisions. Cast iron plaques mounted on fluted posts, the battery tablets, denote the positions of the basic artillery fighting units. Today more than 1,320 markers, tablets, and monuments from the War Department Era continue to document the battle.

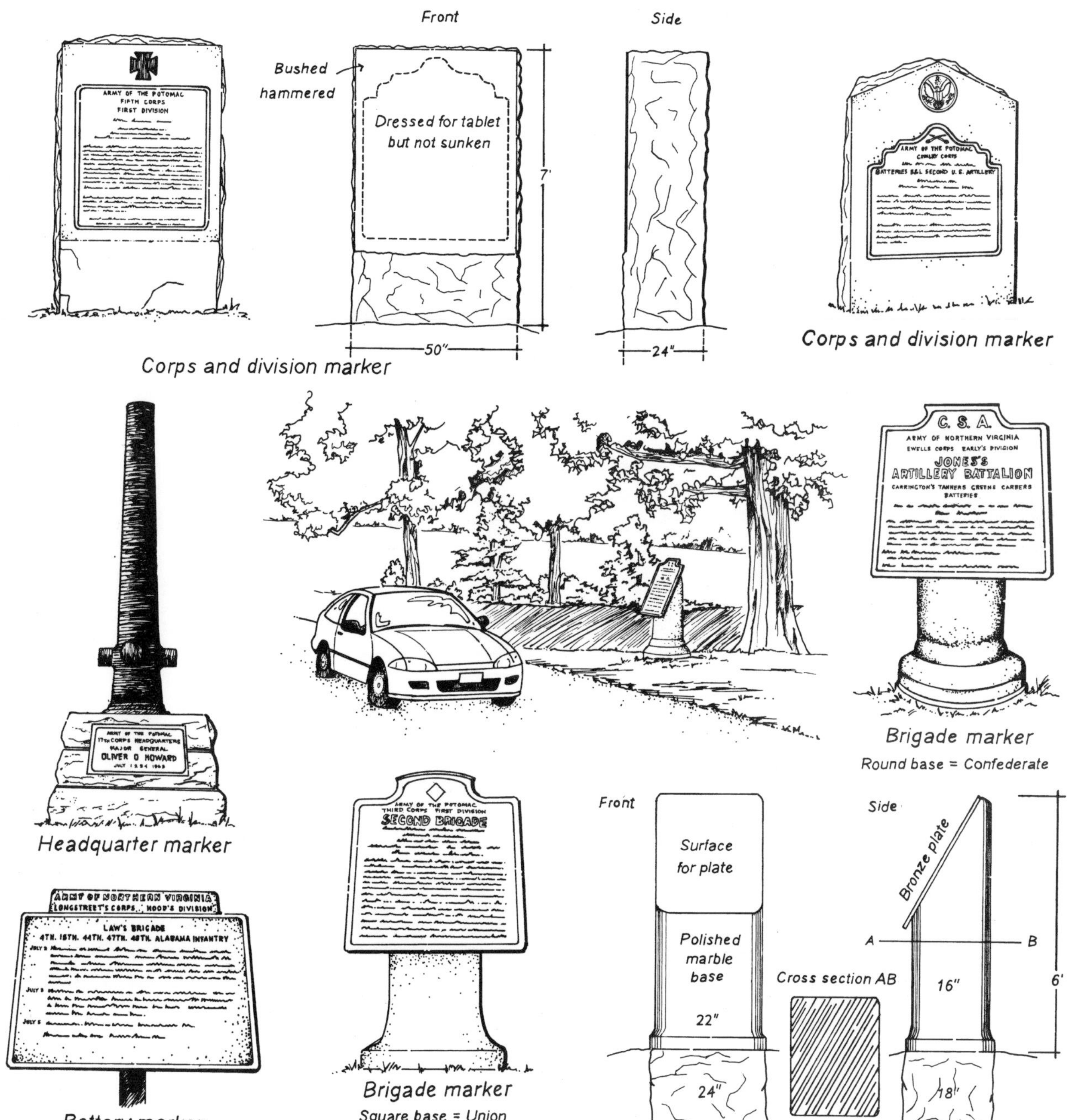

DELINEATED BY: Nicki Yung, 1998

NATIONAL PARK SERVICE
ROADS & BRIDGES RECORDING PROJECT
NATIONAL PARK SERVICE
UNITED STATES DEPARTMENT OF THE INTERIOR

GETTYSBURG NATIONAL MILITARY PARK TOUR ROADS
GETTYSBURG VICINITY ADAMS COUNTY PENNSYLVANIA

SHEET 12 of 14

HISTORIC AMERICAN ENGINEERING RECORD PA-485

IF REPRODUCED, PLEASE CREDIT: HISTORIC AMERICAN ENGINEERING RECORD, NATIONAL PARK SERVICE, NAME OF DELINEATOR, DATE OF THE DRAWING

BOUNDARY DEMARCATIONS

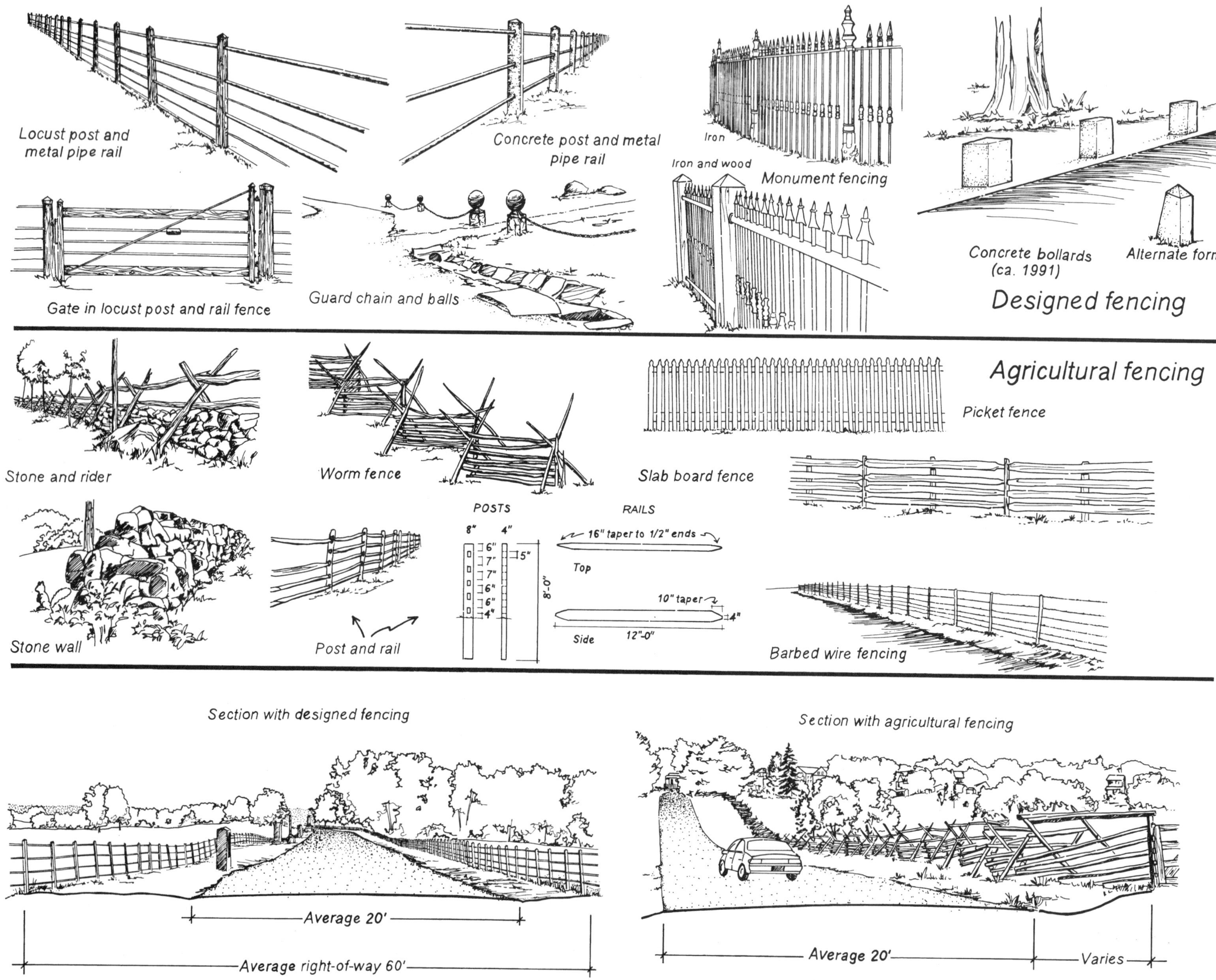

Note: Drawings based on field photographs, notes, historical photographs and original drawings.

Visitors to the Gettysburg National Military Park see many boundary demarcations, including a variety of fence types. Some of the fences reflect the war-time appearance of the battlefield, while others are more formal and were added to control traffic and pedestrian patterns. The fences that one sees today and many that no longer appear on the battlefield fall into two categories: designed, more formal fencing, and agricultural fencing. The more prevalent use of agricultural fencing today reflects the park's recent emphasis on an 1863 landscape appearance.

At the time that the Union and Confederate Armies converged upon Gettysburg they encountered an active agricultural, wooded, and rock region. During the battle, fence lines became defensive battle lines. These fence types included picket, post-and-rail, Virginia worm, stone and rider, and stone walls. Some designated property line, pastures, yards and gardens, but all reflected the available materials.

After the war new boundary systems appeared as soon as battlefield land was purchased for commemorative purposes. The GBMA incorporated the existing fencing patterns into their system, but added twisted wire fencing and barbed wire to offset the newly constructed avenues and to protect the monuments. Some of the most elaborate iron fencing also dates from this period. The War Department formalized the avenues even more by replacing the wire fencing with locust posts, iron caps, and four iron rails, and adding smaller iron features to protect the curves and fragile edges of the road surface.

DELINEATED BY: *Nicki Yung, 1998*

NATIONAL PARK SERVICE
ROADS & BRIDGES RECORDING PROJECT
NATIONAL PARK SERVICE
UNITED STATES DEPARTMENT OF THE INTERIOR

GETTYSBURG NATIONAL MILITARY PARK TOUR ROADS
GETTYSBURG VICINITY ADAMS COUNTY PENNSYLVANIA.

SHEET 13 OF 14

HISTORIC AMERICAN ENGINEERING RECORD
PA-485

LIBRARY OF CONGRESS INDEX NUMBER

IF REPRODUCED, PLEASE CREDIT: HISTORIC AMERICAN ENGINEERING RECORD, NATIONAL PARK SERVICE, NAME OF DELINEATOR, DATE OF THE DRAWING

EXPERIENCING THE BATTLE

Seminary Ridge

When the Gettysburg Battlefield Commission designed the avenue system, its main concern was to convey the lines of battle prior to the beginning of each day's fighting, leaving the battlefields undisturbed for study. The battles took place on varied terrain and it is this variety which adds interest and understanding to the motorists' experience of the battlefield today. The avenues encompass the crests of gently elevated ridges, rocky hillsides, boulder strewn valleys, shaded woodlots, and open expanses of rolling farmland.

Slyder's Woods

1. West Confederate Avenue

West Confederate Avenue follows the gentle rise of Seminary Ridge, a key position for the Confederate troops on the second and third days of fighting. The tree-line of the ridge offered the Confederate infantry and artillery protection as they advanced and retreated from their battles in the open spaces of the Wheatfield, the Peach Orchard, and Little Round Top, and for the final assault on Cemetery Ridge.

Devil's Den

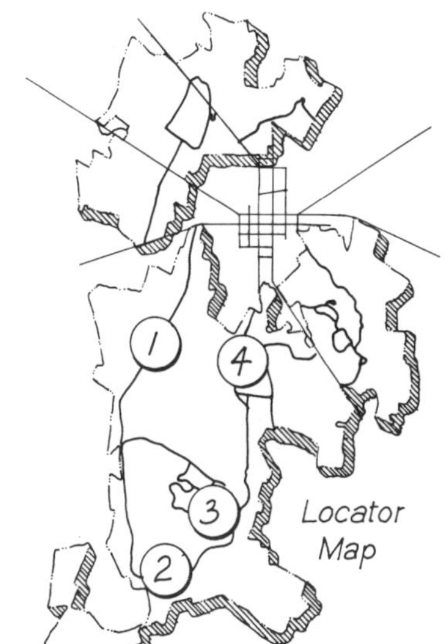

High Water Mark

2. South Confederate Avenue

Motorists on South Confederate Avenue pass through Bushman's and Slyder's Woods, where the Northern and Southern troops lined up on what was then the edge of well-tended woodlots. More densely vegetated today than they were on 2 July 1863, the visitor can still feel the rise of the topography towards Little Round Top, the focus of much of the day's fighting.

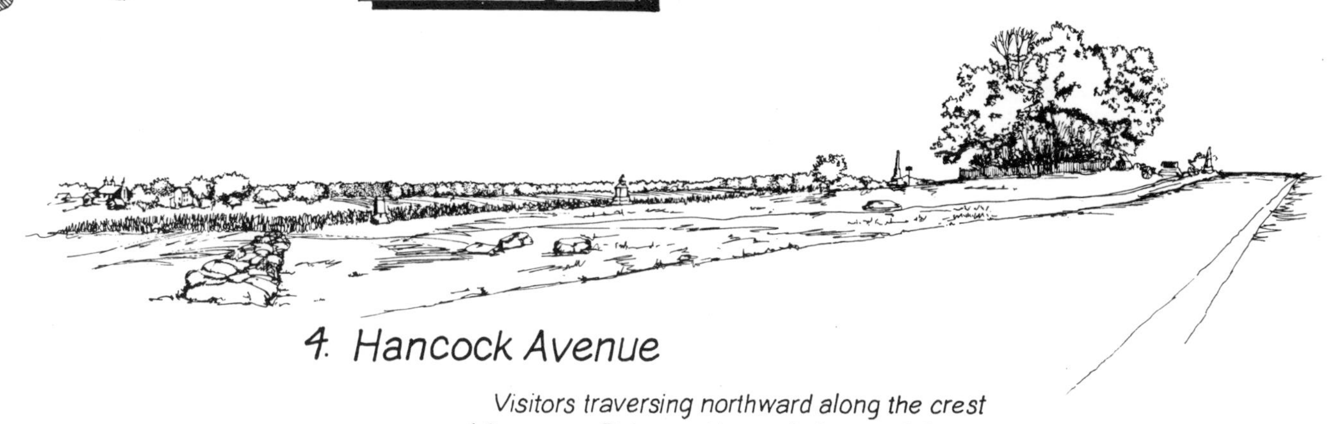

3. Sickles Avenue

The exfoliated eroded rock outcrops of Devil's Den provided the perfect combination of ledges and protective cover to the Confederate sharp shooters. As Sickles Avenue nears Plum Run at the base of Devil's Den and then winds to the top of Devil's Den, motorists get a view, first of what it would have been like to assault the craggy rocks, and then of the strategic view the Confederate soldiers had of Little Round Top.

4. Hancock Avenue

Visitors traversing northward along the crest of Cemetery Ridge on Hancock Avenue follow lines of defense held by Union troops on the 2nd and 3rd days of fighting. At the High water Mark and The Angle it becomes apparent how exposed the troops were to the opposing force's fire and that the advantage offered by the ridge and stone walls was minimal, yet vital.

Edward J. Lupyak, 1998

DELINEATED BY: NATIONAL PARK SERVICE ROADS & BRIDGES RECORDING PROJECT
NATIONAL PARK SERVICE
UNITED STATES DEPARTMENT OF THE INTERIOR

GETTYSBURG NATIONAL MILITARY PARK TOUR ROADS
GETTYSBURG VICINITY ADAMS COUNTY PENNSYLVANIA

SHEET 14 of 14

HISTORIC AMERICAN ENGINEERING RECORD PA-485

SHILOH NATIONAL MILITARY PARK TOUR ROADS

SHILOH, TENNESSEE

Shiloh National Military Park Locator Map

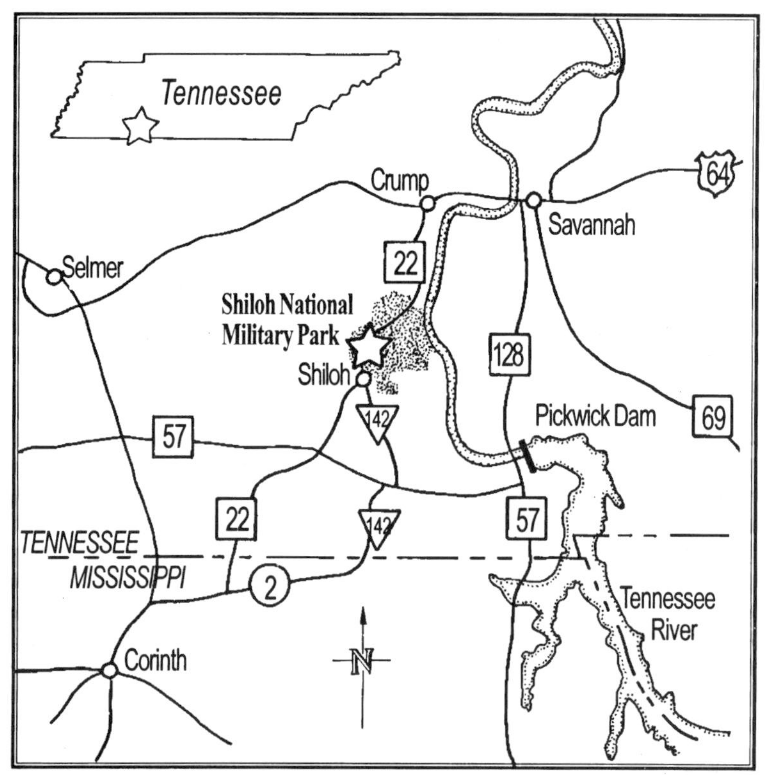

Shiloh National Military Park (SNMP) commemorates one of the first major battles of the western campaign of the Civil War, which occurred on April 6 and 7, 1862 in Shiloh, Tennessee. The Battle of Shiloh was considered the largest and bloodiest Civil War engagement to that date. Some historians claim that the great number of casualties at Shiloh brought the horrors of the war home to the American populace for the first time. The park roads facilitate the access, documentation, and commemoration of the battle. Their history and development reflects varying interpretations of the historic conflict, the ideologies of battlefield preservation, and the changing expectations of a national park experience. Most are a legacy of the historic traces that existed at the time of the battle and of park roadway design. As with other battlefield parks, the first efforts to establish SNMP were made by a group of veterans. They reunited at Shiloh in April 1893 and formed the Shiloh Battlefield Association. In spite of opposition within the U.S. War Department and the popular press because of the site's remoteness, the battlefield was designated a national military park under the auspices of the War Department in December 1894. It was the second military park created by the federal government to memorialize significant military engagements of the Civil War.

Located along the Tennessee River in rural Hardin County, the 3972-acre park was designed to reconstruct and document the site of the historic conflict and to commemorate its participants. The purpose of acquiring the entire battle ground instead of a simple memorial plot was to protect the unrecovered bodies of fallen soldiers and to recreate the historic landscape conditions so that casual visitors and students of military history could thoroughly visualize, appreciate, and learn from the battle events. The park's undulating terrain contains a web of historic roads that wind through open fields and dense forests. Plaques, monuments, and markers that explicate the events of the battle are scattered throughout the site. In 1933, battlefield parks were transferred from the War Department to the National Park Service in the U.S. Department of Interior. NPS officials expanded the parks focus beyond the representation and commemoration of military history to include more recreational goals. There was greater concern for the image of the park as a natural sanctuary. The War Department created a relatively authentic battlefield road system but, NPS administrators did not consider it a good historical park transportation network.

The untidy arrangement of historic roads was altered to fit the broader NPS goals of providing a "smooth, dustless roadway, over which visitors may ride in comfort as they enjoy the beauty of the Park." Modern materials and engineering features were combined with naturalistic design elements to enhance the structural forms and to change the aesthetic qualities of the park roads and landscapes. Roads were realigned to accommodate faster automobiles and roadside interpretation was increased. Since the roads were designed for sight-seeing, greater attention was given to the patterns of scenic views along the tour. The park's roads were also a significant component of local transportation systems. The region's first paved roads were in the park. It is not gated so there is public access twenty-four hours a day. The park's intended recreational and commemorative uses of the SNMP roads and the local expectations and desires continue to be an area of contention and accommodation. Digging below Shiloh's road surfaces reveals traces of more than just old road beds. The transportation networks vividly chronicle battle events, Civil War history, changing technologies, and ideas regarding national park experiences and aesthetics. Touring Shiloh Park roads does not simply take visitors back in time. It also brings them in contact with complex artifacts of national heritage and identity.

This project is part of the Historic American Engineering Record (HAER), a long range program to document historically significant engineering and industrial works in the United States. The HAER program is administrated by the Historic American Buildings Survey/Historic American Engineering Record Division (HABS/HAER) of the National Park Service, U.S. Department of Interior. Shiloh National Military Park (SNMP) Tour Roads Recording Project was cosponsored during the summer of 1998 by HAER (Eric DeLony, Chief) and SNMP (Woody Harrell, Superintendent). The project was funded by the Federal Lands Highway Program through the National Park Service Park Roads and Parkways program. The fieldwork, measured drawings, historical reports, and photographs were prepared under the direction of NPS Park Roads and Bridges Recording Program Manager Todd Croteau and Program Historian Tim Davis. Cynthia Ott prepared both the historical report and the drawings. Formal large-format photography was completed by Jet Lowe.

Delineated by: CYNTHIA OTT, 1999

NATIONAL PARK SERVICE ROADS AND BRIDGES RECORDING PROGRAM
NATIONAL PARK SERVICE
UNITED STATES DEPARTMENT OF THE INTERIOR

SHILOH VICINITY

SHILOH NATIONAL MILITARY PARK TOUR ROADS
SHILOH NATIONAL MILITARY PARK
HARDIN COUNTY

TENNESSEE

SHEET 1of 4

HISTORIC AMERICAN ENGINEERING RECORD
TN-37

Library of Congress Index Number

BATTLEFIELD PARK ROAD TYPES

Horse-drawn road roller at work in the park, c. 1900.
(Drawing based on historic photo, SNMP Archives.)

Concrete slab-raising on roads in the park, c.1930s.
(Drawing based on historic photo, SNMP Archives.)

Shiloh's roads are essential to the park's design and mission. They help recreate the battle narrative. They provide the main access to the site. They are an important aesthetic component of the park landscape. Rather than adding new avenues, like at Gettysburg, Shiloh relied on historic road traces to form the main circulation system in the park. A few new roads were created after the battle, but most of the local dirt roads that existed in 1862 were present when the park was established in 1894. Although park administrators strove to recreate historic road patterns, they frequently abandoned authenticity for practicality. Most of the old road surfaces were continually updated with new materials and engineering features in order to accommodate modern motorists' needs. The park contains a variety of road types that provide different touring experiences. Each category of road forms an integral part of the cultural landscape by shaping the educational and recreational experiences of park visitors.

Dirt Traces

A few roads in the park, including the famous Sunken Road, are maintained as dirt traces. They are no longer accessible to vehicles and are designated as hiking trails. The Sunken Road was not improved because it was reportedly a strategically important battle line on the first day of fighting. It is kept in an antiquated form in order to communicate more fully the conditions at the time of the battle. Other historic roads were not improved because of tour route changes.

Park Roads

The NPS realigned many historic traces in order to enhance visitors' driving experiences in the park. The NPS created gracefully serpentine roads that generally follow the contours of the land so that they appear to blend in with the surrounding landscape. Park roads are 18-20 feet wide and resurfaced with concrete. Most have superelevated curves, inconspicuous turf gutters and no curbs. Grassy borders with clumps of shade trees provide a naturalistic foreground to distant views of the battlefield. Corinth-Pittsburg Road is an example of the park-road type.

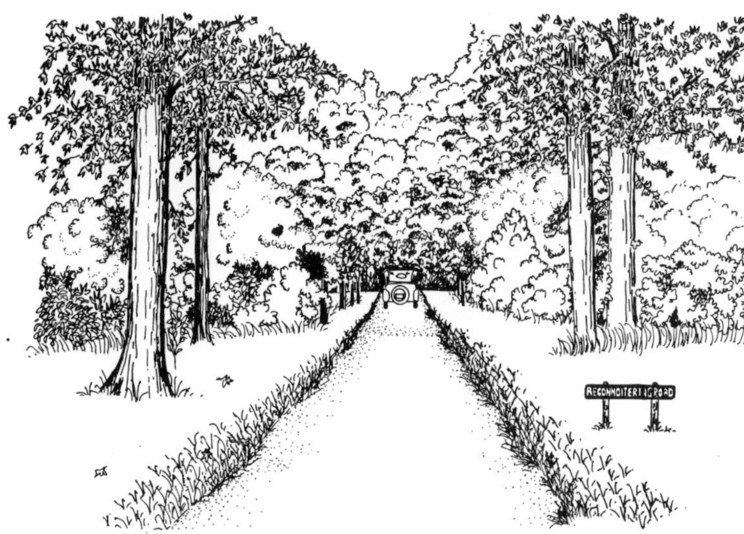

Country Lanes

Country lanes are one-way, 13-foot wide asphalt roads. They are the narrowest vehicular roads in the park. Unlike most park roads, they have few modern engineering features and vegetation is allowed to grow near the edge of the pavement. These factors give this road type the feel of a common rural lane. Reconnoitering Road, one of the most prominent of this type, is not a pre-battle trace, but was cut by troops during the engagement.

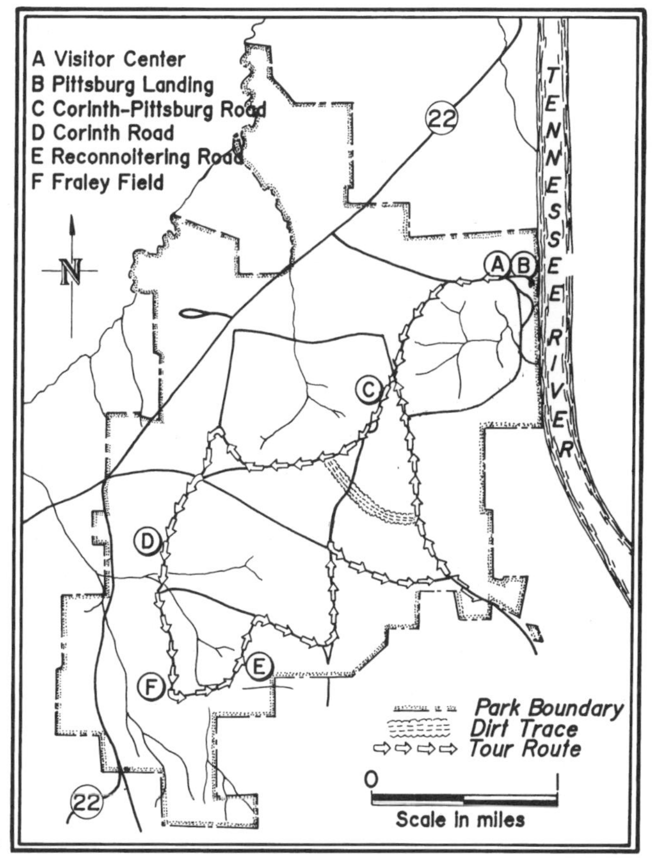

Tour Route

Because the main access to the park was originally via steamboat, the park circulation system starts near the river, not at Fraley Field, where the fighting began in 1862. The general organization of the park tour is a circular route that moves counterclockwise from the visitor center and back again. The tour does not follow a chronological sequence of battle events, but provides the most circuitous and comprehensive route to points of historic interest.

State Highway

Route 22, also known as the belt or bypass road, was constructed along the west side of the park in 1962. It is a two-lane center-striped 40-foot-wide asphalt road designed more for speed than for recreation. The highway was created to move local vehicles and commercial traffic quickly and smoothly through the park without interfering with slower paced tourist traffic. Before it was built, Corinth Road, located in the historic center of the park, was the main local transportation corridor.

HISTORIC AMERICAN ENGINEERING RECORD TN-37

SHEET 2 of 4

SHILOH NATIONAL MILITARY PARK TOUR ROADS
SHILOH NATIONAL MILITARY PARK
HARDIN COUNTY
SHILOH VICINITY TENNESSEE

DELINEATED BY: *Cynthia Ott, 1999*

NATIONAL PARK SERVICE ROADS & BRIDGES RECORDING PROGRAM
NATIONAL PARK SERVICE
UNITED STATES DEPARTMENT OF THE INTERIOR

TOUR ROAD INTERPRETIVE FEATURES

Commemoration & Monuments

Most of the commemorative features were placed on the grounds between 1900 and 1910, but additions continue up to the present. The park was responsible for the design and construction of the standardized monuments made from artillery parts, including the mortuary and headquarters monuments and the cannons. The cannons, and less commonly, the pyramidal head quarters monuments, were also placed in other battlefield parks. The state and regimental monuments were designed and paid for by patriotic and veterans organizations. The War Department required that these monu ments be made of granite and/or bronze and that the inscriptions "be purely his torical [and] be compiled without praise or censure." There are 5 mortuary mon uments, 15 headquarters monuments, over 200 cannon, and 126 state and regimental monuments in the park.

Monument to a fallen soldier, c. 1900

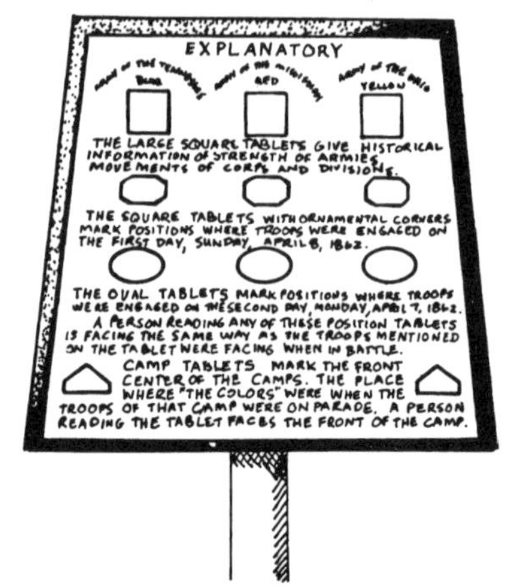

War Department
Explanatory Tablet of Battle Guides, c. 1900

War Department
Iron Markers, c. 1900

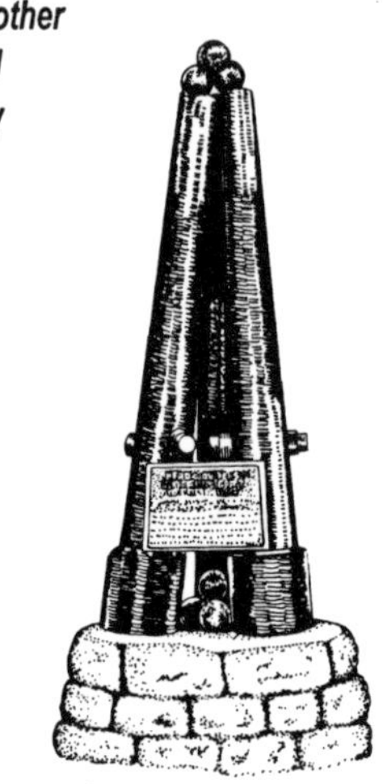

Grant's Headquarters Monument, c. 1900

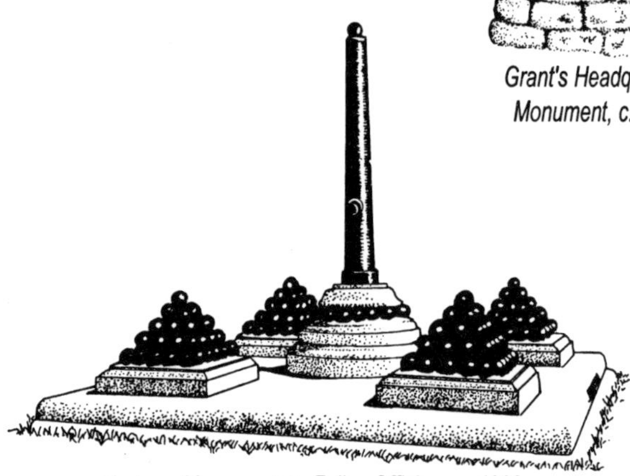

Mortuary Monuments to Fallen Officiers, c.1900

Regimental Monument, c. 1900

National Park Service
Information Station and Road Signs, 1930 to present

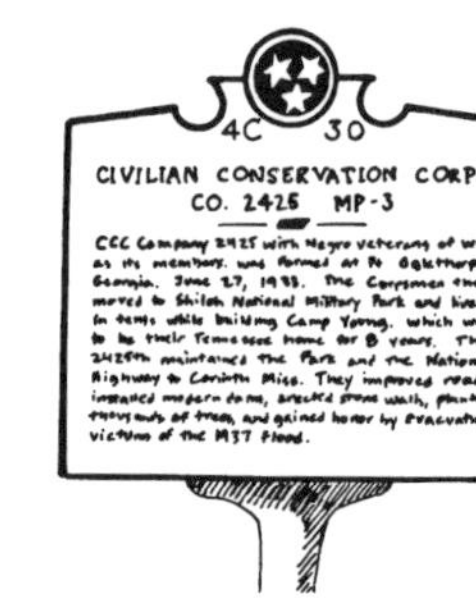

State Highway Historical Marker
Commemorating African-American CCC workers, c. 1990

State Monument, c. 1900

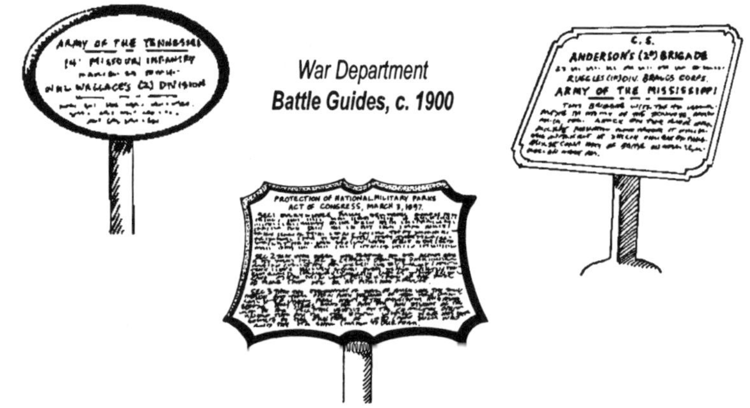

War Department
Battle Guides, c. 1900

Cannon represent artillery positions, c. 1900

Headquarters Monument, c. 1900

Oxen hauling a stone monument from the steamboat landing for placement in the battlefield, ca. 1900. (Drawing based on SNMP Archvies historic photograph.)

Markers and Tablets

The landscape contains hundreds of monuments, markers and tablets that were designed both for instruction and commemoration. Although most are accessible or at least visible from the park roads, the War Department situated them to document historic events at their exact locations, rather than to be roadside attractions, as at other battlefield parks. The War Department erected iron markers and tablets that describe battle events and troop movements. They are coded by size, shape and color. Standardization of designs provides clear clues of the events or scenes being described, creates visual unity through the battlefield landscape, and facilitates self-guided tours. Since the 1930s, the NPS expanded interpretation by adding many educational stations along the tour route. It also erected rustic wooden road signs and tour route markers.

Delineated by: CYNTHIA OTT, 1999

NATIONAL PARK SERVICE ROADS AND BRIDGES RECORDING PROGRAM
NATIONAL PARK SERVICE
UNITED STATES DEPARTMENT OF THE INTERIOR

SHILOH VICINITY

SHILOH NATIONAL MILITARY PARK TOUR ROADS
SHILOH NATIONAL MILITARY PARK
HARDIN COUNTY

TENNESSEE

SHEET 3 of 4

HISTORIC AMERICAN ENGINEERING RECORD
TN-37

Library of Congress Index Number

IF REPRODUCED, PLEASE CREDIT HISTORIC AMERICAN ENGINEERING RECORD, NATIONAL PARK SERVICE, NAME OF DELINEATOR, DATE OF THE DRAWING

BRIDGES & CULVERTS

Shiloh National Military Park is crossed by many tributaries of the Tennessee River. Bridges and culverts were constructed over the years to prevent the waterways from impeding tourist traffic. All of the original wood bridges were replaced by concrete and stone masonry bridges and culverts after a tornado hit the park in 1909. The War Department erected decorative structures that reflected its mission to commemorate the battle. Beginning in the 1930s, the National Park Service constructed more rustic but technologically sophisticated stone and concrete culverts that attempted to conceal human artifice and make the road appear more like a natural part of the landscape.

A War Department-era decorative concrete bridge spanned a creek in the park before it was replaced by a more naturalistic culvert in the 1930s. Abutting stone walls and drainage ditches helped control the flow of water.

Another even more elaborate War Department-era concrete bridge that stood from 1909 to the 1930s.

Evolution of the Dill Creek Crossing

The historic transformations of the Dill Creek crossing on Riverside Drive show the development of the park road's engineering and aesthetic features.

A simple wood bridge was built before or soon after the park's creation.

The War Department relaced it in 1909 with an ornate 60-foot long concrete bridge that memorialized the battle with decorative cannonballs.

Between 1940 and 1941, Civilian Conservation Corps crews, under the auspices of the National Park Service, erected a large concrete culvert that was hidden beneath the road in order to make the drive appear more natural.

1930s-Era Stone Culverts Extant in the Park

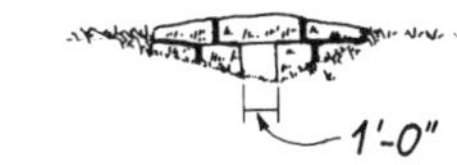

Corinth-Pittsburg Landing Road, south side, next to Ruggle's Battery

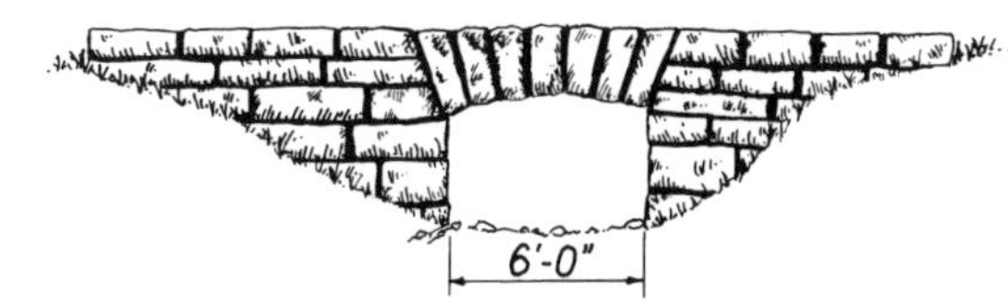

Corinth Road, north side, just west of the United Daughters of the Confederacy Monument

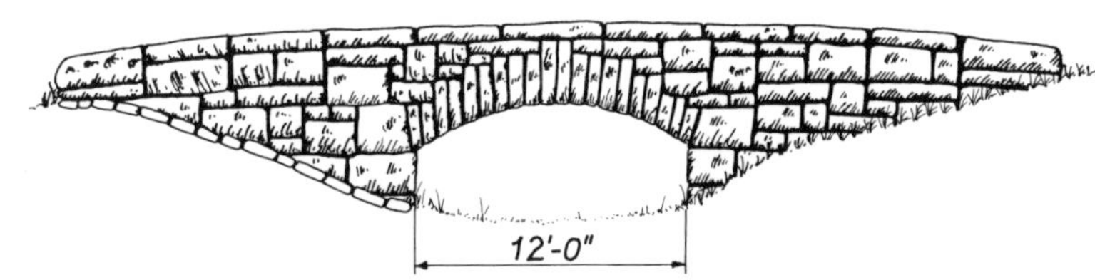

Peabody Road, south side, Shiloh Branch

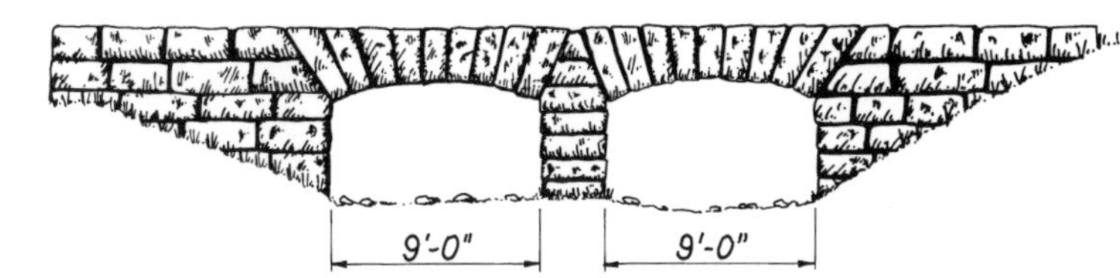

Corinth Road, west side, Shiloh Branch, north culvert

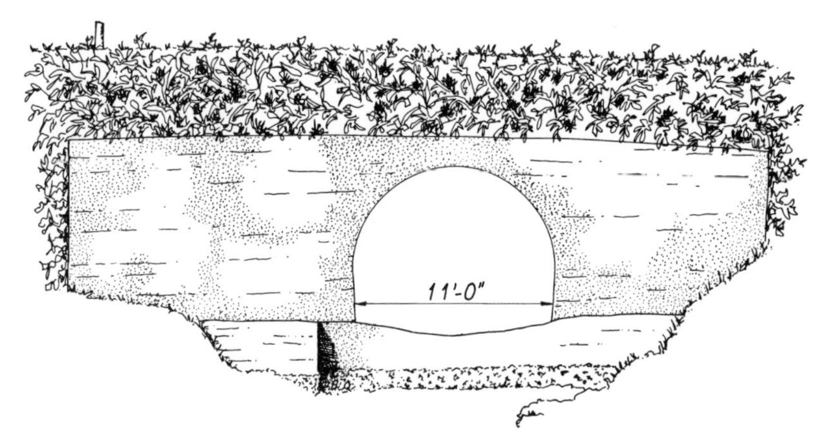

Riverside Drive, west side, Dill Branch

HISTORIC AMERICAN ENGINEERING RECORD TN-37

SHEET 4 OF 4

SHILOH NATIONAL MILITARY PARK TOUR ROADS
SHILOH NATIONAL MILITARY PARK
HARDIN COUNTY

SHILOH VICINITY

TENNESSEE

DELINEATED BY: Cynthia Ott, 1999

NPS ROADS & BRIDGES RECORDING PROGRAM
NATIONAL PARK SERVICE
UNITED STATES DEPARTMENT OF THE INTERIOR

Vicksburg National Military Park Tour Roads

Vicksburg, Mississippi

Entrance Arch near park visitor center, 1997. Scanned from photo by Bill Faust.

Vicksburg National Military Park at Vicksburg, Mississippi was established by the U.S. War Department in 1899 to preserve the site of a pivotal battle of the American Civil War. The park was designed to provide visitors with access to and interpretation of historically significant areas, relating to the 1863 campaign for the city of Vicksburg. After the battle, various veterans groups, states and the federal government placed monuments, memorials and markers throughout the Confederate and Union troop lines to commemorate those who fought there. To facilitate visitors, it was necessary to develop a road system that would complement the site and provide quick and easy access to a monumental landscape with many points of interest.

Park commissioners and engineers involved with planning the tour roads, led by Resident Commissioner William T. Rigby, wanted to make the park as convenient as possible for its visitors. In their opinion, the park was intended for the average citizen who would spend a few hours touring the grounds, not the military or historical expert who would devote days or weeks to the study of siege operations. The result was a park tour designed to be seen from the road via horse, wagon, or carriage...and later automobiles. The tour roads, nearly a century old, have evolved from gravel roadways to an elaborate network of primarily one-way routes paved with concrete and asphalt. The tour roads were laid in curvilinear alignments over and around the rolling landscape passing by the most significant trenches, encampments and battle sites along the siege lines, thus presenting a variety of views of the restored battlefield with monuments, interpretive signs, artillery placements and other historical points of interest. Maintained vistas, mowed bays and other landscape concerns along the roads complement the visitor experience of the park while elaborate drainage systems protect the roadway from erosion of the loess soil during heavy rains.

This project is part of the Historic American Engineering Record (HAER), a long-range program to document historically significant engineering and industrial works in the United States. The HAER program is administered by the Historic American Buildings Survey/Historic American Engineering Record Division (HABS/HAER) of the National Park Service, U.S. Department of the Interior. The Vicksburg National Military Park (VNMP) Roads and Bridges Recording Project was cosponsored by HAER (E. Blaine Cliver, Chief) and VNMP (William Nichols, Superintendent). The project was funded by the Federal Lands Highway Program (Thomas Edick, Administrator) through the National Park Service Park Roads and Parkways Program (Mark Hartsoe, Manager). The field work, measured drawings, historical reports, and photographs were prepared under the direction of Program Manager Todd A. Croteau and Richard Quin, Program Historian. The recording team consisted of William Pete Brooks, field supervisor; landscape architect Deb James, and architect Gregory Seale. Historical reports were prepared by project historian Courtney Youngblood. Formal large-format photography was done by William Foust. The team was assisted by VNMP Historian Terrence Winschel.

Vicksburg National Military Park was transferred to the National Park Service in 1933 and has been managed by the NPS since. Though most of the tour road features remain the same, several major changes have occurred. During the "Mission 66" period of the park service a large portion of the park and road system was ceded to the City of Vicksburg. This included the southern section of the crescent that formed the siege line. Although maintained as a city park, there are noticeable differences in the aesthetic quality. The presence of utility lines, fire hydrants and street lamps distracts from viewing the monuments. Within the national park such amenities are not visible.

The many bridges in the park are a particularly interesting feature of this road system due to their ornamental detailing and early construction technologies. The steel arch bridge on Confederate Avenue over Jackson Road is the only bridge of its kind in the state of Mississippi. Bypassed by another bridge in 1972, the structure remains and has been used as a pedestrian bridge. Nine Melan-type concrete arch bridges were constructed along Union Avenue in 1903. These reinforced concrete bridges have held up remarkably over time, however, geologic forces, erosion, and heavy traffic have taken their toll. The brick-faced concrete arch bridge on Confederate Avenue, spanning Halls Ferry Road, was designed by the National Park Service and Bureau of Public Roads and is also the only bridge of its kind in the state.

This documentation defines the unique characteristics of the Vicksburg National Military Park Tour Roads and interprets the evolution of design, construction and maintenance of the monumental landscape as related to the tour road experience. Today, the tour roads remain heavily used and the older structures along the routes are being considered for replacement. This record will preserve on paper those details that make the park unique.

NATIONAL PARK SERVICE
PARK ROADS AND PARKWAYS RECORDING PROGRAM
NATIONAL PARK SERVICE
UNITED STATES DEPARTMENT OF THE INTERIOR

VICKSBURG VICINITY

VICKSBURG NATIONAL MILITARY PARK TOUR ROADS
VICKSBURG NATIONAL MILITARY PARK
WARREN COUNTY

MISSISSIPPI

SHEET 1 of 7

HISTORIC AMERICAN ENGINEERING RECORD
MS-14

Library of Congress Index Number

IF REPRODUCED, PLEASE CREDIT HISTORIC AMERICAN ENGINEERING RECORD, NATIONAL PARK SERVICE, NAME OF DELINEATOR, DATE OF THE DRAWING

REFERENCE MAP 1

BASED ON: 1962 USGS 7.5 MINUTE MAPS: VICKSBURG EAST, VICKSBURG WEST & REDWOOD QUADRANGLES; "VICKSBURG NATIONAL MILITARY PARK, BY THE PARK COMMISSION..." MAP NOS. 1-4; AND PHOTOGRAPHS FOUND IN THE ARCHIVES OF THE VICKSBURG NATIONAL MILITARY PARK

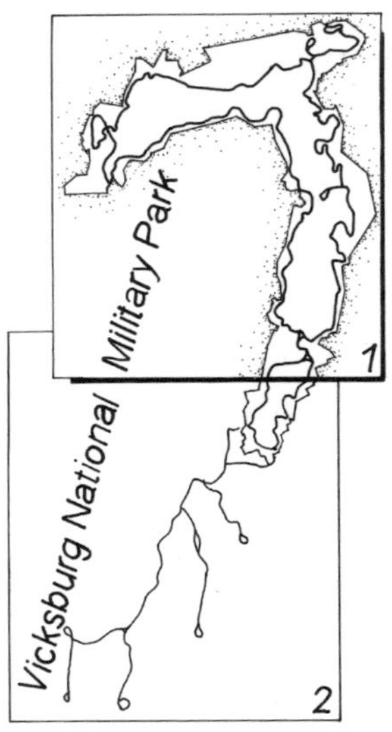

LOCATOR MAP

LEGEND

- Park Tour Road
- Obliterated Road
- Connecting Road
- Historic Bridge
- Major City Street
- Boundary 1963 to present
- Parkland ceded to City, 1963
- Significant Military Site
- State Monument
- Railroad Line

INDEX TO HISTORIC BRIDGES

Bridge	HAER	Mile Post	See Sheet
1 Melan Arch	MS-14A	0.2	8 of 12
2 Melan Arch	MS-14A	1.2	8 of 12
3 Melan Arch	MS-14A	6.4	8 of 12
4 Melan Arch	MS-14A	6.7	8 of 12
5 Melan Arch	MS-14A	6.9	9 of 12
6 Steel	MS-12	11.4	1 of 2

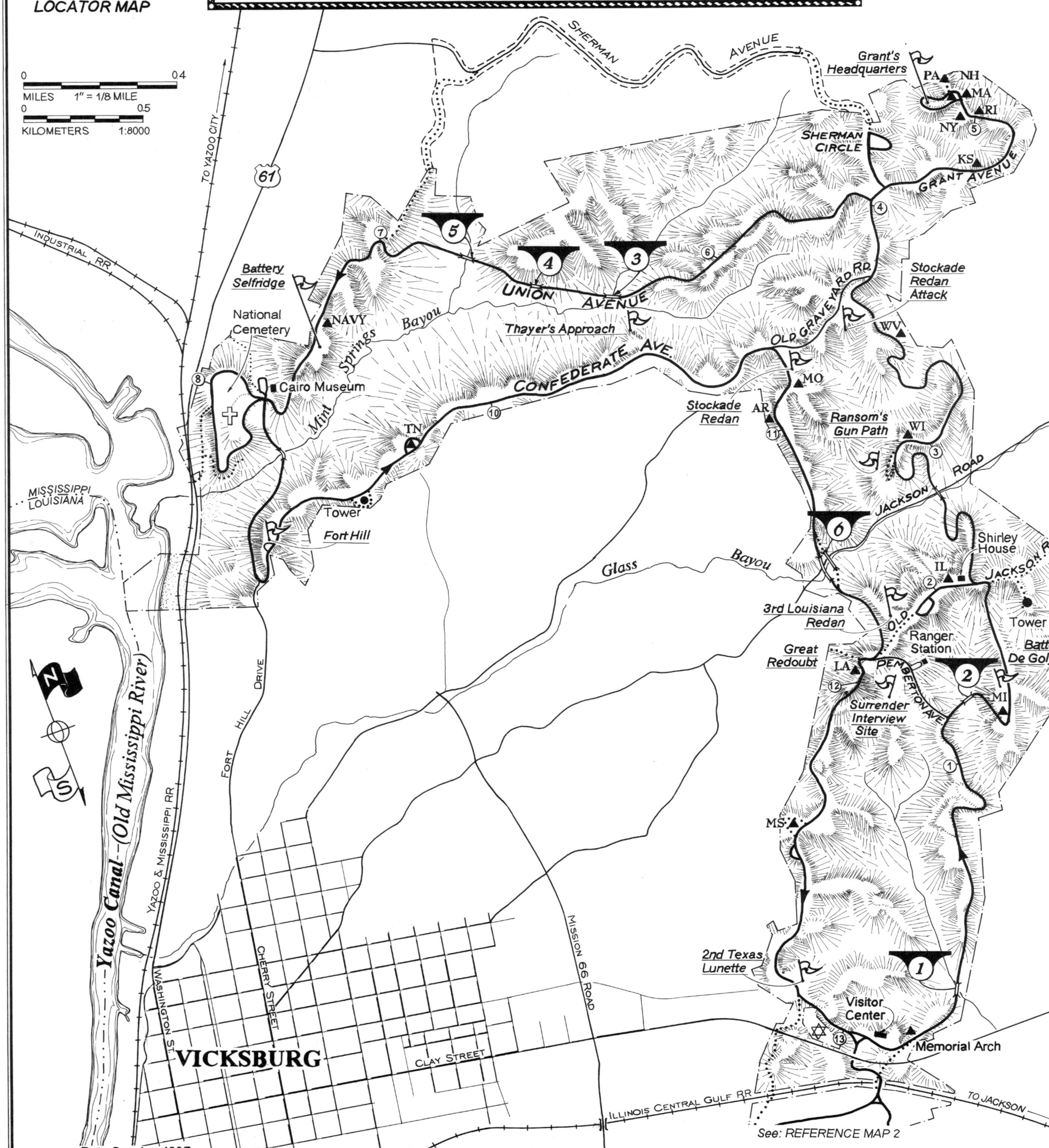

DELINEATED BY: Pete Brooks, 1997

NATIONAL PARK SERVICE ROADS & BRIDGES RECORDING PROGRAM NATIONAL PARK SERVICE UNITED STATES DEPARTMENT OF THE INTERIOR	VICKSBURG NATIONAL MILITARY PARK TOUR ROADS VICKSBURG — WARREN COUNTY — MISSISSIPPI	SHEET 2 of 7	HISTORIC AMERICAN ENGINEERING RECORD MS-14	

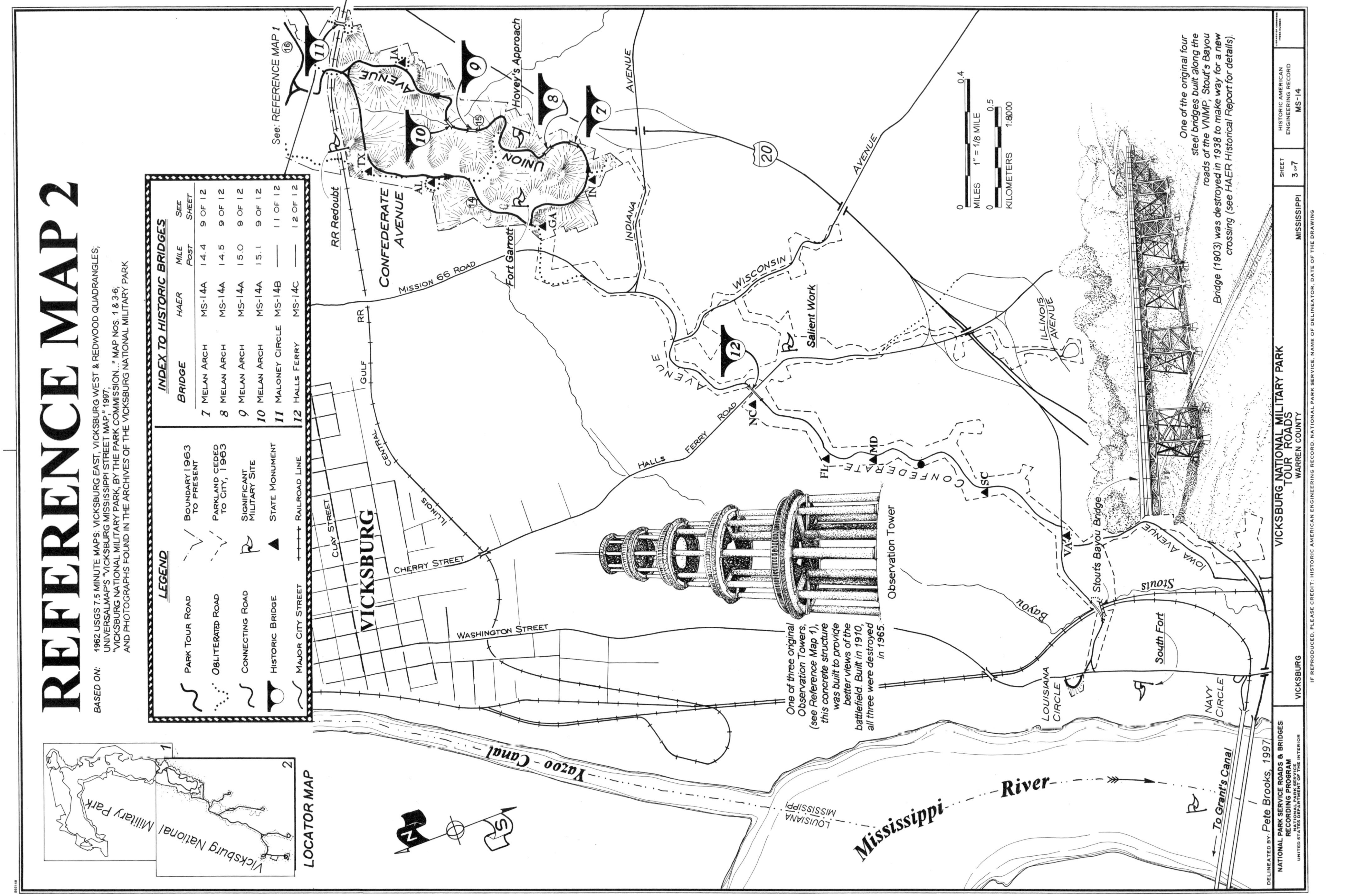

REFERENCE MAP 2
BASED ON: 1962 USGS 7.5 MINUTE MAPS: VICKSBURG EAST, VICKSBURG WEST & REDWOOD QUADRANGLES;
UNIVERSALMAP'S "VICKSBURG MISSISSIPPI STREET MAP," 1997;
"VICKSBURG NATIONAL MILITARY PARK, BY THE PARK COMMISSION..." MAP NOS. 1 & 3-6;
AND PHOTOGRAPHS FOUND IN THE ARCHIVES OF THE VICKSBURG NATIONAL MILITARY PARK
LEGEND
PARK TOUR ROAD
OBLITERATED ROAD
CONNECTING ROAD
HISTORIC BRIDGE
MAJOR CITY STREET
BOUNDARY 1963 TO PRESENT
PARKLAND CEDED TO CITY, 1963
SIGNIFICANT MILITARY SITE
STATE MONUMENT
RAILROAD LINE
INDEX TO HISTORIC BRIDGES
BRIDGE | HAER | MILE POST | SEE SHEET
7 MELAN ARCH | MS-14A | 14.4 | 9 OF 12
8 MELAN ARCH | MS-14A | 14.5 | 9 OF 12
9 MELAN ARCH | MS-14A | 15.0 | 9 OF 12
10 MELAN ARCH | MS-14A | 15.1 | 9 OF 12
11 MALONEY CIRCLE | MS-14B | — | 11 OF 12
12 HALLS FERRY | MS-14C | — | 12 OF 12
LOCATOR MAP
Vicksburg National Military Park
See: REFERENCE MAP 1
Hovey's Approach
UNION AVENUE
RR Redoubt
CONFEDERATE AVENUE
Fort Garrott
MISSION 66 ROAD
INDIANA AVENUE
AVENUE
WISCONSIN
Salient Work
ILLINOIS AVENUE
GULF RR
CENTRAL
ILLINOIS
HALLS FERRY ROAD
CONFEDERATE AVENUE
CLAY STREET
VICKSBURG
CHERRY STREET
WASHINGTON STREET
Observation Tower
One of three original Observation Towers, (see Reference Map 1), this concrete structure was built to provide better views of the battlefield. Built in 1910, all three were destroyed in 1965.
Stouts Bayou Bridge
Stouts Bayou
IOWA AVENUE
South Fort
LOUISIANA CIRCLE
NAVY CIRCLE
To Grant's Canal
Yazoo Canal
Mississippi River
LOUISIANA
MISSISSIPPI
MILES 1" = 1/8 MILE
KILOMETERS 1:8000
One of the original four steel bridges built along the roads of the VNMP, Stout's Bayou Bridge (1903) was destroyed in 1938 to make way for a new crossing (see HAER Historical Report for details).
DELINEATED BY: Pete Brooks, 1997
NATIONAL PARK SERVICE ROADS & BRIDGES RECORDING PROGRAM
NATIONAL PARK SERVICE
UNITED STATES DEPARTMENT OF THE INTERIOR
VICKSBURG
VICKSBURG NATIONAL MILITARY PARK TOUR ROADS
WARREN COUNTY
MISSISSIPPI
SHEET 3 OF 7
HISTORIC AMERICAN ENGINEERING RECORD MS-14
IF REPRODUCED, PLEASE CREDIT: HISTORIC AMERICAN ENGINEERING RECORD, NATIONAL PARK SERVICE, NAME OF DELINEATOR, DATE OF THE DRAWING

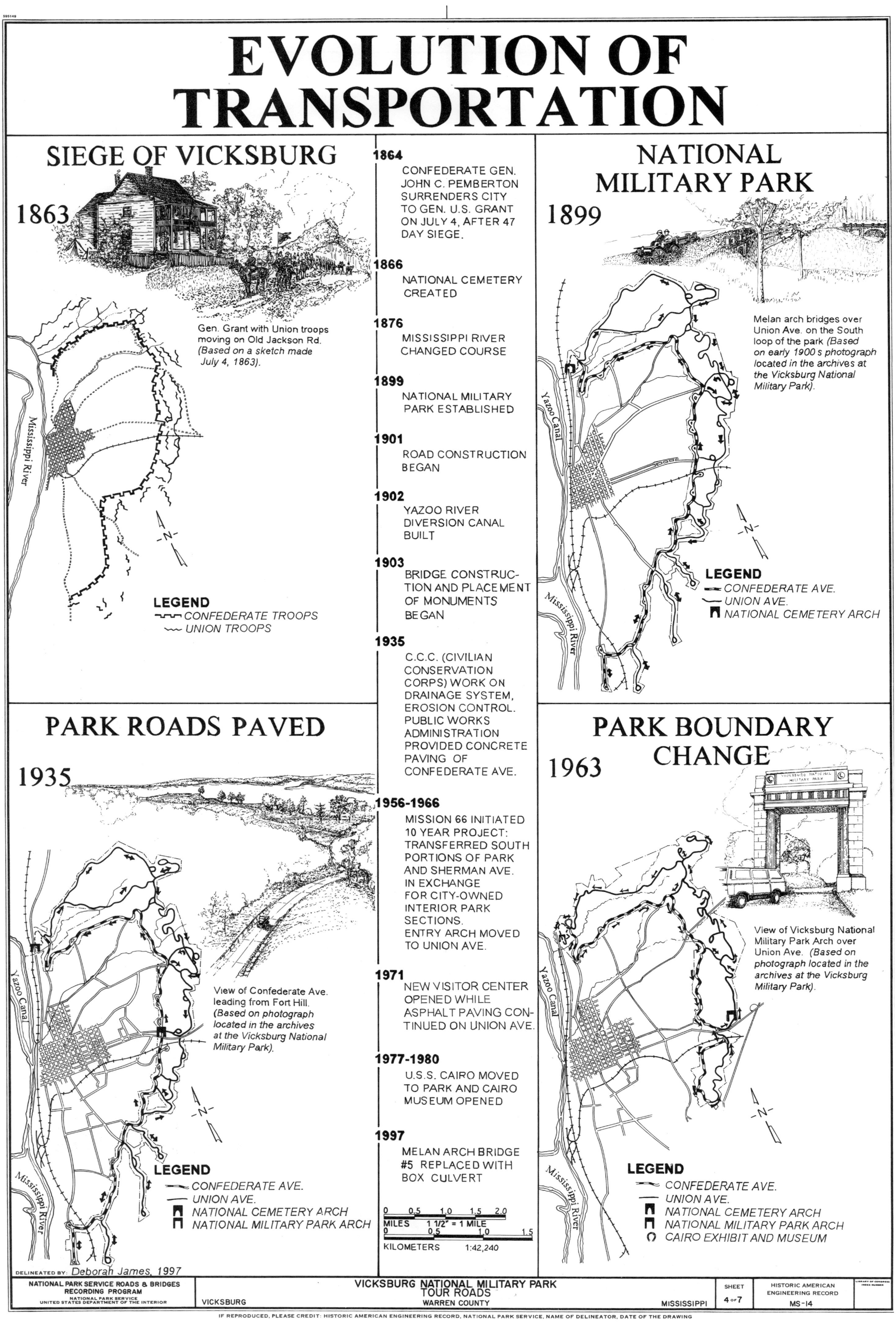
EVOLUTION OF TRANSPORTATION
SIEGE OF VICKSBURG
1863
Gen. Grant with Union troops moving on Old Jackson Rd. (Based on a sketch made July 4, 1863).
Mississippi River
N
LEGEND
CONFEDERATE TROOPS
UNION TROOPS
1864
CONFEDERATE GEN. JOHN C. PEMBERTON SURRENDERS CITY TO GEN. U.S. GRANT ON JULY 4, AFTER 47 DAY SIEGE.
1866
NATIONAL CEMETERY CREATED
1876
MISSISSIPPI RIVER CHANGED COURSE
1899
NATIONAL MILITARY PARK ESTABLISHED
1901
ROAD CONSTRUCTION BEGAN
1902
YAZOO RIVER DIVERSION CANAL BUILT
1903
BRIDGE CONSTRUCTION AND PLACEMENT OF MONUMENTS BEGAN
1935
C.C.C. (CIVILIAN CONSERVATION CORPS) WORK ON DRAINAGE SYSTEM, EROSION CONTROL. PUBLIC WORKS ADMINISTRATION PROVIDED CONCRETE PAVING OF CONFEDERATE AVE.
1956-1966
MISSION 66 INITIATED 10 YEAR PROJECT: TRANSFERRED SOUTH PORTIONS OF PARK AND SHERMAN AVE. IN EXCHANGE FOR CITY-OWNED INTERIOR PARK SECTIONS. ENTRY ARCH MOVED TO UNION AVE.
1971
NEW VISITOR CENTER OPENED WHILE ASPHALT PAVING CONTINUED ON UNION AVE.
1977-1980
U.S.S. CAIRO MOVED TO PARK AND CAIRO MUSEUM OPENED
1997
MELAN ARCH BRIDGE #5 REPLACED WITH BOX CULVERT
NATIONAL MILITARY PARK
1899
Melan arch bridges over Union Ave. on the South loop of the park (Based on early 1900's photograph located in the archives at the Vicksburg National Military Park).
Yazoo Canal
Mississippi River
N
LEGEND
CONFEDERATE AVE.
UNION AVE.
NATIONAL CEMETERY ARCH
PARK ROADS PAVED
1935
View of Confederate Ave. leading from Fort Hill. (Based on photograph located in the archives at the Vicksburg National Military Park).
Yazoo Canal
Mississippi River
N
LEGEND
CONFEDERATE AVE.
UNION AVE.
NATIONAL CEMETERY ARCH
NATIONAL MILITARY PARK ARCH
0 0.5 1.0 1.5 2.0
MILES 1 1/2" = 1 MILE
0 0.5 1.0 1.5
KILOMETERS 1:42,240
PARK BOUNDARY CHANGE
1963
View of Vicksburg National Military Park Arch over Union Ave. (Based on photograph located in the archives at the Vicksburg Military Park).
Yazoo Canal
Mississippi River
N
LEGEND
CONFEDERATE AVE.
UNION AVE.
NATIONAL CEMETERY ARCH
NATIONAL MILITARY PARK ARCH
CAIRO EXHIBIT AND MUSEUM
DELINEATED BY: Deborah James, 1997
NATIONAL PARK SERVICE ROADS & BRIDGES RECORDING PROGRAM
NATIONAL PARK SERVICE
UNITED STATES DEPARTMENT OF THE INTERIOR
VICKSBURG
VICKSBURG NATIONAL MILITARY PARK
TOUR ROADS
WARREN COUNTY
MISSISSIPPI
SHEET 4 OF 7
HISTORIC AMERICAN ENGINEERING RECORD
MS-14
IF REPRODUCED, PLEASE CREDIT: HISTORIC AMERICAN ENGINEERING RECORD, NATIONAL PARK SERVICE, NAME OF DELINEATOR, DATE OF THE DRAWING

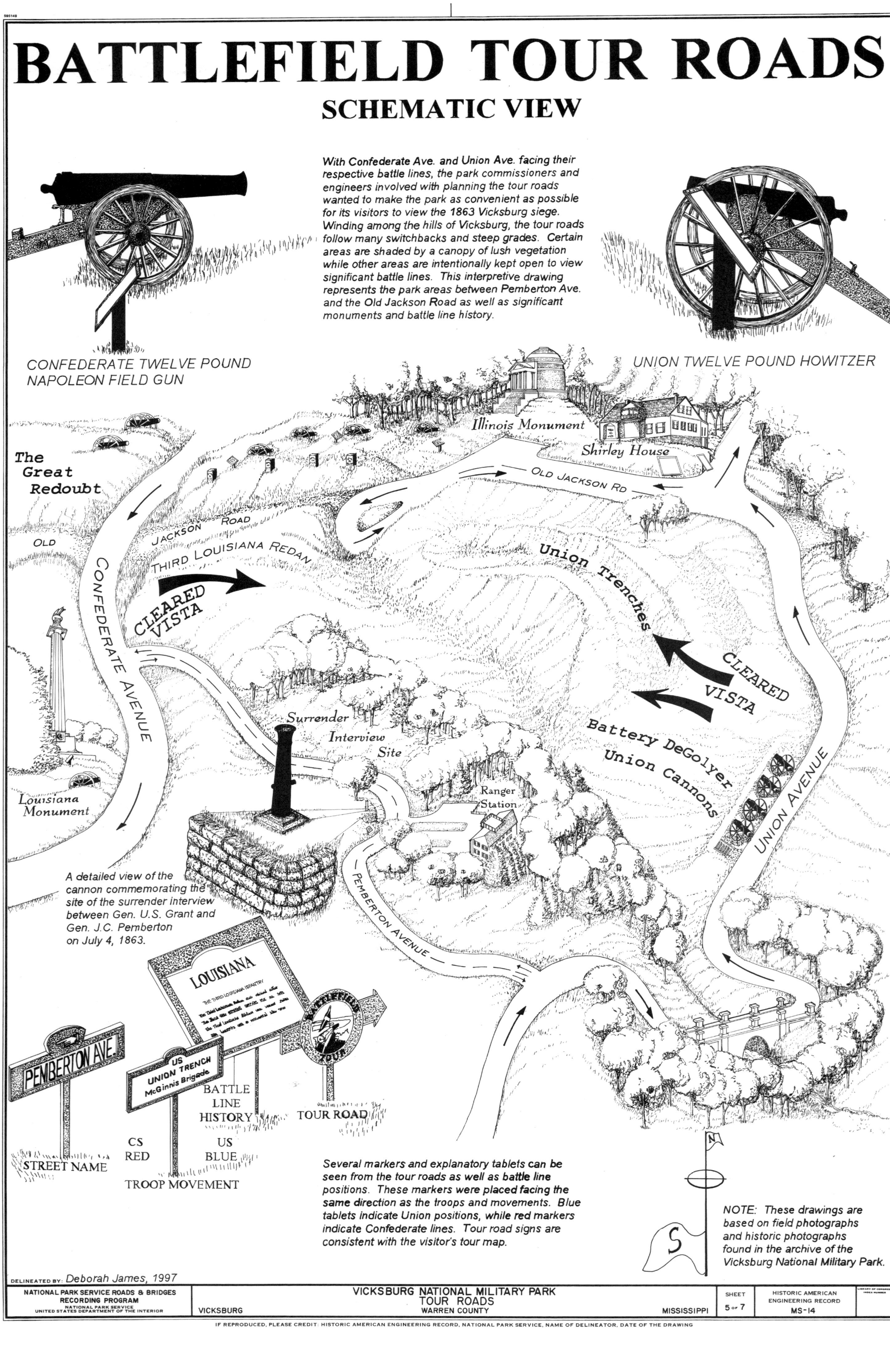
BATTLEFIELD TOUR ROADS
SCHEMATIC VIEW
With Confederate Ave. and Union Ave. facing their respective battle lines, the park commissioners and engineers involved with planning the tour roads wanted to make the park as convenient as possible for its visitors to view the 1863 Vicksburg siege. Winding among the hills of Vicksburg, the tour roads follow many switchbacks and steep grades. Certain areas are shaded by a canopy of lush vegetation while other areas are intentionally kept open to view significant battle lines. This interpretive drawing represents the park areas between Pemberton Ave. and the Old Jackson Road as well as significant monuments and battle line history.
CONFEDERATE TWELVE POUND NAPOLEON FIELD GUN
UNION TWELVE POUND HOWITZER
Illinois Monument
Shirley House
OLD JACKSON RD
The Great Redoubt
OLD JACKSON ROAD
THIRD LOUISIANA REDAN
CLEARED VISTA
Union Trenches
CONFEDERATE AVENUE
CLEARED VISTA
Battery DeGolyer Union Cannons
Surrender Interview Site
Ranger Station
UNION AVENUE
Louisiana Monument
A detailed view of the cannon commemorating the site of the surrender interview between Gen. U.S. Grant and Gen. J.C. Pemberton on July 4, 1863.
PEMBERTON AVENUE
LOUISIANA
PEMBERTON AVE
US UNION TRENCH McGinnis Brigade
BATTLE LINE HISTORY
TOUR ROAD
CS RED
US BLUE
STREET NAME
TROOP MOVEMENT
Several markers and explanatory tablets can be seen from the tour roads as well as battle line positions. These markers were placed facing the same direction as the troops and movements. Blue tablets indicate Union positions, while red markers indicate Confederate lines. Tour road signs are consistent with the visitor's tour map.
NOTE: These drawings are based on field photographs and historic photographs found in the archive of the Vicksburg National Military Park.
DELINEATED BY: Deborah James, 1997
NATIONAL PARK SERVICE ROADS & BRIDGES RECORDING PROGRAM
NATIONAL PARK SERVICE
UNITED STATES DEPARTMENT OF THE INTERIOR
VICKSBURG NATIONAL MILITARY PARK
TOUR ROADS
VICKSBURG
WARREN COUNTY
MISSISSIPPI
SHEET 5 OF 7
HISTORIC AMERICAN ENGINEERING RECORD
MS-14
IF REPRODUCED, PLEASE CREDIT: HISTORIC AMERICAN ENGINEERING RECORD, NATIONAL PARK SERVICE, NAME OF DELINEATOR, DATE OF THE DRAWING

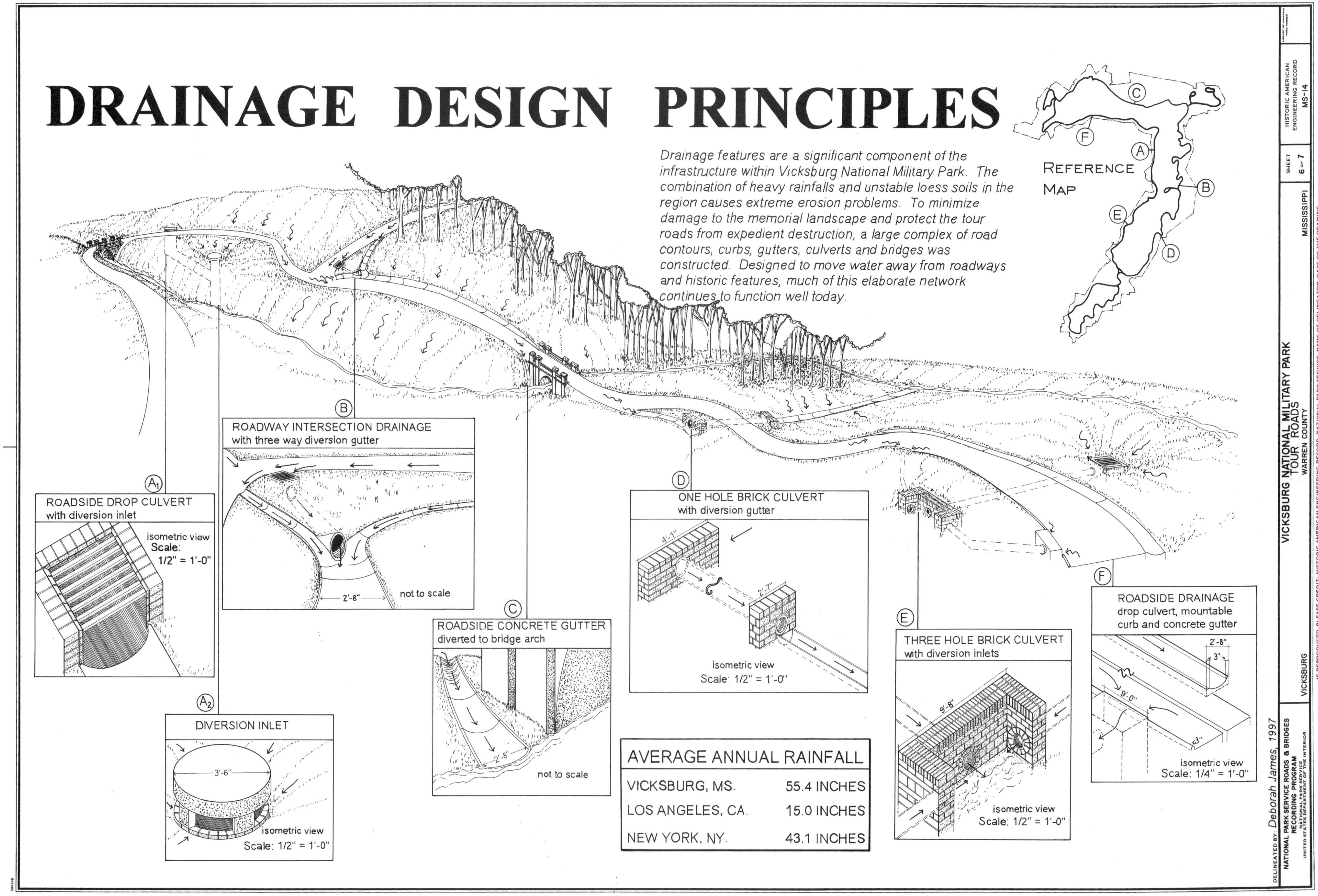
DRAINAGE DESIGN PRINCIPLES
Drainage features are a significant component of the infrastructure within Vicksburg National Military Park. The combination of heavy rainfalls and unstable loess soils in the region causes extreme erosion problems. To minimize damage to the memorial landscape and protect the tour roads from expedient destruction, a large complex of road contours, curbs, gutters, culverts and bridges was constructed. Designed to move water away from roadways and historic features, much of this elaborate network continues to function well today.
REFERENCE MAP
A
B
C
D
E
F
A1
ROADSIDE DROP CULVERT
with diversion inlet
isometric view
Scale: 1/2" = 1'-0"
A2
DIVERSION INLET
3'-6"
isometric view
Scale: 1/2" = 1'-0"
B
ROADWAY INTERSECTION DRAINAGE
with three way diversion gutter
2'-8"
not to scale
C
ROADSIDE CONCRETE GUTTER
diverted to bridge arch
2'-8"
not to scale
D
ONE HOLE BRICK CULVERT
with diversion gutter
4'-1"
2'-7"
isometric view
Scale: 1/2" = 1'-0"
AVERAGE ANNUAL RAINFALL
VICKSBURG, MS. 55.4 INCHES
LOS ANGELES, CA. 15.0 INCHES
NEW YORK, NY. 43.1 INCHES
E
THREE HOLE BRICK CULVERT
with diversion inlets
9'-8"
isometric view
Scale: 1/2" = 1'-0"
F
ROADSIDE DRAINAGE
drop culvert, mountable curb and concrete gutter
2'-8"
3"
9'-0"
3"
isometric view
Scale: 1/4" = 1'-0"
DELINEATED BY: Deborah James, 1997
NATIONAL PARK SERVICE ROADS & BRIDGES RECORDING PROGRAM
NATIONAL PARK SERVICE
UNITED STATES DEPARTMENT OF THE INTERIOR
VICKSBURG NATIONAL MILITARY PARK
TOUR ROADS
WARREN COUNTY
VICKSBURG
MISSISSIPPI
SHEET 6 of 7
HISTORIC AMERICAN ENGINEERING RECORD
MS-14
IF REPRODUCED, PLEASE CREDIT: HISTORIC AMERICAN ENGINEERING RECORD, NATIONAL PARK SERVICE, NAME OF DELINEATOR, DATE OF THE DRAWING

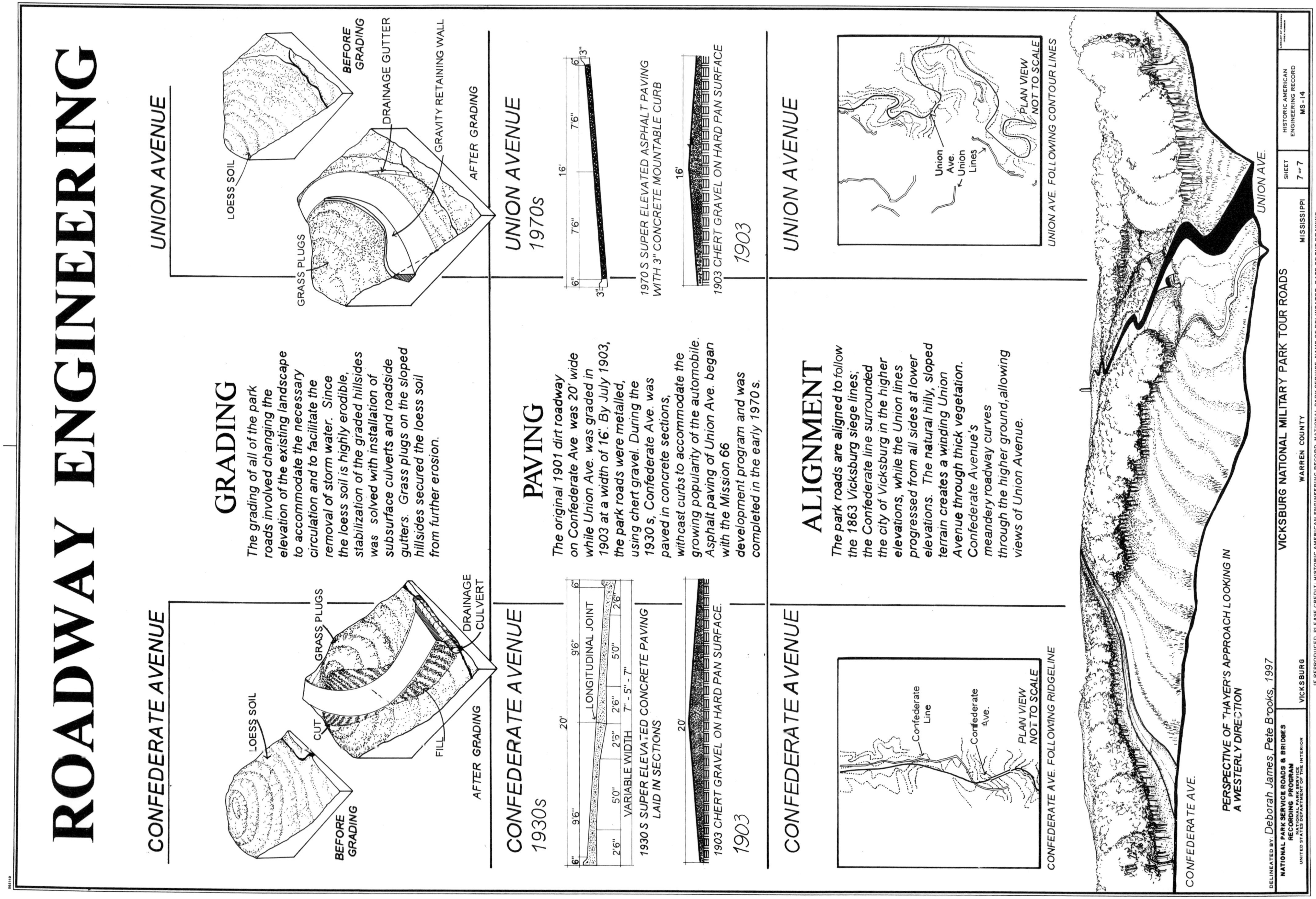
ROADWAY ENGINEERING
CONFEDERATE AVENUE
LOESS SOIL
BEFORE GRADING
CUT
GRASS PLUGS
FILL
DRAINAGE CULVERT
AFTER GRADING
GRADING
The grading of all of the park roads involved changing the elevation of the existing landscape to accommodate the necessary circulation and to facilitate the removal of storm water. Since the loess soil is highly erodible, stabilization of the graded hillsides was solved with installation of subsurface culverts and roadside gutters. Grass plugs on the sloped hillsides secured the loess soil from further erosion.
UNION AVENUE
LOESS SOIL
BEFORE GRADING
DRAINAGE GUTTER
GRAVITY RETAINING WALL
GRASS PLUGS
AFTER GRADING
CONFEDERATE AVENUE 1930s
LONGITUDINAL JOINT
VARIABLE WIDTH 7' - 5' - 7'
1930 S SUPER ELEVATED CONCRETE PAVING LAID IN SECTIONS
1903 CHERT GRAVEL ON HARD PAN SURFACE.
1903
PAVING
The original 1901 dirt roadway on Confederate Ave was 20' wide while Union Ave. was graded in 1903 at a width of 16'. By July 1903, the park roads were metalled, using chert gravel. During the 1930s, Confederate Ave. was paved in concrete sections, withcast curbs to accommodate the growing popularity of the automobile. Asphalt paving of Union Ave. began with the Mission 66 development program and was completed in the early 1970s.
UNION AVENUE 1970s
1970 S SUPER ELEVATED ASPHALT PAVING WITH 3" CONCRETE MOUNTABLE CURB
1903 CHERT GRAVEL ON HARD PAN SURFACE
1903
CONFEDERATE AVENUE
Confederate Line
Confederate Ave.
PLAN VIEW NOT TO SCALE
CONFEDERATE AVE. FOLLOWING RIDGELINE
ALIGNMENT
The park roads are aligned to follow the 1863 Vicksburg siege lines; the Confederate line surrounded the city of Vicksburg in the higher elevations, while the Union lines progressed from all sides at lower elevations. The natural hilly, sloped terrain creates a winding Union Avenue through thick vegetation. Confederate Avenue's meandery roadway curves through the higher ground, allowing views of Union Avenue.
UNION AVENUE
Union Ave.
Union Lines
PLAN VIEW NOT TO SCALE
UNION AVE. FOLLOWING CONTOUR LINES
CONFEDERATE AVE.
UNION AVE.
PERSPECTIVE OF "HAYER'S APPROACH LOOKING IN A WESTERLY DIRECTION
DELINEATED BY: Deborah James, Pete Brooks, 1997
NATIONAL PARK SERVICE ROADS & BRIDGES RECORDING PROGRAM
NATIONAL PARK SERVICE UNITED STATES DEPARTMENT OF THE INTERIOR
VICKSBURG NATIONAL MILITARY PARK TOUR ROADS
VICKSBURG
WARREN COUNTY
MISSISSIPPI
SHEET 7 OF 7
HISTORIC AMERICAN ENGINEERING RECORD MS-14
IF REPRODUCED, PLEASE CREDIT: HISTORIC AMERICAN ENGINEERING RECORD, NATIONAL PARK SERVICE, NAME OF DELINEATOR, DATE OF THE DRAWING

MELAN ARCH BRIDGES 1
1903

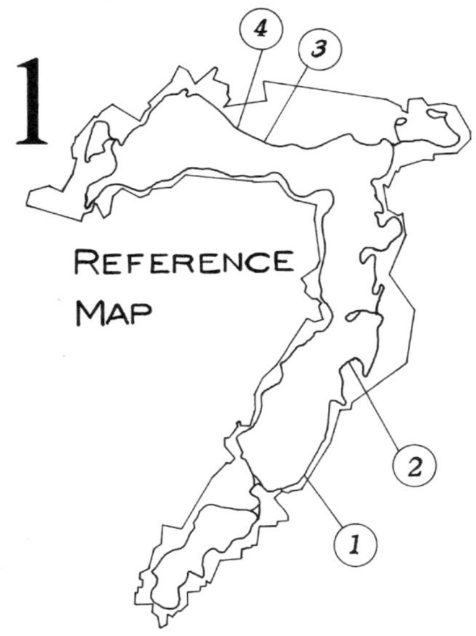

The construction of the Melan Arch Bridges began in July 1903 and was completed by December. The William T. Young Bridge Company won the bid for all nine of the Melan Arch Bridges to be built in the Vicksburg National Military Park. These small bridges were chosen not only to cross small streams and valleys through the Vicksburg National Military Park but also for the beauty they would bring to the park as a whole. Eight of the nine Melan Arch Bridges are still in use today. In 1997 the largest one was replaced with a box culvert after being declared unsafe for vehicular traffic.

Note: Drawings based on field measurements, field photographs, and historic photographs found in the archives of the Vicksburg National Military Park. For more information see H.A.E.R. field notebook.

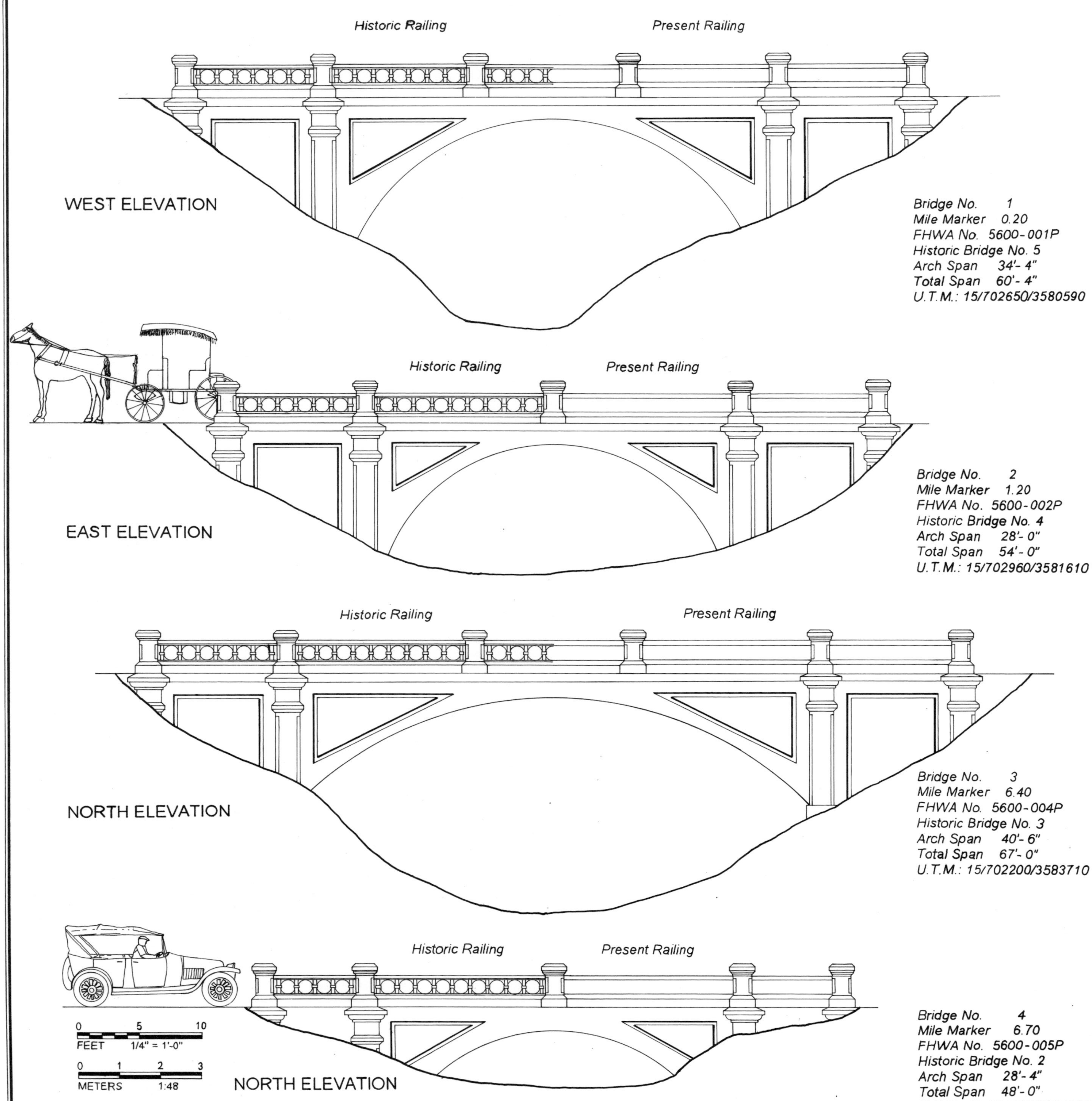

DELINEATED BY: Gregory Seale, 1997

NATIONAL PARK SERVICE ROADS & BRIDGES RECORDING PROGRAM NATIONAL PARK SERVICE UNITED STATES DEPARTMENT OF THE INTERIOR	VICKSBURG NATIONAL MILITARY PARK TOUR ROADS, MELAN ARCH BRIDGES VICKSBURG · WARREN COUNTY · MISSISSIPPI	SHEET 1 OF 3	HISTORIC AMERICAN ENGINEERING RECORD MS-14A

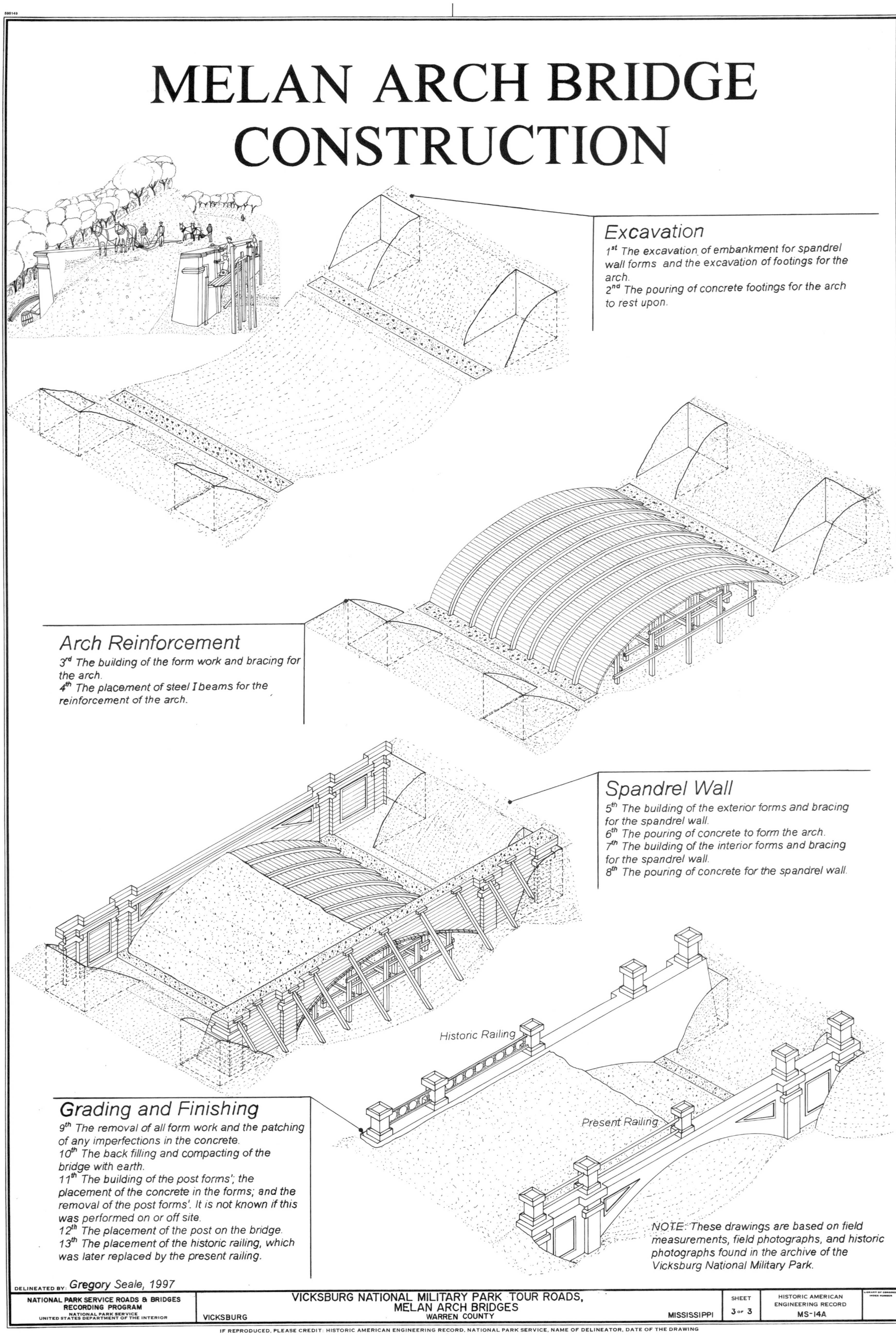
MELAN ARCH BRIDGE
CONSTRUCTION
Excavation
1st The excavation of embankment for spandrel wall forms and the excavation of footings for the arch.
2nd The pouring of concrete footings for the arch to rest upon.
Arch Reinforcement
3rd The building of the form work and bracing for the arch.
4th The placement of steel I beams for the reinforcement of the arch.
Spandrel Wall
5th The building of the exterior forms and bracing for the spandrel wall.
6th The pouring of concrete to form the arch.
7th The building of the interior forms and bracing for the spandrel wall.
8th The pouring of concrete for the spandrel wall.
Historic Railing
Present Railing
Grading and Finishing
9th The removal of all form work and the patching of any imperfections in the concrete.
10th The back filling and compacting of the bridge with earth.
11th The building of the post forms'; the placement of the concrete in the forms; and the removal of the post forms'. It is not known if this was performed on or off site.
12th The placement of the post on the bridge.
13th The placement of the historic railing, which was later replaced by the present railing.
NOTE: These drawings are based on field measurements, field photographs, and historic photographs found in the archive of the Vicksburg National Military Park.
DELINEATED BY: Gregory Seale, 1997
NATIONAL PARK SERVICE ROADS & BRIDGES RECORDING PROGRAM
NATIONAL PARK SERVICE
UNITED STATES DEPARTMENT OF THE INTERIOR
VICKSBURG NATIONAL MILITARY PARK TOUR ROADS,
MELAN ARCH BRIDGES
WARREN COUNTY
VICKSBURG
MISSISSIPPI
SHEET
3 of 3
HISTORIC AMERICAN ENGINEERING RECORD
MS-14A
IF REPRODUCED, PLEASE CREDIT: HISTORIC AMERICAN ENGINEERING RECORD, NATIONAL PARK SERVICE, NAME OF DELINEATOR, DATE OF THE DRAWING

CONFEDERATE AVENUE
STEEL BRIDGE
1903

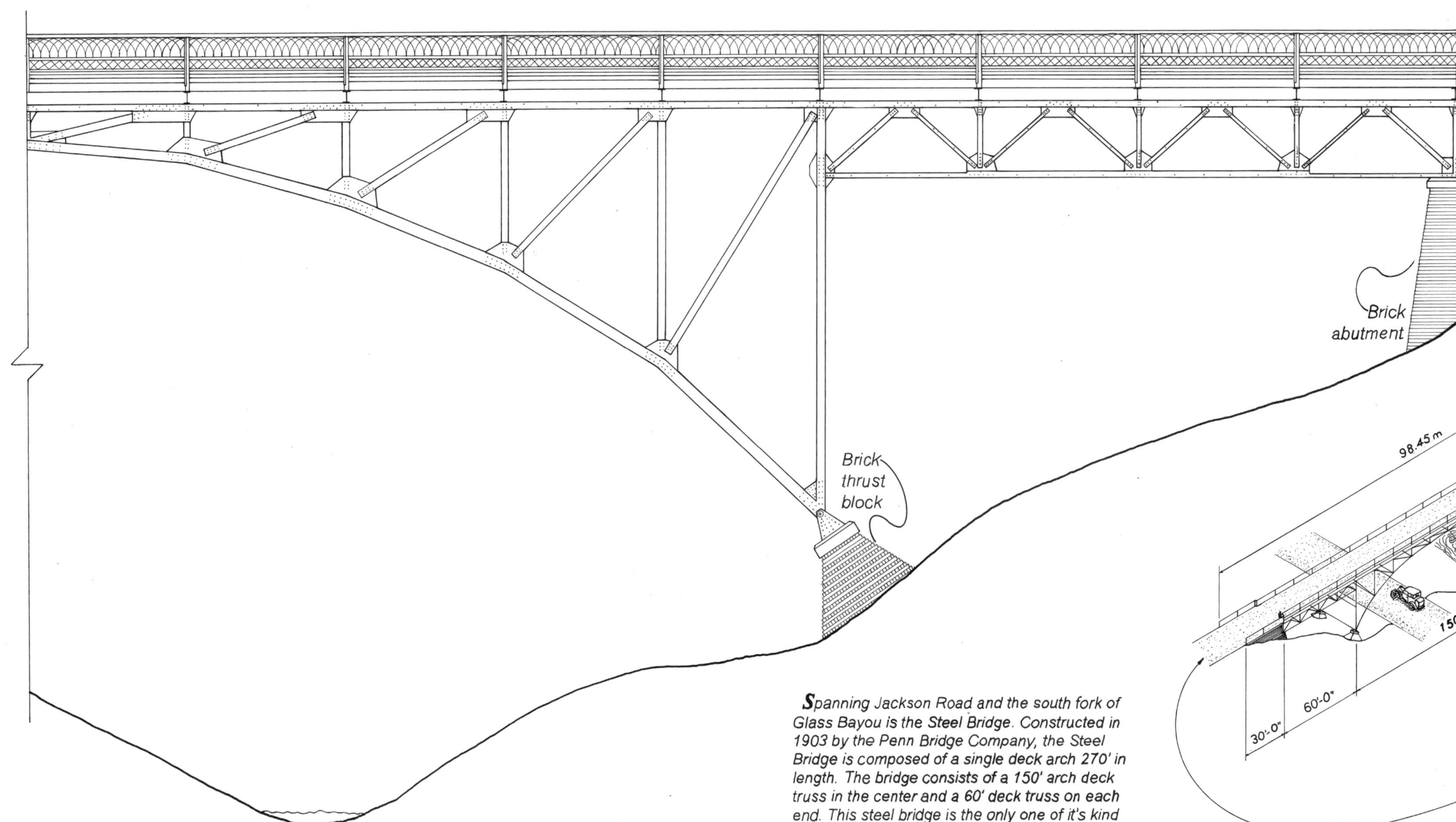

0 5 10 15 20
FEET 3/16" = 1'-0"
0 1 2 3 4 5
METERS 1:64

HALF WEST ELEVATION

U.T.M.: 15/702662/3582400

S*panning Jackson Road and the south fork of Glass Bayou is the Steel Bridge. Constructed in 1903 by the Penn Bridge Company, the Steel Bridge is composed of a single deck arch 270' in length. The bridge consists of a 150' arch deck truss in the center and a 60' deck truss on each end. This steel bridge is the only one of it's kind existing in the state of Mississippi.*

Note: Drawings based on field measurements, field photographs, and historic photographs found in the archives of the Vicksburg National Military Park. For more information see H.A.E.R. field notebook.

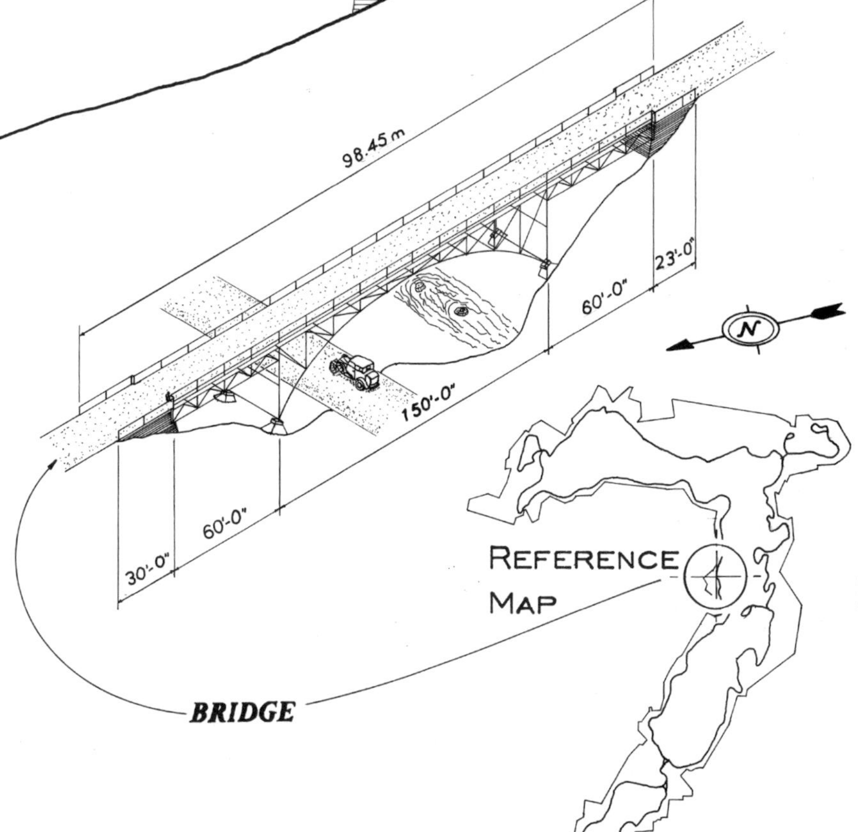

DELINEATED BY: Gregory Seale, 1997

NATIONAL PARK SERVICE ROADS & BRIDGES RECORDING PROGRAM
NATIONAL PARK SERVICE
UNITED STATES DEPARTMENT OF THE INTERIOR

STEEL BRIDGE - 1903
CONFEDERATE AVENUE OVER JACKSON ROAD AND NORTH FORK GLASS BAYOU
VICKSBURG WARREN COUNTY MISSISSIPPI

SHEET 1 OF 2

HISTORIC AMERICAN ENGINEERING RECORD
MS-12

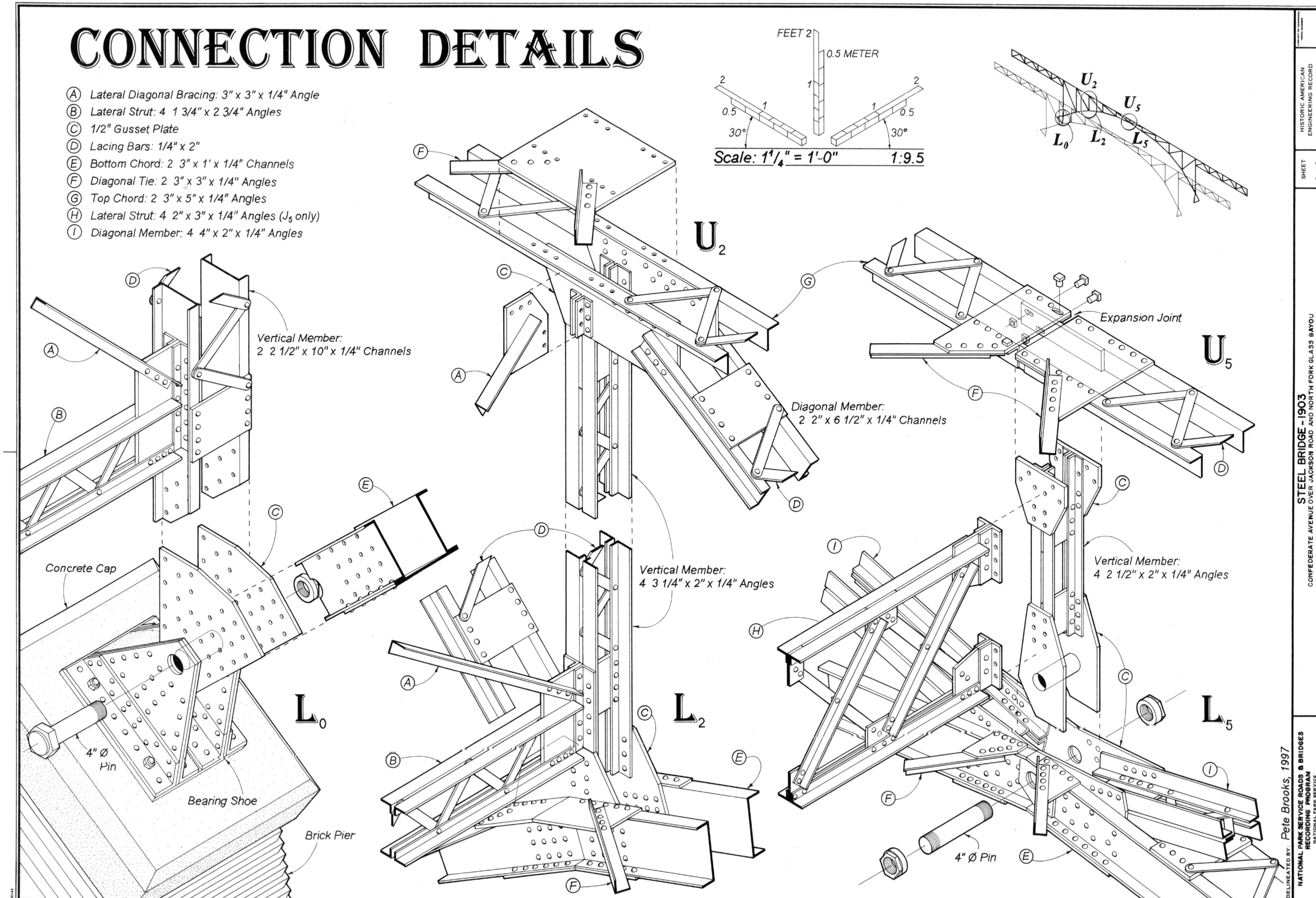
CONNECTION DETAILS
(A) Lateral Diagonal Bracing: 3" x 3" x 1/4" Angle
(B) Lateral Strut: 4 1 3/4" x 2 3/4" Angles
(C) 1/2" Gusset Plate
(D) Lacing Bars: 1/4" x 2"
(E) Bottom Chord: 2 3" x 1' x 1/4" Channels
(F) Diagonal Tie: 2 3" x 3" x 1/4" Angles
(G) Top Chord: 2 3" x 5" x 1/4" Angles
(H) Lateral Strut: 4 2" x 3" x 1/4" Angles (J5 only)
(I) Diagonal Member: 4 4" x 2" x 1/4" Angles
FEET 2
0.5 METER
30°
Scale: 1 1/4" = 1'-0"
1:9.5
U2
U5
L0
L2
L5
Vertical Member: 2 2 1/2" x 10" x 1/4" Channels
Diagonal Member: 2 2" x 6 1/2" x 1/4" Channels
Expansion Joint
Vertical Member: 4 3 1/4" x 2" x 1/4" Angles
Vertical Member: 4 2 1/2" x 2" x 1/4" Angles
Concrete Cap
4" Ø Pin
Bearing Shoe
Brick Pier
4" Ø Pin
DELINEATED BY: Pete Brooks, 1997
NATIONAL PARK SERVICE ROADS & BRIDGES RECORDING PROGRAM
NATIONAL PARK SERVICE
UNITED STATES DEPARTMENT OF THE INTERIOR
STEEL BRIDGE - 1903
CONFEDERATE AVENUE OVER JACKSON ROAD AND NORTH FORK GLASS BAYOU
WARREN COUNTY
VICKSBURG
MISSISSIPPI
SHEET 2 OF 2
HISTORIC AMERICAN ENGINEERING RECORD
MS-12
IF REPRODUCED, PLEASE CREDIT: HISTORIC AMERICAN ENGINEERING RECORD, NATIONAL PARK SERVICE, NAME OF DELINEATOR, DATE OF THE DRAWING

Park Road Precedents

National park roads and parkways both influenced and were influenced by developments outside the National Park System. The National Park Service drew on a long tradition of scenic road building, while precedents set in national park road construction spread to utilitarian and recreational road development at the local, state, and national levels.

The primary challenge facing recreational road builders at the beginning of the twentieth century involved finding ways to accommodate the demands of automobile traffic without compromising traditional standards of aesthetic quality. Two projects that wielded enormous influence in this regard were Oregon's Columbia River Highway and New York's Bronx River Parkway.

The Columbia River Highway was one of the first lengthy scenic drives built expressly for automobiles. It was widely praised for the aesthetic appeal of its bridges, visitor facilities, and vistas, as well as for technical advancements that included modest grades, gently winding alignment, and a patented pavement that was smoother and more durable than conventional surfaces.

New York's Bronx River Parkway was hailed as an even more significant breakthrough. By updating traditional parkway design techniques to meet the needs of recreational motorists, the project shaped the course of twentieth-century parkway development. The Park Service relied heavily on Bronx River Parkway design precedents, hiring key personnel from the design team to help guide the development of national parkways.

HAER documented the Columbia River Highway and Bronx River Parkway with techniques developed through the NPS Park Roads and Bridges Program. Other significant scenic roads recorded during the course of the program included the Taconic State Parkway, which extended the Bronx River Parkway north into upstate New York; Connecticut's Merritt Parkway, which helped popularize the commuter parkway concept; and the Arroyo Seco Parkway, which played a pivotal role in the transition from parkways to freeways. Funding for these projects was provided by various state and local sources.

Historic view of Bronx River Parkway, 1925

HAER and Westchester County Records and Archives Center

Park Road Precedents

Bronx River Parkway Reservation

Cover Sheet	340
Site Map	341
Reclaiming the River	342
The Evolving Parkway Landscape	343
Transforming the Land	344
Landscape Development	345
Parkway Principles	346
Parkway Sections	347
More Than a Motorway	348
Characteristic Road Details	349
The Rigid Frame Concrete Arch	350
Woodland Viaduct	351
Crane Road Bridge	352
Parkway Bridges, Sheet 1	353
Parkway Bridges, Sheet 2	354
Parkway Bridges, Sheet 3	355
Managing the Parkway	356
General Development Plan (1918)	357

Historic Columbia River Highway

Cover Sheet	362
Evolution of Transportation	363
Experiencing the Landscape	364
Scenic Inspiration	365
Grading and Alignment	366
Paving and Drainage	367
Railings	368
Masonry	369
Viaducts	370
Tunnels	371

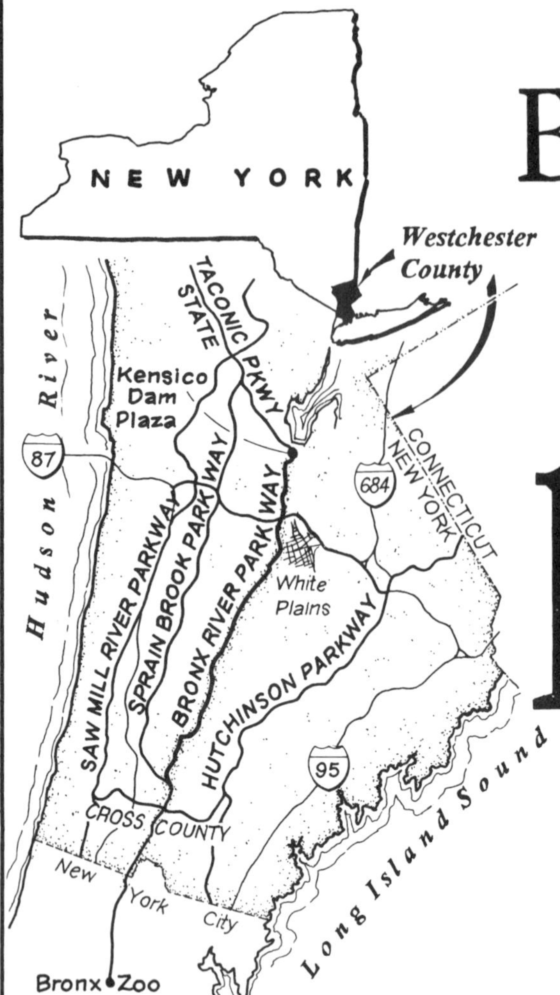

BRONX RIVER PARKWAY RESERVATION

1907 1925

Broad Street Viaduct, ca. 1924, replaced by the Cross County Parkway. From "Report of the Westchester County Park Commission, 1925"

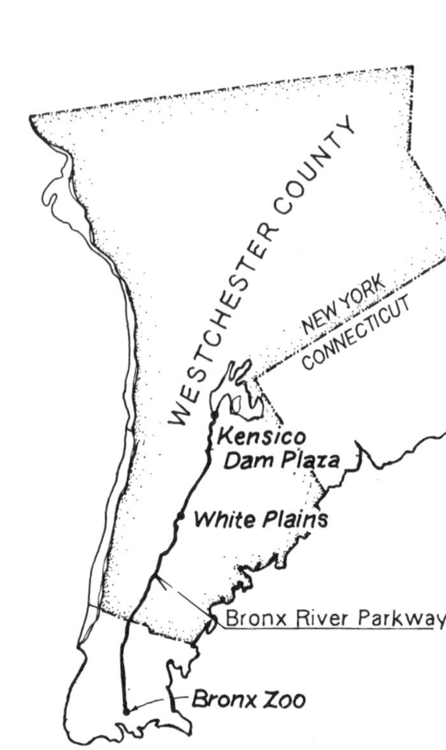

The Bronx River Parkway Reservation was the first public parkway designed explicitly for automobile use. The project began as an environmental restoration and park development initiative, however, aimed at transforming the heavily polluted Bronx River into an attractive linear park. The Bronx Parkway Commission was appointed to undertake this task in 1906. The project was completed in 1925, when the parkway was integrated into the rapidly growing Westchester County park system.

Parkway elements included rejuvenated woodlands, open meadows, water features, recreational pathways, and an attractive scenic drive that highlighted the Bronx River as the leading landscape feature. Designers emphasized a naturalistic appearance with carefully planned scenic vistas, clearings, and artistically composed tree plantings. The parkway drive became a pioneering example of modern motorway development. Its designers combined beauty, safety, and efficiency by reducing the number of dangerous intersections, limiting access from surrounding streets and businesses, and surrounding motorists in a broad swath of landscaped greenery.

The Bronx River Parkway was designed and developed by an exemplary team of engineers, landscape architects, architects, and dedicated amateur park supporters. The Bronx Parkway Commission consisted of Madison Grant, James Cannon, William Niles, and Frank Bethell. Jay Downer served as chief engineer. Consulting landscape architect Hermann Merkell was largely responsible for the parkway's overall design. Gilmore Clarke was the field supervisor for landscape development. Engineers Leslie Holloran and Arthur Hayden also made important contributions. Hayden's rigid-frame bridge design revolutionized the construction of motorway grade-separations. Architectural treatment of parkway bridges was contracted to a number of independent designers including Charles Stoughton, William Delano, and Thomas Hastings.

During recent decades, the parkway drive was widened and modern safety elements were installed in many locations. Much of the Westchester County portion of the parkway continues to fulfill its original function as a multi-purpose recreation, transportation, and environmental protection development, but south of Bronxville the six-lane road retains little historic integrity. The Bronx River Parkway Reservation north of Bronxville was placed on the National Register of Historic Places in 1991.

The Bronx River Parkway Reservation Recording Project was undertaken during the summer of 2001 and is part of the Historic American Engineering Record (HAER), a long range program to document historically significant engineering and industrial works in the United States. HAER (Eric DeLony, Chief) is administered by the Historic American Buildings Survey/Historic American Engineering Record (E. Blaine Cliver, Chief), a division of the National Park Service, U.S. Department of the Interior. The project was funded by Westchester County (Andrew J. Spano, County Executive) and was administered by the Dept. of Parks, Recreation and Conservation (Joseph Stout, Commissioner; Dr. Jack Robbins, Deputy Commissioner of Conservation; David DeLucia, Dir. of Park Facilities). The Dept. of Information Technology (Norman Jacknis, Chief Info. Officer; Patty Dohrenwend, Dir., Westchester County Archives) and the Bronx River Parkway Reservation Conservancy (Lowell Tooley, President; Betsy Dolan, Executive Director) provided additional support.

The drawings, historical report, and photography were completed under the direction of Christopher Marston, Project Leader, and Tim Davis, Program Historian. The field team consisted of Field Supervisor Tanya Folger (SUNY College of Environmental Science and Forestry); architects Brandon Andow (Rhode Island School of Design) and Kamalahasan Ramaswamy (US/ICOMOS, India); landscape architect Karolina Buczek (US/ICOMOS, Poland); and historian Dawn Duensing. Large-format photography was produced by David Haas.

DELINEATED BY: *Tanya Folger, Karolina Buczek, Pete Brooks, Christopher Marston, 2001*

NATIONAL PARK SERVICE
UNITED STATES DEPARTMENT OF THE INTERIOR

BRONX RIVER PARKWAY RESERVATION
THE BRONX TO KENSICO DAM
WESTCHESTER COUNTY

WHITE PLAINS VICINITY

NEW YORK

SHEET 1 of 22

HISTORIC AMERICAN ENGINEERING RECORD
NY-327

IF REPRODUCED, PLEASE CREDIT: HISTORIC AMERICAN ENGINEERING RECORD, NATIONAL PARK SERVICE, NAME OF DELINEATOR, DATE OF THE DRAWING

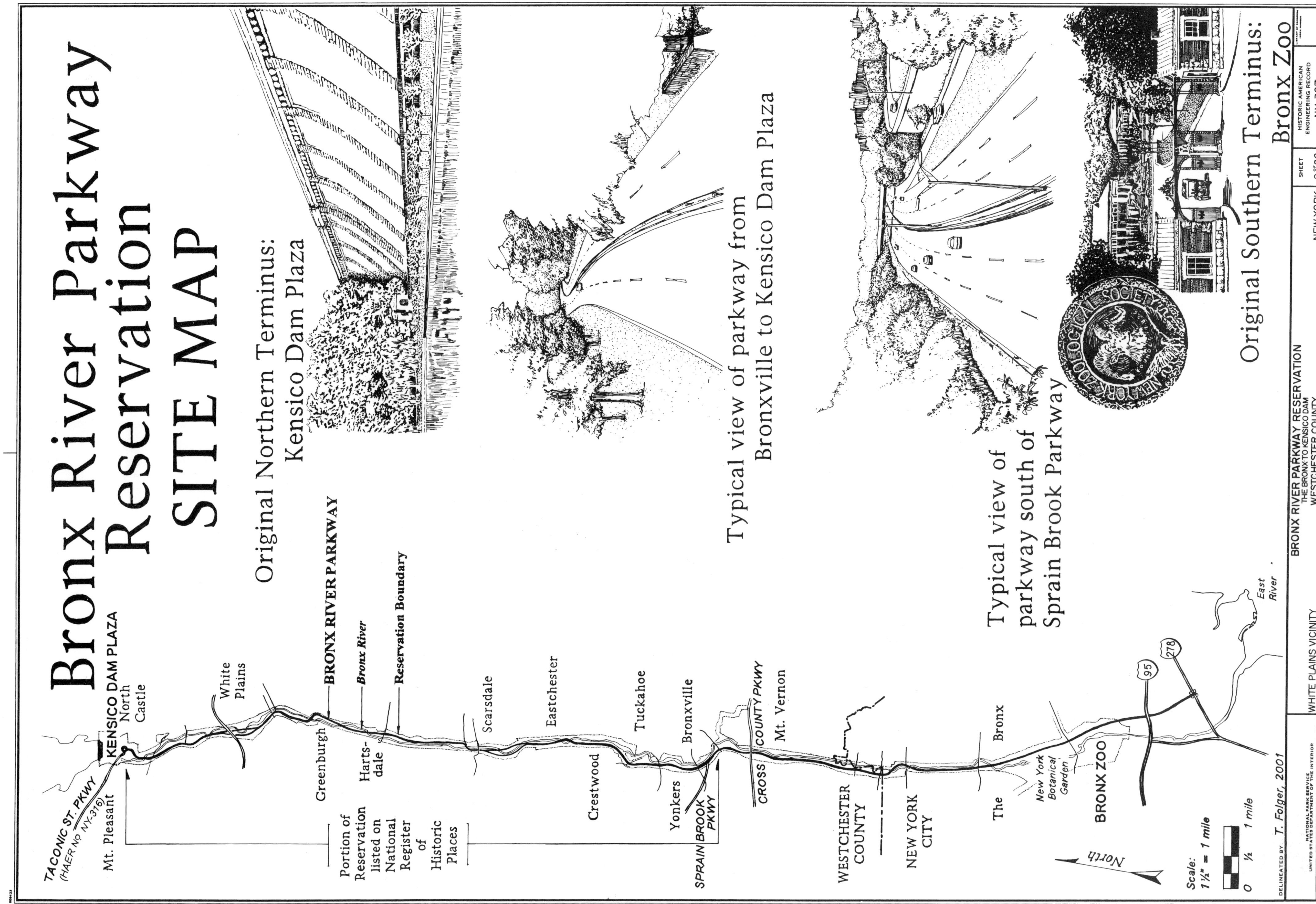
Bronx River Parkway
Reservation
SITE MAP
Original Northern Terminus:
Kensico Dam Plaza
Typical view of parkway from
Bronxville to Kensico Dam Plaza
Typical view of
parkway south of
Sprain Brook Parkway
Original Southern Terminus:
Bronx Zoo
TACONIC ST. PKWY
(HAER NO. NY-316)
Mt. Pleasant
KENSICO DAM PLAZA
North
Castle
White
Plains
BRONX RIVER PARKWAY
Bronx River
Reservation Boundary
Greenburgh
Harts-
dale
Scarsdale
Eastchester
Crestwood
Tuckahoe
Yonkers
Bronxville
SPRAIN BROOK
PKWY
CROSS COUNTY PKWY
Mt. Vernon
WESTCHESTER
COUNTY
NEW YORK
CITY
The Bronx
New York
Botanical
Garden
BRONX ZOO
95
278
East
River
Portion of
Reservation
listed on
National
Register
of
Historic
Places
North
Scale:
1½" = 1 mile
0 ½ 1 mile
DELINEATED BY: T. Folger, 2001
WHITE PLAINS VICINITY
BRONX RIVER PARKWAY RESERVATION
THE BRONX TO KENSICO DAM
WESTCHESTER COUNTY
NEW YORK
SHEET
2 of 22
HISTORIC AMERICAN
ENGINEERING RECORD
NY-327

RECLAIMING THE RIVER

The initial motivation for creating the Bronx River Parkway Reservation was to restore the Bronx River valley to a semblance of its original natural beauty. At the beginning of the 20th century the upper river valley retained much of its scenic charm, but most of the stream south of White Plains was severely degraded by unconstrained development. The Bronx Parkway Commission was created in 1906 to identify the primary sources of pollution and undertake the ambitious project of reclaiming the river.

Raw Sewage

Like most urban streams of its day, the Bronx River served as an open sewer, collecting residential waste from the surrounding neighborhoods. Human sewage emptied directly into the stream, creating foul conditions that prevented people from enjoying the river and threatened animals at the nearby Bronx Zoo. An enclosed trunk sewer was built to protect the river around 1905, but many residences and businesses continued to pollute until the parkway commission compelled them to stop.

Dumping

Urban streams often served as public dumping grounds. The parkway commission spent several years removing refuse and debris from the Bronx River to create a clean unobstructed stream. A 1916 report observed: "Buggies, tea-kettles, bicycles, wagon wheels, boxes, bedsprings, automobile bodies, wash boilers, stoves, utensils of every description, and particularly hot-water heating tanks, were brought forth in astonishing numbers."

Factories

A wide variety of industrial and commercial enterprises discharged pollution directly into the Bronx River. A tannery, coal yards, gas works, and trolley car barns were among the worst polluters but many smaller businesses contributed to the problem. Industrial facilities and the railroad that paralleled the river were also prominent sources of visual pollution, disfiguring the scenery throughout the proposed reservation.

Household Waste & Agricultural Run-off

Numerous small farms, stables, and low-income residences were located right next to the river. Manure from the farms and stables washed directly into the Bronx River. Pollutants from privies leached through the soil, posing additional health risks. The parkway commission viewed the cheap houses, stables, and utilitarian gardens as visual blight that should be replaced by attractive naturalistic scenery.

Billboards & Flooding

Billboard removal was a top priority. By 1907 over seven miles of billboards stretched along the river, advertising to passengers on the nearby railroad. During spring thaws and periods of heavy rainfall the normally shallow Bronx River routinely overflowed its banks. The floodwaters spread sewage over adjacent lowlands and accumulated in stagnant pools that offered fertile breeding grounds for disease-carrying mosquitoes. The parkway commission rechanneled and dredged the river to minimize flooding and create a more free-flowing stream.

Natural Beauty

The Bronx River valley retained appealing vestiges of attractive natural scenery, which the parkway commission sought to preserve and enhance. Butler Woods was one of the region's most charming surviving forests. Located north of Scarsdale, the grove contained brooks and waterfalls and featured large stands of hemlocks, mountain laurel, and northern hardwoods. Prominent resident Emily Butler donated the property to the parkway reservation in 1913.

DELINEATED BY: *Karolina Buczek 2001*

NATIONAL PARK SERVICE
UNITED STATES DEPARTMENT OF THE INTERIOR

BRONX RIVER PARKWAY RESERVATION
THE BRONX TO KENSICO DAM
WESTCHESTER COUNTY

WHITE PLAINS VICINITY

NEW YORK

SHEET 3 OF 22

HISTORIC AMERICAN ENGINEERING RECORD
NY-327

THE EVOLVING PARKWAY LANDSCAPE

The Bronx Parkway Commission (BPC) compiled a comprehensive series of photographs to document the evolution of the parkway landscape. Appearing in annual reports, magazine and newspaper articles, these images celebrated the transformation of the Bronx River from a polluted eyesore into an attractive recreational amenity. The parkway landscape has continued to evolve since the project's completion in 1925, but not always in the manner envisioned by the original designers.

The BPC received widespread acclaim for cleaning up the polluted Bronx River and transforming an unsightly region of factories and low-income housing into an attractive environment for leisure and recreation. Subsequent growth has obscured much of the original composition, where the carefully sculpted river wound through luxuriant meadows dotted with artfully placed trees and shrubs.

Before

After

Current (2001)

Before the parkway was developed, billboards stretched along most of the Bronx River, often lining both sides of the railroad tracks that carried their intended audience. The parkway commission removed over seven miles of signs, replacing the garish commercial display with picturesque scenery composed in the tradition of Anglo-American landscape painting and park design. Many of the original plantings are fully mature, creating a soothing woodland retreat from the surrounding city and suburbs.

The cheap, flood-prone land along the lower Bronx River valley had developed into a mixture of loosely regulated commercial and industrial enterprises, tenements, and small homes served by a confusing array of local streets. Parkway construction transformed this haphazard urban sprawl into a comprehensively designed linear park. The free-flowing parkway drive wound in graceful S-curves through a broad expanse of greenery, providing an attractive and restful experience for commuters and recreational drivers. The roadway was significantly changed in the 1950s-60s to accommodate increasingly heavy traffic. New lanes were added and portions of the roadway were significantly straightened, making it more efficient but less park-like.

DELINEATED BY: Karolina Buczek 2001

NATIONAL PARK SERVICE
UNITED STATES DEPARTMENT OF THE INTERIOR

WHITE PLAINS VICINITY

BRONX RIVER PARKWAY RESERVATION
THE BRONX TO KENSICO DAM
WESTCHESTER COUNTY

NEW YORK

SHEET 4 OF 22

HISTORIC AMERICAN ENGINEERING RECORD
NY-327

TRANSFORMING THE LAND

Existing neighborhoods

Over 370 buildings were removed or demolished to make room for the parkway. Many of these structures were shacks, tenements, stables and other utilitarian buildings that the parkway commission cast as unsightly and socially undesirable. A number of substantial residences and commercial establishments were also purchased or condemned. One of the most densely settled areas was in White Plains, where a large Italian-American population was displaced to further parkway development.

Moving houses

Buildings on land acquired for the parkway were either demolished or sold to the highest bidder. Many were moved to new locations. Houses were jacked up onto log rollers and then pulled away by teams of horses. The Bronx Parkway Commission stipulated that all buildings be moved at least 300′ from the parkway borders so they would not disrupt the desired naturalistic effect.

Cleaning

Once the buildings were removed, crews broke down the foundations and buried them or carted them away. Cleanup crews also removed curbs, pavement, and assorted trash and cleared away unwanted trees, stumps, bushes, and weeds, creating a clean slate for the development of the new parkway landscape.

Grading

The parkway grounds were extensively regraded to eliminate signs of previous uses and create an inviting landscape of smooth fields and gently rolling terrain. Topsoil was spread on areas destined for planting and the future roadbed was prepared for paving with a solid, well-drained subgrade.

Dredging and Excavation

Dredging a deeper watercourse helped control flooding by lowering the water table and containing heavy rains. The huge amount of soil and gravel excavated to improve the river was used to help fill in surrounding wetlands and provide a solid base for the parkway drive. Topsoil was carefully conserved and redistributed prior to final grading and planting. Two million tons of earth were moved during the course of these operations. A special low-profile electric-powered dredging machine allowed the excavation to proceed under existing bridges and low-hanging trees. Surplus military equipment from World War I contributed significantly to the excavation process.

The Bronx River Parkway appears to be a natural landscape, but most of what the viewer sees today is a man-made environment. The parkway commission went to great lengths to transform a region of low-cost housing, mixed agricultural and industrial uses dumps, and swamps into an attractive landscape devoted to leisure and recreation. The transformed landscape combined environmental restoration with carefully calculated aesthetic enhancements to create an idealized version of what the area might have looked like prior to urban development. Tenements factories, and seriously degraded riverbanks gave way to green meadows and shady groves flanking an attractive but extensively reconfigured stream.

Transforming the debris-choked Bronx River into an attractive naturalistic stream and reclaiming the surrounding wetlands were monumental undertakings. Powerful dredges and draglines were employed to create a more free-flowing watercourse that was both attractive and environmentally stable. At least five miles of new channel were excavated to separate the river from the railroad and create a more appealingly "naturalistic" curvilinear effect.

DELINEATED BY: *Karolina Buczek 2001*

NATIONAL PARK SERVICE
UNITED STATES DEPARTMENT OF THE INTERIOR

WHITE PLAINS VICINITY

BRONX RIVER PARKWAY RESERVATION
THE BRONX TO KENSICO DAM
WESTCHESTER COUNTY

NEW YORK

SHEET 5 OF 22

HISTORIC AMERICAN ENGINEERING RECORD
NY-327

LANDSCAPE DEVELOPMENT

"**a humanized naturalness...**
sufficiently diversified
to create woodland groupings
and vistas of all types"

Hermann W. Merkel
Landscape Architect, 1918

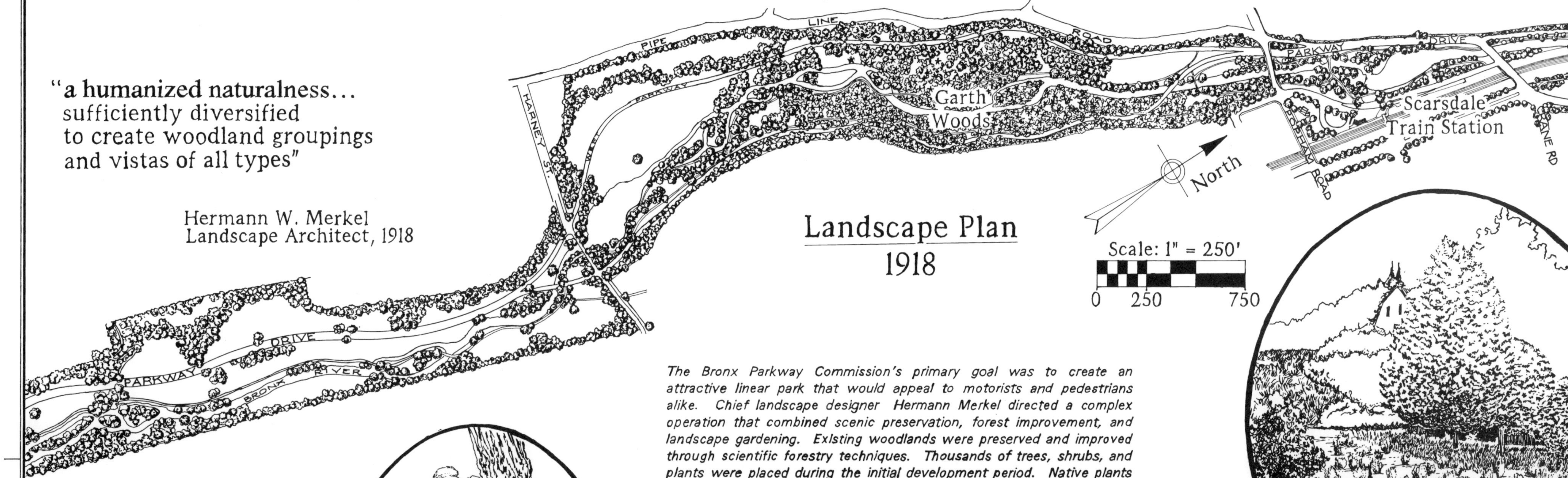

Landscape Plan
1918

The Bronx Parkway Commission's primary goal was to create an attractive linear park that would appeal to motorists and pedestrians alike. Chief landscape designer Hermann Merkel directed a complex operation that combined scenic preservation, forest improvement, and landscape gardening. Existing woodlands were preserved and improved through scientific forestry techniques. Thousands of trees, shrubs, and plants were placed during the initial development period. Native plants predominated, but the parkway was not meant to recreate the primeval forest. It was an artistically composed "naturalistic" landscape designed to produce a pleasing diversity of scenic effects. Meadows, lakes, and trees were artfully interspersed to provide contrast and variety. Both pedestrian and motorist experiences were considered, yielding intimate walks and broad sweeping vistas.

Completed Landscape

Pruning & Trimming

17,000 dead trees removed
140,000 shrubs planted
57,000 trees trimmed
30,000 trees planted
21,000 trees treated

Pesticide Treatment

Mitigating Decay

Planting

DELINEATED BY: *T. Folger, 2001*

NATIONAL PARK SERVICE
UNITED STATES DEPARTMENT OF THE INTERIOR

WHITE PLAINS VICINITY

BRONX RIVER PARKWAY RESERVATION
THE BRONX TO KENSICO DAM
WESTCHESTER COUNTY

NEW YORK

SHEET 6 OF 22

HISTORIC AMERICAN ENGINEERING RECORD
NY-327

PARKWAY PRINCIPLES

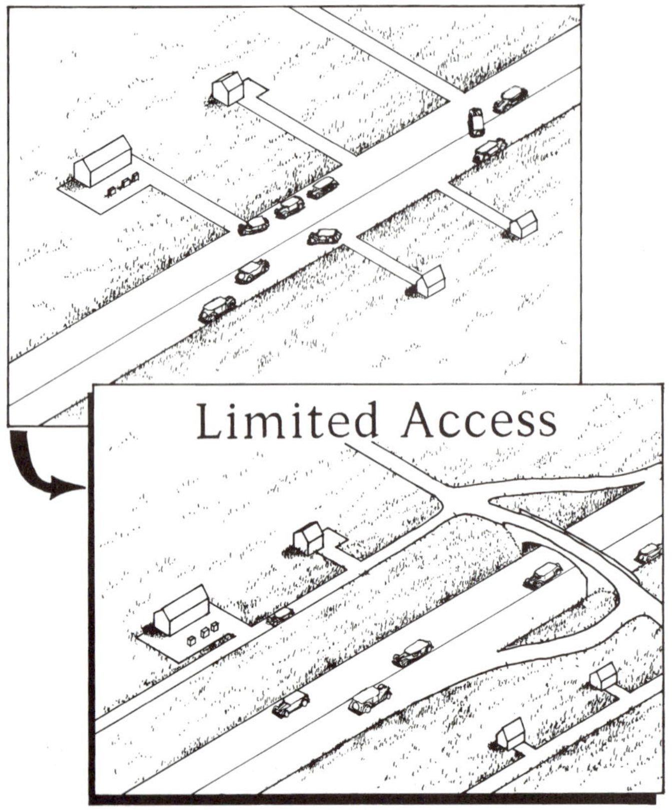

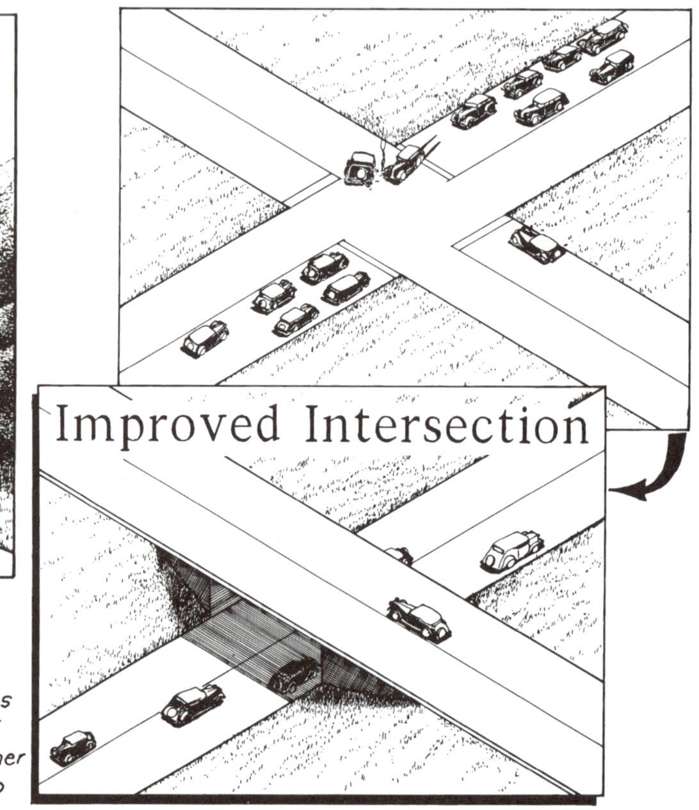

The Bronx River Parkway employed a series of design features that became hallmarks of parkway development throughout the nation. Since most of these enhancements improved safety and efficiency as well as aesthetics, the basic design principles embodied in the Bronx River Parkway were adopted by mainstream highway engineers and played a prominent role in the development of the modern high-speed motorway.

Grade-separated interchanges dramatically improved safety at intersections. Limiting access from surrounding streets and properties further enhanced safety and efficiency while eliminating the unsightly commercial establishments found along many roadways. A broad right-of-way provided additional protection against billboards and other visual intrusions while offering designers more freedom in placing the roadway and enhancing the surrounding landscape. The roadway was laid out in graceful curves that followed the contours of the land, offering greater driving ease and safety at higher speeds. Medians improved safety and could be used to add visual interest or preserve attractive landscape features. The parkway landscape was comprehensively designed to offer modern motorist an updated version of traditional park scenery.

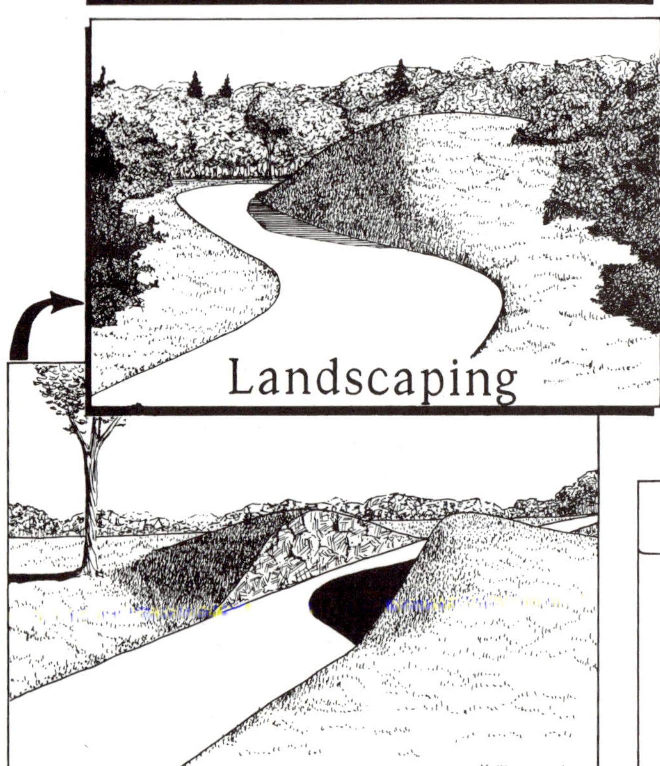

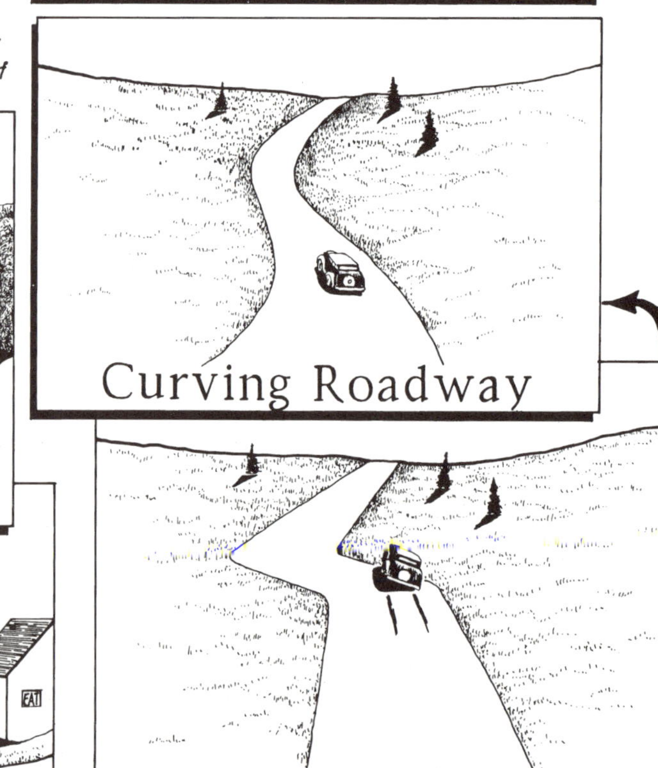

DELINEATED BY: Brandon Andow, 2001

NATIONAL PARK SERVICE
UNITED STATES DEPARTMENT OF THE INTERIOR

WHITE PLAINS VICINITY

BRONX RIVER PARKWAY RESERVATION
THE BRONX TO KENSICO DAM
WESTCHESTER COUNTY

NEW YORK

SHEET
7 OF 22

HISTORIC AMERICAN
ENGINEERING RECORD
NY-327

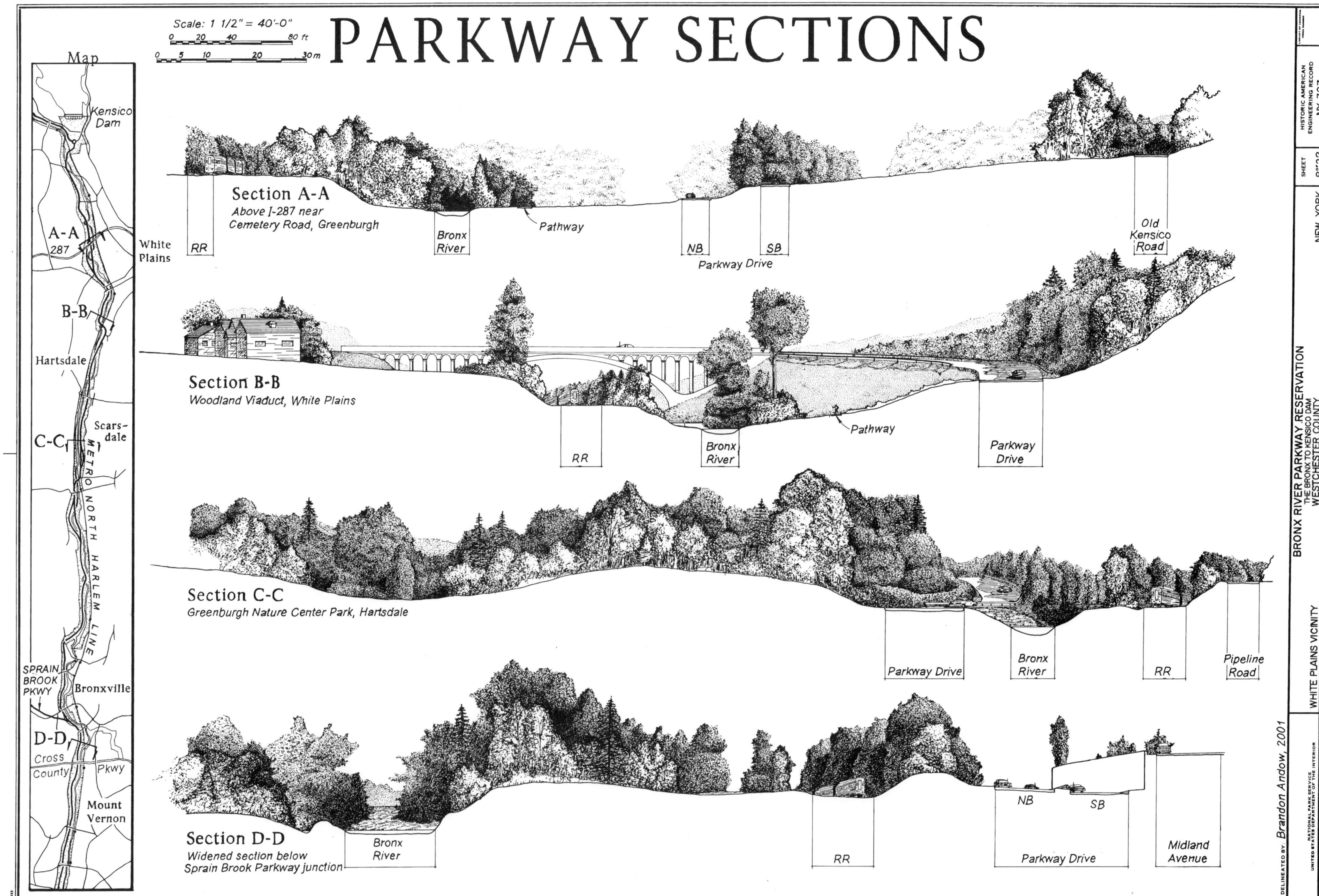
PARKWAY SECTIONS
Scale: 1 1/2" = 40'-0"
0 20 40 80 ft
0 5 10 20 30m
Map
Kensico Dam
A-A
287
White Plains
B-B
Hartsdale
Scars-dale
C-C
METRO NORTH HARLEM LINE
SPRAIN BROOK PKWY
Bronxville
D-D
Cross County Pkwy
Mount Vernon
Section A-A
Above I-287 near Cemetery Road, Greenburgh
RR
Bronx River
Pathway
NB
SB
Parkway Drive
Old Kensico Road
Section B-B
Woodland Viaduct, White Plains
RR
Bronx River
Pathway
Parkway Drive
Section C-C
Greenburgh Nature Center Park, Hartsdale
Parkway Drive
Bronx River
RR
Pipeline Road
Section D-D
Widened section below Sprain Brook Parkway junction
Bronx River
RR
NB
SB
Parkway Drive
Midland Avenue
DELINEATED BY: Brandon Andow, 2001
NATIONAL PARK SERVICE
UNITED STATES DEPARTMENT OF THE INTERIOR
BRONX RIVER PARKWAY RESERVATION
THE BRONX TO KENSICO DAM
WESTCHESTER COUNTY
WHITE PLAINS VICINITY
NEW YORK
SHEET 8 OF 22
HISTORIC AMERICAN ENGINEERING RECORD
NY-327
IF REPRODUCED, PLEASE CREDIT: HISTORIC AMERICAN ENGINEERING RECORD, NATIONAL PARK SERVICE, NAME OF DELINEATOR, DATE OF THE DRAWING

MORE THAN A MOTORWAY

The Bronx River Parkway has always been more than a motorway. In the eyes of its original creators, the parkway's primary purpose was to transform a polluted river valley into an attractive linear park full of recreational opportunities and attractive scenery. The Bronx Parkway Commission developed a variety of recreational facilities aimed at appealing to a wide range of tastes and social preferences.

Many of these activities centered around the Bronx River, which also provided the primary source of scenic inspiration. The commission developed a number of swimming pools along the river and encouraged canoeing, skating, and related activities. Ball fields were created in newly cleared areas to offer mass recreation and spectator sports. Tennis courts catered to a more elite audience. Holiday events were also held in the parkway's open areas and the Boy Scouts were allowed to build a rustic cabin to support their mission. A varied array of bridle trails, footpaths, and picnic areas afforded opportunities for quieter and more solitary or family-oriented leisure pursuits.

The Bronx River Parkway continues to serve as a vital recreational resource. Local residents and visitors enjoy a similarly diverse range of activities today, though the mixture of sports has changed over time. Horseback riding may not be as popular and the Bronx River no longer serves as a primary location for swimming or canoeing, but other activities have appeared or grown in popularity. Bicycling is more popular than ever; in-line skating has gained many adherents, and jogging and other forms of active outdoor recreation are on the rise. Strolling, picnicking, bird watching, and the leisurely enjoyment of nature continue to draw eager enthusiasts to the parkway's scenic splendor.

In 1974 The Westchester County Department of Parks, Recreation and Conservation began a policy of closing portions of the main parkway drive to automobiles on selected Sundays. The happy crowds of bicyclists, skaters, and walkers enjoying the smooth winding pavement underscore the parkway's fundamental role as a broad-based recreational resource.

DELINEATED BY: *Karolina Buczek 2001*

NATIONAL PARK SERVICE
UNITED STATES DEPARTMENT OF THE INTERIOR

WHITE PLAINS VICINITY

BRONX RIVER PARKWAY RESERVATION
THE BRONX TO KENSICO DAM
WESTCHESTER COUNTY

NEW YORK

SHEET
9 OF 22

HISTORIC AMERICAN
ENGINEERING RECORD
NY-327

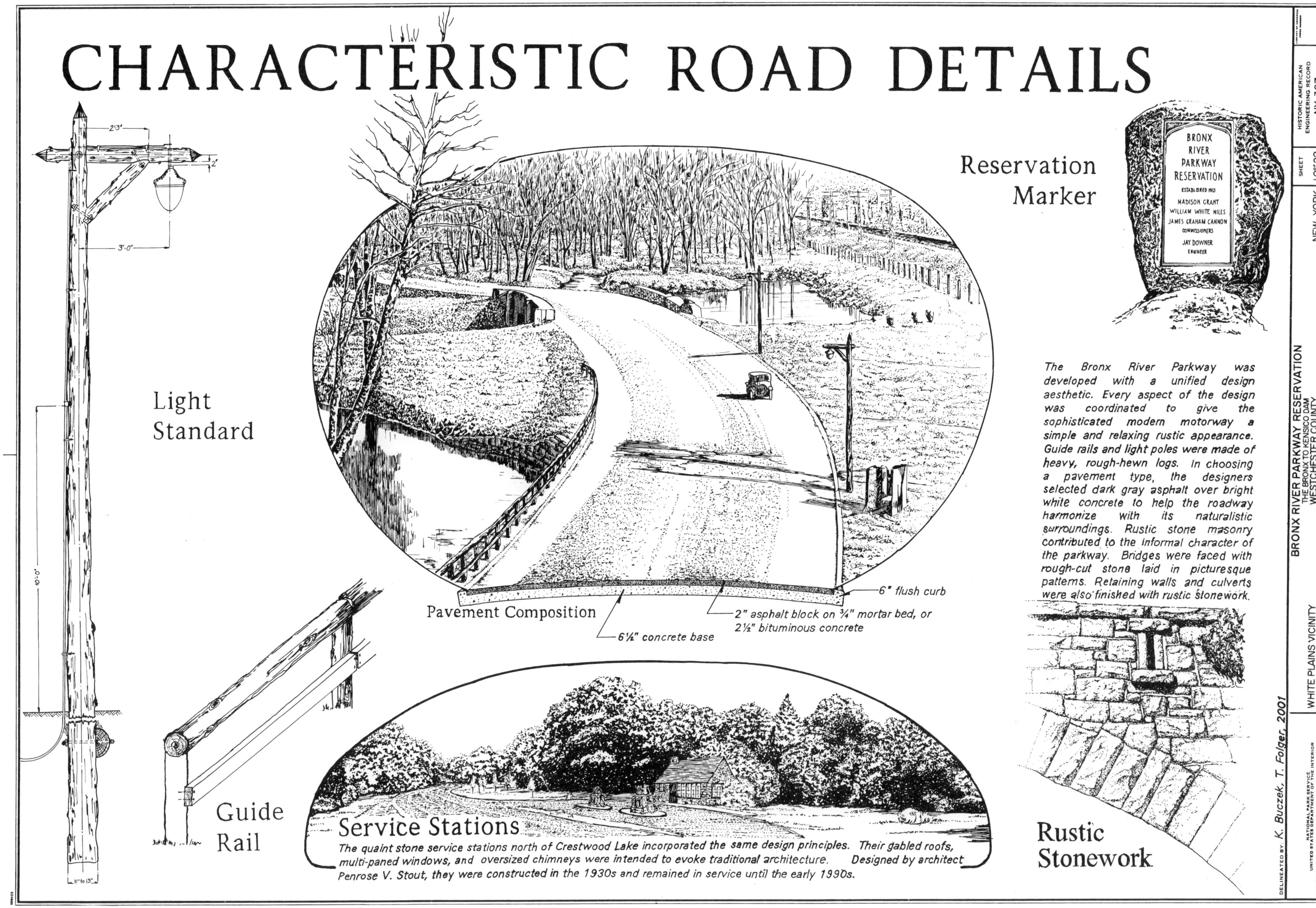
CHARACTERISTIC ROAD DETAILS
2'3"
2'
3'-0"
Light Standard
10'-0"
11" to 13"
Guide Rail
Reservation Marker
BRONX
RIVER
PARKWAY
RESERVATION
ESTABLISHED 1913
MADISON GRANT
WILLIAM WHITE NILES
JAMES GRAHAM CANNON
COMMISSIONERS
JAY DOWNER
ENGINEER
The Bronx River Parkway was developed with a unified design aesthetic. Every aspect of the design was coordinated to give the sophisticated modern motorway a simple and relaxing rustic appearance. Guide rails and light poles were made of heavy, rough-hewn logs. In choosing a pavement type, the designers selected dark gray asphalt over bright white concrete to help the roadway harmonize with its naturalistic surroundings. Rustic stone masonry contributed to the informal character of the parkway. Bridges were faced with rough-cut stone laid in picturesque patterns. Retaining walls and culverts were also finished with rustic stonework.
Pavement Composition
6" flush curb
2" asphalt block on ¾" mortar bed, or 2½" bituminous concrete
6½" concrete base
Service Stations
The quaint stone service stations north of Crestwood Lake incorporated the same design principles. Their gabled roofs, multi-paned windows, and oversized chimneys were intended to evoke traditional architecture. Designed by architect Penrose V. Stout, they were constructed in the 1930s and remained in service until the early 1990s.
Rustic Stonework
DELINEATED BY: K. Buczek, T. Folger, 2001
NATIONAL PARK SERVICE
UNITED STATES DEPARTMENT OF THE INTERIOR
WHITE PLAINS VICINITY
BRONX RIVER PARKWAY RESERVATION
THE BRONX TO KENSICO DAM
WESTCHESTER COUNTY
NEW YORK
SHEET
10 OF 22
HISTORIC AMERICAN ENGINEERING RECORD
NY-327
IF REPRODUCED, PLEASE CREDIT: HISTORIC AMERICAN ENGINEERING RECORD, NATIONAL PARK SERVICE, NAME OF DELINEATOR, DATE OF THE DRAWING

THE RIGID FRAME CONCRETE ARCH

Frame Bridge Elevation

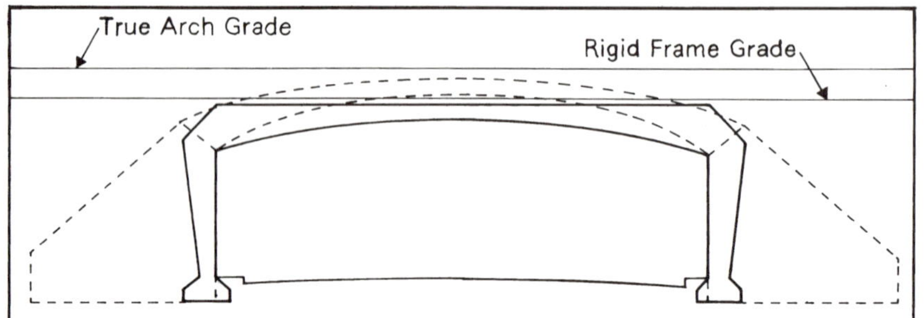

Comparison to True Arch Bridge

The rigid-frame concrete arch bridge was developed by Bronx Parkway Commission engineer Arthur G. Hayden in 1922. This innovative design derived from the need to construct a large number of attractive spans for bridges and grade-separated interchanges along the Bronx River Parkway.

Prior to Hayden's innovation, most grade-separations were constructed with conventional arches, which required much larger structures to provide the same amount of useable vertical clearance, or with steel or concrete girders, which offered uniform clearance but were considered unattractive for parkway use. The rigid-frame arch retained a suggestion of the true arch's graceful curvature but offered a broader span of maximum clearance. Its strength derived from a rigid connection between horizontal and vertical members that spread the load more evenly throughout the entire structure. The rigid-frame arch was also relatively cheap to construct. Its greater structural strength permitted a thinner cross section and required considerably less concrete to stabilize the abutments. The more efficient profile also produced significant savings in related construction costs since the overall height difference between the crossing roadways was minimized and less fill was required to construct suitable approaches to each grade-separation.

The rigid-frame bridge was ideal for parkway use because it was economically and structurally efficient, had a pleasing profile, and was readily adaptable toa wide variety of architectural treatments. It was widely used in parkway development throughout the country and spread rapidly to utilitarian roadways, where its practical qualities made it ideal for large-scale production.

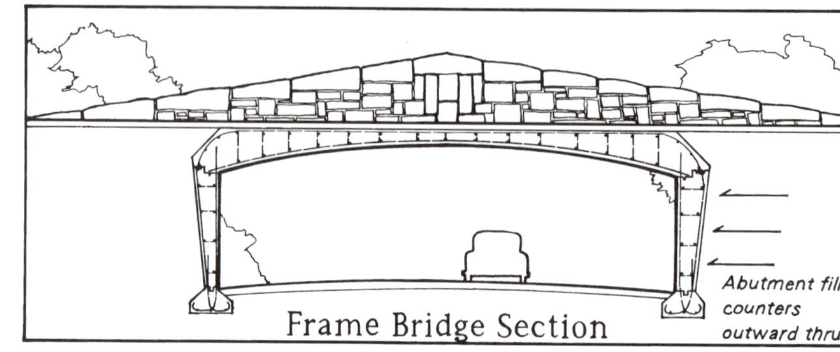

Frame Bridge Section

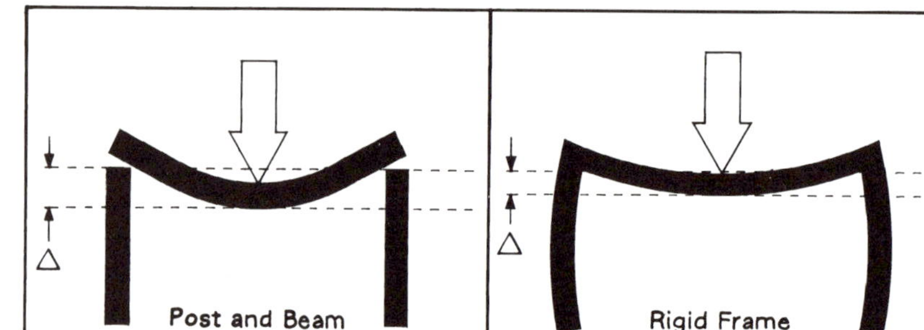

Structural Principles

Construction Process

1.

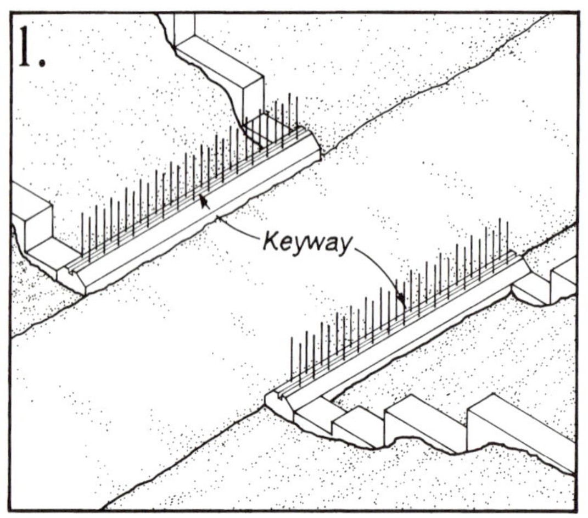

1. Footings

Footings for the bridge and abutment wing walls are formed and poured. Reinforcing bars are carried 3'-4' above footings so they can be spliced to reinforcement of vertical legs. Keyways cleaned to accept succeeding pours.

2.

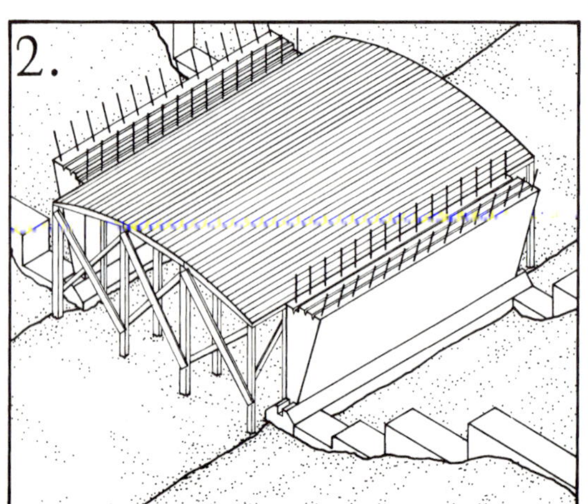

2. Falsework and Vertical Legs

Falsework for concrete arch and stone arch rings is erected. Vertical legs are formed and poured. Any architectural detailing on the inside face of vertical legs is cast with the forms.

3.

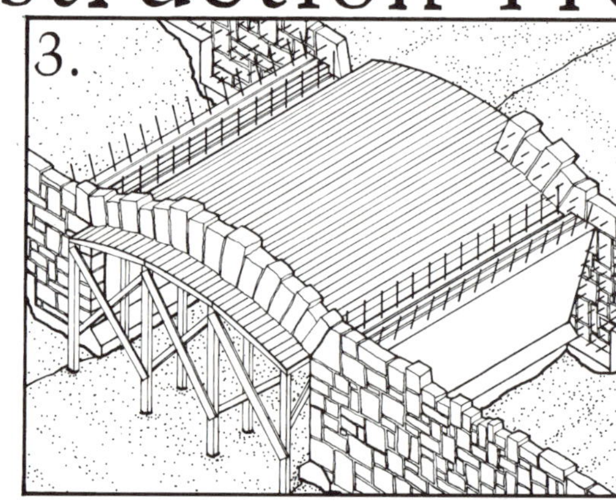

3. Stone Arch and Wing Walls

Stone arch rings and wing wall facings are laid. Stones are secured to concrete by steel anchors embedded in joints and hooked into drill holes on back side of stones.

4.

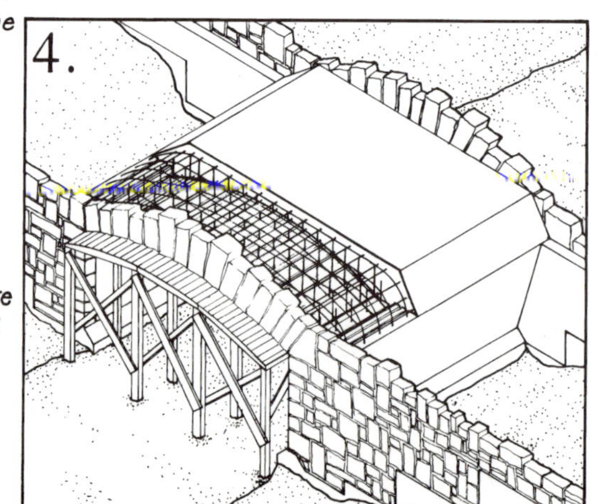

4. Concrete Arch and Wing Walls

The concrete arch and abutment wing walls are cast against the stones. The concrete arch can be cast in one piece or in multiple longitudinal bays.

5.

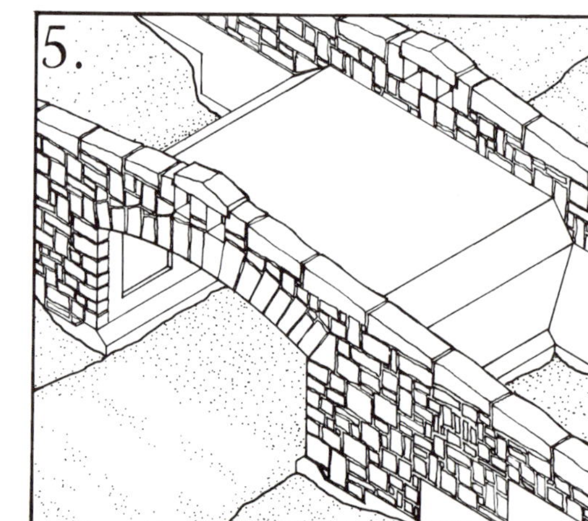

5. Finish Masonry

Masonry parapets are completed and finished with coping stones.

6.

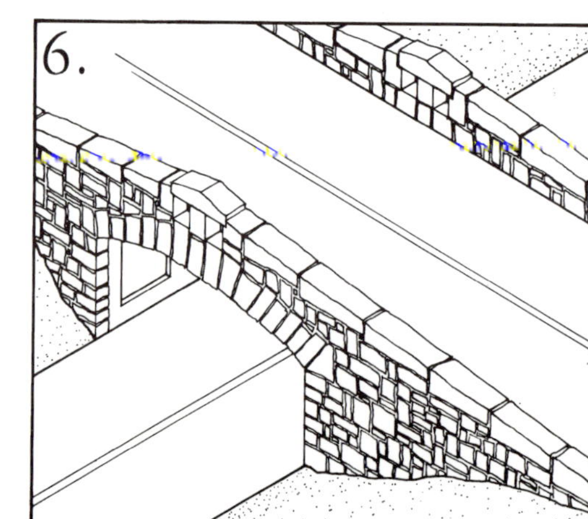

6. Grading and Paving

Abutments are backfilled when roadway approaches are filled to the desired grade. Roadways are then paved with a bituminous wearing surface.

DELINEATED BY: *Brandon Andow, 2001*

NATIONAL PARK SERVICE
UNITED STATES DEPARTMENT OF THE INTERIOR

WHITE PLAINS VICINITY

BRONX RIVER PARKWAY RESERVATION
THE BRONX TO KENSICO DAM
WESTCHESTER COUNTY

NEW YORK

SHEET
11 OF 22

HISTORIC AMERICAN ENGINEERING RECORD
NY-327

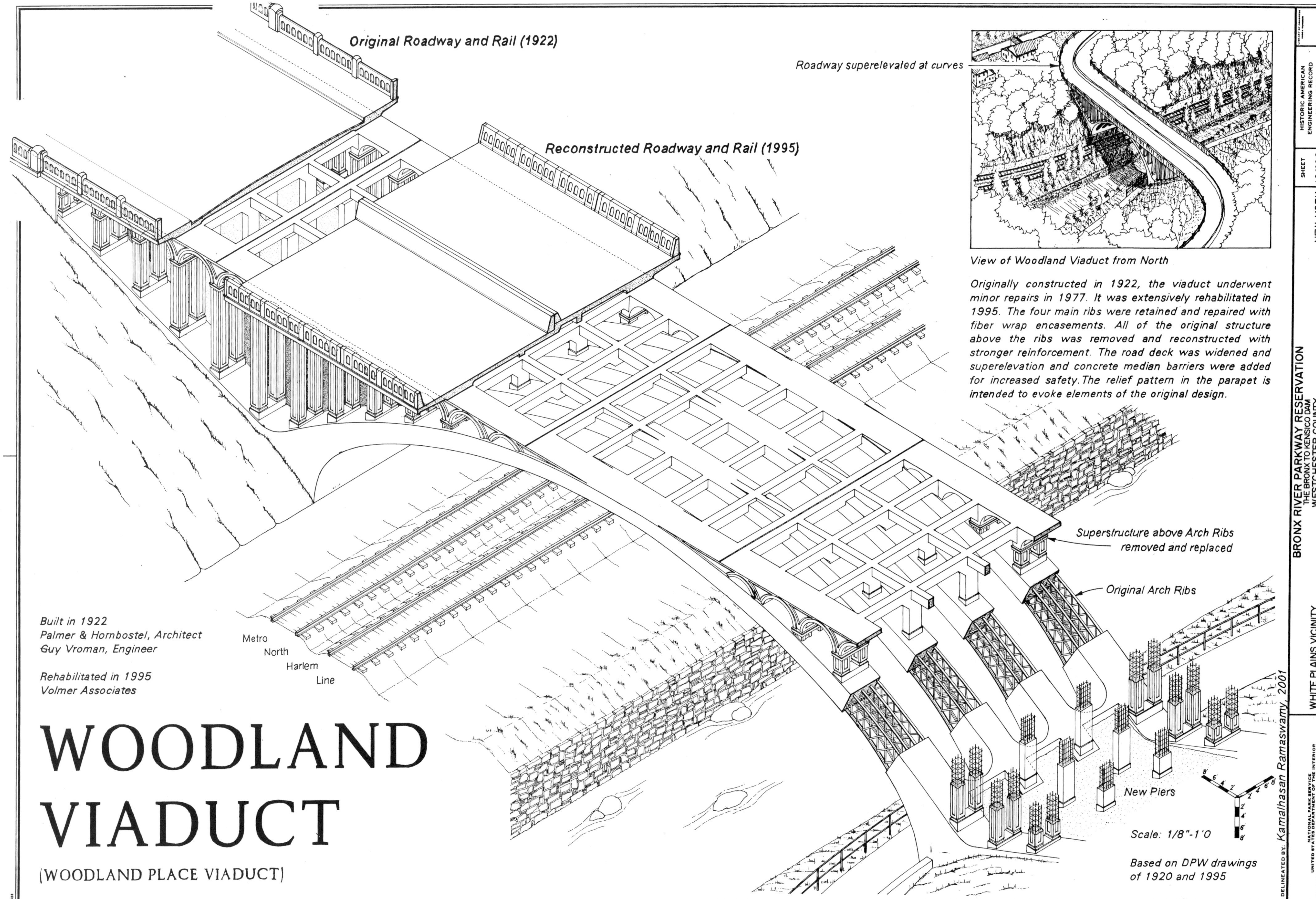
Original Roadway and Rail (1922)
Reconstructed Roadway and Rail (1995)
Roadway superelevated at curves
View of Woodland Viaduct from North
Originally constructed in 1922, the viaduct underwent minor repairs in 1977. It was extensively rehabilitated in 1995. The four main ribs were retained and repaired with fiber wrap encasements. All of the original structure above the ribs was removed and reconstructed with stronger reinforcement. The road deck was widened and superelevation and concrete median barriers were added for increased safety. The relief pattern in the parapet is intended to evoke elements of the original design.
Superstructure above Arch Ribs removed and replaced
Original Arch Ribs
New Piers
Scale: 1/8"-1'0
Based on DPW drawings of 1920 and 1995
Built in 1922
Palmer & Hornbostel, Architect
Guy Vroman, Engineer
Rehabilitated in 1995
Volmer Associates
Metro North Harlem Line
WOODLAND VIADUCT
(WOODLAND PLACE VIADUCT)
DELINEATED BY: Kamalhasan Ramaswamy, 2001
NATIONAL PARK SERVICE
UNITED STATES DEPARTMENT OF THE INTERIOR
WHITE PLAINS VICINITY
BRONX RIVER PARKWAY RESERVATION
THE BRONX TO KENSICO DAM
WESTCHESTER COUNTY
NEW YORK
SHEET 12 OF 22
HISTORIC AMERICAN ENGINEERING RECORD
NY-327
IF REPRODUCED, PLEASE CREDIT: HISTORIC AMERICAN ENGINEERING RECORD, NATIONAL PARK SERVICE, NAME OF DELINEATOR, DATE OF THE DRAWING

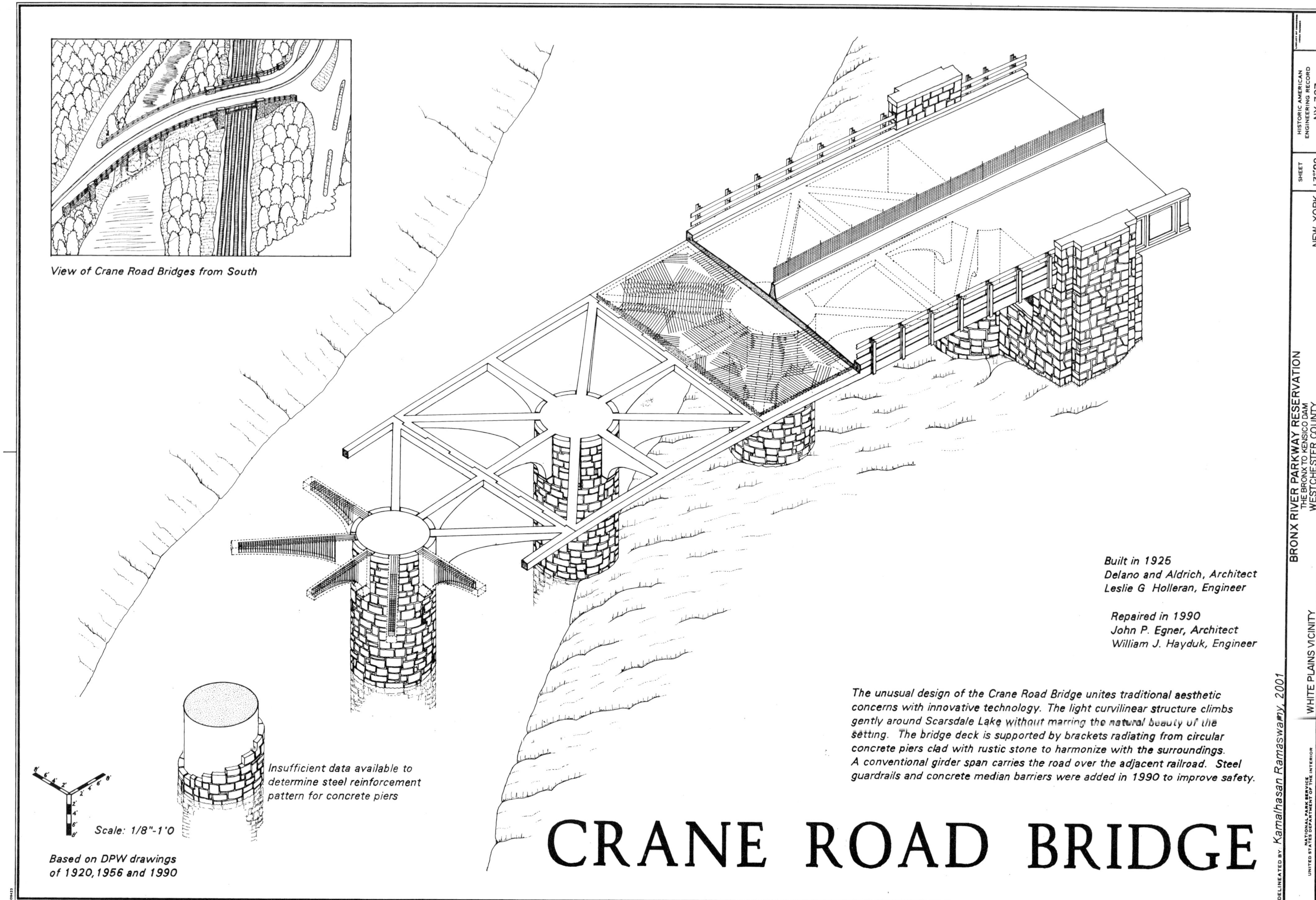

View of Crane Road Bridges from South
Built in 1925
Delano and Aldrich, Architect
Leslie G Holleran, Engineer
Repaired in 1990
John P. Egner, Architect
William J. Hayduk, Engineer
The unusual design of the Crane Road Bridge unites traditional aesthetic concerns with innovative technology. The light curvilinear structure climbs gently around Scarsdale Lake without marring the natural beauty of the setting. The bridge deck is supported by brackets radiating from circular concrete piers clad with rustic stone to harmonize with the surroundings. A conventional girder span carries the road over the adjacent railroad. Steel guardrails and concrete median barriers were added in 1990 to improve safety.
Insufficient data available to determine steel reinforcement pattern for concrete piers
Scale: 1/8"-1'0
Based on DPW drawings of 1920, 1956 and 1990
CRANE ROAD BRIDGE
DELINEATED BY: Kamalhasan Ramaswamy, 2001
NATIONAL PARK SERVICE
UNITED STATES DEPARTMENT OF THE INTERIOR
BRONX RIVER PARKWAY RESERVATION
THE BRONX TO KENSICO DAM
WESTCHESTER COUNTY
WHITE PLAINS VICINITY
NEW YORK
SHEET
13 of 22
HISTORIC AMERICAN ENGINEERING RECORD
NY-327
IF REPRODUCED, PLEASE CREDIT: HISTORIC AMERICAN ENGINEERING RECORD, NATIONAL PARK SERVICE, NAME OF DELINEATOR, DATE OF THE DRAWING

Scale: 3/32" = 1'-0"

0 5 10 15 20 25 30 ft

0 1 2 3 4 5 6 7 8 9 10 m

PARKWAY BRIDGES

Sheet 1 of 3

Based on original DPW drawings
and field measurements

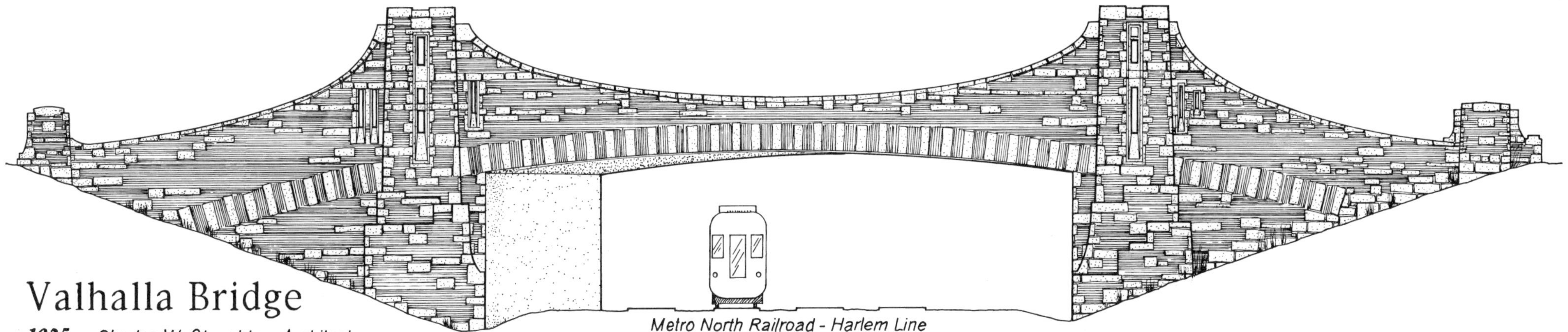

Valhalla Bridge

1925 *Charles W. Stoughton, Architect*
Arthur G. Hayden, Engineer
Modified in 1992 *to carry only northbound traffic.*

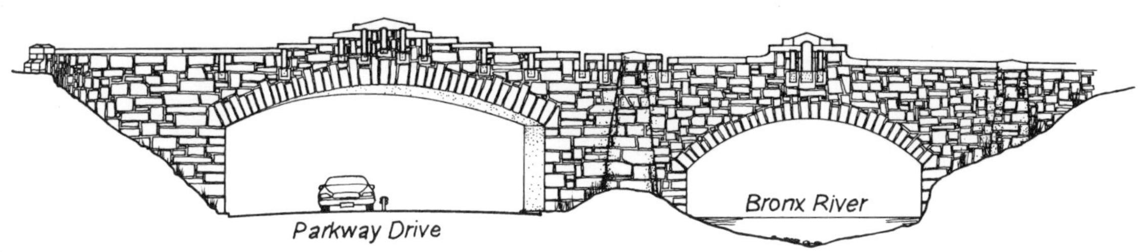

Main Street Bridges

1924 *Charles W. Stoughton, Architect* *Arthur G. Hayden, Engineer*

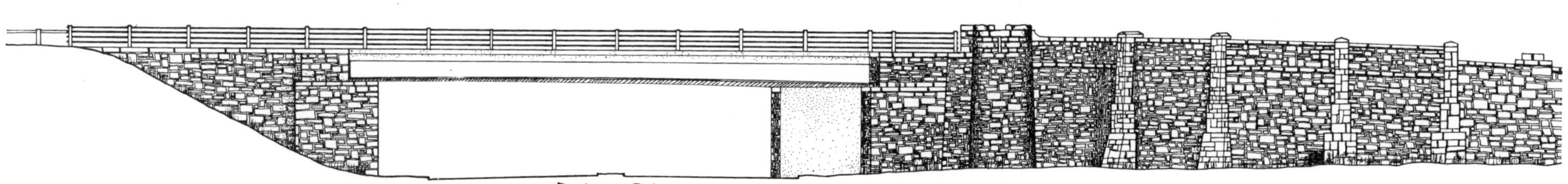

Fenimore Road Bridge

1924 *Gilmore Clarke, Architect* *Arthur G. Hayden, Engineer*

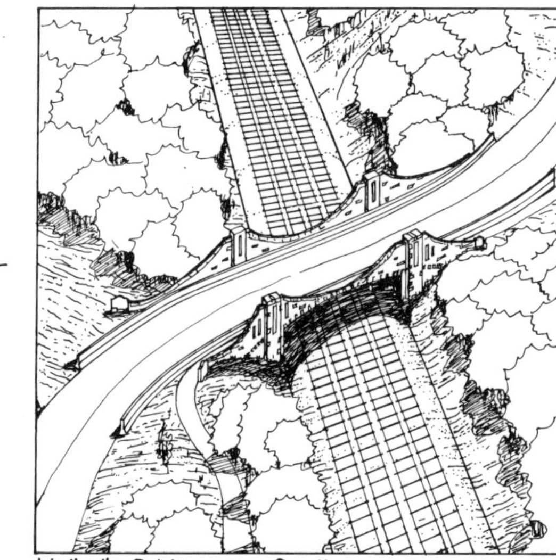

Valhalla Bridge, view South

Main Street Bridges into downtown White Plains

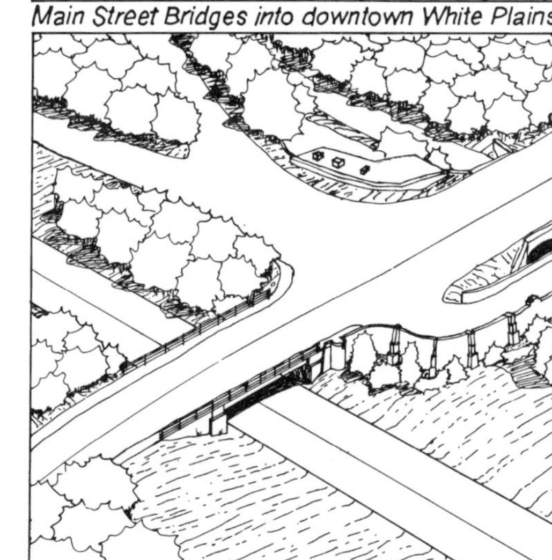

Fenimore Road Bridge, view South

DELINEATED BY: *Brandon Andow, 2001*

NATIONAL PARK SERVICE
UNITED STATES DEPARTMENT OF THE INTERIOR

BRONX RIVER PARKWAY RESERVATION
THE BRONX TO KENSICO DAM
WESTCHESTER COUNTY

WHITE PLAINS VICINITY

NEW YORK

SHEET 14 of 22

HISTORIC AMERICAN ENGINEERING RECORD
NY-327

IF REPRODUCED, PLEASE CREDIT: HISTORIC AMERICAN ENGINEERING RECORD, NATIONAL PARK SERVICE, NAME OF DELINEATOR, DATE OF THE DRAWING

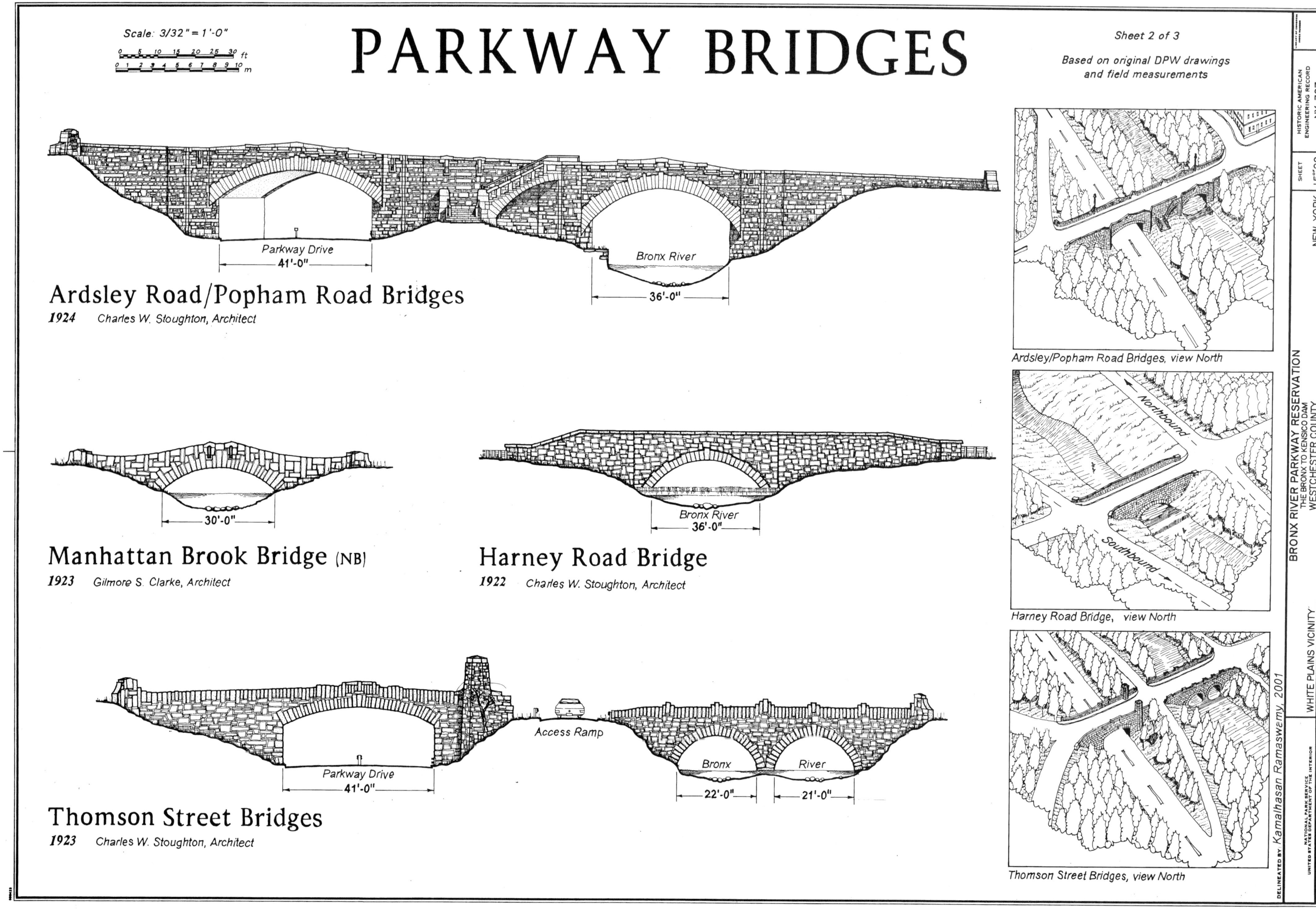

PARKWAY BRIDGES
Scale: 3/32" = 1'-0"
ft
m
Sheet 2 of 3
Based on original DPW drawings
and field measurements
Parkway Drive
41'-0"
Bronx River
36'-0"
Ardsley Road/Popham Road Bridges
1924 Charles W. Stoughton, Architect
Ardsley/Popham Road Bridges, view North
30'-0"
Manhattan Brook Bridge (NB)
1923 Gilmore S. Clarke, Architect
Bronx River
36'-0"
Harney Road Bridge
1922 Charles W. Stoughton, Architect
Northbound
Southbound
Harney Road Bridge, view North
Access Ramp
Parkway Drive
41'-0"
Bronx
River
22'-0"
21'-0"
Thomson Street Bridges
1923 Charles W. Stoughton, Architect
Thomson Street Bridges, view North
DELINEATED BY: Kamalhasan Ramaswamy, 2001
NATIONAL PARK SERVICE
UNITED STATES DEPARTMENT OF THE INTERIOR
WHITE PLAINS VICINITY
BRONX RIVER PARKWAY RESERVATION
THE BRONX TO KENSICO DAM
WESTCHESTER COUNTY
NEW YORK
SHEET
15 of 22
HISTORIC AMERICAN
ENGINEERING RECORD
NY-327
IF REPRODUCED, PLEASE CREDIT: HISTORIC AMERICAN ENGINEERING RECORD, NATIONAL PARK SERVICE, NAME OF DELINEATOR, DATE OF THE DRAWING

PARKWAY BRIDGES

Scale: 3/32" = 1'-0"

0 5 10 15 20 25 30 ft

0 1 2 3 4 5 6 7 8 9 10 m

Sheet 3 of 3

Based on original DPW drawings and field measurements

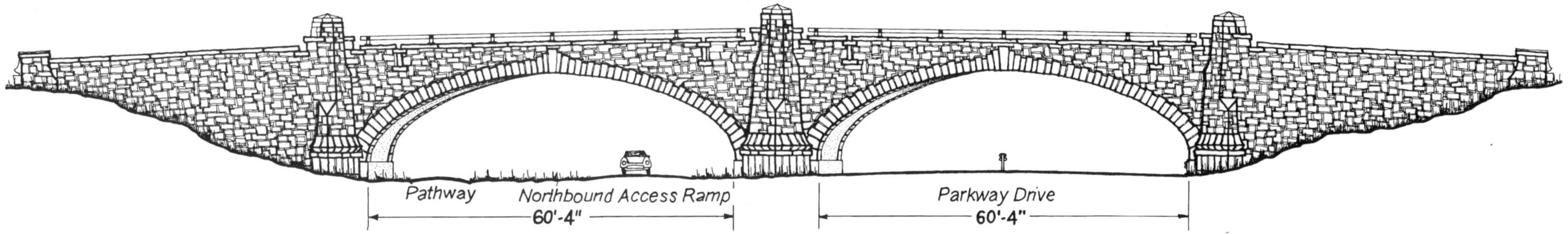

Parkway Viaduct (Park Avenue Bridge)

1923 *Bowdoin & Webster, Architect* *Arthur G. Hayden, Engineer*

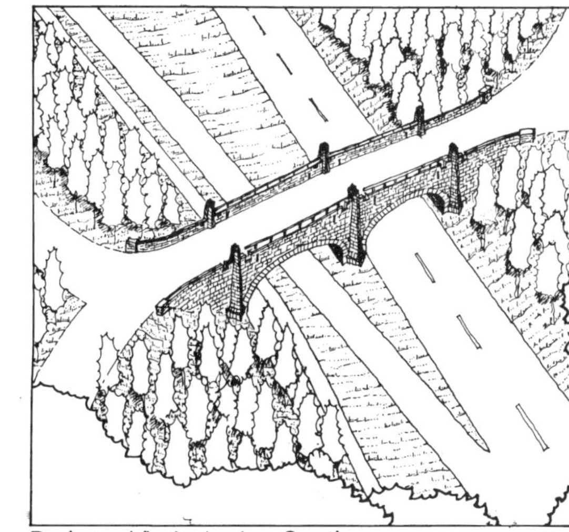

Parkway Viaduct, view South

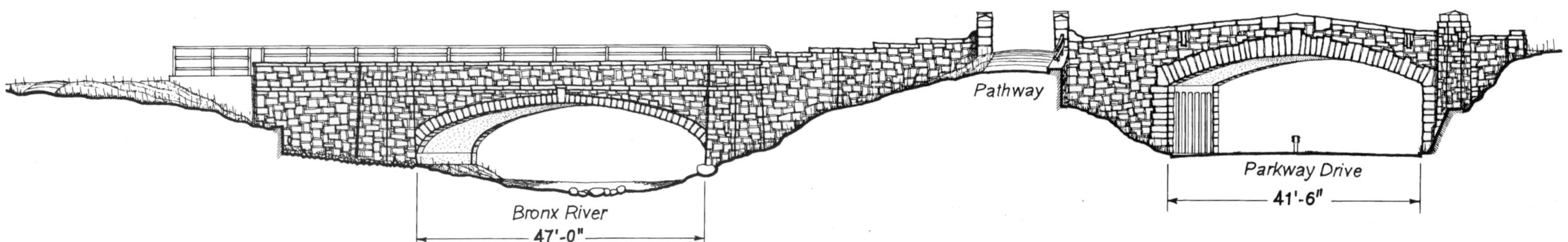

Palmer Road Bridges

1922 *Charles W. Stoughton, Architect* *Arthur G. Hayden, Engineer*

Palmer Road Bridges, view South

Footbridge *Replacement wood deck bridge over original stone foundations, Hartsdale vicinity.*

Footbridge *Locust posts & rails, built 1925 in Butler Woods, non-extant.*

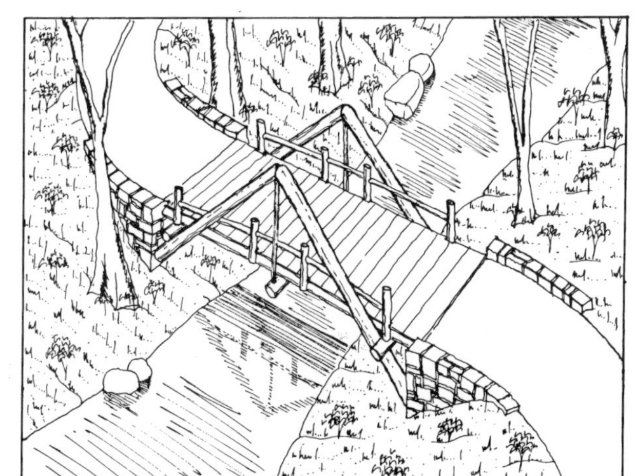

Footbridge *A-frame built 1925 in Garth Woods; extant but in poor condition.*

Footbridge *Pre-fabricated steel truss bridge, on concrete piers; below Harney Road.*

HISTORIC AMERICAN ENGINEERING RECORD NY-327

SHEET 16 of 22

NEW YORK

BRONX RIVER PARKWAY RESERVATION
THE BRONX TO KENSICO DAM
WESTCHESTER COUNTY

WHITE PLAINS VICINITY

DELINEATED BY: *Kamalhasan Ramaswamy, 2001*

NATIONAL PARK SERVICE
UNITED STATES DEPARTMENT OF THE INTERIOR

MANAGING THE PARKWAY

Modernization

Efforts to make the parkway conform to modern engineering standards have resulted in the elimination or alteration of many historic features. Steel barriers have replaced the original log guide rails. Concrete and steel median dividers have been added throughout most of the parkway. Modern steel fixtures have supplanted the rustic wood light poles. Large, high-visibility signs have been posted. These changes detract from the parkway's historic integrity and naturalistic appearance.

Realignment

Much of the Bronx River Parkway's pavement has been realigned since the original roadway was completed in 1925. Most of these modifications were intended to improve traffic flow and increase motorist safety. These changes have significantly changed the character of the winding historic road.

Revegetation

Maintaining the artfully diversified landscape compositions envisioned by the parkway's designers requires considerable time and expense. Many of the original meadows, clearings, and vistas have been obscured by rapidly growing shrubs and trees, significantly altering the open park-like landscape that once extended throughout a much larger portion of the parkway reservation.

Siltation

The Bronx River still serves a major local drainage function. The heavy runoff caused by surrounding commercial and residential development compounds the natural process of siltation, where eroded soil is deposited in slow-moving sections of the river. Parkway managers formerly engaged in a policy of remedial dredging, but this approach has not been practiced for several years. Consequently, some of the carefully designed water features are being filled in with small islands and mudflats.

Invasive Plants and Animals

Non-native plants are displacing many of the native species favored by the parkway's designers. Exotic species have changed the parkway's appearance, reduced biodiversity, and upset the natural balance required for healthy wetlands, forests, and meadows. Most of these exotics are plants and trees, but the overpopulation of Canada geese imposes similar threats. Geese drive out native birds and foul lakes and lawns with their droppings.

DELINEATED BY: *T. Folger, 2001*

NATIONAL PARK SERVICE
UNITED STATES DEPARTMENT OF THE INTERIOR

WHITE PLAINS VICINITY

BRONX RIVER PARKWAY RESERVATION
THE BRONX TO KENSICO DAM
WESTCHESTER COUNTY

NEW YORK

SHEET 17 of 22

HISTORIC AMERICAN ENGINEERING RECORD
NY-327

BRPR GENERAL DEVELOPMENT PLAN (1918) Sheet 1 of 5

The 1918 development plan articulated the general design scheme for the parkway. Consulting landscape architect Hermann Merkel was primarily responsible for these drawings, but every member of the Bronx Parkway Commission made significant contributions to the design. Refinements continued to be made throughout the course of construction and the treatment of some areas changed significantly in some locations.

The 1918 development plans represent the clearest and most comprehensive articulation of the Bronx Parkway Commission's original vision for a linear landscape park.

These reproductions of the 1918 Bronx River Parkway Reservation development plan were plotted by the Historic American Engineering Record in 2001 based on scans of the original drawings located in the Westchester County Records and Archives Center.

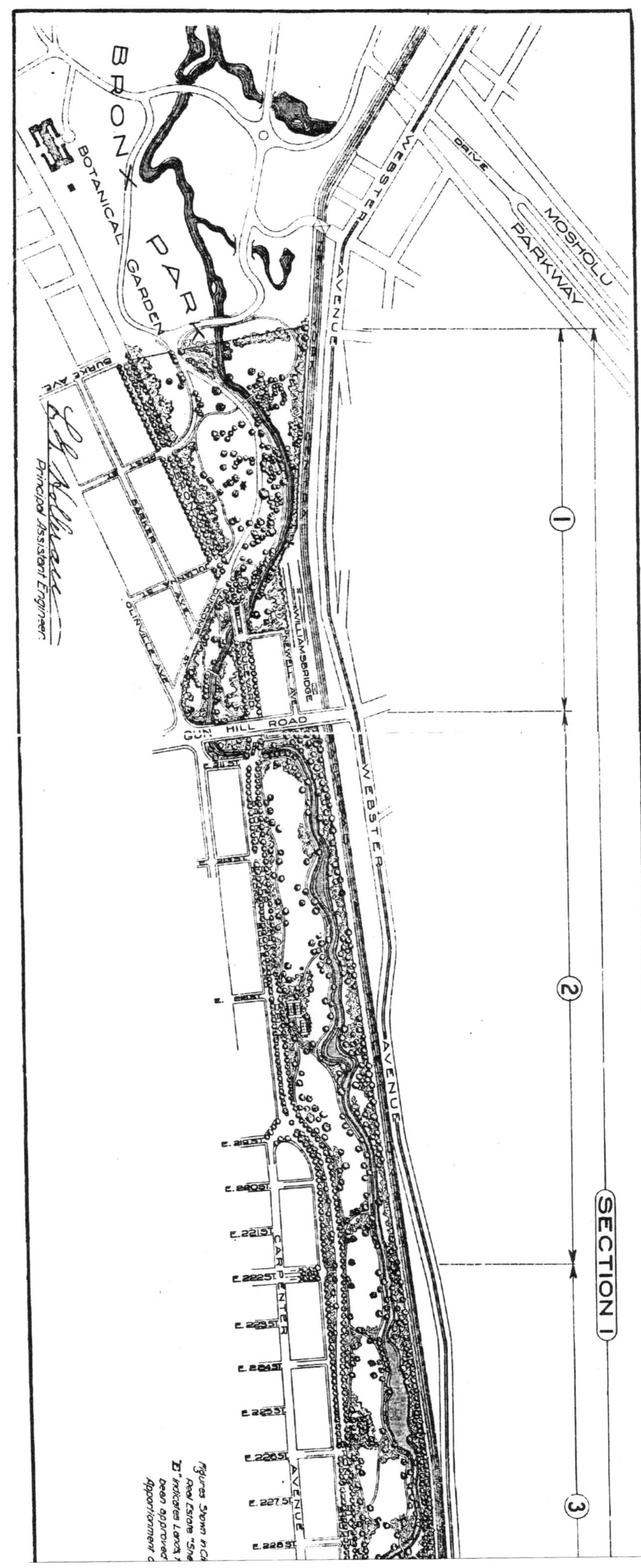

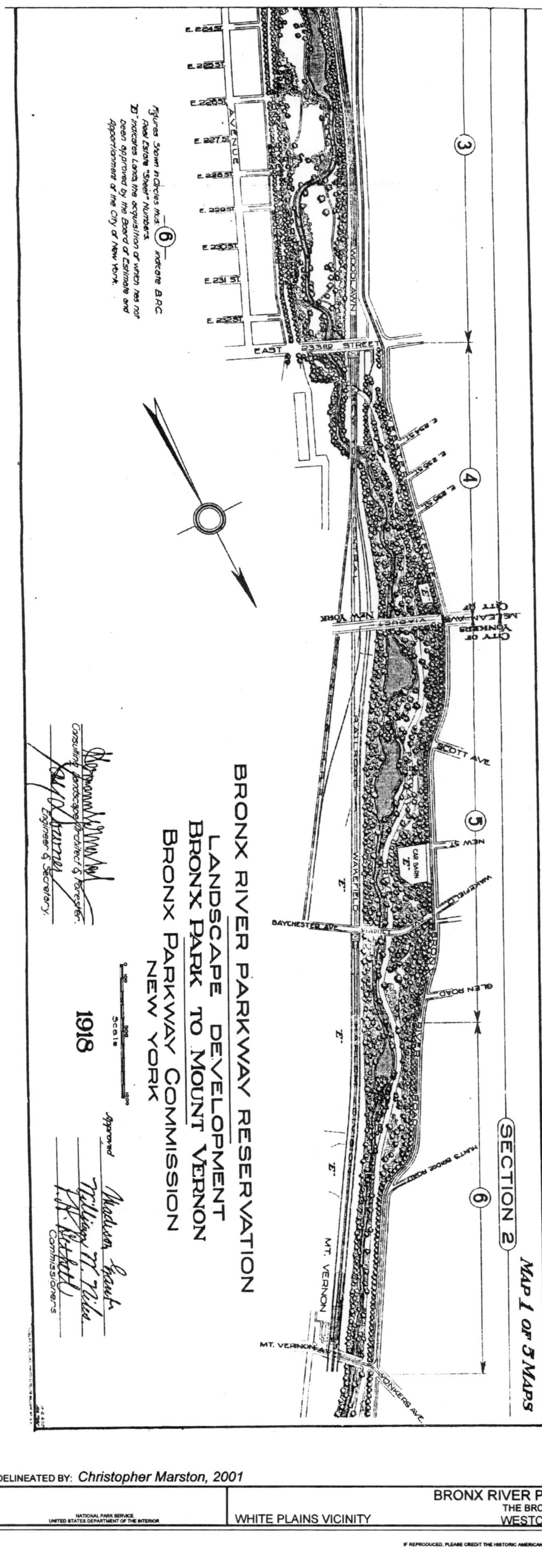

DELINEATED BY: *Christopher Marston, 2001*

NATIONAL PARK SERVICE UNITED STATES DEPARTMENT OF THE INTERIOR	BRONX RIVER PARKWAY RESERVATION THE BRONX TO KENSICO DAM WHITE PLAINS VICINITY — WESTCHESTER COUNTY — NEW YORK	SHEET 18 of 22	HISTORIC AMERICAN ENGINEERING RECORD NY-327	LIBRARY OF CONGRESS

IF REPRODUCED, PLEASE CREDIT THE HISTORIC AMERICAN ENGINEERING RECORD, NATIONAL PARK SERVICE, NAME OF DELINEATOR, DATE OF DRAWING

BRPR GENERAL DEVELOPMENT PLAN (1918) Sheet 2 of 5

The 1918 development plan articulated the general design scheme for the parkway. Consulting landscape architect Hermann Merkel was primarily responsible for these drawings, but every member of the Bronx Parkway Commission made significant contributions to the design. Refinements continued to be made throughout the course of construction and the treatment of some areas changed significantly in some locations.

The 1918 development plans represent the clearest and most comprehensive articulation of the Bronx Parkway Commission's original vision for a linear landscape park.

These reproductions of the 1918 Bronx River Parkway Reservation development plan were plotted by the Historic American Engineering Record in 2001 based on scans of the original drawings located in the Westchester County Records and Archives Center.

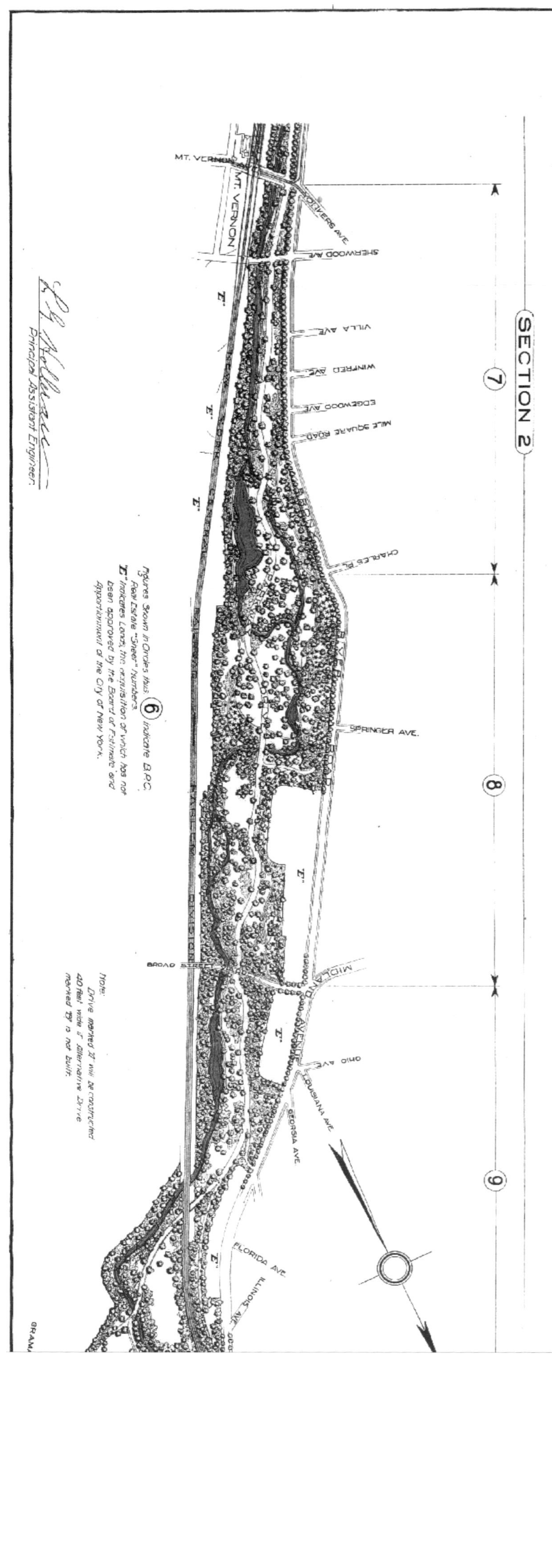

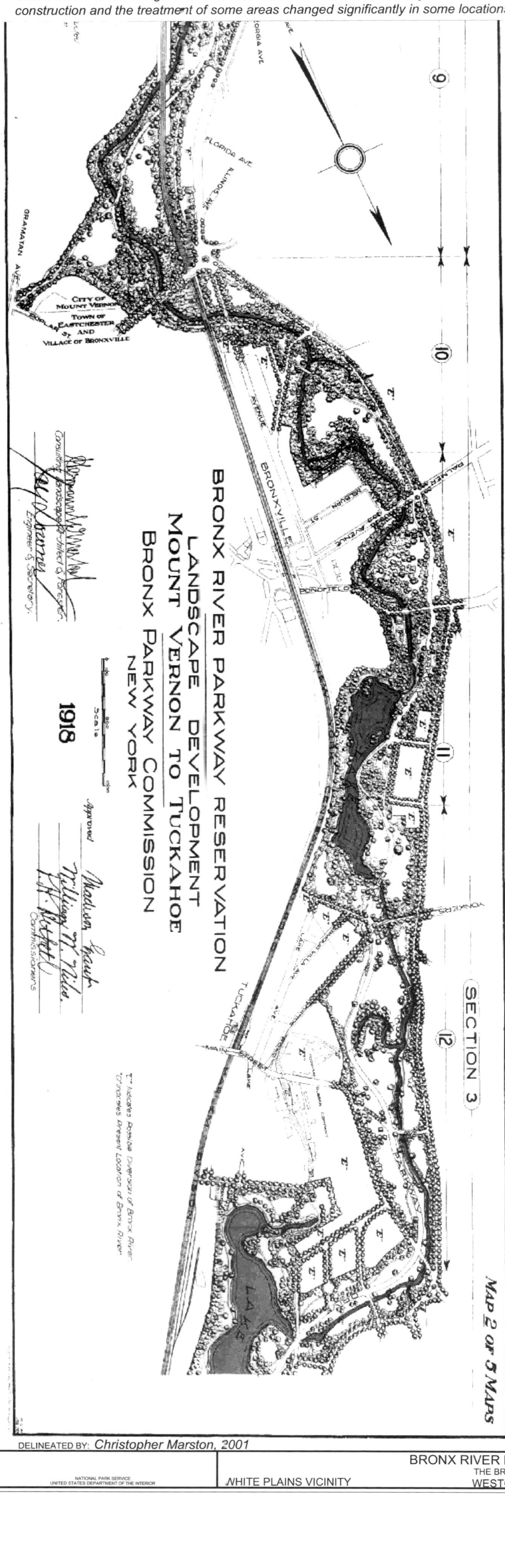

DELINEATED BY: *Christopher Marston, 2001*

NATIONAL PARK SERVICE UNITED STATES DEPARTMENT OF THE INTERIOR	WHITE PLAINS VICINITY	BRONX RIVER PARKWAY RESERVATION THE BRONX TO KENSICO DAM WESTCHESTER COUNTY	NEW YORK	SHEET 19 OF 22	HISTORIC AMERICAN ENGINEERING RECORD NY-327

BRPR GENERAL DEVELOPMENT PLAN (1918) Sheet 3 of 5

The 1918 development plan articulated the general design scheme for the parkway. Consulting landscape architect Hermann Merkel was primarily responsible for these drawings, but every member of the Bronx Parkway Commission made significant contributions to the design. Refinements continued to be made throughout the course of construction and the treatment of some areas changed significantly in some locations.

The 1918 development plans represent the clearest and most comprehensive articulation of the Bronx Parkway Commission's original vision for a linear landscape park.

These reproductions of the 1918 Bronx River Parkway Reservation development plan were plotted by the Historic American Engineering Record in 2001 based on scans of the original drawings located in the Westchester County Records and Archives Center.

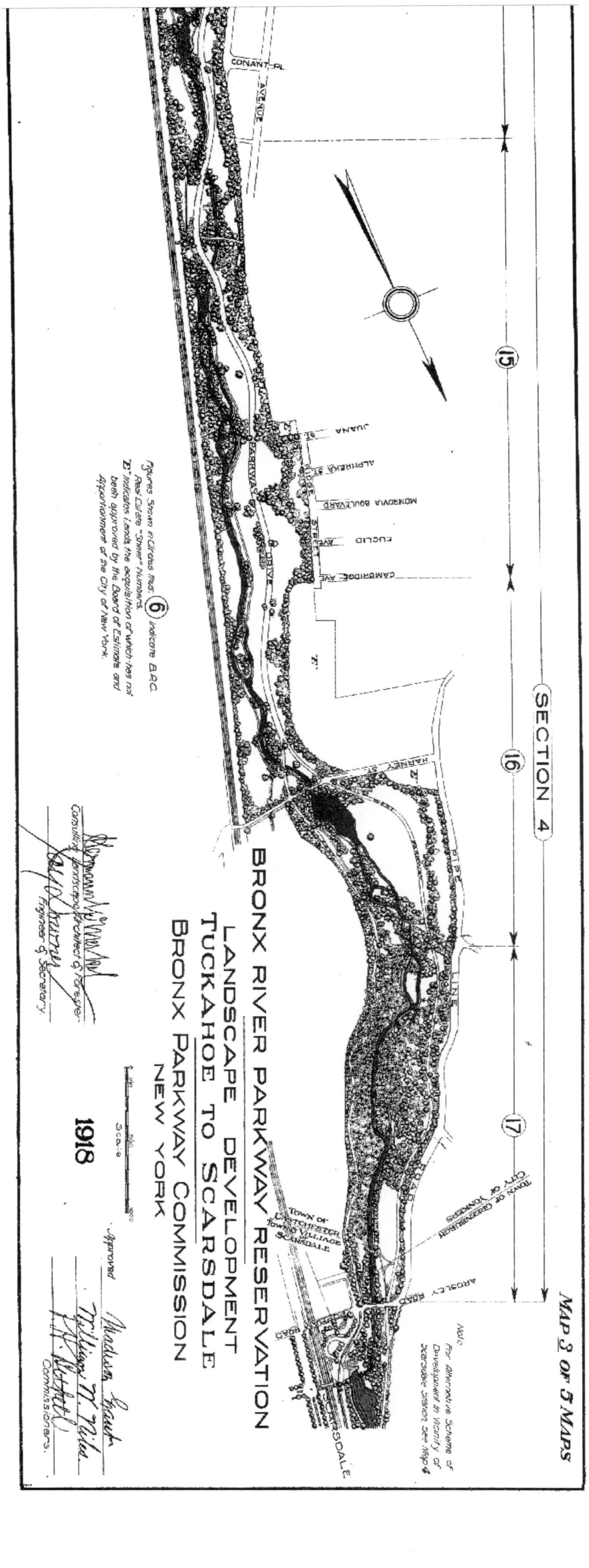

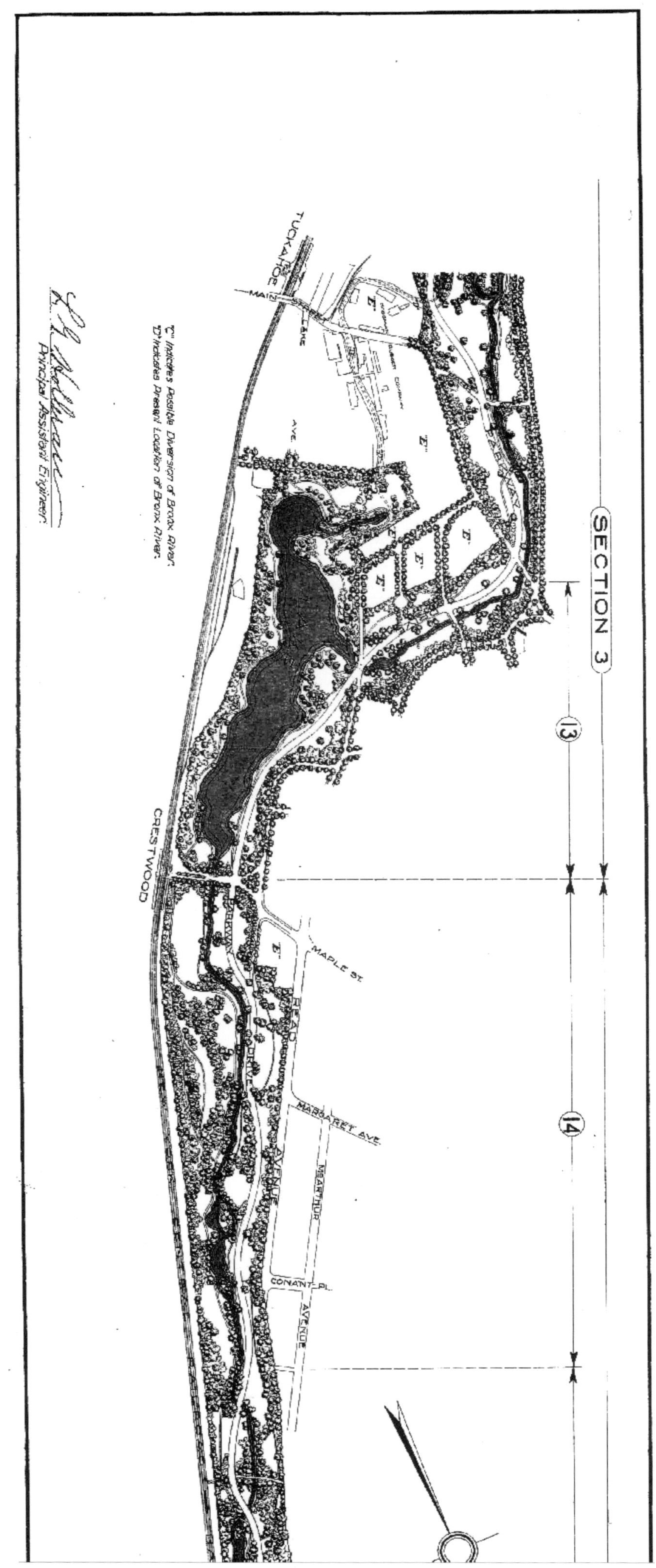

DELINEATED BY: *Christopher Marston, 2001*

NATIONAL PARK SERVICE UNITED STATES DEPARTMENT OF THE INTERIOR	WHITE PLAINS VICINITY	BRONX RIVER PARKWAY RESERVATION THE BRONX TO KENSICO DAM WESTCHESTER COUNTY	NEW YORK	SHEET 20 OF 22	HISTORIC AMERICAN ENGINEERING RECORD NY-327

BRPR GENERAL DEVELOPMENT PLAN (1918) Sheet 4 of 5

The 1918 development plan articulated the general design scheme for the parkway. Consulting landscape architect Hermann Merkel was primarily responsible for these drawings, but every member of the Bronx Parkway Commission made significant contributions to the design. Refinements continued to be made throughout the course of construction and the treatment of some areas changed significantly in some locations.

The 1918 development plans represent the clearest and most comprehensive articulation of the Bronx Parkway Commission's original vision for a linear landscape park.

These reproductions of the 1918 Bronx River Parkway Reservation development plan were plotted by the Historic American Engineering Record in 2001 based on scans of the original drawings located in the Westchester County Records and Archives Center.

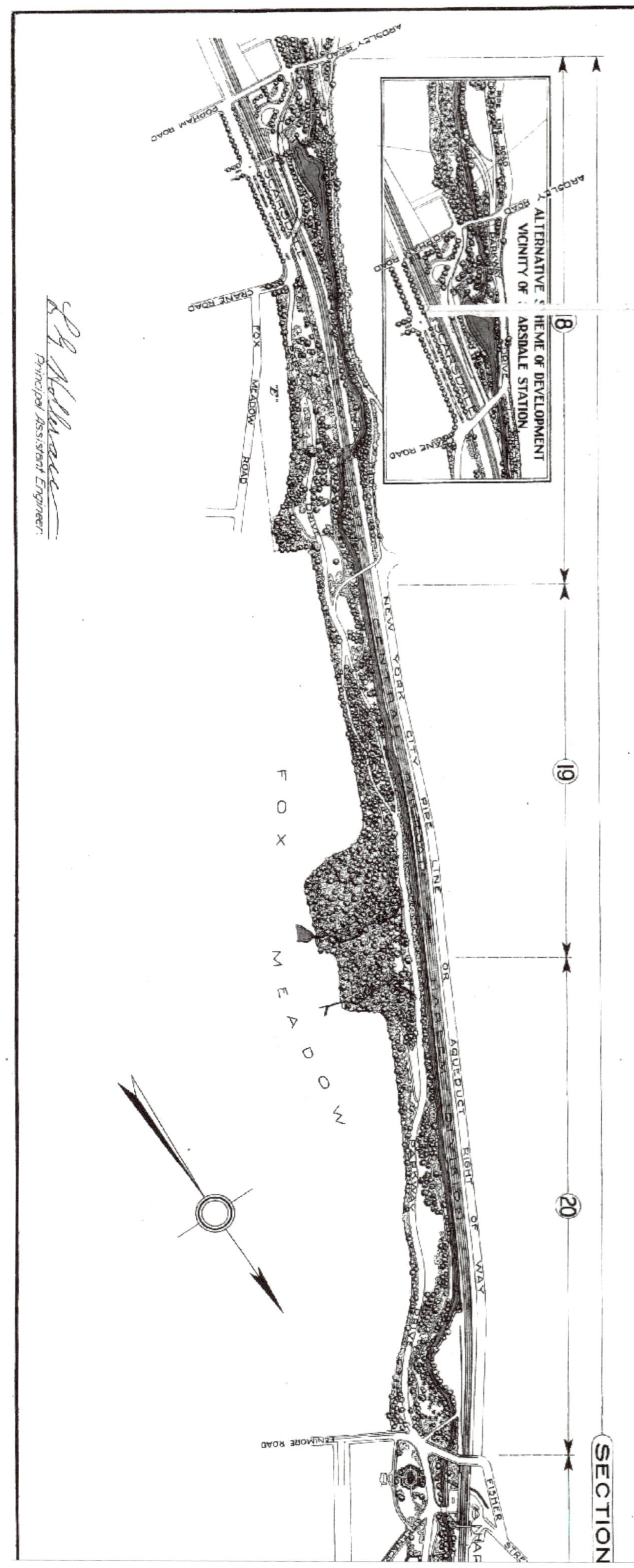

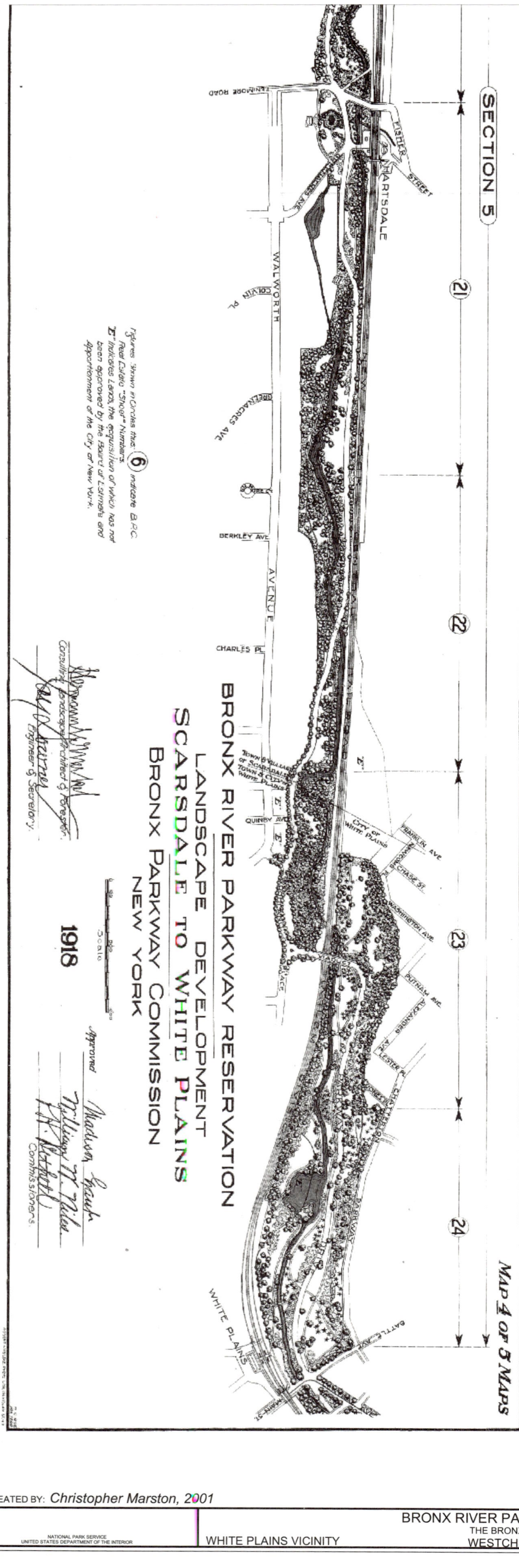

DELINEATED BY: *Christopher Marston, 2001*

NATIONAL PARK SERVICE UNITED STATES DEPARTMENT OF THE INTERIOR	WHITE PLAINS VICINITY	BRONX RIVER PARKWAY RESERVATION THE BRONX TO KENSICO DAM WESTCHESTER COUNTY	NEW YORK	SHEET 21 OF 22	HISTORIC AMERICAN ENGINEERING RECORD NY-327

BRPR GENERAL DEVELOPMENT PLAN (1918) Sheet 5 of 5

The 1918 development plan articulated the general design scheme for the parkway. Consulting landscape architect Hermann Merkel was primarily responsible for these drawings, but every member of the Bronx Parkway Commission made significant contributions to the design. Refinements continued to be made throughout the course of construction and the treatment of some areas changed significantly in some locations.

Nevertheless, the 1918 development plans represent the clearest and most comprehensive articulation of the Bronx Parkway Commission's original vision for a linear landscape park.

These reproductions of the 1918 Bronx River Parkway Reservation development plan were plotted by the Historic American Engineering Record in 2001 based on scans of the original drawings located in the Westchester County Records and Archives Center.

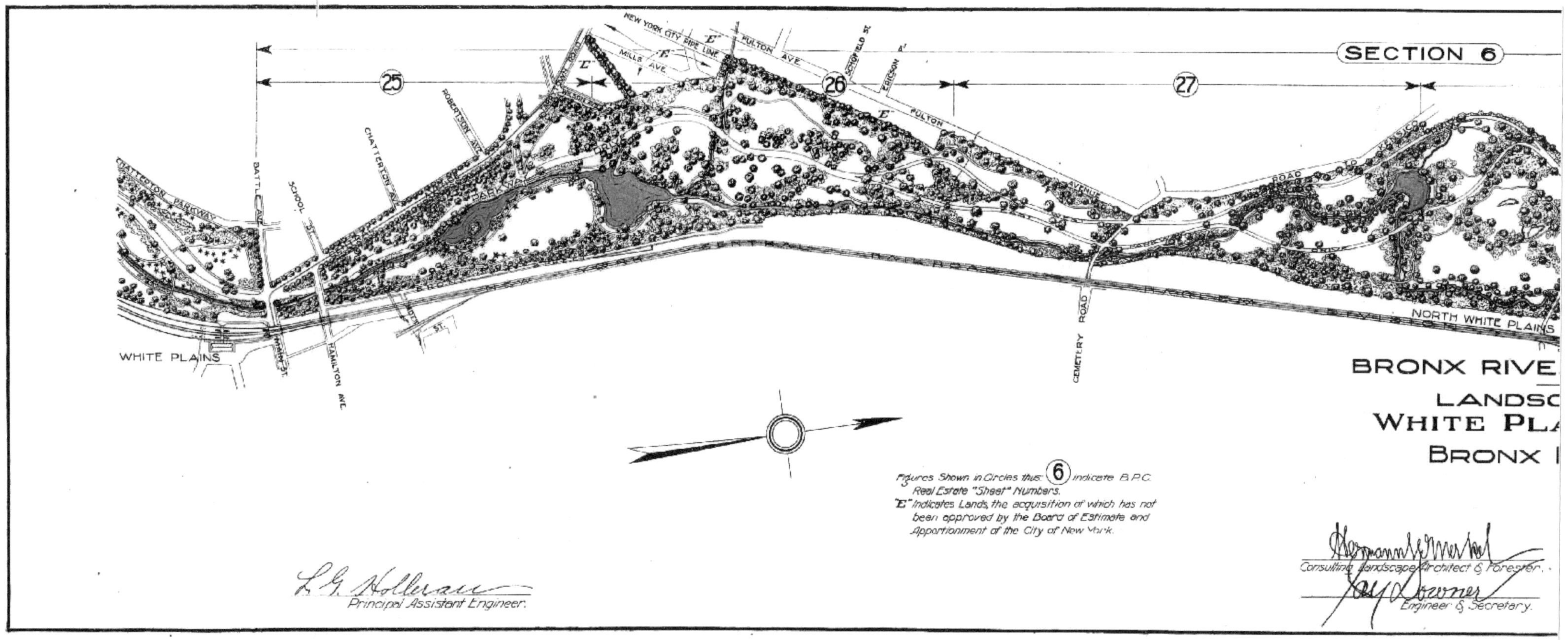

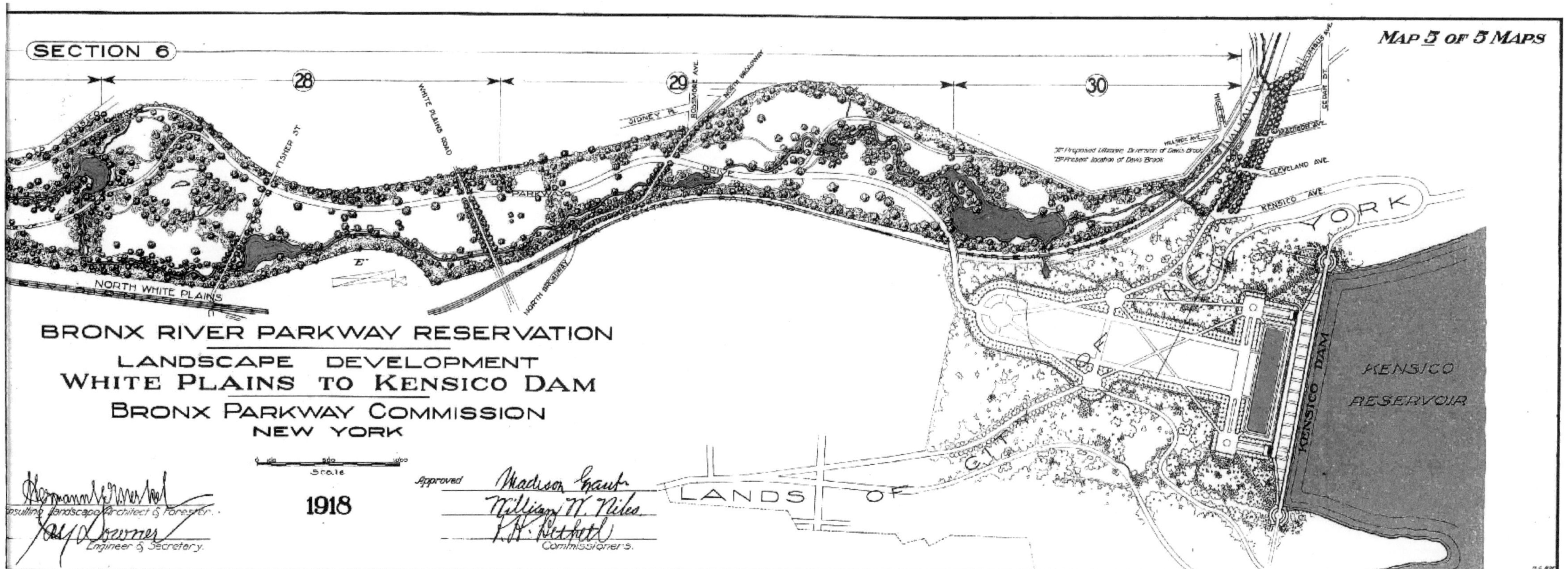

BRONX RIVER PARKWAY RESERVATION
THE BRONX TO KENSICO DAM
WESTCHESTER COUNTY

WHITE PLAINS VICINITY

NEW YORK

HISTORIC AMERICAN ENGINEERING RECORD
NY-327

SHEET 22 of 22

DELINEATED BY: Christopher Marston, 2001

NATIONAL PARK SERVICE
UNITED STATES DEPARTMENT OF THE INTERIOR

"Tourists want three things; a good road to drive on, something worthwhile to see and something worthwhile to eat. . . . We will cash in, year after year, on our crop of scenic beauty, without depleting it in any way."

Samuel Hill

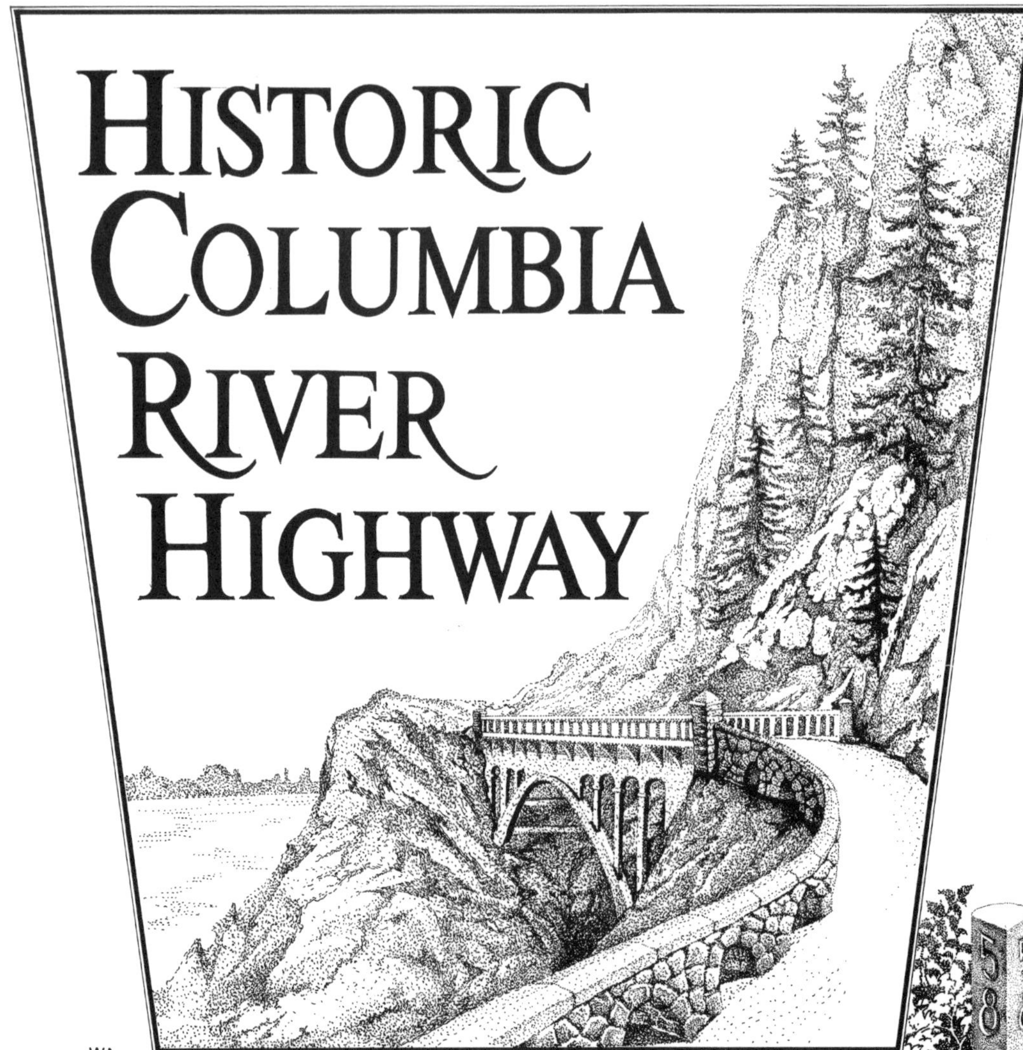

HISTORIC COLUMBIA RIVER HIGHWAY

1913 - 1922

"When I made my preliminary survey here and found myself standing waist-deep in ferns, I remembered my mother's long ago warning, 'Oh, Samuel, do be careful of my Boston fern.' And I then pledged myself that none of this wild beauty should be marred where it could be prevented."

Samuel C. Lancaster

The Pacific Northwest's Historic Columbia River Highway, constructed between 1913 and 1922 in Oregon's Multnomah, Hood River, and Wasco counties, is one of the oldest scenic highways in the United States. Its design and execution were the products of two visionaries: the lawyer, entrepreneur, and good roads' promoter Samuel Hill, and the engineer and landscape architect Samuel C. Lancaster. Several top road and bridge designers assisted them and many Portland businessmen strongly advocated construction of what they saw as the "King of the Roads."

The Historic Columbia River Highway was an early-twentieth-century technical and civic achievement, successfully mixing sensitivity to a magnificent natural landscape with ambitious engineering. The route has gained national significance because it represents one of the earliest applications of cliff-face road building as applied to modern highway construction, but the foremost reason for its construction was Hill's and Lancaster's vision of building a scenic highway above the Columbia River to rival the great roads of Europe.

It consists of many types of construction, including bridges and viaducts, tunnels, loop roads, masonry walls, and buildings and camps, integrated well with its surroundings. The highway served as the link between the many natural phenomena of the landscape—its magnificent waterfalls, rock outcroppings, creeks and dells, and scenic river vistas. Lancaster and his associates believed that they were accomplishing a worthwhile task and they saw the Historic Columbia River Highway as one without rivals, a road superior to all others in the world.

In a more practical light, many observers envisioned the road as a lifeline connecting Portland with the many commercial and agricultural areas along the Columbia River. Some saw it as part of a linear web of similarly constructed routes throughout the Pacific Northwest, in Oregon, Washington, and northern Idaho, and connecting with other highways around the nation.

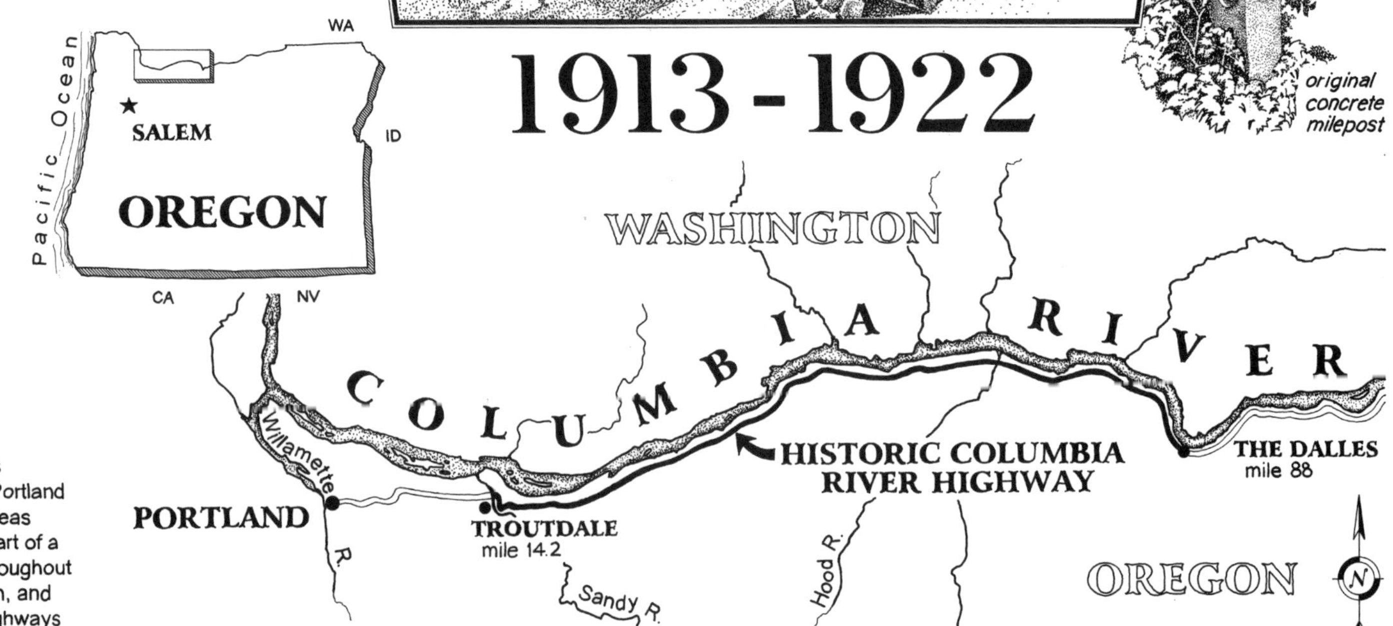

Site Map based on USDA Forest Service Map of the Columbia River Gorge National Scenic Area   Scale: 1½" = 10 miles

The Historic Columbia River Highway served Oregon as part of its principal east-west trunk route, U.S. 30, until a wider, gracefully curved water-level road replaced it in the early 1950s. Much of the original highway remains as a scenic route for less hurried travelers who look forward to experiencing the many natural wonders of the Columbia River Gorge.

This recording project is part of the Historic American Engineering Record (HAER), a long-range program to document historically significant engineering and industrial works in the United States. The HAER program is administered by the Historic American Buildings Survey/Historic American Engineering Record (HABS/HAER) Division of the National Park Service, U.S. Department of the Interior.

The Historic Columbia River Highway Recording Project was cosponsored in 1995 by HABS/HAER, under the general direction of Robert J. Kapsch, Ph.D., Chief, and by the Oregon Department of Transportation (ODOT), Bruce Warner, Region One Manager; in cooperation with the US/International Committee on Monuments and Sites (ICOMOS), the American Society of Civil Engineers (ASCE), and the Historic Columbia River Highway Advisory Committee.

Fieldwork, measured drawings, historical reports, and photographs were prepared under the direction of Eric N. DeLony, Chief of HAER; Todd A. Croteau, HAER Architect, and Dean A. Herrin, Ph.D., HAER Historian. The recording team consisted of Elaine G. Pierce (Chattanooga, Tennessee), Architect and Field Supervisor; Vladimir V. Simonenko (ICOMOS/ Academy of Fine Arts, Kiev, Ukraine), Architect; Christine Rumi (University of Oregon) and Pete Brooks (Yale University), Architectural Technicians; Helen I. Selph (California State Polytechnic University, Pomona) and Jodi C. Zeller (University of Illinois, Urbana-Champaign), Landscape Architectural Technicians; Robert W. Hadlow, Ph.D. (ASCE / Pullman, Washington), Historian; and Jet Lowe (Washington, DC), HAER Photographer. Jeanette B. Kloos, ODOT Region One Scenic Area Coordinator; and Dwight A. Smith, ODOT Cultural Resources Specialist, served as department liaisons.

DELINEATED BY: V. V. Simonenko & Elaine G. Pierce 1995

HISTORIC COLUMBIA RIVER HIGHWAY RECORDING PROJECT
NATIONAL PARK SERVICE
UNITED STATES DEPARTMENT OF THE INTERIOR

TROUTDALE VIC.

HISTORIC COLUMBIA RIVER HIGHWAY 1913-22
LOCATED BETWEEN TROUTDALE AND THE DALLES ALONG THE COLUMBIA RIVER
MULTNOMAH CO. OREGON

SHEET 1 of 27

HISTORIC AMERICAN ENGINEERING RECORD
OR - 36

1915 1926 1957 1987

EVOLUTION OF TRANSPORTATION

(MOFFETT CREEK BRIDGES)

The Columbia River Gorge has long been a transportation corridor. The ancient river, shaped by the volcanic outpourings of the Columbia Basalts, was stripped clean of soil and vegetation as recently as the last Bretz Flood about 13,000 years ago. In time, however, the gorge restocked and fish runs drew American Indians to the area.

American sea trader Robert Gray *entered* the ***River of the West*** in May 1792, and named it for his ship, *Columbia*. Lewis and Clark came down river in 1805. In the 1840s westward-bound pioneers could travel by wagon as far as The Dalles, but had to take to the river in log rafts to make it past the Cascades. It proved a perilous journey, if not an impossible one. In 1845, Samuel Barlow and Joel Palmer located the Barlow Road around Mt. Hood and the first wagons came through the following year. By 1850 steamboats were running on the Columbia and Willamette rivers and reached The Dalles in 1851. A wagon road portage on the Oregon side followed in 1855.

In 1862, the first locomotive, the Oregon Pony, appeared on the Gorge courtesy of the Oregon Steam Navigation Company. The railroad expanded rapidly and by the 1880s the Oregon Railway & Navigation Company ran along the Columbia's south bank from Portland to points east of The Dalles. The wagon road was eliminated and river boat travel was drastically reduced.

The Columbia River Highway opened in 1915 and by 1922 was the first paved road to connect Portland to The Dalles. The highway skirted the southern wall of the gorge and heralded the end of the railroad's 30-year reign. Within twenty years, however, the highway could no longer support the ever increasing traffic and by 1949 a water level route, much of it built on fill, but using some of the original highway, was opened. By 1969, a four-lane highway to The Dalles was built.

About 50 miles of the 1915 road remain in use, reminding us of the changing pace of our lives. Today, jets soar high above and fiber optic cables race below. In the march of technology, the Gorge continues to be a corridor for energy, information and leisure as well as transportation.

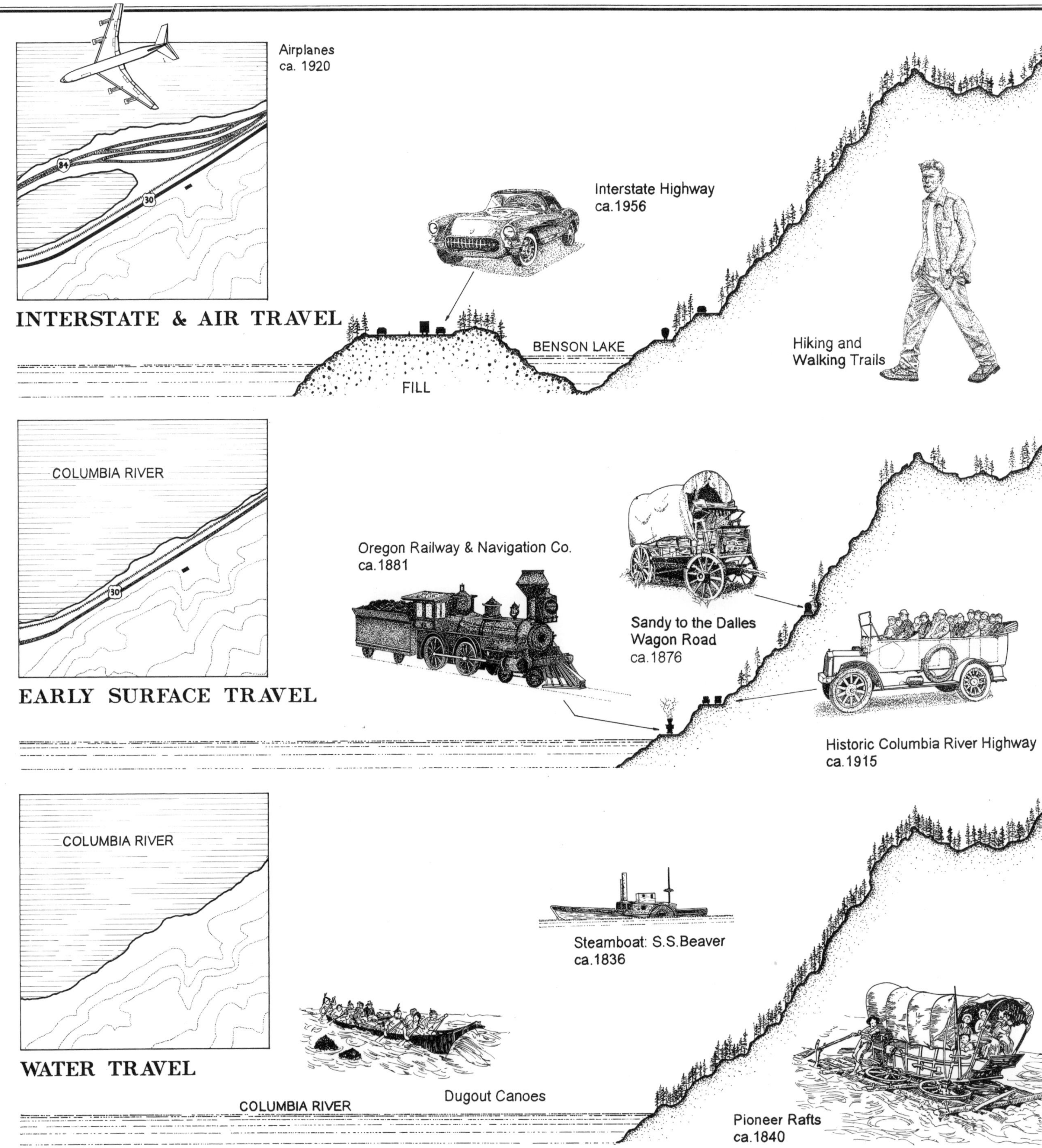

HISTORIC AMERICAN ENGINEERING RECORD OR-36

SHEET 5 of 27

HISTORIC COLUMBIA RIVER HIGHWAY 1913-22
LOCATED BETWEEN TROUTDALE AND THE DALLES ALONG THE COLUMBIA RIVER
MULTNOMAH CO.

OREGON

TROUTDALE VIC.

DELINEATED BY: *Christine Rumi 1995*
HISTORIC COLUMBIA RIVER HIGHWAY RECORDING PROJECT
NATIONAL PARK SERVICE
UNITED STATES DEPARTMENT OF THE INTERIOR

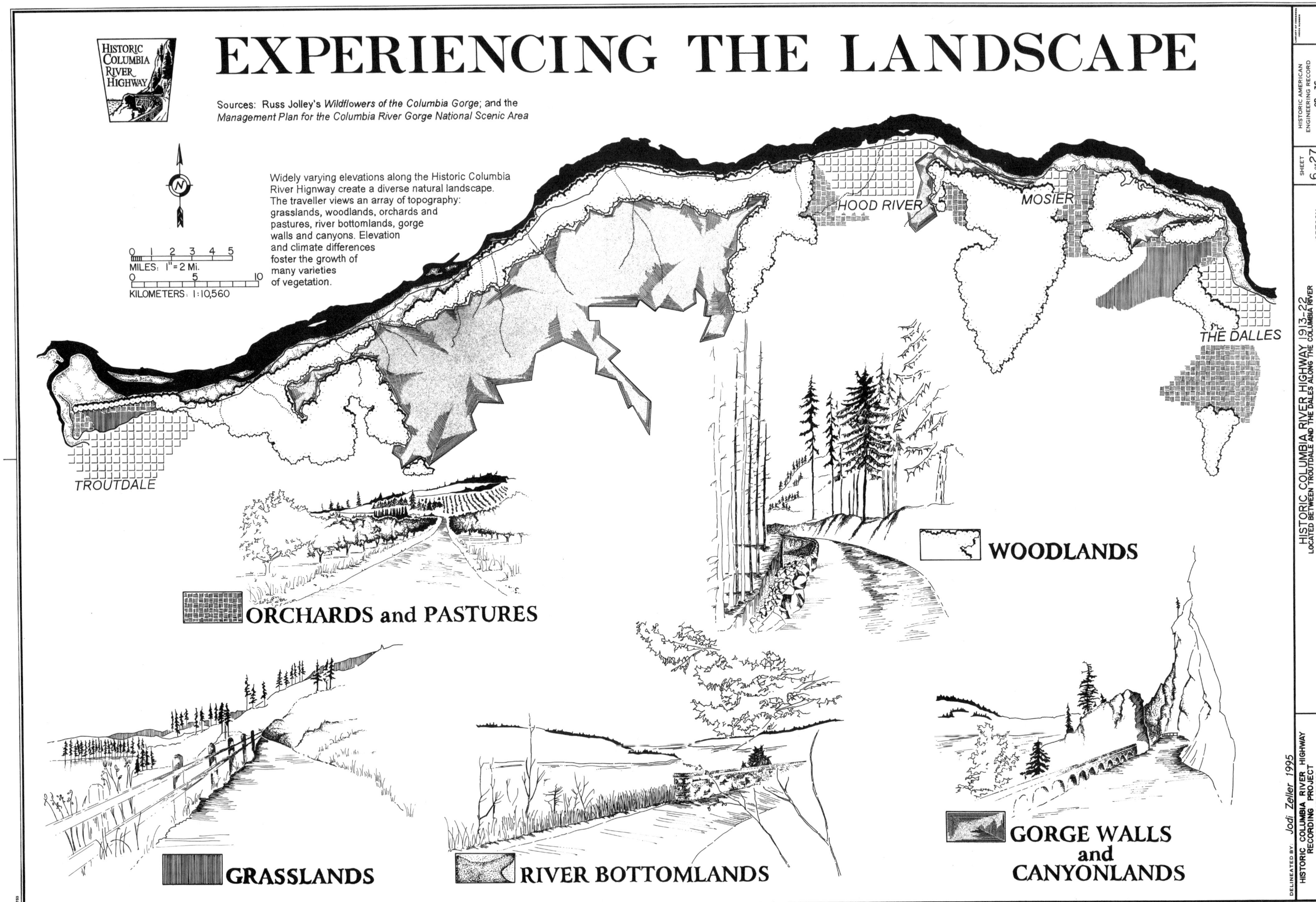
HISTORIC COLUMBIA RIVER HIGHWAY
EXPERIENCING THE LANDSCAPE
Sources: Russ Jolley's Wildflowers of the Columbia Gorge; and the Management Plan for the Columbia River Gorge National Scenic Area
N
Widely varying elevations along the Historic Columbia River Highway create a diverse natural landscape. The traveller views an array of topography: grasslands, woodlands, orchards and pastures, river bottomlands, gorge walls and canyons. Elevation and climate differences foster the growth of many varieties of vegetation.
0 1 2 3 4 5
MILES: 1" = 2 Mi.
0 5 10
KILOMETERS: 1:10,560
TROUTDALE
HOOD RIVER
MOSIER
THE DALLES
WOODLANDS
ORCHARDS and PASTURES
GRASSLANDS
RIVER BOTTOMLANDS
GORGE WALLS and CANYONLANDS
HISTORIC AMERICAN ENGINEERING RECORD
OR - 36
SHEET 6 OF 27
OREGON
HISTORIC COLUMBIA RIVER HIGHWAY 1913-22
LOCATED BETWEEN TROUTDALE AND THE DALLES ALONG THE COLUMBIA RIVER
MULTNOMAH CO.
TROUTDALE VIC.
IF REPRODUCED, PLEASE CREDIT: HISTORIC AMERICAN ENGINEERING RECORD, NATIONAL PARK SERVICE, NAME OF DELINEATOR, DATE OF THE DRAWING
DELINEATED BY: Jodi Zeller 1995
HISTORIC COLUMBIA RIVER HIGHWAY RECORDING PROJECT
NATIONAL PARK SERVICE
UNITED STATES DEPARTMENT OF THE INTERIOR

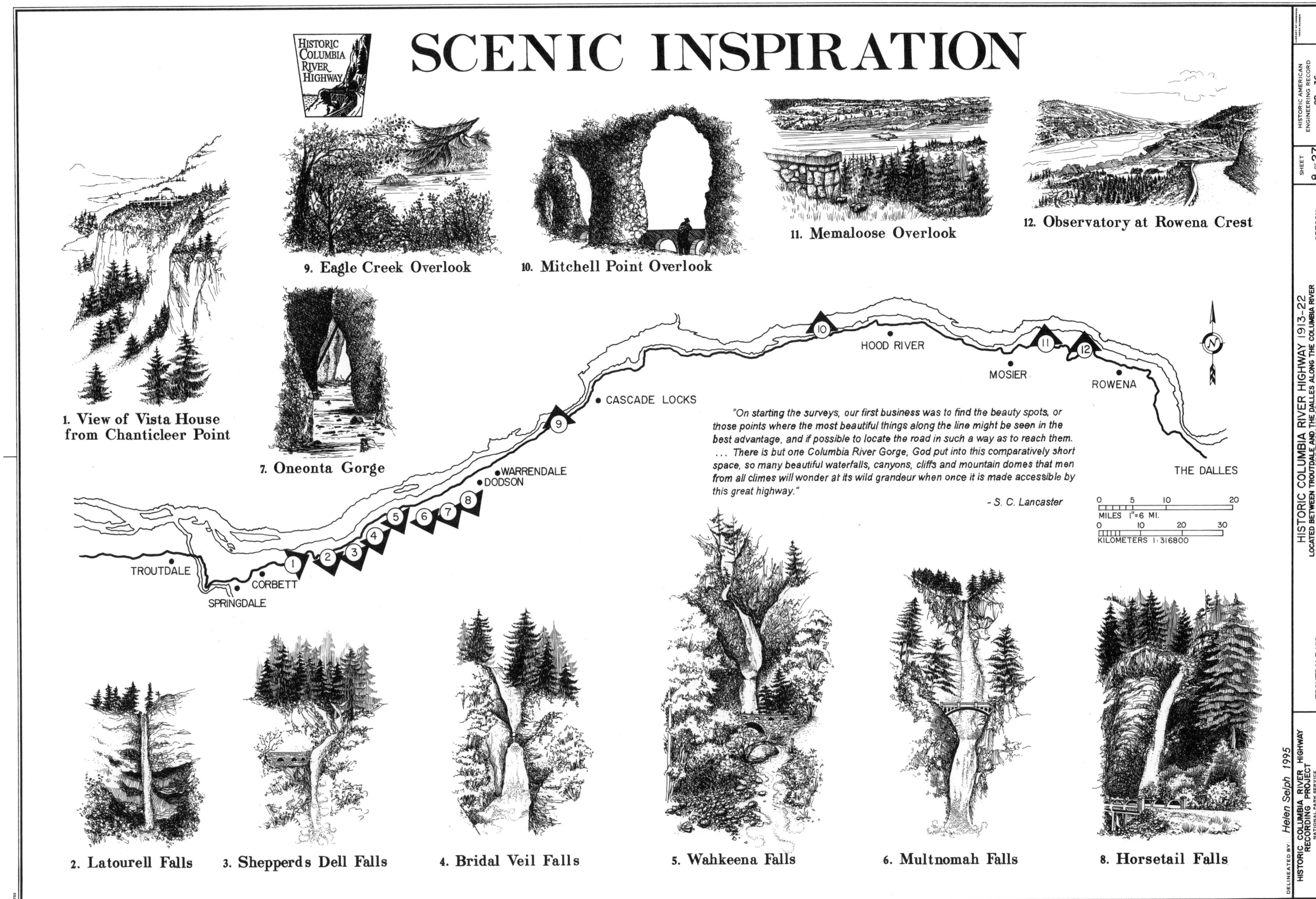
HISTORIC COLUMBIA RIVER HIGHWAY
SCENIC INSPIRATION
1. View of Vista House from Chanticleer Point
9. Eagle Creek Overlook
10. Mitchell Point Overlook
11. Memaloose Overlook
12. Observatory at Rowena Crest
7. Oneonta Gorge
CASCADE LOCKS
HOOD RIVER
MOSIER
ROWENA
THE DALLES
WARRENDALE
DODSON
TROUTDALE
CORBETT
SPRINGDALE
N
"On starting the surveys, our first business was to find the beauty spots, or those points where the most beautiful things along the line might be seen in the best advantage, and if possible to locate the road in such a way as to reach them. . . . There is but one Columbia River Gorge, God put into this comparatively short space, so many beautiful waterfalls, canyons, cliffs and mountain domes that men from all climes will wonder at its wild grandeur when once it is made accessible by this great highway."
- S. C. Lancaster
0 5 10 20
MILES 1"=6 MI.
0 10 20 30
KILOMETERS 1:316800
2. Latourell Falls
3. Shepperds Dell Falls
4. Bridal Veil Falls
5. Wahkeena Falls
6. Multnomah Falls
8. Horsetail Falls
DELINEATED BY: Helen Selph 1995
HISTORIC COLUMBIA RIVER HIGHWAY RECORDING PROJECT
NATIONAL PARK SERVICE
UNITED STATES DEPARTMENT OF THE INTERIOR
TROUTDALE VIC.
HISTORIC COLUMBIA RIVER HIGHWAY 1913-22
LOCATED BETWEEN TROUTDALE AND THE DALLES ALONG THE COLUMBIA RIVER
MULTNOMAH CO.
OREGON
SHEET 8 OF 27
HISTORIC AMERICAN ENGINEERING RECORD
OR - 36
IF REPRODUCED, PLEASE CREDIT: HISTORIC AMERICAN ENGINEERING RECORD, NATIONAL PARK SERVICE, NAME OF DELINEATOR, DATE OF THE DRAWING

GRADING & ALIGNMENT

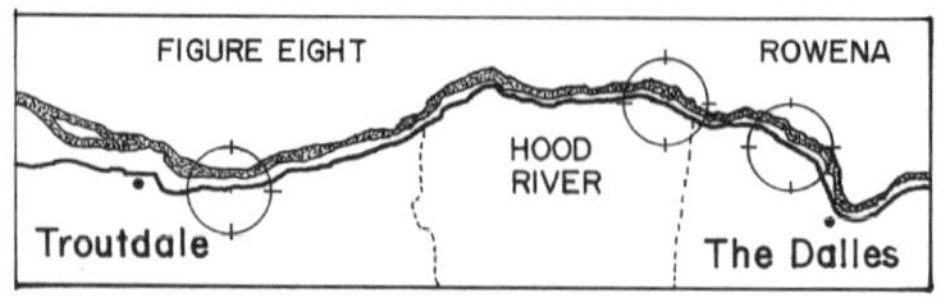

REFERENCE MAP

Maximum grades

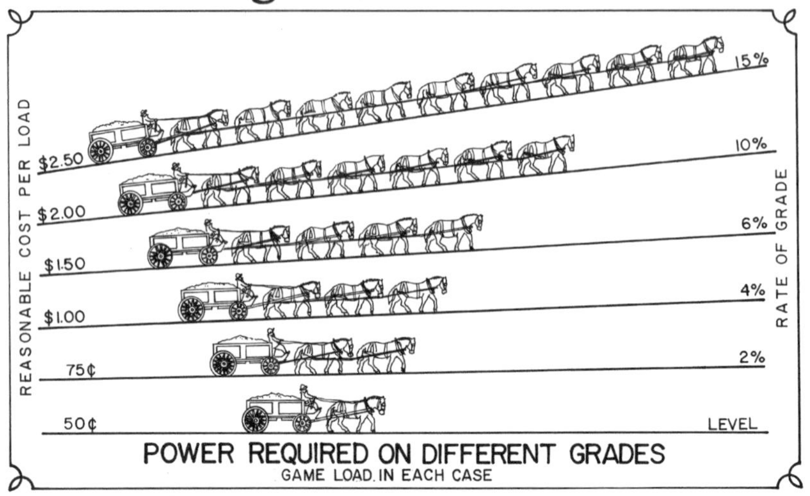

To negotiate the Columbia River Gorge's rugged terrain, wagon roads typically had grades between 10 and 20 percent. In his experiments at Maryhill, Sam Lancaster determined that 5 percent grades make for safer descents and more leisurely inclines. He used the above figure to illustrate that gentler slopes also lowered freight transport costs. (Illustration courtesy of Maryhill Museum, Goldendale, Washington)

Minimum radius & Curve widening

Design standards also included a minimum turning radius of 200 feet to eliminate the dangerous hairpin turns common to mountain roads. While some exceptions were made (see below), in no case was there a radius of less than 100 feet. To facilitate car handling, all curves were widened from 2 feet to 6 feet midway between point of curvature and point of tangency on the inside and tapered.

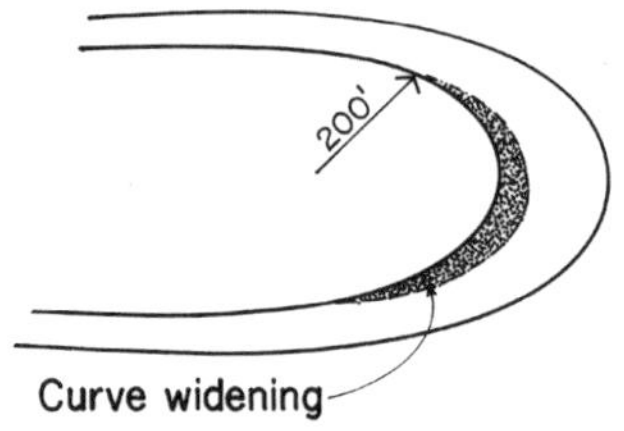

Crowning & Superelevation

Crowning is the standard grading practice in which the road surface is sloped so that water drains away from a high spot in the middle instead of collecting dangerously on the surface. In curves, however, these slopes combine with curve vectors to pull vehicles off the road. Curve banking, sloping the road towards the inside of the curve, was developed to take advantage of the curves centifugal forces. Today, this superelevation permits drivers to maintain highway speeds in curves that would normally require slowing down to keep a vehicle on the road. The designers of the Historic Columbia River Highway decided that "any curve necessitating a super-elevation was dangerous and speed should be slackened and in any case should be enough to prevent side-lash to passengers in a machine at 25 miles an hour; but that the general rule should be nearer a speed of 15 miles per hour." In the end, superevelation was used primarily for drainage purposes.

Exaggerated crowning — Crowned curve — Superelevated curve

Continuous curves

The Historic Columbia River Highway is a route of continuous curves. In addition to providing an everchanging scenic view, these reversing curves served several purposes. Their meandering paths were longer in length than straight shots, allowing more leisurely grades between the same points. Continuous curves could also follow the natural terrain better than straight courses, reducing the need for the extreme cut and fill that characterizes today's faster highways and freeways. Perhaps most significantly, connecting consecutive curves eliminated potential flat spots in between, where crowning might remain level until the next curve. Water collecting there could be potentially hazardous to drivers and also lead to pavement breakdown.

Equipment

Contractors routinely used men, horses and heavy machinery to excavate the roadbed for the Historic Columbia River Highway. In one instance, 30 men, equipped with a steam shovel with a 2□ cubic yard dipper, 4 Ingersoll-Rand air drills, two 15-ton dinkeys, and thirty 4-yard ore cars could move 1,600 cubic yards of overburden and rock each 8-hour day.

Horse-drawn scraping grader

Steam shovel

Loops

In some locations, the steep terrain of the available land couldn't accommodate the standard minimum radius or maximum grade. For example, from Crown Point the road wound around the bluff to Latourell Falls making it necessary to descend 600 feet in a 40 acre tract. To "develop distance" between these points and preserve the maximum grade of 5 percent, the engineers looped the road back and forth on itself down the mountainside, turning several times on curves with not less than a 100 foot turning radius. To compensate for this deviation from the standard 200 foot radius, engineers reduced the grade 1 percent for every 50 foot reduction in turning radius.

Figure Eight Loops
(Troutdale Historical Society)

Hood River Loops
(ODOT)

Rowena Loops
(Steve Lehl)

DELINEATED BY: Helen Selph & Elaine G. Pierce 1995

NATIONAL PARK SERVICE
UNITED STATES DEPARTMENT OF THE INTERIOR

TROUTDALE VIC.

HISTORIC COLUMBIA RIVER HIGHWAY 1913-22
LOCATED BETWEEN TROUTDALE AND THE DALLES ALONG THE COLUMBIA RIVER
MULTNOMAH CO.

OREGON

SHEET 9 of 27

HISTORIC AMERICAN ENGINEERING RECORD
OR - 36

PAVING & DRAINAGE

The Maryhill Experiments

Paving Mix Plant

From 1909 to 1913 Good Roads promoter Samuel Hill financed and built the Maryhill Loops at his Washington estate on the Columbia River. Hill spent over $100,000 of his own money to show the public that blacktop was the road surface of the future.

Under the direction of Samuel C. Lancaster, the roads at Maryhill were built as an experiment to test seven different paving methods for their durability. In three cases, a standard "waterbound" asphalt-macadam road, made by layering courses of stone and dust, was surfaced with various mixtures of hot asphaltic oil and screened stone. These thick mixtures never adequately bound the road surface. In the fourth method, no stone was added to the oil and dust was not included in the layering; this allowed the thinner oil to penetrate the stone layers more thoroughly, but left an insufficient surface. In the next two cases a standard macadam road was sealed with a mixture of oil and sand. The resulting mixture was then rolled, but the water bound nature of the layering still prevent penetration of the surface treatment. Finally, Lancaster built an ordinary dust bound macadam road with no asphaltic treatment.

Ultimately, Hill and Lancaster paved the Historic Columbia River Highway with Warrenite, a patented bitulithic pavement of crushed rock and asphalt. The engineers attributed its durability to the use of stones as large as one inch even though 1/2 inch was the norm. It consisted of a 2 inch layer of dense asphaltic concrete, which was laid while hot on a subbase of broken stone or old macadam road. After rolling, the surface was given a flush coat of asphalt.

Concrete drain cover 32" square

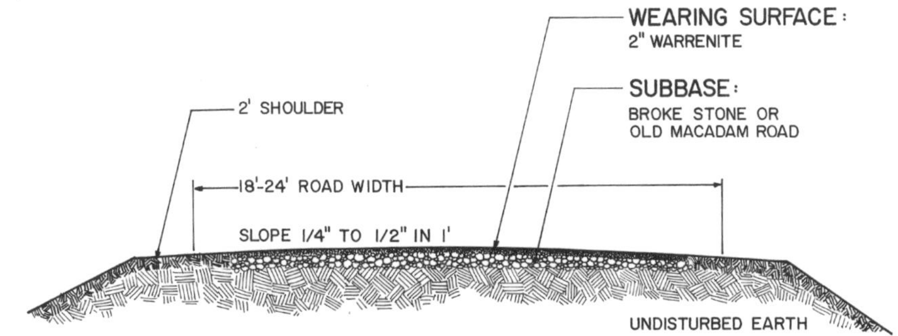

Typical Road Section

Drainage Structures

A road is only as good as its subbase and water is its natural enemy. The designers of the Historic Columbia River Highway were familiar with the legendary rains of the Pacific Northwest and knew the difficulties of road maintenance in the Columbia River Gorge. Their built-in preservation efforts represent one of the earliest rural uses of crowning, drains, curbs and gutters to channel water away from the subbase.

French Drains were the most commonly used method on the Highway for intercepting water before it reached the subbase. In this system, trenches of not less than eighteen inches were dug alongside the road and filled with broken stone or gravel backfill. Drain tiles were added for larger water volumes.

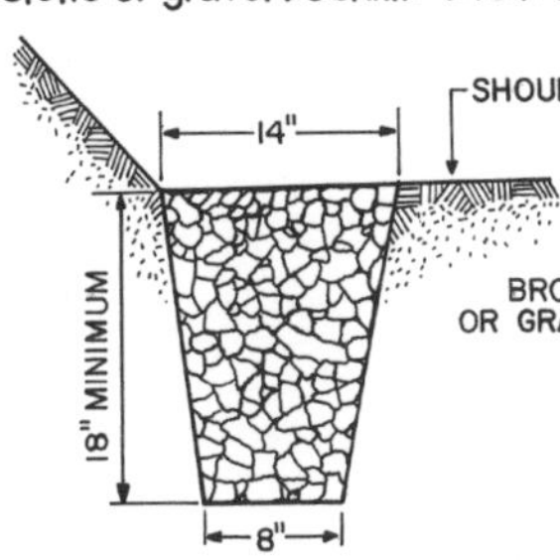

Blind Stone Drain

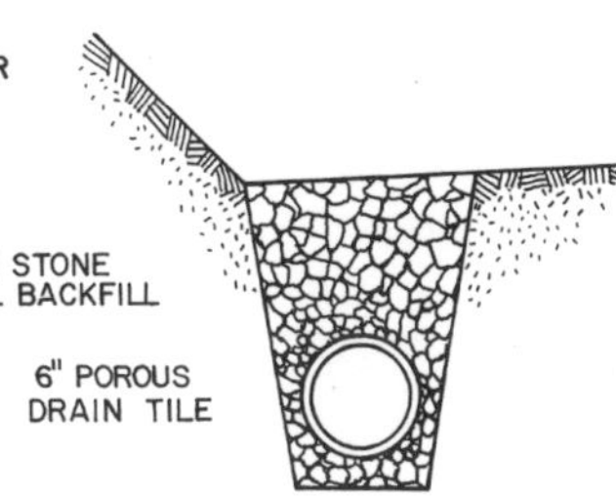

Porous Tile Drain

Gutters were used in several locations. These shallow paved drainage channels carried runoff from the road to inlets where it was directed away from the subbase. Side drains were gutters built on the outside edge of the curb.

Curbs served several functions on the Historic Columbia River Highway. They controlled water on the road surface, preventing it from running off into the subbase, and also helped to stabilize the road edge and present a more finished appearance. In addition, curbs acted as a deterrent to vehicles leaving the pavement at hazardous points.

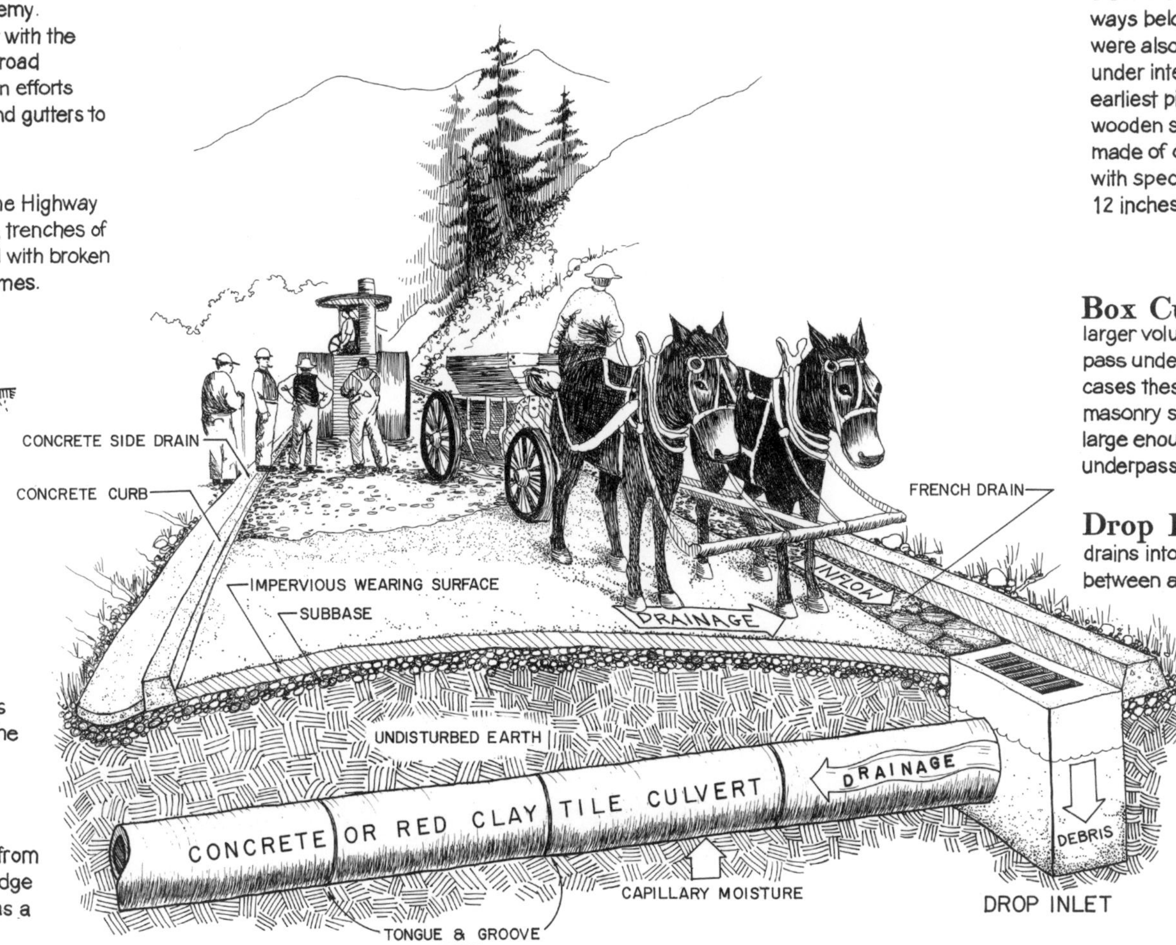

Pipe Culverts carried cross drainage and small waterways below the subbase. They were also used along the Highway under intersecting roads. The earliest pipes were constructed of wooden staves. They were later made of concrete or red clay tile with specified diameters between 12 inches and 36 inches.

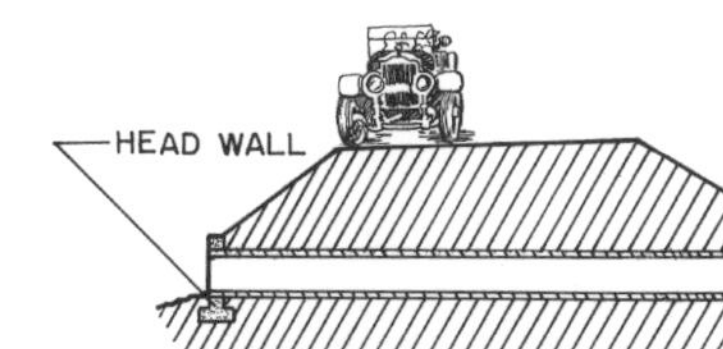

Box Culverts permitted larger volumes of water to safely pass under the roadway. In some cases these large rectangular masonry structures were made large enough to serve as livestock underpasses or cattle crossings.

Drop Inlets are small catch basins that direct runoff from the gutters and drains into cross culverts. In the drawing below, large removable rocks were placed between a French Drain and the inlet to permit easy access to accumulated debris.

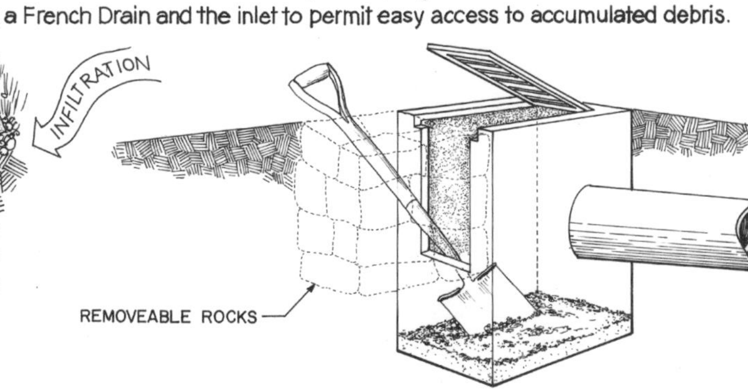

DELINEATED BY: *Helen Selph & Elaine G. Pierce 1995*

HISTORIC COLUMBIA RIVER HIGHWAY RECORDING PROJECT
NATIONAL PARK SERVICE
UNITED STATES DEPARTMENT OF THE INTERIOR

TROUTDALE VIC.

HISTORIC COLUMBIA RIVER HIGHWAY 1913-22
LOCATED BETWEEN TROUTDALE AND THE DALLES ALONG THE COLUMBIA RIVER
MULNOMAH CO.

OREGON

SHEET 10 OF 27

HISTORIC AMERICAN ENGINEERING RECORD
OR - 36

HISTORIC COLUMBIA RIVER HIGHWAY

RAILINGS

Guard rails, or guard fences, are safety devices placed along the edges of roadways to protect traffic on bridge approaches, at curves, and other danger points. It acts as a warning sign and also prevents vehicles from leaving the road. They take many styles, often dictated by suitability to topography, traffic needs, and availability of construction materials. At least seven types of guard fences have been used on the Historic Columbia River Highway.

In the 1930s, an approach to the Sandy River (Stark Street) Bridge was widened to better accommodate traffic. New "rustic" masonry railings replaced older rubble barriers. Their mortared basalt wall design was popularized in national park road construction and became a U.S. Bureau of Public Roads' standard.

The Oregon State Highway Department began erecting its "Standard Guard Fence" in the 1910s on the Historic Columbia River Highway and throughout the state. Its post and plank design offered inexpensive construction, high visibility, and aesthetic appeal.

The department also used a "Standard Arched Masonry Rail" throughout the Historic Columbia River Highway. There were many variations to its design, but they all consisted of slip-face mortared walls of random rubble with arched drainage openings and a screeded concrete cap. Masonry railings were more expensive than other styles but their maintenance costs were very low.

The "Concrete Arch and Cap" is seen along the Historic Columbia River Highway only on two bridges and two viaducts. It was constructed of reinforced-concrete rail caps and posts and concrete plaster and lath arches. Its design continued the arch theme seen in masonry railings and bridges throughout the highway.

The "Post and Cable Fence," seen near Hood River, dates from the 1930s or 1940s. Experience revealed that light wooden fences alone could not contain vehicles. Substituting a twisted cable for the top rail offered added protection for motorists.

The "Guard Fence," employed by the Oregon Department of Transportation since the early 1990s, is reminiscent of the Standard Guard Fence. However, with its steel backing and larger dimensions, it also meets current crash standards.

The "C" Rail, seen throughout the highway, became a popular alternative in the 1930s and 1940s to the Standard Guard Fence. It provided better resistance to wandering automobiles. ODOT is replacing this rail with its modern Guard Fence as it rehabilitates the highway.

The "W" Rail was a popular alternative from the 1940s through the 1980s to replace the old Standard Guard Fence or "C" Rail. ODOT is also replacing this rail with its modern Guard Fence.

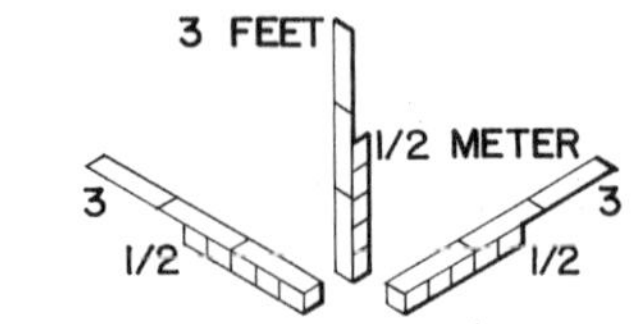

SCALE: 1/2" = 1'-0"   1:24

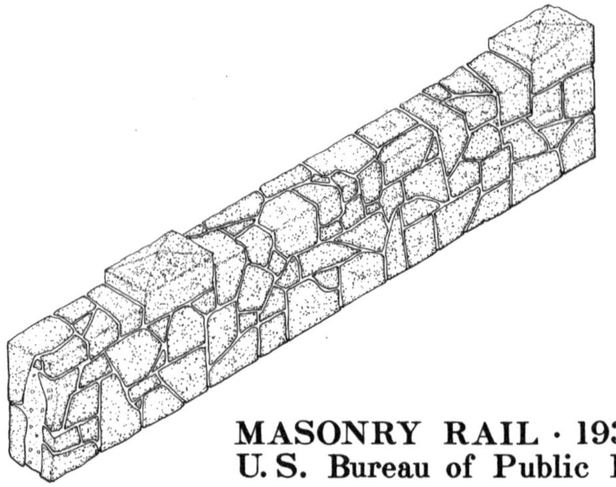

MASONRY RAIL · 1930s
U. S. Bureau of Public Roads
FOUND AT SANDY RIVER (STARK ST.) BRIDGE
(SOLID STONE)

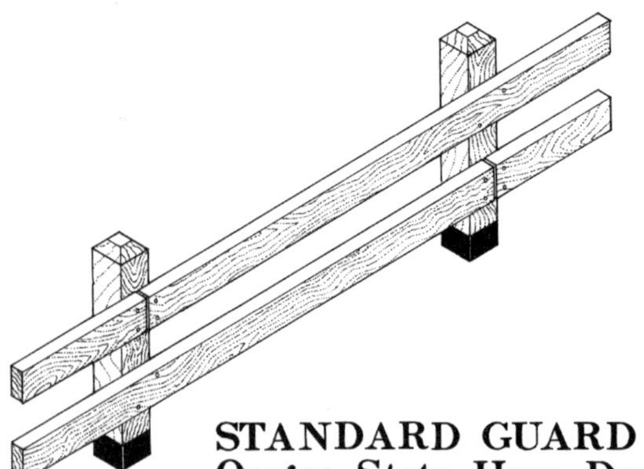

STANDARD GUARD FENCE
Oregon State Hwy. Dept. · 1915 *ca.*
ORIGINALLY FOUND THROUGHOUT HIGHWAY
NO LONGER EXISTS
(NO STEEL)

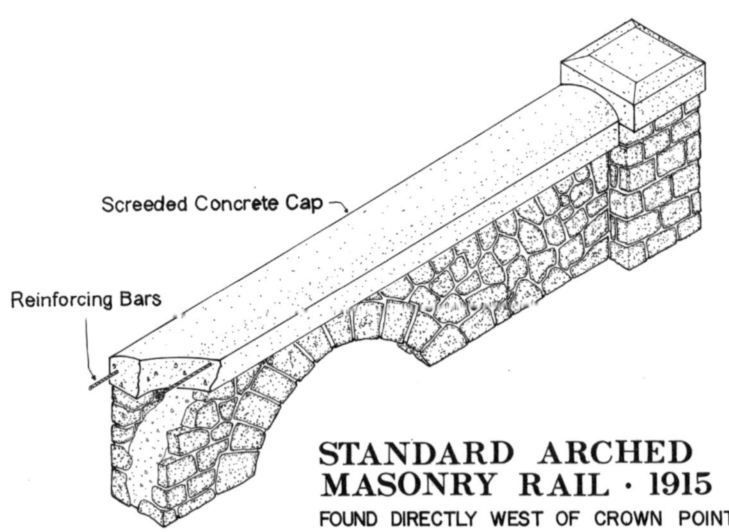

STANDARD ARCHED MASONRY RAIL · 1915
FOUND DIRECTLY WEST OF CROWN POINT
(GROUT CORE)

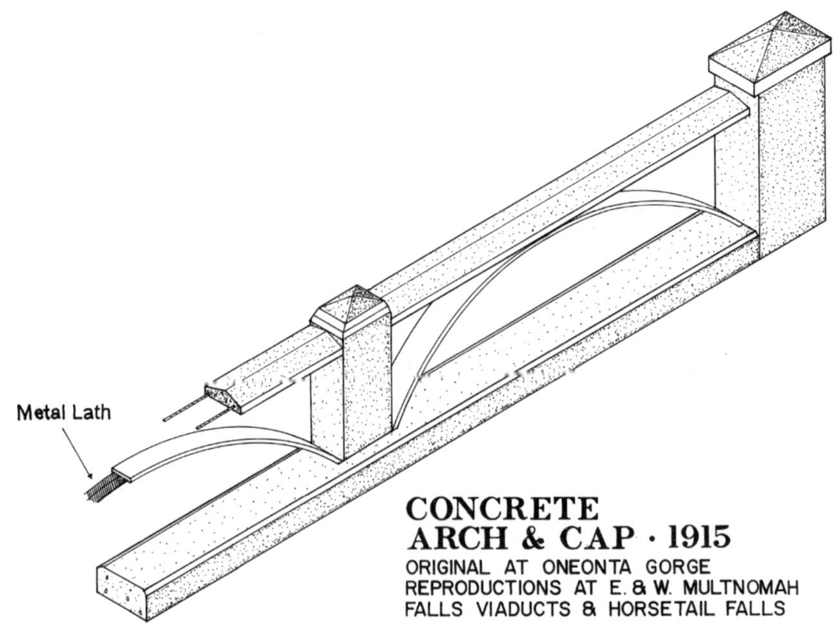

CONCRETE ARCH & CAP · 1915
ORIGINAL AT ONEONTA GORGE
REPRODUCTIONS AT E. & W. MULTNOMAH FALLS VIADUCTS & HORSETAIL FALLS

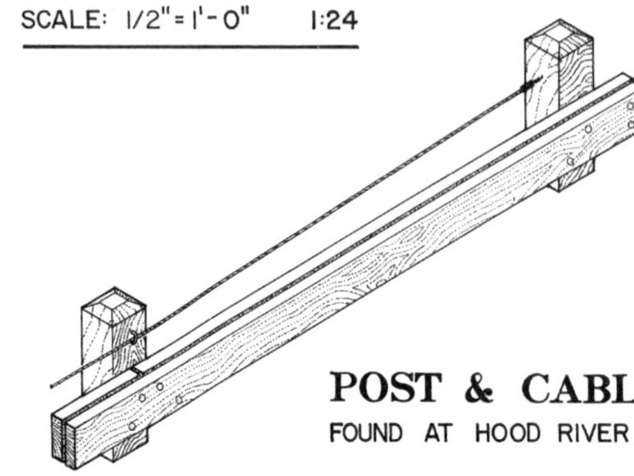

POST & CABLE · 1930s
FOUND AT HOOD RIVER LOOPS

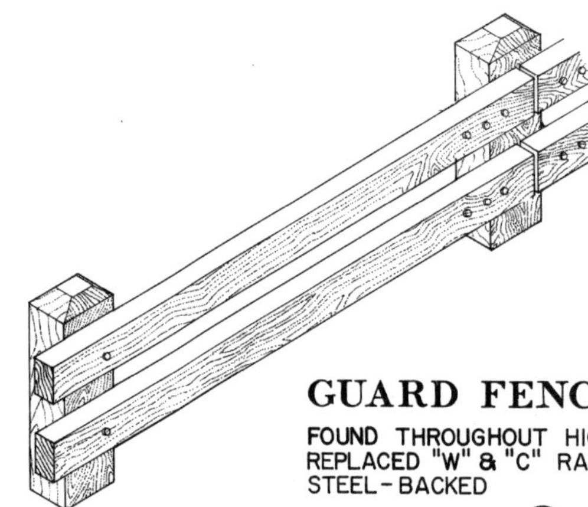

GUARD FENCE · 1987
FOUND THROUGHOUT HIGHWAY
REPLACED "W" & "C" RAILS
STEEL-BACKED

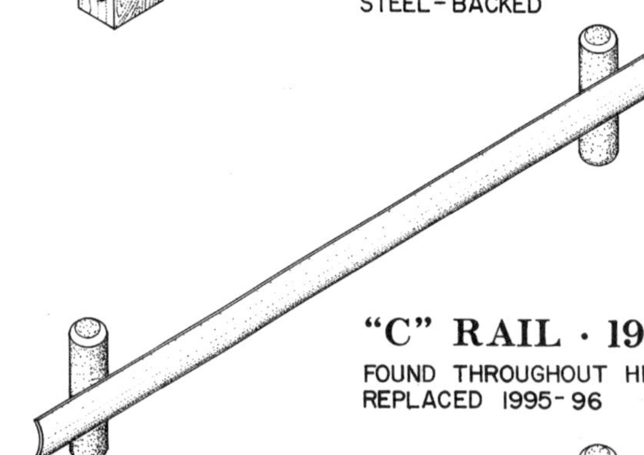

"C" RAIL · 1950s
FOUND THROUGHOUT HIGHWAY
REPLACED 1995-96

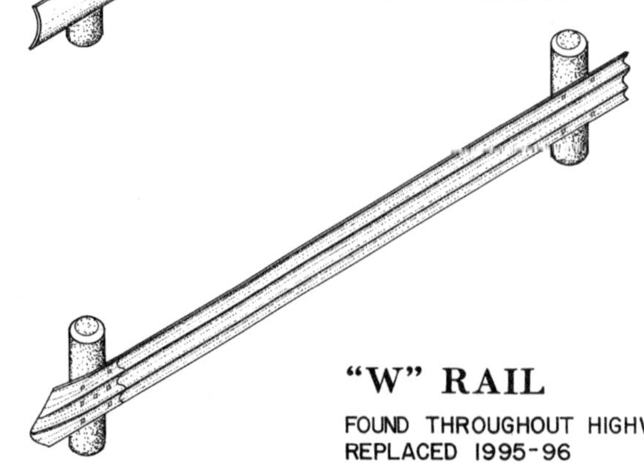

"W" RAIL
FOUND THROUGHOUT HIGHWAY
REPLACED 1995-96

DELINEATED BY: *Christine Rumi 1995*

NATIONAL PARK SERVICE
UNITED STATES DEPARTMENT OF THE INTERIOR

TROUTDALE VIC.

HISTORIC COLUMBIA RIVER HIGHWAY 1913-22
LOCATED BETWEEN TROUTDALE AND THE DALLES ALONG THE COLUMBIA RIVER
MULTNOMAH CO.

OREGON

SHEET 11 of 27

HISTORIC AMERICAN ENGINEERING RECORD
OR-36

HISTORIC COLUMBIA RIVER HIGHWAY

MASONRY

(Information based on 1995 interviews with Richard Fix, master mason)

One of the most distinctive features of the Historic Columbia River Highway is its masonry. Sam Lancaster hired Italian craftsmen to build its many retaining walls, arched guard walls, water stops, and culverts like the one shown above. (Image provided by Maryhill Museum, Goldendale, Washington)

Dressing

While some of the rock came from as far away as a mile, most was taken from the site itself. Moving larger rocks required using hand star drills to make holes into which pins were inserted. Workers strung cables through these pins to move the rock into the work area where they could be broken up using feathers and wedges. In this technique, feathers are dropped into holes drilled 4 - 6 inches apart along the break line. To control breakage, masons gently tapped each wedge along the line. The sequence was repeated until the rock split. Smaller rocks were then shaped by hand with stone hammers and bushing hammers. This dressing was best done while the rock was green, usually within a month of being cut.

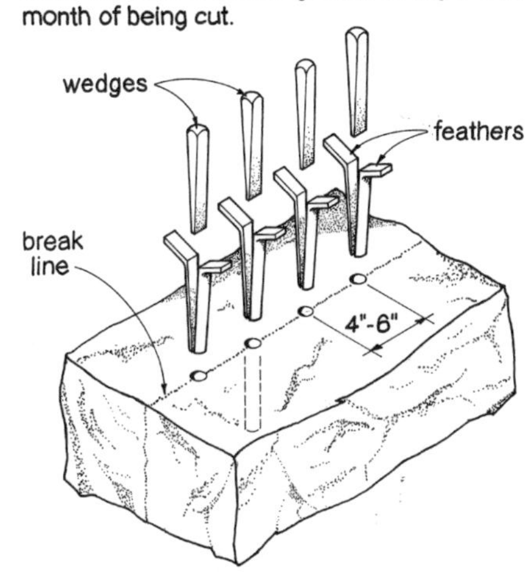

Patterns

Random rubble, as shown in step 3 below, is characterized by irregularly shaped rocks laid in no particular pattern. This method reveals the individual style of each mason. Some favored smaller rocks over larger ones while others chose to spend more time dressing the stones or tooling their joints.

Ashlar patterns, as seen in the retaining wall at right, are laid in courses with attention given to uniformity in height. Frequently a wall with ashlar facing on the front was faced in random rubble on the back.

Arches

While picturesque in appearance, the many arched wall openings on the highway served a more practical function, drainage. Water pooling on the road behind a solid wall could undermine the road surface as well as the wall itself. In a few locations, these openings were semi-circular, but the semi-elliptical arch, as shown below, was more commonly used for several reasons. Semi-elliptical arches spanned greater distances under the same 30" rail height than semi-circular arches, and fewer arches in a given length of wall meant less rock was used. More importantly, semi-elliptical arches were less labor intensive. In a semi-circular arch, the arch stones are roughly identical in shape, requiring masons to dress each stone to a set pattern. In contrast, semi-elliptical arches are flatter; once the curved stones at the bottom of the arch were shaped, the remaining stones were more rectilinear in shape and thus more easily dressed.

Mortar joints

After a wall was built, the concrete grout joints were raked, or recessed, 1" to 1½" from the rock face. The mason then over filled this channel with a finish mortar of fine masonry sand applied with a small grout bag. The excess was then "feathered to the stone" or evenly distributed to camoflage irregularities and create more uniform joint widths. In several locations masons used a bead joiner to tool this smooth joint into a bead or a concave joint.

Ideally, this sand mortar finishing was done while the coarser wall grout was still wet and a good bond could form. On the Historic Columbia River Highway, however, the finishing crew often arrived days after the grout had set. The resulting "cold joints" led to the flaking seen today.

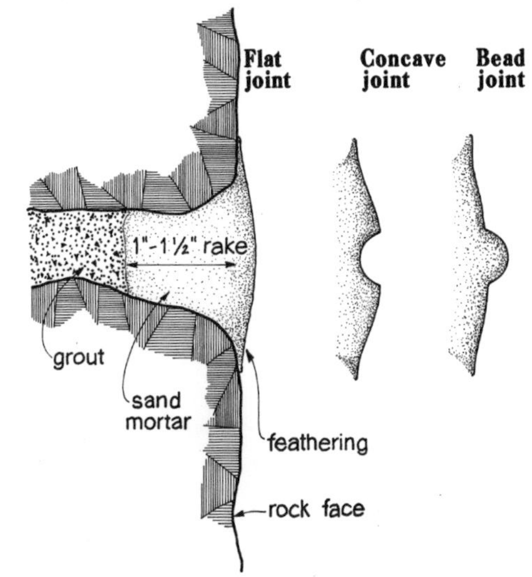

Retaining walls & Guard rocks

Several miles of retaining wall were built along the highway. These walls were dry laid, allowing water to drain away from the roadbed instead of undermining it. Masons used batter-boards and jacklines to maintain the consistent slope of these walls. Workers stood on wooden planks supported by timbers built into the wall as it went up. When the wall was finished, these timbers were left in place and cut off at the wall face. A number of retaining walls were topped with guard rocks that continued the slope of the batter and served to keep traffic on the road. Masons shaped their semi-circular inside face with bush-hammers. (Above image provided by the Oregon Historical Society)

Process: Building a slip-form grout-lock wall on the HCRH

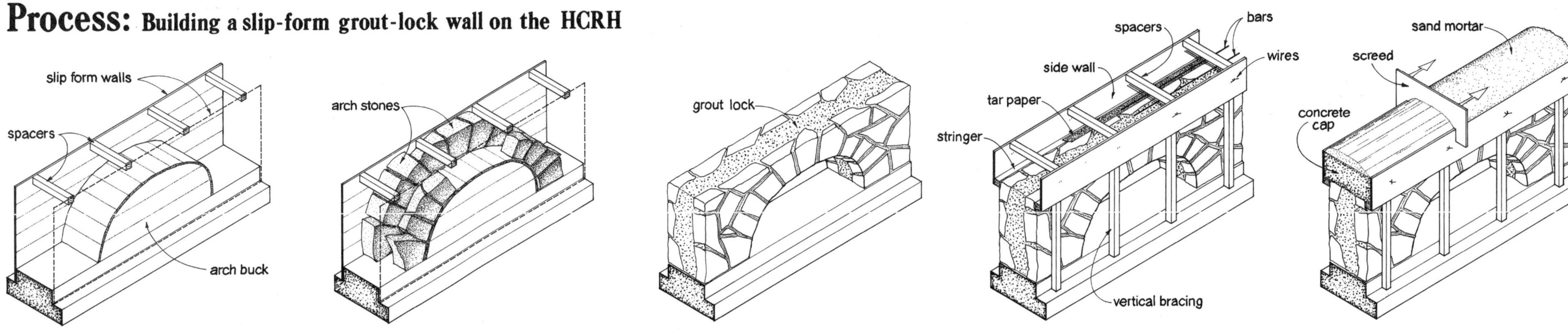

1. Set up the form
This is done on top of a stable surface, like a retaining wall or a concrete foundation.

2. Lay the rock
Arch stones are grouted in first and then the remaining stones are laid and grouted until "laid up to height."

3. Strip the forms & finish the joints
(see Mortar joints above)

4. Build the cap form
Set up vertical bracing, stringers, side walls.
Tie wires, lay bars and insert spacers.
Line the bottom joints with tar paper.

5. Pour & finish the cap
Pour the concrete and screed flush.
Apply sand mortar and screed to shape.
Cut wires and remove forms when set.

DELINEATED BY: Elaine G. Pierce 1995

NATIONAL PARK SERVICE
UNITED STATES DEPARTMENT OF THE INTERIOR

HISTORIC COLUMBIA RIVER HIGHWAY 1913-22
LOCATED BETWEEN TROUTDALE AND THE DALLES ALONG THE COLUMBIA RIVER
MULTNOMAH CO.
TROUTDALE VIC.
OREGON

HISTORIC AMERICAN ENGINEERING RECORD
OR - 36

SHEET 12 of 27

VIADUCTS

(Photograph sources appear beneath each sketch)

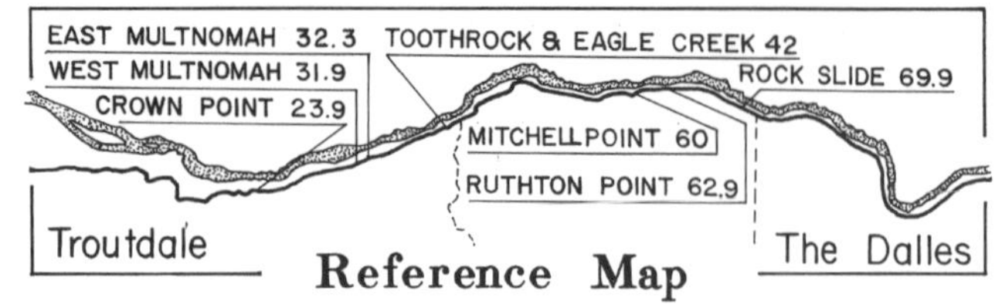

Reference Map

Spanning a cleft

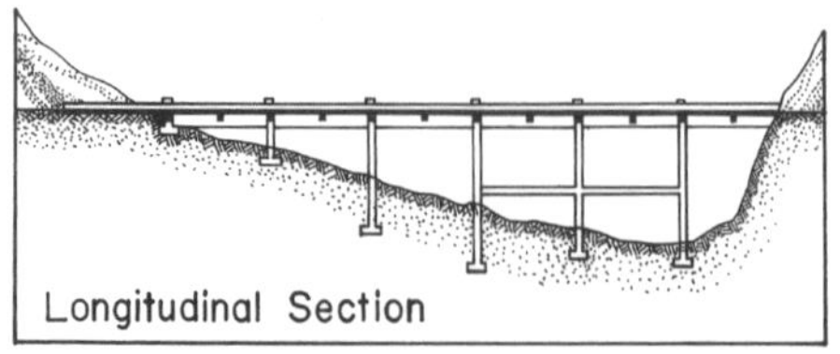

At the west approach to Mitchell Point Tunnel, it was necessary to carry the road alignment from a rock ledge across a talus sloped concavity to the tunnel portal. Engineers spanned this cleft with a 192' viaduct, a series of short reinforced concrete slab bridges with piers at the ends of each individual span.

Mitchell Point 1915
(M. Sherman, et al., *Columbia River Gorge*)

Raising the roadbed

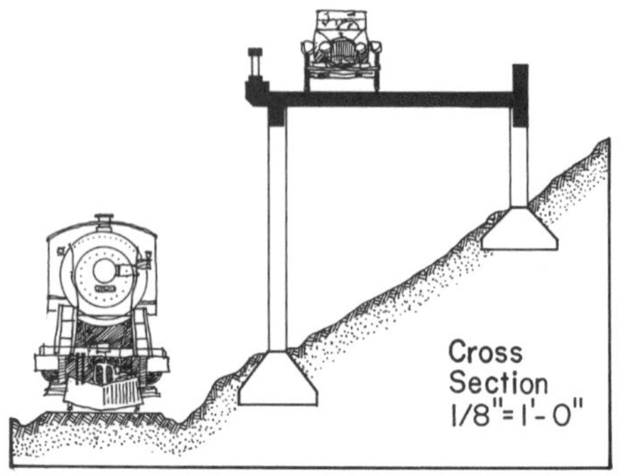

The road immediately west and east of Multnomah Falls runs between a steep mountainside and the Oregon-Washington Railroad and Navigation Company main line, now Union Pacific Railroad, at the river's edge. The topography of the area made construction of a typical retaining wall difficult. Even minimal cut and fill at the toe of these mountainsides, held together by underbrush and timber, might cause rock and debris avalanches to cover the road and block the railroad's main line.

The solution was to employ viaducts, resting on unequal length columns and anchored to the slopes. They rose up the hillsides, nearly overhanging the tracks to maintain vertical clearances for rail traffic when they came dangerously close to the main line's right-of-way. As the result of a storm in 1921, the East Viaduct required stabilization and support walls were poured behind every other column.

West Multnomah 1914
(*First Annual Report of the Oregon Highway Engineer*, 1914)

East Multnomah 1914
(S. C. Lancaster, *The Columbia, America's Great Highway*)

Skirting hillsides

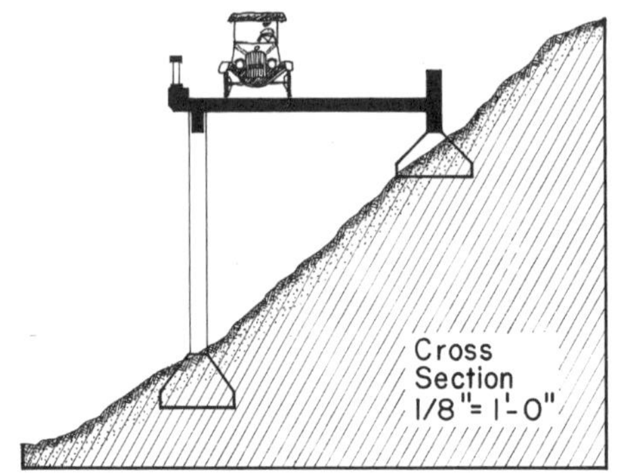

Engineers designed half-viaducts for several locations on the highway to skirt hillsides. They were constructed much like viaducts with unequal-length columns, except that the inside bents consisted only of footings and the inside elevations were anchored into the hillsides or masonry walls. Because of the half-viaducts' inconspicuous design, motorists often didn't realize they were travelling over engineered structures and not surface road.

The 560' Crown Point Viaduct (28 20' reinforced-concrete deck slabs) prevented unnecessary excavation or fill to establish room for a 7' sidewalk and curb. Its design also includes a 4' concrete outer railing and concrete light standards to illuminate the point at night.

High above the river, Toothrock and Eagle Creek viaducts (224') carry the highway around Toothrock, a tall basalt cliff, before dropping down to Eagle Creek. Their designs differ only in their railing treatment; where Toothrock Viaduct uses a concrete spindle and cap design, Eagle Creek Viaduct uses a masonry rail and concrete cap design.

Ruthton Point Viaduct (50') consists of three reinforced-concrete deck girder spans and carried the highway near a promontory west of Hood River. It uses a simple standardized concrete railing panel and cap.

The 34' Rock Slide Viaduct lies a short distance west of the Mosier Twin Tunnels. The uninterrupted roadway surface and the continuous arched rubble parapet railing across the viaduct make it difficult to identify the bridge span from the road.

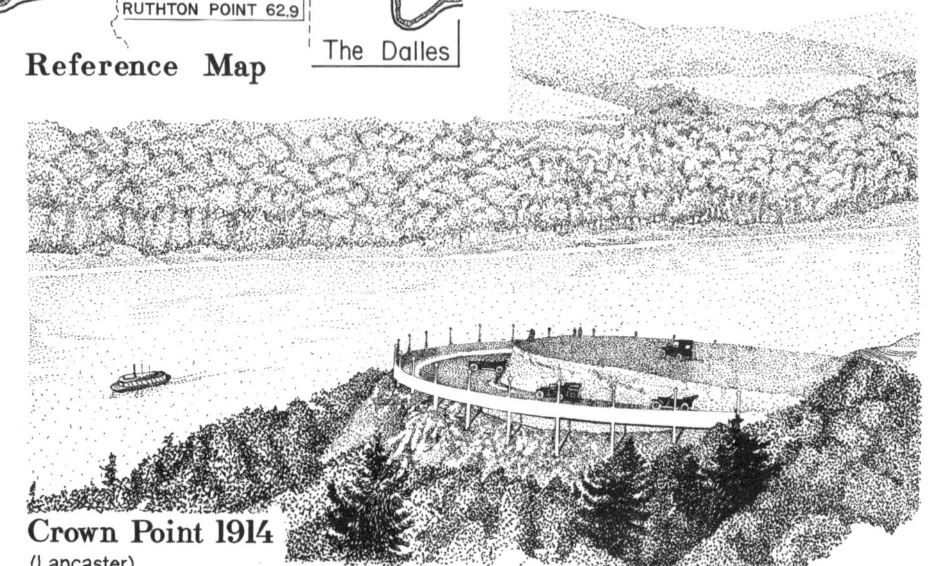

Crown Point 1914
(Lancaster)

Toothrock 1915
(Lancaster)

Eagle Creek 1915
(Maryhill Museum of Art)

Ruthton Point 1918
(D. A. Smith, et al., *Historic Highway Bridges of Oregon*)

Rock Slide 1920
(Smith)

HISTORIC COLUMBIA RIVER HIGHWAY 1913-22
LOCATED BETWEEN TROUTDALE AND THE DALLES ALONG THE COLUMBIA RIVER
MULTNOMAH CO. OREGON
TROUTDALE VIC.
HISTORIC AMERICAN ENGINEERING RECORD OR-36
SHEET 13 of 27
DELINEATED BY: V. V. Simonenko & Elaine G. Pierce 1995
NATIONAL PARK SERVICE
UNITED STATES DEPARTMENT OF THE INTERIOR

TUNNELS

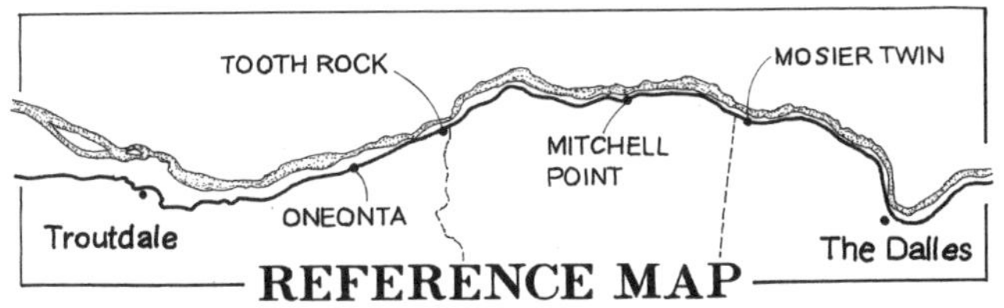

With the railroad to the north and sheer basalt cliffs to the south, tunnels were often the only alternative to longer, more expensive routes on the Historic Columbia River Highway. As a result, three tunnels were constructed along the highway between 1914 and 1921. Like the road itself, these tunnels were soon unable to handle the burgeoning traffic. Their narrow road widths coupled with the threat of rockfall made for hazardous driving.

Oneonta (1914) was the first and shortest tunnel on the original highway. At the west approach, an 80' slab bridge carried the road to the tunnel portal. In 1948, the railroad moved its right-of-way to water level, giving highway engineers a prime opportunity to realign the road to bypass the tunnel. A new 48' deck girder bridge was built parallel to the old bridge and the tunnel was backfilled. The old bridge now serves as an overlook and parking while the old road vanishes into the hillside.

Mitchell Point (1915) and the Mosier Twin Tunnels (1921) were similarly fated. Mitchell Point provided spectacular Gorge views through its five famous windows until the highway was relocated in the 1950's. Ultimately, Mitchell Point was completely destroyed in 1966. (See Sheet 17) The Mosier Twin tunnels, featuring two adits and a cliffwalk, was backfilled by 1958 but is currently being reopened for recreation purposes. (See Sheet 18)

Bonneville Dam's construction forced the realignment of the Historic Columbia River Highway with Tooth Rock Tunnel in 1937. At 827', Tooth Rock became the longest tunnel on the highway, and with a 26' wide roadway could safely accommodate larger vehicles. A lighting system and 4' sidewalks set Tooth Rock apart from the original tunnels on the highway as an example of modern engineering. In 1969, Tooth Rock's roadbed was lowered and sidewalks eliminated to allow even larger vehicles safe passage through the tunnel. Today, it is still in use by high speed traffic on Interstate 84's eastbound lanes.

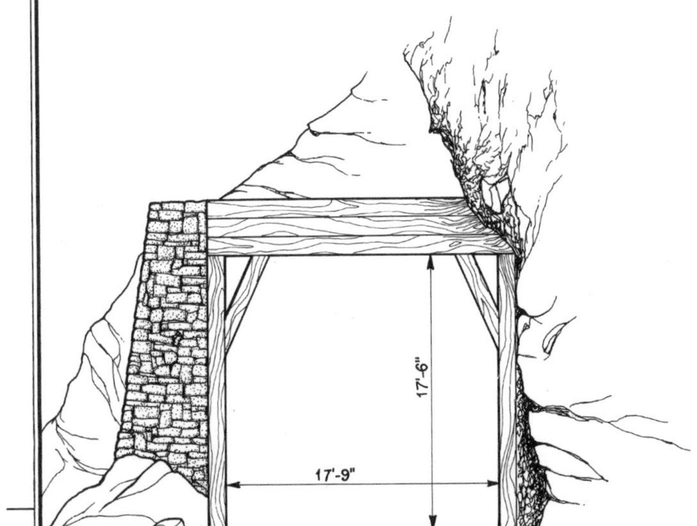

West portal

Oneonta 1914

LENGTH COMPARISON

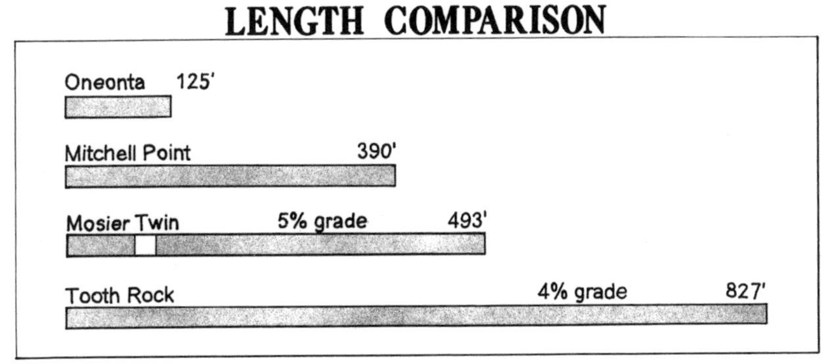

Lantern Detail

Scale: 1/2" = 1' - 0"

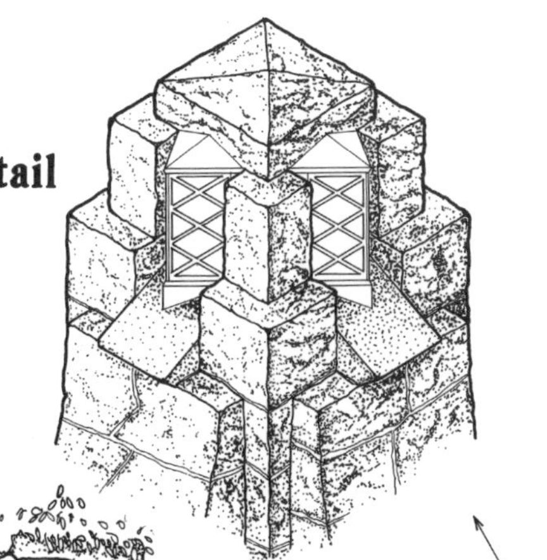

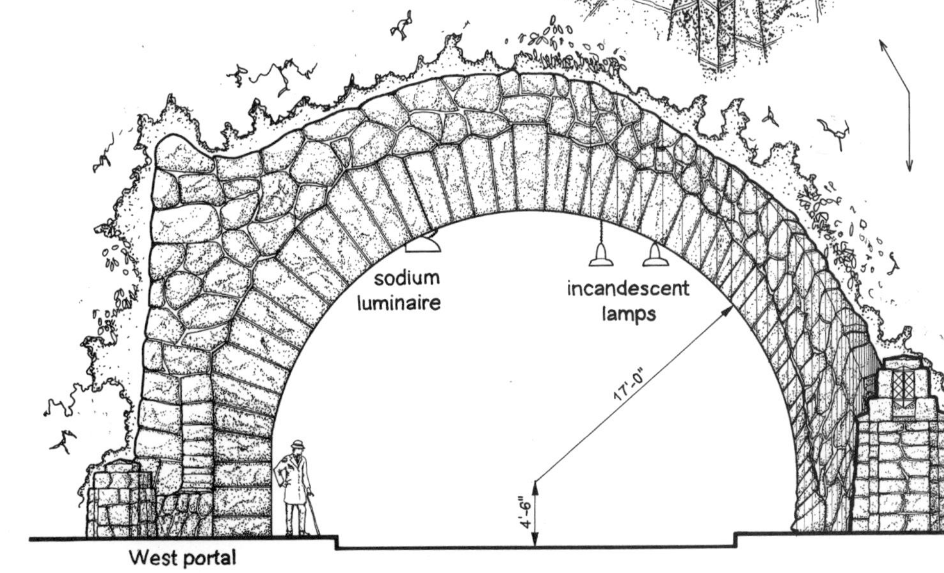

West portal

Tooth Rock 1937

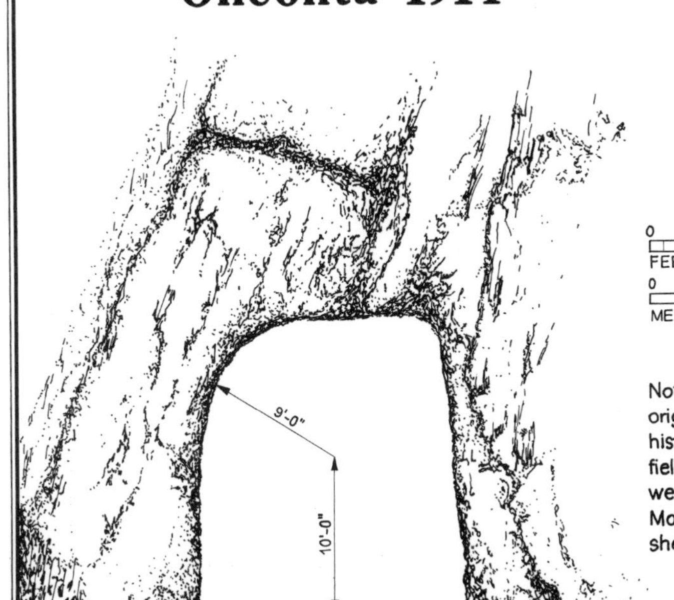

West portal

Mitchell Point 1915

0 5 10 15
FEET 3/16"=1'-0"
0 1 2 3 4 5
METERS 1:64

Note: Drawings based on original design documents, historic photographs, and field measurements. For west portal, tunnel no. 1 of Mosier Twin Tunnels, see sheet 18.

Tunnel no. 1 – East portal

Tunnel no. 2 – West portal

Tunnel no. 2 – East portal

Mosier Twin 1921

DELINEATED BY Pete Brooks 1995

HISTORIC COLUMBIA RIVER HIGHWAY RECORDING PROJECT
NATIONAL PARK SERVICE
UNITED STATES DEPARTMENT OF THE INTERIOR

HISTORIC COLUMBIA RIVER HIGHWAY 1913-22
LOCATED BETWEEN TROUTDALE AND THE DALLES ALONG THE COLUMBIA RIVER
MULTNOMAH CO.
TROUTDALE VIC.
OREGON

SHEET 16 OF 27

HISTORIC AMERICAN ENGINEERING RECORD
OR - 36

Appendix

The Historic American Engineering Record National Park Service Park Roads and Bridges Projects

The preceding portfolio represents a significant but by no means comprehensive compendium of the drawings produced by the Historic American Engineering Record (HAER) National Park Service (NPS) Park Roads and Bridges Program. The complete set of drawings, along with associated historical narratives and large-format photographs, is permanently archived in the Historic American Buildings Survey (HABS)/HAER collection of the Library of Congress's Prints and Photographs Division. Researchers can examine this material on site or access it online through the Library of Congress's American Memory website. A complete list of HABS/HAER NPS Park Roads and Bridges projects is provided below. For administrative purposes, many park road systems were divided into discrete components so that the documentation for specific roads and bridges could be accessed individually. Not all the HAER NPS Park Roads and Bridges documentation was online by the date of this publication, but web users will eventually be able to access the entire collection by searching under the park or project name or even with more general terms such as *park road, park bridge,* and *parkway*. Several of the following projects consist of written histories and large-format photographs with no formal drawing components.

Acadia National Park

- Acadia National Park Motor Roads
- Acadia National Park Roads and Bridges
- Rockefeller Carriage Roads
- Paradise Hills Road
- Cadillac Mountain Road
- Multiple bridges listed individually by name

Arches National Park

- Main Entrance Road
- Moab Canyon Wash Culvert

Baltimore-Washington Parkway

- Baltimore-Washington Parkway

Blue Ridge Parkway

- Blue Ridge Parkway
- Linn Cove Viaduct

Capitol Reef National Park

- Capitol Reef National Park Roads and Bridges

Chickamauga and Chattanooga National Military Park

- Chattanooga National Military Park Tour Roads
- Chickamauga National Military Park Tour Roads

Colonial National Historical Park

- Colonial Parkway
- Colonial National Historical Park Roads and Bridges
- Jamestown Island Loop Road
- Yorktown Battlefield Tour Road
- Williamsburg Tunnel
- Multiple bridges listed individually by name

Colorado National Monument

- Trail of the Serpent
- Rim Rock Drive

Crater Lake National Park

- Crater Lake National Park Roads
- Goodbye Creek Bridge

Death Valley National Park

- Death Valley National Park Roads

Fort Laramie National Historic Site

- North Platte River Bowstring Truss Bridge

George Washington Memorial Parkway

- George Washington Memorial Parkway
- Multiple bridges listed individually by name

Gettysburg National Military Park

- Gettysburg Tour Roads

Glacier National Park

- Going-to-the-Sun Road
- West Side Tunnel
- East Side Tunnel
- Multiple bridges listed individually by name

Glen Canyon National Recreation Area

- Navajo Bridge

Grand Canyon National Park

- Grand Canyon National Park Roads
- Cape Royal Road
- Village Loop Drive
- West Rim
- North Entrance Road
- East Rim Drive
- South Entrance Road
- Multiple bridges listed individually by name

Great Smoky Mountains National Park

- Great Smoky Mountains National Park Roads and Bridges
- Newfound Gap Road
- Clingman's Dome Road
- Little River Road
- Cades Cove
- Foothills Parkway
- Cataloochee Valley
- Roaring Fork Motor Nature Trail
- Big Creek Road
- Northshore Road
- Cosley Park Road
- Deep Creek Road
- Greenbrier Road
- Heintooga Round Bottom Road and Balsam Mountain Road
- Cataloochee Trail and Turnpike
- Rich Mountain Road
- Elkmont Road
- Multiple bridges listed individually by name

Haleakala National Park Roads

- Haleakala National Park Roads

Hawaii Volcanoes National Park

- Hawaii Volcanoes National Park Roads
- Crater Rim Drive
- Chain of Craters Road
- Mauna Loa Road
- Hilina Pali Road

Lassen Volcanic National Park

- Lassen Park Road

Marsh-Billings-Rockefeller National Historical Park

- Marsh-Billings-Rockefeller Carriage Roads

Mesa Verde National Park

- Mesa Verde National Park Main Entrance Road
- Prater-Morefield Tunnel

Mount Rainier National Park

- Mount Rainier National Park Roads and Bridges
- Nisqually Road
- Carbon River Road
- Mowich Lake Road
- West Side Road
- Stevens Canyon Highway
- East Side Highway
- Mather Memorial Parkway
- Yakima Park Highway
- Box Canyon Tunnel
- Stevens Canyon Tunnel
- East Side Highway Tunnel
- Multiple bridges listed individually by name

Natchez Trace Parkway

- Natchez Trace Parkway

National Capital Region

- Potomac Aqueduct
- Arlington Memorial Bridge
- Tidal Reservoir
- Boundary Channel Bridge
- Quantico Creek Bridges
- Stone Bridge

Olympic National Park
- Olympic National Park Road System

Petrified Forest National Park
- Long Logs Road
- Jim Camp Wash Bridge

Redwood National and State Parks
- Redwood National and State Parks Roads

Rock Creek and Potomac Parkway
- Rock Creek and Potomac Parkway

Rock Creek Park
- Rock Creek Park Road System
- Multiple bridges listed individually by name

Rocky Mountain National Park
- Rocky Mountain National Park Roads and Bridges
- Trail Ridge Road
- Fall River Road

Scotts Bluff National Monument
- Scotts Bluff Summit Road

Sequoia National Park
- Generals Highway
- Multiple bridges listed individually by name

Shenandoah National Park
- Skyline Drive

Shiloh National Military Park
- Shiloh Tour Roads

Vanderbilt Mansion National Historic Site
- Vanderbilt Mansion Roads and Bridges
- Multiple bridges listed individually by name

Vicksburg National Military Park
- Vicksburg Tour Road
- Multiple bridges listed individually by name

Wind Cave National Park
- Wind Cave Roads and Bridges
- Beaver Creek Bridge
- Pigtail Bridge

Yellowstone National Park
- Yellowstone Roads and Bridges
- Grand Loop Road
- West Entrance Road
- Gallatin Entrance Road
- North Entrance Road
- Multiple bridges listed individually by name

Yosemite National Park
- Yosemite National Park Roads and Bridges
- All Year Highway
- Glacier Point Road
- Coulterville Road
- Big Oak Flat Road
- Wawona Road
- Tioga Road
- Hetch Hetchy Road
- Big Oak Flat Road Tunnels 1, 2, and 3
- Wawona Tunnel
- Multiple bridges listed individually by name

Zion National Park
- Zion–Mount Carmel Highway
- Zion National Park Roads and Bridges
- Floor of the Valley Road
- Multiple bridges listed individually by name

Associated Non-NPS Road Documentation Projects

Arroyo Seco Parkway
- Arroyo Seco Parkway
- Figueroa Street Tunnels
- Multiple bridges listed individually by name

Bronx River Parkway Reservation
- Bronx River Parkway Reservation

Historic Columbia River Highway
- Historic Columbia River Highway
- Oneonta Tunnel
- Toothrock Tunnel
- Mitchell Point Tunnel
- Mosier Twin Tunnels
- Multiple bridges listed individually by name

Merritt Parkway
- Merritt Parkway
- Multiple bridges listed individually by name

Taconic State Parkway
- Taconic State Parkway

Selected References

General Park and Road History

Abbott, Stanley. "Parkways—Past, Present, and Future." *Parks and Recreation* (December 1948): 681–91.

Albright, Horace, and Robert Cahn. *The Birth of the National Park Service: The Founding Years, 1913–1933.* Salt Lake City: Howe Brothers, 1985.

Albright, Horace, and Frank Taylor. *Oh Ranger! A Book about the National Parks.* New York: Dodd & Mead, 1936.

Bayliss, Dudley. "Parkway Development under the National Park Service." *Parks and Recreation* (February 1937): 255–59.

———. "Planning Our National Park Roads and Our National Parkways." *Traffic Quarterly* 11 (July 1957): 417–40.

Belasco, Warren. *Americans on the Road: From Autocamp to Motel, 1910–1945.* Cambridge: MIT Press, 1979. Reprint, Baltimore: Johns Hopkins University Press, 1997.

Birnbaum, Charles, and Robin Karson, eds. *Pioneers of American Landscape Design.* New York: McGraw-Hill, 2000.

Carr, Ethan. *Wilderness by Design: Landscape Architecture and the National Park Service.* Lincoln: University of Nebraska Press, 1998.

Chittenden, Hiram. *The Yellowstone National Park: Historical and Descriptive.* Cincinnati: Robert Clarke Co., 1895.

———. *The Yellowstone National Park: Historical and Descriptive,* "new" edition. Cincinnati: Stewart & Kidd, 1920.

Culpin, Mary Shivers. *The History of the Construction of the Road System in Yellowstone National Park, 1872–1966: Historic Resource Study,* vol. 1. National Park Service, Rocky Mountain Region, Division of Cultural Resources, 1994.

Cutler, Phoebe. *The Public Landscape of the New Deal.* New Haven: Yale University Press, 1985.

Davis, Timothy. "Mount Vernon Memorial Highway and the Evolution of the American Parkway." Ph.D. diss., University of Texas at

Austin, 1997.
———. "Rock Creek and Potomac Parkway, Washington, D.C.: The Evolution of a Contested Urban Landscape," *Studies in the History of Gardens and Designed Landscapes* 19 (April–June 1999): 123–237.
———. "Mount Vernon Memorial Highway: Changing Conceptions of an American Commemorative Landscape." In *Places of Commemoration: Search for Identity and Landscape Design,* ed. Joachim Wolschke-Bulmahn, 123–77. Washington, D.C.: Dumbarton Oaks, 2001.
———. "'A Pleasing Illusion of Unspoiled Countryside': The American Parkway and the Problematics of an Institutional Vernacular." In *Perspectives in Vernacular Architecture,* ed. Kenneth Breisch and Kim Hoagland, 9:228–46. Knoxville: University of Tennessee Press, 2003.
Demars, Stanford E. *The Tourist in Yosemite, 1855–1985.* Salt Lake City: University of Utah Press, 1991.
Foresta, Ronald. *America's National Parks and Their Keepers.* Washington, D.C.: Resources for the Future, 1984.
Good, Albert H. *Park and Recreation Structures.* 3 vols. Washington, D.C.: Government Printing Office, 1938; reprint, New York: Princeton Architectural Press, 1989.
Haynes, Aubrey L. *The Yellowstone Story: A History of Our First National Park,* vol. 2, 2d edition. Niwot: University Press of Colorado, 1996.
Hewes, Laurence. *American Highway Practice.* 2 vols. New York: John Wiley & Sons, 1942.
Huth, Hans. *Nature and the American: Three Centuries of Changing Attitudes,* 2d edition, with introduction by Douglas Strong. Lincoln: University of Nebraska Press, 1990.
Johnston, Hank. *Yosemite's Yesterdays,* 2d edition. Yosemite, Calif.: Flying Spur Press, 1989.
———. *Yosemite Yesterdays,* vol. 2. Yosemite, Calif.: Flying Spur Press, 1989.
Jolley, Harley E. *The Blue Ridge Parkway.* Knoxville: University of Tennessee Press, 1969.
Marriott, Paul Daniel. *Saving Historic Roads: Design and Policy Guidelines.* New York: John Wiley & Sons, 1998.
Mather, Stephen. "Engineering Applied to National Parks." *Transactions of the American Society of Civil Engineers* 94 (1930): 1181–93.
McClelland, Linda. *Presenting Nature: The Historic Landscape Design of the National Park Service, 1916–1942.* U.S. Department of the Interior, National Park Service, Cultural Resources, Interagency Resources Division, National Register of Historic Places, 1993.
———. *Building the National Parks: Historic Landscape Design and Construction.* Baltimore: John Hopkins University Press, 1998.
Newton, Norman T. *Design on the Land: The Development of Landscape Architecture.* Cambridge: Harvard University Press, 1971.
Pomeroy, Earl. *In Search of the Golden West: The Tourist in Western America.* New York: Alfred A. Knopf, 1957.
Roberts, Ann Rockefeller. *Mr. Rockefeller's Roads: The Untold Story of Acadia's Carriage Roads and Their Creator.* Camden, Maine: Down East Books, 1990.
Rose, Albert. *Historic American Roads, from Frontier Trails to Superhighways.* New York: Crown Publishers, 1976.
Runte, Al. *National Parks: The American Experience,* 2d edition. Lincoln: University of Nebraska Press, 1987.
———. *Yosemite: The Embattled Wilderness.* Lincoln: University of Nebraska Press, 1990.
Sears, John. *Sacred Places: American Tourist Attractions in the Nineteenth Century.* New York: Oxford University Press, 1989.
Sellers, Richard. *Preserving Nature in the National Parks: A History.* New Haven: Yale University Press, 1997.
Shankland, Robert. *Steve Mather of the National Parks,* 3rd edition. New York: Alfred A. Knopf, 1970.
Snow, W. Brewster, ed. *The Highway and the Landscape.* New Brunswick, N.J.: Rutgers University Press, 1959.
Sutter, Paul. *Driven Wild: How the Fight against Automobiles Launched the Modern Wilderness Movement.* Seattle: University of Washington Press, 2002.
Tunnard, Christopher, and Boris Pushkarev. *Man-made America: Chaos or Control?* New Haven: Yale University Press, 1963.
U.S. Department of Agriculture. *Roadside Improvement.* U.S. Department of Agriculture Miscellaneous Publication No. 191. Washington, D.C.: Government Printing Office, 1934.
U.S. Department of the Interior. *Proceedings of the National Park Conference Held at Berkeley, California, March 11, 12, and 13, 1915.* Washington: Government Printing Office, 1915.
U.S. Department of the Interior. National Park Service. *Reports of the Director of the National Park Service to the Secretary of the Interior for the Fiscal Year Ended June 30, 1917–1932.* 16 vols. Washington, D.C.: Government Printing Office.
-. *Park Road Standards.* Washington, D.C.: Government Printing Office, 1968, 1984.
U.S. Department of the Interior. National Park Service. Branch of Planning. *Park Structures and Facilities.* Washington, D.C.: Government Printing Office, 1935.
U.S. Department of Transportation. Federal Highway Administration. *America's Highways, 1776–1976: A History of the Federal-Aid Program.* Washington, D.C.: Government Printing Office, 1976.
———. *Federal Lands Highway Program: Activities, Accomplishments, and Trend Analyses for 1991–1996,* 6 vols. Washington, D.C.: Federal Highway Administration, 1991–96.
———. *Flexibility in Highway Design.* Publication No. FHWA-PD-97-062. n.d. [1997].
Wirth, Conrad. *Parks, Politics, and the People.* Norman: University of Oklahoma Press, 1980.

Documenting Historic Roads, Structures, and Cultural Landscapes

America Preserved: A Checklist of Historic Buildings, Structures, and Sites Recorded by the Historic American Buildings Survey/ Historic American Engineering Record. Washington, D.C.: Library of Congress, 1995.
Appleyard, Donald, Kevin Lynch, and John R. Myer. *The View from the Road.* Cambridge: MIT Press, 1964.
Burns, John A., and the Staff of HABS/HAER, eds. *Recording Historic Structures: Historic American Buildings Survey/Historic American Engineering Record.* Washington, D.C.: American Institute of Architects Press, 1989.
Croteau, Todd. "Recording NPS Roads and Bridges." *CRM* 16, no. 3 (1993): 3–4.
Davis, Timothy. "HAER Documents America's Park Roads and Parkways," *CRM* 23, no. 4 (2000): 12–14.
DeLony, Eric. *Landmark American Bridges.* New York: American Society of Civil Engineers, 1993.
Groth, Paul, and Chris Wilson, eds. *Everyday America: J. B. Jackson and Recent Cultural Landscape Studies.* Berkeley and Los Angeles: University of California Press, 2003.
Hokanson, Drake. *The Lincoln Highway: Main Street across America.* Ames: Iowa University Press, 1988.
Jackson, John Brinckerhoff. *Landscape in Sight: Looking at America.* Edited by Helen Horowitz. New Haven: Yale University Press, 1997.
Leach, Sara Amy. "Documenting Rock Creek and Potomac Parkway." *CRM* 16, no. 3 (1993): 5–7.
Mayall, R. Newton. "Recording Historic American Landscape Architecture." *Landscape Architecture* 26 (October 1935): 1–11.
Page, Robert, Cathy Gilbert, and Susan Dolan. *A Guide to Cultural Landscape Reports: Content, Process, and Techniques.* U.S. Department of the Interior, National Park Service, Park Historic Structures and Cultural Landscape Program, Washington, D.C., 1998.

Peatross, Ford, ed. *Historic America: Buildings, Structures, and Sites.* Washington, D.C.: Library of Congress, 1983.
Raitz, Karl B. *The National Road.* Baltimore: Johns Hopkins University Press, 1996.
———. *Guide to the National Road.* Baltimore: Johns Hopkins University Press, 1996.
Schlereth, Thomas J. *Reading the Road: U.S. 40 and the American Landscape.* Knoxville: University of Tennessee Press, 1997.
U.S. Department of the Interior. National Park Service. *Recording Historic Buildings.* Compiled by Harley J. McKee. Washington, D.C.: Government Printing Office, 1970.
———. *Visual Character of the Blue Ridge Parkway: Virginia and North Carolina.* Washington, D.C.: Government Printing Office, 1997.
U.S. Department of the Interior. National Park Service. Interagency Resources Division. *National Register Bulletin 18: How to Evaluate and Nominate Historic Designed Landscapes.* Washington, D.C.: Government Printing Office, 1994.
U.S. Department of the Interior. National Park Service. National Register, History, and Education. *National Register Bulletin: How to Improve the Quality of Photographs for National Register Nominations,* revised edition. Washington, D.C.: Government Printing Office, 1998.
———. *National Register Bulletin: Researching a Historic Property,* revised edition. Washington, D.C.: Government Printing Office, 1998.

The Four Guardsmen, Generals Highway,
Sequoia National Park, 1993

Photographer: Brian Grogan, HAER

Timothy Davis is the lead historian for the National Park Service's Park Historic Structures and Cultural Landscape Program. After working on several HAER documentation teams, he served as senior historian for the NPS Park Roads and Bridges Program. He received a Ph.D. in American Civilization from the University of Texas at Austin and has written extensively about park roads, parkways, and other aspects of the American landscape.

Todd A. Croteau was the HAER NPS Park Roads and Bridges Program Manager. Serving as project architect on several summer documentation teams, he helped guide the program from traditional recording techniques to more interpretive graphics encompassing broader themes of landscape design. He received a degree in Industrial Design from the Rhode Island School of Design.

Christopher H. Marston has led HAER documentation teams on a wide variety of industrial and engineering sites around the country since 1989. He served as a project leader for the HAER NPS Park Roads and Bridges Program for seven years. He has a Master of Architecture degree from Carnegie Mellon University and a B.S. in Architecture from the University of Virginia, which is a short drive from both Blue Ridge Parkway and Skyline Drive.